Human BIOLOGY

Thirteenth Edition

Sylvia S. Mader

Michael Windelspecht
Appalachian State University

With contributions by

David Cox
Lincoln Land Community College

The McGraw-Hill Companies

HUMAN BIOLOGY, THIRTEENTH EDITION

Published by McGraw-Hill, a business unit of The McGraw-Hill Companies, Inc., 1221 Avenue of the Americas, New York, NY 10020.

This book is printed on acid-free paper.

1 2 3 4 5 6 7 8 9 0 RMN/RMN 10 9 8 7 6 5 4 3

ISBN 978-0-07-352548-8
MHID 0-07-352548-0

Senior Vice President, Products & Markets: *Kurt L. Strand*
Vice President, General Manager, Products & Markets: *Marty Lange*
Vice President, Content Production & Technology Services: *Kimberly Meriwether David*
Managing Director: *Michael S. Hackett*
Director: *Lynn Breithaupt*
Brand Manager: *Eric Weber*
Director of Development: *Rose Koos*
Senior Development Editor: *Anne L. Winch*
Marketing Manager: *Chris Loewenberg*
Senior Project Manager: *Jayne L. Klein*
Senior Buyer: *Sandy Ludovissy*
Media Project Manager: *Janean A. Utley*
Senior Designer: *Laurie B. Janssen*
Cover Designer: *Ron Bissell*
Cover Image: *© Corey Rich, Getty Images*
Content Licensing Specialist: *Lori Hancock*
Photo Research: *Evelyn Jo Johnson*
Compositor and Art Studio: *Electronic Publishing Services Inc., NYC*
Typeface: *10/12 Utopia Std*
Printer: *R. R. Donnelley*

All credits appearing on page or at the end of the book are considered to be an extension of the copyright page.

Library of Congress Cataloging-in-Publication Data

Mader, Sylvia S.
Human biology / Sylvia S. Mader, Michael Windelspecht ; contributions by David Cox. — 13th ed.
p. cm.
Includes index.
ISBN 978-0-07-352548-8 — ISBN 0-07-352548-0 (hard copy : alk. paper) 1. Human biology—Textbooks. I. Windelspecht, Michael, 1963- II. Title.
QP36.M2 2014
612—dc23

2012031773

The Internet addresses listed in the text were accurate at the time of publication. The inclusion of a website does not indicate an endorsement by the authors or McGraw-Hill, and McGraw-Hill does not guarantee the accuracy of the information presented at these sites.

www.mhhe.com

Brief Contents

About the Authors

Sylvia S. Mader Sylvia S. Mader has authored several nationally recognized biology texts published by McGraw-Hill. Educated at Bryn Mawr College, Harvard University, Tufts University, and Nova Southeastern University, she holds degrees in both Biology and Education. Over the years she has taught at University of Massachusetts, Lowell; Massachusetts Bay Community College; Suffolk University; and Nathan Mayhew Seminars. Her ability to reach out to science-shy students led to the writing of her first text, *Inquiry into Life,* that is now in its fourteenth edition. Highly acclaimed for her crisp and entertaining writing style, her books have become models for others who write in the field of biology.

Although her writing schedule is always quite demanding, Dr. Mader enjoys taking time to visit and explore the various ecosystems of the biosphere. Her several trips to the Florida Everglades and Caribbean coral reefs resulted in talks she has given to various groups around the country. She has visited the tundra in Alaska, the taiga in the Canadian Rockies, the Sonoran Desert in Arizona, and tropical rain forests in South America and Australia. A photo safari to the Serengeti in Kenya resulted in a number of photographs for her texts. She was thrilled to think of walking in Darwin's steps when she journeyed to the Galápagos Islands with a group of biology educators. Dr. Mader was also a member of a group of biology educators who traveled to China to meet with their Chinese counterparts and exchange ideas about the teaching of modern-day biology.

Michael Windelspecht As an educator, Michael Windelspecht has taught introductory biology, genetics, and human genetics in the online, traditional, and hybrid environments at community colleges, comprehensive universities, and military institutions. For over a decade he served as the Introductory Biology Coordinator at Appalachian State University, where he directed a program that enrolled over 4,500 students annually. His educational background includes degrees from the University of Maryland, Michigan State University, and the University of South Florida. Dr. Windelspecht is also active in promoting the scientific literacy of secondary school educators. He has received funding for workshops integrating water quality research into the science curriculum, and has spent several summers working with Fulbright programs involving Pakistani middle school teachers.

As an author, Dr. Windelspecht has published five reference textbooks and numerous print and online lab manuals. He served as the series editor for a ten-volume work on the human body. This is his fifth textbook in the Mader series. For years Dr. Windelspecht has been active in the development of multimedia resources for online and hybrid science classrooms. Along with his wife, Sandra, he owns a multimedia production company, Ricochet Creative Productions, which actively develops and assesses new technologies for the science classroom.

CONTRIBUTOR:

Dave Cox Dave Cox serves as Associate Professor of Biology at Lincoln Land Community College, in Springfield, Illinois. He was educated at Illinois College and Western Illinois University. As an educator, Professor Cox teaches introductory biology for non-majors in the traditional classroom format as well as in a hybrid format. He also teaches biology for majors, and marine biology and biological field studies as study-abroad courses in Belize. He serves as the Educational Director for the Sibun Education and Adventure Lodge, located in Belmopan, Belize. Dave served as a contributor to the 11th edition of *Biology,* and this thirteenth edition of *Human Biology.*

Preface

Goals of the Thirteenth Edition

We are a naturally inquisitive species. As children, we become fascinated with life at a very early age. We want to know how our bodies work, why there are differences, and similarities, between ourselves and the other children around us. In other words, at a very early age, children are acting like biologists.

In many ways, today's students in the science classroom face some of the same challenges as their parents did decades ago. The abundance of new terms often overwhelms even the best prepared student, and the study of biological processes and methods of scientific thinking may convince some students that "science isn't their thing." The study of human biology creates an opportunity for teachers to instruct their students using the ultimate model organism—their own bodies. Whether this is their last science class or the first in a long career in allied health, the study of human biology is pertinent to everyone.

There are also challenges that are unique to the modern classroom. Students in the 21st century are being exposed, almost on a daily basis, to exciting new discoveries and insights that, in many cases, were beyond our predictions even a few short years ago. It is our task, as instructors, not only to make these findings available to our students, but to enlighten students as to why these discoveries are important to their lives and society. At the same time, we must provide students with a firm foundation in those core principles on which biology is founded, and in doing so, provide them with the background to keep up with the many discoveries still to come.

This edition of *Human Biology* builds on the strengths of the Mader series of textbooks. To accomplish the goal of improving scientific literacy, while establishing a foundation of knowledge in human biology and physiology, we integrate a tested, traditional learning system with modern digital and pedagogical approaches designed to stimulate and engage today's student.

The authors of the text identified several goals that guided them through the revision of *Human Biology*, Thirteenth Edition:

1. build upon the strengths of the previous editions of the text,
2. enhance the learning process by integrating content that appeals to today's students.
3. deploy new pedagogical elements, including multimedia assets, to increase student interaction with the text,
4. develop a new series of digital assets designed to engage the modern student and provide assessment of learning outcomes.

New Chapter Openers and Featured Readings

To focus on the relevancy of biology to our everyday lives, the authors developed a series of new feature readings for this text. Each of these readings has a human focus, and in many cases represent topics that have recently been in the news. Each feature is supplemented by a series of questions that instructors may use to promote conversations in the classroom, or as the basis for one-minute opinion papers.

Media Integration

Students can improve the effectiveness of their learning by integrating the digital assets of today's courses into their study habits. LearnSmart™ and Connect®, McGraw-Hill's flagship digital tools, can increase student preparedness for class as well as increase retention of difficult topics. These assets may easily be uploaded into any course management system to provide your students with useful study tutorials.

As educators, the authors recognize that today's students are digital learners. Therefore, a significant new feature of this edition is the integration of media assets into the chapter content. Virtually every section of the textbook is now linked to MP3 files, animations of biological processes, and National Geographic and ScienCentral videos. In addition, McGraw-Hill's new 3D animations are integrated into the more difficult chapters of the text.

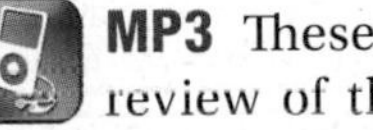

MP3 These three- to five-minute audio files serve as a review of the material in the chapter, and they also assist the student in the pronunciation of scientific terms.

Animation Drawing on McGraw-Hill's vast library of animations, the authors have selected animations that will enhance the student's understanding of complex biological processes.

3D Animation For topics such as photosynthesis and cellular respiration, McGraw-Hill has produced a series of dynamic 3D animations that may be used both as presentation tools in the classroom, and as mini-tutorials that can be assigned within Connect or your course management system.

Video Two different types of movies are integrated into this edition of the text. The ScienCentral videos are short news clips on advances in the sciences. The National Geographic videos provide students with a glimpse of the complexity of life that normally would not be possible in the classroom.

Virtual Lab These simulated experiments serve as excellent tutorials, allowing students to explore the topics covered in select chapters of the text.

Guided Tutorials

In addition to the assets listed above, the authors of the textbook have prepared a series of 2-minute guided tutorials of some of the more difficult topics in the text. A complete list of these tutorials is provided on the inside back cover of the text.

A Student's Guide to Using This Textbook

Case Study The opening case-study is designed to demonstrate how the chapter content is relevant to your life. The authors have not only chosen current-event topics, but have also included a series of two or three questions that you should be thinking of as you progress through the chapter.

Chapter Outline Lists the major sections that will be discussed in the chapter.

Before You Begin Links the content of the chapter with material from earlier in the text. The questions designate important topics that you should understand before proceeding into the chapter.

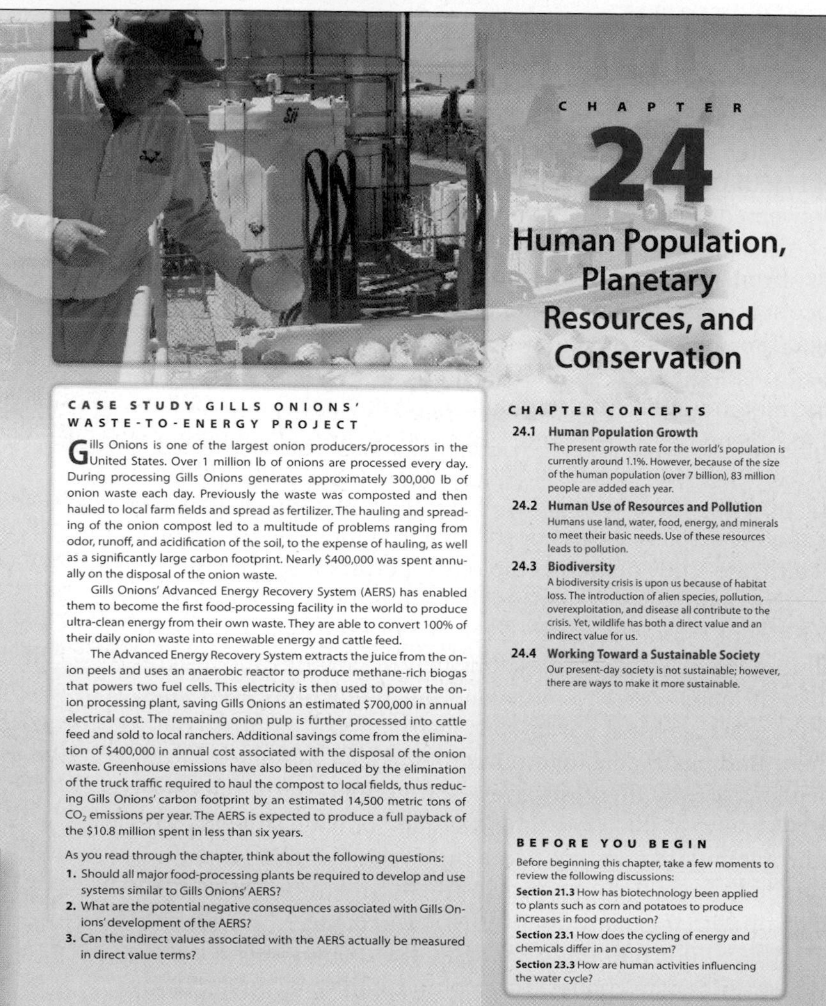

CHAPTER

24

Human Population, Planetary Resources, and Conservation

CASE STUDY GILLS ONIONS' WASTE-TO-ENERGY PROJECT

Gills Onions is one of the largest onion producers/processors in the United States. Over 1 million lb of onions are processed every day. During processing Gills Onions generates approximately 300,000 lb of onion waste each day. Previously the waste was composted and then hauled to local farm fields and spread as fertilizer. The hauling and spreading of the onion compost led to a multitude of problems ranging from odor, runoff, and acidification of the soil, to the expense of hauling, as well as a significantly large carbon footprint. Nearly $400,000 was spent annually on the disposal of the onion waste.

Gills Onions' Advanced Energy Recovery System (AERS) has enabled them to become the first food-processing facility in the world to produce ultra-clean energy from their own waste. They are able to convert 100% of their daily onion waste into renewable energy and cattle feed.

The Advanced Energy Recovery System extracts the juice from the onion peels and uses an anaerobic reactor to produce methane-rich biogas that powers two fuel cells. This electricity is then used to power the onion processing plant, saving Gills Onions an estimated $700,000 in annual electrical cost. The remaining onion pulp is further processed into cattle feed and sold to local ranchers. Additional savings come from the elimination of $400,000 in annual cost associated with the disposal of the onion waste. Greenhouse emissions have also been reduced by the elimination of the truck traffic required to haul the compost to local fields, thus reducing Gills Onions' carbon footprint by an estimated 14,500 metric tons of CO_2 emissions per year. The AERS is expected to produce a full payback of the $10.8 million spent in less than six years.

As you read through the chapter, think about the following questions:

1. Should all major food-processing plants be required to develop and use systems similar to Gills Onions' AERS?
2. What are the potential negative consequences associated with Gills Onions' development of the AERS?
3. Can the indirect values associated with the AERS actually be measured in direct value terms?

CHAPTER CONCEPTS

24.1 Human Population Growth
The present growth rate for the world's population is currently around 1.1%. However, because of the size of the human population (over 7 billion), 83 million people are added each year.

24.2 Human Use of Resources and Pollution
Humans use land, water, food, energy, and minerals to meet their basic needs. Use of these resources leads to pollution.

24.3 Biodiversity
A biodiversity crisis is upon us because of habitat loss. The introduction of alien species, pollution, overexploitation, and disease all contribute to the crisis. Yet, wildlife has both a direct value and an indirect value for us.

24.4 Working Toward a Sustainable Society
Our present-day society is not sustainable; however, there are ways to make it more sustainable.

BEFORE YOU BEGIN

Before beginning this chapter, take a few moments to review the following discussions:

Section 21.3 How has biotechnology been applied to plants such as corn and potatoes to produce increases in food production?

Section 23.1 How does the cycling of energy and chemicals differ in an ecosystem?

Section 23.3 How are human activities influencing the water cycle?

557

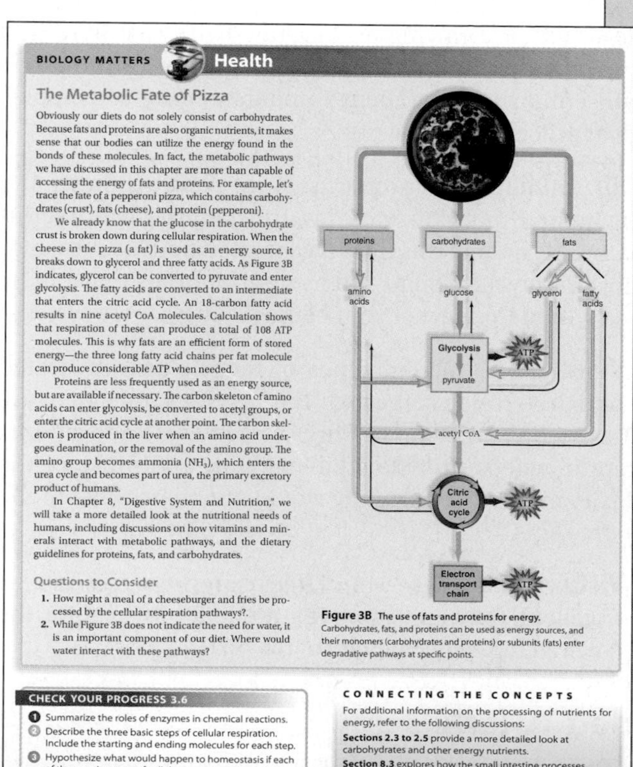

BIOLOGY MATTERS Health

The Metabolic Fate of Pizza

Obviously our diets do not solely consist of carbohydrates. Because fats and proteins are also organic nutrients, it makes sense that our bodies can utilize the energy found in the bonds of these molecules. In fact, the metabolic pathways we have discussed in this chapter are more than capable of accessing the energy of fats and proteins. For example, let's trace the fate of a pepperoni pizza, which contains carbohydrates (crust), fats (cheese), and protein (pepperoni).

We already know that the glucose in the carbohydrate crust is broken down during cellular respiration. When the cheese in the pizza (a fat) is used as an energy source, it breaks down to glycerol and three fatty acids. As Figure 3B indicates, glycerol can be converted to pyruvate and enter glycolysis. The fatty acids are converted to an intermediate that enters the citric acid cycle. An 18-carbon fatty acid results in nine acetyl CoA molecules. Calculation shows that respiration of these can produce a total of 108 ATP molecules. This is why fats are an efficient form of stored energy—the three long fatty acid chains per fat molecule can produce considerable ATP when needed.

Proteins are less frequently used as an energy source, but are available if necessary. The carbon skeleton of amino acids can enter glycolysis, be converted to acetyl groups, or enter the citric acid cycle at another point. The carbon skeleton is produced in the liver when an amino acid undergoes deamination, or the removal of the amino group. The amino group becomes ammonia (NH_3), which enters the urea cycle and becomes part of urea, the primary excretory product of humans.

In Chapter 8, "Digestive System and Nutrition," we will take a more detailed look at the nutritional needs of humans, including discussions on how vitamins and minerals interact with metabolic pathways, and the dietary guidelines for proteins, fats, and carbohydrates.

Questions to Consider

1. How might a meal of a cheeseburger and fries be processed by the cellular respiration pathways?.
2. While Figure 3B does not indicate the need for water, it is an important component of our diet. Where would water interact with these pathways?

Figure 3B The use of fats and proteins for energy. Carbohydrates, fats, and proteins can be used as energy sources, and their monomers (carbohydrates and proteins) or subunits (fats) enter degradative pathways at specific points.

CHECK YOUR PROGRESS 3.6

1. Summarize the roles of enzymes in chemical reactions.
2. Describe the three basic steps of cellular respiration. Include the starting and ending molecules for each step.
3. Hypothesize what would happen to homeostasis if each of three major steps of cellular respiration were missing.

CONNECTING THE CONCEPTS

For additional information on the processing of nutrients for energy, refer to the following discussions:

Sections 2.3 to 2.5 provide a more detailed look at carbohydrates and other energy nutrients.

Section 8.3 explores how the small intestine processes nutrients for absorption.

Section 8.6 describes the importance of carbohydrates, fats, and proteins in the diet.

62

Learning Outcomes

Provide you with an overview of what you need to know in the section. Your instructor can assign activities through Connect™ to help you achieve these outcomes.

Media Integration

Ask your instructor about related quizzes that are available through Connect™ Biology.

Check Your Progress

Questions at the end of each section help you assess and/or apply your understanding of the material in the section. The color codes (red, yellow, and green) are designed to give you an indication of how well you comprehend these topics.

Case Study Conclusion The case study conclusion at the end of the chapter explains how the material you just learned relates to the scenario presented on the opening pages.

Media Integration and Media Study Tools

To enhance your study of the content, multimedia, including MP3 files, animations, and videos, are integrated into every chapter. At the end of each chapter, a media table lists each of the media assets. Go to **www.mhhe.com/maderhuman13e** to access the animations, videos, and MP3 files referenced throughout this book. The website also contains practice tests, animations, and videos organized and integrated by chapter to help you succeed in your study of biology. The ConnectPlus™ platform provides a media-rich eBook, interactive learning tools, and access to the LearnSmart™ system for enhanced student performance.

CASE STUDY CONCLUSION

Following his exam, Steven spent time researching the causes and effects of hypertension on his body. He learned what a blood pressure value of 132/84 mm Hg really meant. The larger number (132) represented the maximum pressure as his heart was beating, while the smaller number (84) was the resting pressure. Ideally, his blood pressure should be less than 120/80 mm Hg, so as his doctor indicated, Steven was prehypertensive. However, the good news was that at his age, prehypertension was largely due to lifestyle choices, namely diet and physical activity. Therefore, Steven resolved to watch his diet more closely, especially the amount of saturated fat and cholesterol, and increase his cardiovascular workouts per week. His doctor also recommended a follow-up visit in a few months to ensure that Steven was making satisfactory progress.

MEDIA STUDY TOOLS

connect plus+ BIOLOGY Enhance your study of this chapter with media! Visit **www.mhhe.com/maderhuman13e** and go to "Media Study Tools" for this chapter to access the following:

Animations	Videos	MP3 Files
5.3 Cardiac Cycle • Conducting System of the Heart • Baroreceptor Reflex Control of Blood Pressure **5.6** Fluid Exchange Across the Walls of Capillaries	**5.3** Heart Stem Cells **5.7** Cardiac Repair • Heart Stem Cells	**5.2** Classification of Blood Vessels **5.3** Cardiac Cycle • Cardiac Conduction System **5.4** Blood Flow and Blood Pressure

SUMMARIZE

5.1 Overview of the Cardiovascular System

The **cardiovascular system** consists of the heart and blood vessels. The heart pumps blood, and blood vessels take blood to and from capillaries, where exchanges of nutrients for wastes occur with tissue cells. Blood is refreshed at the lungs, where gas exchange occurs; at the digestive tract, where nutrients enter the blood; and at the kidney, where wastes are removed from blood.

The **lymphatic system** removes excess fluid from around the tissue and returns it to the cardiovascular system.

5.2 The Types of Blood Vessels

Arteries: **Arteries** and **arterioles** move blood away from the heart. Arteries have the thickest walls, which allows them to withstand blood pressure.

Capillaries: Exchange of substances occurs in the capillaries. **Precapillary sphincters** and **arteriovenous shunts** help control the flow of blood within the capillaries.

Veins: **Veins** and **venules** move blood toward the heart. Veins have relatively weak walls with valves that keep the blood flowing in one direction.

Summarize Provides an excellent overview of the chapter concepts using concise, bulleted summaries, summary tables, and key illustrations. The boldfaced terms represent the key terms from each section of the chapter. These are placed within the summary to help you understand how the term relates to the content of the chapter. An expanded definition of each boldfaced term is provided in the Glossary.

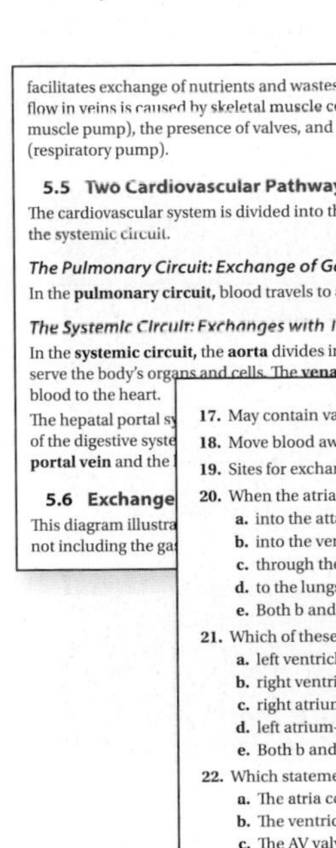

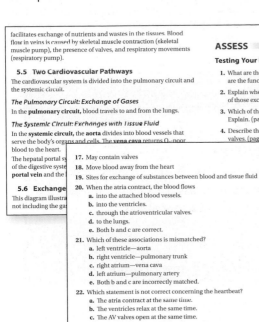

facilitates exchange of nutrients and wastes in the tissues. Blood flow in veins is caused by skeletal muscle contraction (skeletal muscle pump), the presence of valves, and respiratory movements (respiratory pump).

5.5 Two Cardiovascular Pathways

The cardiovascular system is divided into the pulmonary circuit and the systemic circuit.

The Pulmonary Circuit: Exchange of Gases

In the **pulmonary circuit,** blood travels to and from the lungs.

The Systemic Circuit: Exchanges with Tissue Fluid

In the **systemic circuit,** the **aorta** divides into blood vessels that serve the body's organs and cells. The **vena cava** returns O_2-poor blood to the heart.

The hepatal portal sy... of the digestive syste... **portal vein** and the ...

5.6 Exchange...

This diagram illustra... not including the ga...

ASSESS

Testing Your Knowledge of the Concepts

1. What are the two parts of the cardiovascular system, and what are the functions of each part? (page 92)
2. Explain where exchanges occur in the body and the importance of those exchanges. (page 92)
3. Which of the three types of blood vessels are most numerous? Explain. (page 93)
4. Describe the structure of the heart, including the chambers and valves. (pages 94–97)

17. May contain valves
18. Move blood away from the heart
19. Sites for exchange of substances between blood and tissue fluid
20. When the atria contract, the blood flows
    a. into the attached blood vessels.
    b. into the ventricles.
    c. through the atrioventricular valves.
    d. to the lungs.
    e. Both b and c are correct.
21. Which of these associations is mismatched?
    a. left ventricle—aorta
    b. right ventricle—pulmonary trunk
    c. right atrium—vena cava
    d. left atrium—pulmonary artery
    e. Both b and c are incorrectly matched.
22. Which statement is not correct concerning the heartbeat?
    a. The atria contract at the same time.
    b. The ventricles relax at the same time.
    c. The AV valves open at the same time.
    d. The semilunar valves open at the same time.
    e. First the right side contracts; then the left side contracts.
23. Accumulation of plaque in an artery wall is
    a. an aneurysm.
    b. angina pectoris.
    c. atherosclerosis.
    d. hypertension.
    e. a thromboembolism.

ENGAGE

Virtual Lab Blood Pressure

The virtual lab "Blood Pressure" provides an interactive look at how factors such as age and gender influence the risk of hypertension.

Thinking Critically About the Concepts

1. You have to stand in front of the class to give a report. You are nervous, and your heart is pounding. What is the specific mechanism behind this reaction? How would your ECG appear?
2. The cardiovascular system is an elegant example of the concept that structure supports function. Each type of blood vessel has a specific job. Each vessel's physical characteristics enable it to do that job. The muscle walls of the right and left ventricle vary in thickness depending on where they pump the blood. When organ structure is damaged or changed (as arteries are in atherosclerosis), the organ's ability to perform its function may be compromised. Homeostatic conditions, such as blood pressure, may be affected as well. Dietary and lifestyle choices can either prevent damage or harm the cardiovascular system.
    a. What do you think the long-term effects of hypertension would be on the heart? What about other organ systems?
    b. Why do you think a combination of hypertension and atherosclerosis is a particularly dangerous combination?
    c. What factors in your life could be changed to reduce the risk of CVD? Be critical!

Assess

Testing Yourself Questions help you review material and prepare for tests. (See Appendix C for answers.)

Engage

Thinking Critically Questions give you an opportunity to reason as a scientist.

Virtual Labs For selected chapters, these online labs can serve as tutorials to help you better understand the content and provide you with the opportunity to investigate topics associated with the chapter from a scientific perspective.

Teaching and Learning Tools

McGraw-Hill Higher Education and Blackboard Have Teamed Up

The Best of Both Worlds

Blackboard®, the Web-based course-management system, has partnered with McGraw-Hill to better allow students and faculty to use online materials and activities to complement face-to-face teaching. Blackboard features exciting social learning and teaching tools that foster more logical, visually impactful, and active learning opportunities for students. You'll transform your closed-door classrooms into communities where students remain connected to their educational experience 24 hours a day.

This partnership allows you and your students access to McGraw-Hill's Connect® and McGraw-Hill Create™ right from within your Blackboard course—all with one single sign-on.

Not only do you get single sign-on with Connect and Create, you also get deep integration of McGraw-Hill content and content engines right in Blackboard. Whether you're choosing a book for your course or building Connect assignments, all the tools you need are right where you want them—inside of Blackboard.

Gradebooks are now seamless. When a student completes an integrated Connect assignment, the grade for that assignment automatically (and instantly) feeds your Blackboard grade center.

McGraw-Hill and Blackboard can now offer you easy access to industry leading technology and content, whether your campus hosts it or we do. Be sure to ask your local McGraw-Hill representative for details.

McGraw-Hill LearnSmart™

McGraw-Hill LearnSmart™ is available as an integrated feature of **McGraw-Hill Connect® Biology** and provides students with a GPS (Guided Path to Success) for your course. Using artificial intelligence, LearnSmart intelligently assesses a student's knowledge of course content through a series of adaptive questions. It pinpoints concepts the student does not understand and maps out a personalized study plan for success. This innovative study tool also has features that allow instructors to see exactly what students have accomplished and a built-in assessment tool for graded assignments. Visit the following site for a demonstration.

www.mhlearnsmart.com

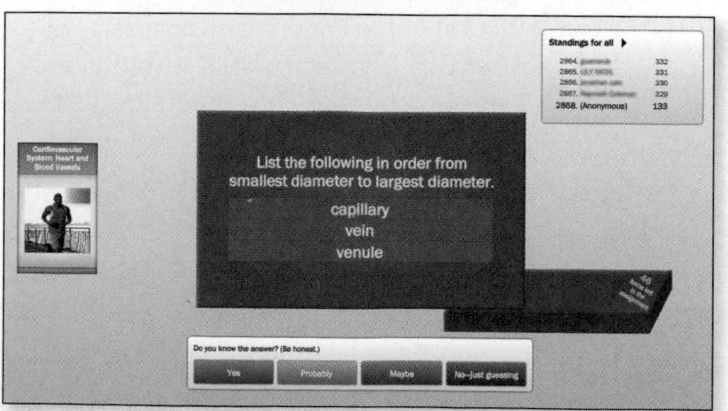

McGraw-Hill LabSmart™ LabSmart

THE Virtual Lab Experience

Based on the same world-class, super-adaptive technology as LearnSmart, **McGraw-Hill LabSmart™** is a must-see, outcomes-based lab simulation. It assesses a student's knowledge and adaptively corrects deficiencies, allowing the student to learn faster and retain more knowledge with greater success.

First, a student's knowledge is adaptively leveled on core learning oucomes: Questioning reveals knowledge deficiencies that are corrected by the delivery of content that is conditional on a student's response. Then, a simulated lab experience requires the student to think and act like a scientist: Recording, interpreting, and analyzing data using simulated equipment found in labs and clinics. The student is allowed to make mistakes—a powerful part of the learning experience! A virtual coach provides subtle hints when needed; asks questions about the student's choices; and allows the student to reflect upon and correct those mistakes. Whether your need is to overcome the logistical challenges of a traditional lab, provide better lab prep, improve student performance, or make your online experience one that rivals the real world, LabSmart accomplishes it all. To learn more, visit

www.mhlabsmart.com

McGraw-Hill Connect® Biology

McGraw-Hill Connect® Biology provides online presentation, assignment, and assessment solutions. It connects your students with the tools and resources they'll need to achieve success.

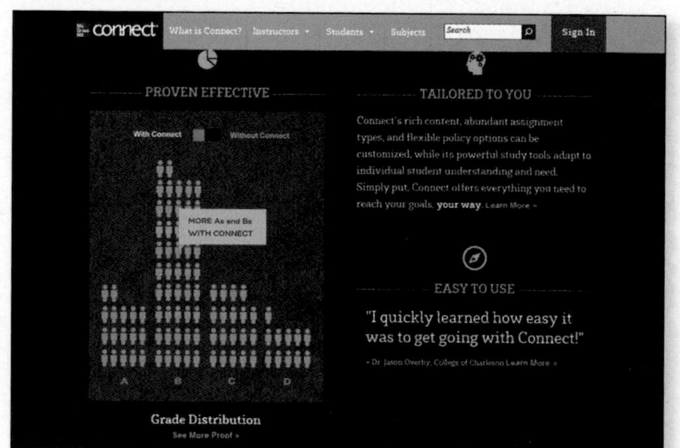

With Connect Biology, you can deliver assignments, quizzes, and tests online. A robust set of questions and activities are presented and aligned with the textbook's learning outcomes. As an instructor, you can edit existing questions and author entirely new problems. Track individual student performance—by question, assignment, or in relation to the class overall—with detailed grade reports. Integrate grade reports easily with Learning Management Systems (LMS), such as WebCT, Blackboard, and much more.

ConnectPlus Biology provides students with all the advantages of Connect Biology, plus 24/7 online access to an eBook. This media-rich version of the book is available through the McGraw-Hill Connect platform and allows seamless integration of text, media, and assessments. To learn more, visit

www.mcgrawhillconnect.com

My Lectures—Tegrity®

McGraw-Hill Tegrity® records and distributes your class lecture with just a click of a button. Students can view anytime/anywhere via computer, iPod, or mobile device. It indexes as it records your PowerPoint® presentations and anything shown on your computer so students can use keywords to find exactly what they want to study. Tegrity is available as an integrated feature of McGraw-Hill Connect Biology and as a standalone.

Animations for a New Generation

Dynamic, 3D animations of key biological processes bring an unprecedented level of control to the classroom. Innovative features keep the emphasis on teaching rather than entertaining.

- An options menu lets you control the animation's level of detail, speed, length, and appearance, so you can create the experience you want.
- Draw on the animation using the whiteboard pen to highlight important areas.
- The scroll bar lets you fast forward and rewind while seeing what happens in the animation, so you can start at the exact moment you want.
- A scene menu lets you instantly jump to a specific point in the animation.
- Pop-ups add detail at important points and help students relate the animation back to concepts from lecture and the textbook.
- A complete visual summary at the end of the animation reminds students of the big picture.
- Animation topics include: Cellular Respiration, Molecular Biology of the Gene, DNA Replication, Cell Cycle and Mitosis, and Membrane Transport.

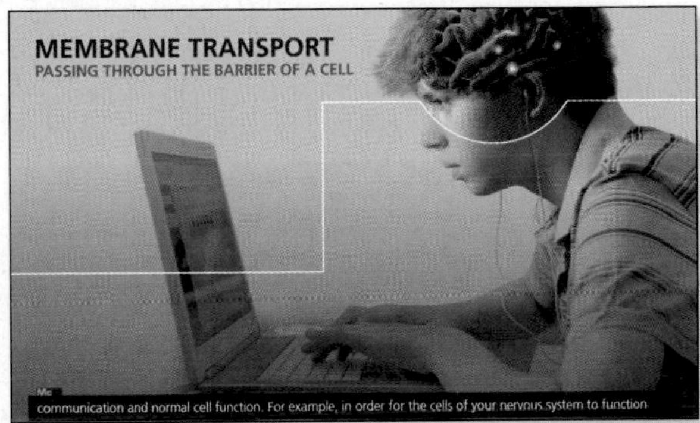

McGraw-Hill Create™

With **McGraw-Hill Create**™, you can easily rearrange chapters, combine material from other content sources, and quickly upload content you have written, like your course syllabus or teaching notes. Find the content you need in Create by searching through thousands of leading McGraw-Hill textbooks. Arrange your book to fit your teaching style. Create even allows you to personalize your book's appearance by selecting the cover and adding your name, school, and course information. Order a Create book and you'll receive a complimentary print review copy in 3–5 business days or a complimentary electronic review copy (eComp) via e-mail in minutes. Go to www.mcgrawhillcreate.com today and register to experience how McGraw-Hill Create empowers you to teach *your* students *your* way. To learn more visit

www.mcgrawhillcreate.com

Presentation Tools

Everything you need for outstanding presentations in one place.

www.mhhe.com/maderhuman13e

- *FlexArt Image Powerpoints*—including every piece of art that has been sized and cropped specifically for superior presentations, as well as labels that can be edited and flexible art that can be picked up and moved on key figures. Also included are tables, photographs, and unlabeled art pieces.
- *Lecture PowerPoints with Animations*—animations illustrating important processes are embedded in the lecture material.
- *Animation PowerPoints*—relevant animations embedded in PowerPoint slides to drop easily into an existing presentation.
- *Labeled JPEG Images*—full-color digital files of all illustrations that can be readily incorporated into presentations, exams, or custom-made classroom materials.
- *Base Art Image Files*—unlabeled digital files of all illustrations.

Presentation Center

In addition to the images from your book, this online digital library contains photos, artwork, animations, and other media from an array of McGraw-Hill textbooks.

Computerized Test Bank

A comprehensive bank of test questions is provided within a computerized test bank powered by McGraw-Hill's flexible electronic testing program, **EZ Test Online.** A new tagging scheme allows you to sort questions by Bloom's difficulty level, learning outcome, topic, and section. With EZ Test Online, instructors can select questions from multiple McGraw-Hill test banks or author their own, and then either print the test for paper distribution or give it online.

Instructor's Manual

The instructor's manual contains chapter outlines, lecture enrichment ideas, and discussion questions.

Laboratory Manual

The *Human Biology Laboratory Manual* is written by Dr. Sylvia Mader. Every laboratory has been written to help students learn the fundamental concepts of biology and the specific content of the chapter to which the lab relates, as well as gain a better understanding of the scientific method.

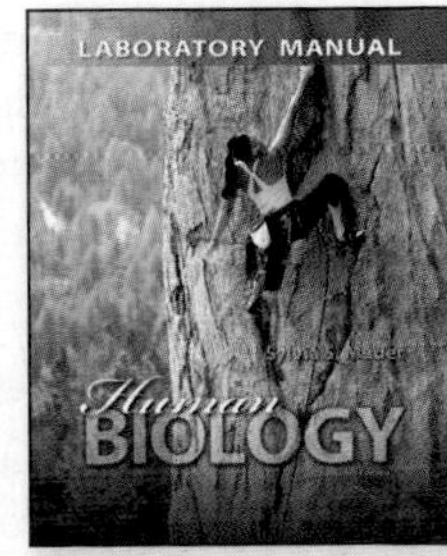

Companion Website

www.mhhe.com/maderhuman13e

The Mader *Human Biology* companion website allows students to access a variety of free digital learning tools that include

- Chapter-level quizzing
- Animations and videos
- Vocabulary flashcards
- Virtual labs

Detailed List of Content Changes in *Human Biology*, Thirteenth Edition

Chapter 1: Exploring Life and Science contains two new feature readings. The first examines how humans have adapted to life at high-elevations. The second examines the science of Robert Koch and the importance of the Koch Postulates. The chapter also contains a link to the virtual lab "Dependent and Independent Variables."

Unit 1: Human Organization

Chapter 2: Chemistry of Life contains an updated periodic table (Fig. 2.1) that includes periods and groups for the elements. **Chapter 3: Cell Structure and Function** includes new images explaining surface-to-volume ration (Fig. 3.2) and the endosymbiotic theory (Fig. 3.5). A new feature, "The Metabolic Fate of Pizza" explores alternate energy sources for glycolysis. Links to the 3D animations "Membrane Transport" and "Cellular Respiration" are included throughout the chapter. The material on the classification of connective tissue (Fig. 4.4) in **Chapter 4: Organization and Regulation of Body Systems** has been updated. A new figure (Fig. 4.7) has been added to the chapter to illustrate the shape and configurations of epithelial cells.

Unit 2: Maintenance of the Human Body

Chapter 5: Cardiovascular System: Heart and Blood Vessels now begins with a new opening article on hypertension in young adults. Table 5.1 has been updated to include information on a hypertensive crisis. A new feature "New Information About Preventing Cardiovascular Disease" explores how dietary factors may reduce cardiovascular disease. **Chapter 6: Cardiovascular System: Blood** contains a new feature reading "Aspirin and Heart Disease." **Chapter 7: The Lymphatic and Immune Systems** includes new figures (Figs. 7.7 and 7.10) that summarize the innate and adaptive immune defenses. The chapter features a new article, "Adult Vaccinations." In addition, the material on monoclonal antibodies has been moved so that it is included in the general discussion of antibodies. The **Infectious Diseases Supplement** has been updated to include more information on both the development of HIV vaccines and influenza viruses. **Chapter 8: Digestive System and Nutrition** now includes references to the USDA's MyPlate program. **Chapter 9: Respiratory System** includes a new article "Artificial Lung Technology" and a new figure (Fig. 9.8) explaining the relationship between air pressure and volume. **Chapter 10: Urinary System** has been reorganized so that a greater emphasis on the role of the kidneys in maintaining water-salt and acid-base balance is placed earlier in the chapter.

Unit 3: Movement and Support in Humans

Chapter 11: Skeletal System has been reorganized so that the content on bone growth and remodeling is included under homeostasis. **Chapter 12: Muscular System** features a new article on botox and wrinkles.

Unit 4: Integration and Coordination in Humans

Chapter 13: Nervous System contains more detailed overview of the types of neuroglia. The section on drug abuse has been updated to include drugs such as K2 and date rape drugs. **Chapter 14: Senses** begins with a more detailed examination of signal transduction and the types of sensory receptors. **Chapter 15: Endocrine System** contains a more detailed discussion of A, B, and D cells in the pancreas and an expanded section on diabetes mellitus. A new feature article "Identifying Insulin as a Chemical Messenger" is included in this chapter.

Unit 5: Reproduction and Development

Chapter 16: Reproductive System contains an expanded discussion on contraceptive injections and vaccines, as well as methods of emergency contraception. **Chapter 17: Human Development and Aging** contains a new feature "Preventing and Testing for Birth Defects" that describes methods of obtaining cells for genetic testing prior to birth. The content regarding the development of the internal sex organs has a greater focus on the role of the *SRY* gene (Fig. 17.9). The section on pregnancy has been expanded to provide an enhanced explanation of changes in a woman's body during pregnancy (Fig. 17.12). The content on aging (section 17.5) has also been expanded to included a more detailed discussion of hypotheses regarding cellular aging.

Unit 6: Human Genetics

Chapter 18: Patterns of Chromosome Inheritance begins with a more detailed look at chromosome structure. The 3D animations "Cell Cycle & Mitosis" and "Meiosis" are integrated throughout the chapter. The content on spermatogenesis and oogenesis has been moved so that it occurs closer to the discussion of meiosis. **Chapter 19: Cancer** includes a new article on how cancer causes death from a physiological perspective. **Chapter 20: Patterns of Genetic Inheritance** includes new examples of genetic disorders in humans. The content of non-Mendelian genetics has been expanded significantly and includes new material on skin color and multifactorial traits. **Chapter 21: DNA Biology and Technology** contains a new feature "Discovering the Structure of DNA" and an increased focus on the molecular mechanisms of DNA replication. The 3D animations "DNA Replication" and "Molecular Biology of the Gene" are integrated into the chapter. A new figure explains the process of a PCR reaction.

Unit 7: Human Evolution and Ecology

Chapter 22: Human Evolution begins with new figure (Fig. 22.1) and discussions regarding early chemical and biological evolution on the planet. The chapter includes a more detailed look at the importance of Charles Darwin's work and an increased focus on the importance of transitional fossils (Fig. 22.3) in understanding evolution. Also included are more detailed discussions of the ardipithicines, Neandertals, and Denisovans. **Chapter 23: Global Ecology and Human Interferences** contains a new feature article "Biomagnification of Mercury." **Chapter 24: Human Population, Planetary Resources, and Conservation** begins with a new article on the Gill Onions Waste to Energy Project. The chapter also contains an updated article on the safety of genetically-engineered foods

Acknowledgments

Dr. Sylvia Mader represents one of the icons of science education. Her dedication to her students, coupled with her clear, concise writing style, has benefited the education of thousands of students over the past three decades. It is an honor to continue her legacy, and to bring her message to the next generation of students. I have been privileged to work with a team of dedicated and talented individuals on this text. My coauthor, David Cox, made significant contributions to the ecology chapters of the book. I am convinced that my abilities as a science educator have been significantly enhanced by our teamwork. Together, we have striven to ensure that the content is presented in the familiar Mader style.

Many dedicated and talented individuals assisted in the development of *Human Biology*. I am very grateful for the help of so many professionals at McGraw-Hill who were involved in bringing this book to fruition. In particular, let me thank my developmental editors Rose Koos and Anne Winch. The talents, project management skills, advice, and most of all, patience, of these individuals are greatly appreciated. The managing director, Michael Hackett, continues to support the development of innovative products. The desire of my editor, Eric Weber, to impact the lives of our students, and assist those who instruct them, is evident throughout this text. The project manager, Jayne Klein, faithfully and carefully steered the book through the publication process. Tamara Maury, the marketing manager, tirelessly promoted the text and educated the sales reps on its message.

The design of the book is the result of the creative talents of Laurie Janssen and many others who assisted in deciding the appearance of each element in the text. I was very lucky to have Bea Sussman as my copyeditor, and Dawnelle Krouse and Mary Swartz as proofreaders. Electronic Publishing Services produced this textbook, in the process emphasizing pedagogy and beauty to arrive at the best presentation on the page. Lori Hancock and Evelyn Jo Johnson did a superb job of finding just the right photographs and micrographs.

Who I am, as an educator and an author, is a direct reflection of what I have learned from my students. Education is a two-way street, and it is my honest opinion that both my professional and personal life have been enriched by my interactions with my students. They have encouraged me to learn more, teach better, and never stop questioning the world around me. Last, but certainly not least, I would also like to acknowledge my wife, Sandra. She has never wavered in her patience and support of my endeavors.

Michael Windelspecht, Ph.D.
Blowing Rock, NC

The thirteenth edition of *Human Biology* would not have the same excellent quality without the input of these contributors and those of the many contributors and reviewers listed below. The authors would especially like to thank the attendees of the Human Biology Symposium in Jackson Hole, Wyoming. The energy and dedication to learning expressed by these individuals served as a major motivating factor for this revision.

McGraw-Hill's 360° Development Process is an ongoing, never-ending, market-oriented approach to building accurate and innovative print and digital products. It is dedicated to continual large-scale and incremental improvement driven by multiple customer feedback loops and checkpoints. This is initiated during the early planning stages of our new products, and intensifies during the development and production stages, then begins again upon publication in anticipation of the next edition.

This process is designed to provide a broad, comprehensive spectrum of feedback for refinement and innovation of our learning tools, for both student and instructor. The 360° Development Process includes market research, content reviews, course- and product-specific symposia, accuracy checks, and art reviews. We appreciate the expertise of the many individuals involved in this process.

Human Biology Symposium Participants

A special thank you to these 2012 Symposium participants for your ideas and enthusiasm.

Brian Berthelsen, *Iowa Western Community College*
David Dorado, *Fullerton College*
Jane Gallagher, *The City College of New York*
Debra Grieneisen, *Harrisburg Area Community College*
Julie Hendriksen, *Pasco-Hernando Community College*
Nola Kelly, *Hudson Valley Community College*
Carol Mack, *Erie Community College*
Terri Pope, *Cuyahoga Community College*
Anil Rao, *Metropolitan State College of Denver*
Mary Ann Sadler, *St. Charles Community College*
Teresa Shakespeare, *Fort Valley State University*
Roy Silcox, *Brigham Young University*
Michael Troyan, *Penn State University*
Gay Williamson, *Mississippi State University*

Ancillary Authors

Appendix Answer Bank: Jeffrey A. Isaacson, *Nebraska Wesleyan University*
Connect Question Bank: Krissy Johnson and Alex James, *Ricochet Creative Productions*
Test Bank: Dave Cox, *Lincoln Land Community College*

Tutorial Development: Krissy Johnson and Sandy Windelspecht, *Ricochet Creative Productions*
Instructor's Manual: Roberta Batorsky, *Middlesex County College*
eBook Quizzes: Ashley Spring, *Brevard Community College*
Companion Website Quizzes: Sylvester Allred, *Northern Arizona University*
Lecture Outlines: Stephanie Songer, *North Georgia College and State University*
LearnSmart™ Authors and Reviewers:
Tammy Atchison, *Pitt Community College*
Megan Berdelman
Dena Berg, *Tarrant County College*
Lisa Bonneau, *Mount Marty College*
Jane Caldwell, *Washington and Jefferson College*
Patrick Galliart, *North Iowa Area Community College*
Dan Matusiak, *St. Charles Community College*
Steven Nilsen
Jerred Seveyka, *Yakima Valley Community College*
MaryJo Witz, *Monroe Community College*

Thirteenth Edition Reviewers

Daniel Aruscavage, *Kutztown University*
Deborah Barker, *Cuesta College*
Shirley Bartido, *New Jersey City University*
Becky Brown, *College of Marin*
Virginia L. Bliss, *Assumption College*
Stephen W. Chordas III, *The Ohio State University*
Michael DeBow, *San Joaquin Delta College*
Elizabeth Jordan, *Santa Monica College*
Marian L. Farrell, *University of Scranton*
Steven Fenster, *Ashland University*
Ferdinand Gomez, *FIU Herbert Wertheim College of Medicine*
Jennifer Grant, *University of Wisconsin-Stout*
Maleka Hashmi, *University of Wisconsin-Stout*
Clare Hays, *Metropolitan State College of Denver*
Jacki Houghton, *Santa Monica College*
Carol Mack, *Erie Community College*
Peter Malo, *Richard J. Daley College*
Nancy Jean Mann, *Cuesta College*
Darcy Medica, *Penn State University*
Joseph Napolitano, *Suffolk County Community College*
Olatunde Okediji, *Albany State University*
Chris Osborne, *Mt. San Jacinto College—Menifee Valley Campus*
Ivan Paul, *John Wood Community College*
Greg Petersen, *Kirkwood Community College*
Darlene Sabio, *Excelsior College*
Tobili Sam-Yellowe, *Cleveland State University*
Jeremy Sellers, *University of North Carolina at Pembroke*
Roy W. Silcox, *Brigham Young University*
Jacqueline Washington, *Nyack College*
Daniel Westholm, *The College of St. Scholastica*

Contents

Readings

BIOLOGY MATTERS Bioethical

BIOLOGY MATTERS Health

BIOLOGY MATTERS Science

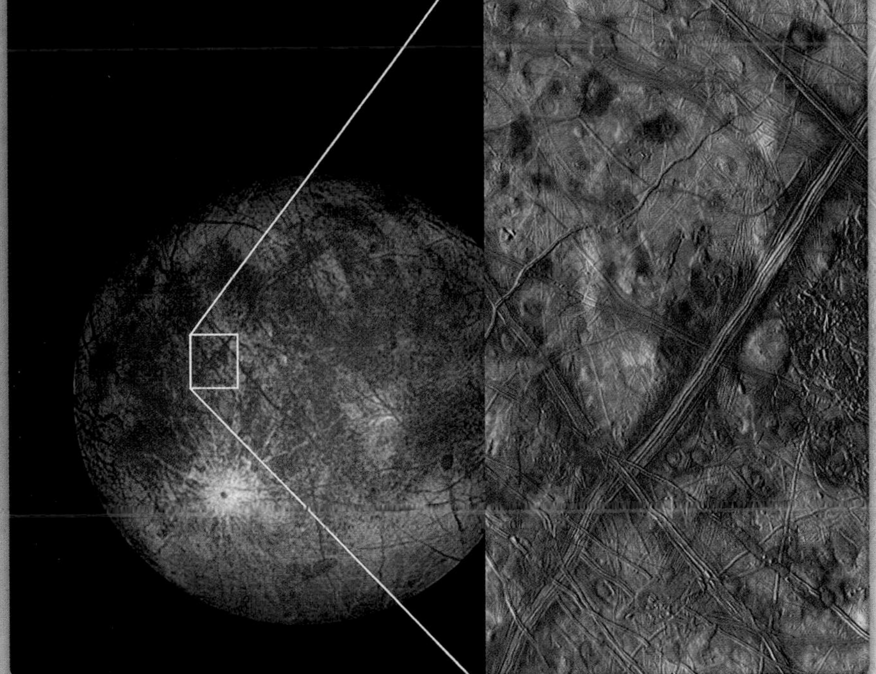

CHAPTER

1

Exploring Life and Science

CASE STUDY THE SEARCH FOR LIFE

What do Europa, Titan, and Earth all have in common? Besides being part of our solar system, they are all at the frontline of our species' effort to understand the nature of life.

Europa (shown above) is one of the larger moons of Jupiter, and it has had held a special fascination for astronomers since Galileo first described it in 1610. Today, Europa is one of the prime candidates to harbor life outside of Earth. Geologists believe that under Europa's icy exterior lies a vast ocean of water. Having analyzed past comet impact sites on the surface of Europa, geologists also believe that this ocean contains the basic ingredients for life, including carbon and possibly even free oxygen. Europa's ocean is warmed by a constant gravitational tug of war between it and Jupiter, and the ocean is protected by an ice casing almost 19 kilometers thick.

Titan is the second-largest satellite in the solar system, larger than even our moon. Although it is in orbit around Saturn, and thus located some distance from the influence of the sun, Titan has become a focal point for the study of extraterrestrial life since the NASA space probe *Cassini–Huygens* first arrived at Saturn in 2004. *Cassini* has detected the presence of the building blocks of life on Titan, including lakes of methane and ammonia, and vast deposits of hydrogen and carbon compounds called hydrocarbons.

On Earth, scientists are exploring the extreme environments near volcanoes and deep-sea thermal vents to get a better picture of what life may have looked like under the inhospitable conditions that dominated at the time we now know life first began on our planet. Already, in the past few years, marine biologists have discovered new forms of life that cannot only live off of hydrogen sulfide, a deadly gas to most life, but also thrive under extreme pressure and temperatures.

As you read through the chapter, think about the following questions:

1. What are the basic characteristics that define life?
2. What evidence would you look for on Europa or Titan that would tell you that life may have existed on these moons in the past?
3. What does it tell us if we discover life on Europa or Titan and it has similar characteristics to life on Earth? What if it is very different?

CHAPTER CONCEPTS

1.1 The Characteristics of Life
The process of evolution accounts for the diversity of living organisms and explains why all life shares the same basic characteristics.

1.2 Humans Are Related to Other Animals
Humans are eukaryotes and are further classified as mammals in the animal kingdom. We differ from other mammals, including apes, by our highly developed brain, upright stance, creative language, and the ability to use a wide variety of tools.

1.3 Science as a Process
Biologists use the scientific process when they study the natural world. A hypothesis is formulated and tested to arrive at a conclusion. Theories explain how the natural world is organized.

1.4 Making Sense of a Scientific Study
Data are more easily understood if results are presented in the form of a graph and are accompanied by a statistical analysis.

1.5 Science and Social Responsibility
Scientific investigations and technology have always been influenced by human values. Everyone has a responsibility to ensure that science and technology are used for the good of all.

1.1 The Characteristics of Life

LEARNING OUTCOMES

Upon completion of this section, you should be able to

1. Explain the basic characteristics that are common to all living organisms.
2. Describe the levels of organization of life.
3. Summarize how the terms homeostasis, metabolism, development, and adaptation all relate to living organisms.
4. Explain why the study of evolution is important in understanding life.

The science of **biology** is the study of living organisms and their environments. All living organisms (Fig. 1.1) share several basic characteristics. They (1) are organized, (2) acquire materials and energy, (3) are homeostatic, (4) respond to stimuli, (5) reproduce and grow, and (6) have an evolutionary history.

MP3 Life Characteristics

Organisms Are Organized

Figure 1.2 illustrates the levels of biological organization. Note that, at the bottom of the figure, **atoms** join together to form the **molecules** that make up a cell. A **cell** is the smallest structural and functional unit of an organism. Some organisms, such as bacteria, are single cells. Humans are *multicellular* because they are composed of many different types of cells. A nerve cell is one of the types of cells in the human body. It has a structure suitable to conducting a nerve impulse.

A **tissue** is a group of similar cells that perform a particular function. Nervous tissue is composed of millions of nerve cells that transmit signals to all parts of the body. Several types of tissues make up an **organ,** and each organ belongs to an **organ system.** The organs of an organ system work together to accomplish a common purpose. The brain works with the spinal cord to send commands to body parts by way of nerves. **Organisms,** such as trees and humans, are a collection of organ systems.

The levels of biological organization extend beyond the individual. All the members of one **species** (a group of interbreeding organisms) in a particular area belong to a **population.** A tropical grassland may have a population of zebras, acacia trees, and humans, for example. The interacting populations of the grasslands make up a **community.** The community of populations interacts with the physical environment to form an **ecosystem.** Finally, all the Earth's ecosystems make up the **biosphere** (Fig. 1.2, *top*).

Figure 1.1 **All life shares common characteristics.**
From the simplest one-celled organisms to complex plants and animals, all life shares several basic characteristics.

Figure 1.2 Levels of biological organization.
Life is connected from the atomic level to the biosphere. While the cell is the basic unit of life, it is comprised of molecules and atoms. The sum of all life on the planet is called the biosphere.

Biosphere
Regions of the Earth's crust, waters, and atmosphere inhabited by living things

Ecosystem
A community plus the physical environment

Community
Interacting populations in a particular area

Population
Organisms of the same species in a particular area

Organism
An individual; complex individuals contain organ systems

human

tree

Organ System
Composed of several organs working together

nervous system

shoot system

Organ
Composed of tissues functioning together for a specific task

the brain

leaves

Tissue
A group of cells with a common structure and function

nervous tissue

epidermal tissue

Cell
The structural and functional unit of all living things

nerve cell

plant cell

Molecule
Union of two or more atoms of the same or different elements

methane

Atom
Smallest unit of an element composed of electrons, protons, and neutrons

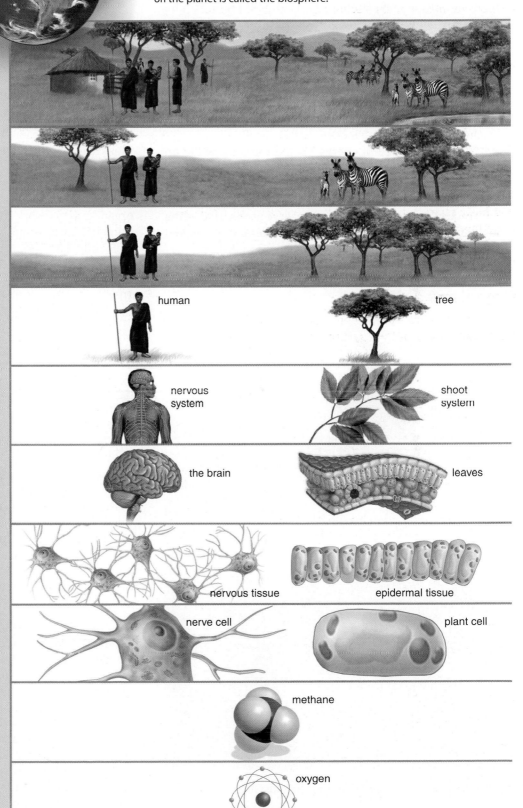

oxygen

Organisms Acquire Materials and Energy

Humans, like all living organisms, cannot maintain their organization or carry on life's activities without an outside source of materials and energy. Humans and other animals acquire materials and energy when they eat food (Fig. 1.3).

Food provides nutrient molecules, which are used as building blocks or for energy. It takes energy (work) to maintain the organization of the cell and of the organism. Some nutrient molecules are broken down completely to provide the necessary energy to convert other nutrient molecules into the parts and products of cells. The term **metabolism** describes all of the chemical reactions that occur within a cell.

The ultimate source of energy for the majority of life on Earth is the sun. Plants, algae, and some bacteria are able to harvest the energy of the sun and convert it to chemical energy by a process called **photosynthesis.** Photosynthesis produces organic molecules, such as sugars, that serve as the basis of the food chain for many other organisms, including humans and all other animals.

a.

b.

Figure 1.3 Humans and other animals must acquire energy. All life, including humans (**a**) and other animals such as this red-tailed hawk (**b**) must acquire energy to survive. The method by which organisms acquire energy is dependent on the species.

APPLICATIONS AND MISCONCEPTIONS

How many cells are in your body?

The number of cells in a human body varies depending on the size of the person and whether cells have been damaged or lost. However, most estimates suggest that there are well over 100 trillion cells in a human body.

Organisms Maintain Homeostasis

For the metabolic pathways within a cell to function correctly, the environmental conditions of the cell must be kept within strict operating limits. The ability of a cell or an organism to maintain an internal environment that operates under specific conditions is called **homeostasis.** In humans, many of our organ systems work to maintain homeostasis. For example, human body temperature normally fluctuates slightly between 36.5 and 37.5°C (97.7 and 99.5°F) during the day. In general, the lowest temperature usually occurs between 2 A.M. and 4 A.M., and the highest usually occurs between 6 P.M. and 10 P.M. However, activity can cause the body temperature to rise, and inactivity can cause it to decline. A number of body systems, including the cardiovascular system and the nervous system, work together to maintain a constant temperature. However, the body's ability to maintain a normal temperature is somewhat dependent on the external temperature. Even though we can shiver when we are cold and perspire when we are hot, we will die if the external temperature becomes overly cold or hot.

This text emphasizes how all the systems of the human body help maintain homeostasis. For example, the digestive system takes in nutrients, and the respiratory system exchanges gases with the environment. The cardiovascular system distributes nutrients and oxygen to the cells and picks up their wastes. The metabolic waste products of cells are excreted by the urinary system. The work of the nervous and endocrine systems is critical because these systems coordinate the functions of the other systems. Throughout the text, the Connecting the Concepts sections will provide you with links to more information on homeostasis.

Organisms Respond to Stimuli

Homeostasis would be impossible without the ability of the body to respond to stimuli. Response to external stimuli is more apparent to us because it involves movement, as when we quickly remove a hand from a hot stove. Certain sensory receptors also detect a change in the internal environment, and then the central nervous system brings about an appropriate response. When you are startled by a loud noise, your heartbeat increases, which causes your blood pressure to increase. If blood pressure rises too high, the brain directs blood vessels to dilate, helping to restore normal blood pressure.

All life responds to external stimuli, often by moving toward or away from a stimulus, such as the sight of food. Organisms may use a variety of mechanisms to move, but movement in humans and other animals is dependent upon their nervous and musculoskeletal systems. The leaves of plants track the passage of the sun during the day; when a houseplant is placed near a window, its stems bend to face the sun. The movement of an animal, whether self-directed or in response to a stimulus, constitutes a large part of its behavior. Some behaviors help us acquire food and reproduce.

Organisms Reproduce and Grow

Reproduction is a fundamental characteristic of life. Cells come into being only from pre-existing cells, and all living things have parents. When organisms **reproduce,** they pass on their genetic information to the next generation. Following the fertilization of the egg by a sperm cell, the resulting zygote undergoes a rapid period of growth and development. This is common in almost all living organisms. Figure 1.4*a* illustrates that an acorn progresses to a seedling before it becomes an adult oak tree. In humans, growth occurs as the fertilized egg develops into a fetus (Fig. 1.4*b*). **Growth,** recognized by an increase in size and often the number of cells, is a part of development. In multicellular organisms, such as humans, the term **development** is used to indicate all the changes that occur from the time the egg is fertilized until death. Therefore, it includes all the changes that occur during childhood, adolescence, and adulthood. Development also includes the repair that takes place following an injury.

The genetic information of all life is deoxyribonucleic acid, or DNA. DNA contains the hereditary information that directs not only the structure of each cell but also its function. The information in the DNA is contained within **genes,** short sequences of hereditary material that specify the instructions for a specific trait. Before reproduction occurs, DNA is replicated so that an exact copy of each gene may be passed on to the offspring. When humans reproduce, a sperm carries genes contributed by a male into the egg, which contains genes contributed by a female. The genes direct both growth and development so that the organism will eventually resemble the parents. Sometimes, **mutations** may cause minor variations in these genes, potentially causing an organism to be better suited for its environment. These mutations are the basis of evolutionary change.

a.

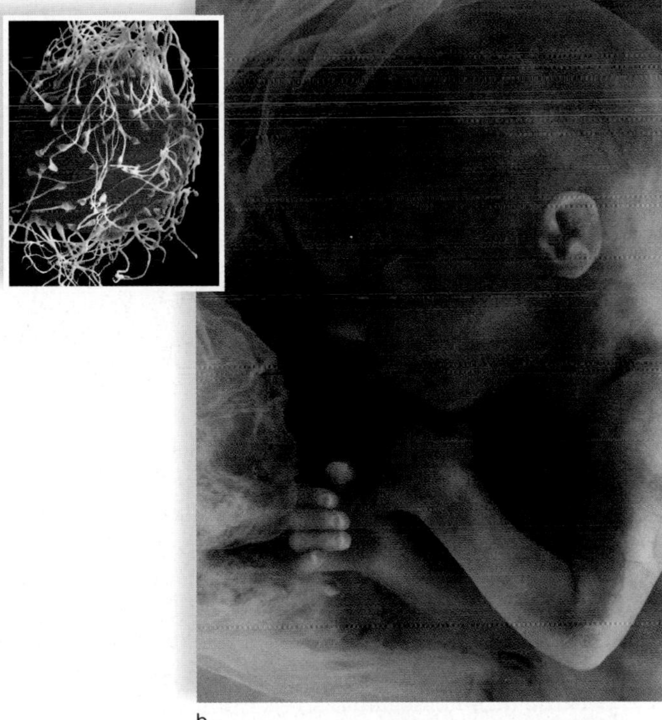

b.

Figure 1.4 Growth and development define life.
a. A small acorn becomes a tree, and (**b**) following fertilization, an embryo becomes a fetus by the process of growth and development.

Organisms Have an Evolutionary History

Evolution is the process by which a population changes over time. When a new variation arises that allows certain members of a population to capture more resources, these members tend to survive and have more offspring than the other, unchanged members. Therefore, each successive generation will include more members with the new variation, which represents an **adaptation** to the environment. Consider, for example, populations of humans that live at high altitudes, such as the cultures living at elevations of over 4,000 meters (m) (14,000 ft) in the Tibetan Plateau. This environment is very low in oxygen. As the Science feature, "Adapting to Life at High Elevations," investigates, these populations have evolved an adaptation that actually reduces the amount of hemoglobin, the oxygen-carrying pigment in the blood. As the feature explains, this adaptation makes life at these altitudes possible. Evolution, which has been going on since the origin of life and which will continue as long as life exists, explains both the unity and the diversity of life. All organisms share the same characteristics of life because their ancestry can be traced to the first cell or cells. Organisms are diverse because they are adapted to different ways of life.

BIOLOGY MATTERS  Science

Adapting to Life at High Elevations

Humans, like all other organisms, have an evolutionary history. This not only means that we share common ancestors with other animals, but that over time, we demonstrate adaptations to changing environmental conditions. One such study of populations living in the high-elevation mountains of Tibet (Fig. 1A) demonstrates how the processes of evolution and adaptation influence humans.

Figure 1A
Individuals living at high elevations, such as these Tibetans, have become adapted to their environment.

Normally, if a person moves to a higher altitude, his or her body responds by making more hemoglobin, the component of blood that carries oxygen, which thickens the blood. For minor elevation changes, this does not present much of a problem. But for people that live at extreme elevations (some people in the Himalayas can live at elevations of over 13,000 ft, or close to 4,000 m), this can present a number of health problems, including chronic mountain sickness, a disease that affects people who live at high altitudes for extended periods of time. The problem is that, as the amount of hemoglobin increases, the blood thickens and becomes more viscous. This can cause problems with elevated blood pressure, or hypertension, and an increase in the formation of blood clots, all of which have negative physiological effects.

Because high hemoglobin levels would be a detriment to people at high elevations, it makes sense that natural selection would favor individuals that produced less hemoglobin at high elevations. Such is the case with the Tibetans in this study. Researchers have identified an allele of a gene that reduces hemoglobin production at high elevations. Comparisons between Tibetans at both high and low elevations strongly suggest that selection has played a role in the prevalence of the high-elevation allele.

The gene is *EPSA1*, a gene on chromosome 2 of humans. *EPSA1* produces a transcription factor, which basically acts as a regulator of which genes are turned on and off in the body, a process called gene expression. The transcription factor produced by *EPSA1* has a number of functions in the body. For example, in addition to controlling the amount of hemoglobin in the blood, this transcription factor also regulates other genes that direct how the body uses oxygen.

When the researchers examined the variations in *EPSA1* in the Tibetan population, they discovered that their version greatly reduces the production of hemoglobin. Therefore, the Tibetan population has lower hemoglobin levels than people living at lower altitudes, thus allowing these individuals to escape the consequences of thick blood.

How long did it take for the original population to adapt to live at higher elevations? Modern genetic analysis allows scientists to model the rate of evolutionary change. By comparing these genes in high-elevation and low-elevation Tibetan populations, the researchers were able to determine that the process probably occurred over less than 3,000 years. If we use a 25-year generation time for humans, that means that the adaptation to the high-elevation environment probably took less than 120 generations for this Tibetan population.

Questions to Consider

1. What other environments do you think could be studied to look for examples of human adaptation?
2. In addition to hemoglobin levels, do you think that people at high elevations may exhibit other adaptations?

CHECK YOUR PROGRESS 1.1

1. Describe the basic characteristics of life.
2. Summarize the levels of biological organization.
3. Explain the relationship between adapations and evolutionary change.

CONNECTING THE CONCEPTS

Both homeostasis and evolution are central themes in the study of biology. For more examples of homeostasis and evolution, refer to the following discussions:

Section 4.8 explains how body temperature is regulated.

Section 10.4 explores the role of the kidneys in fluid and salt homeostasis.

Section 22.3 examines the evolutionary history of humans.

1.2 Humans Are Related to Other Animals

LEARNING OUTCOMES

Upon completion of this section, you should be able to

1. Summarize the place of humans in the overall classification of living organisms.
2. Describe the relationship between humans and the biosphere, and the role of culture in shaping that relationship.

Biologists classify all life as belonging to one of three **domains.** The evolutionary relationships of these domains are presented in Figure 1.5. Two of these domains, domain Bacteria and domain Archaea, contain prokaryotes, one-celled organisms that lack a nucleus. Organisms in the third domain, Eukarya, are classified as being members of one of four

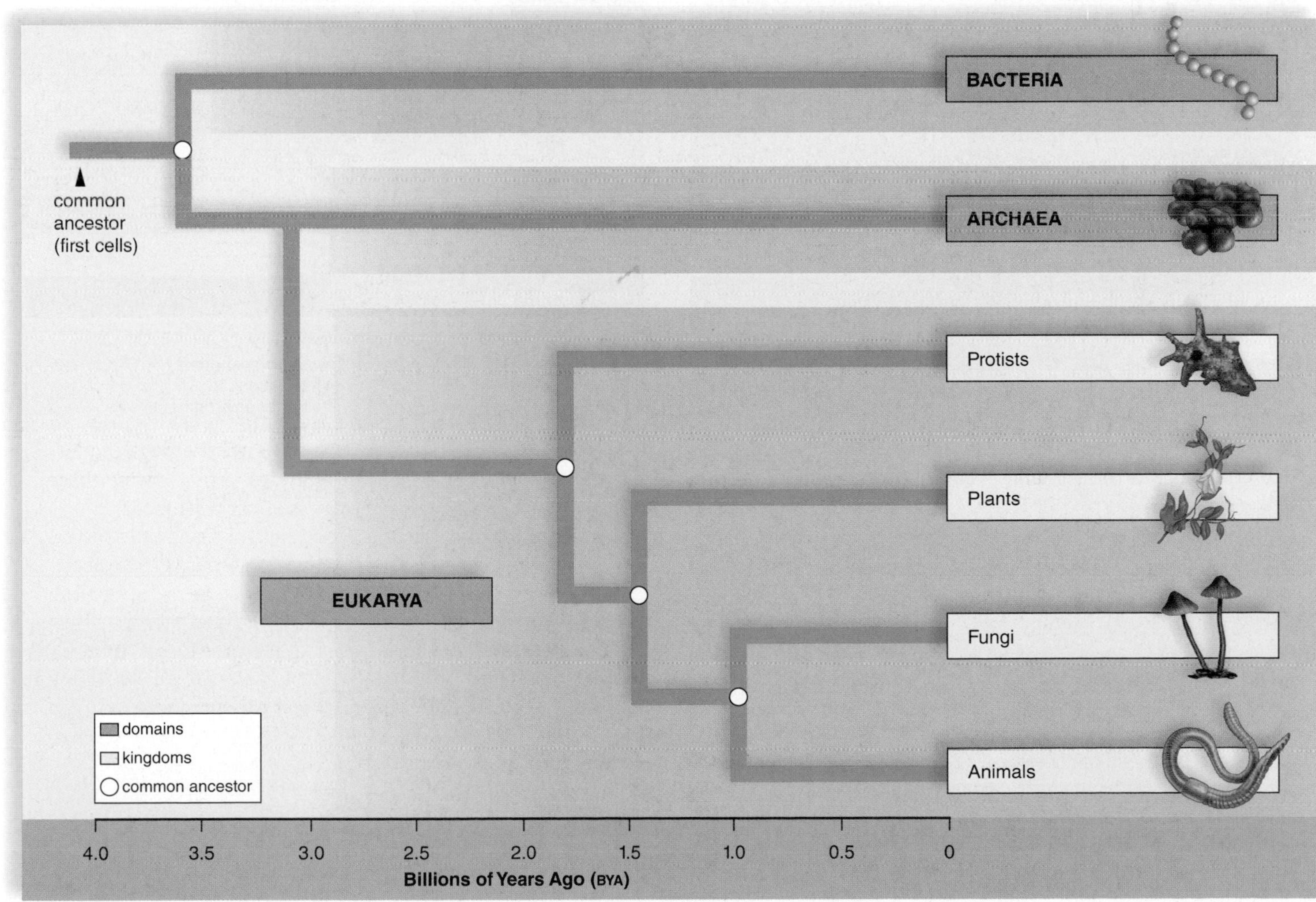

Figure 1.5 **The evolutionary relationships of the three domains of life.**
Living organisms are classified into three domains: Bacteria, Archaea, and Eukarya. The Eukarya are further divided into kingdoms (see Fig. 1.6). A geologic timescale is provided on the bottom for reference.

kingdoms (Fig. 1.6)—plants (Plantae), fungi (Fungi), animals (Animalia), and protists (Protista). Most organisms in kingdom Animalia are *invertebrates,* such as earthworms, insects, and mollusks. **Vertebrates** are animals that have a nerve cord protected by a vertebral column, which gives them their name. Fish, reptiles, amphibians, and birds are all vertebrates. Vertebrates with hair or fur and mammary glands are classified as mammals. Humans, raccoons, seals, and meerkats are examples of mammals.

Humans are most closely related to apes. We are distinguished from apes by our (1) highly developed brains, (2) completely upright stance, (3) creative language, and (4) ability to use a wide variety of tools. Humans did not evolve from apes; apes and humans share a common, apelike ancestor.

Today's apes are our evolutionary cousins. Our relationship to apes is analogous to you and your first cousin being descended from your grandparents. We could not have evolved from our cousins because we are contemporaries—living on Earth at the same time.

Humans Have a Cultural Heritage

Humans have a cultural heritage in addition to a biological heritage. *Culture* encompasses human activities and products passed on from one generation to the next outside of direct biological inheritance. Among animals, only humans have a language that allows us to communicate information and experiences symbolically. We are born without knowledge of an accepted way to behave, but we gradually acquire this knowledge by adult instruction and imitation of role models. Members of the previous generation pass on their beliefs, values, and skills to the next generation. Many of the skills involve

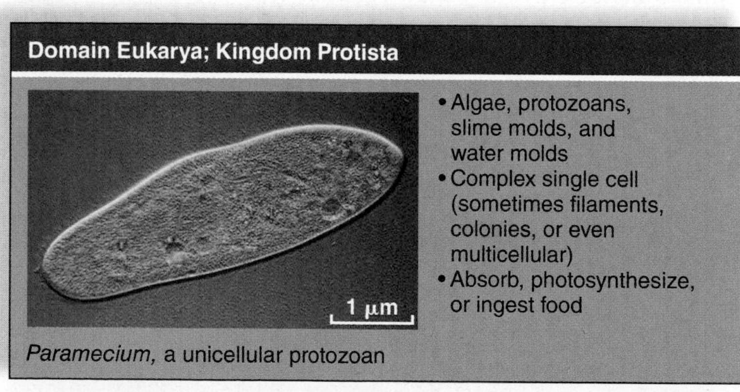

Domain Eukarya; Kingdom Protista
1 μm
• Algae, protozoans, slime molds, and water molds
• Complex single cell (sometimes filaments, colonies, or even multicellular)
• Absorb, photosynthesize, or ingest food
Paramecium, a unicellular protozoan

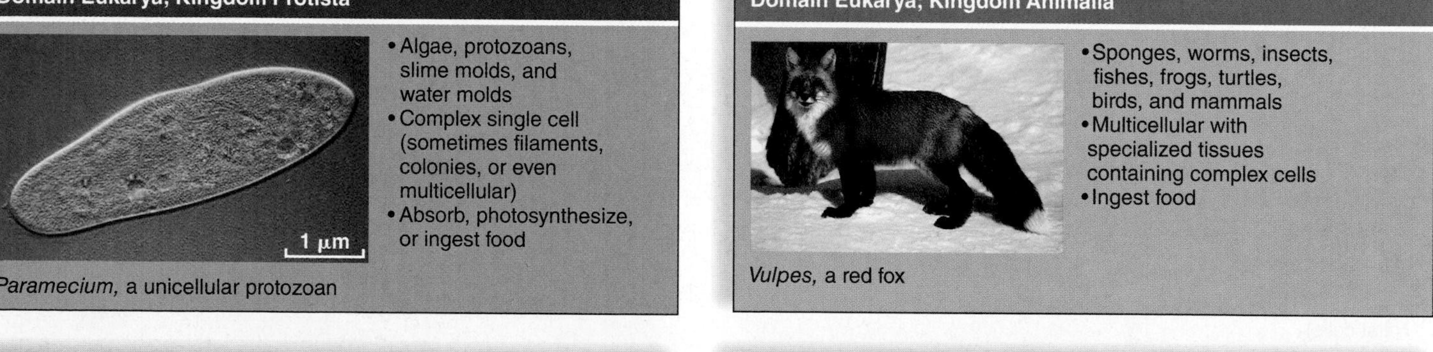

Domain Eukarya; Kingdom Animalia
• Sponges, worms, insects, fishes, frogs, turtles, birds, and mammals
• Multicellular with specialized tissues containing complex cells
• Ingest food
Vulpes, a red fox

Domain Eukarya; Kingdom Fungi
• Molds, mushrooms, yeasts, and ringworms
• Mostly multicellular filaments with specialized, complex cells
• Absorb food
Cantharellula, a club fungi

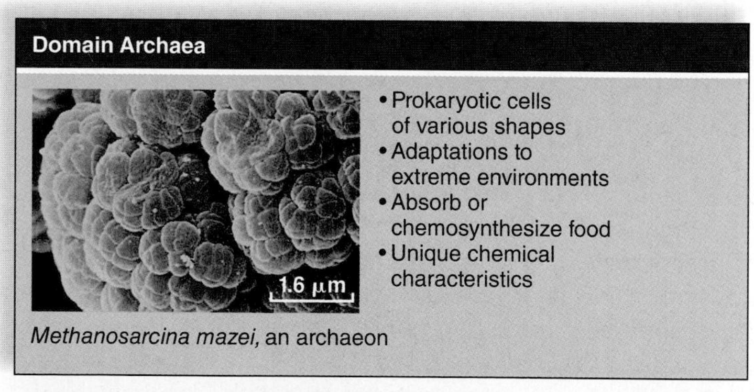

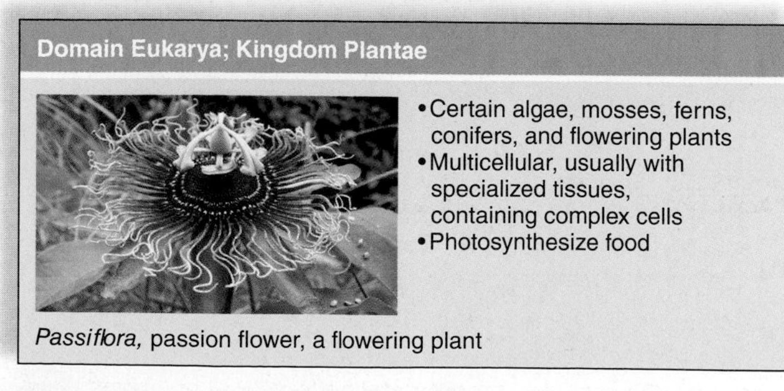

Domain Eukarya; Kingdom Plantae
• Certain algae, mosses, ferns, conifers, and flowering plants
• Multicellular, usually with specialized tissues, containing complex cells
• Photosynthesize food
Passiflora, passion flower, a flowering plant

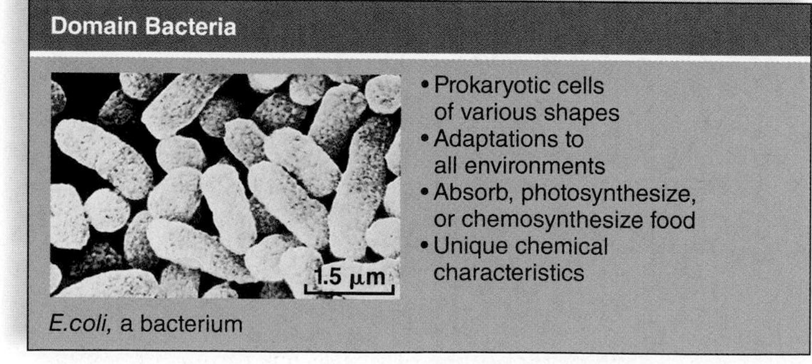

Domain Bacteria
1.5 μm
• Prokaryotic cells of various shapes
• Adaptations to all environments
• Absorb, photosynthesize, or chemosynthesize food
• Unique chemical characteristics
E.coli, a bacterium

Figure 1.6 **The classification of life.**
This figure provides some of the characteristics of the organisms of each of the major domains and kingdoms of life. Humans belong to the domain Eukarya and kingdom Animalia.

tool use, which can vary from how to hunt in the wild to how to use a computer. Human skills have also produced a rich heritage in the arts and sciences. However, a society highly dependent on science and technology has its drawbacks as well. Unfortunately, this cultural development may mislead us into believing that humans are somehow not part of the natural world surrounding us.

Humans Are Members of the Biosphere

All life on Earth is part of the biosphere, a living network that spans the surface of the Earth into the atmosphere and down into the soil and seas. Although humans can raise animals and crops for food, we depend on the environment for many services. Without microorganisms that decompose, the waste we create would soon cover the Earth's surface. Some species of bacteria can clean up pollutants like heavy metals and pesticides.

Freshwater ecosystems, such as rivers and lakes, provide fish to eat, drinking water, and water to irrigate crops. Many of our crops and prescription drugs were originally derived from plants that grew naturally in an ecosystem. Some human populations around the globe still depend on wild animals as a food source. The water-holding capacity of forests prevents flooding, and the ability of forests and other ecosystems to retain soil prevents soil erosion. For many people, these same forests provide a place for recreational activities like hiking and camping.

Humans Threaten the Biosphere

The human population tends to modify existing ecosystems for its own purposes. Humans clear forests and grasslands to grow crops. Later, houses are built on what was once farmland. Clusters of houses become small towns that often grow into cities (Fig. 1.7*a*). The overuse of water supplies by large human populations can result in desertification, or the expansion of desert regions (Fig. 1.7*b*). Human activities have altered almost all ecosystems and reduced **biodiversity,** or the number of different species present in a given area. The present biodiversity of our planet has been estimated to be as high as 15 million species. So far, fewer than 2 million have been identified and named. It is estimated that we are now losing as many as 20,000 to 30,0000 species per year due to human activities. Many biologists are alarmed about the present rate of **extinction** (death of a species). Studies are suggesting that the rate of species loss is now greater than the rates of the five mass extinctions that occurred earlier in our planet's history. Unlike previous extinctions, which were caused by asteroid impacts or changes in the chemistry of the atmosphere, the current extinction crisis has been conclusively linked to human activity.

One of the major bioethical issues of our time is preservation of the biosphere and biodiversity. If we adopt a conservation ethic that preserves the biosphere and biodiversity, we will ensure the continued existence of our species.

a.

b.

Figure 1.7 Humans negatively influence many ecosystems.
a. When humans build cities, diversity is lost. Notice the absence of a variety of plants/trees. **b.** An overuse of water resources can lead to desertification.

APPLICATIONS AND MISCONCEPTIONS

How many humans are there?

As of the end of 2011, it was estimated that there were over 7 billion humans on the planet. Each of those humans needs food, shelter, clean water and air, and materials to maintain a healthy lifestyle. We add an additional 78 million people per year—that is like adding ten New York Cities per year! This makes human population growth one of the greatest threats to the biosphere.

CHECK YOUR PROGRESS 1.2

1. Define the term *biosphere.*
2. Explain why it is important to know the evolutionary relationships between organisms.
3. Summarize how the increase in the human population affects our biosphere.

CONNECTING THE CONCEPTS

To learn more about the preceding material, refer to the following discussions:

Chapter 22 examines recent developments in the study of human evolution.

Chapter 23 provides a more detailed look at ecosystems.

Chapter 24 details some of the emerging threats that humans pose to the biosphere.

1.3 Science as a Process

LEARNING OUTCOMES

Upon completion of this section, you should be able to

1. Describe the general process of the scientific method.
2. Distinguish between a control group and an experimental group in a scientific test.
3. Recognize the importance of scientific journals in the reporting of scientific information.

Science is a way of knowing about the natural world. When scientists study the natural world, they aim to be objective, rather than subjective. Objective observations are supported by factual information, whereas subjective observations involve personal judgment. For example, the fat content of a particular food would be an objective observation of a nutritional study. Reporting about the good or bad taste of the food would be a subjective observation. It is difficult to make objective observations and conclusions because we are often influenced by our prejudices. Scientists must keep in mind that scientific conclusions can change because of new findings. New findings are often made because of recent advances in techniques or equipment.

Importance of Scientific Theories in Biology

Science is not just a pile of facts. The ultimate goal of science is to understand the natural world in terms of scientific theories. **Scientific theories** are concepts that tell us about the order and the patterns within the natural world—in other words, how the natural world is organized. For example, following are some of the basic theories of biology:

Theory	Concept
Cell	All organisms are composed of cells, and new cells only come from pre-existing cells.
Homeostasis	The internal environment of an organism stays relatively constant.
Genes	Organisms contain coded information that dictates their form, function, and behavior.
Ecosystem	Populations of organisms interact with each other and the physical environment.
Evolution	All organisms have a common ancestor, but each is adapted to a particular way of life.

Evolution is the unifying concept of biology because it makes sense of what we know about the nature of life. For example, the theory of evolution enables scientists to understand the tremendous diversity of living things and their relationships. It explains common structural features, physiology, patterns of development, and behaviors. The theory of evolution has been supported by so many observations and experiments for over a hundred years that some biologists refer to it as the **principle** of evolution. This term is the preferred terminology for theories generally accepted as valid by an overwhelming number of scientists.

The Scientific Method Has Steps

Unlike other types of information available to us, scientific information is acquired by a process known as the **scientific method.** The approach of individual scientists to their work is as varied as the scientists. For the sake of discussion, it is possible to speak of the scientific method as consisting of certain steps (Fig. 1.8).

After making initial observations, a scientist will, most likely, study any previous **data,** results, and conclusions reported by previous research. Imagination and creative thinking also help a scientist formulate a **hypothesis.** The hypothesis becomes the basis for more observation and/or experimentation. The new data help a scientist come to a **conclusion** that

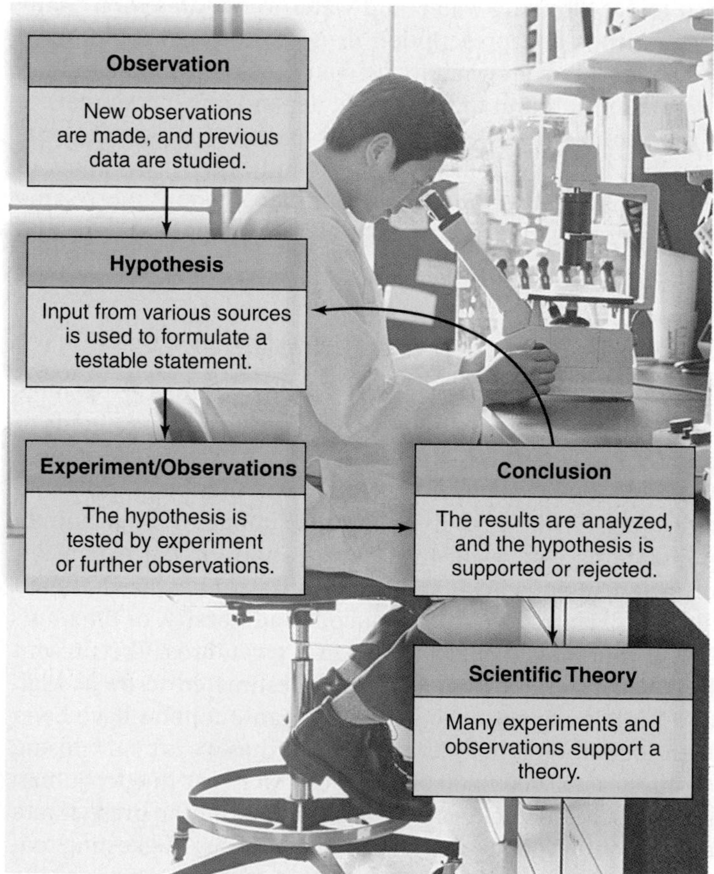

Figure 1.8 **The scientific method.**
On the basis of new and/or previous observations, a scientist formulates a hypothesis. The hypothesis is tested by further observations and/or experiments, and new data either support or do not support the hypothesis. The return arrow indicates that a scientist often chooses to retest the same hypothesis or to test a related hypothesis. Conclusions from many different but related experiments may lead to the development of a scientific theory. For example, studies pertaining to development, anatomy, and fossil remains all support the theory of evolution.

either supports or does not support the hypothesis. Hypotheses are always subject to modification, so they can never be proven true; however, they can be proven untrue. When the hypothesis is not supported by the data, it must be rejected; therefore, some think of the body of science as what is left after alternative hypotheses have been rejected.

Science is different from other ways of knowing by its use of the scientific method to examine a phenomenon. Any suggestions about the natural world not based on data gathered by employing the scientific method cannot be accepted as within the realm of science. Scientific theories are concepts based on a wide range of observations and experiments.

How the Cause of Ulcers Was Discovered

Let's take a look at how the cause of ulcers was discovered so we can get a better idea of how the scientific method works. In 1974, Barry James Marshall was a young resident physician at Queen Elizabeth II Medical Center in Perth, Australia. There he saw many patients who had bleeding stomach ulcers. A pathologist at the hospital, Dr. J. Robin Warren, told him about finding a particular bacterium, now called *Helicobacter pylori,* near the site of peptic ulcers (open sores in the stomach). Using the computer networks available at that time, Marshall compiled much data showing a possible correlation between the presence of *H. pylori* and the occurrence of both gastritis (inflammation of the stomach) and stomach ulcers. On the basis of these data, Marshall formulated a hypothesis: *H. pylori* is the cause of gastritis and ulcers.

Marshall decided to make use of Koch's postulates, the standard criteria that must be fulfilled to show that a pathogen (bacterium or virus) causes a disease (see the Science feature, "Robert Koch," below).

The First Two Criteria

By 1983, Marshall had fulfilled the first and second of Koch's criteria. He was able to isolate *H. pylori* from ulcer patients and grow it in the laboratory. (Success was achieved only after a petri dish was inadvertently left in the incubator for six, instead of two, days.) Further, he had determined that bismuth, the active ingredient in Pepto-Bismol, could destroy the bacteria in a petri dish.

Despite presentation of these findings to the scientific community, most physicians continued to believe that stomach acidity and stress were the causes of stomach ulcers. In those days, patients were usually advised to make drastic changes in their lifestyle or seek psychiatric counseling to "cure" their ulcers. Many scientists believed that no bacterium would be able to survive the normal acidity of the stomach.

The Last Two Criteria

Marshall had a problem in fulfilling the third and fourth of Koch's criteria. He had been unable to infect guinea pigs and rats with the bacteria because the bacteria just did not flourish in the intestinal tract of those animals. Marshall was not able to use human subjects because our society does not condone the use of humans as experimental subjects in dangerous or life-threatening research. Marshall was so determined to support his hypothesis that, in 1985, he decided to perform the experiment on himself! To the disbelief of those in the lab that day, he and another volunteer swallowed a foul-smelling and

BIOLOGY MATTERS  **Science**

Robert Koch

Robert Koch (1843–1910) was a German microbiologist who helped verify the germ theory of disease and established the standard as to whether an organism causes a particular disease. Koch was a scientist who studied anthrax, a type of soil-living bacterium that causes disease in animals and humans. Anthrax was used as a biological weapon in the United States by unknown persons following the terrorist attacks in 2001. Koch expanded on Louis Pasteur's work by establishing a series of steps, now known as Koch's postulates, to identify whether a specific bacterium is responsible for a disease.

Koch's Postulates

- The suspected pathogen (virus or bacterium) must be present in every case of the disease;
- The pathogen must be isolated from the host and grown in a lab dish;
- The disease must be reproduced when a pure culture of the pathogen is inoculated into a healthy susceptible host; and
- The same pathogen must be recovered again from the experimentally infected host.

This procedure is still widely used by scientists and medical professionals. Koch is credited not only with his work on the germ theory but also with the development of several important scientific instruments. He was one of the first to use an incubator, an instrument that maintains a steady temperature for growing bacteria outside of an organism. He also was the first to use gelatin as a medium for growing bacteria in dishes.

Questions to Consider

1. How does Koch's procedure relate to the scientific method?
2. When might these procedures not necessarily be followed?

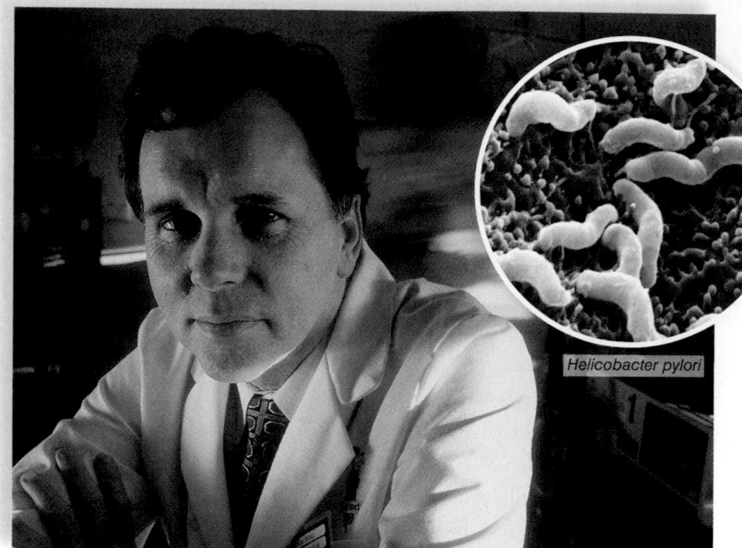

Figure 1.9 **Dr. Barry Marshall and the cause of stomach ulcers.**
Dr. Barry Marshall, pictured here, fulfilled Koch's postulates to show that *Helicobacter pylori* is the cause of peptic ulcers. The inset shows the presence of the bacterium in the stomach.

foul-tasting solution of *H. pylori.* Within the week, they felt lousy and were vomiting up their stomach contents. Examination by endoscopy showed that their stomachs were now inflamed, and biopsies of the stomach lining contained the suspected bacterium (Fig. 1.9). Their symptoms abated without need for medication, and they never developed an ulcer. At Marshall's next talk, he challenged his audience to refute his hypothesis. Many tried, but ultimately the investigators supported his findings.

The Conclusion

In science, many experiments, often involving a considerable number of subjects, are required before a conclusion can be reached. By the early 1990s, at least three independent studies involving hundreds of patients had been published showing that antibiotic therapy could eliminate *H. pylori* from the intestinal tract and cure patients of ulcers wherever they occurred in the tract.

Dr. Marshall received all sorts of prizes and awards, but he and Dr. Warren were especially gratified to receive a Nobel Prize in Medicine in 2005. The Nobel committee reportedly thanked Marshall and Warren for their "pioneering discovery," stating that peptic ulcer disease now could be cured with antibiotics and acid-secretion inhibitors rather than becoming a "chronic, frequently disabling condition."

How to Do a Controlled Study

The work that Marshall and Warren did was largely observational. Often, scientists perform an **experiment,** a series of procedures to test a hypothesis. As an example, let's say investigators want to determine which of two antibiotics best treats an ulcer. When scientists do an experiment, they try to vary just the **experimental variables,** in this case, the medications being tested. A **control group** is not given the medications, but one or more **test groups** are given the medications. If, by chance, the control group shows the same results as a test group, the investigators immediately know the results of their study are invalid because it would mean the medications may have nothing to do with the results.

The study depicted in Figure 1.10 shows how investigators may study this hypothesis.

Hypothesis: Newly discovered antibiotic B is a better treatment for ulcers than antibiotic A, which is in current use.

Investigators who perform clinical research must obtain informed consent from their subjects before proceeding with the research. The informed consent ensures that subjects know details about the research and that their participation is voluntary. The risks and benefits involved in participating in the research are all outlined. It is important to reduce the number of possible variables (differences) such as sex, weight, other illnesses, and so forth between the groups. Therefore, the investigators *randomly* divide a very large group of volunteers (Fig. 1.10*a*) equally into the three groups. The hope is that any differences will be distributed evenly among the three groups. This is more likely to occur if the investigators have a large number of volunteers.

The three groups will be treated (Fig. 1.10*b*) as follows:

Control group: Subjects with ulcers are not treated with either antibiotic.
Test group 1: Subjects with ulcers are treated with antibiotic A.
Test group 2: Subjects with ulcers are treated with antibiotic B.

After the investigators have determined that all volunteers do suffer from ulcers, they will want the subjects to think they are all receiving the *same* treatment. This is an additional way to protect the results from any influence other than the medication. To achieve this end, the subjects in the control group can receive a **placebo,** a treatment that appears to be the same as that administered to the other two groups but contains no medication. In this study, the use of a *placebo* would help ensure the same dedication by all subjects to the study.

The Results

After two weeks of administering the same amount of medication (or placebo) in the same way, the intestinal tract of each subject is examined to determine if ulcers are still present. Endoscopy, depicted in the photograph in Figure 1.10*c,* is one possible way to examine a patient for the presence of ulcers. This procedure is performed under sedation and involves inserting an endoscope—a small, flexible tube with a tiny camera on the end—down the throat and into the stomach. It allows the doctor to see the lining of the stomach and check for possible ulcers. Tests performed during an endoscopy can also determine if *H. pylori* is present.

Endoscopy is somewhat subjective so it is probably best if the examiner is not aware of which group the subject is in. Otherwise, the prejudice of the examiner may influence the examination. When neither the patient nor the examiner is aware of the specific treatment, it is called a *double-blind* study.

In this study, the investigators may decide to determine effectiveness of the medication by the percentage of people who no longer have ulcers. So, if 20 people out of 100 still have ulcers, the medication is 80% effective. The difference in effectiveness is easily read in the graph portion of Figure 1.10*d*.

Conclusion: On the basis of their data, the investigators conclude that their hypothesis has been supported.

Publication of Scientific Studies

Scientific studies are customarily published in a scientific journal, so that all aspects of a study are available to the scientific community. Before information is published in scientific journals, it is typically reviewed by experts. These people ensure that the research is credible, accurate, unbiased, and well executed.

Another scientist should be able to read about an experiment in a scientific journal, repeat the experiment in a different location, and get the same (or very similar) results. Each article begins with a short synopsis of the study so scientists can quickly find the articles of greatest interest to them. The materials and methods of performing each study are clearly outlined so that researchers can more easily repeat the work. Some articles are rejected for publication by reviewers when they believe there is something questionable about the design or manner in which an experiment was conducted.

Further Study

As mentioned previously, the conclusion of one experiment often leads to another experiment. Scientists reading the study described in Figure 1.10 may decide that it would be important to test the difference in the ability of antibiotic A and B to kill *H. pylori* in a petri dish. Or, they may want to test which medication is more effective in women than men, and so forth. The need for scientists to expand on findings explains why science changes and the findings of yesterday may be improved upon tomorrow.

a.

State Hypothesis:
Antibiotic B is a better treatment for ulcers than antibiotic A.

Large number of subjects were selected.

Subjects were divided into three groups.

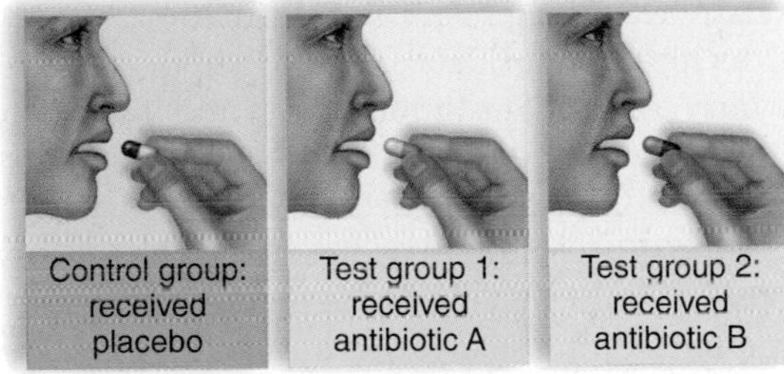

b.

Perform Experiment:
Groups were treated the same except as noted.

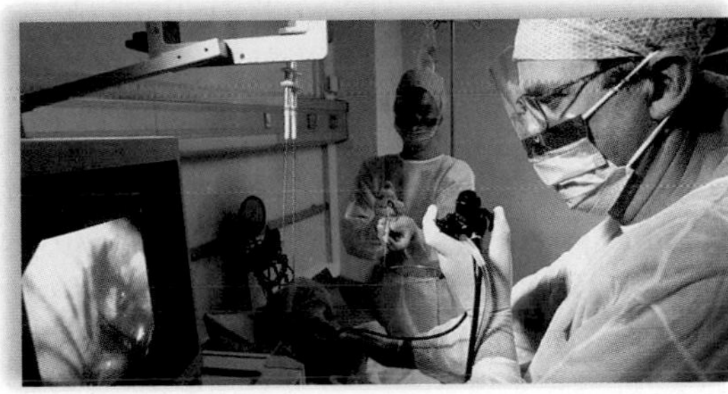

c.

Collect Data:
Each subject was examined for the presence of ulcers.

d.

Treatment of Ulcers by Antibiotics

	Control Group	Test Group 1	Test Group 2
Effectiveness of Treatment (%)	10	60	80

Conclusion: Hypothesis is supported: Antibiotic B is a better treatment for ulcers than antibiotic A.

Figure 1.10 Design of a scientific study.
In this controlled laboratory experiment to test the effectiveness of a medication in humans, subjects were divided into three groups. The control group received a placebo and no medication. Test group 1 received antibiotic A and test group 2 received antibiotic B. The results are depicted in a graph, and it shows that antibiotic B was found to be a more effective treatment than antibiotic A for the treatment of ulcers.

Scientific Journals Versus Other Sources of Information

The information in many scientific journals is highly regarded by scientists because of the review process and because it is "straight from the horse's mouth," so to speak. The investigator who did the research is generally the primary author of a published study. Reading the actual results of the experiment tends to prevent the possibility of misinformation and/or bias. Do you remember playing "pass the message" when you were young? When someone started a message and it was passed around a circle of people, the last person to hear the message rarely received the original message. As each person passed along the message, information was added to or deleted from the original. That same thing may happen to scientific information when it is published in magazines or books or reported by someone other than the original investigator.

Unfortunately, the studies in scientific journals may be technical and difficult for a layperson to read and understand. The general public typically relies on secondary sources of information for its science news. Sometimes, the information may be out of context or misunderstood by the reporter, and the result is transmission of misinformation. Ideally, a reference to the original source (scientific journal article) will be provided so that information can be verified. Remember also that it often takes years to do enough experiments for the scientific community to accept findings as well founded. Be wary of claims that have only limited data to support them and any information that may not be supported by repeated experimentation.

People should be especially careful about scientific information available on the Internet, which is not well regulated. Reliable, credible scientific information can often be found at websites with URLs containing .edu (for educational institution), .gov (for government sites such as the National Institutes of Health or the Centers for Disease Control and Prevention), and .org (for nonprofit organizations such as the American Lung Association or the National Multiple Sclerosis Society). Unfortunately, quite a bit of scientific information on the Internet is intended to entice people into purchasing some sort of product for weight loss, prevention of hair loss, or similar maladies. These websites usually have URLs ending with .com or .net. It pays to question and verify the information from these websites with another source (primary, if possible).

CHECK YOUR PROGRESS 1.3

1. Describe each step of the scientific method.
2. Explain why a controlled study is an important part of the experimental design.
3. List a few pros and cons of using a scientific journal versus other sources of information.

CONNECTING THE CONCEPTS

For more information on the topics presented in this section, refer to the following discussions:

Section S.3 discusses how resistance to antibiotics occurs.
Section 8.3 provides more information on ulcers.

1.4 Making Sense of a Scientific Study

LEARNING OUTCOMES

Upon completion of this section, you should be able to

1. Explain the difference between anecdotal and testimonial data.
2. Interpret information that is presented in a scientific graph.
3. Recognize the importance of statistical analysis to the study of science.

When evaluating scientific information, it is important to consider the type of data given to support it. Anecdotal data—which consist of testimonials by individuals rather than results from a controlled, clinical study—are never considered reliable data. An example of anecdotal data would be claims from people that a particular diet helped them lose weight. Obviously, this doesn't mean the diet will work for everyone. Testimonial data are suspect because the effect of whatever is under discussion may not have been studied with a large number of subjects or a control group.

We must also keep in mind that just because two events occur at the same time, one factor may not be the cause of the other. Dr. Marshall had this problem when his data largely depended on finding *H. pylori* at the site of ulcers. More data were needed before the scientific community could conclude that *H. pylori* was the cause of an ulcer. Similarly, that a human papillomavirus (HPV) infection usually precedes cervical cancer could be viewed only as limited evidence that HPV causes cervical cancer. In this instance, HPV has turned out to be a cause of cervical cancer, but not all correlations (relationships) turn out to be causations. For example, scientific studies do not support the well-entrenched belief that exposure to cold temperatures results in colds. Instead, we now know that viruses cause colds.

What to Look For

Although most everyone who examines a scientific paper is tempted to first read the abstract (synopsis) at the beginning and then skip to the conclusion at the end of the study, it is a good idea to also examine the investigators' methodology and results. The methodology tells us how the scientists conducted their study, and the results tell us what facts (data) were discovered. Always keep in mind that the conclusion is not the same as the data. The conclusion is an interpretation of the data by the individuals who authored the paper. Although scientific papers are reviewed by other scientists, ultimately it is up to us to decide if the conclusion is justified by the data.

Graphs

Data are often depicted in the form of a bar graph (see Fig. 1.10*d*) or a line graph (Fig. 1.11). A graph shows the relationship between two quantities, such as the taking of an antibiotic and the disappearance of an ulcer. As in Figure 1.10, the experimental variable (study groups) is plotted on the *x*-axis (horizontal), and the result (effectiveness) is plotted along the *y*-axis (vertical). Graphs are useful tools to summarize data in a clear and simplified manner. For example, Figure 1.10*d* immediately shows that antibiotic B produced the best results.

The title and labels can assist you in reading a graph; therefore, when looking at a graph, first check the two axes to determine what the graph pertains to. For example, in Figure 1.11, we can see that the investigators were studying blood cholesterol levels over a four-week period. By looking at this graph, we know that the blood cholesterol levels were highest during week 2, and we can also see to what degree the values varied over the course of the study.

Statistical Data

Most authors who publish research articles use statistics to help them evaluate their experimental data. In statistics, the **standard error** tells us how uncertain a particular value is. Suppose you predict how many hurricanes Florida will have next year by calculating the average number during the past ten years. If the number of hurricanes per year varies widely, your standard error will be larger than if the number per year is usually about the same. In other words, the standard error tells you how far off the average could be. If the average number of hurricanes is four and the standard error is $\pm$ 2, then your prediction of four hurricanes is between two and six hurricanes. In Figure 1.11, the standard error is represented by the bars above and below each data point. This provides a visual indication of the statistical analysis of the data.

Statistical Significance

When scientists conduct an experiment, there is always the possibility that the results are due to chance or due to some factor other than the experimental variable. Investigators take into account several factors when they calculate the probability value (p) that their results were due to chance alone. If the probability value is low, researchers describe the results as statistically significant. A probability value of less than 5% (usually written as $p < 0.05$) is acceptable, but, even so, keep in mind that the lower the p value, the *less* likely that results are due to chance. Therefore, the lower the p value, the *greater* the confidence the investigators and you can have in the results. Depending on the type of study, most scientists like to have a p value of <0.05, but p values of <0.001 are common in many studies.

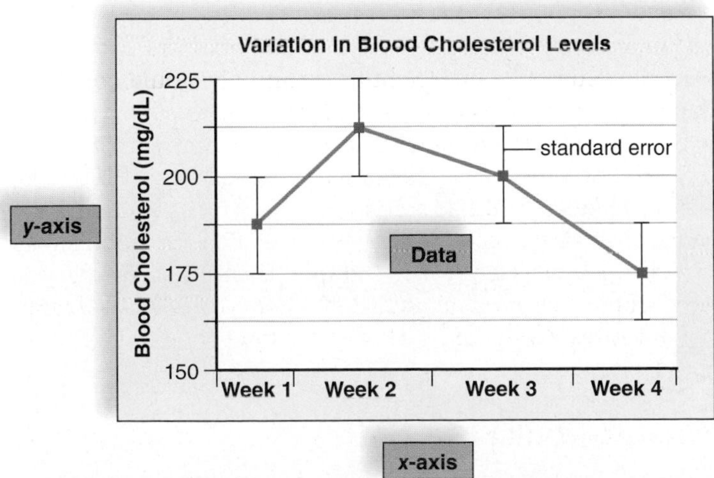

Figure 1.11 The presentation of scientific data.
This line graph shows the variation in the concentration of blood cholesterol over a four-week study. The bars above each data point represent the variation, or standard error, in the results.

CHECK YOUR PROGRESS 1.4

1. Explain why anecdotal data are not used in controlled, clinical studies.
2. Explain why it is important to study an investigator's methodology and results.
3. Summarize how the use of graphs and statistics aids in data analysis.

CONNECTING THE CONCEPTS

For more examples of how graphs can be used to present scientific data, refer to the following discussions:

Figure 7.16 demonstrates how immunizations increase antibody concentration in the blood.

Figure 13.3 shows the relationship between an action potential and voltage across a cell membrane.

Figure 24.1 illustrates human population growth.

1.5 Science and Social Responsibility

LEARNING OUTCOMES

Upon completion of this section, you should be able to

1. Recognize the importance of ethics in scientific studies.
2. Discuss the need for the general public to have a general understanding of science and its relationship to society.

As we have learned in this chapter, science is a systematic way of acquiring knowledge about the natural world. Biologists are scientists who study life, from the tiniest microbes to the tallest trees (Fig. 1.12). Religion, aesthetics, and ethics are other ways in which humans seek order in the natural world. Science differs from these other ways of knowing and learning by its process, based on the scientific method. Science considers hypotheses that can be tested only by experimentation and observation. Only after an immense amount of data has been gathered do scientists arrive at a scientific theory, a well-founded concept about the natural world. Knowing this should help you realize that all scientific theories have merit.

Science is a slightly different endeavor from technology, but science is the driving force behind technology. **Technology** is the application of scientific knowledge to the interests of humans. As members of Western civilization, we have always believed that science and technology offer us ways to improve our lives. Building houses, paving roads, growing crops, and curing illnesses all depend on technology. Just think of how many things you use each day made of plastic and you will know how much technology means to your daily life. Even the field of nuclear physics is of direct benefit to human beings. The findings of nuclear physics play a role in cancer therapy, medical imaging, and homeland security. It has also given us nuclear power and the atomic bomb. In other words, science and technology are not risk-free; uncontrolled technology can result in unanticipated side effects.

Science and Technology: Benefits Versus Risks

Investigations into cell structure and genes led to the current biotechnology revolution. We now know how to manipulate genes. If you have diabetes and are using insulin, it was produced by genetically modified (GM) bacteria. Despite its benefits, there are risks associated with biotechnology. Ecologists are concerned that GM crops could endanger the biosphere. For example, many farmers are now planting GM cotton that produces an insect-killing toxin. Though insect pests are their intended target, toxins may also kill the natural predators that feed on insect pests. It may not be wise to kill off friendly predatory insects. People are now eating GM foods and/or foods that have been manufactured by using GM products. Some people are concerned about the possible effects of GM foods on human health.

In medicine, gene technology raises even more difficult ethical issues. These include whether humans should be cloned or whether gene therapy should be used to modify the inheritance of people, even before they are born. A current debate centers around the use of stem cells to cure human illnesses, such as spinal cord injuries or Alzheimer disease. You may not appreciate this debate until you know that early human embryos are composed only of cells called embryonic

a.

b.

Figure 1.12 Biologists work in many environments.
Data collection can be done (**a**) in the laboratory or (**b**) in the field. Biologists discover basic information about the natural world, including the effects of technology on human health and the environment.

stem cells. Embryonic stem cells are genetically capable of becoming any type of tissue needed to cure a human illness. Should human embryos be dismantled and used for this purpose? It means they will never have the opportunity to become humans.

Everyone Is Responsible

Science, technology, and society have interacted throughout human history, and scientific investigation and technology have always been affected by human values. Studying science, such as human biology, can give citizens the background they need to fully participate in ethical debates. All citizens should assume this responsibility because everyone, not just scientists, needs to be involved in making value judgments about the proper use of technology.

Should windmills be placed in Nantucket Sound or other bodies of water to reduce our use of oil, which causes global warming? To participate in this debate, you have to know what windmills might do to the ecology of Nantucket Sound, what global warming is, and what the evidence for global warming is. In other words, you need to review the data of scientists and only then participate in the debate about how technology should be used in this particular instance.

Should the rise in human population be curtailed to help preserve biodiversity? You might hear on the news that we are in a biodiversity crisis—the number of extinctions expected to occur in the near future is unparalleled in the history of the Earth. To help answer this question, you will want to know how large the human population is, what the benefits of biodiversity are, how it is threatened, and what the threat is to humankind if biodiversity is not preserved.

Scientists can inform and educate us, but they need not bear the burden of making these decisions alone because science does not make value judgments. This is the job of all of us.

CHECK YOUR PROGRESS 1.5

1. List examples that demonstrate both the risks and benefits of technology.
2. Discuss how human values affect science and technology.
3. Discuss the following: In your opinion, which members of society should be responsible for deciding how technology should be used?

CONNECTING THE CONCEPTS

The topic of bioethics is covered in several of the boxed readings in this text. For some examples of bioethical discussions that involve biotechnology, refer to the following discussions:

Chapter 17 contains the Bioethical feature, "Reproductive and Therapeutic Cloning," on whether humans should be cloned.

Chapter 21 contains a bioethical feature examining the use of DNA profiling in the justice system.

CASE STUDY CONCLUSION

In this chapter, you have explored some of the basic characteristics of life as we know it. The question is, How can we apply our knowledge of life on Earth to detect life on other planets? Most likely, the life on moons such as Europa and Titan is not highly organized. Most scientists believe that simple multicellular organisms may be the only life-forms that can survive this distance from the sun. Thus, future missions to Europa and Titan will likely look for evidence of life-acquiring materials and using energy. When you eat food, you produce carbon dioxide and other waste products. Living organisms on other planets should do the same. By studying the extreme environments of Europa, Titan, and our own planet, we may be better able to define the basic properties of life and what it really means when we say that something "is alive."

MEDIA STUDY TOOLS

Enhance your study of this chapter with media!
Visit **www.mhhe.com/maderhuman13e** and go to "Media Study Tools" for this chapter to access the following:

MP3 File

1.1 Life Characteristics

SUMMARIZE

1.1 The Characteristics of Life

Biology is the study of life. All living **organisms** share common characteristics; they

- have levels of organization—**atoms, molecules, cells, tissues, organs, organ systems, organisms, populations, community, ecosystem,** and **biosphere;**
- acquire materials and energy from the environment. **Metabolism** is the sum of the reactions involved in these processes. **Photosynthesis,** which occurs in organisms such as plants, is responsible for producing the organic molecules that serve as food for most organisms.

- **reproduce;** and experience **growth,** and in many cases **development;**
- maintain **homeostasis** to maintain the conditions of an internal environment;
- respond to stimuli; and
- have an evolutionary history and are adapted to a way of life.

1.2 Humans Are Related to Other Animals

The classification of living organisms mirrors their evolutionary relationships. Humans are mammals, a type of **vertebrate** in the animal **kingdom** of the **domain** Eukarya. Humans differ from other mammals, including apes, by their

- highly developed brains;
- completely upright stance;
- creative language; and
- ability to use a wide variety of tools.

Humans Have a Cultural Heritage

Language, tool use, values, and information are passed on from one generation to the next.

Humans Are Members of the Biosphere

Humans depend on the biosphere for its many services, such as absorption of pollutants, sources of water and food, prevention of soil erosion, and natural beauty. Human population growth and use of resources threaten the **biodiversity** of the biosphere and is leading to the **extinction** of many species.

1.3 Science as a Process

Science is a way of knowing about the natural world. The **scientific method** consists of

- making an observation;
- formulating a **hypothesis;**
- carrying out experiments and observations to collect **data;**
- coming to a **conclusion;** and
- presenting results of the study for peer review.

Over time, widely accepted concepts and ideas that explain patterns in the natural world may become principles of **scientific theories.**

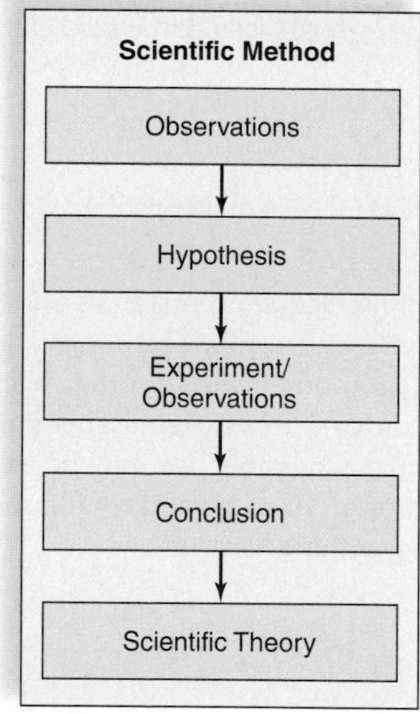

How the Cause of Ulcers Was Discovered

Dr. Marshall followed Koch's postulates to show that *H. pylori* causes ulcers.

How to Do a Controlled Study

- In an **experiment,** the **test group** is exposed to the **experimental variable,** while the **control group** is not. The control group may be given a **placebo** to ensure that the participants all behave the same during the experiment.
- All groups are otherwise treated the same, and it is best if the subjects and the technicians do not know what group they are in.
- The results and conclusion are published in a scientific journal.

Scientific Journals Versus Other Sources of Information

Primary sources of information are best; if secondhand sources are used, the reader needs to carefully evaluate the source. The Internet is not regulated, although URLs that end in .edu, .gov, and .org most likely provide reliable scientific information.

1.4 Making Sense of a Scientific Study

In general, when assessing a scientific study you should beware of anecdotal and correlational data because they need further study to be substantiated. When examining graphs, ensure that they clearly show the relationship between quantities.

Statistical Data

The **standard error** tells us how uncertain a particular value is. The statistical significance tells us how trustworthy the results are. The lower the statistical significance, the less likely the results are due to chance and the more likely they are due to the experimental variable.

1.5 Science and Social Responsibility

Scientific information is based on observation and experimentation. Therefore, scientists need not make value judgments for us.The use of modern **technology** has its risks, and all citizens need to be able to make informed decisions regarding how and when technology should be used.

ASSESS

Testing Your Knowledge of the Concepts

1. Name and describe the basic characteristics of life. (pages 2–6)
2. What is homeostasis, and why is it important? Give some examples that show how systems work together to maintain homeostasis. (page 4)
3. What are the major domains and kingdoms of life? (pages 7–8)
4. How do human activities threaten the biosphere? (page 9)
5. Discuss the importance of a scientific theory, and describe several theories basic to understanding biological principles. (page 10)
6. With reference to the steps of the scientific method, explain how scientists arrive at a theory. (page 10)
7. What are Koch's postulates, and what are they used for? (page 11)
8. What is a control group, and what is the importance of a control group in a controlled study? (page 12)
9. How do science and technology improve our lives? (pages 16–17)

In questions 10–13, match each description with the correct characteristic of life from the key.

Key:

a. Life is organized.
b. Living organisms reproduce and grow.
c. Living organisms respond to stimuli.
d. Living organisms have an evolutionary history.
e. Living organisms acquire materials and energy.

10. Human heart rate increases when the person is scared.

11. Humans produce only humans.

12. Humans need to eat for building blocks and energy.

13. Similar cells form tissues in the human body.

14. The level of organization that includes two or more tissues that work together is a(an)
a. organ.
b. tissue.
c. organ system.
d. organism.

15. The level of organization most responsible for the maintenance of homeostasis is the ______________ level.
a. cellular
b. organ system
c. organ
d. tissue

16. In an experiment, the ______________ is exposed to the experimental variable.
a. test group
b. control group
c. Both a and b are correct.
d. Neither a nor b is correct.

17. Humans belong to all of the following groups, except
a. the animal kingdom.
b. domain Eukarya.
c. invertebrates.
d. mammals.

ENGAGE

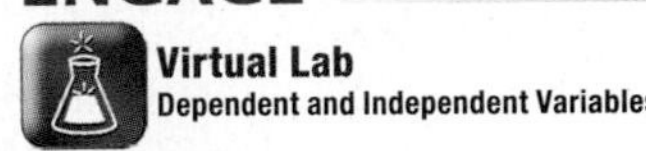

The virtual lab "Dependent and Independent Variables" provides an interactive tutorial for how scientists construct scientific experiments.

Thinking Critically About the Concepts

1. Viruses are generally lumped into a "germs" grouping with bacteria. Viruses are composed of a small amount of genetic material (DNA or RNA) wrapped in a protein coat. Can something so simple be considered a living organism? Why aren't viruses mentioned in the system of classification covered in section 1.2? (*Hint:* Consider the shared characteristics of all living organisms.)
2. Can anecdotal data ever be a considered a reliable data source for scientific research? Why or why not? If not, what purpose can anecdotal data serve?
3. In the case of Europa and Titan, if life were found to exist there, would that change our definition of the basic characteristics of life? Would that change our definition of a biosphere?

CHAPTER

2

Chemistry of Life

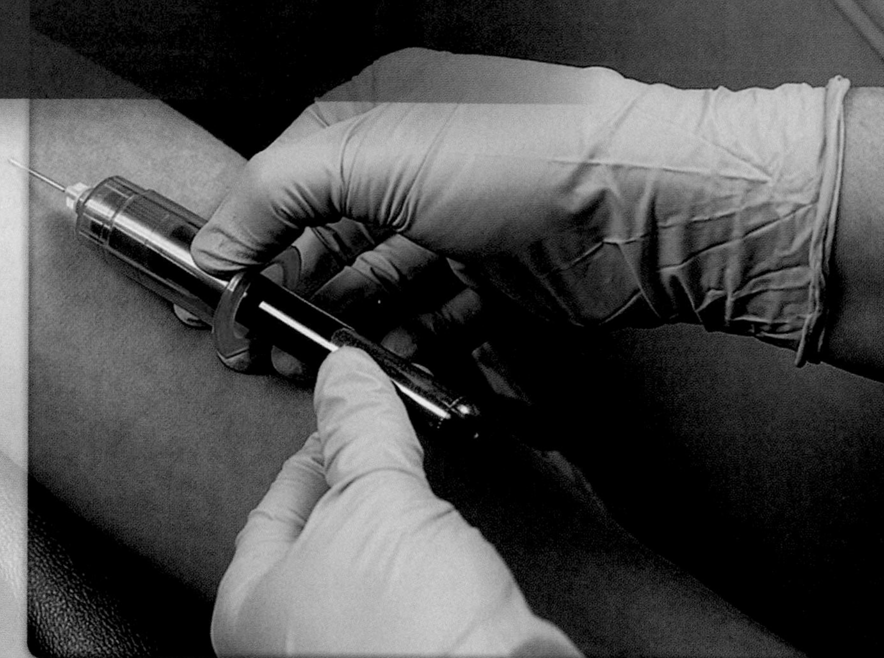

CHAPTER CONCEPTS

2.1 From Atoms to Molecules
All matter is composed of atoms, which react with one another to form molecules.

2.2 Water and Life
The properties of water make life, as we know it, possible. Living organisms are affected adversely by water that is too acidic or too basic.

2.3 Molecules of Life
Carbohydrates, lipids, proteins, and nucleic acids are macromolecules with specific functions in cells.

2.4 Carbohydrates
Blood sugar is glucose, and humans store glucose as glycogen. Cellulose is plant material that is a source of fiber in the diet.

2.5 Lipids
Fats and oils are long-term energy storage molecules. The amount in the diet has an effect on our health. Other lipids, such as the sterols and phospholipids, function differently in the body.

2.6 Proteins
Proteins have numerous and varied functions in cells. The structure of a protein determines its function.

2.7 Nucleic Acids
DNA is the genetic material of life. RNA serves as a helper to DNA. ATP is an energy molecule that is used by the cell to do metabolic work.

BEFORE YOU BEGIN

Before beginning this chapter, take a few moments to review the following discussions:

Section 1.1 What are the basic characteristics of all living organisms?

Figure 1.2 What is the difference between an atom and a molecule?

CASE STUDY BLOOD CHEMISTRY AND NUTRITION

As a 26-year-old male, David knew that he was slightly overweight and definitely not in the physical condition that he had been five years earlier. Still, he considered himself to be healthy; therefore, he only reluctantly agreed to the blood test at his annual physical exam. A few weeks later, David's doctor called and asked David to come in for a review of the results.

In the office, David's doctor explained that he had some real concerns about the blood test results. The doctor explained that David had a total cholesterol value of 218 and a blood triglyceride value of 150 milligrams per deciliter (mg/dl). Furthermore, his good cholesterol (HDL) was low (40 mg/dl) and his bad cholesterol (LDL) was high (130 mg/dl). If these values remained uncorrected, there would be an increased risk of heart disease and atherosclerosis and, potentially, diseases such as diabetes and cancer in David's future. David's doctor recommended that he reduce his dietary fat intake, increase his exercise, and come back in three months for a follow-up visit. If his blood lipids did not come into acceptable ranges, David's doctor was going to put him on atorvastatin (Lipitor), a cholesterol-reducing medication.

David realized that he understood very little about the chemistry that made his body work, and he was curious about some of the terms the doctor mentioned during the visit. For example, what is the difference between good and bad cholesterol? And what is a triglyceride? In doing some online research, he discovered that actually there is no such thing as "good" and "bad" cholesterol. The LDL and HDL actually refer to a type of protein that is involved in lipid transport in the body. David was confused, because his doctor had referred to these as being cholesterol. David realized that he not only knew very little about his body's chemistry, but he also was in the dark as to the function of many nutrients in his body.

As you read through the chapter, think about the following questions:

1. What is it about cholesterol that would make it increase the risk for the diseases the doctor mentioned?
2. Is all cholesterol bad for the body?
3. What are the roles of proteins in the body, and how do they relate to David's cholesterol values?

2.1 From Atoms to Molecules

LEARNING OUTCOMES

Upon completion of this section, you should be able to

1. Distinguish between atoms and elements.
2. Describe the structure of an atom.
3. Define an isotope and summarize its application in both medicine and biology.
4. Distinguish between ionic and covalent bonds.

Matter refers to anything that takes up space and has mass. It is helpful to remember that matter can exist as a solid, a liquid, or a gas. Not only are humans composed of matter, but so is the food we eat, water we drink, and the air we breathe.

Elements

An **element** is one of the basic building blocks of matter; an element cannot be broken down by chemical means. Considering the variety of living and nonliving things in the world, it's remarkable that there are only 92 naturally occurring elements. It is even more surprising that over 90% of the human body is composed of just four elements: carbon, nitrogen, oxygen, and hydrogen. Even so, other elements, such as iron, are important to our health. Iron-deficiency anemia results when the diet doesn't contain enough iron for the making of hemoglobin. Hemoglobin serves an important function in the body because it transports oxygen to our cells.

Each element has a name and a symbol. For example, carbon has been assigned the atomic symbol C, and iron has been assigned the symbol Fe. Some of the symbols we use for elements are derived from Latin. For example, the symbol for sodium is Na because *natrium,* in Latin, means "sodium." Likewise, the symbol for iron is Fe because *ferrum* means "iron." Chemists arrange the elements in a *periodic* table, so named because all the elements in a column show *periodicity,* meaning that each column of elements behaves similarly during chemical reactions. For example, all the elements in column VII (7) undergo the same type of chemical reactions, for reasons we will soon explore. Figure 2.1 shows only some of the first 36 elements in the periodic table, but a complete table is available in Appendix A.

Periods	I	II	III	IV	V	VI	VII	VIII
1	1 **H** 1.008							2 **He** 4.003
2	3 **Li** 6.941	4 **Be** 9.012	5 **B** 10.81	6 **C** 12.01	7 **N** 14.01	8 **O** 16.00	9 **F** 19.00	10 **Ne** 20.18
3	11 **Na** 22.99	12 **Mg** 24.31	13 **Al** 26.98	14 **Si** 28.09	15 **P** 30.97	16 **S** 32.07	17 **Cl** 35.45	18 **Ar** 39.95
4	19 **K** 39.10	20 **Ca** 40.08	31 **Ga** 69.72	32 **Ge** 72.59	33 **As** 74.92	34 **Se** 78.96	35 **Br** 79.90	36 **Kr** 83.60

atomic number — atomic symbol — atomic mass; Groups

Figure 2.1 **A portion of the periodic table of elements.** The number on the top of each square is the atomic number, which increases from left to right. The letter symbols represent each element and are sometimes abbreviations of Greek or Latin names. Below the symbol is the value for atomic mass. A complete periodic table can be found in Appendix A.

APPLICATIONS AND MISCONCEPTIONS

Where do elements come from?

We are all familiar with elements. Iron, sodium, oxygen, and carbon are all common terms in our lives, but where do they originate from?

Normal chemical reactions do not produce elements. The majority of the heavier elements, such as iron, are produced only by the intense chemical reactions within stars. When these stars reach the end of their life, they explode, producing a supernova. Supernovas scatter the heavier elements into space, where they eventually are involved in the formation of planets. The late astronomer and philosopher Carl Sagan (1934–1996) frequently referred to humans as "star stuff." In many ways, we—all living organisms—are formed from elements that originated within the stars.

Atoms

An **atom** is the smallest unit of an element that still retains the chemical and physical properties of the element. The same name is given to the element and the atoms of the element. Though it is possible to split an atom, an atom is the smallest unit to enter into chemical reactions. Physicists have identified a number of subatomic particles that make up atoms. The three best known subatomic particles include positively charged **protons,** uncharged **neutrons,** and negatively charged **electrons.** Protons and neutrons are located within the nucleus of an atom, and electrons move about the nucleus. Figure 2.2 shows the arrangement of the subatomic particles of some common elements. In Figure 2.2, the circle around the nucleus of the atom represents an **electron shell,** the average location of electrons. Notice that most of an atom is empty space. If we could draw an atom the size of a football stadium, the nucleus would be like a gumball in the center of the field, and the electrons would be tiny specks whirling about in the upper stands.

Animation Atomic Structure

The Periodic Table

Atoms not only have an atomic symbol, they also have an atomic number and mass number. All atoms of an element have the same number of protons housed in the nucleus. This

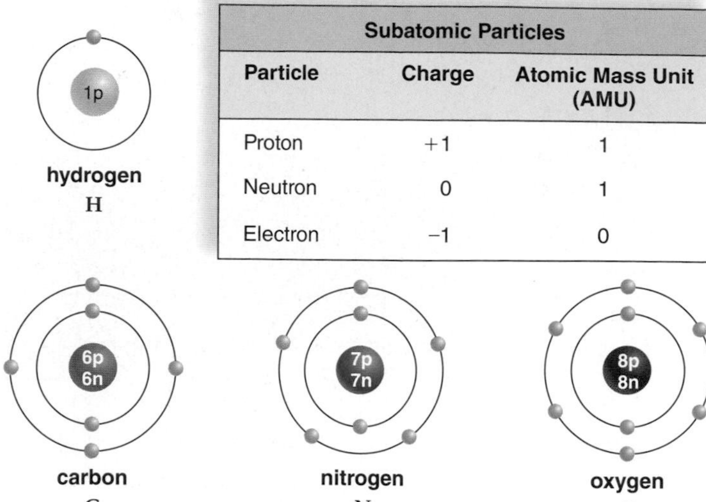

Subatomic Particles		
Particle	**Charge**	**Atomic Mass Unit (AMU)**
Proton	+1	1
Neutron	0	1
Electron	−1	0

Figure 2.2 The atomic structure of select elements. Notice how the protons (p) and neutrons (n) are located in the nucleus and the electrons (blue dots) are found in energy oribitals around the nucleus.

is called the **atomic number,** which accounts for the unique properties of this type of atom.

Each atom also has its own mass number dependent on the number of subatomic particles in that atom. Protons and neutrons are assigned one atomic mass unit (AMU) each. Electrons are so small that their AMU is considered zero in most calculations (Fig. 2.2). Therefore, the **mass number** of an atom is the sum of protons and neutrons in the nucleus.

By convention, when an atom stands alone (and not in the periodic table, discussed next), the atomic number is written as a subscript to the lower left of the atomic symbol. The mass number is written as a superscript to the upper left of the atomic symbol. Regardless of position, the smaller number is always the atomic number, as shown here for carbon:

mass number — 12, atomic number — 6, atomic symbol — C

$$^{12}_{6}\mathrm{C}$$

The atoms shown in the periodic table (see Fig. 2.1) are assumed to be electrically neutral. Therefore, the atomic number not only tells you the number of protons, it also tells you the number of electrons. The **atomic mass** (the number below the atomic symbol on the periodic table) is the average of the AMU for all the isotopes (discussed next) of that atom. To determine the number of neutrons, subtract the number of protons from the atomic mass, and take the closest whole number.

Isotopes

Isotopes of the same type of atom have the same number of protons (thus, the same atomic number) but different numbers of neutrons. Therefore, their mass numbers are different. For example, the element carbon 12 (^{12}C) has six neutrons, carbon 13 (^{13}C) has seven neutrons, and carbon 14 (^{14}C) has eight neutrons. You can determine the number of neutrons for an isotope by subtracting the atomic number (carbon's atomic number is 6; see Fig. 2.1) from the mass number.

Unlike the other two isotopes of carbon, carbon 14 is unstable and breaks down over time. As carbon 14 decays, it releases various types of energy in the form of rays and subatomic particles; therefore, it is a **radioisotope.** The radiation given off by radioisotopes can be detected in various ways. You may be familiar with the use of a Geiger counter to detect radiation.

Animation Half-Life

Low Levels of Radiation

The importance of chemistry to biology and medicine is nowhere more evident than in the many uses of radioisotopes. A radioisotope behaves the same chemically as the stable isotopes of an element. This means that you can put a small amount of radioisotope in a sample and it becomes a *tracer* by which to detect molecular changes.

Specific tracers are used in imaging the body's organs and tissues and can be used to diagnose the presence of tumors. For example, after a patient drinks a solution containing a minute amount of iodine 131, it becomes concentrated in the thyroid, which uses iodine to make the hormone thyroxine. After the thyroid has had some time to accumulate the iodine 131 an image may indicate whether it is healthy in structure and function (Fig. 2.3*a*). Another example is in the use of positron-emission tomography (PET) to determine the comparative activity of tissues. Radioactively labeled glucose, which emits

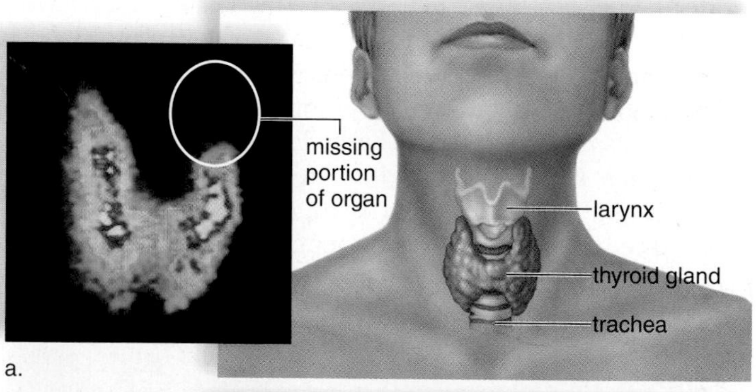

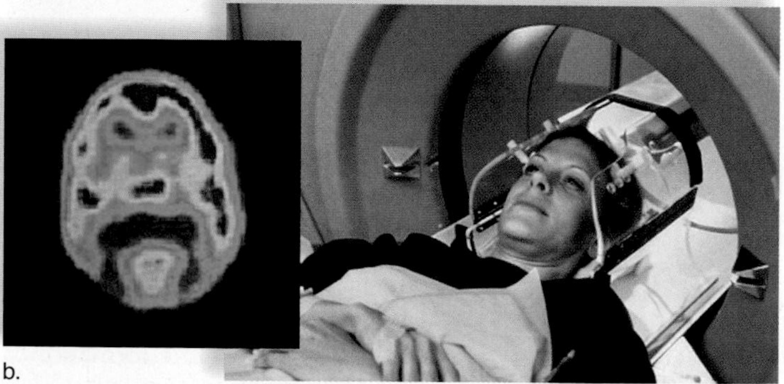

Figure 2.3 Medical uses for low-level radiation. **a.** The missing area (indicated by the circle) in this thyroid scan indicates the presence of a tumor that does not take up radioactive iodine. **b.** A PET (positron-emission tomography) scan reveals which portions of the brain are most active (red surrounded by light green).

a subatomic particle known as a positron, can be injected into the body. The radiation given off is detected by sensors and analyzed by a computer. The result is a color image that shows which tissues took up glucose and are metabolically active (red areas in Fig. 2.3*b*). A PET scan of the brain can help diagnose a brain tumor, Alzheimer disease, epilepsy, and can determine whether a stroke has occurred.

Video Nuclear Medicine

High Levels of Radiation

Radioactive substances in the environment can harm cells, damage DNA, and cause cancer. The release of radioactive particles following a nuclear power plant accident, such as occurred at the Japanese Fukushima nuclear plant in 2011, can have far-reaching and long-lasting effects on human health. However, the effects of radiation can also be put to good use. Radiation from radioisotopes has been used for many years to kill bacteria and viruses and to sterilize medical and dental products. Increasingly, the technology is being used to increase the safety of our food supply. By using specific types of radiation, food can be sterilized without irradiating or damaging the food (Fig. 2.4*a*).

Radiation can be used to ensure public safety against bacterial infection. It is used to sterilize the U.S. mail and other packages to free them of possible pathogens, such as anthrax spores.

The ability of radiation to kill cells is often applied to cancer cells. Radioisotopes can be introduced into the body in a way that allows radiation to destroy only cancer cells, with little risk to the rest of the body (Fig. 2.4*b*). Another form of high-energy radiation, X-rays, can be used for medical diagnosis and cancer therapy.

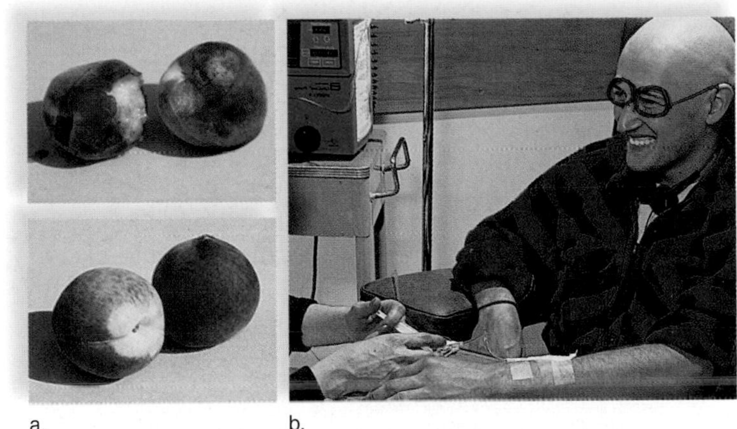

Figure 2.4 Uses of high-level radiation.
a. Radiation can be used to sterilize food by killing bacteria and fungi.
b. Physicians can use radiation therapy to kill cancer cells.

Molecules and Compounds

Atoms often bond with one another to form a chemical unit called a **molecule.** A molecule can contain atoms of the same type, as when an oxygen atom joins with another oxygen atom to form oxygen gas. Or, the atoms can be different, as when an oxygen atom joins with two hydrogen atoms to form water. When the atoms are different, a **compound** is formed. Two types of bonds join atoms: the ionic bond and the covalent bond.

MP3 Chemical Bonding

Ionic Bonding

Atoms with more than one shell are most stable when the outer shell contains eight electrons. During an ionic reaction, atoms give up or take on an electron or electrons to achieve a stable outer shell.

Figure 2.5 depicts a reaction between a sodium (Na) atom and a chlorine (Cl) atom. Sodium, with one electron in the

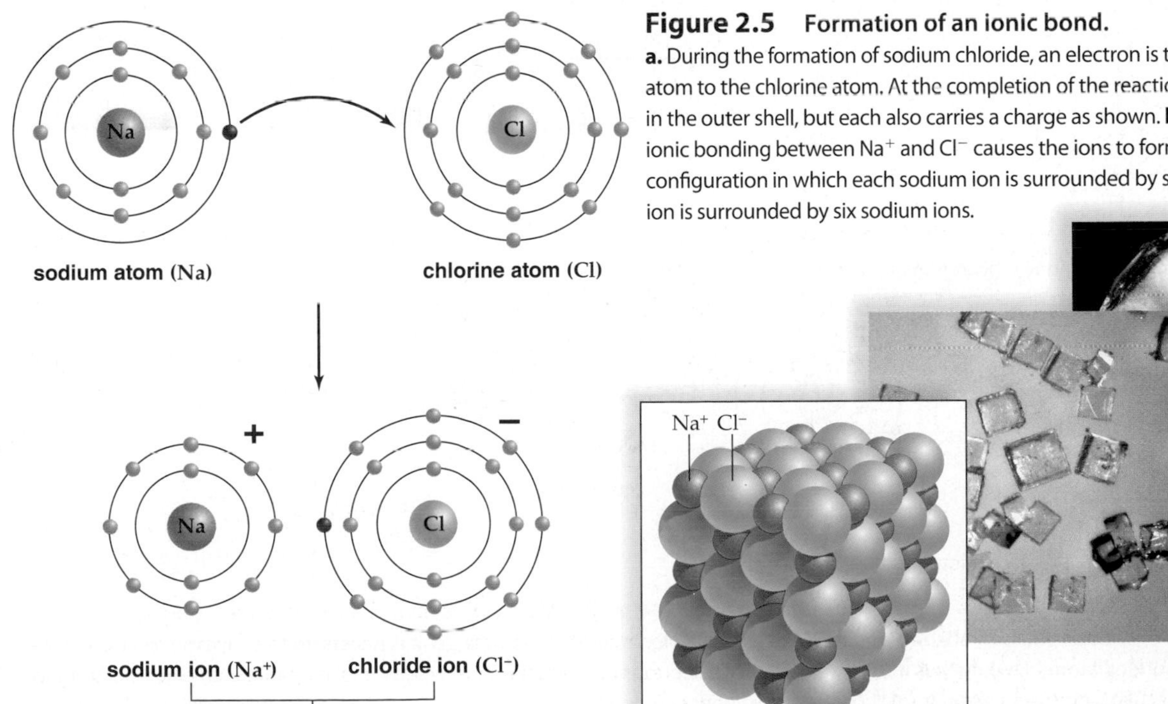

Figure 2.5 Formation of an ionic bond.
a. During the formation of sodium chloride, an electron is transferred from the sodium atom to the chlorine atom. At the completion of the reaction, each atom has eight electrons in the outer shell, but each also carries a charge as shown. **b.** In a sodium chloride crystal, ionic bonding between Na^+ and Cl^- causes the ions to form a three-dimensional lattice configuration in which each sodium ion is surrounded by six chloride ions and each chloride ion is surrounded by six sodium ions.

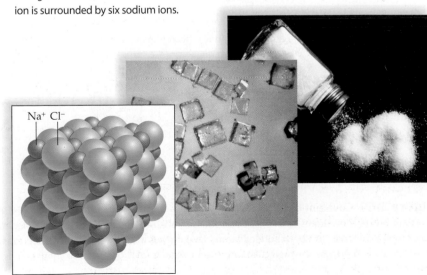

outer shell, reacts with a single chlorine atom. Why? Once the reaction is finished and sodium loses one electron to chlorine, sodium's outer shell will have eight electrons. Similarly, a chlorine atom, which has seven electrons already, needs to acquire only one more electron to have a stable outer shell.

Ions are particles that carry either a positive (+) or negative (−) charge. When the reaction between sodium and chlorine is finished, the sodium ion carries a positive charge because it now has one less electron than protons, and the chloride ion carries a negative charge because it now has one more electron than protons:

Sodium Ion (Na^+)	Chloride Ion (Cl^-)
11 protons (+)	17 protons (+)
10 electrons (−)	18 electrons (−)
One (+) charge	One (−) charge

The attraction between oppositely charged sodium ions and chloride ions forms an **ionic bond.** The resulting compound, sodium chloride, is table salt, which we use to enliven the taste of foods.

Animation
Ionic Bonds

In contrast to sodium, why would calcium, with two electrons in the outer shell, react with two chlorine atoms? Whereas calcium needs to lose two electrons, each chlorine, with seven electrons already, requires only one more electron to have a stable outer shell. The resulting salt ($CaCl_2$) is called calcium chloride. Calcium chloride is used as a deicer in northern climates.

The balance of various ions in the body is important to our health. Too much sodium in the blood can contribute to high blood pressure. Calcium deficiency leads to rickets (a bowing of the legs) in children. Too much or too little potassium results in heartbeat irregularities and can be fatal. Bicarbonate, hydrogen, and hydroxide ions are all involved in maintaining the acid–base balance of the body (see pages 28–29).

Covalent Bonding

Atoms share electrons in **covalent bonds.** The overlapping, outermost shells in Figure 2.6 indicate that the atoms are sharing electrons. Just as two hands participate in a handshake, each atom contributes one electron to the shared pair. These electrons spend part of their time in the outer shell of each atom; therefore, they are counted as belonging to both bonded atoms.

Double and Triple Bonds Besides a single bond, in which atoms share only a pair of electrons, a double or a triple bond can form. In a double bond, atoms share two pairs of electrons; in a triple bond, atoms share three pairs of electrons. For example, in Figure 2.6*b*, each oxygen atom (O) requires two more electrons to achieve a total of eight electrons in the outer shell. Four electrons are placed in the outer, overlapping shells in the diagram.

Animation
Ionic and Covalent Bonds

Structural and Molecular Formulas Covalent bonds can be represented in a number of ways. In contrast to the diagrams in Figure 2.6, structural formulas use straight lines to show

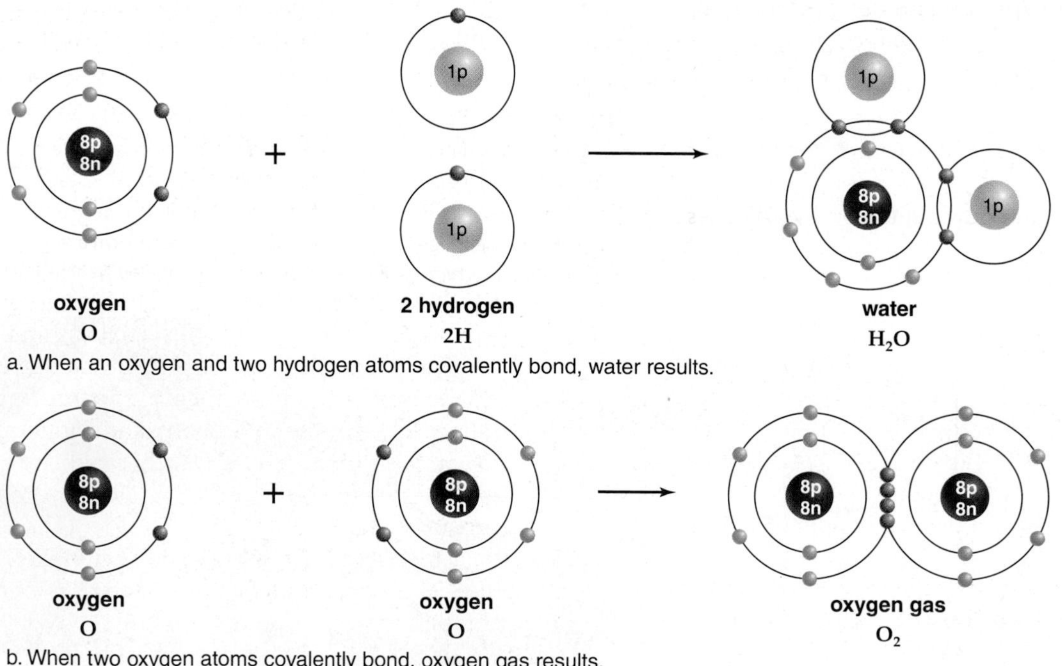

Figure 2.6 Covalent bonds.
Covalent bonds allow atoms to fill their outer shells by sharing electrons. Because the electrons are being shared, it is necessary to count the electrons in the outer shell as belonging to both bonded atoms. Hydrogen is most stable with two electrons in the outer shell; oxygen is most stable with eight electrons in the outer shell. Therefore, the molecular formula for water is (**a**) H_2O, and for oxygen gas it is (**b**) O_2.

the covalent bonds between the atoms. Each line represents a pair of shared electrons. Molecular formulas indicate only the number of each type of atom making up a molecule. A comparison follows:

Structural formula: H—O—H, O═O

Molecular formula: H_2O, O_2

What would be the structural and molecular formulas for carbon dioxide? Carbon, with four electrons in the outer shell, requires four more electrons to complete its outer shell. Each oxygen, with six electrons in the outer shell, needs only two electrons to complete its outer shell. Therefore, carbon shares two pairs of electrons with each oxygen atom, and the formulas are as follows:

Structural formula: O═C═O

Molecular formula: CO_2

CHECK YOUR PROGRESS 2.1

1. Explain the organization of an atom of carbon.
2. Determine the number of neutrons found in the isotopes of calcium ^{40}Ca and ^{48}Ca. (*Hint:* The atomic number of calcium is 20.)
3. Explain the role of radioisotopes in modern society.
4. Distinguish between the structure of ionic and colvalent bonds and give an example of each.

CONNECTING THE CONCEPTS

The study of biology has a firm foundation in chemistry. To see this relationship in more detail, refer to the following discussions:

Section 3.3 describes how cells move ions across cell membranes.

Section 10.3 explains how our urinary system uses ions to maintain homeostasis.

2.2 Water and Life

LEARNING OUTCOMES

Upon completion of this section, you should be able to

1. List the properties of water.
2. Explain the role of hydrogen bonds in the properties of water.
3. Summarize the structure of the pH scale and the importance of buffers to biological systems.

Water is the most abundant molecule in living organisms, usually making up about 60–70% of the total body weight. Furthermore, the physical and chemical properties of water make life as we know it possible.

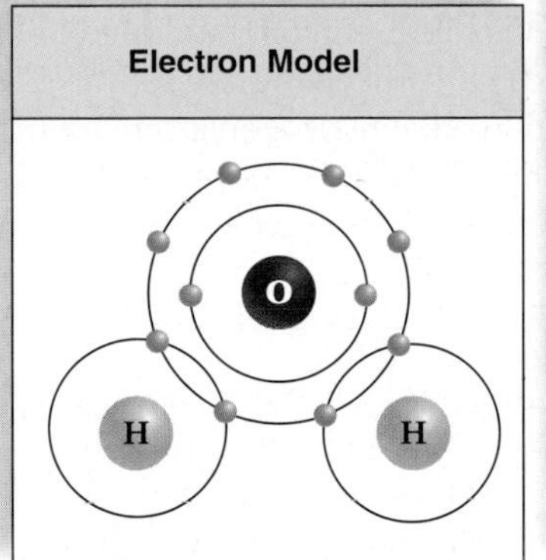

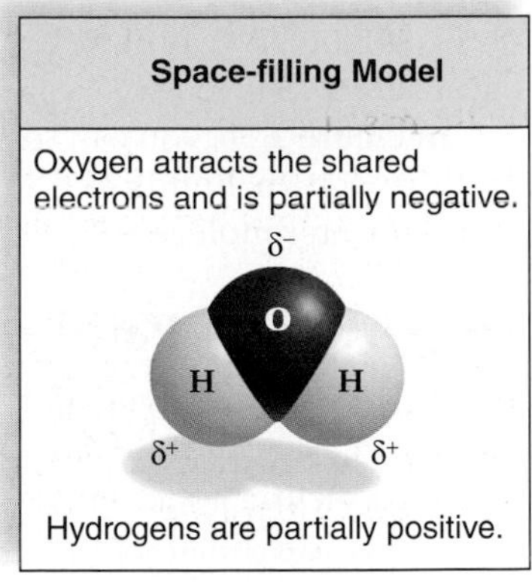

Space-filling Model
Oxygen attracts the shared electrons and is partially negative.
δ⁻
O
H
H
δ⁺
δ⁺
Hydrogens are partially positive.

a. Water (H_2O)

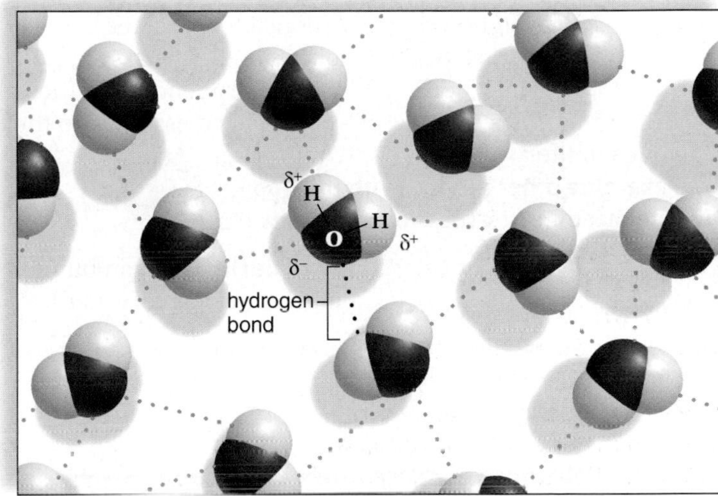

b. Hydrogen bonding between water molecules

Figure 2.7 **Hydrogen bonds and water molecules.**
a. Two models for the structure of water. The diagram on the left shows the sharing of electrons between the oxygen and hydrogen atoms. The diagram on the right illustrates that water is a polar molecule because electrons are not equally shared. Electrons move closer to oxygen, creating a partial negative charge δ^-, whereas hydrogen has a partial positive charge δ^+. **b.** The partial charges allow hydrogen bonds (dotted lines) to form temporarily between water molecules.

In water, the electrons spend more time circling the oxygen (O) atom than the hydrogens because oxygen, the larger atom, has a greater ability to attract electrons than do the smaller hydrogen (H) atoms. The negatively charged electrons are closer to the oxygen atom, so the oxygen atom becomes slightly negative. In turn, the hydrogens are slightly positive. Therefore, water is a **polar** molecule; the oxygen end of the molecule has a slight negative charge (δ^-), and the hydrogen end has a slight positive charge (δ^+).

In Figure 2.7*a*, the diagram on the left shows a structural model of water, and the one on the right is called a space-filling model.

Hydrogen Bonds

A **hydrogen bond** is the attraction of a slightly positive, covalently bonded hydrogen to a slightly negative atom in the vicinity. These usually occur between a hydrogen and either an oxygen or nitrogen atom. A hydrogen bond is represented by a dotted line because it is relatively weak and can be broken rather easily.

In Figure 2.7*b*, you can see that each hydrogen atom, being slightly positive, bonds to the slightly negative oxygen atom of another water molecule.

Properties of Water

The first cell(s) evolved in water, and all living organisms are 70–90% water. Because of hydrogen bonding, water molecules cling together, and this association gives water its unique chemical properties. Without hydrogen bonding between molecules, water would freeze at −100°C and boil at −91°C, making most of the water on Earth steam, and life unlikely. Hydrogen bonding is responsible for water being a liquid at temperatures typically found on the Earth's surface. It freezes at 0°C and boils at 100°C (see Appendix B). These and other unique properties of water make it essential to the existence of life as we know it. When scientists examine other planets with the hope of finding life, they first look for signs of water.

Animation Water Properties

Water Has a High Heat Capacity A **calorie**[1] is the amount of heat energy needed to raise the temperature of 1 g of water 1°C. In comparison, other covalently bonded liquids require input of only about half this amount of energy to rise in temperature 1°C. The many hydrogen bonds that link water molecules together help water absorb heat without a great change in temperature. Converting 1 g of the coldest liquid water to ice requires the loss of 80 calories of heat energy (Fig. 2.8*a*). Water holds onto its heat, and its temperature falls more slowly than that of other liquids. Because the temperature of water rises and falls slowly, we are better able to maintain our normal internal temperature and are protected from rapid temperature changes.

Water Has a High Heat of Evaporation When water boils, it evaporates—that is, vaporizes into the environment. Converting 1 g of the hottest water to a gas requires an input of 540 calories of energy. Water has a high heat of evaporation because hydrogen bonds must be broken before water boils. Water's high heat of vaporization gives our bodies an efficient way to release excess body heat in a hot environment. When we sweat, or get splashed with water, body heat is used to vaporize water, thus cooling us down (Fig. 2.8*b*). Because of water's high heat of vaporization and ability to hold onto its heat, temperatures along the coasts are moderate. During the summer, the ocean absorbs and stores solar heat, and during the winter, the ocean releases it slowly. In contrast, the interior regions of continents experience abrupt changes in temperatures.

[1]Be aware that a calorie—spelled with a lowercase *c*—is not the same as a Calorie (capital *C*), which is used to measure the heat energy of food. A Calorie is 1,000 calories, or one kilocalorie.

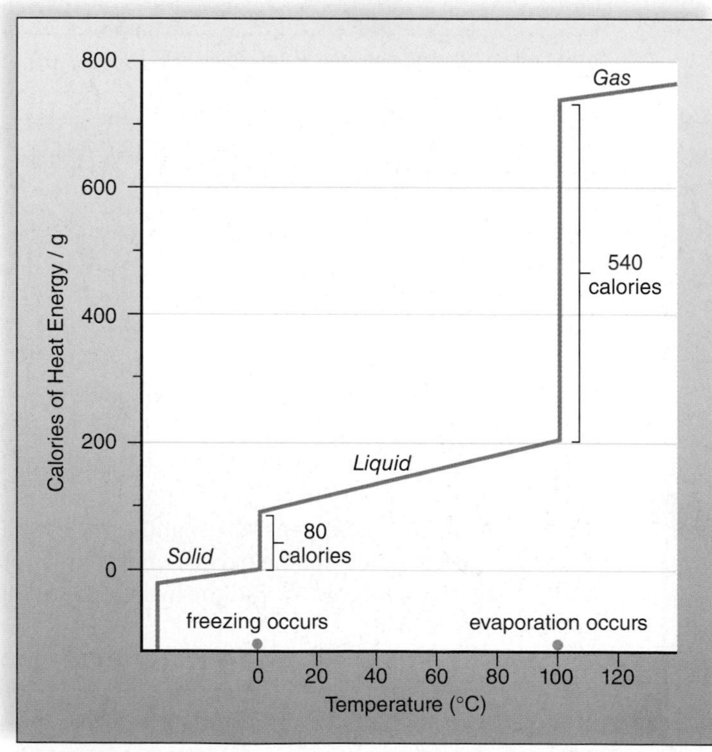

a. Calories lost when 1 g of liquid water freezes and calories required when 1 g of liquid water evaporates.

b. Water's high heat of evaporation is useful for cooling the body

Figure 2.8 Temperature and Water.
a. The high heat capacity of water resists a change from a liquid to gaseous state. If it was a gas at a lower temperature, life could not exist. **b.** Body heat vaporizes sweat. In this way, bodies cool when the temperature rises.

Water Is a Solvent Due to its polarity, water facilitates chemical reactions, both outside and within living systems. As a solvent, it dissolves a great number of substances, especially those that are also polar. A *solution* contains dissolved substances, which are then called *solutes.* When ionic compounds—for example, sodium chloride (NaCl)—are put into water, the negative ends of the water molecules are attracted to the sodium ions, and the positive ends of the water molecules are attracted to the chloride ions. This attraction causes the sodium ions and the chloride ions to separate, or dissociate, in water.

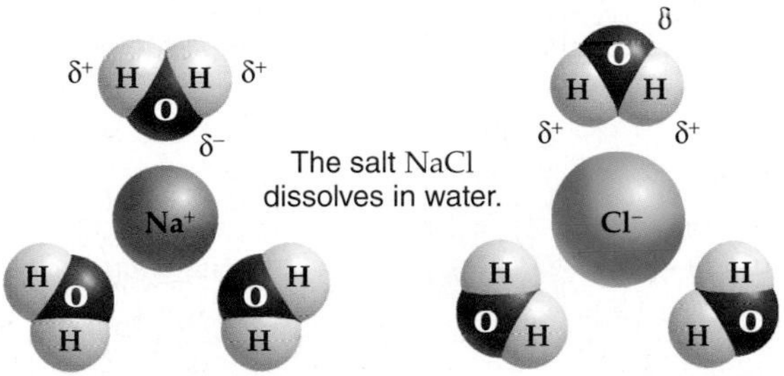

Molecules that can attract water are said to be **hydrophilic.** When ions and molecules disperse in water, they move about and collide, allowing reactions to occur. Molecules that cannot attract water, also known as *nonpolar* molecules, are said to be **hydrophobic.** Hydrophilic molecules tend to attract other polar molecules; similarly, hydrophobic substances usually associate with other nonpolar molecules. Many oils, such as vegetable oils, are nonpolar, and therefore hydrophobic. This is why oil does not mix well with water.

Water Molecules Are Cohesive and Adhesive Cohesion refers to the ability of water molecules to cling to each other due to hydrogen bonding. At any moment in time, a water molecule can form hydrogen bonds with at most four other water molecules. Because of cohesion, water exists as a liquid under the conditions of temperature and pressure present at the Earth's surface. The strong cohesion of water molecules is apparent because water flows freely, yet water molecules do not separate from each other.

Adhesion refers to the ability of water molecules to cling to other polar surfaces. This is a result of water's polarity. Many animals, including humans, contain internal vessels in which water assists the transport of nutrients and wastes because the cohesion and adhesion of water allows blood to fill the tubular vessels of the cardiovascular system. For example, the liquid portion of our blood, which transports dissolved and suspended substances about the body, is 92% water. The water in our blood assists in the transport of nutrients and oxygen to our cells, and the removal of waste material from the cells.

APPLICATIONS AND MISCONCEPTIONS

How do lungs stay open and keep from collapsing?

Our lives depend on water's cohesive property. A thin film of water coats the surface of the lungs and the inner chest wall. This film allows the lungs to stick to the chest wall, keeping the lungs open so we can breathe.

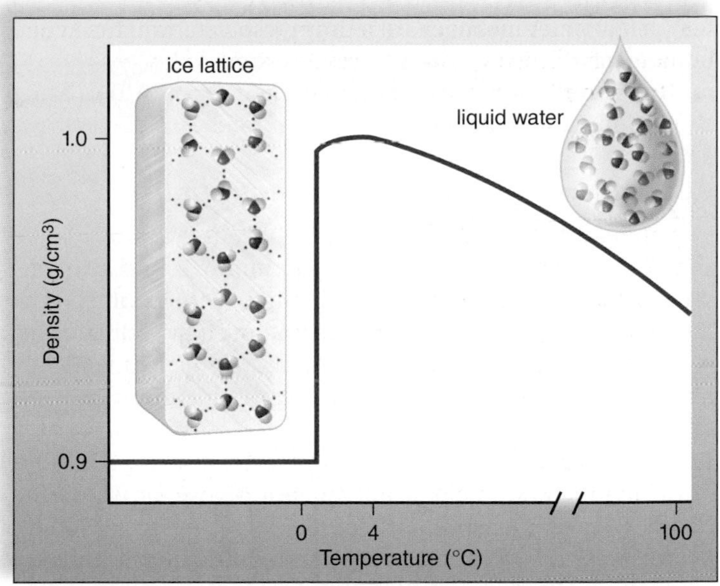

Figure 2.9 **Ice is Less Dense than Water.**
Ice is less dense than water because the hydrogen bonds in ice are farther apart than the hydrogen bonds of liquid water.

Frozen Water Is Less Dense Than Liquid Water As liquid water cools, the molecules come closer together. Water is most dense at 4°C, but the water molecules are still moving about (Fig. 2.9). At temperatures below 4°C, only vibrational movement occurs, and hydrogen bonding becomes more rigid but also more open. This means that water expands as it reaches 0°C and freezes, which is why cans of soda burst when placed in a freezer, or why frost heaves make northern roads bumpy in the winter. It also means that ice is less dense than liquid water, and therefore ice floats on liquid water.

This property of water plays an important role in many aquatic ecosystems. If ice did not float on water, it would sink to the bottom, and ponds, lakes, and perhaps even the ocean would freeze solid, making life impossible in the water and also on land. Instead, bodies of water always freeze from the top down. When a body of water freezes on the surface, the ice acts as an insulator to prevent the water below it from freezing. This allows aquatic organisms to survive the winter. As ice melts in the spring, it draws heat from the environment, helping to prevent a sudden change in temperature that might be harmful to life.

Acids and Bases

When water molecules dissociate (break up), they release an equal number of hydrogen ions (H^+) and hydroxide ions (OH^-):

$$\underset{\text{water}}{H{-}O{-}H} \rightleftharpoons \underset{\text{hydrogen ion}}{H^+} + \underset{\text{hydroxide ion}}{OH^-}$$

Only a few water molecules at a time dissociate, and the actual number of H^+ or OH^- is 10^{-7} moles/liter. A **mole** is a unit of scientific measurement for atoms, ions, and molecules.

MP3 Water and pH

Acidic Solutions (High H^+ Concentrations)

Lemon juice, vinegar, tomatoes, and coffee are all acidic solutions. What do they have in common? **Acids** are substances that dissociate in water, releasing hydrogen ions (H^+). The acidity of a substance depends on how fully it dissociates in water. For example, an important inorganic acid is hydrochloric acid (HCl), which dissociates in this manner:

$$HCl \longrightarrow H^+ + Cl^-$$

If hydrochloric acid is added to a beaker of water, the number of hydrogen ions (H^+) increases greatly. In our bodies, hydrochloric acid is produced by the stomach and aids in food digestion.

Basic Solutions (Low H^+ Concentrations)

Milk of magnesia and ammonia are commonly known basic substances. **Bases** are substances that either take up hydrogen ions (H^+) or release hydroxide ions (OH^-). For example, an important base is sodium hydroxide (NaOH), which dissociates almost completely in this manner:

$$NaOH \longrightarrow Na^+ + OH^-$$

If sodium hydroxide is added to a beaker of water, the number of hydroxide ions increases. Sodium hydroxide is also called *lye* and is contained in many drain-cleaning products.

Some acids and bases are strong, meaning that they donate a large number of H^+ or OH^- ions. You should not taste strong acids or bases because they are destructive to cells. Many household cleansers, such as ammonia or bleach, have poison symbols and carry an important warning not to ingest the product.

pH Scale

The **pH scale** is used to indicate the acidity or basicity (alkalinity) of a solution. The pH scale (Fig. 2.10) ranges from 0 to 14. A pH of 7 represents a neutral state in which the hydrogen ion and hydroxide ion concentrations are equal. A pH below 7 is an acidic solution because the hydrogen ion concentration is greater than the hydroxide concentration. A pH above 7 is basic because the $[OH^-]$ is greater than the $[H^+]$. Further, as we move down the pH scale from pH 14 to pH 0, each unit is 10 times more acidic than the previous unit.[2] As we move up the scale from 0 to 14, each unit is 10 times more basic than the previous unit. Therefore, pH 5 is 100 times more acidic than pH 7 and 100 times more basic than pH 3.

The pH scale was devised to eliminate the use of cumbersome numbers. For example, the possible hydrogen ion concentrations of a solution are on the left of this listing and the pH is on the right:

	$[H^+]$ (moles per liter)	pH
0.000001	$= 1 \times 10^{-6}$	6
0.0000001	$= 1 \times 10^{-7}$	7
0.00000001	$= 1 \times 10^{-8}$	8

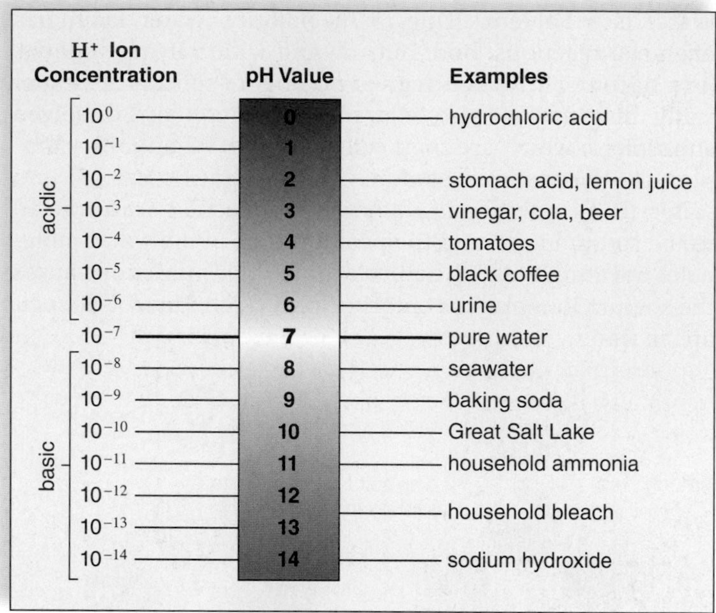

Figure 2.10 **The pH scale.**
The pH scale ranges from 0 to 14, with 0 being the most acidic and 14 being the most basic. A solution at pH 7 (neutral pH) has equal amounts of hydrogen ions (H^+) and hydroxide ions (OH^-). An acidic pH has more H^+ than OH^- and a basic pH has more OH^- than H^+. The pH of familiar solutions can be seen on the scale.

To further illustrate the relationship between hydrogen ion concentration and pH, consider the following question. Which of the pH values listed indicates a higher hydrogen ion concentration $[H^+]$ than pH 7, and therefore would be an acidic solution? A number with a smaller negative exponent indicates a greater quantity of hydrogen ions than one with a larger negative exponent. Therefore, pH 6 is an acidic solution.

MP3 Acid-Base Balances

Buffers

From organisms to ecosystems, the pH needs to be maintained within a narrow range to prevent negative consequences. Normally, pH stability is possible because the body and the environment have **buffers** to prevent pH changes. At the ecosystem level, a problem arises when precipitation in the form of rain or snow becomes so acidic that the environment runs out of natural buffers in the soil or water. Rain normally has a pH of about 5.7, but very acidic rain—sometimes

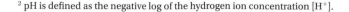

[2] pH is defined as the negative log of the hydrogen ion concentration $[H^+]$.

with a pH as low as 2.1—is recorded every year across the Appalachian Mountains. Why is the pH of the rain so low? Rain becomes acidic because the burning of fossil fuels emits sulfur dioxides (SO_2) and nitrogen oxides (NO_2) into the atmosphere. They combine with water to produce sulfuric acid (H_2SO_4) and nitric acid (HNO_3). Acid deposition destroys statues and kills forests (see Chapter 23). It also leads to fish kills in lakes and streams.

Buffers help keep the pH within normal limits because they are chemicals or combinations of chemicals that take up excess hydrogen ions (H^+) or hydroxide ions (OH^-). For example, carbonic acid (H_2CO_3) is a weak acid that minimally dissociates and then re-forms in the following manner:

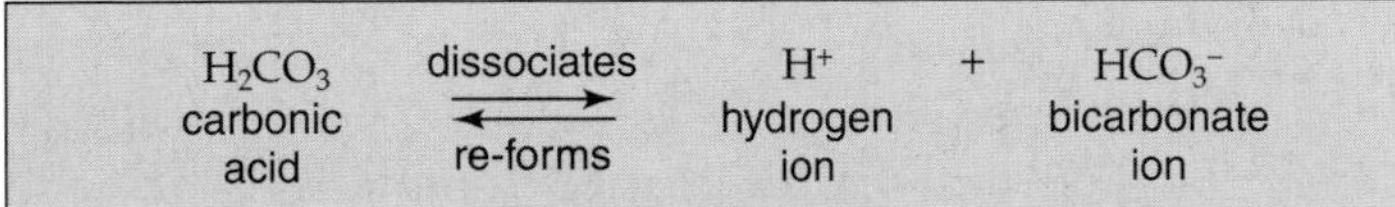

The pH of our blood when we are healthy is always about 7.4—just slightly basic (alkaline). Blood always contains a combination of some carbonic acid and some bicarbonate ions. When hydrogen ions (H^+) are added to blood, the following reaction occurs:

$$H^+ + HCO_3^- \longrightarrow H_2CO_3$$

When hydroxide ions (OH^-) are added to blood, this reaction occurs:

$$OH^- + H_2CO_3 \longrightarrow HCO_3^- + H_2O$$

These reactions prevent any significant change in blood pH.

CHECK YOUR PROGRESS 2.2

1. List the characteristics of water that help support life.
2. Contrast the hydrogen ion concentration of acids and bases.
3. Summarize why the pH of the environment (soil and water) and the body rarely change.

CONNECTING THE CONCEPTS

The properties of water play an important role in human physiology and maintaining homeostasis. For a few examples, refer to the following discussions:

Section 5.6 provides a description of how the body moves nutrients using the circulatory system.

Section 9.6 explores how the respiratory system helps regulate the pH of the blood.

Section 10.4 examines how the kidneys maintain water and pH homeostasis in the body.

2.3 Molecules of Life

LEARNING OUTCOMES

Upon completion of this section, you should be able to

1. List the four classes of organic molecules that are found in cells.
2. Describe the processes by which the organic molecules are assembled and disassembled.

Four categories of **organic molecules**—carbohydrates, lipids, proteins, and nucleic acids—are unique to cells. In biology, **organic** doesn't refer to how food is grown; it refers to a molecule that contains carbon (C) and hydrogen (H) and is usually associated with living organisms.

MP3 Organic Molecules

Each type of organic molecule in cells is composed of subunits. When a cell constructs a **macromolecule,** a molecule that contains many subunits, it uses a **dehydration reaction,** a type of synthesis reaction. During a dehydration reaction, an —OH (hydroxyl group) and an —H (hydrogen atom), the equivalent of a water molecule, are removed as the molecule forms (Fig. 2.11*a*). The reaction is reminiscent of a train whose length is determined by how many boxcars it has hitched together. To break down macromolecules, the cell uses a **hydrolysis reaction** in which the components of water are added during the breaking of the bond between the molecules (Fig. 2.11*b*).

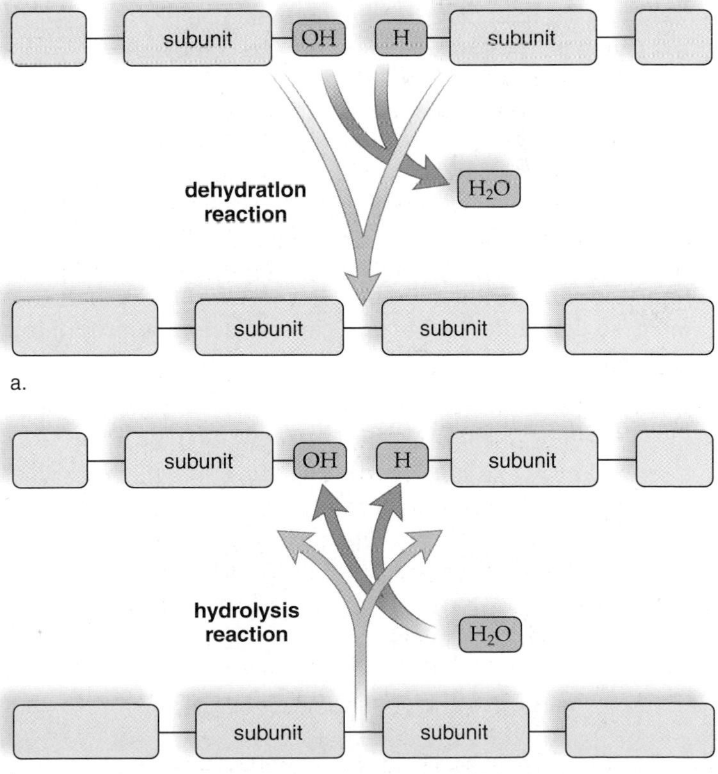

Figure 2.11 The breakdown and synthesis of macromolecules. **a.** Subunits bond in a dehydration reaction. Water is given off and a macromolecule forms. **b.** In the reverse reaction, macromolecules are divided into subunits by hydrolysis.

CHECK YOUR PROGRESS 2.3

1. Explain the difference between an organic and an inorganic molecule.
2. List the four classes of molecules that are unique to cells.
3. Describe the type of reaction that occurs during the synthesis of a macromolecule.

CONNECTING THE CONCEPTS

The organic nutrients not only play an important role in the building of cells but also serve as sources of energy. For more information, refer to the following discussions:

Section 3.6 explains how the body uses organic nutrients as energy sources.

Sections 8.1–8.4 explore how the digestive system processes the organic nutrients and prepares them for transport.

2.4 Carbohydrates

LEARNING OUTCOMES

Upon completion of this section, you should be able to

1. Summarize the basic chemical properties of a carbohydrate.
2. State the roles of carbohydrates in human physiology.
3. Compare the structure of simple and complex carbohydrates.
4. Explain the importance of fiber in the diet.

Carbohydrate molecules are characterized by the presence of the atomic grouping H—C—OH, in which the ratio of hydrogen atoms (H) to oxygen atoms (O) is approximately 2:1. This ratio is the same as the ratio in water (*hydros* in Greek means "water," so the name "hydrates of carbon" seems appropriate). **Carbohydrates,** first and foremost, function for quick and short-term energy storage in all organisms, including humans.

Simple Carbohydrates

If a carbohydrate is made up of just one ring and its number of carbon atoms is low (from five to seven), it is called a simple sugar, or **monosaccharide.** The designation *pentose* means a 5-carbon sugar, and the designation *hexose* means a 6-carbon sugar. **Glucose,** the hexose our bodies use as an immediate source of energy, can be written in any one of these ways:

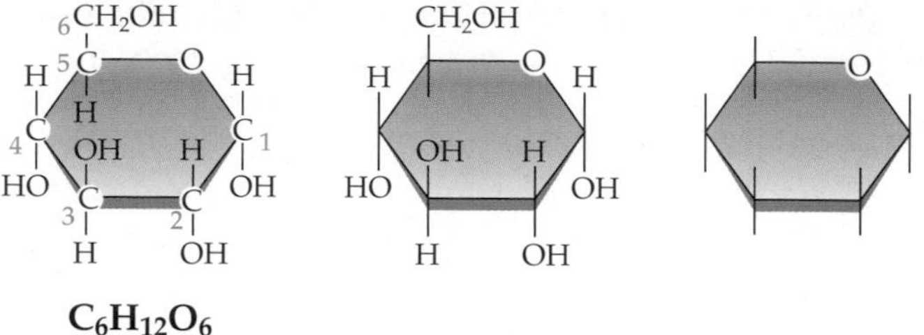

Other common hexoses are fructose, found in fruits, and galactose, a constituent of milk.

A **disaccharide** (*di,* "two"; *saccharide,* "sugar") is made by joining only two monosaccharides together by a dehydration reaction. Maltose is a disaccharide that is formed by a dehydration reaction between two glucose molecules (Fig. 2.12).When our hydrolytic digestive juices break down maltose, the result is two glucose molecules. When glucose and fructose join, the disaccharide sucrose forms. Sucrose, ordinarily derived from sugarcane and sugar beets, is commonly known as table sugar.

Complex Carbohydrates (Polysaccharides)

Macromolecules such as starch, glycogen, and cellulose are **polysaccharides** that contain many glucose units.

Starch and **glycogen** are readily stored forms of glucose in plants and animals, respectively. Some of the macromolecules in starch are long chains of up to 4,000 glucose units. Starch and glycogen have slightly different structures. Starch has fewer side branches, or chains, than does glycogen. A side chain of glucose branches off from the main chain, as shown in Figures 2.13 and 2.14. Flour, used for baking and typically obtained from grinding wheat, is high in starch. Pasta and potatoes are also high-starch foods.

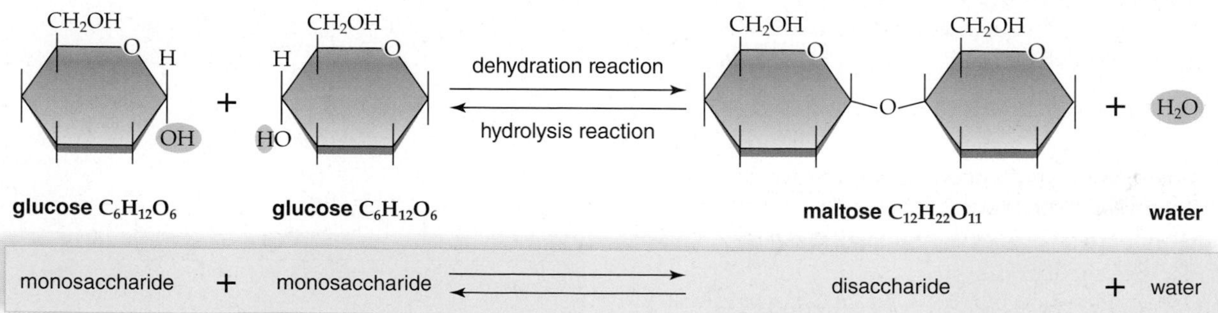

Figure 2.12 The synthesis and breakdown of a dissacharide.
Maltose, a dissacharide, is formed by a dehydration reaction between two glucose molecules. The breakdown of maltose occurs following a hydrolysis reaction and the addition of water.

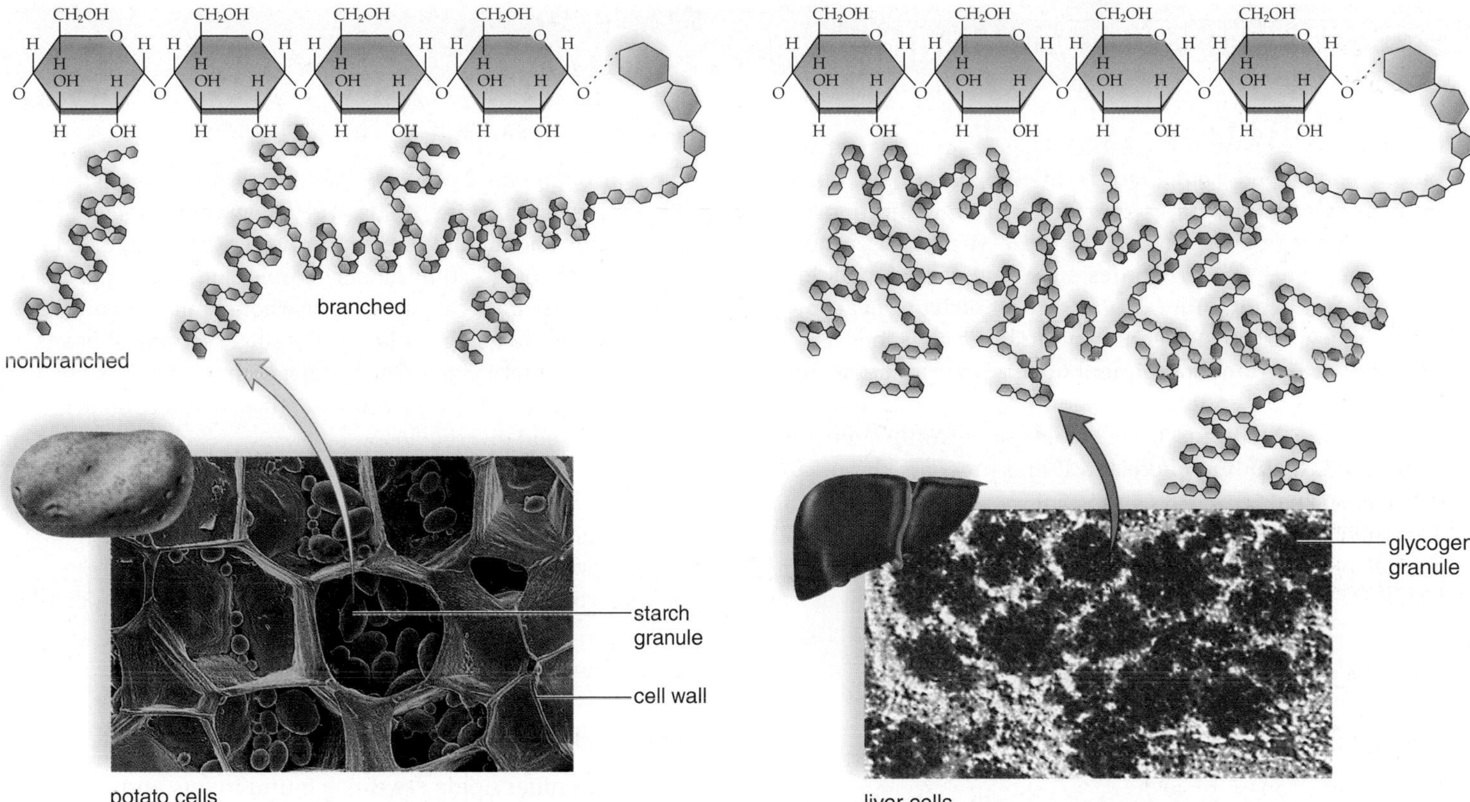

Figure 2.13 **Starch is a plant complex carbohydrate.** Starch has straight chains of glucose molecules. Some chains are also branched, as indicated. The electron micrograph shows starch granules in potato cells.

Figure 2.14 **Glycogen is an animal complex carbohydrate.** Glycogen is more branched than starch. The electron micrograph shows glycogen granules in liver cells.

After we eat starchy foods, such as potatoes, bread, and cake, glucose enters the bloodstream and, normally, the liver stores glucose as glycogen. The release of the hormone insulin from the pancreas promotes the storage of glucose as glycogen. In between eating, the liver releases glucose so that, normally, the blood glucose concentration is always about 0.1%.

The polysaccharide **cellulose,** commonly called fiber, is found in plant cell walls. In cellulose, the glucose units are joined by a slightly different type of linkage than that in starch or glycogen (Fig. 2.15). Though this might seem to be a technicality, it is important because we are unable to digest foods containing this type of linkage; therefore, cellulose largely passes through our digestive tract as fiber, or roughage.

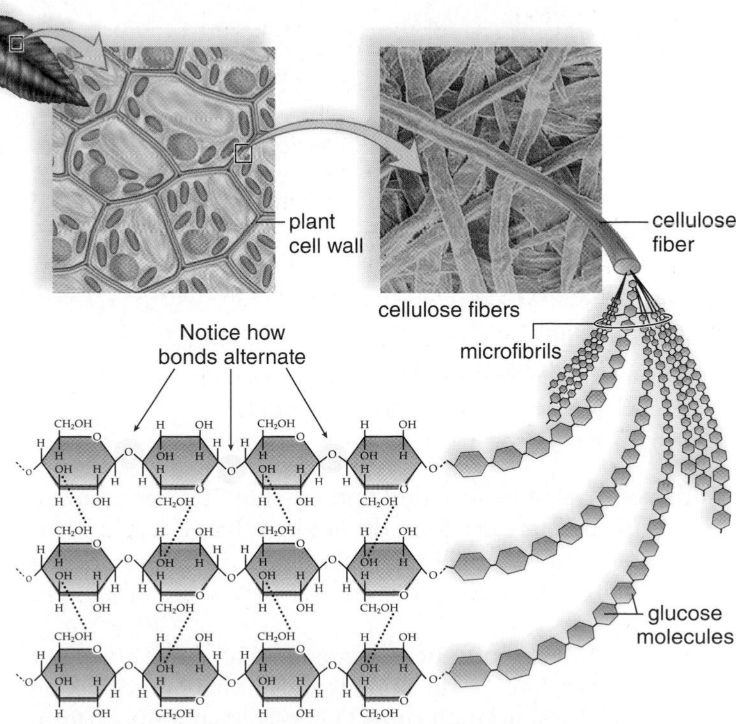

Figure 2.15 **Fiber is a plant complex carbohydrate.** Unlike starch, we don't get energy directly from fiber. This is because of the arrangements of the chemical bonds in the fiber chains. However, fiber is an important component of our diet and helps keep our digestive system healthy.

CHECK YOUR PROGRESS 2.4

1. Explain the function of simple carbohydrates and polysaccharides in humans.
2. Describe the difference in structure between a simple carbohydrate and the various complex carbohydrates.
3. Explain why some complex carbohydrates may be used for energy, but others may not.

BIOLOGY MATTERS  Health

Fiber in the Diet

Fiber, also called *roughage,* is mainly composed of the undigested carbohydrates that pass through the digestive system. Most fiber is derived from the structural carbohydrates of plants. This includes such material as cellulose, pectins, and lignin. Fiber is not truly a nutrient because we do not use it directly for energy or building cells, but it is an extremely important component of our diet. Fiber adds bulk to material in the intestines, keeping the colon functioning normally, and it binds many types of harmful chemicals in the diet, including cholesterol, and prevents them from being absorbed.

There are two basic types of fiber—insoluble and soluble. Soluble fiber dissolves in water and acts in the binding of cholesterol. Soluble fiber is found in many fruits, as well as oat grains. Insoluble fiber provides the bulk to the fecal material and is found in bran, nuts, seeds, and whole wheat foods.

An adult male should consume around 38 g per day of fiber, whereas the average adult female should consume approximately 25 g per day. One serving of whole-grain bread (one slice) provides about 3 g of fiber, and a single serving of beans (1/2 cup) contains 4–5 g of fiber. A diet high in fiber has been shown to reduce the risk of cardiovascular disease, diabetes, colon cancer, and diverticulosis.

Questions to Consider

1. How might a diet that is high in fiber also prevent a person from overeating?
2. Why would you want to vary the types of fiber in your diet?

CONNECTING THE CONCEPTS

In humans, carbohydrates primarily serve as energy molecules, although fiber does act to ensure the health of the digestive system. For more information on the interaction of the human body with carbohydrates, refer to the following discussions:

Section 3.6 explores the use of carbohydrates for energy at the cellular level.

Section 8.3 examines how the digestive system processes carbohydrates.

Section 8.6 explains how fiber promotes the health of the digestive system.

2.5 Lipids

LEARNING OUTCOMES

Upon completion of this section, you should be able to

1. Compare the structure of fats, phospholipids, and steroids.
2. State the function of each class of lipids.

Lipids are diverse in structure and function, but they have a common characteristic: They do not dissolve in water. Their low solubility in water is due to an absence of polar groups. They contain little oxygen and consist mostly of carbon and hydrogen atoms.

Lipids contain more energy per gram than other biological molecules; therefore, fats in animals and oils in plants function well as energy-storage molecules. Others (phospholipids) form a membrane so that the cell is separated from its environment and has inner compartments as well. Steroids are a large class of lipids that includes, among other molecules, the sex hormones.

MP3 Lipids

Fats and Oils

The most familiar lipids are those found in fats and oils. **Fats,** usually of animal origin (e.g., lard and butter), are solid at room temperature. **Oils,** usually of plant origin (e.g., corn oil and soybean oil), are liquid at room temperature. Fat has several functions in the body. It is used for long-term energy storage, insulates against heat loss, and forms a protective cushion around major organs. Steroids are formed from smaller lipid molecules and function as chemical messengers.

Emulsifiers can cause fats to mix with water. They contain molecules with a nonpolar end and a polar end. The molecules position themselves about an oil droplet so that their polar ends project outward. The droplet *disperses* in water, which means that **emulsification** has occurred. In our bodies, emulsification occurs during the digestion of fatty foods. To assist in the breakdown of fats, the liver manufactures bile. Bile is stored by the gallbladder and released into the small intestine following a meal. Bile emulsifies the fats in the food, allowing greater access by the digestive enzymes. Fats and oils form when one glycerol molecule reacts with three fatty acid molecules (Fig. 2.16). A fat is sometimes called a **triglyceride** because of its three-part structure, or the term *neutral fat* can be used because the molecule is nonpolar and carries no charges.

Waxes are molecules made up of one fatty acid combined with another single organic molecule, usually an alcohol (chemists refer to "alcohols" as an entire group of molecules that includes drinking alcohol and rubbing alcohol). Waxes prevent loss of moisture from body surfaces. Cerumen, or ear wax, is a very thick wax produced by glands lining the outer ear canal (see Chapter 14). It protects the ear canal from irritation and infection by trapping particles, bacteria, and viruses. When ear wax is completely washed away by swimming or diving, the result is a painful "swimmer's ear."

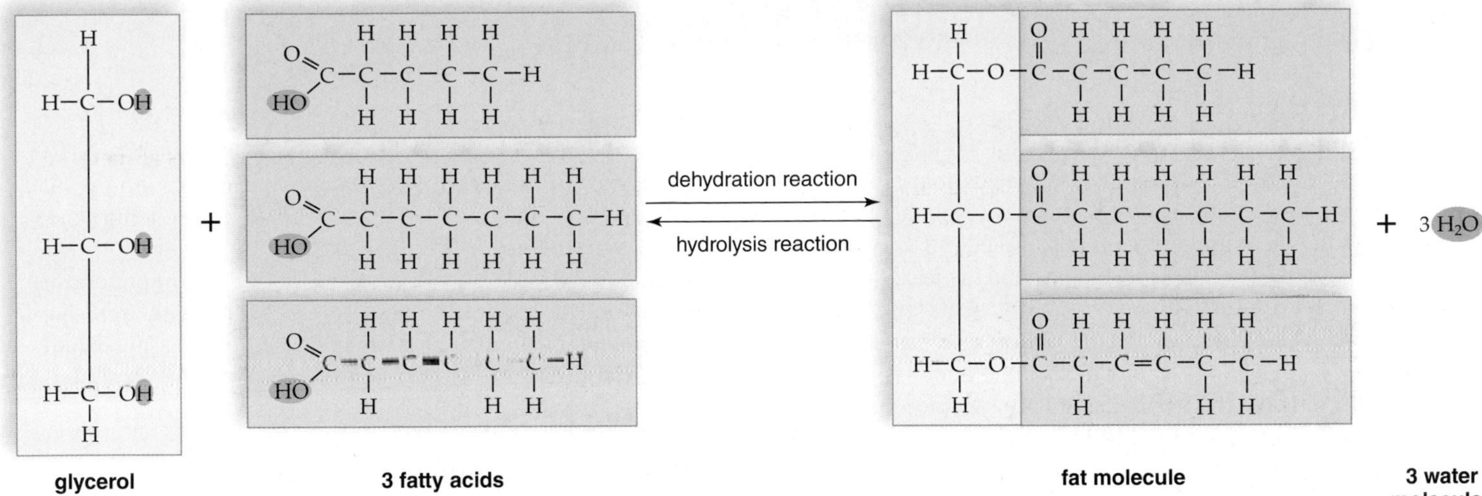

Figure 2.16 **Structure of a triglyceride.**
Triglycerides are formed when three fatty acids combine with glycerol by dehydration synthesis reactions. The reverse reaction starts digestion of fat; hydrolysis introduces water, and fatty acid–glycerol bonds are broken.

Saturated, Unsaturated, and Trans Fatty Acids

A **fatty acid** is a carbon-hydrogen chain that ends with the acidic group —COOH (Fig. 2.16, *left*). Most of the fatty acids in cells contain 16 or 18 carbon atoms per molecule, although smaller ones with fewer carbons are also known. Fatty acids are either saturated or unsaturated. **Saturated fatty acids** have no double bonds between the carbon atoms. The chain is saturated, so to speak, with all the hydrogens it can hold. **Unsaturated fatty acids** have double bonds in the carbon chain wherever the number of hydrogens is less than two per carbon (Fig. 2.17).

H H / —C=C— — **Unsaturated cis fats** (oils)

H H / —C—C— / H H — **Saturated** (butter)

H / —C=C— / H — **Unsaturated trans fats** (hydrogenated oils)

Figure 2.17 **Comparison of saturated, unsaturated, and trans fats.**
Saturated fats have no double bonds between carbons in the fatty acid. Unsaturated fats have one or more double bonds in the fatty acid. For a fat to be a trans fat, the hydrogens need to be on opposite sides of the carbon–carbon double bond.

BIOLOGY MATTERS  **Health**

Good and Bad Cholesterol

Blood tests to analyze your lipid profile are part of many annual medical exams. After your annual exam, your doctor tells you that your total cholesterol is 210, your HDL value is low (34), and your LDL value is high (110). You know that you need to get your total cholesterol below 200, the threshold of a healthy diet; but like most people, you have no idea what the other two numbers mean. Then you remember that LDL is commonly referred to as "bad" cholesterol, and HDL is called the "good" cholesterol. In actuality, these molecules are not forms of cholesterol; they are types of proteins. The lipoproteins in the body serve as a form of fat and cholesterol carrier, moving these nutrients around as needed. An LDL is a lipoprotein that is full of triglycerides and cholesterol, whereas an HDL is basically empty. Thus, a high LDL value indicates that your carriers were usually full, meaning that your diet must be providing too many of these nutrients. After additional research, you find out that other factors, such as the amount of dietary fiber, daily exercise, and even genetics, can play a role in regulating "good" and "bad" levels of these lipoproteins. Furthermore, the numbers that the doctor gave you were actually concentrations of these molecules in your blood (in milligrams per deciliter, or mg/dl). In today's world, it is important that we all understand the terminology associated with our own medical history.

Questions to Consider

1. Why might physicians resort to calling HDLs and LDLs "good" and "bad"?
2. Why would a high HDL level result in a low LDL level?

BIOLOGY MATTERS  Health

The Omega-3 Fatty Acids

Not all fats are bad. In fact, some of them are essential to our health. A special class of unsaturated fatty acids, the omega-3 fatty acids, is considered both an essential and developmentally important nutrient. The name *omega-3* (also called *n-3 fatty acids*) is derived from the location of the double bond in the carbon chain.

The three important omega-3 fatty acids are linolenic acid (ALA), docosahexaenoic acid (DHA), and eicosapentaenoic acid (EPA). Omega-3 fatty acids are a major component of the fatty acids in the brain, and adequate amounts of them appear to be important in children and young adults. A diet that is rich in these fatty acids also offers protection against cardiovascular disease, and research is ongoing with regard to other health benefits. DHA may reduce the risk of Alzheimer disease. DHA and EPA may be manufactured from APA in small amounts within our bodies. Some of the best sources of omega-3 fatty acids are cold-water fish such as salmon and sardines. Flax oil, also called *linseed oil,* is an excellent plant-based source of omega-3 fatty acids.

Although the fatty acids are an important component of the diet, nutritionists warn not to overdo the diet with excessive supplements as the omega-3s may cause health issues when taken in large doses.

Questions to Consider

1. Why might a diet high in omega-3 fatty acids also be low in cholesterol?
2. Why would omega-3 fatty acids reduce the risk of cardiovascular disease?

In general, oils, present in cooking oils and bottle margarines, are liquids at room temperature because the presence of a double bond creates a bend in the fatty acid chain. Such kinks prevent close packing between the hydrocarbon chains and account for the fluidity of oils. On the other hand, butter, which contains saturated fatty acids and no double bonds, is a solid at room temperature.

Saturated fats, in particular, contribute to the disease *atherosclerosis.* Atherosclerosis is caused by formation of lesions, or *atherosclerotic plaques,* on the inside of blood vessels. The plaques narrow blood vessel diameter, choking off the blood and oxygen supply to tissues. Atherosclerosis is the primary cause of cardiovascular disease (heart attack and stroke) in the United States. Even more harmful than naturally occurring saturated fats are the **trans fats** (*trans* in Latin means "across"; Fig. 2.17), which are created artificially from vegetable oils. Trans fats may be partially hydrogenated to make them semisolid. Complete hydrogenation of oils causes all double bonds to become saturated. Partial hydrogenation does not saturate all bonds. It reconfigures some double bonds, but some of the hydrogen atoms end up on different sides of the chain. This configuration makes these bonds difficult to break, thus allowing trans fats to accumulate in the circulatory system. Trans fats are found in shortenings, solid margarines, and many processed foods (snack foods, baked goods, and fried foods).

Current dietary guidelines from the American Heart Association (AHA) advise replacing trans fats with unsaturated oils. In particular, monounsaturated oils (like olive oil, with one double bond in the carbon chain) are recommended. Polyunsaturated oils (many double bonds in the carbon chain) such as corn oil, canola oil, and safflower oil also fit in the AHA guidelines.

Dietary Fat

For good health, the diet should include some fat; but, for the reasons stated, the first thing to do when looking at a nutrition label is to check the total amount of fat per serving. The total recommended amount of fat in a 2,000-calorie diet is 65 grams (g). That information results in the % Daily Value (DV) given in the sample nutrition label for macaroni and cheese in Figure 2.18. As of January 2006, food manufacturers are required to list the amount of trans fats greater than 0.5 g per serving in the nutrition label for a food.

Animation
Anatomy of a Food Label

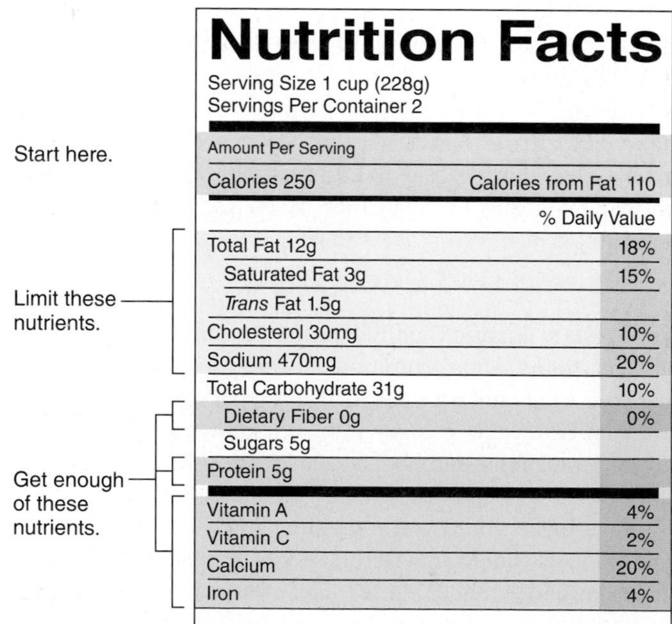

Figure 2.18 Understanding a food label.
Food labels provide some important information about the product. Each of the items listed on the label is referenced to the % Daily Value, which is based on a 2,000-calorie diet. In general, total fat, cholesterol, and sodium should be limited in the diet.

Phospholipids

Phospholipids have a phosphate group (Fig. 2.19). They are constructed like fats, except that in place of the third fatty acid, there is a phosphate group or a grouping that contains both phosphate and nitrogen. These molecules are not electrically neutral, as are fats, because the phosphate and nitrogen-containing groups are ionized. They form the polar (hydrophilic) head of the molecule, and the rest of the molecule becomes the nonpolar (hydrophobic) tails. (Remember that *hydrophilic* means "water-loving" and *hydrophobic* is "water-fearing.")

Phospholipids are the primary components of cellular membranes. They spontaneously form a *bilayer* (a sort of molecular "sandwich") in which the hydrophilic heads (the sandwich "bread") face outward toward watery solutions, and the tails (the sandwich "filling") form the hydrophobic interior (Fig. 2.19*b*).

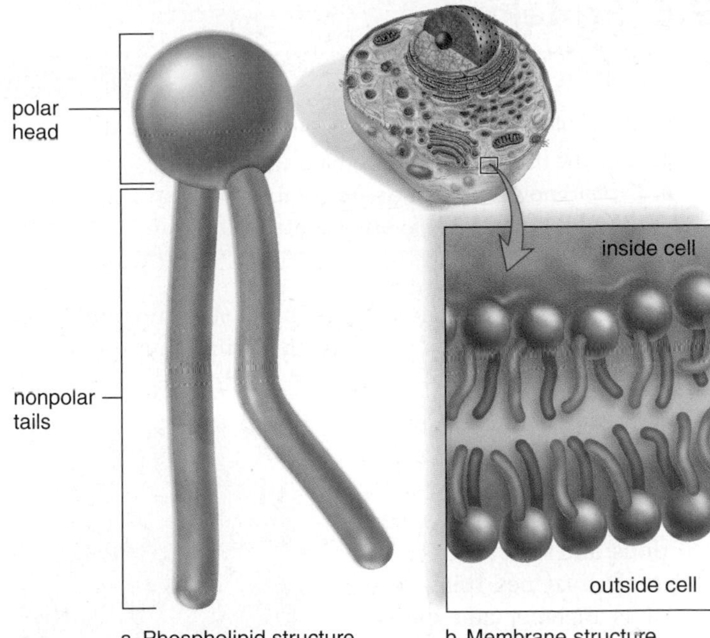

Figure 2.19 **Structure of a phospholipid.**
a. Phospholipids are structured like fats with one fatty acid replaced by a polar phosphate group. Therefore, the head is polar, whereas the tails are nonpolar. **b.** This causes the molecules to arrange themselves in a "sandwich" arrangement when exposed to water—polar phosphate groups on the outside of the layer, nonpolar lipid tails on the inside of the layer.

Steroids

Steroids are lipids that have an entirely different structure from those of fats. Steroid molecules have a backbone of four fused carbon rings. Each one differs primarily by the attached molecules, called *functional groups,* attached to the rings. Cholesterol is a component of an animal cell's plasma membrane and is the precursor of several other steroids, such as the sex hormones estrogen and testosterone. The liver usually makes all the cholesterol the body needs. Dietary sources should be restricted because elevated levels of cholesterol, saturated fats, and trans fats are linked to atherosclerosis, a disease of the blood vessels in which fatty plaques accumulate inside blood vessel linings and reduce blood flow.

The male sex hormone, testosterone, is formed primarily in the testes; the female sex hormone, estrogen, is formed primarily in the ovaries. Testosterone and estrogen differ only by the functional groups attached to the same carbon backbone. However, they have a profound effect on the body and the sexuality of humans and other animals (Fig. 2.20). The taking of anabolic steroids, usually to build muscle strength, is illegal because the side effects are harmful to the body (see the Health feature, "Anabolic Steroid Use," in Chapter 12).

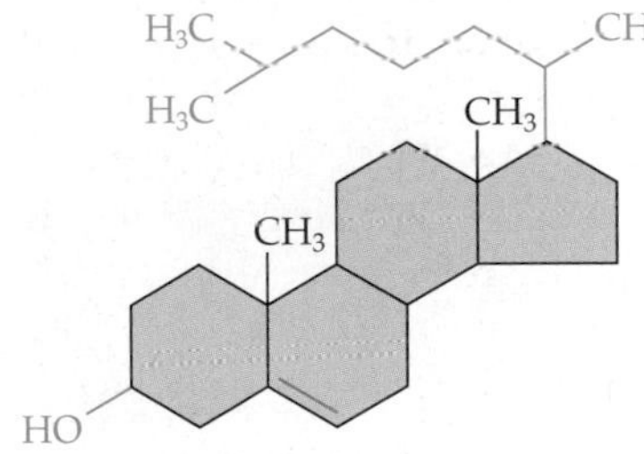

Figure 2.20
Examples of steroids.
a. All steroids are made from cholesterol and have four carbon rings. Compare the structure of (**b**) testosterone and (**c**) estrogen, and notice the slight changes in their attached groups (shown in blue).

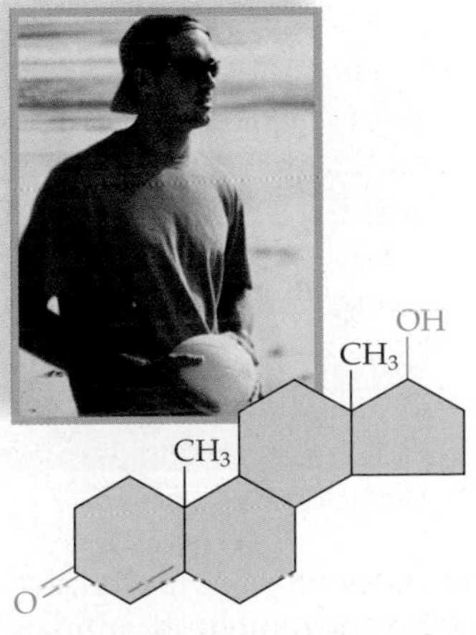

CHECK YOUR PROGRESS 2.5

1. State the function of fats and oils in the human body.
2. List the uses of phospholipids and steroids in the body.
3. Explain the health consequences of a diet high intrans fats.

CONNECTING THE CONCEPTS

Fats and lipids have a variety of uses in the human body. To see how they interact with the various systems of the body, refer to the following discussions:

Section 5.7 provides more information on atherosclerosis.
Section 8.3 explores the digestion and absorption of fats.
Section 15.1 examines how some lipids act as hormones in the body.

2.6 Proteins

LEARNING OUTCOMES

Upon completion of this section, you should be able to

1. Describe the structure of an amino acid.
2. Explain how amino acids are combined to form proteins.
3. Summarize the four levels of protein structure.

Proteins are of primary importance in the structure and function of cells. Some of their many functions in humans follow.

Support: Some proteins are structural proteins. Keratin, for example, makes up hair and nails. Collagen lends support to ligaments, tendons, and skin.

Hair is a protein.

Enzymes: Enzymes bring reactants together and thereby speed chemical reactions in cells. They are specific for one particular type of reaction and only function at body temperature.

Transport: Channel and carrier proteins in the plasma membrane allow substances to enter and exit cells. Some other proteins transport molecules in the blood of animals; **hemoglobin** in red blood cells is a complex protein that transports oxygen.

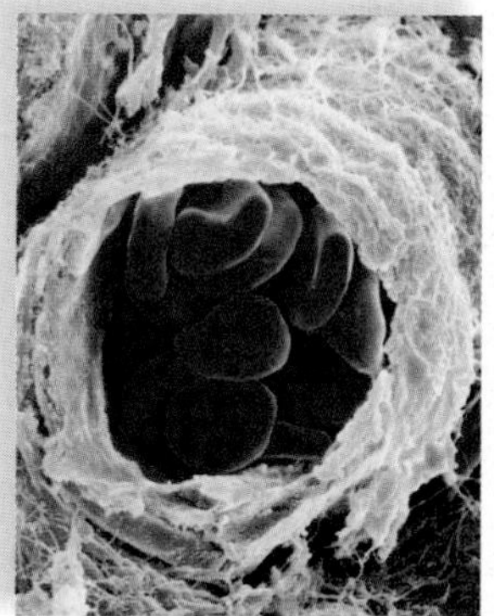

Hemoglobin is a protein.

Defense: Antibodies are proteins. They combine with foreign substances, called antigens. In this way, they prevent antigens from destroying cells and upsetting homeostasis.

Hormones: Hormones are regulatory proteins. They serve as intercellular messengers that influence the metabolism of cells. The hormone insulin regulates the content of glucose in the blood and in cells. The presence of growth hormone determines the height of an individual.

Muscle contains protein.

Motion: The contractile proteins actin and myosin allow parts of cells to move and cause muscles to contract. Muscle contraction facilitates the movement of animals from place to place.

The structures and functions of vertebrate cells and tissues differ according to the type of proteins they contain. For example, muscle cells contain actin and myosin, red blood cells contain hemoglobin, and support tissues contain collagen.

Amino Acids: Subunits of Proteins

Proteins are macromolecules with **amino acid** subunits. The central carbon atom in an amino acid bonds to a hydrogen atom and also to three other groups of atoms. The name *amino acid* is appropriate because one of these groups is an $—NH_2$ (amino group) and another is a —COOH (carboxyl group, an acid). The third group is the *R* group for an amino acid:

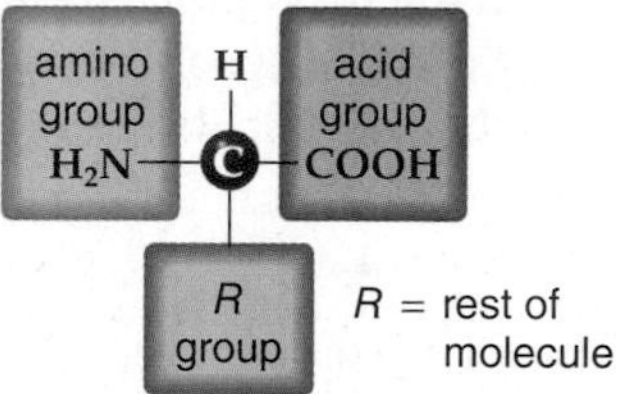

Amino acid

Amino acids differ according to their particular *R* group. The *R* groups range in complexity from a single hydrogen atom to a complicated ring compound. Some *R* groups are polar and some are not. Also, the amino acid cysteine ends with an —SH group, which often serves to connect one chain of amino acids to another by a disulfide bond, —S—S—. Several amino acids commonly found in cells are shown in Figure 2.21.

MP3 Proteins

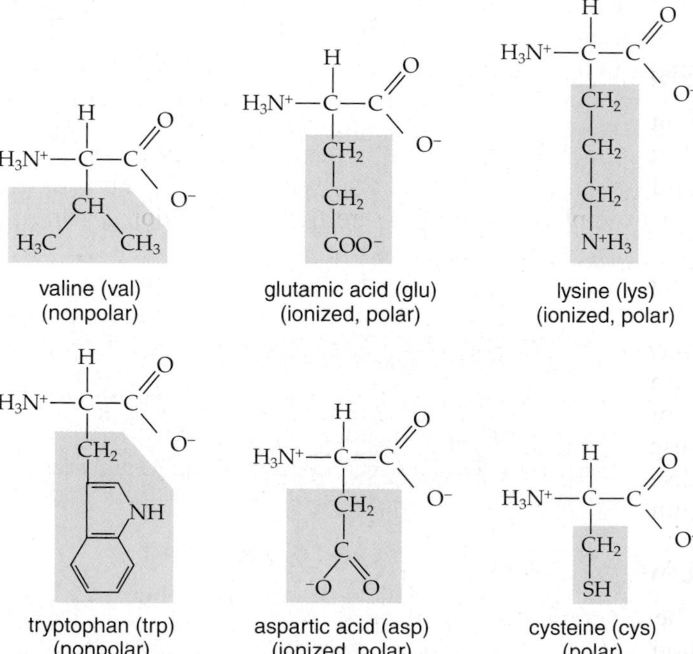

Figure 2.21 The structure of a few amino acids.
Amino acids all have an amine group (H_3N^+), an acid group (COO^-), and an *R* group, all attached to the central carbon atom. The *R* groups (screened in blue) are all different. Some *R* groups are nonpolar and hydrophobic; others are polar and hydrophilic. Still others are polar and ionized.

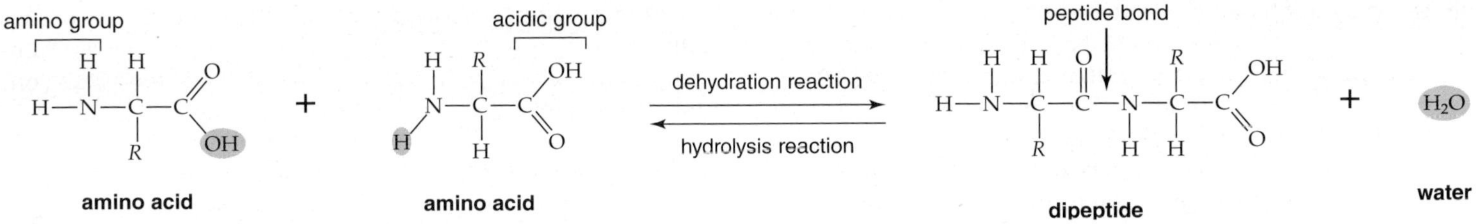

amino group · acidic group · peptide bond

amino acid + amino acid ⇄ (dehydration reaction / hydrolysis reaction) dipeptide + H_2O water

Figure 2.22 Synthesis and breakdown of a protein.
Amino acids join by peptide bonds using a dehydration reaction, and a water molecule is given off. In the reverse reaction, peptide bonds are broken by hydrolysis, and a water molecule is introduced.

Peptides

Figure 2.22 shows how two amino acids join by a dehydration reaction between the carboxyl group of one and the amino group of another. The covalent bond between two amino acids is called a **peptide bond.** When three or more amino acids are linked by peptide bonds, the chain that results is called a **polypeptide.** The atoms associated with the peptide bond share the electrons unevenly because oxygen attracts electrons more than nitrogen. Therefore, the hydrogen attached to the nitrogen has a slightly positive charge (δ^+), whereas the oxygen has a slightly negative charge (δ^-):

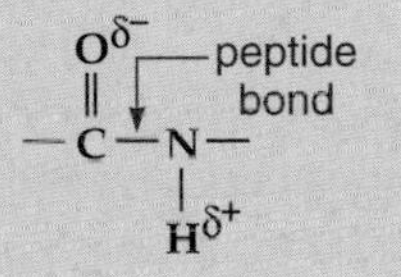

$O^{\delta-}$ — peptide bond — $-C-N-$ — $H^{\delta+}$

δ^- = slightly negative
δ^+ = slightly positive

Shape of Proteins

Proteins cannot function unless they have a specific shape. When proteins are exposed to extremes in heat and pH, they undergo an irreversible change in shape called **denaturation.** For example, we are all aware that the addition of vinegar (an acid) to milk causes curdling. Similarly, heating causes coagulation of egg whites, which contain a protein called albumin. Denaturation occurs because the normal bonding between the *R* groups has been disturbed. Once a protein loses its normal shape, it is no longer able to perform its usual function. Researchers recognize a change in protein shape is responsible for diseases such as Alzheimer disease and Creutzfeldt-Jakob disease (the human form of mad cow disease).

Animation Protein Denaturation

Levels of Protein Organization

The structure of a protein has at least three levels of organization and can have four levels (Fig. 2.23). The first level, called the *primary structure,* is the linear sequence of the amino acids joined by peptide bonds. Each particular polypeptide has its own sequence of amino acids.

The *secondary structure* of a protein comes about when the polypeptide takes on a certain orientation in space. Once

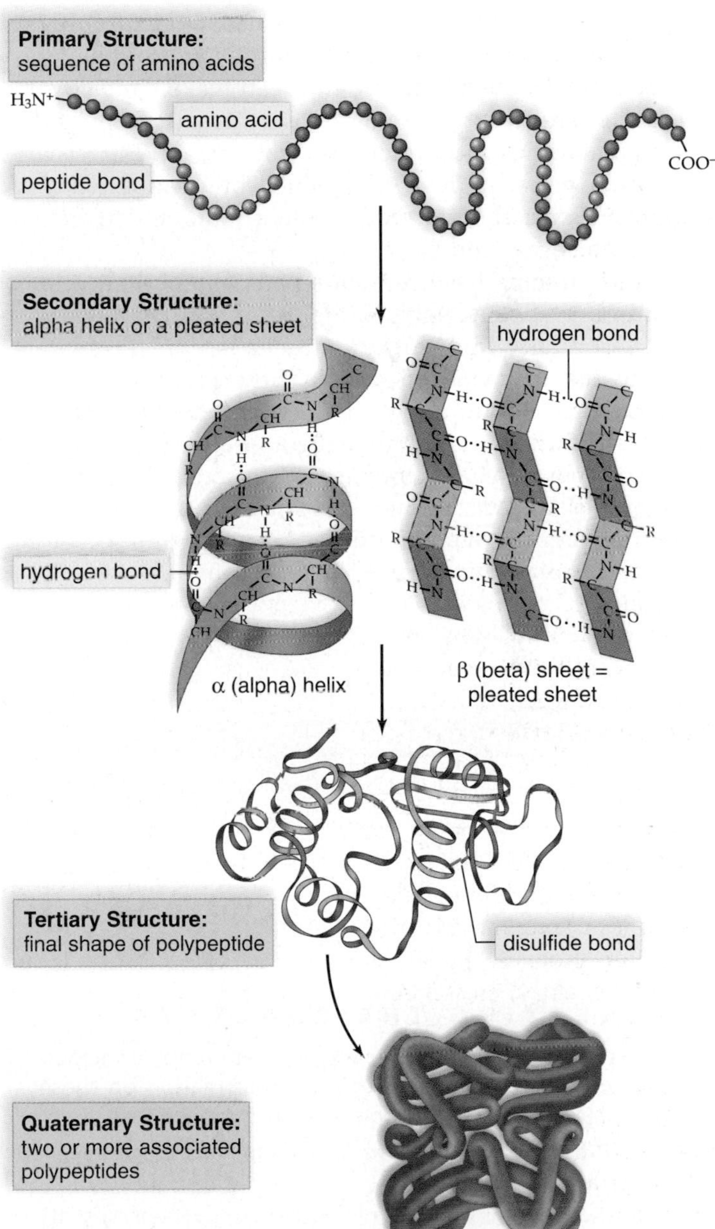

Figure 2.23 Levels of protein structure.
The structure of proteins can differ significantly. Primary structure, the sequence of amino acids, determines secondary and tertiary structure. Quaternary structure is created by assembling smaller proteins into a large structure.

amino acids are assembled into a polypeptide, the resulting C=O section between amino acids in the chain is polar, having a partially negative charge. (Remember that oxygen holds on to electrons longer than carbon, and that's what causes the partially negative charge.) Hydrogen bonding is possible between the C=O of one amino acid and the N—H of another amino acid in a polypeptide. Coiling of the chain results in an α (alpha) helix, or a right-handed spiral, and a folding of the chain results in a pleated sheet. Hydrogen bonding between peptide bonds holds the shape in place.

The *tertiary structure* of a protein is its final, three-dimensional shape. In enzymes, the polypeptide bends and twists in different ways. In most enzymes, the hydrophobic portions are packed on the inside and the hydrophilic portions are on the outside where they can make contact with water. The tertiary structure of the enzymes determines the types of molecules with which they will interact. The tertiary shape of a polypeptide is maintained by various types of bonding between the *R* groups; covalent, ionic, and hydrogen bonding all occur.

Some proteins have only one polypeptide, and others have more than one polypeptide, each with its own primary, secondary, and tertiary structures. These separate polypeptides are arranged to give these proteins a fourth level of structure, termed the *quaternary structure.* Hemoglobin is a complex protein having a quaternary structure; many enzymes also have a quaternary structure. Each of four polypeptides in hemoglobin is tightly associated with a nonprotein *heme* group. A heme group contains an iron (Fe) atom that binds to oxygen; in that way, hemoglobin transports O_2 to the tissues.

CHECK YOUR PROGRESS 2.6

1. Describe the major functions of proteins in organisms.
2. Explain the structure of an amino acid.
3. Describe how the shape of a protein relates to its function.

CONNECTING THE CONCEPTS

Almost every function of the body is somehow connected to the activity of a protein. For more information on these processes, refer to the following discussions:

Section 7.1 gives more information on how misfolded proteins may cause disease.

Section 8.3 explains how the digestive system processes proteins.

Section 13.1 explores how some proteins are used as neurotransmitters in the nervous system.

Section 21.2 examines the process of protein synthesis in a cell.

2.7 Nucleic Acids

LEARNING OUTCOMES

Upon completion of this section, you should be able to

1. Explain the differences between RNA and DNA.
2. Summarize the role of ATP in cellular reactions.

Each cell has a storehouse of information that specifies how a cell should behave, respond to the environment, and divide to make new cells. **Nucleic acids,** which are polymers of nucleotides, store information, include instructions for life, and conduct chemical reactions (Fig. 2.24). **DNA** (**deoxyribonucleic acid**) is one type of nucleic acid that not only stores information about how to copy, or replicate, itself, but also specifies the order in which amino acids are to be joined to make a protein.

RNA (**ribonucleic acid**) is another diverse type of nucleic acid that has multiple uses. Messenger RNA (mRNA) is a temporary copy of a gene in the DNA that specifies what the amino acid sequence will be during the process of protein synthesis. Transfer RNA (tRNA) is also necessary in synthesizing proteins, and helps to translate the sequence of nucleic acids in a gene into the correct sequence of amino acid during protein synthesis. Ribosomal RNA (rRNA) works as an enzyme to form the peptide bonds between amino acids in a polypeptide. A wide range of other RNA molecules also perform important functions within the cell.

Not all nucleotides are made into DNA or RNA polymers. Some nucleotides are directly involved in metabolic functions in cells. For example, some are components of **coenzymes,** nonprotein organic molecules that help regulate enzymatic reactions. **ATP** (**adenosine triphosphate**) is a special nucleotide that stores large amounts of energy needed for synthetic reactions and for various other energy-requiring processes in cells.

MP3 Nucleic Acids

How the Structure of DNA and RNA Differs

Though both DNA and RNA are polymers of nucleotides, there are some small differences in the types of subunits each contains and in their final structure. These differences give DNA and RNA their unique functions in the body.

Nucleotide Structure

Each **nucleotide** is a molecular complex of three types of subunit molecules—phosphate (phosphoric acid), a pentose (5-carbon) sugar, and a nitrogen-containing base:

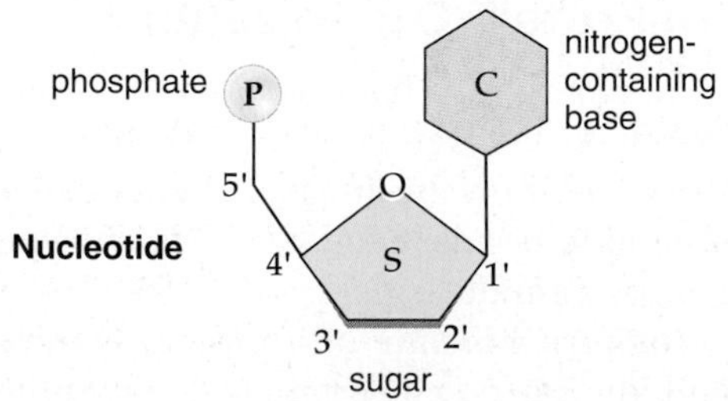

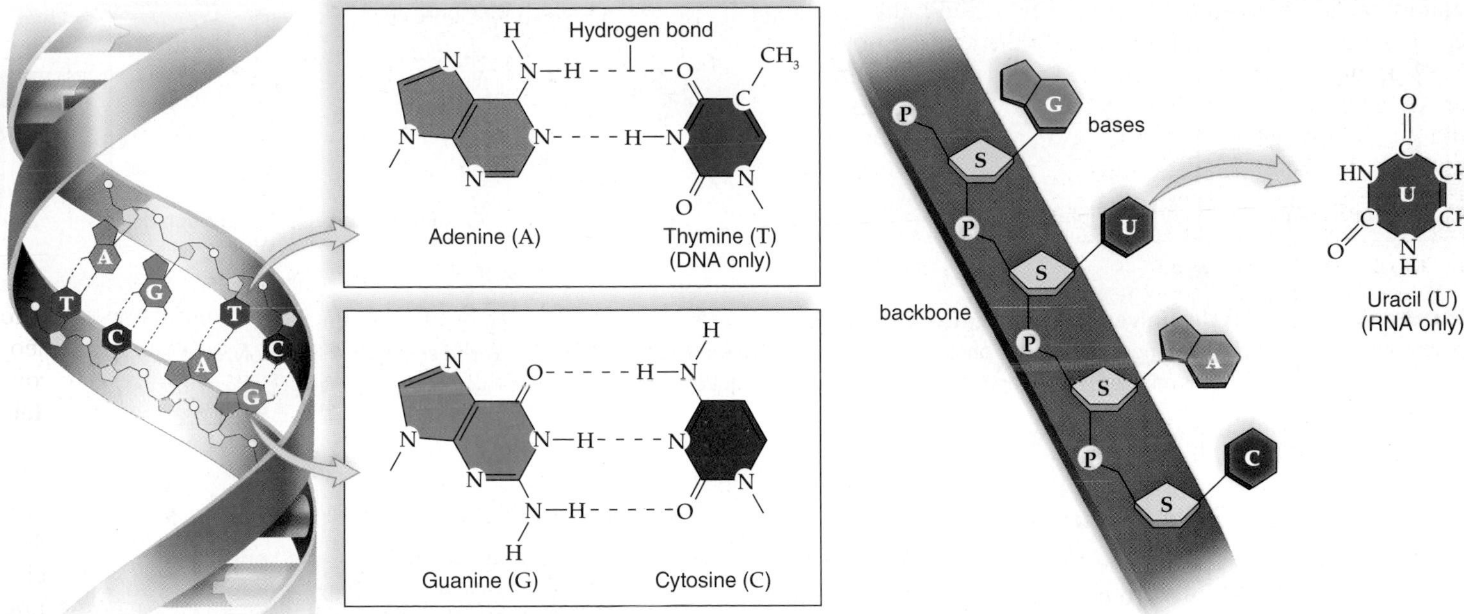

a. DNA structure with base pairs: A with T and G with C

b. RNA structure with bases G, U, A, C

Figure 2.24 The structure of DNA and RNA.
a. In DNA, adenine and thymine are a complementary base pair. Note the hydrogen bonds that join them (like the steps in a spiral staircase). Likewise, guanine and cytosine can pair. **b.** RNA has uracil instead of thymine, so complementary base pairing isn't possible.

Table 2.1 DNA Structure Compared to RNA Structure

	DNA	RNA
Sugar	Deoxyribose	Ribose
Bases	Adenine, guanine, thymine, cytosine	Adenine, guanine, uracil, cytosine
Strands	Double-stranded with base pairing	Single-stranded
Helix	Yes	No

The nucleotides in DNA contain the sugar deoxyribose, and the nucleotides in RNA contain the sugar ribose; this difference accounts for their respective names. There are four different types of bases in DNA: **adenine (A)**, **thymine (T)**, **guanine (G)**, and **cytosine (C)** (Table 2.1). The base can have two rings (adenine or guanine) or one ring (thymine or cytosine). In RNA, the base **uracil (U)** replaces the base thymine. These structures are called *bases* because their presence raises the pH of a solution.

Polynucleotide Structure

The nucleotides link to make a polynucleotide called a *strand,* which has a backbone made up of phosphate-sugar-phosphate-sugar. The bases project to one side of the backbone. The nucleotides of a gene occur in a definite order, and so do the bases. After many years of work, researchers now know the sequence of the bases in human DNA—the human genome. This breakthrough has already led to improvements in genetic counseling, gene therapy, and medicines to treat the causes of many human illnesses.

DNA is double-stranded, with the two strands twisted about each other in the form of a *double helix* (see Fig. 2.24*a*). In DNA, the two strands are held together by hydrogen bonds between the bases. When coiled, DNA resembles a spiral staircase. When unwound, it resembles a stepladder. The uprights (sides) of the ladder are made entirely of phosphate and sugar molecules, and the rungs of the ladder exhibit **complementary base pairing.** Thymine (T) always pairs with adenine (A), and guanine (G) always pairs with cytosine (C). Complementary bases have shapes that fit together.

Complementary base pairing allows DNA to replicate in a way that ensures that the sequence of bases will remain the same. This is important because it is the sequence of bases that determines the sequence of amino acids in a protein. RNA is single-stranded. When RNA forms, complementary base pairing with one DNA strand passes the correct sequence of bases to RNA (Fig. 2.24*b*). RNA is the nucleic acid directly involved in protein synthesis.

Animation DNA Structure

ATP: An Energy Carrier

In addition to being the subunits of nucleic acids, nucleotides have metabolic functions. When adenosine (adenine plus ribose) is modified by the addition of three phosphate groups instead of one, it becomes ATP (adenosine triphosphate), which is an energy carrier in cells.

MP3 ATP

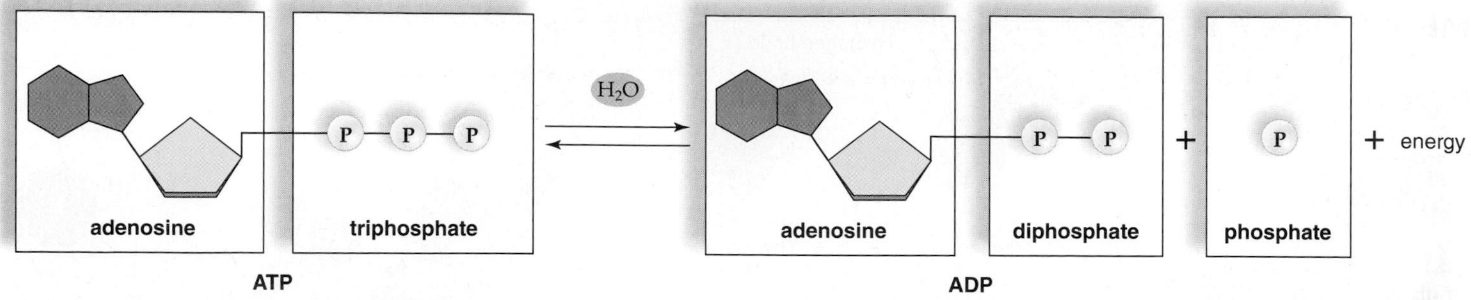

Figure 2.25 **ATP is the universal energy currency of cells.**
ATP is composed of the base adenosine and three phosphate groups (called a triphosphate). When cells need energy, ATP is hydrolyzed (water is added) forming ADP and Ⓟ. Energy is released. To recycle ATP, energy from food is required and the reverse reaction occurs: ADP and Ⓟ join to form ATP, and water is given off.

Structure of ATP Suits Its Function

ATP is a high-energy molecule because the last two phosphate bonds are unstable and easily broken. Usually in cells, the last phosphate bond is hydrolyzed, leaving the molecule **ADP** (**adenosine diphosphate**) and a molecule of inorganic phosphate Ⓟ (Fig. 2.25). The energy released by ATP breakdown is used by the cell to synthesize macromolecules, such as carbohydrates and proteins. In muscle cells, the energy is used for muscle contraction; and in nerve cells, it is used for the conduction of nerve impulses. After ATP breaks down, it can be recycled by adding Ⓟ to ADP. Notice in Figure 2.25 that an input of energy is required to re-form ATP.

CHECK YOUR PROGRESS 2.7

1. Describe the structure of ATP.
2. State the type of bond that joins the bases within a DNA double helix. Why are these bonds used?
3. Compare the structure of DNA and RNA. What impact do these differences have on their function?

CONNECTING THE CONCEPTS

As the information-carrying and energy molecules of the body, the nucleic acids play an important role in how our cells, tissues, and organs function. For more information on this class of molecules, refer to the following discussions:

Section 3.6 examines the metabolic pathways that generate ATP in a cell.

Section 21.1 provides a more detailed look at the structure of DNA and RNA.

Section 21.2 explores how DNA contains the information to make proteins.

Section 21.4 examines how advances in biotechnology are giving scientists the ability to manipulate DNA in the laboratory.

CASE STUDY CONCLUSION

After three months of work, David felt more prepared for his visit with his physician. Not only had he made some important adjustments to his diet by limiting the amount of dietary fat and watching the cholesterol content of food but he had also increased his weekly exercise regime. More importantly, he now had a better understanding of how the chemistry of certain molecules related to his health. David now recognized that *cholesterol* was an important molecule in his body; however, because it was also hydrophobic, it could cause problems with his circulatory system. Furthermore, he also understood what his doctor meant by the terms *good* and *bad cholesterol.* David's doctor was actually referring to lipoproteins, a form of protein that transports lipids and cholesterol in the blood. A high level of LDL—the "bad cholesterol"—meant that his body had an excess of fat to be transported, and HDL represented empty transport proteins. Ideally, low LDL and high HDL values signified a healthy cardiovascular system and a reduced risk for a number of diet-related diseases.

MEDIA STUDY TOOLS

Enhance your study of this chapter with media! Visit **www.mhhe.com/maderhuman13e** and go to "Media Study Tools" for this chapter to access the following:

Animations	Video	MP3 Files
2.1 Atomic Structure • Half-Life • Ionic Bonds • Ionic and Covalent Bonds **2.2** Water Properties **2.5** Anatomy of a Food Label **2.6** Protein Denaturation **2.7** DNA Structure	**2.1** Nuclear Medicine	**2.1** Chemical Bonding **2.2** Water and pH • Acid-Base Balances **2.3** Organic Molecules **2.4** Carbohydrates **2.5** Lipids **2.6** Proteins **2.7** Nucleic Acids • ATP

SUMMARIZE

2.1 From Atoms to Molecules

- **Matter** is composed of elements; each **element** is made up of just one type of **atom.** Elements are identified by an atomic symbol and an **atomic number,** which indicates the number of protons in an atom of an element. The **mass number** of an atom is based on the number of **protons** and **neutrons** in the nucleus.
- **Electrons** orbit the nucleus of an atom in **electron shells.** An atom's chemical properties depend on the number of electrons in its outer electron shell.
- **Isotopes** of an element vary in the number of neutrons. The **atomic mass** on the periodic table reflects the average mass of these isotopes. **Radioisotopes** are unstable isotopes that are useful in scientific studies and medicine.
- Atoms may bind with one another to form **molecules** and **compounds. Ionic bonds** are formed between atoms that have gained or lost electron(s) to form **ions. Covalent bonds** are formed by a sharing of electrons.

2.2 Water and Life

- The properties of water occur because water is a **polar** compound, resulting in the formation of **hydrogen bonds** between water molecules.
- Water is a liquid, instead of a gas, at room temperature. The energy needed to change the state of water is called a **calorie.**
- Water heats and freezes slowly, moderating temperatures and allowing bodies to cool by vaporizing water.
- Frozen water is less dense than liquid water, so ice floats on water.
- Water is cohesive and fills tubular vessels, such as blood vessels. A thin film of water allows the lungs to adhere to the chest wall.
- Water is the universal solvent because of its polarity. **Hydrophilic** molecules interact easily with water. **Hydrophobic** molecules do not interact well with water.
- pH is determined by the hydrogen ion concentration $[H^+]$. **Acids** increase H^+ but decrease the pH of water, and **bases** decrease H^+ but increase the pH of water. The **pH scale** reflects whether a solution is acid or basic (alkaline). **Buffers** help cells and organisms maintain a constant pH.

2.3 Molecules of Life

- Carbohydrates, lipids, proteins, and nucleic acids are **organic molecules** with specific functions in cells.
- **Dehydration reactions** form macromolecules from their building blocks. **Hydrolysis reactions** break down **macromolecules.**

2.4 Carbohydrates

- **Carbohydrates** are short-term energy storage molecules.
- Simple carbohydrates are **monosaccharides** or **disaccharides. Glucose** is a monosaccharide used by cells for quick energy.
- Complex carbohydrates are **polysaccharides. Starch, glycogen,** and **cellulose** (fiber) are polysaccharides containing many glucose units.
- Plants store glucose as starch, whereas animals store glucose as glycogen. Cellulose forms plant cell walls. Cellulose is dietary fiber. Fiber plays an important role in digestive system health.

2.5 Lipids

- **Lipids** are nonpolar molecules that do not dissolve in water. **Fats,** also called **triglycerides,** and **oils** are lipids that act as long-term energy storage molecules.
- **Fatty acids** can be **saturated** or **unsaturated. Trans fats** are unsaturated fatty acids that have adverse effects on your health.
- Plasma membranes contain **phospholipids.**
- **Steroids** are complex lipids composed of three interlocking rings. Testosterone and estrogen are steroids.
- Cholesterol is a steroid that is transported by proteins called lipoproteins (LDLs and HDLs).

2.6 Proteins

- **Proteins** may be structural proteins (keratin, collagen), hormones, or enzymes that speed chemical reactions. Proteins account for cell movement (actin, myosin), enable muscle contraction (actin, myosin), or transport molecules in blood (**hemoglobin**).
- Proteins are macromolecules with **amino acid** subunits. A peptide is composed of two amino acids linked by a **peptide bond,** and a **polypeptide** contains many amino acids.
- A protein has levels of structure: A primary structure is determined by the sequence of amino acids that forms a polypeptide. A secondary structure is an α (alpha) helix or pleated sheet. A tertiary structure occurs when the secondary structure forms a three-dimensional, globular shape. A quaternary structure occurs when two or more polypeptides join to form a single protein. **Denaturation** represents an irreversible change in the shape of a protein.

2.7 Nucleic Acids

- **Nucleic acids** are macromolecules composed of nucleotides. **Nucleotides** are composed of a sugar, a base, and a phosphate. DNA and RNA are polymers of nucleotides.
- **DNA** (**deoxyribonucleic acid**) contains the sugar deoxyribose; contains the bases **adenine** (**A**), **guanine** (**G**), **thymine** (**T**), and **cytosine** (**C**); is double-stranded; and forms a helix. The helix exhibits **complementary base pairing** between the strands of DNA.
- **RNA** (**ribonucleic acid**) contains the sugar ribose; contains the bases adenine, guanine, **uracil** (**U**), and cytosine; and does not form a helix.
- **ATP** (**adenosine triphosphate**) is a high-energy molecule because its bonds are unstable.
- ATP undergoes hydrolysis to **ADP** (**adenosine diphosphate**) + Ⓟ, which releases energy used by cells to do metabolic work.

Organic molecules	Examples	Monomers	Functions
Carbohydrates	Monosaccharides, disaccharides, polysaccharides	CH_2OH, H, O, H, OH, H, HO, OH, H, OH **Glucose**	Immediate energy and stored energy; structural molecules
Lipids	Fats, oils, phospholipids, steroids	H–C–OH, H–C–OH, H–C–OH **Glycerol** HO–C(=O)–C–C–C–C–C–R **Fatty acid**	Long-term energy storage; membrane components
Proteins	Structural, enzymatic, carrier, hormonal, contractile	amino group H_2N – C (H) – acid group COOH; *R* group **Amino acid**	Support, metabolic, transport, regulation, motion
Nucleic acids	DNA, RNA	phosphate P, base C, S **Nucleotide**	Storage of genetic information

ASSESS

Testing Your Knowledge of the Concepts

1. Name the subatomic particles of the atom. Describe their charge, atomic mass, and location in the atom. (pages 21–22)
2. Why can a radioisotope be used as a tracer in the human body? Give an example. (pages 22–23)
3. Explain the difference between an ionic bond and a covalent bond. (pages 23–25)
4. Relate the properties of water to its polarity and hydrogen bonding between water molecules. (pages 26–27)
5. On the pH scale, which numbers indicate a basic solution? An acidic solution? A neutral solution? What makes a solution basic, acidic, or neutral? (pages 27–28)
6. What are buffers, and why are they important to life? (pages 28–29)
7. Name, describe, and give an example of each class of carbohydrate. What is the main function of each of the three classes? (pages 30–31)
8. What are the subunits of a triglyceride? What is the difference between a saturated and an unsaturated fatty acid? What are the functions of fats in the body? (pages 32–33)
9. How does the structure of a phospholipid differ from a triglyceride? Describe the arrangement of phospholipids in the plasma membrane. (page 35)
10. What are the building blocks of proteins? How do these components bond together to make a protein? (pages 36–37)
11. Discuss the primary, secondary, and tertiary structures of proteins. Why are these structures so important? (pages 37–38)
12. What structural differences are there between DNA and RNA? (pages 38–39)
13. What type of reaction releases the energy of an ATP molecule? Explain. (page 40)
14. The atomic number gives the
    - **a.** number of neutrons in the nucleus.
    - **b.** number of protons in the nucleus.
    - **c.** weight of the atom.
    - **d.** number of protons in the outer shell.
15. Isotopes differ in their
    - **a.** number of protons.
    - **b.** atomic number.
    - **c.** number of neutrons.
    - **d.** number of electrons.
16. Which type of bond results from the complete transfer of electrons from one atom to another?
    - **a.** covalent
    - **b.** ionic
    - **c.** hydrogen
    - **d.** neutral
17. What is true of a solution that goes from pH 5 to pH 8?
    - **a.** The H^+ concentration decreases as the solution becomes more basic.
    - **b.** The H^+ concentration increases as the solution becomes more acidic.
    - **c.** The H^+ concentration decreases as the solution becomes more acidic.

18. An example of a polysaccharide used for energy storage in humans is

a. cellulose. **c.** cholesterol.
b. glycogen. **d.** starch.

19. Saturated and unsaturated fatty acids differ in the

a. number of carbon-to-carbon double bonds.
b. consistency at room temperature.
c. number of hydrogen atoms present.
d. All of these are correct.

20. An RNA nucleotide differs from a DNA molecule in that RNA has

a. a ribose sugar. **c.** a uracil base.
b. a phosphate molecule. **d.** Both a and c are correct.

ENGAGE

The virtual lab "Nutrition" provides a more detailed look at how your diet provides you with daily amounts of nutrients such as protein, carbohydrates, calories, and fats. The lab allows you to select a variety of foods for breakfast, lunch, dinner, and snacks; and it charts how these selections influence your % Daily Value (DV) of these nutrients.

Thinking Critically About the Concepts

The case study in this chapter included information about healthy cholesterol levels in the blood. There are several biological molecules essential to life, and the body needs them in certain amounts in order to function properly. Cholesterol is one of these molecules. Sometimes, the body can make these molecules for us; other times, we need to get them through our diet. Either way, the balancing of "just enough" versus "too much" is the role of homeostasis.

1. If a doctor were to prescribe cholesterol-lowering medicine for David, why would he also have to change his diet and exercise program? Why would the medication alone not be enough?
2. Cholesterol is a needed substance for the human body; we use it as a base for certain hormones and as a support structure in our plasma membranes. If it is needed in the body, how can it also be harmful?
3. Name another substance that is needed in the body but can be detrimental to homeostasis in large quantities.
4. The presence and absence of certain molecules in the blood can alter the pH of the blood. Why would it be important to maintain a steady blood pH?
5. Substances that the body makes or we ingest through the diet, like carbohydrates and lipids, can be stored in different areas for future use. Why is storage an important part of homeostasis?

C H A P T E R

3

Cell Structure and Function

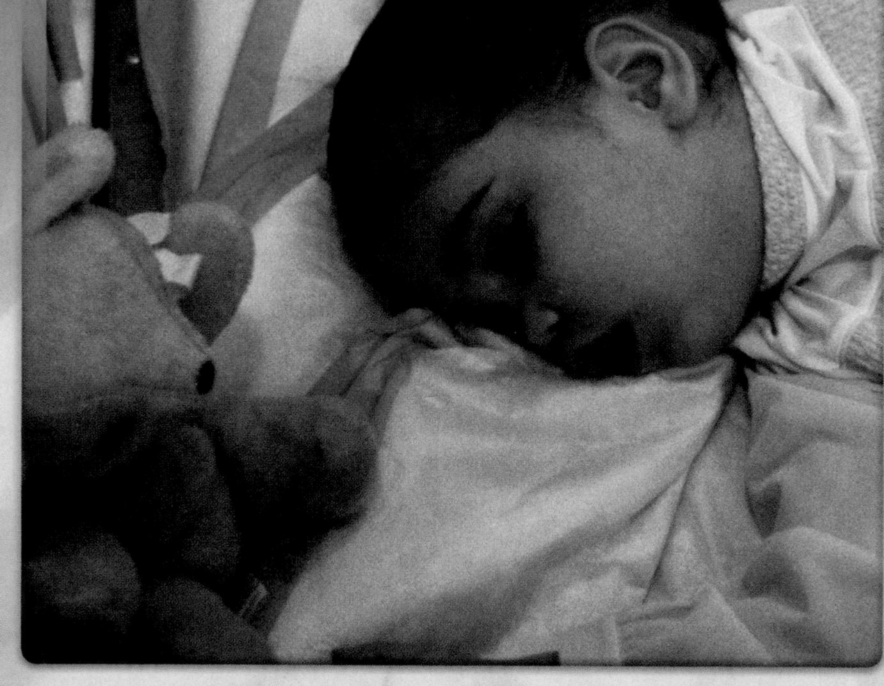

CHAPTER CONCEPTS

3.1 What Is a Cell?
Cells are the basic units of life. Cell size is limited by the surface area-to-volume ratio.

3.2 How Cells Are Organized
Human cells are eukaryotic cells, with a plasma membrane, cytoplasm, and a nucleus. Within the cytoplasm are a variety of organelles that carry out specific functions.

3.3 The Plasma Membrane and How Substances Cross It
The structure of the plasma membrane influences its permeability. Passive and active transport mechanisms, diffusion, transport by carriers, and the use of vesicles allow substances to cross the plasma membrane.

3.4 The Nucleus and Endomembrane System
The nucleus stores the genetic material. Ribosomes act as sites for protein synthesis. The endomembrane system acts as a series of interchangeable organelles that manufacture and modify proteins—and other organic molecules—for use by the cell.

3.5 The Cytoskeleton, Cell Movement, and Cell Junctions
The cytoskeleton is composed of fibers that maintain the shape of the cell and assist the movement of organelles. Cilia and flagella contain microtubules and can move using ATP energy. In tissues, cells are connected by junctions that allow for coordinated activities.

3.6 Mitochondria and Cellular Metabolism
Mitochondria are the sites of cellular respiration, an aerobic process that produces the majority of ATP for a cell. Fermentation, an aerobic process, produces only two ATP per glucose molecule.

BEFORE YOU BEGIN

Before beginning this chapter, take a few moments to review the following discussions:

Section 2.2 What properties of water make it a crucial molecule for life as we know it?

Sections 2.3 to 2.7 What are the basic roles of carbohydrates, fats, proteins, and nucleic acids in the cell?

Section 2.7 What is the role of ATP in a cell?

CASE STUDY WHEN CELLS MALFUNCTION

Mary and Kevin first noticed that something was wrong with their newborn about four months after birth. Whereas most newborns rapidly strengthen and are developing the ability to hold their head up and are demonstrating hand–eye coordination, their baby seemed to be weakening. In addition, Mary began to sense that something was wrong when their baby started having trouble swallowing his formula. After consulting with their pediatrician, Mary and Kevin decided to bring their child to a local pediatric research hospital to talk with physicians trained in newborn developmental disorders.

After a series of tests that included blood work and a complete physical examination, the specialists at the research center informed Kevin and Mary that the symptoms their newborn was exhibiting were characteristic of a condition called Tay–Sachs disease. This condition is a rare metabolic disorder that causes one of the internal components of the cell, the lysosome, to malfunction. Because of this malfunction, fatty acids were accumulating in the cells of their child. These accumulations were causing the neurons to degrade, producing the symptoms noted by the parents.

What puzzled the research team was the fact that neither Kevin nor Mary were of Eastern European descent. Populations from this area are known to have a higher rate of the mutation that causes Tay–Sachs disease. However, genetic testing of both Kevin and Mary indicated that they were carriers for the trait, meaning that though they each had one normal copy of the gene associated with Tay–Sachs disease, each carried a defective copy as well. Only one good copy of the gene is needed for the lysosome to function correctly. Unfortunately, each had passed on a copy of the defective gene to their child.

Despite the poor prognosis for their child, both Kevin and Mary were determined to learn more about how this defect caused the lysosome to malfunction and about what treatments were being developed to prolong the life span of a child with Tay–Sachs disease.

As you read through the chapter, think about the following questions:

1. What organelle produces the lysosomes?
2. What is the role of the lysosome in a normally functioning cell?
3. Why would a malfunction in the lysosome cause an accumulation of fatty acids in the cell?

3.1 What Is a Cell?

LEARNING OUTCOMES

Upon completion of this section, you should be able to

1. State the basic principles of the cell theory.
2. Explain how the surface area-to-volume ratio limits cell size.
3. Summarize the role of microscopy in the study of cells.

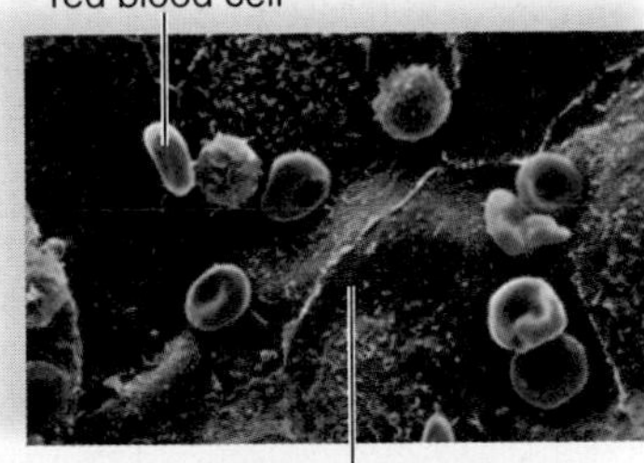

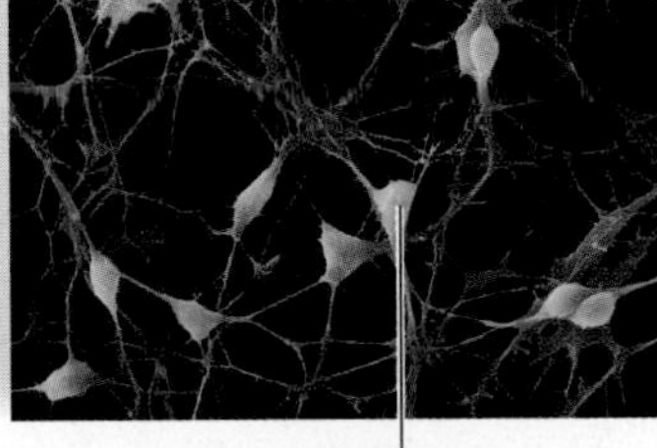

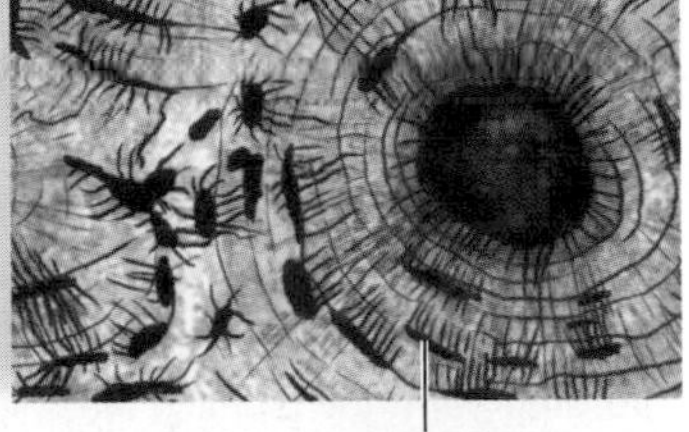

Figure 3.1 Cells vary in structure and function. A cell's structure is related to its function. Despite differences in appearance, all exchange substances with their environment.

All organisms, including humans, are composed of cells. From the single-celled bacteria to plants and complex animals such as ourselves, the cell represents the fundamental unit of life. Despite their importance, most cells are small and can be seen only under a microscope. The small size of cells means that they are measured using the smaller units of the metric system, such as the *micrometer* (μm). A micrometer is 1/1,000 millimeter (mm). The micrometer is the common unit of measurement for people who use microscopes professionally (see Appendix B for a complete list of metric units). Most human cells are about 100 μm in diameter, about the width of a human hair. The internal contents of a cell are even smaller and, in most cases, may only be viewed using powerful microscopes. Because of this small size, the **cell theory,** one of the fundamental principles of modern biology, was not formulated until after the invention of the microscope in the seventeenth century.

The Cell Theory

A cell is the basic unit of life. According to the cell theory, nothing smaller than a cell is alive. A single-celled organism exhibits the basic characteristics of life that were presented in Chapter 1. There is no smaller unit of life that is able to reproduce and grow, respond to stimuli, remain homeostatic, take in and use materials from the environment, and become adapted to the environment. In short, life has a cellular nature.

All living organisms are made up of cells. Although it may be apparent that a unicellular organism is necessarily a cell, what about multicellular ones? Humans are multicellular. Is there any tissue in the human body not composed of cells? At first, you might be inclined to say that bone is not composed of cells. However, if you were to examine bone tissue under the microscope (Fig. 3.1), you would be able to see that it, too, is composed of cells surrounded by material they have deposited. Cells may differ in their appearance, as is shown in the comparison of several cell types in Figure 3.1. However, despite these differences, they all have certain structures in common. In general, it is important to recognize that the structure of a cell is directly related to its function.

New cells arise only from pre-existing cells. Until the nineteenth century, most people believed in spontaneous generation, that is, that nonliving objects could give rise to living organisms. For example, maggots were thought to arise from meat hung in the butcher shop. Maggots often appeared in meat to which flies had access. However, people did not realize that the living maggots did not spontaneously generate from the nonliving meat. A series of experiments by Francesco Redi in the seventeenth century demonstrated that meat that was placed within sealed containers did not generate maggots. In other words, life did not generate spontaneously. In 1864, the French scientist Louis Pasteur conducted a now-classic set of experiments using bacterial cells. His experiments proved conclusively that spontaneous generation of life from nonlife was not possible.

When mice or humans reproduce, a sperm cell joins with an egg cell to form a zygote. This is the first cell of a new multicellular organism. By reproducing, parents pass a copy of their genes to their offspring. The genes contain the instructions that allow the zygote to grow and develop into the complete organism.

Cell Size

A few cells, such as the egg of a chicken or frog, are large enough to be seen by the naked eye. In comparison, a human egg cell is around 100 μm in size, placing it right at the limit of what can be viewed by our eyes. However, most cells are much smaller. The small size of cells is explained by considering the *surface area-to-volume ratio* of cells. Nutrients enter a cell—and waste exits a cell—at its surface. Therefore, the greater the amount of surface, the greater the ability to get material in and out of the cell. A large cell requires more nutrients and produces more waste than a small cell. Yet, as cells become smaller in volume, the proportionate amount of surface area actually increases. You can see this by comparing the cubes in Figure 3.2.

We would expect, then, that there would be a limit to how large an actively metabolizing cell can become. Once a chicken's egg is fertilized and starts metabolizing, it divides repeatedly without increasing in size. Cell division increases the amount of surface area needed for adequate exchange of materials.

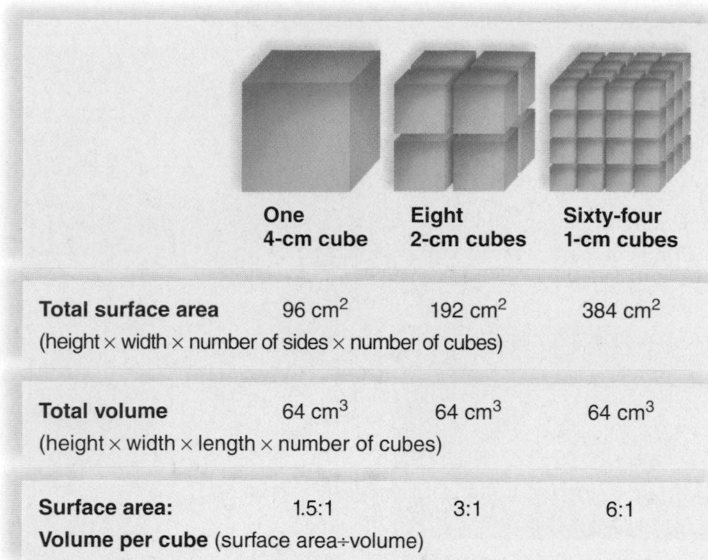

Figure 3.2 Surface area-to-volume ratio limits cell size. As cell size decreases from 4 cm^3 to 1 cm^3, the ratio of the surface area to volume increases.

Table 3.1 Resolving Power of the Eye and Common Microscopes

	Magnification	Resolving Power
Eye	N/A	0.1 mm (100 μm)
Light microscope	1,000×	0.0001 mm (0.1 μm)
Transmission electron microscope	50,000×	0.000001 mm (0.01 μm)

Microscopy

Microscopes provide scientists with a deeper look into how cells function. There are many different types of microscopes, from compound light microscopes to powerful electron microscopes. The *magnification,* or the ratio between the observed size of an image and its actual size, varies with the type of microscope. In addition, the resolution of the image varies between microscopes (Table 3.1). *Resolution* is the ability to distinguish between two adjacent points, and it represents the minimum distance between two objects that allows them to be seen as two different objects. Usually, the more powerful the microscope, the greater the resolution. Figure 3.3 illustrates images of a red blood cell taken by three different types of microscopes. A *compound light microscope* (Fig. 3.3*a*) uses a set of glass lenses and light rays passing through the object to magnify objects. The image can be viewed directly by the human eye.

The *transmission electron microscope* makes use of a stream of electrons to produce magnified images (Fig. 3.3*b*). The human eye cannot see the image. Therefore, it is projected onto a fluorescent screen or photographic film to produce an image (or *micrograph*) that can be viewed. The magnification and resolution produced by a transmission electron microscope is much higher than that of a light microscope. Therefore, this microscope has the ability to produce enlarged images with greater detail.

A *scanning electron microscope* provides a three-dimensional view of the surface of an object (Fig. 3.3*c*). A narrow beam of electrons is scanned over the surface of the specimen, which is coated with a thin layer of metal. The metal gives off secondary electrons, which are collected to produce a television-type picture of the specimen's surface on a screen.

In the laboratory, the light microscope is often used to view live specimens. However, this is not the case for the electron microscopes. Because electrons cannot travel very far in air, a strong vacuum must be maintained along the entire path of the electron beam. Often, cells are treated before

Figure 3.3 Micrographs of human red blood cells. **a.** Light micrograph (LM) of many cells in a large vessel (stained). **b.** Transmission electron micrograph (TEM) of just three cells in a small vessel (colored). **c.** Scanning electron micrograph (SEM) gives a three-dimensional view of cells and vessels (colored).

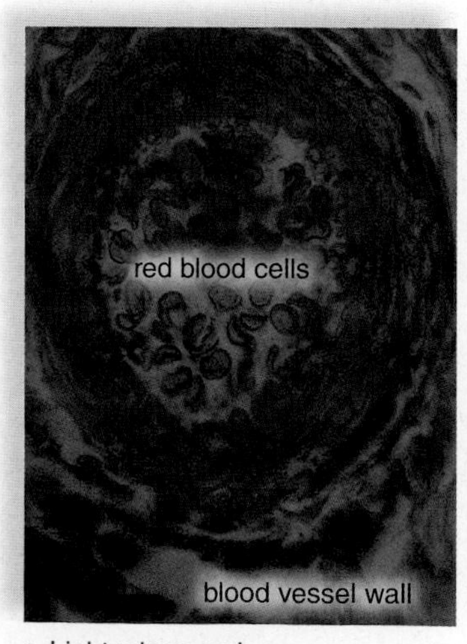

a. Light micrograph

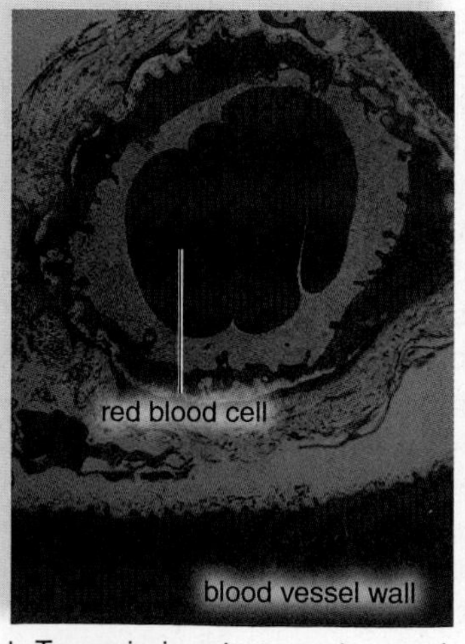

b. Transmission electron micrograph

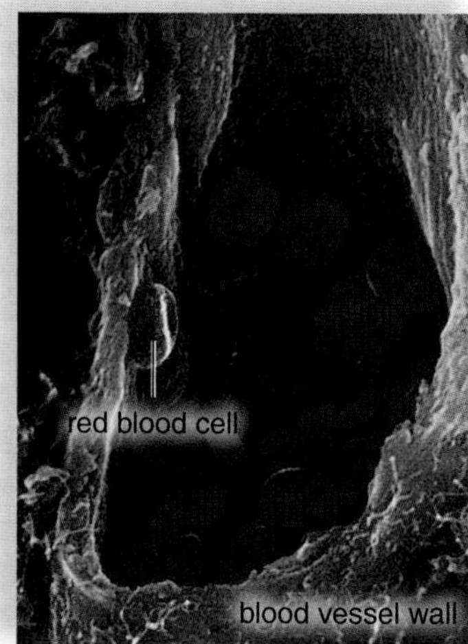

c. Scanning electron micrograph

BIOLOGY MATTERS  Science

Coloring Organisms Green: Green Fluorescent Proteins and Cells

Most cells lack any significant pigmentation. Thus, cell biologists frequently rely on dyes to produce enough contrast to resolve organelles and other cellular structures. The first of these dyes were developed in the nineteenth century from chemicals used to stain clothes in the textile industry. Since then, significant advances have occurred in the development of cellular stains.

In 2008, three scientists—Martin Chalfie, Roger Y. Tsien, and Osamu Shimomura—earned the Nobel Prize in Chemistry or Medicine for their work with a protein called *green fluorescent protein,* or GFP. GFP is a bioluminescent protein found in the jellyfish *Aequorea victoria,* commonly called the crystal jelly (Fig. 3A*a*). The crystal jelly is a native of the West Coast of the United States. Normally, this jellyfish is transparent. However, when disturbed, special cells in the jellyfish release a fluorescent protein called aequorin. Aequorin fluoresces with a green color. The research teams of Chalfie, Tsien, and Shimomura were able to isolate the fluorescent protein from the jellyfish and develop it as a molecular tag. These tags can be generated for almost any protein within the cell, revealing not only its cellular location but also how its distribution within the cell may change as a result of a response to its environment. Figure 3A*b* shows how a GFP-labeled antibody can be used to identify the cellular location of the actin proteins in a human cell. Actin is one of the prime components of the cell's microfilaments, which in turn are part of the cytoskeleton of the cell. This image shows the distribution of actin in a human cell.

Questions to Consider

1. Discuss how a researcher might use a GFP-labeled protein in a study of a disease, such as cancer.
2. How do studies such as these support the idea that preserving the diversity of life on the planet is important?

a. jellyfish

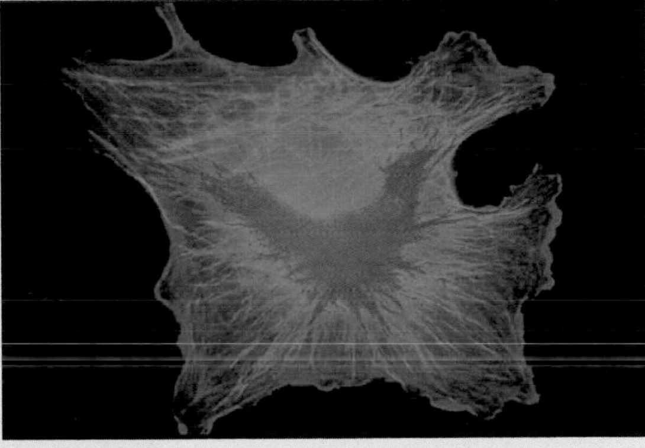

b. actin filaments

Figure 3A **GFP shows details of the interior of cells.** **a.** The jellyfish *Aequorea victoria* and (**b**) the GFP stain of a human cell. This illustration shows a human cell tagged with a GFP-labeled antibody to the actin protein.

being viewed under a microscope. Because most cells are transparent, they are often stained with colored dyes before being viewed under a light microscope. Certain cellular components take up the dye more than other components; therefore, contrast is enhanced. A similar approach is used in electron microscopy, except in this case the sample is treated with electron-dense metals (such as gold) to provide contrast. The metals do not provide color, so electron micrographs may be colored after the micrograph is obtained. The expression "falsely colored" means that the original micrograph was colored after it was produced. In addition, during electron microscopy, cells are treated so that they do not decompose in the vacuum. Frequently they are also embedded into a matrix, which allows a researcher to slice the cell into very thin pieces, providing cross sections of the interior of the cell.

CHECK YOUR PROGRESS 3.1

1. Summarize the cell theory and state its importance to the study of biology.
2. Explain how a cell's size relates to its function.
3. Compare and contrast the information that may be obtained from a light microscope and an electron microscope.

CONNECTING THE CONCEPTS

For more on the cells mentioned in this section, refer to the following discussions:

Section 6.2 discusses how red blood cells transport gases within the circulatory system.

Section 6.6 provides an overview of how red blood cells help maintain homeostasis in the body.

Section 17.1 examines the complex structure of a human egg cell.

3.2 How Cells Are Organized

LEARNING OUTCOMES

Upon completion of this section, you should be able to

1. Identify the components of a human cell and state the function of each.
2. Distinguish between the structure of a prokaryotic cell and that of a eukaryotic cell.
3. Summarize how eukaryotic cells evolved from prokaryotic cells.

Biologists classify cells into two broad categories—the prokaryotes and eukaryotes. The prokaryotic group includes two groups of bacteria, the eubacteria and the archaebacteria. The structure of the bacteria is discussed in more detail in Chapter 7. Within the eukaryotic group are the animals, plants, fungi, and some single-celled organisms called protists. The general structure of a human eukaryotic cell is shown in Figure 3.4. Despite their differences, both types of cells have a **plasma membrane,** an outer membrane

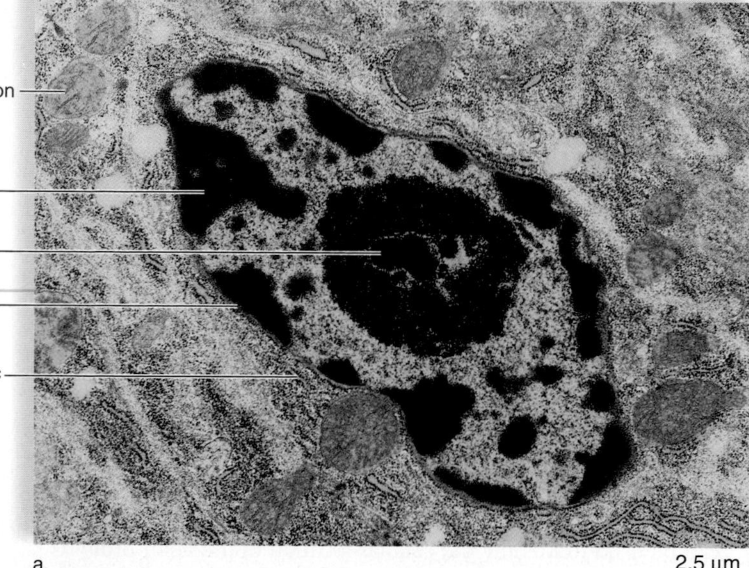

Figure 3.4 The structure of a typical eukaryotic cell.

a. A transmission electron micrograph of the interior structures of a cell. **b.** The structure and function of the components of a eukaryotic cell.

Plasma membrane: outer surface that regulates entrance and exit of molecules

protein

phospholipid

CYTOSKELETON: maintains cell shape and assists movement of cell parts:

Microtubules: cylinders of protein molecules present in cytoplasm, centrioles, cilia, and flagella

Intermediate filaments: protein fibers that provide support and strength

Actin filaments: protein fibers that play a role in movement of cell and organelles

Centrioles: short, cylinders of microtubules .

Centrosome: microtubule organizing center that contains a pair of centrioles

Lysosome: vesicle that digests macromolecules and even cell parts

Vesicle: membrane-bounded sac that stores and transports substances

Cytoplasm: semifluid matrix outside nucleus that contains organelles

b.

NUCLEUS:

Nuclear envelope: double membrane with nuclear pores that encloses nucleus

Chromatin: diffuse threads containing DNA and protein

Nucleolus: region that produces subunits of ribosomes

ENDOPLASMIC RETICULUM:

Rough ER: studded with ribosomes, processes proteins

Smooth ER: lacks ribosomes, synthesizes lipid molecules

Ribosomes: particles that carry out protein synthesis

Mitochondrion: organelle that carries out cellular respiration, producing ATP molecules

Polyribosome: string of ribosomes simultaneously synthesizing same protein

Golgi apparatus: processes, packages, and secretes modified cell products

that regulates what enters and exits a cell. The plasma membrane is a phospholipid bilayer. This bilayer is a "sandwich" made of two layers of phospholipids. Their polar phosphate molecules form the top and bottom surfaces of the bilayer, and the nonpolar lipid lies in between. The phospholipid bilayer is **selectively permeable,** which means it allows certain molecules—but not others—to enter the cell. Proteins scattered throughout the plasma membrane play important roles in allowing substances to enter the cell. All types of cells also contain **cytoplasm,** which is a semifluid medium that contains water and various types of molecules suspended or dissolved in the medium. The presence of proteins accounts for the semifluid nature of the cytoplasm. The cytoplasm contains **organelles.** Originally, the term *organelle* referred to only membranous structures, but we will use it to include any well-defined subcellular structure. Eukaryotic cells have many different types of organelles.

Internal Structure of Eukaryotic Cells

The most prominent organelle within the **eukaryotic cell** is a **nucleus,** a membrane-enclosed structure in which DNA is found. **Prokaryotic cells** (such as bacterial cells) lack a nucleus. Although the DNA of prokaryotic cells is centrally placed within the cell, it is not surrounded by a membrane.

Eukaryotic cells also possess organelles, each type of which has a specific function (Fig. 3.4). Many organelles are surrounded by a membrane, which allows compartmentalization of the cell. This keeps the various cellular activities separated from one another.

MP3 Cellular Organelles

Evolutionary History of the Eukaryotic Cell

Figure 3.5 shows that the first cells to evolve were prokaryotic cells. Prokaryotic cells today are represented by the bacteria and archaea, which differ mainly by their chemistry. Bacteria are well known for causing diseases in humans, but they also have great environmental and commercial importance. The archaea are known for living in extreme environments that may mirror the first environments on Earth. These environments are too hot, too salty, and/or too acidic for the survival of most cells. Evidence widely supports the hypothesis that eukaryotic cells evolved from the archaea.

The internal structure of eukaryotic cells is believed to have evolved as the series of events shown in Figure 3.5. The nucleus could have formed by *invagination* of the plasma membrane, a process whereby a pocket is formed in the plasma membrane. The pocket would have enclosed the DNA of the cell, thus forming its nucleus. Surprisingly, some of the organelles in eukaryotic cells may have arisen by engulfing prokaryotic cells. The engulfed prokaryotic cells were not digested; rather, they then evolved into different organelles. One of these events would have given the eukaryotic cell a mitochondrion. Mitochondria are organelles that carry on cellular respiration. Another such event may have produced the chloroplast. Chloroplasts are found in cells that carry out photosynthesis. This process is often called *endosymbiosis.*

Animation Endosymbiosis

Early prokaryotic organisms, such as the archaeans, were well adapted to life on the early Earth. The environment that

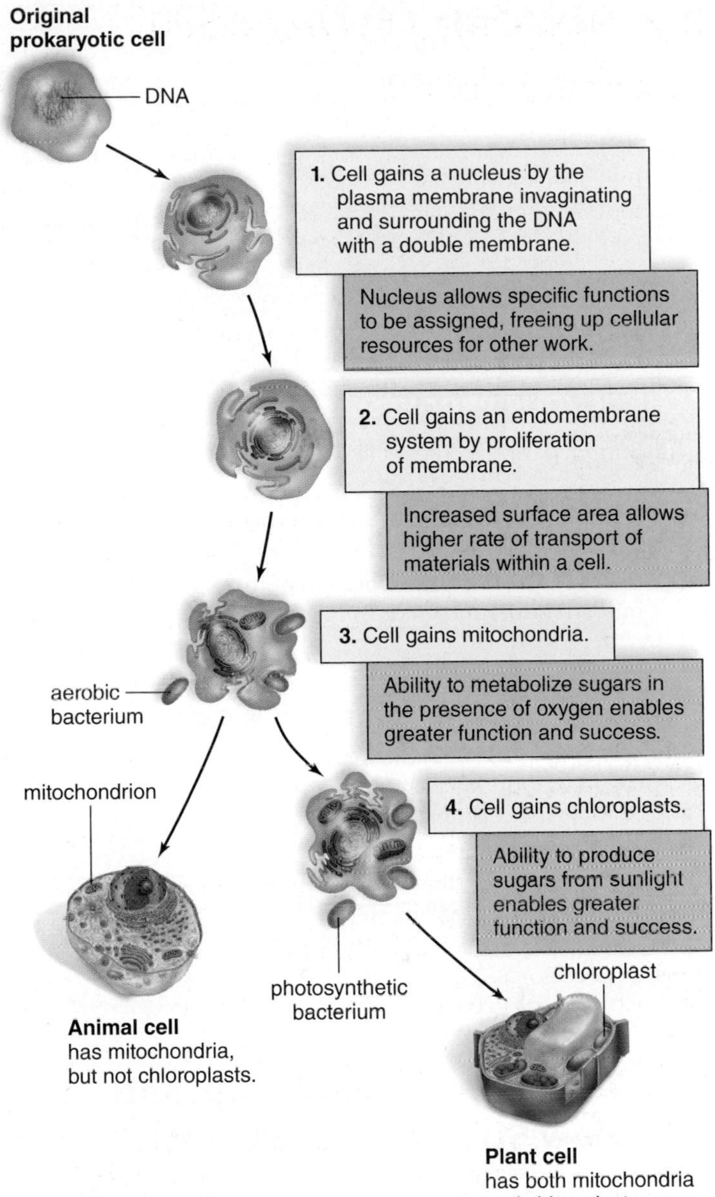

Figure 3.5 The evolution of eukaryotic cells.
Invagination of the plasma membrane of a prokaryotic cell could have created the nucleus. Later, the cell gained organelles, some of which may have been independent prokaryotes.

they evolved in contained conditions that would be instantly lethal to life today. The atmosphere contained no oxygen; instead, it was filled with carbon monoxide and other poisonous gases; the temperature of the planet was greater than 200°F; and there was no ozone layer to protect organisms from damaging radiation from the sun.

Despite these conditions, prokaryotic life survived and in doing so gradually adapted to Earth's environment. In the process, most of the archaea bacteria went extinct. However, we now know that some are still around and can be found in some of the most inhospitable places on the planet, such as thermal vents and salty seas. The study of these ancient bacteria is still shedding light on the early origins of life.

APPLICATIONS AND MISCONCEPTIONS

How old are the bacteria?

Scientists now recognize that the first cells on Earth were the prokaryotes. This ancient group of organisms first appeared on the planet over 3.5 billion years ago. Sometimes that amount of time can be hard to visualize, so a geologic timescale has been provided in the animation "Geologic History of Earth."

Animation
Geologic History of Earth

CHECK YOUR PROGRESS 3.2

1. Summarize the three main components of a eukaryotic cell.
2. Describe the main differences between a eukaryotic and a prokaryotic cell.
3. Describe the possible evolution of the nucleus, mitochondria, and chloroplast.

CONNECTING THE CONCEPTS

The material in this section summarizes some previous concepts of eukaryotic and prokaryotic cells and the role of phospholipids in the cell membrane. For more information, refer to the following discussions:

Section 1.2 illustrates the difference in the classification of eukaryotic and prokaryotic cells.

Section 7.1 provides more information on the structure of bacterial cells.

3.3 The Plasma Membrane and How Substances Cross It

LEARNING OUTCOMES

Upon completion of this section, you should be able to

1. Describe the structure of the cell membrane and list the type of molecules found in the membrane.
2. Distinguish between diffusion, osmosis, and facilitated transport, and state the role of each in the cell.
3. Explain how tonicity relates to the direction of water movement across a membrane.
4. Compare passive-transport and active-transport mechanisms.
5. Summarize how eukaryotic cells move large molecules across membranes.

Like all cells, a human cell is surrounded by an outer plasma membrane (Fig. 3.6). The plasma membrane marks the boundary between the outside and the inside of the cell. The integrity and function of the plasma membrane are necessary to the life of the cell.

The plasma membrane is a phospholipid bilayer with attached or embedded proteins. A phospholipid molecule has a polar head and nonpolar tails (see Fig. 2.19). When phospholipids are placed in water, they naturally form a spherical bilayer. The polar heads, being charged, are *hydrophilic* (attracted to water). They position themselves to face toward the watery environment outside and inside the cell. The nonpolar tails are *hydrophobic* (not attracted to water). They turn inward toward one another, where there is no water.

At body temperature, the phospholipid bilayer is a liquid. It has the consistency of olive oil. The proteins are able to change their position by moving laterally. The **fluid-mosaic model** is a working description of membrane structure. It states that the protein molecules form a shifting pattern within the fluid phospholipid bilayer. Cholesterol lends support to the membrane.

Short chains of sugars are attached to the outer surface of some protein and lipid molecules. These are called *glycoproteins* and *glycolipids,* respectively. These carbohydrate chains, specific to each cell, help mark the cell as belonging to a particular individual. They account for why people have different blood types, for example. Other glycoproteins have a special configuration that allows them to act as a receptor for a chemical messenger, such as a hormone. Some plasma membrane proteins form channels through which certain substances can enter cells. Others are either enzymes that catalyze reactions or carriers involved in the passage of molecules through the membrane.

Plasma Membrane Functions

The plasma membrane isolates the interior of the cell from the external environment. In doing so, it allows only certain molecules and ions to enter and exit the cytoplasm freely. Therefore, the plasma membrane is said to be selectively permeable (Fig. 3.7). Small, lipid-soluble molecules, such as oxygen and carbon dioxide, can pass through the membrane easily. The small size of water molecules allows them to freely cross the membrane by using protein channels called *aquaporins.* Ions and large molecules cannot cross the membrane without more direct assistance, which will be discussed later.

Diffusion

Diffusion is the random movement of molecules from an area of higher concentration to an area of lower concentration, until they are equally distributed. Diffusion is a passive way for molecules to enter or exit a cell. No cellular energy is needed to bring it about.

Certain molecules can freely cross the plasma membrane by diffusion. When molecules can cross a plasma membrane, which way will they go? The molecules will move in both directions. But the *net movement* will be from the region of higher concentration to the region of lower concentration, until

Figure 3.6 Organization of the plasma membrane. A plasma membrane is composed of a phospholipid bilayer in which proteins are embedded. The hydrophilic heads of phospholipids are a part of the outside surface and the inside surface of the membrane. The hydrophobic tails make up the interior of the membrane. Note the plasma membrane's asymmetry—carbohydrate chains are attached to the outside surface, and cytoskeleton filaments are attached to the inside surface. Cholesterol lends support to the membrane.

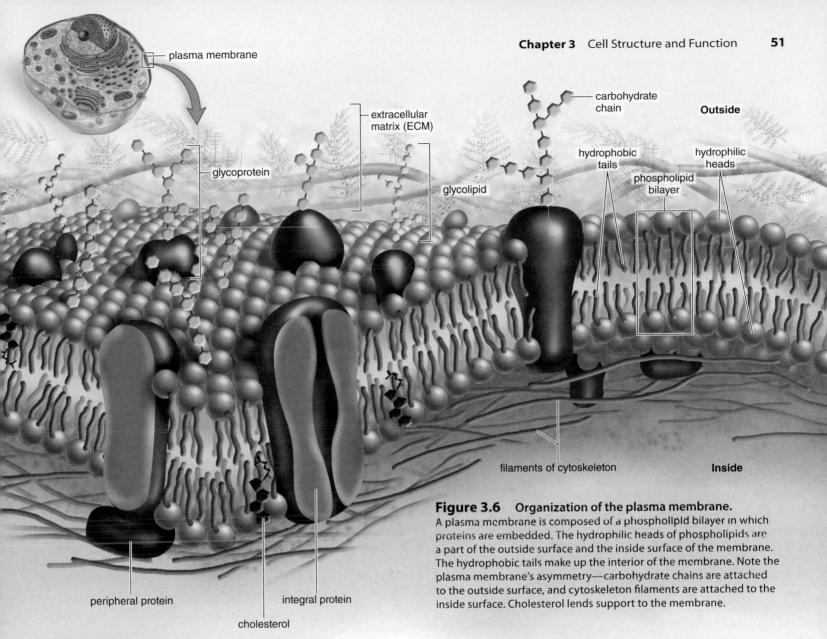

equilibrium is achieved. At equilibrium, as many molecules of the substance will be entering as leaving the cell (Fig. 3.8). Oxygen diffuses across the plasma membrane, and the net movement is toward the inside of the cell. This is because a cell uses oxygen when it produces ATP molecules for energy purposes.

Osmosis

Osmosis is the net movement of water across a semipermeable membrane, from an area of higher concentration to an area of lower concentration. The membrane separates the two areas, and solute is unable to pass through the membrane. Water will tend to flow from the area that has less solute (and therefore more water) to the area with more solute (and therefore less water). **Tonicity** refers to the osmotic characteristics of a solution across a particular membrane, such as a red blood cell membrane.

Figure 3.7 Selective permeability of the plasma membrane. Small, uncharged molecules are able to cross the membrane, whereas large or charged molecules cannot. Water travels freely across membranes through aquaporins.

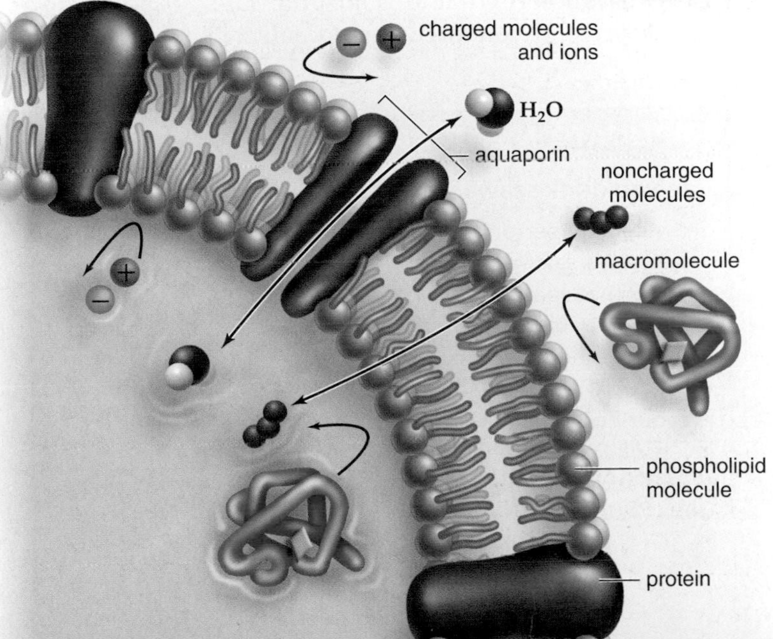

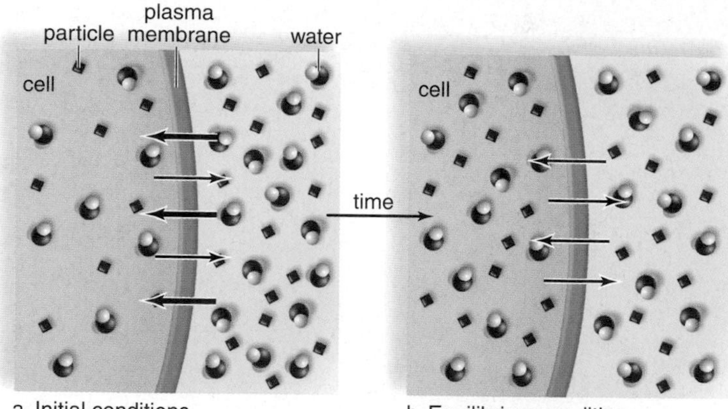

a. Initial conditions b. Equilibrium conditions

Figure 3.8 **Diffusion across the plasma membrane.** **a.** When a substance can diffuse across the plasma membrane, it will move back and forth across the membrane, but the net movement will be toward the region of lower concentration. **b.** At equilibrium, equal numbers of particles and water have crossed in both directions, and there is no net movement.

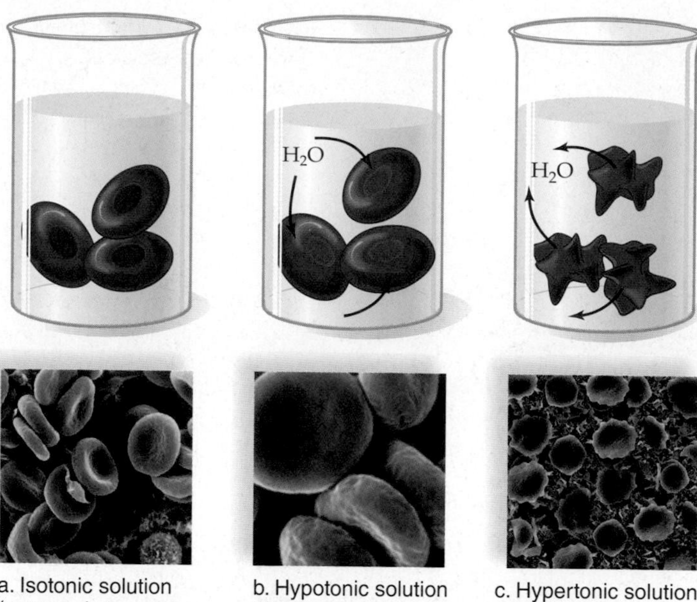

a. Isotonic solution (same solute concentration as in cell)

b. Hypotonic solution (lower solute concentration than in cell)

c. Hypertonic solution (higher solute concentration than in cell)

Figure 3.9 **Effects of changes in tonicity on red blood cells.** **a.** In an isotonic solution, cells remain the same. **b.** In a hypotonic solution, cells gain water and may burst (lysis). **c.** In a hypertonic solution, cells lose water and shrink (crenation).

Normally, body fluids are *isotonic* to cells (Fig. 3.9*a*). There is the same concentration of nondiffusible solutes and water on both sides of the plasma membrane. Therefore, cells maintain their normal size and shape. Intravenous solutions given in medical situations are usually isotonic.

Solutions that cause cells to swell or even to burst due to an intake of water are said to be *hypotonic.* A hypotonic solution has a lower concentration of solute and a higher concentration of water than the cells. If red blood cells are placed in a hypotonic solution, water enters the cells. They swell to bursting (Fig. 3.9*b*). *Lysis* is used to refer to the process of bursting cells. Bursting of red blood cells is termed *hemolysis.*

Solutions that cause cells to shrink or shrivel due to loss of water are said to be *hypertonic.* A hypertonic solution has a higher concentration of solute and a lower concentration of water than do the cells. If red blood cells are placed in a hypertonic solution, water leaves the cells; they shrink (Fig. 3.9*c*). The term *crenation* refers to red blood cells in this condition. These changes have occurred due to osmotic pressure. **Osmotic pressure** controls water movement in our bodies. For example, in the small and large intestines, osmotic pressure allows us to absorb the water in food and drink. In the kidneys, osmotic pressure controls water absorption as well.

3D Animation Membrane Transport: Osmosis

Animation How Osmosis Works

APPLICATIONS AND MISCONCEPTIONS

Can you drink seawater?

Seawater is hypertonic to our cells. Seawater contains approximately 3.5% salt, whereas our cells contain 0.9%. Once the salt entered the blood, your cells would shrivel up and die as they lost water trying to dilute the excess salt. Your kidneys can only produce urine that is slightly less salty than seawater, so you would dehydrate providing the amount of water necessary to rid your body of the salt. In addition, salt water contains high levels of magnesium ions, which cause diarrhea and further dehydration.

Facilitated Transport

Many solutes do not simply diffuse across a plasma membrane. They are transported by means of protein carriers within the membrane. During **facilitated transport,** a molecule is transported across the plasma membrane from the side of higher concentration to the side of lower concentration (Fig. 3.10). This is a passive means of transport because the cell does not need to expend energy to move a substance down its concentration gradient. Each protein carrier, sometimes called a *transporter,* binds only to a particular molecule, such as glucose. Type 2 diabetes results when cells lack a sufficient number of glucose transporters.

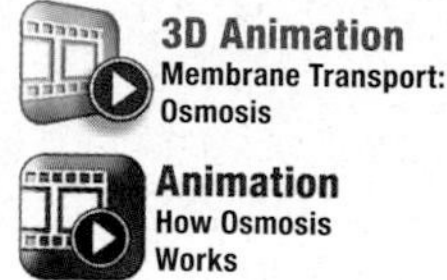

Active Transport

During **active transport,** a molecule is moving from a *lower* to *higher* concentration. One example is the concentration of iodine ions in the cells of the thyroid gland. In the digestive tract, sugar is completely absorbed from the gut by cells that line the intestines. In another example, water homeostasis is maintained by the kidneys by the active transport of sodium ions (Na^+) by cells lining kidney tubules.

Active transport requires a protein carrier and the use of cellular energy obtained from the breakdown of ATP. When

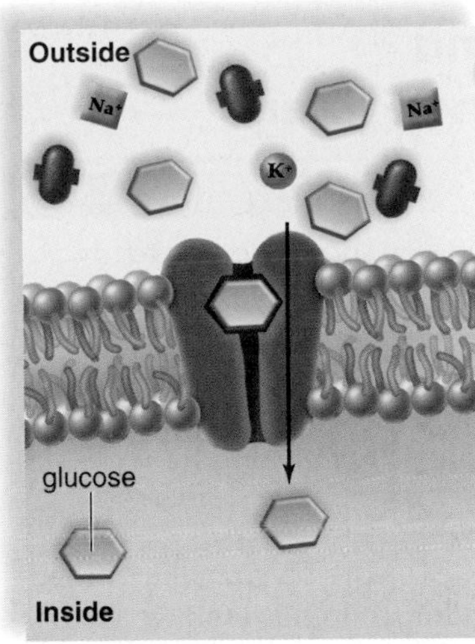

Figure 3.10 **Facilitated transport across a cell membrane.**
This is a passive form of transport in which substances move down their concentration gradient through a protein carrier. In this example, glucose (green) moves into the cell by facilitated transport. The end result will be an equal distribution of glucose on both sides of the membrane.

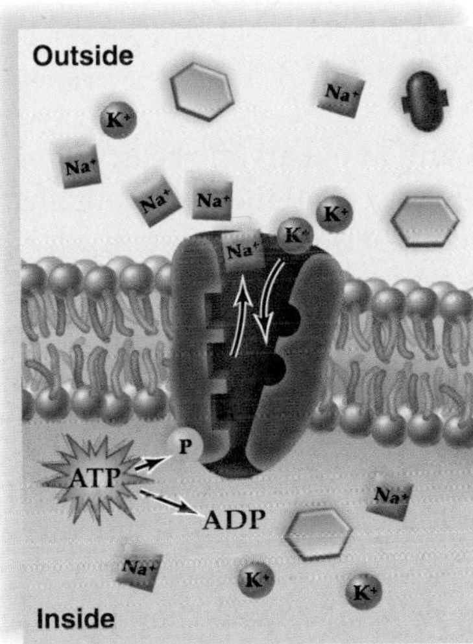

Figure 3.11 **Active transport and the sodium–potassium pump.**
This is a form of transport in which a molecule moves from low concentration to high concentration. It requires a protein carrier and energy. Na^+ exits and K^+ enters the cell by active transport, so Na^+ will be concentrated outside and K^+ will be concentrated inside the cell.

ATP is broken down, energy is released. In this case, the energy is used to carry out active transport. Proteins involved in active transport often are called *pumps.* Just as a water pump uses energy to move water against the force of gravity, energy is used to move substances against their concentration gradients. One type of pump active in all cells moves sodium ions (Na^+) to the outside and potassium ions (K^+) to the inside of the cell (Fig. 3.11). This type of pump is associated especially with nerve and muscle cells.

The passage of salt (NaCl) across a plasma membrane is of primary importance in cells. First, sodium ions are pumped across a membrane. Then, chloride ions diffuse through channels that allow their passage. In cystic fibrosis, a mutation in these chloride ion channels causes them to malfunction. This leads to the symptoms of this inherited (genetic) disorder.

Endocytosis and Exocytosis

During *endocytosis,* a portion of the plasma membrane invaginates, or forms a pouch, to envelop a substance and fluid. Then the membrane pinches off to form an endocytic vesicle inside the cell (Fig. 3.12*a*). Some white blood cells are able to take up pathogens (disease-causing agents) by endocytosis. Here the process is given a special name: **phagocytosis.** Usually, cells take up small molecules and fluid, and then the process is called *pinocytosis* (Fig.3.12*b*).

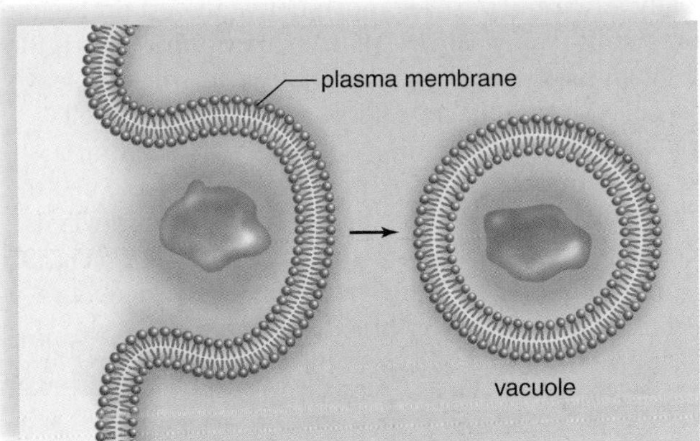

a. Phagocytosis

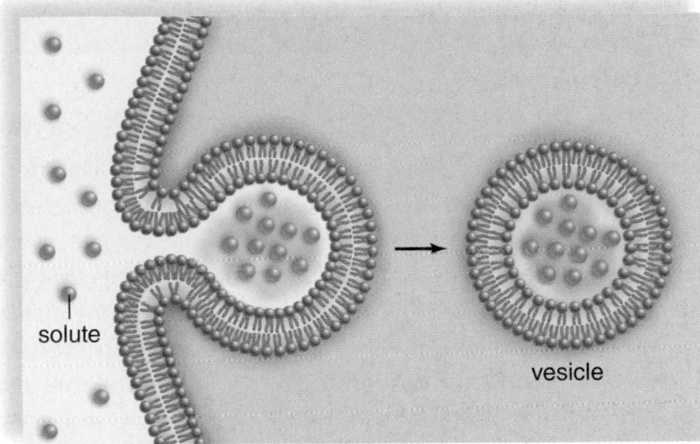

b. Pinocytosis

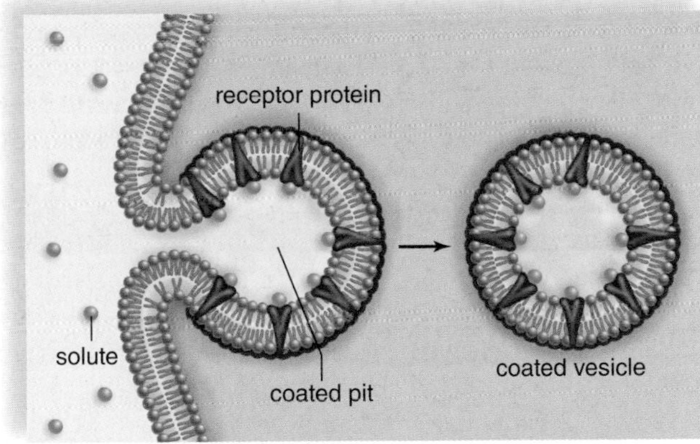

c. Receptor-mediated endocytosis

Figure 3.12 **Movement of large molecules across the membrane.**
a. Large substances enter a cell by endocytosis (phagocytosis). **b.** Small molecules, and fluids enter a cell by pinocytosis. **c.** In receptor-mediated endocytosis, molecules first bind to specific receptors and are then brought into the cell by endocytosis.

During *exocytosis,* a vesicle fuses with the plasma membrane as secretion occurs. Later in the chapter, we will see that a steady stream of vesicles move between certain organelles, before finally fusing with the plasma membrane. This is the way that signaling molecules, called *neurotransmitters,* leave one nerve cell to excite the next nerve cell or a muscle cell.

A special form of endocytosis uses a receptor, a special form of membrane protein, on the surface of the cell to concentrate specific molecules of interest for endocytosis. This process is called *receptor-mediated endocytosis* (Fig. 3.12*c*). An inherited form of cardiovascular disease occurs when cells fail to take up a combined lipoprotein and cholesterol molecule from the blood by receptor-mediated endocytosis.

Animation Endocytosis and Exocytosis

APPLICATIONS AND MISCONCEPTIONS

What causes cystic fibrosis?

In 1989, scientists determined that defects in a gene on chromosome 7 were the cause of cystic fibrosis (CF). This gene, called *CFTR* (cystic fibrosis conductance transmembrane regulator), codes for a protein that is responsible for the movement of chloride ions across the membranes of cells that produce mucus, sweat, and saliva. Defects in this gene cause an improper water–salt balance in the excretions of these cells, which in turn leads to the symptoms of CF. To date, there are over 1,400 known mutations in the CF gene. This tremendous amount of variation in this gene accounts for the differences in the severity of the disease in CF patients.

By knowing the precise gene that causes the disease, scientists have been able to develop new treatment options for people with CF. At one time, an individual with CF rarely saw his or her twentieth birthday; now it is routine for people to live into their thirties and forties. New treatments, such as gene therapy, are being explored for sufferers of CF.

Video Good Poison

CHECK YOUR PROGRESS 3.3

1. Describe the structure and overall function of the plasma membrane.
2. Compare and contrast diffusion,osmosis, facilitated transport, and active transport.
3. Discuss the various ways materials can enter and leave cells.

CONNECTING THE CONCEPTS

The movement of materials across a cell membrane is crucial to the maintenance of homeostasis for many organ systems in humans. For some examples, refer to the following discussions:

Section 8.3 examines how nutrients, including glucose, are moved into the cells of the digestive system.

Section 10.4 investigates how the movement of salts by the urinary system maintains blood homeostasis.

Section 20.2 explains the patterns of inheritance associated with cystic fibrosis.

3.4 The Nucleus and Endomembrane System

LEARNING OUTCOMES

Upon completion of this section, you should be able to

1. Describe the structure of the nucleus and explain its role as the storage place of the genetic information.
2. Summarize the function of the organelles of the endomembrane system.
3. Explain the role and location of the ribosomes.

The nucleus and several organelles are involved in the production and processing of proteins. The endomembrane system is a series of membrane organelles that function in the processing of materials for the cell.

The Nucleus

The nucleus, a prominent structure in eukaryotic cells, stores genetic information (Fig. 3.13). Every cell in the body contains the same genes. Genes are segments of DNA that contain information for the production of specific proteins. Each type of cell has certain genes turned on and others turned off. DNA, with RNA acting as an intermediary, specifies the proteins in a cell. Proteins have many functions in cells, and they help determine a cell's specificity.

Chromatin is the combination of DNA molecules and proteins that make up the **chromosomes.** Chromatin can coil tightly to form visible chromosomes during *meiosis* (cell division that forms reproductive cells in humans) and *mitosis* (cell division that duplicates cells). Most of the time, however, the chromatin is uncoiled. Individual chromosomes cannot be distinguished and the chromatin appears grainy in electron micrographs of the nucleus. Chromatin is immersed in a semifluid medium called the nucleoplasm. A difference in pH suggests that nucleoplasm has a different composition from cytoplasm.

Micrographs of a nucleus often show a dark region (or sometimes more than one) of chromatin. This is the **nucleolus,** where ribosomal RNA (rRNA) is produced. This is also where rRNA joins with proteins to form the subunits of ribosomes.

The nucleus is separated from the cytoplasm by a double membrane known as the **nuclear envelope.** This is continuous with the **endoplasmic reticulum (ER)**, a membranous system of saccules and channels discussed in the next section. The nuclear envelope has **nuclear pores** of sufficient size to permit the passage of ribosomal subunits out of the nucleus and proteins into the nucleus.

Ribosomes

Ribosomes are organelles composed of proteins and rRNA. Protein synthesis occurs at the ribosomes. Ribosomes are often attached to the endoplasmic reticulum; but they also may occur free within the cytoplasm, either singly or in groups called

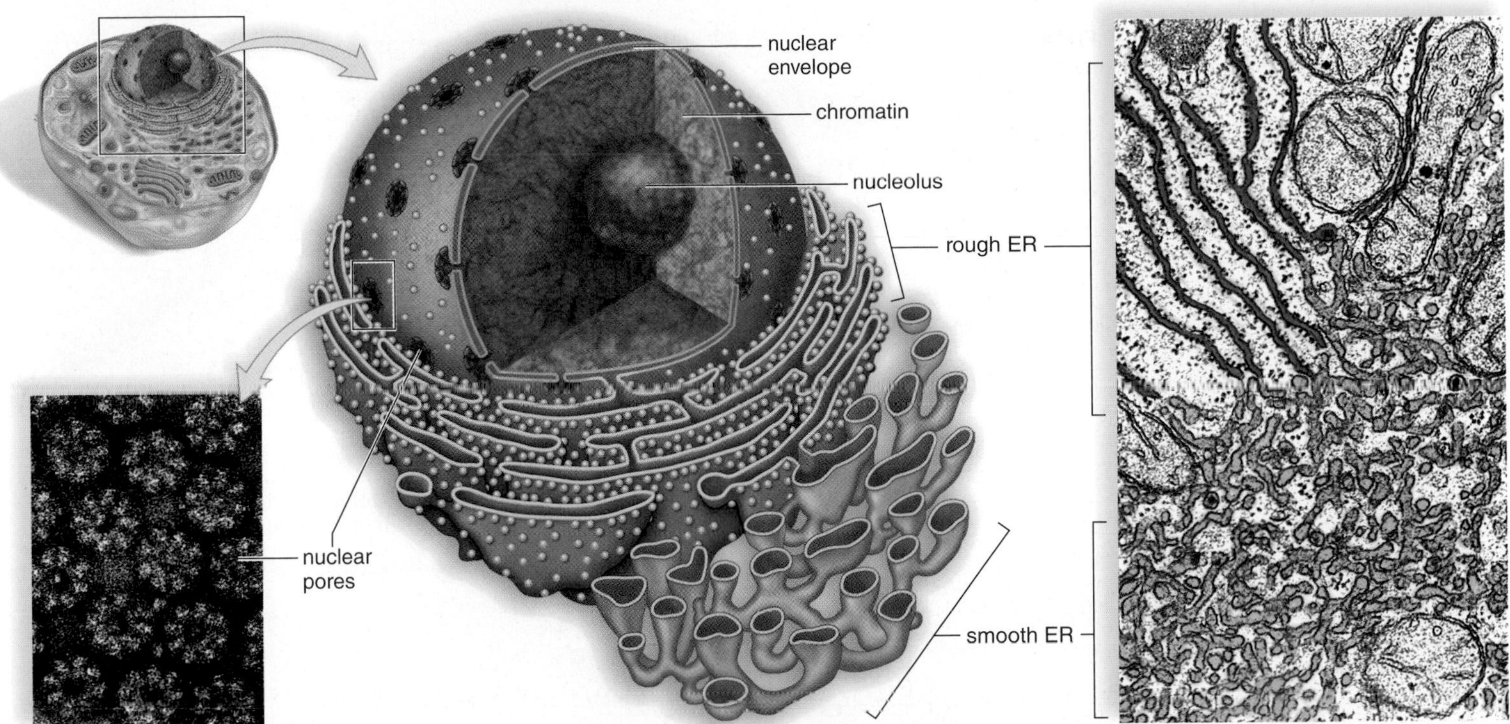

Figure 3.13 The nucleus and endoplasmic reticulum.
The nucleus contains chromatin. Chromatin has a special region called the nucleolus, where rRNA is produced and ribosome subunits are assembled. The nuclear envelope contains pores (TEM, *left*) that allow substances to enter and exit the nucleus to and from the cytoplasm. The nuclear envelope is attached to the endoplasmic reticulum (TEM, *right*), which often has attached ribosomes, where protein synthesis occurs.

polyribosomes. Proteins synthesized at ribosomes attached to the endoplasmic reticulum have a different destination from that of proteins manufactured at ribosomes free in the cytoplasm.

The Endomembrane System

The **endomembrane system** consists of the nuclear envelope, the endoplasmic reticulum, the Golgi apparatus, lysosomes, and **vesicles** (tiny membranous sacs) (Fig. 3.14). This system compartmentalizes the cell so that chemical reactions are restricted to specific regions. The vesicles transport molecules from one part of the system to another.

The Endoplasmic Reticulum

The endoplasmic reticulum (ER) has two portions. Rough ER is studded with ribosomes on the side of the membrane that faces the cytoplasm. Here, proteins are synthesized and enter the ER interior, where processing and modification begin. Some of these proteins are incorporated into the membrane, and some are for export. Smooth ER, continuous with rough ER, does not have attached ribosomes. Smooth ER synthesizes the phospholipids that occur in membranes and has various other functions, depending on the particular cell. In the testes, it produces testosterone. In the liver, it helps detoxify drugs.

The ER forms transport vesicles in which large molecules are transported to other parts of the cell. Often, these vesicles are on their way to the plasma membrane or the Golgi apparatus.

The Golgi Apparatus

The **Golgi apparatus** is named for Camillo Golgi, who discovered its presence in cells in 1898. The Golgi apparatus consists of a stack of slightly curved saccules, whose appearance can be compared to a stack of pancakes. Here, proteins and lipids received from the ER are modified. For example, a chain of sugars may be added to them. This makes them glycoproteins and glycolipids, molecules often found in the plasma membrane.

The vesicles that leave the Golgi apparatus move to other parts of the cell. Some vesicles proceed to the plasma membrane, where they discharge their contents. In all, the Golgi apparatus is involved in processing, packaging, and secretion.

Lysosomes

Lysosomes, membranous sacs produced by the Golgi apparatus, contain *hydrolytic enzymes.* Lysosomes are found in all cells of the body but are particularly numerous in white blood cells that engulf disease-causing microbes. When a lysosome fuses with such an endocytic vesicle, its contents are digested by lysosomal enzymes into simpler subunits that then enter the cytoplasm. In a process called autodigestion, parts of a cell may be broken down by the lysosomes. Some human diseases are caused by the lack of a particular lysosome enzyme. Tay–Sachs disease, as discussed in the chapter opener, occurs when an undigested substance collects in nerve cells, leading to developmental problems and death in early childhood.

Animation Lysosomes

Figure 3.14 **The endomembrane system.** The organelles in the endomembrane system work together to produce, modify, secrete, and digest proteins and lipids.

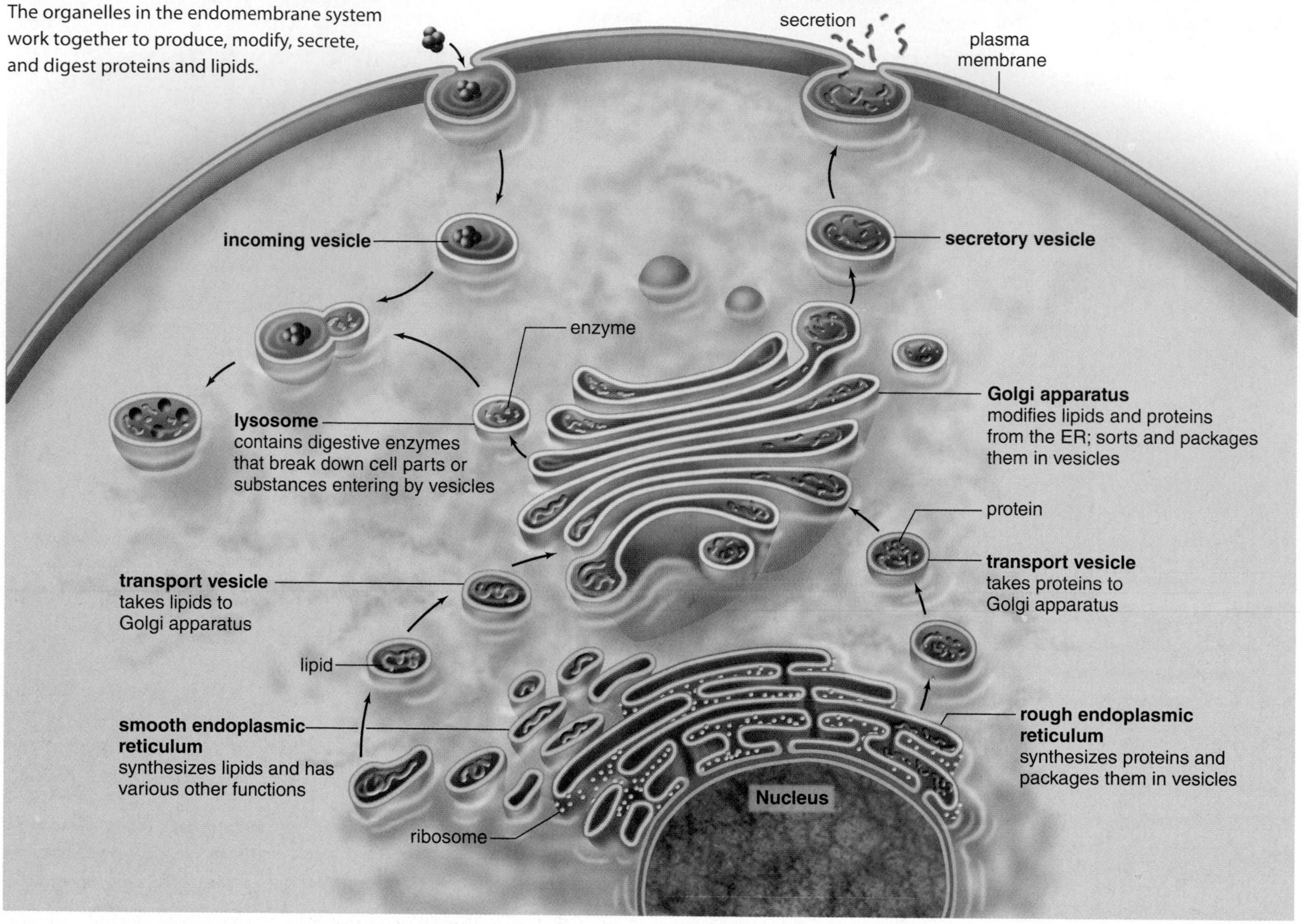

CHECK YOUR PROGRESS 3.4

1. Describe the functions of the following organelles: endoplasmic reticulum, Golgi apparatus, and lysosomes.
2. Explain how the nucleus, ribosomes, and rough endoplasmic reticulum contribute to protein synthesis.
3. Describe the function of the endomembrane system, including the formation and actions of transport vesicles.

CONNECTING THE CONCEPTS

For a more detailed look at how the organelles of the endomembrane system function, refer to the following discussions:

Section 17.5 contains information on how aging is related to the breakdown of cellular organelles.

Section 20.2 explores the patterns of inheritance associated with Tay–Sachs disease.

Section 21.2 provides a more detailed look at how ribosomes produce proteins.

3.5 The Cytoskeleton, Cell Movement, and Cell Junctions

LEARNING OUTCOMES

Upon completion of this section, you should be able to

1. Explain the role of the cytoskeleton in the cell.
2. Summarize the major protein fibers in the cytoskeleton.
3. Describe the role of flagella and cilia in human cells.
4. Compare the function of adhesion junctions, gap junctions, and tight junctions in human cells.

It took a high-powered electron microscope to discover that the cytoplasm of the cell is crisscrossed by several types of protein fibers collectively called the **cytoskeleton** (see Fig. 3.4). The cytoskeleton helps maintain a cell's shape and either anchors the organelles or assists their movement, as appropriate.

In the cytoskeleton, **microtubules** are much larger than actin filaments. Each is a cylinder that contains rows of a protein called tubulin. The regulation of microtubule assembly is under the control of a microtubule organizing center called the **centrosome** (see Fig. 3.4). Microtubules help maintain the shape of the cell and act as tracks along which organelles move. During cell division, microtubules form spindle fibers, which assist the movement of chromosomes. **Actin filaments,** made of a protein called *actin,* are long, extremely thin fibers that usually occur in bundles or other groupings. Actin filaments are involved in movement. Microvilli, which project from certain cells and can shorten and extend, contain actin filaments. **Intermediate filaments,** as their name implies, are intermediate in size between microtubules and actin filaments. Their structure and function differ according to the type of cell.

Cilia and Flagella

Cilia (sing., cilium) and **flagella** (sing., flagellum) are involved in movement. The ciliated cells that line our respiratory tract sweep debris trapped within mucus back up the throat. This helps keep the lungs clean. Similarly, ciliated cells move an egg along the oviduct, where it will be fertilized by a flagellated sperm cell (Fig. 3.15). Motor molecules, powered by ATP, allow the microtubules in cilia and flagella to interact and bend and, thereby, move.

The importance of normal cilia and flagella is illustrated by the occurrence of a genetic disorder. Some individuals have an inherited genetic defect that leads to malformed microtubules in cilia and flagella. Not surprisingly, these individuals suffer from recurrent and severe respiratory infections. The ciliated cells lining respiratory passages fail to keep their lungs clean. They are also unable to reproduce naturally due to the lack of ciliary action to move the egg in a female or the lack of flagella action by sperm in a male.

APPLICATIONS AND MISCONCEPTIONS

How fast does a human sperm swim?

Individual sperm speeds vary considerably and are greatly influenced by environmental conditions. However, in recent studies, researchers found that some human sperm could travel at top speeds of approximately 20 cm/hour. This means that these sperm could reach the female ovum in less than an hour. Scientists are interested in sperm speed so that they can design new contraceptive methods.

Video Human Sperm

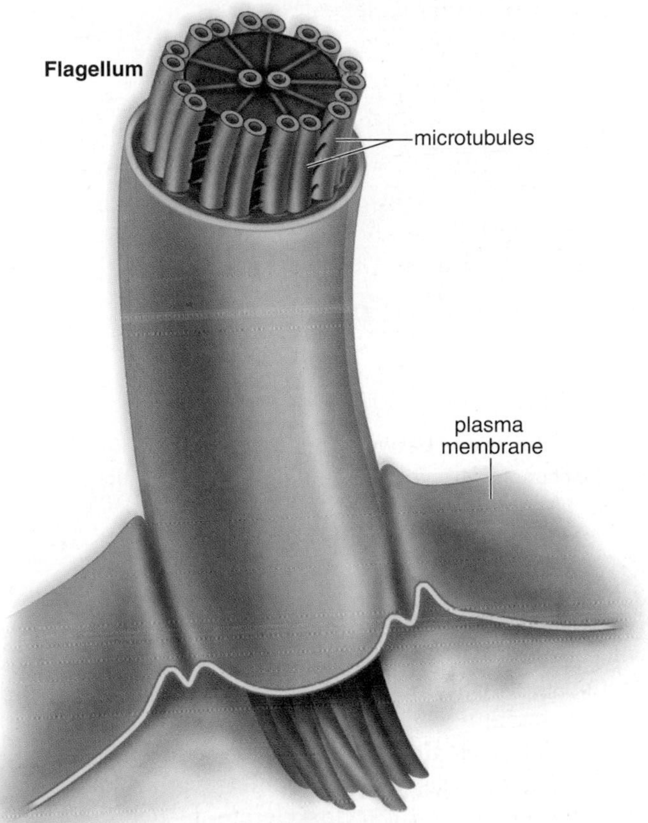

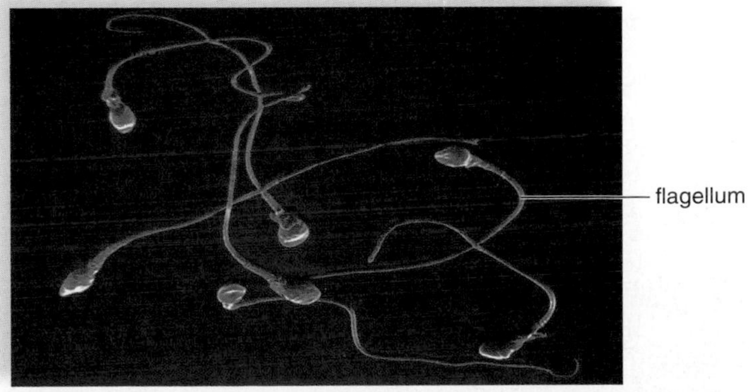

Figure 3.15 Structure and function of the flagella and cilia. Human reproduction is dependent on the normal activity of cilia and flagella. **a.** Both cilia and flagella have an inner core of microtubules within a covering of plasma membrane. **b.** Cilia within the oviduct move the egg to where it is fertilized by a flagellated sperm. **c.** Sperm have very long flagella.

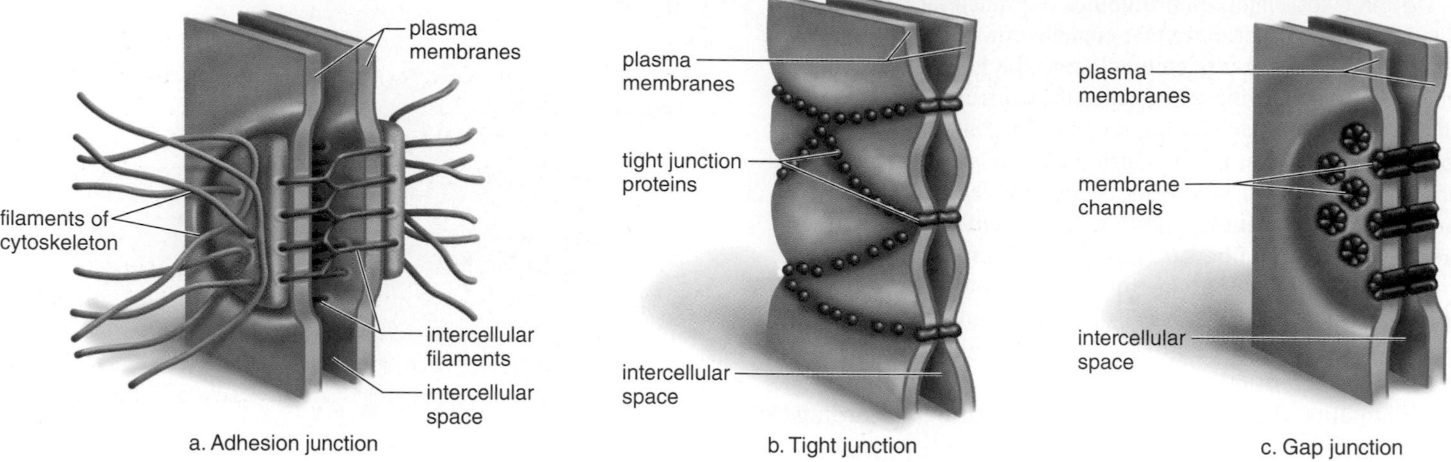

Figure 3.16 Junctions between cells.
a. Adhesion junctions mechanically connect cells. **b.** Tight junctions form barriers with the external environment. **c.** Gap junctions allow for communication between cells.

Junctions Between Cells

As we will see in the next chapter, human tissues are known to have junctions between their cells that allow them to function in a coordinated manner. Figure 3.16 illustrates the three main types of cell junctions in human cells.

Adhesion junctions serve to mechanically attach adjacent cells. In these junctions, the cytoskeletons of two adjacent cells are interconnected. They are a common type of junction between skin cells. In *tight junctions,* connections between the plasma membrane proteins of neighboring cells produce a zipperlike barrier. These types of junctions are common in the digestive system and the kidney, where it is necessary to contain fluids (digestive juices and urine) within a specific area. *Gap junctions* serve as communication portals between cells. In these junctions, channel proteins of the plasma membrane fuse, allowing easy movement between adjacent cells.

CHECK YOUR PROGRESS 3.5

1. List the three types of fibers found in the cytoskeleton.
2. Describe the structure of cilia and flagella and state the function of each.
3. List the types of junctions found in animal cells and provide a function for each.

CONNECTING THE CONCEPTS

The cytoskeleton of the cell plays an important role in many aspects of our physiology. To explore this further, refer to the following discussions:

Section 9.1 investigates how the ciliated cells of the respiratory system function.

Section 16.2 explains the role of the flagellated sperm cell in reproduction.

Section 18.1 explores how the cytoskeleton is involved in cell division.

3.6 Mitochondria and Cellular Metabolism

LEARNING OUTCOMES

Upon completion of this section, you should be able to

1. Identify the key structures of a mitochondrion.
2. Summarize the relationship between the mitochondria and energy-generating pathways of the cell.
3. Summarize the roles of glycolysis, citric acid cycle, electron transport chain, and fermentation in energy generation.
4. Illustrate the stages of the ATP cycle.

Mitochondria (sing., mitochondrion) are often called the powerhouses of the cell. Just as a powerhouse burns fuel to produce electricity, the mitochondria convert the chemical energy of glucose products into the chemical energy of ATP molecules. In the process, mitochondria use up oxygen and give off carbon dioxide. Therefore, the process of producing ATP is called **cellular respiration.** The structure of mitochondria is appropriate to the task. The inner membrane is folded to form little shelves called *cristae.* These project into the *matrix,* an inner space filled with a gel-like fluid (Fig. 3.17). The matrix of a mitochondrion contains enzymes for breaking down glucose products. ATP production then occurs at the cristae. Protein complexes that aid in the conversion of energy are located in an assembly-line fashion on these membranous shelves.

The structure of a mitochondrion supports the hypothesis that they were originally prokaryotes engulfed by a cell. Mitochondria are bounded by a double membrane, as a prokaryote would be if taken into a cell by endocytosis. Even more interesting is the observation that mitochondria have their own genes—and they reproduce themselves!

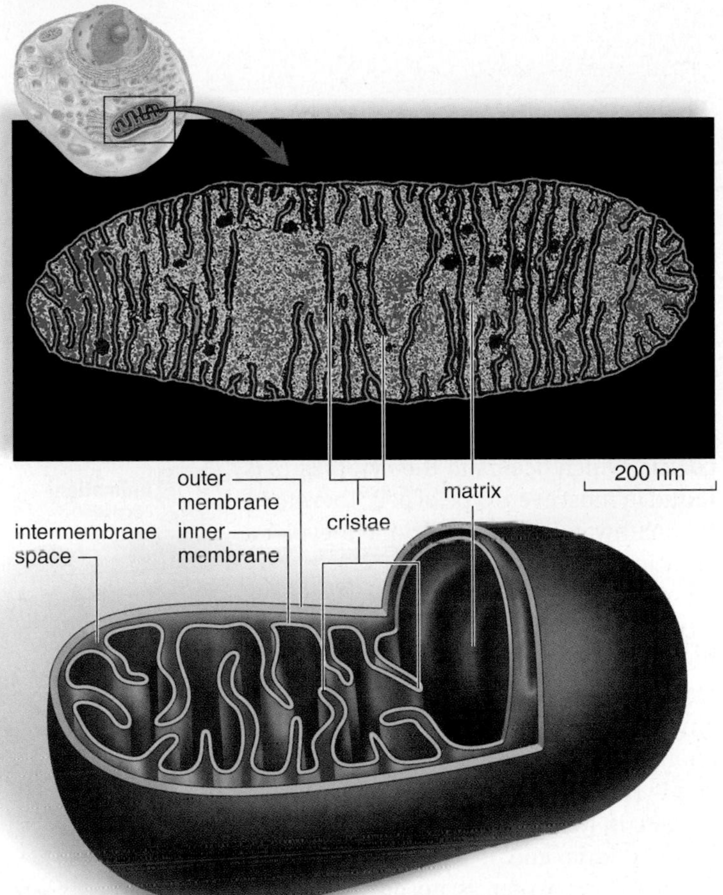

Figure 3.17 **The structure of a mitochondrion.**
A mitochondrion (TEM, *above*) is bounded by a double membrane, and the inner membrane folds into projections called cristae. The cristae project into a semifluid matrix that contains many enzymes.

Cellular Respiration and Metabolism

Cellular respiration is an important component of **metabolism,** which includes all the chemical reactions that occur in a cell. Often, metabolism requires metabolic pathways and is carried out by enzymes sequentially arranged in cells:

$$A \xrightarrow{1} B \xrightarrow{2} C \xrightarrow{3} D \xrightarrow{4} E \xrightarrow{5} F \xrightarrow{6} G$$

The letters, except *A* and *G*, are **products** of the previous reaction and the **reactants** for the next reaction. *A* represents the beginning reactant, and *G* represents the final product. The numbers in the pathway refer to different enzymes. *Each reaction in a metabolic pathway requires a specific enzyme.* The mechanism of action of enzymes has been studied extensively because enzymes are so necessary in cells.

Animation Biochemical Pathways

Metabolic pathways are highly regulated by the cell. One type of regulation is *feedback inhibition.* In feedback inhibition, one of the end products of the metabolic pathway interacts with an enzyme early in the pathway. In most cases, this feedback slows down the pathway so that the cell does not produce more product than it needs.

Animation Feedback Inhibition of Biochemical Pathways

Enzymes

Enzymes are metabolic assistants that speed up the rate of a chemical reaction. The reactant(s) that participate(s) in the reaction is/are called the enzyme's **substrate(s).** Enzymes are often named for their substrates. For example, lipids are broken down by lipase, maltose by maltase, and lactose by lactase.

Enzymes have a specific region, called an **active site,** where the substrates are brought together so they can react. An enzyme's specificity is caused by the shape of the active site. Here the enzyme and its substrate(s) fit together in a specific way, much as the pieces of a jigsaw puzzle fit together (Fig. 3.18). After one reaction is complete, the product or products are

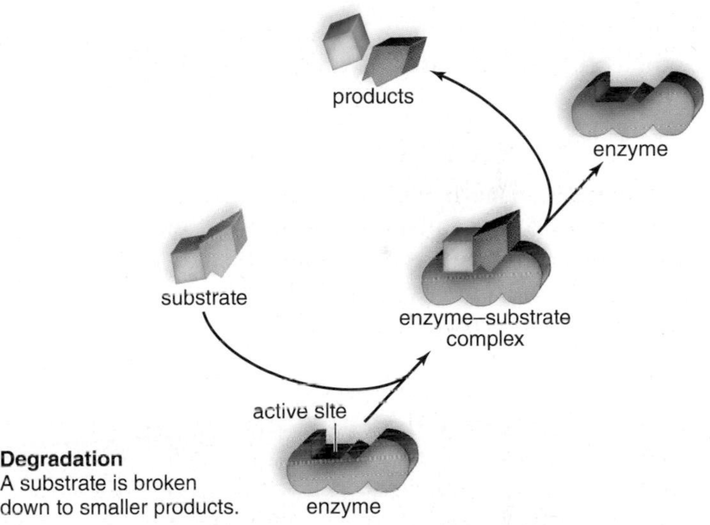

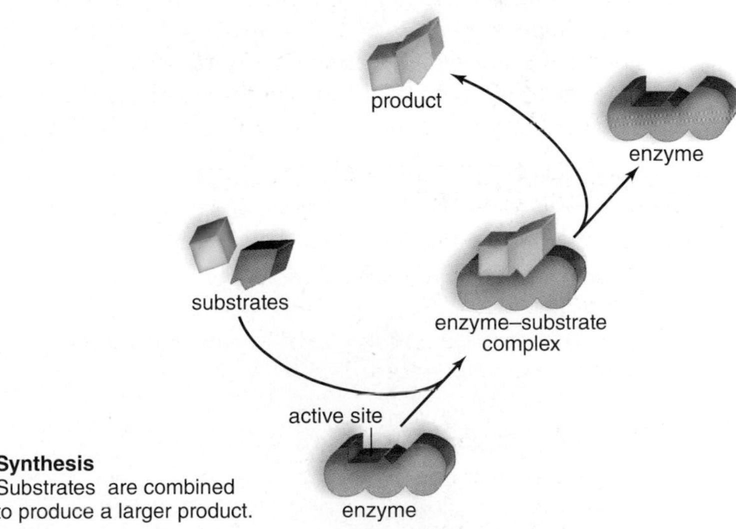

Figure 3.18 **Action of an enzyme.**
An enzyme has an active site, where the substrates and enzyme fit together in such a way that the substrates are oriented to react. Following the reaction, the products are released and the enzyme is free to act again. Some enzymes carry out degradation, in which the substrate is broken down to smaller products. Other enzymes carry out synthesis, in which the substrates are combined to produce a larger product.

released. The enzyme is ready to be used again. Therefore, a cell requires only a small amount of a particular enzyme to carry out a reaction. A chemical reaction can be summarized in the following manner:

$$E + S \longrightarrow ES \longrightarrow E + P$$

where E = enzyme, S = substrate, ES = enzyme–substrate complex, and P = product. An enzyme can be used over and over again.

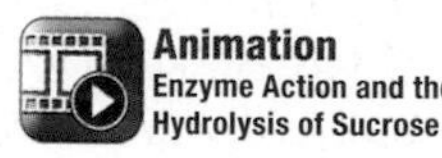

Coenzymes are nonprotein molecules that assist the activity of an enzyme and may even accept or contribute atoms to the reaction. It is interesting that vitamins are often components of coenzymes. The vitamin niacin is a part of the coenzyme **NAD^+ (nicotinamide adenine dinucleotide)**, which carries hydrogen (H) and electrons.

Cellular Respiration

After blood transports glucose and oxygen to cells, cellular respiration begins. Cellular respiration breaks down glucose to carbon dioxide and water. Three pathways are involved in the breakdown of glucose—glycolysis, the citric acid cycle, and the electron transport chain (Fig. 3.19). These metabolic pathways allow the energy within a glucose molecule to be slowly released so that ATP can be gradually produced. Cells would lose a tremendous amount of energy, in the form of heat, if glucose breakdown occurred all at once. When humans burn wood or coal, the energy escapes all at once as heat. But a cell gradually "burns" glucose, and energy is captured as ATP.

MP3
Cellular Respiration

Glycolysis **Glycolysis** means "sugar splitting." During glycolysis, glucose, a six-carbon (C_6) molecule, is split so that the result is two three-carbon (C_3) molecules of *pyruvate*. Glycolysis, which occurs in the cytoplasm, is found in most every type of cell. Therefore, this pathway is believed to have evolved early in the history of life.

Animation
Glycolysis

Glycolysis is an **anaerobic** pathway, because it does not require oxygen. This pathway can occur in microbes that live in bogs or swamps or our intestinal tract, where there is no oxygen. During glycolysis, hydrogens and electrons are removed from glucose, and NADH results. The breaking of bonds releases enough energy for a net yield of two ATP molecules.

Pyruvate is a pivotal molecule in cellular respiration. When oxygen is available, the molecule enters mitochondria and is completely broken down. When oxygen is not available, fermentation occurs (discussion follows).

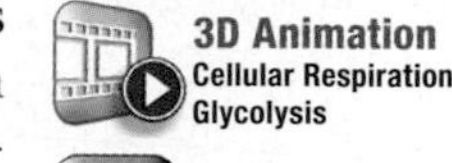

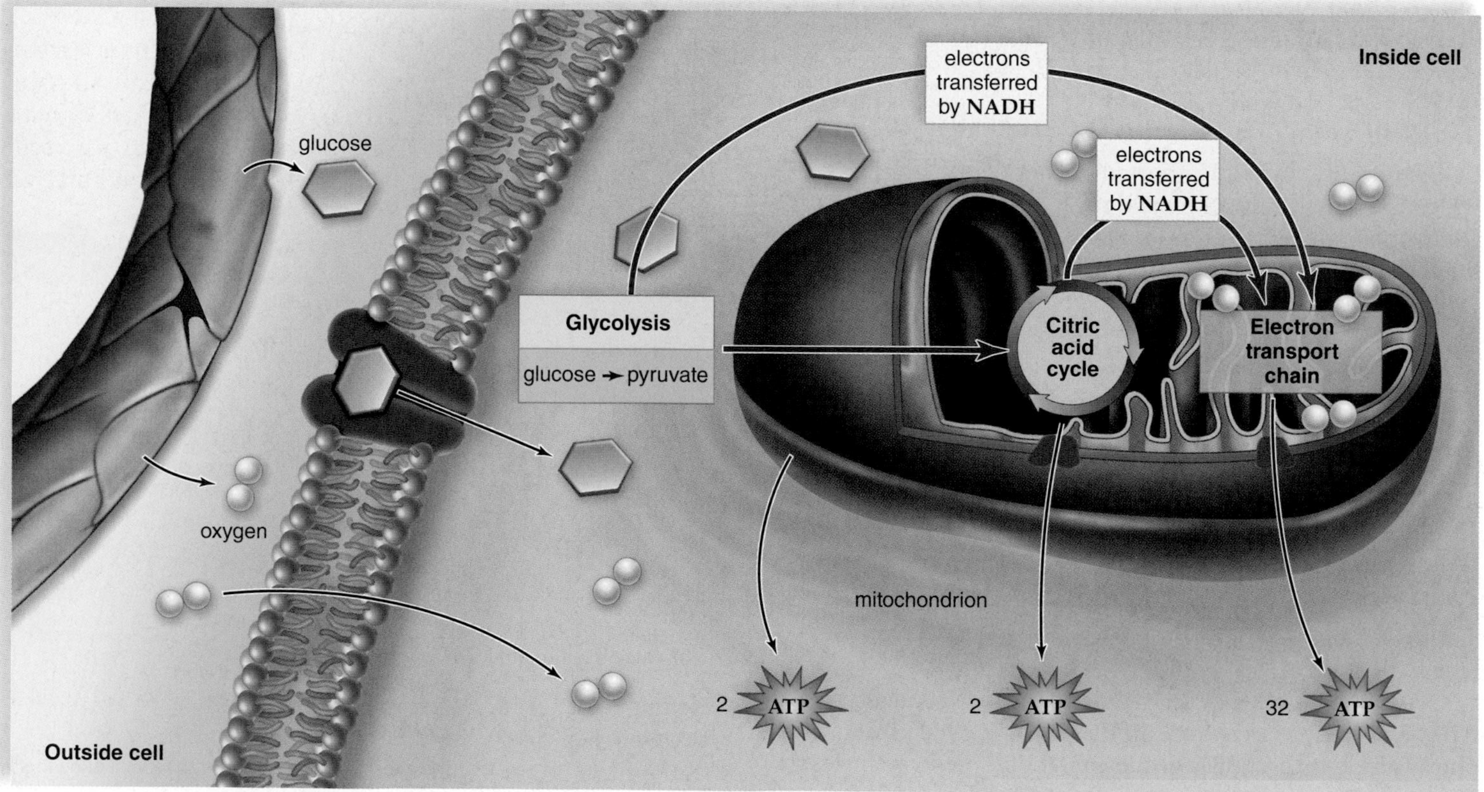

Figure 3.19 **Production of ATP.**
Glucose enters a cell from the bloodstream by facilitated transport. The three main pathways of cellular respiration (glycolysis, citric acid cycle, and electron transport chain) all produce ATP, but most is produced by the electron transport chain. NADH carries electrons to the electron transport chain from glycolysis and the citric acid cycle. ATP exits a mitochondrion by facilitated transport.

Citric Acid Cycle Each of the pyruvate molecules, after a brief modification, enters the citric acid cycle as acetyl CoA. The **citric acid cycle,** also called the *Krebs cycle,* is a cyclical series of enzymatic reactions that occurs in the matrix of mitochondria. The purpose of this pathway is to complete the breakdown of glucose by breaking the remaining C—C bonds. As the reactions progress, carbon dioxide is released, a small amount of ATP (two per glucose) is produced, and the remaining hydrogen and electrons are carried away by NADH. The cellular respiration pathways have the ability to use organic molecules other than carbohydrates as an energy source. Both fats and proteins may be converted to compounds that enter the citric acid cycle. More information on these processes is provided in the Health feature, "The Metabolic Fate of Pizza," on page 62.

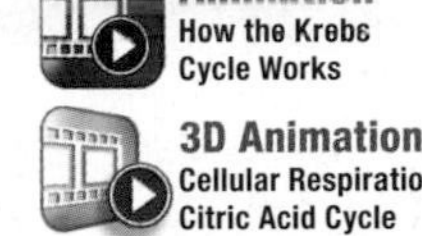

Electron Transport Chain NADH molecules from glycolysis and the citric acid cycle deliver electrons to the **electron transport chain.** The members of the electron transport chain are carrier proteins grouped into complexes. These complexes are embedded in the cristae of a mitochondrion. Each carrier of the electron transport chain accepts two electrons and passes them on to the next carrier. The hydrogens carried by NADH molecules will be used later.

High-energy electrons enter the chain and, as they are passed from carrier to carrier, the electrons lose energy. Low-energy electrons emerge from the chain. Oxygen serves as the final acceptor of the electrons at the end of the chain. After oxygen receives the electrons, it combines with hydrogens and becomes water.

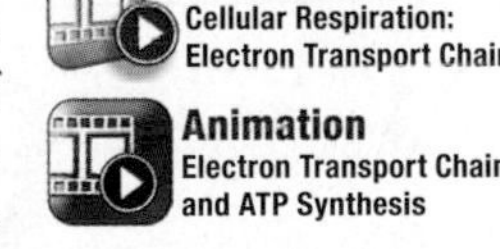

The presence of oxygen makes the electron transport chain **aerobic.** Oxygen does not combine with any substrates during cellular respiration. Breathing is necessary to our existence, and the sole purpose of oxygen is to receive electrons at the end of the electron transport chain.

The energy, released as electrons pass from carrier to carrier, is used for ATP production. It took many years for investigators to determine exactly how this occurs, and the details are beyond the scope of this text. Suffice it to say that the inner mitochondrial membrane contains an ATP-synthase complex that combines ADP + Ⓟ to produce ATP. The ATP-synthase complex produces about 32 ATP per glucose molecule. Overall, the reactions of cellular respiration produce between 36 and 38 ATP molecules.

ATP-ADP Cycle Each cell produces ATP within its mitochondria; therefore, each cell uses ATP for its own purposes. Figure 3.20 shows the ATP cycle. Glucose breakdown leads to ATP buildup, and then ATP is used for the metabolic work of the cell. Muscle cells use ATP for contraction, and nerve cells use it for conduction of nerve impulses. ATP breakdown releases heat.

Fermentation

Fermentation is an anaerobic process, meaning that it does not require oxygen. When oxygen is not available to cells, the electron transport chain soon becomes inoperative. This is because oxygen is not present to accept electrons. In this case, most cells have a safety valve so that some ATP can still be produced. Glycolysis operates as long as it is supplied with "free" NAD^+—NAD^+ that can pick up hydrogens and electrons. Normally, NADH takes electrons to the electron transport chain and, thereby, is recycled to become NAD^+. However, if the system is not working due to a lack of oxygen, NADH passes its hydrogens and electrons to pyruvate molecules, as shown in the following reaction:

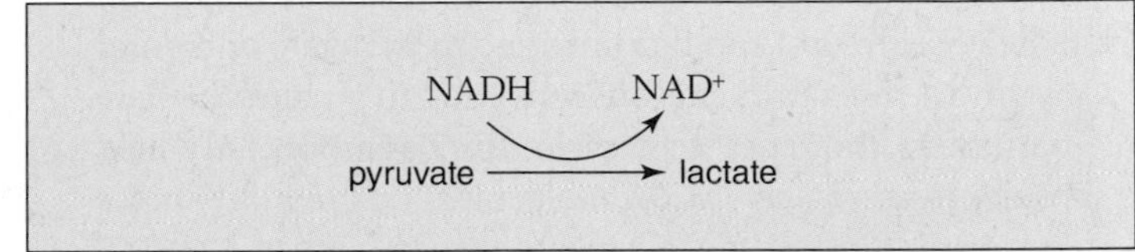

This means that the citric acid cycle and the electron transport chain do not function as part of fermentation. When oxygen is available again, lactate can be converted back to pyruvate and metabolism can proceed as usual.

Fermentation can give us a burst of energy for a short time, but it produces only two ATP per glucose molecule. Also, fermentation results in the buildup of lactate. Lactate is toxic to cells and causes muscles to cramp and fatigue. If fermentation continues for any length of time, death follows.

Fermentation takes its name from yeast fermentation. Yeast fermentation produces alcohol and carbon dioxide (instead of lactate). When yeast is used to leaven bread, carbon dioxide production makes the bread rise. When yeast is used to produce alcoholic beverages, it is the alcohol that humans make use of.

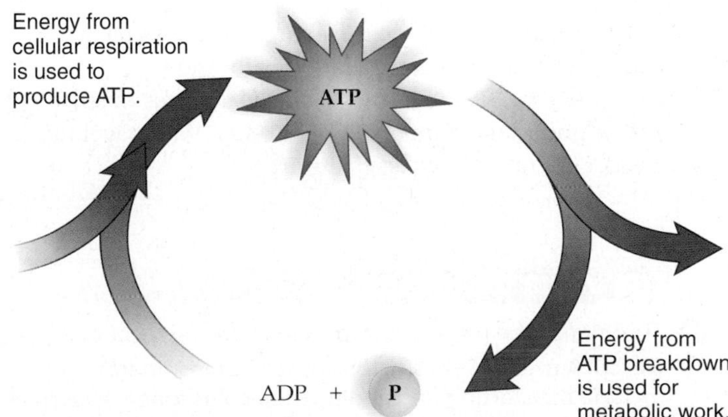

Figure 3.20 **The ATP cycle.**
The breakdown of organic nutrients, such as glucose, by cellular respiration transfers energy to form ATP. ATP is used for energy-requiring reactions, such as muscle contraction. ATP breakdown also gives off heat. Additional food energy rejoins ADP and Ⓟ to form ATP again.

BIOLOGY MATTERS  Health

The Metabolic Fate of Pizza

Obviously our diets do not solely consist of carbohydrates. Because fats and proteins are also organic nutrients, it makes sense that our bodies can utilize the energy found in the bonds of these molecules. In fact, the metabolic pathways we have discussed in this chapter are more than capable of accessing the energy of fats and proteins. For example, let's trace the fate of a pepperoni pizza, which contains carbohydrates (crust), fats (cheese), and protein (pepperoni).

We already know that the glucose in the carbohydrate crust is broken down during cellular respiration. When the cheese in the pizza (a fat) is used as an energy source, it breaks down to glycerol and three fatty acids. As Figure 3B indicates, glycerol can be converted to pyruvate and enter glycolysis. The fatty acids are converted to an intermediate that enters the citric acid cycle. An 18-carbon fatty acid results in nine acetyl CoA molecules. Calculation shows that respiration of these can produce a total of 108 ATP molecules. This is why fats are an efficient form of stored energy—the three long fatty acid chains per fat molecule can produce considerable ATP when needed.

Proteins are less frequently used as an energy source, but are available if necessary. The carbon skeleton of amino acids can enter glycolysis, be converted to acetyl groups, or enter the citric acid cycle at another point. The carbon skeleton is produced in the liver when an amino acid undergoes deamination, or the removal of the amino group. The amino group becomes ammonia (NH_3), which enters the urea cycle and becomes part of urea, the primary excretory product of humans.

In Chapter 8, "Digestive System and Nutrition," we will take a more detailed look at the nutritional needs of humans, including discussions on how vitamins and minerals interact with metabolic pathways, and the dietary guidelines for proteins, fats, and carbohydrates.

Questions to Consider

1. How might a meal of a cheeseburger and fries be processed by the cellular respiration pathways?.
2. While Figure 3B does not indicate the need for water, it is an important component of our diet. Where would water interact with these pathways?

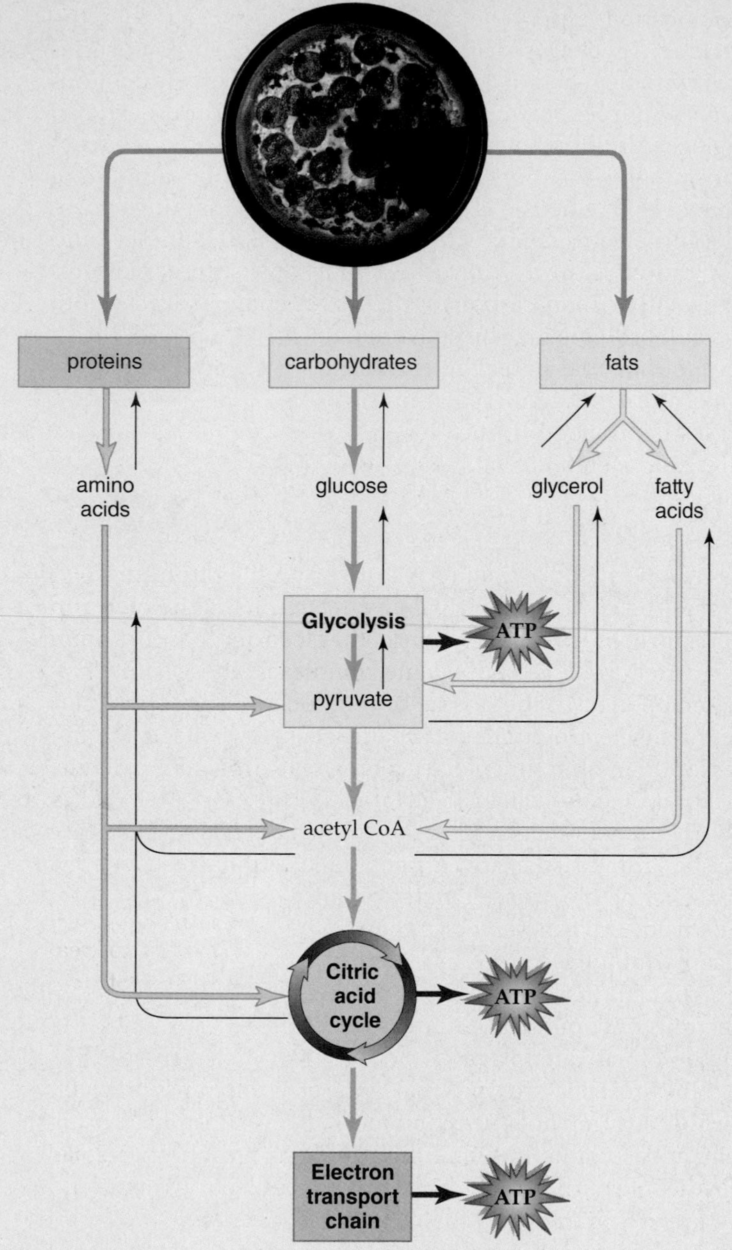

Figure 3B The use of fats and proteins for energy. Carbohydrates, fats, and proteins can be used as energy sources, and their monomers (carbohydrates and proteins) or subunits (fats) enter degradative pathways at specific points.

CHECK YOUR PROGRESS 3.6

1. Summarize the roles of enzymes in chemical reactions.
2. Describe the three basic steps of cellular respiration. Include the starting and ending molecules for each step.
3. Hypothesize what would happen to homeostasis if each of three major steps of cellular respiration were missing.

CONNECTING THE CONCEPTS

For additional information on the processing of nutrients for energy, refer to the following discussions:

Sections 2.3 to 2.5 provide a more detailed look at carbohydrates and other energy nutrients.

Section 8.3 explores how the small intestine processes nutrients for absorption.

Section 8.6 describes the importance of carbohydrates, fats, and proteins in the diet.

BIOLOGY MATTERS  **Science**

Stem-Cell Research

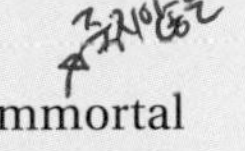

In the human body, stem cells are analogous to immortal "parents." Their "offspring," called *daughter cells,* can remain as stem cells and potentially divide indefinitely. However, most daughter cells differentiate further, forming mature cells called *end cells.* Research using stem cells has remained a source of controversy since 1998, when scientists discovered how to isolate and grow human stem cells in the laboratory.

There are primarily two different types of stem cells: embryonic and adult. Advantages and disadvantages exist for each type. *Embryonic stem cells* are derived from fertilized embryos at various stages of development. Fertilized human ova stored in infertility clinics are often used as the source of embryonic stem cells. The use of these cells for research has sparked tremendous controversy, because many people believe these cells have the potential to become a human. *Adult stem cells* are undifferentiated cells found in various body tissues, whose purpose is to repair or replace damaged tissues. The use of adult stem cells is generally accepted. However, adult stem cells lack the flexibility of embryonic stem cells, because adult stem cells form far fewer types of end cells.

With all the time and money spent on stem-cell research, how close are we to using stem cells for the cure of disease? Advances in stem-cell therapy are being announced all the time. For example, in 2012, researchers announced that they had successfully used stem cells to produce a neuron cell that could be used as a model for producing drugs to treat Alzheimer disease. Some of the most successful uses of stem cells have involved Parkinson disease. Parkinson disease is a progressive motor control disorder, triggered by the death of certain neurons in the brain. These neurons are responsible for releasing the neurotransmitter dopamine onto specific brain cells that control movement. (This is why Parkinson patients are often treated with L-dopa, which is converted into dopamine.) It is now possible to cause stem cells in the laboratory to differentiate into neurons that produce dopamine. To be used for transplant purposes, however, the stem cells must produce enough end cells for transplant. Further, the cells must survive after the transplant and function correctly for the remainder of the patient's life. Finally, transplanted cells must not harm the patient. The usual risks of surgery would still exist for the transplant recipient: damage to healthy tissue, bleeding, infection.

A possible solution was introduced in 2008 when researchers first developed the use of *induced pluripotent stem cells,* or *iPS cells.* These cells are normal cells of the body that have been chemically "convinced" to return to an undifferentiated state. In other words, it is now possible to induce adult cells of the body to form stem cells. By doing so, researchers hope to be able to bypass some of the controversies surrounding the use of embryonic stem cells and, in the process, develop a more rapid and effective method of obtaining stem cells to fight specific diseases. For this groundbreaking work, *Science Magazine* was awarded its 2008 Breakthrough of the Year Award.*

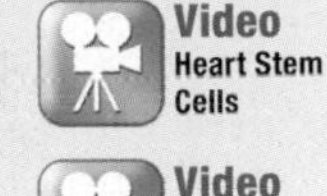

Questions to Consider

1. How much time and money should be spent on a therapy that may work only after "years of intensive research"? Would this money be better spent on therapies that have a higher likelihood of success?
2. Should the president remove the ban on certain types of stem cells so that this research can proceed faster?
3. What criteria and ethical considerations should be used to select Parkinson patients for stem-cell therapy?

* "Breakthrough of the Year: Reprogramming Cells," *Science* 322, no. 5909 (2008), http://www.sciencemag.org/cgi/content/full/322/5909/1766 (accessed February 22, 2012).

CASE STUDY CONCLUSION

Over the next few months, both Kevin and Mary dedicated hours to understanding the causes and treatments of Tay–Sachs disease. They learned that the disease is caused by a recessive mutation that limits the production of an enzyme called beta-hexosaminidase A. This enzyme is loaded into a newly formed lysosome by the Golgi apparatus. The enzyme's function is to break down a specific type of fatty acid chain called *gangliosides.* Gangliosides play an important role in the early formation of the neurons in the brain. Tay–Sachs disease occurs when the gangliosides overaccumulate in the neurons.

Though the prognosis for their child was initially poor—very few children with Tay–Sachs live beyond the age of four, the parents were encouraged to find out what advances in a form of medicine called gene therapy might be able to prolong the life of their child. In gene therapy, a correct version of the gene is introduced into specific cells in an attempt to regain lost function. Some initial studies using mice as a model had demonstrated an ability to reduce ganglioside concentrations by providing a working version of the gene that produced beta-hexosaminidase A to the neurons of the brain. Though research was still ongoing, it was a promising piece of information for both Kevin and Mary.

MEDIA STUDY TOOLS

Enhance your study of this chapter with media! Visit **www.mhhe.com/maderhuman13e** and go to "Media Study Tools" for this chapter to access the following:

Animations	Videos	MP3 Files
3.2 Endosymbiosis • Geologic History of Earth **3.3** Diffusion Through Cell Membranes • Hemolysis and Crenation • How Osmosis Works • How Facilitated Diffusion Works • How the Sodium-Potassium Pump Works • Endocytosis and Exocytosis **3.4** Lysosomes **3.6** Biochemical Pathways • Feedback Inhibition of Biochemical Pathways • Enzyme Action and the Hydrolysis of Sucrose • How the NAD+ Works • How Glycolysis Works • How the Krebs Cycle Works • Electron Transport Chain and ATP Synthesis	**3.3** Good Poison **3.5** Human Sperm **3.6** Heart Stem Cells • Making Brain Cells	**3.2** Cellular Organelles **3.3** Membrane Structure • Simple Diffusion • Osmosis

3D Animations

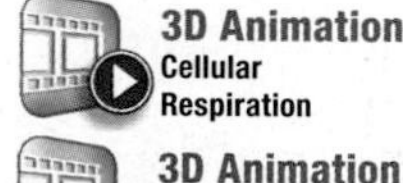

McGraw-Hill's 3D animations, "Cellular Respiration" and "Membrane Transport" provide a dynamic exploration of the key concepts of this chapter and are available through McGraw-Hill Connect®.

SUMMARIZE

3.1 What Is a Cell?

The **cell theory** states that cells are the basic units of life and that all life comes from pre-existing cells. Microscopes are used to view cells, which must remain small to have a favorable surface area-to-volume ratio.

3.2 How Cells Are Organized

The human cell is a **eukaryotic cell** with a **nucleus** that contains the genetic material. The cell is surrounded by a **plasma membrane,** a **selectively permeable** barrier that limits the movement of materials in and out of the cell. Between the plasma membrane and the nucleus is the **cytoplasm,** which contains various **organelles.** Organelles in the cytoplasm have specific functions. **Prokaryotic cells,** such as the bacteria, are smaller than eukaryotic cells, and lack a nucleus.

3.3 The Plasma Membrane and How Substances Cross It

The **fluid-mosaic model** describes the structure of the plasma membrane. The plasma membrane contains

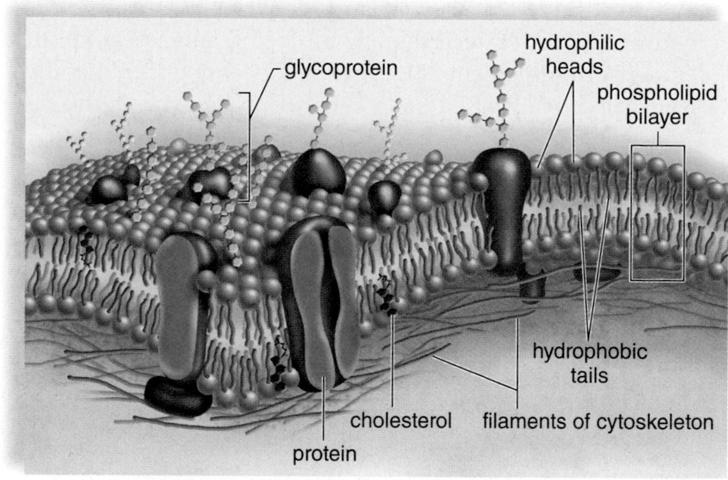

- a phospholipid bilayer that selectively regulates the passage of molecules and ions into and out of the cell; and
- embedded proteins, which allow certain substances to cross the plasma membrane.

Passage of molecules into or out of cells can be passive or active.

- Passive mechanisms do not require energy. Examples are are **diffusion, osmosis,** and **facilitated transport. Tonicity** and **osmotic pressure** control the process of osmosis.
- Active mechanisms require an input of energy. Examples are **active transport** (sodium-potassium pump), endocytosis (**phagocytosis** and pinocytosis), receptor-mediated endocytosis, and exocytosis.

3.4 The Nucleus and the Endomembrane System

- The nucleus houses DNA, which specifies the order of amino acids in proteins. It is surrounded by a **nuclear envelope** that contains **nuclear pores** for communication and the movement of materials.
- **Chromatin** is a combination of DNA molecules and proteins that make up **chromosomes.**
- The **nucleolus** produces ribosomal RNA (rRNA).
- Protein synthesis occurs in **ribosomes,** small organelles composed of proteins and rRNA.

The Endomembrane System

The **endomembrane system** consists of the nuclear envelope, **endoplasmic reticulum (ER)**, Golgi apparatus, lysosomes, and **vesicles**.

- The rough ER has ribosomes, where protein synthesis occurs.
- Smooth ER has no ribosomes and has various functions, including lipid synthesis.
- The **Golgi apparatus** processes and packages proteins and lipids into vesicles for secretion or movement into other parts of the cell.
- **Lysosomes** are specialized vesicles produced by the Golgi apparatus. They fuse with incoming vesicles to digest enclosed material, and they autodigest old cell parts.

3.5 The Cytoskeleton, Cell Movement, and Cell Junctions

- The **cytoskeleton** consists of **microtubules, actin filaments,** and **intermediate filaments** that give cells their shape; and it allows organelles to move about the cell. Microtubules are organized by **centrosomes. Cilia** and **flagella,** which contain microtubules, allow a cell to move.
- Cell junctions connect cells to form tissues and to facilitate communication between cells.

3.6 Mitochondria and Cellular Metabolism

- **Mitochondria** have an inner membrane that forms cristae, which project into the matrix.
- Mitochondria are involved in cellular respiration, which uses oxygen and releases carbon dioxide.
- During **cellular respiration,** mitochondria convert the energy of glucose into the energy of ATP molecules.

Cellular Respiration and Metabolism

- **Metabolism** represents all of the chemical reactions that occur in a cell. A metabolic pathway is a series of reactions, each of which has its own **enzyme.** The materials entering these reactions are called **reactants,** and the materials leaving the pathway are called **products.**
- Enzymes bind their **substrates** in the **active site.**
- Sometimes enzymes require **coenzymes**—such as **NAD$^+$** (**nicotinamide adenine dinucleotide**)—nonprotein molecules that participate in the reaction.
- Cellular respiration is the enzymatic breakdown of glucose to carbon dioxide and water.
- Cellular respiration includes three pathways: **glycolysis** (occurs in the cytoplasm and is **anaerobic**), the **citric acid cycle** (releases carbon dioxide), and the **electron transport chain** (passes electrons to oxygen). Reactions that occur within the mitochondria (citric acid cycle and electron transport chain) are **aerobic.**

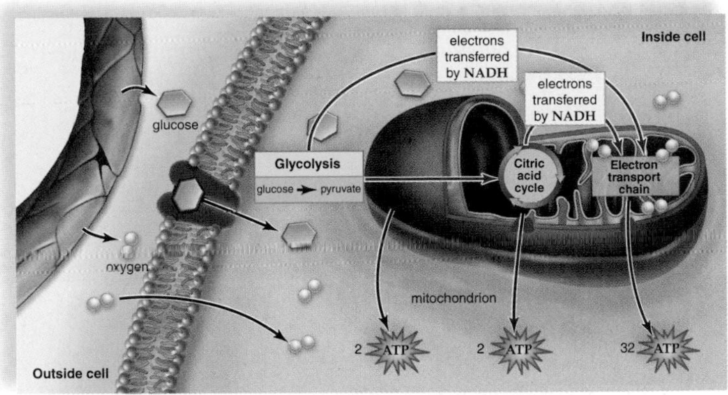

Fermentation

- If oxygen is not available in cells, the electron transport chain is inoperative, and **fermentation** (which does not require oxygen) occurs. Fermentation serves to recycle NAD$^+$ molecules so that the cell can produce a small amount of ATP by glycolysis.

ASSESS

Testing Your Knowledge of the Concepts

Complete the following questions.

1. Explain the three key concepts of the cell theory. (page 45)
2. Which type of microscope would you use to observe the swimming behavior of a flagellated protozoan? Explain. (page 46)
3. Describe how the eukaryotic cell gained mitochondria and chloroplasts. (page 49)
4. Invagination of plasma membrane produced what structures in eukaryotic cells not present in prokaryotic cells? (page 49)
5. How does the organization of the plasma membrane relate to its function? (pages 50–54)
6. Describe four different ways materials can enter and/or leave a cell. (pages 50–54)
7. For the following cell organelles, describe the structure and function of each: nucleus, nucleolus, ribosomes, endoplasmic reticulum (rough and smooth), Golgi apparatus, lysosomes, centrioles, and mitochondria. (pages 54–56)
8. Explain the purpose of the cytoskeleton in a cell. (pages 56–57)
9. Describe an enzyme and coenzyme. Explain the mechanism of enzyme function, particularly the relationship of shape to its activity. (pages 58–59)
10. Which stage of cellular respiration produces the most ATP? Explain. (pages 58–60)

11. The small size of cells is best correlated with
- **a.** the fact that they are self-reproducing.
- **b.** an adequate surface area for exchange of materials.
- **c.** their vast versatility.
- **d.** All of these are correct.

12. Which of the following is not part of the fluid-mosaic model?
- **a.** phospholipids
- **b.** proteins
- **c.** cholesterol
- **d.** chromatin

13. Facilitated transport differs from diffusion in that facilitated diffusion
- **a.** involves the passive use of a carrier protein.
- **b.** involves the active use of a carrier protein.
- **c.** moves a molecule from a low to a high concentration.
- **d.** involves the use of ATP molecules.

14. When a cell is placed in a hypotonic solution,
- **a.** solute exits the cell to equalize the concentration on both sides of the membrane.
- **b.** water exits the cell toward the area of lower solute concentration.
- **c.** water enters the cell toward the area of higher solute concentration.
- **d.** solute exits and water enters the cell.

In questions 15–18, match each function to the proper organelle in the key.

Key:

a. mitochondrion	**c.** Golgi apparatus
b. nucleus	**d.** rough ER

15. Packaging and secretion

16. ATP production (powerhouse of cell)

17. Protein synthesis

18. Control center for the cell

19. Vesicles carrying proteins for secretion move from the ER to the
- **a.** smooth ER.
- **b.** lysosomes.
- **c.** Golgi apparatus.
- **d.** nucleolus.

20. The active site of an enzyme
- **a.** is identical to that of any other enzyme.
- **b.** is the part of the enzyme where the substrate can fit.
- **c.** can be used over and over.
- **d.** is where the coenzyme binds.
- **e.** Both b and c are correct.

21. Which of the following pathways is anaerobic?
- **a.** electron transport chain
- **b.** citric acid cycle
- **c.** glycolysis
- **d.** All of these pathways are aerobic.

22. Use these terms to label the following diagram of the plasma membrane: carbohydrate chain, filaments of the cytoskeleton, hydrophilic heads, hydrophobic tails, membrane protein (used twice), phospholipid, and phospholipid bilayer.

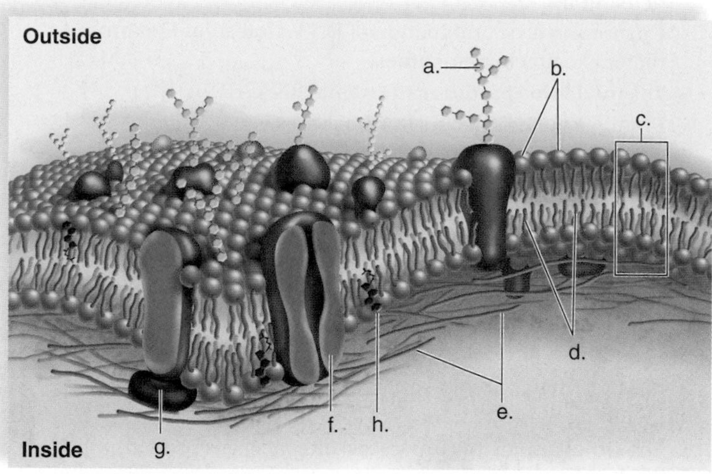

ENGAGE

The virtual lab "Enzyme-Controlled Reactions" provides an interactive investigation of how environmental conditions regulate enzyme activity.

Thinking Critically About the Concepts

In the case study at the beginning of the chapter, the child had malfunctioning lysosomes, which caused an accumulation of fatty acid in the system. Each part of a cell plays an important role in the homeostasis of the entire body.

1. What might occur if the cells of the body contained malfunctioning mitochondria?
2. What would happen to homeostasis if enzymes were no longer produced in the body?
3. Knowing what you know about the function of a lysosome, what might occur if the cells' lysosomes were overproductive instead of malfunctioning?

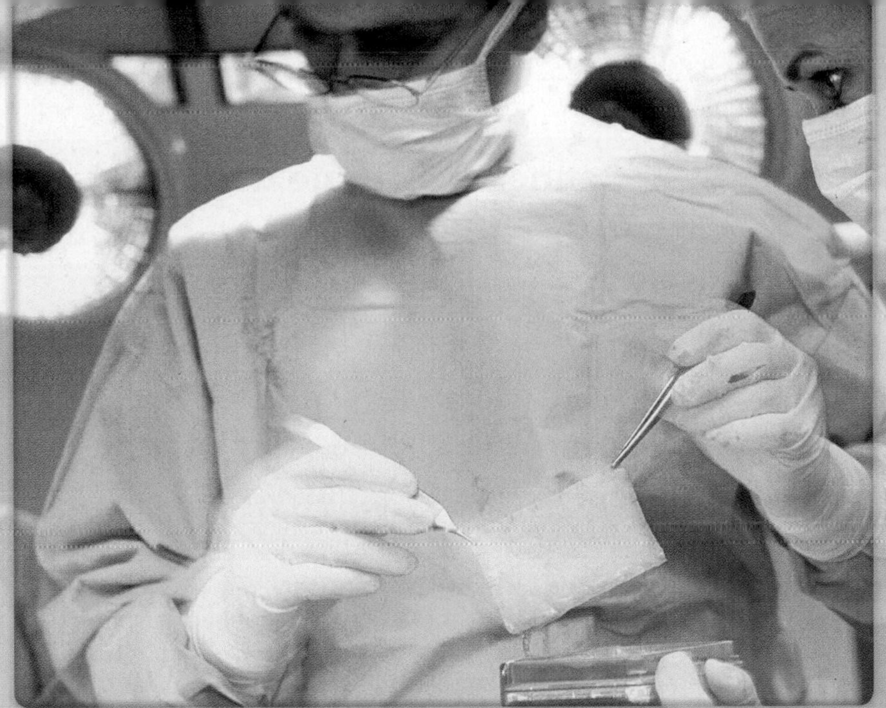

CHAPTER

4

Organization and Regulation of Body Systems

CASE STUDY ARTIFICIAL SKIN

When Kristen awoke in the hospital, she discovered that the fire in her home had produced third-degree burns over a large portion of her legs. The doctors informed Kristen that the burns on her legs covered too much of an area for autografting, the traditional grafting technique that removes skin from other parts of her body to cover the burn areas. Another available option was allografting, where skin is removed from another person, or a cadaver, and used to cover the burn areas. However, Kristen's specialists were not eager to take that route because complications often develop due to rejection of the foreign tissue or infections.

Instead, the doctors recommended a relatively new technique—artificial skin. Just a decade ago, the concept of artificial skin may have been found only in a science-fiction movie, but advances in medical technology have made the use of artificial skin a reality. The purpose of using artificial skin is not to permanently replace the damaged tissue; rather, the procedure is designed to protect the damaged tissue and allow time for the patient's skin to heal itself.

The first step of the procedure, after the removal of the burned tissue, is to cover the wound with the artificial skin. This skin contains collagen, a connective tissue, and an adhesive-like carbohydrate that allows the artificial skin to bind to the underlying tissue. Initially, the artificial skin contains a plastic wrapping that simulates the epidermis and protects the tissue from water loss and infection. The next step is to remove a small sample of epidermal cells from an unburned area of skin on the patient's body. These are taken to a laboratory and placed in incubators to grow sheets of skin. Once ready, the plastic covering on the patient is replaced by the sheets of epidermal cells. Over time, the laboratory-grown artificial skin is integrated into the newly growing skin.

As you read through the chapter, think about the following questions:

1. What types of tissue are normally found in skin?
2. Why would burn damage to the skin represent such a serious challenge to Kristen's health?
3. Why would it be more difficult to produce new dermis in the laboratory than new epidermis?

CHAPTER CONCEPTS

4.1 Types of Tissues
The body contains four types of tissues: connective, muscular, nervous, and epithelial.

4.2 Connective Tissue Connects and Supports
Connective tissues bind and support body parts.

4.3 Muscular Tissue Moves the Body
Muscular tissue moves the body and its parts.

4.4 Nervous Tissue Communicates
Nervous tissue transmits information throughout the body.

4.5 Epithelial Tissue Protects
Epithelial tissues line cavities and cover surfaces.

4.6 Integumentary System
The skin is the largest organ and plays an important role in maintaining homeostasis.

4.7 Organ Systems, Body Cavities, and Body Membranes
Organ systems contain multiple organs that interact to carry out a process.

4.8 Homeostasis
Homeostasis maintains the internal environment and is made possible by feedback mechanisms.

BEFORE YOU BEGIN

Before beginning this chapter, take a few moments to review the following discussions:

Section 1.1 How do tissues and organs fit into the levels of biological organization?

Section 3.2 How are cells structured?

Section 3.5 How are cells linked together to form tissues?

4.1 Types of Tissues

LEARNING OUTCOME

Upon completion of this section, you should be able to

1. Describe the four types of tissues and provide a general function for each.

Recall from the material on the levels of biological organization (see Fig. 1.2) that cells are composed of molecules; a tissue is a group of similar cells; an organ contains several types of tissues; and several organs are found in an organ system. In this chapter, we will further explore the tissue, organ, and organ system levels of organization.

MP3 Tissues Overview

A **tissue** is composed of specialized cells of the same type that perform a common function in the body. The tissues of the human body can be categorized into four major types:

Connective tissue binds and supports body parts.
Muscular tissue moves the body and its parts.
Nervous tissue receives stimuli and conducts nerve impulses.
Epithelial tissue covers body surfaces and lines body cavities.

Cancers are classified according to the type of tissue from which they arise. Sarcomas are cancers arising in muscular or connective tissue (especially bone or cartilage). Leukemias are cancers of the blood. Lymphomas are cancers of lymphoid tissue. Carcinomas, the most common type, are cancers of epithelial tissue. The chance of developing cancer in a particular tissue is related to the rate of cell division. Both epithelial cells and blood cells reproduce at a high rate. Thus, carcinomas and leukemias are common types of cancer.

CHECK YOUR PROGRESS 4.1

1. List the four major tissue types found in the human body.
2. Describe how cancers are classified.
3. Explain why carcinomas and leukemias are the most common cancer types.

CONNECTING THE CONCEPTS

For more information on cancer, refer to the following discussions:

Section 6.3 examines leukemia as a form of blood cancer.

Section 19.1 provides additional information on the most common types of cancer.

4.2 Connective Tissue Connects and Supports

LEARNING OUTCOMES

Upon completion of this section, you should be able to

1. Describe the primary types of connective tissue and provide a function for each.
2. Compare the structure and function of bone and cartilage.
3. Differentiate between blood and lymph.

Connective tissue is diverse in structure and function. Despite these apparent differences all types of connective tissue have three similar components: specialized cells, ground substance, and protein fibers. These components are shown in Figure 4.1, a diagrammatic representation of loose fibrous connective tissue. The ground substance is a noncellular material that separates the cells. It varies in consistency from solid (bone) to semifluid (cartilage) to fluid (blood).

MP3 Connective Tissue

The fibers are of three possible types. White **collagen fibers** contain collagen, a protein that gives them flexibility and strength. **Reticular fibers** are very thin collagen fibers, highly branched proteins that form delicate supporting networks. Yellow **elastic fibers** contain elastin, a protein that is not as strong as collagen but is more elastic (meaning that it can return to its original shape; elastic fibers may stretch over 100 times their relaxed size without damaging the proteins). Inherited connective tissue disorders arise when people inherit genes that lead to malformed fibers. For example, in Marfan syndrome, there are mutations in the fibrillin gene (*FBN1*). Fibrillin is a component of elastic fibers and these mutations cause a decrease in the

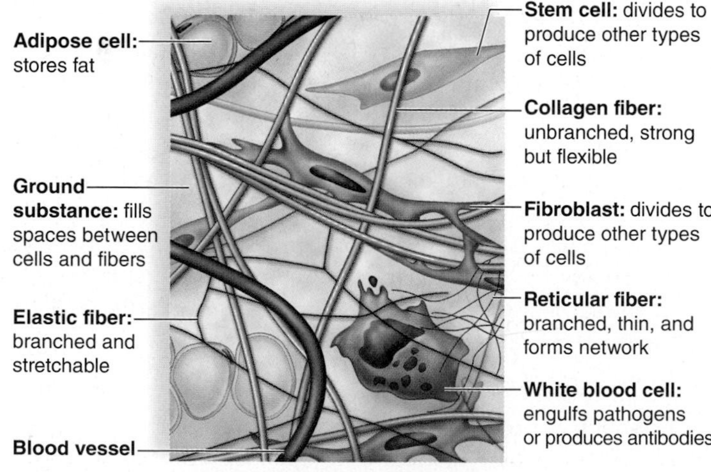

Figure 4.1 **Components of connective tissue.** All connective tissues have three components: specialized cells, ground substance, and protein fibers. Loose fibrous connective tissue is shown here.

elasticity of the connective tissues that are normally rich in elastic fibers, such as the aorta. Individuals with this disease often die from aortic rupture, which occurs when the aorta cannot expand in response to increased blood pressure.

Fibrous Connective Tissue

Fibrous tissue exists in two forms: loose fibrous tissue and dense fibrous tissue. Both loose fibrous and dense fibrous connective tissues have cells called **fibroblasts** located some distance from one another and separated by a jellylike ground substance containing white collagen fibers and yellow elastic fibers (Fig. 4.2). **Matrix** is a term that includes ground substance and fibers.

Loose fibrous connective tissue, also called areolar tissue, supports epithelium and many internal organs. Its presence in lungs, arteries, and the urinary bladder allows these organs to expand. It forms a protective covering enclosing many internal organs, such as muscles, blood vessels, and nerves.

Adipose tissue is a special type of loose connective tissue in which the cells enlarge and store fat. Adipose tissue has little extracellular matrix. Its cells, which are called **adipocytes,** are crowded, and each is filled with liquid fat. The body uses this stored fat for energy, insulation, and organ protection. Adipose tissue also releases a hormone called *leptin,* which regulates appetite-control centers in the brain. Adipose tissue is primarily found beneath the skin, around the kidneys, and on the surface of the heart.

> **APPLICATIONS AND MISCONCEPTIONS**
>
> **Does dieting get rid of fat cells?**
>
> Unfortunately, no. Dieting can shrink the size of the fat cells (adipocytes), but their number stays the same. This means that it is easier to regain the weight if diet and exercise are not maintained. Liposuction is the only way to remove fat cells from the body.

Dense fibrous connective tissue contains many collagen fibers packed together. This type of tissue has more specific functions than does loose connective tissue. For example, dense fibrous connective tissue is found in **tendons,** which connect muscles to bones, and in **ligaments,** which connect bones to other bones at joints.

Supportive Connective Tissue

Cartilage and bone are the two main supportive connective tissues. Each provides structure, shape, protection, and leverage for movement. Generally, cartilage is more flexible than bone because it lacks mineralization of the matrix. The supportive connective tissues are covered in more detail in section 11.1.

Cartilage

In **cartilage,** the cells lie in small chambers called **lacunae** (sing., lacuna), separated by a solid, yet flexible, matrix. This

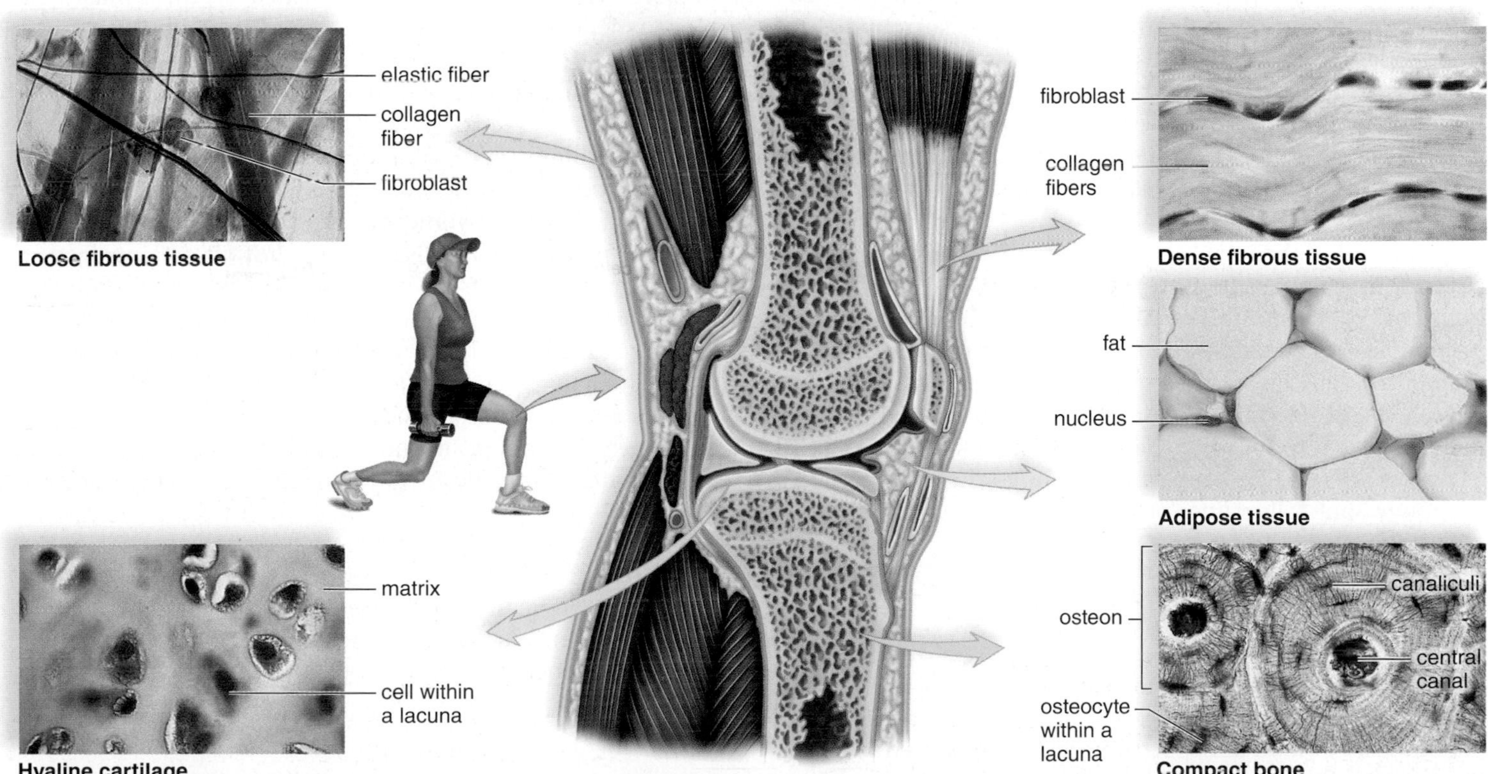

Figure 4.2 Connective tissues in the knee.
Most types of connective tissue may be found in the knee.

matrix is formed by cells called *chondroblasts* and *chondrocytes.* Because this tissue lacks a direct blood supply, it often heals slowly. The three types of cartilage are distinguished by the type of fiber found in the matrix.

Hyaline cartilage (Fig. 4.2), the most common type of cartilage, contains only fine collagen fibers. The matrix has a glassy, translucent appearance. Hyaline cartilage is found in the nose and at the ends of the long bones and the ribs, and it forms rings in the walls of respiratory passages. The fetal skeleton also is made of this type of cartilage. Later, the cartilaginous fetal skeleton is replaced by bone.

Elastic cartilage has more elastic fibers than hyaline cartilage. For this reason, it is more flexible and is found, for example, in the framework of the outer ear.

Fibrocartilage has a matrix containing strong collagen fibers. Fibrocartilage is found in structures that withstand tension and pressure, such as the disks between the vertebrae in the backbone and the cushions in the knee joint.

Bone

Bone is the most rigid connective tissue. It consists of an extremely hard matrix of inorganic salts, notably calcium salts. These salts are deposited around protein fibers, especially collagen fibers. The inorganic salts give bone rigidity. The protein fibers provide elasticity and strength, much as steel rods do in reinforced concrete. Cells called *osteoblasts* and *osteoclasts* are responsible for forming the matrix in bone tissue.

Compact bone makes up the shaft of a long bone (Fig. 4.2). It consists of cylindrical structural units called *osteons* (see section 11.1). The central canal of each osteon is surrounded by rings of hard matrix. Bone cells are located in lacunae between the rings of matrix. In the central canal, nerve fibers carry nerve impulses, and blood vessels carry nutrients that allow bone to renew itself. Thin extensions of bone cells within canaliculi (minute canals) connect the cells to each other and to the central canal.

The ends of the long bones are composed of spongy bone covered by compact bone. Spongy bone also surrounds the bone marrow cavity. This, in turn, is covered by compact bone forming a "sandwich" structure. **Spongy bone** appears as an open, bony latticework with numerous bony bars and plates. These are separated by irregular spaces. Although lighter than compact bone, spongy bone is still designed for strength. Just as braces are used for support in buildings, the solid portions of spongy bone follow lines of stress.

Fluid Connective Tissue

Blood

Blood represents a fluid connective tissue. **Blood,** which consists of formed elements (Fig. 4.3) and plasma, is located in blood vessels. Blood transports nutrients and oxygen to **tissue fluid.** Tissue fluid bathes the body's cells and removes carbon dioxide and other wastes. Blood helps distribute heat and also plays a role in fluid, ion, and pH balance. The systems of the body help keep blood composition and chemistry within normal limits. The structure and function of blood is covered in more detail in Chapter 6.

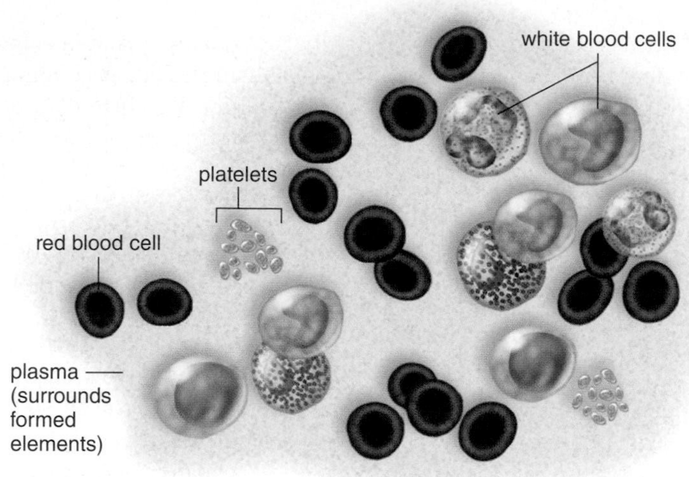

Figure 4.3 The formed elements of blood.
Red blood cells, which lack a nucleus, transport oxygen. Each type of white blood cell has a particular way to fight infections. Platelets, fragments of a particular cell, function in helping to seal injured blood vessels.

The formed elements of blood each have specific functions. The **red blood cells** (**erythrocytes**) are small, biconcave, disk-shaped cells without nuclei. The presence of the red pigment hemoglobin makes the cells red, which, in turn, makes the blood red. Hemoglobin is composed of four units. Each unit is composed of the protein globin and a complex iron-containing structure called *heme.* The iron forms a loose association with oxygen; therefore, red blood cells transport oxygen.

White blood cells (**leukocytes**) may be distinguished from red blood cells because they have a nucleus. Without staining, leukocytes would be translucent. There are many different types of white blood cells, but all are involved in protecting the body from infection. Some white blood cells are *generalists,* meaning that they will respond to any foreign invader in the body. These are phagocytic cells, because they engulf infectious agents such as bacteria. Others are more specific and may either produce antibodies (molecules that combine with foreign substances to inactivate them) or may directly attack specific invading agents or infected cells in the body.

Platelets (**thrombocytes**) are not complete cells. Rather, they are fragments of giant cells present only in bone marrow. When a blood vessel is damaged, platelets form a plug that seals the vessel, and injured tissues release molecules that help the clotting process.

Lymph

Lymph is also a fluid connective tissue. Lymph is a clear (sometimes faintly yellow) fluid derived from the fluids surrounding

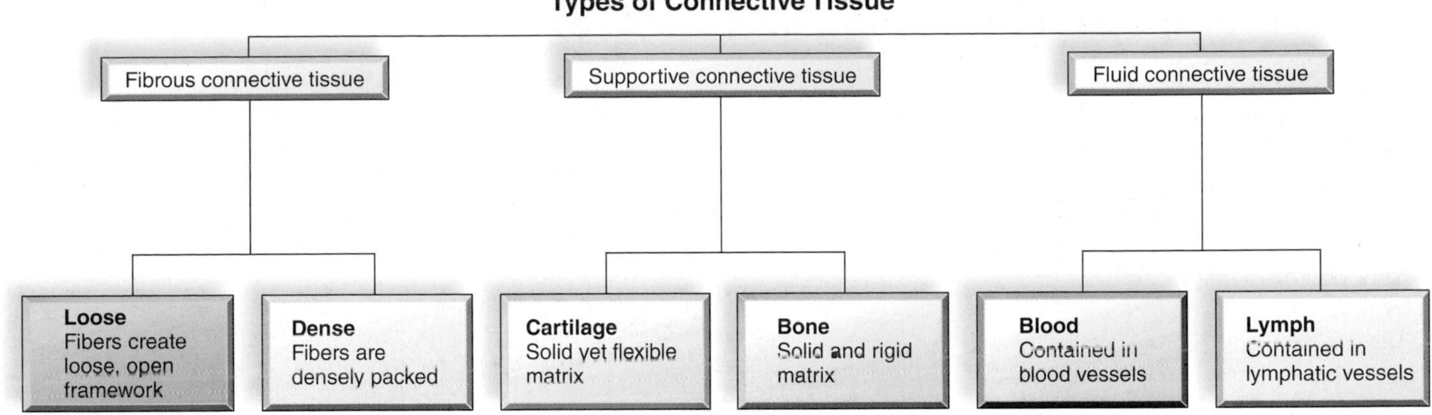

Figure 4.4 **Types of connective tissue.**
Connective tissue is divided into three general categories—fibrous, supportive, and fluid.

the tissues. It contains white blood cells. Lymphatic vessels absorb excess tissue fluid and various dissolved solutes in the tissues. They transport lymph to particular vessels of the cardiovascular system. Lymphatic vessels absorb fat molecules from the small intestine. Lymph nodes, composed of fibrous connective tissue, occur along the length of lymphatic vessels. Lymph is cleansed as it passes through lymph nodes, in particular, because white blood cells congregate there. Lymph nodes enlarge when you have an infection.

Figure 4.4 provides a summary of the classification of each of the major types of connective tissue.

CHECK YOUR PROGRESS 4.2

1. Describe the three general categories of connective tissue, and provide some examples of each type.
2. Explain the difference in the composition of the matrix in each of the three classes of connective tissue.
3. Describe how each of the two fluid connective tissues are important to homeostasis.

CONNECTING THE CONCEPTS

The tissue types discussed in this section are examined in greater detail later in the book. For more information, refer to the following discussions:

Section 6.1 details the types of formed elements found in blood.

Section 7.2 explains the role of the lymphatic system in moving lymph through the body.

Section 11.1 provides a more detailed examination of cartilage and bone.

4.3 Muscular Tissue Moves the Body

LEARNING OUTCOME

Upon completion of this section, you should be able to

1. Distinguish between the three types of muscles with regard to location and function in the body.

Muscular tissue is specialized to contract. It is composed of cells called *muscle fibers,* which contain actin and myosin filaments. The interaction of these filaments accounts for movement. These interactions, and the structure of the muscles of the body, are covered in greater detail in Chapter 12. The three types of vertebrate muscular tissue are skeletal, smooth, and cardiac.

MP3 Muscular Tissue

Skeletal muscle is also called *voluntary muscle* (Fig. 4.5*a*). It is attached by tendons to the bones of the skeleton. When it contracts, body parts move. Contraction of skeletal muscle is under voluntary control and occurs faster than in the other muscle types. Skeletal muscle fibers are cylindrical and long—sometimes they run the length of the muscle. They arise during development when several cells fuse, resulting in one fiber with multiple nuclei. The nuclei are located at the periphery of the cell, just inside the plasma membrane. The fibers have alternating light and dark bands that give them a **striated,** or striped, appearance. These bands are due to the placement of actin filaments and myosin filaments in the cell.

Smooth muscle is so named because the cells lack striations (Fig. 4.5*b*). The spindle-shaped cells each have a single nucleus. These cells form layers in which the thick middle portion of one cell is opposite the thin ends of adjacent cells. Consequently, the nuclei form an irregular pattern in the tissue. Smooth muscle is involuntary, meaning that it is not under conscious or voluntary control. Smooth muscle is found in the walls of viscera (intestine, bladder, and other internal

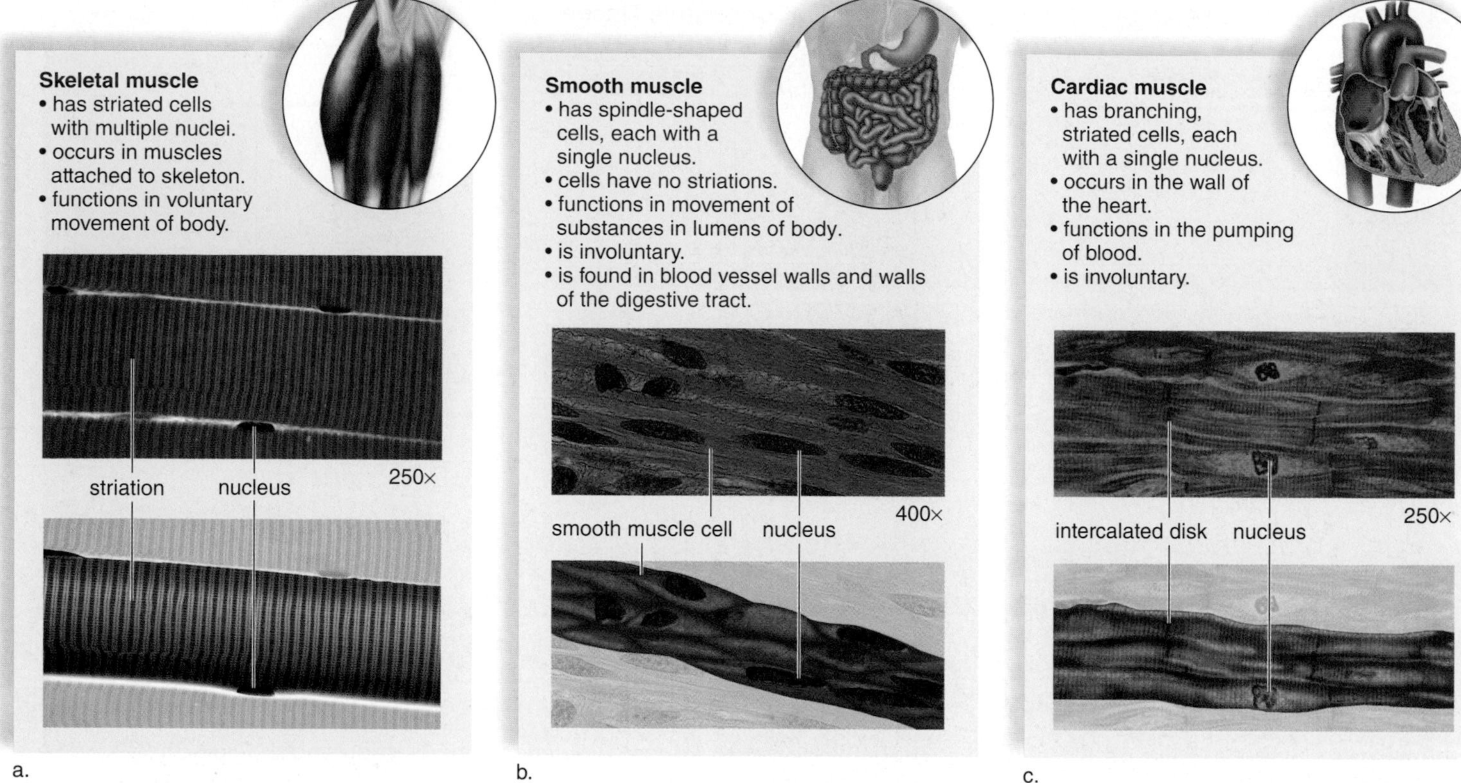

Skeletal muscle
• has striated cells with multiple nuclei.
• occurs in muscles attached to skeleton.
• functions in voluntary movement of body.
striation
nucleus
250×
a.
Smooth muscle
• has spindle-shaped cells, each with a single nucleus.
• cells have no striations.
• functions in movement of substances in lumens of body.
• is involuntary.
• is found in blood vessel walls and walls of the digestive tract.
smooth muscle cell
nucleus
400×
b.
Cardiac muscle
• has branching, striated cells, each with a single nucleus.
• occurs in the wall of the heart.
• functions in the pumping of blood.
• is involuntary.
intercalated disk
nucleus
250×
c.

Figure 4.5 The three types of muscle tissue.
a. Skeletal muscle is voluntary and striated. **b.** Smooth muscle is involuntary and nonstriated. **c.** Cardiac muscle is involuntary and striated.

organs) and blood vessels. For this reason, it is sometimes referred to as *visceral muscle.* Smooth muscle contracts more slowly than skeletal muscle but can remain contracted for a longer time. When the smooth muscle of the bladder contracts, urine is sent into a tube called the urethra, which takes it to the outside. When the smooth muscle of the blood vessels contracts, blood vessels constrict, helping to raise blood pressure.

Cardiac muscle (Fig. 4.5*c*) is found only in the walls of the heart. Its contraction pumps blood and accounts for the heartbeat. Cardiac muscle combines features of both smooth and skeletal muscle. Like skeletal muscle, it has striations, but the contraction of the heart is involuntary for the most part. Cardiac muscle cells also differ from skeletal muscle cells in that they usually have a single, centrally placed nucleus. The cells are branched and seemingly fused, one with another. The heart appears to be composed of one large interconnecting mass of muscle cells. Cardiac muscle cells are separate and individual, but they are bound end to end at *intercalated disks.* These are areas where folded plasma membranes between two cells contain adhesion junctions and gap junctions (see section 3.5).

CHECK YOUR PROGRESS 4.3

1. Explain the difference in the structure and function of skeletal, smooth, and cardiac muscle.
2. Describe where each type of muscle fiber is found in the body.
3. Explain why smooth muscle and cardiac muscle are involuntary and summarize what advantage this provides to homeostasis.

CONNECTING THE CONCEPTS

Muscle tissue plays an important role in our physiology. For more information on each of the three types of muscles, refer to the following discussions:

Section 5.3 provides a more detailed look at how the heartbeat is generated.

Figure 8.2 illustrates how smooth muscle lines the digestive tract.

Section 12.2 examines the structure and function of skeletal muscle.

4.4 Nervous Tissue Communicates

LEARNING OUTCOMES

Upon completion of this section, you should be able to

1. Distinguish between neurons and neuroglia cells.
2. Describe the structure of a neuron.

Nervous tissue consists of nerve cells, called neurons, and neuroglia, the cells that support and nourish the neurons. Nervous tissue is the central component of the nervous system (see Chapter 13), which serves three primary functions in the body: sensory input, integration of data, and motor output.

MP3 Nervous Tissue

Neurons

A **neuron** is a specialized cell that has three parts: dendrites, a cell body, and an axon (Fig. 4.6). A *dendrite* is an extension that receives signals from sensory receptors or other neurons. The *cell body* contains most of the cell's cytoplasm and the nucleus. An *axon* is an extension that conducts nerve impulses. Long axons are covered by myelin, a white fatty substance. The term *fiber*[1] is used here to refer to an axon along with its myelin sheath, if it has one. Outside the brain and spinal cord, fibers bound by connective tissue form nerves. **Nerves** conduct signals from sensory receptors to the spinal cord and the brain, where integration, or processing, occurs. However, the phenomenon called sensation occurs only in the brain. Nerves also conduct signals from the spinal cord and brain to muscles and glands and other organs. This triggers a characteristic response from each tissue. For example, muscles contract and glands secrete. In this way, a coordinated response to the original sensory input is achieved.

Neuroglia

In addition to neurons, nervous tissue contains neuroglia. **Neuroglia** are cells that outnumber neurons nine to one and take up more than half the volume of the brain. Although the primary function of neuroglia is to support and nourish neurons, research is being conducted to determine how much they directly contribute to brain function. Neuroglia do not have long extensions (axons or dendrites). However, researchers are gathering evidence that neuroglia do communicate among themselves and with neurons, even without these extensions. Examples of neuroglia found in the brain are microglia, astrocytes, and oligodendrocytes (Fig. 4.6). *Microglia,* in addition to supporting neurons, engulf bacterial and cellular debris. *Astrocytes* provide nutrients to neurons and produce a hormone known as glial-derived neurotrophic factor (GDNF). This growth factor is currently undergoing clinical trials as a therapy for Parkinson disease and other diseases caused by neuron degeneration. *Oligodendrocytes* form the myelin sheaths around fibers in the brain and spinal cord. Outside the brain, Schwann cells are the type of neuroglia that encircle long nerve fibers and form a myelin sheath.

[1] In connective tissue, a fiber is a component of the matrix; in muscular tissue, a fiber is a muscle cell; in nervous tissue, a fiber is an axon.

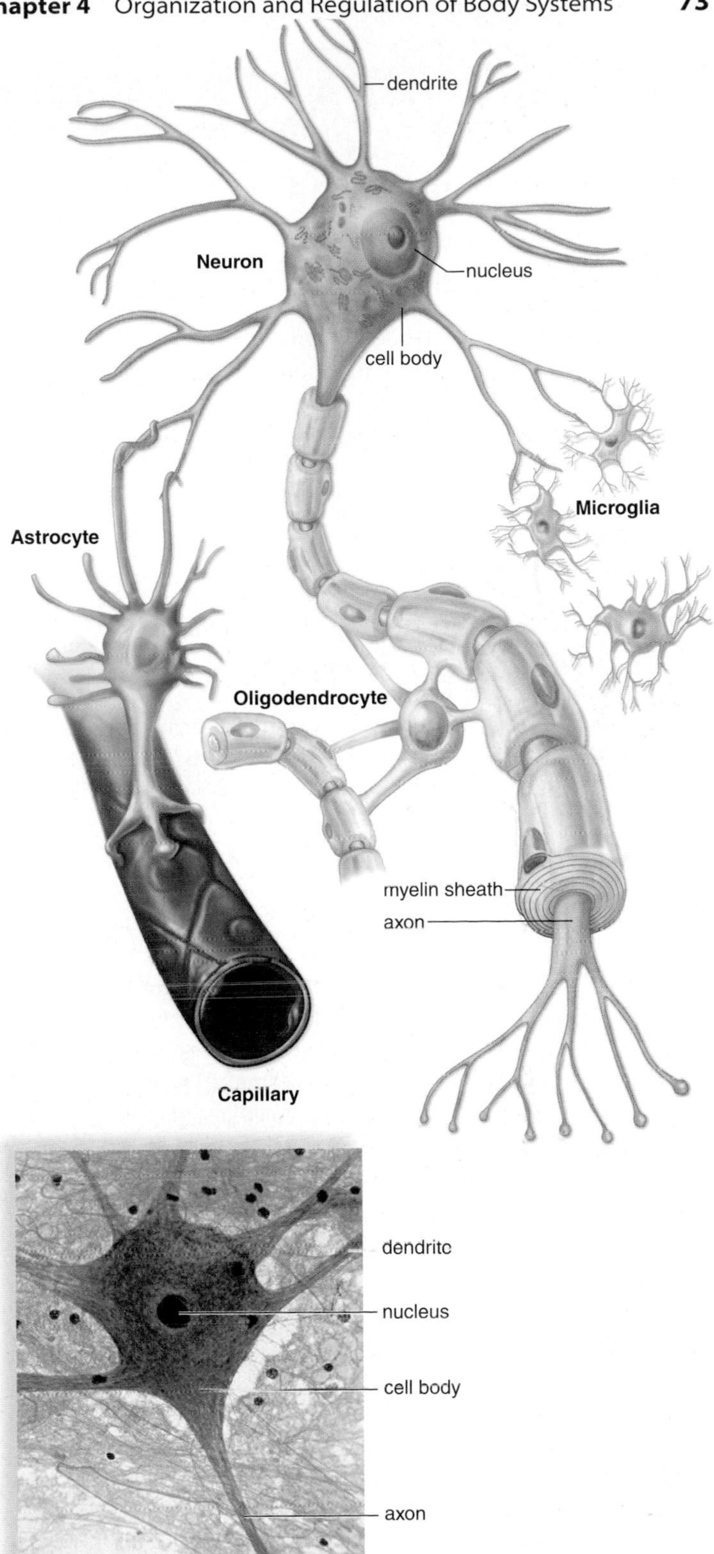

Figure 4.6 A neuron and examples of supporting neuroglia cells.

Neurons conduct nerve impulses. Neuroglia support and service neurons. Microglia are a type of neuroglia that become mobile in response to inflammation and phagocytize debris. Astrocytes lie between neurons and a capillary. Therefore, substances entering neurons from the blood must first pass through astrocytes. Oligodendrocytes form the myelin sheaths around fibers in the brain and spinal cord.

BIOLOGY MATTERS  Science

Nerve Regeneration and Stem Cells

In humans, axons outside the brain and spinal cord can regenerate—but not those inside these organs (Fig. 4A). After injury, axons in the human central nervous system (CNS) degenerate, resulting in permanent loss of nervous function. Interestingly, about 90% of the cells in the brain and the spinal cord are not even neurons. They are glial cells. In nerves outside the brain and spinal cord, the glial cells are Schwann cells that help axons regenerate. The glial cells in the CNS are oligodendrocytes and astrocytes, and they inhibit axon regeneration.

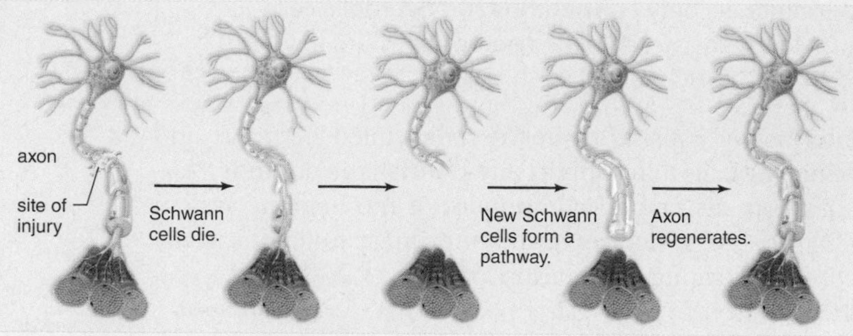

Figure 4A Regeneration of nerve cells.
Outside the CNS, nerves regenerate because new neuroglia called Schwann cells form a pathway for axons to reach a muscle. In the CNS, comparable neuroglia called oligodendrocytes do not have this function.

The spinal cord does contain its own stem cells. When the spinal cord is injured in experimental animals, these stem cells do proliferate. But instead of becoming functional neurons, they become glial cells. Researchers are trying to understand the process that triggers the stem cells to become glial cells. In the future, this understanding would allow manipulation of stem cells into neurons.

In early experiments with neural stem cells in the laboratory, scientists at Johns Hopkins University caused embryonic stem (ES) cells to differentiate into spinal cord motor neurons, the type of nerve cell that causes muscles to contract. The motor neurons then produced axons. When grown in the same dish with muscle cells, the motor neurons formed neuromuscular junctions and even caused muscle contractions. The cells were then transplanted into the spinal cords of rats with spinal cord injuries. Some of the transplanted cells survived for longer than a month within the spinal cord. However, no improvement in symptoms was seen and no functional neuron connections were made.

In later experiments by the same research group paralyzed rats were first treated with drugs and nerve growth factors to overcome inhibition from the central nervous system. These techniques significantly increased the success of the transplanted neurons. Amazingly, axons of transplanted neurons reached the muscles, formed neuromuscular junctions, and provided partial relief from the paralysis. Research is being done on the use of both the body's own stem cells and laboratory-grown stem cells to repair damaged CNS neurons. While many questions remain, the current results are promising.

Questions to Consider

1. Why do you think that neurons cannot simply be transplanted from other areas of the body?
2. How might this research also help patients who suffer from neurodegenerative diseases such as Parkinson disease?

APPLICATIONS AND MISCONCEPTIONS

How fast is a reflex?

A reflex is a built-in pathway that allows the body to react quickly to a response. One example, the knee-jerk, or patellar reflex, is tested by tapping just below the kneecap. The lower leg will then involuntarily kick forward. The reaction is designed to protect the thigh muscle from excessive stretch. The knee-jerk reflex is an example of a simple stretch reflex. There is only one pathway required: the stretch sensation (caused by tapping the knee), to the spinal cord, to the leg muscle. The whole circuit is complete within a millisecond—or 1/1,000 second!

CHECK YOUR PROGRESS 4.4

1. Describe the structure and function of a neuron.
2. Discuss the different types of neuroglial cells and the function of each.
3. Explain how the neurons and neuroglial cells work together to make nervous tissue function.

CONNECTING THE CONCEPTS

Nervous tissue plays an important role in transmitting the signals needed to maintain homeostasis. For more information on how neurons work, refer to the following discussions:

Section 13.1 provides a more detailed examination of how neurons function.

Section 14.1 discusses how neurons are involved in sensation.

4.5 Epithelial Tissue Protects

LEARNING OUTCOMES

Upon completion of this section, you should be able to

1. State the role of epithelial cells in the body.
2. Distinguish between the different forms of epithelial tissue with regard to location and function.

Epithelial tissue, also called epithelium (pl., epithelia), consists of tightly packed cells that form a continuous layer. Epithelial tissue covers surfaces and lines body cavities. Usually, it has a protective function. It can also be modified to carry out secretion, absorption, excretion, and filtration.

Epithelial cells are named based on their appearance (Fig. 4.7). All epithelial cells are exposed to the environment on one side. On the other side, they are bounded by a **basement membrane.** The basement membrane should not be confused with the plasma membrane (in the cell) or the body membranes that line the cavities of the body. Instead, the basement membrane is a thin layer of various types of carbohydrates and proteins that anchors the epithelium to underlying connective tissue.

MP3 Epithelial Tissue

Simple Epithelia

Epithelial tissue is either simple or stratified. Simple epithelia have only a single layer of cells (Fig. 4.8) and are classified according to cell type. **Squamous epithelium,** composed of flattened cells, is found lining the air sacs of lungs and walls of blood vessels. Its shape and arrangement permit exchanges of substances in these locations. Oxygen–carbon-dioxide exchange occurs in the lungs, and nutrient–waste exchange occurs across blood vessels in the tissues.

Cuboidal epithelium consists of a single layer of cube-shaped cells. This type of epithelium is frequently found in glands, such as the salivary glands, the thyroid gland, and the pancreas. Simple cuboidal epithelium also covers the ovaries and lines kidney tubules, the portions of the kidney in which urine is formed. When cuboidal cells are involved in absorption, they have microvilli (minute cellular extensions of the plasma membrane). These increase the surface area of the cells. When cuboidal cells function in active transport, they contain many mitochondria.

Columnar epithelium has cells resembling rectangular pillars or columns, with nuclei usually located near the bottom of each cell. This epithelium is found lining the digestive tract, where microvilli expand the surface area and aid in absorbing the products of digestion. Ciliated columnar epithelium is found lining the oviducts, where it propels the egg toward the uterus, or womb.

Pseudostratified columnar epithelium (Fig. 4.8) is so named because it appears to be layered (*pseudo,* "false"; *stratified,* "layers"). However, true layers do not exist because each cell touches the basement membrane. In particular, the irregular placement of the nuclei creates the appearance of

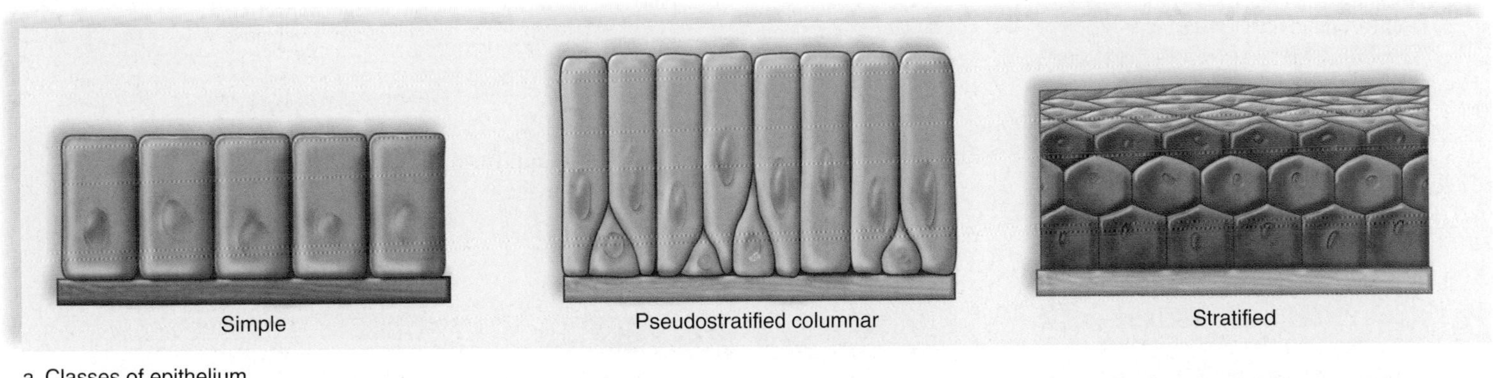

a. Classes of epithelium

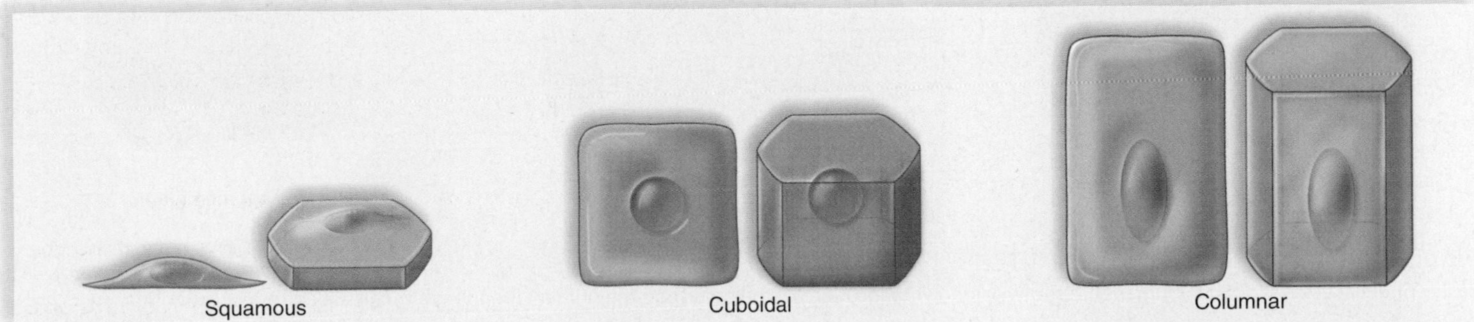

b. Cell shapes

Figure 4.7 Shapes of epithelial cells.
Epithelial cells have a variety of shapes and configurations in tissues.

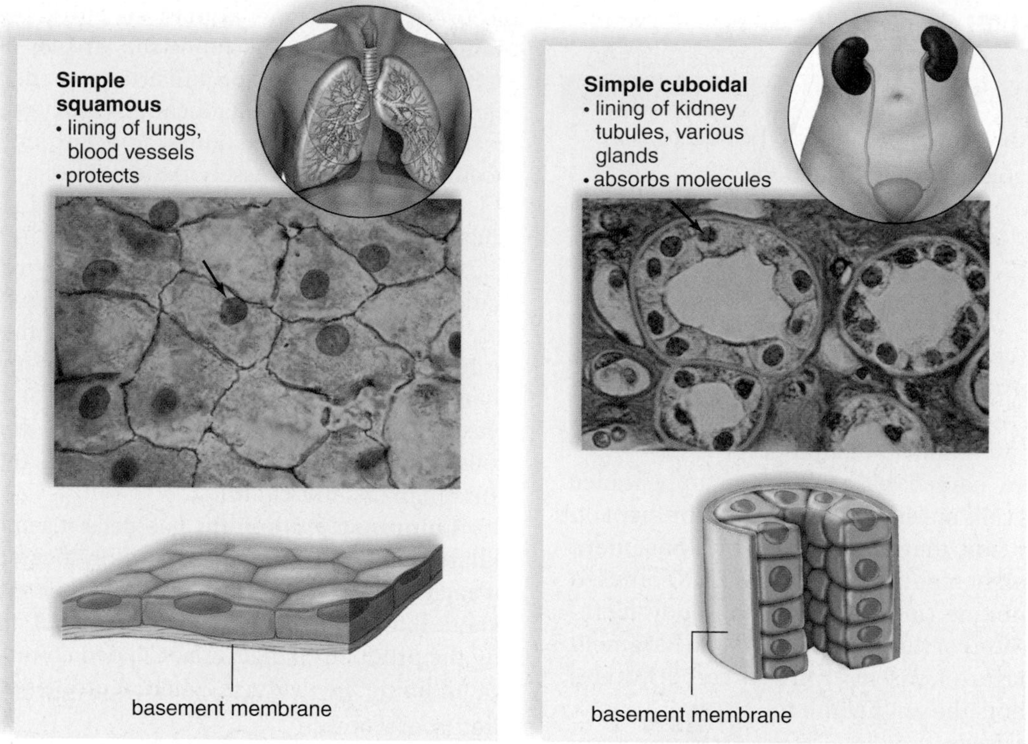

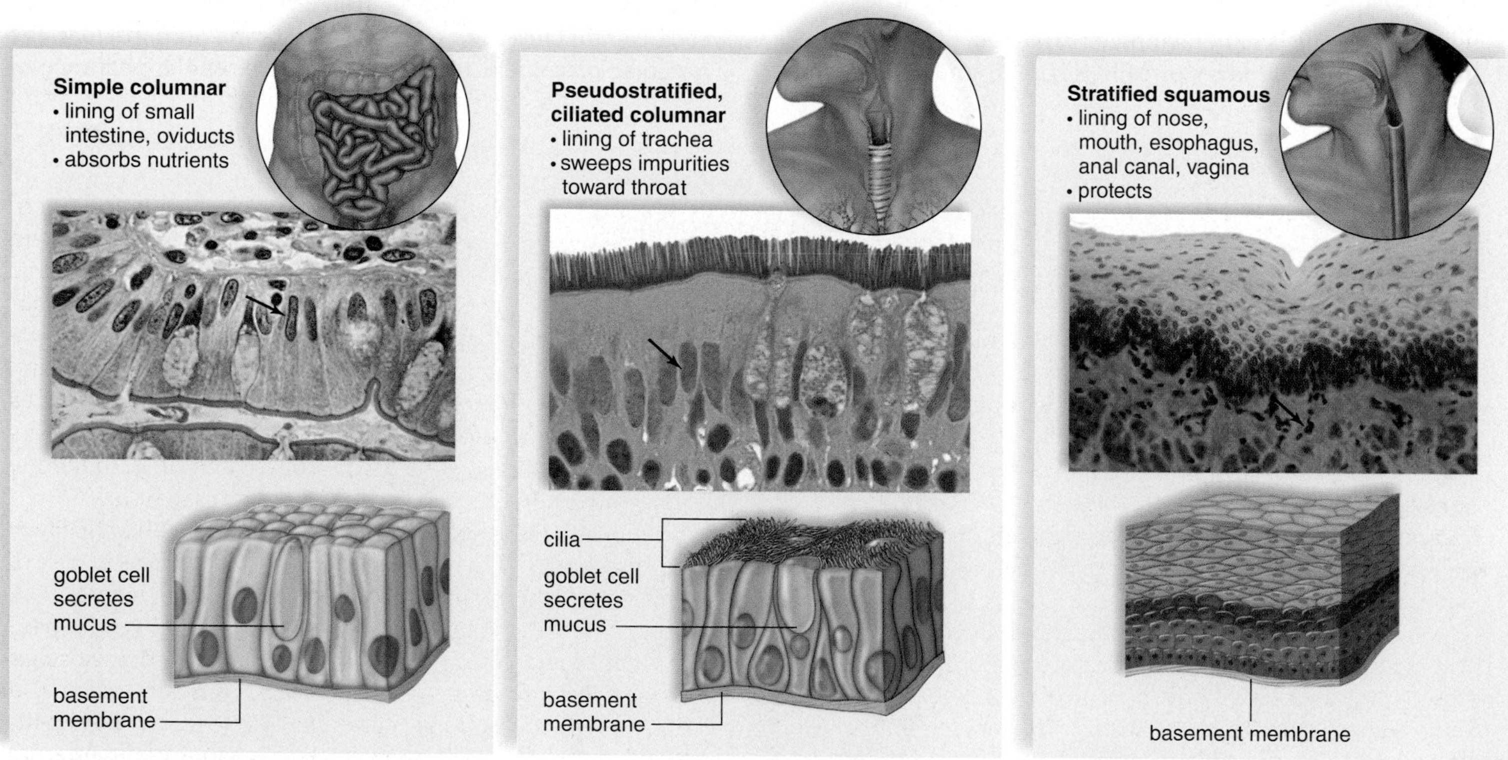

Figure 4.8 **The basic types of epithelial cells.**
Basic epithelial tissues found in humans are shown, along with locations of the tissue and the primary function of the tissue at these locations.

several layers, whereas only one exists. The lining of the windpipe, or trachea, is pseudostratified ciliated columnar epithelium. A secreted covering of mucus traps foreign particles. The upward motion of the cilia carries the mucus to the back of the throat, where it may either be swallowed or expectorated (spit out). Smoking can cause a change in the secretion of mucus and can inhibit ciliary action, resulting in a chronic inflammatory condition called *bronchitis.*

In some cases, columnar and pseudostratified columnar epithelium secretes a product. In this case, it is said to be glandular. A **gland** can be a single epithelial cell, as in the case of mucus-secreting goblet cells, or a gland can contain many cells. Glands with ducts that secrete their product onto the outer surface (e.g., sweat glands and mammary glands) or into a cavity (e.g., salivary glands) are called **exocrine glands.** Ducts can be simple or compound:

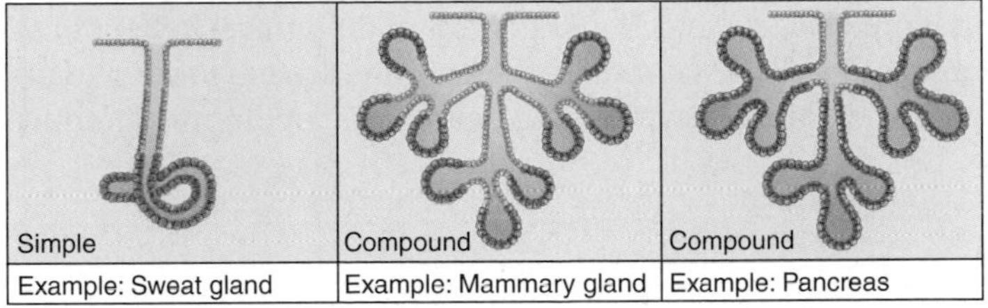

Glands that have no ducts are appropriately known as the ductless glands, or endocrine glands. **Endocrine glands** (e.g., pituitary gland and thyroid) secrete hormones directly into the bloodstream.

Stratified Epithelia

Stratified epithelia have layers of cells piled one on top of the other (Fig. 4.8). Only the bottom layer touches the basement membrane. The nose, mouth, esophagus, anal canal, the outer portion of the cervix (adjacent to the vagina), and vagina are lined with stratified squamous epithelium. Cancer of the cervix is detectable by doing a *pap smear.* Cells lining the cervix are smeared onto a slide later examined to detect any abnormalities.

As we shall see, the outer layer of skin is also stratified squamous epithelium, but the cells have been reinforced by keratin, a protein that provides strength. Stratified cuboidal and stratified columnar epithelia also are found in the body.

Transitional epithelium was originally named because it was thought to be an intermediate form of epithelial cells. Now the term is used to imply changeability, because tissue changes in response to tension. It forms the lining of the urinary bladder, the ureters (tubes that carry urine from the kidneys to the bladder), and part of the urethra (the single tube that carries urine to the outside). All are organs that may need to stretch. When the bladder is distended, this epithelium stretches, and the outer cells take on a squamous appearance.

CHECK YOUR PROGRESS 4.5

1. List the functions of epithelial tissue.
2. Describe the structure of each major type of epithelial tissue.
3. Summarize how the structure of some epithelial tissue relates to its function. Give some specific examples.

CONNECTING THE CONCEPTS

Epithelial tissue is involved in the operation of most organs of the body. For more information, refer to the following discussions:

Section 8.3 describes how specialized epithelial cells in the stomach secrete hydrochloric acid.

Section 9.6 examines how gas exchange occurs across the epithelial cells of the lungs.

Section 15.1 provides more information on the endocrine and exocrine glands of the body.

4.6 Integumentary System

LEARNING OUTCOMES

Upon completion of this section, you should be able to

1. Describe the structure and function of human skin.
2. Compare the function of the epidermis and dermis.
3. Identify the function of the accessory organs associated with the skin.

In some cases, a tissue is associated with a particular organ. For example, nervous tissue is typically associated with the brain. But an **organ** is composed of two or more types of tissues working together to perform particular functions. The skin is an organ comprised of all four tissue types: epithelial, connective, muscular, and nervous tissue. An **organ system** contains many different organs that cooperate to carry out a process, such as the digestion of food. The skin has several accessory organs (hair, nails, sweat glands, and sebaceous glands) and, therefore, is sometimes referred to as the **integumentary system.**

Skin is the most conspicuous system in the body because it covers the body. In an adult, the skin has a surface area of about 1.8 square meters (m^2) (over 19.5 square feet [ft^2]). It accounts for nearly 15% of the weight of an average human. The skin has numerous functions. It protects underlying tissues from physical trauma, pathogen invasion, and water loss. It also helps regulate body temperature. Therefore, skin plays a significant role in homeostasis, the relative constancy of the internal environment. The skin even synthesizes certain chemicals that affect the rest of the body. Skin contains sensory receptors such as touch and temperature receptors. Thus it helps us to be aware of our surroundings and communicate with others.

Regions of the Skin

The skin has two regions: the epidermis and the dermis (Fig. 4.9). A **subcutaneous layer,** sometimes called the *hypodermis,* is found between the skin and any underlying structures, such as muscle or bone.

The Epidermis

The **epidermis** is made up of stratified squamous epithelium. New epidermal cells for the renewal of skin are derived from stem (basal) cells. The importance of these stem cells is observed when there is an injury to the skin. If an injury, such as a burn, is deep enough to destroy stem cells, then the skin can no longer replace itself. As soon as possible, the damaged tissue is removed and skin grafting begins. The skin needed for grafting is usually taken from other parts of the patient's body. This is called *autografting,* as opposed to allografting. In *allografting,* the graft is received from another person and is sometimes obtained from cadavers. Autografting is preferred because rejection rates are low. If the damaged area is extensive, it may be difficult to acquire enough skin for autografting. In that case, small amounts of epidermis are removed and cultured in the laboratory. This produces thin sheets of skin that can be transplanted back to the patient (see the chapter opener).

Newly generated skin cells become flattened and hardened as they push to the surface (Fig. 4.10). Hardening takes place because the cells produce *keratin,* a waterproof protein. These cells are also called keratinocytes. Outer skin cells are dead and keratinized, so the skin is waterproof. This prevents water loss and helps maintain water homeostasis. The skin's waterproofing also prevents water from entering the body when the skin is immersed. Dandruff occurs when the rate of keratinization in the skin of the scalp is two or three times the normal rate. A thick layer of dead keratinized cells, arranged in spiral and concentric patterns, forms fingerprints and footprints that are genetically unique.

Two types of specialized cells are located deep in the epidermis. **Langerhans cells** are macrophages, white blood cells that phagocytize infectious agents and then travel to lymphatic organs. There they stimulate the immune system to react to the pathogen. **Melanocytes,** lying deep in the epidermis, produce

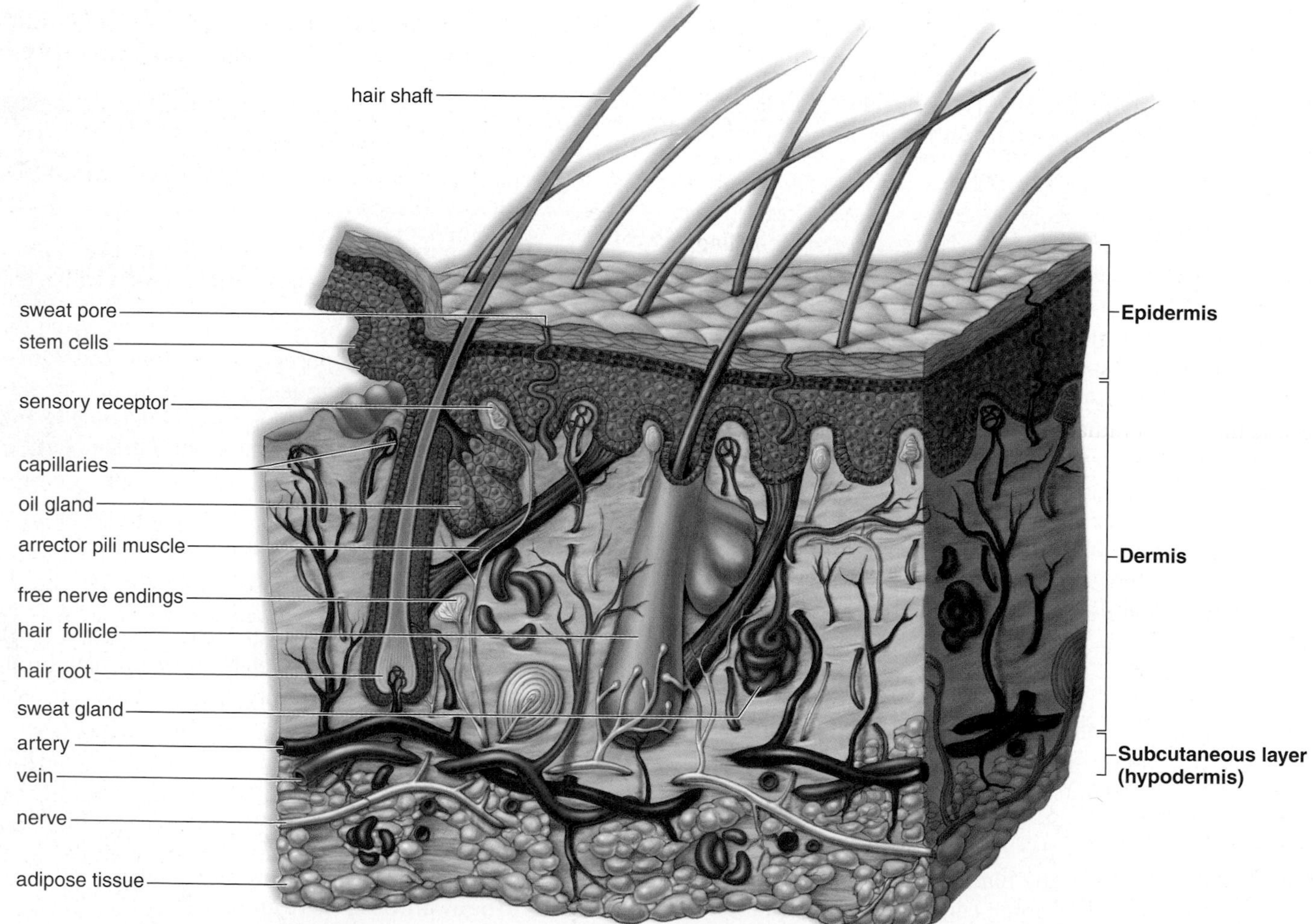

Figure 4.9 **Anatomy of human skin.**
Skin consists of two regions: the epidermis and the dermis. A subcutaneous layer (the hypodermis) lies below the dermis.

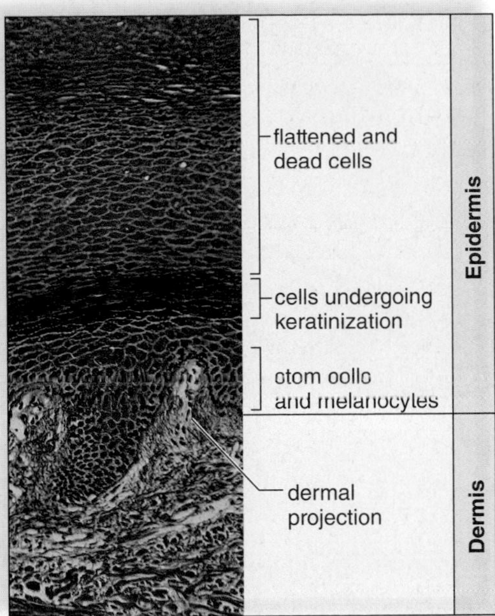

Figure 4.10 A light micrograph of human skin.
The keratinization of cells is shown in this image.

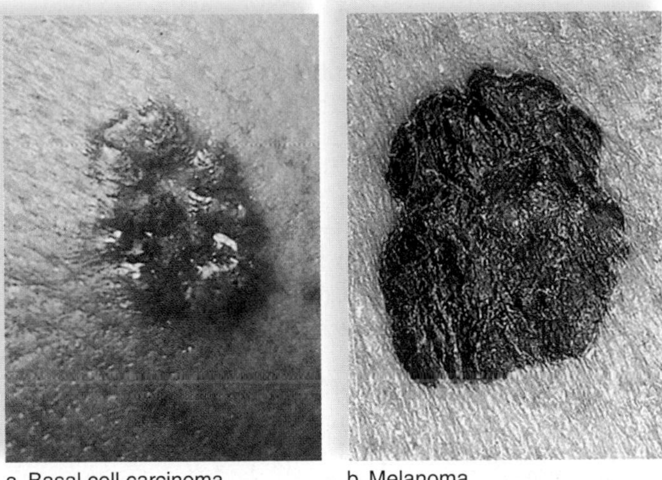

Figure 4.11 Cancers of the skin.
a. Basal cell carcinoma derived from stem cells and (**b**) melanoma derived from melanocytes are types of skin cancer.

melanin. This is the main pigment responsible for skin color. The number of melanocytes is about the same in all individuals, so variation in skin color is due to the amount of melanin produced and its distribution. When skin is exposed to the sun, melanocytes produce more melanin. This protects the skin from the damaging effects of the ultraviolet (UV) radiation in sunlight. The melanin is passed to other epidermal cells, and the result is tanning. In some people, this results in the formation of patches of melanin called freckles. Another pigment, called carotene, is present in epidermal cells and in the dermis. It gives the skin of some Asian populations its yellowish hue. The pinkish color of fair-skinned people is due to the pigment hemoglobin in the red blood cells in the blood vessels of the dermis.

Video
Skin Color Gene

Some ultraviolet radiation does serve a purpose, however. Certain cells in the epidermis convert a steroid related to cholesterol into **vitamin D** with the aid of UV radiation. However, only a small amount of UV radiation is needed. Vitamin D leaves the skin and helps regulate both calcium and phosphorus metabolism in the body. Calcium and phosphorus have a variety of roles and are important in the proper development and mineralization of the bones.

Skin Cancer Whereas we tend to associate a tan with good health, it signifies that the body is trying to protect itself from the dangerous rays of the sun. Too much ultraviolet radiation is dangerous and can lead to skin cancer. Basal cell carcinoma (Fig. 4.11*a*), derived from stem cells gone awry, is the more common type of skin cancer and is the most curable. Melanoma (Fig. 4.11*b*), the type of skin cancer derived from melanocytes, is extremely serious. To prevent skin cancer, you should stay out of the sun between the hours of 10 A.M. and 3 P.M. When you are in the sun, follow these guidelines:

Video
Melanoma Marker

- Use a broad-spectrum sunscreen that protects from both UVA (long-wave) and UVB (short-wave) radiation and has a sun protection factor (SPF) of at least 15. (This means that if you usually burn, for example, after a 20-minute exposure, it will take 15 times longer, or 5 hours, before you will burn.)
- Wear protective clothing. Choose fabrics with a tight weave and wear a wide-brimmed hat.
- Wear sunglasses that have been treated to absorb UVA and UVB radiation.

Also, avoid tanning machines; even if they use only high levels of UVA radiation, the deep layers of the skin will become more vulnerable to UVB radiation.

APPLICATIONS AND MISCONCEPTIONS

Are tanning beds safe?

According to a study sponsored by the World Health Organization (WHO),* individuals who use tanning beds prior to the age of 30 increase their chances of developing skin cancer by 75%. Though exposure to UV radiation does increase vitamin D levels slightly, it also results in an increased formation of cataracts (clouding of the cornea of the eyes) and a suppressed immune system. The UV radiation from tanning beds is known to damage the DNA of cells deep in the skin and can reduce the amount of collagen in the connective tissues, resulting in premature aging of the skin. For these reasons, the WHO has issued a request for increased government monitoring of tanning beds and has classified them as "known to be carcinogenic to humans."

* "Sunbeds, Tanning, and UV Exposure," World Health Organization, http://www.who.int/mediacentre/factsheets/fs287/en (accessed July 18, 2011).

The Dermis

The **dermis** is a region of dense fibrous connective tissue beneath the epidermis. (*Dermatology* is a branch of medicine that specializes in diagnosing and treating skin disorders.) The dermis contains collagen and elastic fibers. The collagen fibers are flexible but offer great resistance to overstretching. They prevent the skin from being torn. The elastic fibers maintain normal skin tension but also stretch to allow movement of underlying muscles and joints. The number of collagen and elastic fibers decreases with age and with exposure to the sun. This causes the skin to become less supple and more prone to wrinkling. The dermis also contains blood vessels that nourish the skin. When blood rushes into these vessels, a person blushes. When blood is minimal in them, a person turns "blue." Blood vessels in the dermis play a role in temperature regulation. If body temperature starts to rise, the blood vessels in the skin will dilate. As a result, more blood is brought to the surface of the skin for cooling. If the outer temperature cools, the blood vessels constrict, so less blood is brought to the skin's surface.

The sensory receptors—primarily in the dermis—are specialized for touch, pressure, pain, hot, and cold. These receptors supply the central nervous system with information about the external environment. The sensory receptors also account for the use of the skin as a means of communication between people. For example, the touch receptors play a major role in sexual arousal.

The Subcutaneous Layer

Technically speaking, the subcutaneous layer beneath the dermis is not a part of skin. It is a common site for injections. This layer is composed of loose connective tissue and adipose tissue, which stores fat. Fat is a stored source of energy in the body. Adipose tissue helps to thermally insulate the body from either gaining heat from the outside or losing heat from the inside. A well-developed subcutaneous layer gives the body a rounded appearance and provides protective padding against external assaults. Excessive development of the subcutaneous layer accompanies obesity.

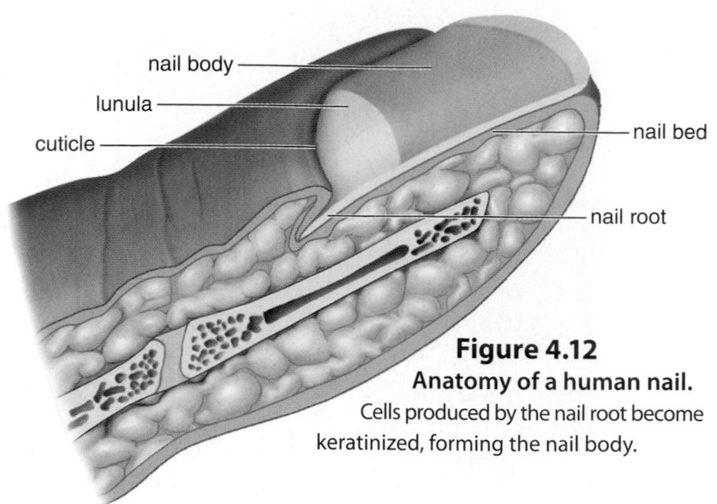

Figure 4.12
Anatomy of a human nail.
Cells produced by the nail root become keratinized, forming the nail body.

Accessory Organs of the Skin

Nails, hair, and glands are structures of epidermal origin, even though some parts of hair and glands are largely found in the dermis. **Nails** are a protective covering of the distal part of fingers and toes, collectively called *digits* (Fig. 4.12). Nails grow from special epithelial cells at the base of the nail in the portion called the *nail root.* The *cuticle* is a fold of skin that hides the nail root. The whitish color of the half-moon shaped base, or *lunula,* results from the thick layer of cells in this area. The cells of a nail become keratinized as they grow out over the nail bed.

Hair follicles begin at a bulb in the dermis and continue through the epidermis where the hair shaft extends beyond the skin (see Fig. 4.9). A dark hair color is largely due to the production of true melanin by melanocytes present in the bulb. If the melanin contains iron and sulfur, hair is blond or red. Graying occurs when melanin cannot be produced, but white hair is due to air trapped in the hair shaft.

Contraction of the *arrector pili muscles* attached to hair follicles causes the hairs to "stand on end" and goosebumps to develop. Epidermal cells form the root of a hair, and their division causes a hair to grow. The cells become keratinized and die as they are pushed farther from the root.

Each hair follicle has one or more **oil glands** (see Fig. 4.9), also called *sebaceous glands,* which secrete sebum. *Sebum* is an oily substance that lubricates the hair within the follicle and the skin. The oil secretions from sebaceous glands are acidic and retard the growth of bacteria. If the sebaceous glands fail to discharge (usually because they are blocked with keratinocytes), the secretions collect and form "whiteheads." Over time, the sebum in a whitehead oxidizes to form a "blackhead." Acne is an inflammation of the sebaceous glands that most often occurs during adolescence due to hormonal changes.

Sweat glands (see Fig. 4.9), also called *sudoriferous glands,* are numerous and present in all regions of skin. A sweat gland

APPLICATIONS AND MISCONCEPTIONS

Are there side effects of Botox treatments?

Botox is a drug used to reduce the appearance of facial wrinkles and lines. Botox is the registered trade name for a derivative of botulinum toxin A, a protein toxin produced by the bacterium *Clostridium botulinum.* Botox stops communication between motor nerves and muscles, causing muscle paralysis. Treatments are direct injections under the skin, where the toxin causes facial muscle paralysis. The injections reduce the appearance of wrinkles and lines that appear as a result of normal facial muscle movement. However, Botox treatment is not without side effects. Excessive drooling or a slight rash around the injection site are among the milder side effects of treatment. Spreading of Botox from the injection site may also paralyze facial muscles unintended for treatment. In a few cases, muscle pain and weakness have resulted. Though rare, more serious side effects, including allergic reactions, may also occur. When performed in a medical facility by a licensed physician, Botox treatment is generally considered safe and effective.

BIOLOGY MATTERS  **Science**

Face Transplantation

In 2005, a French surgical team led by Professors Bernard Devauchelle and Jean Michel Dubernard was able to perform the world's first partial face transplant. The recipient was a woman, Isabelle Dinoire, severely disfigured by a dog mauling. Muscles, veins, arteries, nerves, and skin were transplanted onto the lower half of Isabelle's face (Fig. 4B, *top left*). The donor's lips, chin, and nose were transplanted. The donor was a brain-dead patient whose family had agreed to donate all their loved one's organs and tissues. The donor shared Isabelle's blood type and was a good tissue match. Eighteen months after the surgery (Fig. 4B, *top right*), Isabelle was able to eat, drink, and smile.

In 2008, a surgical team at Henri-Mondor Hospital in France was able to perform the first full face transplant. The patient Pascal Coler suffered from a condition called neurofibromatosis, which caused tumors to grow on his face, producing severe disfiguration (Fig. 4B, *bottom left*). As was the case with Isabelle, tissue from a donor's face was used to reconstruct Pascal's appearance (Fig. 4B, *bottom right*).

Although the ability to do these types of transplant has existed for some time, doctors remain concerned about the ethical aspects of the procedure. Organ transplantation has always involved some moral concerns, because the donor must still be alive when the organs are harvested. Historically, face transplants have been a "quality-of-life" issue and not a "life-or-death" surgery. However, this attitude is changing as injured soldiers returning from wars in Afghanistan and Iraq are being considered for face transplants in the United States. In addition, recipients of face transplants must frequently undergo extensive counseling to prepare themselves emotionally for the "new face," and all recipients must spend the remainder of their lives on immunosuppressive drugs.

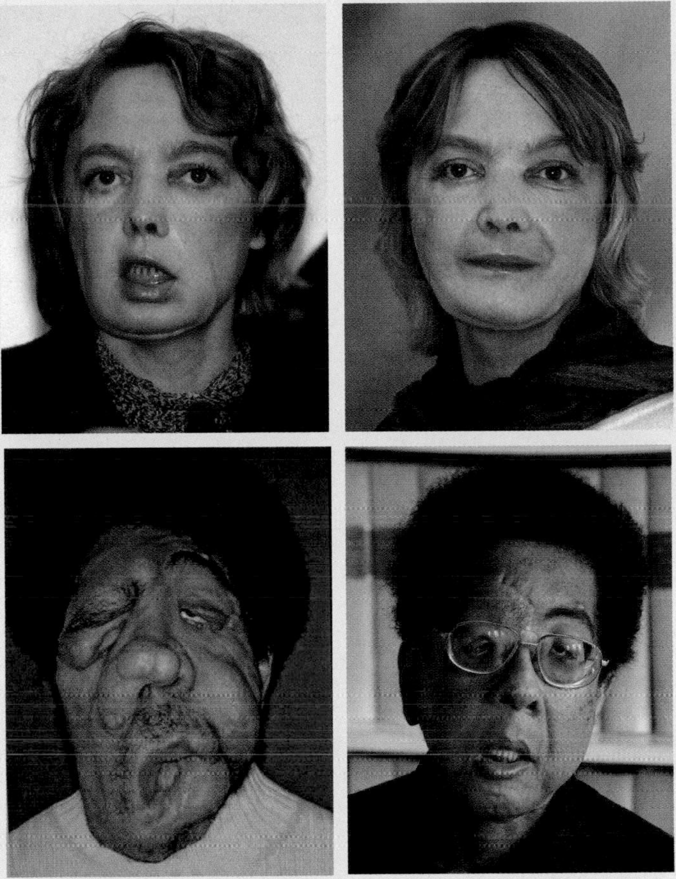

Figure 4B Face transplant recipients. Isabelle Dinoire (*top, left* and *right*) and Pascal Coler (*bottom, left* and *right*) are both examples of successful face transplants.

Questions to Consider

1. What area of the skin do you think would be the hardest for the surgeon to reattach? Why?
2. What functions of the skin might be impaired in the recipient of a face transplant?

is a tubule that begins in the dermis and either opens into a hair follicle or, more often, opens onto the surface of the skin. Sweat glands play a role in modifying body temperature. When body temperature starts to rise, sweat glands become active. Sweat absorbs body heat as it evaporates. Once the body temperature lowers, sweat glands are no longer active.

CHECK YOUR PROGRESS 4.6

1. Compare the structure and function of the epidermis and dermis.
2. Explain how each accessory organ of the skin aids in homeostasis.
3. Hypothesize why structures like sensory receptors and substances for skin color are found in the dermis and not the epidermis.

CONNECTING THE CONCEPTS

For more information on the role of the skin in human physiology, refer to the following discussions:

Section 8.6 provides more information on the relationship between vitamin D and bone calcium homeostasis.

Sections 19.1 and 19.2 explain how cells develop into cancer cells and present some of the different forms of cancer.

Section 22.5 explores the evolutionary reasons for variations in skin color.

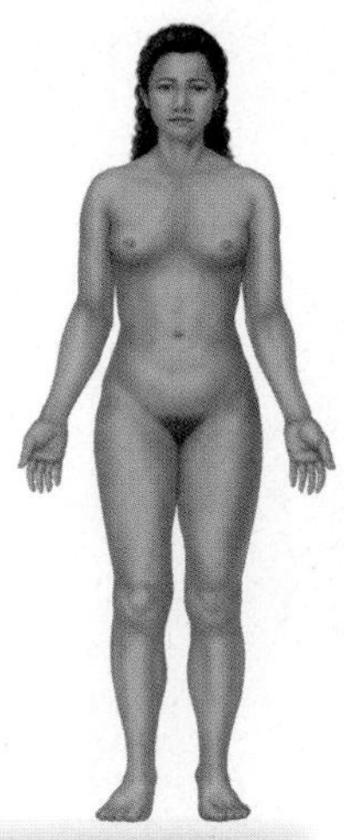

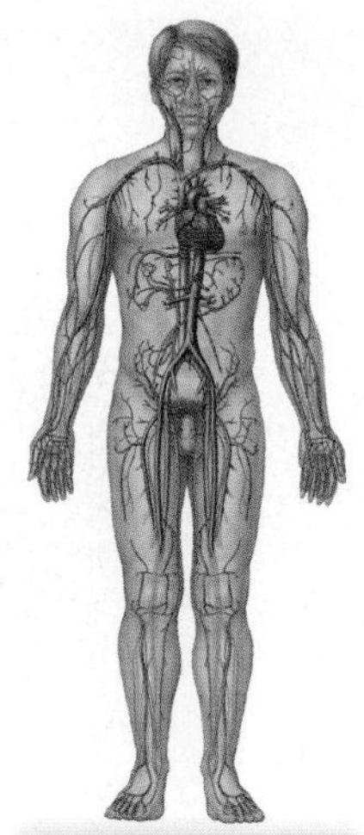

Cardiovascular system
• transport system for nutrients, waste
• provides temperature, pH, and fluid homeostasis
Organ: Heart

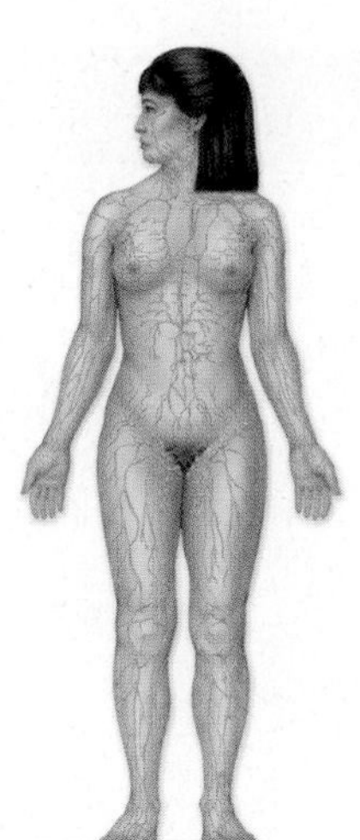

Lymphatic and immune systems
• defends against infectious diseases
• provides fluid homeostasis
• assists in absorption and transport of fats
Organs: Lymphatic vessels, lymph nodes, spleen

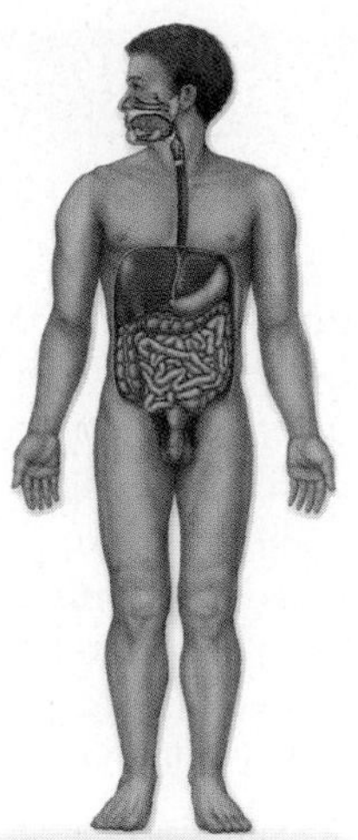

Digestive system
• ingests, digests, and processes food
• absorbs nutrients and eliminates waste
• involved in fluid homeostasis
Organs: Oral cavity, esophagus, stomach, small intestine, large intestine, salivary glands, liver, gallbladder, pancreas

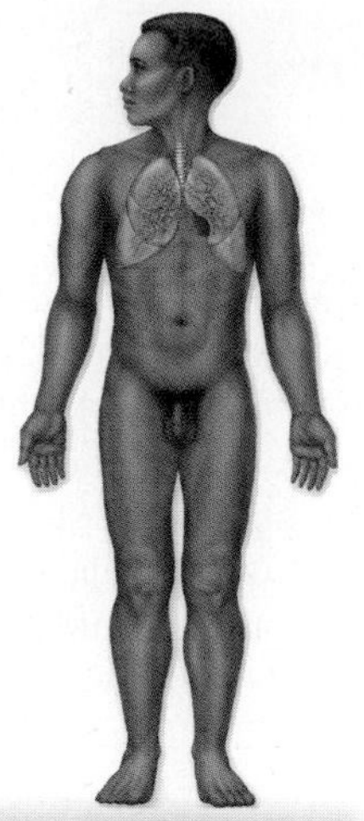

Respiratory system
• exchanges gases at both lungs and tissues
• assists in pH homeostasis
Organs: Lungs

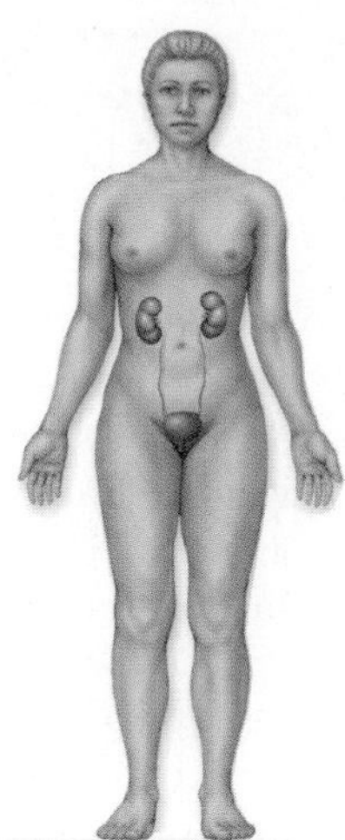

Figure 4.13 Organ systems of the body.

4.7 Organ Systems, Body Cavities, and Body Membranes

LEARNING OUTCOMES

Upon completion of this section, you should be able to

1. Summarize the function of each organ system in the human body.
2. Identify the major cavities of the human body.
3. Name the body membranes and provide a function for each.

Recall from our discussion of the skin in section 4.6 that a group of tissues performing a common function is called an organ. Organs with a similar function, in turn, form organ systems. Some of these organ systems, such as the respiratory system, occupy specific cavities of the body, and others, such as the muscular and circulatory systems, are found throughout the body. The organs and cavities of the body are lined with membranes that often secrete fluid to aid in the physiology of the organ or organ system.

Organ Systems

Figure 4.13 illustrates the organ systems of the human body. It should be emphasized that just as organs work together in an organ system, so do organ systems work together in the body. In some cases, it is arbitrary to assign a particular organ to one system when it also assists the functioning of many other systems. In addition, the organs listed in Figure 4.13 represent the major structures in the body. Often, other structures and glands contribute to the operation of the organ system.

Body Cavities

The human body is divided into two main cavities: the ventral cavity and the dorsal cavity (Fig. 4.14*a*). Called the *coelom* in early development, the *ventral cavity* later becomes the thoracic, abdominal, and pelvic cavities. The thoracic cavity contains the lungs and the heart. The thoracic cavity is separated from the abdominal cavity by a horizontal muscle called the *diaphragm.* The stomach, liver, spleen, pancreas, gallbladder, and most of the small and large intestines are in the abdominal cavity. The pelvic cavity contains the rectum, the urinary bladder, the internal reproductive organs, and the rest of the small and large intestine. Males have an external extension of the abdominal wall called the *scrotum,* which contains the testes.

The *dorsal cavity* has two parts: (1) The cranial cavity within the skull contains the brain; (2) The vertebral canal, formed by the vertebrae, contains the spinal cord.

Body Membranes

Body membranes line cavities and the internal spaces of organs and tubes that open to the outside. The body membranes are of four types: mucous, serous, and synovial membranes and the meninges.

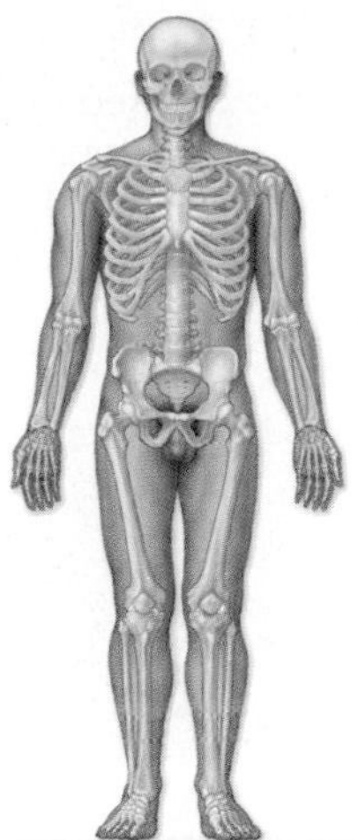

Skeletal system
- provides support and protection
- assists in movement
- stores minerals
- produces blood cells

Organs: Bones

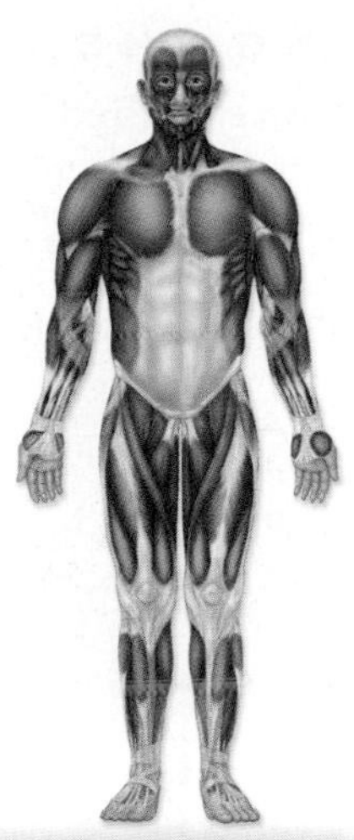

Muscular system
- assists in movement and posture
- produces heat

Organs: Muscles

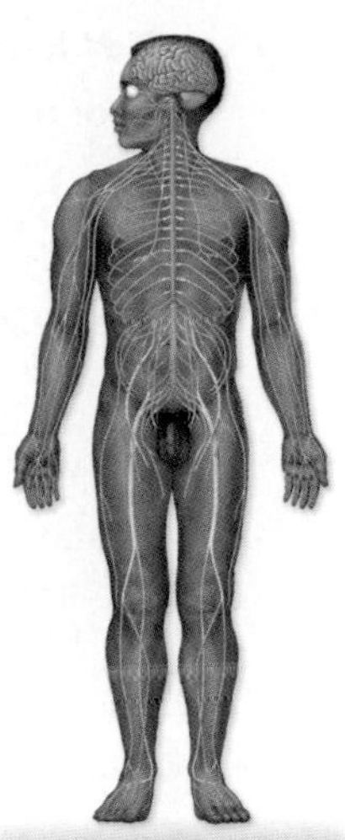

Nervous system
- receives, processes, and stores sensory input
- provides motor output
- coordinates organ systems

Organs: Brain, spinal cord

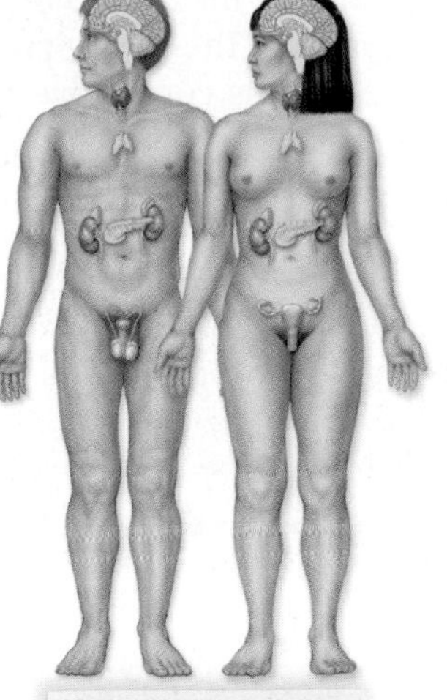

Endocrine system
- produces hormones
- coordinates organ systems
- regulates metabolism and stress responses
- involved in fluid and pH homeostasis

Organs: Testes, ovaries, adrenal glands, pancreas, thymus, thyroid, pineal gland

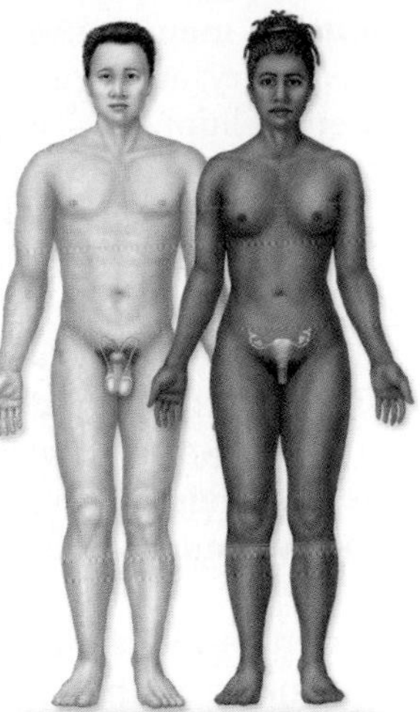

Reproductive system
- produces and transports gametes
- nurtures and gives birth to offspring in females

Organs: Testes, penis, ovaries, uterus, vagina

Figure 4.13—*continued.*

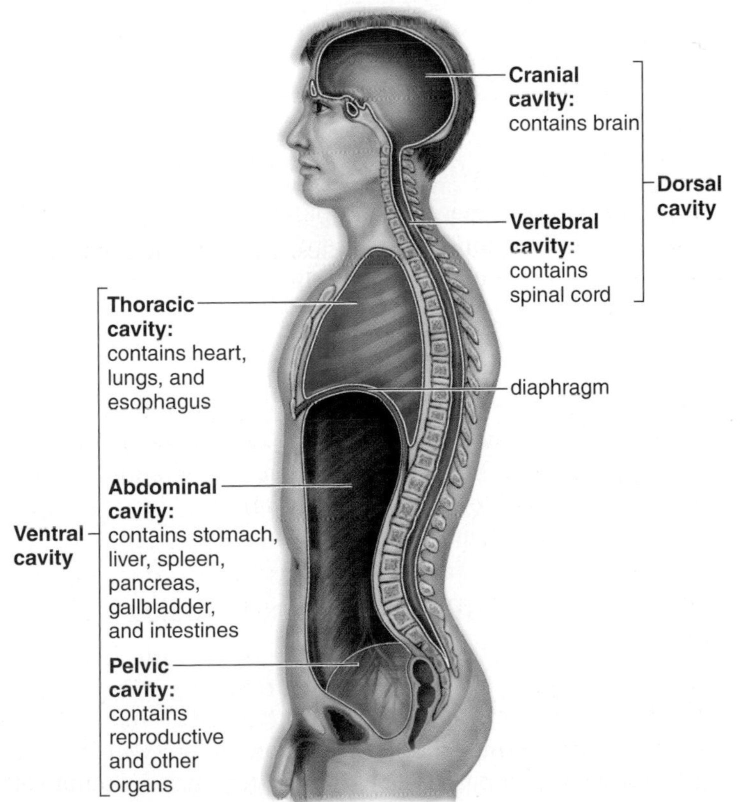

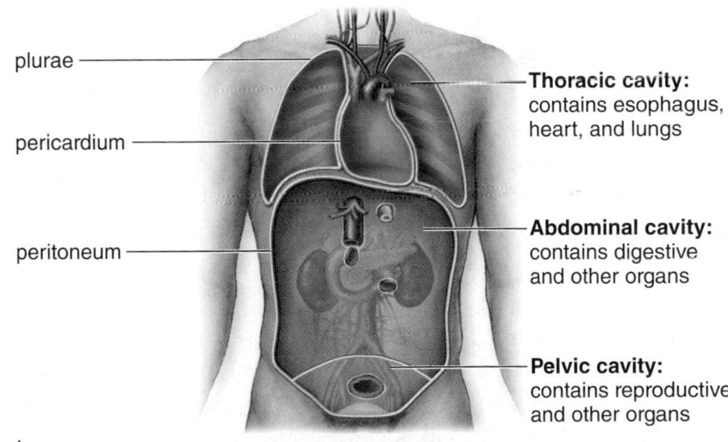

Figure 4.14 Body cavities of humans.
a. Side view. The posterior or dorsal (toward the back) cavity contains the cranial cavity and the vertebral canal. The brain is in the cranial cavity, and the spinal cord is in the vertebral canal. In the anterior or ventral (toward the front) cavity, the diaphragm separates the thoracic cavity from the abdominal cavity. The heart and lungs are in the thoracic cavity; the other internal organs are in either the abdominal cavity or the pelvic cavity.
b. Frontal view of the thoracic cavity, showing serous membranes.

Mucous membranes line the tubes of the digestive, respiratory, urinary, and reproductive systems. They are composed of an epithelium overlying a loose fibrous connective tissue layer. The epithelium contains specialized cells that secrete mucus. This mucus ordinarily protects the body from invasion by bacteria and viruses. Hence, more mucus is secreted and expelled when a person has a cold and has to blow her/his nose. In addition, mucus usually protects the walls of the stomach and small intestine from digestive juices. This protection breaks down when a person develops an ulcer.

Serous membranes line and support the lungs, the heart, and the abdominal cavity and its internal organs (Fig. 4.14*b*). They secrete a watery fluid that keeps the membranes lubricated. Serous membranes support the internal organs and compartmentalize the large thoracic and abdominal cavities.

Serous membranes have specific names according to their location. The pleurae (sing., **pleura**) line the thoracic cavity and cover the lungs. The pericardium forms the pericardial sac and covers the heart. The peritoneum lines the abdominal cavity and covers its organs. A double layer of peritoneum, called mesentery, supports the abdominal organs and attaches them to the abdominal wall. Peritonitis is a life-threatening infection of the peritoneum.

Synovial membranes composed only of loose connective tissue line the cavities of freely movable joints. They secrete synovial fluid into the joint cavity. This fluid lubricates the ends of the bones so that they can move freely. In rheumatoid arthritis, the synovial membrane becomes inflamed and grows thicker, restricting movement.

The **meninges** (sing., meninx) are membranes found within the dorsal cavity. They are composed only of connective tissue and serve as a protective covering for the brain and spinal cord. Meningitis is a life-threatening infection of the meninges.

APPLICATIONS AND MISCONCEPTIONS

What causes meningitis?

Meningitis is caused by an infection of the meninges by either a virus or a bacterium. Viral meningitis is less severe than bacterial meningitis, which in some cases can result in brain damage and death. Bacterial meningitis is usually caused by one of three species of bacteria: *Haemophilus influenzae* type b (Hib), *Streptococcus pneumoniae,* and *Neisseria meningitidis.* Vaccines are available for Hib bacteria and some forms of *S. pneumoniae* and *N. meningitidis.* The Centers for Disease Control and Prevention (CDC) recommends that individuals between the ages of 11 and 18 be vaccinated against bacterial meningitis.

CHECK YOUR PROGRESS 4.7

1. Summarize the overall function of each of the body systems and explain how each aids in homeostasis.
2. Describe the location of the two major body cavities.
3. List the four types of body membranes and describe the structure and function of each.

CONNECTING THE CONCEPTS

Each of the organ systems listed in this section is covered in greater detail in later chapters of the text. For more information on body cavities and body membranes, refer to the following discussions:

Section 7.3 describes how mucous membranes assist the immune response of the body.

Section 9.4 illustrates how the thoracic cavity is involved in breathing.

Section 11.5 explains how synovial joints (and their associated membranes) allow movement.

4.8 Homeostasis

LEARNING OUTCOMES

Upon completion of this section, you should be able to

1. Define *homeostasis* and provide an example.
2. Distinguish between positive and negative feedback mechanisms.

Homeostasis is the body's ability to maintain a relative constancy of its internal environment by adjusting its physiological processes. Even though external conditions may change dramatically, we have physiological mechanisms that respond to disturbances and limit the amount of internal change. Conditions usually stay within a narrow range of normalcy. For example, blood glucose, pH levels, and body temperature typically fluctuate during the day, but not greatly. If internal conditions change to any great degree, illness results.

MP3 Homeostasis

The Internal Environment

The internal environment has two parts: blood and tissue fluid. Blood delivers oxygen and nutrients to the tissues and carries carbon dioxide and wastes away. Tissue fluid, not blood, bathes the body's cells. Therefore, tissue fluid is the medium through which substances are exchanged between cells and blood. Oxygen and nutrients pass through tissue fluid on their way to tissue cells from the blood. Then carbon dioxide and wastes are carried away from the tissue cells by the tissue fluid and are brought back into the blood. The cooperation of body systems is required to keep these substances within the range of normalcy in blood and tissue fluid.

The Body Systems and Homeostasis

The nervous and endocrine systems are particularly important in coordinating the activities of all the other organ systems as they function to maintain homeostasis (Fig. 4.15). The nervous system is able to bring about rapid responses to any changes in the internal environment. The nervous system issues commands by electrochemical signals rapidly transmitted to effector organs, which can be muscles, such as skeletal muscles, or glands, such as sweat and salivary glands. The endocrine system brings about

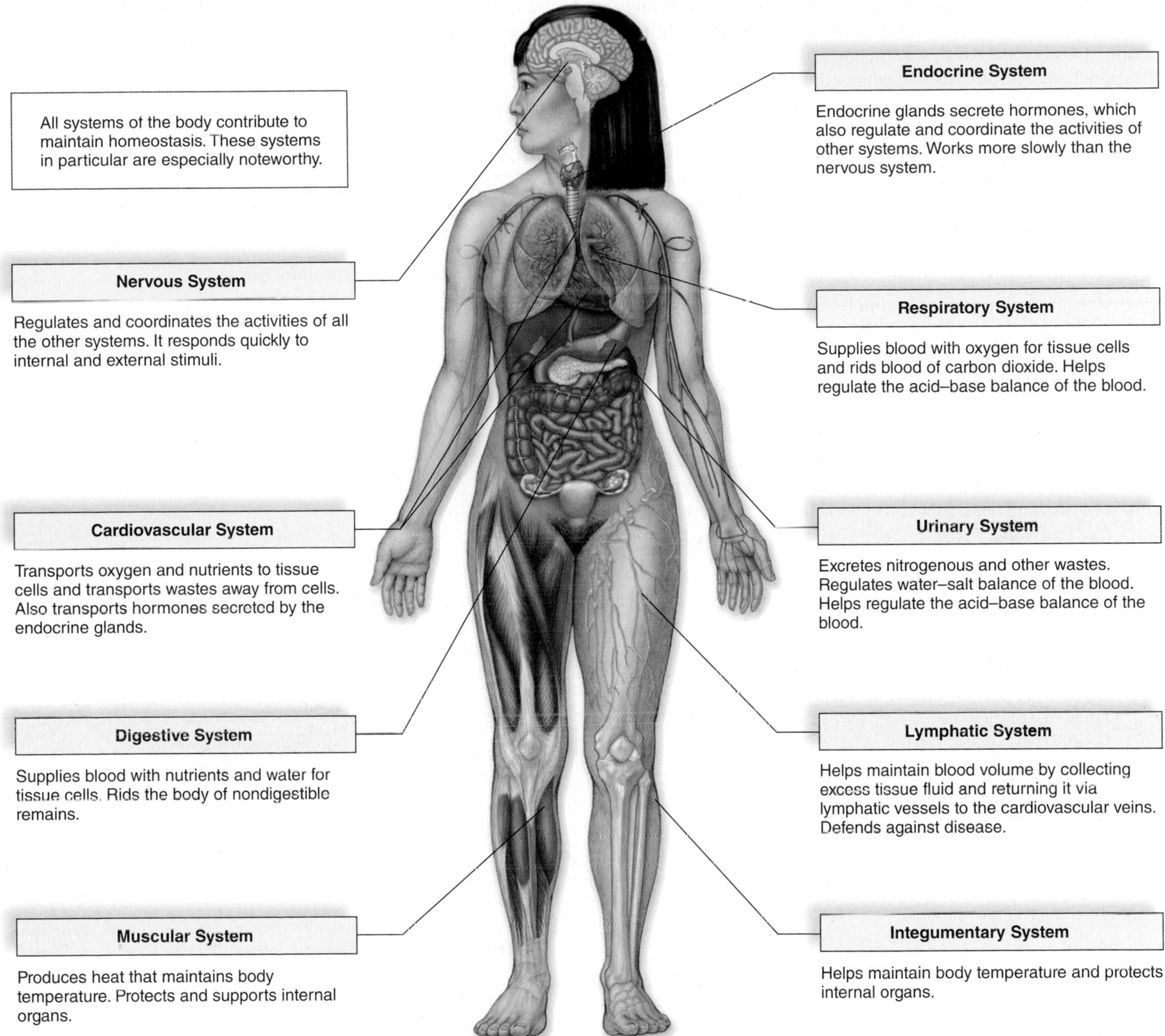

All systems of the body contribute to maintain homeostasis. These systems in particular are especially noteworthy.
Nervous System
Regulates and coordinates the activities of all the other systems. It responds quickly to internal and external stimuli.
Cardiovascular System
Transports oxygen and nutrients to tissue cells and transports wastes away from cells. Also transports hormones secreted by the endocrine glands.
Digestive System
Supplies blood with nutrients and water for tissue cells. Rids the body of nondigestible remains.
Muscular System
Produces heat that maintains body temperature. Protects and supports internal organs.
Endocrine System
Endocrine glands secrete hormones, which also regulate and coordinate the activities of other systems. Works more slowly than the nervous system.
Respiratory System
Supplies blood with oxygen for tissue cells and rids blood of carbon dioxide. Helps regulate the acid–base balance of the blood.
Urinary System
Excretes nitrogenous and other wastes. Regulates water–salt balance of the blood. Helps regulate the acid–base balance of the blood.
Lymphatic System
Helps maintain blood volume by collecting excess tissue fluid and returning it via lymphatic vessels to the cardiovascular veins. Defends against disease.
Integumentary System
Helps maintain body temperature and protects internal organs.

Figure 4.15 **Homeostasis by the organ systems of the human body.**
All the organ systems contribute to homeostasis in many ways. Some of the main contributions of each system are given in this illustration.

slower responses, but they generally have more lasting effects. Glands of the endocrine system, such as the pancreas or the thyroid, release hormones. *Hormones,* such as insulin from the pancreas, are chemical messengers that must travel through the blood and tissue fluid to reach their targets.

The nervous and endocrine systems together direct numerous activities that maintain homeostasis, but all the organ systems must do their part to keep us alive and healthy. Picture what would happen if any component of the cardiovascular, respiratory, digestive, or urinary system failed (Fig. 4.15). If someone is having a heart attack, the heart is unable to pump the blood to supply cells with oxygen. Or think of a person who is choking. The trachea (or windpipe) is blocked, so no air can reach the lungs for uptake by the blood. Unless the obstruction is removed quickly, cells will begin to die as the blood's supply of oxygen is depleted. When the lining of the digestive tract is damaged, as in a severe bacterial infection, nutrient absorption is impaired and cells face an energy crisis. It is important not only to maintain adequate nutrient levels in the blood but also to eliminate wastes and toxins. The liver makes urea, a nitrogenous end product of protein metabolism. Urea and other metabolic wastes are excreted by the kidneys, the urine-producing organs of the body. The kidneys rid the body of nitrogenous wastes and also help to adjust the blood's water-salt and acid-base balances.

A closer examination of how the blood glucose level is maintained helps us understand homeostatic mechanisms. When a healthy person consumes a meal and glucose enters the blood, the pancreas secretes the hormone insulin. Then glucose is removed from the blood as cells take it up. In the liver, glucose is stored in the form of glycogen. This storage is beneficial because later, if blood glucose levels drop, glycogen can be broken down to ensure that the blood level remains constant. Homeostatic mechanisms can fail. In diabetes mellitus, the pancreas cannot produce enough insulin or the body cells cannot respond appropriately to it. Therefore, glucose does not enter the cells and they must turn to other molecules, such as fats and proteins, to survive. This, along with too much glucose in the blood, leads to the numerous complications of diabetes mellitus.

Another example of homeostasis is the ability of the body to regulate the acid–base balance of the body. When carbon dioxide enters the blood, it combines with water to form carbonic acid. However, the blood is buffered, and pH stays within normal range as long as the lungs are busy excreting carbon dioxide. These two mechanisms are backed up by the kidneys, which can rid the body of a wide range of acidic and basic substances and, therefore, adjust the pH.

Negative Feedback

Negative feedback is the primary homeostatic mechanism that keeps a variable, such as the blood glucose level, close to a particular value, or set point. A homeostatic mechanism has at least two components: a sensor and a control center (Fig. 4.16). The sensor detects a change in the internal environment. The control center then brings about an effect to bring conditions back to normal. Then the sensor is no longer activated. In other words, a **negative feedback** mechanism is present when the output of the system resolves or corrects the original stimulus. For example, when blood pressure rises, sensory receptors signal a control center in the brain. The center stops sending nerve signals to muscle in the arterial walls. The arteries can then relax. Once the blood pressure drops, signals no longer go to the control center.

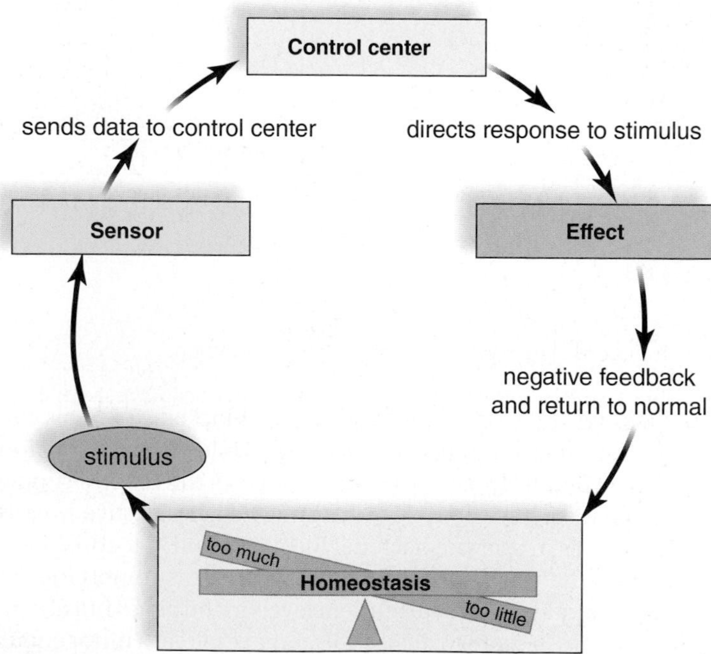

Figure 4.16 Negative feedback mechanisms. This diagram shows how the basic elements of a feedback mechanism work. A sensor detects the stimulus, and a control center brings about an effect that resolves, or corrects, the stimulus.

Mechanical Example

A home heating system is often used to illustrate how a more complicated negative feedback mechanism works (Fig. 4.17).

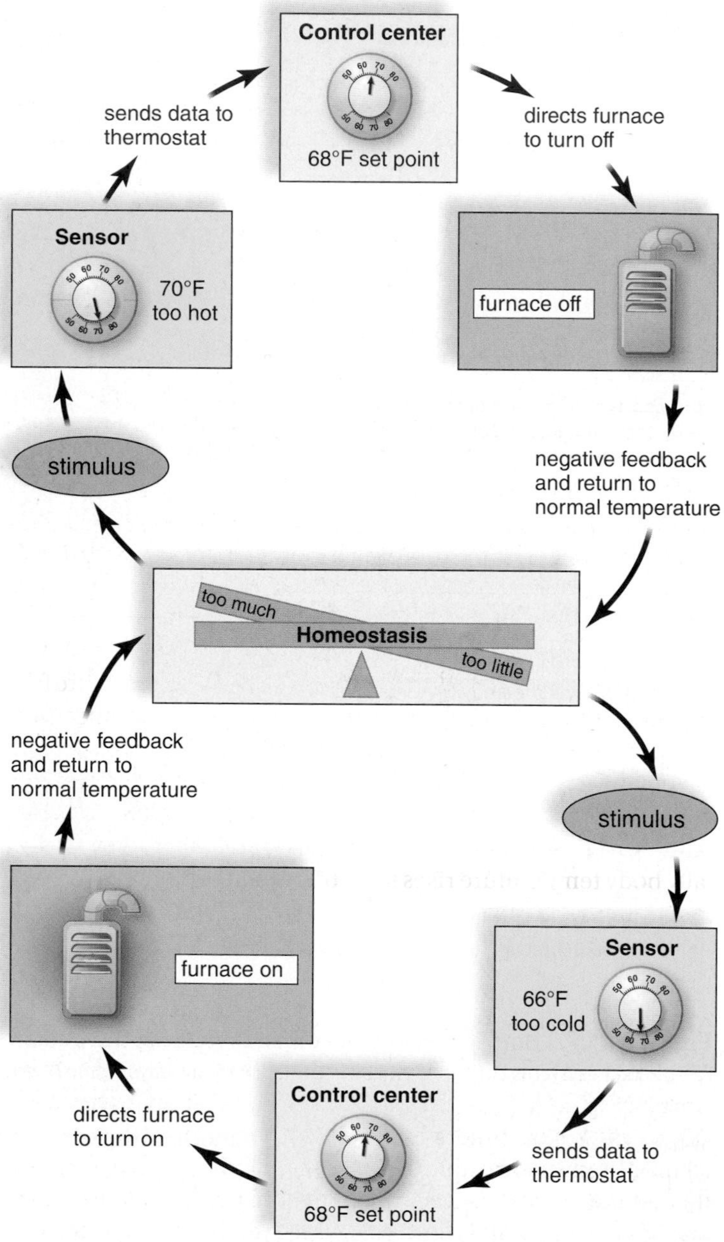

Figure 4.17 Action of a complex negative feedback mechanism. This diagram shows how room temperature is returned to normal when the room becomes too hot (*above*) or too cold (*below*). The thermostat contains both the sensor and the control center. *Above:* The sensor detects that the room is too hot, and the control center turns the furnace off. The stimulus is resolved, or corrected, when the temperature returns to normal. *Below:* The sensor detects that the room is too cold, and the control center turns the furnace on. Once again, the stimulus is resolved, or corrected, when the temperature returns to normal.

You set the thermostat at 68°F. This is the *set point.* The thermostat contains a thermometer, a sensor that detects when the room temperature is above or below the set point. The thermostat also contains a control center. It turns the furnace off when the room is warm and turns it on when the room is cool. When the furnace is off, the room cools a bit. When the furnace is on, the room warms a bit. In other words, typical of negative feedback mechanisms, there is a fluctuation above and below normal.

Human Example: Regulation of Body Temperature

The sensor and control center for body temperature are located in a part of the brain called the *hypothalamus.* A negative feedback mechanism prevents change in the same direction. Body temperature does not get warmer and warmer because warmth brings about a change toward a lower body temperature. Likewise, body temperature does not get continuously colder. A body temperature below normal brings about a change toward a warmer body temperature.

Above-Normal Temperature When the body temperature is above normal, the control center directs the blood vessels of the skin to dilate (Fig. 4.18, *above*). This allows more blood to flow near the surface of the body, where heat can be lost to the environment. In addition, the nervous system activates the sweat glands, and the evaporation of sweat helps lower body temperature. Gradually, body temperature decreases to 98.6°F.

Below-Normal Temperature When the body temperature falls below normal, the control center directs (via nerve impulses) the blood vessels of the skin to constrict (Fig. 4.18, *below*). This conserves heat. If body temperature falls even lower, the control center sends nerve impulses to the skeletal muscles and shivering occurs. Shivering generates heat, and gradually body temperature rises to 98.6°F. When the temperature rises to normal, the control center is inactivated.

MP3 Temperature Regulation

Positive Feedback

Positive feedback is a mechanism that brings about an increasing change in the same direction. When a woman is giving birth, the head of the baby begins to press against the cervix (entrance to the womb), stimulating sensory receptors there. When nerve signals reach the brain, the brain causes the pituitary gland to secrete the hormone oxytocin. Oxytocin travels in the blood and causes the uterus to contract. As labor continues, the cervix is increasingly stimulated, and uterine contractions become stronger until birth occurs.

A positive feedback mechanism can be harmful, as when a fever causes metabolic changes that push the fever higher. Death occurs at a body temperature of 113°F because cellular proteins denature at this temperature and metabolism stops.

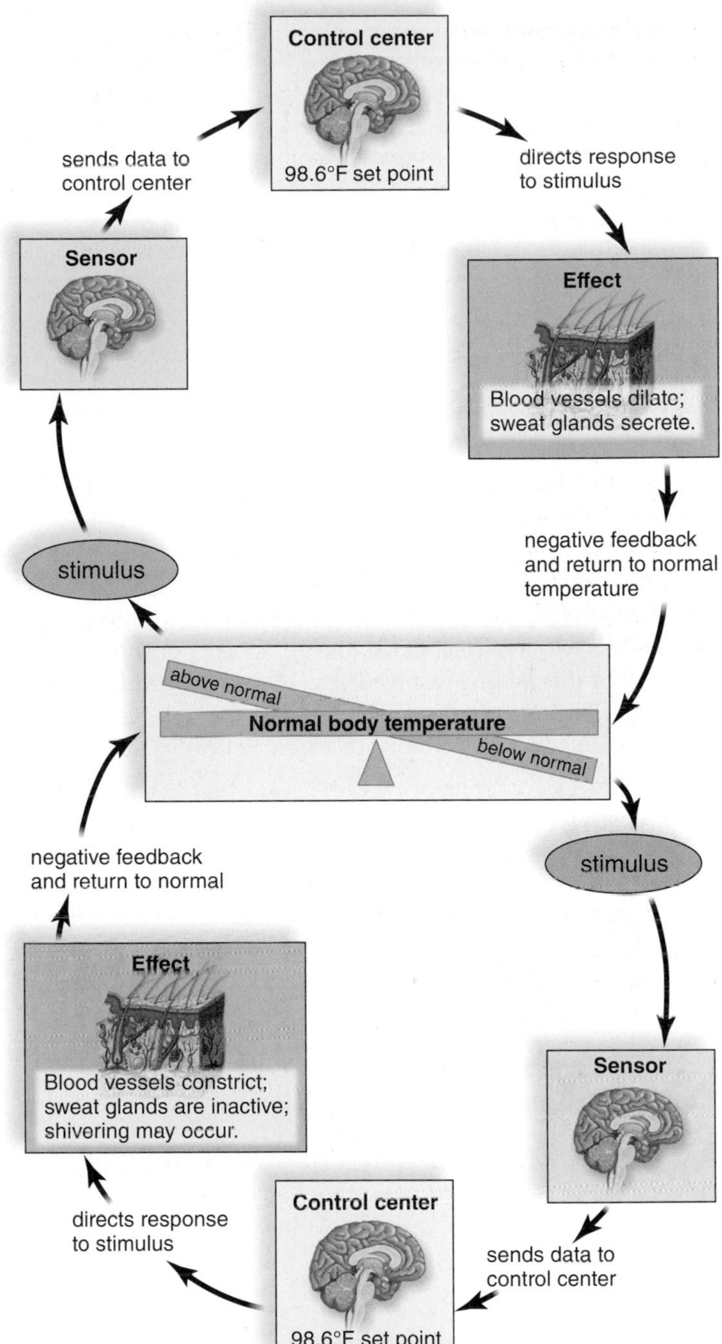

Figure 4.18 Body temperature homeostasis.
Above: When body temperature rises above normal, the hypothalamus senses the change and causes blood vessels to dilate and sweat glands to secrete so that temperature returns to normal. *Below:* When body temperature falls below normal, the hypothalamus senses the change and causes blood vessels to constrict. In addition, shivering may occur to bring temperature back to normal. In this way, the original stimulus is resolved, or corrected.

However, positive feedback loops such as those involved in childbirth, blood clotting, and the stomach's digestion of protein assist the body in completing a process that has a definite cutoff point.

Animation Feedback Mechanisms

CHECK YOUR PROGRESS 4.8

1. Define *homeostasis* and explain why it is important to body function.
2. Summarize how each body system contributes to homeostasis.
3. Distinguish between negative feedback and positive feedback mechanisms in regard to homeostasis.

CONNECTING THE CONCEPTS

The maintenance of homeostasis is an important function of all organ systems. For more information on the organ systems, refer to the following discussions:

Section 6.6 describes how the cardiovascular system helps maintain homeostasis.

Section 10.4 explores the role of the urinary system in maintaining homeostasis.

Section 13.2 describes the role of the hypothalamus as part of the central nervous svstem.

CASE STUDY CONCLUSION

Skin is a very complex organ, consisting of all of the four types of tissues. By developing an understanding of how tissues interact to form organs, it is now possible to develop synthetic tissues such as artificial skin. Though scientists are now expanding these techniques to include the development of artificial dermal tissue, the complexity of the interactions of the various tissues (nervous, muscle, connective) in this layer has presented some obstacles. However, there have been some important advances. Recently, scientists have developed newer forms of artificial skin that release antibiotics directly onto the healing tissue, further protecting the patient against life-threatening infections. Research on artificial tissues is not confined to epithelial tissue. Scientists are actively exploring the development of artificial cardiac tissue and artificial replacement organs such as the kidney, liver, and lungs. Like Kristen's artificial skin, these will not be plastic substitutes but, rather, organic material that integrates with the patient's living cells to replace damaged tissue.

MEDIA STUDY TOOLS

Enhance your study of this chapter with media! Visit **www.mhhe.com/maderhuman13e** and go to "Media Study Tools" for this chapter to access the following:

Animation	Videos	MP3 Files
4.8 Feedback Mechanisms	**4.6** Skin Color Gene • Melanoma Marker	**4.1** Tissues Overview **4.2** Connective Tissue **4.3** Muscle Tissue **4.4** Nervous Tissue **4.5** Epithelial Tissue **4.6** Skin and Its Tissues **4.7** Body Cavities **4.8** Homeostasis • Temperature Regulation

SUMMARIZE

4.1 Types of Tissues

Tissues are made of specialized cells of the same type that perform a common function. Human tissues are categorized into four groups: connective, muscular, nervous, and epithelial.

4.2 Connective Tissue Connects and Supports

Connective tissues have cells separated by a **matrix** that contains ground substance and fibers. Examples of fibers include **collagen fibers, reticular fibers,** and **elastic fibers.** There are three general classes of connective tissue:

- Fibrous connective tissue contains cells called **fibroblasts.** An example of **loose connective tissue** is **adipose tissue,** which contains cells called **adipocytes. Dense fibrous connective tissue** is found in **tendons** and **ligaments.**
- Supportive connective tissue consists of **cartilage** and bone. The matrix for cartilage is solid, yet flexible. The cells are found in chambers called **lacunae.** Examples are **hyaline cartilage,**

elastic cartilage, and **fibrocartilage.** The matrix for bone is solid and rigid. Examples are **compact bone** and **spongy bone.**

- Fluid connective tissue is found in the **blood** and **lymph.** The cells of blood include **red blood cells** (**erythrocytes**), **white blood cells** (**leukocytes**), and **platelets** (**thrombocytes**) in a **tissue fluid** called plasma.

4.3 Muscular Tissue Moves the Body

Muscular tissue is of three types: skeletal, smooth, and cardiac.

- **Skeletal muscle** and **cardiac muscle** are **striated.**
- Cardiac and smooth muscle are involuntary.
- Skeletal muscle is found in muscles attached to bones.
- **Smooth muscle** is found in internal organs.
- Cardiac muscle makes up the heart.

4.4 Nervous Tissue Communicates

- **Nervous tissue** is composed of **neurons** and several types of **neuroglia.**
- Each neuron has dendrites, a cell body, and an axon. Axons conduct nerve impulses.
- Neurons may be organized into **nerves,** which are surrounded by connective tissue.

4.5 Epithelial Tissue Protects

Epithelial tissue covers the body and lines its cavities.

- Types of simple epithelia are **squamous, cuboidal, columnar,** and **pseudostratified columnar.**
- Certain epithelial tissues may have cilia or microvilli.
- Stratified epithelia have many layers of cells, with only the bottom layer touching the **basement membrane.**
- Epithelia may be structured as a **gland,** which secretes a product either into ducts (**exocrine glands**) or into the blood (**endocrine glands**).

4.6 Integumentary System

Organs are comprised of two or more tissues working together for a common function. An **organ system** involves multiple organs cooperating for a specific function. **Skin** and its accessory organs comprise the **integumentary system.** Skin has two regions:

- The **epidermis** contains stem cells, which produce new epithelial cells, and **Langerhans cells** that provide protection against infectious agents. **Melanocytes** produce the coloration of the skin. Within the epidermis, **vitamin D** may be synthesized from cholesterol.
- The **dermis** contains epidermally derived glands and hair follicles, nerve endings, blood vessels, and sensory receptors.
- A **subcutaneous layer** (hypodermis) lies beneath the skin.
- Accessory organs of the skin include the **nails, hair follicles, oil glands, and sweat glands.**

4.7 Organ Systems, Body Cavities, and Body Membranes

Organs make up organ systems, summarized in the table on this page. Some organs are found in particular body cavities. Body cavities are lined by membranes, such as the **mucous membranes, serous membranes, synovial membranes,** and **meninges.**

Table 4.1 Organ Systems

Transport	**Integumentary**
Cardiovascular (heart and blood vessels)	Skin and accessory organs
Lymphatic and Immune (lymphatic vessels)	
Maintenance	**Motor**
Digestive (e.g., stomach, intestines)	Skeletal (bones and cartilage)
Respiratory (tubes and lungs)	Muscular (muscles)
Urinary (tubes and kidneys)	
Control	**Reproduction**
Nervous (brain, spinal cord, and nerves)	Reproductive (tubes and testes in males; tubes and ovaries in females)
Endocrine (glands)	

4.8 Homeostasis

Homeostasis is the relative constancy of the internal environment, tissue fluid, and blood. All organ systems contribute to homeostasis.

- The cardiovascular, respiratory, digestive, and urinary systems directly regulate the amount of gases, nutrients, and wastes in the blood, keeping tissue fluid constant.
- The lymphatic system absorbs excess tissue fluid and functions in immunity.
- The nervous system and endocrine system regulate the other systems.

Negative Feedback

Negative feedback mechanisms keep the environment relatively stable. When a sensor detects a change above or below a set point, a control center brings about an effect that reverses the change and returns conditions to normal. Examples include the following:

- regulation of blood glucose level by insulin,
- regulation of room temperature by a thermostat and furnace, and
- regulation of body temperature by the brain and sweat glands.

Positive Feedback

In contrast to negative feedback, a **positive feedback** mechanism brings about rapid change in the same direction as the stimulus and does not achieve relative stability. These mechanisms are useful under certain conditions, such as during birth.

ASSESS

Testing Your Knowledge of the Concepts

1. What are the functions of the four major tissue types? (page 69)
2. What features do all connective tissues have in common? Why? (page 68)
3. Explain why skeletal muscle is voluntary and cardiac muscle and smooth muscle are involuntary. (pages 71–72)
4. What are the two types of cells found in nervous tissue? Briefly describe the structure and function of each. (pages 73–74)

5. How are epithelial tissues classified? Describe each major type, and give at least one location for each type and the reason why it is found in that location. (pages 75–77)
6. Explain why the skin is sometimes referred to as the integumentary system. (page 77)
7. Referring to Figure 4.13, list each organ system, the major organs, major functions of each, and how each aids in homeostasis. (pages 82–84)
8. What organs of the body are found in the thoracic cavity? The abdominal cavity? (pages 83–84)
9. List the types of membranes found in the body, their functions, and their locations. (pages 82–84)
10. Why is homeostasis defined as the "*relative* constancy of the internal environment"? Does negative feedback or positive feedback tend to promote homeostasis? Explain. (pages 84–86)
11. Which of the following is not a type of fibrous connective tissue?
    a. hyaline cartilage **c.** tendons and ligaments
    b. areolar tissue **d.** adipose tissue
12. Blood is a(n) ________ tissue because it has a ________.
    a. connective; gap junction
    b. muscular; ground substance
    c. epithelial; gap junction
    d. connective; ground substance
13. This type of muscle contains striations.
    a. smooth muscle
    b. skeletal muscle
    c. cardiac muscle
    d. Both b and c are correct.
    e. All of these are correct.
14. Which of the following forms the myelin sheath around nerve fibers outside the brain and spinal cord?
    a. microglia **c.** Schwann cells
    b. neurons **d.** astrocytes
15. Which of these is not a type of epithelial tissue?
    a. simple cuboidal and stratified columnar
    b. bone and cartilage
    c. stratified squamous and simple squamous
    d. pseudostratified and transitional
    e. All of these are epithelial tissues.
16. What type of epithelial tissue is found in the walls of the urinary bladder to provide it with the ability to distend?
    a. simple cuboidal epithelium
    b. transitional epithelium
    c. pseudostratified columnar epithelium
    d. stratified squamous epithelium
17. Without melanocytes, skin would
    a. be too thin. **c.** not tan.
    b. lack nerves. **d.** not be waterproof.
18. Which of the following is a function of skin?
    a. temperature regulation
    b. manufacture of vitamin D
    c. protection from invading pathogens
    d. All of these are correct.
19. The skeletal system functions in
    a. blood cell production. **c.** movement.
    b. mineral storage. **d.** All of these are correct.
20. Which system helps control pH balance?
    a. digestive **c.** respiratory
    b. urinary **d.** Both b and c are correct.
21. Which type of membrane is found lining systems open to the outside environment, such as the respiratory system?
    a. serous **c.** synovial
    b. mucous **d.** meningeal
22. Which allows rapid change in one direction and does not achieve stability?
    a. homeostasis **c.** negative feedback
    b. positive feedback **d.** All of these are correct.
23. Which of the following is an example of negative feedback?
    a. Uterine contractions increase as labor progresses.
    b. Insulin decreases blood sugar levels after a meal is eaten.
    c. Sweating increases as body temperature drops.
    d. Platelets continue to plug an opening in a blood vessel until blood flow stops.

ENGAGE

Thinking Critically About the Concepts

In the hierarchy of biological organization, you have learned that groups of cells make tissues and two or more tissue types compose an organ. In this chapter, the four types of tissues (connective, muscular, nervous, and epithelial) have been discussed in detail. The skin is an organ system referred to as the integumentary system that contains all four tissue types. In addition to the epidermis and dermis, the integumentary system also includes accessory structures such as nails, sweat glands, sebaceous glands, and hair follicles. Each of these components of the skin aids in the various functions of the integumentary system. In the case study, Kristen has burns severe enough to need artificial skin treatment. This treatment will help Kristen's skin repair itself while mimicking some of the functions the integumentary system does for homeostasis.

1. The doctor diagnosed Kristen's burn as severe. Which of the following best describes a severe burn versus a superficial burn?
    a. Superficial burns include the epidermis, dermis, and hypodermis layers.
    b. Severe burns only include the layers of the epidermis.
    c. Severe burns include anything including and below the dermis.
    d. Superficial burns occur on the limbs only.
2. What accessory structures and tissues are damaged in a severe burn? Why?
3. What types of functions will the artificial skin perform while Kristen's own skin is repairing itself?
4. What effects can a severe burn have on overall homeostasis of the body? Give a few examples.
5. Which structures in the dermis will have the slowest repair time compared to others? Which might never repair themselves fully? Why?
6. Without the integumentary system, what might happen to the functions of the cardiovascular system? The nervous system?

CHAPTER

5

Cardiovascular System: Heart and Blood Vessels

CASE STUDY HYPERTENSION IN YOUNG ADULTS

As Steven entered his junior year of college, he decided that he finally had to get back into shape. While he had always been active in high school, the demands of working and studying, and his social life, had taken their toll on him over the past few years. He weighed more and exercised less than ever before. However, before playing on the intramural soccer team, he went to see his local physician for a checkup.

At the physician's office, the nurse measured Steven's weight and blood pressure. He had put on 15 pounds since his last visit a few years ago, and his blood pressure was now 132/84 mm Hg. Steven was not concerned until his doctor informed him that high blood pressure, or hypertension, is defined as a blood pressure measurement of 140/90 mm Hg or higher. Steven's blood pressure value qualified him as "prehypertensive," meaning that he was at an increased risk of developing hypertension in the future.

Hypertension is a disease of the cardiovascular system. Long-term elevated blood pressure is correlated with an increase in heart disease, stroke, kidney disease, and vision problems. Most young adults consider hypertension to be a disease of middle age, but recent studies suggest that almost one in five young adults may have high blood pressure, making it one of the major health risks facing young adults.

After his exam, Steven realized that he knew very little about systolic and diastolic blood pressure, and in fact had never really considered how his lifestyle influenced the health of his circulatory system. He was determined to find some answers to his questions.

As you read through the chapter, think about the following questions:

1. What is the difference between the diastolic and systolic values?
2. How is blood pressure normally regulated in the body?
3. What factors may cause an individual to develop hypertension?

CHAPTER CONCEPTS

5.1 Overview of the Cardiovascular System
In the cardiovascular system, the heart pumps blood, and blood vessels transport blood about the body. Exchanges between the cardiovascular system and tissues occur at the capillaries.

5.2 The Types of Blood Vessels
Arteries, which branch into arterioles, move blood from the heart to capillaries. Capillaries empty into venules. Venules join to form veins that return blood to the heart.

5.3 The Heart Is a Double Pump
The heart's right side pumps blood to the lungs, and the left side pumps blood to the rest of the body.

5.4 Features of the Cardiovascular System
Blood pressure decreases as blood moves from the arteries to the capillaries. Skeletal muscle contraction largely accounts for movement of blood in the veins.

5.5 Two Cardiovascular Pathways
The pulmonary circuit takes blood from the heart to the lungs and back to the heart. The systemic circuit carries blood from the heart to all other organs, and then returns blood to the heart.

5.6 Exchange at the Capillaries
At tissue capillaries, nutrients and oxygen are exchanged for carbon dioxide and other wastes, permitting body cells to remain alive and healthy.

5.7 Cardiovascular Disorders
Hypertension and atherosclerosis are disorders that can lead to stroke, heart attack, and aneurysm. Ways to prevent and treat cardiovascular disease are constantly being improved.

BEFORE YOU BEGIN

Before beginning this chapter, take a few moments to review the following discussions:

Section 3.3 How do substances cross plasma membranes?

Section 4.1 What are the four types of tissues?

Section 4.7 What are the role(s) of the cardiovascular system?

5.1 Overview of the Cardiovascular System

LEARNING OUTCOMES

Upon completion of this section, you should be able to

1. Identify the two components of the cardiovascular system.
2. Summarize the functions of the cardiovascular system.
3. Explain the purpose of the lymphatic system in circulation.

The **cardiovascular system** consists of (1) the heart, which pumps blood, and (2) the blood vessels, through which the blood flows. The beating of the heart sends blood into the blood vessels. In humans, blood is always contained within blood vessels.

Circulation Performs Exchanges

Even though circulation of blood depends on the beating of the heart, the overall purpose of circulation is to service the cells of the body. Remember that cells are surrounded by tissue fluid that is used to exchange substances between the blood and cells. Blood removes waste products from tissue fluid. Blood also brings tissue fluid the oxygen and nutrients cells require to continue their existence. Blood would not be able to continue to perform this function if it did not become oxygenated in the lungs (exchanging carbon dioxide waste for needed oxygen) and cleansed by the kidneys (Fig. 5.1). At the lungs, blood drops off carbon dioxide and picks up oxygen, as indicated by the two arrows in and out of the lungs in Figure 5.1.

Gas exchange is not the only function of blood. Nutrients enter the bloodstream at the intestines and transport much-needed substances to the body's cells. Blood is purified of its wastes at the kidneys, and water and salts are retained as needed. The liver is important, because it takes up amino acids from the blood and returns needed proteins. Liver proteins transport substances such as fats in the blood. The liver also removes **toxins** and chemicals that may have entered the blood at the intestines, and its colonies of white blood cells destroy bacteria and other pathogens. Thousands of miles of blood vessels, which form an intricate circuit reaching almost every cell of the body, move the blood and its contents through the body to and from all the body's organs.

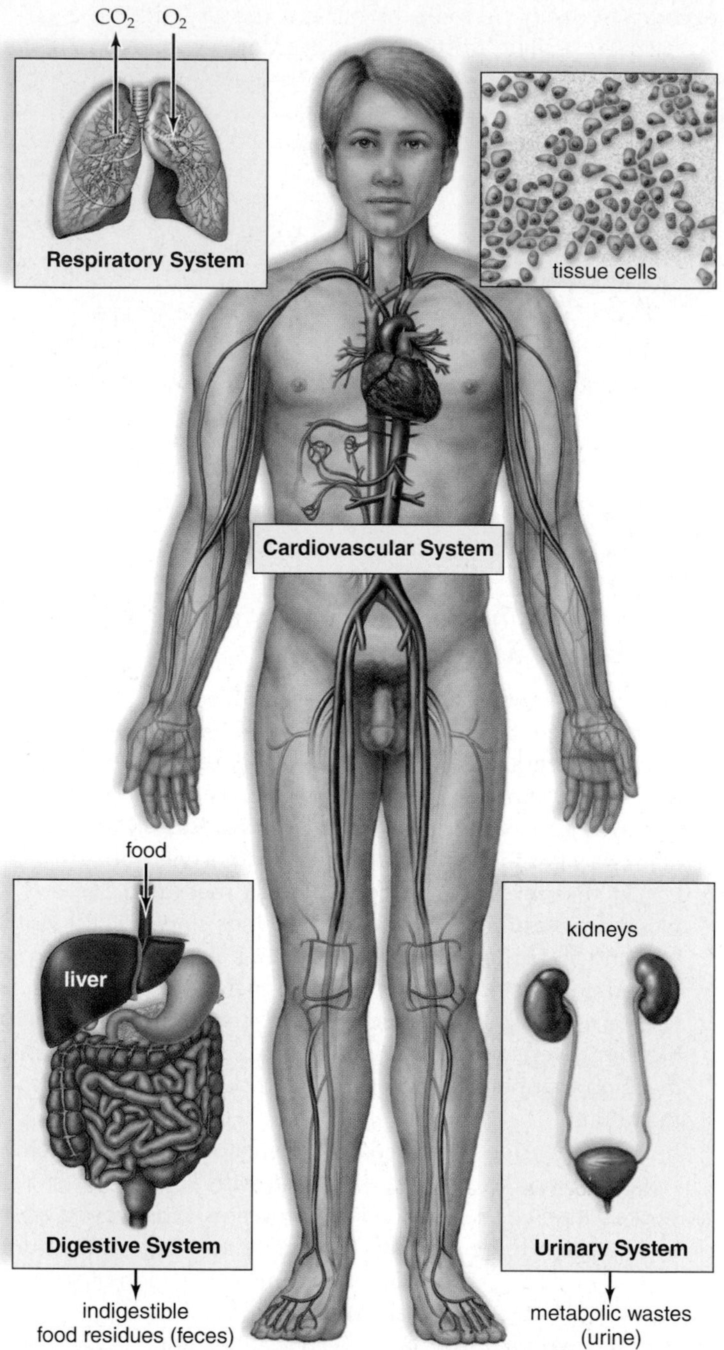

Figure 5.1 The cardiovascular system and homeostasis. The cardiovascular system transports blood throughout the body—and, with the help of other systems in the body, it maintains favorable conditions for the cells of the body.

Functions of the Cardiovascular System

In this chapter, we will see that

1. contractions of the heart generate blood pressure, which moves blood through blood vessels;
2. blood vessels transport the blood from the heart into arteries, capillaries, and veins; and blood then returns to the heart so the circuit can be completed;
3. gas exchange (pickup of carbon dioxide waste and drop-off of oxygen for the cells) occurs at the smallest-diameter vessels, the capillaries; and
4. the heart and blood vessels regulate blood flow, according to the needs of the body.

The **lymphatic system** (see Chapter 7) assists the cardiovascular system because lymphatic vessels collect excess tissue fluid and return it to the cardiovascular system. When exchanges occur between blood and tissue fluid, fluid collects in the tissues. This excess fluid enters lymphatic vessels, which start in the tissues and end at cardiovascular veins in the shoulders. As soon as fluid enters lymphatic vessels, it is called *lymph.* Lymph, you will recall, is a fluid connective tissue, as is blood (see Chapter 4).

CHECK YOUR PROGRESS 5.1

1. Describe the two parts of the cardiovascular system.
2. Summarize the functions of the cardiovascular system.
3. Explain how the lymphatic system interacts with the cardiovascular system.

CONNECTING THE CONCEPTS

As the primary transport system in the human body, the cardiovascular system interacts with all of the major organ systems. In doing so, it plays an important role in the homeostasis of a variety of body functions. For more information on these interactions, refer to the following discussions:

Section 4.8 investigates the role of the cardiovascular system in homeostasis.

Section 7.2 takes a greater look at how the lymphatic system interacts with the cardiovascular system.

Section 8.3 examines how the cardiovascular system moves nutrients that are absorbed by the small intestine.

Section 10.2 explores the interaction of the cardiovascular system with the urinary system.

5.2 The Types of Blood Vessels

LEARNING OUTCOMES

Upon completion of this section, you should be able to

1. Describe the structure and function of the three types of blood vessels.
2. Explain how blood flow is regulated in each of the three types of blood vessels.

The structure of the three types of blood vessels (arteries, capillaries, and veins) is appropriate to their function (Fig. 5.2).

MP3 Classification of Blood Vessels

The Arteries: From the Heart

An **artery** is a blood vessel that transports blood away from the heart. The structure of an artery is well suited for the transport of leaving the heart under pressure. The arterial wall has three layers, the innermost of which is a thin layer of cells called *endothelium.* Endothelium is surrounded by a relatively thick middle layer of smooth muscle and elastic tissue. The artery's outer layer is connective tissue. The strong walls of an artery give it support when blood enters under pressure; the elastic tissue allows an artery to expand to absorb the pressure. **Arterioles** are small arteries barely visible to the naked eye. Whereas the middle layer of arterioles has some elastic tissue, it is composed mostly of smooth muscle. These muscle fibers encircle the arteriole. When the fibers contract, the vessel constricts; when these muscle fibers relax, the vessel dilates.

Figure 5.2 **Structure of a capillary bed.**
A capillary bed is a maze of tiny vessels that lies between an arteriole and a venule. When precapillary sphincter muscles are relaxed, the capillary bed opens and blood flows through the capillaries. Otherwise, blood flows through a shunt directly from the arteriole to the venule. As blood passes through a capillary in the tissues, it gives up its oxygen (O_2). Therefore, blood goes from being O_2-rich in the arteriole to being O_2-poor in the vein. (Note: Blood vessels are colored red if they carry O_2-rich blood and blue if they carry O_2-poor blood. Gas exchange is reversed in the pulmonary circuit, as described in section 5.5.)

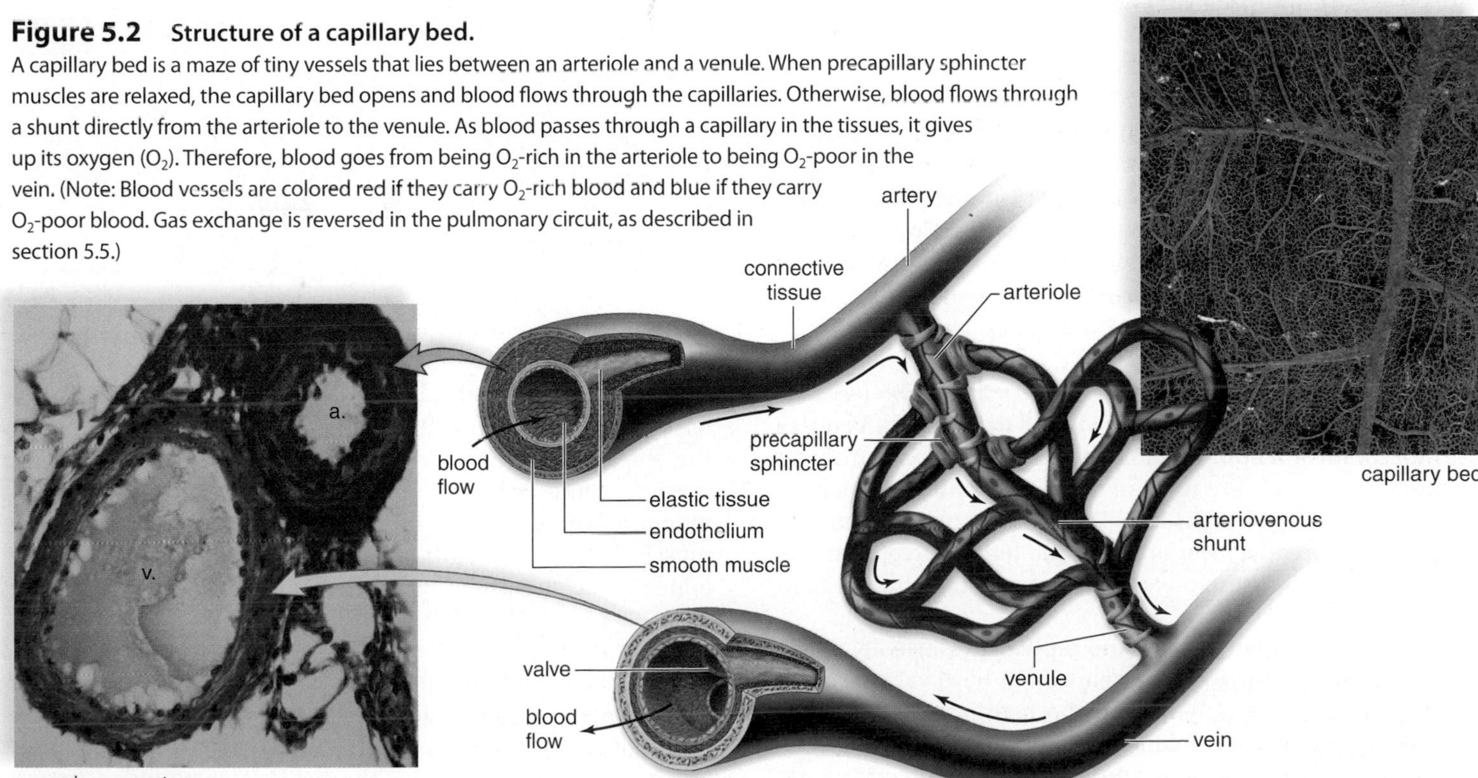

The constriction or dilation of arterioles controls blood pressure. When arterioles constrict, blood pressure rises. Dilation of arterioles causes blood pressure to fall.

The Capillaries: Exchange

Arterioles branch into capillaries. The structure of a capillary is adapted for the exchange of materials with the cells of the body. Each capillary is an extremely narrow, microscopic tube with a wall composed only of endothelium. Capillary endothelium is formed by a single layer of epithelial cells with a basement membrane. Although capillaries are small, their total surface area in humans is about 6,300 square meters (m^2). Capillary beds (networks of many capillaries) are present in all regions of the body, so no cell is far from a capillary and thus not far from gas exchange with blood. In the tissues, only certain capillaries are open at any given time. For example, after eating, the capillaries supplying the digestive system are open, whereas most serving the muscles are closed. Rings of muscle called **precapillary sphincters** control the blood flow through a capillary bed (Fig. 5.2). Constriction of the sphincters closes the capillary bed. When a capillary bed is closed, the blood moves to an area where gas exchange is needed, going directly from arteriole to venule through a pathway called an **arteriovenous shunt.**

APPLICATIONS AND MISCONCEPTIONS

If I have been standing all day, why do shoes that are loose in the morning feel so tight by evening?

Capillaries are the small, thin vessels used for tissue gas exchange, but they can also leak due to these characteristics. When you stand for long periods, gravity increases the pressure inside the capillaries. Tissue fluid leaks into tissue spaces—and your shoes get too snug. When you lie down at night, your feet are level with your heart. Excess tissue fluid that caused swelling now filters back into lymphatic vessels, which empty into shoulder veins, and fluid returns to the cardiovascular system. By morning, your feet have lost their excess water, and those shoes are loose again.

The Veins: To the Heart

Veins are blood vessels that return blood to the heart**. Venules** are small veins that drain blood from the capillaries and then join to form a vein. The walls of both veins and venules have the same three layers as arteries. However, there is less smooth muscle in the middle layer of a vein and less connective tissue in the outer layer. Therefore, the wall of a vein is thinner than that of an artery.

Because the blood leaving the capillaries is usually under low pressure (see section 5.4), veins often have valves**,** which allow blood to flow only toward the heart when open and prevent backward flow of blood when closed. Valves are extensions of the inner-wall layer and are found in the veins that carry blood against the force of gravity, especially the veins of the lower extremities. The walls of veins are thinner, so they can expand to a greater extent. At any one time, about 70% of the blood is in the veins. In this way, the veins act as a blood reservoir. If blood is lost due to hemorrhaging, nervous stimulation causes the veins to constrict, providing more blood to the rest of the body.

CHECK YOUR PROGRESS 5.2

1. List and detail the different types of blood vessels.
2. Describe the differences in function of all the blood vessels.
3. Hypothesize what areas of the body would have more precapillary sphincters than others and why.

CONNECTING THE CONCEPTS

Blood vessels act as conduits for the movement of blood, and in doing so, blood vessels interact with all of the tissues and cells of the body. For more information on blood and blood vessels, refer to the following discussions:

Section 6.1 examines the composition of blood and the function of its various components.

Figure 9.11 illustrates the major arteries and veins involved in moving gases within the body.

Figure 10.6 demonstrates how the capillaries in the kidney assist in the filtration of the blood.

5.3 The Heart Is a Double Pump

LEARNING OUTCOMES

Upon completion of this section, you should be able to

1. Identify the structures and chambers of the human heart.
2. Describe the flow of blood through the human heart.
3. Explain the internal and external controls of the heartbeat.

The **heart** is a cone-shaped, muscular organ located between the lungs, directly behind the sternum (breastbone). The heart is tilted so that the apex (the pointed end) is oriented to the left (Fig. 5.3). To approximate the size of your heart, make a fist, then clasp the fist with your opposite hand. The major portion of the heart is the interior wall of tissue called the **myocardium,** consisting largely of cardiac muscle tissue. The muscle fibers of myocardium are branched. Each fiber is tightly joined to neighboring fibers by structures called *intercalated disks* (see Fig. 5.4*b*). The intercalated disks also include cell junctions like gap junctions and desmosomes. **Gap junctions** are used to aid in simultaneous contractions of the cardiac fibers. **Desmosomes** include arrangements of protein fibers that tightly hold the membranes of adjacent cells together and prevent overstretching. The heart is surrounded by the

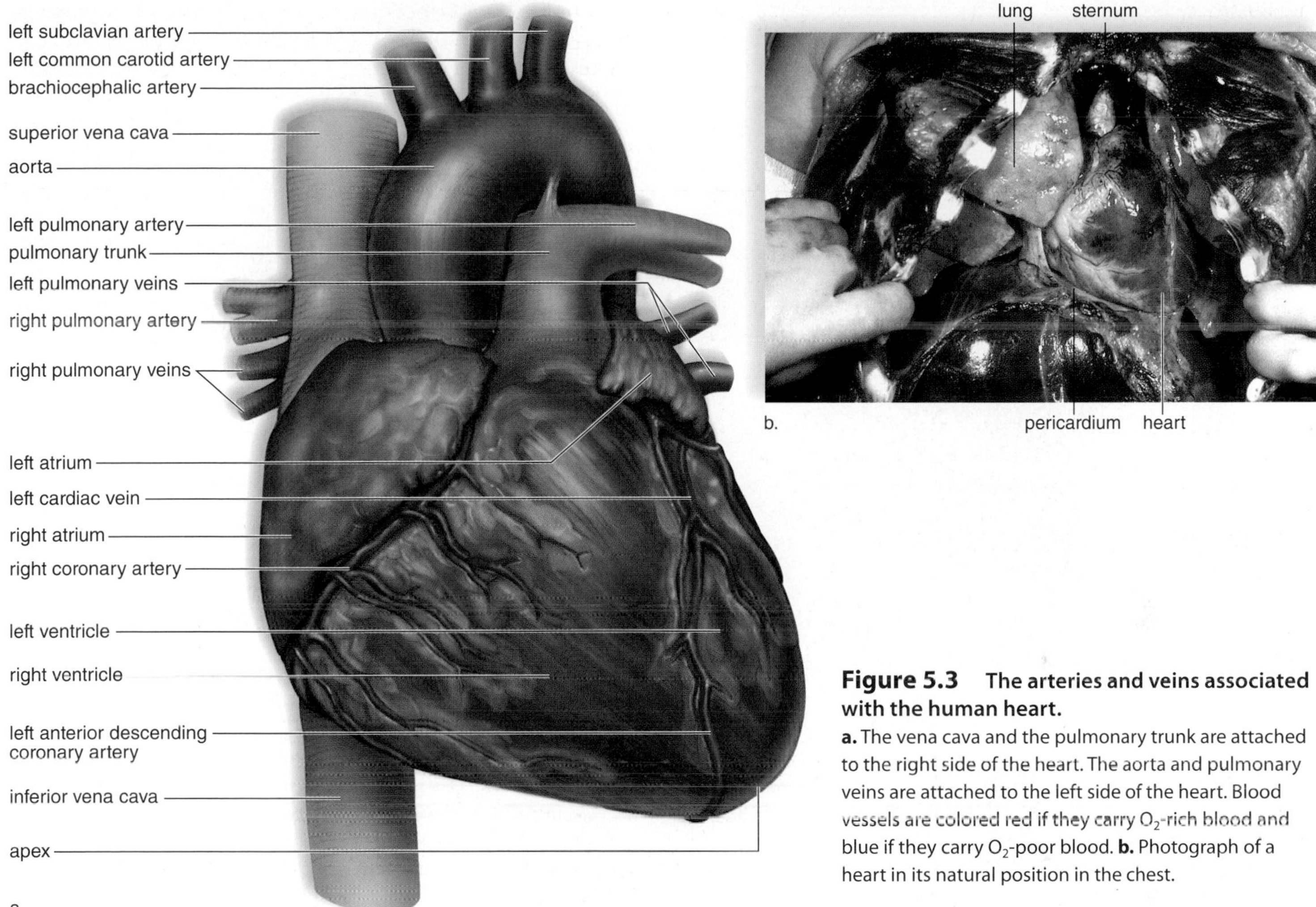

Figure 5.3 The arteries and veins associated with the human heart.
a. The vena cava and the pulmonary trunk are attached to the right side of the heart. The aorta and pulmonary veins are attached to the left side of the heart. Blood vessels are colored red if they carry O_2-rich blood and blue if they carry O_2-poor blood. **b.** Photograph of a heart in its natural position in the chest.

pericardium, a thick, membranous sac that supports and protects the heart. The inside of the pericardium secretes pericardial fluid (a lubrication fluid), and the pericardium slides smoothly over the heart's surface as it pumps the blood.

Internally, a wall called the **septum** separates the heart into a right side and a left side (Fig. 5.4*a*). The heart has four chambers. The two upper, thin-walled atria (sing., **atrium**) are called the right atrium and the left atrium. Each atrium has a wrinkled, earlike flap on the outer surface called an *auricle.* The two lower chambers are the thick-walled **ventricles,** called the right ventricle and the left ventricle (Fig. 5.4*a*).

Heart valves keep blood flowing in the right direction and prevent its backward movement. The valves that lie between the atria and the ventricles are called the **atrioventricular (AV) valves.** These valves are supported by strong fibrous strings called *chordae tendineae,* which are attached to papillary muscles that project from the ventricular walls. Chordae anchor the valves, preventing them from inverting when the heart contracts. The AV valve on the right side is called the *tricuspid* valve because it has three flaps, or cusps. The AV valve on the left side is called the *bicuspid* valve because it has two flaps. The bicuspid valve is commonly referred to as the *mitral* valve, because it has a shape like a bishop's hat, or miter. The remaining two valves are the **semilunar valves,** with flaps shaped like half-moons. These valves lie between the ventricles and their attached vessels. The semilunar valves are named for their attached vessels: The pulmonary semilunar valve lies between the right ventricle and the pulmonary trunk. The aortic semilunar valve lies between the left ventricle and the aorta.

Coronary Circulation: The Heart's Blood Supply

The myocardium—the middle, muscular layer of the three layers of the walls of the heart—receives oxygen and nutrients from the coronary arteries. Likewise, wastes are removed by the cardiac veins. The blood that flows through the heart contributes little to either nutrient supply or waste removal.

The **coronary arteries** (see Fig. 5.3) serve the heart muscle itself. These arteries are the first branches off the aorta. They originate just above the aortic semilunar valve. They lie on the exterior surface of the heart, where they divide into

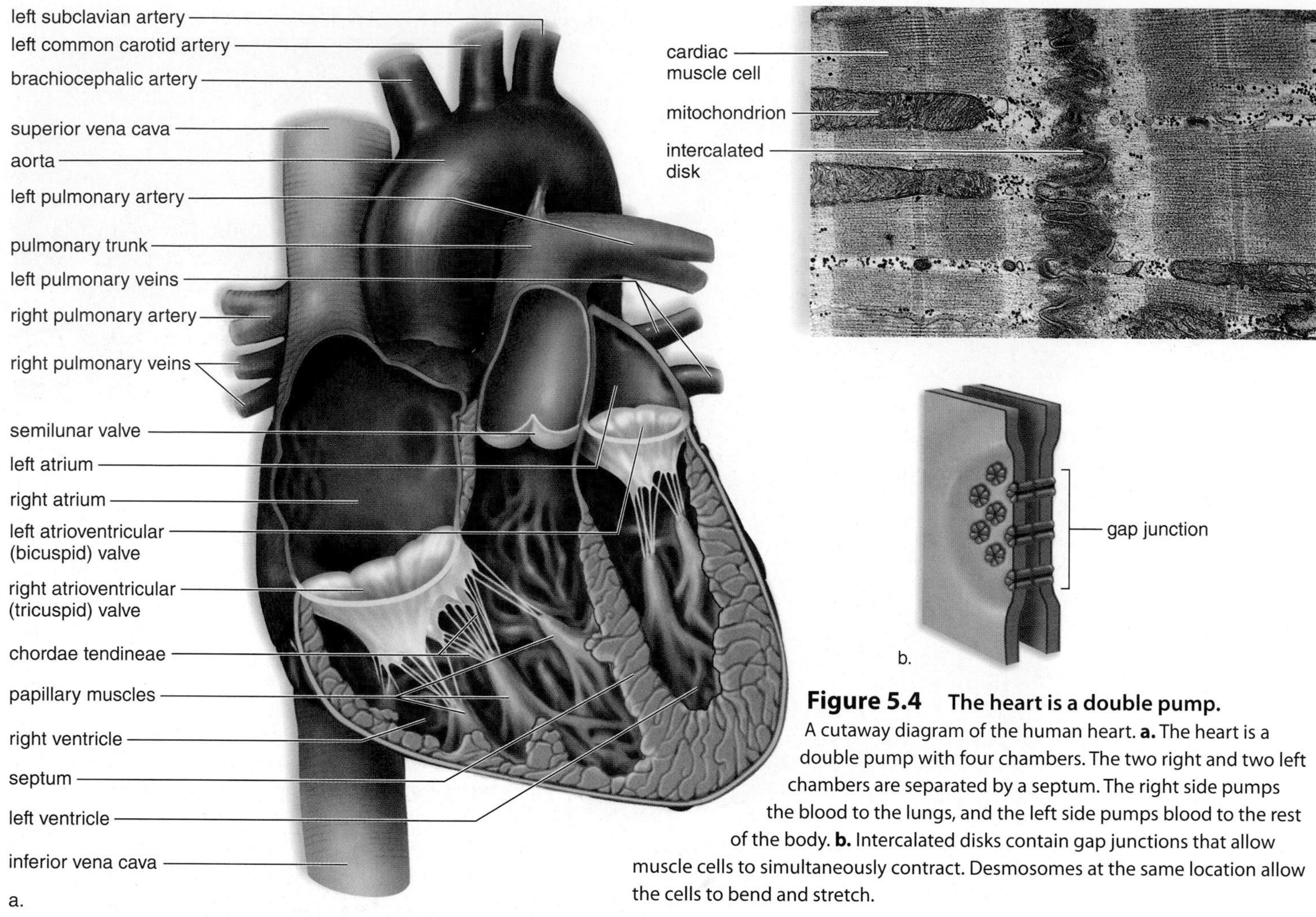

Figure 5.4 The heart is a double pump. A cutaway diagram of the human heart. **a.** The heart is a double pump with four chambers. The two right and two left chambers are separated by a septum. The right side pumps the blood to the lungs, and the left side pumps blood to the rest of the body. **b.** Intercalated disks contain gap junctions that allow muscle cells to simultaneously contract. Desmosomes at the same location allow the cells to bend and stretch.

diverse arterioles. The coronary capillary beds join to form venules, which converge to form the cardiac veins, which empty into the right atrium. Because they have a very small diameter, they can become easily clogged leading to coronary artery disease.

If these vessels become completely clogged, oxygen and nutrients such as glucose will not reach the muscles of the heart. This may result in a *myocardial infarction,* or heart attack. Coronary artery disease may be treated with medication, coronary bypass surgery, angioplasty, or stem-cell therapies (for more information, see page 108).

Video Heart Stem Cells

Passage of Blood Through the Heart

Recall that intercalated disks (see Fig. 5.4*b*) join fibers of cardiac muscle cells, allowing them to communicate with each other. By sending electrical signals between cells, both atria and then both ventricles contract simultaneously. We can trace the path of blood through the heart and body in the following manner.

- The superior vena cava and the inferior vena cava (see Fig. 5.4*a*) carry oxygen-poor blood from body veins to the right atrium.
- The right atrium contracts (simultaneously with the left ventricle), sending blood through an atrioventricular valve (the tricuspid valve) to the right ventricle.
- The right ventricle contracts, pumping blood through the pulmonary semilunar valve into the pulmonary trunk. The pulmonary trunk, which carries oxygen-poor blood, divides into two **pulmonary arteries,** which go to the lungs.
- Pulmonary capillaries within the lungs allow gas exchange. Oxygen enters the blood; carbon dioxide waste is excreted from the blood.
- Four **pulmonary veins,** which carry oxygen-rich blood, enter the left atrium.
- The left atrium pumps blood through an atrioventricular valve (the bicuspid valve) to the left ventricle.
- The left ventricle contracts (at the same time as the right ventricle), sending blood through the aortic semilunar valve into the aorta.

- Arteries and arterioles supply tissue capillaries. Tissue capillaries drain into increasingly larger veins. Veins drain into the superior and inferior vena cava, and the cycle starts again.

From this description, you can see that oxygen-poor blood never mixes with oxygen-rich blood and that blood must go through the lungs to pass from the right side to the left side of the heart. The heart is a double pump because the right ventricle of the heart sends blood through the lungs, and the left ventricle sends blood throughout the body. Thus, the left ventricle has the harder job of pumping blood.

The atria have thin walls, and each pumps blood into the ventricle right below it. The ventricles are thicker, and they pump blood into arteries (pulmonary artery and aorta) that travel to other parts of the body. The thinner myocardium of the right ventricle pumps blood to the lungs nearby in the thoracic cavity. The left ventricle has a thicker wall with more cardiac muscle cells than the right ventricle, and this enables it to pump blood out of the heart with enough force to send it through the body.

The pumping of the heart sends blood under pressure out into the arteries. The left side of the heart pumps with greater force, so blood pressure is greatest in the aorta. Blood pressure then decreases as the total cross-sectional area of arteries and then arterioles increases (see Fig. 5.8). A different mechanism, aside from blood pressure, is used to move blood in the veins.

The Heartbeat Is Controlled

Each heartbeat is called a **cardiac cycle** (Fig. 5.5). Recall that when the heart beats, first the two atria contract at the same time. Next, the two ventricles contract at the same time. Then, all chambers relax. **Systole,** the working phase, refers to contraction of the chambers, and **diastole,** the resting phase, refers to relaxation of the chambers (Fig. 5.5). The heart contracts, or beats, about 70 times a minute on average in a healthy adult, with each heartbeat lasting about 0.85 second with a normal resting rate varying from 60 to 80 beats per minute.

There are two audible heartbeat sounds referred to as "lub-dub." The first sound, "lub," occurs when increasing pressure of blood inside a ventricle forces the cusps of the AV valves to slam shut (Fig. 5.5*b*). In contrast, the pressure of blood inside a ventricle causes the semilunar valves (pulmonary and aortic) to open. The "dub" occurs when the ventricles relax, and

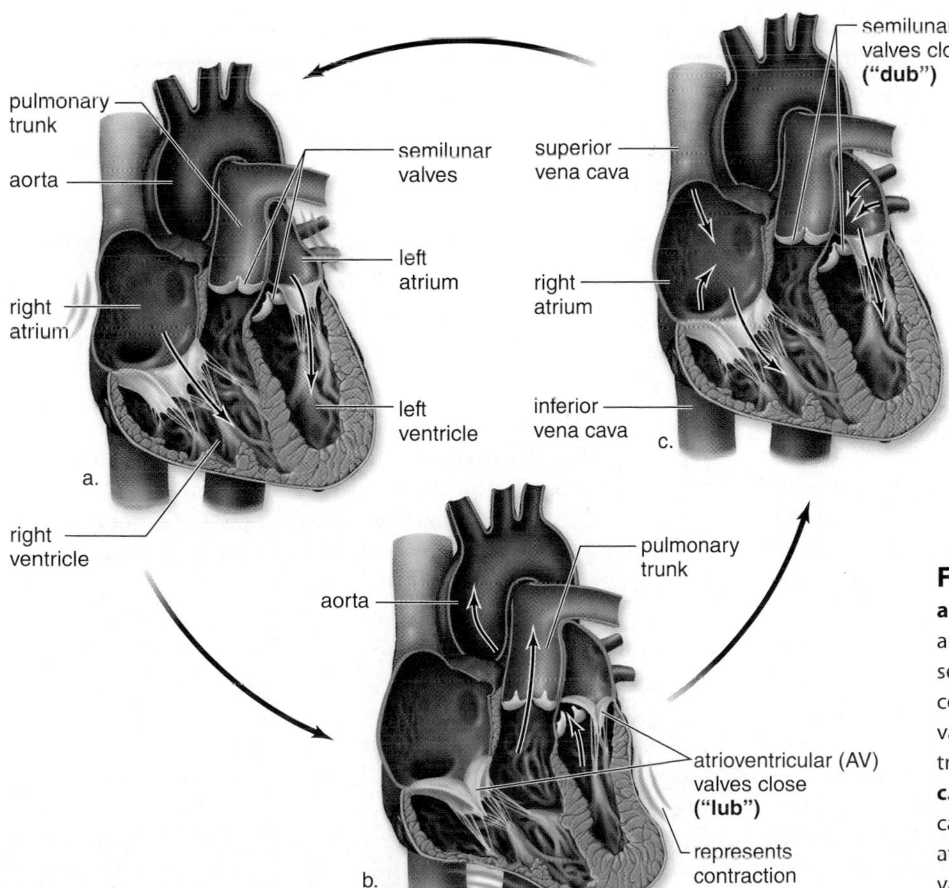

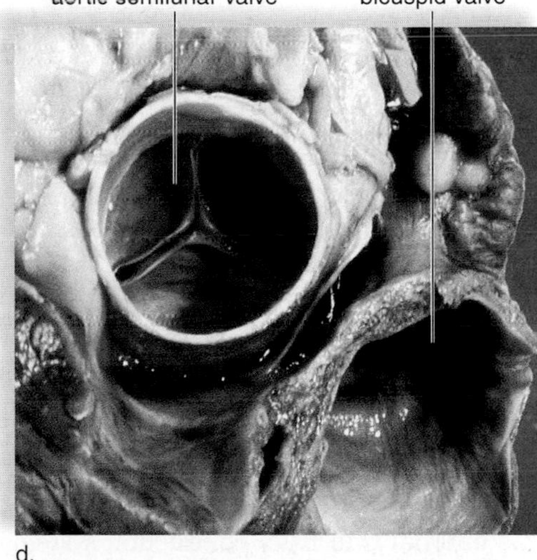

Figure 5.5 The stages of the cardiac cycle. **a.** When the atria contract, the ventricles are relaxed and filling with blood. Atrioventricular valves are open; semilunar valves are closed. **b.** When the ventricles contract, atrioventricular valves are closed; semilunar valves are open. Blood is pumped into the pulmonary trunk and aorta; corresponds to the "lub" sound. **c.** Backward flow of blood against the semilunar valves causes the "dub" sound. When the heart is relaxed, both atria and ventricles fill with blood. The atrioventricular valves are open; semilunar valves are closed. **d.** Aortic semilunar valve and bicuspid valve.

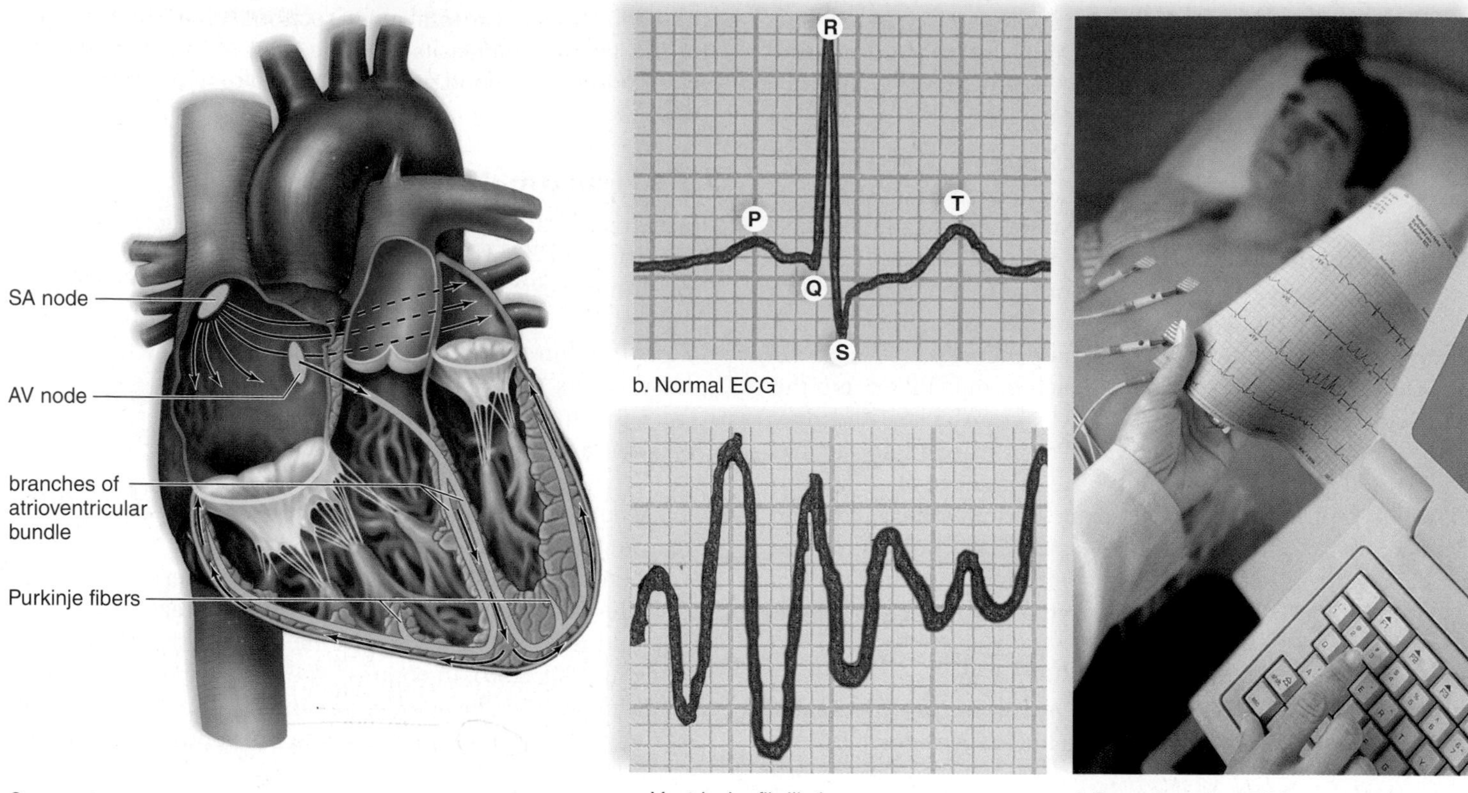

Figure 5.6 An electrical signal pathway through the heart.
a. The SA node sends out a stimulus (black arrows), which causes the atria to contract. When this stimulus reaches the AV node, it signals the ventricles to contract. The electrical signal passes down the two branches of the atrioventricular bundle to the Purkinje fibers. Thereafter, the ventricles contract. **b.** A normal ECG indicates that the heart is functioning properly. The P wave occurs just prior to atrial contraction; the QRS complex occurs just prior to ventricular contraction; and the T wave occurs when the ventricles are recovering from contraction. **c.** Ventricular fibrillation produces an irregular ECG due to irregular stimulation of the ventricles. **d.** The recording of an ECG.

blood in the arteries flows backward momentarily, causing the semilunar valves to close (Fig. 5.5*c*). A heart murmur, or a slight swishing sound after the "lub," is often due to leaky valves, which allow blood to pass back into the atria after the AV valves have closed. Faulty valves can be surgically corrected.

Internal Control of Heartbeat

The rhythmic contraction of the atria and ventricles is due to the internal (intrinsic) conduction system of the heart. Nodal tissue is a unique type of cardiac muscle located in two regions of the heart. Nodal tissue has both muscular and nervous characteristics. The **SA (sinoatrial) node** is located in the upper dorsal wall of the right atrium. The **AV (atrioventricular) node** is located in the base of the right atrium very near the septum (Fig. 5.6*a*). The SA node initiates the heartbeat and automatically sends out an excitation signal every 0.85 second. This causes the atria to contract. When signal impulses reach the AV node, there is a slight delay that allows the atria to finish their contraction before the ventricles begin their contraction. The signal for the ventricles to contract travels from the AV node through the two branches of the *atrioventricular (AV) bundle* before reaching the numerous and smaller *Purkinje fibers.* The AV bundle, its branches, and the Purkinje fibers work efficiently because gap junctions (see

APPLICATIONS AND MISCONCEPTIONS

Can the automatic external defibrillators (AEDs) found in airports and other public places be used by anyone in an emergency?

When an emergency happens in a public place, anyone can use an AED. If a person collapses, perhaps suffering from ventricular fibrillation, the computerized device will make the decisions. It will explain, step-by-step, how to first check for breathing and pulse. Next, it will explain how to apply pads to the chest to deliver the shock, if necessary.

Once the chest pads are attached, the computer analyzes heart activity to determine if a shock is needed. Strong electrical current is applied for a short time to defibrillate the heart. The rescuer moves back, pushing a button when prompted. Voice instructions also explain how to do cardiopulmonary resuscitation (CPR) until paramedics arrive.

However, it's a good idea to first be familiar with CPR and AED use. The Red Cross and many hospitals regularly offer introductory and refresher classes. With training, you might be able to save a person's life!

section 3.5) allow electrical current to flow from cell to cell (see Fig. 5.4*b*). The SA node is also called the **pacemaker** because it regulates heartbeat. If the SA node fails to work properly, the heart still beats due to signals generated by the AV node. But the beat is slower (40 to 60 beats per minute). To correct this condition, it is possible to implant an artificial pacemaker, which automatically gives an electrical stimulus to the heart every 0.85 second.

External Control of Heartbeat

The body has an external (extrinsic) way to regulate the heartbeat. A cardiac control center in the medulla oblongata, a portion of the brain that controls internal organs, can alter the beat of the heart by way of the parasympathetic and sympathetic portions of the nervous system. As will be presented in Chapter 13, the parasympathetic division promotes those functions associated with a resting state. The sympathetic division brings about those responses associated with fight or flight. It makes sense that the parasympathetic division decreases SA and AV nodal activity when we are inactive. By contrast, the sympathetic division increases SA and AV nodal activity when we are active or excited.

The hormones epinephrine and norepinephrine, released by the adrenal medulla, also stimulate the heart. During exercise, for example, the heart pumps faster and stronger due to sympathetic stimulation and the release of epinephrine and norepinephrine (for more information, see Chapters 13 and 15).

Animation
Baroreceptor Reflex Control of Blood Pressure

The Electrocardiogram Is a Record of the Heartbeat

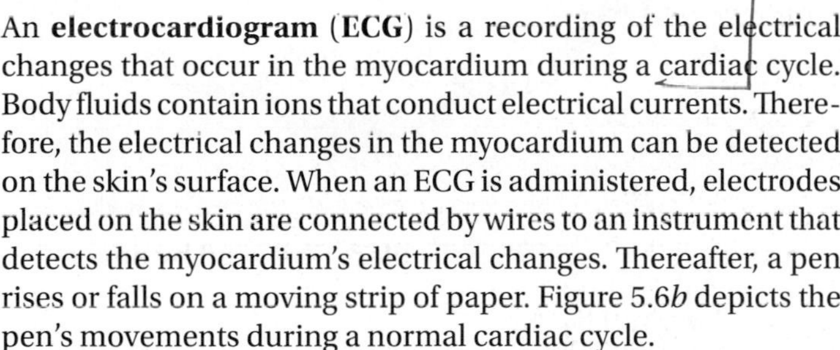

An **electrocardiogram (ECG)** is a recording of the electrical changes that occur in the myocardium during a cardiac cycle. Body fluids contain ions that conduct electrical currents. Therefore, the electrical changes in the myocardium can be detected on the skin's surface. When an ECG is administered, electrodes placed on the skin are connected by wires to an instrument that detects the myocardium's electrical changes. Thereafter, a pen rises or falls on a moving strip of paper. Figure 5.6*b* depicts the pen's movements during a normal cardiac cycle.

When the SA node triggers an impulse, the atrial fibers produce an electrical change called the P wave. The P wave indicates the atria are about to contract. After that, the QRS complex signals that the ventricles are about to contract. The electrical changes that occur as the ventricular muscle fibers recover produce the T wave.

Various types of abnormalities can be detected by an ECG. One of these, called *ventricular fibrillation,* is caused by uncoordinated, irregular electrical activity in the ventricles. Compare Figure 5.6*b* with Figure 5.6*c* and note the irregular line illustrated in Figure 5.6*c*. Ventricular fibrillation is of special interest because it can be caused by an injury, heart attack, or drug overdose. It is the most common cause of sudden cardiac death in a seemingly healthy person over age 35. Once the ventricles are fibrillating, coordinated pumping of the heart ceases and body tissues quickly become oxygen starved. Normal electrical conduction must be reestablished as quickly as possible or the person will die. A strong electrical current is applied to the chest for a short time in a process called *defibrillation.* In response, all heart cells discharge their electricity at once. Then, the SA node may be able to reestablish a coordinated beat.

MP3
Cardiac Conduction System

CHECK YOUR PROGRESS 5.3

1. Describe the flow of blood through the heart.
2. Explain what causes the "lub" and the "dub" sounds of a heartbeat.
3. Summarize the internal and external controls of the heartbeat.

CONNECTING THE CONCEPTS

For more information on the heart as a muscle and to gain a greater understanding of how our heart develops and ages, refer to the following discussions:

Figure 12.1 compares the structure of cardiac muscle to other muscle tissues.

Figure 17.7 illustrates the fetal circulation pathway.

Section 17.5 explores some of the changes to the cardiovascular system as we age.

5.4 Features of the Cardiovascular System

LEARNING OUTCOMES

Upon completion of this section, you should be able to

1. Explain how blood pressure differs in veins, arteries, and capillaries.
2. Distinguish between systolic and diastolic pressure.
3. Analyze systolic and diastolic pressures to assess the cardiovascular health of an individual.

When the left ventricle contracts, blood is sent out into the aorta under pressure. A progressive decrease in pressure occurs as blood moves through the arteries, arterioles, capillaries, venules, and finally the veins. Blood pressure is highest in the aorta. By contrast, pressure is lowest in the superior and inferior venae cavae, which enter the right atrium (see Fig. 5.4*a*).

Pulse Rate Equals Heart Rate

The surge of blood entering the arteries causes their elastic walls to stretch, but then they almost immediately recoil. This rhythmic expansion and recoil of an arterial wall can be felt as a **pulse** in any artery that runs close to the body's surface. It is customary to feel the pulse by placing several fingers on either the radial artery (near the outer border of the palm side of the wrist) or the carotid artery (located on either side of the trachea in the neck).

Normally, the pulse rate indicates the heart rate because the arterial walls pulse whenever the left ventricle contracts. The pulse rate is usually 70 beats per minute in a healthy adult but can vary between 60 and 80 beats per minute.

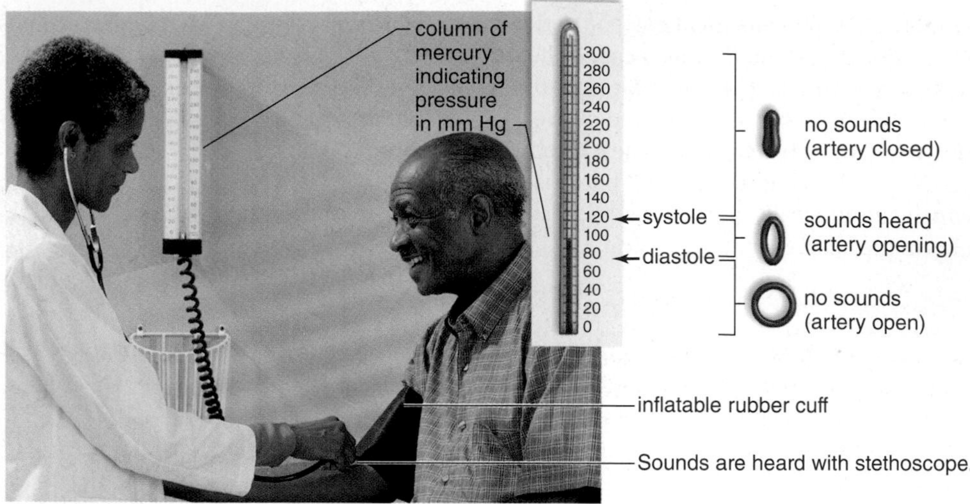

Figure 5.7 Sphygmomanometers measure blood pressure.
The sphygmomanometer cuff is first inflated until the artery is completely closed by its pressure. Next, the pressure is gradually reduced. The clinician listens with a stethoscope for the first sound, indicating that blood is moving past the cuff in an artery. This is systolic blood pressure. The pressure in the cuff is further reduced until no sound is heard, indicating that blood is flowing freely through the artery. This is diastolic pressure. The procedure can be done manually or by the computerized sphygmomanometers commonly found in stores.

Blood Flow Is Regulated

The beating of the heart is necessary to homeostasis because it creates the pressure that propels blood in the arteries and the arterioles. Arterioles lead to the capillaries where exchange with tissue fluid takes place, thus supplying cells with nutrients and removing waste material.

Blood Pressure Moves Blood in Arteries

Blood pressure is the pressure of blood against the wall of a blood vessel. A sphygmomanometer (blood pressure instrument) can be used to measure blood pressure, usually in the brachial artery of the arm (Fig. 5.7). The highest arterial pressure, called the **systolic pressure,** is reached during ejection of blood from the heart. The lowest arterial pressure, called the **diastolic pressure,** occurs while the heart ventricles are relaxing. Blood pressure is measured in millimeters mercury (mm Hg). Normal resting blood pressure for a young adult should be slightly lower than 120 mm Hg over 80 mm Hg, or 120/80, but these values can vary somewhat and still be within the range of normal blood pressure (Table 5.1). The number 120 represents the systolic pressure, and 80 represents the diastolic pressure. High blood pressure is called *hypertension,* and low blood pressure is called *hypotension.*

Both systolic and diastolic blood pressure decrease with distance from the left ventricle because the total cross-sectional area of the blood vessels increases (Fig. 5.8)—there are more arterioles than arteries. The decrease in blood pressure causes the blood velocity to gradually decrease as it flows toward the capillaries.

Table 5.1 Normal Values for Adult Blood Pressure*

	Top Number (Systolic)	**Bottom Number (Diastolic)**
Hypotension	Less than 95	Less than 50
Normal	Below 120	Below 80
Prehypertension	120–139	80–89
Stage 1 hypertension	140–159	90–99
Stage 2 hypertension	160 or more	100 or more
Hypertensive crisis (emergency care needed)	Higher than 180	Higher than 110

*Blood pressure values established by the American Heart Association (www.heart.org).

Blood Flow Is Slow in the Capillaries

There are many more capillaries than arterioles, and blood moves slowly through the capillaries (Fig. 5.8). This is important because the slow progress allows time for the exchange of

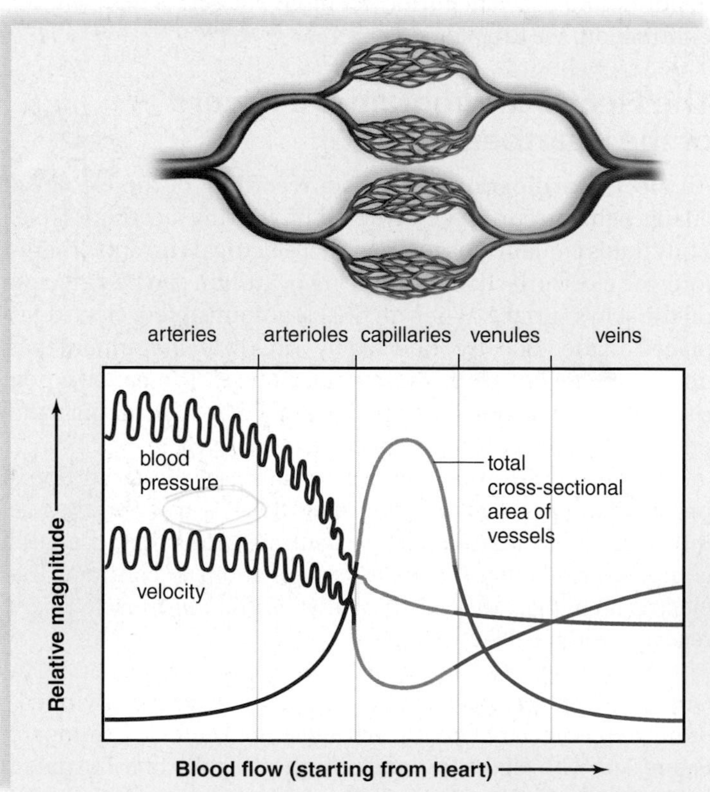

Figure 5.8 Blood velocity and pressure in the blood vessels.
This diagram shows the velocity (speed) and relative blood pressure as the blood flows from the heart (on left) to the veins (on right). The cross-sectional line indicates the total area of the blood vessels. Notice how, as the area increases in the capillaries, there is a marked decrease in blood velocity and pressure.

substances between the blood in the capillaries and the surrounding tissues. Any needed changes in flow rate are adjusted by the opening and closing of the precapillary sphincters.

Blood Flow in Veins Returns Blood to Heart

By studying Figure 5.8, it can be seen that the velocity of blood flow increases from capillaries to veins. As an analogy, imagine a single narrow street emptying its cars into a fast multilane highway. Likewise, as capillaries empty into veins, blood can travel faster. However, it's also apparent from Figure 5.8 that blood pressure is minimal in venules and veins. Blood pressure thus plays only a small role in returning venous blood to the heart. Venous return is dependent upon three additional factors:

1. the *skeletal muscle pump,* dependent on skeletal muscle contraction;
2. the *respiratory pump,* dependent on breathing; and
3. valves in veins.

The skeletal muscle pump functions every time a muscle contracts. When skeletal muscles contract, they compress the weak walls of the veins. This causes the blood to move past a valve (Fig. 5.9). Once past the valve, blood cannot flow backward. The importance of the skeletal muscle pump in moving blood in the veins can be demonstrated by forcing a person to stand rigidly still for an hour or so. Fainting may occur because lack of muscle contraction causes blood to collect in the limbs. Poor venous return deprives the brain of needed oxygen. In this case, fainting is beneficial because the resulting horizontal position aids in getting blood to the head.

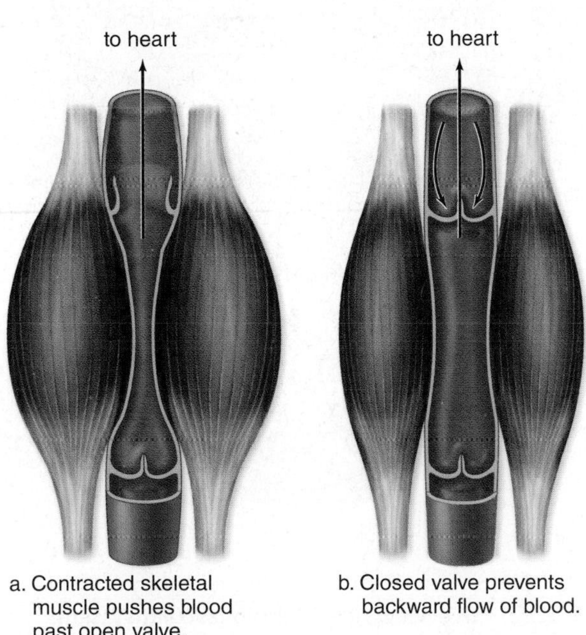
to heart
to heart
a. Contracted skeletal muscle pushes blood past open valve.
b. Closed valve prevents backward flow of blood.

Figure 5.9 The skeletal muscle pump.
a. Pressure on the walls of a vein, exerted by skeletal muscles, increases blood pressure within the vein and forces a valve open. **b.** When external pressure is no longer applied to the vein, blood pressure decreases, and back pressure forces the valve closed. Closure of the valves prevents the blood from flowing backward. Therefore, veins return blood to the heart.

The respiratory pump works much like an eyedropper. When the dropper's suction bulb is released, the bulb expands. Fluid travels from higher pressure in the bottom of the glass tube, moving up the tube toward the lower pressure at the top. When we inhale, the chest expands, and this reduces pressure in the thoracic cavity. Blood will flow from higher pressure (in the abdominal cavity) to lower pressure (in the thoracic cavity). When we exhale, the pressure reverses, but again the valves in the veins prevent backward flow.

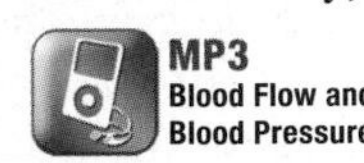
MP3
Blood Flow and Blood Pressure

CHECK YOUR PROGRESS 5.4

1. Explain what the pulse rate of a person indicates.
2. Compare and contrast the pressure of blood flow in the veins, arteries, and capillaries.
3. Explain why valves are needed in the veins.

CONNECTING THE CONCEPTS

The maintenance of blood pressure is an important aspect of homeostasis. Blood pressure is influenced by a variety of factors, including diet and internal controls. For examples, refer to the following discussions:

Section 8.6 explores how lipids and minerals—namely, sodium—in the diet influence blood pressure.

Section 10.4 explains how the water–salt balance and an enzyme called renin influence blood pressure.

APPLICATIONS AND MISCONCEPTIONS

What causes varicose veins to develop and why are they more common in pregnant women?

Veins are thin-walled tubes divided into many separate chambers by vein valves. Excessive stretching occurs if veins are overfilled with blood. For example, if a person stands in one place for a long time, leg veins can't drain properly and blood pools in them. As the vein expands, vein valves become distended and fail to function. These two mechanisms cause the veins to bulge and be visible on the skin's surface. Obesity, sedentary lifestyle, female gender, genetic predisposition, and increasing age are risk factors for varicose veins. Hemorrhoids are varicose veins in the rectum.

Pregnancy in particular can cause leg veins to become distended, because the fetus in the abdomen compresses the large abdominal veins. Once again, leg veins can't drain properly and the valves malfunction. Women who don't develop varicose veins during pregnancy may just be genetically "lucky."

Most varicose veins are only a cosmetic problem. Occasionally they will cause pain, muscle cramps, or itching and become a medical problem. There are safe and effective ways to get rid of them, so see your doctor.

5.5 Two Cardiovascular Pathways

LEARNING OUTCOMES

Upon completion of this section, you should be able to

1. Compare blood flow in the pulmonary and systemic circuits.
2. Identify the major arteries and veins of both the pulmonary and systemic circuits.
3. Compare the oxygen content of the blood in the arteries and veins of the pulmonary and systemic circuits.
4. Explain the location and purpose of the hepatic portal system.

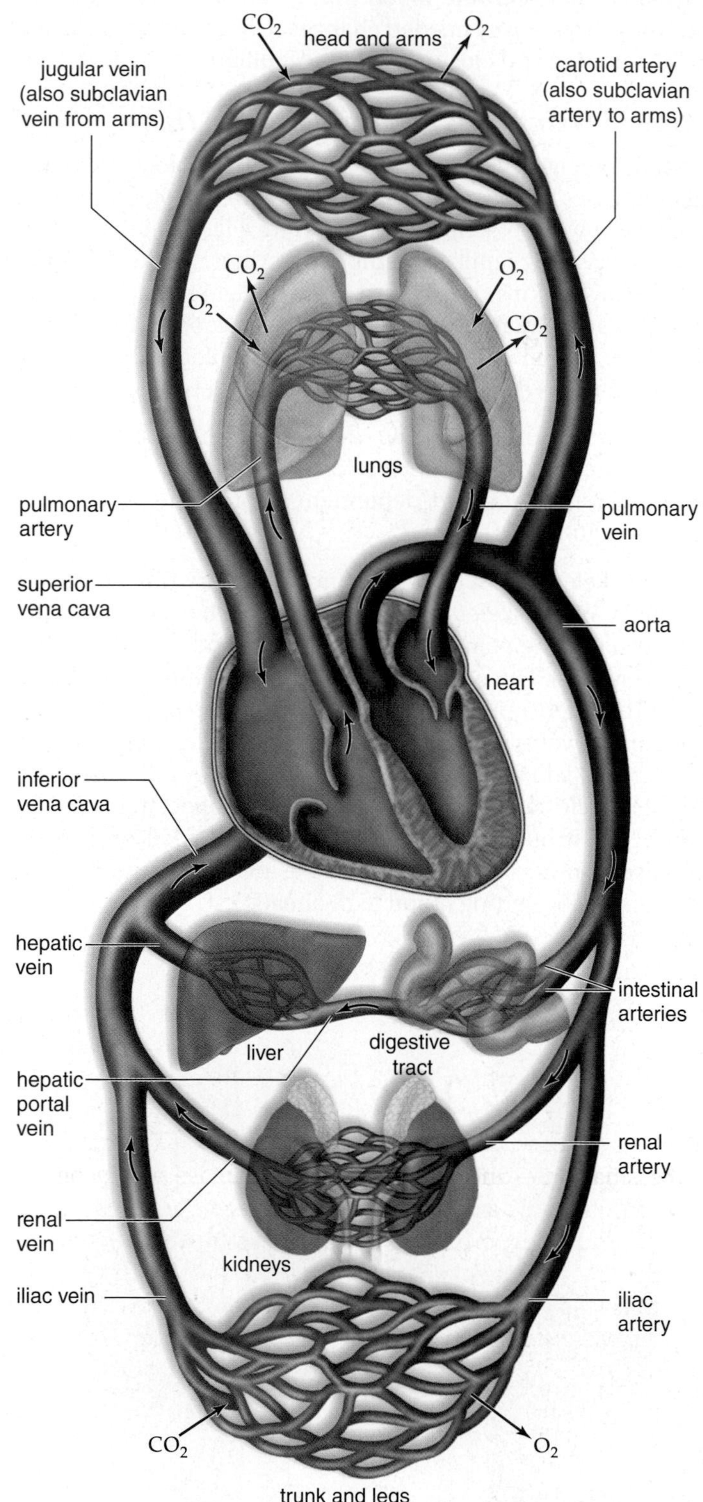

Figure 5.10 Overview of the cardiovascular system. The blue-colored vessels carry blood high in carbon dioxide, and the red-colored vessels carry blood high in oxygen; the arrows indicate the flow of blood. Compare this diagram, useful for learning to trace the path of blood, with Figure 5.11 to realize that arteries and veins go to all parts of the body. Also, there are capillaries in all parts of the body. No cell is located far from a capillary.

The blood flows in two circuits: the **pulmonary circuit,** which circulates blood through the lungs, and the **systemic circuit,** which serves the needs of body tissues (Fig. 5.10). While it is common to discuss these circuits relative to the movement of oxygen and carbon dioxide, it is important to note that these circuits also deliver a variety of nutrients, hormones, and other factors. For this reason, both circuits are important in the maintenance of homeostasis in the body.

The Pulmonary Circuit: Exchange of Gases

The path of blood through the lungs can be traced as follows: Blood from all regions of the body first collects in the right atrium and then passes into the right ventricle, which pumps it into the pulmonary trunk. The pulmonary trunk divides into the right and left pulmonary arteries, which branch as they approach the lungs. The arterioles take blood to the pulmonary capillaries, where carbon dioxide is given off and oxygen is picked up. Blood then passes through the pulmonary venules, which lead to the four pulmonary veins that enter the left atrium. Blood in the pulmonary arteries is oxygen-poor but blood in the pulmonary veins is oxygen-rich, so it is not correct to say that all arteries carry blood high in oxygen and all veins carry blood low in oxygen (as people tend to believe). Just the reverse is true in the pulmonary circuit.

The Systemic Circuit: Exchanges with Tissue Fluid

The systemic circuit includes all of the arteries and veins shown in Figure 5.10. (For simplicity, some blood vessels are not shown.) The heart pumps blood through 60,000 miles of blood vessels to deliver nutrients and oxygen and remove wastes from all body cells.

The largest artery in the systemic circuit, the **aorta,** receives blood from the heart; the largest veins, the superior and inferior venae cavae (sing., **vena cava**), return blood to the heart. The superior vena cava collects blood from the head, the chest, and the arms, and the inferior vena cava collects blood from the lower body regions. Both enter the right atrium.

Tracing the Path of Blood

It's easy to trace the path of blood in the systemic circuit by beginning with the left ventricle, which pumps blood into the aorta. Branches from the aorta go to the organs and major body regions. For example, this is the path of blood to and from the lower legs (Fig 5.11):

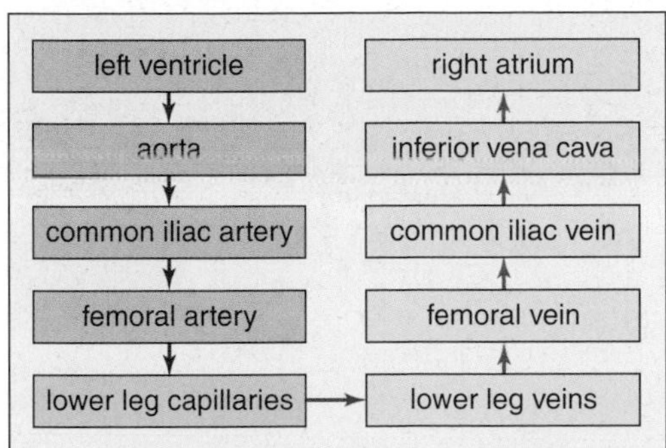

When tracing blood, first mention the aorta, then the artery branching from the aorta. Then list the region of the body where the capillaries are found, followed by the vein returning blood to the vena cava. In many instances, the artery and the vein that serve the same region are given the same name, as is the case for the femoral artery and femoral vein (Fig. 5.11). What happens in between the artery and the vein? Arterioles from the artery branch into capillaries, where exchange takes place, and then venules join into the vein that enters a vena cava.

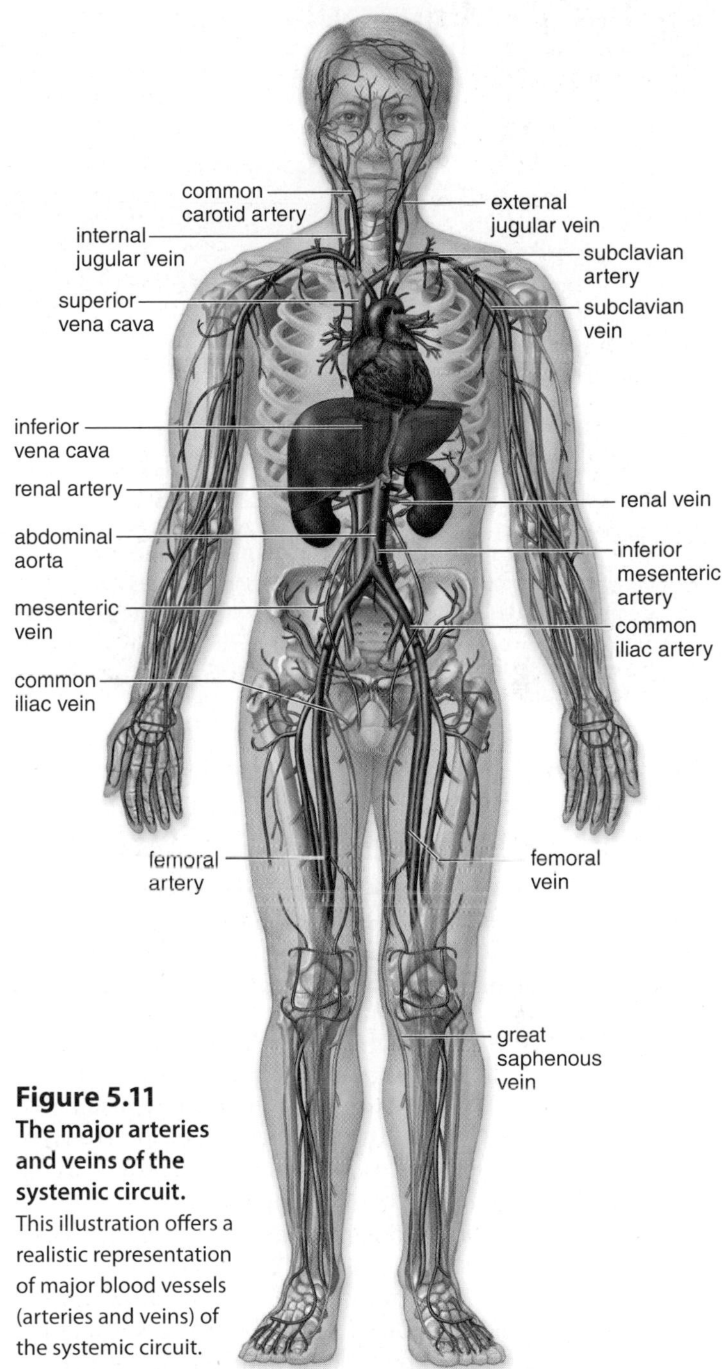

Figure 5.11
The major arteries and veins of the systemic circuit.
This illustration offers a realistic representation of major blood vessels (arteries and veins) of the systemic circuit.

Hepatic Portal System: Specialized for Blood Filtration

The **hepatic portal vein** (see Fig. 5.10) drains blood from the capillary beds of the digestive tract to a capillary bed in the liver. (The term *portal* indicates that it lies between two capillary beds.) The blood in the hepatic portal vein is oxygen-poor, but rich in glucose, amino acids, and other nutrients absorbed by the small intestine. The liver stores glucose as glycogen, and it either stores amino acids or uses them immediately to manufacture blood proteins. The liver also purifies the blood of toxins and pathogens that have entered the body by way of the intestinal capillaries. After blood has filtered slowly through the liver, it is collected by the **hepatic vein** and returned to the inferior vena cava.

CHECK YOUR PROGRESS 5.5

1. Describe the flow of blood in the pulmonary circuit.
2. Describe the path of blood from the heart to the digestive tract and back to the heart by way of the hepatic portal vein.
3. Compare the relative oxygen content of the blood flowing in the pulmonary artery with that in the pulmonary vein.

CONNECTING THE CONCEPTS

During fetal development, minor changes in the flow of blood occur due to a bypassing of the lungs. For more on this and additional information on the hepatic portal system and pulmonary circuits, refer to the following discussions:

Figure 8.8 illustrates how the hepatic portal vein brings nutrients to the liver.

Figure 9.11 diagrams the transport of oxygen and carbon dioxide in the pulmonary and systemic circuits.

Figure 17.7 shows how the systemic circuit interacts with the placenta during development.

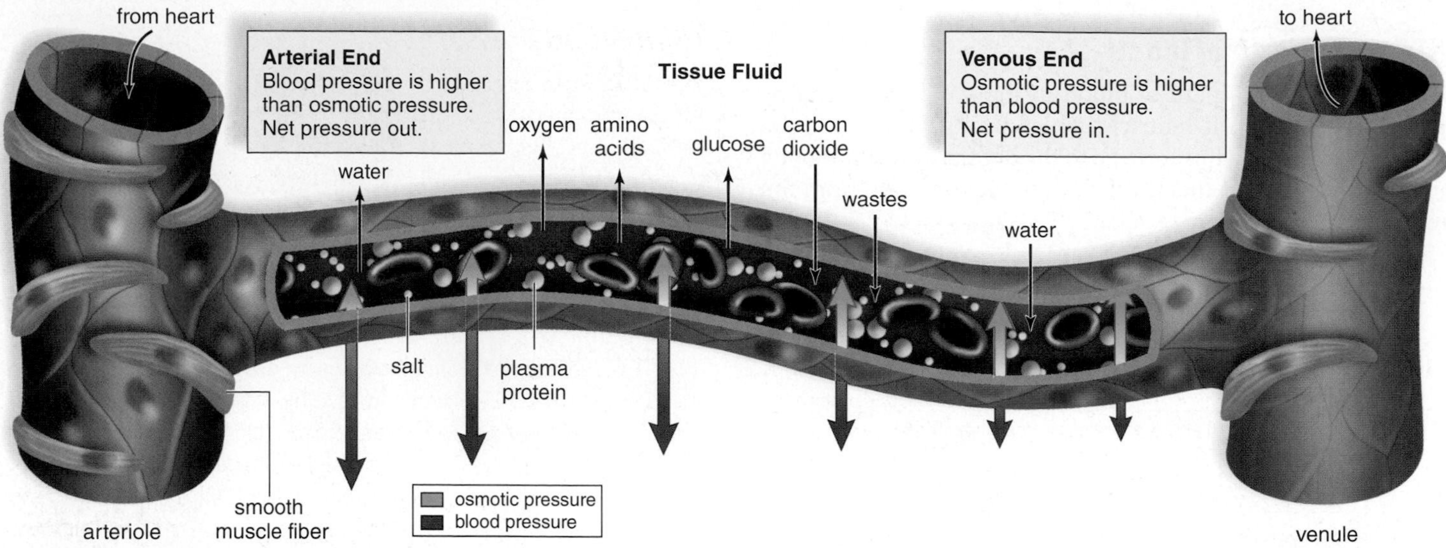

Figure 5.12 The movement of solutes in a capillary bed.
The figure shows the process of capillary exchange and the forces that aid the process. At the arterial end of a capillary, the blood pressure is higher than the osmotic pressure. Tissue fluid tends to leave the bloodstream. In the midsection, solutes, including oxygen (O_2) and carbon dioxide (CO_2), diffuse from high to low concentration. Carbon dioxide and wastes diffuse into the capillary while nutrients and oxygen enter the tissues. At the venous end of a capillary, the osmotic pressure is higher than the blood pressure. Tissue fluid tends to reenter the bloodstream. The red blood cells and the plasma proteins are too large to exit a capillary.

5.6 Exchange at the Capillaries

LEARNING OUTCOMES

Upon completion of this section, you should be able to

1. Describe the processes that move materials across the walls of a capillary.
2. Explain what happens to the excess fluid that leaves the capillaries.

Two forces control movement of fluid through the capillary wall: blood pressure, which tends to cause fluids in the blood to move from capillary to tissue spaces, and osmotic pressure, which tends to cause water to move in the opposite direction. At the arterial end of a capillary, blood pressure (30 mm Hg) is higher than the osmotic pressure of blood (21 mm Hg) (Fig. 5.12). Osmotic pressure is created by the presence of solutes dissolved in plasma, the liquid fraction of the blood. Dissolved plasma proteins are of particular importance in maintaining the osmotic pressure. Most plasma proteins are manufactured by the liver. Blood pressure is higher than osmotic pressure at the arterial end of a capillary, so water exits a capillary at the arterial end.

Midway along the capillary, where blood pressure is lower, the two forces essentially cancel each other, and there is no net movement of fluid. Solutes now diffuse according to their concentration gradient: Oxygen and nutrients (glucose and amino acids) diffuse out of the capillary; carbon dioxide and wastes diffuse into the capillary. Red blood cells and almost all plasma proteins remain in the capillaries. The substances that leave a capillary contribute to tissue fluid, the fluid between the body's cells. Plasma proteins are too large to readily pass out of the capillary. Thus, tissue fluid tends to contain all components of plasma, except much lower amounts of protein.

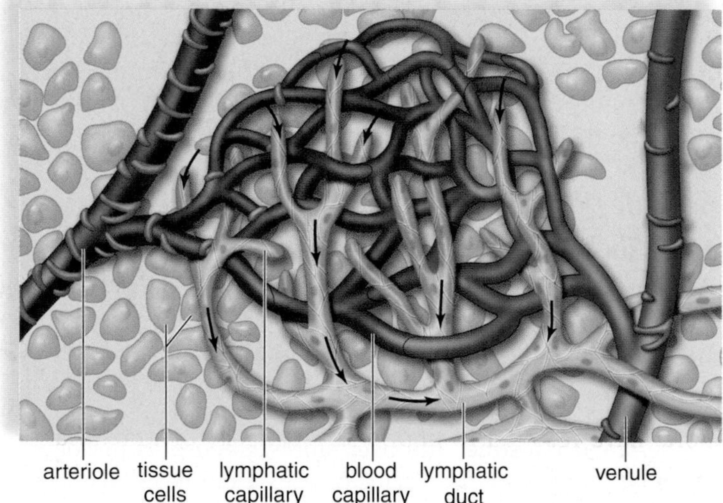

Figure 5.13 Interaction of lymphatic and capillary beds.
Lymphatic capillary beds lie alongside blood capillary beds. When lymphatic capillaries take up excess tissue fluid, it becomes lymph. Lymph returns to the cardiovascular system through cardiovascular veins in the chest. Precapillary sphincters can shut down a blood capillary, and blood then flows through the shunt.

At the venule end of a capillary, blood pressure has fallen even more. Osmotic pressure is greater than blood pressure, and fluid tends to move back. Almost the same amount of fluid that left the capillary returns to it, although some excess tissue fluid is always collected by the lymphatic capillaries (Fig. 5.13). Tissue fluid contained within lymphatic vessels is called **lymph.** Lymph is returned to the systemic venous blood when the major lymphatic vessels enter the subclavian veins in the shoulder region.

Animation Fluid Exchange Across the Walls of Capillaries

CHECK YOUR PROGRESS 5.6

1. Explain what happens to the excess fluid created during capillary exchange.
2. Describe the exchange of materials across the walls of a capillary.
3. Summarize what occurs when blood and osmotic pressure change at the venous end of a capillary.

CONNECTING THE CONCEPTS

The movement of materials in and out of the capillaries plays an important role in the function of a number of other body systems. For examples of the importance of capillaries, refer to the following discussions:

Section 7.2 provides additional information on the role of the lymphatic system.

Section 9.6 explains how gas exchange occurs across the walls of the capillaries in the lungs.

Section 10.3 explores how the urinary system establishes gradients to move waste materials out of the capillaries for excretion.

5.7 Cardiovascular Disorders

LEARNING OUTCOMES

Upon completion of this chapter, you should be able to

1. Explain the underlying causes of cardiovascular disease in humans.
2. Summarize how advances in medicine can treat cardiovascular disorders.

Cardiovascular disease (CVD) is the leading cause of early death in the Western countries. In the United States, CVD is responsible for more than 33% of all deaths. Modern research efforts have resulted in improved diagnosis, treatment, and prevention. This section discusses the range of advances that have been made, first in correcting vascular disorders and then in correcting heart disorders. The Bioethical feature, "Cardiovascular Disease Prevention: Who Pays for an Unhealthy Lifestyle," on page 106 includes risk factors that one should avoid to possibly prevent CVD from developing in the first place, while the Health feature, "New Information About Preventing Cardiovascular Disease," on page 107 examines new evidence on the role of other factors, such as omega-3 fatty acids, in reducing CVD risk.

Disorders of the Blood Vessels

Hypertension and atherosclerosis often lead to stroke or heart attack, due to an artery blocked by a blood clot or clogged by plaque. Treatment involves removing the blood clot or prying open the affected artery. Another possible outcome is an **aneurysm,** a burst blood vessel. An aneurysm can be prevented by replacing a blood vessel that is about to rupture with an artificial one.

High Blood Pressure

Hypertension occurs when blood moves through the arteries at a higher pressure than normal. Also known as high blood pressure, hypertension is sometimes called a silent killer. It may not be detected until it has caused a heart attack, stroke, or even kidney failure. Hypertension is present when the systolic blood pressure is 140 or greater or the diastolic blood pressure is 90 or greater (see Table 5.1). Though systolic and diastolic pressures are both important, diastolic pressure is emphasized when medical treatment is being considered.

The best safeguard against developing hypertension is regular blood pressure checks and a lifestyle that lowers the risk of CVD. If hypertension is present, prescription drugs can help lower blood pressure. Diuretics cause the kidneys to excrete more urine, ridding the body of excess fluid. In addition, hormones (the body's chemical messengers) that raise blood pressure can be inactivated. Drugs called beta-blockers and ACE (angiotensin-converting enzyme) inhibitors help to control hypertension caused by hormones.

Hypertension is often seen in individuals who have **atherosclerosis.** Atherosclerosis is caused by formation of lesions, or *atherosclerotic plaques,* on the inside of blood vessels. The **plaques** narrow blood vessel diameter, choking off blood and oxygen supply to the tissues (see Fig. 5B). In most instances, atherosclerosis begins in early adulthood and develops progressively through middle age, but symptoms may not appear until an individual is 50 or older. To prevent the onset and development of atherosclerosis, the American Heart Association and other organizations recommend a diet low in saturated fat and cholesterol but rich in omega-3 polyunsaturated fatty acids.

Atherosclerotic plaques can cause a clot to form on the irregular, roughened arterial wall. As long as the clot remains stationary, it is called a thrombus. If the thrombus dislodges and moves along with the blood, it is called an embolus. A **thromboembolism** consists of a clot first carried in the bloodstream that then becomes completely stationary when it lodges in a small blood vessel. If a thromboembolism is not treated, the life-threatening complications described in the next section can result.

Research has suggested several possible causes for atherosclerosis aside from hypertension. Chief among these are smoking and a diet rich in lipids and cholesterol. Research also indicates that a low-level bacterial or viral infection that spreads to the blood may cause an injury that starts the process of atherosclerosis. Surprisingly, such an infection may originate with gum diseases or be due to *Helicobacter pylori* (see Chapter 1). People who have high levels of C-reactive protein, which occurs in the blood following a cold or injury, are more likely to have a heart attack.

Stroke, Heart Attack, and Aneurysm

Stroke, heart attack, and aneurysm are associated with hypertension and atherosclerosis. A cerebrovascular accident (CVA), also called a **stroke,** often results when a small cranial arteriole bursts or is blocked by an embolus. Lack of oxygen

BIOLOGY MATTERS  **Bioethical**

Cardiovascular Disease Prevention: Who Pays for an Unhealthy Lifestyle?

Cardiovascular disease (CVD) is not only the number one killer in the United States today; it is also one of the most expensive. More than $444 billion is spent annually for health care and lost productivity, according to the American Heart Association. Family history of heart attack under age 55, male gender, and ethnicity (people of African-American descent are at great risk) are unalterable risk factors for CVD. However, most cases of CVD are preventable (Fig. 5A). Preventable risk factors include:

- Use of tobacco: Nicotine constricts arterioles, increasing blood pressure. As a result, the heart must pump harder to propel blood. Carbon monoxide in smoke decreases the blood's oxygen-carrying ability. (Tobacco use also causes many different cancers, as will be discussed in Chapter 19.)
- Drug and alcohol abuse: Stimulants, like cocaine and amphetamines, can cause irregular heartbeat and lead to heart attacks. Intravenous (IV) drug use may cause cerebral blood clots and stroke. Alcohol abuse can destroy body organs, including the heart. (However, low to moderate alcohol consumption actually decreases CVD risk.)
- Obesity and a sedentary lifestyle: Extra body tissue requires an additional blood supply, increasing the heart's workload. Hypertension develops as the heart works harder to pump blood. Obesity also increases the risk of type 2 diabetes. Diabetes causes blood vessel damage and atherosclerosis. Lack of exercise contributes to obesity.
- Poor diet: A diet high in saturated fats and cholesterol is a risk for CVD. Cholesterol is ferried in the blood by two proteins: LDL (low-density lipoprotein, the "bad" lipoprotein) and HDL (high-density lipoprotein, the "good" lipoprotein). LDL carries its cholesterol to deposit in tissues, but HDL carries cholesterol to the liver where it can be metabolized. Elevated blood LDL and/or low blood HDL levels can contribute to cardiovascular disease. A diet low in saturated fat and cholesterol can help to restore LDL and HDL to recommended levels. Drugs called statins can further lower the LDL level if necessary.
- Stress: A stress-filled lifestyle contributes directly to CVD and may also cause a person to overeat, avoid exercise, start or increase smoking, or abuse drugs and alcohol.
- Poor dental hygiene: Avoiding the dentist or dental clinic results in gum inflammation. Microbes from infected gums can enter the bloodstream and trigger formation of atherosclerotic plaques.

With few exceptions, these risk factors are personal choices.

Figure 5A **Risk factors for CVD that are under an individual's control.**
Unhealthy lifestyle choices—smoking, drinking alcohol to excess, obesity, and a sedentary lifestyle—are factors that a person has the power to change.

Who Pays for Treatment?

People who practice risky behaviors already pay more: for example, they pay extra for health or life insurance. Tobacco and alcohol taxes also help to defray health-care costs. However, insurance companies may not cover pre-existing conditions, including CVD. And what about the uninsured? In these cases, treatment costs are borne by taxpayers. Ultimately, we all pay for CVD treatment—and we all have an incentive to ensure that individuals adopt healthy lifestyles—or pay more for treatment if they refuse.

Is Legislation Needed?

Several organizations, such as the World Health Organization, have advocated legislation to help pay for CVD and its consequences. One proposed example is the so-called "fat tax" on foods with high fat and/or poor nutritional value. The tax would pay for treatment of obesity-related disease, including CVD. Other initiatives stress education of both adults and children regarding healthy lifestyle choices. However, critics oppose this type of legislation, insisting that rewarding healthy behaviors and penalizing risky behaviors are beyond the scope of government.

Questions to Consider

1. Would you support charging higher insurance premiums or taxes to people who don't practice a healthy lifestyle?
2. Should public funding be used for prevention programs if money could be saved in the future?
3. The public does subsidize health care that disproportionately benefits the less healthy, so do you believe that financial interest trumps personal freedom in this matter? Why or why not?

BIOLOGY MATTERS **Health**

New Information About Preventing Cardiovascular Disease

For decades, several factors have been associated with an increased risk of cardiovascular disease (CVD), especially atherosclerosis (Fig. 5B). Some of these cannot be avoided, such as increasing age, male gender, family history of heart disease, and belonging to certain races, including African American, Mexican American, and Native American. Other risk factors—smoking, obesity, high cholesterol, hypertension, diabetes, physical inactivity—can be avoided or at least affected by changing one's behavior or taking medications. In recent years, however, other factors have been under consideration; we discuss a few of them here.

- **Alcohol** Alcohol abuse can destroy just about every organ in the body, the heart included. But recent research suggests that a moderate level of alcohol intake can improve cardiovascular health by improving the blood cholesterol profile, decreasing unwanted clot formation, increasing blood flow in the heart, and reducing blood pressure. According to the American Heart Association, people who consume one or two drinks per day have a 30–50% reduction in cardiovascular disease compared to nondrinkers.

  Please note, however, that the maximum protective effect is achieved with only one or two drinks per day—consuming more than that increases the risk of many alcohol-related problems. The American Heart Association does not recommend that nondrinkers start using alcohol, nor that drinkers increase their consumption, based on these findings.
- **Resveratrol** The "red wine effect" or "French paradox" refers to the observation that levels of CVD in France are relatively low, despite the consumption of a high-fat diet. One possible explanation is that wine is frequently consumed with meals.

  In addition to its alcohol content, red wine contains especially high levels of antioxidants, including resveratrol. Resveratrol is mainly produced in the skin of grapes, so it is also found in grape juice. Resveratrol supplements are also available at health food stores.

  The benefits of resveratrol alone are questionable, however, and most controlled studies to date have demonstrated no beneficial effects. The lower incidence of CVD in the French may be due to multiple factors, including lifestyle and genetic differences.
- **Omega-3 Fatty Acids** The influence that diet has on blood cholesterol levels has been well studied. It is generally beneficial to minimize our intake of foods high in saturated fat (red meat, cream, and butter) and trans fats (most margarines, commercially baked goods, and deep-fried foods). Replacing these harmful fats with healthier ones, such as monounsaturated fats (olive and canola oils) and polyunsaturated fats (corn, safflower, and soybean oils), is beneficial.

  In addition, the American Heart Association now recommends eating at least two servings of fish a week, especially fish like salmon, mackerel, herring, lake trout, sardines, and albacore tuna, which are high in omega-3 fatty acids. These essential fatty acids can decrease triglyceride levels, slow the growth rate of atherosclerotic plaque, and lower blood pressure.

  However, children and pregnant women are advised to limit their fish consumption because of the high levels of mercury contamination in some fishes. For middle-aged and older men and postmenopausal women, the benefits of fish consumption far outweigh the potential risks.

Questions to Consider

1. Would you be helping your health if you decided to eat mackerel every day and drink two glasses of red wine with it? Why or why not?
2. What would be some difficulties in trying to determine the true cause of the "French paradox"?

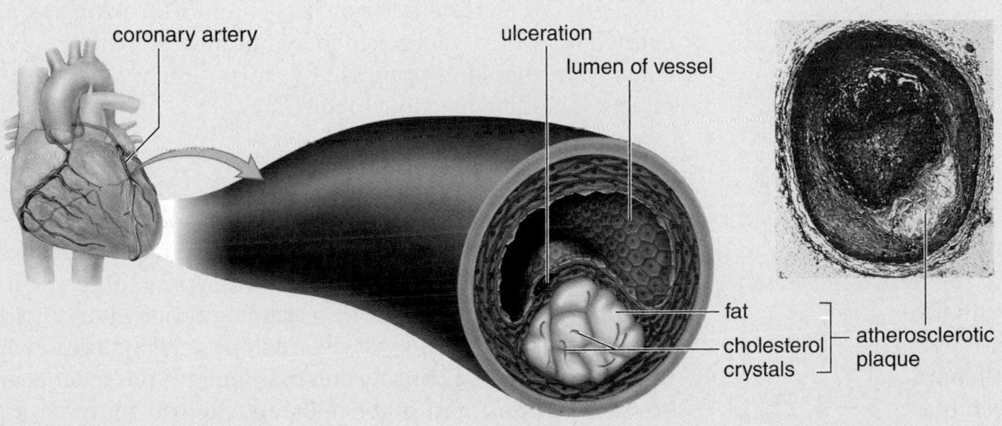

Figure 5B Coronary arteries and plaque. Atherosclerotic plaque is an irregular accumulation of cholesterol and fat. When fat is present in a coronary artery, a heart attack is more likely to occur because of restricted blood flow.

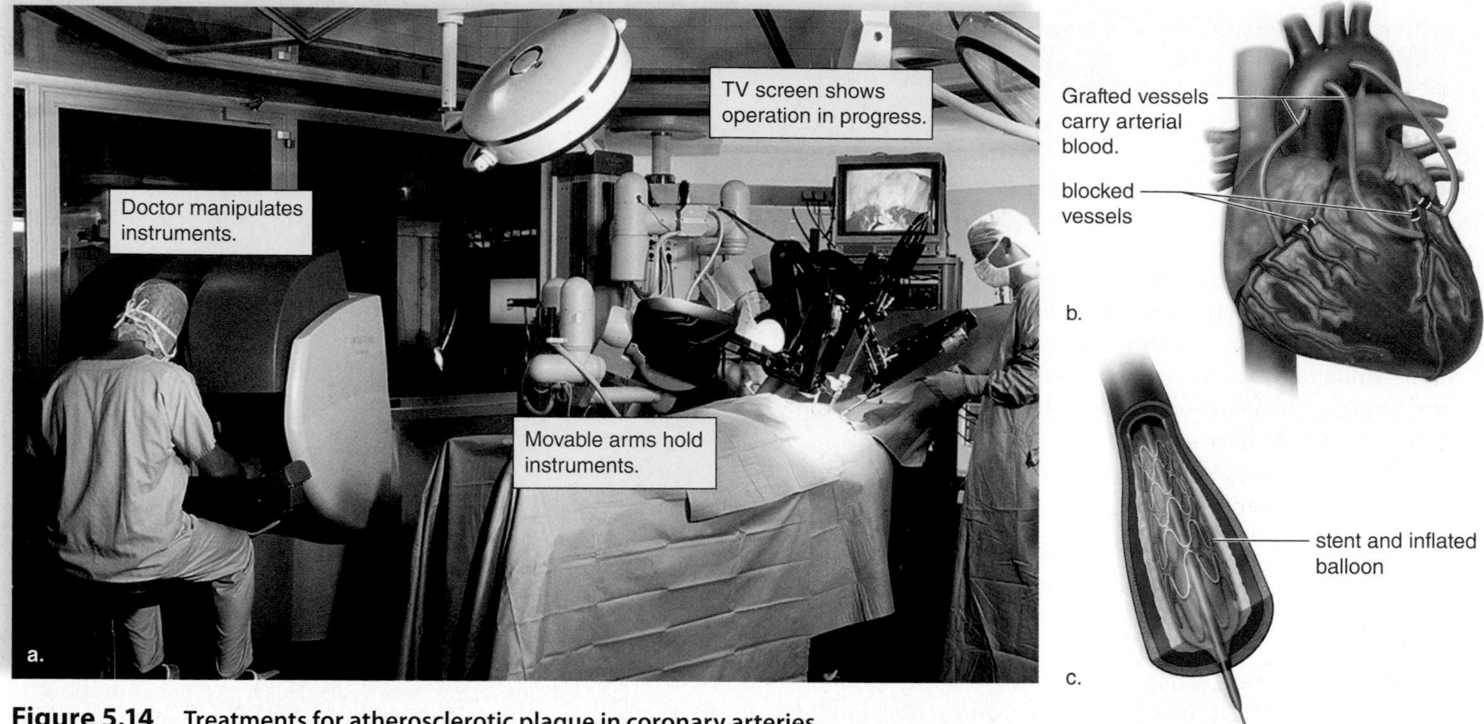

Figure 5.14 **Treatments for atherosclerotic plaque in coronary arteries.**
a. Robotic surgery techniques can assist in coronary bypass operations. **b.** In this procedure, blood vessels (leg veins or arteries placed in the chest) are stitched to coronary arteries, bypassing the obstruction. **c.** In a stent procedure, a balloon-tipped catheter guides the stent (a cylinder of expandable metal mesh) to the obstructed area. When the balloon is inflated, the stent expands and opens the artery.

causes a portion of the brain to die, and paralysis or death can result. A person is sometimes forewarned of a stroke by a feeling of numbness in the hands or the face, difficulty in speaking, or temporary blindness in one eye.

A myocardial infarction (MI), also called a **heart attack,** occurs when a portion of the heart muscle dies due to lack of oxygen. If a coronary artery becomes partially blocked, the individual may then suffer from **angina pectoris.** Characteristic symptoms of angina pectoris include a feeling of pressure, squeezing, or pain in the chest. Pressure and pain can extend to the left arm, neck, jaw, shoulder, or back. Nausea and vomiting, anxiety, dizziness, and shortness of breath may accompany the chest discomfort. Nitroglycerin or related drugs dilate blood vessels and help relieve the pain. When a coronary artery is completely blocked, perhaps because of a thromboembolism, a heart attack occurs.

An aneurysm is a ballooning of a blood vessel, most often the abdominal artery or the arteries leading to the brain. Atherosclerosis and hypertension can weaken the wall of an artery to the point that an aneurysm develops. If a major vessel such as the aorta bursts, death is likely. It is possible to replace a damaged or diseased portion of a vessel, such as an artery, with a plastic tube. Cardiovascular function is preserved because exchange with tissue cells can still take place at the capillaries. In the future, it may be possible to use vessels made in the laboratory by injecting a patient's cells inside an inert mold.

Dissolving Blood Clots

Medical treatment for a thromboembolism includes the use of tissue plasminogen activator (t-PA), a biotechnology drug that converts plasminogen, a protein found in blood, into plasmin, an enzyme that dissolves blood clots. Tissue plasminogen activator is also being used for stroke patients, but with less success. Some patients experience life-threatening bleeding in the brain. A better treatment might be new biotechnology drugs that act on the plasma membrane to prevent brain cells from releasing and/or receiving toxic chemicals caused by the stroke.

If a person has symptoms of angina or a stroke, aspirin may be prescribed. Aspirin lowers the probability of clot formation. There is evidence that aspirin protects against first heart attacks, but there is no clear support for taking aspirin every day to prevent strokes in symptom-free people. Physicians warn that long-term use of aspirin might have harmful effects, including bleeding in the brain.

Treating Clogged Arteries

Cardiovascular disease used to require open-heart surgery and, therefore, a long recuperation time and a long unsightly scar that could occasionally ache. Now, bypass surgery can be accomplished by using robotic technology (Fig. 5.14*a*). A video camera and instruments are inserted through small cuts, while the surgeon sits at a console and manipulates interchangeable grippers, cutters, and other tools attached to movable arms above the operating table. Looking through two eyepieces, the surgeon gets a three-dimensional view of the operating field. Robotic surgery is also used in valve repairs and other heart procedures.

A *coronary bypass operation* is one way to treat an artery clogged with plaque. A surgeon takes a blood vessel—usually a vein from the leg—and stitches one end to the aorta and the other end to a coronary artery located past the point of obstruction. Figure 5.14*b* shows a triple bypass in which three blood vessel segments have been used to allow blood to flow freely from the aorta to cardiac muscle by way of the coronary artery. Since 1997, gene therapy has been used instead of a coronary bypass to grow new blood vessels that will carry blood to cardiac muscle. The surgeon needs only make a small incision and inject many copies of the gene that codes for vascular endothelial growth factor (VEGF) between the ribs directly into the area of the heart that most needs improved blood flow. The VEGF encourages new blood vessels to sprout out of an artery. If collateral blood vessels do form, they transport blood past clogged arteries, making bypass surgery unnecessary. About 60% of patients who undergo the procedure do show signs of vessel growth within two to four weeks.

Another alternative to bypass surgery is available, namely, the stent (Fig. 5.14*c*). A stent is a small metal mesh cylinder that holds a coronary artery open after a blockage has been cleared. Until a couple of years ago, stenting was a second step following angioplasty. During **angioplasty,** a plastic tube was inserted into an artery of an arm or leg and then guided through a major blood vessel toward the heart. When the tube reached the region of plaque in an artery, a balloon attached to the end of the tube was inflated, forcing the vessel open. Now, instead, a stent with an inner balloon is pushed into the blocked area. When the balloon inside the stent is inflated, it expands, locking itself in place. Some patients go home the same day or, at most, after an overnight stay. The stent is more successful when it is coated with a drug that seeps into the artery lining and discourages cell growth. Uncoated stents can close back up in a few months, but the drug used in the coated ones successfully discourages closure. Blood clotting might occur after stent placement, so recipients have to take anticlotting medications.

Heart Failure

When a person has *heart failure,* the heart no longer pumps as it should. Heart failure is a growing problem because people who used to die from heart attacks now survive but are left with damaged hearts. Often the heart is oversized, not because the cardiac wall is stronger but because it is sagging and swollen. One idea is to wrap the heart in a fabric sheath to prevent it from getting too big. This might allow better pumping, similar to the way a weight lifter's belt restricts and reinforces stomach muscles. But a failing heart can have other problems, such as an abnormal heart rhythm. To counter that condition, it's possible to place an implantable cardioverter-defibrillator (ICD) just beneath the skin of the chest. This device can sense both an abnormally slow and an abnormally fast heartbeat. If the former, the ICD generates the missing beat like a pacemaker does. If the latter, it sends the heart a sharp jolt of electricity to slow it down. If the heart rhythm becomes erratic, the ICD sends an even stronger shock—like a defibrillator does.

Heart Transplants

Although heart transplants are now generally successful, many more people are waiting for new hearts than there are organs available. Today, only about 2,200 heart transplants are done annually, though many thousands of people could use them. Genetically altered pigs may one day be used as a source of hearts because of the shortage of human hearts. Also, bone marrow stem cells have been injected into the heart, and researchers report that they apparently become cardiac muscle!

Today, a left ventricular assist device (LVAD), implanted in the abdomen, is an alternative to a heart transplant. A tube passes blood from the left ventricle to the device, which pumps it on to the aorta. A cable passes from the device through the skin to an external battery, which the patient must tote around. Soon, as many as 20,000 people a year may be outfitted with an LVAD. Instead of an LVAD, a few pioneers have volunteered to receive a Jarvik 2000, which is a pump inserted inside the left ventricle. The Jarvik is powered by an external battery no larger than a C battery.

Even fewer patients have received a device called a **total artificial heart** (**TAH**), such as that shown in Figure 5.15. An internal battery and controller regulate the pumping speed, and an external battery powers the device by passing electricity through the skin via external and internal coils. A rotating centrifugal pump moves silicon hydraulic fluid between left and right sacs to force blood out of the heart into the pulmonary trunk and the aorta. Initially, recipients were near death, and

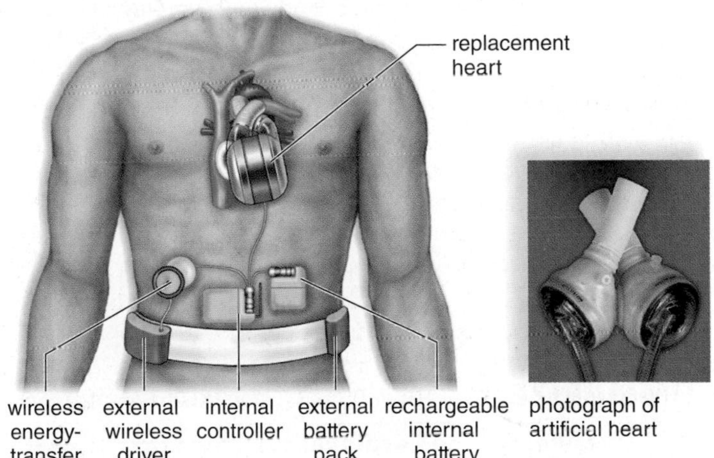

Figure 5.15 **An artificial heart.**
The CardioWest temporary total artificial heart moves blood in the same manner as a natural heart. The controller is implanted into the patient's abdomen. The heart is powered by an internal rechargeable battery, with an external battery as a backup system.
Courtesy of SynCardia Systems, Inc.

most survived for only a few days. However, medical advances have increased the survival time to months for many recipients, making the TAH a temporary treatment for people who are experiencing heart failure. To date, over 1,000 TAHs have been implanted into patients.

CHECK YOUR PROGRESS 5.7

1. Summarize the cardiovascular disorders that are common in humans.
2. Summarize the treatments that are available for cardiovascular disorders.
3. Discuss why CVD is the leading source of death in Western countries.

CONNECTING THE CONCEPTS

Many cardiovascular disorders are caused by incorrect diet or a sedentary lifestyle. Others are the result of genetic abnormalities. For more on the relationships between diet and cardiovascular help and how scientists identify genes associated with cardiovascular disease, refer to the following discussions:

Section 8.6 explores how each class of nutrient relates to a healthy lifestyle.

Table 8.4 lists methods of lowering saturated fats and cholesterol in the diet.

Sections 21.3 and 21.4 explain how biotechnology and genomics are identifying new genes associated with disease.

CASE STUDY CONCLUSION

Following his exam, Steven spent time researching the causes and effects of hypertension on his body. He learned what a blood pressure value of 132/84 mm Hg really meant. The larger number (132) represented the maximum pressure as his heart was beating, while the smaller number (84) was the resting pressure. Ideally, his blood pressure should be less than 120/80 mm Hg, so as his doctor indicated, Steven was prehypertensive. However, the good news was that at his age, prehypertension was largely due to lifestyle choices, namely diet and physical activity. Therefore, Steven resolved to watch his diet more closely, especially the amount of saturated fat and cholesterol, and increase his cardiovascular workouts per week. His doctor also recommended a follow-up visit in a few months to ensure that Steven was making satisfactory progress.

MEDIA STUDY TOOLS

Enhance your study of this chapter with media! Visit **www.mhhe.com/maderhuman13e** and go to "Media Study Tools" for this chapter to access the following:

 **Animations**	 **Videos**	 **MP3 Files**
5.3 Cardiac Cycle • Conducting System of the Heart • Baroreceptor Reflex Control of Blood Pressure **5.6** Fluid Exchange Across the Walls of Capillaries	**5.3** Heart Stem Cells **5.7** Cardiac Repair • Heart Stem Cells	**5.2** Classification of Blood Vessels **5.3** Cardiac Cycle • Cardiac Conduction System **5.4** Blood Flow and Blood Pressure

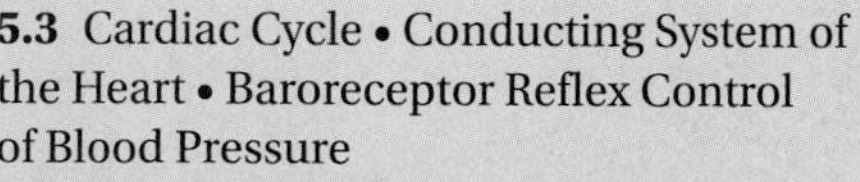

SUMMARIZE

5.1 Overview of the Cardiovascular System

The **cardiovascular system** consists of the heart and blood vessels. The heart pumps blood, and blood vessels take blood to and from capillaries, where exchanges of nutrients for wastes occur with tissue cells. Blood is refreshed at the lungs, where gas exchange occurs; at the digestive tract, where nutrients enter the blood; and at the kidney, where wastes are removed from blood.

The **lymphatic system** removes excess fluid from around the tissue and returns it to the cardiovascular system.

5.2 The Types of Blood Vessels

Arteries: **Arteries** and **arterioles** move blood away from the heart. Arteries have the thickest walls, which allows them to withstand blood pressure.

Capillaries: Exchange of substances occurs in the capillaries. **Precapillary sphincters** and **arteriovenous shunts** help control the flow of blood within the capillaries.

Veins: **Veins** and **venules** move blood toward the heart. Veins have relatively weak walls with valves that keep the blood flowing in one direction.

5.3 The Heart Is a Double Pump

The **heart** is the pump of the circulatory system and consists of a right and left side separated by a **septum.** Each side has an **atrium** and a **ventricle.** Valves, such as the **atrioventricular (AV) valves** and **semilunar valves,** keep the blood moving in the correct direction. The tissue of the heart, the **myocardium,** is contained within a sac called the **pericardium.** At the cellular level, the tissues interact using **gap junctions** and **desmosomes.** The heart supplies itself with blood using the **coronary arteries.**

Passage of Blood Through the Heart

- The right atrium receives O_2-poor blood from the body, and the right ventricle pumps it into the pulmonary circuit (**pulmonary arteries**).
- The left atrium receives O_2-rich blood from the lungs (**pulmonary veins**), and the left ventricle pumps it into the systemic circuit.

The Heartbeat Is Controlled

During the **cardiac cycle,** the **SA (sinoatrial) node** (or **pacemaker**) initiates the heartbeat by causing the atria to contract. The **AV (atrioventricular) node** conveys the stimulus to the ventricles, causing them to contract (**systole**). The heart sounds, "lub-dub," are due to the closing of the atrioventricular valves, followed by the closing of the semilunar valves. The muscles of the heart relax (**diastole**) between contractions. An **electrocardiogram (ECG)** may be used to measure the activity of the heart.

5.4 Features of the Cardiovascular System

The **pulse** indicates the heartbeat rate. **Blood pressure** caused by the beating of the heart accounts for the flow of blood in the arteries. The **systolic pressure** is the maximum, while the **diastolic pressure** is the minimum. The reduced velocity of blood flow in capillaries facilitates exchange of nutrients and wastes in the tissues. Blood flow in veins is caused by skeletal muscle contraction (skeletal muscle pump), the presence of valves, and respiratory movements (respiratory pump).

5.5 Two Cardiovascular Pathways

The cardiovascular system is divided into the pulmonary circuit and the systemic circuit.

The Pulmonary Circuit: Exchange of Gases

In the **pulmonary circuit,** blood travels to and from the lungs.

The Systemic Circuit: Exchanges with Tissue Fluid

In the **systemic circuit,** the **aorta** divides into blood vessels that serve the body's organs and cells. The **vena cava** returns O_2-poor blood to the heart.

The hepatal portal system moves blood between the capillary beds of the digestive system and liver. This system consists of the **hepatic portal vein** and the **hepatic vein.**

5.6 Exchange at the Capillaries

This diagram illustrates capillary exchange in tissues of the body—not including the gas-exchanging surfaces of the lungs.

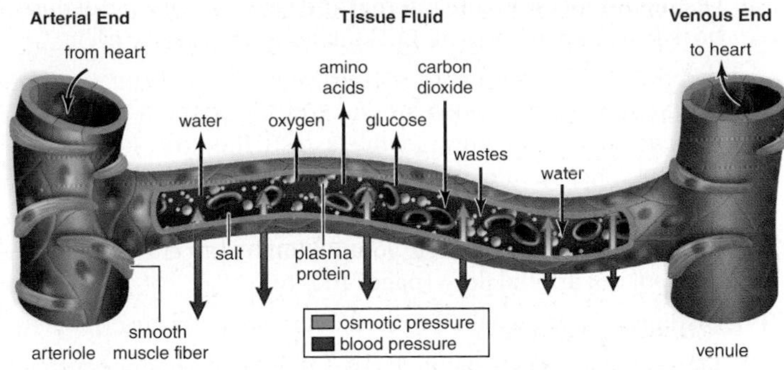

- At the arterial end of a cardiovascular capillary, blood pressure is greater than osmotic pressure; therefore, fluid leaves the capillary.
- In the midsection, oxygen and nutrients diffuse out of the capillary, and carbon dioxide and other wastes diffuse into the capillary.
- At the venous end, osmotic pressure created by the presence of proteins exceeds blood pressure, causing most of the fluid to reenter the capillary. Some fluid remains as interstitial (tissue) fluid.

Excess fluid not picked up at the venous end of the cardiovascular capillary enters the lymphatic capillaries.

- **Lymph** is tissue fluid contained within lymphatic vessels.
- The lymphatic system is a one-way system. Its fluid is returned to blood by way of a cardiovascular vein.

5.7 Cardiovascular Disorders

Cardiovascular disease is the leading cause of death in the Western countries.

- **Hypertension** and **atherosclerosis** can lead to **stroke, heart attack, angina pectoris** (chest pain), or an **aneurysm.** Atherosclerotic **plaques** increase the risk of these conditions. If these plaques dislodge in the circulatory system, a **thromboembolism** may result.
- Following a heart-healthy diet, getting regular exercise, maintaining a proper weight, and not smoking reduce cardiovascular disease risk.

ASSESS

Testing Your Knowledge of the Concepts

1. What are the two parts of the cardiovascular system, and what are the functions of each part? (page 92)
2. Explain where exchanges occur in the body and the importance of those exchanges. (page 92)
3. Which of the three types of blood vessels are most numerous? Explain. (page 93)
4. Describe the structure of the heart, including the chambers and valves. (pages 94–97)
5. What is the function of the septum in the heart? What would happen if the heart had no septum? (page 94)
6. Trace the path of blood through the heart, including chambers, valves, and vessels the blood travels through. (pages 96–97)
7. Describe the cardiac cycle, using the terms *systole* and *diastole.* What are the roles of the SA node and the AV node in the cardiac cycle? (pages 97–98)

8. Distinguish between the internal and external controls of the heartbeat. Explain how an ECG relates to the cardiac cycle. (pages 98–99)
9. In what vessel is the blood pressure the highest? The lowest? In what vessel is blood flow rate the fastest? The slowest? Why are the pressure and rate in the capillaries important to capillary function? (pages 99–100)
10. Explain why skeletal muscle contraction has an effect on venous flow but not arterial flow. (pages 100–101)
11. Distinguish between the two cardiovascular pathways. (page 102)
12. Trace the pathway of blood to and from the brain in the systemic circuit. (page 103)
13. Describe the process by which nutrients are exchanged for wastes across a capillary, using glucose and carbon dioxide as examples. (pages 104–105)
14. What is the most probable association between high blood pressure and a heart attack? With this association in mind, what type of diet might help prevent a heart attack? (page 105)

In questions 15–19, match the descriptions to the blood vessel in the key. Answers may be used more than once.

Key:

a. venules
b. veins
c. capillaries
d. arteries
e. arterioles

15. Drain blood from capillaries
16. Empty into capillaries
17. May contain valves
18. Move blood away from the heart
19. Sites for exchange of substances between blood and tissue fluid
20. When the atria contract, the blood flows
    a. into the attached blood vessels.
    b. into the ventricles.
    c. through the atrioventricular valves.
    d. to the lungs.
    e. Both b and c are correct.
21. Which of these associations is mismatched?
    a. left ventricle—aorta
    b. right ventricle—pulmonary trunk
    c. right atrium—vena cava
    d. left atrium—pulmonary artery
    e. Both b and c are incorrectly matched.
22. Which statement is not correct concerning the heartbeat?
    a. The atria contract at the same time.
    b. The ventricles relax at the same time.
    c. The AV valves open at the same time.
    d. The semilunar valves open at the same time.
    e. First the right side contracts; then the left side contracts.
23. Accumulation of plaque in an artery wall is
    a. an aneurysm.
    b. angina pectoris.
    c. atherosclerosis.
    d. hypertension.
    e. a thromboembolism.
24. Label the following diagram of the cardiovascular system using the alphabetized list below the figure:

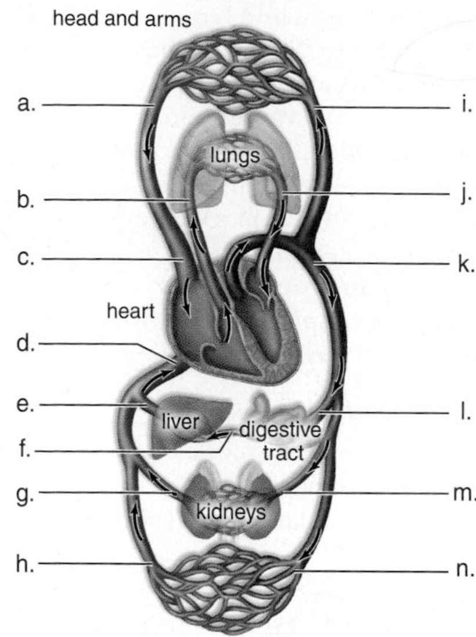

aorta	hepatic portal vein
carotid artery	hepatic vein
iliac artery	pulmonary artery
iliac vein	pulmonary vein
inferior vena cava	renal artery
jugular vein	renal vein
interstitial arteries	superior vena cava

ENGAGE

The virtual lab "Blood Pressure" provides an interactive look at how factors such as age and gender influence the risk of hypertension.

Thinking Critically About the Concepts

1. You have to stand in front of the class to give a report. You are nervous, and your heart is pounding. What is the specific mechanism behind this reaction? How would your ECG appear?
2. The cardiovascular system is an elegant example of the concept that structure supports function. Each type of blood vessel has a specific job. Each vessel's physical characteristics enable it to do that job. The muscle walls of the right and left ventricle vary in thickness depending on where they pump the blood. When organ structure is damaged or changed (as arteries are in atherosclerosis), the organ's ability to perform its function may be compromised. Homeostatic conditions, such as blood pressure, may be affected as well. Dietary and lifestyle choices can either prevent damage or harm the cardiovascular system.
    a. What do you think the long-term effects of hypertension would be on the heart? What about other organ systems?
    b. Why do you think a combination of hypertension and atherosclerosis is a particularly dangerous combination?
    c. What factors in your life could be changed to reduce the risk of CVD? Be critical!

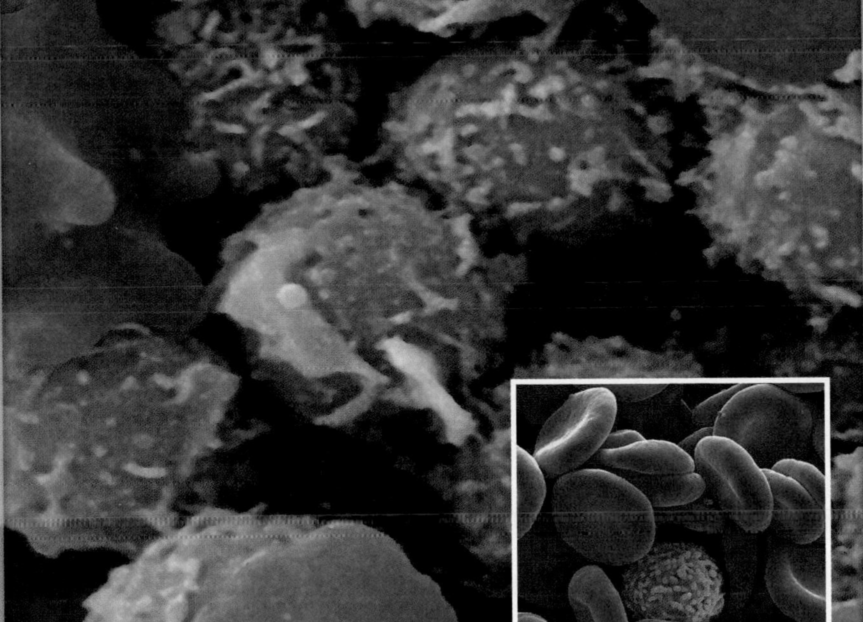

CHAPTER

6

Cardiovascular System: Blood

CASE STUDY CANCER OF THE BLOOD—LEUKEMIA

Ben was a 20-year-old placekicker on the football team at the university. He had been active all of his life, from playing baseball and football as a teenager to participating in intense intramural activities in college. He considered himself to be in fine health, and he worked hard to maintain his position on the team.

At the start of his second season, Ben noticed that he did not have as much energy as he was accustomed to. Practices seemed to take a lot more out of him than they normally did. He was frequently tired and often went to bed right after his practice. During the football drills, he often experienced a severe shortness of breath. Rather than risk his place on the team, Ben tried to cover up these symptoms; but over the next few weeks, his health worsened and he noticed swelling in his neck, armpits, and groin. In addition, bruises that he received in practice seemed to be taking a lot longer to heal. Ben became concerned, and when he approached his trainers, they immediately sent him to the team physician.

After completing a physical exam, Ben's physician ordered a complete blood count, or CBC test. The CBC test determines the total quantity of red and white blood cells, the amount of hemoglobin in the blood, and the number of platelets. The results of these tests indicated that Ben had a very high white blood cell count, but that these cells were not functioning correctly. The team physician immediately referred Ben to a specialist. A bone marrow biopsy confirmed what the doctors suspected; Ben had a type of cancer called acute lymphocytic leukemia (ALL).

As you read through the chapter, think about the following questions:

1. What are the roles of white blood cells in the body?
2. Why would the doctors order a bone marrow biopsy if they suspected a blood disorder?
3. Why is leukemia considered to be a form of cancer?

CHAPTER CONCEPTS

6.1 Blood: An Overview
Blood, a liquid connective tissue, is a transport medium with a wide variety of functions in the body.

6.2 Red Blood Cells and Transport of Oxygen
Red blood cells contain hemoglobin and are involved in the transport of gases in the body.

6.3 White Blood Cells and Defense Against Disease
A variety of white blood cells, or leukocytes, exists. Collectively, these cells help the body fight infection.

6.4 Platelets and Blood Clotting
Platelets are cell fragments that maintain homeostasis by repairing breaks in blood vessels and preventing loss of blood.

6.5 Blood Typing and Transfusions
Blood typing is based on specific antigens found on the surface of red blood cells.

6.6 Homeostasis
Blood and the cardiovascular system play an important role in the maintenance of homeostasis in the body.

BEFORE YOU BEGIN

Before beginning this chapter, take a few moments to review the following discussions:

Section 4.2 Why is blood considered to be a connective tissue?

Section 4.8 How does the circulatory system contribute to homeostasis?

Section 5.6 How does the blood exchange nutrients and wastes with the tissues of the body?

6.1 Blood: An Overview

LEARNING OUTCOMES

Upon completion of this section, you should be able to

1. List the functions of blood in the human body.
2. Compare the composition of formed elements and plasma in the blood.

In Chapter 5, we learned that the cardiovascular system consists of the heart, which pumps the blood, and the blood vessels, which conduct blood around the body. In this chapter, we will learn about the functions and composition of blood.

Functions of Blood

The human heart is an amazing muscular pump. With each beat, the human heart pumps approximately 75 ml of blood. On average, the heart beats 70 times per minute. Thus, the heart pumps roughly 5,250 ml per minute—75 ml/beat × 70 beats/minute—circulating the body's entire blood supply once each minute! If needed (when exercising, for example), the heart can cycle blood throughout the body even faster.

The functions of blood fall into three general categories: transport, defense, and regulation.

Transport. Blood is the primary transport medium. Blood acquires oxygen in the lungs and distributes it to tissue cells. Similarly, blood picks up nutrients from the digestive tract for delivery to the tissues. In its return trip to the lungs, blood transports carbon dioxide. Every time a person exhales, the carbon dioxide waste is eliminated. Blood also transports other wastes, such as excess nitrogen from the breakdown of proteins, to the kidneys for elimination. Blood exchanges nutrients and wastes with tissues by capillary exchange (see Fig. 5.12). In doing so, it maintains homeostasis by keeping the composition of tissue fluid within normal limits.

In addition to the transport of nutrients and wastes, various organs and tissues secrete hormones into the blood. Blood transports these to other organs and tissues, where they serve as signals that influence cellular metabolism. Blood is well suited for its role in transporting substances. Proteins in the blood help transfer hormones to the tissues. Special proteins called lipoproteins (also known as HDL and LDL) carry lipids, or fats, throughout the body. Most important, hemoglobin (found in red blood cells) is specialized to combine with oxygen and deliver it to cells. Hemoglobin also assists in transferring waste carbon dioxide back to the lungs.

Carbon monoxide (chemical formula CO) is a colorless, odorless gas, which can also bind to hemoglobin. Each year, hundreds of people die accidentally from CO poisoning, caused by malfunction of a fuel-burning appliance or by improper venting of CO fumes. When CO is present, it takes the place of oxygen (O_2) in hemoglobin. As a result, cells are starved of oxygen. Tragically, treatment may be delayed, because symptoms of poisoning—headache, body ache, nausea, dizziness, and drowsiness—can be mistaken for the "flu." Government guidelines recommend that *any* equipment producing CO be checked regularly to ensure proper ventilation. Finally, CO detectors, like smoke detectors, should be installed and used properly.

Defense. Blood defends the body against pathogen invasion and blood loss. Certain blood cells are capable of engulfing and destroying pathogens by a process called *phagocytosis* (see Chapter 3).

Other white blood cells produce and secrete antibodies into the blood. An **antibody** is a protein that combines with and disables specific pathogens. Disabled pathogens can then be destroyed by the phagocytic white blood cells.

When an injury occurs, blood clots and defends against blood loss. Blood clotting involves platelets (described in section 6.4) and proteins. For example, prothrombin and fibrinogen are two inactive blood proteins. They circulate constantly, ready to be activated to form a clot if needed. Without blood clotting, we could bleed to death even from a small cut.

Regulation. Blood plays an important role in regulating the body's homeostasis. Blood helps regulate body temperature by picking up heat, mostly from active muscles, and transporting it about the body. If the body becomes too warm, blood is transported to dilated blood vessels in the skin. Heat disperses to the environment, and the body cools to a normal temperature.

The liquid portion of blood, the plasma**,** contains dissolved salts and proteins. These solutes create blood's *osmotic pressure,* which keeps the liquid content of the blood high (see section 3.3, for a review of osmosis). In this way, blood plays a role in helping to maintain its own water-salt balance.

The chemical buffers, body chemicals that stabilize blood pH, in blood help regulate the body's acid-base balance and keep it at a relatively constant pH of 7.4.

Composition of Blood

Blood is a tissue, and, like any tissue, it contains cells and cell fragments (Fig. 6.1). Collectively, the cells and cell fragments are called the **formed elements.** The cell and cell fragments are suspended in a liquid called **plasma.** Therefore, blood is classified as a liquid connective tissue.

The Formed Elements

The formed elements are red blood cells, white blood cells, and platelets. These are produced in red bone marrow, which can be found in most bones of a child but only in certain bones of an adult. Red bone marrow contains *pluripotent stem cells,* the parent cells that divide and give rise to all of the various types of blood cells (Fig. 6.1). Stem cells are the focus of a tremendous amount of research in the medical community. Scientists are discovering that under the right conditions in the laboratory, stem cells can be "coaxed" into becoming a greater variety of cell types—cells that might have the potential to treat human

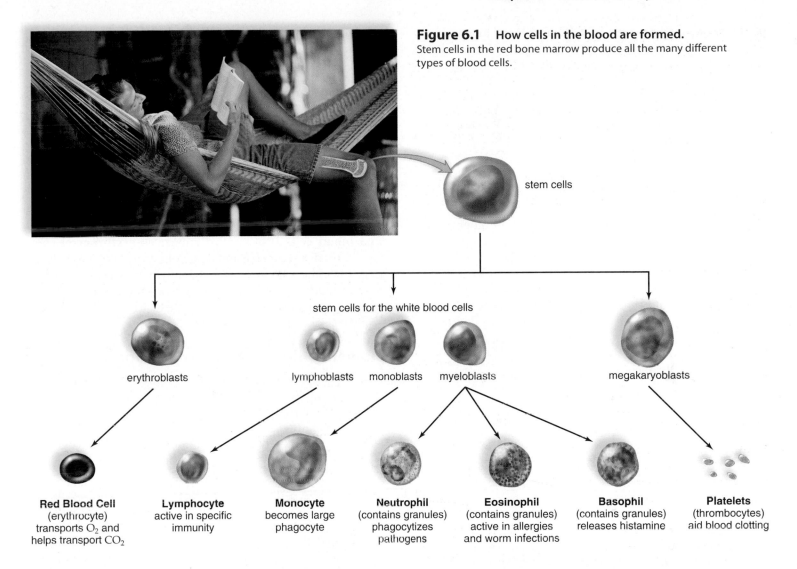

Figure 6.1 How cells in the blood are formed.
Stem cells in the red bone marrow produce all the many different types of blood cells.

stem cells

stem cells for the white blood cells

erythroblasts

lymphoblasts

monoblasts

myeloblasts

megakaryoblasts

Red Blood Cell (erythrocyte) transports O_2 and helps transport CO_2

Lymphocyte active in specific immunity

Monocyte becomes large phagocyte

Neutrophil (contains granules) phagocytizes pathogens

Eosinophil (contains granules) active in allergies and worm infections

Basophil (contains granules) releases histamine

Platelets (thrombocytes) aid blood clotting

diseases such as diabetes and Alzheimer disease, among many others (see Chapter 21 for a discussion of stem-cell research).

Red blood cells are two to three times smaller than white blood cells, but there are many more of them. There are millions of red blood cells and only thousands of white blood cells in a mm^3 of blood, about this size: ■

Plasma

Plasma is the liquid medium for carrying various substances in the blood. It also distributes the heat generated as a by-product of metabolism, particularly muscle contraction. About 91% of plasma is water (Fig. 6.2). The remaining 9% of plasma consists of various salts (ions) and organic molecules. Salts are dissolved in plasma. As mentioned previously, salts and plasma proteins maintain the osmotic pressure of blood. Salts also function as buffers that help maintain blood pH. Small organic molecules such as glucose and amino acids are nutrients for cells; urea is a nitrogenous waste product on its way to the kidneys for excretion.

The most abundant organic molecules in blood are called the **plasma proteins.** The liver produces the majority of the

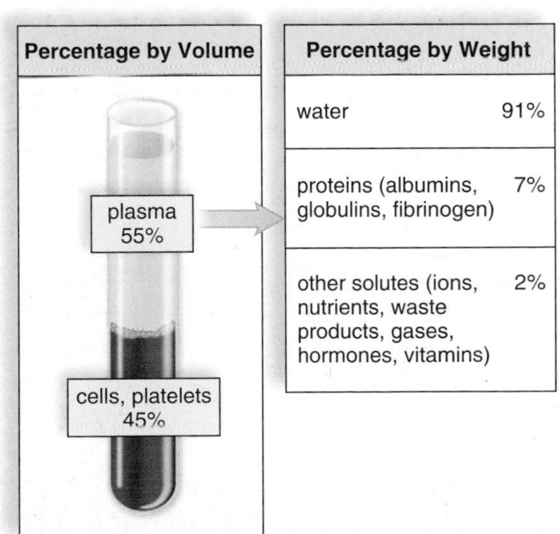

Figure 6.2 The composition of blood plasma.
Plasma, the liquid portion of blood, is mainly water and proteins. Solutes such as nutrients, vitamins, and hormones are transported in plasma.

plasma proteins. The plasma proteins have many functions that help maintain homeostasis. Like salts, they are able to take up and release hydrogen ions. Therefore, they help keep blood pH around 7.4. Plasma proteins are too large to pass through capillary walls. They remain in the blood, establishing an osmotic gradient between blood and tissue fluid. This **osmotic pressure** is a force that prevents excessive loss of plasma from the capillaries into tissue fluid.

Animation Fluid Exchange Across the Walls of Capillaries

Three major types of plasma proteins are the **albumins, globulins,** and **fibrinogen.** Albumins are the most abundant plasma proteins and contribute most to plasma's osmotic pressure. They also combine with and help transport other organic molecules. The globulins are of three types called alpha, beta, and gamma globulins. Alpha and beta globulins also combine with and help transport substances in the blood such as hormones, cholesterol, and iron. Gamma globulins are also known as antibodies and are produced by white blood cells called lymphocytes, not by the liver. Gamma globulins are important in fighting disease-causing pathogens (see section 6.3). Fibrinogen is an inactive plasma protein. Once activated, fibrinogen forms a blood clot (see section 6.4).

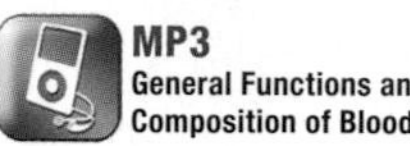
MP3 General Functions and Composition of Blood

CHECK YOUR PROGRESS 6.1

1. Summarize the functions of blood.
2. List the different types of plasma proteins and explain why each is important.
3. Describe why blood is a connective tissue.

CONNECTING THE CONCEPTS

For more information on the topics presented in this section, refer to the following discussions:

Section 2.5 contains a Health feature, "Good and Bad Cholesterol," on page 33, on the differences between HDLs and LDLs in the blood.

Section 4.2 examines why blood is classified as a connective tissue.

Section 7.4 explores how white blood cells are involved in the immune response.

APPLICATIONS AND MISCONCEPTIONS

Are stem cells only present in embryos?

Actually, this is a common misconception brought on by a media focus on embryonic stem (ES) cells. In reality, any actively dividing tissue requires a form of stem cell to act as the original source of cells. The difference is that many ES cells are totipotent—meaning that they retain the ability to form almost any type of cell—though most adult stem cells are pluripotent. Pluripotent cells have undergone an additional stage of specialization and, therefore, can only produce a limited variety of new cell types.

Video Heart Stem Cells

6.2 Red Blood Cells and Transport of Oxygen

LEARNING OUTCOMES

Upon completion of this section, you should be able to

1. Explain the role of hemoglobin in gas transport.
2. Compare the transport of oxygen and carbon dioxide by red blood cells.
3. Summarize the role of erythropoietin in red blood cell production.

Red blood cells (RBCs), also known as **erythrocytes,** are small (usually between 6 and 8 micrometers), biconcave disks (Fig. 6.3*a*). The presence of hemoglobin in these cells, and their unique internal structure, makes them a unique cell type in the body. They are also very abundant; there are 4 to 6 million red blood cells per mm^3 of whole blood.

How Red Blood Cells Carry Oxygen

Red blood cells are highly specialized for oxygen (O_2) transport. RBCs contain **hemoglobin (Hb)**, a pigment with a high affinity (or attraction) for oxygen. It is also responsible for the red coloration of RBCs and the blood. The globin portion of hemoglobin is a protein that contains four highly folded polypeptide chains. The heme part of hemoglobin is an iron-containing group in the center of each polypeptide chain (Fig. 6.3*b*). The iron combines reversibly with oxygen. This means that heme accepts O_2 in the lungs and then lets go of it in the tissues. By contrast, carbon monoxide (CO) combines with the iron of heme and then will not easily let go.

Each hemoglobin molecule can transport four molecules of O_2, and each RBC contains about 280 million hemoglobin molecules. This means that each red blood cell can carry over a billion molecules of oxygen.

Red blood cells are an excellent example of structure suiting function. They have no nucleus; their biconcave shape comes about because they lose their nucleus during maturation (see Fig. 6.1). The biconcave shape of RBCs gives them a greater surface area for the diffusion of gases into and out of the cell. All of the internal space of RBCs is used for transport of oxygen. Aside from having no nucleus, RBCs also lack most organelles, including mitochondria. RBCs anaerobically produce ATP, and they do not consume any of the oxygen they transport.

When oxygen binds to heme in the lungs, hemoglobin assumes a slightly different shape and is called **oxyhemoglobin.** In the tissues, heme gives up this oxygen, and hemoglobin resumes its former shape, called **deoxyhemoglobin.** The released oxygen diffuses out of the blood into tissue fluid and then into cells.

How Red Blood Cells Help Transport Carbon Dioxide

After blood picks up carbon dioxide (CO_2) in the tissues, about 7% is dissolved in plasma. If the percentage of plasma CO_2

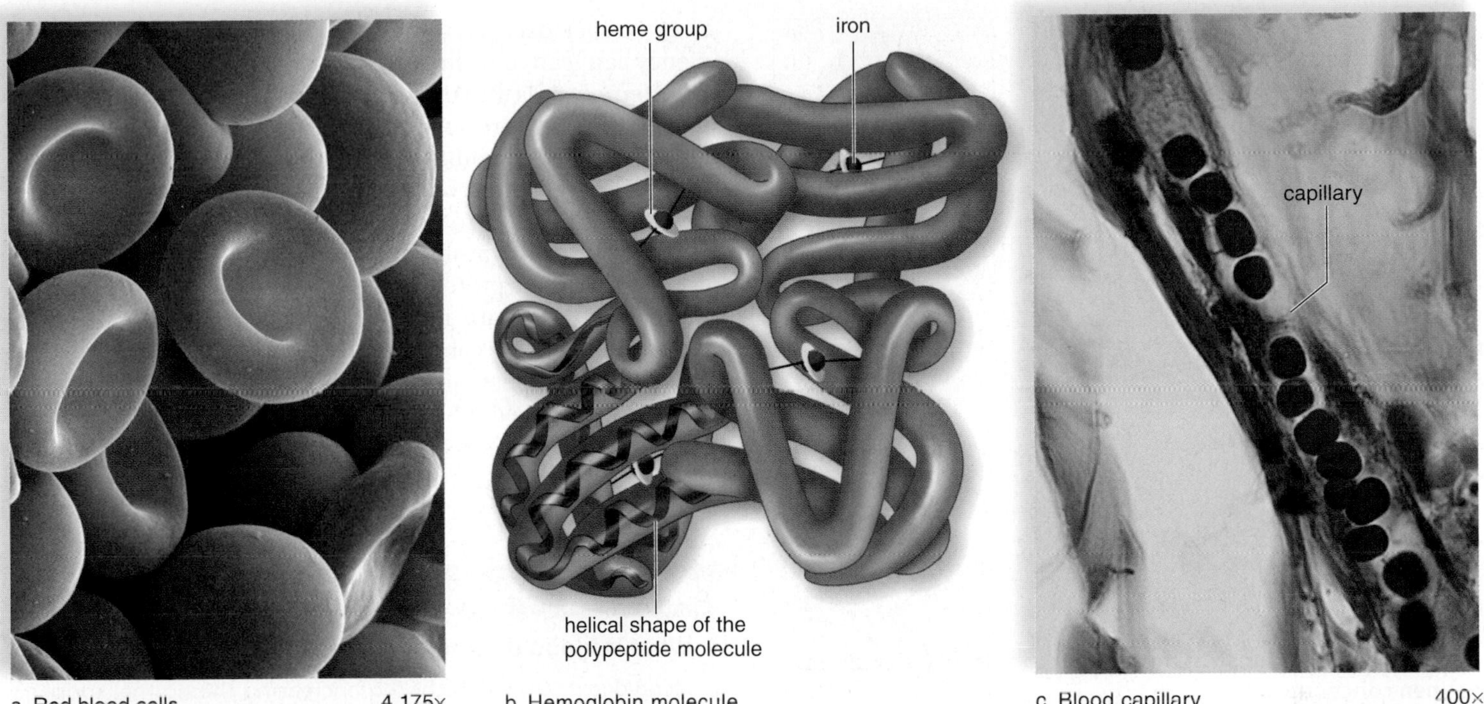

Figure 6.3 Red blood cells and the structure of hemoglobin.
a. Red blood cells are biconcave disks containing many molecules of hemoglobin. **b.** Hemoglobin contains two types of polypeptide chains (blue, purple), forming the molecule's globin portion. An iron-containing heme group is in the center of each chain. Oxygen combines loosely with iron when hemoglobin is oxygenated. **c.** Red blood cells move single file through the capillaries.

were higher than 7%, plasma would be carbonated and bubble like a soda. Instead, hemoglobin directly transports about 25% of CO_2, combining it with the globin protein. Hemoglobin-carrying CO_2 is termed *carbaminohemoglobin.*

The remaining CO_2 (about 68%) is transported as the bicarbonate ion (HCO_3^-) in the plasma. Consider this equation:

CO_2	+	H_2O	$\rightleftharpoons$	H_2CO_3	$\rightleftharpoons$	H^+	+	HCO_3^-
carbon dioxide		water		carbonic acid		hydrogen ion		bicarbonate ion

Carbon dioxide moves into RBCs, combining with cellular water to form carbonic acid (arrows pointing left to right in the preceding equation illustrate this part of the reaction). An enzyme found inside RBCs, called *carbonic anhydrase,* speeds the reaction. Carbonic acid quickly separates, or dissociates, to form hydrogen ions (H^+) and bicarbonate ions (HCO_3^-). The bicarbonate ions diffuse out of the RBCs to be carried in the plasma. The H^+ from this equation binds to globin, the protein portion of hemoglobin. Thus, hemoglobin assists plasma proteins and salts in keeping the blood pH constant. When blood reaches the lungs, the reaction is reversed (arrows pointing right to left). Hydrogen ions and bicarbonate ions reunite to re-form carbonic acid. The carbonic anhydrase enzyme also speeds this reverse reaction. Carbon dioxide diffuses out of the blood and into the airways of the lungs, to be exhaled from the body (also see Fig. 9.11).

Red Blood Cells Are Produced in Bone Marrow

The RBC stem cells in the bone marrow divide and produce new cells that differentiate into mature RBCs (see Fig. 6.1). As red blood cells mature, they acquire hemoglobin and lose their nucleus and other internal organelles. The lack of internal organelles means that RBCs are incapable of conducting many of the functions of other eukaryotic cells in the body. For example, because they lack mitochondria, RBCs are not able to carry out cellular respiration, and instead rely upon the processes of glycolysis and fermentation. In addition, the cytoskeleton of RBCs contains proteins that provide their unique shape, thus allowing them to move through narrow capillary beds. Also, due to their lack of a nucleus, RBCs are unable to replenish important proteins and repair cellular damage. Therefore, red blood cells only live about 120 days. When they age, red blood cells are phagocytized by white blood cells (macrophages) in the liver and spleen.

It is estimated that about 2 million RBCs are destroyed per second, and therefore, an equal number must be produced to keep the red blood cell count in balance. When red blood cells are broken down, hemoglobin is released. The globin portion of hemoglobin is broken down into its component amino acids, which are recycled by the body. The majority of the iron is recovered and returned to the bone marrow for reuse, although a small amount is lost and must be replaced in the diet. The rest of the heme portion of the molecule undergoes chemical degradation and is excreted by the liver

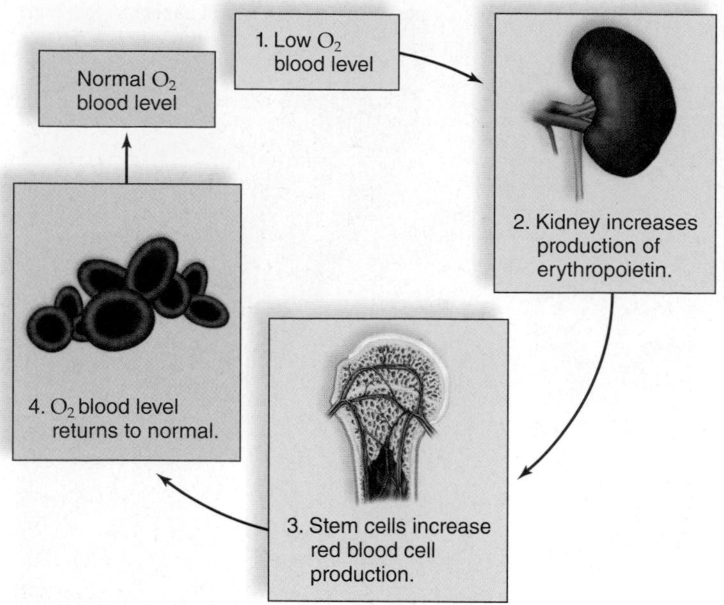

Figure 6.4 **Response of the kidneys to a decrease in blood oxygen concentration.**
The kidneys release increased amounts of erythropoietin whenever the oxygen capacity of the blood is reduced. Erythropoietin stimulates the red bone marrow to speed up its production of red blood cells, which carry oxygen.

and kidneys. The body has a way to boost the number of RBCs when insufficient oxygen is being delivered to the cells. The kidneys release a hormone called **erythropoietin (EPO)**, which stimulates the stem cells in bone marrow to produce more red blood cells (Fig. 6.4). The liver and other tissues also produce EPO for the same purpose.

Animation Hemoglobin Breakdown

Disorders Involving Red Blood Cells

If the liver fails to excrete heme, it accumulates in tissues, causing a condition called *jaundice.* In jaundice, the skin and whites of the eyes turn yellow. Likewise, when skin is bruised, the chemical breakdown of heme causes the skin to change color from red/purple to blue to green to yellow. When there is an insufficient number of red blood cells or the cells do not have enough hemoglobin, the individual suffers from **anemia** and has a tired, run-down feeling. Iron, vitamin B_{12}, and the B vitamin folic acid are necessary for the production of red blood cells. *Iron-deficiency anemia* is the most common form. It results from inadequate intake of dietary iron, which causes insufficient hemoglobin synthesis. A lack of vitamin B_{12} causes *pernicious anemia,* in which stem-cell activity is reduced due to inadequate DNA production. As a consequence, fewer red blood cells are produced. *Folic-acid-deficiency anemia* also leads to a reduced number of RBCs, particularly during pregnancy. Pregnant women should consult with their health-care provider about the need to increase their intake of folic acid, because a deficiency can lead to birth defects in the newborn.

Hemolysis is the rupturing of red blood cells. In hemolytic anemia, the rate of red blood cell destruction increases. **Sickle-cell disease** is a hereditary condition in which the individual has sickle-shaped red blood cells that tend to rupture as they pass through the narrow capillaries. The RBCs look like those to the right. The problem arises because the protein in two of the four chains making up hemoglobin is abnormal. The life expectancy of sickle-shaped red blood cells is about 90 days instead of 120 days.

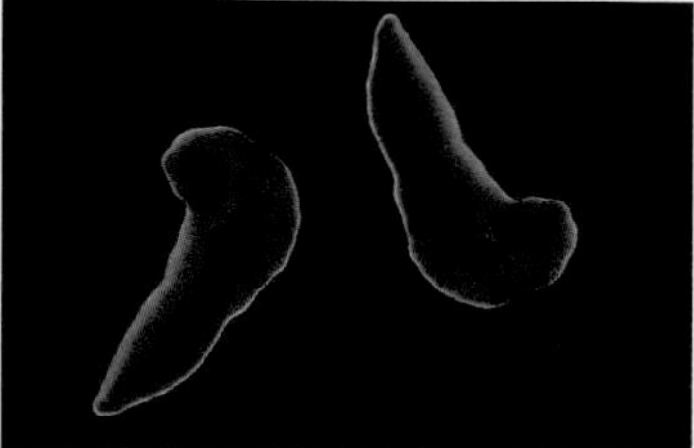

1,600×, colorized SEM

APPLICATIONS AND MISCONCEPTIONS

What is blood doping and is it safe?

Blood doping is any method of increasing the normal supply of RBCs for the purpose of delivering oxygen more efficiently, reducing fatigue, and giving athletes a competitive edge. To accomplish blood doping, athletes can inject themselves with EPO some months before the competition. These injections will increase the number of RBCs in their blood. Several weeks later, four units of their blood are removed and centrifuged to concentrate the RBCs. The concentrated RBCs are reinfused shortly before the athletic event. Blood doping is a dangerous, illegal practice. Several cyclists died in the 1990s from heart failure, probably due to blood that was too thick with cells for the heart to pump.

CHECK YOUR PROGRESS 6.2

1. Describe the anatomy of an RBC and how that anatomy leads to the cell's ability to transport oxygen.
2. Describe the pros and cons of a mature RBC not containing a nucleus.
3. Detail a few blood disorders, including their causes and symptoms.

CONNECTING THE CONCEPTS

For additional information on the topics presented in this section, refer to the following discussions:

Section 9.6 details the process of gas exchange in the lungs.

Section 10.4 explains how the kidneys assist in maintaining red blood cell homeostasis.

Section 20.2 describes the patterns of inheritance associated with sickle-cell disease.

6.3 White Blood Cells and Defense Against Disease

LEARNING OUTCOMES

Upon completion of this section, you should be able to

1. Explain the function of white blood cells in the body.
2. Distinguish between granular and agranular leukocytes.
3. Describe some of the disorders associated with white blood cells.

White blood cells (**leukocytes**) differ from red blood cells in that they are usually larger, have a nucleus, lack hemoglobin, and are translucent unless stained. White blood cells are not as numerous as red blood cells. There are only 5,000–11,000 per mm^3 of blood. White blood cells are derived from stem cells in the red bone marrow, where most types mature. There are several types of white blood cells (Fig. 6.5), and the production of each type is regulated by a protein called a *colony-stimulating factor* (*CSF*). In a person with normally functioning bone marrow, the numbers of white blood cells can double within hours, if needed. White blood cells are able to squeeze through pores in the capillary wall; therefore, they are also found in tissue fluid and lymph (Fig. 6.6).

White Blood Cells	Function
Granular leukocytes	
• Neutrophils 	Phagocytize pathogens and cellular debris.
• Eosinophils 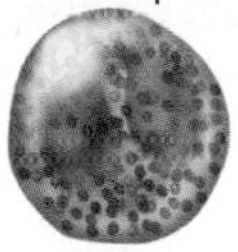	Use granule contents to digest large pathogens, such as worms, and reduce inflammation.
• Basophils 	Promote blood flow to injured tissues and the inflammatory response.
Agranular leukocytes	
• Lymphocytes 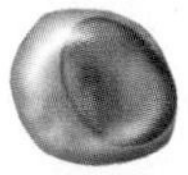	Responsible for specific immunity; B cells produce antibodies; T cells destroy cancer and virus-infected cells.
• Monocytes 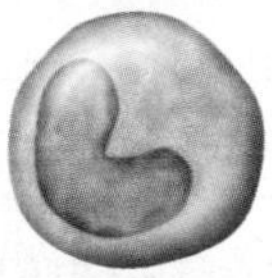	Become macrophages that phagocytize pathogens and cellular debris.

Figure 6.5 **Some examples of white blood cells.** Neutrophils, eosinophils, and basophils are granular leukocytes. Lymphocytes and monocytes have few, if any, granules (agranular).

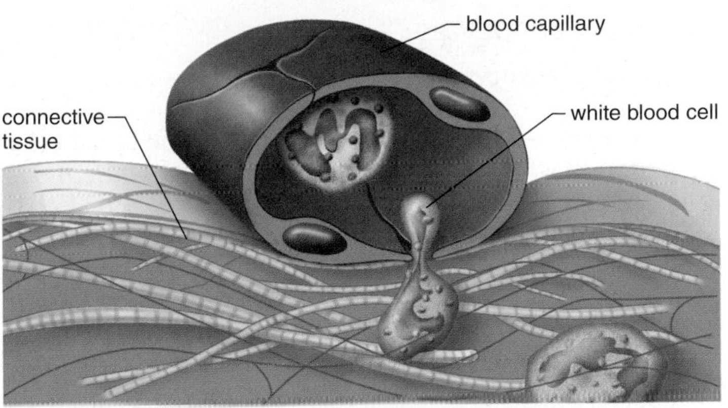

Figure 6.6 **Movement of white blood cells into the tissue.** White blood cells can squeeze between the cells of a capillary wall and enter body tissues.

White blood cells fight infection; thus, they are an important part of the immune system. The **immune system,** discussed in Chapter 7, consists of a variety of cells, tissues, and organs that defend the body against pathogens, cancer cells, and foreign proteins. Many white blood cells live only a few days; often they die while fighting pathogens. Others live for months or even years.

White blood cells have various ways to fight infection. Certain ones are very good at phagocytosis. During phagocytosis, a projection from the cell surrounds a pathogen and literally engulfs it. A vesicle containing the pathogen is formed inside the cell. Lysosomes attach and empty their digestive enzymes into the vesicle. The enzymes break down the pathogen. Other white blood cells produce antibodies, proteins that combine with antigens. An **antigen** is a cell or other substance foreign to the individual that invokes an immune response. The antibody–antigen pair is then marked for destruction, again by phagocytosis. We will be describing antigens and antibodies involved in blood typing and coagulation in section 6.5.

Animation Phagocytosis

Types of White Blood Cells

White blood cells are classified into the **granular leukocytes** and the **agranular leukocytes** (Fig. 6.5). Granulocytes have noticeable cytoplasmic granules, which can be easily seen when the cells are stained and examined with a microscope. Granules, like lysosomes, contain various enzymes and proteins. Agranulocytes contain only sparse, fine granules, which are not easily viewed under a microscope.

Granular Leukocytes

The granular leukocytes include neutrophils, eosinophils, and basophils.

Neutrophils account for 50–70% of all white blood cells. Therefore, they are the most abundant of the white blood cells. They have a multilobed nucleus, so they are called *polymorphonuclear leukocytes,* or "polys." Compared to other granular leukocytes, the granules of neutrophils are neither easily stained with acidic red dye nor with basic purple dye. This accounts for their name, neutrophil. Neutrophils are usually first responders to bacterial infection, and their intense phagocytic activity is essential to overcoming an invasion by a pathogen.

Eosinophils have a bilobed nucleus. Their large, abundant granules take up eosin and become a red color. This accounts for the name eosinophil. They are involved in the protection of the body against large parasites (namely the parasitic worms) and also in the phagocytosis of allergens and proteins associated with the inflammatory response.

Basophils are the rarest of the white blood cells, but play an important role in the immune response. They have a U-shaped or lobed nucleus. Their granules take up the basic stain and become a dark-blue color. This accounts for the name, basophil. In the connective tissues, basophils, and similar cells called **mast cells,** release histamine associated with allergic reactions. Histamine dilates blood vessels but constricts the air tubes that lead to the lungs, which is what happens during an asthma attack when someone has difficulty breathing.

Agranular Leukocytes

The agranular leukocytes include the lymphocytes and the monocytes. Lymphocytes and monocytes do not have granules and have nonlobular nuclei. They are sometimes called the mononuclear leukocytes.

Lymphocytes account for 25–35% of all white blood cells. Therefore, they are the second most abundant type of white blood cell. Lymphocytes are responsible for specific immunity to particular pathogens and toxins (poisonous substances). The lymphocytes are of two types: B cells and T cells. Mature B cells called plasma cells produce antibodies, the proteins that combine with target pathogens and mark them for destruction. Some T cells (cytotoxic T cells) directly destroy pathogens. The AIDS virus attacks one of several types of T cells. In this way, the virus causes immune deficiency, an inability to defend the body against pathogens. B lymphocytes and T lymphocytes are discussed more fully in Chapter 7.

Monocytes are the largest of the white blood cells. After taking up residence in the tissues, they differentiate into even larger macrophages. In the skin, they become dendritic cells. Like the neutrophils, macrophages and dendritic cells are active phagocytes, destroying pathogens, old cells, and cellular debris. Macrophages and dendritic cells also stimulate other white blood cells, including lymphocytes, to defend the body.

Disorders Involving White Blood Cells

Immune deficiencies are sometimes inherited. Children have **severe combined immunodeficiency (SCID)** when the stem cells of white blood cells lack an enzyme called adenosine deaminase. Without this enzyme, B and T lymphocytes do not develop and the body cannot fight infections. About 100 children are born with the disease each year. Injections of the missing enzyme can be given twice weekly, but a bone marrow transplant from a compatible donor is the best way to cure the disease.

Cancer is due to uncontrolled cell growth. **Leukemia,** which means "white blood," refers to a group of cancers that involve uncontrolled white blood cell proliferation. Most of these white blood cells are abnormal or immature. Therefore, they are incapable of performing their normal defense functions. Each type of leukemia is named for the type of cell dividing out of control. For example, lymphocytic leukemia involves abnormal lymphocyte proliferation.

An Epstein–Barr virus (EBV) infection of lymphocytes is the cause of **infectious mononucleosis,** so named because the lymphocytes are mononuclear. EBV, a member of the herpes virus family, is one of the most common human viruses. Symptoms of infectious mononucleosis are fever, sore throat, and swollen lymph glands. Although symptoms usually disappear in one or two months without medication, EBV remains dormant and hidden in a few cells in the throat and blood for the rest of a person's life. Stress can reactivate the virus. Reactivation means that a person's saliva can pass on the infection to someone else, as with intimate kissing. This is why mononucleosis is called the "kissing disease."

CHECK YOUR PROGRESS 6.3

1. Explain why there are different types of white blood cells.
2. Describe the structure and function of each type of white blood cell.
3. Summarize three disorders of white blood cells.

CONNECTING THE CONCEPTS

White blood cells play an important role in the defense against disease. For more information on these cells, refer to the following discussions:

Section 7.3 describes the role of the monocytes in the innate immune response.

Section 7.4 details how the lymphocytes are involved in the adaptive defenses.

Section S.1 explores how the HIV virus infects white blood cells, resulting in the disease called AIDS.

6.4 Platelets and Blood Clotting

LEARNING OUTCOMES

Upon completion of this section, you should be able to

1. Explain how blood clotting relates to homeostasis.
2. List the steps in the formation of a blood clot.
3. Describe disorders associated with blood clotting.

Platelets (**thrombocytes**) result from fragmentation of large cells, called **megakaryocytes,** in the red bone marrow. Platelets are produced at a rate of 200 billion a day, and the blood contains 150,000–300,000 per mm^3. These formed elements are involved in the process of blood **clotting,** or **coagulation.** Also involved are the plasma proteins **prothrombin** and **fibrinogen,** manufactured in the liver and deposited in the blood. Vitamin K is necessary to the production of prothrombin.

Blood Clotting

The blood-clotting process helps the body maintain homeostasis in the cardiovascular system by ensuring that the plasma and formed elements remain within the blood vessels. At least 12 clotting factors and calcium ions (Ca^{2+}) participate in the formation of a blood clot.

When a blood vessel in the body is damaged, platelets clump at the site of the puncture and partially seal the leak (Fig. 6.7). Platelets and the injured tissues release a clotting factor called prothrombin activator that converts prothrombin in the plasma to **thrombin.** This reaction requires calcium ions (Ca^{2+}). Thrombin, in turn, acts as an enzyme that severs two short amino acid chains from a fibrinogen molecule, one of the proteins in plasma. These activated fragments then join end to end, forming long threads of fibrin.

Fibrin threads wind around the platelet plug in the damaged area of the blood vessel and provide the framework for the clot. Red blood cells trapped within the fibrin threads make the clot appear red. A fibrin clot is temporary. Once blood vessel repair starts, an enzyme called plasmin destroys the fibrin network so tissue cells can grow.

After blood clots, a yellowish fluid called **serum** escapes from the clot. It contains all the components of plasma except fibrinogen and prothrombin.

1. Blood vessel is punctured.

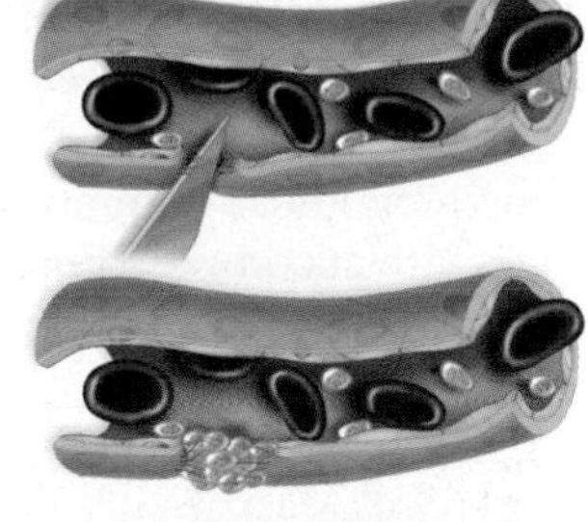

2. Platelets congregate and form a plug.

3. Platelets and damaged tissue cells release prothrombin activator, which initiates a cascade of enzymatic reactions.

prothrombin activator

prothrombin —Ca^{2+}→ thrombin

fibrinogen —Ca^{2+}→ fibrin threads

4. Fibrin threads form and trap red blood cells.

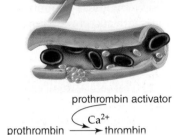

a. Blood-clotting process

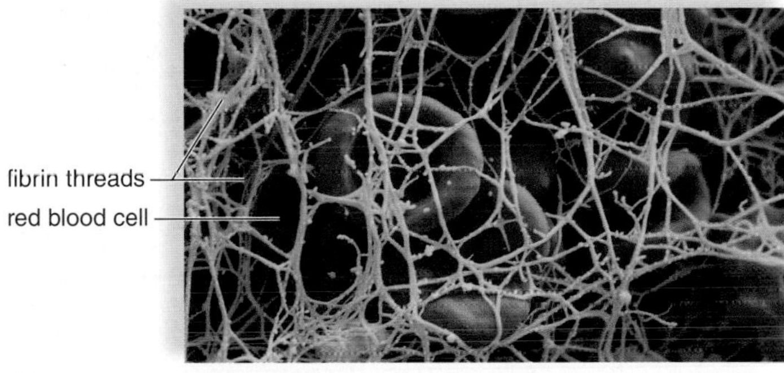

b. Blood clot 4,400×

Figure 6.7 **The steps in the formation of a blood clot.** **a.** Platelets and damaged tissue cells release prothrombin activator, which acts on prothrombin in the presence of Ca^{2+} (calcium ions) to produce thrombin. Thrombin acts on fibrinogen in the presence of Ca^{2+} to form fibrin threads. **b.** A scanning electron micrograph of a blood clot shows red blood cells caught in the fibrin threads.

Disorders Related to Blood Clotting

An insufficient number of platelets is called **thrombocytopenia.** Thrombocytopenia is either due to low platelet production in bone marrow or increased breakdown of platelets outside the marrow. A number of conditions, including leukemia, can lead to thrombocytopenia. It can also be drug-induced. Symptoms include bruising, rash, and nosebleeds or bleeding in the mouth. Gastrointestinal bleeding or bleeding in the brain are possible complications.

If the lining of a blood vessel becomes roughened, a clot can form spontaneously inside an unbroken blood vessel. Most often, roughening occurs because an atherosclerotic plaque has formed (see Chapter 5). Rarely, the vessel lining is damaged during placement of an intravenous tube. The spontaneous clot is called a *thrombus* (pl., thrombi) if it remains stationary inside the blood vessel. Sitting for long periods, as when traveling, can also cause thrombus formation. If the clot dislodges and travels in the blood, it is called an embolus. If a **thromboembolism** is not treated, blood flow to the tissues can stop completely. Heart attack or stroke can result, as discussed in Chapter 5.

Hemophilia is an inherited clotting disorder that causes a deficiency in a clotting factor. There are many forms of the disorder. Hemophilia A, caused by deficiency of clotting factor VIII, is more likely to occur in boys than in girls. Hemophilia A is caused by an abnormal copy of the factor VIII production gene, found on the X chromosome. This hemophilia arises when a boy has an abnormal gene on his single X chromosome. Girls only need one normal gene between their two X chromosomes to make the normal amounts of clotting factor VIII (see Chapter 20). In hemophilia, the slightest bump can cause bleeding into the joints. Cartilage degeneration in the joints and absorption of underlying bone can follow. The most frequent cause of death is bleeding into the brain with accompanying neurological damage. Regular injections of factor VIII can successfully treat the disease.

CHECK YOUR PROGRESS 6.4

1. List the components of the blood that are involved in the formation of a blood clot.
2. Describe the stages of blood clotting.
3. Summarize a few of the blood-clotting disorders.

CONNECTING THE CONCEPTS

For more information on the disorders discussed in this section, refer to the following discussions:

Section 5.7 describes how blood clots relate to disorders of the cardiovascular system.

Section 20.4 explores the inheritance of hemophilia in the royal families of Europe.

APPLICATIONS AND MISCONCEPTIONS

What are some other forms of clotting disorders?

In addition to hemophilia A, there are several other forms of clotting disorders. Hemophilia B, also called *Christmas disease* after the first person to be diagnosed with the disease (Stephen Christmas in 1952), is caused by a deficiency in clotting factor IX. Von Willebrand disease is caused by a deficiency in the gluelike protein that holds the platelets together at the site of the damaged blood vessel. All of these disorders may be partially treated with injections of the missing proteins.

BIOLOGY MATTERS  **Health**

Aspirin and Heart Disease

Aspirin (Fig. 6A), or acetyl salicylic acid, has long been recognized as treatment for pain. However, low-dose use of aspirin is now known to reduce the chances of some forms of heart disease, including strokes and heart attacks. It does this by interfering with the normal cascade of events in the blood clotting-pathway, as shown in Figure 6.7a.

When located free in the blood, platelets (thrombocytes) exist as individual cell fragments. However, following an injury, the damaged cells release chemicals that cause the platelets to become "sticky" and clump together into larger groups called aggregates.

The conversion of platelets to the stickier version is due to a chemical called thromboxane. Thromboxane is released by a platelet in response to an injury or wound in the area. As platelets aggregate, they release more thromboxane. This is an example of a positive feedback mechanism (see section 4.8) and it allows for a rapid response to an injury, thus reducing the loss of blood.

A problem occurs when platelets start to stick together in the absence of a wound. These unintended clots can form blockages in the heart and brain, causing heart attacks and strokes. Aspirin inhibits the activity of thromboxane, effectively making the platelets less likely to stick together. Several studies have indicated that the use of aspirin by individuals with heart disease, such as atherosclerosis, reduces the chances of a heart attack or stroke.

It is important to note that the actual dosage of aspirin that produces this effect has not been conclusively determined by the medical community, although most agree that doses around 75–80 milligrams (mg) per day are sufficient. There is no evidence that higher doses produce an increased effect, and in fact, can cause side effects such as ulcers and abdominal bleeding. While many drugstores sell low-dose aspirin (81 mg "baby" aspirin) over the counter, it is highly recommended that you consult with your physician before taking a daily dose of aspirin.

Figure 6A Aspirin and heart disease. Low doses of aspirin may prevent some forms of heart disease.

Questions to Consider

1. As a medical researcher, what other areas of the blood-clotting pathway would you target for inhibition to decrease the risks of stroke?
2. Under what medical conditions do you think that aspirin should not be used? Why?

BIOLOGY MATTERS **Health**

What To Know When Giving Blood

According to the American Red Cross, over 9.5 million people donate blood each year. Yet, despite that number, donated blood is often in short supply. Every 2 seconds in the United States a person needs blood, resulting in a need for over 38,000 donations a day. Here are some facts regarding the procedure.

The Procedure

Before the procedure, you will be asked a series of private and confidential questions about your health and lifestyle. Your temperature, blood pressure, and pulse will be recorded and a drop of your blood will be tested to ensure that you're not anemic.

The supplies used for your donation are sterile and are used only for you. You can't be infected with a disease when donating blood. When the actual donation is started, you may feel a brief "sting." The procedure takes about 10 minutes, and you will have given about a pint of blood (Fig. 6B). Your body replaces the liquid part (plasma) in hours and the cells in a few weeks.

You will have several opportunities both prior to and after giving blood to let Red Cross officials know whether you consider your blood to be safe. Immediately after you donate, you are given a number to call if you decide that your blood may not be safe to give to another person. Donated blood is tested for syphilis bacteria and AIDS antibodies, as well as hepatitis and other viruses. You are notified if tests are positive, and your blood won't be used if it could make someone ill. However, you should *never* use the process of a blood donation to get tested for any medical condition, especially AIDS. It is possible to have a negative result for AIDS antibodies and yet still spread the virus, because forming antibodies takes several weeks after exposure.

The Cautions

Some medications and medical conditions have waiting periods before you can donate blood. Before giving blood, you should inform the medical staff if you meet any of the conditions in the following list:

- You have recently had an infection or a fever.
- You have taken or are taking drugs that slow blood clotting. You should wait 48 hours after taking aspirin or aspirin-related drugs.
- You have had malaria, have taken drugs for malaria prevention, or have traveled to malaria-prone countries.
- You have a medical history of hepatitis or tuberculosis.
- You have been treated for syphilis or gonorrhea in the last 12 months.
- You have AIDS, have had a positive HIV test, or are at risk for getting an HIV infection due to one of the following:
    - You have ever injected illegal drugs.
    - You have taken clotting factor concentrates for hemophilia.
    - You have been given money or drugs for sex since 1977.
    - You had a sexual partner within the last year who did any of the above things.

  For men:
    - You had sex *even once* with another man since 1977, *or* had sex with a female prostitute within the last year.

  For women:
    - You had sex with a male or female prostitute within the last year, *or* had a male sexual partner who had sex with another man *even once* since 1977.

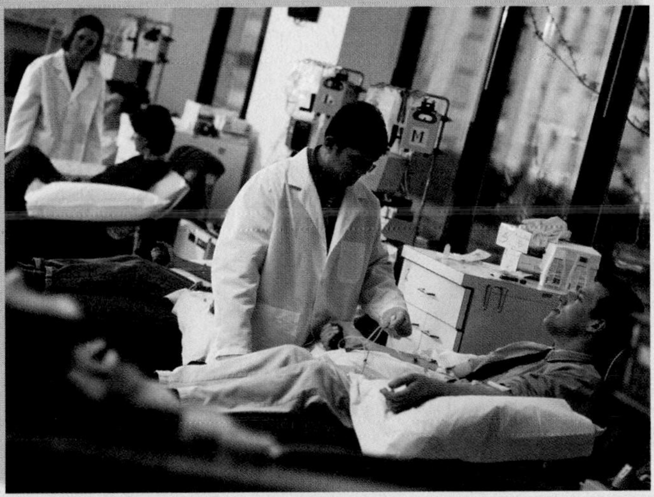

Figure 6B Donating blood can help save a life.

After the Procedure

Most people feel fine while they give blood and afterward, but a few donors have an upset stomach or feel faint or dizzy after donation. Resting, drinking fluid, and eating a snack usually help. Occasionally, bruising, redness, and pain occur at your donation site, so avoid strenuous exercise and lifting for a day or so. Very rarely, a person may have muscle spasms and/or suffer nerve damage. You should contact your physician if you have any concerns about potential side effects of donating blood.

For more information about blood donation, including eligibility requirements, visit the Red Cross website at www.redcrossblood.org.

Questions to Consider

1. Why would individuals be restricted from donating if they have had a sexually-transmitted disease in the past 12 months, even if they have been treated?
2. Are there other risk factors that you think should be considered that would exclude individuals from donating blood?

6.5 Blood Typing and Transfusions

LEARNING OUTCOMES

Upon completion of this section, you should be able to

1. Explain what determines blood types in humans.
2. Predict the compatibility of blood types for a transfusion.
3. Summarize the role of Rh factor in hemolytic disease of the newborn.

A **blood transfusion** is the transfer of blood from one individual to another. For transfusions to be done safely, blood must be typed so that **agglutination** (clumping of red blood cells) does not occur when blood from different people is mixed. Blood typing usually involves determining the ABO blood group and whether the individual is Rh-negative (Rh^-) or Rh-positive (Rh^+).

ABO Blood Groups

Only certain types of blood transfusions are safe because the plasma membranes of red blood cells carry glycoproteins that can be antigens to other individuals. An antigen is any substance that is foreign to a person's body. ABO blood typing is based on the presence or absence of two possible antigens, called type A antigen and type B antigen. Whether these antigens are present or not depends on the particular inheritance of the individual.

MP3 Blood Groupings

In Figure 6.8, type A antigen and type B antigen are given different shapes and colors. Study each figure to see the type(s) of antigens found on the red blood cell plasma membrane. As you might expect, type A blood has the A antigen (blue spheres) and type B blood has the B antigen (violet triangles). If you guessed that type AB blood has both A and B antigens, you're correct. Finally, notice that type O blood has neither antigen on its red blood cells.

Each blood type has antibodies (yellow Y-shaped molecules) that correspond to the *opposite* blood type. Thus, an individual with type A blood has anti-B antibodies in the plasma, and a person with type B blood has anti-A antibodies in the plasma. Further, a person with type O blood has both antibodies in the plasma and someone with type AB blood lacks both antibodies (Fig. 6.8). Anti-A and/or anti-B antibodies are not present at birth, but they appear over the course of several months.

Finally, observe that each antibody has a *binding site* that will combine with its corresponding antigen in a tight lock-and-key fit. The anti-B antibodies have a triangular binding site on the top of each Y-shaped molecule. This binding site fits snugly with the purple, triangular B antigen. Similarly, the anti-A antibodies have a spherical binding site shaped to form a perfect fit with the A antigen. The presence of these antibodies can cause agglutination.

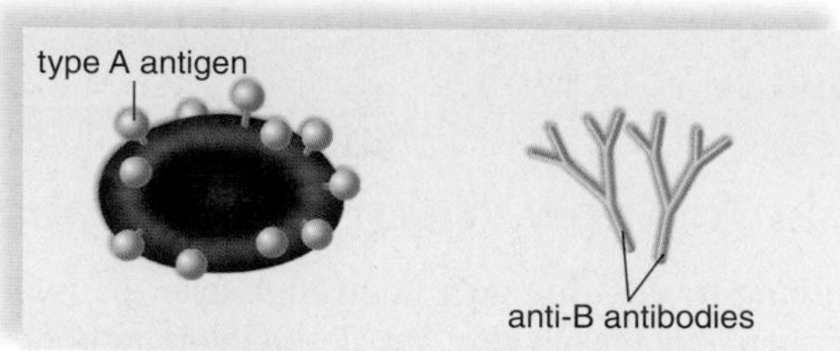

Type A blood. Red blood cells have type A surface antigens. Plasma has anti-B antibodies.

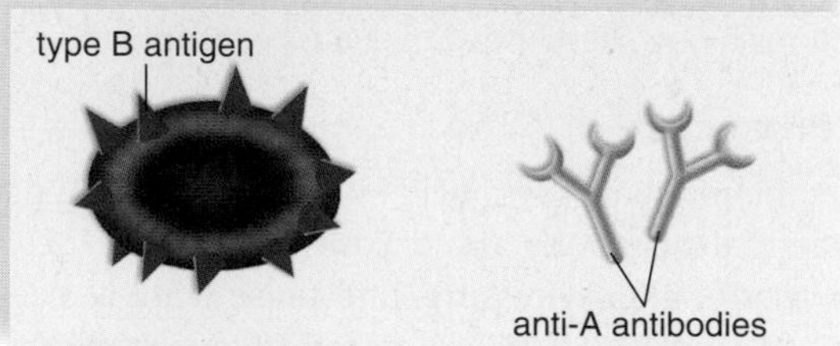

Type B blood. Red blood cells have type B surface antigens. Plasma has anti-A antibodies.

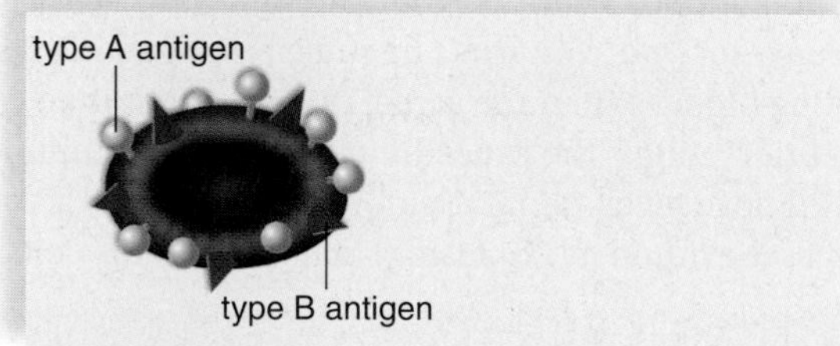

Type AB blood. Red blood cells have type A and type B surface antigens. Plasma has neither anti-A nor anti-B antibodies.

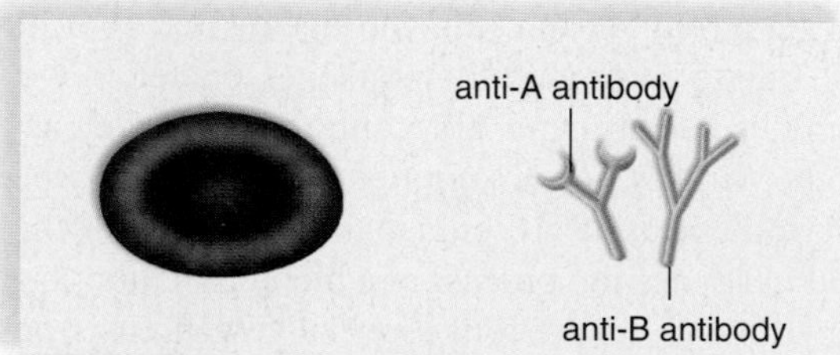

Type O blood. Red blood cells have neither type A nor type B surface antigens. Plasma has both anti-A and anti-B antibodies.

Figure 6.8 **The ABO blood type system.**
In the ABO system, blood type depends on the presence or absence of antigens A and B on the surface of red blood cells. In these drawings, A and B antigens are represented by different shapes on the red blood cells. The possible anti-A and anti-B antibodies in the plasma are shown for each blood type. An anti-B antibody cannot bind to an A antigen, and vice versa.

APPLICATIONS AND MISCONCEPTIONS

What is the most common ABO blood type? What is the least common ABO blood type?

It depends on your genetic inheritance. In the United States and other countries with a large immigrant population, the most common ABO type is type O. However, type A blood is most common in several European countries, including Portugal, Armenia, Switzerland, and Norway. Worldwide, type B blood is common in central Asia, India, and the Far East. Worldwide, the least common ABO type is type AB.

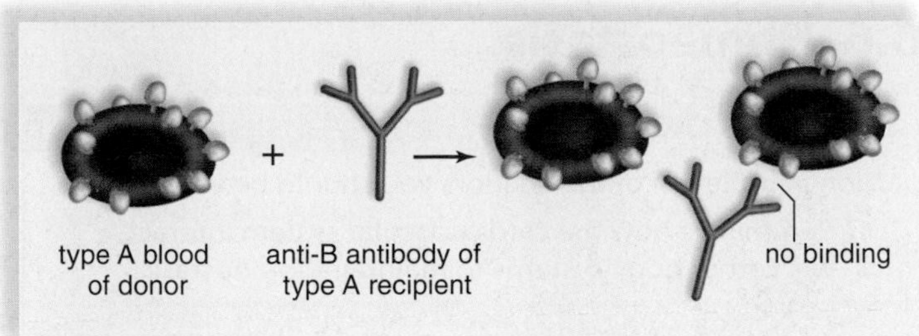

a. No agglutination

b. Agglutination

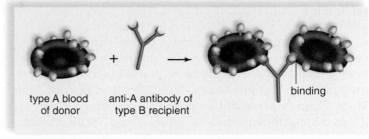

Figure 6.9 Blood compatibility and agglutination.
No agglutination occurs in (**a**) because anti-B antibodies cannot combine with the A antigen. Agglutination occurs in (**b**) when anti-A antibodies in the recipient combine with A antigen on donor red blood cells.

Blood Compatibility

Blood compatibility is very important when transfusions are done. The antibodies in the plasma must not combine with the antigens on the surface of the red blood cells or else agglutination occurs. With agglutination, anti-A antibodies have combined with type A antigens, and anti-B antibodies have combined with type B antigens. Therefore, agglutination is expected if the donor has type A blood and the recipient has type B blood (Fig. 6.9). What about other combinations of blood types? Try out all other possible donors and recipients to see if agglutination will occur.

Type O blood is sometimes called the *universal donor* because the red blood cells of type O blood lack A and B antigens. Type O donor blood should not agglutinate with any other type of recipient blood. Likewise, type AB blood is sometimes called *universal recipient* blood because the plasma lacks A and B antibodies. Type AB recipient blood should not agglutinate with any other type of donor blood. In practice, however, there are other possible blood groups, aside from ABO blood groups. Before blood can be safely transfused from one person to another, it is necessary to physically combine donor blood with recipient blood on a glass slide, then observe whether agglutination occurs. This procedure, called blood-type crossmatching, takes only a few minutes. It is done before every blood transfusion is performed. Type O blood donation without crossmatching is performed only in an emergency, when blood loss is severe and the patient's survival is at stake.

Rh Blood Groups

The designation of blood type usually also includes whether the person has or does not have the Rh factor on the red blood cell. Rh-negative individuals normally do not have antibodies to the Rh factor, but they make them when exposed to the Rh factor.

If a mother is Rh-negative and the father is Rh-positive, a child can be Rh-positive. During a pregnancy, Rh-positive antigens can leak across the placenta into the mother's bloodstream. The presence of these Rh-positive antigens causes the mother to produce anti-Rh antibodies (Fig. 6.10). Usually, in a subsequent pregnancy with another Rh-positive baby, the anti-Rh antibodies may cross the placenta and destroy

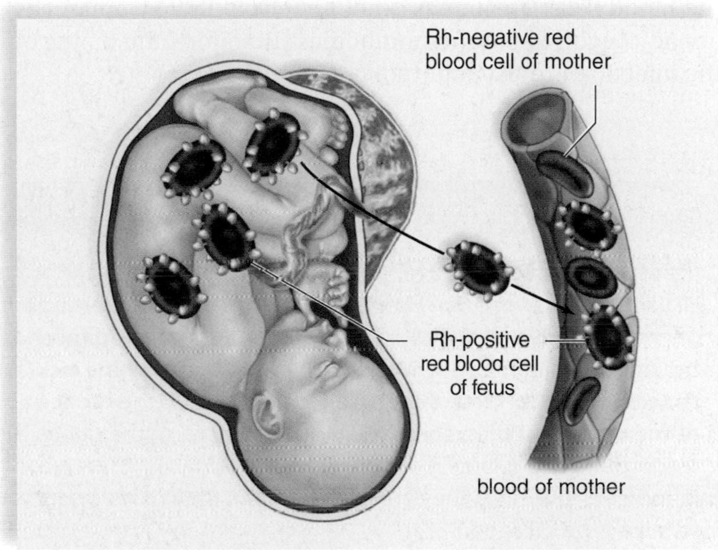

a. Fetal Rh-positive red blood cells leak across placenta into mother's bloodstream.

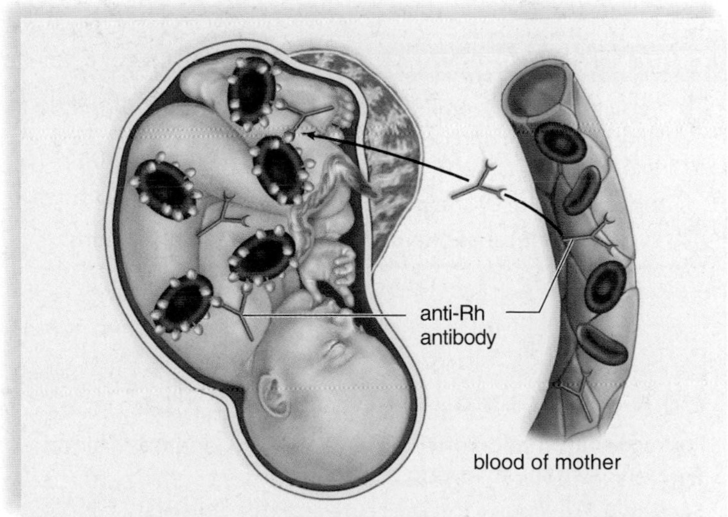

b. Mother forms anti-Rh antibodies that cross the placenta and attack fetal Rh-positive red blood cells.

Figure 6.10 Rh factor disease (hemolytic disease of the newborn).
a. Due to a pregnancy in which the child is Rh-positive, an Rh-negative mother can begin to produce antibodies against Rh-positive red blood cells. **b.** Usually in a subsequent pregnancy, these antibodies can cross the placenta and cause hemolysis of an Rh-positive child's red blood cells.

the unborn child's red blood cells. This is called *hemolytic disease of the newborn (HDN)* because hemolysis starts in the womb and continues after the baby is born. Due to red blood cell destruction, the baby will be severely anemic. Excess hemoglobin breakdown products in the blood can lead to brain damage and intellecutal disability, or even death. The Rh problem is prevented by giving Rh-negative women an Rh immunoglobulin injection no later than 72 hours after giving birth to an Rh-positive child. This injection contains anti-Rh antibodies that attack any of the baby's red blood cells in the mother's blood before these cells can stimulate her immune system to produce her own antibodies. This treatment is termed RhoGAM, and does not harm the newborn's red blood cells. Treatment is not beneficial if the woman has already begun to produce antibodies. Therefore, the timing of the injection is most important.

APPLICATIONS AND MISCONCEPTIONS

Is ABO blood typing accurate?

Usually, the answer is yes. However, there are some interesting genetic disorders that may complicate traditional ABO blood typing. For example, individuals with Bombay syndrome lack the enzyme to correctly attach A and B antigens to the surface of the red blood cells. These individuals may carry the genes to produce the A and B antigens on the surface of the red blood cells, but since they are not attached, they may appear to have type O blood.

CHECK YOUR PROGRESS 6.5

1. Explain what determines blood type and list the four different types of blood.
2. Solve the following: If a person has type A blood, to whom may he or she donate blood?
3. Explain what causes hemolytic disease of the newborn.

CONNECTING THE CONCEPTS

For more information on the chemistry and genetics of blood types, refer to the following discussions:

Section 3.3 describes the structure and function of glycoproteins.

Section 20.3 examines patterns of inheritance associated with human ABO blood types.

6.6 Homeostasis

LEARNING OUTCOME

Upon completion of this section, you should be able to

1. Summarize how the cardiovascular system interacts with other body systems to maintain homeostasis.

Overall, the body systems work together to promote a stable internal environment that enables our bodies to function at peak efficiency. The process of regulating this environment is called homeostasis. To promote homeostasis, all of the organ systems must work together. For example, after studying all night for a final exam, you oversleep the next morning. You wake up just a few minutes before class begins. Immediately, your heart starts beating faster and your blood pressure rises, due to sympathetic *nervous system* stimulation (see Chapter 13). Hormones, such as epinephrine (adrenaline) from the adrenal glands, pour into the bloodstream to prolong your body's physical response to stress. The *endocrine system* is at work (see Chapter 15).

Frantic, you throw on your clothes and sprint to the bus. Your *muscular* and *respiratory systems* meet the demand (see Chapters 9 and 12). As you run, your muscles require more oxygen. Breathing and heart rate increase to meet the demand. Carbon dioxide generated by hardworking muscles is carried in the blood to the lungs so it can be expelled. To eliminate the heat produced by your muscles, you sweat and blood flow to your skin increases.

The nightmare continues. In your mad dash up the steps to class, you fall, skinning your knee. At first, blood flows from the wound, but a clot soon seals off the injured area. As a scab forms, healing has already begun. Any injury that breaks the skin can allow harmful bacteria into the body, but the immune system usually deals quickly with invaders. Later, you might notice that your knee is not only skinned but also black and blue. Bruises form when blood leaks out of damaged vessels underneath the skin and then clots.

You make it to class just in time. As you settle down to take the exam, your heartbeat and breathing gradually slow, because the muscle need is not as great. As you concentrate, blood flow increases in certain highly active parts of your brain. The brain is a very demanding organ in terms of both oxygen and glucose.

You start to feel hungry. There is little blood glucose to provide cellular fuel, so glycogen stored in the liver is broken down to make glucose available to blood and body cells. Finally, you return the exam and eat breakfast afterward. When most nutrients from your meal are absorbed by the *digestive system,* they enter the bloodstream (see Chapter 8). However, fats enter lacteals, vessels of the *lymphatic system,* for transport to the blood (see Chapter 7).

Notice in this scenario how many organ systems interacted with the cardiovascular system. The cardiovascular system functions as the major transport system in the human body. It is involved not only in the transport of gases such as oxygen, but also the movement of nutrients and the removal of waste material. Thus, it plays a critical role in the regulation of homeostasis.

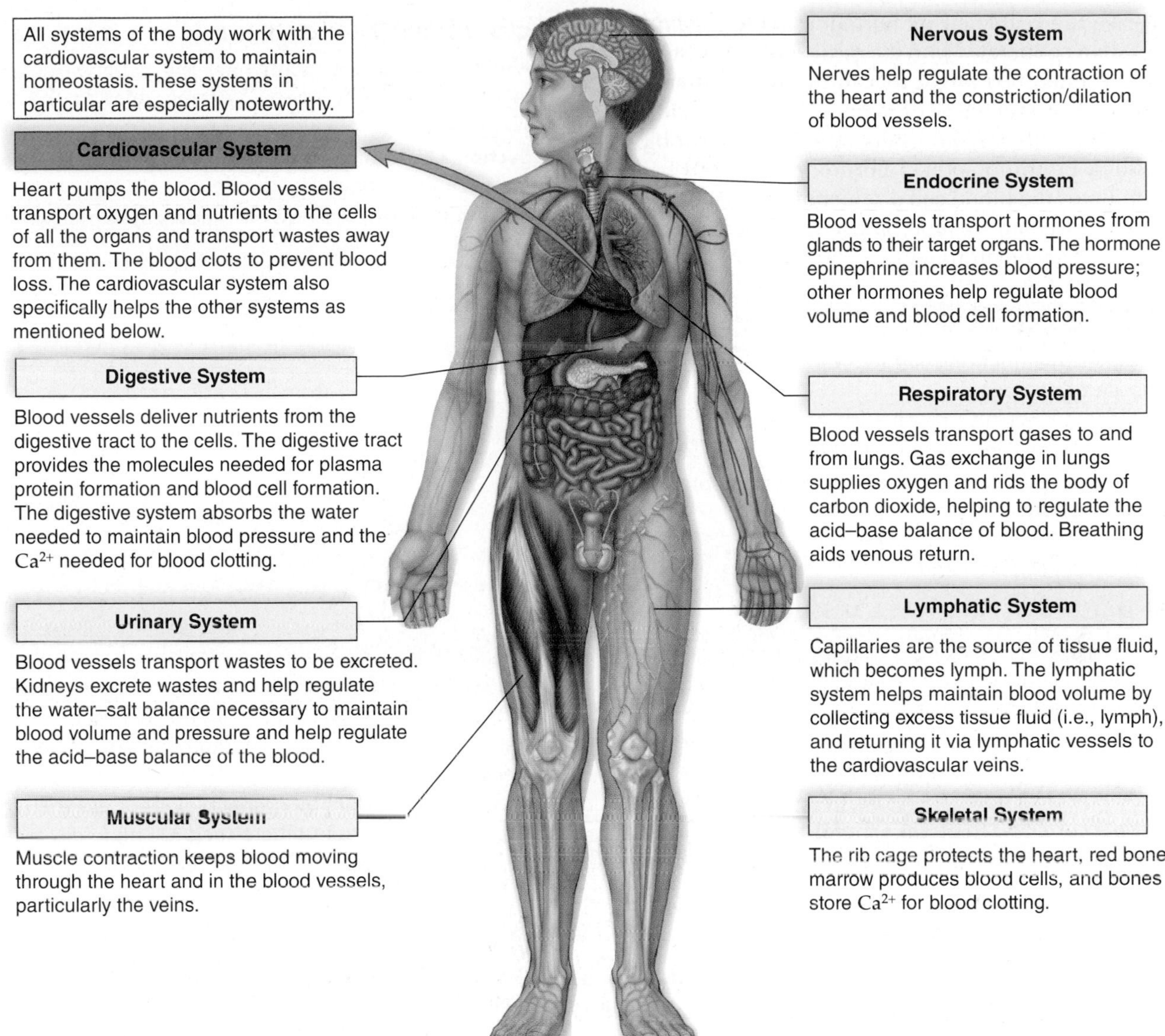

Figure 6.11 How body systems cooperate to ensure homeostasis.
Each of these systems makes critical contributions to the functioning of the cardiovascular system and, therefore, to homeostasis. See if you can suggest contributions by each system before looking at the information given.

How Body Systems Work Together

Figure 6.11 summarizes how these and other systems cooperate with the cardiovascular system. You previously learned that the body's internal environment contains blood and tissue fluid. Tissue fluid originates from blood plasma but contains very little plasma protein. Tissue fluid is absorbed by lymphatic capillaries, after which it is referred to as lymph. The lymph courses through lymphatic vessels, eventually returning to the venous system. Thus, the cardiovascular and lymphatic systems are intimately linked.

Homeostasis is possible only if the cardiovascular system delivers oxygen from the lungs, as well as nutrients from the digestive system, to the tissue fluid surrounding cells. Simultaneously, the cardiovascular system also removes metabolic wastes, delivering waste to excretory organs.

The three components of the muscular system make essential contributions to blood movement. Cardiac muscle contractions circulate blood throughout the body. Contraction or relaxation of the smooth muscle in blood vessel walls changes vessel diameter and helps to maintain the correct blood pressure. Further, skeletal muscle contraction compresses both cardiovascular and lymphatic veins. Lymph returns to cardiovascular veins, and blood in the cardiovascular veins drains back to the heart. The circulation of tissue fluid and blood is then complete.

The skeletal and endocrine systems are vital to cardiovascular homeostasis. Red bone marrow produces blood cells. Without the needed blood cells, the person becomes anemic and lacks an immune response. In addition, bones contribute calcium ions (Ca^{2+}) to the process of blood clotting. Without blood clotting, bleeding from an injury (even something as simple as skinning a knee) could be fatal. Both blood cell production and bone calcium release are regulated by hormones. Once again, the endocrine system cooperates with the cardiovascular system.

Finally, we must not forget that the urinary system has functions besides producing and excreting urine. The kidneys help regulate the acid-base and salt-water balances of blood and tissue fluid. Erythropoietin, a hormone produced by the kidneys, stimulates red blood cell production. The urinary system joins the muscular, skeletal, and endocrine systems to maintain the internal environment.

CHECK YOUR PROGRESS 6.6

1. Explain how the functions of the cardiovascular system contribute to homeostasis.
2. Summarize how each system listed in Figure 6.11 interacts with the cardiovascular system.
3. List a few examples of disorders that affect homeostasis and describe why.

CONNECTING THE CONCEPTS

For more information on how the cardiovascular system interacts with other body systems, refer to the following discussions:

Section 7.2 examines the interaction of the cardiovascular system and lymphatic system.

Section 10.3 explores how the urinary system removes wastes from the blood.

CASE STUDY CONCLUSION

Leukemia is a cancer of the blood. Like any cancer, leukemia is caused by uncontrolled cell growth. In Ben's case, his symptoms could be explained by the fact that the white blood cells in his bone marrow were multiplying uncontrollably and blocking out the ability of the red blood cells and platelets to complete their normal tasks. Therefore, he felt tired and was unable to heal correctly. The treatment for leukemia almost always involves chemotherapy—the use of chemicals to kill the cells that are growing uncontrollably and restore the normal balance of formed elements in the blood. Increasingly, people with leukemia are using bone marrow stem-cell transplants. In this case, either chemotherapy or radiation treatment (sometimes both) is used to kill all of the bone marrow cells. Stem cells from a compatible donor are then inserted into the bone marrow, where they will hopefully establish a new population of healthy blood cells and platelets.

MEDIA STUDY TOOLS

Enhance your study of this chapter with media! Visit **www.mhhe.com/maderhuman13e** and go to "Media Study Tools" for this chapter to access the following:

 **Animations**

6.1 Fluid Exchange Across the Walls of Capillaries
6.2 Hemoglobin Breakdown
6.3 Phagocytosis

 **Video**

6.1 Heart Stem Cells

 **MP3 Files**

6.1 General Functions and Composition of Blood
6.5 Blood Groupings

SUMMARIZE

6.1 Blood: An Overview

Blood functions to

- transport hormones, oxygen, and nutrients to cells;
- transport carbon dioxide and other wastes from cells;
- fight infections by transporting **antibodies** and cells of the immune system;
- maintain blood pressure and regulate body temperature; and
- keep the pH of body fluids within normal limits.

These functions help maintain homeostasis.

Blood has two main components: **plasma** and **formed elements** (red blood cells, white blood cells, and platelets).

Plasma

Plasma is a fluid connective tissue, 91% of which is water. **Plasma proteins** (albumins, globulins, and fibrinogen) are mostly produced by the liver. These proteins maintain **osmotic pressure** and help regulate pH. **Albumins** transport other molecules, **globulins** function in immunity, and prothrombin and **fibrinogen** enable blood clotting.

6.2 Red Blood Cells and Transport of Oxygen

Red blood cells (**erythrocytes**) lack a nucleus and other organelles. They contain **hemoglobin** (**Hb**), which combines with oxygen and transports it to the tissues. Hemoglobin assists in carbon dioxide transport, as well.

Red blood cell (RBC) production is controlled by the blood oxygen concentration. When oxygen concentration decreases, the kidneys increase production of the hormone **erythropoietin (EPO)**. In response, more red blood cells are produced by the bone marrow.

Diseases involving RBCs include **anemia** (not enough hemoglobin to transport oxygen), **hemolysis** (rupturing of RBCs), and **sickle-cell disease** (malformed RBCs).

6.3 White Blood Cells and Defense Against Disease

White blood cells (leukocytes) are larger than red blood cells. They have a nucleus and are translucent unless stained. White blood cells are either **granular leukocytes** or **agranular leukocytes**. White blood cells are an important part of the **immune system,** which protects the body from infection. They often use phagocytosis to ingest foreign compounds or cells called **antigens.**

- The granular leukocytes are eosinophils, basophils (and mast cells), and neutrophils. Neutrophils are abundant, respond first to infections, and phagocytize pathogens.
- The agranular leukocytes include monocytes and lymphocytes. **Monocytes** are the largest white blood cells. They can become macrophages that phagocytize pathogens and cellular debris. **Lymphocytes** (B cells and T cells) are responsible for specific immunity.

Diseases associated with white blood cells include **severe combined immunodeficiency (SCID**; inability to fight infections), **leukemia** (white blood cell cancer), and **infectious mononucleosis** (produced by infection with the EBV virus).

6.4 Platelets and Blood Clotting

Platelets (thrombocytes) result from fragmentation of **megakaryocytes** in the red bone marrow and function in blood clotting.

Blood Clotting

Platelets and the plasma proteins, **prothrombin** (and **thrombin**) and **fibrinogen**, function in blood clotting (**coagulation**), an enzymatic process. Fibrin threads that trap red blood cells result from the enzymatic reaction.

Diseases associated with improper blood clotting include **thrombocytopenia** (insufficient number of platelets), a **thromboembolism** (movement of the blood clot into the heart, lungs, or brain), and **hemophilia** (loss of a specific clotting factor).

6.5 Blood Typing and Transfusions

Blood typing usually involves determining the ABO blood group and whether the person is Rh– or Rh+. Determining blood type is necessary for transfusions so that **agglutination** (clumping) of red blood cells does not occur.

ABO Blood Groups

ABO blood typing determines the presence or absence of type A antigen and type B antigen on the surface of red blood cells.

- *Type A Blood* Type A surface antigens; plasma has anti-B antibodies.
- *Type B Blood* Type B surface antigens; plasma has anti-A antibodies.
- *Type AB Blood* Both type A and type B surface antigens; plasma has neither anti-A nor anti-B antibodies (universal recipient).
- *Type O Blood* Neither type A nor type B surface antigens. Plasma has both anti-A and anti-B antigens (universal donor).
- *Agglutination* Agglutination occurs if the corresponding antigen and antibody are mixed (i.e., if the donor has type A blood and the recipient has type B blood).

Rh Blood Groups

The Rh antigen must also be considered when transfusing blood. It is very important during pregnancy because an Rh-negative mother may form antibodies to the Rh antigen while carrying or after the birth of an Rh-positive child. These antibodies can cross the placenta to destroy the red blood cells of an Rh-positive child.

6.6 Homeostasis

Homeostasis depends upon the cardiovascular system because it serves the needs of the cells. Other body systems are also critical to cardiovascular system function:

- The digestive system supplies nutrients.
- The respiratory system supplies oxygen and removes carbon dioxide from the blood.
- The nervous and endocrine systems help maintain blood pressure. Endocrine hormones regulate red blood cell formation and calcium balance.
- The lymphatic system returns tissue fluid to the veins.
- Skeletal muscle contraction (skeletal system) and breathing movements (respiratory system) propel blood in the veins.

ASSESS

Testing Your Knowledge of the Concepts

1. The transport function of blood is dependent on what components? The defense function of blood is dependent on what components? The regulatory functions of blood are dependent on what components? (pages 114–115)
2. Why is the osmotic pressure of blood important to overall homeostasis? (page 116)
3. Describe the structure and functions of a red blood cell, including the molecule hemoglobin. What is the role of red blood cells in the blood? How is the production of red blood cells regulated? (pages 116–118)
4. Briefly describe the different types of white blood cells by structure and function. (pages 119–120)
5. What formed element is crucial to blood clotting? What other substances are necessary for clotting? Explain the steps that take place when blood clots. (page 121)
6. Distinguish between disorders involving erythrocytes, leukocytes, and platelets. (pages 118, 120, and 121)
7. For each type of ABO blood, give the antigen or antigens and antibody or antibodies present. List which blood type each can receive and to which type each can be given. (pages 124–125)
8. Explain what occurs in hemolytic disease of the newborn. (page 126)
9. Choose five body systems, and briefly explain how they interact with cardiovascular system function. (pages 126–127)

10. Which of the following is not a formed element in the blood?
  a. red blood cells
  b. platelets
  c. white blood cells
  d. organic molecules and salts

In questions 11–15, match each description with a white blood cell in the key.

Key:

  a. lymphocytes
  b. neutrophils
  c. basophils
  d. monocytes
  e. eosinophils

11. U-shaped nucleus; blue-stained granules that release histamine

12. Includes B cells and T cells that provide specific immunity

13. Bilobed nucleus, red-stained granules, allergic reactions, and parasitic worms

14. Largest; no granules; become macrophages

15. Most abundant; multilobed nucleus; first responders to invasion

16. Which of the plasma proteins contribute(s) most to osmotic pressure?
  a. albumin
  b. globulins
  c. erythrocytes
  d. fibrinogen

17. Stem cells are responsible for
  a. red blood cell production.
  b. white blood cell production.
  c. platelet production.
  d. the production of all formed elements.

18. Which hemoglobin component is recovered for reuse following red blood cell destruction?
  a. heme
  b. globin
  c. iron
  d. Both b and c are correct.

19. When the oxygen capacity of the blood is reduced,
  a. the liver produces more bile.
  b. the kidneys release erythropoietin.
  c. the thymus produces more red blood cells.
  d. sickle-cell disease occurs.
  e. All of the choices are correct.

20. Which of the following is not true of white blood cells?
  a. formed in red bone marrow
  b. carry oxygen and carbon dioxide
  c. can leave the bloodstream and enter tissues
  d. can fight disease and infection

21. Which of the following is in the correct sequence for blood clotting?
  a. prothrombin activator, prothrombin, thrombin
  b. fibrin threads, prothrombin activator, thrombin
  c. thrombin, fibrinogen, fibrin threads
  d. prothrombin, clotting factors, fibrinogen
  e. Both a and c are correct.

22. Theoretically, a person with type AB blood should be able to receive
  a. type B and type AB blood.
  b. type O and type B blood.
  c. type A and type O blood.
  d. All of the choices are correct.

23. Blood is associated with which of the following forms of homeostasis?
  a. nutrient supply to the body
  b. supply of gases such as oxygen and carbon dioxide
  c. removal of waste material
  d. transport of hormones
  e. All of the choices are correct.

ENGAGE

Thinking Critically About the Concepts

After reading the homeostasis section of this chapter, you should better understand the interactions between different organ systems that are required for homeostasis. No system ever operates alone. Keep in mind what you've learned about the cardiovascular system as you continue your studies. You may be surprised by how often you hear about concepts from Chapters 5 and 6 in upcoming chapters.

Ben, from the case study, suffers from leukemia. By causing huge numbers of abnormal white blood cells to be produced, the disease disrupts homeostasis in all organ systems. Thrombocytopenia, or platelet deficiency, may result in a fatal hemorrhage. Diseased white blood cells can't battle infections. Most important, when bone marrow produces leukemia cells instead of red blood cells, tissues cannot receive the oxygen they require. Without aggressive treatment, Ben's disease would be fatal.

1. Carbon monoxide (CO) is a deadly, odorless, colorless gas. Hemoglobin in red blood cells binds much more closely to CO than it does to oxygen. Hemoglobin's ability to transport oxygen is severely compromised in the presence of CO.
  a. A malfunctioning furnace is one potential cause for CO poisoning. What are some other situations in which you could be exposed to carbon monoxide? What safeguards should be in place when around these situations?
  b. If cells are deprived of oxygen, what is the effect on the production of cellular energy? Be specific.

2. There are three specific nutrients mentioned in this chapter that are necessary for the body to form red blood cells.
  a. Can you name all three?
  b. What are good food sources for these nutrients?

3. What type of organic molecule is hemoglobin? What nutrient is required to form hemoglobin?

4. The hormone erythropoietin is produced by the kidneys whenever more red blood cells are necessary. Think of several reasons for the body needing more red blood cells.

5. Athletes who abuse erythropoietin have many more red blood cells than usual. After examining Figure 6.3 and imagining many more red blood cells than usual in this capillary, explain why an athlete might die from having too many red blood cells.

6. For several years, researchers have attempted to produce an artificial blood for transfusions. Artificial blood would most likely be safer and more readily available than human blood. While artificial blood might not have all the characteristics of human blood, it would be useful on the battlefield and in emergency situations. Which characteristics of normal blood must artificial blood have to be useful, and which would probably be too difficult to reproduce?

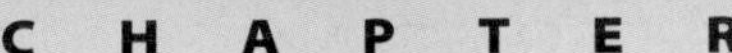

C H A P T E R

7

The Lymphatic and Immune Systems

CASE STUDY LUPUS

Abigail is a healthy and active 12-year-old girl. She has a pretty unremarkable patient history that included nothing but the normal childhood diseases—croup when she was an infant, chicken pox a few years ago, several ear infections, and a couple of bad bouts of the flu. Over the last few months, she has been complaining to her mother that she was tired, her arms and legs ached, and her knees and elbows were bothering her. Her mother immediately thought it was "growing pains" and dismissed the symptoms. Within a few weeks, Abigail developed a rash on her cheeks and across the bridge of her nose, which resembled a butterfly shape. Her mother believed it was due to the new face soap she had switched to. Thinking Abigail might be sensitive to an ingredient in the soap, her mother quickly switched back to the old brand. The rash did not subside. Abigail suddenly began developing ulcers in her mouth, which interfered with eating and drinking. Within a few more weeks, Abigail also began experiencing some digestive issues—stomachaches after eating, periodic bouts of diarrhea, and a noticeable weight loss. Within another few weeks, Abigail began losing handfuls of hair. Her mother took her to the pediatrician.

Dr. Koos did a full exam on Abigail and ran a battery of tests over the next few days. The tests included a complete blood count (CBC), urinalysis, various protein assays, and an ANA (antinuclear antibody) test, which is a test commonly used to aid in the diagnosis of many different autoimmune disorders. Once the results were in and a diagnosis was finally obtained, Dr. Koos explained that Abigail had lupus, an autoimmune disease.

As you read through the chapter, think about the following questions:

1. What is an autoimmune disease?
2. Are the symptoms Abigail experienced common to autoimmune diseases?
3. How is an autoimmune disease acquired?

CHAPTER CONCEPTS

BEFORE YOU BEGIN

Before beginning this chapter, take a few moments to review the following discussions:

Section 2.6 How is a protein's structure related to its function?

Section 3.2 What are some differences between prokaryotic and eukaryotic cells?

Section 6.3 What is the role of white blood cells in defense against pathogens?

7.1 Microbes, Pathogens, and You

LEARNING OUTCOMES

Upon completion of this section, you should be able to

1. Distinguish between a prokaryotic and eukaryotic cell.
2. Identify the structures of a prokaryotic cell.
3. Describe the structure of a general virus.

The term *microbe* applies to microscopic organisms, such as bacteria, viruses, and protists that are widely distributed in the environment. They may be found on both inanimate objects and on the surfaces and interiors of plants and animals. While the term often is associated with diseases, in fact many of the activities of microbes are useful to humans. We eat foods produced by bacteria every day. They contribute to the production of yogurt, cheese, bread, beer, wine, and many pickled foods. Today, drugs available through biotechnology are produced by bacteria. Microbes help us in still another way. Without the activity of decomposers, the biosphere (including ourselves) would cease to exist. When a tree falls to the forest floor, it eventually rots because decomposers, including bacteria and fungi, break down the remains of dead organisms to inorganic nutrients. Plants need these inorganic nutrients to make the many molecules that become food for us.

Despite all these benefits, there are certain bacteria and viruses known as **pathogens,** or disease-causing agents. The body has three lines of defense against invasion:

1. Barriers to entry, such as the skin and mucous membranes of body cavities, act to prevent pathogens from gaining entrance into the body.
2. First responders, such as the phagocytic white blood cells, act to prevent an infection after an invasion has occurred due to a pathogen getting past a barrier and into the body.
3. Acquired defenses overcome an infection by killing the particular disease-causing agent that has entered the body. Acquired defenses also protect us against cancer.

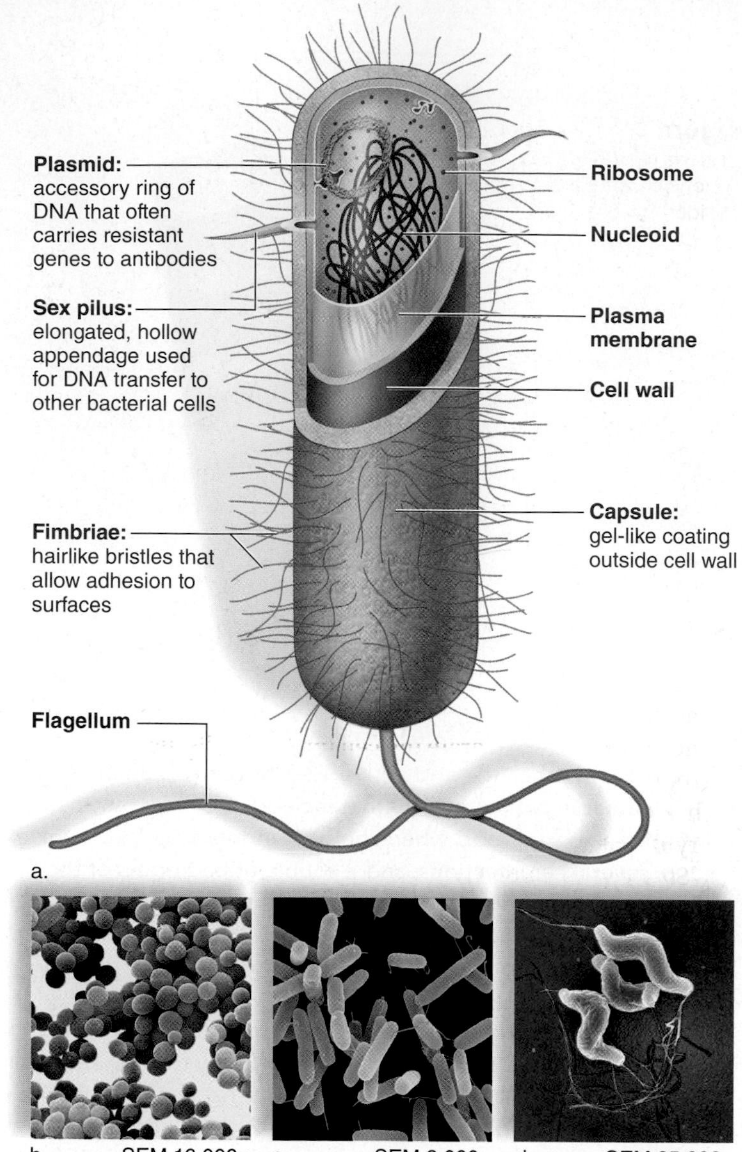

Figure 7.1 Typical shapes of bacteria.
a. Bacterial features, especially the four with definitions, contribute to the ability of bacteria to cause disease. Bacteria occur in three shapes. **b.** *Staphylococcus aureus* is a sphere-shaped bacterium that causes toxic shock syndrome. **c.** *Pseudomonas aeruginosa* is a rod-shaped bacterium that causes urinary tract infections. **d.** *Campylobacter jejuni* is a curve-shaped bacterium that causes food poisoning.

Bacteria

Bacteria are single-celled prokaryotes that do not have a nucleus. Figure 7.1 illustrates the main features of bacterial anatomy and shows the three common shapes: *coccus* (spherical-shaped), *bacillus* (rod-shaped), and *spirillum* (curved, sometimes spiral-shaped). Bacteria have a cell wall that contains a unique amino disaccharide. The "cillin" antibiotics, such as penicillin, interfere with the production of the cell wall. The cell wall of some bacteria is surrounded by a **capsule** that has a thick, gelatinous consistency. Capsules often allow bacteria to stick to surfaces such as teeth. They also prevent phagocytic white blood cells from taking them up and destroying them.

Motile bacteria usually have long, very thin appendages called **flagella** (sing., flagellum). The flagella rotate 360° and cause the bacterium to move backward. Some bacteria have **fimbriae,** stiff fibers that allow the bacteria to adhere to surfaces such as host cells. Fimbriae allow a bacterium to cling to and gain access to the body. In contrast, a **pilus** is an elongated hollow appendage used to transfer DNA from one cell to another. Genes that allow bacteria to be resistant to antibiotics can be passed from one to the other through a pilus by a process called *conjugation.*

Bacteria are independent cells that are capable of performing many diverse functions. Their DNA is packaged in a chromosome that occupies the center of the cell. Many bacteria also have small circular pieces of DNA called **plasmids.** Genes that allow bacteria to be resistant to antibiotics are often located in a plasmid. Abuse of antibiotic therapy increases the number of resistant bacterial strains that are difficult to kill, even with antibiotics.

Animation Antibiotic Resistance

Bacteria reproduce by a process called *binary fission* (Fig. 7.2). The single, circular chromosome attached to the plasma membrane is copied. Then the chromosomes are

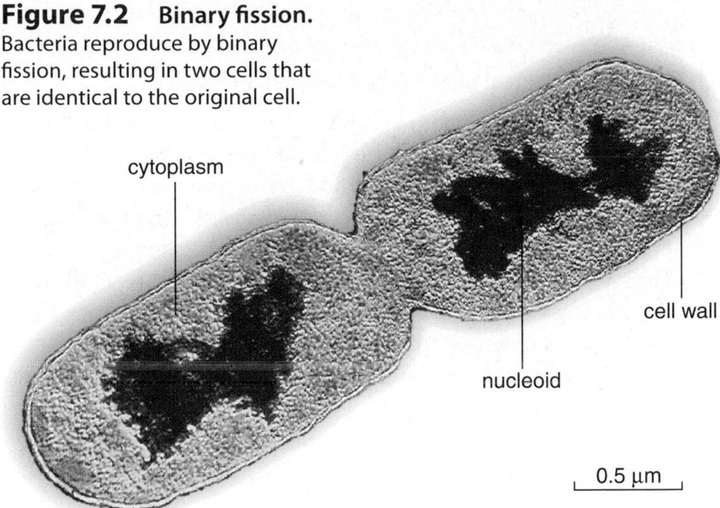

Figure 7.2 **Binary fission.** Bacteria reproduce by binary fission, resulting in two cells that are identical to the original cell.

separated as the cell enlarges. The newly formed plasma membrane and cell wall separate the cell into two cells. Bacteria can reproduce rapidly under favorable conditions, with some species doubling their numbers every 12 minutes.

Animation: Binary Fission

Strep throat, tuberculosis, gangrene, gonorrhea, and syphilis are well-known bacterial diseases. Not only does growth of bacteria cause disease but some bacteria release molecules called **toxins** that inhibit cellular metabolism. For example, it is important to have a tetanus shot because the bacteria that causes this disease, *Clostridium tetani,* produces a toxin that prevents relaxation of muscles. In time, the body contorts because all the muscles have contracted; if no medical treatment is available, suffocation can occur. More on pathogenic bacteria is provided in the Infectious Disease Supplement following this chapter.

Viruses

Viruses bridge the gap between the living and the nonliving. Outside a host, viruses are essentially chemicals that can be stored on a shelf. But when the opportunity arises, viruses replicate inside cells, and during this period, they clearly appear to be alive.

However, viruses are acellular (not composed of cells) and they lack the metabolic machinery needed to acquire and use nutrients. In effect, they are cellular parasites. Viruses cause diseases such as colds, flu, measles, chicken pox, polio, rabies, AIDS, genital warts, and genital herpes. The Infectious Disease Supplement following this chapter takes a more detailed look at some important viral pathogens.

The largest viruses are only about one-quarter the size of a bacterium, about one-hundredth the size of a eukaryotic cell (Fig. 7.3). However, most are much smaller and may only be viewed using the most powerful microscopes. A virus always has two parts: an outer capsid composed of protein units and an inner core of nucleic acid (Fig. 7.4). A virus carries the genetic information needed to reproduce itself. In contrast to cellular organisms, the viral genetic material need not be double-stranded DNA, nor even DNA. Indeed, some viruses, such as HIV and influenza, have RNA as their genetic material. A virus may also contain various enzymes that help it reproduce.

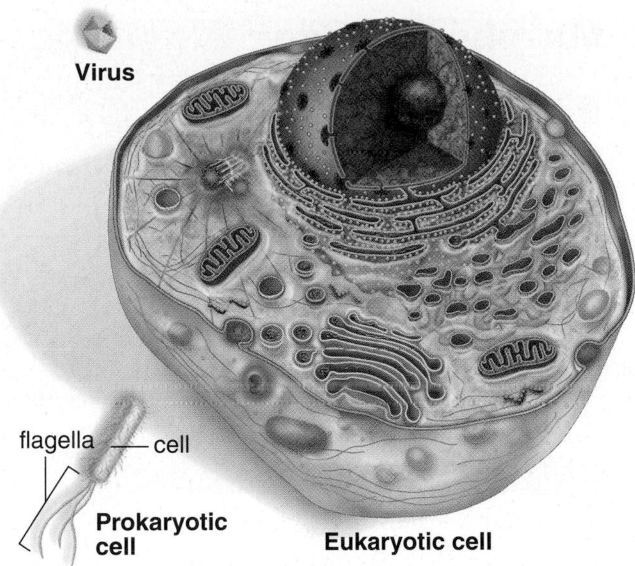

Figure 7.3 **Comparative sizes of viruses, bacteria, and eukaryotic cells.**
Viruses are tiny noncellular particles, whereas bacteria are small independent cells. Eukaryotic cells are more complex and larger because they contain a nucleus and many organelles.

Viruses are microscopic pirates, commandeering the metabolic machinery of a host cell. Viruses gain entry into and are specific to a particular host cell because portions of the virus are specific for a receptor on the host cell's outer surface. Once the virus is attached, the viral genetic material (DNA or RNA) enters the cell. Inside the cell, the nucleic acid codes for the protein units in the capsid. In addition, the virus may have genes for special enzymes needed for the virus to reproduce and exit from the host cell. In large measure, however, a virus relies on the host's enzymes and ribosomes for its own reproduction.

APPLICATIONS AND MISCONCEPTIONS

Does refrigeration kill bacteria?

The answer is no. The speed at which bacteria reproduce depends on a number of factors, including moisture in the environment and temperature. At the temperatures found in most refrigerators and freezers, bacterial growth is slowed but the bacteria are not killed. Once the temperature returns to a favorable level, the bacteria resume normal cell division. The only way to kill most food-related bacteria is by using high temperatures, such as those used in boiling or thorough cooking.

Animation: Food Pathogens & Temperature

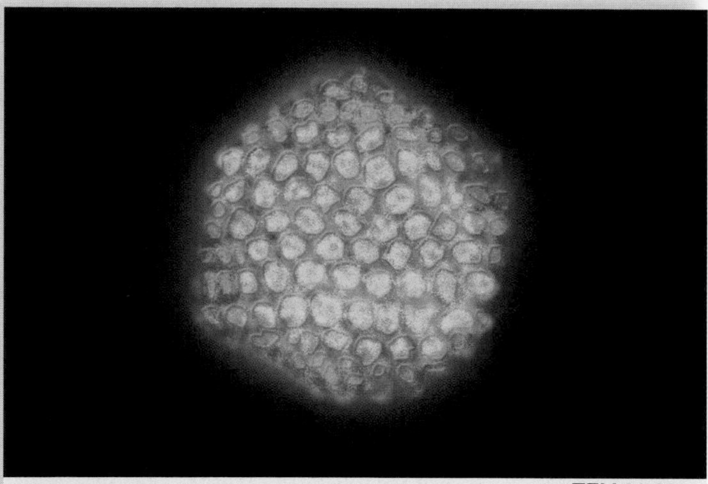

Adenovirus: DNA virus with a polyhedral capsid and a fiber at each corner.

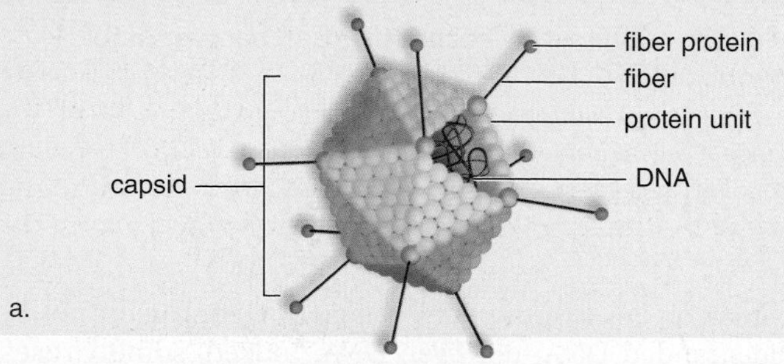

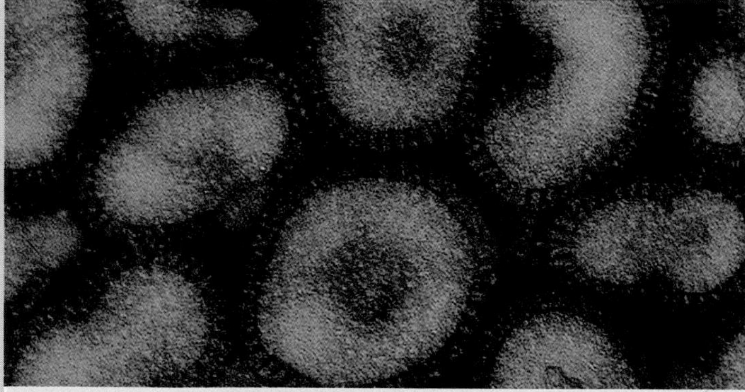

Influenza virus: RNA virus with a spherical capsid surrounded by an envelope with spikes.

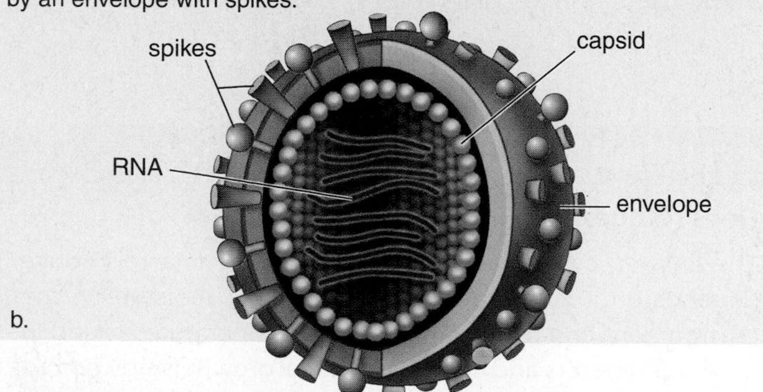

Figure 7.4 Typical virus structures.
Despite their diversity, all viruses have an outer capsid, composed of protein subunits, and a nucleic acid core, composed of either DNA or RNA but not both. **a.** Adenoviruses cause colds, and (**b**) influenza viruses cause the flu.

APPLICATIONS AND MISCONCEPTIONS

Are antibacterial cleansers really effective?

Studies show that using antibacterial soap when washing your hands is no more effective in preventing the spread of bacteria than using pure soap and water. In some cases, the main ingredient in most antibacterial soaps, triclosan, may actually cause some bacteria to become resistant to certain drugs like amoxicillin, which is used to kill bacteria. Waterless hand sanitizers (which are alchohol-based) are recommended to be used in conjunction with, not as a replacement for, regular hand washing. Proper hand washing and sanitizing can prevent the spread of colds, flus, and certain diarrhea disorders when done often and properly.

Prions

Prions, infectious particles made strictly of proteins, cause a group of degenerative diseases of the nervous system, also called wasting diseases. Originally thought to be viral diseases, prions cause **Creutzfeldt-Jakob disease** (**CJD**) in humans; scrapie in sheep; and bovine spongiform encephalopathy (BSE), commonly called mad cow disease**,** in cattle. These infections are apparently transmitted by ingestion of brain and nerve tissues from infected animals. Prion proteins are believed to play the role of a "housekeeper" in the brains of healthy individuals. However, in some people, a "rogue" form of the prion protein folds into a new shape and in the process loses its original function. The "rogue" protein is able to refold normal prion proteins into the new shape and thus cause disease. Nervous tissue is lost, and calcified plaques show up in the brain due to activity of prions. Thankfully, the incidence of prion diseases in humans is very low.

CHECK YOUR PROGRESS 7.1

1. Describe what makes a virus or bacteria a pathogen.
2. Detail the structures in bacteria that can be associated with virulence, the ability to cause disease.
3. Summarize why viruses are considered parasites that always cause disease.

CONNECTING THE CONCEPTS

For more information on the topics presented in this section, refer to the following discussions:

Section S.3 explores the problem of antibiotic resistance in medicine.

Section 16.6 explains the link between bacteria and some sexually transmitted diseases.

Section 19.2 examines how some viruses may cause cancer.

7.2 The Lymphatic System

LEARNING OUTCOMES

Upon completion of this section, you should be able to

1. Describe the structure of the lymphatic system.
2. Summarize how the lymphatic system contributes to homeostasis.
3. Explain how the lymphatic system interacts with the circulatory system.

The **lymphatic system** consists of lymphatic vessels and the lymphatic organs. This system, closely associated with the cardiovascular system, has four main functions that contribute to homeostasis: (1) Lymphatic capillaries absorb excess tissue fluid and return it to the bloodstream; (2) in the small intestines, lymphatic capillaries called lacteals absorb fats in the form of lipoproteins and transport them to the bloodstream; (3) the lymphatic system is responsible for the production, maintenance, and distribution of lymphocytes; and (4) the lymphatic system helps defend the body against pathogens.

MP3 Lymphatic System

Lymphatic Vessels

Lymphatic vessels form a one-way system of capillaries to vessels and, finally, to ducts. These vessels take lymph to cardiovascular veins in the shoulders (Fig. 7.5). As mentioned in Chapter 6, lymphatic capillaries take up excess tissue fluid. Tissue fluid is mostly water, but it also contains solutes (i.e., nutrients, electrolytes, and oxygen) derived from plasma. Tissue fluid also contains cellular products (i.e., hormones, enzymes, and wastes) secreted by cells. The fluid inside lymphatic vessels is called **lymph.** Lymph is usually a colorless liquid, but after a meal, it appears creamy because of its lipid content.

The lymphatic capillaries join to form lymphatic vessels that merge before entering either the thoracic duct or the right lymphatic duct. The larger thoracic duct returns lymph collected from the body below the thorax, the left arm, and left side of the head and neck into the left subclavian vein. The

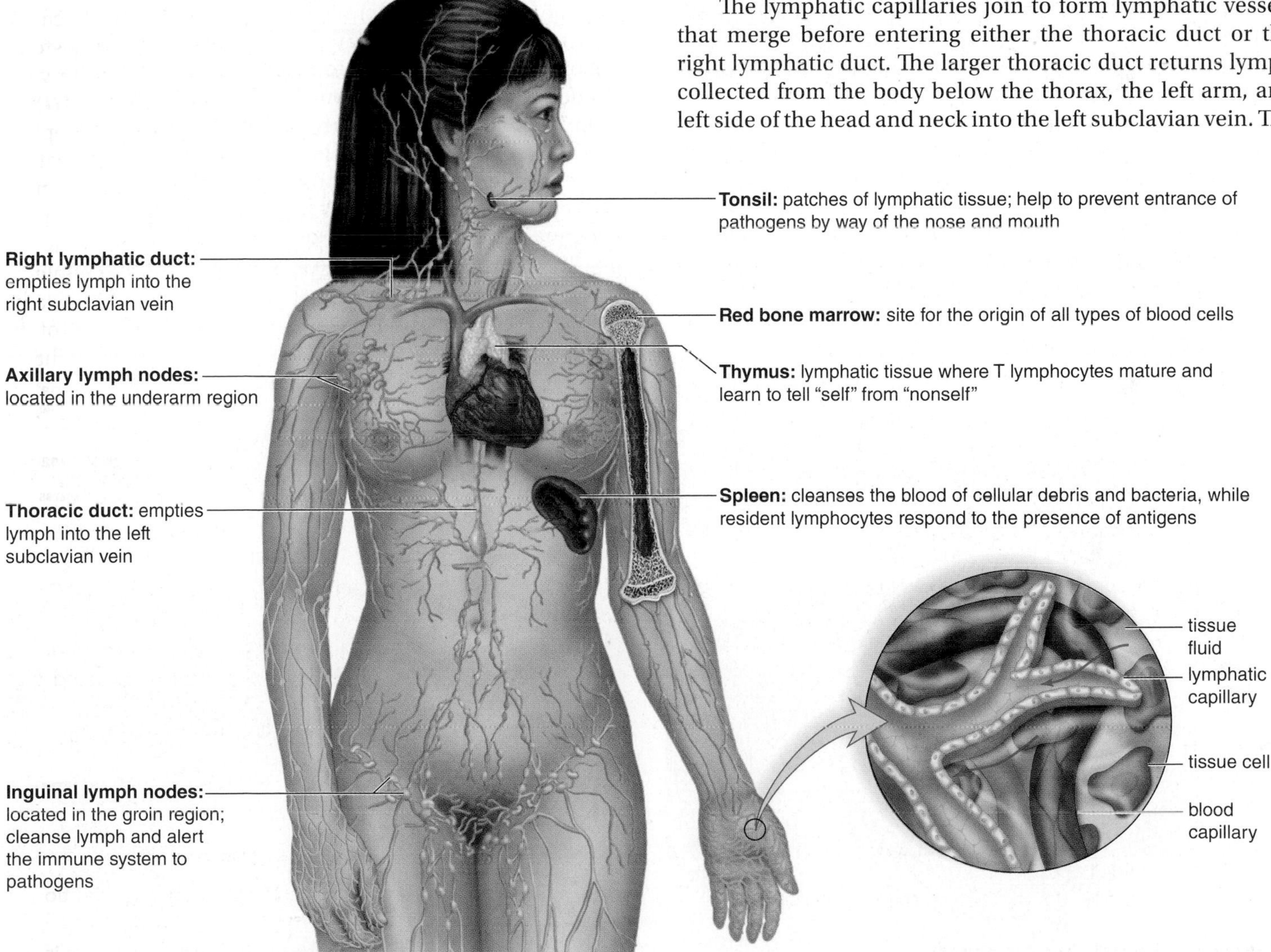
Tonsil: patches of lymphatic tissue; help to prevent entrance of pathogens by way of the nose and mouth
Right lymphatic duct: empties lymph into the right subclavian vein
Red bone marrow: site for the origin of all types of blood cells
Axillary lymph nodes: located in the underarm region
Thymus: lymphatic tissue where T lymphocytes mature and learn to tell "self" from "nonself"
Thoracic duct: empties lymph into the left subclavian vein
Spleen: cleanses the blood of cellular debris and bacteria, while resident lymphocytes respond to the presence of antigens
tissue fluid
lymphatic capillary
tissue cell
blood capillary
Inguinal lymph nodes: located in the groin region; cleanse lymph and alert the immune system to pathogens

Figure 7.5 Functions of the lymphatic system components.
Lymphatic vessels drain excess fluid from the tissues and return it to the cardiovascular system. The enlargement shows that lymphatic vessels, like cardiovascular veins, have valves to prevent backward flow. The lymph nodes, spleen, thymus, and red bone marrow are the main lymphatic organs that assist immunity.

right lymphatic duct returns lymph from the right arm and right side of the head and neck into the right subclavian vein.

The construction of the larger lymphatic vessels is similar to that of cardiovascular veins, including the presence of valves. The movement of lymph within lymphatic capillaries is largely dependent upon skeletal muscle contraction. Lymph forced through lymphatic vessels as a result of muscular compression is prevented from flowing backward by one-way valves.

Animation Lymphatic System

Lymphatic Organs

The lymphatic organs are divided into two categories. The primary lymphatic organs include the red bone marrow and the thymus, whereas the lymph nodes and spleen represent the secondary lymphatic organs. Figure 7.6 shows tissue samples taken from the primary and secondary lymphatic organs.

The Primary Lymphatic Organs

Red bone marrow (Fig 7.6*a*) produces all types of blood cells. In a child, most bones have red bone marrow; in an adult, it is limited to the sternum, vertebrae, ribs, part of the pelvic girdle, and the upper ends of the humerus and femur. In addition to the red blood cells, bone marrow produces the various types of white blood cells: neutrophils, eosinophils, basophils, lymphocytes, and monocytes. Lymphocytes are either **B cells** (**B lymphocytes**) or **T cells** (**T lymphocytes**). B cells mature in the bone marrow, but the T cells mature in the thymus. Any B cell that reacts with cells of the body is removed in the bone marrow and does not enter the circulation. This ensures that the B cells do not harm normal cells of the body.

The soft, bilobed **thymus** (Fig. 7.6*b*) is located in the thoracic cavity between the trachea and the sternum, superior to the heart. The thymus will begin shrinking in size before puberty and is noticeably smaller in an adult than in a child.

The thymus has two functions: (1) The thymus produces thymic hormones, such as thymosin, thought to aid in the maturation of T lymphocytes. Thymosin may also have other functions in immunity. (2) Immature T lymphocytes migrate from the bone marrow through the bloodstream to the thymus, where they mature. Only about 5% of these cells ever leave the thymus. These T lymphocytes have survived a critical test: If any show the ability to react with the individual's cells, they die in the thymus. If they have potential to attack a pathogen, they can leave the thymus. The thymus is absolutely critical to immunity because without mature, properly functioning T cells, the body's response to specific pathogens is poor or absent.

Secondary Lymphatic Organs

The secondary lymphatic organs are the spleen, the lymph nodes, and other organs containing lymphoid tissue. These include the tonsils, Peyer's patches, and the appendix.

The **spleen** (Fig. 7.6*c*) filters blood. The spleen, the largest lymphatic organ, is located in the upper left region of the abdominal cavity posterior to the stomach. Connective tissue divides the spleen into regions known as white pulp and red pulp. The red pulp, which surrounds venous sinuses (cavities), is involved in filtering the blood. Blood entering the spleen must pass through the sinuses before exiting. Here, macrophages that are like powerful vacuum cleaners engulf pathogens and debris, such as worn-out red blood cells.

The spleen's outer capsule is relatively thin, and an infection or a blow can cause the spleen to burst. Although the spleen's functions are replaced by other organs, a person without a spleen is often slightly more susceptible to infections and may have to receive antibiotic therapy indefinitely.

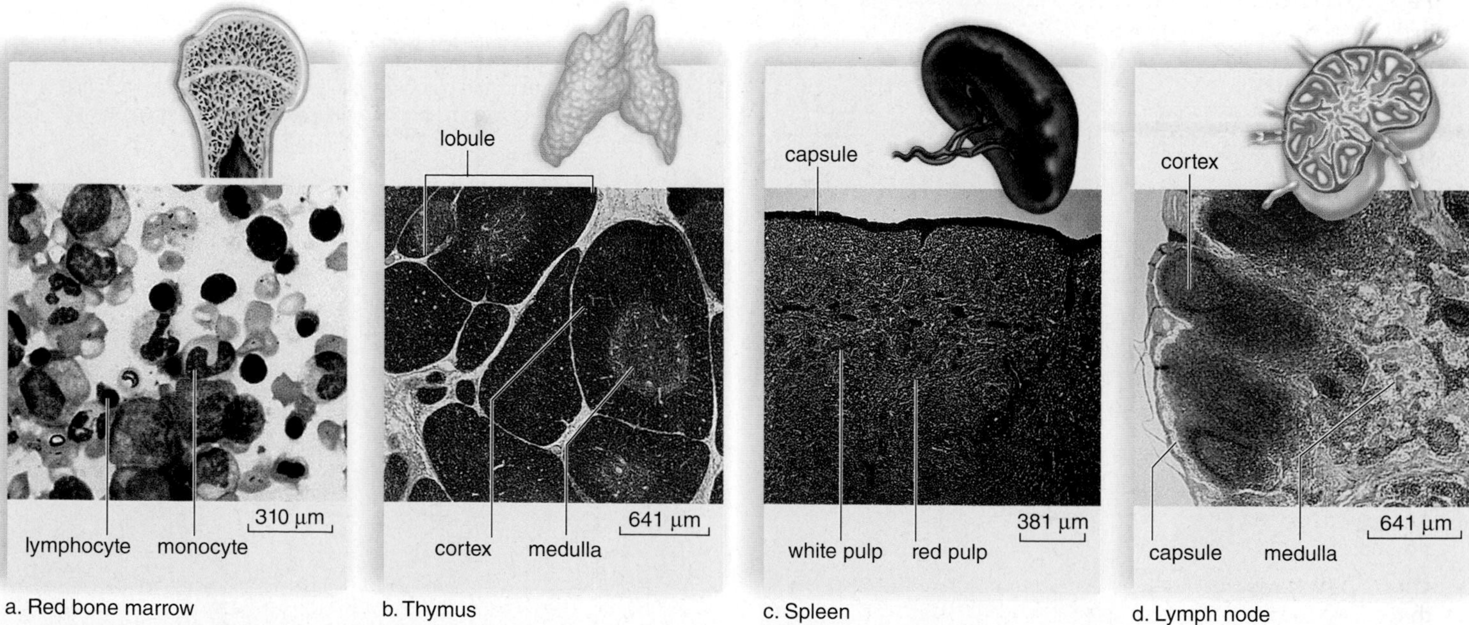

Figure 7.6 Tissue samples from primary and secondary lymphatic organs.
Red bone marrow (**a**) and the thymus (**b**) are the primary lymphatic organs. Blood cells, including lymphocytes, are produced in red bone marrow. B cells mature in the bone marrow, but T cells mature in the thymus. The spleen (**c**) and the lymph nodes (**d**) are secondary lymphatic organs. Lymph is cleansed in lymph nodes, and blood is cleansed in the spleen.

APPLICATIONS AND MISCONCEPTIONS

What does it mean when my lymph nodes are swollen?

Lymph nodes will swell when they are fighting an infection. When the body is invaded by a bacteria or virus, individual nodes can swell from 1/2 to 2 inches in diameter. Cuts, burns, bites, rashes, and any break in the skin can cause an infection and thus the swelling of your lymph nodes. Swollen nodes near the groin mean infection on a leg or lower abdomen; in your armpits mean arms or chest; and swollen nodes on the front of your neck mean an infection in the face, ears, nose, or throat. Some diseases, like chicken pox, can cause all your nodes to swell.

Lymph nodes (Fig. 7.6*d*), which occur along lymphatic vessels, filter lymph. Connective tissue forms a capsule and also divides a lymph node into compartments. Each compartment contains a sinus that increases in size toward the center of the node. As lymph courses through the sinuses, it is exposed to macrophages, which engulf pathogens and debris. Lymphocytes, also present in sinuses, fight infections and attack cancer cells.

Lymph nodes are named for their location. For example, inguinal nodes are in the groin, and axillary nodes are in the armpits. Physicians often feel for the presence of swollen, tender lymph nodes in the neck as evidence that the body is fighting an infection. This is a noninvasive, preliminary way to help make such a diagnosis.

Lymphatic nodules are concentrations of lymphatic tissue not surrounded by a capsule. The *tonsils* are patches of lymphatic tissue located in a ring about the pharynx. The tonsils perform the same functions as lymph nodes; but because of their location, they are the first to encounter pathogens and antigens that enter the body by way of the nose and mouth.

Peyer's patches are located in the intestinal wall and tissues within the appendix, a small extension of the large intestine, and encounter pathogens that enter the body by way of the intestinal tract.

CHECK YOUR PROGRESS 7.2

1. Describe how the lymphatic system contributes to fluid homeostasis in the body.
2. Detail the differences between a primary and a secondary lymphatic organ and give an example of each.
3. Predict what might happen to the body if the lymphatic ducts did not allow lymph to drain.

CONNECTING THE CONCEPTS

As noted, the lymphatic system plays a role in the movement of fats and the return of excess fluid to the circulatory system. For more information on these functions, refer to the following discussions:

Section 5.1 examines how the lymphatic system interacts with the circulatory system.

Section 8.3 explores how the lymphatic system is involved in the processing of fat in the diet.

Figure 8.6 diagrams the location of the lymphatic vessels in the small intestine.

7.3 Innate Immune Defenses

LEARNING OUTCOMES

Upon completion of this section, you should be able to

1. List examples of the body's innate defenses.
2. Summarize the events in the inflammatory response.
3. Explain the role of the complement system.

We are constantly exposed to microbes such as viruses, bacteria, and fungi in our environment. Immunity is the capability of killing or removing foreign substances, pathogens, and cancer cells from the body. Mechanisms of innate immunity are fully functional without previous exposure to these invaders, whereas adaptive immunity (see section 7.4) is initiated and amplified by exposure. As summarized in Figure 7.7, innate immune defenses include: physical and chemical barriers; the inflammatory response; phagocytes and natural killer cells; and protective proteins such as complement and interferons.

Innate defenses occur immediately or very shortly after infection occurs. With innate immunity, there is no recognition that an intruder has attacked before, and therefore no immunological "memory" is present for the attacker.

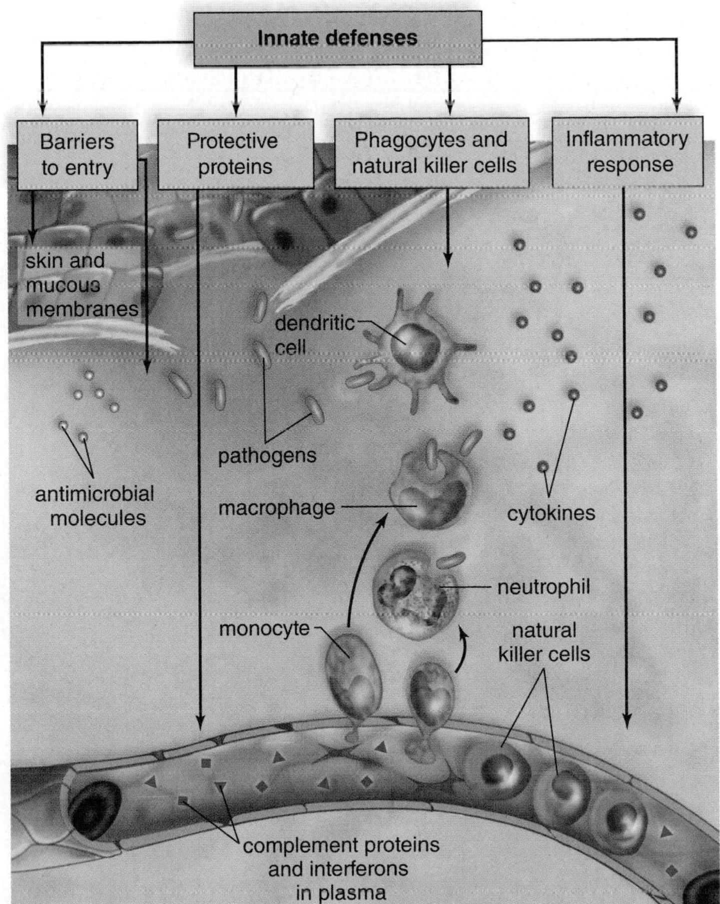

Figure 7.7 Overview of innate immune defenses. Most innate defenses act rapidly to detect and respond to specific molecules expressed by pathogens.

Physical and Chemical Barriers to Entry

The body has built-in barriers, both physical and chemical, that serve as the first line of defense against an infection by pathogens.

The intact skin is generally an effective physical barrier that prevents infection. Mucous membranes lining the respiratory, digestive, reproductive, and urinary tracts are also physical barriers to entry by pathogens. For example, the ciliated cells that line the upper respiratory tract sweep mucus and trapped particles up into the throat, where they can be swallowed, spit, or coughed out.

The chemical barriers to infection include the secretions of sebaceous (oil) glands of the skin. These secretions contain chemicals that weaken or kill certain bacteria on the skin.

Perspiration, saliva, and tears contain an antibacterial enzyme called **lysozyme.** Saliva also helps to wash microbes off the teeth and tongue, and tears wash the eyes. Similarly, as urine is voided from the body, it flushes bacteria from the urinary tract.

The acid pH of the stomach inhibits growth or kills many types of bacteria. At one time, it was thought that no bacterium could survive the acidity of the stomach. But now we know that ulcers are caused by the bacterium *Helicobacter pylori* (see Chapter 1). Similarly, the acidity of the vagina and its thick walls discourage the presence of pathogens.

Finally, a significant chemical barrier to infection is created by the normal flora, microbes that usually reside in the mouth, intestine, and other areas. By using available nutrients and releasing their own waste, these resident bacteria prevent potential pathogens from taking up residence. For this reason, chronic use of antibiotics can make a person susceptible to pathogenic infection by killing off the normal flora.

Inflammatory Response

The inflammatory response exemplifies the second line of defense against invasion by a pathogen. Inflammation employs mainly neutrophils and macrophages to surround and kill (engulf by phagocytosis) pathogens trying to get a foothold inside the body. Protective proteins are also involved. Inflammation is usually recognized by its four hallmark symptoms: redness, heat, swelling, and pain (Fig. 7.8).

The four signs of the inflammatory response are due to capillary changes in the damaged area, and all serve to protect the body. Chemical mediators, such as **histamine,** released by damaged tissue cells and mast cells, cause the capillaries to dilate and become more permeable. Excess blood flow due to enlarged capillaries causes the skin to redden and become warm. Increased temperature in an inflamed area tends to inhibit growth of some pathogens. Increased blood flow brings white blood cells to the area. Increased permeability of capillaries allows fluids and proteins, including blood-clotting factors, to escape into the tissues. Clot formation in the injured area prevents blood loss. The excess fluid in the area presses on nerve endings, causing the familiar pain associated with swelling. Together, these events summon white blood cells to the area.

As soon as the white blood cells arrive, they move out of the bloodstream into the surrounding tissue. The neutrophils are first and actively phagocytize debris, dead cells, and bacteria they encounter. The many neutrophils attracted to the area can usually localize any infection and keep it from spreading.

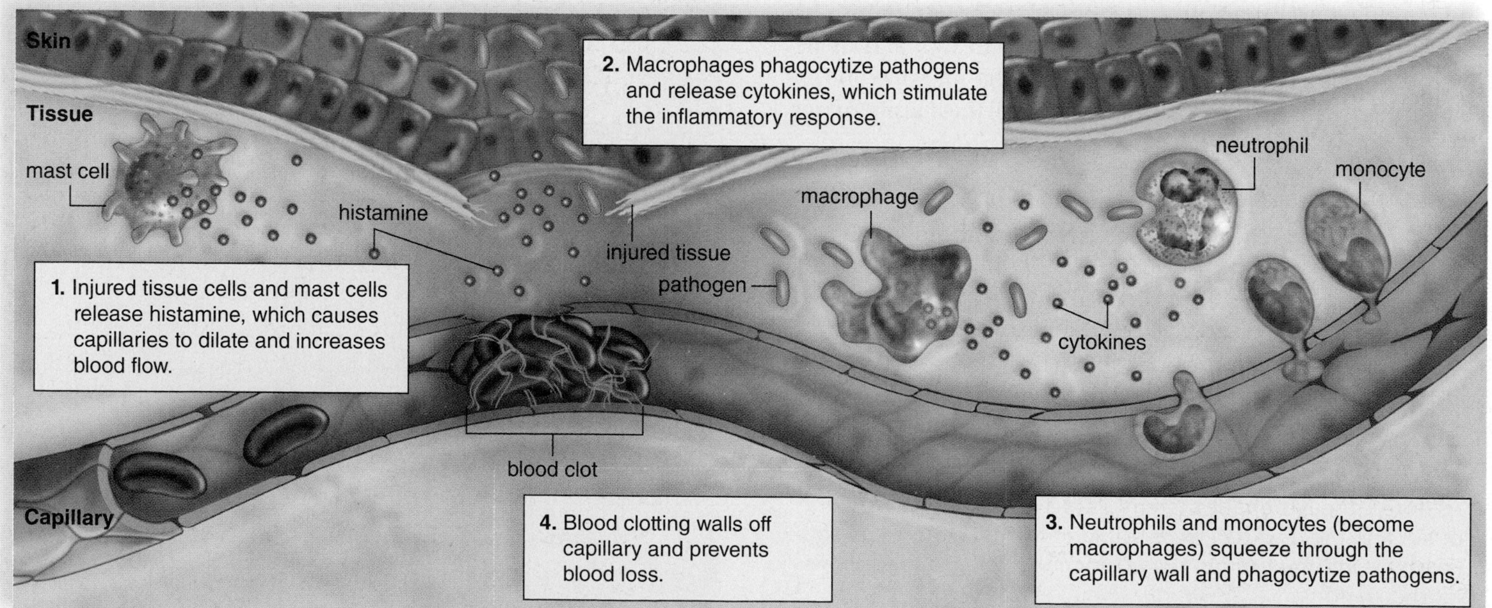

Figure 7.8 **Steps of the inflammatory response.**
1. Due to capillary changes in a damaged area and the release of chemical mediators, such as histamine by mast cells, an inflamed area exhibits redness, heat, swelling, and pain. **2.** Macrophages release cytokines, which stimulate the inflammatory and other immune responses. **3.** Monocytes and neutrophils squeeze through capillary walls from the blood and phagocytize pathogens. **4.** A blood clot can form a seal in a break in a blood vessel.

APPLICATIONS AND MISCONCEPTIONS

How do antihistamines work?

Once histamine is released from mast cells, it binds to receptors on other body cells. There, the histamine causes the familiar symptoms associated with infections and allergies: sneezing, itching, runny nose, and watery eyes. Antihistamines work by blocking the receptors on the cells so that histamine can no longer bind. For allergy relief, antihistamines are most effective when taken before exposure to the allergen.

If neutrophils die off in great quantity, they become a yellow-white substance called pus.

When an injury is not serious, the inflammatory response is short-lived and the healing process will quickly return the affected area to a normal state. Nearby cells secrete chemical factors to ensure the growth (and repair) of blood vessels and new cells to fill in the damaged area.

If, on the other hand, the neutrophils are overwhelmed, they call for reinforcements by secreting chemical mediators called **cytokines.** Cytokines attract more white blood cells to the area, including monocytes. Monocytes are longer-lived cells that become **macrophages,** even more powerful phagocytes than neutrophils. Macrophages can enlist the help of lymphocytes to carry out specific defense mechanisms.

Animation Inflammatory Response

Inflammation is the body's natural response to an irritation or injury and serves an important role. Once the healing process has begun, inflammation rapidly subsides. However, in some cases, chronic inflammation may last for weeks, months, or even years if an irritation or infection cannot be overcome. Inflammatory chemicals may cause collateral damage to the body, in addition to killing the invaders. Should an inflammation persist, anti-inflammatory medications, such as aspirin, ibuprofen, or cortisone, can minimize the effects of various chemical mediators.

Protective Proteins

The **complement system,** often simply called complement, is composed of a number of blood plasma proteins designated by the letter *C* and a number. The complement proteins "complement" certain immune responses, which accounts for their name. For example, they are involved in and amplify the inflammatory response because certain complement proteins can bind to mast cells and trigger histamine release. Others can attract phagocytes to the scene. Some complement proteins bind to the surface of pathogens already coated with antibodies, which ensures that the pathogens will be phagocytized by a neutrophil or macrophage.

Certain other complement proteins join to form a membrane attack complex that produces holes in the surface of bacteria. Fluids then enter the bacterial cell to the point that they burst (Fig. 7.9).

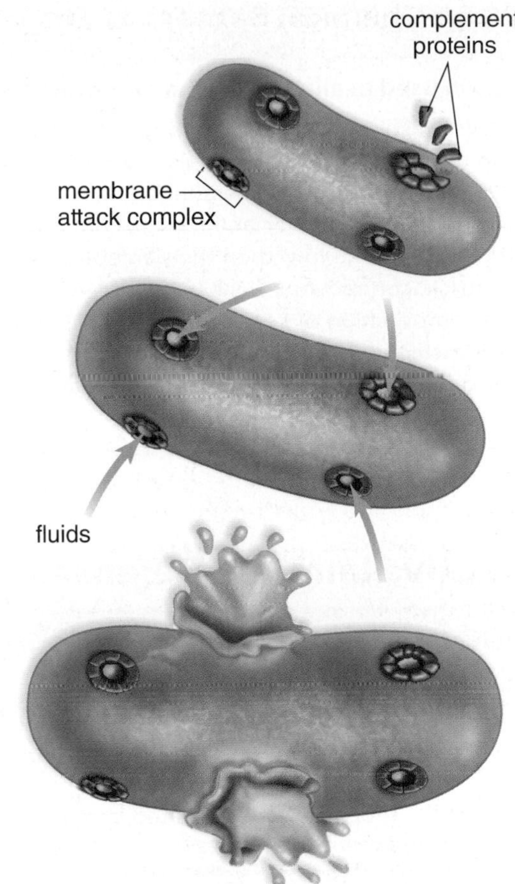

Figure 7.9 **Action of the complement system.**
When the immune response activates complement proteins in the blood plasma, they form a membrane attack complex that makes holes in bacterial cell walls and plasma membranes, allowing fluids to enter, causing the cell to burst.

Interferons are proteins produced by virus-infected cells as a warning to noninfected cells in the area. Interferons bind to receptors of noninfected cells, causing them to prepare for possible attack by producing substances that interfere with viral replication. Interferons are used as treatment in certain viral infections, such as hepatitis C.

Animation Antiviral Activity of Interferon

CHECK YOUR PROGRESS 7.3

1. List some examples of the body's innate defenses.
2. Describe the blood cells associated with innate defenses, and detail how they function.
3. Discuss how the complement proteins got their name.

CONNECTING THE CONCEPTS

For more information on the topics presented in this section, refer to the following discussions:

Section 4.7 examines how the skin forms a physical barrier to the exterior environment.

Section 6.3 provides additional details on white blood cells.

Section S.3 discusses the problems with the overuse of antibiotics.

APPLICATIONS AND MISCONCEPTIONS

Why is aspirin used to alleviate so many symptoms?

The chemicals in aspirin decrease the body's ability to make prostaglandins. Prostaglandins, made by most of the body's tissues, are substances used as messengers in the perception and response to pain, fever, and muscle contractions. If the prostaglandin level is low, then the perception of pain, fever, and muscle contractions will be lowered. Aspirin also decreases the production of certain substances needed in the beginning stages of blood clotting, which is why it is prescribed to patients with certain cardiovascular clotting disorders.

7.4 Adaptive Immune Defenses

LEARNING OUTCOMES

Upon completion of this section, you should be able to

1. Explain the role of an antigen in the acquired defenses.
2. Summarize the process of antibody-mediated immunity and list the cells involved in the process.
3. Summarize the process of cell-mediated immunity and list the cells involved in the process.

When innate defenses have failed to prevent an infection, adaptive defenses come into play. Adaptive defenses overcome an infection by doing away with the particular disease-causing agent that has entered the body. Adaptive defenses also provide some protection against cancer.

How Adaptive Defenses Work

Adaptive defenses respond to large molecules, normally protein structures, called **antigens** that the immune system recognizes as foreign to the body. Fragments of bacteria, viruses, molds, or parasitic worms can all be antigenic. Further, abnormal plasma membrane proteins produced by cancer cells may also be antigens. We do not ordinarily become immune to our own normal cells, so it is said that the immune system is able to distinguish self from nonself.

Adaptive defenses primarily depend on the action of lymphocytes, which differentiate as either B cells (B lymphocytes) or T cells (T lymphocytes). B cells and T cells are capable of recognizing antigens because they have specific antigen receptors. These antigen receptors are plasma membrane proteins whose shape allows them to combine with particular antigens. Each lymphocyte has only one type of receptor. It is often said that the receptor and the antigen fit together like a lock and key. We encounter millions of different antigens during our lifetime, so we need a diversity of B cells and T cells to protect us against them. Remarkably, this diversification occurs during the maturation process. Millions of specific B cells and/or T cells are formed, increasing the likelihood that at least one will recognize any possible antigen.

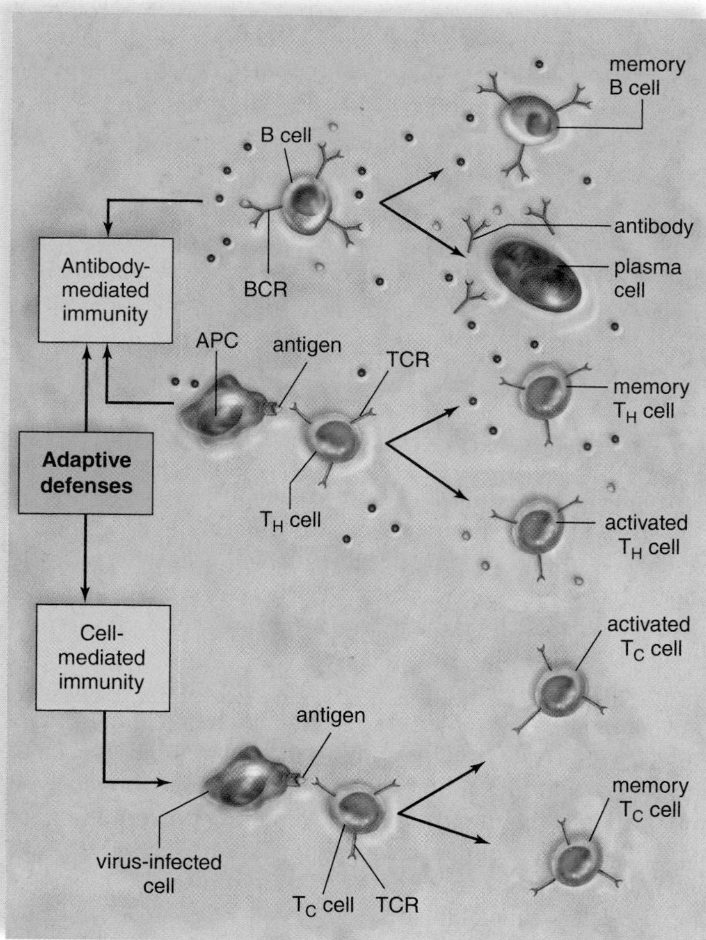

Figure 7.10 **Overview of adaptive immune defenses.** B cells, helper T (T_H) cells, and cytotoxic T (T_C) cells respond to specific antigens by dividing and differentiating. The BCRs of B cells bind to whole, intact antigens, while the TCRs of T_H and T_C cells only bind to antigens that are processed and presented by MHC proteins on the surface of other cells. When activated by antigens, B cells differentiate into antibody-secreting plasma cells, T_H cells become cytokine-secreting cells, and T_C cells are able to destroy virus-infected or cancer cells. Each cell type also produces memory cells that can respond more quickly to a subsequent exposure to the same antigen.

B cells differentiate into other types of cells, and so do T cells. The two major pathways of adaptive immunity that we will be discussing, and the cells associated with each pathway, are presented in Figure 7.10.

B Cells and Antibody-Mediated Immunity

The receptor on a B cell is called a B-cell receptor (BCR). The clonal selection model (Fig. 7.11) states that an antigen selects then binds to the BCR of only one type B cell. Then this B cell produces multiple copies of itself. The resulting group of identical cells is called a clone. Similarly, an antigen can bind to a T-cell receptor (TCR) and this T cell will clone.

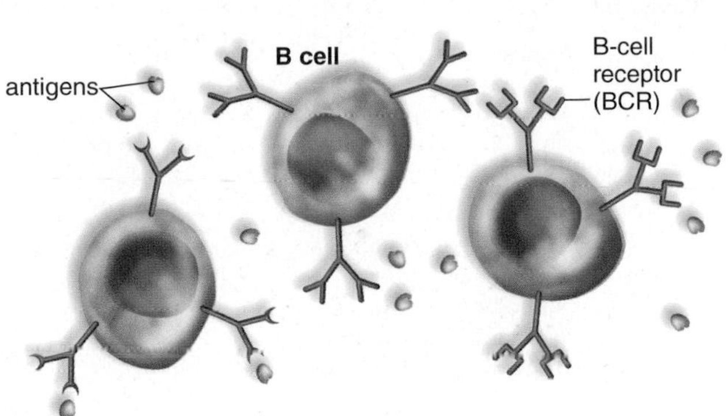

a. **Activation**–When a B-cell receptor binds to an antigen activation occurs.

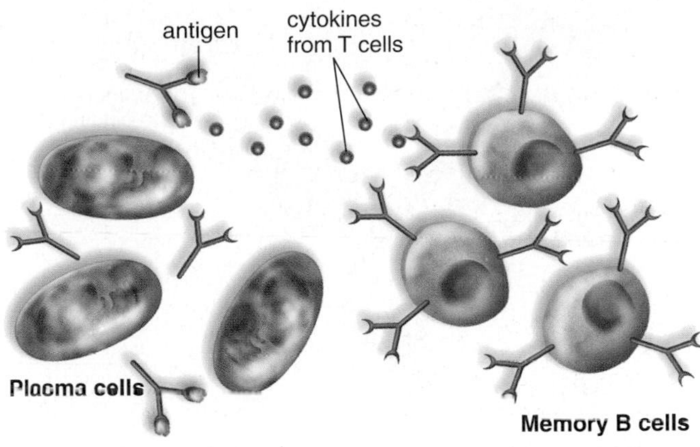

b. **Clonal expansion**–During clonal expansion, cytokines secreted by helper T cells stimulate B cells to clone mostly into plasma cells or memory cells.

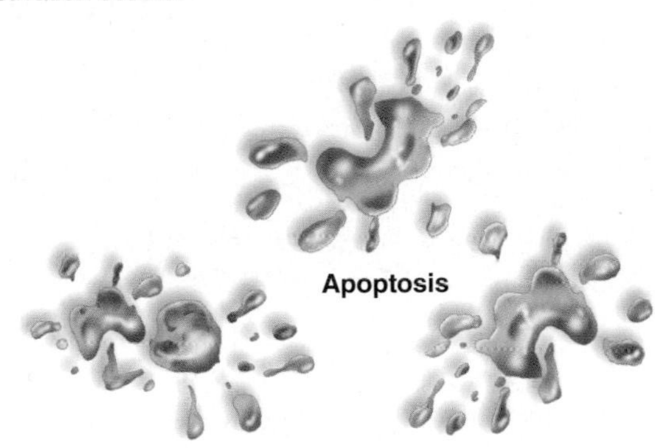

c. **Apoptosis**–Apoptosis, or programmed cell death, occurs to plasma cells left in the system after the infection has passed.

Figure 7.11 Antibody-mediated immunity, or the clonal selection model for B cells.
a. Activation of a B cell occurs when its B-cell receptor (BCR) combines with an antigen (colored green). In the presence of cytokines produced by T cells, the B cell undergoes clonal expansion (**b**), producing many plasma cells that secrete antibodies specific to the antigen. **c.** After the infection, plasma cells undergo apoptosis.

B Cells Become Plasma Cells and Memory B Cells Note in Figure 7.11 that each B cell has a specific BCR represented by shape. Only the B cell with a BCR that has a shape that fits the antigen (green circle) undergoes clonal expansion. During clonal expansion, cytokines secreted by helper T (T_H) cells (see page 144) stimulate B cells to clone. Most of the cloned B cells become **plasma cells,** which circulate in the blood and lymph. Plasma cells are larger than regular B cells because they have extensive rough endoplasmic reticulum. This is for the mass production and secretion of antibodies to a specific antigen. Antibodies identical to the BCR of the activated B cell are secreted from the plasma B cell. Some cloned B cells become memory cells, the means by which long-term immunity is possible. If the same antigen enters the system again, memory B cells quickly divide and transform into plasma cells. The correct type of antibody can quickly be produced by these plasma cells.

Once the threat of an infection has passed, the development of new plasma cells ceases and those present undergo apoptosis. **Apoptosis** is the process of programmed cell death. It involves a cascade of specific cellular events leading to the death and destruction of the cell and removal of the cell remnants from the body as a waste product.

Defense by B cells is called humoral immunity or **antibody-mediated immunity** because activated B cells become plasma cells that produce antibodies. Collectively, plasma cells probably produce as many as 2 million different antibodies. A human doesn't have 2 million genes, so there cannot be a separate gene for each type of antibody. It has been found that scattered DNA segments can be shuffled and combined in various ways to produce the DNA sequence coding for the BCR unique to each type of B cell.

Animation Antibody Diversity

Characteristics of B Cells

- Antibody-mediated immunity against pathogens
- Produced and mature in bone marrow
- Directly recognize antigen and then undergo clonal selection
- Clonal expansion produces antibody-secreting plasma cells as well as memory B cells

Structure of an Antibody The basic unit that composes antibody molecules is a Y-shaped protein molecule with two arms. Each arm has a "heavy" (long) polypeptide chain and a "light" (short) polypeptide chain (Fig. 7.12). These chains have constant regions, located at the trunk of the Y, where the sequence of amino acids is fixed. The class of antibody of each individual molecule is determined by the structure of the antibody's constant region. The variable regions form an antigen--binding site. Their shape is specific to a particular antigen. The

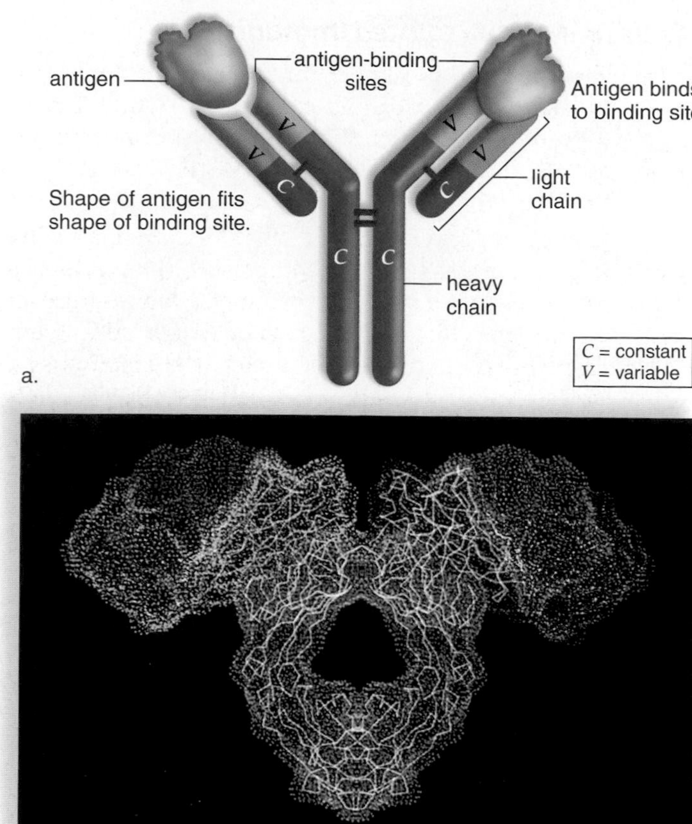

Figure 7.12 The structure of an antibody.
a. An antibody contains two heavy (long) polypeptide chains and two light (short) chains arranged so there are two variable regions where a particular antigen is capable of binding with an antibody. **b.** Computer model of an antibody molecule. The antigen combines with the two side branches.

antigen combines with the antibody at the antigen-binding site in a lock-and-key manner. Antibodies may consist of single Y-shaped molecules, called *monomers,* or may be paired together in a molecule termed a *dimer.* Very large antibodies belonging to the M class are pentamers—clusters of five Y-shaped molecules linked together.

Antigens can be part of a pathogen, such as a virus or a toxin like that produced by tetanus bacteria. Antibodies sometimes react with viruses and toxins by coating them completely, a process called *neutralization.* Often, the reaction produces a clump of antigens combined with antibodies, termed an *immune complex.* The antibodies in an immune complex are like a beacon that attracts white blood cells.

Classes of Antibodies There are five different classes of circulating antibodies (Table 7.1). IgG antibodies are the major type in blood, and lesser amounts are found in lymph and tissue fluid. IgG antibodies bind to pathogens and their toxins. IgG antibodies can cross the placenta from a mother to her fetus, so the newborn has temporary, partial immune protection. As mentioned, IgM antibodies are pentamers. IgM antibodies are the first antibodies produced by a newborn's body.

Table 7.1 Antibodies

Class	Presence	Function
IgG	Main antibody type in circulation; crosses the placenta from mother to fetus	Binds to pathogens, activates complement, and enhances phagocytosis by white blood cells
IgM	Antibody type found in circulation; largest antibody; first antibody formed by a newborn; first antibody formed with any new infection	Activates complement; clumps cells
IgA	Main antibody type in secretions such as saliva and milk	Prevents pathogens from attaching to epithelial cells in digestive and respiratory tract
IgD	Antibody type found on surface of immature B cells	Presence signifies readiness of B cell
IgE	Antibody type found as antigen receptors on mast cells in tissues	Responsible for immediate allergic response and protection against certain parasitic worms

IgM antibodies are the first to appear in blood soon after an infection begins and the first to disappear before the infection is over. They are good activators of the complement system. IgA antibodies are monomers or dimers containing two Y-shaped structures. They are the main type of antibody found in body secretions: saliva, tears, mucus, and breast milk. IgA molecules bind to pathogens and prevent them from reaching the bloodstream. The main function of IgD molecules seems to be to serve as antigen receptors on immature B cells. IgE antibodies are responsible for prevention of parasitic worm infections, but they can also cause immediate allergic responses.

Animation Antibodies

Monoclonal Antibodies

Every plasma cell derived from the same B cell secretes antibodies against a specific antigen. These are **monoclonal antibodies** because all of them are the same type and because they are produced by plasma cells derived from the same B cell. One method of producing monoclonal antibodies in vitro (outside the body) is depicted in Figure 7.13. B lymphocytes are removed from an animal, normally a lab mouse, and are exposed to a particular antigen. The resulting plasma cells are fused with myeloma cells (malignant plasma cells that live and divide indefinitely; they are immortal cells). The fused cells are called hybridomas—*hybrid-* because they result from the fusion of two different cells, and *-oma* because one of the cells is a cancer cell.

Animation Monoclonal Antibody Production

At present, monoclonal antibodies are being used for quick and certain diagnosis of various conditions. For example, the hormone human chorionic gonadotropin (HCG) is present in the urine of a pregnant woman. A monoclonal antibody can be used to detect this hormone. Monoclonal antibodies are also

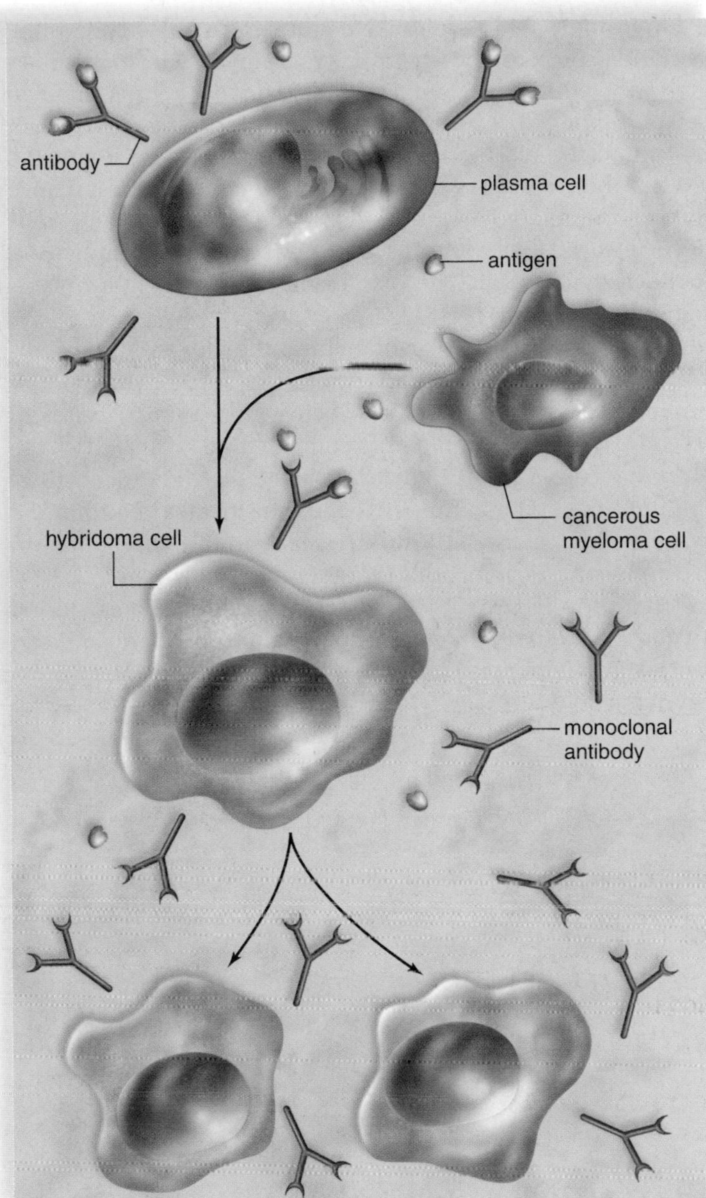

Figure 7.13 The production of monoclonal antibodies. Plasma cells of the same type (derived from immunized mice) are fused with myeloma (cancerous) cells, producing hybridoma cells that are "immortal." Hybridoma cells divide and continue to produce the same type of antibody, called monoclonal antibodies.

used to identify infections like H1N1 flu, HIV, and RSV, a respiratory virus infection common in young children. Because they can be used to distinguish between cancerous and normal tissue cells, they are also used to carry radioisotopes or toxic drugs to tumors, which can then be selectively destroyed. Trastuzumab (Herceptin) is a monoclonal antibody used in the treatment of breast cancer. It binds to a protein receptor on breast cancer cells and prevents the cancer cells from dividing as quickly. Antibodies that bind to cancer cells also can activate the complement system and can increase phagocytosis by macrophages and neutrophils.

T Cells and Cell-Mediated Immunity

Cell-mediated immunity is named for the action of T cells that directly attack diseased cells and cancer cells. Other T cells, however, release cytokines that stimulate both nonspecific and specific defenses.

How T Cells Recognize an Antigen When a T cell leaves the thymus, it has a unique **T-cell receptor (TCR)**, just as B cells have. Unlike B cells, however, T cells are unable to recognize an antigen without help. The antigen must be displayed to them by an **antigen presenting cell (APC)**, such as a macrophage. After phagocytizing a pathogen, such as a bacterium, APCs travel to a lymph node or the spleen, where T cells also congregate. In the meantime, the APC has broken the pathogen apart in a lysosome. A piece of the pathogen is then displayed in the groove of a **major histocompatibility complex (MHC)** protein on the cell's surface.

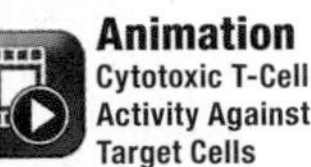

Human MHC proteins are called **human leukocyte antigens (HLAs)**. These proteins are found on all of our body cells. More than 50 different proteins have been identified, and each human has a unique combination. One exception are the HLAs of identical twins. Because identical twins arise from division of a single zygote, their HLA proteins are identical. MHC antigens are self proteins because they mark the cell as belonging to a particular individual. The importance of self proteins in plasma membranes was first recognized when it was discovered that they contribute to the specificity of tissues and make it difficult to transplant tissue from one human to another. Comparison studies of the more than 50 MHC antigens must always be carried out before a transplant is attempted. The greater the number of these proteins that match, the more likely the transplant will be successful.

When an antigen-presenting cell links a foreign antigen to the self protein on its plasma membrane, it carries out an important safeguard for the rest of the body. The T cell to be activated can compare the antigen and self protein side by side. The activated T cell and all of the daughter cells, which it will form, can recognize foreign from self. These T cells go on to destroy cells carrying foreign antigens, while leaving normal body cells unharmed.

Clonal Expansion In Figure 7.14, the T cells have specific TCRs, represented by their different shapes. A macrophage is presenting an antigen to a T cell that has the specific TCR that will combine with this particular antigen, represented by a green circle. The T cell is activated and undergoes clonal expansion. Many copies of the activated T cell are produced during clonal expansion. The two classes of MHC proteins are called MHC I and MHC II. A subgroup of T cells recognizes APCs that display an antigen within the groove of an MHC I protein. These T cells will activate and become cytotoxic T (T_C) cells. A subgroup of T cells recognizes APCs that display an antigen within the groove of an MHC II protein. These T cells will activate and become helper T (T_H) cells. Helper T cells are necessary for regulating B cells.

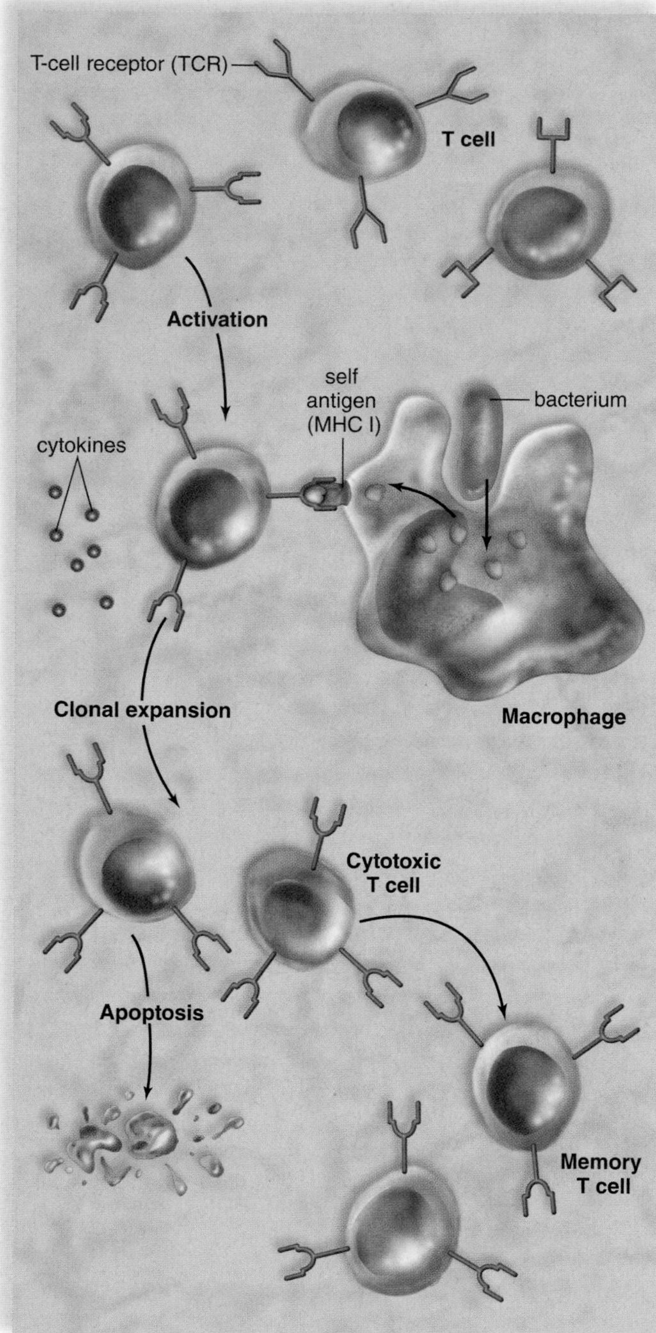

Figure 7.14 The clonal selection model for T cells. Activation of a T cell occurs when its T-cell receptor (TCR) can combine with an antigen presented by a macrophage. In this example, cytotoxic T cells are produced. When the immune response is finished, they undergo apoptosis. A small number of memory T cells remain.

As the illness disappears, the immune reaction wanes. Activated T cells become susceptible to apoptosis. As mentioned previously, apoptosis contributes to homeostasis by regulating the number of cells present in an organ, or in this case, in the immune system. When apoptosis does not occur as it should, the potential exists for an autoimmune response (see page 149) or for T-cell cancers (i.e., lymphomas and leukemias).

Cytotoxic T (T_C) cells have storage vacuoles containing perforins and storage vacuoles containing enzymes called granzymes. After a cytotoxic T cell binds to a virus-infected cell or tumor cell, it releases perforin molecules, which punch holes into the plasma membrane, forming a pore. Cytotoxic T cells then deliver granzymes into the pore. These cause the cell to undergo apoptosis and die. Once cytotoxic T cells have released the perforins and granzymes, they move on to the next target cell. Cytotoxic T cells are responsible for **cell-mediated immunity** (Fig. 7.15).

Helper T (T_H) cells regulate immunity by secreting cytokines, the chemicals that enhance the response of all types of immune cells. B cells cannot be activated without T-cell help (see Fig. 7.11). The human immunodeficiency virus (HIV), the virus that causes AIDS, infects helper T cells and other cells of the immune system. The virus thus inactivates the immune response and makes HIV-infected individuals susceptible to the opportunistic infections that eventually kill them.

Notice in Figure 7.14 that a few of the clonally expanded T cells are **memory T cells.** They remain in the body and can jump-start an immune reaction to an antigen previously present in the body.

Characteristics of T Cells

- Cell-mediated immunity against virus-infected cells and cancer cells
- Produced in bone marrow, mature in thymus
- Antigen must be presented in groove of an HLA (MHC) molecule
- Cytotoxic T cells destroy nonself antigen-bearing cells
- Helper T cells secrete cytokines that control the immune response

CHECK YOUR PROGRESS 7.4

1. Detail how innate defense differs from adaptive defense.
2. Describe how adaptive defense responds against a virus-infected cell.
3. Differentiate which blood cells are mainly responsible for adaptive defense, and state how they function.

CONNECTING THE CONCEPTS

For more information on the topics presented in this section, refer to the following discussions:

Section 6.3 summarizes the formation and specialization of lymphocytes.

Section S.1 provides additional information on HIV and the AIDS epidemic.

Section 19.4 examines how immunotherapy using cytotoxic T cells may be used to treat cancer.

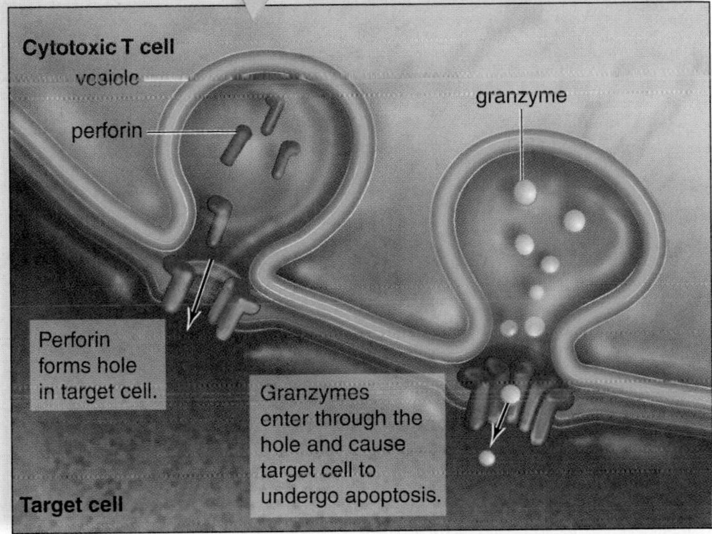

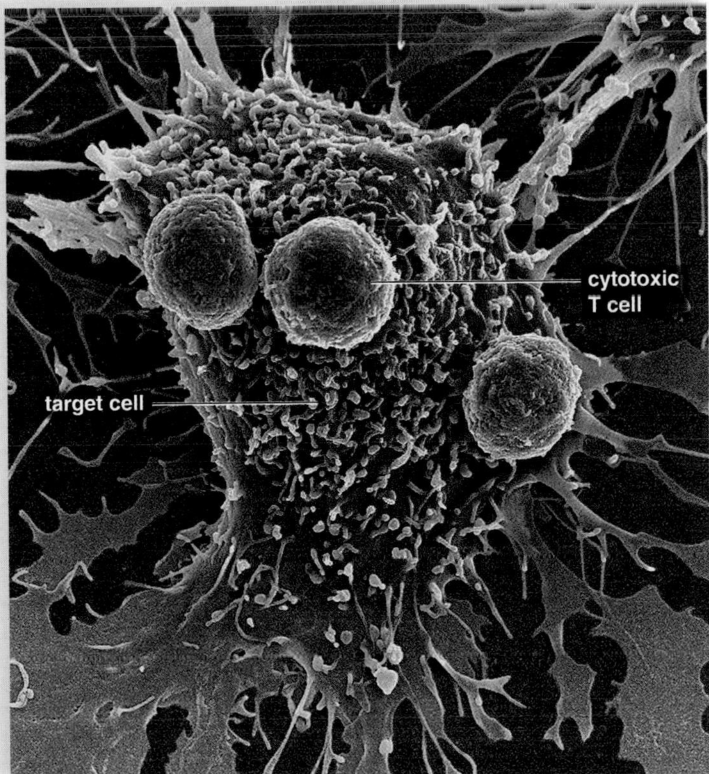

Figure 7.15 How cytotoxic T cells kill infected cells.
Cytotoxic T cells bind to the target cell, punch holes in its membrane, and inject chemicals that cause the cell to die.

7.5 Acquired Immunity

LEARNING OUTCOMES

Upon completion of this section, you should be able to

1. Distinguish between active and passive immunity.
2. Recognize the importance of cytokines in immunity.

Immunity occurs naturally through infection or is brought about artificially by medical intervention. The two types of acquired immunity are active and passive. In the process of **active immunity,** the individual alone produces antibodies against an antigen. In **passive immunity,** the individual is given prepared antibodies via an injection.

Active Immunity

Active immunity sometimes develops naturally after a person is infected with a pathogen. However, active immunity is often induced when a person is well to prevent future infection. Artificial exposure to an antigen through immunization can prevent future disease. The United States is committed to immunizing all children against the common types of childhood disease. A full list of recommended childhood vaccinations is available from the Centers for Disease Control and Prevention's (CDC) website at www.cdc.gov/vaccines. Information on recommended adult vaccinations is provided in the Health feature, "Adult Vaccinations," on page 148.

Immunization involves the use of **vaccines,** substances that contain an antigen to which the immune system responds. Traditionally, vaccines are the pathogens themselves, or their products, that have been treated to be no longer virulent (able to cause disease). Today, it is possible to genetically engineer bacteria to mass-produce a protein from pathogens, and this protein can be used as a vaccine. This method is used to produce the vaccine for the viral-induced disease hepatitis B, and a vaccine for malaria made by the same method is currently going through FDA approval.

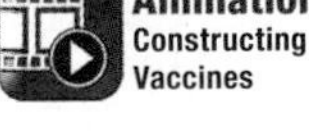

After a vaccine is given, it is possible to follow an immune response by determining the amount of antibody present in a sample of plasma—this is called the *antibody titer.* After the first exposure to a vaccine, a primary response occurs. For a period of several days, no antibodies are present. Then the titer rises slowly, levels off, and gradually declines as the antibodies bind to the antigen or simply break down (Fig. 7.16). After a second exposure to the vaccine, a secondary response is expected. The titer rises rapidly to a level much greater than before. Then it slowly declines. The second exposure is called a "booster" because it boosts the antibody titer to a high level. The high antibody titer now is expected to help prevent disease symptoms, even if the individual is exposed to the disease-causing antigen.

Active immunity depends upon the presence of memory B cells and memory T cells capable of responding to lower doses of antigen. Active immunity is usually long lasting, although a booster may be required after many years for certain vaccines.

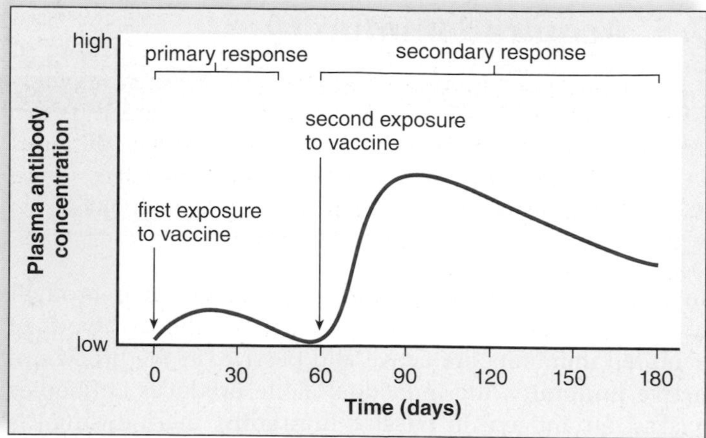

Figure 7.16 **How immunizations cause active immunity.** During immunization, the primary response, after the first exposure to a vaccine, is minimal, but the secondary response, which may occur after the second exposure, shows a dramatic rise in the amount of antibody present in plasma.

Passive Immunity

Passive immunity occurs when an individual is given prepared antibodies or immune cells to combat a disease. These antibodies are not produced by the individual's plasma cells, so passive immunity is temporary. For example, newborn infants are passively immune to some diseases because IgG antibodies have crossed the placenta from the mother's blood (Fig. 7.17*a*). These antibodies soon disappear, and within a few months, infants become more susceptible to infections. Breast-feeding prolongs the natural passive immunity an infant receives from the mother because IgG and IgA antibodies are present in the mother's milk (Fig. 7.17*b*).

Even though passive immunity does not last, it is sometimes used to prevent illness in a patient who has been unexpectedly exposed to an infectious disease. Usually, the patient receives a gamma globulin injection of serum that contains antibodies, in some cases, taken from individuals who have recovered from the illness (Fig. 7.17*c*). For example, a health-care worker who suffers an accidental needlestick may come into contact with the blood from a patient infected with hepatitis virus. Immediate treatment with a gamma globulin injection (along with simultaneous vaccination against the virus) can typically prevent the viruses from causing infection.

Cytokines and Immunity

Cytokines are signaling molecules produced by T lymphocytes, macrophages, and other cells. Cytokines regulate white blood cell formation and/or function, so they are being investigated as a possible adjunct therapy (treatments used in conjunction with the primary treatment) for cancer and AIDS. Both interferon, produced by virus-infected cells, and **interleukins,** produced by various white blood cells, have been used as immunotherapeutic drugs. These are used particularly to enhance the ability of the individual's T cells to fight cancer.

Most cancer cells carry an altered protein on their cell surface, so they should be attacked and destroyed by cytotoxic T cells. Whenever cancer develops, it is possible that cytotoxic T cells have not been activated. In that case, cytokines might awaken the immune system and lead to the destruction of the cancer. In one technique, researchers withdrew T cells from the patient and presented cancer cell antigens to them. They then activated the cells by culturing them in the presence of an interleukin. The T cells were reinjected into the patient, who was given doses of interleukin to maintain the killer activity of the T cells.

Scientists actively engaged in interleukin research believe that interleukins soon will be used as adjuncts for vaccines. They are currently used as treatment adjuncts for chronic diseases like psoriasis, rheumatoid arthritis, and irritable bowel syndrome and are sometimes used for the treatment of chronic infectious diseases and the treatment of cancer. Interleukin antagonists also may prove helpful in preventing skin and organ rejection, autoimmune diseases like lupus and Crohn's disease, and allergies.

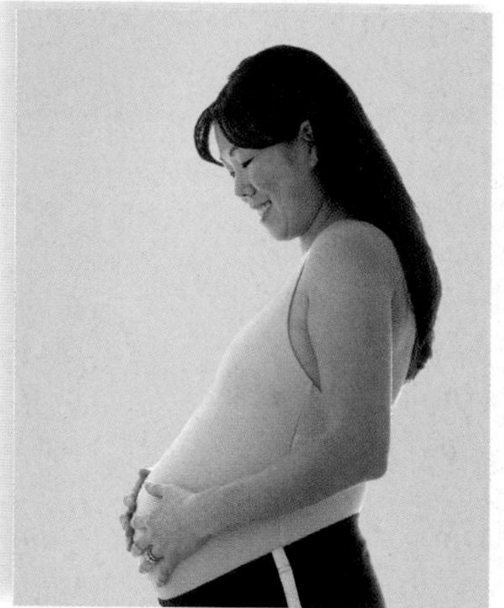

a. Antibodies (IgG) cross the placenta.

b. Antibodies (IgG, IgA) are secreted into breast milk.

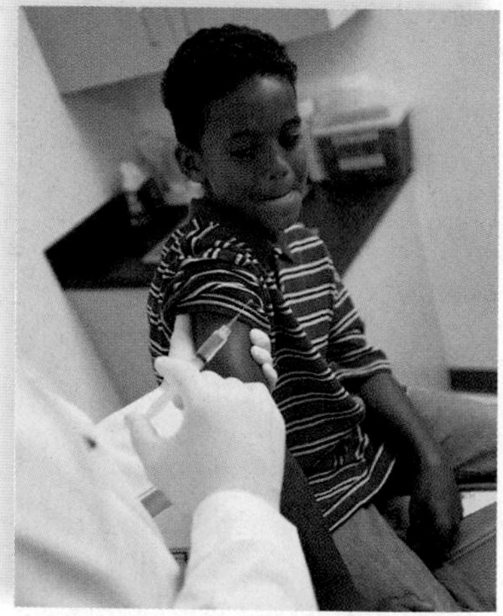

c. Antibodies can be injected by a physician.

Figure 7.17 **Delivery mechanisms of passive immunity.** During passive immunity, antibodies are received (**a**) by crossing the placenta, (**b**) in breast milk, or (**c**) by injection. The body is not producing the antibodies, so passive immunity is short-lived.

CHECK YOUR PROGRESS 7.5

1. Define *acquired immunity* and give some examples.
2. Describe the two different types of acquired immunity.
3. Compare the two types of immune therapies that can assist passive immunity.

CONNECTING THE CONCEPTS

For more information on the topics presented in this section, refer to the following discussion:

Section 19.4 examines how cytokines may be used to treat cancer.

7.6 Hypersensitivity Reactions

LEARNING OUTCOMES

Upon completion of this section, you should be able to

1. Explain what causes an allergic reaction.
2. Identify the causes of select autoimmune diseases.

Sometimes, the immune system responds in a manner that harms the body, as when individuals develop allergies, receive an incompatible blood type, suffer tissue rejection, or have an autoimmune disease.

Allergies

Allergies are hypersensitivities to substances, such as pollen, food, or animal hair, that ordinarily would do no harm to the body. The responses to these antigens, called **allergens,** usually include some degree of tissue damage.

An **immediate allergic response** can occur within seconds of contact with the antigen. The response is caused by antibodies known as IgE (see Table 7.1). IgE antibodies are attached to receptors on the plasma membrane of mast cells in the tissues and also to basophils in the blood. When an allergen attaches to the IgE antibodies on these cells, the cells release histamine and other substances that bring about the allergic symptoms. When pollen is an allergen, histamine stimulates the mucous membranes of the nose and eyes to release fluid. This causes the runny nose and watery eyes typical of hay fever. If a person has asthma, the airways leading to the lungs constrict, resulting in difficult breathing accompanied by wheezing. When food contains an allergen, nausea, vomiting, and diarrhea often result.

Anaphylactic shock is an immediate allergic response that occurs because the allergen has entered the bloodstream. Bee stings and penicillin shots are known to cause this reaction because both inject the allergen into the blood. Anaphylactic shock is characterized by a sudden and life-threatening drop in blood pressure due to increased permeability of the capillaries by histamine. Taking epinephrine can counteract this reaction until medical help is available.

People with allergies produce ten times more IgE than those without allergies. A new treatment using injections of monoclonal IgG antibodies for IgEs is being tested in individuals with severe food allergies. More routinely, injections of the allergen are given so that the body will build up high quantities of IgG antibodies. The hope is that IgG antibodies will combine with allergens received from the environment before they have a chance to reach the IgE antibodies located in the membrane of mast cells and basophils.

A **delayed allergic response** is initiated by memory T cells at the site of allergen contact in the body. The allergic response is regulated by the cytokines secreted by both T cells and macrophages. A classic example of a delayed allergic response is the skin test for tuberculosis (TB). When the test result is positive, the tissue where the antigen was injected becomes red and hardened. This shows that there was prior exposure to the bacterium that causes TB. Contact dermatitis, which occurs when a person is allergic to poison ivy, jewelry, cosmetics, and many other substances that touch the skin, is also an example of a delayed allergic response.

Other Immune Problems

Certain organs, such as the skin, the heart, and the kidneys, could be transplanted easily from one person to another if the body did not attempt to reject them. Rejection of transplanted tissue results because the recipient's immune system recognizes that the transplanted tissue is not "self." Cytotoxic T cells respond by attacking the cells of the transplanted tissue.

Organ rejection can be controlled by carefully selecting the organ to be transplanted and administering **immunosuppressive** drugs. It is best if the transplanted organ has the same type of MHC antigens as those of the recipient, because cytotoxic T cells recognize foreign MHC antigens. Two well-known immunosuppressive drugs, cyclosporine and tacrolimus, act by inhibiting the production of certain T-cell cytokines.

Xenotransplantation is the use of animal organs instead of human organs in transplant patients. Scientists have chosen to use the pig because animal husbandry has long included the raising of pigs as a meat source, and pigs are prolific. Genetic engineering can make pig organs less antigenic. The ultimate goal is to make pig organs as widely accepted as type O blood.

An alternative to xenotransplantation exists because tissue engineering is making organs in the laboratory. Scientists have transplanted lab-grown urinary bladders into human patients. They hope that production of organs lacking HLA antigens will one day do away with the problem of rejection.

Immune system disorders occur when a patient has an immune deficiency or when the immune system attacks the body's own cells.

When a person has an immune deficiency, the immune system is unable to protect the body against disease. Infrequently, a child may be born with an impaired immune system. For example, in **severe combined immunodeficiency (SCID)**, both antibody- and cell-mediated immunity are lacking or

BIOLOGY MATTERS  Health

Adult Vaccinations

Many people mistakenly believe that you receive your full complement of vaccinations by the time you leave high school. In reality, being vaccinated is a lifelong activity. The Centers for Disease Control and Prevention (CDC) has identified a series of vaccinations that are recommended after the age of 18 (Table 7A). In many cases, vaccinations are recommended when an individual is determined to be at risk for a specific disease or condition. For example, while the vaccination for hepatitis B (HepB) may not be required, it is recommended for individuals who have had more than one sexual partner during a six-month period, have been diagnosed with a sexually transmitted disease (STD; see Chapter 16), use injection drugs, or may have been exposed to blood or infected body fluid. In most cases, vaccinations are recommended as a protective measure even if you are not in an at-risk category.

Table 7A Recommended Vaccination Schedule for Adults

Vaccine	19-49 Years Old	50-64 Years Old	65 Years & Older
Influenza	Annual vaccination every fall or winter		
Pneumococcal polysaccharide (PPSV)	One to two vaccinations if you smoke or have chronic medical conditions.*		Vaccination at age 65 if no prior vaccination history.
Tetanus, diphtheria, pertussis (Td, Tdap)	One Tdap vaccine if younger than 65 years of age. Then Td booster every 10 years**		Tdap vaccine if you are a health-care worker, in contact with infants, or are pregnant. Td booster every 10 years**
Hepatitis B (HepB)	Vaccination needed if you are at risk for hepatitis B infection or desire protection. Vaccine is administered as 3 doses over a 6 month period.		
Hepatitis A (HepA)	Vaccination needed if you are at risk for hepatitis A infection or desire protection. Vaccine is administered as 2 doses 6–18 months apart.		
Human papillomavirus (HPV)	Vaccination needed for females younger than 26 and males younger than 21. Males 22 to 26 may receive vaccination if risk conditions exist.* Vaccination is administered as 3 doses over 6 months.		
Measles, mumps, Rubella (MMR)	One vaccination needed if born after 1957. Second vaccination may be necessary for some people.*		
Varicella (Chickenpox)	Vaccination needed if you have never had chickenpox, or were vaccinated with only one dose.*		
Meningococcal	Ages 19–21 and a first-year college student living in a resident hall or have at-risk medical conditions.* Additional boosters may be needed.*		
Zoster		Vaccination for individuals 60 years or older.	

*Consult with your health-care provider to assess your medical conditions and risk of infection.
**Consult your health-care provider if you have not been vaccinated with at least 3 tenatus- and diphtheria-containing vaccinations during your life (Td or Tdap) or have a deep or dirty wound.

As always, if you have questions regarding any of these diseases, or about your personal need for vaccinations, consult with your health-care provider. For more information on vaccination schedules from birth through adulthood, visit the CDC's website at www.cdc.gov/vaccines, or the Immunization Action Coalition (www.immunize.org).

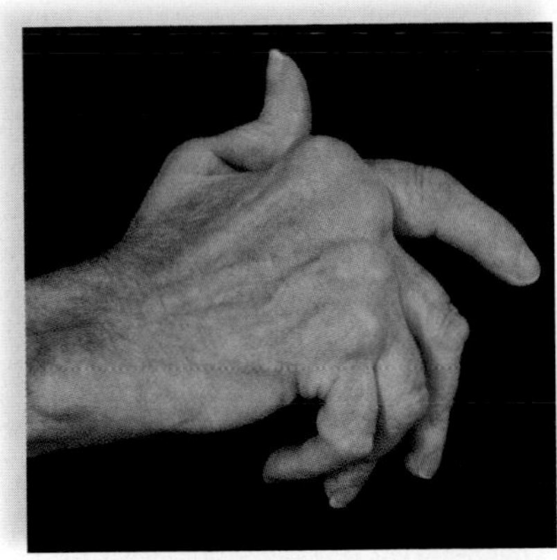

Figure 7.18 Rheumatoid arthritis.
Rheumatoid arthritis is due to recurring inflammation in skeletal joints. Complement proteins, T cells, and B cells all participate in deterioration of the joints, which eventually become immobile.

inadequate. Without treatment, even common infections can be fatal. Bone marrow transplants and gene therapy have been successful in SCID patients. Acquired immune deficiencies can be caused by infections, chemical exposure, or radiation. Acquired immunodeficiency syndrome (AIDS) is a result of an infection with the human immunodeficiency virus (HIV). As a result of a weakened immune system, AIDS patients show a greater susceptibility to infections and also have a higher risk of cancer (see the Infectious Diseases Supplement that follows this chapter).

When cytotoxic T cells or antibodies mistakenly attack the body's own cells, the person has an **autoimmune disease.** The exact cause of autoimmune diseases is not known, although it appears to involve both genetic and environmental factors. People with certain HLA antigens are more susceptible. Women are more likely than men to develop an autoimmune disease.

Sometimes the autoimmune disease follows an infection. For example, in **rheumatic fever,** antibodies induced by a streptococcal (bacterial) infection of the throat also react with heart muscle. This causes an inflammatory response with damage to the heart muscle and valves. **Rheumatoid arthritis** (Fig. 7.18) is an autoimmune disease in which the joints are chronically inflamed. It is thought that antigen–antibody complexes, complement, neutrophils, activated T cells, and macrophages are all involved in the destruction of cartilage in the joints. A person with **systemic lupus erythematosus** (SLE), commonly just called *lupus,* has various symptoms including a facial rash, fever, and joint pain. In these patients, damage to the central nervous system, heart, and kidneys can be fatal. SLE patients produce high levels of anti-DNA antibodies. All human cells (except red blood cells) contain DNA, so the symptoms of lupus interfere with tissues throughout the body. **Myasthenia gravis** develops when antibodies attach to and interfere with the function of neuromuscular junctions. The result is severe muscle weakness, eventually resulting in death from respiratory failure. In **multiple sclerosis** (**MS**), T cells attack the myelin sheath covering nerve fibers, causing CNS dysfunction, double vision, and muscular weakness. Treatments for autoimmune diseases usually involve drugs designed to decrease the immune response.

CHECK YOUR PROGRESS 7.6

1. Define the types of complications and disorders associated with the functioning of the immune system.
2. Detail how an antibody works during an allergic reaction.
3. Hypothesize why an autoimmune disorder sometimes develops after an infection.

CONNECTING THE CONCEPTS

For more information on the topics presented in this section, refer to the following discussions:

Section S.1 examines how tuberculosis has become a worldwide epidemic.

Section 21.3 explores how gene therapy may be used to treat diseases such as SCID.

CASE STUDY CONCLUSION

Lupus is an autoimmune disease. The immune system normally makes antibodies that attack foreign cells to keep the body healthy. In an autoimmune disease, the person's immune system makes antibodies that attack the healthy cells of the body, instead of invading pathogen cells. Abigail's immune system was attacking her, in addition to attacking any bacteria and viruses that got into her system. Lupus is also considered a rheumatic disease, or a disorder that affects the muscles, joints, and connective tissue. That explained the pain in her arms and legs.

Dr. Koos went on to detail that she believed Abigail had a type of lupus called systemic lupus erythematosus (SLE). SLE is the most common form of lupus and affects multiple organ systems. It is most commonly seen developing in people in their twenties and thirties, although it is not uncommon for symptoms to develop in people as young as ten years old. The exact cause of lupus is unclear; It is suggested that there is a genetic predisposition to lupus that can be activated by infection, stress, and even increasing levels of estrogen. Dr. Koos explained that 90% of the 1.5 million Americans living with lupus (an estimated 10,000 of them are children under 18) are female.

Dr. Koos explained that there was no cure for lupus and current treatments were aimed at managing the symptoms. She recommended that Abigail have a team of health-care professionals help her manage her lupus. Abigail had appointments with a rheumatologist to manage her muscle and joint pain; a dermatologist to help with the periodic rashes; and a nephrologist, or kidney specialist, because lupus patients tend to develop kidney problems. Abigail is currently on a daily NSAID (nonsteroidal anti-inflammatory drug) to control muscle and joint pain. During a flare-up, she can also take corticosteroids to control inflammation. With proper care and caution, Abigail can keep her lupus in check, decrease the severity of flare-ups, and live a healthy and productive life.

MEDIA STUDY TOOLS

Enhance your study of this chapter with media! Visit **www.mhhe.com/maderhuman13e** and go to "Media Study Tools" for this chapter to access the following:

 Animations	 Videos	 MP3 Files
7.1 Antibiotic Resistance • Binary Fission • Viral Entry into Host Cells • Food Pathogens & Temperature • How Prions Arise • Prion Diseases **7.2** Lymphatic System **7.3** Inflammatory Response • Activation of Complement • Antiviral Activity of Interferon **7.4** Antibody Diversity • Antibodies • Cytotoxic T Cell Activity Against Target Cells • Monoclonal Antibody Production **7.5** Constructing Vaccines	**7.1** *E. coli* Wars • Cranberries vs. Bacteria • How Viruses Attack **7.3** Neutrophils **7.5** Potato Vaccine	**7.2** Lymphatic System **7.3** Barriers and Nonspecific Defenses

SUMMARIZE

7.1 Microbes, Pathogens, and You

Microbes perform valuable services, but they also cause disease. Disease-causing microbes are collectively called **pathogens.**

Bacteria

- **Bacteria** are prokaryotic cells that consist of a plasma membrane, a cell wall containing unique polysaccharides, and cytoplasm. Some bacteria have a gel-like **capsule** surrounding the cell wall. Some bacteria possess **fimbriae** to attach to surfaces, **flagella** for movement, and a **pilus** for the transfer of DNA between cells. Within the cell, **plasmids** sometimes contain genes that allow for resistance to antibiotics.
- Bacteria may cause disease by multiplying in hosts and also by producing toxins.

Viruses

- **Viruses** are noncellular particles consisting of a protein coat and a nucleic acid core that take over the machinery of the host to reproduce.

Prions

- **Prions** are misfolded proteins that have the ability to infect cells. **Creutzfeldt-Jakob disease** (**CJD**) is a prion disease in humans.

7.2 The Lymphatic System

The lymphatic system consists of lymphatic vessels that return lymph to cardiovascular veins.

The primary lymphatic organs are

- the **red bone marrow,** where all blood cells are made and the **B cells** (**B lymphocytes**) mature; and
- the **thymus,** where **T cells** (**T lymphocytes**) mature.

The secondary lymphatic organs are

- the **spleen; lymph nodes;** and other organs containing lymphoid tissue, such as the tonsils, Peyer's patches, and the appendix. Blood is cleansed of pathogens and debris in the spleen. Lymph is cleansed of pathogens and debris in the nodes.

7.3 Innate Immune Defenses

Immunity involves innate and adaptive defenses. The innate defenses include the following:

- chemical barriers such as **lysozyme** enzymes;
- physical barriers to entry; and
- the **inflammatory response,** which involves the action of phagocytic neutrophils and **macrophages.** Chemicals such as **histamine** and **cytokines** act as chemical signals. The **complement system** utilizes protective proteins and **interferons.**

7.4 Adaptive Immune Defenses

Adaptive defenses require B cells and T cells, also called B lymphocytes and T lymphocytes. The adaptive defenses respond to **antigens,** or foreign objects, in the body.

B Cells and Antibody-Mediated Immunity

- The **clonal selection model** explains how activated B cells undergo clonal selection with production of plasma cells and memory B cells, after their **B-cell receptor** (**BCR**) combines with a specific antigen.
- **Plasma cells** secrete antibodies and eventually undergo **apoptosis.** Plasma cells are responsible for **antibody-mediated immunity.**
- An antibody is usually a Y-shaped molecule that has two binding sites for a specific antigen.
- Memory B cells remain in the body and produce antibodies if the same antigen enters the body at a later date.
- **Monoclonal antibodies,** produced by the same plasma cell, have various functions, from detecting infections to treating cancer.

T Cells and Cell-Mediated Immunity

- T cells possess a unique **T-cell receptor** (**TCR**). For a T cell to recognize an antigen, the antigen must be presented by an **antigen-presenting cell** (**APC**) such as a macrophage. Once digested within a lysosome, the antigen is presented on the **major histocompatibility complex** (**MHC**) of the cell. These MHC proteins belong to a class of molecules called **human leukocyte antigens** (**HLAs**).

- Activated T cells undergo clonal expansion until the illness has been stemmed. Then, most of the activated T cells undergo apoptosis. A few cells remain, however, as memory T cells.
- The two main types of T cells are cytotoxic T cells and helper T cells.
- **Cytotoxic T cells** kill virus-infected cells or cancer cells on contact because they bear a nonself protein. They are involved in the process of **cell-mediated immunity.**
- **Helper T cells** produce cytokines and stimulate other immune cells.
- Some activated T cells remain as **memory T cells** to combat future infections by the same pathogen.

7.5 Acquired Immunity

- **Active immunity** can be induced by **immunization** using **vaccines** when a person is well and in no immediate danger of contracting an infectious disease. Active immunity depends upon the presence of memory cells in the body.
- **Passive immunity** is needed when an individual is in immediate danger of succumbing to an infectious disease. Passive immunity is short-lived because the antibodies are administered to—and not made by—the individual.
- Cytokines, including **interleukins,** are a form of passive immunity used to treat AIDS and to promote the body's ability to recover from cancer.

7.6 Hypersensitivity Reactions

Allergies occur when the immune system reacts vigorously to **allergens,** substances not normally recognized as foreign.

- **Immediate allergic responses,** usually consisting of coldlike symptoms, are due to the activity of antibodies. One example is **anaphylactic shock.**
- **Delayed allergic responses,** such as contact dermatitis, are due to the activity of T cells.
- Tissue rejection occurs when the immune system recognizes a tissue as foreign. **Immunosuppressive** drugs may inhibit tissue rejection. Xenotransplantation is the use of animal tissue in place of human tissue.
- Immune deficiencies can be inherited or can be caused by infection, chemical exposure, or radiation. One example is **severe combined immunodeficiency (SCID),** in which adaptive responses are inoperative.
- **Autoimmune diseases** occur when the immune system reacts to tissues/organs of the individual as if they were foreign. Examples are **rheumatic fever, rheumatoid arthritis, systemic lupus erythematosus, myasthenia gravis,** and **multiple sclerosis (MS).**

ASSESS

Testing Your Knowledge of the Concepts

1. Describe the basic characteristics of bacteria. Explain how five particular features contribute to the ability of bacteria to cause disease. (pages 132–133)
2. What is the structure of a virus? Is a virus living? Explain how a virus is able to reproduce. (pages 133–134)
3. What are prions, and how do they cause disease? (page 134)
4. Why are the red bone marrow and the thymus termed primary lymphatic organs? (page 135)
5. In what ways are the spleen and lymph nodes similar, and in what ways are they different? (pages 136–137)
6. How do chemical and physical barriers protect the body? (page 138)
7. How do innate defenses differ from adaptive defenses? (pages 137, 140)
8. Describe the steps that occur during an inflammatory response. (pages 138–139)
9. How do the roles of B cells and T cells differ in the adaptive defenses? (page 140)
10. What is the clonal selection model as it applies to B cells? What becomes of the clones that are produced? (pages 140–141)
11. Describe the structure of an antibody. What are the five main classes, where are they found, and what are their functions? (page 142)
12. Discuss the production of monoclonal antibodies and their applications. (pages 142–143)
13. Explain how T cells recognize an antigen. What are the types of T cells, and how do they function in immunity? (pages 143–144)
14. How is active immunity achieved? How is passive immunity achieved? (pages 145–146)
15. Discuss allergies, tissue rejection, and autoimmune diseases as they relate to the immune system. (pages 147–149)
16. Which of the following is a function of the spleen?
    a. produces T cells
    b. removes worn-out red blood cells
    c. produces immunoglobulins
    d. produces macrophages
    e. regulates the immune system
17. Which of the following is a function of the thymus?
    a. production of red blood cells
    b. secretion of antibodies
    c. production and maintenance of stem cells
    d. site for the maturation of T lymphocytes
18. Which of the following is a function of the secondary lymphatic organs?
    a. transport of lymph
    b. clonal selection of B cells
    c. located where lymphocytes encounter antigens
    d. All of the choices are correct.
19. Defense mechanisms that function to protect the body against many infectious agents are called
    a. adaptive.
    b. innate.
    c. barriers to entry.
    d. immunity.
20. Which of the following is most directly responsible for the increase in capillary permeability during the inflammatory reaction?
    a. pain
    b. white blood cells
    c. histamine
    d. tissue damage

21. Which of the following is *not* a goal of the inflammatory reaction?
    a. brings more oxygen to damaged tissues
    b. decreases blood loss from a wound
    c. decreases the number of white blood cells in the damaged tissues
    d. prevents entry of pathogens into damaged tissues
22. Which of the following is *not* correct concerning interferon?
    a. Interferon is a protective protein.
    b. Virus-infected cells produce interferon.
    c. Interferon has no effect on viruses.
    d. Interferon can be used to treat certain viral infections.
23. Which one of these does *not* pertain to B cells?
    a. have passed through the thymus
    b. have specific receptors
    c. are responsible for antibody-mediated immunity
    d. synthesize and liberate antibodies
24. Which of these pertain(s) to T cells?
    a. have specific receptors
    b. are of more than one type
    c. are responsible for cell-mediated immunity
    d. stimulate antibody production by B cells
    e. All of the choices are correct.
25. During a secondary immune response,
    a. antibodies are made quickly and in great amounts.
    b. antibody production lasts longer than in a primary response.
    c. B cells become plasma cells.
    d. All of the choices are correct.
26. Active immunity may be produced by
    a. having a disease.
    b. receiving a vaccine.
    c. receiving gamma globulin injections.
    d. Both a and b are correct.
    e. Both b and c are correct.
27. Which of the following is *not* an innate body defense?
    a. complement
    b. the skin and mucous membranes
    c. antibodies
    d. lysozyme

ENGAGE

Thinking Critically About the Concepts

An allergic response is an overreaction of the immune system in response to an antigen. Such responses always require a prior exposure to the antigen. The reaction can be immediate (within seconds to minutes) or delayed (within hours). Insect venom reactions often involve the development of hives and itching. Asthmalike symptoms include shortness of breath and wheezing. Decreased blood pressure will eventually cause loss of consciousness. This immediate, severe allergic effect, called anaphylaxis, can be fatal if untreated. In addition to insect venom, food allergens such as those in milk, peanut butter, and shellfish may also elicit life-threatening symptoms. In these cases, IgE antibodies, histamine, and other inflammatory chemicals are the culprits involved in making the reaction so severe.

1. How are B cells involved in immediate allergic responses?
2. An allergist is a doctor that treats people with known allergies. What type of treatments do you think are used? Why?
3. Why do you think certain allergens affect some people more than others? Why are some people asymptomatic to a particular antigen whereas others are affected?
4. Think of an analogy (something you're already familiar with) for the barrier defenses like your skin and mucous membranes.
5. Someone bitten by a poisonous snake should be given some antivenom (antibodies) to prevent them from dying. If they are bitten by the same type of snake three years after the initial bite, will they have immunity to the venom, or should they get another shot of antivenom? Justify your response with an explanation of the type of immunity someone gains from a shot of antibodies.

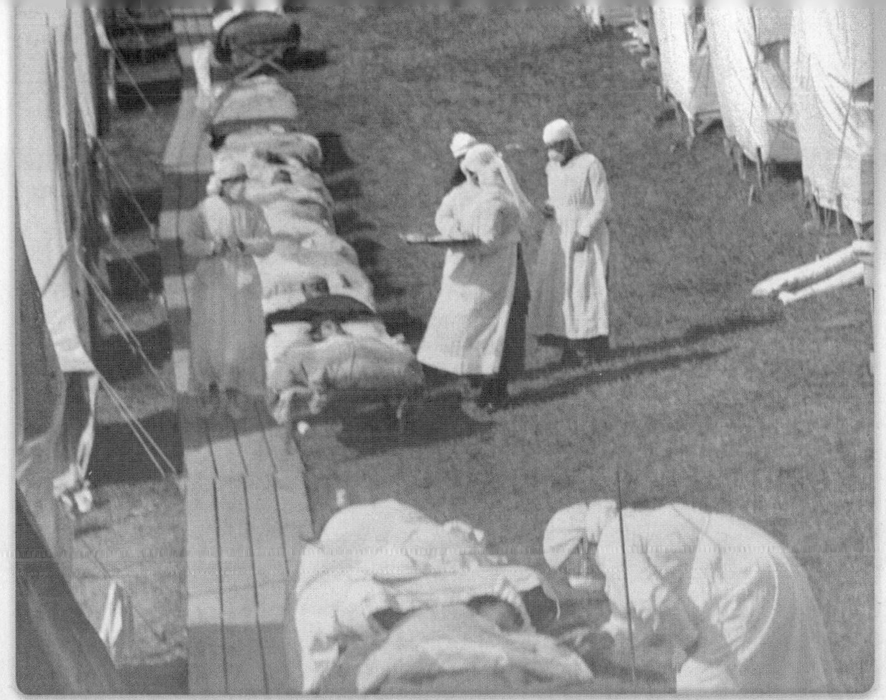

Infectious Diseases Supplement

THE SCIENCE OF EPIDEMIOLOGY

Epidemiology is the study of diseases in populations, including the causes, distribution, and control of these diseases. Humans have a long history of struggle with infectious diseases. Infectious diseases are caused by pathogens (from the Greek word for "suffering"), which include certain bacteria, viruses, fungi, parasites, protozoans, and prions.

Infectious disease outbreaks are described in the earliest written documents. Smallpox outbreaks were described 10,000 years ago in the Nile Valley. Malaria was described 4,000 years ago in China. The Chinese treated it with the Qinghao plant—the source of the drug currently used to treat malaria. Outbreaks of bubonic plague ("Black Death") in Europe decimated the continent's population. The spread of the disease, as well as the public response to it, literally determined European history.

In more modern times, the battle between infectious agents and humans includes the development of vaccinations. The Chinese first developed vaccines around 200 BCE. In 1796, the smallpox vaccine was introduced to the Western world. Death rates from this virus dropped dramatically. The disease was eradicated from the human population in 1977. A rabies vaccine was developed in 1885, and vaccines for diphtheria, typhus, tetanus, and polio soon followed. Child mortality rates plummeted and remain low in countries where vaccination is routine.

The death rates from infectious diseases dropped still further in the 1940s with the development of antibiotics. With the developments of vaccines and antibiotics, scientists hoped they could eradicate other pathogens or, at least, significantly limit the suffering caused by infectious diseases. However, infectious diseases still kill large numbers of people. According to the World Health Organization (WHO), six diseases—acute respiratory infections, HIV/AIDS, diarrheal diseases, malaria, measles, and tuberculosis—account for 90% of all deaths attributed to infectious diseases. This supplement examines the causes, preventions, and treatments for several of the important infectious diseases.

CHAPTER CONCEPTS

S.1 HIV/AIDS, Tuberculosis, and Malaria

Epidemiology is the study of diseases in populations. The terms *epidemic* and *pandemic* are used to describe disease outbreaks. HIV/AIDS, tuberculosis, malaria, and influenza are examples of pandemic diseases.

S.2 Emerging Diseases

Emerging diseases include diseases that have never before been seen, as well as those previously recognized in a small number of people in isolated settings. Diseases that have been present throughout history, but not known to be caused by a pathogen, are also considered to be emerging diseases. Reemerging diseases are previously known diseases undergoing resurgence, often due to human carelessness.

S.3 Antibiotic Resistance

Misuse of antibiotics has led to the selection of antibiotic-resistant organisms. Some organisms have developed multidrug resistances, and these organisms are very difficult to treat.

BEFORE YOU BEGIN

Before beginning this material, take a few moments to review the following discussions:

Section 7.1 What is the difference in the structure of a bacterium and a virus?

Section 7.4 What is the role of a T cell in the immune response?

Section 7.5 How do immunizations protect an individual against disease?

S.1 HIV/AIDS, Tuberculosis, and Malaria

LEARNING OUTCOMES

Upon completion of this section, you should be able to

1. Distinguish between an outbreak, epidemic, and pandemic.
2. Summarize the stages of the HIV life cycle.
3. Summarize the barriers in the development of an effective HIV vaccine.
4. Describe the causes of tuberculosis and malaria.

A disease is classified as an **epidemic** if there are more cases of the disease than expected in a certain area for a certain period. The number of cases that constitute an epidemic depends on what is expected. For example, a few cases of a very rare disease may constitute an epidemic, whereas a larger number of a very common disease may not be. If the epidemic is confined to a local area, it is usually called an **outbreak.** Global epidemics are called **pandemics** (Fig. S.1). HIV/AIDS, tuberculosis, malaria, and influenza are all examples of current pandemics. Organizations such as the Centers for Disease Control and Prevention (CDC) and the WHO monitor and respond to the threats of **infectious diseases.** These organizations are primarily responsible for determining whether an outbreak has reached epidemic or pandemic levels.

HIV/AIDS

Acquired immunodeficiency syndrome (**AIDS**) is caused by a virus known as the **human immunodeficiency virus** (**HIV**). There are two main types of HIV: HIV-1 and HIV-2. HIV-1 is the more widespread, virulent form of HIV. Of the two types of HIV, HIV-2 corresponds to a type of immunodeficiency virus found in the green monkey, which lives in western Africa. In addition, researchers have found a virus identical to HIV-1 in a subgroup of chimpanzees once common in west-central Africa. Perhaps HIV viruses were originally found only in nonhuman primates. They could have mutated to HIV after humans ate nonhuman primates for meat.

HIV can infect cells with particular surface receptors. Most importantly, HIV infects and destroys cells of the immune system, particularly helper T cells and macrophages. As the number of helper T cells declines, the body's ability to fight an infection also declines. As a result, the person becomes ill with various diseases. AIDS is the advanced stage of HIV infection, in which a person develops one or more of a number of opportunistic infections. An **opportunistic infection** is one that has the *opportunity* to occur only because the immune system is severely weakened.

Origin of and Prevalence of HIV

It is generally accepted that HIV originated in Africa and then spread to the United States and Europe by way of the Caribbean. However, the exact dates of the first human cases of HIV are still being investigated. Recent molecular analyses of the HIV virus suggest that the virus may have first infected humans sometime between 1884 and 1924. Direct evidence of HIV in humans has been obtained from tissue and blood samples taken in the 1950s and 1960s. HIV has been found in a preserved 1959 blood sample taken from a man who lived in an African country now called the Democratic Republic of the Congo. British scientists have been able to show that AIDS came to their country perhaps as early

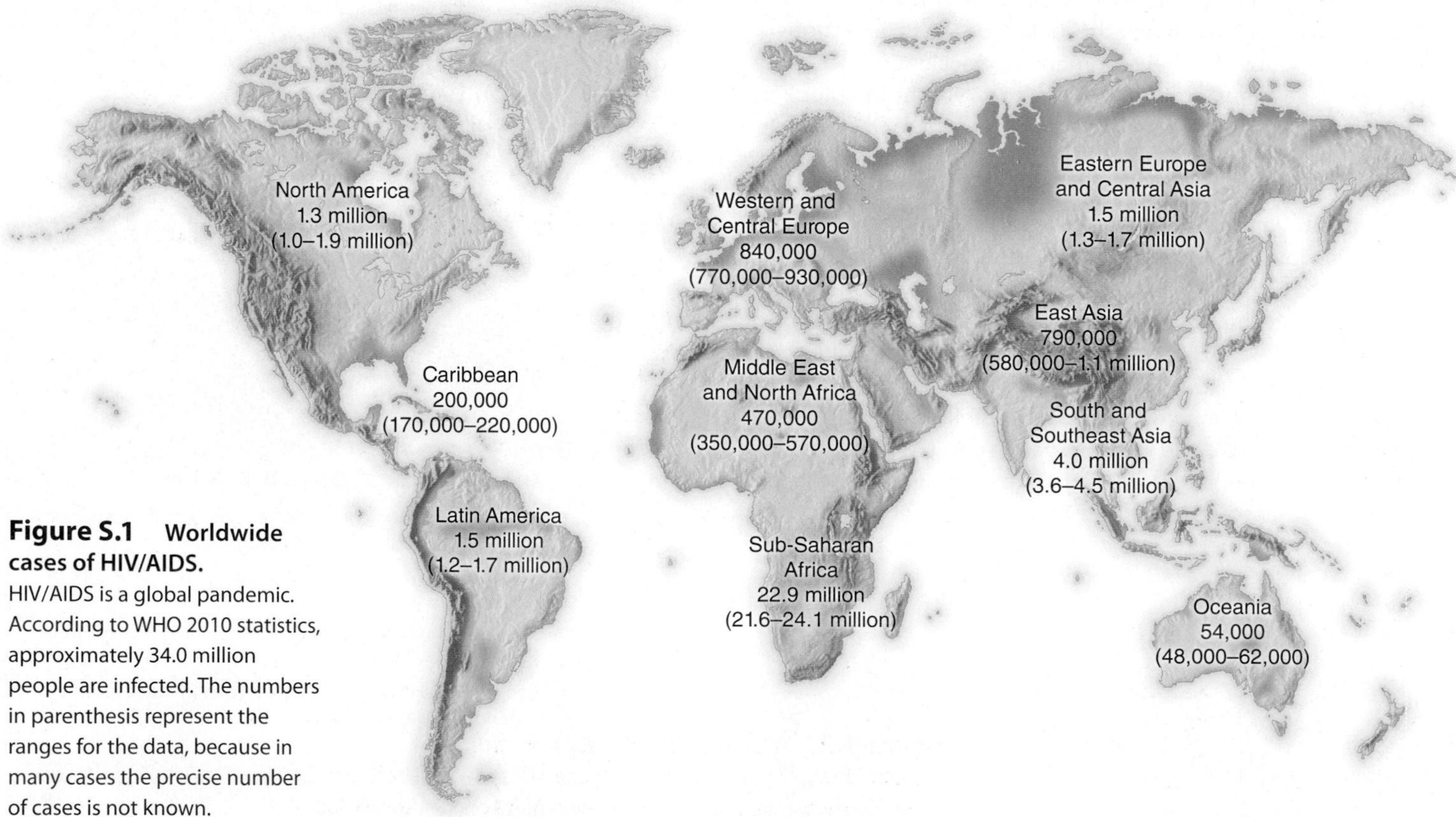

Figure S.1 Worldwide cases of HIV/AIDS. HIV/AIDS is a global pandemic. According to WHO 2010 statistics, approximately 34.0 million people are infected. The numbers in parenthesis represent the ranges for the data, because in many cases the precise number of cases is not known.

as 1959. They examined the preserved tissues of a Manchester seaman who died that year and concluded that he most likely died of AIDS. Similarly, it is thought that HIV entered the United States on numerous occasions as early as the 1950s. But the first documented case is a 15-year-old male who died in Missouri in 1969, with skin lesions now known to be characteristic of an AIDS-related cancer. Doctors froze some of his tissues because they could not identify the cause of death. Researchers also want to test the preserved tissue samples of a 49-year-old Haitian who died in New York in 1959 of the type of pneumonia now known to be AIDS-related.

Throughout the 1960s, it was customary in the United States to list leukemia as the cause of death in immunodeficient patients. Most likely, some of these people actually died of AIDS. HIV is not extremely infectious. Thus, it took several decades for the number of AIDS cases to increase to the point that AIDS became recognizable as a specific and separate disease. The name *AIDS* was coined in 1982, and HIV was found to be the cause of AIDS in 1983–84.

Worldwide, estimates of HIV/AIDS infection rates and deaths are updated every two to three years. As of 2010, (Table S.1), an estimated 34.0 million people were living with HIV infection. Among the 2.7 million new HIV infections, nearly 20% are in people under the age of 15. Although the number of deaths due to HIV/AIDS is declining, in 2010 the disease still claimed 1.8 million lives, bringing the total number of deaths attributed to HIV/AIDS to over 30 million. As of 2010, at least 0.8% of the adults in the world have an HIV infection.

As we can deduce from studying Figure S.1 and Table S.1, most people infected with HIV live in the developing (poor, low- to middle-income) countries. The hardest-hit regions are shown in Figure S.2.

Table S.1 HIV Global Statistics, 2010

	People Living with HIV	New Infections	AIDS Deaths	Adult Prevalence (Ages 15–49)
Sub-Saharan Africa	22.9 million	1.9 million	1.2 million	5.0%
Asia	4.8 million	358,000	306,000	0.3%
Latin America	1.5 million	100,000	67,000	0.4%
Caribbean	200,000	12,000	9,000	0.9%
North America	1.3 million	58,000	20,000	0.6%
Western and Central Europe	840,000	30,000	9,900	0.2%
Eastern Europe and Central Asia	1.5 million	160,000	90,000	0.9%
North Africa and the Middle East	470,000	59,000	35,000	0.2%
Oceania	54,000	3,300	1,600	0.3%
Total	**34.0 million**	**2.7 million**	**1.8 million**	**0.8%**

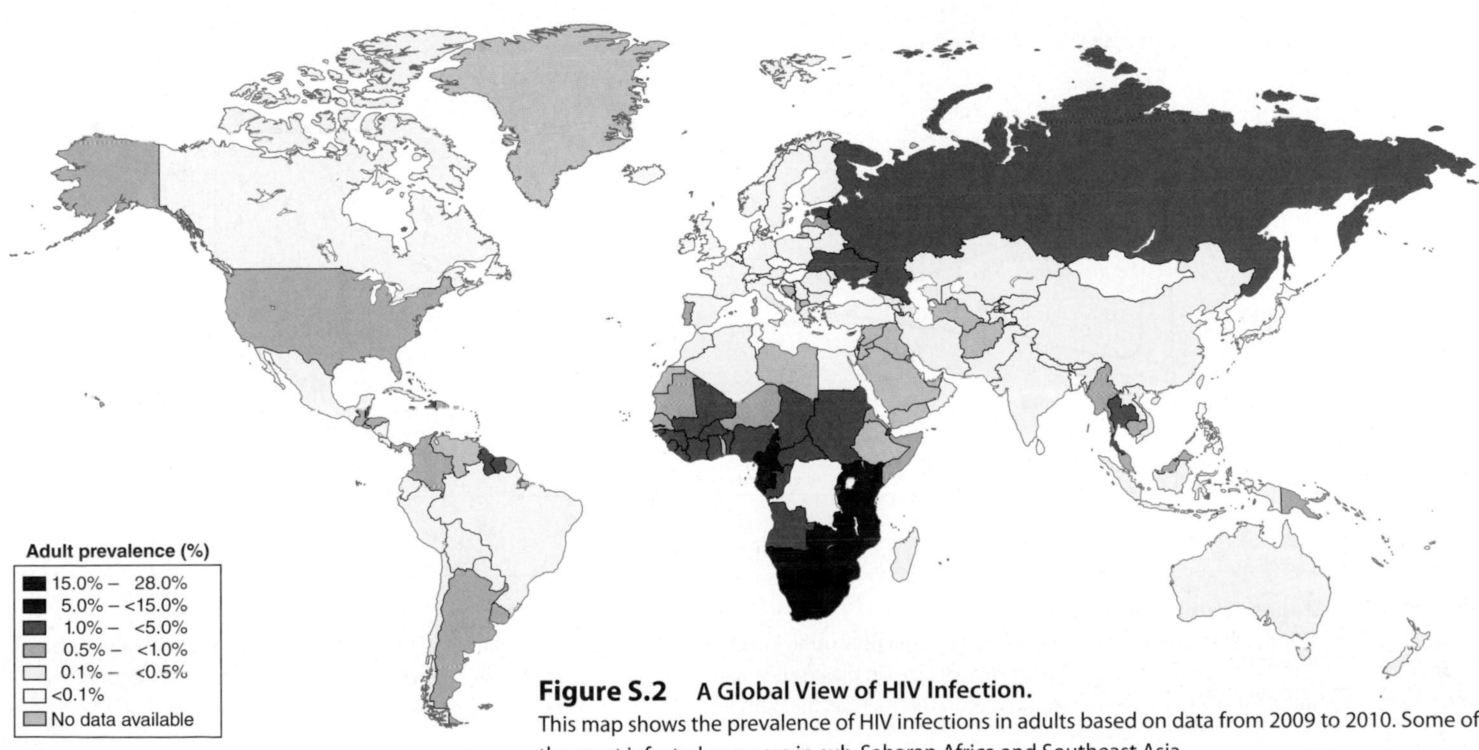

Figure S.2 A Global View of HIV Infection.
This map shows the prevalence of HIV infections in adults based on data from 2009 to 2010. Some of the most infected areas are in sub-Saharan Africa and Southeast Asia.

Phases of an HIV Infection

HIV occurs as several subtypes. HIV-1C is prominent in Africa, and HIV-1B causes most infections in the United States. The following description of the phases of HIV infection pertains to an HIV-1B infection. The helper T cells and macrophages infected by HIV are called *CD4 cells* because they display a molecule called CD4 on their surface. With the destruction of CD4 cells, the immune system is significantly impaired. After all, macrophages present the antigen to helper T cells. In turn, helper T lymphocytes coordinate the immune response. B lymphocytes are stimulated to produce antibodies, and cytotoxic T cells destroy cells infected with a virus. In the United States, one of the most common causes of AIDS deaths is from *Pneumocystis jiroveci* pneumonia (PCP); in Africa, tuberculosis kills more HIV-infected people than any other AIDS-related illness.

In 1993, the CDC issued clinical guidelines for the classification of HIV to help clinicians track the status, progression, and phases of HIV infection. The classification of HIV infection will be discussed in three categories (or phases); the system is based on two aspects of a person's health—the CD4 T-cell count and the history of AIDS-defining illnesses.

Category A: Acute Phase A person in category A typically has no apparent symptoms (asymptomatic), is highly infectious, and has a CD4 T-cell count that has never fallen below 500 cells per cubic millimeter (cells/mm^3) of blood, which is sufficient for the immune system to function normally (Fig. S.3). A normal CD4 T-cell count is at least 800 cells/mm^3.

Today, investigators are able to track not only the blood level of CD4 T cells but also the viral load. The viral load is the number of HIV particles in the blood. At the start of an HIV-1B infection, the virus replicates ferociously, and the killing of CD4 T cells is evident because the blood level of these cells drops dramatically. During the first few weeks of infection, some people (1–2%) develop flulike symptoms (fever, chills, aches, swollen lymph nodes) that may last an average of two weeks. After this, a person may remain "symptom free" for years. At the beginning of this acute phase of infection, an HIV antibody test is usually negative because it generally takes an average of 25 days before there are detectable levels of HIV antibodies in body fluids.

After time, the body responds to the infection by increased activity of immune cells, and the HIV blood test becomes positive. During this phase, the number of CD4 T cells is greater than the viral load (Fig. S.3); but some investigators believe that a great unseen battle is going on. The body is staying ahead of the hordes of viruses entering the blood by producing as many as 1 to 2 billion new helper T lymphocytes each day. This is called the "kitchen-sink model" for CD4 T-cell loss. The sink's faucet (production of new CD4 T cells) and the sink's drain (destruction of CD4 T cells) are wide open. As long as the body can produce enough new CD4 T cells to keep pace with the destruction of these cells by HIV and by cytotoxic T cells, the person has a healthy immune system that can deal with the infection. In other words, a person in category A would have no history of conditions listed in categories B and C.

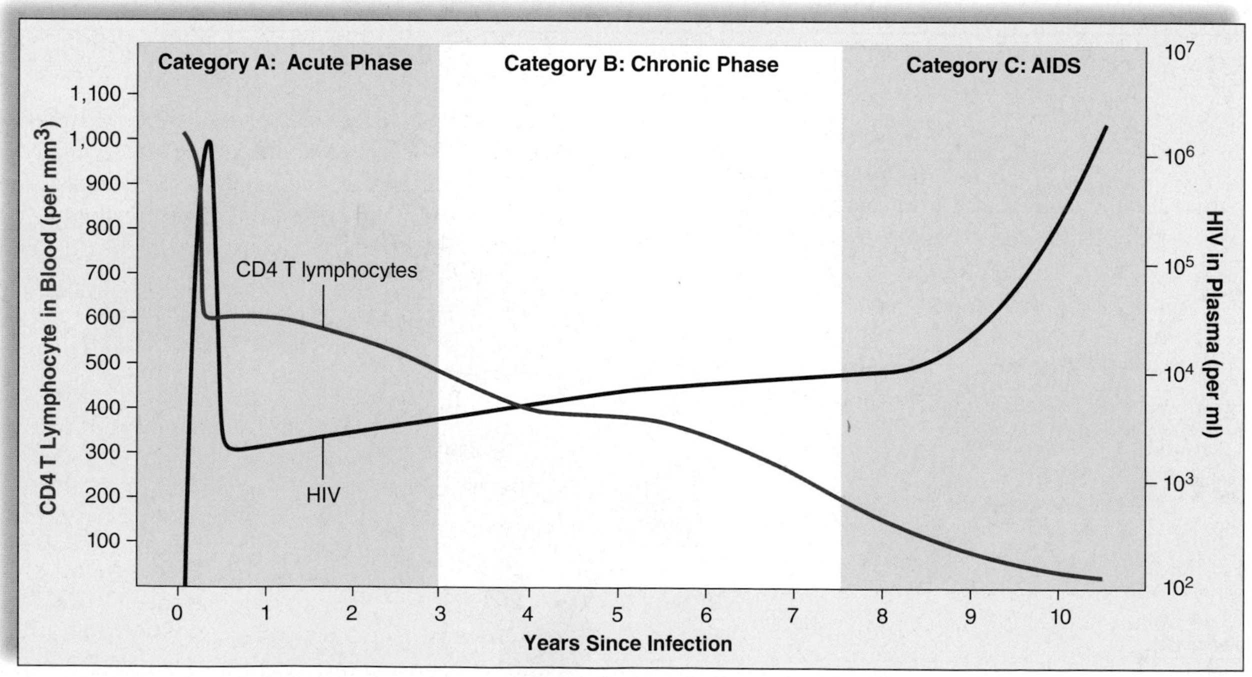

Figure S.3 Stages of an HIV infection.
In category A individuals, the number of HIV particles in plasma rises upon infection and then falls. The number of CD4 T lymphocytes falls but stays above 500/mm^3. In category B individuals, the number of HIV particles in plasma is slowly rising and the number of T lymphocytes is decreasing. In category C individuals, the number of HIV particles in plasma rises dramatically as the number of T lymphocytes falls below 200/mm^3.

Category B: Chronic Phase A person in category B would have a CD4 T-cell count between 499 and 200 cells/mm^3 and one or more of a variety of symptoms related to an impaired immune system. The symptoms might include yeast infections of the mouth or vagina, cervical dysplasia (precancerous abnormal growth), prolonged diarrhea, thick sores on the tongue (hairy leukoplakia), or shingles (to list a few). Swollen lymph nodes, unexplained persistent or recurrent fevers, fatigue, coughs, or diarrhea are often seen as well. During this chronic stage of infection, the number of HIV particles is on the rise (Fig. S.3). However, they do not as yet have any of the conditions listed for category C.

Category C: AIDS A person in category C is diagnosed with AIDS. When a person has AIDS, the CD4 T-cell count has fallen below 200 cells/mm^3 or the person has developed one or more of the 25 AIDS-defining illnesses (or opportunistic infections) described by the CDC's list of conditions in the 1993 AIDS surveillance case definition. Persons with AIDS die from one or more opportunistic diseases rather than from the HIV infection. Recall that an opportunistic illness occurs only when the immune system is weakened. These diseases include the following:

- *Pneumocystis jiroveci* pneumonia—a fungal infection of the lungs
- *Mycobacterium tuberculosis*—a bacterial infection usually of lymph nodes or lungs but may be spread to other organs
- Toxoplasmic encephalitis—a protozoan parasitic infection, often seen in the brain of AIDS patients
- Kaposi's sarcoma—an unusual cancer of the blood vessels, which gives rise to reddish-purple, coin-sized spots and lesions on the skin
- Invasive cervical cancer—a cancer of the cervix, which spreads to nearby tissues

Once one or more of these opportunistic infections have occurred, the person will remain in category C. Newly developed drugs can treat opportunistic diseases. Still, most AIDS patients are repeatedly hospitalized due to weight loss, constant fatigue, and multiple infections. Death usually follows in two to four years. Although there is still no cure for AIDS, many people with HIV infection are living longer, healthier lives due to the expanding use of antiretroviral therapy.

HIV Structure

HIV consists of two single strands of RNA (its nucleic acid genome); various proteins; and an envelope, which it acquires from its host cell (Fig. S.4). The virus's genetic material is protected by a series of three protein coats: the nucleocapsid, capsid, and matrix. Within the matrix are the following three very important enzymes: reverse transcriptase, integrase, and protease:

- *Reverse transcriptase* catalyzes reverse transcription, the conversion of the viral RNA to viral DNA.

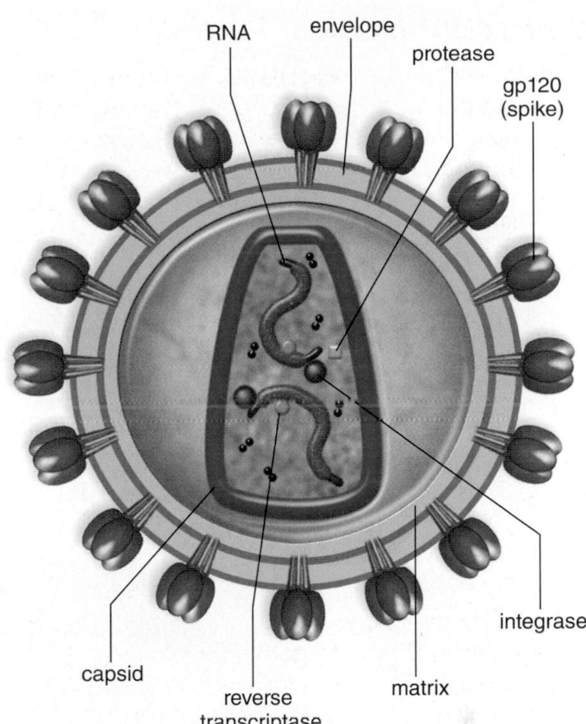

Figure S.4 **The structure of the human immunodeficiency virus.** HIV is composed of two strands of RNA, a capsid, and an envelope containing spike proteins.

- *Integrase* catalyzes the integration of viral DNA into the DNA of the host cell.
- *Protease* catalyzes the breakdown of the newly synthesized viral polypeptides into functional viral proteins.

Embedded in HIV's envelope are protein spikes referred to as *gp120s*. These spikes must be present for HIV to gain entry into its target immune cells. The genome for HIV consists of RNA instead of DNA, which classifies HIV as a retrovirus. A **retrovirus** must use reverse transcription to convert its RNA into viral DNA. Then, it can insert its genome into the host's genome (DNA).

Animation Replication Cycle of a Retrovirus

HIV Life Cycle

The events that occur in the reproductive cycle of an HIV virus (Fig. S.5) are listed here as follows:

1. *Attachment.* During attachment, HIV binds to the plasma membrane of its target cell. The spike located on the surface of HIV, gp120, binds to a CD4 receptor on the surface of a helper T cell or macrophage.
2. *Fusion.* After attachment occurs, HIV fuses with the plasma membrane and the virus enters the cell.
3. *Entry.* During a process called *uncoating,* the capsid and protein coats are removed, releasing RNA and viral proteins into the cytoplasm of the host cell.

Animation Entry of a Virus into a Host Cell

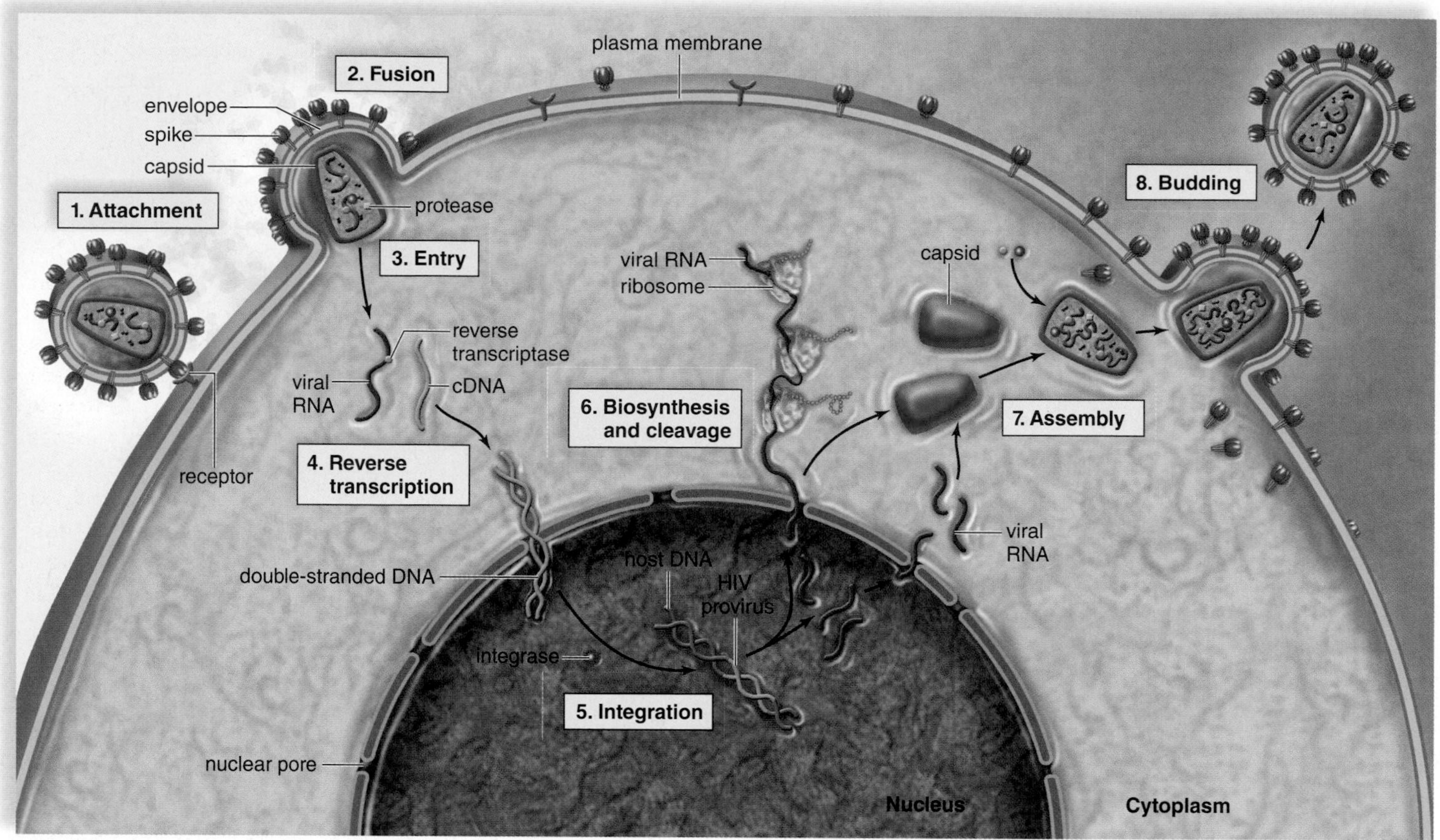

Figure S.5 **HIV replication in a host cell.**
HIV is a retrovirus that uses reverse transcription to produce viral DNA. Viral DNA integrates into the cell's chromosomes before it reproduces and buds from the cell.

4. *Reverse transcription.* This event in the reproductive cycle is unique to retroviruses. During this phase, an enzyme called reverse transcriptase catalyzes the conversion of HIV's single-stranded RNA into double-stranded viral DNA. Usually in cells, DNA is transcribed into RNA. Retroviruses can do the opposite only because they have a unique enzyme from which they take their name (*retro* in Latin means "reverse").
5. *Integration.* The newly synthesized viral DNA, along with the viral enzyme integrase, migrates into the nucleus of the host cell. Then, with the help of integrase, the host cell's DNA is spliced. Double-stranded viral DNA is then integrated into the host cell's DNA (chromosome). Once viral DNA has integrated into the host cell's DNA, HIV is referred to as a **provirus,** meaning it is now a part of the cell's genetic material. HIV is usually transmitted to another person by means of cells that contain proviruses. Also, proviruses serve as a latent reservoir for HIV during drug treatment. Even if drug therapy results in an undetectable viral load, investigators know that there are still proviruses inside infected lymphocytes.
6. *Biosynthesis and cleavage.* When the provirus is activated, perhaps by a new and different infection, the normal cell machinery directs the production of more viral RNA. Some of this RNA becomes the genetic material for new viral particles. The rest of viral RNA brings about the synthesis of very long polypeptides. These polypeptides have to be cut up into smaller pieces. This cutting process, called cleavage, is catalyzed by the HIV protease enzyme.
7. *Assembly.* Capsid proteins, viral enzymes, and RNA can now be assembled to form new viral particles.
8. *Budding.* During budding, the virus gets its envelope and envelope marker coded for by the viral genetic material. The envelope is actually host plasma membrane.

The life cycle of an HIV virus includes transmission to a new host. Body secretions, such as semen from an infected male, contain proviruses inside CD4 T cells. When this semen is discharged into the vagina, rectum, or mouth, infected CD4 T cells migrate through the organ's lining and enter the body. The receptive partner in anal–rectal intercourse appears to be most at risk, because the lining of the rectum is apparently prone to penetration. CD4 macrophages present in tissues are believed to be the first infected when proviruses enter the body. When these macrophages move to the lymph nodes, HIV begins to infect CD4 T cells. HIV can hide out in local lymph nodes for some time, but eventually the lymph nodes degenerate. Large numbers of HIV particles can then enter the bloodstream. Now the viral load begins to increase; when it exceeds the CD4 T-cell count, the individual progresses to the final phase of an HIV infection.

Animation HIV Replication

Transmission and Prevention of HIV

HIV is transmitted by sexual contact with an infected person, including vaginal or rectal intercourse and oral-genital contact. Also, needle sharing among intravenous drug users is high-risk behavior. A less common mode of transmission (and now rare in countries where blood is screened for HIV) is through transfusions of infected blood or blood-clotting factors. Babies born to HIV-infected women may become infected before or during birth or through breast-feeding after birth. From a global perspective, heterosexual sex is the main mode of HIV transmission. In some nations, however, men who have sex with men, IV drug abusers, and sex industry (prostitution) workers are the most common transmitters of HIV. Differences in cultures, sexual practices, and belief systems around the world influence the type of HIV prevention strategies needed to fight the spread of the disease.

Blood, semen, vaginal fluid, and breast milk are the body fluids known to have the highest concentrations of HIV. However, HIV is not transmitted through casual contact in the workplace, schools, or social settings. Casual kissing, hugging, or shaking hands doesn't spread the virus. Likewise, you can't be infected by touching toilet seats, doorknobs, dishes, drinking glasses, food, or pets. The general message of HIV prevention across the globe is either abstinence, sex with only one uninfected partner, or consistent use of a condom during sexual encounters.

HIV Testing and Treatment for HIV

Generally, an HIV test does not test for the virus. Instead, initial HIV tests are designed to detect the presence of HIV antibodies in the body. Most people will develop antibodies to HIV within two to eight weeks (average of 25 days), but it can take from three to six months. Test results for conventional blood, oral, and urine HIV antibody tests may not be available for weeks. Rapid HIV antibody tests are now being used, with results available within 20 minutes.

At one time, an HIV infection almost invariably led to AIDS and an early death because there were no drugs for controlling the progression of HIV disease in infected people. But since late 1995, scientists have gained a much better understanding of the structure of HIV and its life cycle. Now, therapy is available that successfully controls HIV replication. Patients remain in the chronic phase of infection for a variable number of years, meaning that the development of AIDS is postponed. However, this is not the case for those in poor nations where funds for HIV drugs are limited. Only 20% of those worldwide who need antiretroviral therapy are receiving this treatment.

APPLICATIONS AND MISCONCEPTIONS

What is the difference between an HIV test, rapid test, or home-use test kit?

The Food and Drug Administration (FDA) regulates the terminology associated with at home testing for diseases such as HIV/AIDS. In general the term test indicates that a sample (blood, urine) is collected and then sent to a registered lab for analysis. A "rapid test" is performed on site by a registered health-care worker. In a "home test," the person purchasing the kit collects the samples, performs the tests, and analyzes the results. For a home test, the data are not necessarily verified by a registered health-care worker. Although an Internet search and a visit to almost any drugstore reveals a number of over-the-counter (OTC) home tests for HIV, the only home test that is currently approved by the FDA is called the OraQuick In-Home HIV Test. This test provides results within 20 minutes, but the individual is still recommended to confirm the presence of the virus using a blood test. Another type of test that is approved by the FDA is called the Home Access HIV-1 Test System (sometimes also called the Express HIV-1 Test System). This is a highly reliable test that also has safeguards in place to protect an individual's identity. This test does not provide immediate results, but instead involves sending a sample to a registed lab. The results are reported in about seven days. As is always the case, it is highly recommended that you talk with your physician regarding the results of any form of medical test.

AIDS in the United States is presently caused by HIV-1B, and drug therapy has brought the condition under control. But the drug therapy has two dangers. Infected people may become lax in their efforts to avoid infection because they know that drug therapy is available. Further, drug use leads to drug-resistant viruses. Even now, some HIV-1B viruses are known to have become drug resistant when patients have failed to adhere to their drug regimens.

Another possible problem is that the present treatment is designed for HIV-1B. It's possible that in 5 to 20 years, the more-developed countries, including the United States, will experience a new epidemic of AIDS caused by HIV-1C. Therefore, it behooves the more-developed countries to do all they can to help African countries aggressively seek a solution to the HIV-1C epidemic.

There is no cure for AIDS, but a treatment called *highly active antiretroviral therapy* (*HAART*) is usually able to stop HIV replication to such an extent that the viral load becomes undetectable. HAART uses a combination of drugs that interfere with the life cycle of HIV. Entry inhibitors stop HIV from entering a cell. The virus is prevented from binding to a receptor in the plasma membrane. Reverse transcriptase inhibitors, such as zidovudine (AZT), interfere with the operation of the reverse transcriptase enzyme. Integrase inhibitors prevent HIV from inserting its genetic material into that of the host cells.

Protease inhibitors prevent protease from cutting up newly created polypeptides. Assembly and budding inhibitors are in the experimental stage, and none are available as yet. Ideally, a drug combination will make the virus less likely to replicate or successfully mutate. If reproduction can be suppressed, viral resistance to this therapy will be less likely to occur and the drugs won't lose their effectiveness. However, investigators have found that when HAART is discontinued, the virus rebounds.

Animation Treatment of HIV Infection

Throughout the world, more than 5 million people are now receiving HIV treatment. In 2009 alone, 1.2 million people received HIV antiretroviral therapy for the first time—a 30% increase in the number of people receiving treatment in a single year. This is largely due to the President's Emergency Plan for AIDS Relief, or PEPFAR program, which began as a commitment from the United States to provide $15 billion in antiretroviral drugs to resource-poor countries from 2003 to 2008. Since then, the program has been continued and expanded. A study published in 2009 showed the PEPFAR program had cut AIDS death rates by more than 10% in targeted countries in Africa, though it had no appreciable effect on the prevalence of the disease in those nations.

An HIV-positive pregnant woman who takes reverse transcriptase inhibitors during her pregnancy reduces the chances of HIV transmission to her newborn. If possible, drug therapy should be delayed until the tenth to twelfth week of pregnancy to minimize any adverse effects of AZT on fetal development. If treatment begins at this time and delivery is by cesarean section, the chance of transmission from mother to infant is very slim (about 1%).

The consensus is that control of the AIDS pandemic will not occur until a vaccine preventing HIV infection is developed. Unfortunately, none of these would be a cure. Instead, the vaccine would be a preventive measure that helps people who are not yet infected escape infection (preventive vaccine). Alternatively, a therapeutic vaccine could slow the progression of the disease upon future infection. Scientists have studied more than 50 different preventive vaccines and over 30 therapeutic vaccines. A large-scale clinical trial (called RV144) of a preventive vaccine concluded in 2009 in Thailand; though there is some evidence that the vaccine may help reduce HIV infection rates, researchers are still analyzing the data and working on follow-up studies. Still, the success of RV144 has encouraged researchers in believing that the development of a preventive vaccine may be possible in the near future. The Science feature, "The Challenges of Developing an AIDS Vaccine," reviews the difficulties in developing an effective AIDS vaccine. Thus, you can understand why we do not have an AIDS vaccine yet and, indeed, why we may never have an ideal one.

Tuberculosis

In 1882, Robert Koch was the first to see the causative agent of tuberculosis (TB) with a microscope. At that time, TB caused 14% of deaths in Europe. The disease was called *consumption,* because it seemed to consume the patients from the inside until they wasted away. In the 1940s, with the advent of effective antibiotics to fight TB, it was thought that the disease could be eliminated. However, control measures were not implemented consistently and tuberculosis cases began to rise in the 1980s. It is estimated that one-third of the world's population has been exposed to TB. Each year, approximately 9 million people are infected, and 2 million die (Fig. S.6). HIV infections are a contributing factor to the increase in TB cases. Tuberculosis is currently the number one cause of death in AIDS patients.

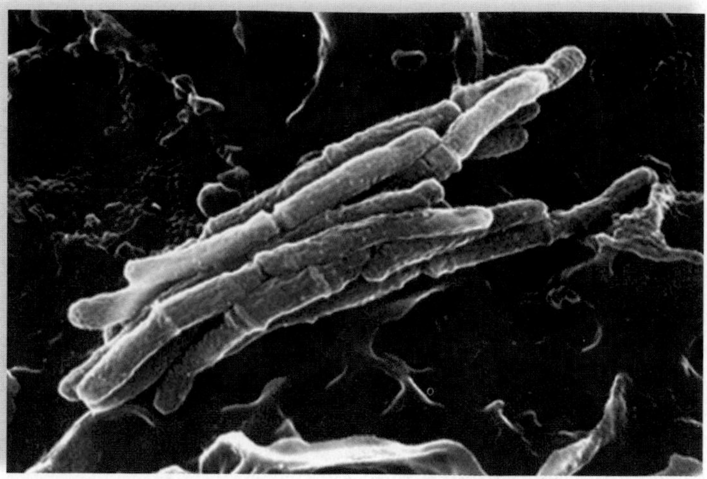

SEM 6,200×

Figure S.6 The causative agent of tuberculosis. Tuberculosis is caused by *Mycobacterium tuberculosis,* shown here from a sputum sample.

Causative Agent and Transmission

Tuberculosis is caused by a species of rod-shaped bacterium called *Mycobacterium tuberculosis.* In nature, it is a very slow-growing bacterium. The cells have a thick, waxy coating and can exist for weeks in a dehydrated state. The organism is spread by airborne droplets, introduced into the air when an infected person coughs, sings, or sneezes. The bacteria can float in the air for several hours and still be infectious. The likelihood of infection increases with the length and frequency of exposure to an individual with active TB. This makes TB especially contagious on airplane flights longer than 8 hours. It also makes the caregivers for those with TB at risk.

Disease

The incubation period is 4 to 12 weeks, and the disease develops very slowly. Once the bacteria reach the lungs, they are consumed by macrophages. Other white blood cells rush to the infected area. Together, they wall off the original infection site, producing small, hard nodules, or *tubercles,* in the lungs. It is these tubercles that give the disease its name. In most patients, the bacteria remain alive within the tubercles, but the disease does not progress. These patients are said to have

BIOLOGY MATTERS  Science

The Challenges of Developing an AIDS Vaccine

An ideal AIDS vaccine would be inexpensive and able to provide lifelong protection against all strains of HIV (Fig. SA). Is this possible? Effective vaccines have been developed against diseases such as hepatitis B, smallpox, polio, tetanus, influenza, and the measles. But what about AIDS? The top obstacles in the development of an AIDS vaccine include the following:

1. The ideal AIDS vaccine is one that would prevent HIV entry into human cells and prevent the progression and transmission of the disease. However, no vaccine has ever proven to be 100% effective at blocking a virus from entry into cells. Instead, most vaccines prevent, modify, or weaken the disease caused by the infection.
2. Due to HIV's high rate of mutation, there are several genetically different types and subtypes of HIV. HIV strains may differ by 10% within one person and by 35% in people across the globe. Viruses that are genetically different may have different surface proteins, and HIV surface proteins are the focus of many AIDS vaccines. The question is, will scientists need a vaccine for each HIV subtype or will one vaccine provide protection for all HIV variants?
3. If a vaccine produces only short-term protection, people would need to continue to get booster shots similar to shots given yearly for the flu.
4. There are concerns that an AIDS vaccine would make people more vulnerable to HIV infection. It is believed that in some diseases, such as yellow fever and Rift Valley fever, the antibodies produced after receiving the vaccine helped the virus infect more cells.
5. HIV can be transmitted as a free virus and in infected cells, so perhaps a vaccine would need to stimulate both cellular and antibody-mediated responses. This is complicated by the fact that HIV infects and destroys immune cells, particularly T cells. In addition, most successful vaccines stimulate only antibody production.
6. Most vaccines in use today against other diseases are prepared from live-attenuated (weakened) forms of the infectious virus. There are concerns that an AIDS vaccine made using weakened forms of the live virus will cause HIV disease (AIDS).
7. HIV inserts its genetic material into human cells, where it can hide from the immune system.

Although there are many difficulties in vaccine development, AIDS vaccine trials are under way. The process can take many years. After a vaccine has been tested in animals, it must pass through three phases of clinical trials before it is marketed or administered to the public. In Phases I and II, the vaccine is tested from one to two years in a small number of HIV-uninfected volunteers. The most effective vaccines move into Phase III. In Phase III, the vaccine is tested three to four years in thousands of HIV-uninfected people. A Phase III trial of the RV144 HIV vaccine was conducted from 2003 to 2009 in Thailand and involved over 16,000 volunteers. Though the results are still being analyzed, the vaccine is reported to reduce the rate of HIV infection by 32%. Whereas this is a much lower percentage than will be needed for a vaccine to be considered effective, most scientists have many reasons to be optimistic that an AIDS vaccine can and will be developed. The RV144 results, and other successes with vaccinating monkeys, suggest that the research is beginning to yield results. But the most compelling reason for optimism is the human body's ability to suppress the infection. The immune system is able to successfully and effectively decrease the HIV viral load in the body, helping to delay the onset of AIDS an average of ten years in 60% of people who are HIV-infected in the United States. Studies have shown that a small number of people remain HIV-uninfected after repeated exposure to the virus, and a few HIV-infected individuals maintain a healthy immune system for over 15 years. It is these stories of the human body's ability to fight HIV infection that keep scientists hopeful that there is a way to help the body fight HIV infection. Most scientists agree that—despite the obstacles or the "setbacks"—an AIDS vaccine is possible somewhere in our future.

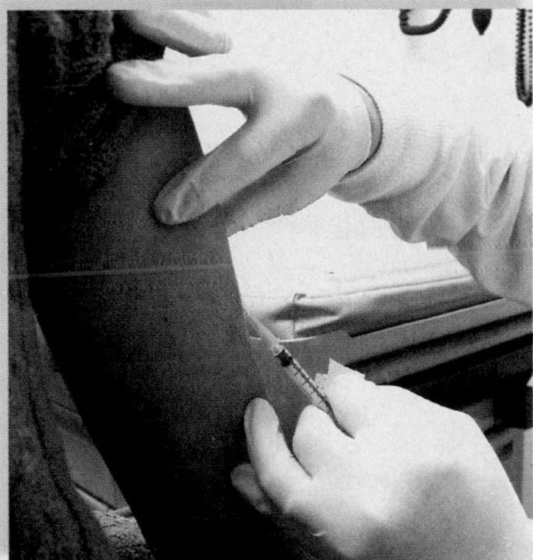

Figure SA A preventive HIV vaccine is not yet available. Different strains of HIV as well as a high viral mutation rate are just two of the reasons it has been difficult to develop a vaccine against HIV.

Questions to Consider

1. Why might an HIV/AIDS vaccine actually give some people a false sense of security?
2. What cultural factors may be inhibiting the development of a vaccine?

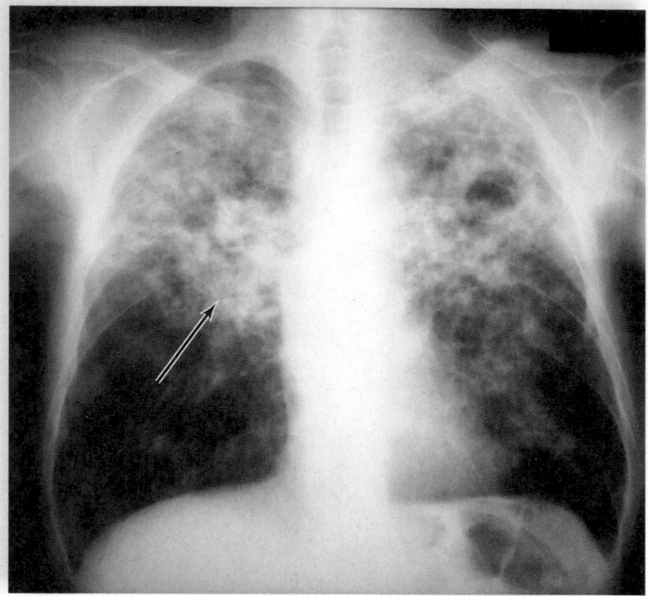

Figure S.7 An X-ray of a TB-infected lung.
TB causes tubercles in the lungs. In the active state, these liquefy and cavities are seen on an X-ray.

latent TB. They do not feel sick, and they are not contagious. A person with latent TB will test positively on a tuberculosis skin test. Tubercles often calcify and can be seen on a chest X-ray (Fig. S.7). The combination of skin testing and X-ray findings confirms the diagnosis of tuberculosis.

If the immune system fails to control *Mycobacterium tuberculosis* in the lungs, active disease may occur. A person with active disease is contagious. The tubercle liquefies and forms a cavity. The bacteria can then spread from these cavities throughout the body, especially to the kidney, spine, and brain. It may be fatal. Symptoms of active TB include a bad cough, chest pain, and coughing up blood or sputum. As the disease progresses, symptoms include fatigue, loss of appetite, chills, fevers, and night sweats. The patient begins to lose weight and wastes away.

APPLICATIONS AND MISCONCEPTIONS

Does the TB skin test expose you to tuberculosis?

The TB skin test does not expose you to the *Mycobacterium tuberculosis* bacteria. Instead, in a lab, a small amount of purified protein derivative (called PPD) is obtained from strains of the bacterium. When the PPD is injected under the skin, it acts as an antigen to the immune system. If you have previously been exposed to TB, the antibody-mediated response will invoke a reaction (swelling, hardness) against the PPD protein. A medical professional familiar with your medical history then measures the degree of the reaction to assess your exposure. Because the PPD protein is derived from the bacterium, it is not capable of causing the disease.

Treatment and Prevention

Due to the resurgence of antibiotic-resistant strains, multiple anti-TB drugs are given simultaneously for 12 to 24 months. The most common drugs are isoniazid (INH), rifampin (RIF), ethambutol, and pyrazinamide. Though it is unlikely that bacteria will develop resistance to multiple drugs at the same time, there are several drug-resistant forms of TB. Examples are multidrug-resistant TB (MDR TB), which is resistant to both isoniazid and rifampin, and extensively drug-resistant TB (XDR TB), a rare form of TB that is resistant to isoniazid and rifampin, plus one additional antibiotic. It takes at least six months to kill all of the *Mycobacterium tuberculosis* in the body, so the length of drug treatment is long.

Public health officials try to prevent the spread of TB by identifying and treating all cases of active TB. Active TB patients are isolated for at least two weeks at the beginning of drug treatment to prevent transmission. Thereafter, they are monitored for drug compliance and reappearance of symptoms. Anyone exposed to an active case of TB is treated.

Malaria

Malaria is called the world's invisible pandemic. Most people in the West do not even consider malaria a major health threat, yet there are 350–500 million cases per year with over 1 million deaths, mostly in sub-Saharan Africa (Fig. S.8). Key to the geographic distribution of malaria is transmission of the disease by a mosquito vector that depends on temperature and rainfall and thus survives well in tropical areas. A **vector** is a living organism—usually an insect or animal—that transfers the pathogen from one host to another.

Causative Agent and Transmission

The parasites that cause malaria belong to the genus *Plasmodium.* These are protists (see Fig. 1.5). There are four species that infect humans: *P. malariae, P. falciparum, P. vivax,* and *P. ovale. P. falciparum* causes more disease and death than the other species. The parasite is spread by the female *Anopheles gambiae* mosquito. Half of the life cycle occurs in the human, and the remainder happens in the mosquito. As the female mosquito feeds on human blood, she injects saliva containing an anticoagulant along with the parasite. The parasites travel to the liver, where they undergo asexual reproduction. The parasites are released from the liver to infect more liver cells and erythrocytes (red blood cells). Inside the erythrocyte, the *Plasmodium* enlarges and divides until it bursts the erythrocytes. This red blood cell stage is cyclic and repeats every 48 to 72 hours. Some parasites within the erythrocytes don't destroy their host cells. Instead, they develop into the sexual form of the parasite. When these are ingested by another

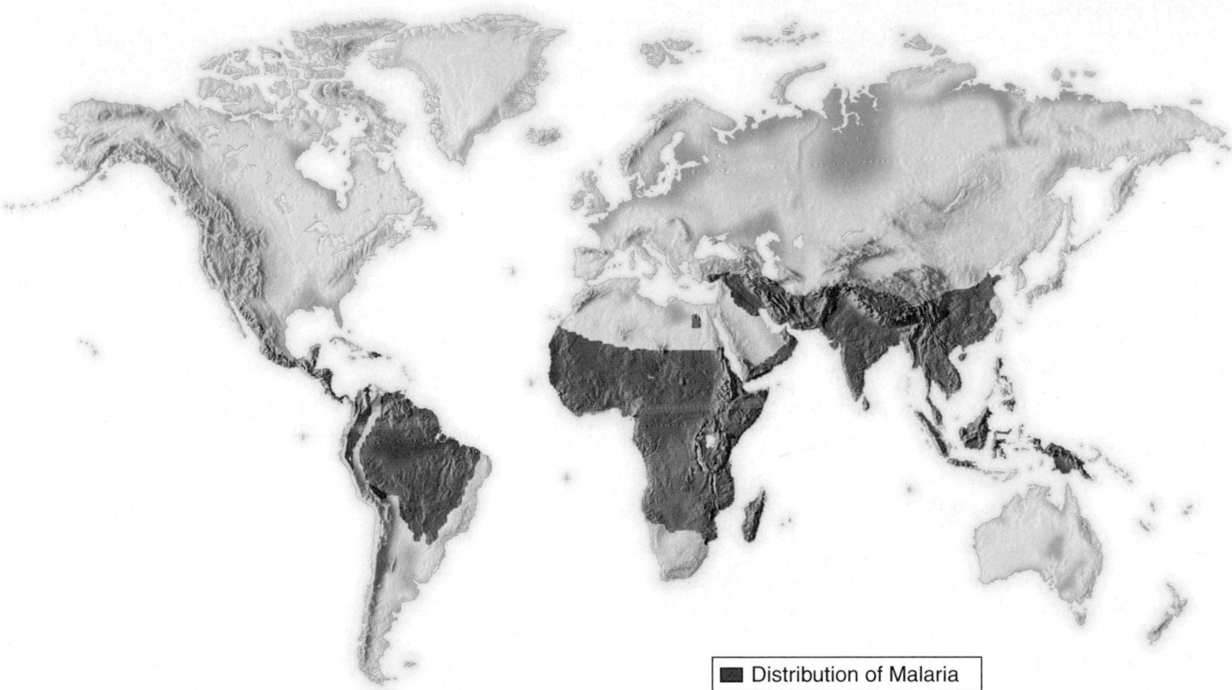

Figure S.8 Areas of the world that are hardest hit by malaria.
Although malaria may occur in many regions of the world, the disease is most often found in subtropical areas where its mosquito vector lives.

mosquito during a blood meal, they develop into male and female gametes within the gut of the mosquito. The gametes fuse, undergo mitosis, and form the parasites that migrate to the salivary glands of the mosquito to continue the cycle.

Disease

People at significant risk for malaria include those who have little or no immunity to the parasite. Children, pregnant women, and travelers are most likely to fall victim to the disease. Diagnosis of malaria depends upon the presence of parasites in the blood. The incubation period from time of bite to onset of symptoms varies from 7 to 30 days. The symptoms of malaria range from very mild to fatal. Most infected people develop a flulike illness with chills and fevers interrupted by sweating. These symptoms exhibit a cyclical pattern every 48 to 72 hours corresponding to bursting of the red blood cells in the body. Milder cases of malaria are often confused with influenza or a cold; therefore, treatment is delayed. More severe cases cause severe anemia due to destruction of red blood cells, cerebral malaria, acute kidney failure, cardiovascular collapse, shock, and death.

Treatment and Prevention

Malaria is a curable disease if it is diagnosed and treated correctly in a short period. A person exhibiting malaria symptoms should be treated within 24 hours of onset. Common antimalarial drugs include quinine and artesunate. Treatment helps reduce symptoms and breaks the transmission pattern for the disease. Antimalarial drugs can also be used prophylactically, or before infection. They do not prevent the initial infection following the mosquito bite. Instead, they prevent the development of the parasites in the blood.

Health organizations are working on the prevention of infection through vector control. Strategies include eliminating the mosquito by removing its breeding sites and by insecticide fogging of large areas. Additional efforts are aimed at preventing humans from being bitten by the mosquito, using simple mosquito nets. The use of insecticide-treated mosquito nets for children has reduced the incidence of malaria.

Drug-resistant *Plasmodium* and insecticide-resistant *Anopheles* are becoming significant problems. *P. falciparum* and *P. vivax* have developed strains that are resistant to the antimalarial drugs. Efforts to develop a malaria vaccine are ongoing.

Influenza

The common name for influenza is the "flu," and each year it affects between 5 and 20% of Americans and causes an estimated 36,000 fatalities. Influenza is a viral infection that causes runny nose, cough, chills, fever, head and body aches, and nausea. You can catch influenza by inhaling virus-laden droplets that have been coughed or sneezed into the air by an infected person, or by contact with contaminated objects, such as door handles or bedding. The viruses then attach to and infect cells of the respiratory tract.

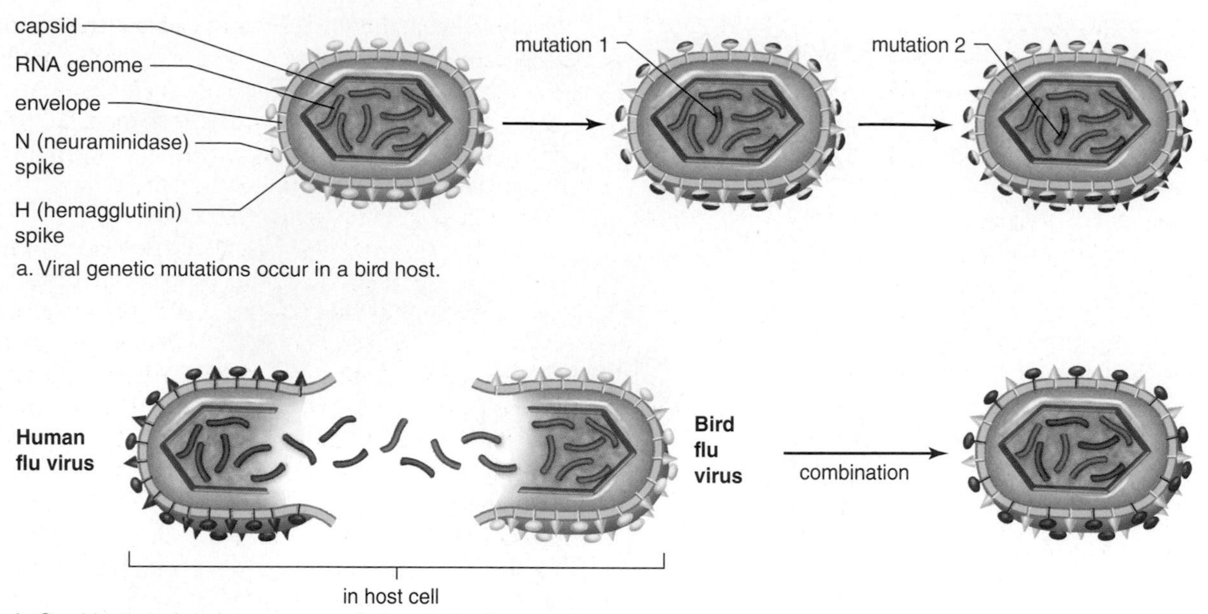

Figure S.9 The bird flu virus.
a. Genetic mutations in bird flu viral spikes could allow the virus to infect the human upper respiratory tract. **b.** Alternatively, a combination of bird flu and human spikes could allow the virus to infect the human upper respiratory tract.

Influenza Viruses

The influenza virus has an H (hemagglutinin) spike and an N (neuraminidase) spike (Fig. S.9*a, left*). Its H spike allows the virus to bind to its receptor, and its N spike attacks host plasma membranes in a way that allows mature viruses to exit the cell. Both H spikes and N spikes have variations in their shapes: 16 types of H and 9 types of N spikes are known. Furthermore, each type of spike can occur in different varieties called subtypes. Many of the influenza viruses are assigned specific codes based on the type of spike. For example, H5N1 virus gets its name from its variety of H5 spikes and its variety of N1 spikes. Our immune system can recognize only the particular variety of H spikes and N spikes it has been exposed to in the past by infection or immunization. When a new influenza virus arises, one for which there is little or no immunity in the human population, a flu pandemic (global outbreak) may occur.

Possible Bird Flu Pandemic of the Future

Currently, the H5N1 subtype of influenza virus is of great concern because of its potential to reach pandemic proportions. An H5N1 is common in wild birds such as waterfowl, and can readily infect domestic poultry such as chickens, which is why it is referred to as an avian influenza or a bird flu virus. An H5N1 virus has infected waterfowl for some time without causing serious illness. A more pathogenic version of H5N1 appeared about a decade ago in China, and promptly started to cause widespread and severe illness in domestic chickens. Scientists are still trying to determine what made H5N1 become so lethal, first to chickens, and then to humans.

Why can the bird flu H5N1 infect humans? Because the virus can attach to both a bird flu receptor and to a human flu receptor. Close contact between domestic poultry and humans is necessary for this to happen. At this time, the virus has rarely been transmitted from one human to another, and only among people who have close contact with one another, such as members of the same household. The concern is that with additional mutations, the H5N1 virus could become capable of sustained human-to-human transmission, and then spread around the world.

How could H5N1 become better at spreading within the human population? At this time, bird flu H5N1 infects mostly the lungs. Most human flu viruses infect the upper respiratory tract, trachea, and bronchi and can be spread by coughing. If a spontaneous mutation in the H spike of H5N1 enabled it to attack the upper respiratory tract, then it could be easily spread from human to human by coughing and sneezing. Or, another possibility is that a combining of spikes could occur in a person who is infected with both the bird flu and the human flu viruses (Fig. S.9*b*). According to the CDC, over the past decade an increasing number of humans infected with an H5N1 virus have been reported in Asia, the Pacific, the Near East, Africa, and Europe. Over half of these people have died. The good news is that the FDA has approved an H5N1 vaccine for individuals between the ages of 18 and 64. While additional mutations in the virus may reduce the overall effectiveness of the vaccine, it is believed to provide a good foundation of protection against this form of influenza.

Video Flu Fighter

CHECK YOUR PROGRESS S.1

1. Describe the differences between an outbreak, epidemic, and pandemic, and give an example of each.
2. Summarize the HIV replication cycle, and list the types of cells this virus infects.
3. Explain the role of the mosquito in the malarial life cycle.
4. Explain how variation may occur in influenza viruses such as H5N1.

CONNECTING THE CONCEPTS

For more information on the topics presented in this section, refer to the following discussions:

Section 6.2 examines the role of erythrocytes in the body.

Section 7.1 describes the structure of a typical virus.

Section 9.3 explores the structure of the lungs.

S.2 Emerging Diseases

LEARNING OUTCOMES

Upon completion of this section, you should be able to

1. Define the term *emerging disease.*
2. List some examples of emerging diseases.

In the past several years, avian influenza (H5N1; see section S.1), swine flu (H1N1), and severe acute respiratory syndrome (SARS) have generated a lot of press. These are considered new or **emerging diseases.** The National Institute of Allergy and Infectious Diseases (NIAID) lists 18 pathogens that are newly recognized in the last two decades and five additional pathogens that are considered to be reemerging. Reemerging diseases are ones that have reappeared after a significant decline in incidence. *Streptococcus,* the bacteria that causes strep throat and other infections, is considered to be a reemerging pathogen due to increasing resistance to antibiotics. Finally, there are diseases that have been known throughout human history but had not been known to be caused by an infectious agent or the pathogen had never been identified. Ulcers caused by *Helicobacter pylori* (recognized in 1983) are an example.

Figure S.10 **A potential source of the SARS epidemic.** It appears that civets may have been the source of the SARS epidemic. Civet meat is considered a delicacy in some parts of China.

Where do emerging diseases come from? Some of these diseases may result from new and/or increased exposure to animals or insect populations that act as vectors for disease. Changes in human behavior and use of technology can result in new diseases. SARS is thought to have arisen in Guandong, China, due to consumption of civets, a type of exotic cat considered a delicacy (Fig. S.10). The civets were possibly infected by exposure to horseshoe bats sold in open markets. Legionnaires' disease emerged in 1976 due to contamination of a large air-conditioning system in a hotel. The bacteria thrived in the cooling tower used as the water source for the air-conditioning system. In addition, globalization results in the transport of diseases all over the world that were previously restricted to isolated communities. The first SARS cases were reported in southern China the week of November 16, 2002. By the end of February 2003, SARS had reached nine countries/provinces, mostly through airline travel. Some pathogens mutate and change hosts, jumping from birds to humans, for example. Before 1997, avian flu was thought to affect only birds. A mutated strain jumped to humans in the 1997 outbreak. To control that epidemic, officials killed 1.5 million chickens to remove the source of the virus.

NIAID also monitors reemerging diseases. Reemerging diseases have been known in the past but were thought to have been controlled. Diseases in this category include known diseases that are spreading from their original geographic location or diseases that have suddenly increased in incidence. A change in geographic location could be due to global warming allowing expansion of habitats for insect vectors. Reemerging diseases can also be due to human carelessness, as in the abuse of antibiotics or poorly implemented vaccination programs. This allows previously controlled diseases to resurge.

CHECK YOUR PROGRESS S.2

1. Define *emerging disease,* and give an example.
2. Explain how emerging diseases arise.
3. Explain what may be done to reduce the threat of emerging and reemerging diseases.

CONNECTING THE CONCEPTS

For more information on the topics in this section, refer to the following discussions:

Section 1.3 examines the research that identified *H. pylori* as a cause of ulcers.

Section 7.1 provides additional information on the H1N1 virus.

S.3 Antibiotic Resistance

LEARNING OUTCOMES

Upon completion of this section, you should be able to

1. Summarize how a pathogen becomes resistant to an antibiotic.
2. Explain the significance of XDR TB and MRSA.

Some well-known pathogens are becoming more difficult to fight due to the advent of **antibiotic resistance.** Just four years after penicillin was introduced in 1943, bacteria began developing resistance to it. The use of antibiotics does not cause humans to become resistant to the drugs. Instead, pathogens become resistant. There are some organisms in a population naturally resistant to the drug (Fig. S.11). They have acquired this resistance through mutations or interactions with other organisms. The drug regimen kills the susceptible ones while leaving naturally resistant ones to multiply and repopulate the patient's body. The new population is then resistant to the drug. Tuberculosis, malaria, gonorrhea, *Staphylococcus aureus,* and enterococci (or group D *Streptococcus*) are a few of the diseases and organisms connected to antibiotic resistance. Unfortunately, more and more organisms are becoming multidrug resistant, leaving health-care facilities with little choice for treatment of infection.

What is being done about this problem? The CDC, U.S. Food and Drug Administration, and U.S. Department of Agriculture have formed a cooperative organization charged with monitoring antibiotic-resistant organisms. Pharmaceutical companies are developing new antibiotics. However, the best way to fight antibiotic resistance is to prevent it from happening in the first place by using antibiotics wisely. Take all of the antibiotics prescribed as directed. Do not skip doses or discontinue treatment when you feel better. Do not expect a doctor to prescribe antibiotics for all infections. For example, antibiotics are ineffective in treating viral infections such as colds. Do not save unused antibiotics or take antibiotics prescribed for a different infection. Antibiotics are important drugs for fighting infection and improper usage lessens their effectiveness for all of us.

XDR TB

XDR TB stands for extensively drug-resistant tuberculosis. It is resistant to almost all of the drugs used to treat TB. This includes the first-line antibiotics, the older and cheaper ones, as well as the second-line antibiotics, the newer and more expensive drugs. Therefore, the treatment options are very limited. Fortunately, this is still relatively rare. There were only 83 cases in the United States between 1993 and 2007. MDR TB, multidrug-resistant TB, is more common. These organisms are resistant to the first-line antibiotics. Eastern Europe and Southeast Asia have the highest rates of MDR TB.

In 2007, a patient allegedly diagnosed with XDR TB traveled to Europe and Canada by air, then tried to enter the United States by car. The CDC put him under federal quarantine in Atlanta. Patients with XDR TB should not be traveling on airlines, and the CDC recommended testing for passengers and crew on those airline flights.

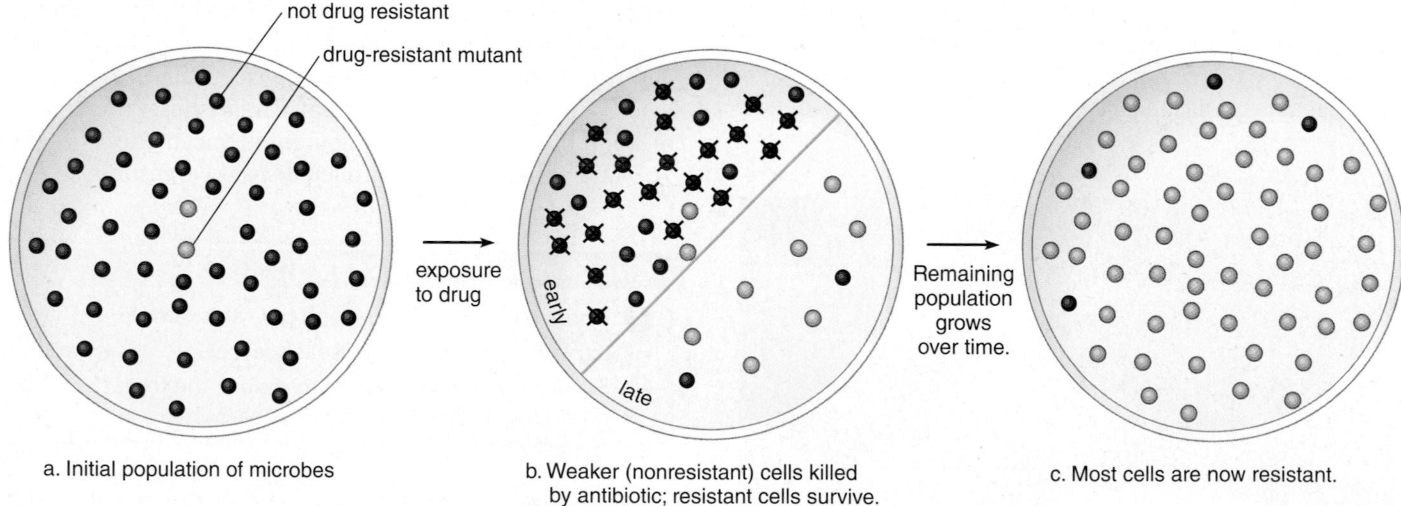
not drug resistant
drug-resistant mutant
exposure to drug
early
late
Remaining population grows over time.
a. Initial population of microbes
b. Weaker (nonresistant) cells killed by antibiotic; resistant cells survive.
c. Most cells are now resistant.

Figure S.11 Development of antibiotic resistance.
a. In a microbe population, random mutation can result in cells with drug resistance. **b.** Drug use kills all nonresistant microbes, leaving only the stronger, resistant cells to survive. **c.** The new microbe population is now mostly resistant to the antibiotic.

MRSA

Methicillin-resistant *Staphylococcus aureus,* or **MRSA,** is another antibiotic-resistant bacterium (Fig. S.12). MRSA is resistant to methicillin and other common antibiotics such as penicillin and amoxicillin. It causes "staph" infections such as boils and infection of the hair follicles. In 1974, only 2% of staph infections were caused by MRSA, but by 2004, the number had risen to 63%. In response, the CDC conducted an aggressive campaign to educate health-care workers on MRSA prevention. The program was very successful, and between 2005 and 2008 the number of MRSA infections in hospitals declined by 28%.

MRSA is especially common in athletes who share equipment. MRSA is also prevalent in nursing homes and hospitals where patients have already reduced immune responses. Patients infected with MRSA generally have longer hospital stays with poorer outcomes. It can be fatal. It is passed from nonsymptomatic carriers to patients, usually through hand contact. Hand washing is critical for preventing transmission.

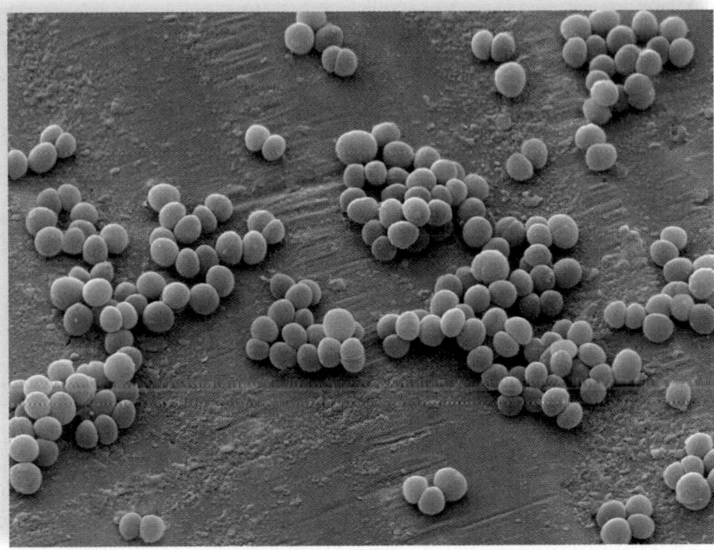

Figure S.12 **Methicillin-resistant *Staphylococcus aureus*.**

APPLICATIONS AND MISCONCEPTIONS

How dangerous is MRSA?

To illustrate how serious a threat MRSA represents, consider the following fact. In 2005, MRSA infections caused more deaths in the United States (18,650) than AIDS (16,000). The good news is that the rate of MRSA infections has been falling consistently since 2005, mostly due to an increased awareness among health professionals.

CHECK YOUR PROGRESS S.3

1. Explain how bacteria become resistant to an antibiotic.
2. Describe the correct procedure for taking antibiotics.
3. Summarize the significance of XDR TB and MRSA.

CONNECTING THE CONCEPTS

For more information on the topics in this section, refer to the following discussion:

Section 7.1 describes the structure and reproductive cycle of bacteria.

CHAPTER

8

Digestive System and Nutrition

CHAPTER CONCEPTS

8.1 Overview of Digestion
The gastrointestinal (GI) tract contains a number of organs that carry out several processes as they digest food.

8.2 The Mouth, Pharynx, and Esophagus
Chemical and mechanical digestion begins in the mouth before food enters the pharynx and the esophagus, which take it to the stomach.

8.3 The Stomach and Small Intestine
The stomach stores food and continues chemical digestion, which is completed in the small intestine. The products of digestion are absorbed by the small intestine into the blood or lymph.

8.4 The Accessory Organs and Regulation of Secretions
The accessory organs—the pancreas, liver, and gallbladder—assist the organs of the GI tract in the processing of food.

8.5 The Large Intestine and Defecation
The large intestine contains bacteria, and serves to absorb water, salts, and vitamins before defecation.

8.6 Nutrition and Weight Control
A healthy diet includes a balance of energy nutrients and adequate levels of vitamins, minerals, and water.

BEFORE YOU BEGIN

Before beginning this chapter, take a few moments to review the following discussions:

Sections 2.4 to 2.7 What are the roles of carbohydrates, lipids, proteins, and nucleic acids?

Section 3.3 What is the role of the facilitated and active transport mechanisms in moving nutrients across a plasma membrane?

Section 3.6 What are the roles of enzymes and coenzymes?

CASE STUDY GI TRACT DILEMMA

Nicole, 32, went to a gastroenterologist complaining about intense heartburn, belching, stomach pain, nausea, and weight loss. She had also recently begun throwing up small amounts of blood. She had used over-the-counter (OTC) heartburn medications like Pepcid AC for years. Recently, it had stopped being as effective for her heartburn and she developed stomachaches after eating. She had also developed a sharp stabbing pain in her upper right abdomen and noticed she was generally fatigued most days. She smoked half a pack of cigarettes a day, drank two to four alcoholic drinks per week, followed a moderately healthy diet, and exercised only occasionally. At 5′6″, she was not heavyset but presently carried about 10 to 20 pounds more than her ideal weight. Dr. Winch ordered an EGD (esophagogastroduodenoscopy) during which a tube with a camera would be placed down her throat and into her stomach to take pictures of the area from her esophagus to her duodenum, the place where her stomach and small intestine meet. The doctor also ordered an upper gastrointestinal (GI) series, commonly called a barium swallow. In that exam, the patient swallows a liquid chemical called barium that lines the digestive tract with a chalky coating. Barium is illuminated on X-rays, thus allowing the digestive tract organs to be evaluated.

As you read through the chapter, think about the following questions:

1. Why would Nicole's height, weight, and lifestyle habits have an impact on her digestion?
2. What is heartburn, how can it develop, and how can lifestyle choices affect it?
3. What are other diagnostic tools used in gastroenterology and what types of disorders can they identify?

8.1 Overview of Digestion

LEARNING OUTCOMES

Upon completion of this section, you should be able to

1. State the function of each organ of the gastrointestinal tract.
2. List the accessory organs and provide a function for each.
3. Describe the structure of the gastrointestinal tract wall.

The organs of the digestive system are located within a tube called the gastrointestinal (GI) tract, which is depicted in Figure 8.1. Food, whether it is a salad or a cheeseburger, consists of the organic macromolecules you studied in Chapter 2: carbohydrates, fats, and proteins. These molecules are too big to cross plasma membranes. The purpose of digestion is to **hydrolyze,** or break down using water, these macromolecules to their subunit molecules. The subunit molecules, mainly monosaccharides, amino acids, fatty acids, and glycerol, can cross plasma membranes using facilitated and active transport (see section 3.3). Our food also contains water, salts, vitamins, and minerals that help the body function normally. The nutrients made available by digestive processes are carried by the blood to our cells. The following processes are necessary to the digestive process.

MP3 Organs of Digestion

- **Ingestion** occurs when the mouth takes in food. Ingestion can be associated with our diet. The expression "You are what you eat" recognizes that our diet is very important to our health. By learning about and developing good nutritional habits, we increase our likelihood of enjoying a longer, more active, and productive life. Unfortunately, poor diet and lack of physical activity now rival smoking as major causes of preventable death in the United States.
- **Digestion** involves the breakdown of larger pieces of food into smaller pieces that can be acted on by the digestive enzymes. Digestion may be either mechanical or chemical. Mechanical digestion occurs primarily in the mouth by chewing and by wavelike contractions of the smooth muscles in the stomach.

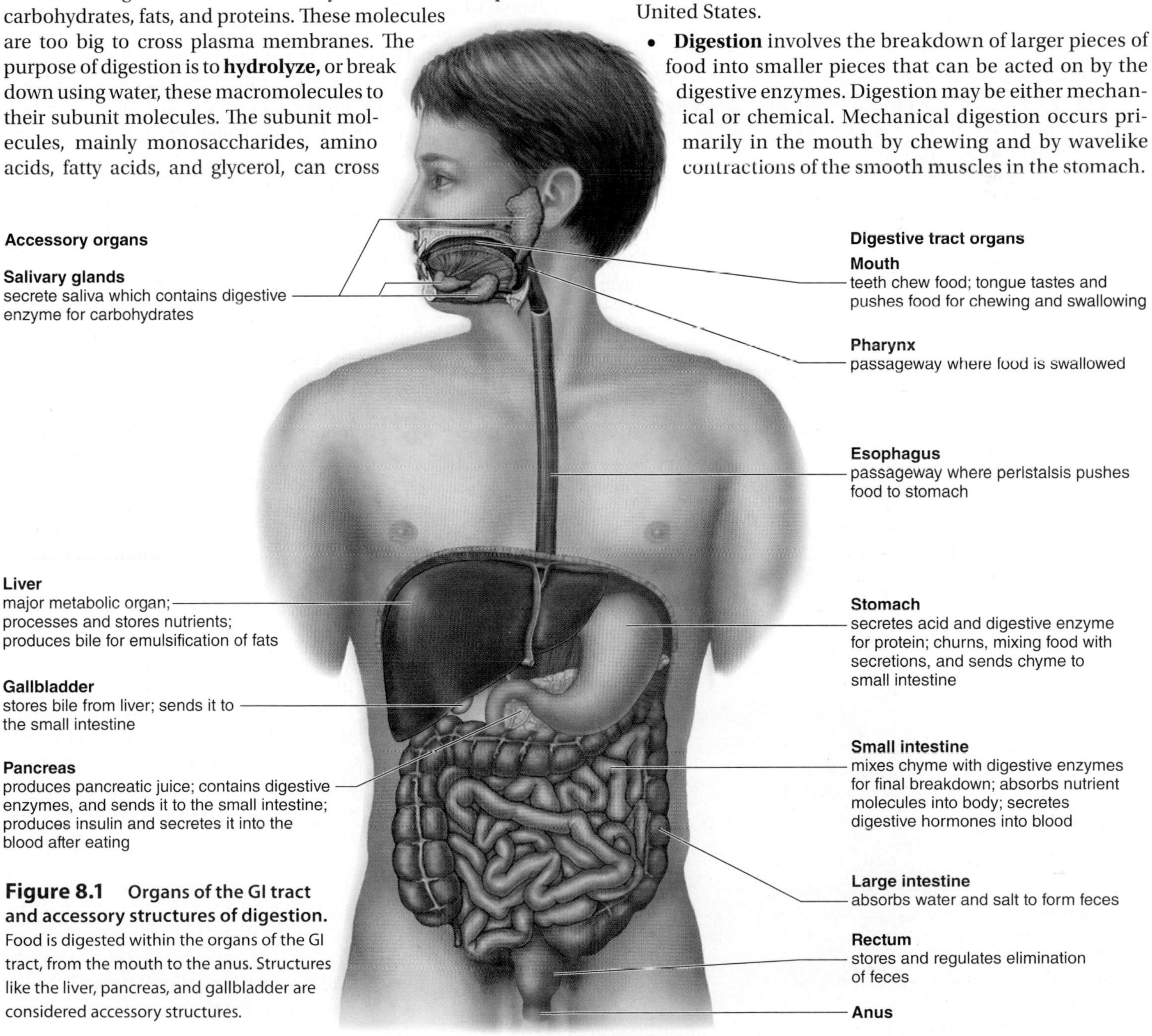

Accessory organs

Salivary glands
secrete saliva which contains digestive enzyme for carbohydrates

Liver
major metabolic organ; processes and stores nutrients; produces bile for emulsification of fats

Gallbladder
stores bile from liver; sends it to the small intestine

Pancreas
produces pancreatic juice; contains digestive enzymes, and sends it to the small intestine; produces insulin and secretes it into the blood after eating

Digestive tract organs

Mouth
teeth chew food; tongue tastes and pushes food for chewing and swallowing

Pharynx
passageway where food is swallowed

Esophagus
passageway where peristalsis pushes food to stomach

Stomach
secretes acid and digestive enzyme for protein; churns, mixing food with secretions, and sends chyme to small intestine

Small intestine
mixes chyme with digestive enzymes for final breakdown; absorbs nutrient molecules into body; secretes digestive hormones into blood

Large intestine
absorbs water and salt to form feces

Rectum
stores and regulates elimination of feces

Anus

Figure 8.1 Organs of the GI tract and accessory structures of digestion. Food is digested within the organs of the GI tract, from the mouth to the anus. Structures like the liver, pancreas, and gallbladder are considered accessory structures.

During chemical digestion, digestive enzymes hydrolyze our food's macromolecules into absorbable subunits. The digestive enzymes all have pH ranges at which they are most effective; the compartmentalization of the digestive tract helps to establish these ideal pH ranges. Chemical digestion begins in the mouth, continues in the stomach, and is completed in the small intestine.

- **Movement** of GI tract contents along the digestive tract is very important for the tract to fulfill its other functions. For example, food must be passed along from one organ to the next, normally by contractions of smooth muscle tissue called **peristalsis**, and indigestible remains must be expelled.
- **Absorption** occurs as subunit molecules produced by chemical digestion (i.e., nutrients) cross the wall of the GI tract and enter the cells lining the tract. From there, the nutrients enter the blood for delivery to the cells.
- **Elimination** removes molecules that cannot be digested and need to be discharged from the body. The removal of indigestible wastes through the anus is termed *defecation*.

MP3 Overview of the Digestive System

Wall of the Digestive Tract

We can compare the GI tract to a garden hose that has a beginning (mouth) and an end (anus). The **lumen** is the open area of a hollow organ or vessel and in the GI tract is the central space that contains food being digested. The wall of the GI tract has four layers (Fig. 8.2). Each layer can be associated with a particular function and disorder.

The inner layer of the wall next to the lumen is called the **mucosa.** The mucosal layer contains cells that produce and secrete mucus used to protect all the layers of the tract from the digestive enzymes inside the lumen. Glands in the mucosa of the mouth, stomach, and small intestine also release digestive enzymes. Hydrochloric acid, an important digestive enzyme, is produced by glands in the mucosa of the stomach.

Diverticulosis is a condition in which portions of the mucosa of any part of the GI tract—but primarily the large intestine—literally have pushed through the other layers and formed pouches, where food can collect. The pouches can be likened to an inner tube that pokes through weak places in a tire. When the pouches become infected or inflamed, the condition is called *diverticulitis.* This happens in 10–25% of people with diverticulosis.

The second layer in the GI wall is called the **submucosa.** The submucosal layer is a broad band of loose connective tissue that contains blood vessels, lymphatic vessels, and nerves. These are the vessels that will carry the nutrients absorbed by the mucosa. Lymph nodules, called Peyer's patches, are also in the submucosa. Like the tonsils, they help protect us from disease. The submucosa contains blood vessels, so it can be the site of an inflammatory response (see Chapter 7) that leads to *inflammatory bowel disease* (*IBD*). Chronic diarrhea, abdominal pain, fever, and weight loss are symptoms of IBD.

The third layer is termed the **muscularis,** and it contains two layers of smooth muscle. The inner, circular layer encircles the tract. The outer, longitudinal layer lies in the same direction as the tract. The contraction of these muscles, under nervous and hormonal control, accounts for peristalsis and subsequent movement of digested food from the esophagus to the anus. The muscularis can be associated with *irritable bowel syndrome* (*IBS*), in which contractions of the wall cause abdominal pain, constipation, and/or diarrhea. The underlying cause of IBS is not known, although some suggest that because this area is under nervous system control, stress may be an underlying cause.

The fourth and outermost layer of the tract is the **serosa** (serous membrane layer), which secretes a lubricating fluid. The serosa is a part of the peritoneum, the internal lining of the abdominal cavity.

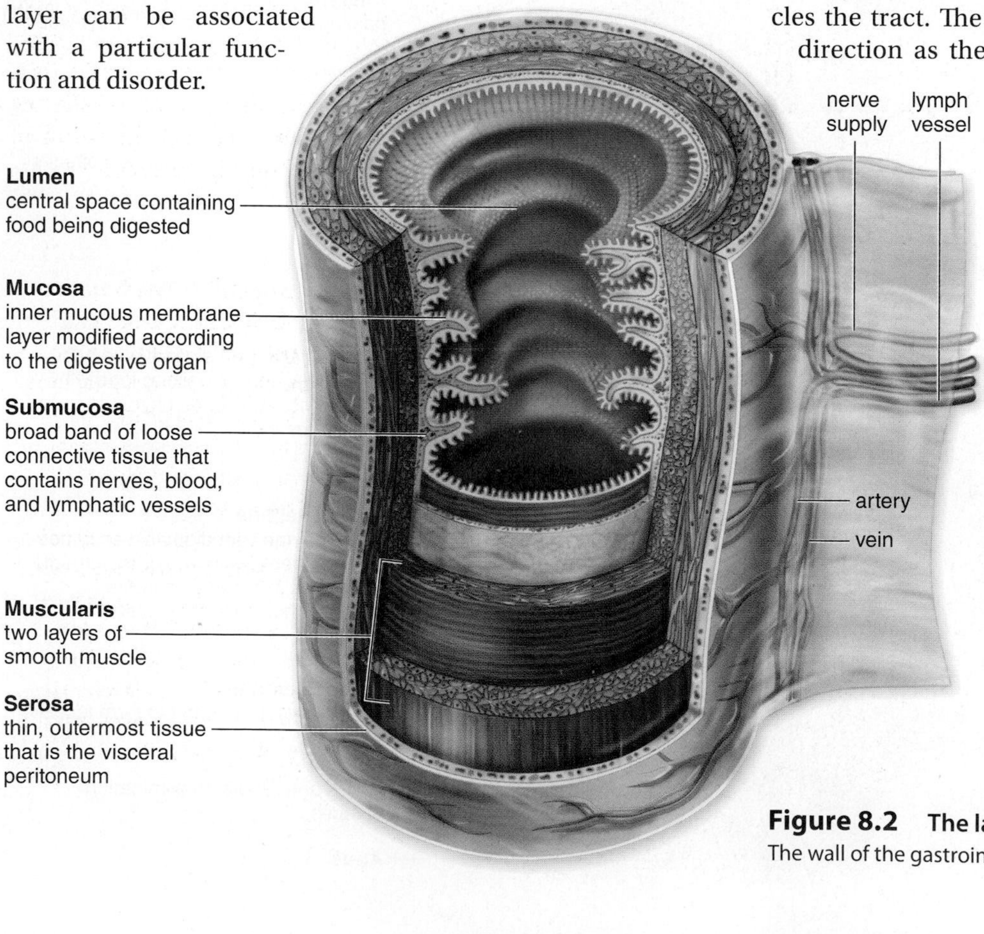

Figure 8.2 The layers of the gastrointestinal tract wall. The wall of the gastrointestinal tract contains the four layers noted.

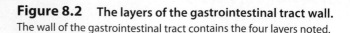

CHECK YOUR PROGRESS 8.1

1. List the organs and all the accessory structures of the digestive system.
2. Describe the processes that occur during the digestive process.
3. Identify the four layers of the GI tract from the lumen outward.

CONNECTING THE CONCEPTS

Several other organ systems interact directly with the digestive system. For more information on these interactions, refer to the following discussions:

Section 5.5 discusses the hepatic portal circulatory system, which moves blood from the intestines to the liver.

Section 7.2 examines the structure of the lymphatic system.

Section 13.4 explores how the nervous system controls the overall activity level of the gastrointestinal tract.

8.2 The Mouth, Pharynx, and Esophagus

LEARNING OUTCOMES

Upon completion of this section, you should be able to

1. Identify the structures of the mouth, pharynx, and esophagus, and provide a function for each.
2. Explain the series of events involved in swallowing.
3. Summarize the diseases and conditions associated with the mouth, pharynx, and esophagus.

The mouth, the pharynx, and the esophagus are in the first part of the GI tract.

The Mouth

The **mouth** (also called the *oral cavity*) receives food and begins the process of mechanical and chemical digestion. The mouth is bounded externally by the lips and cheeks. The lips extend from the base of the nose to the start of the chin. The red portion of the lips is poorly keratinized, and this allows blood to show through.

The roof of the mouth separates the nasal cavities from the oral cavity. The roof has two parts: an anterior (toward the front) *hard palate* and a posterior (toward the back) *soft palate* (Fig. 8.3*a*). The hard palate contains several bones, but the soft palate is composed entirely of muscle. The soft palate ends in a finger-shaped projection called the *uvula*. The *tonsils* are also in the back of the mouth on either side of the tongue. Tonsils are lymphatic tissue and help protect us from disease. In the *nasopharynx*, the location where the nasal cavity opens above the soft palate, there is a single pharyngeal tonsil. This is commonly called the *adenoids*.

Three pairs of **salivary glands** (see Fig. 8.1) secrete saliva by way of ducts to the mouth. One pair of salivary glands lies at the side of the face immediately below and in front of the ears. The ducts of these salivary glands open on the inner surface of the cheek just above the second upper molar. This pair of glands swells when a person has the mumps, a viral disease. The measles, mumps, and rubella (MMR) vaccination you likely had as a child prevents the mumps. Another pair of salivary glands lies beneath the tongue, and still another pair lies beneath the floor of the oral cavity. The ducts from these salivary glands open under the tongue. You can locate the openings if you use your tongue to feel for small flaps on the inside of your cheek and under your tongue. Saliva is a solution of mucus and water. Saliva also contains **salivary amylase,** an enzyme that begins the chemical digestion of starch, as well as bicarbonate and the antimicrobial compound called lysozyme.

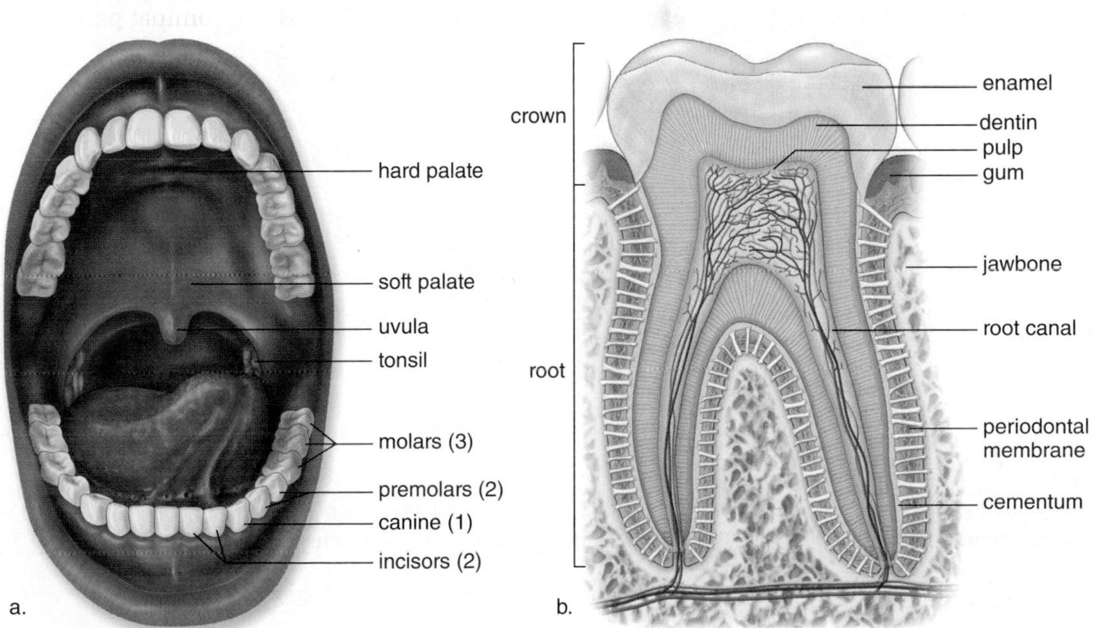

Figure 8.3 The functions of the different teeth and a cross section of a tooth showing the crown and root. **a.** The chisel-shaped incisors bite; the pointed canines tear; the fairly flat premolars grind; and the flattened molars crush food. **b.** Longitudinal section of a tooth. The crown is the portion that projects above the gumline and can be replaced by a dentist if damaged. When a "root canal" is done, the nerves are removed. When the periodontal membrane is inflamed, the teeth can loosen.

The Teeth and Tongue

Mechanical digestion occurs when our teeth chew food into pieces convenient for swallowing. During the first two years of life, the 20 smaller deciduous, or baby, teeth appear. These are eventually replaced by 32 adult teeth (see Fig. 8.3*a*). The "wisdom teeth," the third pair of molars, sometimes fail to erupt. If they push on the other teeth and/or cause pain, they can be removed by a dentist or oral surgeon. Each tooth has two main divisions: a crown, the portion of the tooth above the gumline, and a root, the portion below the gumline (Fig. 8.3*b*). The crown has a layer of enamel, an extremely hard outer covering of calcium compounds; dentin, a thick layer of bonelike material; and an inner pulp, which contains the nerves and the blood vessels. Dentin and pulp also make up a portion of the root, which includes periodontal membranes to anchor the tooth into the jawbone.

Tooth decay, called **dental caries** or cavities, occurs when bacteria within the mouth metabolize sugar. Acids produced during this metabolism erode the teeth. Tooth decay can be painful when it is severe enough to reach the nerves of the inner pulp. Two measures can prevent tooth decay: eating a limited amount of sweets and daily brushing and flossing of teeth. Fluoride treatments, particularly in children, can make the enamel stronger and more resistant to decay. Gum disease, known to be linked to cardiovascular disease, is more apt to occur with aging. Inflammation of the gums, called **gingivitis,** can spread to the periodontal membrane, which lines the tooth socket. A person then has **periodontitis,** characterized by a loss of bone and loosening of the teeth. Extensive dental work may be required or teeth will be completely lost. Stimulation of the gums in a manner advised by a dentist is helpful in controlling this condition. Medications are also available.

The tongue is covered by a mucous membrane, which contains the sensory receptors called taste buds (see Chapter 14). When taste buds are activated by the presence of food, nerve impulses travel by way of nerves to the brain. The tongue is composed of skeletal muscle, and it assists the teeth in carrying out mechanical digestion by moving food around in the mouth. In preparation for swallowing, the tongue forms chewed food into a mass called a **bolus,** which it pushes toward the pharynx.

The Pharynx and Esophagus

Both the mouth and the nasal passages lead to the **pharynx,** a hollow space at the back of the throat (Fig. 8.4). In turn, the pharynx opens into both the food passage (esophagus) and air passage (trachea, or windpipe). These two tubes are parallel to each other, with the trachea anterior to (in front of) the esophagus.

Swallowing

Swallowing has a voluntary phase; however, once food or drink is pushed back far enough into the pharynx, swallowing becomes a reflex action performed automatically or involuntarily. During swallowing, food normally enters the **esophagus,** a muscular tube that moves food into the stomach, because other possible avenues are blocked. The soft palate moves back to close off the nasal passages, and the trachea moves up under the **epiglottis** to cover the glottis. The **glottis** is the opening to the larynx (voice box) and, therefore, the air passage. We do not breathe when we swallow. The up-and-down movement of the Adam's apple, the front part of the larynx, is easy to observe when a person swallows (Fig. 8.4*a*). Sometimes, the

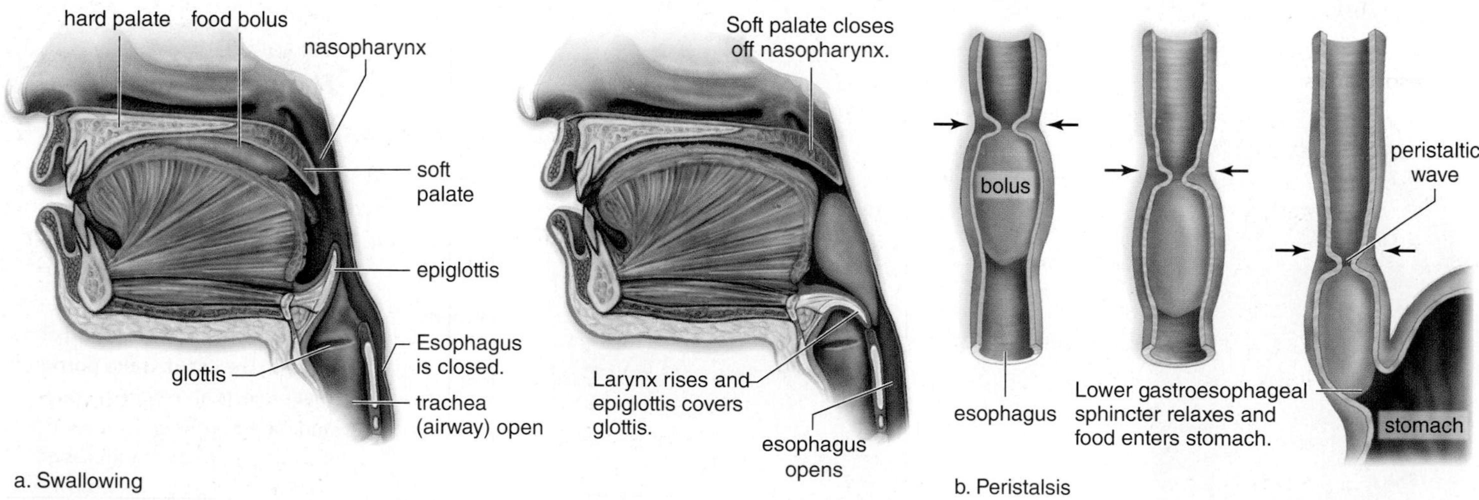

Figure 8.4 The process of swallowing.
a. When food is swallowed, the tongue pushes a bolus of food up against the soft palate (*left*). Then, the soft palate closes off the nasal cavities, and the epiglottis closes off the larynx so the bolus of food enters the esophagus (*right*). **b.** Peristalsis moves food through a sphincter into the stomach.

epiglottis does not cover the glottis fast enough or completely enough and food or liquid can end up in the trachea instead of the esophagus. When this occurs, the muscles around the lungs contract and force a cough that will bring the food back up the trachea and into the pharynx.

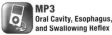

MP3 Oral Cavity, Esophagus, and Swallowing Reflex

Peristalsis

Peristalsis pushes food through the esophagus. The peristaltic contractions continue in the stomach and intestines. The esophagus plays no role in the chemical digestion of food. Its sole purpose is to move the food bolus from the mouth to the stomach. A constriction called the *lower gastroesophageal sphincter* marks the entrance of the esophagus to the stomach. **Sphincters** are muscles that encircle tubes and act as valves. The tubes close when the sphincters contract and they open when the sphincters relax. When food or saliva is swallowed, the sphincter relaxes for a moment to allow the food or saliva to enter the stomach (Fig. 8.4*b*). The sphincter then contracts, preventing the acidic stomach contents from backing up into the esophagus.

When the lower esophageal sphincter fails to open and allow food into the stomach, or when the sphincter is opened and food moves from the stomach back to the esophagus, **heartburn** occurs. As discussed in the Health feature, "Heartburn (GERD)," on page 174, this condition can lead to damage of the esophagus and lower esophageal sphincter. Vomiting occurs when strong contractions of the abdominal muscles and the **diaphragm** (the muscle separating the thoracic and abdominal cavities) force the contents of the stomach into the esophagus and into the oral cavity.

CHECK YOUR PROGRESS 8.2

1. Describe the anatomy of the mouth, teeth, pharynx, and esophagus.
2. Detail how mechanical digestion and chemical digestion occur in the mouth.
3. Explain what ordinarily prevents food from entering the nose or entering the trachea when you swallow and why this occurs.

CONNECTING THE CONCEPTS

For more information on the interaction of the oral cavity and pharynx with other body systems, refer to the following discussions:

Section 7.2 describes the function of lymphatic tissue, such as that found in the tonsils.

Section 9.4 defines the role of the diaphragm in breathing.

Section 13.4 explains a reflex response by the nervous system.

8.3 The Stomach and Small Intestine

LEARNING OUTCOMES

Upon completion of this section, you should be able to

1. Describe the structure of the stomach, and explain its role in digestion.
2. Describe the structure of the small intestine, and explain its role in digestion.
3. Explain how carbohydrates, lipids, and proteins are processed by the small intestine.

The stomach and small intestine complete the digestion of food, which began in the mouth.

The Stomach

The **stomach** (Fig. 8.5) is a thick-walled, J-shaped organ that lies on the left side of the body beneath the diaphragm. The stomach is continuous with the esophagus above and the duodenum of the small intestine below. The stomach stores food, initiates the digestion of protein, and controls the movement of food into the small intestine. Nutrients are not absorbed by the stomach. However, it does absorb alcohol, because alcohol is fat soluble and can pass through membranes easily.

The stomach wall has the usual four layers (see Fig. 8.2), but two of them are modified for particular functions. The muscularis contains three layers of smooth muscle (Fig. 8.5*a*). In addition to the circular and longitudinal layers, the stomach also contains a layer of smooth muscle that runs obliquely to the other two. The **oblique layer** also allows the stomach to stretch and to mechanically break down food into smaller fragments that are mixed with gastric juice.

APPLICATIONS AND MISCONCEPTIONS

Why does my stomach "growl"?

As you digest food and liquids you are also moving gas and air through your GI tract. When pockets of gas and air get squeezed by peristalsis in your stomach and small intestine, it makes a noise or "growl." So why do you "growl" when your stomach is empty? The process of digestion begins long before you eat. When your stomach is empty, the brain will tell the stomach muscles to begin peristalsis to aid in stimulating hunger. Those muscle contractions around an empty stomach vibrate and echo causing the "growling" sound.

BIOLOGY MATTERS  Health

Heartburn (GERD)

In the absence of other symptoms, that burning sensation in your chest may have nothing to do with your heart. Instead, it is likely due to acid reflux. Almost everyone has had acid reflux and heartburn at some time. The burning sensation occurs in the area of the esophagus that lies behind the heart, and that is why it is termed *heartburn.* We might think it means we have a heart problem. Heartburn occurs because the contents of the stomach are more acidic than those of the esophagus. When the stomach contents pass upward into the esophagus (Fig. 8A), the acidity begins to erode the lining of the esophagus, producing the burning sensation associated with heartburn.

Some people experience heartburn after eating a large meal that overfills their stomach. Women sometimes get heartburn during pregnancy when the developing fetus pushes the internal organs upward. Pressure applied to the abdominal wall because of obesity also causes heartburn.

When heartburn or acid reflux becomes a chronic condition, the patient is diagnosed with gastroesophageal reflux disease (GERD). The term signifies that the stomach (*gastric* refers to the stomach) and the esophagus are involved in the disease. In GERD, patients' reflux is more frequent, remains in the esophagus longer, and often contains higher levels of acid than in a patient with typical heartburn or acid reflux. People diagnosed with GERD may also experience pain in their chest, feel like they are choking, and have trouble swallowing.

Patients with weak or abnormal esophageal contractions have difficulty pushing food into the stomach. These weak contractions can also prevent reflux from being pushed back into the stomach after it has entered the esophagus, thus producing GERD. When patients are lying down, the effects of abnormal esophageal contractions become more severe. This is because gravity is not helping to return reflux to the stomach. Some people with GERD may have weaker than normal lower gastroesophageal sphincters. Their sphincters don't fully close after food is pushed into the stomach. Surgery to tighten the sphincter may alleviate GERD.

Over-the-counter medications act as treatments for acid reflux because they have a basic pH, which neutralizes stomach acid. Other drugs, such as Nexium and Prilosec, reduce acid production. Overall, persons with acid reflux are recommended to first try modifying their eating habits.

Diet and Exercise

It has been found that diet and weight control can help control acid reflux. Some tips to reduce acid reflux include:

- Eating several small meals a day instead of three large meals. Avoiding foods that lead to stomach acidity, such as tomato sauces, citrus fruits, alcohol and caffeinated beverages.
- Reducing consumption of high-fat meals (fast-food) and foods high in refined sugar (cakes and candy).
- Increasing the amount of complex carbohydrates (such as multigrain bread, brown rice, and pasta) in your diet.
- Participating in light exercise (bike riding, walking, yoga) and light weight lifting to help control weight.

Questions to Consider

1. From the perspective of pH, how do antacids help control heartburn and GERD?
2. Considering that the GI tract consists of layers of muscles (see Fig. 8.2), why might exercise help control GERD?

Figure 8A Preventing acid reflux. In GERD, the lower gastroesophageal sphincter does not close correctly, allowing the contents of the stomach to enter the esophagus.

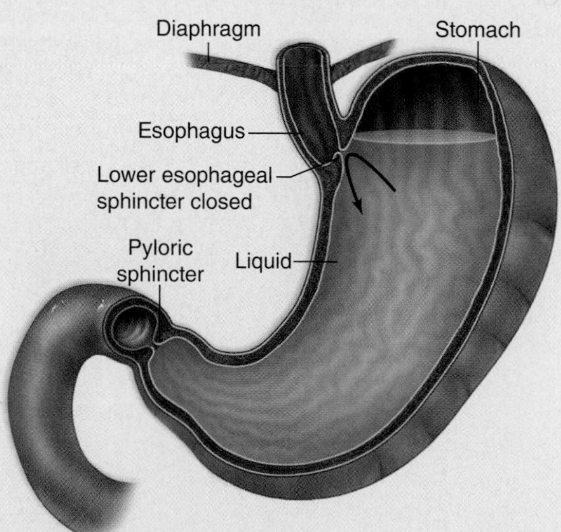

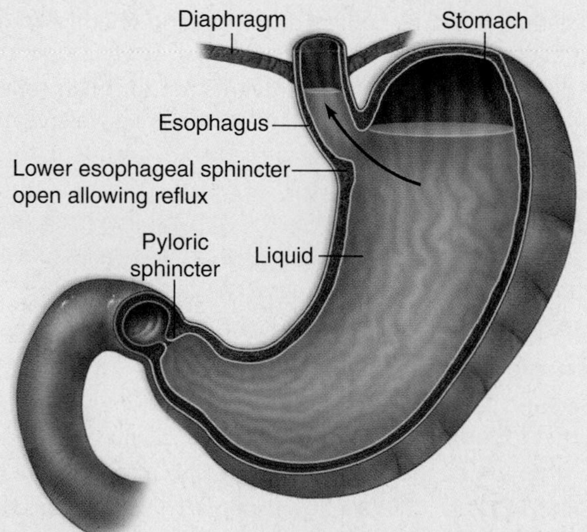

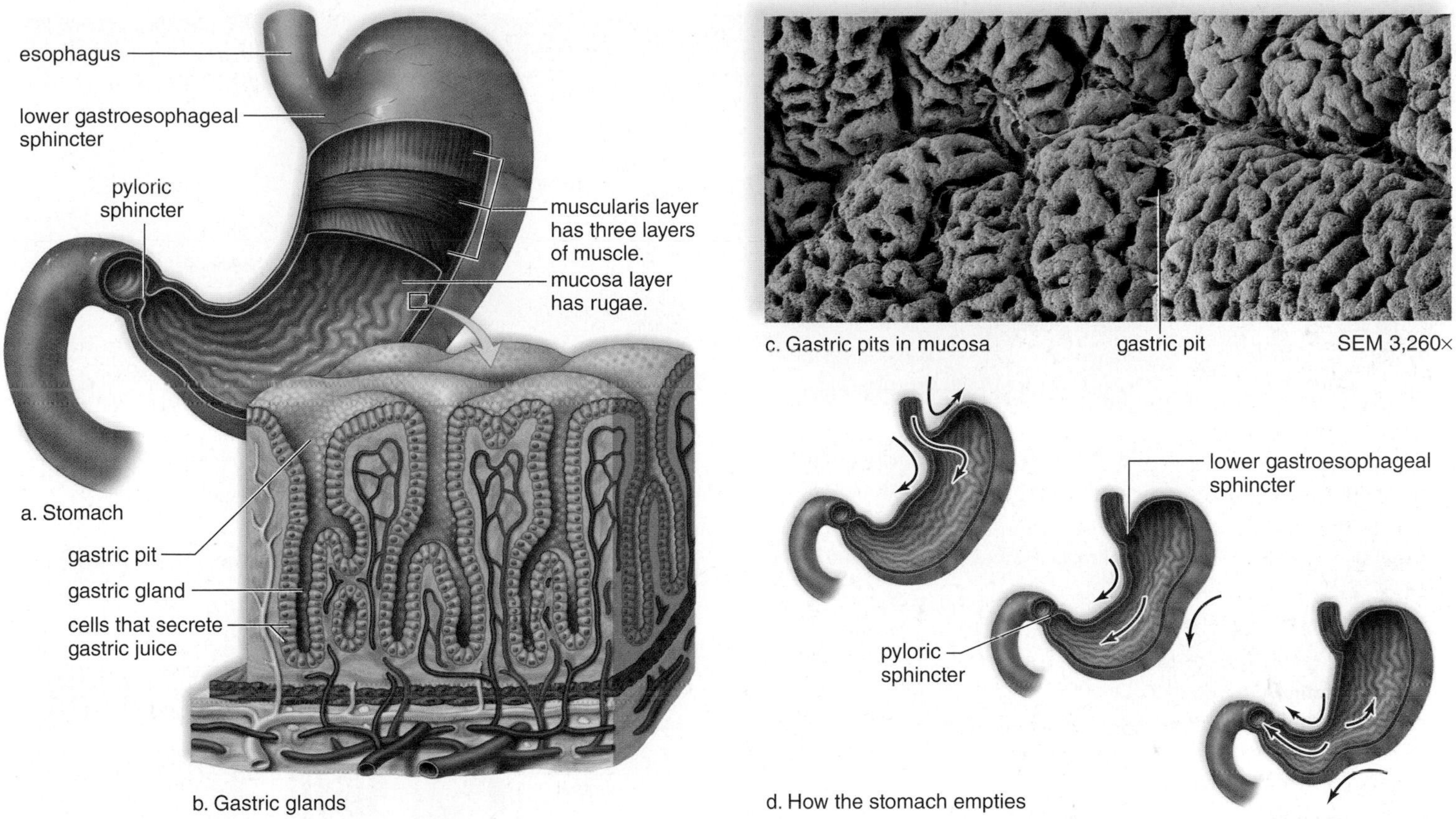

Figure 8.5 The layers of the stomach.
a. Structure of the stomach showing the three layers of the muscularis and the folds called rugae. **b, c.** Gastric glands present in the mucosa secrete mucus, HCl, and pepsin, an enzyme that digests protein. **d.** Peristalsis in the stomach controls the secretion of chyme into the small intestine at the pyloric sphincter.

The mucosa of the stomach has deep folds called **rugae.** These disappear as the stomach fills to an approximate capacity of 1 liter. The mucosa of the stomach has millions of gastric pits, which lead into **gastric glands** (Fig. 8.5*b, c*). The gastric glands produce gastric juice. Gastric juice contains an enzyme called **pepsin,** which digests protein, plus hydrochloric acid (HCl) and mucus. HCl causes the stomach to be very acidic with a pH of about 2. This acidity is beneficial because it kills most bacteria present in food. Although HCl does not digest food, it does break down the connective tissue of meat and activates pepsin.

Normally, the stomach empties in about 2 to 6 hours. When food leaves the stomach, it is a thick, soupy liquid of partially digested food called **chyme.** Chyme's entry into the small intestine is regulated so that small amounts enter at intervals. Peristaltic waves move the chyme toward the pyloric sphincter, which closes and squeezes most of the chyme back, allowing only a small amount to enter the small intestine at one time (Fig. 8.5*d*).

The Small Intestine

The **small intestine** is named for its small diameter compared with that of the large intestine. The small intestine is very long, averaging about 6 m (18 ft) in length, compared with the large intestine, which is about 1.5 m (4.5 ft) in length.

Digestion Is Completed in the Small Intestine

The small intestine contains a wide range of enzymes to digest the carbohydrate, protein, and fat content of food (Table 8.1). Most of these enzymes are secreted by the pancreas and enter via a duct at the **duodenum,** the name for the first 25 cm of the small intestine. Another duct brings bile from the liver and gallbladder into the duodenum (see Fig. 8.8). Bile emulsifies fat. Emulsification is a form of mechanical digestion that causes fat droplets to disperse in water. After emulsification, the **lipase** enzyme, produced by the pancreas, hydrolyzes fats to form the monoglycerides and fatty acids. Pancreatic **amylase** begins the digestion of carbohydrates. An intestinal enzyme completes the digestion of carbohydrates to glucose. Similarly, pancreatic trypsin, a **protease** enzyme, begins and intestinal enzymes finish the digestion of proteins to amino acids. The intestine has a slightly basic pH because pancreatic juice contains sodium bicarbonate ($NaHCO_3$), which neutralizes chyme.

Table 8.1 Major Digestive Enzymes

Enzyme	Produced By	Site of Action	Optimum pH	Digestion
Carbohydrate Digestion				
Salivary amylase	Salivary glands	Mouth	Neutral	Starch + H_2O → maltose
Pancreatic amylase	Pancreas	Small intestine	Basic	Starch + H_2O → maltose
Maltase	Small intestine	Small intestine	Basic	Maltose + H_2O → glucose + glucose
Lactase	Small intestine	Small intestine	Basic	Lactose + H_2O → glucose + galactose
Protein Digestion				
Pepsin	Gastric glands	Stomach	Acidic	Protein + H_2O → peptides
Trypsin	Pancreas	Small intestine	Basic	Protein + H_2O → peptides
Peptidases	Small intestine	Small intestine	Basic	Peptide + H_2O → amino acids
Nucleic Acid Digestion				
Nuclease	Pancreas	Small intestine	Basic	RNA and DNA + H_2O → nucleotides
Nucleosidases	Small intestine	Small intestine	Basic	Nucleotide + H_2O → base + sugar + phosphate
Fat Digestion				
Lipase	Pancreas	Small intestine	Basic	Fat droplet + H_2O → monoglycerides + fatty acids

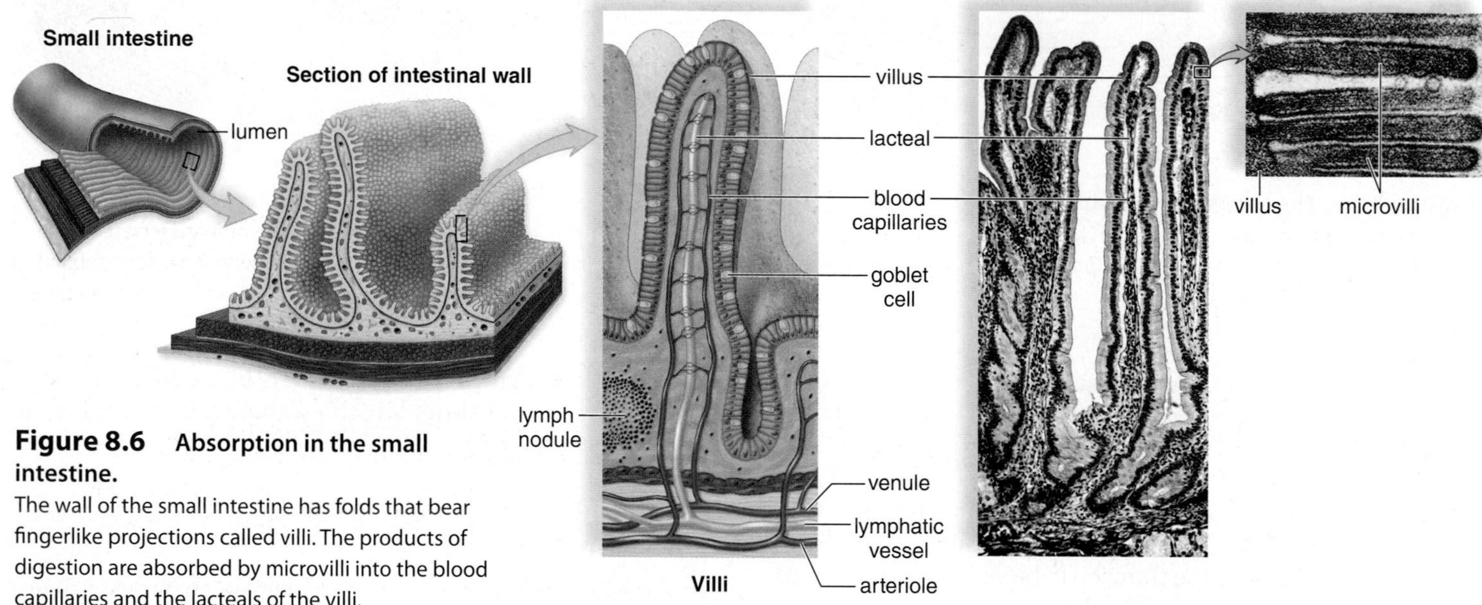

Figure 8.6 Absorption in the small intestine.
The wall of the small intestine has folds that bear fingerlike projections called villi. The products of digestion are absorbed by microvilli into the blood capillaries and the lacteals of the villi.

Nutrients Are Absorbed in the Small Intestine

The wall of the small intestine absorbs the sugar, amino acid, glycerol, and fatty acids molecules that were the products of the digestive process. The mucosa of the small intestine is modified for absorption. It has been suggested that the surface area of the small intestine is approximately that of a tennis court. This great surface area functions to absorb more nutrients than a smaller area would. The mucosa of the small intestine contains fingerlike projections called **villi** (sing., villus), which give the intestinal wall a soft, velvety appearance (Fig. 8.6). A villus has an outer layer of columnar epithelial cells, and each of these cells has thousands of microscopic extensions called *microvilli.* Collectively, in electron micrographs, microvilli give the villi a fuzzy border known as the brush border. The microvilli contain enzymes, called brush border enzymes, that complete the digestive process. The microvilli greatly increase the surface area of the villus for the absorption of nutrients.

Nutrients are absorbed into the vessels of a villus (Fig. 8.7). A villus contains blood capillaries and a small lymphatic capillary called a **lacteal.** As you know, the lymphatic system is an adjunct to the cardiovascular system. Lymphatic vessels carry a fluid called lymph to the cardiovascular veins. Sugars (monosaccharides) and amino acids enter the blood capillaries of a villus. Single molecules of glycerol, called monoglycerides, and fatty acids enter the epithelial cells of the villi. Lipoprotein droplets, called *chylomicrons,* are formed

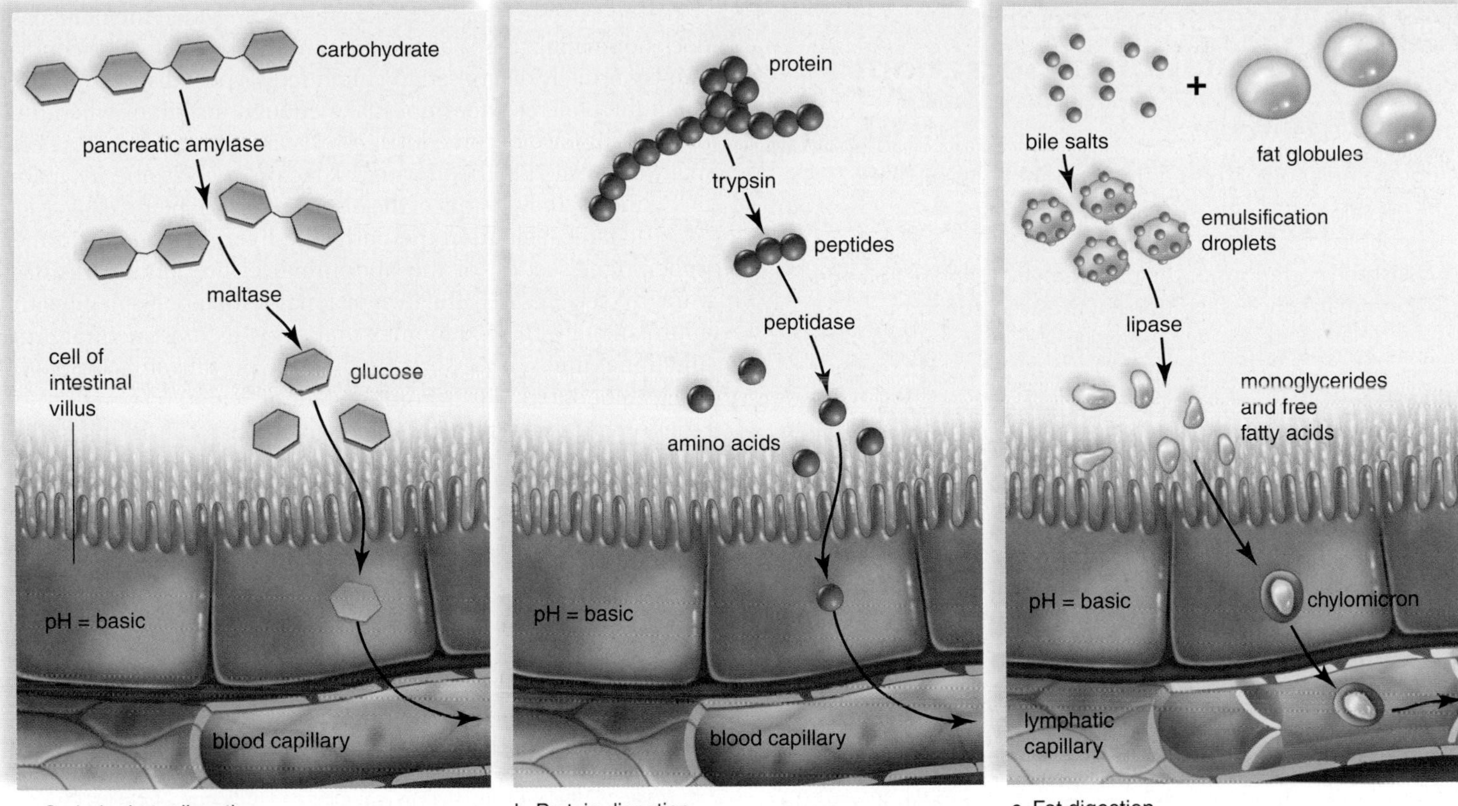

Figure 8.7 **Digestion and absorption of organic nutrients.**
a. Carbohydrate is digested to glucose, which is actively transported into the cells of intestinal villi. From there, glucose moves into the bloodstream.
b. Proteins are digested to amino acids, which are actively transported into the cells of intestinal villi. From there, amino acids move into the bloodstream.
c. Fats are emulsified by bile and digested to monoglycerides and fatty acids. These diffuse into cells, where they recombine and join with proteins. These lipoproteins, called chylomicrons, enter a lacteal.

when monoglycerides and fatty acids are rejoined in the villi epithelia cells. Chylomicrons then enter a lacteal. After nutrients are absorbed, they are eventually carried to all the cells of the body by the bloodstream.

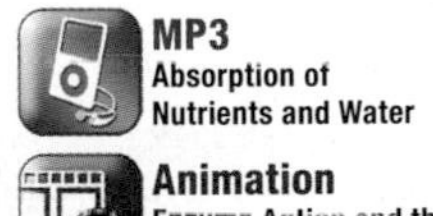

Lactose Intolerance

Lactose is the primary sugar in milk. People who do not have the brush border enzyme lactase cannot digest lactose. The result is a condition called **lactose intolerance,** characterized by diarrhea, gas, bloating, and abdominal cramps after ingesting milk and other dairy products. Diarrhea occurs because the indigestible lactose causes fluid retention in the small intestine. Gas, bloating, and cramps occur when bacteria break down the lactose anaerobically.

Persons with lactose intolerance can consume dairy products that are lactose-free or in which lactose has already been digested. These include lactose-free milk, cheese, and yogurt. A dietary supplement that aids in the digestion of lactose is available also.

CHECK YOUR PROGRESS 8.3

1. Describe the functions of the stomach and how the wall of the stomach is modified to perform these functions.
2. Detail the functions of the small intestine and how the wall of the small intestine is modified to perform these functions.
3. Discuss why absorption of most molecules and nutrients is done in the small intestine and not the stomach.

CONNECTING THE CONCEPTS

To better understand the function of the stomach and small intestine, refer to the following discussions:

Section 2.2 explains how to interpret pH values.

Sections 2.4 to 2.6 provide additional details on the structure of carbohydrates, lipids, and proteins.

Section 4.5 describes the structure of the epithelial tissue that lines the small intestine.

8.4 The Accessory Organs and Regulation of Secretions

LEARNING OUTCOMES

Upon completion of this section, you should be able to

1. Explain the function of the pancreas, liver, and gallbladder during digestion.
2. List the secretions of the pancreas, liver, and gallbladder.
3. Summarize how secretions of the accessory organs are regulated.

We first take a look at the roles of the pancreas, liver, and gallbladder in digestion before considering how the secretions of these organs and those of the GI tract are regulated.

The Accessory Organs

The **pancreas** is a fish-shaped, spongy, grayish-pink organ that stretches across the back of the abdomen behind the stomach. Most pancreatic cells produce pancreatic juice, which enters the duodenum via the pancreatic duct (Fig. 8.8*a*). Pancreatic juice contains sodium bicarbonate ($NaHCO_3$) and digestive enzymes for all types of food. Sodium bicarbonate neutralizes acid chyme from the stomach. Pancreatic amylase digests starch, trypsin digests protein, and **pancreatic** lipase digests fat.

The pancreas is also an endocrine gland that secretes the hormone insulin into the blood. A **hormone** is a protein or steroid produced by a cell that affects the function of a different (target) cell. When the blood glucose level rises rapidly, the pancreas produces an overload of insulin to bring the level under control and back to homeostasis. Type 1 diabetes occurs when the pancreas does not manufacture sufficient amounts of insulin. This condition is normally diagnosed in childhood. Type 2 diabetes occurs when the pancreas does not make enough insulin or when the body's cells have become insulin-resistant. Type 2 diabetes normally occurs in adulthood with risk factors such as obesity, inactivity, and a family history of the disease (see Chapter 15).

The largest gland in the body, the **liver,** lies mainly in the upper right section of the abdominal cavity, under the diaphragm (see Fig. 8.1). The liver is a major metabolic gland with approximately 100,000 lobules that serve as its structural and functional units (Fig. 8.8*b*). The hepatic portal vein (Fig. 8.8*b*) brings blood to the liver from the GI tract capillary bed. Capillaries of the lobules filter this blood. In a sense, the liver acts like a sewage treatment plant when it removes poisonous substances from the blood and detoxifies them (Table 8.2).

Table 8.2 Functions of the Liver

1. Destroys old red blood cells; excretes bilirubin, a breakdown product of hemoglobin in bile, a liver product
2. Detoxifies blood by removing and metabolizing poisonous substances
3. Stores iron (Fe^{2+}), the water-soluble vitamin B_{12}, and the fat-soluble vitamins A, D, E, and K
4. Makes plasma proteins, such as albumins and fibrinogen, from amino acids
5. Stores glucose as glycogen after a meal, and breaks down glycogen to glucose to maintain the glucose concentration of blood between eating periods
6. Produces urea after breaking down amino acids
7. Helps regulate blood cholesterol level, converting some to bile salts

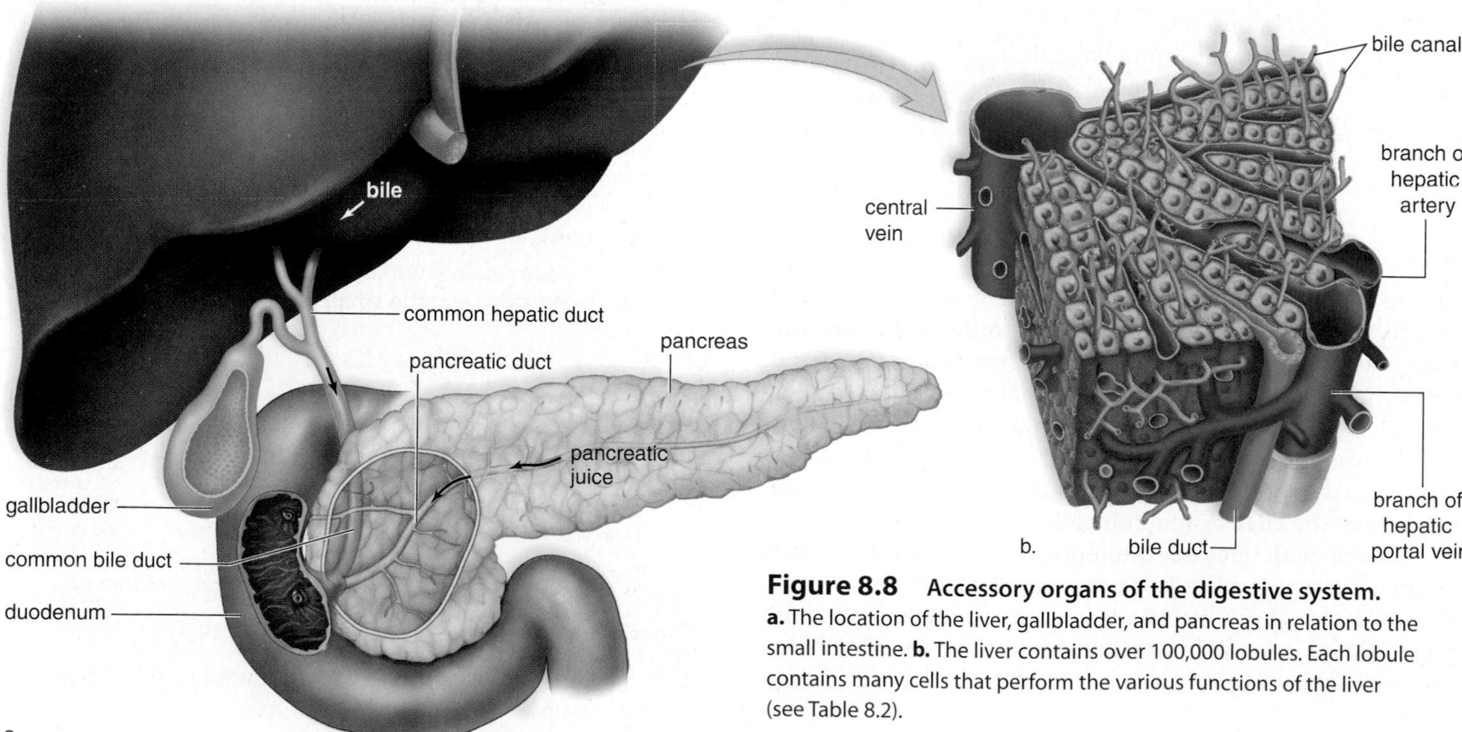

Figure 8.8 Accessory organs of the digestive system. **a.** The location of the liver, gallbladder, and pancreas in relation to the small intestine. **b.** The liver contains over 100,000 lobules. Each lobule contains many cells that perform the various functions of the liver (see Table 8.2).

The liver is also a storage organ. It removes iron and the vitamins A, D, E, K, and B_{12} from blood and stores them. The liver is also involved in blood glucose homeostasis. In the presence of insulin, the liver stores glucose as glycogen. When blood glucose becomes low, the liver releases glucose by breaking down glycogen. If need be, the liver converts glycerol (from fats) and amino acids to glucose molecules. As amino acids are converted to glucose, the liver combines their amino groups with carbon dioxide to form **urea,** the usual nitrogenous waste product in humans. Plasma proteins needed in the blood (see section 6.1) are also made by the liver.

The liver helps regulate blood **cholesterol** levels as well. Some cholesterol is converted to bile salts by the liver. **Bile** is a solution of bile salts, water, cholesterol, and bicarbonate. It has a yellowish-green color because it also contains bilirubin, a pigment protein formed during the breakdown of hemoglobin, which is a process also performed by the liver. Bile is stored in the **gallbladder,** a pear-shaped organ just below the liver, until it is sent via the bile ducts to the duodenum. **Gallstones** form when liquid stored in the gallbladder hardens into pieces of stonelike material. In the small intestine, bile salts emulsify fat. When fat is emulsified, it breaks up into droplets. The droplets provide a large surface area that can be acted upon by digestive enzymes.

Liver Disorders

Hepatitis and cirrhosis are two serious diseases that affect the entire liver and hinder its ability to repair itself. Therefore, they are life-threatening diseases. When a person has a liver ailment, bile pigments may leak into the blood causing **jaundice.** Jaundice is a yellowish tint to the whites of the eyes and also to the skin of light-pigmented persons. Jaundice can result from **hepatitis,** inflammation of the liver. Viral hepatitis occurs in several forms. Hepatitis A is usually acquired from sewage-contaminated drinking water and food. Hepatitis B, which is usually spread by sexual contact, can also be spread by blood transfusions or contaminated needles. The hepatitis B virus is more contagious than the AIDS virus, spread in the same way. Vaccines are available for hepatitis A and hepatitis B. Hepatitis C is usually acquired by contact with infected blood and can lead to chronic hepatitis, liver cancer, and death. There is no vaccine for hepatitis C.

Cirrhosis is another chronic disease of the liver. First, the organ becomes fatty, and then liver tissue is replaced by inactive fibrous scar tissue. Cirrhosis of the liver is often seen in alcoholics, due to malnutrition and to the excessive amounts of alcohol (a toxin) the liver is forced to break down. Physicians are also beginning to see cirrhosis of the liver in obese people, who are overweight due to a diet high in fatty foods.

The liver has amazing regenerative powers and can recover if the rate of regeneration exceeds the rate of damage. During liver failure, however, there may not be enough time to let the liver heal itself and liver transplantation is usually the preferred treatment. The liver is a vital organ, and its failure leads to death.

Regulation of Digestive Secretions

The secretions of digestive juices are controlled by the nervous system and by digestive hormones. When you look at or smell food, the parasympathetic nervous system automatically stimulates gastric secretion. Also, when a person has eaten a meal particularly rich in protein, the stomach produces the hormone gastrin. Gastrin enters the bloodstream, and soon the secretory activity of gastric glands increases.

Cells of the duodenal wall produce two other hormones of particular interest—secretin and cholecystokinin (CCK). Secretin release is stimulated by acid, especially the HCl present in chyme. Partially digested proteins and fat stimulate the release of CCK. Soon after these hormones enter the bloodstream, the pancreas increases its output of pancreatic juice. Pancreatic juice buffers the acidic chyme entering the intestine from the stomach and helps digest food. CCK also causes the liver to increase its production of bile and causes the gallbladder to contract and release stored bile. The bile then aids the digestion of fats that stimulated the release of CCK. Figure 8.9 summarizes the actions of gastrin, secretin, and CCK.

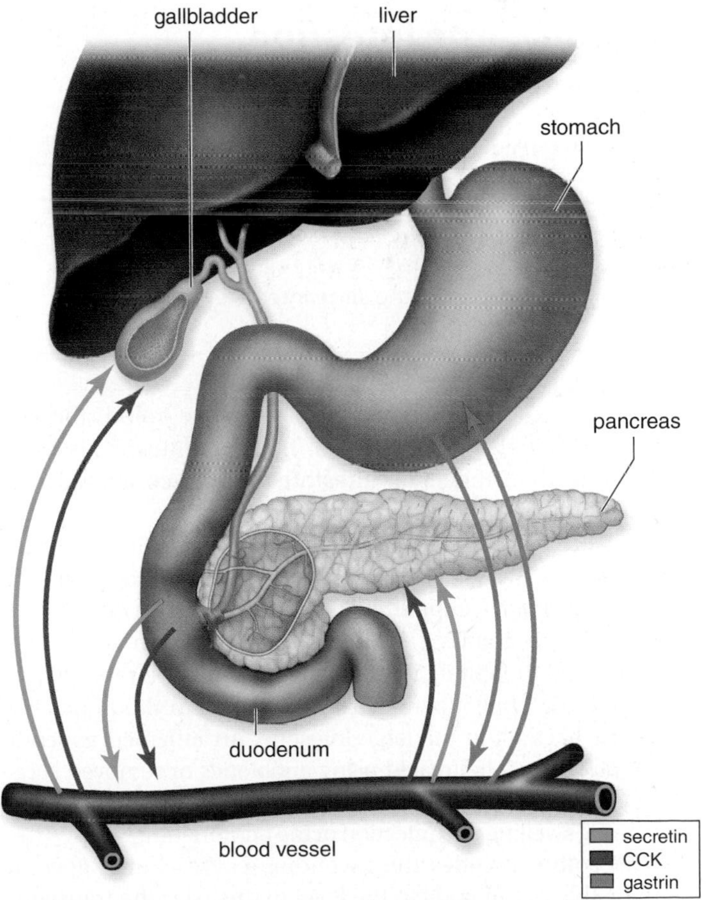

Figure 8.9 Hormonal control and regulation of digestion. Gastrin (blue) from the lower stomach feeds back to stimulate the upper part of the stomach to produce digestive juice. Secretin (green) and CCK (purple) from the duodenal wall stimulate the pancreas to secrete digestive juice and the gallbladder to release bile.

CHECK YOUR PROGRESS 8.4

1. Name and describe the functions of three main accessory organs that assist with the digestive process.
2. Discuss what might occur if each of the accessory organs of digestion did not function properly. Give an example for each.
3. Detail why the regulation of digestive secretions is important to the overall process of digestion. Describe what might occur without regulation.

CONNECTING THE CONCEPTS

For more information on the accessory organs, refer to the following discussions:

Section 5.5 describes the movement of blood and nutrients from the intestines to the liver via the hepatic portal system.

Section 13.4 outlines how the nervous system interacts with the liver and pancreas to maintain homeostasis.

Section 15.5 explains the role of the pancreas in maintaining blood glucose homeostasis.

8.5 The Large Intestine and Defecation

LEARNING OUTCOMES

Upon completion of this section, you should be able to

1. Describe the structure and function of the large intestine.
2. List the disorders of the large intestine and provide a cause for each.

The **large intestine** includes the cecum, the colon, the rectum, and the anal canal (Fig. 8.10). The large intestine is larger in diameter than the small intestine (6.5 cm compared with 2.5 cm), but it is shorter in length (see Fig. 8.1).

The **cecum** is the first portion of the large intestine joining the end of the small intestine. The cecum usually has a small projection called the vermiform **appendix** (the term vermiform describes its "wormlike" appearance) (Fig. 8.10). In humans, the appendix is thought to aid in fighting infections. Scientists have recently proposed that the appendix may also contribute to the population of needed bacteria in the large intestine. An inflamed appendix (*appendicitis*) may be treated using antibiotics or removed surgically. Should the appendix burst, the result can be *peritonitis,* a life-threatening swelling and infection of the peritoneum.

The **colon** includes the ascending colon, which goes up the right side of the body to the level of the liver; the transverse colon, which crosses the abdominal cavity just below the liver and the stomach; the descending colon, which passes down the left side of the body; and the sigmoid colon, which enters the **rectum,** the last 20 cm of the large intestine. The rectum opens at the **anus,** where **defecation,** the expulsion of feces, occurs.

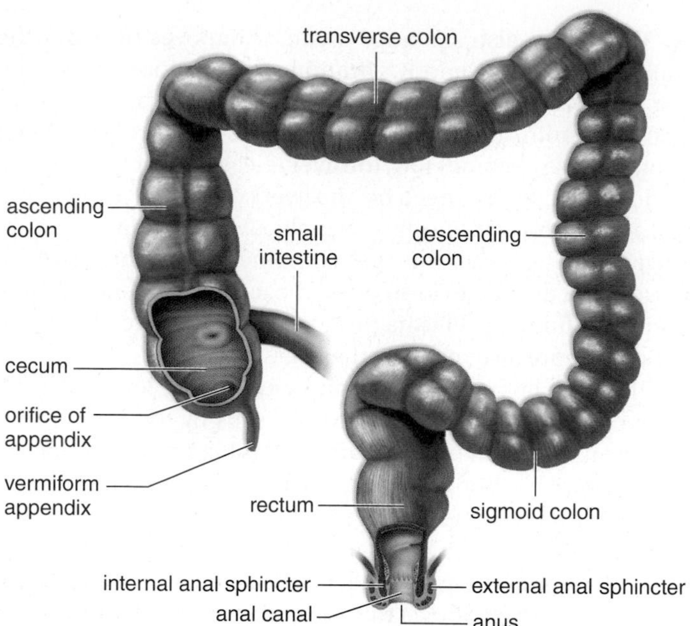

Figure 8.10 **The regions of the large intestine.**
The regions of the large intestine are the cecum, colon, rectum, and anal canal.

Functions of the Large Intestine

The large intestine does not produce any digestive enzymes, and it does not absorb any nutrients. The function of the large intestine is to absorb water, an important process to prevent dehydration of the body and maintain homeostasis.

The large intestine can absorb vitamins produced by intestinal flora, the bacteria that inhabit the intestine and aid in keeping us healthy. For many years, it was believed that *Escherichia coli* were the major inhabitants of the colon, but culture methods now show that over 99% of the colon bacteria are other types of bacteria. The bacteria in the large intestine break down indigestible material and also produce B-complex vitamins and most of the vitamin K needed by our bodies.

Video Fat Microbes

The large intestine forms feces. The consistency of normal feces is usually three-fourths water and one-fourth solid wastes. Bacteria, dietary **fiber** (indigestible remains), and other indigestible materials make up the solid wastes. Bacterial action on indigestible materials causes the odor of feces and also accounts for the presence of gas. Stercobilin, which is a breakdown product of **bilirubin** (the yellow-orange bile pigment produced from the breakdown of hemoglobin), and oxidized iron cause the brown color of feces.

Defecation, ridding the body of feces, is also a function of the large intestine. Peristalsis occurs infrequently in the large intestine, but when it does, feces are forced into the rectum. Feces collect in the rectum until it is appropriate to defecate. At that time, stretching of the rectal wall initiates nerve impulses to the spinal cord. Shortly thereafter, the rectal muscles contract and the anal sphincters relax. This allows the feces to exit the body through the anus (Fig. 8.10). A person can inhibit defecation by contracting the external anal sphincter made of skeletal muscle. Ridding the body of indigestible remains is another way the digestive system helps maintain homeostasis.

APPLICATIONS AND MISCONCEPTIONS

Are coliform bacteria dangerous in the water?

Water is considered unsafe for swimming when the coliform (nonpathogenic intestinal) bacterial count reaches a certain number. A high count indicates that a significant amount of feces has entered the water. The more feces present, the greater the possibility that disease-causing bacteria are also present. For most states, the fecal coliform count has to be under 200 fecal bacteria per every 100 milliliters (ml) of water to be considered safe for swimming. The national standard for safe drinking water is 0 fecal bacteria per 100 ml of water.

Disorders of the Colon and Rectum

The large intestine is subject to a number of disorders. Many of these can be prevented or minimized by a good diet and proper hygiene.

Diarrhea

Diarrhea is characterized by bowel movements that are loose or watery. The major causes of **diarrhea** are infection of the lower intestinal tract and nervous stimulation. The intestinal wall becomes irritated, and peristalsis increases when an infection occurs. Water is not absorbed, and the diarrhea that results rids the body of the infectious organisms. In nervous diarrhea, the nervous system stimulates the intestinal wall and diarrhea results. Most people have several occurrences of diarrhea each year without suffering any health consequences. However, prolonged diarrhea can lead to dehydration because of water loss, which can lead to an imbalance of salts in the blood that can affect heart muscle contraction and potentially lead to death.

Constipation

When a person is constipated, the feces are dry and hard, making it difficult for them to be expelled. Diets that lack whole-grain foods, as well as ignoring the urge to defecate, are often the causes of **constipation.** When feces are not expelled regularly, additional water is absorbed from them. The material becomes drier, harder, and more difficult to eliminate. Adequate water and fiber intake can help regularity of defecation. The frequent use of laxatives is discouraged because it can result on a dependence of their use for normal bowel movements. If, however, it is necessary to take a laxative, the most natural is a bulk laxative. Like fiber, it produces a soft mass of cellulose in the colon. Lubricants such as mineral oil make the colon slippery; saline laxatives such as milk of magnesia act osmotically by preventing water from being absorbed. Some laxatives are irritants that increase peristalsis.

Chronic constipation is associated with the development of **hemorrhoids,** enlarged and inflamed blood vessels at the anus. Other contributing factors for hemorrhoid development include pregnancy, aging, and anal intercourse.

Diverticulosis

As mentioned previously (see page 170), diverticulosis is the occurrence of little pouches of mucosa where food can collect. The pouches form when the mucosa pushes through weak spots in the muscularis. A frequent site is the last part of the descending colon.

Irritable Bowel Syndrome

Also mentioned previously (see page 170), irritable bowel syndrome (IBS), or spastic colon, is a condition in which the muscularis contracts powerfully but without its normal coordination. The symptoms are abdominal cramps; gas; constipation; and urgent, explosive stools (feces discharge).

Inflammatory Bowel Disease

Inflammatory bowel disease (IBD) is a collective term for a number of inflammatory disorders. Ulcerative colitis and Crohn's disease are the most common of these. Ulcerative colitis affects the large intestine and rectum and results in diarrhea, rectal bleeding, abdominal cramps, and urgency in defecation. Crohn's disease is normally isolated to the small intestine but can affect any area of the digestive tract, including the colon and rectum. It is characterized by the breakdown of the lining of the affected area resulting in ulcers. Ulcers are painful and cause bleeding because they erode the submucosal layer, where there are nerves and blood vessels. This also results in the inability to absorb nutrients at the affected sites. Symptoms of Crohn's disease include diarrhea, weight loss, abdominal cramping, anemia, bleeding, and malnutrition.

Polyps and Cancer

The colon is subject to the development of **polyps,** small growths arising from the epithelial lining. Polyps, whether benign or cancerous, can be removed surgically. If colon cancer is detected while still confined to a polyp, the expected outcome is a complete cure. The National Cancer Institute estimates over 143,000 new cases of colon and rectal (colorectal) cancer are diagnosed per year in the United States. Some investigators believe that dietary fat increases the likelihood of colon cancer, because dietary fat causes an increase in bile secretion. It could be that intestinal bacteria convert bile salts to substances that promote the development of cancer. Fiber in the diet seems to inhibit the development of colon cancer, and regular elimination reduces the time that the colon wall is exposed to any cancer-promoting agents in feces.

One diagnostic tool for all of these disorders is an endoscopic exam called a colonoscopy. In a colonoscopy a flexible tube containing a camera is inserted into the GI tract, usually from the anus. The doctor may then examine the length of the colon, as well as take tissue samples (biopsies) for additional tests. However, as discussed in the Health feature, "Swallowing a Camera," on page 182, this procedure is gradually being replaced by the PillCam, a camera you swallow.

BIOLOGY MATTERS  Health

Swallowing a Camera

During a traditional endoscopy procedure, the doctor uses an endoscope (a retractable, tubelike instrument with an embedded camera) to examine the patient's GI tract. The PillCam has become a viable alternative to traditional endoscopy. With a gulp of water, PillCam is swallowed, and it travels through the digestive system. Instead of spending an uncomfortable half day or more at the doctor's office, a patient visits the doctor in the morning, swallows the camera, puts on the recording device, and goes about his or her daily routine.

Propelled by the normal muscular movement of the digestive system, PillCam embarks on a 4- to 8-hour journey through the digestive system. As it travels through the stomach, the twists and turns of the small intestine, and the large intestine, PillCam continuously captures high-quality, wide-angle film footage (Fig. 8B) of its journey and beams this information to the recording device worn by the patient.

At the end of the day, PillCam reaches the end of its journey, and it is defecated. Later, the recording device is returned to the doctor's office so that the data may be retrieved. The doctor can view PillCam's journey as a 90-minute movie.

Like the food that we eat, PillCam traverses the numerous twists and turns of the intestine with ease. This provides a more accurate diagnosis, because a larger portion of the GI tract can be examined. PillCam does not create discomfort in the patient, so the considerable risks involved with using anaesthetics and painkillers are eliminated. And finally, the doctor does not need to be present for the entire procedure, saving valuable time and money for both doctor and patient.

However, the PillCam does have some limitations. For example, during a colonoscopy, a colonoscope, an instrument similar to an endoscope, is inserted through the anus into the large intestine. During the examination, the physician is able to remove precancerous tissue, usually polyps, for additional studies. Because a PillCam is disposable, it cannot be used to remove tissue samples for analysis, so its use as a replacement for colonoscopy is somewhat limited.

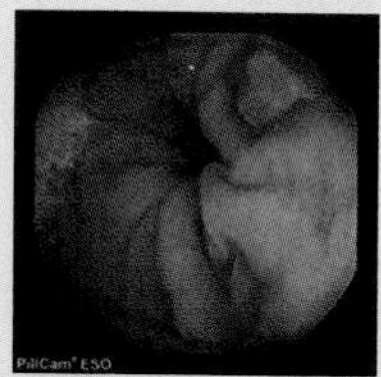
esophagus

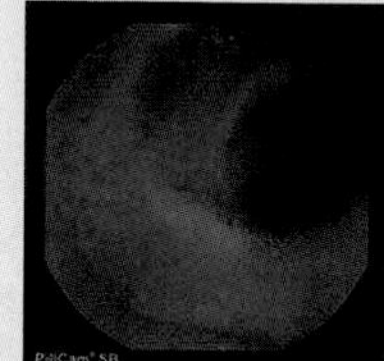
normal GI tract

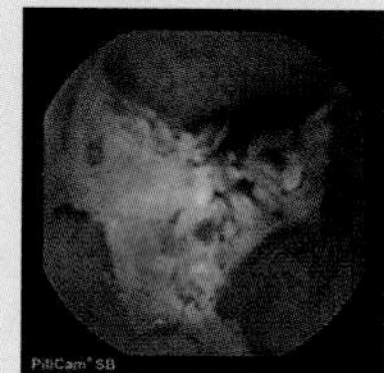
Crohn's disease

Figure 8B An endoscopy procedure using a PillCam. The PillCam traverses through the entire GI tract, taking pictures after it is swallowed.

Questions to Consider

1. How might a PillCam be used to make a traditional endoscopic exam more effective?
2. What factors might regulate how long the PillCam takes to move through the GI tract?

CHECK YOUR PROGRESS 8.5

1. Describe the different parts of the large intestine and provide the function for each.
2. Detail how the functions of the large intestine contribute to homeostasis. Give some examples.
3. Discuss how a disorder of the large intestine can affect homeostasis overall. Give a few examples.

CONNECTING THE CONCEPTS

For more information on the large intestine, refer to the following discussions:

Section 2.4 describes the role of fiber as a complex carbohydrate.

Section 7.3 examines how the bacteria in the large intestine act as an innate defense against disease.

Section 19.2 outlines some of the more common causes of cancer.

8.6 Nutrition and Weight Control

LEARNING OUTCOMES

Upon completion of this section, you should be able to

1. Calculate a BMI value, and interpret its relationship to your overall health.
2. Identify the role of each class of nutrient in the human body.

Obesity, or being significantly overweight, has become one of the greatest health problems in the United States. According to the Centers for Disease Control and Prevention (CDC), almost 36% of adults, and 17% of children, are classified as obese. These statistics are of great concern because excess body fat is associated with a higher risk for premature death, type 2 diabetes, hypertension, cardiovascular disease, stroke, gallbladder disease, respiratory dysfunction, osteoarthritis, and certain types of cancer.

The Health feature, "Searching for the Magic Weight-Loss Bullet," explains the various ways people have tried to keep their weight under control. The conclusion is that achieving or maintaining a healthy weight requires not only eating a variety of healthy foods but also exercising. In other words, to reverse a trend toward obesity, eat fewer calories by making wiser food choices and be more active.

Defining Obesity

Today, obesity is often defined as having a **body mass index (BMI)** of 30 or greater. The body mass index uses a person's height and weight (Fig. 8.11) to calculate a general approximation of their percent body fat. The BMI can also be calculated by dividing weight in pounds (lb) by height in inches (in.) squared and multiplying by a conversion factor of 703:

$$\text{weight (lbs)}/\text{height (in)}^2 \times 703$$

Most people find that it is easier to use the table (Fig. 8.11), or one of the variety of online calculators. As a general rule,

a healthy BMI = 18.5 to 24.9;
an overweight BMI = 25 to 29.9;
an obese BMI = 30 or higher; and
a morbidly obese BMI = 40 or higher.

Your BMI gives you an idea of how much of your weight is due to adipose tissue, commonly known as fat. In general, the

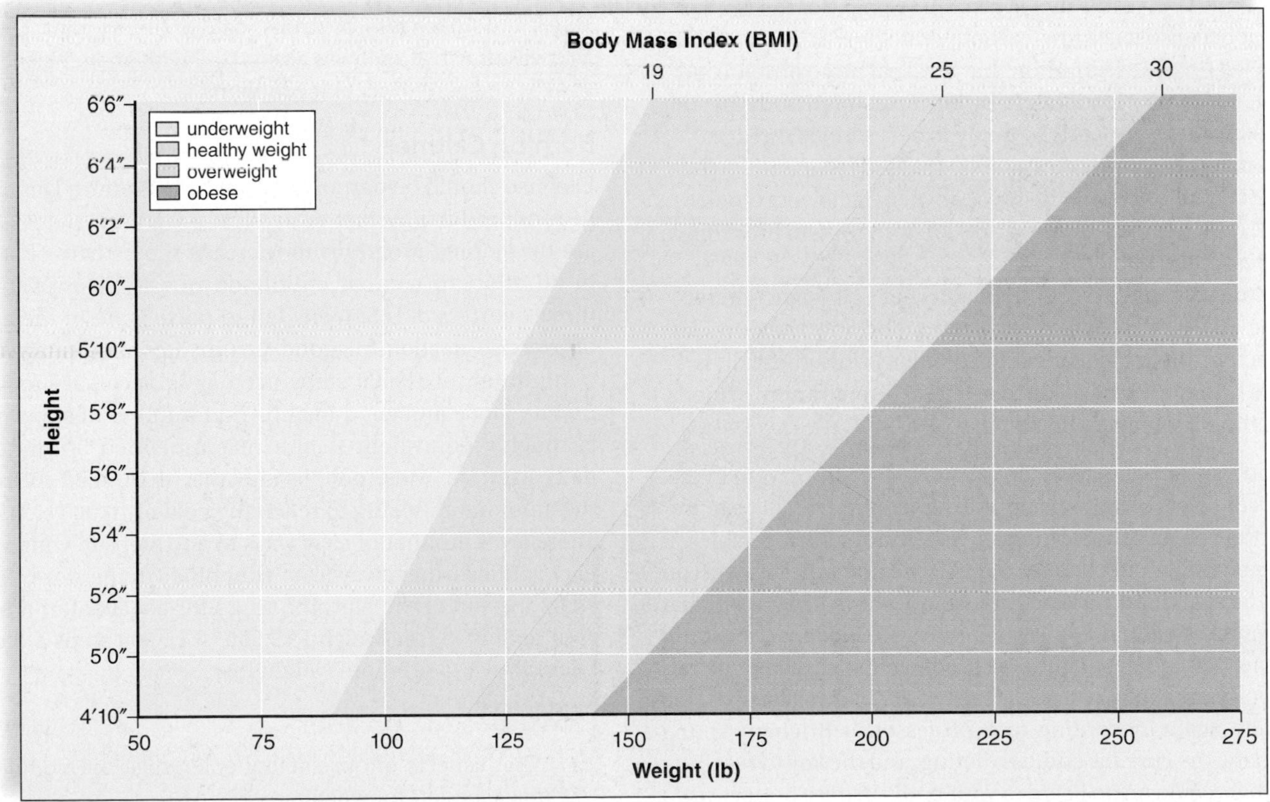

Source: U.S. Department of Agriculture: *Dietary Guidelines for Americans, 2005*

Figure 8.11 The body mass index chart.
Match your weight with your height, then determine your body mass index (BMI). Healthy BMI = 18.5 to 24.9; overweight BMI = 25 to 29.9; obese BMI = 30 to 39.9; morbidly obese BMI = 40 or more.

BIOLOGY MATTERS  Health

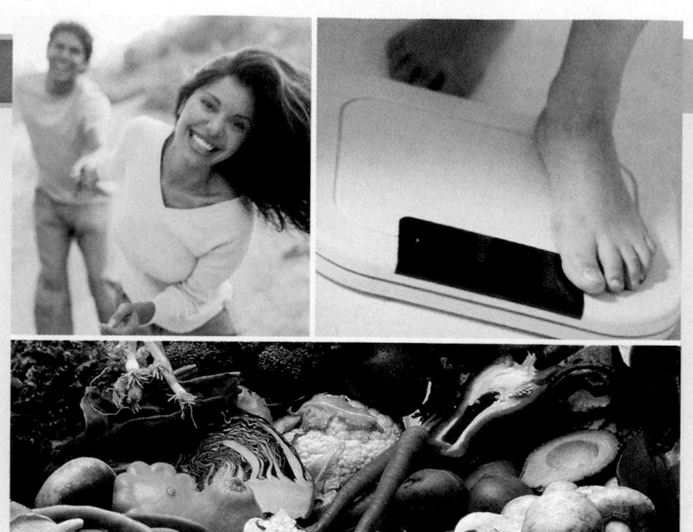

Searching for the Magic Weight-Loss Bullet

"Eat a variety of foods, watch your weight, and exercise" doesn't sound like a very glamorous way to lose weight. Besides, you can't sell the message to the public and make a lot of money. No wonder the public, always looking for the magic weight-loss bullet, is offered so many solutions to being overweight, most of which are not healthy. The solutions involve trendy diet programs, new prescription medications, and even surgery. The latter two options are for people who have tried a low-calorie diet and regular physical activity but have been unsuccessful in losing weight. Prescription medications should only be taken under a physician's supervision; and, of course, surgeries are only done by physicians.

Trendy Diet Programs

Various diets for the overweight have been around for many years, and here are some recently touted:

The Pritikin Diet This diet encourages the consumption of large amounts of carbohydrates and fiber in the form of whole grains and vegetables. The diet is so low fat that the dieter may not be able to consume a sufficient amount of "healthy" fats.

The Atkins Diet This diet is just the opposite of the Pritikin diet because it is a low-carbohydrate (low-carb) diet. It is based on the assumption that if we eat more protein and fat, our bodies lose weight by burning stored body fat. The Atkins diet is thought by many to be a serious threat to homeostasis. It puts a strain on the body to maintain the blood glucose level, the breakdown of fat lowers blood pH, and the excretion of nitrogen from protein breakdown stresses the kidneys.

The Zone Diet and the South Beach Diet As a reaction to the Atkins diet, these diets recommend only "healthy" fats and permit low-sugar carbs. In other words, these diets are bringing us back, once again, to "Eat a variety of foods, maintain your weight, and exercise."

You may have heard of the caveman diet. The caveman diet mimics the diets of humans prior to agricultural pursuits. It promotes the consumption of meats, fish, fruits, and vegetables. If you would rather not change your lifestyle dramatically, you can "flush" fat away, according to some nutritionists. They assert that by consuming certain foods, such as cayenne pepper, mustard, cinnamon, green vegetables, and omega-3-rich fish, you can boost your metabolic rate and cleanse your body of fat.

Then again, according to a professor at Brigham Young University, the cure for endless dieting, and the key to reaching a healthy weight, are to listen to your body. Using a hunger scale of 1 to 10, with 1 being starving and 10 being very overfull, keep around 3 to 5 and you will eat less. Unfortunately, such sensible advice doesn't seem to have caught on yet.

Prescription Drugs

Because obesity is officially classified as a disease, the pharmaceutical industry is actively developing products to assist with weight loss. Despite lawsuits against the manufacturers of the prescription drug fen-phen (Pondimin) for causing heart problems, new drugs continue to enter clinical trials. Not all are successful. For example, rimonabant (Acomplia) was supposed to block pleasure receptors in the brain, but due to suspected neurological complications never reached the market in the United States. Another class of drugs, called sibutramine (Meridia) was withdrawn after it was associated with an increased rate of strokes and cardiovascular problems.

Burning Calories

Exercise should be part of any weight-loss effort. Despite cuts in funding for physical education programs, many schools are developing programs to increase the activity of students in an effort to combat childhood obesity. However, about three-quarters of U.S. teens fail to participate in the amount of exercise needed for good health. Research suggests that a minimum of 10,000 steps per day is necessary for weight maintenance and good health. That number of steps per day is roughly equivalent to the recommended 30 minutes of daily exercise. Most people find that they need to increase the amount of walking to reach the goal of 10,000 steps a day. There are a number of easy ways to add steps to your routine. Park a little farther away from your office or the store. Take the stairs instead of the elevator, or go for a walk after a meal. If your goal is to lose weight, 12,000 to 15,000 steps a day have been shown to promote weight loss.

Questions to Consider

1. What aspects of our society encourage the popularity of magic bullets for weight loss?
2. Why do you think that so many of the weight-loss drugs are removed from the market for unanticipated side effects?

taller you are, the more you could weigh without it being due to fat. Using BMI in this way works for most people, especially if they tend to be sedentary. But your BMI number should be used only as a general guide. It does not take into account fitness, bone structure, or gender. For example, a weight lifter might have an obese BMI, not because of the amount of body fat, but because of increased bone and muscle weight.

Animation BMI

Classes of Nutrients

A **nutrient** can be defined as a required component of food that performs a physiological function in the body. Nutrients provide us with energy, promote growth and development, and regulate cellular metabolism.

Carbohydrates

Carbohydrates are either simple or complex (see section 2.4). Glucose is a simple sugar preferred by the body as an energy source. Complex carbohydrates, consisting of multiple sugar units, are digested to glucose. Although body cells can use fatty acids as an energy source, brain cells require glucose. For this reason alone, it is necessary to include carbohydrates in the diet because the body is unable to convert fatty acids to glucose.

Any product made from refined grains, such as white bread, cake, and cookies, should be minimized in the diet. During refinement, fiber is removed from the grains, along with vitamins and minerals, in order for the final product to be mainly starch. In contrast, sources of complex carbohydrates, such as beans, peas, nuts, fruits, and whole-grain products, are recommended as good sources of vitamins, minerals, and also fiber (Fig. 8.12). Insoluble fiber adds bulk to fecal material and stimulates movements of the large intestine, preventing constipation. Soluble fiber combines with bile salts and cholesterol in the small intestine and prevents them from being absorbed.

Figure 8.12 Foods rich in fiber.
Plants provide a good source of fiber in the diet. They also provide a good source of vitamins and minerals when they are not processed (refined).

Table 8.3 Reducing High-Glycemic-Index Carbohydrates

To Reduce Dietary Sugar
1. Eat fewer sweets, such as candy, soft drinks, ice cream, and pastries.
2. Eat fresh fruits or fruits canned without heavy syrup.
3. Use less sugar—white, brown, or raw—and less honey and syrups.
4. Avoid sweetened breakfast cereals.
5. Eat less jelly, jam, and preserves.
6. Eat fresh fruit; especially avoid artificial fruit juices.
7. When cooking, use spices, such as cinnamon, instead of sugar to flavor foods.
8. Do not put sugar in tea or coffee.
9. Avoid processed foods made from refined carbohydrates, such as white bread, rice, and pasta, and limit potato intake.

Can Carbohydrates Be Harmful? Nutritionists now recognize that the high intake of refined carbohydrates and fructose sweeteners processed from cornstarch is contributing to the increase in obesity in the United States. In addition, these foods are said to have a high **glycemic index,** because they quickly increase blood glucose. When the blood glucose level rises rapidly, the pancreas produces an overload of insulin to bring the level under control. Investigators tell us that a chronically high insulin level may lead to insulin resistance, type 2 diabetes, and increased fat deposition. Deposition of fat is associated with coronary heart disease, liver ailments, and several types of cancer.

Table 8.3 gives suggestions on how to reduce your intake of dietary sugars.

Carbohydrates are the preferred energy source for the body, but select complex carbohydrates in whole grains, beans, nuts, and fruits contain fiber, vitamins, and minerals in addition to carbohydrates.

MP3 Carbohydrates

Proteins

Dietary proteins are digested to amino acids (see section 2.6), which cells use to synthesize hundreds of cellular proteins. Of the 20 different amino acids, eight are **essential amino acids** that must be present in the diet because the body cannot make them, and two others are ones the body makes insufficiently. Eggs, milk products, meat, poultry, and most other foods derived from animals contain all eight essential amino acids and are "complete" or "high-quality" protein sources. Legumes (beans and peas) (Fig. 8.13), other types of vegetables, seeds and nuts, and grains supply us with amino acids. However, each of these alone is an incomplete protein source because of a deficiency in at least one of the essential amino acids. Absence of one essential amino acid prevents use of the other 19 amino acids. Therefore, vegetarians are counseled to combine two or more incomplete types of plant products to acquire all the essential amino acids. Tofu, soy milk, and other foods made from processed soybeans are complete protein sources. A balanced vegetarian diet is possible with a little knowledge and planning.

Figure 8.13 Foods rich in proteins and complex carbohydrates. Beans are a good source of complex carbohydrates and protein. But beans don't supply all the essential amino acids. To ensure a complete source of protein in the diet, beans should be eaten in combination with a grain, such as rice.

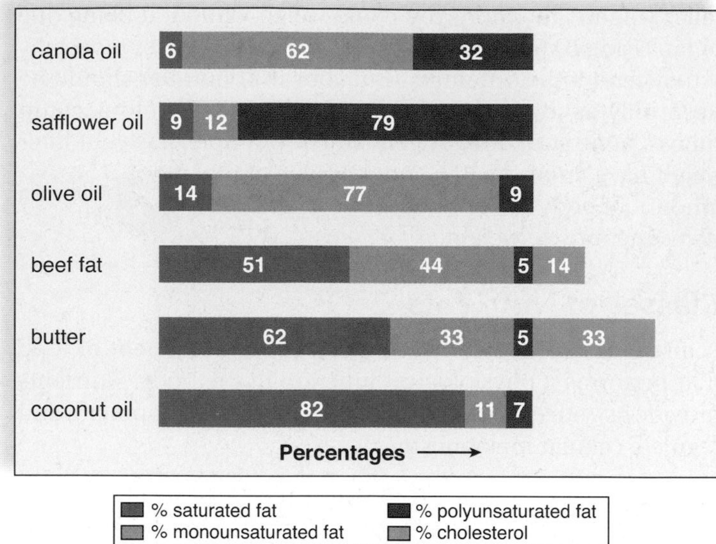

Figure 8.14 Saturated and unsaturated fatty acids. This illustration gives the percentages of cholesterol and saturated and unsaturated fatty acids in selected fats and oils.

A daily supply of essential amino acids is needed because they are not stored in the body, unlike the other amino acids that can be stored as proteins and metabolized for cells' needs. However, it does not take very much protein to meet the daily requirement. Two servings of meat a day (one serving is equal in size to a deck of cards) is usually plenty.

Animation Protein and Muscle Production

Can Proteins Be Harmful? The liver removes the nitrogen-containing compound from an amino acid. By converting this portion to urea, the liver enables potentially toxic nitrogen to be removed from our bodies. However, large amounts of water are needed to properly excrete urea. Therefore, dehydration may occur if protein consumption is excessive. High-protein diets, especially those rich in animal proteins, can also increase calcium loss in urine. Excretion of calcium may lead to kidney stones and bone loss.

Certain types of meat, especially red meat, are known to be high in saturated fats; other sources of protein, such as chicken, fish, and eggs, are more likely to be low in saturated fats. As you recall from section 5.7, excessive dietary saturated fat is a risk factor for cardiovascular disease.

Sufficient proteins are needed to supply the essential amino acids. Meat and dairy sources of protein may supply unwanted saturated fat, but vegetable sources do not.

MP3 Proteins

Lipids

Fats, oils, and cholesterol are lipids (see section 2.5). Saturated fats, which are solids at room temperature, usually have an animal origin. Two well-known exceptions are palm oil and coconut oil, which contain mostly saturated fats and come from the plants mentioned (Fig. 8.14). Butter and fats associated with meats (like the fat on steak and bacon) contain saturated fats.

Oils contain unsaturated fatty acids, which do not promote cardiovascular disease. Corn oil and safflower oil are high in polyunsaturated fatty acids. Polyunsaturated oils are the only type of fat that contains linoleic acid and linolenic acid, two fatty acids the body cannot make. These fatty acids must be supplied by diet, so they are called **essential fatty acids.**

Olive oil and canola oil are well known to contain a larger percentage of monounsaturated fatty acids than other types of cooking oils. Omega-3 fatty acids—with a double bond in the third position—are believed to preserve brain function and protect against heart disease. Flaxseed contains abundant omega-3 fatty acids. Cold-water fish like salmon, sardines, and trout are also an excellent source.

Can Lipids Be Harmful? The risk for cardiovascular disease is increased by a diet high in saturated fats and cholesterol. Saturated fats contribute to the formation of lesions associated with atherosclerosis inside the blood vessels. These lesions, called *atherosclerotic plaques,* limit the flow of blood through these vessels (see section 5.7). Cholesterol is carried in the blood by the two transport proteins: high-density lipoprotein (HDL) and low-density lipoprotein (LDL). Cholesterol transported by HDL (the "good" lipoprotein) ends up in the liver where the cholesterol is metabolized. Cholesterol carried by LDL (the "bad" lipoprotein) ends up being deposited in the tissues. Atherosclerotic plaques form when levels of HDL are low and/or when levels of LDL are high. Recommended levels of HDL and LDL can be reestablished by a diet low in saturated fats and cholesterol.

Trans-fatty acids (trans fats) arise when unsaturated fatty acids are hydrogenated to produce a solid fat. The function of the cell membrane receptors that clear cholesterol from the bloodstream may be reduced by trans fats resulting in higher blood cholesterol level. Trans fats are found in commercially packaged goods, such as cookies and crackers. Unfortunately, other snacks such as microwave popcorn may be sources as well. Be aware that any packaged goods containing *partially hydrogenated* vegetable oils or shortening contain trans fats. Some margarines used for home cooking or baking incorporate hydrogenated vegetable oil. Commercially fried foods, such as french fries from some fast-food chains, should be strictly limited in a healthy diet. Though tasty, these are often full of trans fats.

Table 8.4 gives suggestions on how to reduce dietary saturated fat and cholesterol. It is not a good idea to rely on commercially produced low-fat foods. In some products, sugars have replaced the fat, and in others, protein is used.

Unsaturated fats such as those in oils do not lead to cardiovascular disease and are preferred. Fats and oils contain many more calories per gram than do carbohydrates and protein.

MP3 Lipids

Table 8.4 Reducing Certain Lipids

To Reduce Saturated Fats and Trans Fats in the Diet

1. Choose poultry, fish, or dry beans and peas as a protein source.
2. Remove skin from poultry, and trim fat from red meats before cooking; place on a rack so that fat drains off.
3. Broil, boil, or bake rather than fry.
4. Limit your intake of butter, cream, trans fats, shortenings, and tropical oils (coconut and palm oils).
5. Use herbs and spices to season vegetables instead of butter, margarine, or sauces. Use lemon juice instead of salad dressing.
6. Drink skim milk instead of whole milk, and use skim milk in cooking and baking.

To Reduce Dietary Cholesterol

1. Avoid cheese, egg yolks, liver, and certain shellfish (shrimp and lobster). Preferably, eat white fish and poultry.
2. Substitute egg whites for egg yolks in both cooking and eating.
3. Include soluble fiber in the diet. Oat bran; oatmeal; beans; corn; and fruits, such as apples, citrus fruits, and cranberries, are high in soluble fiber.

BIOLOGY MATTERS  Health

Trans Fats and Food Labels

When the Food and Drug Administration required the addition of trans fat information to food labels in 2006, many food companies created labels touting their products as "trans fat free." A check of the label details (Fig. 8C) would list 0 grams of trans fat in the area where fat grams are listed. But a more thorough check of the list of ingredients might reveal a bit of trans fat lurking in the food.

If you see "partially hydrogenated oil" listed with the ingredients, there *are* some trans fats in that particular food. Trans fats only have to be listed with the breakdown of fat grams when there are 0.5 grams or more per serving. Limiting trans fats to 1% of daily calories is recommended by the American Heart Association. Unfortunately, eating more than one serving of a food with "hidden" trans fats might push some people over the recommended daily intake of trans fats.

Trans fats were commonly used in foods to extend the food products' shelf life. Many companies have started to discontinue the use of trans fats, given the health risks associated with them. Several fast food companies have made very public announcements of their goal to eliminate the use of trans fats in their preparation of food. Many cities, such as New York and Philadelphia, have banned the use of trans fats in restaurant and bakery foods, and other cities are considering similar action.

Nutrition Facts
Serving Size 17 Crackers (31g)
Servings Per Container About 7

Amount Per Serving
Calories 130 Calories from Fat 30

	% Daily Value*
Total Fat 3.5g	5%
Saturated Fat 0.5g	1%
Trans Fat 0g	
Cholesterol 0mg	0%
Sodium 180mg	8%
Total Carbohydrate 23g	8%
Dietary Fiber 2g	10%
Sugars 3g	
Protein 4g	

Vitamin A 0% • Vitamin C 0%
Calcium 0% • Iron 4%

*Percent Daily Values are based on a 2,000 calorie diet. Your daily values may be higher or lower depending on your calorie needs:

	Calories:	2,000	2,500
Total Fat	Less than	65g	80g
Sat Fat	Less than	20g	25g
Cholesterol	Less than	300mg	300mg
Sodium	Less than	2,400mg	2,400mg
Total Carbohydrate		300g	375g
Dietary Fiber		25g	30g

Fat content reduced from 7g to 3.5g per serving.

INGREDIENTS: ENRICHED FLOUR (WHEAT FLOUR, NIACIN, REDUCED IRON, THIAMIN MONONITRATE, RIBOFLAVIN, FOLIC ACID), WATER, WHOLE WHEAT FLOUR, CRACKED WHEAT FLOUR, VEGETABLE SHORTENING (COTTONSEED OIL, PARTIALLY HYDROGENATED SOYBEAN OIL, CITRIC ACID, TBHQ [ANTIOXIDANT]), SUGAR, RYE FLOUR, MALTED BARLEY FLOUR, MALTED CORN FLOUR, CORN SYRUP, SOY LECITHIN, SALT, CORN FLOUR, EXTRACTIVES OF PAPRIKA AND TURMERIC (FOR COLOR), SODIUM SULFITE, WHEY.
CONTAINS: WHEAT, SOY AND MILK.

Figure 8C A typical food label.

Animation Anatomy of a Food Label

Questions to Consider

1. How could the food label be changed to more accurately reflect the trans fat content of a food?
2. What are some other areas of the current food label that are misleading?

Minerals

Minerals are divided into *major minerals* and *trace minerals.* Major minerals are needed at quantities greater than 100 milligrams (mg) per day. Trace minerals are needed at levels less than 100 mg per day. Table 8.5 lists the selected minerals and gives their functions and food sources.

The major minerals are constituents of cells and body fluids and are structural components of tissues. The trace minerals are often part of larger molecules. For example, iron (Fe^{2+}) is present in hemoglobin, and iodine (I^-) is a part of hormones produced by the thyroid gland. Zinc (Zn^{2+}), copper (Cu^{2+}), and manganese (Mn^{2+}) are present in enzymes that catalyze a variety of reactions. As research continues, more and more elements are added to the list of trace minerals considered essential. During the past three decades, for example, very small amounts of selenium, molybdenum, chromium, nickel, vanadium, silicon, and even arsenic have been found to be essential to good health. Table 8.5 also provides signs of deficiency and toxicity for the selected minerals.

Occasionally, individuals do not receive enough iron (especially women), calcium, magnesium, or zinc in their diets. Adult females need more iron in their diet than males (15 mg compared with 10 mg) because they lose hemoglobin each month during menstruation. Stress can bring on a magnesium deficiency, and due to its high-fiber content, a vegetarian diet may make zinc less available to the body. However, a varied and complete diet usually supplies enough of each type of mineral.

Calcium

Calcium (Ca^{2+}) is a major mineral needed for the construction of bones and teeth. It is also necessary for nerve conduction, muscle contraction, and blood clotting. Many people take calcium supplements to prevent or counteract **osteoporosis,** a degenerative bone disease that afflicts an estimated one-fourth of older men and one-half of older women in the United States. Osteoporosis develops because bone-eating cells called osteoclasts are more active than bone-forming cells called osteoblasts. Therefore, the bones are porous, and they break easily because they lack sufficient calcium. Recommended calcium intakes vary by age, but in general 1,000 mg a day is recommended for men and women. After age 50 in women, and 70 in men, this increases to 1,200 mg a day. For many people, calcium supplements are needed to obtain these levels.

Animation Osteoporosis

Table 8.5 Minerals

Mineral	Functions	Food Sources	Health Concerns	
			Deficiency	Toxicity
Major (More than 100 mg/Day Needed)				
Calcium (Ca^{2+})	Strong bones and teeth, nerve conduction, muscle contraction, blood clotting	Dairy products, leafy green vegetables	Stunted growth in children, low bone density in adults	Kidney stones; interferes with iron and zinc absorption
Phosphorus (PO_4^{3-})	Bone and soft tissue growth; part of phospholipids, ATP, and nucleic acids	Meat, dairy products, sunflower seeds, food additives	Weakness, confusion, pain in bones and joints	Low blood and bone calcium levels
Potassium (K^+)	Nerve conduction, muscle contraction	Many fruits and vegetables, bran	Paralysis, irregular heartbeat, eventual death	Vomiting, heart attack, death
Sulfur (S^{2-})	Stabilizes protein shape, neutralizes toxic substances	Meat, dairy products, legumes	Not likely	In animals, depresses growth
Sodium (Na^+)	Nerve conduction, pH and water balance	Table salt	Lethargy, muscle cramps, loss of appetite	Edema, high blood pressure
Chloride (Cl^-)	Water balance	Table salt	Not likely	Vomiting, dehydration
Magnesium (Mg^{2+})	Part of various enzymes for nerve and muscle contraction, protein synthesis	Whole grains, leafy green vegetables	Muscle spasm, irregular heartbeat, convulsions, confusion, personality changes	Diarrhea
Trace (Less than 100 mg/Day Needed)				
Zinc (Zn^{2+})	Protein synthesis, wound healing, fetal development and growth, immune function	Meats, legumes, whole grains	Delayed wound healing, stunted growth, diarrhea, mental lethargy	Anemia, diarrhea, vomiting, renal failure, abnormal cholesterol levels
Iron (Fe^{2+})	Hemoglobin synthesis	Whole grains, meats, prune juice	Anemia, physical and mental sluggishness	Iron toxicity disease, organ failure, eventual death
Copper (Cu^{2+})	Hemoglobin synthesis	Meat, nuts, legumes	Anemia, stunted growth in children	Damage to internal organs if not excreted
Iodine (I^-)	Thyroid hormone synthesis	Iodized table salt, seafood	Thyroid deficiency	Depressed thyroid function, anxiety
Selenium (SeO_4^{2-})	Part of antioxidant enzyme	Seafood, meats, eggs	Vascular collapse, possible cancer development	Hair and fingernail loss, discolored skin
Manganese (Mn^{2+})	Part of enzymes	Nuts, legumes, green vegetables	Weakness and confusion	Confusion, coma, death

Small-framed, Caucasian women with a family history of osteoporosis are at greatest risk of developing the disease. Smoking and drinking more than nine cups of caffeinated drinks daily may also contribute. Vitamin D is an essential companion to calcium in preventing osteoporosis. Other vitamins may also be helpful; for example, magnesium has been found to suppress the cycle that leads to bone loss. In addition to adequate calcium and vitamin intake, exercise helps prevent osteoporosis. Medications are also available that slow bone loss while increasing skeletal mass.

Sodium

Sodium plays a major role in regulating the body's water balance, as does chloride (Cl^-). Sodium plays an important role in the movement of materials across the plasma membrane (see section 3.3) as well as the conduction of a nerve impulse (see section 13.1). The recommended amount of sodium intake per day is 500 mg, although the average American takes in more than 3,400 mg every day. This imbalance has caused concern because sodium in the form of salt intensifies hypertension (high blood pressure). About one-third of the sodium we consume occurs naturally in foods. Another third is added during commercial processing, and we add the last third either during home cooking or at the table in the form of table salt.

Clearly, it is possible to cut down on the amount of sodium in the diet. Table 8.6 gives recommendations for doing so.

APPLICATIONS AND MISCONCEPTIONS

Where does most of the sodium in the diet come from?

Contrary to popular belief, the majority of the sodium in your diet does not come from the salt you put on your food when you are eating. Instead, most of our sodium (over three-quarters!) comes from processed foods and condiments. Sodium is used in these items both to preserve the food or condiment and to make it taste better. But sometimes the amount of sodium is phenomenal. A single teaspoon of soy sauce contains almost 1,000 mg of sodium, and a half-cup of prepared tomato sauce typically has over 400 mg of sodium. Websites such as nutritiondata.com can help you track your daily sodium intake.

Table 8.6 Reducing Dietary Sodium

To Reduce Dietary Sodium

1. Use spices instead of salt to flavor foods.
2. Add little or no salt to foods at the table, and add only small amounts of salt when you cook.
3. Eat unsalted crackers, pretzels, potato chips, nuts, and popcorn.
4. Avoid hot dogs, ham, bacon, luncheon meats, smoked salmon, sardines, and anchovies.
5. Avoid processed cheese and canned or dehydrated soups.
6. Avoid brine-soaked foods, such as pickles or olives.
7. Read nutrition labels to avoid high-salt products.

Vitamins

Vitamins are organic compounds (other than carbohydrate, fat, and protein) that the body uses for metabolic purposes but is unable to produce in adequate quantity. Many vitamins are portions of coenzymes, enzyme helpers. For example, niacin is part of the coenzyme NAD, and riboflavin is part of another dehydrogenase, FAD, discussed in Chapter 3. Coenzymes are needed in only small amounts because each can be used over and over. Not all vitamins are coenzymes. Vitamin A, for example, is a precursor for the visual pigment that prevents night blindness. If vitamins are lacking in the diet, various symptoms develop. There are 13 vitamins, divided into those that are *fat soluble* (Table 8.7) and those that are *water soluble* (Table 8.8). The differences between fat and water-soluble vitamins have to do with how the compound gets absorbed into the body, how it is transported by the body, its interaction with body tissues, and how it is stored within the cells.

Table 8.7 Fat-Soluble Vitamins

Mineral	Functions	Food Sources	Health Concerns	
			Deficiency	Toxicity
Vitamin A	Antioxidant synthesized from beta-carotene; needed for healthy eyes, skin, hair, and mucous membranes, and for proper bone growth	Deep yellow/orange and leafy, dark green vegetables; fruits; cheese; whole milk; butter; eggs	Night blindness, impaired growth of bones and teeth	Headache, dizziness, nausea, hair loss, abnormal development of fetus
Vitamin D	A group of steroids needed for development and maintenance of bones and teeth and for absorption of calcium	Milk fortified with vitamin D, fish liver oil; also made in the skin when exposed to sunlight	Rickets, decalcification and weakening of bones	Calcification of soft tissues, diarrhea, possible renal damage
Vitamin E	Antioxidant that prevents oxidation of vitamin A and polyunsaturated fatty acids	Leafy green vegetables, fruits, vegetable oils, nuts, whole-grain breads and cereals	Unknown	Diarrhea, nausea, headaches, fatigue, muscle weakness
Vitamin K	Needed for synthesis of substances active in clotting of blood	Leafy green vegetables, cabbage, cauliflower	Easy bruising and bleeding	Can interfere with anticoagulant medication

Table 8.8 Water-Soluble Vitamins

Vitamin	Functions	Food Sources	Health Concerns	
			Deficiency	Toxicity
Vitamin C	Antioxidant; needed for forming collagen; helps maintain capillaries, bones, and teeth	Citrus fruits, leafy green vegetables, tomatoes, potatoes, cabbage	Scurvy, delayed wound healing, infections	Gout, kidney stones, diarrhea, decreased copper
Thiamine (vitamin B_1)	Part of coenzyme needed for cellular respiration; also promotes activity of the nervous system	Whole-grain cereals, dried beans and peas, sunflower seeds, nuts	Beriberi, muscular weakness, enlarged heart	Can interfere with absorption of other vitamins
Riboflavin (vitamin B_2)	Part of coenzymes, such as FAD*; aids cellular respiration, including oxidation of protein and fat	Nuts, dairy products, whole-grain cereals, poultry, leafy green vegetables	Dermatitis, blurred vision, growth failure	Unknown
Niacin (nicotinic acid)	Part of coenzyme NAD†; needed for cellular respiration, including oxidation of protein and fat	Peanuts, poultry, whole-grain cereals, leafy green vegetables, beans	Pellagra, diarrhea, mental disorders	High blood sugar and uric acid, vasodilation, etc.
Folacin (folic acid)	Coenzyme needed for production of hemoglobin and formation of DNA	Dark, leafy green vegetables; nuts; beans; whole-grain cereals	Megaloblastic anemia, spina bifida	May mask B_{12} deficiency
Vitamin B_6	Coenzyme needed for synthesis of hormones and hemoglobin; CNS control	Whole-grain cereals, bananas, beans, poultry, nuts, leafy green vegetables	Rarely, convulsions, vomiting, seborrhea, muscular weakness	Insomnia, neuropathy
Pantothenic acid	Part of coenzyme A needed for oxidation of carbohydrates and fats; aids in the formation of hormones and certain neurotransmitters	Nuts, beans, dark green vegetables, poultry, fruits, milk	Rarely, loss of appetite, mental depression, numbness	Unknown
Vitamin B_{12}	Complex, cobalt-containing compound; part of the coenzyme needed for synthesis of nucleic acids and myelin	Dairy products, fish, poultry, eggs, fortified cereals	Pernicious anemia	Unknown
Biotin	Coenzyme needed for metabolism of amino acids and fatty acids	Generally in foods, especially eggs	Skin rash, nausea, fatigue	Unknown

*FAD = flavin adenine dinucleotide
†NAD = nicotinamide adenine dinucleotide

Antioxidants

Over the past several decades, numerous statistical studies have determined that a diet rich in fruits and vegetables can protect against cancer. Cellular metabolism generates free radicals, unstable molecules that carry an extra electron. The most common free radicals in cells are superoxide (O_2^-) and hydroxide (OH^-). To stabilize themselves, free radicals donate an electron to DNA, to proteins (including enzymes), or to lipids, which can be found in plasma membranes. Such donations often damage these cellular molecules and thereby may lead to disruptions of cellular functions, including even cancer.

Vitamins C, E, and A are believed to defend the body against free radicals and are therefore termed antioxidants. These vitamins are especially abundant in fruits and vegetables. Dietary guidelines (Fig. 8.15) suggest that we increase our consumption of fruits and vegetables each day. To achieve this goal, think in terms of salad greens, raw or cooked vegetables, dried fruit, and fruit juice, in addition to apples and oranges and other fresh fruits.

Dietary supplements may provide a potential safeguard against cancer and cardiovascular disease. Nutritionists do not think people should take supplements instead of improving their intake of fruits and vegetables. There are many beneficial compounds in these foods that cannot be obtained from a vitamin pill. These compounds enhance one another's absorption or action and also perform independent biological functions.

Vitamin D

Skin cells contain a precursor cholesterol molecule converted to vitamin D after UV exposure. Vitamin D leaves the skin and is modified first in the kidneys and then in the liver until finally it becomes calcitriol. Calcitriol promotes the absorption of calcium by the intestines. When taking a calcium supplement, it is a good idea to get one with added vitamin D. The lack of vitamin D leads to rickets in children. Rickets, characterized by bowing of the legs, is caused by defective mineralization of the skeleton. Most milk is fortified with vitamin D, which helps prevent the occurrence of rickets.

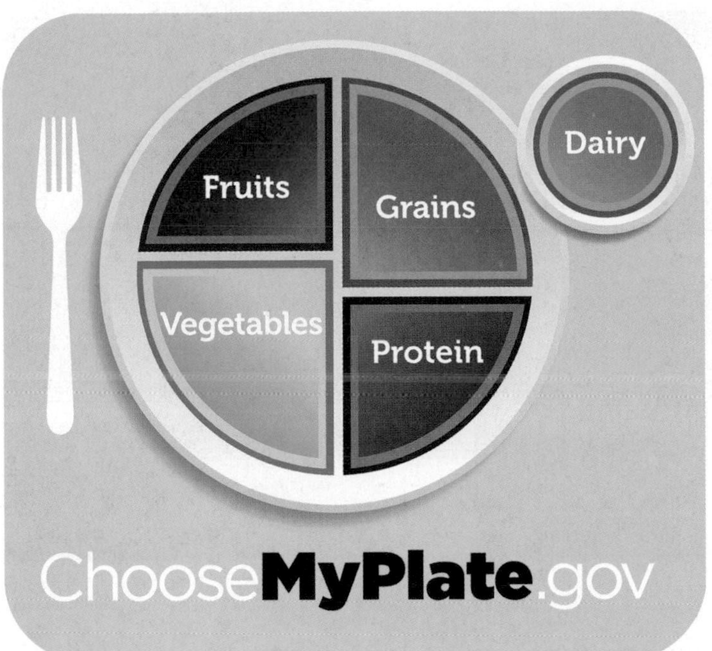

Figure 8.15 MyPlate.
The U.S. Department of Agriculture (USDA) developed this visual representation of a food plate as a guide to better health. The size differences on the plate for each food group suggest what portion of your meal should consist of each category. The five different colors illustrate that foods, in correct proportions , are needed each day for good health.

How to Plan Nutritious Meals

Many serious disorders in Americans are linked to a diet that results in excess body fat. Whereas genetics is a factor in being overweight, a person cannot become fat without taking in more food energy (calories) than is needed. People need calories (energy) for their basal metabolism. Basal metabolism is the number of calories a person's body burns at rest to maintain normal body functions. A person also needs calories for exercise. The less exercise, the fewer calories needed beyond the basal metabolic rate. So, the first step in planning a diet is to limit the number of calories to an amount that a person will use each day. Let's say you do all the necessary calculations (beyond the scope of this book). You discover that, as a woman, the maximum number of calories you can take in each day is 2,000. If you're a man, you can afford 2,500 calories without gaining weight.

The new guidelines developed by the U.S. Department of Agriculture (USDA) are called MyPlate (Fig. 8.15). This graphical representation replaced the older pyramids because most people found it easier to interpret. It can be used to help you decide how those calories should be distributed among the foods to eat. MyPlate emphasizes the proportions of each food group that should be consumed daily. In addition, the USDA provides recommendations concerning the minimum quantity of foods in each group that should be eaten daily (see ChooseMyPlate.gov). In general, we should

- eat a variety of foods. Foods from all food groups should be included in the diet.
- eat more of these foods: fruits, vegetables, whole grains, and fat-free or low-fat milk products. Choose dark green vegetables, orange vegetables, and leafy vegetables. Dry beans and peas are good sources for fiber and a great protein source as well. Limit potatoes and corn. When eating grains, choose whole grains such as brown rice, oatmeal, and whole-wheat bread. Choose fruit as a snack or a topping for foods, instead of sugar.
- choose lean meats such as poultry; fish high in omega-3 fatty acids, such as salmon, trout, and herring, in moderate-sized portions. Include oils rich in monounsaturated and polyunsaturated fatty acids in the diet.
- eat less of foods high in saturated or trans fats, added sugars, cholesterol, salt, and alcohol.
- be physically active every day. If weight loss is needed, decrease calorie intake slowly, while maintaining adequate nutrient intake and increasing physical activity.

Animation Calories in a Sandwich

Eating Disorders

People with eating disorders are dissatisfied with their body image. Social, cultural, emotional, and biological factors all contribute to the development of an eating disorder. Serious conditions such as obesity, anorexia nervosa, and bulimia nervosa can lead to malnutrition, disability, and death. Regardless of the eating disorder, early recognition and treatment are crucial.

Anorexia nervosa is a severe psychological disorder characterized by an irrational fear of getting fat. Victims refuse to eat enough food to maintain a healthy body weight (Fig. 8.16*a*). A self-imposed starvation diet is often accompanied by occasional binge eating, followed by purging and extreme physical activity to avoid weight gain. Binges usually include large amounts of high-calorie foods, and purging episodes involve self-induced vomiting and laxative abuse. About 90% of people suffering from anorexia nervosa are young women; an estimated 1 in 200 teenage girls is affected.

A person with **bulimia nervosa** binge-eats and then purges to avoid gaining weight (Fig. 8.16*b*). The binge–purge cycle behavior can occur several times a day. People with bulimia nervosa can be difficult to identify because their body weights are often normal and they tend to conceal their binging and purging practices. Women are more likely than men to develop bulimia; an estimated 4% of young women suffer from this condition.

Other abnormal eating practices include binge-eating disorder and muscle dysmorphia. Many obese people suffer from **binge-eating disorder,** a condition characterized by episodes of overeating without purging. Stress, anxiety, anger, and depression can trigger food binges. A person suffering from **muscle dysmorphia** (Fig. 8.16*c*) thinks his or her body is underdeveloped. Body-building activities and a preoccupation with diet and body form accompany this condition. Each day, the person may spend hours in the gym working out on muscle-strengthening equipment. Unlike anorexia nervosa and bulimia, muscle dysmorphia affects more men than women.

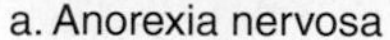
a. Anorexia nervosa

b. Bulimia nervosa

c. Muscle dysmorphia

Figure 8.16 The characteristics of different eating disorders.
a. People with anorexia nervosa have a mistaken body image and think they are fat, even though they are thin. **b.** Those with bulimia nervosa overeat and then purge their bodies of the food they have eaten. **c.** People with muscle dysmorphia think their muscles are underdeveloped. They spend hours at the gym and are preoccupied with diet as a way to gain muscle mass.

CHECK YOUR PROGRESS 8.6

1. Briefly describe and give an example of each class of nutrients.
2. Discuss why carbohydrates and fats might be the cause of the obesity epidemic today.
3. Detail why it is important to overall homeostasis to have a balanced diet.

CONNECTING THE CONCEPTS

For more information on how the body utilizes nutrients, refer to the following discussions:

Section 11.1 explains how vitamin D acts as a hormone that influences bone growth.

Section 14.4 examines the role of vitamin A in vision.

Section 19.2 examines how a diet with adequate levels of vitamins A and C may help prevent cancer.

CASE STUDY CONCLUSION

Nicole's test results indicated that she had gastroesophageal reflux disease, or GERD, as well as a duodenal ulcer. GERD is a condition in which food and liquid travel backward up the esophagus from the stomach (see the Health feature, "Heartburn (GERD)," on page 174). After you swallow food, it mixes with very acidic gastric juices to chemically digest the food so it can enter the intestine, where nutrients can be removed. When this mixture travels up the esophagus, the acidity of the digestive juices irritates the lining of the esophagus, causing heartburn and lesions. If not treated, the esophagus can begin to deteriorate. The duodenal ulcer Nicole had was a raw area on the duodenum. The ulcer was caused by the same acidic digestive juices affecting this area that affected her esophagus. Additionally, Nicole was tested for levels of *Helicobacter pylori*—a type of bacteria that commonly lives in the digestive tracts of humans. Some people have elevated levels of *H. pylori*, which can cause heartburn and reflux as well as ulcerations of the stomach and duodenum. Her levels of *H. pylori* were significant. Dr. Winch gave Nicole a few medications: an antibiotic to decrease the *H. pylori* level; a sucralfate (a medication that coats the area of the ulceration so it can heal); and pantoprazole (Protonix), a medication to reduce the amount of acid the stomach makes. The doctor also suggested a few lifestyle changes that could help. Drinking and smoking have been shown to worsen GERD and ulcers. Dr. Winch also showed Nicole studies that suggested that a healthy diet and moderate exercise aided in managing the levels of not only acidic gastric juices produced but also *H. pylori*. Dr. Winch also cautioned Nicole against taking OTC medications that could cause stomach inflammations, such as aspirin, ibuprofen, and naproxen. By managing her diet, lifestyle, and medications, Nicole can get her GI tract back in working order.

MEDIA STUDY TOOLS

Enhance your study of this chapter with media! Visit **www.mhhe.com/maderhuman13e** and go to "Media Study Tools" for this chapter to access the following:

Animations	Video	MP3 Files
8.3 Three Phases of Gastric Secretion • Enzyme Action and the Hydrolysis of Sucrose **8.6** BMI • Protein and Muscle Production • Osteoporosis • Calories in a Sandwich • Anatomy of a Food Label • B Vitamins	**8.5** Fat Microbes	**8.1** Organs of Digestion • Overview of the Digestive System **8.2** Oral Cavity, Esophagus, and Swallowing Reflex **8.3** The Stomach • Absorption of Nutrients and Water **8.6** Carbohydrates • Proteins • Lipids

SUMMARIZE

8.1 Overview of Digestion

- The purpose of the digestive system is to **hydrolyze** macromolecules to their smallest subunits. The organs of the digestive system are located within the GI tract.
- The processes of digestion require **ingestion, digestion, movement** (**peristalsis**), **absorption,** and **elimination.**
- All parts of the tract have four layers, called the **mucosa, submucosa, serosa,** and **muscularis.** These layers surround the **lumen,** or interior space of the GI tract.
- A disorder of the muscularis called **diverticulosis** can affect any organ of the GI tract, but primarily occurs in the large intestine.

8.2 The Mouth, Pharynx, and Esophagus

- In the **mouth** (or oral cavity), teeth chew the food. The **salivary glands** produce saliva, which contains **salivary amylase** for digesting starch, and the tongue forms a **bolus** for swallowing.
- Both the mouth and the nose lead into the **pharynx.** The pharynx opens into both the esophagus (food passage) and trachea (air passage). During swallowing, the opening into the nose is blocked by the soft palate, and the **epiglottis** covers the opening to the trachea (**glottis**). Food enters the esophagus, and peristalsis begins. The esophagus moves food to the stomach by peristalsis. **Sphincters** control the movement of the bolus.
- Disorders of the mouth include **dental caries** (cavities) of the teeth and **gingivitis** and **periodontitis** of the gums.
- **Heartburn** occurs when the contents of the stomach enter the **esophagus.** During vomiting, the abdominal muscles and **diaphragm** propel the food through the esophagus and out the mouth.

8.3 The Stomach and Small Intestine

- The **stomach** expands and stores food and also churns, mixing food with the acidic gastric juices. The stomach contains an **oblique layer** of smooth muscle,and folds called **rugae,** to assist in mixing the food. **Gastric glands** produce gastric juice, which contains **pepsin,** an enzyme that digests protein. The material leaving the stomach is called **chyme.**
- The **duodenum** of the **small intestine** receives bile from the liver and pancreatic juice from the pancreas. Bile emulsifies fat and readies it for digestion by lipase.
- The pancreas produces enzymes that digest starch (**amylase**), protein (**proteases** such as trypsin), and fat (**lipase**). The intestinal enzymes finish the process of chemical digestion.
- Brush border enzymes of the small intestine complete the digestive process. Small nutrient molecules are absorbed at the **villi** in the walls of the small intestine. The nutrients enter the capillaries of the circulatory system and **lacteals** of the lymphatic system. **Lactose intolerance** occurs when individuals are missing the lactase enzyme in the small intestine.

8.4 The Accessory Organs and Regulation of Secretions

Three accessory organs of digestion send secretions to the duodenum via ducts. These organs are the pancreas, liver, and gallbladder.

- The **pancreas** produces pancreatic juice, which contains digestive enzymes for carbohydrate, protein, and fat. The pancreas also releases the **hormone** insulin, which regulates blood glucose levels.
- The **liver** produces **bile,** destroys old blood cells, detoxifies blood, stores iron, makes plasma proteins, stores glucose as glycogen, breaks down glycogen to glucose, produces **urea,** and helps regulate blood **cholesterol** levels.
- The **gallbladder** stores bile, produced by the liver. The secretions of digestive juices are controlled by the nervous system and by hormones.
- Gastrin produced by the lower part of the stomach stimulates the upper part of the stomach to secrete pepsin. Secretin and CCK produced by the duodenal wall stimulate the pancreas to secrete its juices and the gallbladder to release bile.
- Disorders of the liver include **jaundice, hepatitis,** and **cirrhosis.** **Gallstones** may affect the operation of the gallbladder.

8.5 The Large Intestine and Defecation

- The **large intestine** consists of the **cecum;** the **colon** (including the ascending, transverse, and descending colon); the **appendix;** and the **rectum,** which ends at the **anus.**
- The large intestine absorbs water, salts, and some vitamins; forms the feces; and carries out **defecation. Fiber** provides bulk to the feces. The color of the feces is the result of **bilirubin** (a waste product).
- Disorders of the large intestine include **diarrhea, constipation, hemorrhoids,** diverticulosis, irritable bowel syndrome, inflammatory bowel disease, **polyps,** and cancer.

8.6 Nutrition and Weight Control

The **nutrients** released by the digestive process should provide us with adequate energy, essential amino acids and fatty acids, and all necessary vitamins and minerals.

- The **body mass index** (**BMI**) may be used to determine the percent body fat. **Obesity** is associated with many illnesses, including type 2 diabetes and cardiovascular disease. The new MyPlate dietary guidelines provide a visual representation of dietary proportions to maintain good health.
- Carbohydrates are necessary in the diet, but simple sugars and refined starches cause a rapid release of insulin that can lead to type 2 diabetes. The **glycemic index** may be used to predict which foods will release carbohydrates quickly into the blood.
- Proteins supply **essential amino acids.**
- Unsaturated fatty acids, particularly the omega-3 fatty acids, are protective against cardiovascular disease. Saturated fatty acids and trans fats contribute to heart disease. **Essential fatty acids** must be supplied by the diet.
- **Minerals** are also required by the body in certain amounts. **Osteoporosis** is an example of a disease caused by a mineral deficiency (calcium).
- **Vitamins** are organic compounds that act as metabolic assistants.
- Eating disorders include **anorexia nervosa, bulimia nervosa, binge-eating disorder,** and **muscle dysmorphia.**

ASSESS

Testing Your Knowledge of the Concepts

1. Argue that absorption is the most important of the five processes of digestion over the other four processes. (pages 169–170)
2. List the main organs of the digestive tract, and state the contribution of each to the digestive process. (page 169)
3. Name the enzymes involved in the digestion of starch, protein, and fat, and tell where these enzymes are active and what they do. (page 176)
4. Discuss the absorption of the products of digestion into the lymphatic and cardiovascular systems. (pages 176–177)
5. Why are the pancreas, liver, and gallbladder considered *accessory* organs of digestion and not organs of digestion? (pages 178–179)
6. Name and state the functions of the hormones that assist the nervous system in regulating digestive secretions. (page 179)
7. What is the chief contribution of each of these in the body: carbohydrates, proteins, fats, fruits, and vegetables? (pages 185–187)
8. Which three eating disorders involve binge eating? How are these three disorders different from one another? (page 191)
9. Tracing the path of food in the following list (a–f), which step is out of order first?
   a. mouth
   b. pharynx
   c. esophagus
   d. small intestine
   e. stomach
   f. large intestine
10. Which association is incorrect?
    a. mouth—starch digestion
    b. esophagus—protein digestion
    c. small intestine—starch, lipid, protein digestion
    d. stomach—food storage
    e. liver—production of bile
11. Which association is incorrect?
    a. pancreas—produces alkaline secretions and enzymes
    b. salivary glands—produce saliva and amylase
    c. gallbladder—produces digestive enzymes
    d. liver—produces bile
12. Peristalsis occurs
    a. from the mouth to the small intestine.
    b. from the beginning of the esophagus to the anus.
    c. only in the stomach.
    d. only in the small and large intestine.
    e. only in the esophagus and stomach.
13. Bile
    a. is an important enzyme for the digestion of fats.
    b. cannot be stored.
    c. is made by the gallbladder.
    d. emulsifies fat.
    e. All of these are correct.
14. Which of the following is not a function of the liver in adults?
    a. produces bile
    b. detoxifies alcohol
    c. stores glucose
    d. produces urea
    e. makes red blood cells
15. The large intestine
    a. digests all types of food.
    b. is the longest part of the intestinal tract.
    c. absorbs water.
    d. is connected to the stomach.
    e. is subject to hepatitis.
16. The amino acids that must be consumed in the diet are called essential. Nonessential amino acids
    a. can be produced by the body.
    b. are only needed occasionally.
    c. are stored in the body until needed.
    d. can only be found in the diet; the body cannot synthesize these amino acids.

In questions 17–20, match each statement to an answer in the key. Answers may be used more than once. Some may have more than one answer.

Key:

a. gastrin
b. secretin
c. CCK
d. All of these are correct.
e. None of these are correct.

17. Stimulates gallbladder to release bile

18. Stimulates the stomach to digest protein

19. Secreted by duodenum

20. Secreted by the stomach

In questions 21–25, match each statement to a vitamin or mineral in the key.

Key:

a. calcium
b. vitamin K
c. sodium
d. iodine
e. vitamin A

21. Needed to make thyroid hormone

22. Needed for night vision

23. Needed for bones, teeth, and muscle contraction

24. Needed for nerve conduction, pH, and water balance

25. Needed for making clotting proteins

ENGAGE

The virtual lab "Nutrition" provides an interactive investigation of how your food choices relate to your daily intake of select nutrients.

Thinking Critically About the Concepts

Bariatric surgery is a type of medical procedure that reduces the size of the stomach and enables food to bypass a section of the small intestine. The surgery is generally done when obese individuals have unsuccessfully tried numerous ways to lose weight and their health is compromised by their weight. There are many risks associated with the surgery, but it helps a number of people lose a considerable amount of weight and ultimately improve their overall health. After people undergo the surgery, there are several lifestyle changes they must make to avoid nutritional deficiencies and to compensate for the small size of their stomach.

1. **a.** Why do some people who have had bariatric surgery process their food in a blender (or have to chew thoroughly) before swallowing their food?
   b. Why should people who have had bariatric surgery drink liquids between meals rather than with meals?
2. What risk is there to the esophagus after bariatric surgery?

C H A P T E R

9

Respiratory System

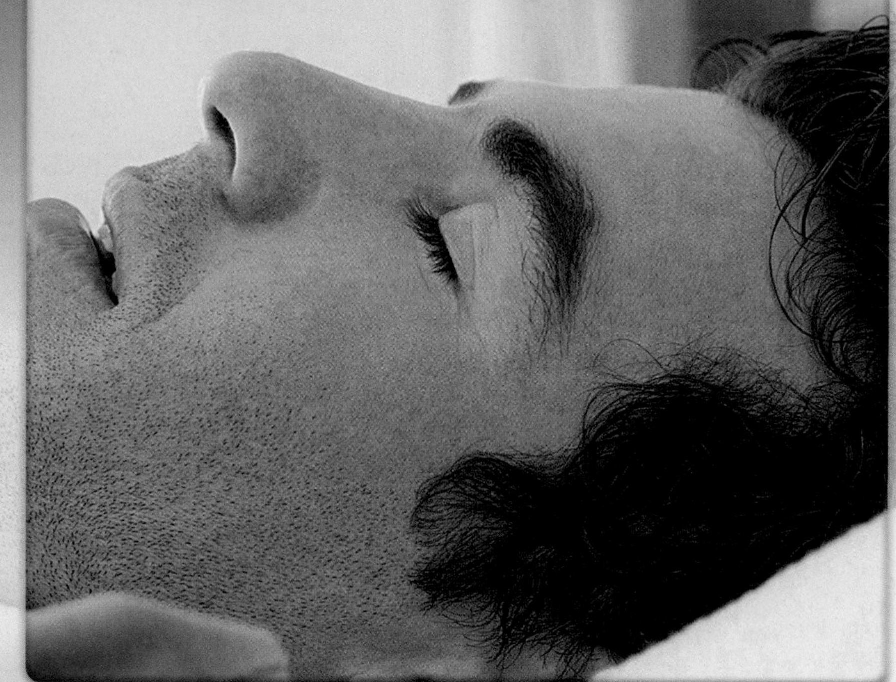

CHAPTER CONCEPTS

9.1 The Respiratory System
The respiratory system is divided into two regions, the upper and lower respiratory tracts.

9.2 The Upper Respiratory Tract
Air is warmed and filtered in the nose, then moves across the pharynx to pass through the glottis into the larynx.

9.3 The Lower Respiratory Tract
The trachea leads to the bronchial tree, which ends in the lungs. Gas exchange occurs in the air sacs called alveoli.

9.4 Mechanism of Breathing
During inspiration, expansion of the chest moves air into the lungs; during expiration, air leaves the lungs.

9.5 Control of Ventilation
The respiratory center in the brain automatically regulates breathing by relying on special receptors to detect changes in the pH of the blood.

9.6 Gas Exchanges in the Body
In the body, oxygen is carried by hemoglobin in red blood cells, and carbon dioxide is transported primarily as bicarbonate ions. These gases are exchanged in the lungs and tissues.

9.7 Respiration and Health
Several illnesses are associated with the upper and lower respiratory tracts.

BEFORE YOU BEGIN

Before beginning this chapter, take a few moments to review the following discussions:

Section 2.2 What do the values of the pH scale indicate?

Figure 4.15 What is the role of the respiratory system in maintaining homeostasis?

Section 6.2 How are red blood cells involved in transporting oxygen and carbon dioxide?

CASE STUDY SLEEP APNEA

For weeks, Justin had felt rundown and tired. He had just moved into the city and started a new job. Even though the past few weeks had been hectic, he had been trying to get at least 7 hours of sleep each night. However, when he woke up each morning, he still felt tired. He was finding it more difficult each day to concentrate on his job, and he found himself forgetting important details of meetings. Concerned that his lack of sleep might cause problems at work, Justin scheduled an appointment with his doctor.

Justin's doctor decided to do a complete physical exam. He checked Justin's weight and blood pressure and asked Justin a series of questions about his diet; medicines he was taking; and, finally, his sleep habits. Though Justin was in general a healthy individual, he reported that he had put on a few pounds since leaving college and starting his job and that his sleep problems seemed to be even worse if he had a few drinks before going to bed. Justin also mentioned that he rarely slept through the night and often awoke suddenly with a feeling of being out of breath.

Given these symptoms, the doctor suggested that Justin might be suffering from sleep apnea. He explained to Justin that there are two types of sleep apnea—obstructive sleep apnea and central sleep apnea. Central sleep apnea is usually caused by an illness or injury to the central nervous system and is associated with neurological problems in the brain. Because Justin did not have a history of any of these conditions, his doctor focused on obstructive sleep apnea. In this condition, the airways in the upper respiratory system become blocked. To test for a sleep disorder, the doctor referred Justin to the local sleep center for a polysomnogram. This test examines blood oxygen levels during sleep, electrical patterns in the brain, eye movement, heart and breathing rate, and the sleeping position of the individual.

As you read through the chapter, think about the following questions:

1. What structures of the upper respiratory system might contribute to obstructive sleep apnea?
2. Why does the polysomnogram monitor brain function as well as breathing and heart rates?
3. Why might Justin's weight gain and occasional alcohol use contribute to the condition?

9.1 The Respiratory System

LEARNING OUTCOMES

Upon completion of this section, you should be able to

1. Summarize the role of the respiratory system in the body.
2. Distinguish between inspiration and expiration.
3. Identify the structures of the human respiratory system.

The organs of the **respiratory system** ensure that oxygen enters the body and carbon dioxide leaves the body (Fig. 9.1). During **inspiration,** or inhalation (breathing in), air is conducted from the atmosphere to the lungs by a series of cavities, tubes, and openings, illustrated in Figure 9.1. During **expiration,** or exhalation (breathing out), air is conducted from the lungs to the atmosphere by way of the same structures.

Ventilation is another term for breathing that includes both inspiration and expiration. Once ventilation has occurred, the respiratory system depends on the cardiovascular system to transport oxygen (O_2) from the lungs to the tissues and carbon dioxide (CO_2) from the tissues to the lungs.

Gas exchange is necessary because the cells of the body carry out cellular respiration to make energy in the form of ATP. During cellular respiration, cells use up O_2 and produce CO_2. The respiratory system provides these cells with O_2 and removes CO_2.

MP3 Respiratory Structure and Function

CHECK YOUR PROGRESS 9.1

1. Trace the path of air from the nasal cavities to the lungs.
2. Explain how the flow of air differs during *inspiration* and *expiration.*
3. Describe the function(s) of the respiratory system.

CONNECTING THE CONCEPTS

The respiratory and circulatory systems cooperate extensively to maintain homeostasis in the body. For more on the interactions of these two systems, refer to the following discussions:

Section 5.5 outlines the circulatory pathways that move gases to and from the lungs.

Section 6.2 describes the role of the red blood cells in the transport of gases.

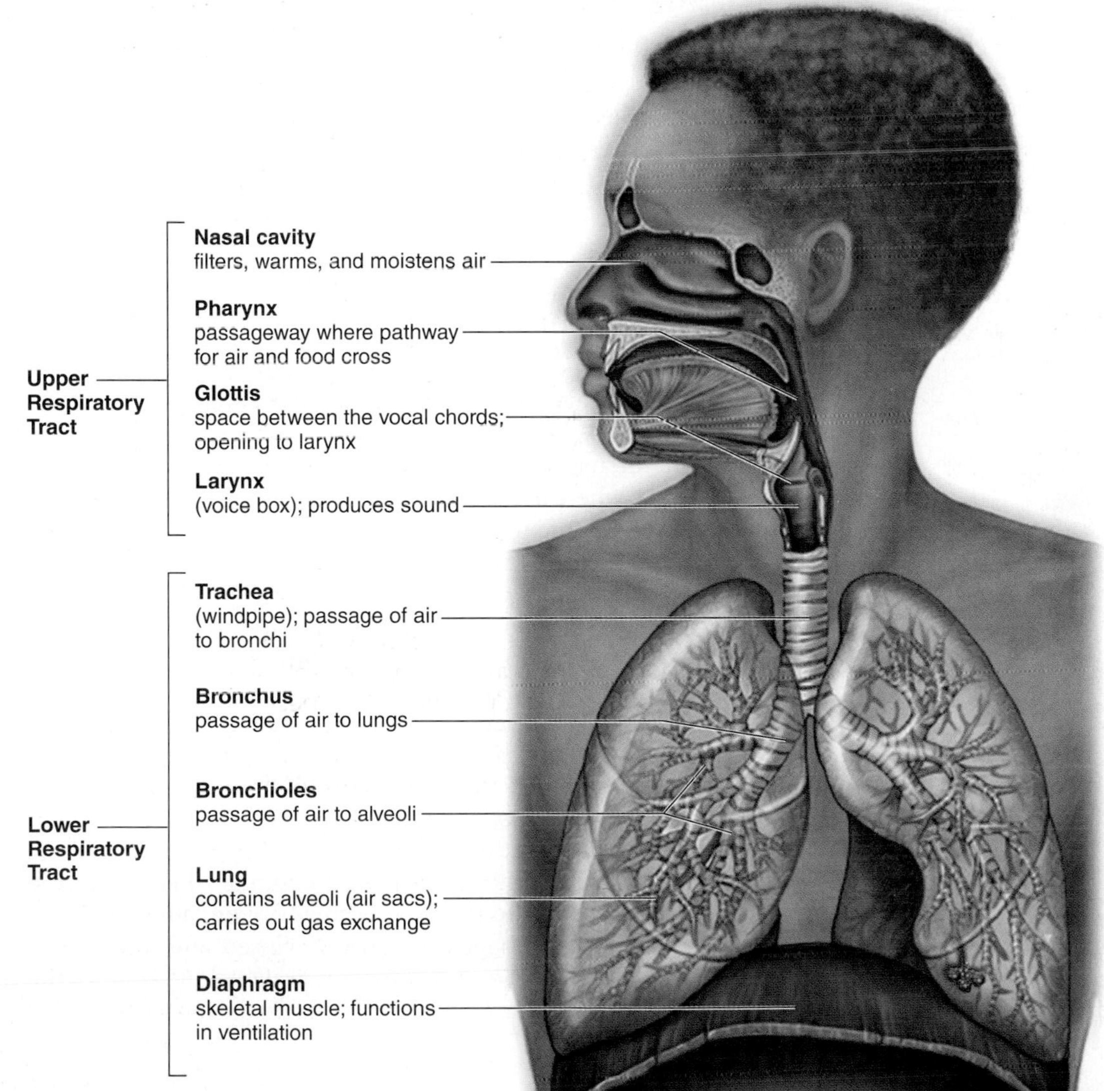

Figure 9.1 **The human respiratory tract.** The respiratory tract extends from the nose to the lungs. Note the organs in the upper respiratory tract and the ones in the lower respiratory tract.

9.2 The Upper Respiratory Tract

LEARNING OUTCOMES

Upon completion of this section, you should be able to

1. Summarize the role of the nose, pharynx, and larynx in respiration.
2. Identify the structures of the upper respiratory system and provide their function.
3. Explain how sound is produced by the larynx.

The nasal cavities, pharynx, and larynx are the organs of the upper respiratory tract (Fig. 9.2).

The Nose

The nose opens at the nares (nostrils) that lead to the **nasal cavities.** The nasal cavities are narrow canals separated from each other by a septum composed of bone and cartilage (Fig. 9.2).

Air entering the nasal cavities is met by large stiff hairs that act as a screening device. The hairs filter the air and trap small particles (dust, mold spores, pollen, etc.) so they don't enter air passages. The rest of the nasal cavities are lined by mucous membrane. The mucus secreted by this membrane helps trap dust and move it to the pharynx, where it can be swallowed or expectorated by coughing or spitting. Under the mucous layer is the submucosa. The submucosa contains a large number of capillaries that help warm and moisten the incoming air. When we breathe out on a cold day, the moisture in the outgoing air condenses so that we can see our breath. The abundance of capillaries in the submucosa also makes us susceptible to nosebleeds if the nose suffers an injury.

In the narrow upper recesses of the nasal cavities are special ciliated cells that act as odor receptors. Nerves lead from these cells to the brain, where the impulses generated by the odor receptors are interpreted as smell (see section 14.3).

The tear (lacrimal) glands drain into the nasal cavities by way of tear ducts. When you cry, your nose runs as tears drain from the eye surface into the nose. The nasal cavities also connect with the sinuses (cavities) of the skull. At times, fluid may accumulate in these sinuses, causing an increase in pressure, resulting in a sinus headache.

Air in the nasal cavities passes into the nasopharynx, the upper portion of the pharynx. Connected to the nasopharynx are tubes called **auditory tubes** (also called **eustachian tubes**) that connect to the middle ear. When air pressure inside the middle ears equalizes with the air pressure in the nasopharynx, the auditory tube openings may create a "popping" sensation. When in a plane taking off or landing, some people chew gum or yawn in an effort to move air from the ears and prevent the popping.

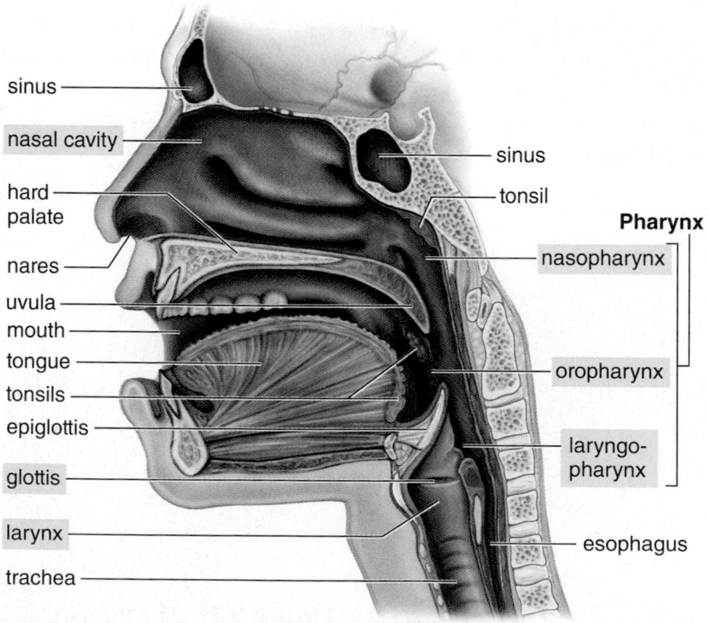

Figure 9.2 The upper respiratory tract.
This drawing shows the path of air from the nasal cavities to the trachea, in the lower respiratory tract. The designated structures are in the upper respiratory tract.

APPLICATIONS AND MISCONCEPTIONS

Is stifling a sneeze harmful?

Scientists have found that air travels at 100 miles an hour during a sneeze. This is enough force to propel sneeze droplets up to 12 ft away from the person sneezing. If a sneeze is stifled, the air is forced into the auditory (eustachian) tube and middle ear, potentially causing damage to the middle ear.

The Pharynx

The **pharynx,** commonly referred to as the throat, is a funnel-shaped passageway that connects the nasal and oral cavities to the larynx. The pharynx has three parts: the nasopharynx, where the nasal cavities open above the soft palate; the oropharynx, where the oral cavity opens; and the laryngopharynx, which opens into the larynx.

The **tonsils** form a protective ring at the junction of the oral cavity and the pharynx. The tonsils contain lymphocytes, which protect against invasion of inhaled foreign antigens. The tonsils are the primary defense during breathing, because inhaled air passes directly over the tonsils. In the tonsils, B cells and T cells are prepared to respond to antigens that may subsequently invade internal tissues and fluids. Therefore, the respiratory tract assists the immune system in maintaining homeostasis.

In the pharynx, the air passage and the food passage lie parallel to each other and share a common opening in the laryngopharynx. The larynx is normally open, allowing air to pass, but the esophagus is normally closed and opens only when a person swallows. If someone swallows and some of the food enters the larynx, coughing occurs in an effort to dislodge

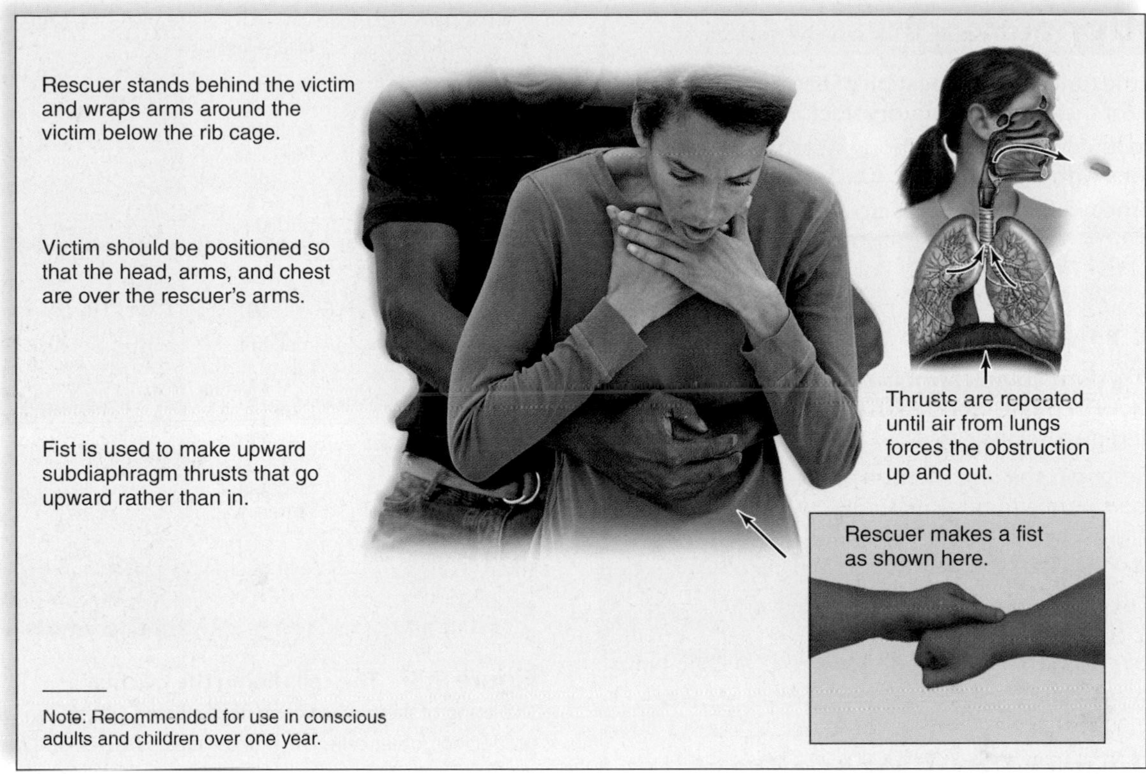

Rescuer stands behind the victim and wraps arms around the victim below the rib cage.
Victim should be positioned so that the head, arms, and chest are over the rescuer's arms.
Fist is used to make upward subdiaphragm thrusts that go upward rather than in.
Thrusts are repeated until air from lungs forces the obstruction up and out.
Rescuer makes a fist as shown here.
Note: Recommended for use in conscious adults and children over one year.

Figure 9.3 **The Heimlich maneuver.**
The Heimlich maneuver is used when someone's ability to breathe is prevented by an obstruction blocking the airway. The steps associated with the Heimlich maneuver are shown in the illustration.

the food. If the passageway remains blocked, the Heimlich maneuver (Fig. 9.3) may be used to dislodge food blocking the **epiglottis** and airway.

The Larynx

The **larynx** is a cartilaginous structure that serves as a passageway for air between the pharynx and the trachea. The larynx can be pictured as a triangular box whose apex, the Adam's apple (or laryngeal prominence), is located at the front of the neck. The larynx is also called the *voice box* because it houses the vocal cords. The **vocal cords** are mucosal folds supported by elastic ligaments, and the slit between the vocal cords is called the **glottis** (Fig. 9.4). When air is expelled through the glottis, the vocal cords vibrate, producing sound. At the time of puberty, the growth of the larynx and the vocal cords is much more rapid and accentuated in the male than in the female, causing the male to have a more prominent Adam's apple and a deeper voice. The voice "breaks" in the young male due to his inability to control the longer vocal cords.

The high or low pitch of the voice is regulated when speaking and singing by changing the tension on the vocal cords. The greater the tension, as when the glottis becomes narrower, the higher the pitch. When the glottis is wider, the pitch is lower (Fig. 9.4, *right*). The loudness or intensity of the voice depends upon the amplitude of the vibrations—the degree to which the vocal cords vibrate.

Ordinarily, when food is swallowed, the larynx moves upward against the epiglottis, a flap of tissue that prevents food from passing into the larynx. You can detect the movement of the larynx by placing your hand gently on your larynx and swallowing.

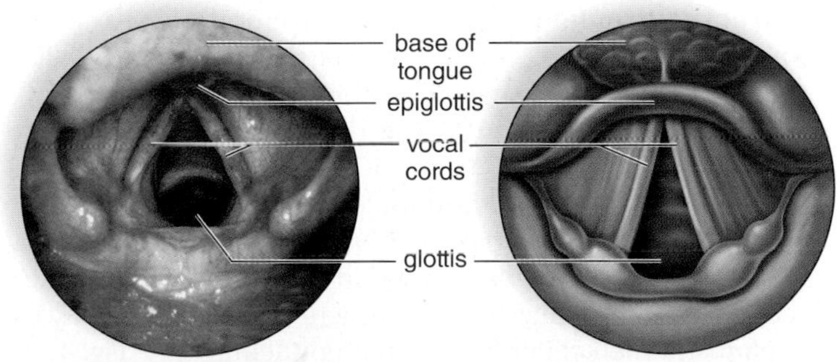

Figure 9.4 **The vocal cords.**
Viewed from above, the vocal cords can be seen to stretch across the glottis, the opening to the trachea. When air is expelled through the glottis, the vocal cords vibrate, producing sound. The glottis is narrow when we produce a high-pitched sound, and it widens as the pitch deepens.

CHECK YOUR PROGRESS 9.2

1. Describe the function and location of each of the structures of the upper respiratory tract.
2. Name and briefly describe the body systems that have connections with the pharynx.
3. Describe the two pathways that cross in the pharynx.

CONNECTING THE CONCEPTS

Several organ systems interact within the region of the upper respiratory tract. For more information on these systems, refer to the following discussions:

Section 7.2 describes the organization of the lymphatic system and the role of lymphatic tissue.

Section 8.2 illustrates the connection between the upper digestive and upper respiratory systems.

Section 14.3 examines how the respiratory system contributes to the sense of smell.

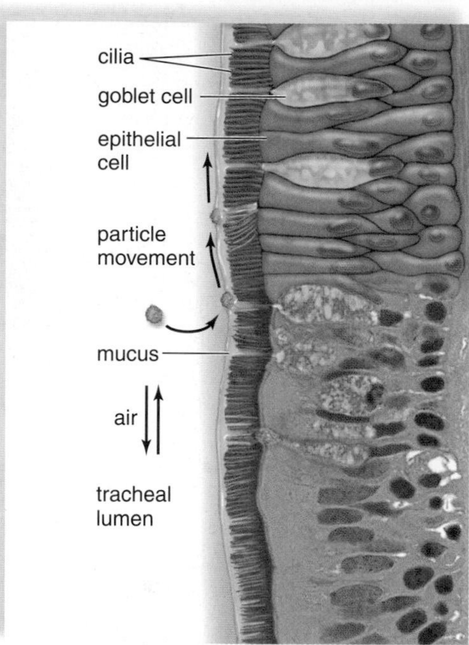

Figure 9.5 The cells lining the trachea. The lining of the trachea consists of ciliated epithelium with mucus-producing goblet cells. The mucus traps particles, and the cilia help move the mucus toward the throat.

9.3 The Lower Respiratory Tract

LEARNING OUTCOMES

Upon completion of this section, you should be able to

1. Summarize the role of the trachea, bronchial tree, and lungs in respiration.
2. Identify the structures of the lower respiratory system and provide their function.
3. Explain how the alveoli increase the efficiency of the respiratory system.

Once the incoming air makes its way past the larynx, it enters the lower respiratory tract. The lower respiratory tract consists of the trachea, the bronchial tree, and the lungs.

The Trachea

The **trachea,** commonly called the windpipe, is a tube connecting the larynx to the primary bronchi. Its walls consist of connective tissue and smooth muscle reinforced by C-shaped cartilaginous rings. The rings prevent the trachea from collapsing.

The trachea lies anterior to the esophagus. It is separated from the esophagus by a flexible muscular wall. This orientation allows the esophagus to expand when swallowing. The mucous membrane that lines the trachea has an outer layer of pseudostratified ciliated columnar epithelium (see Fig. 4.8) and goblet cells. The goblet cells produce mucus, which traps debris in the air as it passes through the trachea. The mucus is then swept toward the pharynx and away from the lungs by the cilia that project from the epithelium (Fig. 9.5).

When one coughs, the tracheal wall contracts, narrowing its diameter. Therefore, coughing causes air to move more rapidly through the trachea, helping to expel mucus and foreign objects. Smoking is known to destroy the cilia; consequently, the soot in cigarette smoke collects in the lungs. Smokers often develop heavy coughs as a result. Smoking is discussed more fully in the Health feature, "Questions About Smoking, Tobacco, and Health," on page 213.

If the trachea is blocked because of illness or the accidental swallowing of a foreign object, a breathing tube can be inserted by way of an incision made in the trachea. This tube acts as an artificial air intake and exhaust duct. The operation is called a **tracheostomy.**

The Bronchial Tree

The trachea divides into right and left primary **bronchi** (sing., bronchus), which lead into the right and left lungs (see Fig. 9.1). The bronchi branch into a few secondary bronchi that also branch, until the branches become about 1 mm in diameter and are called **bronchioles.** The bronchi resemble the trachea in structure. As the bronchial tubes divide and subdivide, their walls become thinner, and the small rings of cartilage are no longer present. During an asthma attack, the smooth muscle of the bronchioles contracts, causing bronchiolar constriction and characteristic wheezing. Each bronchiole leads to an elongated space enclosed by a multitude of air pockets or sacs called **alveoli** (sing., alveolus) (Fig. 9.6).

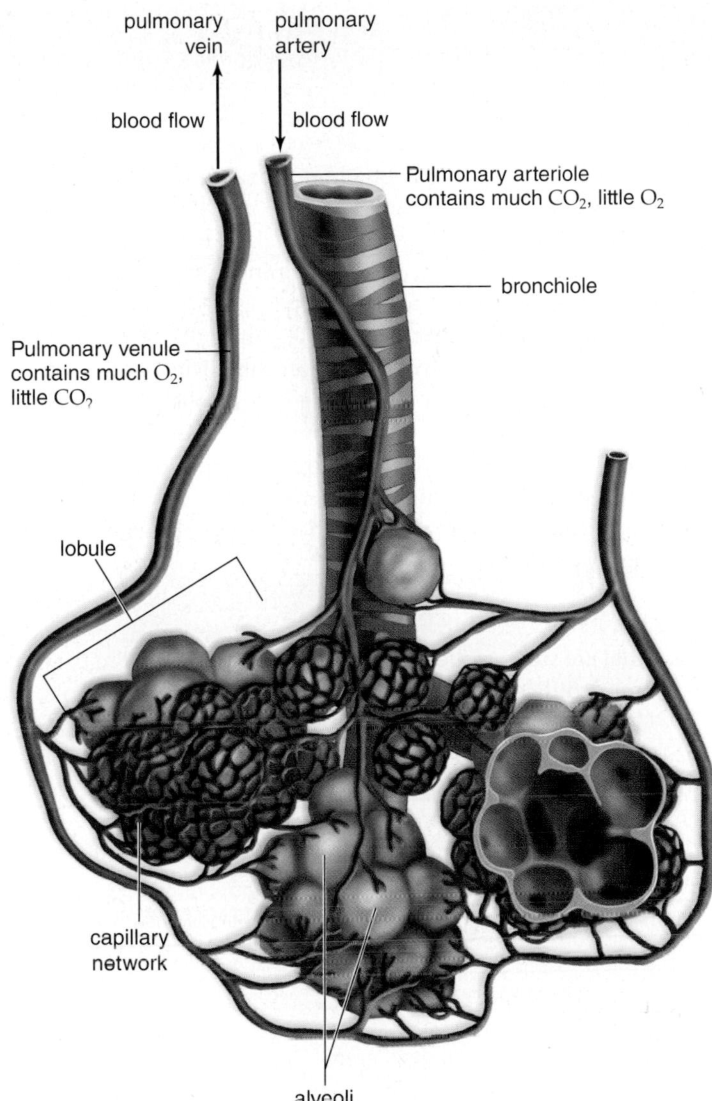

Figure 9.6 Pulmonary circulation to and from the lungs. The lungs consist of alveoli surrounded by an extensive capillary network. The pulmonary artery carries O_2-poor (CO_2-rich) blood (colored blue), and the pulmonary vein carries O_2-rich blood (colored red).

The Lungs

The **lungs** are paired, cone-shaped organs in the thoracic cavity. In the center of the thoracic cavity are the trachea, heart, thymus, and esophagus. The lungs are on either side of the trachea. The right lung has three lobes, and the left lung has two lobes, allowing room for the heart, which points left. Each lobe is further divided into lobules, and each lobule has a bronchiole serving many alveoli.

The lungs follow the contours of the thoracic cavity, including the diaphragm, the muscle that separates the thoracic cavity from the abdominal cavity. Each lung is enclosed by pleurae (sing., **pleura**), two layers of serous membrane that produce serous fluid. The parietal pleura adheres to the thoracic cavity wall, and the visceral pleura adheres to the surface of the lung. *Surface tension* is the tendency for water molecules to cling to one another due to hydrogen bonding between molecules. Surface tension holds the two pleural layers together. Therefore, the lungs must follow the movement of the thorax when breathing occurs. Pleurisy occurs when these layers become inflamed. Breathing, sneezing, and coughing are quite painful because the layers rub against each other. Causes of plueurisy include viral infections (such as the flu), tuberculosis, and pneumonia.

The Alveoli

The lungs have about 300 million alveoli, with a total cross-sectional area of 50–70 m^2. That's about the size of a tennis court. Each alveolar sac is surrounded by blood capillaries. The walls of the sac and the capillaries are largely simple squamous epithelium (see Fig. 4.8). Gas exchange occurs between air in the alveoli and blood in the capillaries. Oxygen diffuses across the alveolar wall and enters the bloodstream, and carbon dioxide diffuses from the blood across the alveolar wall to enter the alveoli (Fig. 9.6). The Science feature, "Artificial Lung Technology," on page 202, reviews how scientists are now able to generate artificial alveoli and lungs to aid in medical research.

The alveoli of human lungs are lined with a **surfactant,** a film of lipoprotein that lowers the surface tension of water and prevents the alveoli from closing. The lungs collapse in some newborn babies—especially premature infants—who lack this film. The condition, called *infant respiratory distress syndrome,* is now treatable by surfactant replacement therapy.

CHECK YOUR PROGRESS 9.3

1. Briefly describe the functions of the organs of the lower respiratory system.
2. Detail the structures of the lower respiratory system that participate in gas exchange.
3. Discuss what might occur to overall homeostasis if the alveoli did not function properly.

CONNECTING THE CONCEPTS

For more information on the structures of the thoracic cavity, refer to the following discussions:

Section 2.2 explains the role of hydrogen bonds in the properties of water.

Section 7.2 describes the function of the thymus as an organ of the lymphatic system.

Section 8.2 summarizes the role of the esophagus in the digestive system.

BIOLOGY MATTERS  Science

Artificial Lung Technology

Some organs, such as the kidneys, are relatively easy to transplant from one well-matched individual to another, with a high rate of success. Lungs are more difficult to transplant, however, with a ten-year survival rate of only 10–20%. Some very recent research is showing how one day it may be possible to replace diseased lungs with laboratory-grown versions.

In early 2010, a group of scientists at Yale University anesthetized a group of adult rats, surgically removed their left lungs, and replaced them with tissue-engineered lungs that had been produced in the laboratory. For periods of up to 2 hours, the implanted lung tissue exchanged oxygen and carbon dioxide at similar rates as the natural lungs.*

In order to build the artificial rat lungs, the Yale researchers began by removing lungs from adult rats, and treating the organs to remove most of the cellular components, while preserving the airways and supporting connective tissue matrix. Next, they placed these decellularized structures, along with various types of cultured cells, into a sterile container designed to mimic some aspects of the fetal environment in which the lungs normally first develop. The scientists found that the cells were able to form much of the lung tissues as well as the blood vessels needed to supply the tissue with blood and transport gases.

Although these results are an important early step toward growing replacement lungs in the lab, there is a long way to go before anyone will contemplate implanting engineered lungs into humans.

In a separate study published on the same day as the Yale study, a Harvard group announced that they had developed a pea-sized device that mimics human lung tissues. Made of human lung cells, a permeable membrane, plus blood capillary cells, all mounted on a microchip (Fig. 9A), the device is able to mimic the function of alveoli. When the researchers placed bacteria on the alveolar side of the device, and white blood cells on the capillary side, the blood cells crossed the membrane, mimicking an immune response.† At a minimum, the researchers hope that their "lung-on-a-chip" device can be used for testing certain drugs or the effects of various toxins on the lungs, which might replace much of the animal testing that is currently performed.

Questions to Consider

1. What are some of the aspects of lung structure that make it a more difficult organ to grow in the lab than, for example, a urinary bladder?
2. Besides increased availability, what are two other potential advantages of laboratory-grown lungs (or other tissues) compared to regular donor tissues?

* Petersen, T. H., et al. "Tissue-Engineered Lungs for in vivo Implantation," *Science* 329:538–41 (2010).

† Huh. D., et al. "Reconstituting Organ-Level Lung Functions on a Chip," *Science* 328:1662–68 (2010).

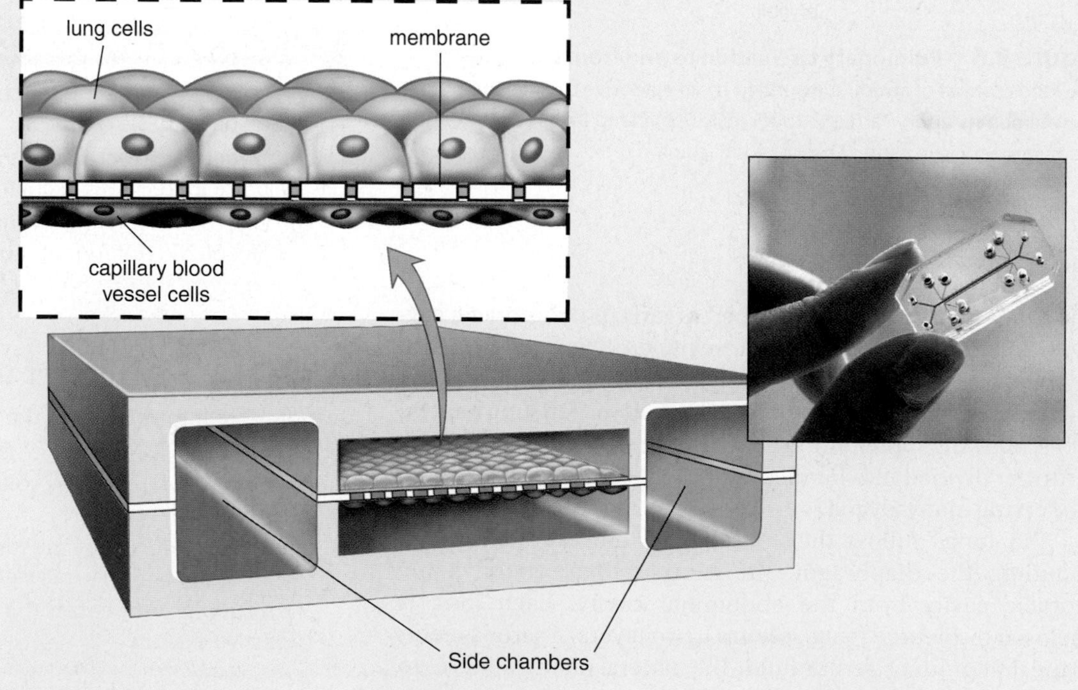

Figure 9A Lung on a chip.
About the size of a credit card, the device consists of a semipermeable membrane with lung cells on one side and blood vessel cells on the other. Liquid medium flows in a channel on the blood cell side, while air flows in and out on the lung cell side.

9.4 Mechanism of Breathing

LEARNING OUTCOMES

Upon completion of this section, you should be able to

1. Contrast the processes of inspiration and expiration during ventilation.
2. Define the terms *tidal volume, vital capacity,* and *residual volume* in relation to ventilation.
3. Summarize the purpose of the inspiratory and expiratory reserve volumes.

Ventilation, or breathing, has two phases. The process of inspiration, also called inhalation, moves air into the lungs; the process of expiration, also called exhalation, moves air out of the lungs. To understand ventilation, the manner in which air enters and exits the lungs, it is necessary to remember the following facts:

1. Normally, there is a continuous column of air from the pharynx to the alveoli of the lungs.
2. The lungs lie within the sealed thoracic cavity. The rib cage, consisting of the ribs joined to the vertebral column posteriorly and to the sternum anteriorly, forms the top and sides of the thoracic cavity. The intercostal muscles lie between the ribs. The diaphragm and connective tissue form the floor of the thoracic cavity.
3. The lungs adhere to the thoracic wall by way of the pleura. Any space between the two pleurae is minimal due to the surface tension of the fluid between them.

Inspiration

Inspiration is the active phase of ventilation because this is the phase in which the diaphragm and the external intercostal muscles contract (Fig. 9.7*a*). In its relaxed state, the diaphragm is dome-shaped. During inspiration, it contracts and becomes a flattened sheet of muscle. Also, the external intercostal muscles contract, causing the rib cage to move upward and outward.

Following contraction of the diaphragm and the external intercostal muscles, the volume of the thoracic cavity is larger than it was before. As the thoracic volume increases, the lungs

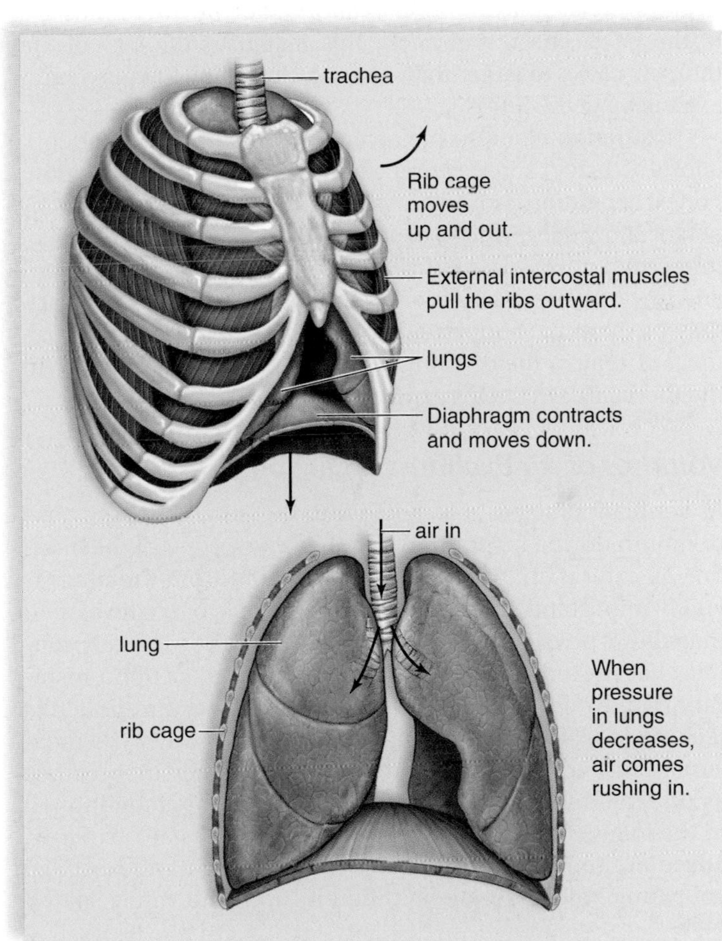

a. Inspiration

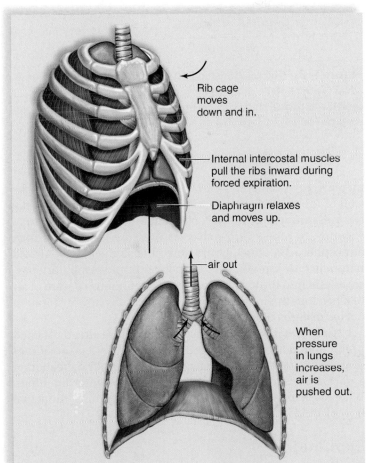

b. Expiration

Figure 9.7 The thoracic cavity during inspiration and expiration.
a. During inspiration, the thoracic cavity and lungs expand so that air is drawn in. **b.** During expiration, the thoracic cavity and lungs resume their original positions and pressures. Now air is forced out.

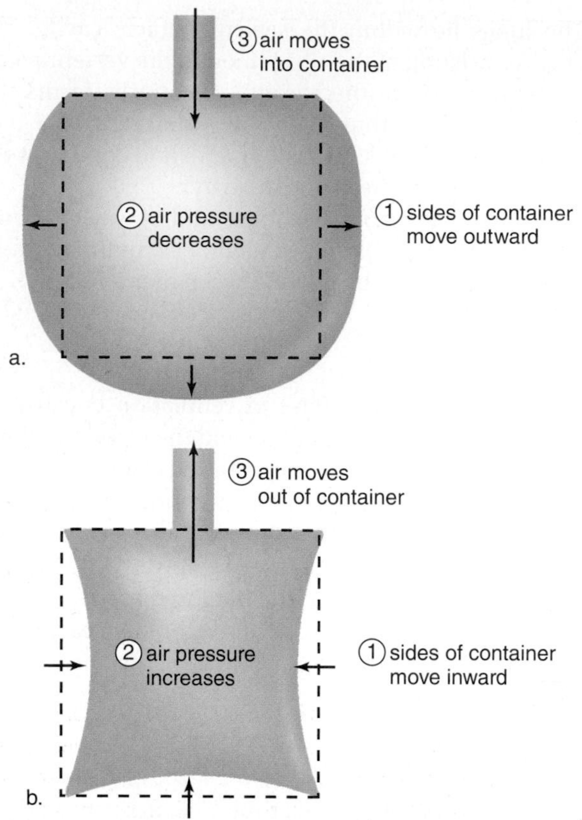

Figure 9.8 The relationship between air pressure and volume. When the sides of the container move outward, the volume in the container increases and the air pressure decreases. When the sides of the container collapse, the volume decreases and air pressure increases.

increase in volume as well because the lung adheres to the wall of the thoracic cavity. As the lung volume increases, the air pressure within the alveoli decreases, creating a partial vacuum. In other words, alveolar pressure is now less than atmospheric pressure (air pressure outside the lungs). Air will naturally flow from outside the body into the respiratory passages and into the alveoli, because a continuous column of air reaches into the lungs. Figure 9.8*a* illustrates how inspiration occurs using a container as a model.

Air comes into the lungs because they have already opened up; air does not force the lungs open. This is why it is sometimes said that *humans inhale by negative pressure.* The creation of a partial vacuum in the alveoli causes air to enter the lungs. Whereas inspiration is the active phase of breathing, the actual flow of air into the alveoli is passive.

Expiration

Usually, expiration is the passive phase of breathing, and no effort is required to bring it about. During expiration, the diaphragm and external intercostal muscles relax. The rib cage returns to its resting position, moving down and inward (see Fig. 9.7*b*). The elastic properties of the thoracic wall and lung tissue help them to recoil. In addition, the lungs recoil because the surface tension of the fluid lining the alveoli tends to draw them closed. Returning to the container model (Fig. 9.8*b*), if the sides and bottom of the container relax, the volume of the container decreases, the air pressure inside increases. As a result, the air flows out.

What keeps the alveoli from collapsing as a part of expiration? Recall that the presence of surfactant lowers the surface tension within the alveoli. Also, as the lungs recoil, the pressure between the pleura decreases, and this tends to make the alveoli stay open. The importance of the reduced intrapleural pressure is demonstrated when in an accident the thoracic cavity is punctured (a "punctured lung"). Air now enters the intrapleural space, causing the lung to collapse.

Maximum Inspiratory Effort and Forced Expiration

If you recall the last time you exercised vigorously—perhaps running in a race, or even just climbing all those stairs to your classroom—you probably remember that you were breathing a lot harder than normal during and immediately after that heavy exercise. Maximum inspiratory effort involves muscles of the back, chest, and neck. This increases the size of the thoracic cavity to larger than normal, thus allowing maximum expansion of the lungs.

Expiration can also be forced. The maximum inspiratory efforts of heavy exercise are accompanied by forced expiration. Forced expiration is also necessary to sing, blow air into a trumpet, or blow out birthday candles. Contraction of the internal intercostal muscles can force the rib cage to move downward and inward. Also, when the abdominal wall muscles contract, they push on the abdominal organs. In turn, the organs push upward against the diaphragm and the increased pressure in the thoracic cavity helps expel air.

Volumes of Air Exchanged During Ventilation

As ventilation occurs, air moves into the lungs from the nose or mouth during inspiration and then moves out of the lungs during expiration. A free flow of air to and from the lungs is vitally important. Therefore, a technique has been developed that allows physicians to determine if there is a medical problem that prevents the lungs from filling with air upon inspiration and releasing it from the body upon expiration. This technique is illustrated in Figure 9.9, which shows the measurements recorded by a spirometer when a person breathes as directed by a technician. The actual numbers mentioned in the following discussion about lung volumes are averages. These numbers are affected by gender, height, and age, so your lung volumes may be different from the values stated here.

Tidal Volume Normally, when we are relaxed, only a small amount of air moves in and out with each breath, similar, perhaps, to the tide at the beach. This amount of air, called the **tidal volume,** is only about 500 ml.

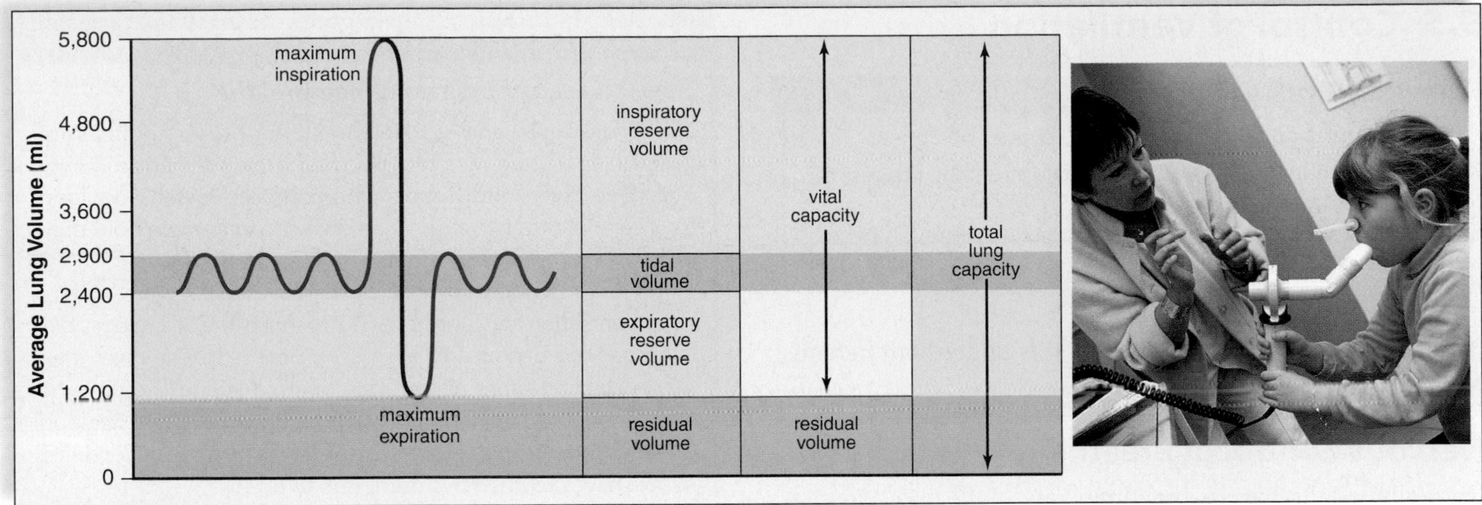

Figure 9.9 Measuring the vital capacity of the lungs.
A spirometer is an instrument that measures the amount of air inhaled and exhaled with each breath. During inspiration, there is an upswing, and during expiration, there is a downswing. Vital capacity (red) is measured by taking the deepest breath and then exhaling as much as possible.

Vital Capacity It is possible to increase the amount of air inhaled and, therefore, the amount exhaled by deep breathing. The maximum volume of air that can be moved in plus the maximum amount that can be moved out during a single breath is called the **vital capacity.** It is called vital capacity because your life depends on breathing, and the more air you can move, the better off you are. A number of different illnesses, such as pulmonary fibrosis (see section 9.7), can decrease vital capacity.

Inspiratory and Expiratory Reserve Volume As noted previously, we can increase inspiration by expanding the chest and also by lowering the diaphragm to the maximum extent possible. Forced inspiration (**inspiratory reserve volume**) usually adds another 2,900 ml of inhaled air, and that's quite a bit more than a tidal volume of only 500 ml!

We can increase expiration by contracting the abdominal and thoracic muscles. This so-called **expiratory reserve volume** is usually about 1,400 ml of air. You can see from Figure 9.9 that vital capacity is the sum of tidal, inspiratory reserve, and expiratory reserve volumes.

APPLICATIONS AND MISCONCEPTIONS

What happens when "the wind gets knocked out of you"?

The "wind may get knocked out of you" following a blow to the upper abdomen in the area of the stomach. There is a network of nerves in that region called the solar plexus. Trauma to this area can cause the diaphragm to experience a sudden, involuntary, and painful contraction called a spasm. It's not possible to breathe while the diaphragm experiences this spasm—hence, the feeling of being breathless. The pain and inability to breathe stop once the diaphragm relaxes.

Residual Volume It is a curious fact that some of the inhaled air never reaches the lungs; instead, it fills the nasal cavities, trachea, bronchi, and bronchioles (see Fig. 9.1). These passages are not used for gas exchange; therefore, they are said to contain **dead air space.** To ensure that newly inhaled air reaches the lungs, it is better to breathe slowly and deeply.

Also, note in Figure 9.9 that even after a very deep exhalation, some air (about 1,000 ml) remains in the lungs. This is called the **residual volume.** The residual volume is the amount of air that can't be exhaled from the lungs. In some lung diseases to be discussed later, the residual volume gradually increases because the individual has difficulty emptying the lungs. Increased residual volume will cause the expiratory reserve volume to be reduced. As a result, vital capacity is decreased as well.

MP3 Ventilation

CHECK YOUR PROGRESS 9.4

1. Explain how the volume (size) of the thoracic cavity affects the pressure in the lungs.
2. Distinguish between the different volumes of air exchanged during ventilation and describe when each is used.
3. Discuss what effect insufficient expiration might have on overall homeostasis.

CONNECTING THE CONCEPTS

To understand how the lungs develop as we age, refer to the following discussions:

Section 17.3 describes the development of the lungs in the fetus.

Section 17.5 examines the effects of aging on the ability of the lungs to exchange gases.

9.5 Control of Ventilation

LEARNING OUTCOMES

Upon completion of this section, you should be able to

1. Explain how the nervous system controls the process of breathing.
2. Explain the role of chemoreceptors and pH levels in regulating breathing rate.

Breathing is controlled in two ways. It is under both nervous and chemical control.

Nervous Control of Breathing

Normally, adults have a breathing rate of 12 to 20 ventilations per minute. The rhythm of ventilation is controlled by a **respiratory control center** located in the medulla oblongata of the brain. The respiratory control center automatically sends out nerve signals to the diaphragm and the external intercostal muscles of the rib cage, causing inspiration to occur (Fig. 9.10). When the respiratory center stops sending nerve signals to the diaphragm and the rib cage, the muscles relax and expiration occurs.

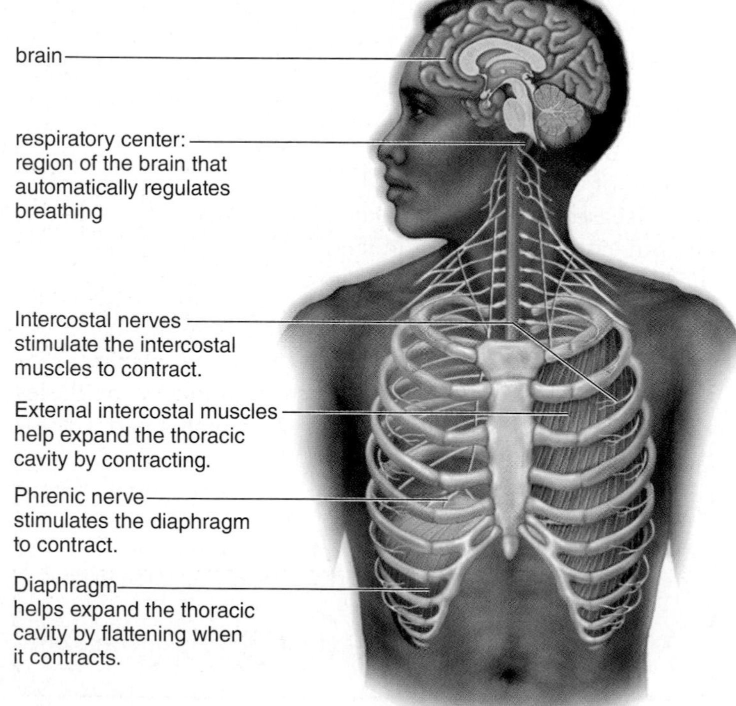
brain
respiratory center: region of the brain that automatically regulates breathing
Intercostal nerves stimulate the intercostal muscles to contract.
External intercostal muscles help expand the thoracic cavity by contracting.
Phrenic nerve stimulates the diaphragm to contract.
Diaphragm helps expand the thoracic cavity by flattening when it contracts.

Figure 9.10 The control of breathing by the respiratory center. During inspiration, the respiratory center, located in the medulla oblongata, stimulates the external intercostal (rib) muscles to contract via the intercostal nerves and stimulates the diaphragm to contract via the phrenic nerve. The thoracic cavity and then the lungs expand and air comes rushing in. Expiration occurs due to a lack of stimulation from the respiratory center to the diaphragm and intercostal muscles. As the thoracic cavity and then the lungs resume their original size, air is pushed out.

APPLICATIONS AND MISCONCEPTIONS

How long can people hold their breath?

Most individuals can hold their breath for 1 to 2 minutes. With practice, many are able to hold their breath for up to 3 minutes. Free divers, individuals who compete to see how long they can hold their breath while diving in water, can hold their breath for 5 minutes or more.

Researchers hope to more closely investigate divers' use of hyperventilation and lung "packing" to make free diving possible. Lung packing begins with very deep breathing. The diver then "packs," adding more air by additional breathing through the mouth. The use of hyperventilation or lung packing to prolong breath holding should never be tried by untrained individuals. People have drowned by attempting to do so.

Sudden infant death syndrome (SIDS), or crib death, claims the life of about 2,500 infants a year in the United States. An infant under one year of age is put to bed seemingly healthy, and sometime while sleeping, the child stops breathing. Though the precise cause of SIDS is not known, scientists have ruled out vaccinations, vomiting, and infections as factors. Most research is focusing on miscommunication between the respiratory center of the brain and the lungs and, possibly, problems with heart function.

Although the respiratory center automatically controls the rate and depth of breathing, its activity can be influenced by nervous input. We can voluntarily change our breathing pattern to accommodate activities such as speaking, singing, eating, swimming under water, and so forth. Following forced inspiration, stretch receptors in the airway walls respond to increased pressure. These receptors initiate inhibitory nerve impulses. The impulses travel from the inflated lungs to the respiratory center. This temporarily stops the respiratory center from sending out nerve signals. In this manner, excessive stretching of the elastic tissue of the lungs is prevented.

Chemical Control of Breathing

As you know, working cells produce carbon dioxide, which enters the blood. There, carbon dioxide combines with water, forming an acid, which breaks down and gives off hydrogen ions (H^+). These hydrogen ions can change the pH of the blood. **Chemoreceptors** are sensory receptors in the body that are sensitive to chemical composition of body fluids. Two sets of chemoreceptors sensitive to pH can cause breathing to speed up. A centrally placed set is located in the medulla oblongata of the brain stem, and a peripherally placed set is in the circulatory system. Carotid bodies, located in the carotid arteries, and aortic bodies, located in the aorta, are sensitive to blood pH. These chemoreceptors are not strongly affected by low oxygen (O_2) levels. Instead, they are stimulated when the carbon dioxide entering the blood is sufficient to change blood pH.

When the pH of the blood becomes more acidic (decreases), the respiratory center increases the rate and depth of breathing. With an increased breathing rate, more carbon dioxide is removed from the blood. The hydrogen ion concentration returns to normal, and the breathing rate returns to normal.

Most people are unable to hold their breath for more than 1 minute. When you hold your breath, metabolically produced carbon dioxide begins accumulating in the blood. As a result, H^+ accumulates and the blood becomes more acidic. The respiratory center, stimulated by the chemoreceptors, is able to override a person's voluntary inhibition of respiration. Breathing resumes, despite attempts to prevent it.

MP3 Control of Respiration

CHECK YOUR PROGRESS 9.5

1. Explain why we automatically breathe 12 to 20 times a minute. Describe what might happen to homeostasis if those numbers were only 2 to 5 times a minute.
2. Describe the nervous system's control of the respiratory system.
3. Discuss why it's not possible to hold your breath for more than a minute or so.

CONNECTING THE CONCEPTS

For more on the structures that regulate breathing, refer to the following discussions:

Section 13.2 examines the location and function of the medulla oblongata.

Section 14.1 describes how chemoreceptors help maintain homeostasis.

9.6 Gas Exchanges in the Body

LEARNING OUTCOMES

Upon completion of this section, you should be able to

1. Distinguish between external and internal respiration.
2. Summarize the chemical processes that are involved in external and internal respiration.
3. Identify the role of carbonic anhydrase and carbaminohemoglobin in respiration.

Gas exchange is critical to homeostasis. Oxygen needed to produce energy must be supplied to all the cells, and carbon dioxide must be removed from the body during gas exchange. As mentioned previously, respiration includes the exchange of gases not only in the lungs but also in the tissues (Fig. 9.11).

The principles of diffusion govern whether O_2 or CO_2 enters or leaves the blood in the lungs and in the tissues. Gases exert pressure, and the amount of pressure each gas exerts is called its partial pressure, symbolized as P_{O_2} and P_{CO_2}. If the partial pressure of oxygen differs across a membrane, oxygen will diffuse from the higher to lower partial pressure.

Animation Changes in the Partial Pressure of Oxygen and Carbon Dioxide

External Respiration

External respiration refers to the exchange of gases between air in the alveoli and blood in the pulmonary capillaries (see Fig. 9.6 and Fig. 9.11*a*). Blood in the pulmonary capillaries has a higher P_{CO_2} than atmospheric air. Therefore, *CO_2 diffuses out of the plasma into the lungs.* Most of the CO_2 is carried in plasma as **bicarbonate ions** (HCO_3^-). In the low-P_{CO_2} environment of the lungs, the reaction proceeds to the right.

$$H^+ + HCO_3^- \longrightarrow H_2CO_3 \xrightarrow{\text{carbonic anhydrase}} H_2O + CO_2$$

hydrogen ion, bicarbonate ion, carbonic acid, water, carbon dioxide

The enzyme **carbonic anhydrase** speeds the breakdown of carbonic acid (H_2CO_3) in red blood cells.

What happens if you hyperventilate (breathe at a high rate) and therefore push this reaction far to the right? The blood will have fewer hydrogen ions, and alkalosis, a high blood pH, results. In that case, breathing is inhibited, and you may suffer from various symptoms ranging from dizziness to continuous contractions of the skeletal muscles. You may have heard that you should inhale and exhale from a paper bag after hyperventilating. Doing so increases the CO_2 in your blood, because you're inhaling the CO_2 you just exhaled into the bag. This restores a normal blood pH. What happens if you hypoventilate (breathe at a low rate) and this reaction does not occur? Hydrogen ions build up in the blood and acidosis occurs. Buffers may compensate for the low pH, and breathing most likely increases. Extreme changes in blood pH affect enzyme function, which may lead to coma and death.

The pressure pattern for O_2 during external respiration is the reverse of that for CO_2. Blood in the pulmonary capillaries is low in oxygen, and alveolar air contains a higher partial pressure of oxygen. Therefore, *O_2 diffuses into plasma and then into red blood cells in the lungs.* Hemoglobin takes up this oxygen and becomes **oxyhemoglobin** (HbO_2).

$$Hb + O_2 \longrightarrow HbO_2$$

deoxyhemoglobin, oxygen, oxyhemoglobin

Internal Respiration

Internal respiration refers to the exchange of gases between the blood in systemic capillaries and the tissue cells. In Figure 9.11*b*, internal respiration is shown at the bottom. Blood entering systemic capillaries is a bright red color because red blood cells contain oxyhemoglobin. The temperature in the tissues is higher and the pH is slightly lower (more acidic),

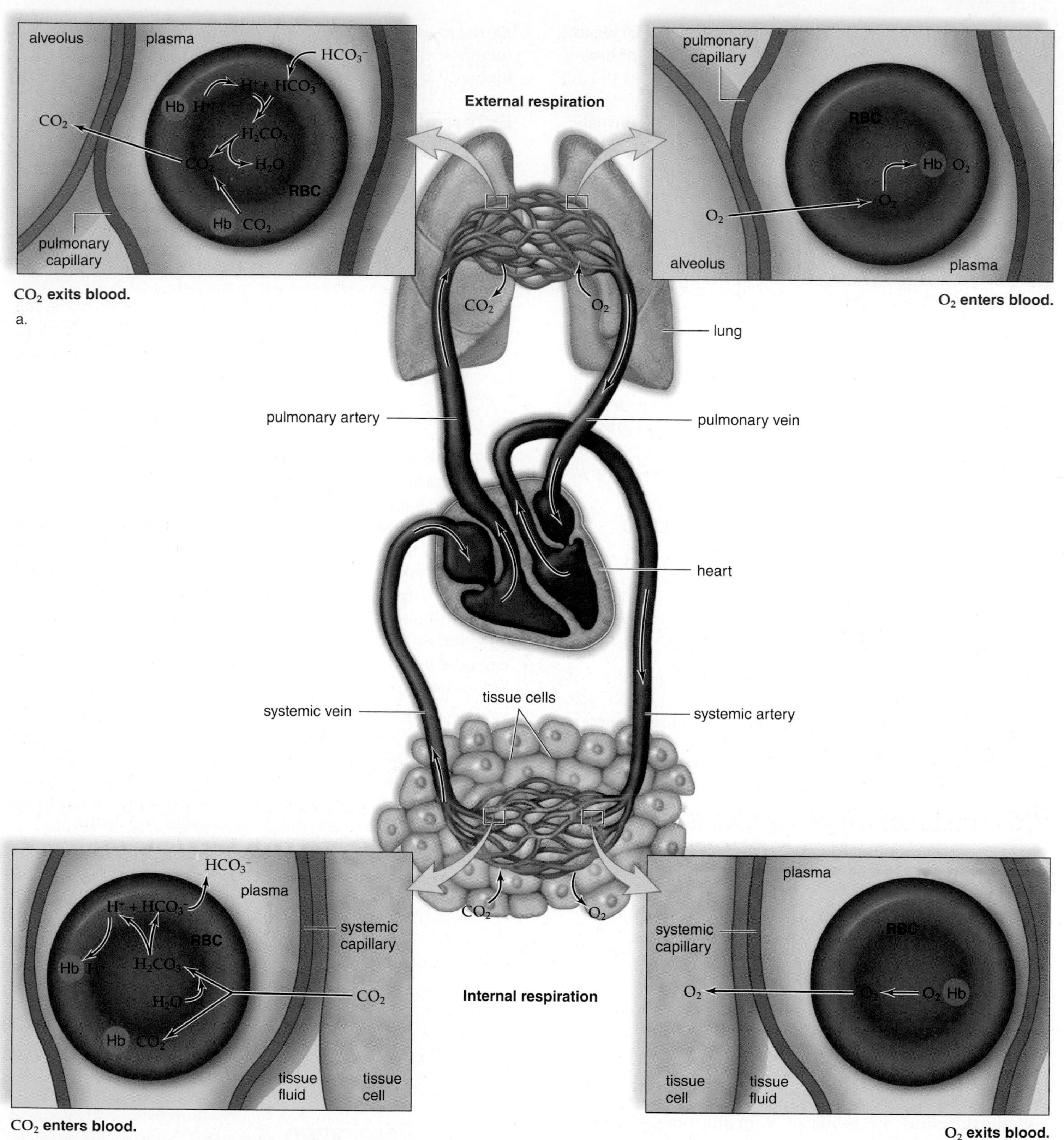

Figure 9.11 Movement of gases during external and internal respiration.
a. During external respiration in the lungs, HCO_3^- is converted to CO_2, which exits the blood. O_2 enters the blood and hemoglobin (Hb) carries O_2 to the tissues. **b.** During internal respiration in the tissues, O_2 exits the blood, and CO_2 enters the blood. Most of the CO_2 enters red blood cells, where it becomes the bicarbonate ion, carried in the plasma. Some hemoglobin combines with CO_2 and some combines with H^+.

so oxyhemoglobin naturally gives up oxygen. After oxyhemoglobin gives up O_2, it diffuses out of the blood into the tissues.

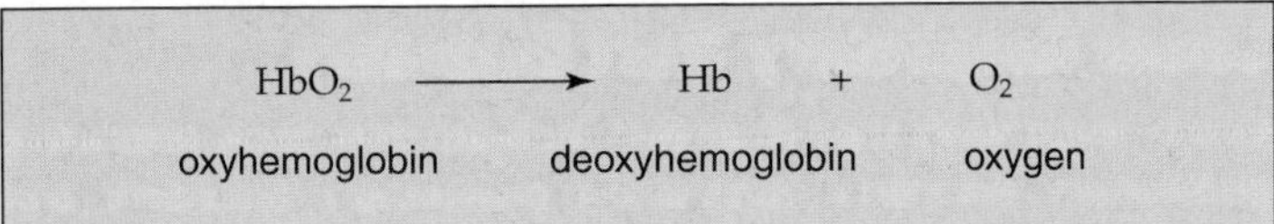

Oxygen diffuses out of the blood into the tissues because the P_{O_2} of tissue fluid is lower than that of blood. The lower P_{O_2} is due to cells continuously using up oxygen in cellular respiration (see Fig. 3.19). *Carbon dioxide diffuses into the blood from the tissues* because the P_{CO_2} of tissue fluid is higher than that of blood. Carbon dioxide is produced during cellular respiration and collects in tissue fluid.

After CO_2 diffuses into the blood, most enters the red blood cells, where a small amount is taken up by hemoglobin, forming **carbaminohemoglobin** ($HbCO_2$). In plasma, CO_2 combines with water, forming carbonic acid (H_2CO_3), which dissociates to hydrogen ions (H^+) and bicarbonate ions (HCO_3^-).

Animation Gas Exchange During Respiration

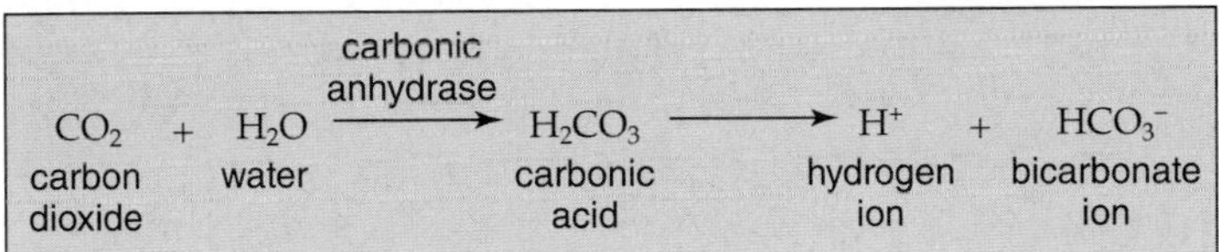

The enzyme carbonic anhydrase, mentioned previously, speeds the reaction in red blood cells. Bicarbonate ions (HCO_3^-) diffuse out of red blood cells and are carried in the plasma. The globin portion of hemoglobin combines with excess hydrogen ions produced by the overall reaction, and Hb becomes HHb, called *reduced hemoglobin*. In this way, the pH of blood remains fairly constant. Blood that leaves the systemic capillaries is a dark maroon color because red blood cells contain reduced hemoglobin.

MP3 Gas Exchange

CHECK YOUR PROGRESS 9.6

1. Describe the differences between external respiration and internal respiration.
2. Describe how hemoglobin functions in the transport of both oxygen and carbon dioxide.
3. Detail the influence of P_{O_2} on both external and internal respiration.

CONNECTING THE CONCEPTS

For more on the mechanisms by which respiration maintains homeostasis, refer to the following discussions:

Section 2.2 explores the relationship between H^+ concentration and pH.

Section 6.1 examines how components of the blood plasma help buffer the pH of the blood.

Section 6.2 provides additional information on hemoglobin and red blood cells.

9.7 Respiration and Health

LEARNING OUTCOMES

Upon completion of this section, you should be able to

1. Identify the symptoms and causes of selected upper respiratory tract infections.
2. Identify the symptoms and causes of selected lower respiratory tract disorders.
3. Summarize the relationship between smoking, cancer, and emphysema.

The respiratory tract is constantly exposed to environmental air. The quality of this air and whether it contains infectious pathogens, such as bacteria and viruses, or harmful chemicals, influences the overall health of the respiratory system.

Upper Respiratory Tract Infections

Upper respiratory infections (URIs) can spread from the nasal cavities to the sinuses, middle ears, and larynx. What we call "strep throat" is a primary bacterial infection caused by *Streptococcus pyogenes* that can lead to a generalized URI and even a systemic (affecting the body as a whole) infection. The symptoms of strep throat are severe sore throat, high fever, and white patches on a dark red throat. Because strep throat is bacterial, it can be treated successfully with antibiotics.

Sinusitis

Sinusitis develops when nasal congestion blocks the tiny openings leading to the sinuses (see Fig. 9.2). Symptoms include postnasal discharge and facial pain that worsens when the patient bends forward. Pain and tenderness usually occur over the lower forehead or over the cheeks (sometimes also producing toothaches). Successful treatment depends on restoring proper drainage of the sinuses. Even a hot shower and sleeping upright can be helpful. Nasal spray decongestants and oral antihistamines also alleviate the symptoms of sinusitis. Sprays may be preferred because they treat the symptoms without the side effects, such as drowsiness, of oral medicine. However, nasal sprays can become habit forming if used for long periods. Persistent sinusitis should be evaluated by a health-care professional.

Otitis Media

Otitis media is an infection of the middle ear. This infection is considered here because it is a complication often seen in children who have a nasal infection. Infection can spread by way of the auditory tube from the nasopharynx to the middle ear. Pain is the primary symptom of a middle-ear infection. A sense of fullness, hearing loss, vertigo (dizziness), and fever may also be present. Antibiotics are prescribed if necessary, but physicians are aware today that overuse of antibiotics can lead to resistance of bacteria to antibiotics (see section S.3). Tubes (called tympanostomy tubes) are sometimes placed in the eardrums of children with multiple recurrences to help prevent the buildup of pressure in the middle ear and reduce the possibility of hearing loss. Normally, the tubes fall out with time.

Tonsillitis

Tonsillitis occurs when the tonsils become inflamed and enlarged. The tonsil in the posterior wall of the nasopharynx is often called the adenoid. If tonsillitis occurs frequently and enlargement makes breathing difficult, the tonsils can be removed surgically in a **tonsillectomy.** Fewer tonsillectomies are performed today than in the past because we now know that the tonsils are lymphatic tissue and serve to trap many of the pathogens that enter the pharynx. Therefore, they are a first line of defense against invasion of the body.

Laryngitis

Laryngitis is an infection of the larynx with accompanying hoarseness, leading to the inability to talk in an audible voice. Usually, laryngitis disappears with treatment of the URI. Persistent hoarseness without the presence of a URI is one of the warning signs of cancer and should be looked into by a physician.

Lower Respiratory Tract Disorders

Lower respiratory tract disorders include infections, restrictive pulmonary disorders, obstructive pulmonary disorders, and lung cancer.

Lower Respiratory Infections

Acute bronchitis is an infection of the primary and secondary bronchi. Usually, it is preceded by a viral URI that has led to a secondary bacterial infection. Most likely, a nonproductive cough has become a deep cough that expectorates mucus and perhaps pus.

Pneumonia is a viral or bacterial infection of the lungs in which the bronchi and alveoli fill with thick fluid (Fig. 9.12). Most often, it is preceded by influenza. High fever and chills with headache and chest pain are symptoms of pneumonia. Rather than being a generalized lung infection, pneumonia may be localized in specific lobules of the lungs.

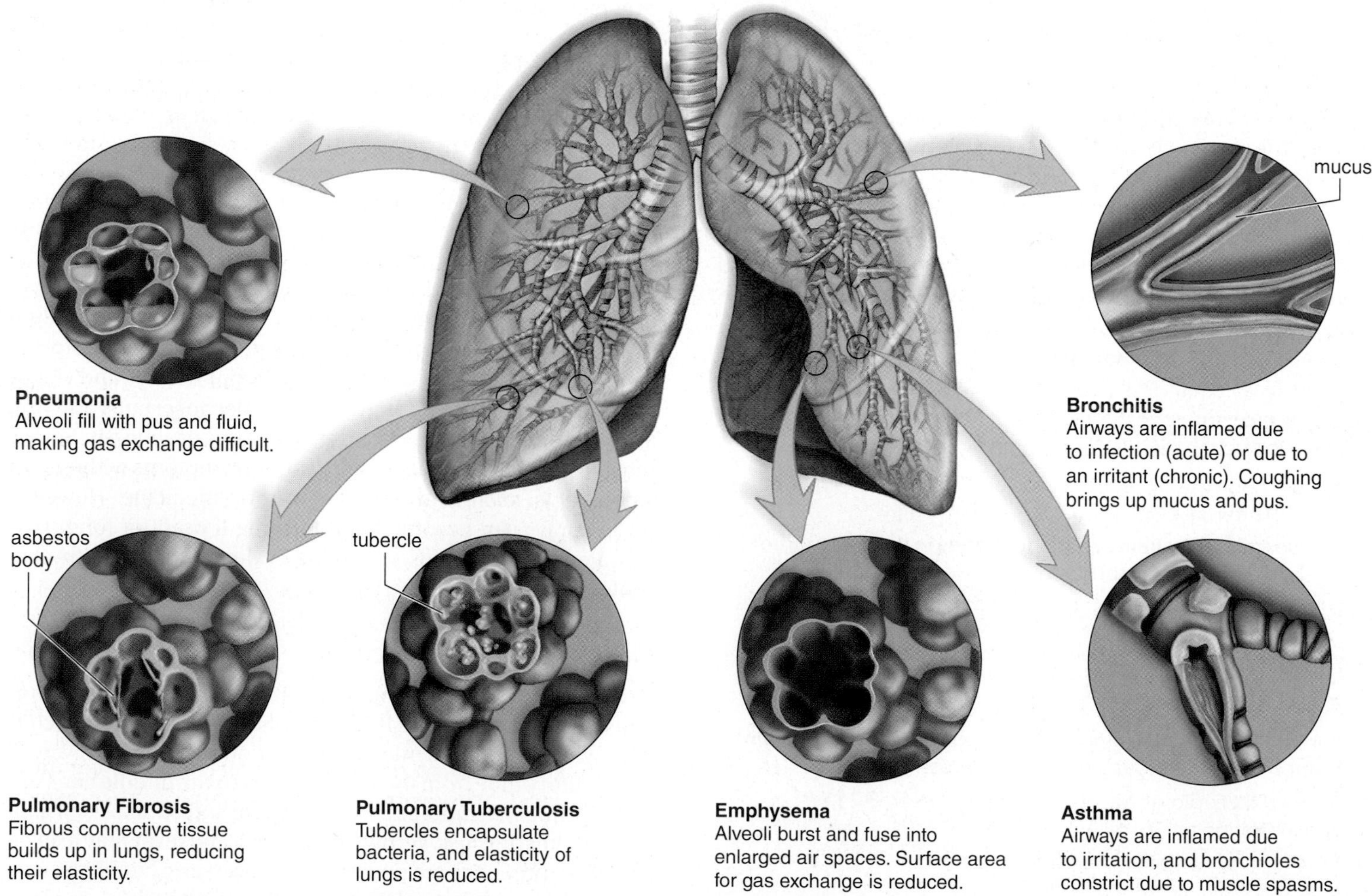

mucus
Pneumonia
Alveoli fill with pus and fluid, making gas exchange difficult.
Bronchitis
Airways are inflamed due to infection (acute) or due to an irritant (chronic). Coughing brings up mucus and pus.
asbestos body
tubercle
Pulmonary Fibrosis
Fibrous connective tissue builds up in lungs, reducing their elasticity.
Pulmonary Tuberculosis
Tubercles encapsulate bacteria, and elasticity of lungs is reduced.
Emphysema
Alveoli burst and fuse into enlarged air spaces. Surface area for gas exchange is reduced.
Asthma
Airways are inflamed due to irritation, and bronchioles constrict due to muscle spasms.

Figure 9.12 Some diseases and disorders of the respiratory system.
Exposure to infectious pathogens and/or polluted air, including tobacco smoke, causes the diseases and disorders shown here.

Obviously, the more lobules involved, the more serious is the infection. Pneumonia can be caused by a bacterium that is usually held in check but has gained the upper hand due to stress and/or reduced immunity. AIDS patients are subject to a particularly rare form of pneumonia caused by a fungus named *Pneumocystis jiroveci* (formerly *Pneumocystis carinii*). Pneumonia of this type is almost never seen in individuals with a healthy immune system.

Pulmonary tuberculosis, commonly called **tuberculosis,** is a bacterial disease that was called *consumption* at one time. When the bacteria (*Mycobacterium tuberculosis*) invade the lung tissue, the cells build a protective capsule around the foreigners, isolating them from the rest of the body. This tiny capsule is called a tubercle. If the resistance of the body is high, the imprisoned organisms die, but if the resistance is low, the organisms eventually can be liberated. If a chest X-ray detects active tubercles, the individual is put on appropriate drug therapy to ensure the localization of the disease and the eventual destruction of any live bacteria. It is possible to tell if a person has ever been exposed to tuberculosis with a *tuberculin* test. This procedure uses a highly diluted bacterial extract injected into the patient's skin. If a tuberculin test is positive, an X-ray will be done to confirm an active disease. A person may have a positive test but no active disease.

Restrictive Pulmonary Disorders

In restrictive pulmonary disorders, vital capacity is reduced because the lungs have lost their elasticity. Inhaling particles such as silica (sand), coal dust, asbestos, and fiberglass can lead to **pulmonary fibrosis,** a condition in which fibrous connective tissue builds up in the lungs. The lungs cannot inflate properly and are always tending toward deflation. Breathing asbestos is also associated with the development of cancer. Asbestos was formerly used widely as a fireproofing and insulating agent, so unwarranted exposure has occurred. It has been projected that 2 million deaths caused by asbestos exposure—mostly in the workplace—will occur in the United States between 1990 and 2020.

APPLICATIONS AND MISCONCEPTIONS

What is cystic fibrosis?

Cystic fibrosis is a genetic disease in which chloride ion transporters in mucus-producing epithelial cells do not function correctly. Because of this, the mucous secretions are often very thick and easily clog small structures such as the alveoli. Though cystic fibrosis is usually considered to be just a disease of the respiratory system, because this is where the symptoms usually first appear, it can also cause problems in organs such as the pancreas. Just a few decades ago, people with cystic fibrosis usually died before the age of 20. New advances in medicine and treatments now enable individuals to live into their thirties and forties.

Video
Good
Poison

Obstructive Pulmonary Disorders

In obstructive pulmonary disorders, air does not flow freely in the airways and the time it takes to inhale or exhale maximally is greatly increased. Several disorders, including chronic bronchitis, emphysema, and asthma, are collectively referred to as chronic obstructive pulmonary disease (COPD) because they tend to recur.

In **chronic bronchitis,** the airways are inflamed and filled with mucus. A cough that brings up mucus is common. The bronchi have undergone degenerative changes, including the loss of cilia and their normal cleansing action. Under these conditions, an infection is more likely to occur. Smoking is the most frequent cause of chronic bronchitis. Exposure to other pollutants can also cause chronic bronchitis.

Emphysema is a chronic and incurable disorder in which the alveoli are distended and their walls damaged. As a result, the surface area available for gas exchange is reduced. Emphysema, most often caused by smoking, is often preceded by chronic bronchitis. Air trapped in the lungs leads to alveolar damage and a noticeable ballooning of the chest. The elastic recoil of the lungs is reduced, so not only are the airways narrowed but the driving force behind expiration is also reduced. The victim is breathless and may have a cough. The surface area for gas exchange is reduced, so less oxygen reaches the heart and the brain. Even so, the heart works furiously to force more blood through the lungs, and an increased workload on the heart can result. Lack of oxygen to the brain can make the person feel depressed, sluggish, and irritable. Exercise, drug therapy, supplemental oxygen, and giving up smoking may relieve the symptoms and possibly slow the progression of emphysema. Severe emphysema may be treated by lung transplantation or lung volume reduction surgery (LVRS). During LVRS, a third of the most diseased lung tissue is removed. The removal enables the remaining tissue to function better. The result is an increase in patients' breathing ability and lung capacity.

Asthma is a disease of the bronchi and bronchioles that is marked by wheezing, breathlessness, and sometimes a cough and expectoration of mucus. The airways are unusually sensitive to specific irritants, which can include a wide range of allergens such as pollen, animal dander, dust, tobacco smoke, and industrial fumes. Even cold air can be an irritant. When exposed to the irritant, the smooth muscle in the bronchioles undergoes spasms. It now appears that chemical mediators given off by immune cells in the bronchioles cause the spasms. Most asthma patients have some degree of bronchial inflammation that further reduces the diameter of the airways and contributes to the seriousness of an attack. Asthma is not curable, but it is treatable. Special inhalers can control the inflammation and possibly prevent an attack, and other types of inhalers can stop the muscle spasms should an attack occur.

Lung Cancer

Lung cancer is more prevalent in men than in women but has surpassed breast cancer as a cause of death in women. The increase in the incidence of lung cancer in women is

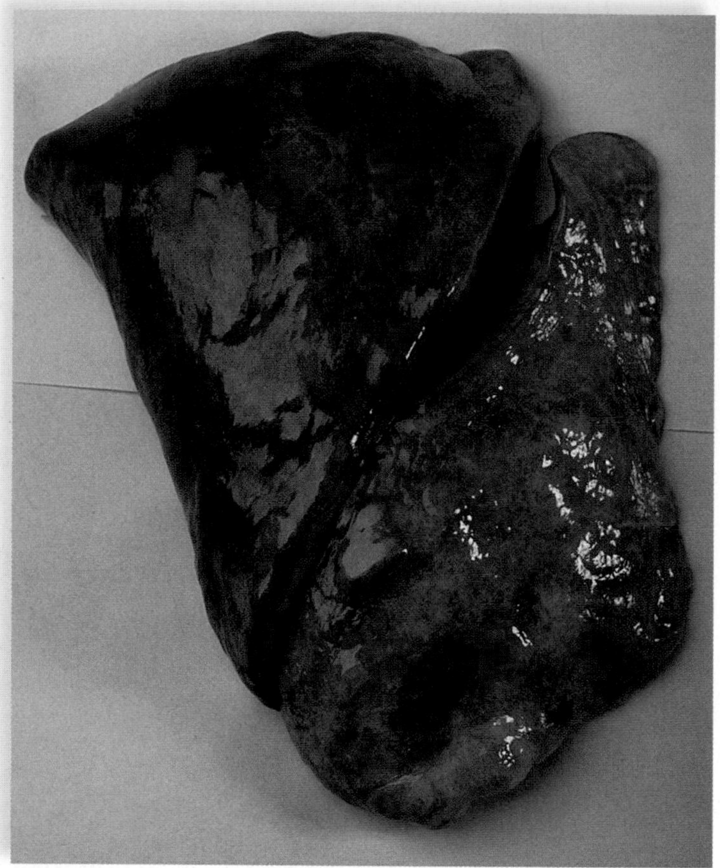

a. Normal lung

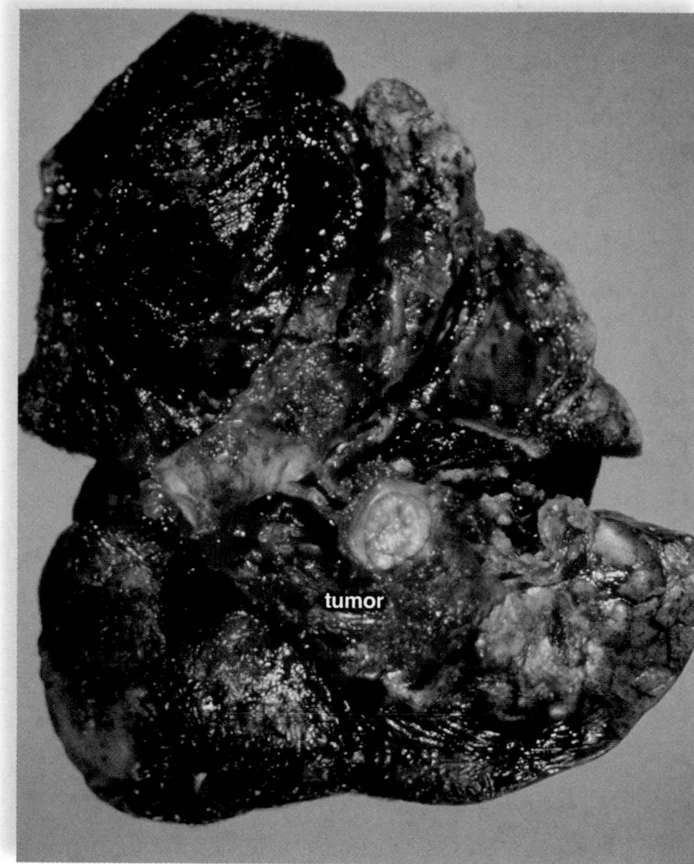

b. Lung cancer

Figure 9.13 **Effect of smoking on a human lung.**
a. Normal lung, note the healthy red color. **b.** Lungs of a heavy smoker. Notice how black the lungs are except where cancerous tumors have formed.

directly related to increased numbers of women who smoke. Autopsies on smokers have revealed the progressive steps by which the most common form of lung cancer develops. The first event appears to be thickening and callusing of the cells lining the bronchi. (Callusing occurs whenever cells are exposed to irritants.) Then cilia are lost, making it impossible to prevent dust and dirt from settling in the lungs. Following this, cells with atypical nuclei appear in the callused lining. A tumor consisting of disordered cells with atypical nuclei is considered cancer in situ (at one location). A normal lung versus a lung with cancerous tumors is shown in Figure 9.13. A final step occurs when some of these cells break loose and penetrate other tissues, a process called metastasis. Now the cancer has spread. The original tumor may grow until a bronchus is blocked, cutting off the supply of air to that lung. The entire lung then collapses, the secretions trapped in the lung spaces become infected, and pneumonia or a lung abscess (localized area of pus) results. The only treatment that offers a possibility of cure is to remove a lobe or the whole lung before metastasis has had time to occur. This operation is called **pneumonectomy.** If the cancer has spread, chemotherapy and radiation are also required.

Research also indicates that exposure to secondhand smoke can also cause lung cancer and other illnesses normally associated with smoking. If a person stops voluntary smoking and avoids secondhand smoke, and if the body tissues are not already cancerous, the lungs may return to normal over time.

CHECK YOUR PROGRESS 9.7

1. Name and describe the symptoms of some common respiratory infections and disorders of the upper respiratory tract and of the lower respiratory tract.
2. Detail how each of the common respiratory infections in the preceding question can be treated.
3. Describe the three respiratory disorders commonly associated with smoking tobacco.

CONNECTING THE CONCEPTS

For more on the diseases that influence the respiratory system, refer to the following discussions:

Section S.1 provides a more detailed look at tuberculosis and influenza epidemics.

Section S.3 explains the causes and consequences of antibiotic resistance.

Sections 19.1 and 19.2 examine the biology of cancer cells and the major causes of cancer.

BIOLOGY MATTERS  Health

Questions About Smoking, Tobacco, and Health

Is cigarette smoking really addictive?

Yes. Nicotine is an addictive drug (just like heroin and cocaine). Small amounts make the smoker want to smoke more. Also, nicotine not only affects the mood and nature of the smoker, it may cause withdrawal symptoms if the smoker attempts to stop. The younger a person is when he or she begins to smoke, the more likely he or she is to develop an addiction to nicotine.

What are some of the short-term and long-term effects of smoking cigarettes?

Short-term effects include shortness of breath and nagging coughs, diminished ability to smell and taste, premature aging of the skin, and increased risk of sexual impotence in men. Smokers tend to tire easily during physical activity. Long-term effects include many types of cancer, heart disease, aneurysms, bronchitis, emphysema, and stroke. Smoking contributes to the severity of pneumonia and asthma.

Does smoking cause cancer?

Yes. Tobacco use accounts for about one-third of cancer deaths in the United States. Smoking causes almost 90% of lung cancers. Smoking also causes cancers of the oral cavity, pharynx, larynx, and esophagus. It contributes to the development of cancers of the bladder, pancreas, cervix, kidney, and stomach. Smoking is also linked to the development of some leukemias.

Why do smokers have "smoker's cough"?

Cigarette smoke contains chemicals that irritate the air passages and lungs. When a smoker inhales these substances, the body tries to protect itself by producing mucus and coughing. The nicotine in smoke decreases the sweeping action of cilia, so some of the poisons in the smoke remain in the lungs.

If you smoke but do not inhale, is there any danger?

Yes. Wherever smoke touches living cells, it does harm. Even if smokers don't inhale, they are breathing the smoke as secondhand smoke and are still at risk for lung cancer. Pipe and cigar smokers, who often do not inhale, are at an increased risk for lip, mouth, tongue, and several other cancers.

Does cigarette smoking affect the heart?

Yes. Smoking increases the risk of heart disease, the number one cause of death in the United States. Cigarette smoking is the biggest risk factor for sudden heart attacks. Smokers who have a heart attack are more likely to die within an hour of the heart attack than nonsmokers. Cigarette smoke at very low levels (much lower than the levels that cause lung disease) can cause damage to the heart.

How does smoking affect pregnant women and their babies?

Smoking during pregnancy is linked with a greater chance of miscarriage, premature delivery, stillbirth, infant death, low birth weight, and sudden infant death syndrome (SIDS). Up to 10% of infant deaths would be prevented if pregnant women did not smoke. When a pregnant woman smokes, the nicotine, carbon monoxide, and other dangerous chemicals in smoke enter her bloodstream and then pass into the baby's body. This prevents the baby from getting essential nutrients and oxygen for growth.

What are the dangers of environmental tobacco smoke?

Environmental tobacco smoke, also called secondhand smoke, increases the rate of heart disease by over 25% in nonsmokers, and causes an estimated 46,000 deaths from heart disease each year. Children whose parents smoke are more likely to suffer from asthma, pneumonia or bronchitis, ear infections, coughing, wheezing, and increased mucus production in the first two years of life.

Are chewing tobacco and snuff safe alternatives to cigarette smoking?

No. The juice from smokeless tobacco is absorbed directly through the lining of the mouth. This creates sores and white patches that often lead to cancer of the mouth and damage to the teeth and gums. Smokeless tobacco users greatly increase their risk of other cancers, including those of the pharynx

How can people stop smoking?

There are a number of organizations, including the American Cancer Society and the American Lung Association, that offer suggestions for how to quit smoking. Both organizations also offer support groups to people interested in quitting. There is even advice about how people can help their friends quit smoking. Nicotine Anonymous offers a 12-step program for kicking a nicotine addiction, modeled after the 12-step Alcoholics Anonymous program. Recovering smokers have reported that a support system is very important. Several smoking-cessation drugs have been approved by the FDA, and others are in development phases. One such drug, varenicline (Chantix), affects areas in the brain stimulated by nicotine. Varenicline may lessen withdrawal symptoms by mimicking the effects of nicotine. In addition, several over-the-counter products are available to replace nicotine. These include nicotine patches, gum, and lozenges. These products are designed to supply enough nicotine to alleviate the withdrawal symptoms people experience during attempts to stop smoking.

Questions to Consider

1. What can you do to reduce your exposure to secondhand smoke?
2. Why does cigarette smoke have a negative effect on organ systems that are not in direct contact with the actual smoke?

CASE STUDY CONCLUSION

The results of the polysomnogram treatment indicated that Justin's condition was most likely due to periods of sleep apnea during the evening. Over the 7-hour test period, Justin experienced an average of five apnea events per hour. His blood oxygen concentration was also low during these periods. The brain wave tests did not indicate anything abnormal, suggesting that the problem was associated with obstructive sleep apnea, not central sleep apnea.

As a first level of treatment, Justin's doctor suggested that he increase his exercise, shed the extra pounds, and reduce his use of alcohol. In addition, he scheduled Justin to be fitted for a device called a CPAP (continuous positive airway pressure) mask. The device delivers a constant flow of air into the upper respiratory tract. This serves to keep the airways open, reducing the frequency of apnea events. Justin would have to use the device nightly, because the CPAP mask was not a cure for obstructive sleep apnea. If diet and exercise did not restore Justin's normal sleep patterns, the other option would be surgery to remove some of the soft tissues in the pharynx region, a procedure called an uvulopalatopharyngoplasty (UPPP). However, Justin's doctor was confident that his changes in lifestyle and use of the CPAP mask would help reduce Justin's occurrences of sleep apnea.

MEDIA STUDY TOOLS

Enhance your study of this chapter with media! Visit **www.mhhe.com/maderhuman13e** and go to "Media Study Tools" for this chapter to access the following:

 **Animations**	 **Video**	 **MP3 Files**
9.6 Changes in the Partial Pressure of Oxygen and Carbon Dioxide • Gas Exchange During Respiration	**9.7** Good Poison	**9.1** Respiratory Structure and Function **9.4** Ventilation **9.5** Control of Respiration **9.6** Gas Exchange

SUMMARIZE

9.1 The Respiratory System

The **respiratory system** is responsible for the process of **ventilation** (breathing), which includes **inspiration** and **expiration.**

The respiratory tract consists of the nose, the pharynx, the larynx, the trachea, the bronchi, the bronchioles, and the lungs.

9.2 The Upper Respiratory Tract

Air from the nose enters the pharynx and passes through:

- the **nasal cavities,** which filter and warm the air;
- the **pharynx,** the opening into parallel air and food passageways; the upper region of the pharynx, the nasopharynx, is connected to the middle ear by **auditory tubes** (also called **eustachian tubes);** the **epiglottis** blocks entry of food into the lower respiratory tract; and
- the **larynx,** the voice box that houses the **vocal cords;** the **glottis** is the small opening between the vocal cords. The **tonsils** are lymphatic tissue that help protect the respiratory system.

9.3 The Lower Respiratory Tract

- The **trachea** (windpipe) is lined with goblet cells and ciliated cells; in a **tracheostomy,** a small incision is made in the trachea to assist breathing.
- The **bronchi** enter the **lungs** and branch into smaller **bronchioles.**
- The lungs consist of the **alveoli,** air sacs surrounded by a capillary network; alveoli are lined with a **surfactant** that prevents them from closing. Each lung is enclosed by a membrane called a **pleura.**

9.4 Mechanism of Breathing

Breathing involves inspiration and expiration of air.

Inspiration

The diaphragm lowers, and the rib cage moves upward and outward; the lungs expand, and air rushes in.

Expiration

The diaphragm relaxes and moves up. The rib cage moves down and in; pressure in the lungs increases; air is pushed out of the lungs.

Respiratory volumes can be measured by

- the **tidal volume,** the amount of air that normally enters and exits with each breath;
- the **vital capacity,** the amount of air that moves in plus the amount that moves out with maximum effort.

- the **inspiratory reserve volume** and **expiratory reserve volume,** the difference between normal amounts and the maximum effort amounts of air moved; and
- the **residual volume,** the amount of air that stays in the lungs when we breathe; also called **dead air space.**

9.5 Control of Ventilation

The **respiratory control center** in the brain automatically causes us to breathe 12 to 20 times a minute. Extra carbon dioxide in the blood can decrease the pH; if so, **chemoreceptors** alert the respiratory center, which increases the rate of breathing.

9.6 Gas Exchanges in the Body

Both **external respiration** and **internal respiration** depend on diffusion. Hemoglobin activity is essential to the transport of gases and, therefore, to external and internal respiration.

External Respiration

- CO_2 diffuses out of plasma into lungs; **carbonic anhydrase** accelerates the breakdown of HCO_3^- ions in red blood cells;
- O_2 diffuses into the plasma and then into red blood cells in the capillaries. O_2 is carried by hemoglobin, forming **oxyhemoglobin.**

Internal Respiration

- O_2 diffuses out of the blood into the tissues.
- CO_2 diffuses into the blood from the tissues. CO_2 is carried in the plasma as **bicarbonate ions** (HCO_3^-); a small amount links with hemoglobin to form **carbaminohemoglobin.**

9.7 Respiration and Health

A number of illnesses are associated with the respiratory tract.

Upper Respiratory Tract Infections

- Infections of the nasal cavities, sinuses, throat, tonsils, and larynx are all upper respiratory tract infections.
- These include **sinusitis, otitis media, laryngitis,** and **tonsillitis;** tonsils, which are lymphatic tissue, may be removed by a **tonsillectomy.**

Lower Respiratory Tract Disorders

- Lower respiratory infections include **acute bronchitis, pneumonia,** and pulmonary **tuberculosis.**
- Restrictive pulmonary disorders are exemplified by **pulmonary fibrosis.**
- Obstructive pulmonary disorders are exemplified by **chronic bronchitis, emphysema,** and **asthma.**
- Smoking can eventually lead to **lung cancer** and other disorders; sections of diseased lung may be removed by a **pneumonectomy.**

ASSESS

Testing Your Knowledge of the Concepts

1. Name the three parts of the pharynx. (page 198)
2. How is the structure of the trachea important for respiration, as well as digestion? (page 200)
3. Describe the structure of an alveolus, and explain how it is suited for gas exchange. (page 201)
4. What are the steps of inspiration and expiration? How is breathing controlled? (pages 203–204, 206–207)
5. Using Figure 9.9, list and define the volumes and capacities of air movement. (pages 204–205)
6. Describe what occurs during external respiration, using two important equations. What is the driving force for the gas exchange? (page 207)
7. Describe what occurs during internal respiration, using two important equations. What is the driving force for the gas exchange? (pages 207–209)
8. Describe several upper and lower respiratory tract disorders (other than cancer). If appropriate, explain why breathing is difficult with these conditions. (pages 209–211)
9. List the steps by which lung cancer develops. (pages 211–212)
10. Which of these is anatomically incorrect?
    - **a.** The nose has two nasal cavities.
    - **b.** The pharynx connects the nasal and oral cavities to the larynx.
    - **c.** The larynx contains the vocal cords.
    - **d.** The trachea enters the lungs.
    - **e.** The lungs contain many alveoli.
11. How is inhaled air modified before it reaches the lungs?
    - **a.** It must be humidified.
    - **b.** It must be warmed.
    - **c.** It must be filtered.
    - **d.** All of these are correct.
12. What is the name of the structure that prevents food from entering the trachea?
    - **a.** glottis
    - **b.** septum
    - **c.** epiglottis
    - **d.** Adam's apple

In questions 13–17, match each description with a structure in the key.

Key:

- **a.** pharynx
- **b.** glottis
- **c.** larynx
- **d.** trachea
- **e.** bronchi
- **f.** bronchioles

13. Branched tubes that lead from bronchi to the alveoli
14. Reinforced tube that connects larynx with bronchi
15. Chamber behind oral cavity and between nasal cavity and larynx
16. Opening into larynx
17. Divisions of the trachea that enter lungs
18. Which of these is incorrect concerning inspiration?
    - **a.** Rib cage moves up and out.
    - **b.** Diaphragm contracts and moves down.
    - **c.** Pressure in lungs decreases, and air comes rushing in.
    - **d.** The lungs expand because air comes rushing in.
19. Air enters the human lungs because
    - **a.** atmospheric pressure is lower than the pressure inside the lungs.
    - **b.** atmospheric pressure is greater than the pressure inside the lungs.
    - **c.** although the pressures are the same inside and outside, the partial pressure of oxygen is lower within the lungs.
    - **d.** the residual air in the lungs causes the partial pressure of oxygen to be lower than it is outside.

20. The maximum volume of air that can be moved in and out during a single breath is called the
    - a. expiratory and inspiratory reserve volume.
    - b. residual volume.
    - c. tidal volume.
    - d. vital capacity.
    - e. functional residual capacity.
21. The enzyme carbonic anhydrase
    - a. causes the blood to be more basic in the tissues.
    - b. speeds up the conversion of carbonic acid to carbon dioxide and water, and the reverse.
    - c. actively transports carbon dioxide out of capillaries.
    - d. is active only at high altitudes.
    - e. All of the choices are correct.
22. Hemoglobin assists transport of gases by
    - a. combining with oxygen.
    - b. combining with CO_2.
    - c. combining with H^+.
    - d. being present in red blood cells.
    - e. All of the choices are correct.
23. In humans, the respiratory center
    - a. is stimulated by carbon dioxide.
    - b. is located in the medulla oblongata.
    - c. controls the rate of breathing.
    - d. All of the choices are correct.
24. Which of the following is not true of obstructive pulmonary disorders?
    - a. Air does not flow freely in the airways.
    - b. Vital capacity is reduced due to loss of lung elasticity.
    - c. Disorders may include chronic bronchitis, emphysema, and asthma.
    - d. Ventilation takes longer to occur.
25. Label this diagram of the human respiratory tract.

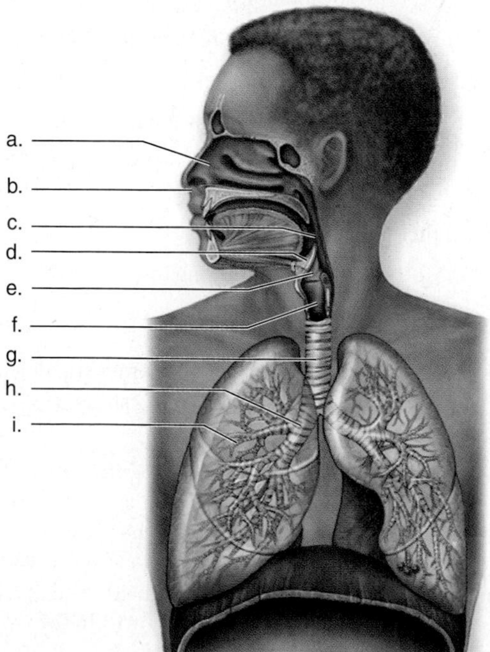

ENGAGE

Thinking Critically About the Concepts

1. Children may also suffer from obstructive sleep apnea (OSA). Symptoms experienced during the daytime include breathing through the mouth and difficulty focusing. Children don't often have the excessive sleepiness during the day that adults like Justin have. Children with OSA often have enlarged tonsils or adenoids. Removal of the tonsils and adenoids often alleviates the problem. Continuous positive airway pressure (CPAP) treatment might be necessary if the OSA continues after surgery. Why would enlarged tonsils or adenoids cause OSA?
2. Explain the expression, "The food went down the wrong tube," by referring to structures along the path of air.
3. Long-term smokers often develop a chronic cough.
    - a. What purpose does smokers' cough serve?
    - b. Why are nonsmokers less likely to develop a chronic cough?
4. Professional singers who need to hold a long note while singing must exert a great deal of control over their breathing. What muscle(s) need(s) conditioning to acquire better control over their breathing?
5. Why would someone who nearly drowned have a blue tint to his or her skin?

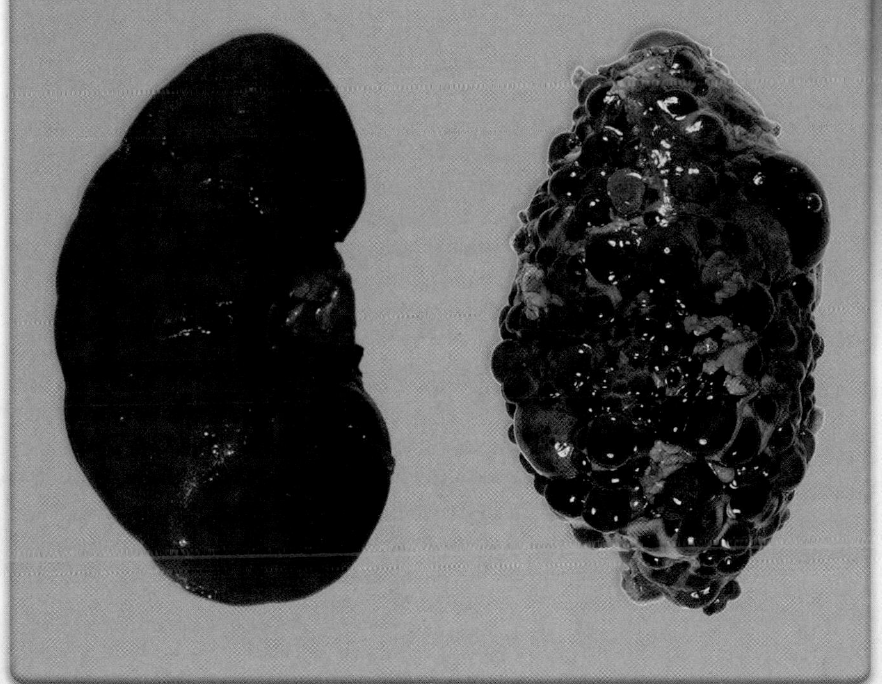

CHAPTER

10

Urinary System

CASE STUDY POLYCYSTIC KIDNEY DISEASE

Michael and Jada were excited about the birth of their child. Married for three years, they already had a healthy daughter, and they were happy that she would have a younger sister to play with. After Aiesha was born, however, it soon became clear that something was wrong. She weighed only 5 lb 4 oz at birth. The first time she urinated, there was an obvious tinge of blood in her urine. In addition, she also seemed to urinate much more frequently than was to be expected for an infant, and her blood pressure was higher than was normal. When her doctors performed ultrasound and magnetic resonance imaging (MRI) scans of her abdominal organs, they found that Aiesha had signs of polycystic kidney disease (PKD). Aiesha's doctors explained that in PKD, cysts (small, fluid-filled sacs) form within the collecting ducts of the nephrons in the interior of the kidneys. The ultrasound results indicated that both of Aiesha's kidneys were covered in cysts (see the kidney above on right) and that this usually meant that the cysts were present inside the kidneys as well. Michael and Jada were informed that the presence of these cysts explained Aiesha's symptoms. The doctors also told Michael and Jada that PKD would most likely cause Aiesha's kidneys to fail and that they should immediately prepare her for a kidney transplant. Because PKD is a genetic disorder, the physicians suggested that both parents undergo genetic tests to see if they were carriers for PKD.

As you read through the chapter, think about the following questions:

1. What is the role of the kidneys in the body?
2. How would problems in the collecting ducts of the nephrons cause kidney failure?
3. Why would problems with the kidneys result in blood in the urine and high blood pressure?

CHAPTER CONCEPTS

10.1 The Urinary System
In the urinary system, kidneys produce urine, which is stored in the bladder before being discharged from the body. The kidneys are major organs of homeostasis.

10.2 Kidney Structure
Microscopically, the kidneys are composed of nephrons. Each nephron filters the blood and produces urine.

10.3 Urine Formation
Urine is composed primarily of nitrogenous waste products, salts, and water. Urine formation is a stepwise process.

10.4 Kidneys and Homeostasis
In addition to excreting waste, the kidneys are involved in the salt–water balance and the acid–base balance of the blood.

10.5 Kidney Function Disorders
Various types of illnesses, including diabetes, kidney stones, and infections, can lead to renal failure. Hemodialysis is needed for the survival of patients with renal failure.

BEFORE YOU BEGIN

Before beginning this chapter, take a few moments to review the following discussions:

Section 2.2 What determines whether a solution is acidic or basic?

Section 3.3 How does water move across a plasma membrane?

Section 4.8 How do feedback mechanisms contribute to the maintenance of homeostasis?

10.1 The Urinary System

LEARNING OUTCOMES

Upon completion of this section, you should be able to

1. Summarize the functions of the urinary system.
2. Identify the organs of the urinary system and state their function.

The **urinary system** is the organ system of the body that plays a major role in maintaining the salt, water, and pH homeostasis of the blood. Collectively, these organs carry out the process of **excretion,** or the removal of metabolic waste from the body. These metabolic waste materials are the by-products of the normal activities of the cells and tissues. In comparison to excretion, defecation—a process of the digestive system (see Chapter 8)—eliminates undigested food and bacteria in the form of feces. Excretion in humans is performed by the formation and discharge of urine from the body.

Functions of the Urinary System

As the urinary system carries out the process of excretion, it performs a number of important functions that contribute to homeostasis. Each of these functions is described in greater detail in section 10.4.

Excretion of Metabolic Wastes

The metabolic waste of humans consists primarily of nitrogenous waste, such as urea, creatinine, ammonium, and uric acid. **Urea,** a waste product of amino acid metabolism, is the primary nitrogenous end product of metabolism in humans. In the liver, the breakdown of amino acids releases ammonia, a compound that is very toxic to cells. The liver rapidly combines the ammonia with carbon dioxide to produce urea, which is much less harmful. Normally, urea levels in the blood are between 10 and 20 milligrams per deciliter (mg/dl). Elevated urea levels in the blood may cause **uremia,** a condition that causes cardiac arrhythmia, vomiting, respiratory problems, and potentially death. Treatments for elevated urea levels are discussed in section 10.5.

In addition to urea, the kidneys secrete creatinine and uric acid. **Creatinine** is a waste product that results from the breakdown of creatine phosphate, a high-energy phosphate reserve molecule in muscles. **Uric acid** is formed from the metabolic processing of nucleotides (such as adenine and thymine). Uric acid is rather insoluble. If too much uric acid is present in blood, crystals form and precipitate out. Crystals of uric acid sometimes collect in the joints, producing a painful ailment called **gout.**

Maintenance of Water–Salt Balance

A principal function of the kidneys is to maintain the appropriate water–salt balance of the blood. As you know, salts, such as NaCl, have the ability to influence the rate and direction of osmosis (see section 3.3). Therefore, the more salts there are in the blood, the greater the blood volume and the greater the blood pressure. By regulating the concentration of certain ions, namely sodium (Na^+) and potassium (K^+), in the blood, the kidneys regulate blood pressure. In addition to ions, the kidneys also maintain the appropriate blood level of ions such as potassium (K^+), bicarbonate (HCO_3^-), and calcium (Ca^{2+}).

Maintenance of Acid–Base Balance

The kidneys regulate the acid–base balance of the blood. For a person to remain healthy, the blood pH should be just about 7.4. The kidneys monitor and help control blood pH, mainly by excreting hydrogen ions (H^+) and reabsorbing the bicarbonate ions (HCO_3^-) as needed to keep blood pH at 7.4. Urine usually has a pH of 6 or lower because our diet often contains acidic foods.

Secretion of Hormones

The kidneys assist the endocrine system in hormone secretion. The kidneys release renin, an enzyme that leads to aldosterone secretion. Aldosterone is a hormone produced by the adrenal glands (Fig. 10.1), which lie atop the kidneys. As described in section 10.4, aldosterone is involved in regulating the water–salt balance of the blood. The kidneys also release **erythropoietin (EPO)**, a hormone that regulates the production of red blood cells.

Additional Functions of the Kidneys

The kidneys also reabsorb filtered nutrients and synthesize vitamin D. Vitamin D is a molecule that promotes calcium ion (Ca^{2+}) absorption from the digestive tract.

Organs of the Urinary System

The urinary system consists of the kidneys, ureters, urinary bladder, and urethra (Fig. 10.1).

Kidneys

The **kidneys** are paired organs located on either side of the vertebral column at the same level as the small of the lower back. They lie in depressions beneath the peritoneum, where they receive some protection from the lower rib cage. Due to the shape of the liver, the right kidney is positioned slightly lower than the left. The kidneys are bean-shaped and reddish-brown in color. The fist-sized organs are covered by a tough capsule of fibrous connective tissue, called a renal capsule. Masses of adipose tissue adhere to each kidney. The concave side of a kidney has a depression where a **renal artery** enters and a **renal vein** and a ureter exit the kidney. The renal artery transports blood to be filtered to the kidneys, and the renal vein carries filtered blood away from the kidneys.

Ureters

The **ureters** conduct urine from the kidneys to the bladder. They are small, muscular tubes about 25 cm long and 5 mm in diameter. The wall of a ureter has three layers: an inner mucosa (mucous membrane), a smooth muscle layer, and an outer fibrous coat of connective tissue. Peristaltic contractions cause urine to enter the bladder even if a person is lying down. Urine enters the bladder in spurts that occur at the rate of one to five per minute.

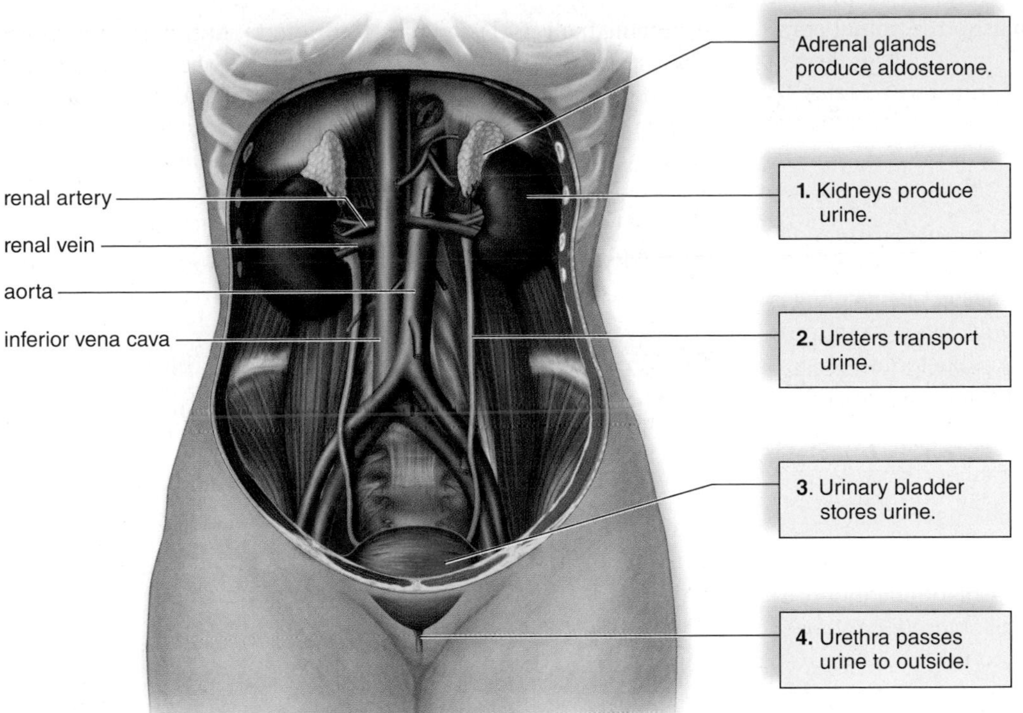

Figure 10.1 The urinary system. The kidneys, ureters, urinary bladder, and urethra. The adrenal glands are part of the endocrine system.

APPLICATIONS AND MISCONCEPTIONS

What is a "floating kidney"?

A floating kidney, a condition also known as nephroptosis, occurs when the kidney becomes detached from its position and moves freely beneath the peritoneum. A floating kidney may develop in people who are very thin, or in someone who has recently received a sharp blow to the back. When the kidney becomes dislodged, it may form a kink in the ureter, causing urine to back up into the kidney. This can potentially result in damage to the structures inside the kidney. Surgery can correct a floating kidney by reattaching it to the abdominal wall.

Urinary Bladder

The **urinary bladder** stores urine until it is expelled from the body. The bladder has three openings: two for the ureters and one for the urethra, which drains the bladder (Fig. 10.2).

The bladder wall is expandable because it contains a middle layer of circular fibers of smooth muscle and two layers of longitudinal smooth muscle. The epithelium of the mucosa becomes thinner, and folds in the mucosa called *rugae* disappear as the bladder enlarges. The bladder's rugae are similar to those of the stomach. A layer of transitional epithelium enables the bladder to stretch and contain an increased volume of urine. The urinary bladder has a maximum capacity of between 700 and 800 ml.

The bladder has other features that allow it to retain urine. After urine enters the bladder from a ureter, small folds of bladder mucosa act like a valve to prevent backward flow. Two sphincters in close proximity are found where the urethra exits the bladder. The internal sphincter occurs around the opening to the urethra. It is composed of smooth muscle and is involuntarily controlled. An external sphincter is composed of skeletal muscle that can be voluntarily controlled.

When the urinary bladder fills to about 250 ml with urine, stretch receptors are activated by the enlargement of the bladder.

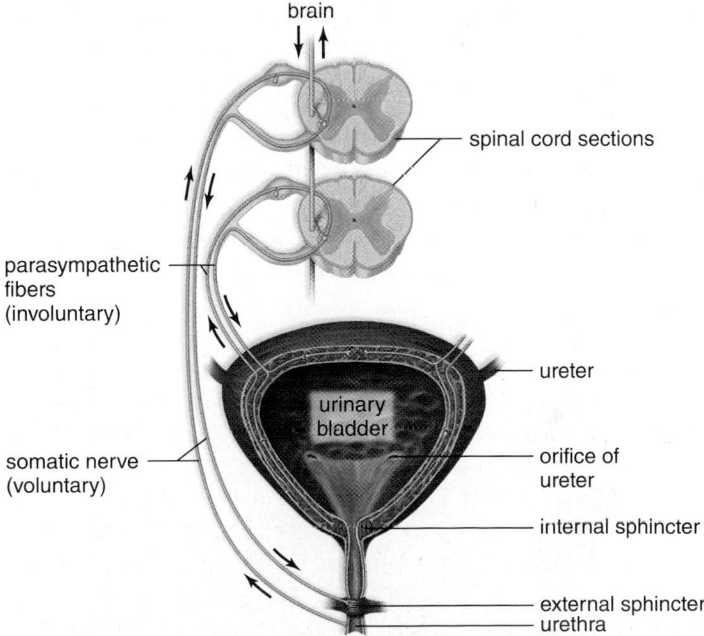

Figure 10.2 Sensory impulses trigger a desire to urinate. As the bladder fills with urine, sensory impulses go to the spinal cord and then to the brain. The brain can override the urge to urinate. When urination occurs, motor nerve impulses cause the bladder to contract and the sphincters to relax.

These receptors send sensory nerve signals to the spinal cord. Subsequently, motor nerve impulses from the spinal cord cause the urinary bladder to contract and the sphincters to relax so that urination, also called *micturition,* is possible (Fig. 10.2).

Urethra

The **urethra** is a small tube that extends from the urinary bladder to an external opening. Therefore, its function is to remove urine from the body. The urethra has a different length in females than in males. In females, the urethra is only about 4 cm long. The short length of the female urethra makes bacterial invasion of the urinary tract easier. In males, the urethra averages 20 cm when the penis is flaccid (limp, nonerect). As the urethra leaves the male urinary bladder, it is encircled by the prostate gland. The prostate sometimes enlarges, restricting the flow of urine in the urethra. The Health feature, "Urinary Difficulties Due to an Enlarged Prostate," discusses this problem in men.

In females, the reproductive and urinary systems are not connected. However, in males, the urethra carries urine during urination and sperm during ejaculation.

APPLICATIONS AND MISCONCEPTIONS

What is an overactive bladder?

In an overactive bladder the muscles of the bladder contract, even though the bladder may not be full. Muscle contraction causes strong feelings of urgency to go to the bathroom. Use of medications, such as tolterodine (Detrol LA) and oxybutynin (Ditropan XL), control the symptoms by blocking nerve signals to the bladder and calming the bladder's muscle contractions. An overactive bladder can also be treated without medication by urinating at set times of the day or by doing exercises to strengthen the muscles that control urination.

CHECK YOUR PROGRESS 10.1

1. List and briefly describe the functions of the organs of the urinary system.
2. Summarize the processes the kidneys perform to maintain homeostasis.
3. Predict what might occur to overall homeostasis if the kidneys could not excrete metabolic waste products from the body.

CONNECTING THE CONCEPTS

For more on the interaction of the urinary system with other body systems, refer to the following discussions:

Section 6.1 describes the composition of blood.

Section 12.3 describes how muscles use creatine phosphate as an energy molecule.

Section 15.4 examines the structure and function of the adrenal glands.

BIOLOGY MATTERS  Science

Lab-Grown Bladders

You're probably familiar with organ transplants done with organs harvested from people who have recently died or even from living donors. Did you know that some organs can now be grown in a lab and used for transplantation?

Figure 10A A lab-grown bladder is readied for transplantation.
There is less risk of rejection when a lab-grown bladder is transplanted.

Bladders grown in a lab have been successfully transplanted into a number of patients (Fig. 10A). Each patient in one particular study contributed cells from his or her own diseased bladder. Those cells were cultured in a lab and encouraged to form a new bladder by growing them on a collagen form shaped like a bladder. Eventually, the new bladders were attached to each patient's diseased bladder. This alleviated the incontinence problems the patients had experienced. The risk of kidney damage was also decreased in these individuals by lowering the pressure inside their enlarged bladders. The cells for the new bladders were taken from each patient, so there was no risk of rejection.

Researchers hope this technique can be used to grow other types of tissues and organs that would be available for transplant. This could be the one potential solution to the limited number of organs available for transplant.

Questions to Consider

1. Why is it possible to grow bladders in the lab, but not a kidney?
2. What do you think are some possible extensions of this technology in other parts of the body?

BIOLOGY MATTERS  Health

Urinary Difficulties Due to an Enlarged Prostate

The prostate gland, part of the male reproductive system, surrounds the urethra at the point where the urethra leaves the urinary bladder (Fig. 10B). The prostate gland produces and adds a fluid to semen as semen passes through the urethra within the penis. At about age 50, the prostate gland often begins to enlarge, growing from the size of a walnut to that of a lime or even a lemon. This condition is called benign prostatic hyperplasia (BPH). As it enlarges, the prostate squeezes the urethra, causing urine to back up—first into the bladder, then into the ureters, and finally, perhaps, into the kidneys. While BPH is technically a disorder of the reproductive system, its symptoms are almost exclusively associated with urination and excretion, and therefore it is often discussed as a urinary system disorder.

Treatment Emphasis Is on Early Detection

The treatment for BPH can involve (1) invasive procedures to reduce the size of the prostate or (2) medications that shrink the prostate and/or improve urine flow. For the former, prostate tissue can be destroyed by applying microwaves to a specific portion of the gland. In some cases, a physician may decide to surgically remove that prostate tissue. This may be accomplished by abdominal surgery, which requires an incision of the abdomen, or access to the prostate via the urethra. This operation, called transurethral resection of the prostate (TURP), requires careful consideration because one study found that the death rate during the five years following TURP is much higher than that following abdominal surgery.

Some drug treatments recognize that prostate enlargement is due to a prostate enzyme (5-alpha-reductase) that acts on the male sex hormone testosterone, converting it into a substance that promotes prostate growth. Growth is fine during puberty, but continued growth in an adult is undesirable. These drugs contain substances that interfere with the action of the enzyme that promotes growth. One of the ingredients is an extract from a plant called the saw palmetto. It is particularly effective during the early stages of prostate enlargement. While it is sold in tablet form as an over-the-counter nutrient supplement, it should not be taken unless the need for it is confirmed by a physician. The prescription drugs finasteride (Propecia) and dutasteride (Avodart) are more powerful inhibitors of the same growth enzyme, but patients complain of erectile dysfunction and loss of libido while on the drugs.

Another common treatment for BPH involves the use of alpha-blockers, such as tamsulosin (Flomax). Alpha-blockers target specific receptors (called α-adrenergic receptors) on the surface of smooth muscle tissue. Tamsulosin inhibits the interaction of the nervous system with the smooth muscle, causing it to relax and promote urine flow. Similarly, drugs such as tadalafil (Cialis) inhibit an enzyme called phosphodiesterase type 5 (PDE5) in smooth muscle tissue, causing it to relax. Like tamsulosin, the use of tadalafil causes the relaxation of the prostate, enhancing the flow of urine.

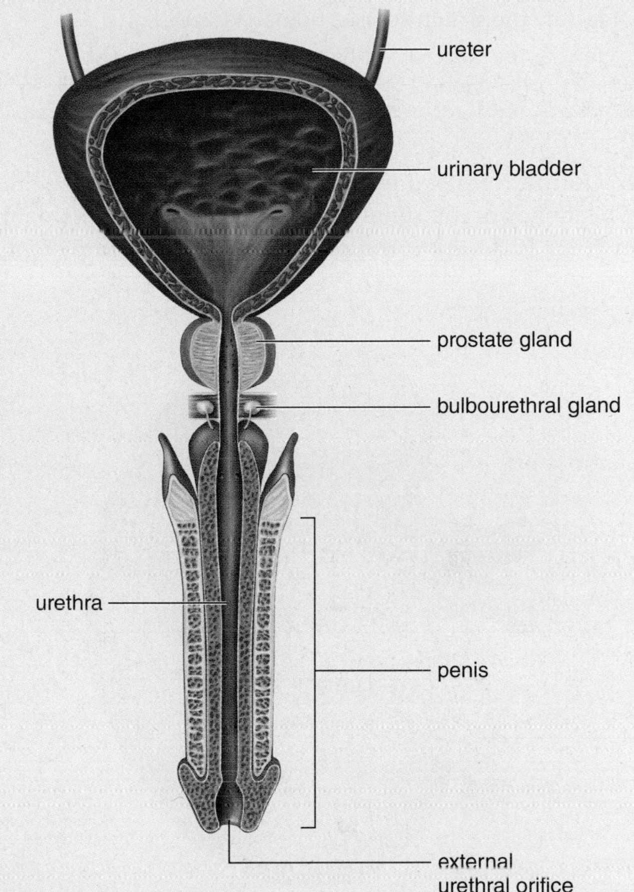

Figure 10B Location of the prostate gland.
Note the position of the prostate gland, which can enlarge to obstruct urine flow.

Many men are concerned that BPH may be associated with prostate cancer, but the two conditions are not necessarily related. BPH occurs in the inner zone of the prostate, whereas cancer tends to develop in the outer area. If prostate cancer is suspected, blood tests and a biopsy, in which a tiny sample of prostate tissue is surgically removed, will confirm the diagnosis.

Enlarged Prostate and Cancer

Although prostate cancer is the second most common cancer in men, it is not a major killer. Typically, prostate cancer is so slow growing that the survival rate is about 98% if the condition is detected early.

Questions to Consider

1. What is the role of the prostate in the male reproductive system?
2. Given how alpha-blockers function, what other applications might they have in humans?

10.2 Kidney Structure

LEARNING OUTCOMES

Upon completion of this section, you should be able to

1. Identify the structures of a human kidney.
2. Identify the structures of a nephron and state the function of each.

When a kidney is sliced lengthwise, it is possible to see that many branches of the renal artery and vein reach inside the kidney (Fig. 10.3*a, b*). If the blood vessels are removed, it is easier to identify the three regions of a kidney. (1) The **renal cortex** is an outer, granulated layer that dips down in between a radially striated inner layer called the renal medulla. (2) The **renal medulla** consists of cone-shaped tissue masses called renal pyramids. (3) The **renal pelvis** is a central space, or cavity, continuous with the ureter (Fig. 10.3*c, d*).

Microscopically, the kidney is composed of over 1 million **nephrons,** sometimes called renal, or kidney, tubules (Fig. 10.3*e*). The nephrons filter the blood and produce urine. Each nephron is positioned so that the urine flows into a collecting duct. Several nephrons enter the same collecting duct. The collecting ducts eventually enter the renal pelvis.

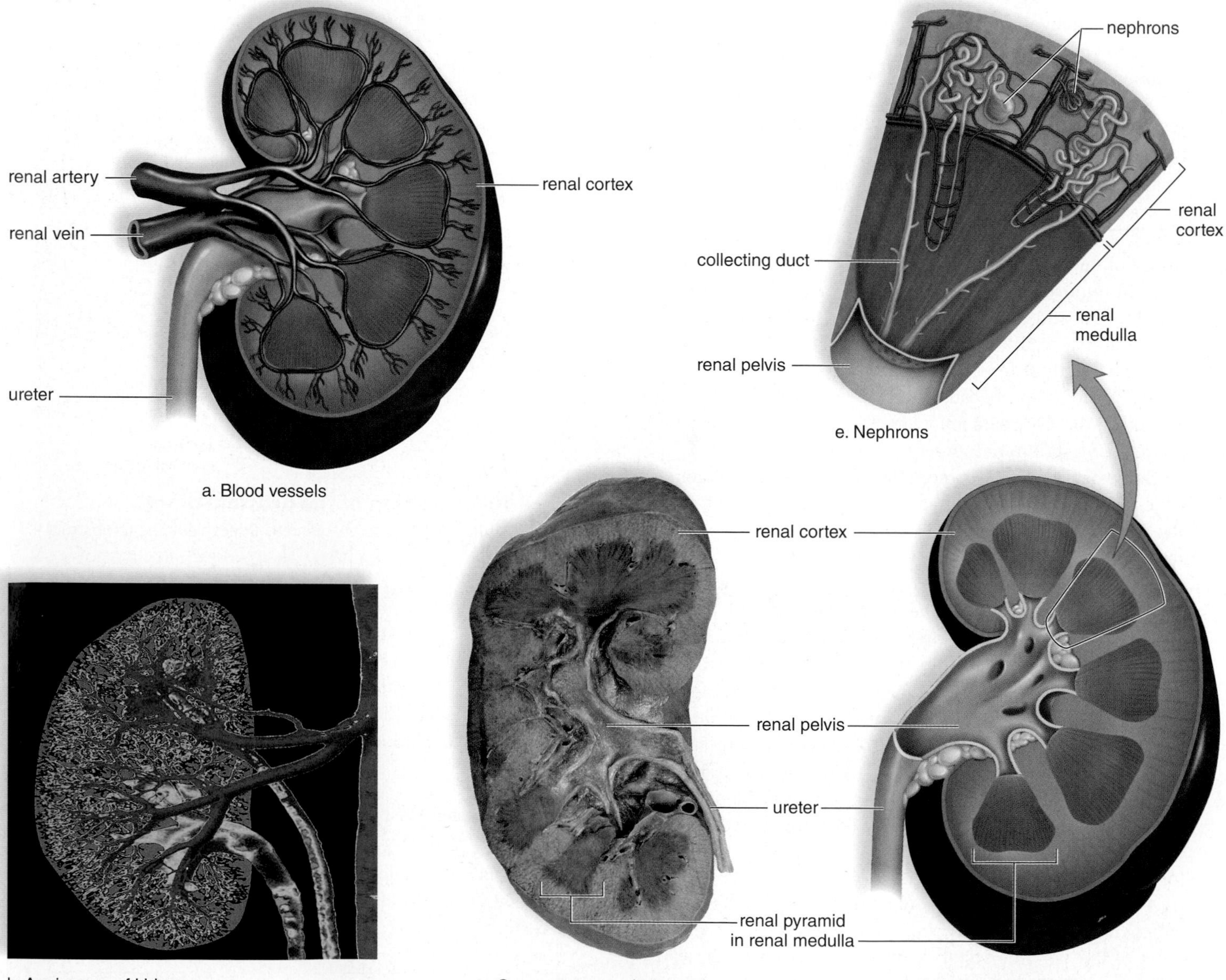

Figure 10.3 The anatomy of a human kidney.
a. A longitudinal section of the kidney showing the blood supply. The renal artery divides into smaller arteries, and these divide into arterioles. Venules join to form small veins, which join to form the renal vein. **b.** A procedure called an angiogram highlights blood vessels by injecting a contrast medium that is opaque to X-rays. **c.** and **d.** The same section without the blood supply. Now it is easier to distinguish the renal cortex; the renal medulla; and the renal pelvis, which connects with the ureter. The renal medulla consists of the renal pyramids. **e.** An enlargement showing the placement of nephrons.

Anatomy of a Nephron

Each nephron has its own blood supply including two capillary regions (Fig. 10.4). From the renal artery, an afferent arteriole transports blood to the **glomerulus,** a knot of capillaries inside the glomerular capsule. Blood leaving the glomerulus is carried away by the efferent arteriole. Blood pressure is higher in the glomerulus because the efferent arteriole is narrower than the afferent arteriole. The efferent arteriole divides and forms the **peritubular capillary network,** which surrounds the rest of the nephron. Blood from the efferent arteriole travels through the peritubular capillary network. Then the blood goes into a venule that carries blood into the renal vein.

Parts of a Nephron

Each nephron is made up of several parts (Fig. 10.4). Some functions are shared by all parts of the nephron. However, the specific structure of each part is especially suited to a particular function.

First, the closed end of the nephron is pushed in on itself to form a cuplike structure called the **glomerular capsule** (Bowman's capsule). The outer layer of the glomerular capsule is composed of squamous epithelial cells. The inner layer is made up of podocytes that have long cytoplasmic extensions. The podocytes cling to the capillary walls of the glomerulus and leave pores that allow easy passage of small molecules

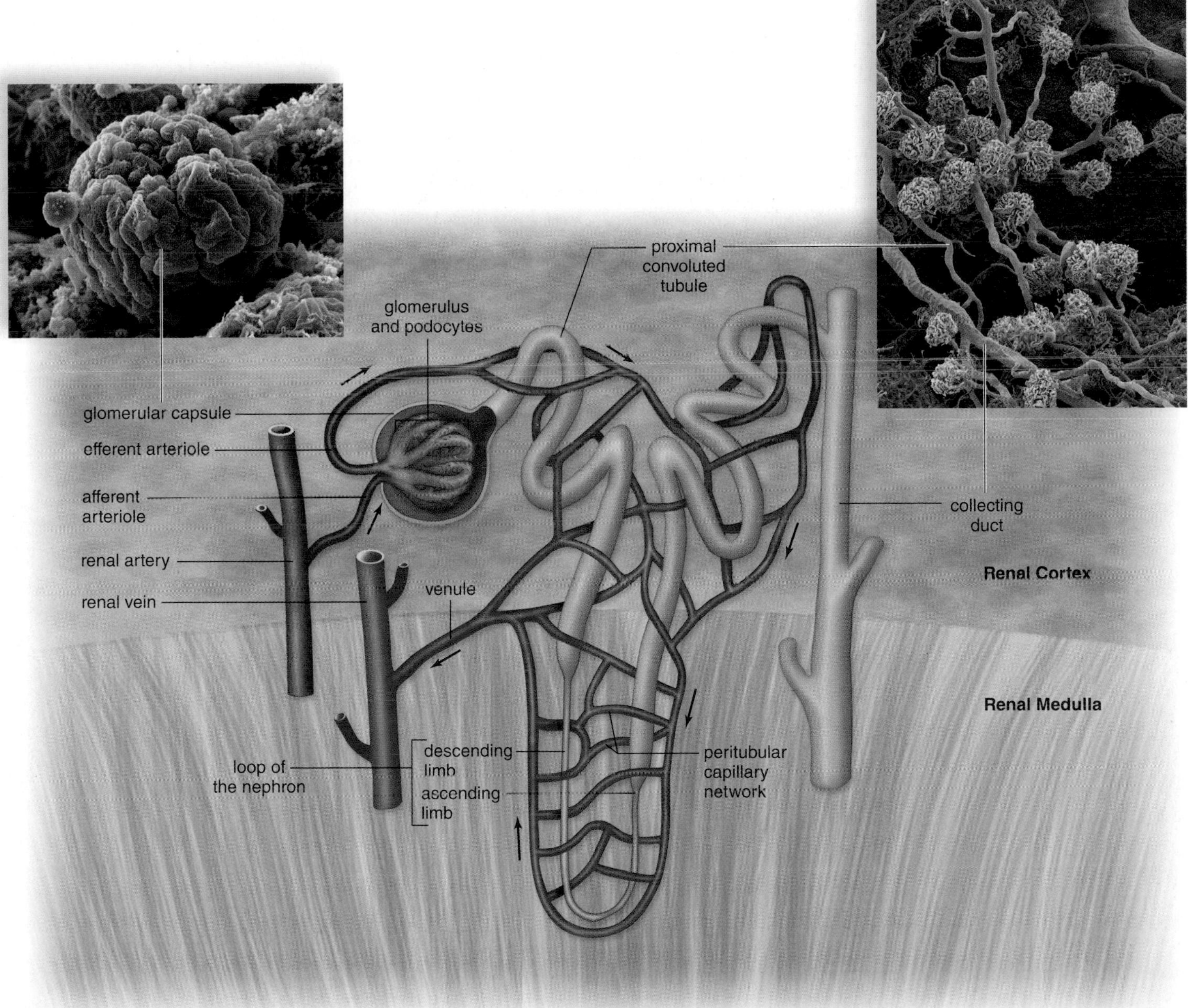

Figure 10.4 The structure of a nephron.
A nephron is made up of a glomerular capsule, the proximal convoluted tubule, the loop of the nephron, the distal convoluted tubule, and the collecting duct. The photomicrographs show the microscopic anatomy of these structures. The arrows indicate the path of blood around the nephron.

from the glomerulus to the inside of the glomerular capsule. This process, called glomerular filtration, produces a filtrate of the blood.

Next, there is a **proximal convoluted tubule.** The cuboidal epithelial cells lining this part of the nephron have numerous microvilli, about 1 micrometer (μm) in length, that are tightly packed and form a brush border (Fig. 10.5). A brush border greatly increases the surface area for the tubular reabsorption of filtrate components. Each cell also has many mitochondria, which can supply energy for active transport of molecules from the lumen to the peritubular capillary network.

Simple squamous epithelium appears as the tube narrows and makes a U-turn called the **loop of the nephron** (loop of Henle). Each loop consists of a descending limb and an ascending limb. The descending limb of the loop allows water to diffuse into tissue surrounding the nephron. The ascending limb actively transports salt from its lumen to interstitial tissue. As we shall see, this activity facilitates the reabsorption of water by the nephron and collecting duct.

The cuboidal epithelial cells of the **distal convoluted tubule** have numerous mitochondria, but they lack microvilli. This means that the distal convoluted tubule is not specialized for reabsorption. Instead, its primary function is ion exchange. During ion exchange, cells reabsorb certain ions, returning them to the blood. Other ions are secreted from the blood into the tubule. The distal convoluted tubules of several nephrons enter one collecting duct. Many **collecting ducts** carry urine to the renal pelvis.

As shown in Figure 10.4, the glomerular capsule and the convoluted tubules always lie within the renal cortex. The loop of the nephron dips down into the renal medulla. A few nephrons have a very long loop of the nephron, which penetrates deep into the renal medulla. Collecting ducts are also located in the renal medulla, and together they give the renal pyramids their appearance.

MP3 The Functional Anatomy of the Urinary System

CHECK YOUR PROGRESS 10.2

1. Name the three major areas of a kidney.
2. Describe the different parts of a nephron with regard to structure and function.
3. Discuss why the nephrons have such an intricate capillary system.

CONNECTING THE CONCEPTS

For more information on the topics presented in this section, refer to the following discussions:

Section 4.5 illustrates the various forms of epithelial tissue and provides a function for each.

Section 5.2 summarizes the function of arterioles, capillaries, and venules in the circulatory system.

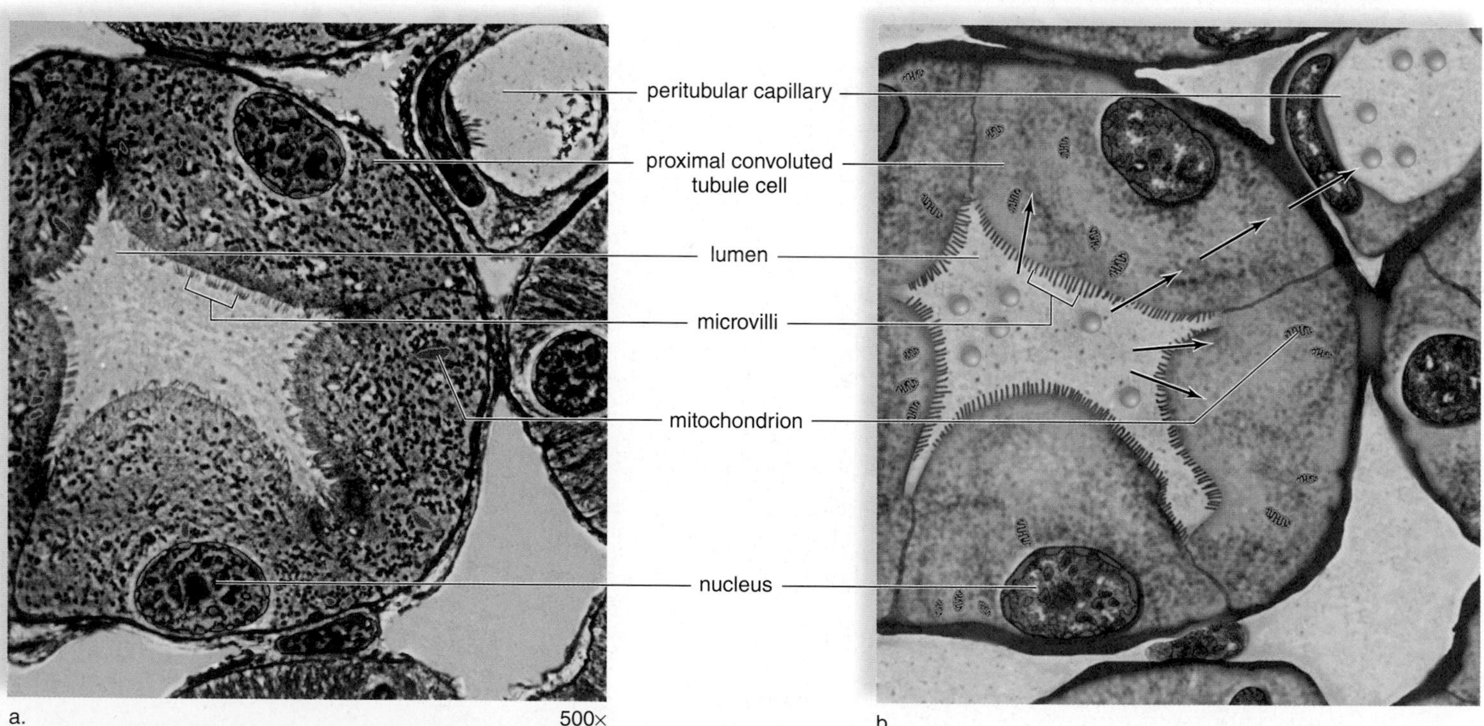

Figure 10.5 The specialized cells of the proximal convoluted tubule.
a. This photomicrograph shows that the cells lining the proximal convoluted tubule have a brushlike border composed of microvilli, which greatly increase the surface area exposed to the lumen. The peritubular capillary network surrounds the cells. **b.** Diagrammatic representation of (**a**) shows that each cell has many mitochondria, which supply the energy needed for active transport, the process that moves molecules (green) from the lumen of the tubule to the capillary, as indicated by the arrows.

10.3 Urine Formation

LEARNING OUTCOMES

Upon completion of this section, you should be able to

1. Summarize the three processes involved in the formation of urine.
2. List the components of the glomerular filtrate.
3. Describe how tubular reabsorption processes nutrient and salt molecules.
4. Explain the substances that are removed from the blood by tubular secretion.

Figure 10.6 gives an overview of urine formation, which is divided into three processes.

Glomerular Filtration

Glomerular filtration occurs when whole blood enters the glomerulus by way of the afferent arteriole. The afferent arteriole (Fig. 10.6, inset) has a larger diameter than the efferent arteriole, resulting in an increase in glomerular blood pressure. Because of this, water and small molecules move from the glomerulus to the inside of the glomerular capsule. This is a filtration process because large molecules and formed elements

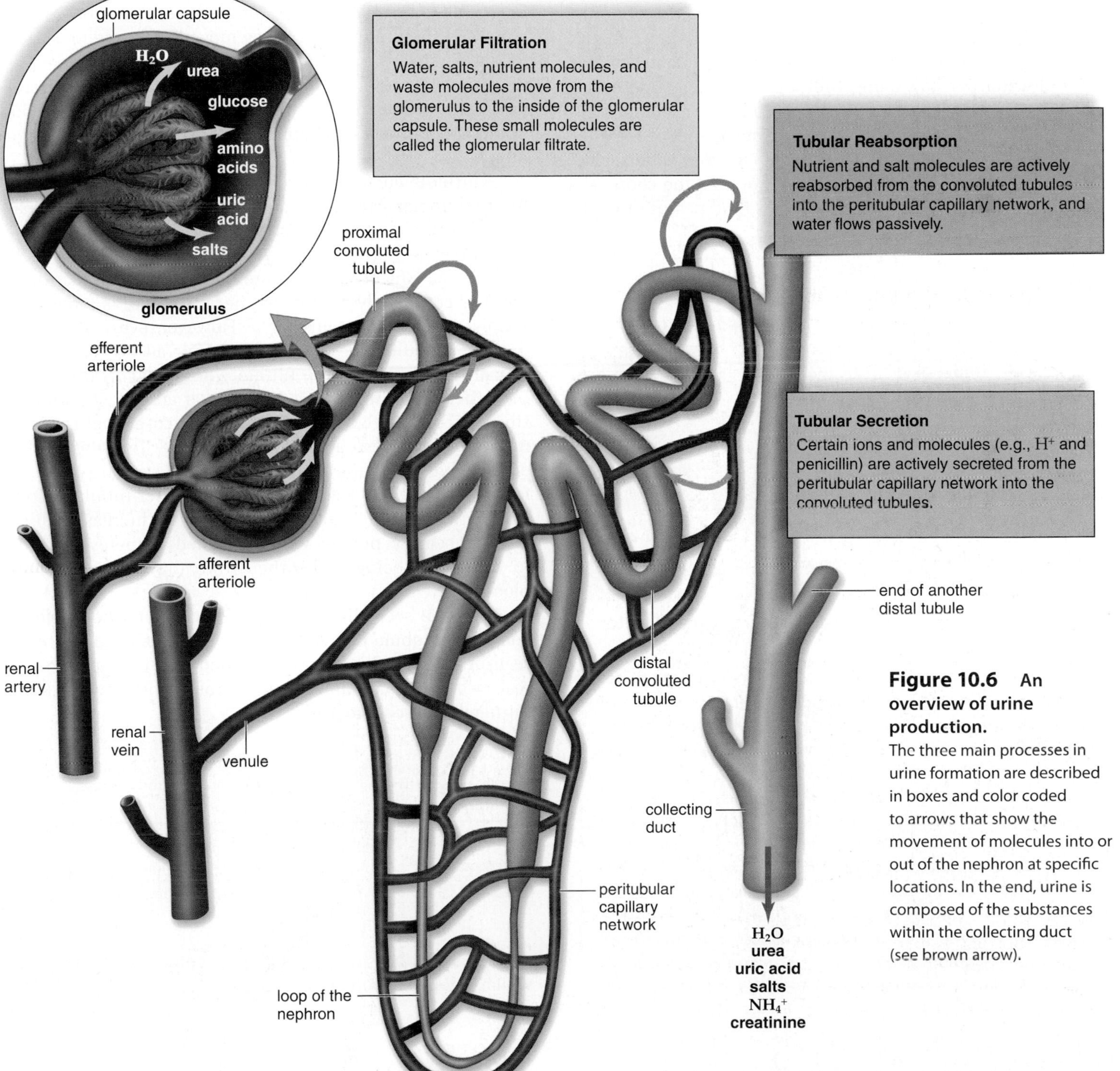
glomerular capsule
H_2O
urea
glucose
amino acids
uric acid
salts
glomerulus
Glomerular Filtration
Water, salts, nutrient molecules, and waste molecules move from the glomerulus to the inside of the glomerular capsule. These small molecules are called the glomerular filtrate.
Tubular Reabsorption
Nutrient and salt molecules are actively reabsorbed from the convoluted tubules into the peritubular capillary network, and water flows passively.
Tubular Secretion
Certain ions and molecules (e.g., H^+ and penicillin) are actively secreted from the peritubular capillary network into the convoluted tubules.
proximal convoluted tubule
efferent arteriole
afferent arteriole
renal artery
renal vein
venule
end of another distal tubule
distal convoluted tubule
collecting duct
peritubular capillary network
loop of the nephron
H_2O
urea
uric acid
salts
NH_4^+
creatinine

Figure 10.6 An overview of urine production. The three main processes in urine formation are described in boxes and color coded to arrows that show the movement of molecules into or out of the nephron at specific locations. In the end, urine is composed of the substances within the collecting duct (see brown arrow).

are unable to pass through the capillary wall. In effect, then, blood in the glomerulus has two portions: the filterable components and the nonfilterable components.

Filterable Blood Components	Nonfilterable Blood Components
Water	Formed elements (blood cells and platelets)
Nitrogenous wastes	Plasma proteins
Nutrients	
Salts (ions)	

The nonfilterable components leave the glomerulus by way of the efferent arteriole. The **glomerular filtrate** inside the glomerular capsule now contains the filterable blood components in approximately the same concentration as plasma.

As indicated in Table 10.1, nephrons in the kidneys filter 180 l of water per day, along with a considerable amount of small molecules (such as glucose) and ions (such as sodium). If the composition of urine were the same as that of the glomerular filtrate, the body would continually lose water, salts, and nutrients. Therefore, we can conclude that the composition of the filtrate must be altered as this fluid passes through the remainder of the tubule.

Table 10.1 Reabsorption from Nephrons

Substance	Amount Filtered (per Day)	Amount Excreted (per Day)	Reabsorption (%)
Water (liters)	180	1.8	99.0
Sodium (g)	630	3.2	99.5
Glucose (g)	180	0.0	100.0
Urea (g)	54	30.0	44.0

g = grams

APPLICATIONS AND MISCONCEPTIONS

Can you drink urine if there is no other source of water available?

You're probably familiar with news reports of trapped individuals who survived for days without water by drinking their urine. If you're a fan of *Man vs. Wild* on the Discovery Channel, you may remember an episode in the Australian Outback when Bear Grylls drank his urine.

In some cultures, people drink urine as an alternative medicine or for cosmetic reasons. This practice is known as urine therapy. Urine from someone who is healthy should be sterile and, therefore, drinkable. However, urine contains any number of other substances besides water, like salts and metabolites of various drugs. If these substances are highly concentrated, you may become dehydrated more quickly. Many survival guides advise against drinking your urine.

Tubular Reabsorption

Tubular reabsorption occurs as molecules and ions are passively and actively reabsorbed from the nephron into the blood of the peritubular capillary network. The osmolarity of the blood is maintained by the presence of plasma proteins and salt. Sodium ions (Na^+) are actively transported by one of two types of transport proteins. First, the movement of sodium may be coupled to the movement of larger solutes, such as amino acids or glucose. This is called a *symport*, because both solutes are being moved in the same direction. The second mechanism involves an *antiport* protein, which moves Na^+ ions into the cell, while transporting H^+ ions out of the cell. This also serves to regulate the pH balance of the blood, because the movement of H^+ ions outward reduces the acidity of the blood. As sodium ions are being moved, chloride ions (Cl^-) follow passively. The reabsorption of salt (NaCl) increases the osmolarity of the blood compared with the filtrate. Therefore, water moves passively from the tubule into the blood. About 65% of Na^+ is reabsorbed at the proximal convoluted tubule.

Nutrients such as glucose and amino acids return to the peritubular capillaries almost exclusively at the proximal convoluted tubule. This is a selective process because only molecules recognized by carrier proteins are actively reabsorbed. Glucose is an example of a molecule that ordinarily is completely reabsorbed because there is a plentiful supply of carrier proteins for it. However, every substance has a maximum rate of transport. After all its carriers are in use, any excess in the filtrate will appear in the urine. In **diabetes mellitus,** because the liver and muscles fail to store glucose as glycogen, the blood glucose level is above normal and glucose appears in the urine. The presence of excess glucose in the filtrate raises its osmolarity. Therefore, less water is reabsorbed into the peritubular capillary network. The frequent urination and increased thirst experienced by people with untreated diabetes are due to less water being reabsorbed from the filtrate into the blood.

We have seen that the filtrate that enters the proximal convoluted tubule is divided into two portions: components reabsorbed from the tubule into blood, and components not reabsorbed that continue to pass through the nephron to be further processed into urine.

Reabsorbed Filtrate Components	Nonreabsorbed Filtrate Components
Most water	Some water
Nutrients	Much nitrogenous waste
Required salts (ions)	Excess salts (ions)

The substances not reabsorbed become the tubular fluid, which enters the loop of the nephron.

BIOLOGY MATTERS  **Health**

Urinalysis

A routine urinalysis can detect abnormalities in urine color, concentration, and content. Significant information about the general state of one's health, particularly in the diagnosis of renal and metabolic diseases, is provided by this analysis also. A complete urinalysis consists of three phases of examination: physical, chemical, and microscopic.

Physical Examination

The physical examination describes urine color, clarity, and odor. The color of normal, fresh urine is usually pale yellow. However, the color may vary from almost colorless (dilute) to dark yellow (concentrated). Pink, red, or smoky brown urine is usually a sign of bleeding that may be due to a kidney, bladder, or urinary tract infection. Liver disorders can also produce dark brown urine. The clarity of normal urine may be clear or cloudy. However, a cloudy urine sample is also characteristic of abnormal levels of bacteria. Finally, urine odor is usually "nutty" or aromatic; but a foul-smelling odor is characteristic of a urinary tract infection. A sweet, fruity odor is characteristic of glucose in the urine or diabetes mellitus. Urine's odor is also affected by the consumption of garlic, curry, asparagus, and vitamin C.

Chemical Examination

The chemical examination is often done with a dipstick, a thin strip of plastic impregnated with chemicals that change color upon reaction with certain substances present in urine (Fig. 10C). The color change on each segment of the dipstick is compared with a standardized color chart. Dipsticks can be used to determine urine's specific gravity; pH; and content of glucose, bilirubin, urobilinogen, ketone, protein, nitrite, blood, and white blood cells (WBCs) in the urine.

1. *Specific gravity* is an indicator of how well the kidneys are able to adjust tonicity in urine. Normal values for urine specific gravity range from 1.002 to 1.035. A high value (concentrated urine) may be a result of dehydration or diabetes mellitus.
2. Normal *urine pH* can be as low as 4.5 and as high as 8.0. In patients with kidney stone disease, urine pH has a direct effect on the type of stones formed.
3. *Glucose* is normally not present in urine. If present, diabetes mellitus is suspected.
4. *Bilirubin* (a by-product of hemoglobin degradation) is not normally present in the urine. *Urobilinogen* (a by-product of bilirubin degradation) is normally present in very small amounts. High levels of bilirubin or urobilinogen may indicate liver disease.
5. *Ketones* are not normally found in urine. Urine ketones are a by-product of fat metabolism. The presence of ketones in the urine may indicate diabetes mellitus or use of a low-carbohydrate diet, such as the Atkins diet.

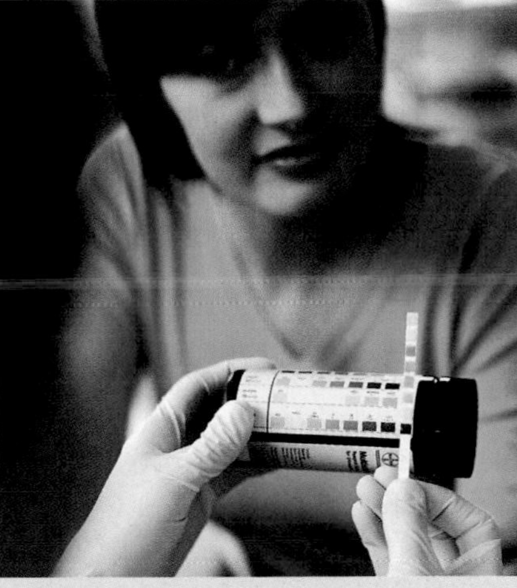

Figure 10C A urinalysis provides an indication of a person's overall health. This patient's urine has been tested and the multiple-test stick is being compared to a reference chart. The dark brown pad shows that the patient has glucose in her urine; this indicates that she has diabetes mellitus.

6. Plasma *proteins* should not be present in urine. A significant amount of urine protein (proteinuria) is usually a sign of kidney damage.
7. Urine typically does not contain *nitrates.* The presence of nitrites is a sign of urinary tract infection.
8. It is not abnormal for blood to show up in the urine when a woman is menstruating. Blood present in the urine at other times may indicate a bacterial infection or kidney damage.
9. The chemical test for *WBCs* is normally negative. A high urine WBC count usually indicates a bacterial infection somewhere in the urinary tract.

Microscopic Examination

For the microscopic examination, the urine is centrifuged, and the sediment (solid material) is examined under a microscope. When renal disease is present, the urine often contains an abnormal amount of cellular material. Urinary casts are sediments formed by the abnormal coagulation of protein material in the distal convoluted tubule or the collecting duct. The presence of crystals in the urine is characteristic of kidney stones, kidney damage, or problems with metabolism.

Forensic Analysis

Urinalysis is also used by the federal government and some businesses to screen potential employees for the use of numerous illegal drugs. Drugs associated with date rape (such as flunitrazepam [Rohypnol], known as "roofies," and GHB, or gamma-hydroxybutyrate) can be detected in the urine of victims to determine if they were drugged.

Questions to Consider

1. What might be the difference between the levels of water-soluble and fat-soluble chemicals in the urine?
2. Why does diabetes result in high levels of glucose in the urine?

Tubular Secretion

Tubular secretion is a second way by which substances are removed from blood and added to the tubular fluid. Hydrogen ions (H^+), creatinine, and drugs such as penicillin are some of the substances moved by active transport from blood into the kidney tubule. In the end, urine contains substances that have undergone glomerular filtration but have not been reabsorbed and substances that have undergone tubular secretion. Tubular secretion is now known to occur along the length of the kidney tubule.

Animation Kidney Function

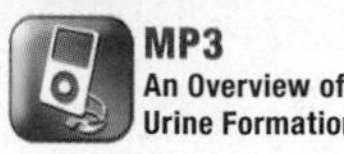
MP3 An Overview of Urine Formation

CHECK YOUR PROGRESS 10.3

1. List the three major processes in urine formation and state the location in the nephron where each occurs.
2. Detail the functions of glomerular filtration, tubular reabsorption, and tubular secretion.
3. Predict what might happen to the overall function of a nephron if glomerular filtration could not occur.

CONNECTING THE CONCEPTS

For more information on the topics presented in this section, refer to the following discussions:

Section 3.3 describes the processes of osmosis, active transport, and the movement of ions across a cell membrane.

Section 6.1 lists the substances found in the plasma component of blood.

10.4 Kidneys and Homeostasis

LEARNING OUTCOMES

Upon completion of this section, you should be able to

1. Summarize how the kidney maintains the water–salt balance of the body.
2. State the purpose of ADH, ANH, and aldosterone in homeostasis.
3. Explain how the kidneys assist in the maintenance of the pH levels of the blood.

The kidneys play a major role in homeostasis, from the maintenance of the water–salt balance in the body to regulating the pH of the blood. In doing so, the kidneys interact with every other organ system of the human body (Fig. 10.7).

Kidneys Excrete Waste Molecules

In Chapter 8, we compared the liver to a sewage treatment plant because it removes poisonous substances from the blood and prepares them for excretion. Similarly, the liver produces urea, the primary nitrogenous end product of humans, which is excreted by the kidneys. If the liver is a sewage treatment plant, the tubules of the kidney are like the trucks that take the sludge, prepared waste, away from the town (the body).

Metabolic waste removal is absolutely necessary for maintaining homeostasis. The blood must constantly be cleansed of the nitrogenous wastes, end products of metabolism. The liver produces urea, and muscles make creatinine. These wastes, and also uric acid from the cells, are carried by the cardiovascular system to the kidneys. The urine-producing kidneys are responsible for the excretion of nitrogenous wastes. They are assisted to a limited degree by the sweat glands in the skin, which excrete perspiration, a mixture of water, salt, and some urea. In times of kidney failure, urea is excreted by the sweat glands and forms a substance called uremic frost on the skin.

Water–Salt Balance

Most of the water found in the filtrate is reabsorbed into the blood before urine leaves the body. All parts of a nephron and the collecting duct participate in the reabsorption of water. The reabsorption of salt always precedes the reabsorption of water. In other words, water is returned to the blood by the process of osmosis. During the process of reabsorption, water passes through water channels, called **aquaporins,** within a plasma membrane protein.

Sodium ions (Na^+) are important in plasma. Usually more than 99% of the Na^+ filtered at the glomerulus is returned to the blood. The kidneys also excrete or reabsorb other ions, such as potassium ions (K^+), bicarbonate ions (HCO_3^-), and magnesium ions (Mg^{2+}) as needed.

Reabsorption of Salt and Water from Cortical Portions of the Nephron

The proximal convoluted tubule, the distal convoluted tubule, and the cortical portion of the collecting ducts are present in the renal cortex. Most of the water (65%) that enters the glomerular capsule is reabsorbed from the nephron into the blood at the proximal convoluted tubule. Na^+ is actively reabsorbed, and Cl^- follows passively. Aquaporins are always open and water is reabsorbed osmotically into the blood.

Hormones regulate the reabsorption of sodium and water in the distal convoluted tubule. **Aldosterone** is a hormone secreted by the adrenal glands, which sit atop the kidneys. This hormone promotes ion exchange at the distal convoluted tubule. Potassium ions (K^+) are excreted, and sodium ions (Na^+) are reabsorbed into the blood. The release of aldosterone is set into motion by the kidneys. The **juxtaglomerular apparatus** is a region of contact between the afferent arteriole and the distal convoluted tubule (Fig. 10.8). When blood volume (and, therefore, blood pressure) falls too low for filtration to occur, the juxtaglomerular apparatus can respond to the decrease by secreting **renin.** Renin is an enzyme that ultimately leads to secretion of aldosterone by the adrenal glands.

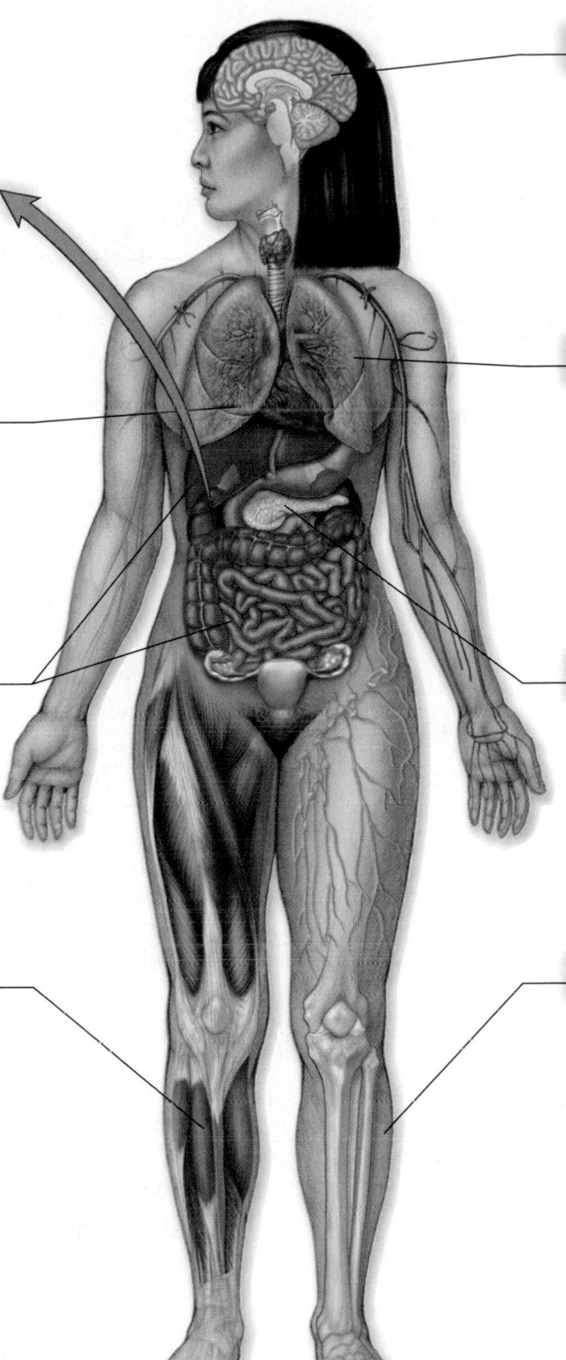

All systems of the body work with the urinary system to maintain homeostasis. These systems are especially noteworthy.
Urinary System
As an aid to all the systems, the kidneys excrete nitrogenous wastes and maintain the water–salt balance and the acid–base balance of the blood. The urinary system also specifically helps the other systems.
Cardiovascular System
Production of renin by the kidneys helps maintain blood pressure. Blood vessels transport nitrogenous wastes to the kidneys and carbon dioxide to the lungs. The buffering system of the blood helps the kidneys maintain the acid–base balance.
Digestive System
The liver produces urea excreted by the kidneys. The yellow pigment found in urine, called urochrome (breakdown product of hemoglobin), is produced by the liver. The digestive system absorbs nutrients, ions, and water. These help the kidneys maintain the proper level of ions and water in the blood.
Muscular System
The kidneys regulate the amount of ions in the blood. These ions are necessary to the contraction of muscles, including those that propel fluids in the ureters and urethra.
Nervous System
The kidneys regulate the amount of ions (e.g., K^+, Na^+, Ca^{2+}) in the blood. These ions are necessary for nerve impulse conduction. The nervous system controls urination.
Respiratory System
The kidneys and the lungs work together to maintain the acid–base balance of the blood.
Endocrine System
The kidneys produce renin, leading to the production of aldosterone, a hormone that helps the kidneys maintain the water–salt balance. The kidneys produce the hormone erythropoietin, and they change vitamin D to a hormone. The posterior pituitary secretes ADH, which regulates water retention by the kidneys.
Integumentary System
Sweat glands excrete perspiration, a solution of water, salt, and some urea.

Figure 10.7 The urinary system and homeostasis.
The urinary system works primarily with these systems to bring about homeostasis.

Research scientists speculate that excessive renin secretion—and thus, reabsorption of excess salt and water—might contribute to high blood pressure.

Aquaporins are not always open in the distal convoluted tubule. Another hormone called **antidiuretic hormone (ADH)** must be present. ADH is produced by the hypothalamus and secreted by the posterior pituitary according to the osmolarity of the blood. If our intake of water has been low, ADH is secreted by the posterior pituitary. Water moves from the distal convoluted tubule and the collecting duct into the blood.

Atrial natriuretic hormone (ANH) is a hormone secreted by the atria of the heart when cardiac cells are stretched due to increased blood volume. ANH inhibits the secretion of renin by the juxtaglomerular apparatus and the secretion of aldosterone by the adrenal glands. Its effect, therefore, is to promote the excretion of sodium ions (Na^+), called *natriuresis.* Normally, salt reabsorption creates an

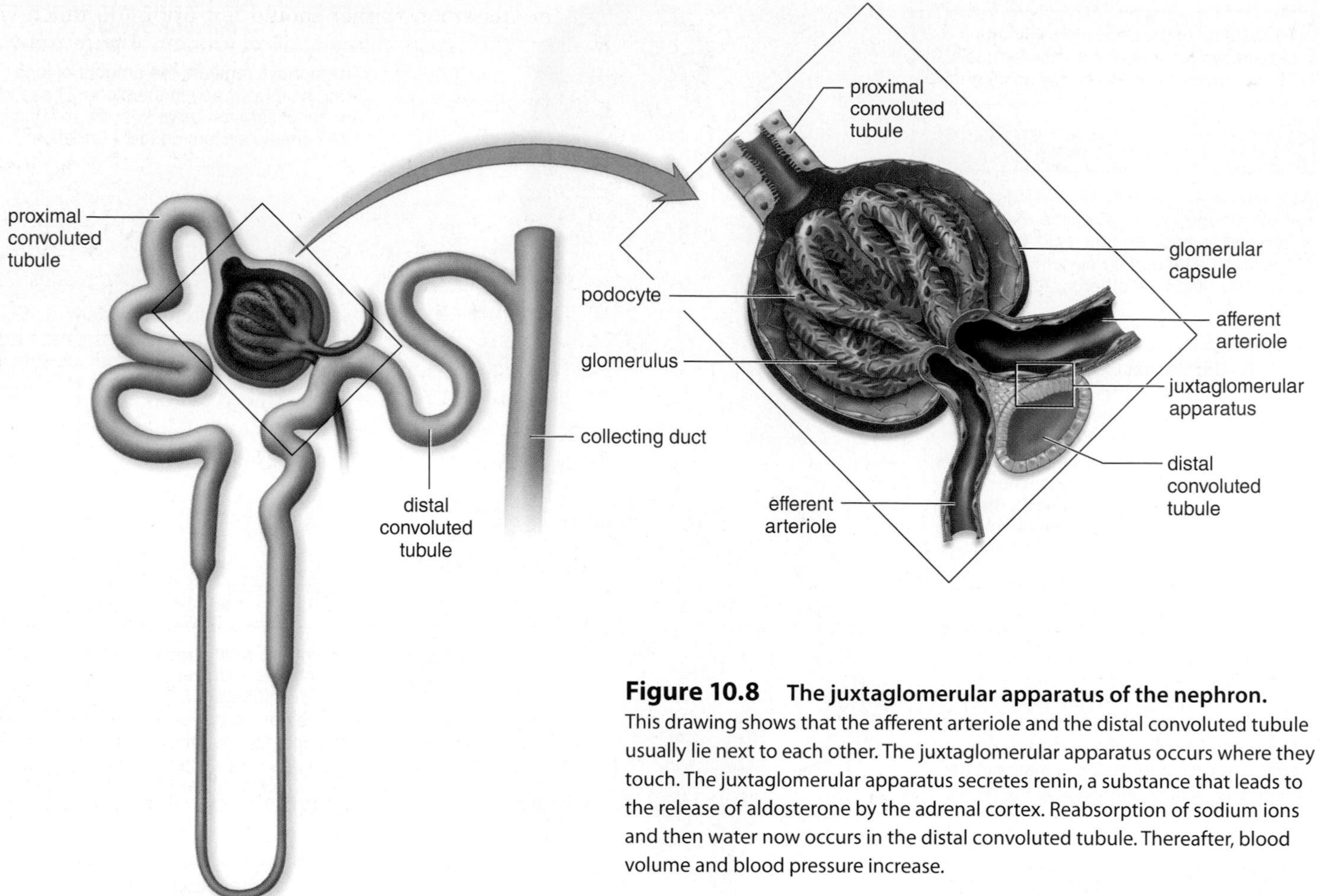

Figure 10.8 **The juxtaglomerular apparatus of the nephron.** This drawing shows that the afferent arteriole and the distal convoluted tubule usually lie next to each other. The juxtaglomerular apparatus occurs where they touch. The juxtaglomerular apparatus secretes renin, a substance that leads to the release of aldosterone by the adrenal cortex. Reabsorption of sodium ions and then water now occurs in the distal convoluted tubule. Thereafter, blood volume and blood pressure increase.

osmotic gradient that causes water to be reabsorbed. Thus, by causing salt excretion, ANH causes water excretion, too. If ANH is present, less water will be reabsorbed, even if ADH is also present.

Reabsorption of Salt and Water from Medullary Portions of the Nephron

The ability of humans to regulate the tonicity of their urine is dependent on the work of the medullary portions of the nephron (loop of the nephron) and the collecting duct.

The Loop of the Nephron A long loop of the nephron (also called the loop of Henle), which typically penetrates deep into the renal medulla, is made up of a descending limb and an ascending limb. Salt (NaCl) passively diffuses out of the lower portion of the *ascending limb.* Any remaining salt is actively transported from the thick upper portion of the limb into the tissue of the outer medulla (Fig. 10.9). In the end, the concentration of salt is greater in the direction of the inner medulla. Surprisingly, however, the inner medulla has an even higher concentration of solutes than expected. It is believed that urea leaks from the lower portion of the collecting duct, contributing to the high solute concentration of the inner medulla.

Water leaves the *descending limb* along its entire length via a countercurrent mechanism because of the osmotic gradient within the medulla. Although water is reabsorbed as soon as fluid enters the descending limb, the remaining fluid within the limb encounters an increasing osmotic concentration of solute. Therefore, water continues to be reabsorbed, even to the bottom of the descending limb. The ascending limb does not reabsorb water primarily because it lacks aquaporins (as indicated by the dark line in Fig. 10.9). Its job is to help establish the solute concentration gradient. Water is returned to the cardiovascular system when it is reabsorbed.

The Collecting Duct Fluid within the collecting duct encounters the same osmotic gradient established by the ascending limb of the nephron. Therefore, water diffuses from the entire length of the collecting duct into the blood if aquaporins are open, as they are if ADH is present.

To understand the action of ADH, consider its name, *antidiuretic hormone.* Diuresis means increased amount of urine, and antidiuresis means decreased amount of urine. When ADH is present, more water is reabsorbed (blood volume and pressure rise) and a decreased amount of urine results. ADH ultimately fine-tunes the tonicity of urine according to

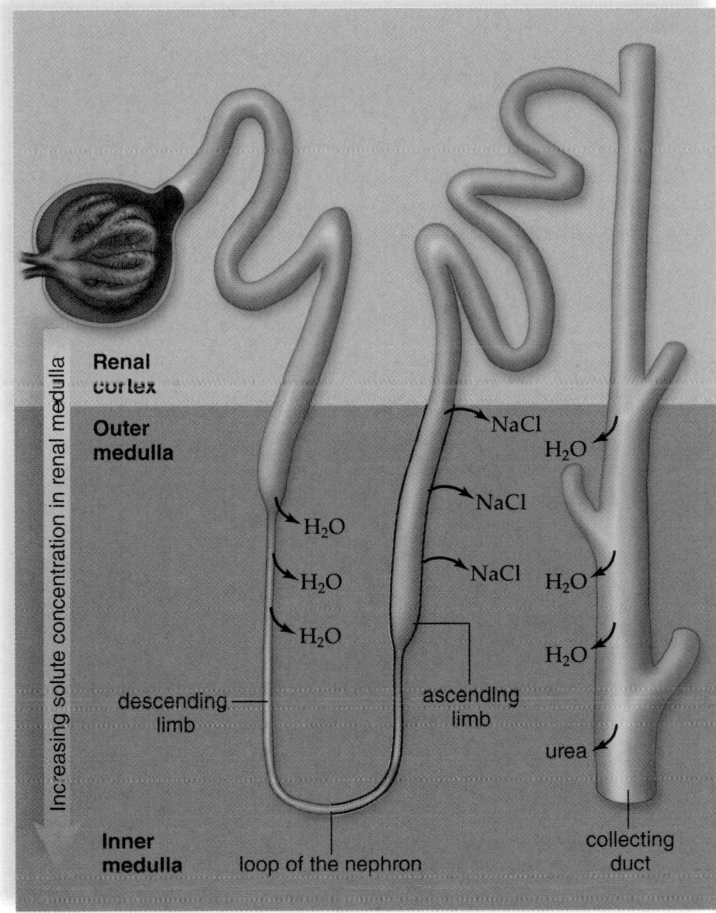

Figure 10.9 **Movement of salt and water within a nephron.** Salt (NaCl) diffuses and is actively transported out of the ascending limb of the loop of the nephron into the renal medulla; also, urea is believed to leak from the collecting duct and to enter the tissues of the renal medulla. This creates a hypertonic environment, which draws water out of the descending limb and the collecting duct. This water is returned to the cardiovascular system. (The thick black outline of the ascending limb means that it is impermeable to water.)

the needs of the body. For example, ADH is secreted at night when we are not drinking water, and this explains why the first urine of the day is more concentrated.

Interaction of Renin, Aldosterone, and ANH

If blood does not have the usual water–salt balance, blood volume and blood pressure are affected. Without adequate blood pressure, exchange across capillary walls cannot take place, nor is glomerular filtration possible in the kidneys.

What happens if you have insufficient sodium ions (Na^+) in your blood and tissue fluid? This can occur due to prolonged heavy sweating, as in athletes running a marathon. When blood Na^+ concentration falls too low, blood pressure falls and the renin-aldosterone sequence begins. Then the kidneys increase Na^+ reabsorption, conserving as much as possible. Subsequently, the osmolarity of the blood and the blood pressure return to normal.

The marathon runner should not drink too much water too fast. Quickly ingesting a large amount of pure water can dilute the body's remaining Na^+ and disrupt water–salt balance. While the sports drinks preferred by athletes contain both sodium and water, they may also be high in calories, and are not recommended for routine exercise lasting less than 90 minutes.

By contrast, think what happens if you eat a big tub of salty popcorn at the movies. When salt (NaCl) is absorbed from the digestive tract, the Na^+ content of the blood increases above normal. This results in increased blood volume. The atria of the heart are stretched by this increased blood volume, and the stretch triggers the release of ANH by the heart. ANH inhibits sodium and water reabsorption by the proximal convoluted tubule and collecting duct. Blood volume then decreases because more sodium and water are excreted in the urine.

Diuretics

Diuretics are chemicals that increase the flow of urine. Drinking alcohol causes diuresis because it inhibits the secretion of ADH. The dehydration that follows is believed to contribute to the symptoms of a hangover. Caffeine is a diuretic because it increases the glomerular filtration rate and decreases the tubular reabsorption of sodium ions (Na^+). Diuretic drugs developed to counteract high blood pressure also decrease the tubular reabsorption of Na^+. A decrease in water reabsorption and a decrease in blood volume and pressure follow.

Acid–Base Balance of Body Fluids

The pH scale, as discussed in Chapter 2, can be used to indicate the basicity (alkalinity) or the acidity of body fluids. A basic solution has a lesser hydrogen ion concentration [H^+] than the neutral pH of 7.0. An acidic solution has a greater [H^+] than neutral pH. The normal pH for body fluids is between 7.35 and 7.45. This is the pH at which our proteins, such as cellular enzymes, function properly. If the blood pH rises above 7.45, a person is said to have **alkalosis,** and if the blood pH decreases below 7.35, a person is said to have **acidosis.** Alkalosis and acidosis are abnormal conditions that may need medical attention.

The foods we eat add basic or acidic substances to the blood, and so does metabolism. For example, cellular respiration adds carbon dioxide that combines with water to form carbonic acid, and fermentation adds lactic acid. The pH of body fluids stays at just about 7.4 via several mechanisms, primarily acid–base buffer systems, the respiratory center, and the kidneys.

MP3 Acid-Base Balance

Acid–Base Buffer Systems

The pH of the blood stays near 7.4 because the blood is buffered. A **buffer** is a chemical or a combination of chemicals that can take up excess hydrogen ions (H^+) or excess hydroxide ions (OH^-). One of the most important buffers in the blood is a combination of carbonic acid (H_2CO_3) and bicarbonate ions (HCO_3^-). When hydrogen ions (H^+) are added to blood, the following reaction occurs:

$$H^+ + HCO_3^- \longrightarrow H_2CO_3$$

When hydroxide ions (OH^-) are added to blood, this reaction occurs:

$$OH^- + H_2CO_3 \longrightarrow HCO_3^- + H_2O$$

These reactions temporarily prevent any significant change in blood pH. A blood buffer, however, can be overwhelmed unless some more permanent adjustment is made. The next adjustment to keep the pH of the blood constant occurs at pulmonary capillaries.

Respiratory Center

As discussed in Chapter 9, the respiratory center in the medulla oblongata increases the breathing rate if the hydrogen ion concentration of the blood rises. Increasing the breathing rate rids the body of hydrogen ions because the following reaction takes place in pulmonary capillaries:

$$H^+ + HCO_3^- \rightleftharpoons H_2CO_3 \rightleftharpoons H_2O + CO_2$$

In other words, when carbon dioxide is exhaled, this reaction shifts to the right and the amount of hydrogen ions is reduced.

It is important to have the correct proportion of carbonic acid and bicarbonate ions in the blood. Breathing readjusts this proportion so that this particular acid–base buffer system can continue to absorb H^+ and OH^- as needed.

The Kidneys

As powerful as the acid–base buffer and the respiratory center mechanisms are, only the kidneys can rid the body of a wide range of acidic and basic substances and, otherwise, adjust the pH. The kidneys are slower acting than the other two mechanisms, but they have a more powerful effect on pH. For the sake of simplicity, we can think of the kidneys as reabsorbing bicarbonate ions and excreting hydrogen ions as needed to maintain the normal pH of the blood (Fig. 10.10). If the blood is acidic, hydrogen ions are excreted and bicarbonate ions are reabsorbed. If the blood is basic, hydrogen ions are not excreted and bicarbonate ions are not reabsorbed. The urine is usually acidic, so it follows that an excess of hydrogen ions is usually excreted. Ammonia (NH_3) provides another means of buffering and removing the hydrogen ions in urine:

$$H^+ + NH_3 \longrightarrow NH_4^+$$

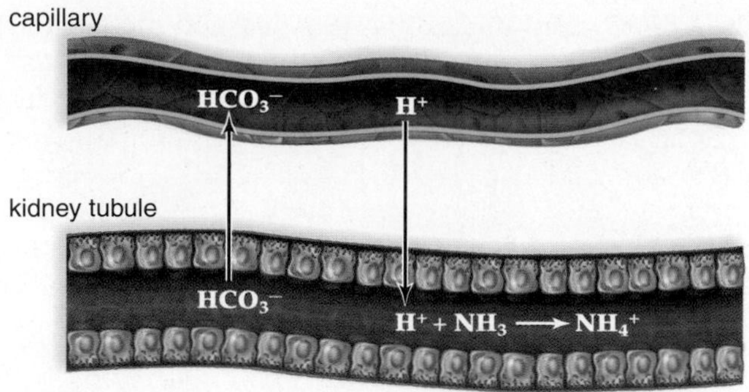

Figure 10.10 Blood pH is maintained by the kidneys. In the kidneys, bicarbonate ions (HCO_3^-) are reabsorbed, and hydrogen ions (H^+) are excreted as needed to maintain the pH of the blood. Excess hydrogen ions are buffered, for example, by ammonia (NH_3), which becomes ammonium (NH_4^+). Ammonia is produced in tubule cells by the deamination of amino acids.

Ammonia (whose presence is obvious in the diaper pail or kitty litter box) is produced in tubule cells by the deamination of amino acids. Phosphate provides another means of buffering hydrogen ions in urine.

The importance of the kidneys' ultimate control over the pH of the blood cannot be overemphasized. As mentioned, the enzymes of cells cannot continue to function if the internal environment does not have near-normal pH.

The Kidneys Assist Other Systems

Aside from producing renin, the kidneys assist the endocrine system and also the cardiovascular system by producing erythropoietin. Erythropoietin (EPO) is a hormone secreted by the kidneys. When blood oxygen decreases, EPO increases red blood cell synthesis by stem cells in the bone marrow (see Chapter 6). When the concentration of red blood cells increases, blood oxygen increases also. During kidney failure, the kidneys may produce less EPO, resulting in fewer red blood cells and symptoms of fatigue. Drugs such as epoetin (Procrit) represent a form of EPO produced by genetic engineering and biotechnology (see Chapter 21). Like normal EPO, epoetin increases red blood cell synthesis and energy levels. This drug is often used to stimulate red bone marrow production in patients in renal failure or recovering from chemotherapy.

APPLICATIONS AND MISCONCEPTIONS

Is it true that urine will neutralize the sting from a jellyfish?

You might remember the scene from the TV series *Survivor* when a player was stung by a jellyfish. The individual called for someone to come urinate on the sting. However, urine has not been shown to help jellyfish stings in scientific studies. Instead, experts recommend the application of vinegar to a sting to alleviate its effects.

The kidneys assist the skeletal, nervous, and muscular systems by helping to regulate the amount of calcium ions (Ca^{2+}) in the blood. The kidneys convert vitamin D to its active form needed for Ca^{2+} absorption by the digestive tract, and they regulate the excretion of electrolytes, including Ca^{2+}. The kidneys also regulate the sodium ion (Na^{+}) and potassium ion (K^{+}) content of the blood. These ions, needed for nerve conduction, are necessary to the contraction of the heart and other muscles in the body.

APPLICATIONS AND MISCONCEPTIONS

Does cranberry juice really prevent or cure a urinary tract infection?

Research has supported the use of cranberry juice to prevent urinary tract infections. It appears to prevent bacteria that would cause infection from adhering to the surfaces of the urinary tract. However, cranberry juice has not been shown to be an effective treatment for an already existing urinary tract infection.

Video Cranberries vs. Bacteria

CHECK YOUR PROGRESS 10.4

1. Detail the differences between dilute and concentrated urine.
2. Describe the action of the three hormones used to influence urine production, and discuss how they work together.
3. Summarize how the action of the kidneys regulates body fluid pH, and clarify why this is important to homeostasis.

CONNECTING THE CONCEPTS

For additional information on water–salt balance and maintenance of the pH of the blood, refer to the following discussions:

Section 2.2 explains how buffers regulate pH levels.

Section 9.5 examines the role of the respiratory center in the medulla oblongata of the brain.

Section 15.4 explores the production of aldosterone by the adrenal glands.

10.5 Kidney Function Disorders

LEARNING OUTCOMES

Upon completion of this section, you should be able to

1. List the major diseases of the urinary system, and summarize their causes.
2. Describe how hemodialysis can help restore homeostasis of the blood in the event of kidney failure.

Many types of illnesses—especially diabetes, hypertension, and inherited conditions—cause progressive renal disease and renal failure. Infections are also contributory. If the infection is localized in the urethra, it is called **urethritis.** If the infection invades the urinary bladder, it is called **cystitis.** Finally, if the kidneys are affected, the infection is called **pyelonephritis.**

Urinary tract infections, an enlarged prostate gland, pH imbalances, or an intake of too much calcium can lead to kidney stones. Kidney stones are hard granules made of calcium, phosphate, uric acid, and protein. Kidney stones form in the renal pelvis and usually pass unnoticed in the urine flow. If they grow to several centimeters and block the renal pelvis or ureter, a reverse pressure builds up and destroys nephrons. When a large kidney stone passes, strong contractions within a ureter can be excruciatingly painful.

One of the first signs of nephron damage is albumin, white blood cells, or even red blood cells in the urine. As described in the Health feature, "Urinalysis," on page 227, a urinalysis can detect urine abnormalities rapidly. If damage is so extensive that more than two-thirds of the nephrons are inoperative, urea and other waste substances accumulate in the blood. This condition is called **uremia.** Although nitrogenous wastes can cause serious damage, the retention of water and salts is of even greater concern. The latter causes edema, fluid accumulation in the body tissues. Imbalance in the ionic composition of body fluids can lead to loss of consciousness and to heart failure.

Hemodialysis

Patients with renal failure can undergo **hemodialysis,** using either an artificial kidney machine or continuous ambulatory peritoneal dialysis (CAPD). *Dialysis* is defined as the diffusion of dissolved molecules through a semipermeable natural or synthetic membrane that has pore sizes that allow only small molecules to pass through. In an artificial kidney machine (Fig. 10.11), the patient's blood is passed through a membranous tube that is in contact with a dialysis solution, or dialysate.

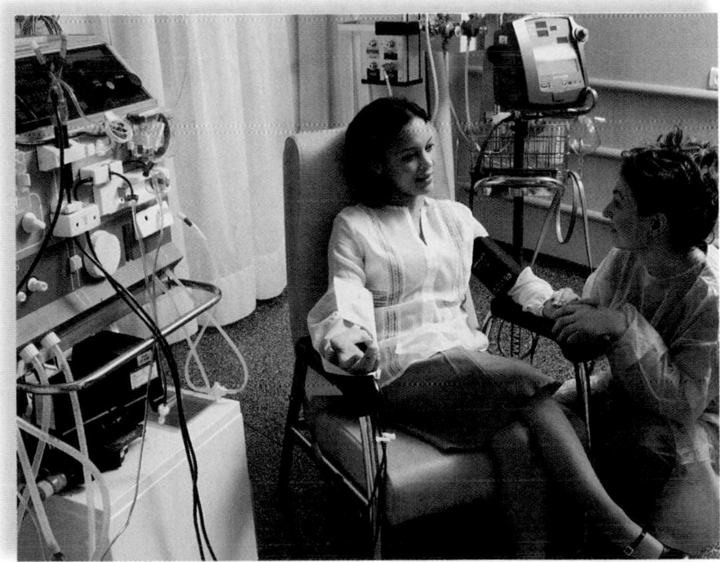

Figure 10.11 **Hemodialysis using an artificial kidney machine.** As the patient's blood is pumped through dialysis tubing, it is exposed to a dialysate (dialysis solution). Wastes exit from blood into the solution because of a preestablished concentration gradient. In this way, not only is blood cleansed but its water–salt and acid–base balances can also be adjusted.

Substances more concentrated in the blood diffuse into the dialysate, and substances more concentrated in the dialysate diffuse into the blood. The dialysate is continuously replaced to maintain favorable concentration gradients. In this way, the artificial kidney can be used either to extract substances from blood, including waste products or toxic chemicals and drugs, or to add substances to blood—for example, bicarbonate ions (HCO_3^-) if the blood is acidic. In the course of a 3- to 6-hour hemodialysis, from 50 to 250 g of urea can be removed from a patient, which greatly exceeds the amount excreted by normal kidneys. Therefore, a patient needs to undergo treatment only about twice a week.

CAPD is so named because the peritoneum is the dialysis membrane. A fresh amount of dialysate is introduced directly into the abdominal cavity from a bag that is temporarily attached to a permanently implanted plastic tube. The dialysate flows into the peritoneal cavity by gravity. Waste and salt molecules pass from the blood vessels in the abdominal wall into the dialysate before the fluid is collected 4 to 8 hours later. The solution is drained into a bag from the abdominal cavity by gravity, and then it is discarded. One advantage of CAPD over an artificial kidney machine is that the individual can go about his or her normal activities during CAPD.

Replacing a Kidney

Patients with renal failure sometimes undergo a kidney transplant operation, during which a functioning kidney from a donor is received. As with all organ transplants, there is the possibility of organ rejection. Receiving a kidney from a close relative has the highest chance of success. The current one-year survival rate is 97% if the kidney is received from a relative and 90% if it is received from a nonrelative. In the future, transplantable kidneys may be created in a laboratory. Another option could be to use kidneys from specially bred pigs whose organs would not be antigenic to humans.

CHECK YOUR PROGRESS 10.5

1. List and detail a few common causes of renal disease.
2. Describe hemodialysis, and relate its function to that of a kidney.
3. Explain why hemodialysis would need to be done frequently in a patient with renal failure.

CONNECTING THE CONCEPTS

For more information on organ transplants, refer to the following discussions:

Section 5.7 describes the options available for individuals experiencing heart failure.

Section 7.6 examines how the immune system potentially interferes with organ transplants.

Section 19.4 provides an overview of how bone marrow transplants can be used to treat cancer.

CASE STUDY CONCLUSION

Michael and Jada learned that there are two forms of polycystic kidney disease (PKD) and that both are genetic disorders. The majority of cases (over 90%) belong to a class called autosomal dominant PKD, named after the type of mutation that causes the disease (see Chapter 20). Individuals with this form of PKD typically develop symptoms in their thirties and forties. The other, rarer form, is called autosomal recessive PKD, and it is caused by a defect in a gene called *PKHD1*. *PKHD1* contains the instructions for the manufacture of a protein called fibrocystin, an important molecule in ensuring that the kidneys function correctly. Unfortunately for Michael and Jada, the genetic test indicated that they were both carriers for a defective copy of *PKHD1*, and that Aiesha had inherited a defective copy from both parents. Without functioning fibrocystin protein, Aiesha's kidneys—and also possibly her liver—would eventually fail as the nephrons inside her kidneys struggled to perform their function and fluid accumulated inside the cysts. The only alternative for Aiesha would be a kidney transplant. Luckily for Aiesha, kidney transplants have a 90%+ success rate, so if a compatible donor could be found, the chances of Aiesha having a healthy childhood were very good.

MEDIA STUDY TOOLS

Enhance your study of this chapter with media! Visit **www.mhhe.com/maderhuman13e** and go to "Media Study Tools" for this chapter to access the following:

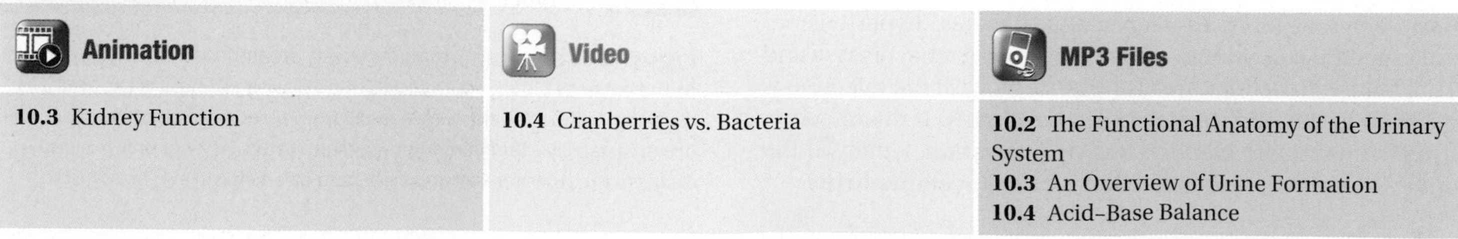

Animation	Video	MP3 Files
10.3 Kidney Function	**10.4** Cranberries vs. Bacteria	**10.2** The Functional Anatomy of the Urinary System **10.3** An Overview of Urine Formation **10.4** Acid–Base Balance

SUMMARIZE

10.1 The Urinary System

The **urinary system** maintains the salt, water, and pH homeostasis of the body through **excretion** of waste materials.

Functions of the Urinary System

The functions of the urinary system include the following:

- excreting nitrogenous wastes, including **urea, uric acid,** and **creatinine.** Elevated levels of urea result in **uremia,** while increased levels of uric acid cause **gout.**
- maintaining the normal water-salt balance of the blood;
- maintaining the acid-base balance of the blood; and
- assisting the endocrine system in hormone secretion by secreting **erythropoietin** (**EPO**) and releasing renin.

Organs of the Urinary System

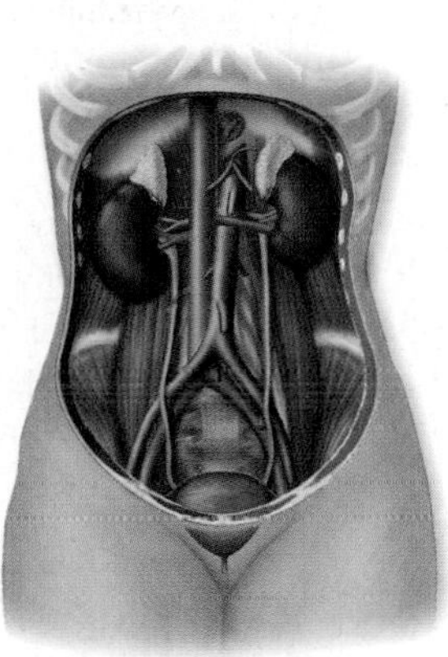

- **Kidneys** produce urine. **Renal arteries** and **renal veins** transport blood to and from the kidneys.
- **Ureters** transport urine to the bladder.
- The **urinary bladder** stores urine.
- The **urethra** releases urine to the outside.

10.2 Kidney Structure

- Macroscopic structures are the **renal cortex, renal medulla,** and **renal pelvis.**
- Microscopic structures are the **nephrons.**

Anatomy of a Nephron

- Each nephron has its own blood supply. The afferent arteriole divides to become the **glomerulus.** The efferent arteriole branches into the **peritubular capillary network.**
- Each region of the nephron is anatomically suited to its task in urine formation. The **glomerular capsule** begins the filtration of the blood. **Proximal convoluted tubules** are involved in the reabsorption of filtrates. The **loop of the nephron** facilitates the reabsorption of water, and the **distal convoluted tubule** is involved in ion exchange.
- The **collecting ducts** of the nephrons move urine to the renal pelvis, where it enters the ureters.

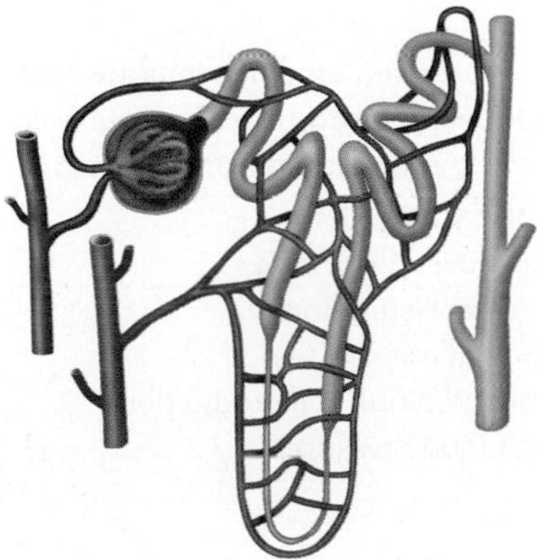

10.3 Urine Formation

Urine is composed primarily of nitrogenous waste products and salts in water. The steps in urine formation include the following:

Glomerular Filtration

Glomerular filtration separates the filterable and nonfilterable components of the blood. Water, salts, nutrients, and wastes move from the glomerulus to the inside of the glomerular capsule. **The glomerular filtrate** contains the filterable components.

Tubular Reabsorption

Tubular reabsorption reabsorbs ions and other molecules from the nephron into the blood. Nutrients and salt molecules are reabsorbed from the convoluted tubules into the peritubular capillary network; water follows. In **diabetes mellitus,** tubular reabsorption is unable to completely reabsorb the excess glucose, and some is excreted in the urine.

Tubular Secretion

Tubular secretion actively transports certain molecules from the peritubular capillary network into the convoluted tubules.

10.4 Kidneys and Homeostasis

- The kidneys maintain the water-salt balance of blood using transport channels called **aquaporins.**
- The kidneys also keep blood pH within normal limits.
- The urinary system works with the other systems of the body to maintain homeostasis.

Water-Salt Balance

- If blood volume falls, the **juxtaglomerular apparatus** releases **renin,** which in turn influences the production of aldosterone.
- In the renal cortex, most of the reabsorption of salts and water occurs in the proximal convoluted tubule. **Aldosterone** controls the reabsorption of sodium, and **antidiuretic hormone** (**ADH**) controls the reabsorption of water in the distal convoluted tubule. **Atrial natriuretic hormone** (**ANH**) acts contrary to aldosterone.
- In the renal medulla, the ascending limb of the loop of the nephron establishes a solute gradient that increases toward the inner medulla. The solute gradient draws water from the descending limb of the loop of the nephron and from the collecting duct. When ADH is present, more water is reabsorbed from the collecting duct and a decreased amount of urine results.
- **Diuretics** in the diet (caffeine) increase the flow of urine from the body.

Acid-Base Balance of Body Fluids

- The kidneys maintain the acid-base balance of blood (blood pH). **Alkalosis** occurs when the blood pH is above 7.45, whereas **acidosis** represents a blood pH below 7.35.
- **Buffers** help maintain blood pH at about 7.4. Ammonia buffers H^+ in the urine.

10.5 Kidney Function Disorders

- Various types of problems, including diabetes, kidney stones, and infections, can lead to renal failure, which necessitates either undergoing hemodialysis—by using a kidney machine or CAPD—or receiving a kidney transplant.
- **Urethritis, cystitis,** and **pyelonephritis** are infections of the urethra, urinary bladder, and kidneys, respectively.
- During kidney failure, **hemodialysis** may be needed to perform the normal functions of the kidneys.

ASSESS

Testing Your Knowledge of the Concepts

1. In what ways do the four functions of the urinary system contribute to homeostasis? (page 218)
2. Describe the path of urine and the structure and function of each organ mentioned. (pages 219–220)
3. Describe the macroscopic structure of the kidney. How does the structure aid in the function? (page 222)
4. Trace the path of blood into and out of the kidney. Why is the kidney so highly vascularized? (page 222)
5. Name the parts of a nephron, and describe the structure of each part. (pages 223–234)
6. Why would it be proper to associate glucose with the first two processes involved in urine formation, glomerular filtration and tubular reabsorption, but not the third process, tubular secretion? (pages 225–228)
7. Explain how hypertonic urine can be formed, detailing where and how salt and water move, and the influence of hormones on the process. (pages 228–231)
8. Describe three ways that the body maintains the acid–base balance of body fluids. (pages 231–232)
9. How do the kidneys assist other body systems? (pages 232–233)
10. Explain how the process of dialysis cleanses the blood. (page 233)
11. Which of these is found in the renal medulla?
    a. loop of the nephron
    b. collecting ducts
    c. peritubular capillaries
    d. All of the choices are correct.
12. Which of these functions of the kidneys are mismatched?
    a. excretes metabolic wastes—rids the body of urea
    b. maintains the water–salt balance—helps regulate blood pressure
    c. maintains the acid–base balance—rids the body of uric acid
    d. secretes hormones—secretes erythropoietin
    e. All of the choices are correct.
13. Which of the following is not correct?
    a. Uric acid is produced from the breakdown of amino acids.
    b. Creatinine is produced from breakdown reactions in the muscles.
    c. Urea is the primary nitrogenous waste of humans.
    d. Ammonia results from the deamination of amino acids.
14. When tracing the path of filtrate, the loop of the nephron follows which structure?
    a. collecting duct
    b. distal convoluted tubule
    c. proximal convoluted tubule
    d. glomerulus
    e. renal pelvis
15. When tracing the path of blood, the blood vessel that follows the renal artery is the
    a. peritubular capillary.
    b. efferent arteriole.
    c. afferent arteriole.
    d. renal vein.
    e. glomerulus.
16. The function of the descending limb of the loop of the nephron in the process of urine formation is
    a. reabsorption of water.
    b. production of filtrate.
    c. reabsorption of solutes.
    d. secretion of solutes.
17. Which of the following materials would not normally be filtered from the blood at the glomerulus?
    a. water
    b. urea
    c. protein
    d. glucose
    e. sodium ions
18. Which of the following materials would not be maximally reabsorbed from the filtrate?
    a. water
    b. glucose
    c. sodium ions
    d. urea
    e. amino acids
19. Reabsorption of the glomerular filtrate occurs primarily at the
    a. proximal convoluted tubule.
    b. distal convoluted tubule.
    c. loop of the nephron.
    d. collecting duct.
20. Sodium is actively extruded from which part of the nephron?
    a. descending portion of the proximal convoluted tubule
    b. ascending portion of the loop of the nephron
    c. ascending portion of the distal convoluted tubule
    d. descending portion of the collecting duct
21. Which of these hormones is most likely to directly cause a drop in blood pressure?
    a. aldosterone
    b. antidiuretic hormone (ADH)
    c. erythropoietin
    d. atrial natriuretic hormone (ANH)
22. The presence of ADH (antidiuretic hormone) causes an individual to excrete
    a. sugars.
    b. less water.
    c. more water.
    d. Both a and c are correct.
23. To lower blood acidity,
    a. hydrogen ions are excreted and bicarbonate ions are reabsorbed.
    b. hydrogen ions are reabsorbed and bicarbonate ions are excreted.
    c. hydrogen ions and bicarbonate ions are reabsorbed.
    d. hydrogen ions and bicarbonate ions are excreted.
    e. urea, uric acid, and ammonia are excreted.
24. The function of erythropoietin is
    a. reabsorption of sodium ions.
    b. excretion of potassium ions.
    c. reabsorption of water.
    d. to stimulate red blood cell production.
    e. to increase blood pressure.

26. Label this diagram of a nephron.

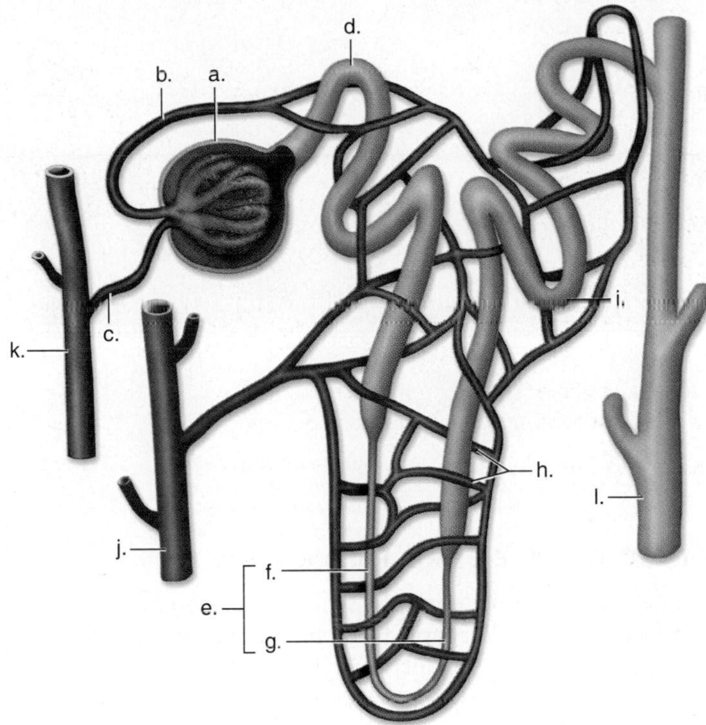

ENGAGE

Thinking Critically About the Concepts

A form of diabetes that may be unfamiliar to most people is diabetes insipidus. When the word "diabetes" is used, everyone tends to think of the type related to insulin. That form of diabetes is diabetes mellitus, which can be type 1 or type 2. In diabetes insipidus, the kidneys are unable to perform their function of water and salt homeostasis due to a lack of ADH in the body. In both cases of diabetes, kidney function is affected.

1. Why would frequent urination be a symptom of both diabetes mellitus and diabetes insipidus?
2. Why would increasing the production of red blood cells in people in renal failure alleviate their symptom of fatigue?
3. A side effect of using medications to calm an overactive bladder is constipation. Look back to Chapter 8 and identify the type of muscle found throughout the digestive system. Speculate as to why constipation might occur after the use of tolterodine (Detrol LA).
4. Men who have prostate cancer often have some or all of the prostate removed. The surgeon tries very hard not to damage the sphincter muscles associated with the bladder. What problem might a man experience if the bladder's sphincter muscles were damaged during the surgery?
5. Recall the anatomical position of the kidneys, and consider the ways they might be damaged in sports. Describe special equipment designed specifically to protect the kidneys.

CHAPTER 11

Skeletal System

CHAPTER CONCEPTS

11.1 Overview of the Skeletal System
Bones are the organs of the skeletal system. The tissues of the system are compact and spongy bone, various types of cartilage, and fibrous connective tissue in the ligaments that hold bones together.

11.2 Bones of the Axial Skeleton
The axial skeleton lies in the midline of the body and consists of the skull, the hyoid bone, the vertebral column, and the rib cage.

11.3 Bones of the Appendicular Skeleton
The appendicular skeleton consists of the bones of the pectoral girdle, upper limbs, pelvic girdle, and lower limbs.

11.4 Articulations
Joints are classified according to their degree of movement. Synovial joints are freely movable.

11.5 Bone Growth and Homeostasis
Bone is a living tissue that grows, remodels, and repairs itself. Bone homeostasis is performed by a variety of cells.

BEFORE YOU BEGIN

Before beginning this chapter, take a few moments to review the following discussions:

Section 4.1 What is the role of connective tissue in the body?

Section 4.8 How does the skeletal system contribute to homeostasis?

Section 8.6 What are the roles of calcium and vitamin D in the body?

CASE STUDY KNEE REPLACEMENT

Jackie was an outstanding athlete in high school, and even now, in her early fifties, she tries to stay in shape. However, during her customary 3-mile jogs, she has been having an increasingly hard time ignoring the pain in her left knee. She had torn some ligaments in her knee playing intramural and intermural volleyball in college, and it had never quite felt the same. In her forties, she was able to control the pain by taking over-the-counter medications; but two years ago she had had arthroscopic surgery to remove some torn cartilage and calcium deposits. Now that the pain was getting worse than before, she knew her best option might be a total knee replacement.

Although it sounds drastic, replacing old, arthritic joints with new artificial ones is becoming increasingly routine. Almost 4 million Americans, and 1 in 20 individuals over the age of 50, have undergone either partial or total knee replacement surgery. During the procedure, a surgeon removes bone from the bottom of the femur and the top of the tibia and replaces each with caps made of metal or ceramic, held in place with bone cement. A plastic plate is installed to allow the femur and tibia to move smoothly against each other, and a smaller plate is attached to the kneecap (patella) so that it can function properly.

As you read through the chapter, think about the following questions:

1. What is the role of cartilage in the knee joint?
2. What specific portions of these long bones are being removed during knee replacement?
3. Why does Jackie's physical condition make her an ideal candidate for knee replacement?

11.1 Overview of the Skeletal System

LEARNING OUTCOMES

Upon completion of this section, you should be able to

1. State the functions of the skeletal system.
2. Describe the structure of a long bone and list the types of tissues it contains.
3. List the three types of cartilage found in the body and provide a function for each.

The **skeletal system** consists of two types of connective tissue: bone and the cartilage found at **joints.** In addition, **ligaments** formed of fibrous connective tissue join the bones.

Functions of the Skeleton

The skeleton does more than merely provide a frame for the body. For example, consider the following functions:

The skeleton supports the body. The bones of the legs support the entire body when we are standing, and bones of the pelvic girdle support the abdominal cavity.

The skeleton protects soft body parts. The bones of the skull protect the brain; the rib cage protects the heart and lungs; and the vertebrae protect the spinal cord, which makes nervous connections to all the muscles of the limbs.

The skeleton produces blood cells. All bones in the fetus have red bone marrow that produces blood cells. In the adult, only certain bones produce blood cells.

The skeleton stores minerals and fat. All bones have a matrix that contains calcium phosphate, a source of calcium ions and phosphate ions in the blood. Fat is stored in yellow bone marrow.

The skeleton, along with the muscles, permits flexible body movement. Whereas articulations (joints) occur between all the bones, we associate body movement in particular with the bones of limbs.

Anatomy of a Long Bone

The bones of the body vary greatly in size and shape. To better understand the anatomy of a bone, we will use a long bone (Fig. 11.1), a type of bone common in the arms and legs. The shaft, or main portion of the bone, is called the *diaphysis.* The diaphysis has a large **medullary cavity,** whose walls are composed of compact bone. The medullary cavity is lined with a thin, vascular membrane (the endosteum) and is filled with yellow bone marrow that stores fat.

The expanded region at the end of a long bone is called an *epiphysis* (pl., epiphyses). The epiphyses are composed largely of spongy bone that contains red bone marrow, where blood cells are made. The epiphyses are coated with a thin layer of hyaline cartilage, which is also called articular cartilage because it occurs at a joint.

Except for the articular cartilage on the bone's ends, a long bone is completely covered by a layer of fibrous connective tissue called the **periosteum.** This covering contains blood vessels, lymphatic vessels, and nerves. Note in Figure 11.1 how a blood vessel penetrates the periosteum and enters the bone. Branches of the blood vessel are found throughout the medullary cavity. Other branches can be found in hollow cylinders called central canals within the bone tissue. The periosteum is continuous with ligaments and tendons connected to a bone.

MP3 Bone Structure

APPLICATIONS AND MISCONCEPTIONS

Does everyone have 206 bones?

A newborn has nearly 300 bones, some of which fuse together as the child grows. The adult human skeleton has approximately 206 bones, but the number varies between individuals. Some people have extra bones, called Wormian bones, that help to fuse skull bones together. Others may have additional small bones in the ankles and feet.

Bone

Compact bone is highly organized and composed of tubular units called osteons. In a cross section of an osteon, bone cells called **osteocytes** lie in **lacunae** (sing., lacuna), tiny chambers arranged in concentric circles around a central canal (Fig. 11.1). Matrix fills the space between the rows of lacunae. Tiny canals called *canaliculi* (sing., canaliculus) run through the matrix. These canaliculi connect the lacunae with one another and with the central canal. The cells stay in contact by strands of cytoplasm that extend into the canaliculi. Osteocytes nearest the center of an osteon exchange nutrients and wastes with the blood vessels in the central canal. These cells then pass on nutrients and collect wastes from the other cells via gap junctions (see Fig. 3.16).

Compared with compact bone, **spongy bone** has an unorganized appearance (Fig. 11.1). It contains numerous thin plates, called *trabeculae,* separated by unequal spaces. Although this makes spongy bone lighter than compact bone, spongy bone is still designed for strength. Just as braces are used for support in buildings, the trabeculae follow lines of stress. The spaces of spongy bone are often filled with **red bone marrow,** a specialized tissue that produces all types of blood cells. The osteocytes of spongy bone are irregularly placed within the trabeculae. Canaliculi bring them nutrients from the red bone marrow.

MP3 Bone Histology

Cartilage

Cartilage is not as strong as bone, but it is more flexible. Its matrix is gel-like and contains many collagenous and elastic fibers. The cells, called **chondrocytes,** lie within lacunae that are irregularly grouped. Cartilage has no nerves, making it well suited for padding joints where the stresses of movement are intense. Cartilage also has no blood vessels and relies on neighboring tissues for nutrient and waste exchange. This makes it slow to heal.

The three types of cartilage differ according to the type and arrangement of fibers in the matrix. *Hyaline cartilage* is firm and somewhat flexible. The matrix appears uniform and glassy, but actually it contains a generous supply of collagen fibers. Hyaline cartilage is found at the ends of long bones, in the nose, at the ends of the ribs, and in the larynx and trachea.

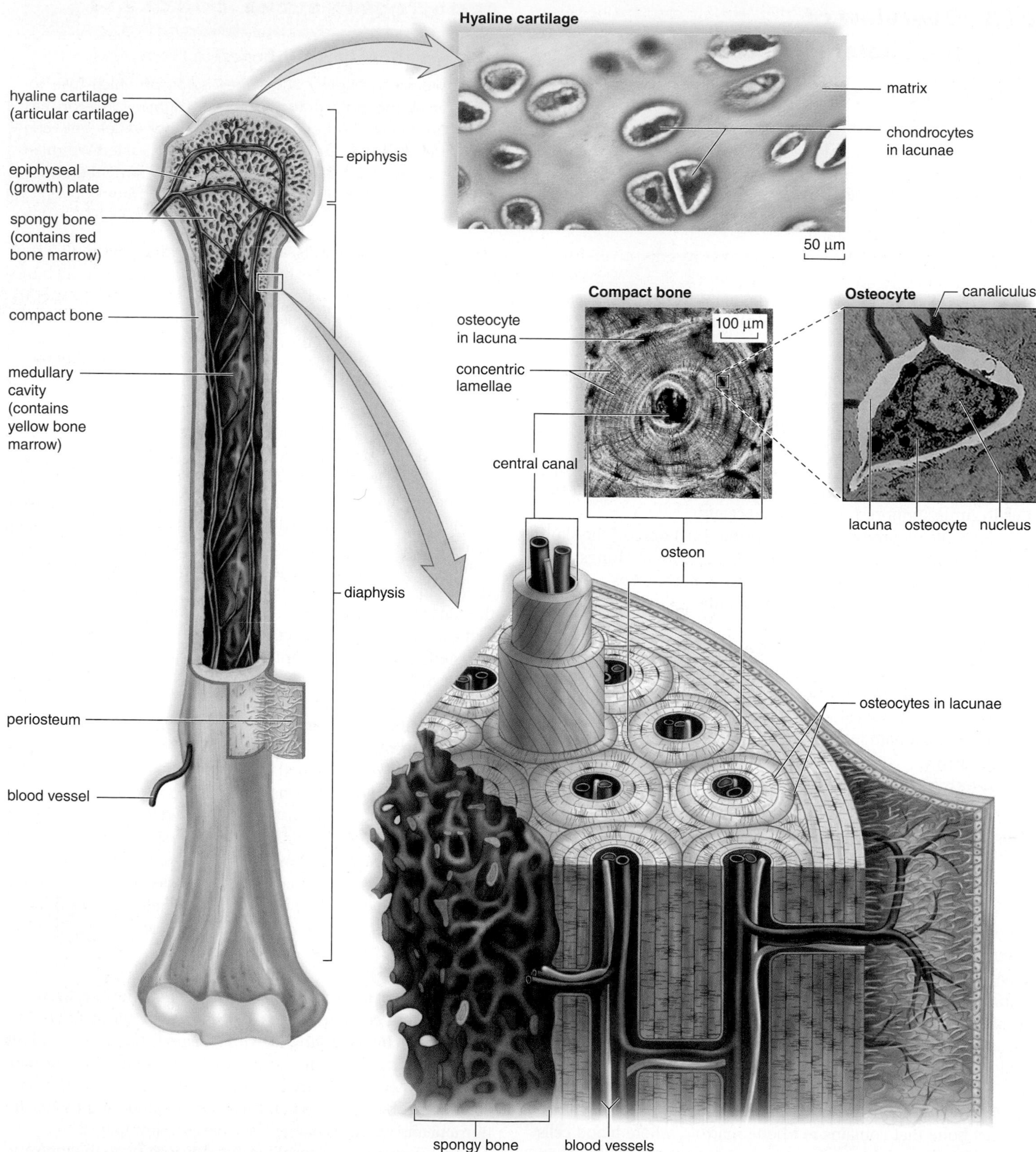

Figure 11.1 The anatomy of a long bone.
A long bone is formed of an outer layer of compact bone. Spongy bone, which lies beneath compact bone, may contain red bone marrow. The central shaft of a long bone contains yellow marrow, a form of stored fat. Periosteum, a fibrous membrane, encases the bone except at its ends. Hyaline cartilage covers the ends of bones.

Fibrocartilage is stronger than hyaline cartilage because the matrix contains wide rows of thick, collagenous fibers. Fibrocartilage is able to withstand both tension and pressure and is found where support is of prime importance—in the disks located between the vertebrae and also in the cartilage of the knee.

Elastic cartilage is more flexible than hyaline cartilage because the matrix contains mostly elastin fibers. This type of cartilage is found in the ear flaps and the epiglottis.

Fibrous Connective Tissue

Fibrous connective tissue contains rows of cells called fibroblasts separated by bundles of collagenous fibers. This tissue makes up ligaments and tendons. Ligaments connect bone to bone. Tendons connect muscle to bone at a joint (also called an articulation).

CHECK YOUR PROGRESS 11.1

1. Describe the typical anatomy of compact and spongy bone.
2. List the three types of cartilage and detail their structure and where each can be found in the body.
3. Predict what might happen to homeostasis if the skeletal system did not exist. Give a few examples.

CONNECTING THE CONCEPTS

For a better understanding of the types of connective tissue presented in this section, refer to the following discussions:

Section 4.2 examines the general structure and function of the connective tissues of the body.

Section 6.1 provides an overview of the blood cells formed in the bone marrow.

Section 12.1 provides additional information on the function of tendons in the body.

11.2 Bones of the Axial Skeleton

LEARNING OUTCOMES

Upon completion of this section, you should be able to

1. Identify the bones of the skull, hyoid, vertebral column, and rib cage.
2. Identify the regions of the vertebral column.
3. Explain the function of the sinuses and intervertebral disks in relation to the axial skeleton.

The 206 bones of the skeleton are classified according to whether they occur in the axial skeleton or the appendicular skeleton (Fig. 11.2). The **axial skeleton** lies in the midline of the body and consists of the skull, hyoid bone, vertebral column, and the rib cage.

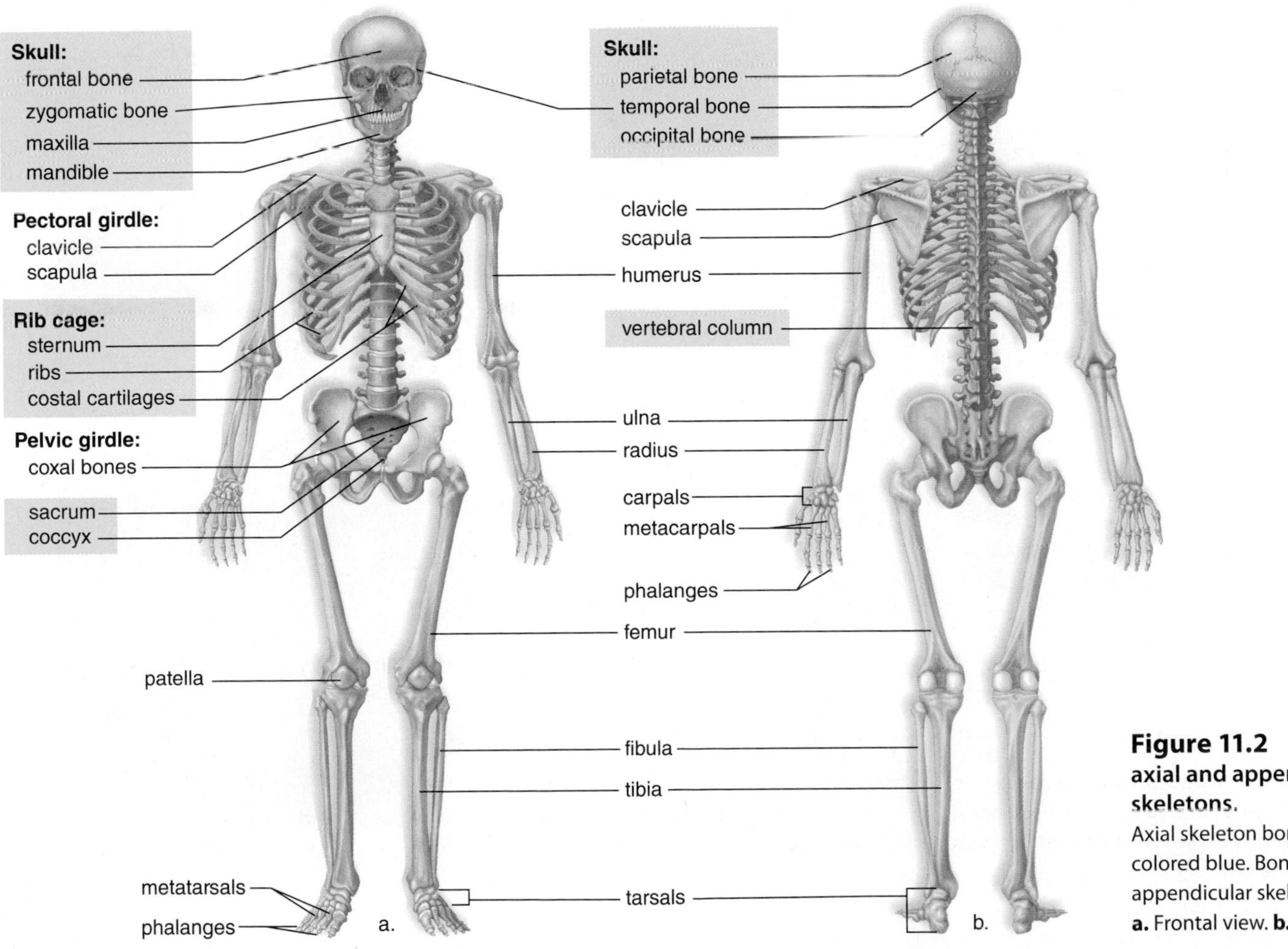

Figure 11.2 The axial and appendicular skeletons. Axial skeleton bones are colored blue. Bones of the appendicular skeleton are tan. **a.** Frontal view. **b.** Dorsal view.

The Skull

The **skull** is formed by the cranium (braincase) and the facial bones. However, some cranial bones contribute to the face.

The Cranium

The cranium protects the brain. In adults, it is composed of eight bones fitted tightly together. In newborns, certain cranial bones are not completely formed. Instead, these bones are joined by membranous regions called **fontanels.** The fontanels usually close by the age of 16 months by the process of intramembranous ossification (see section 11.5).

The major bones of the cranium have the same names as the lobes of the brain: frontal, parietal, occipital, and temporal. On the top of the cranium (Fig. 11.3*a*), the *frontal bone* forms the forehead, the *parietal bones* extend to the sides, and the *occipital bone* curves to form the base of the skull. Here there is a large opening, the **foramen magnum** (Fig. 11.3*b*), through which the spinal cord passes and becomes the brain stem. Below the much larger parietal bones, each *temporal bone* has an opening (external auditory canal) that leads to the middle ear.

The *sphenoid bone,* shaped like a bat with outstretched wings, extends across the floor of the cranium from one side to the other. The sphenoid is the keystone of the cranial bones because all the other bones articulate with it. The sphenoid completes the sides of the skull and also contributes to forming the orbits (eye sockets). The *ethmoid bone,* which lies in front of the sphenoid, also helps form the orbits and the nasal septum. The orbits are completed by various facial bones. The eye sockets are called orbits because we can rotate our eyes.

Some of the bones of the cranium contain the **sinuses,** air spaces lined by mucous membrane. The sinuses reduce the weight of the skull and give a resonant sound to the voice. Sinuses are named according to the bones in which they are located. The major sinuses are the frontal, sphenoid, ethmoid, and maxillary. A smaller set of sinuses, called the mastoid sinuses, drain into the middle ear. *Mastoiditis,* a condition that can lead to deafness, is an inflammation of these sinuses.

The Facial Bones

The most prominent of the facial bones are the mandible, the maxillae (sing., maxilla), the zygomatic bones, and the nasal bones.

The mandible, or lower jaw, is the only movable portion of the skull, and it also forms the chin (Figs. 11.3 and 11.4). The maxillae form the upper jaw and a portion of the eye socket. Further, the hard palate and the floor of the nose are formed by the maxillae (anterior) joined to the palatine bones (posterior). Tooth sockets are located on the mandible and on the maxillae. The grinding action of the mandible and maxillae allows us to chew our food.

The lips and cheeks have a core of skeletal muscle. The *zygomatic bones* are the cheekbone prominences, and the *nasal bones* form the bridge of the nose. Other bones (e.g., ethmoid and vomer) are a part of the nasal septum, which divides

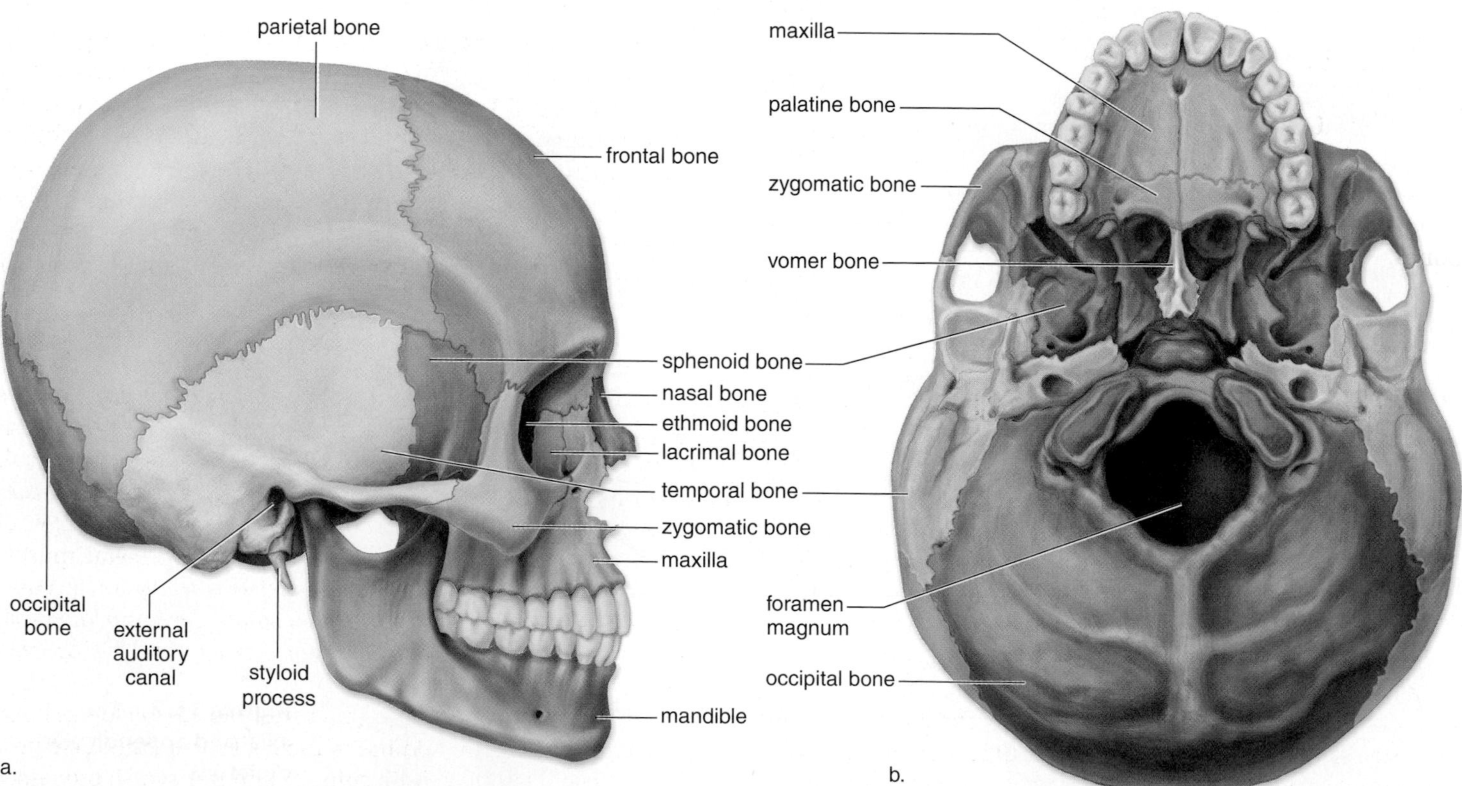

Figure 11.3 **The bones of the skull.**
a. Lateral view. **b.** Inferior view.

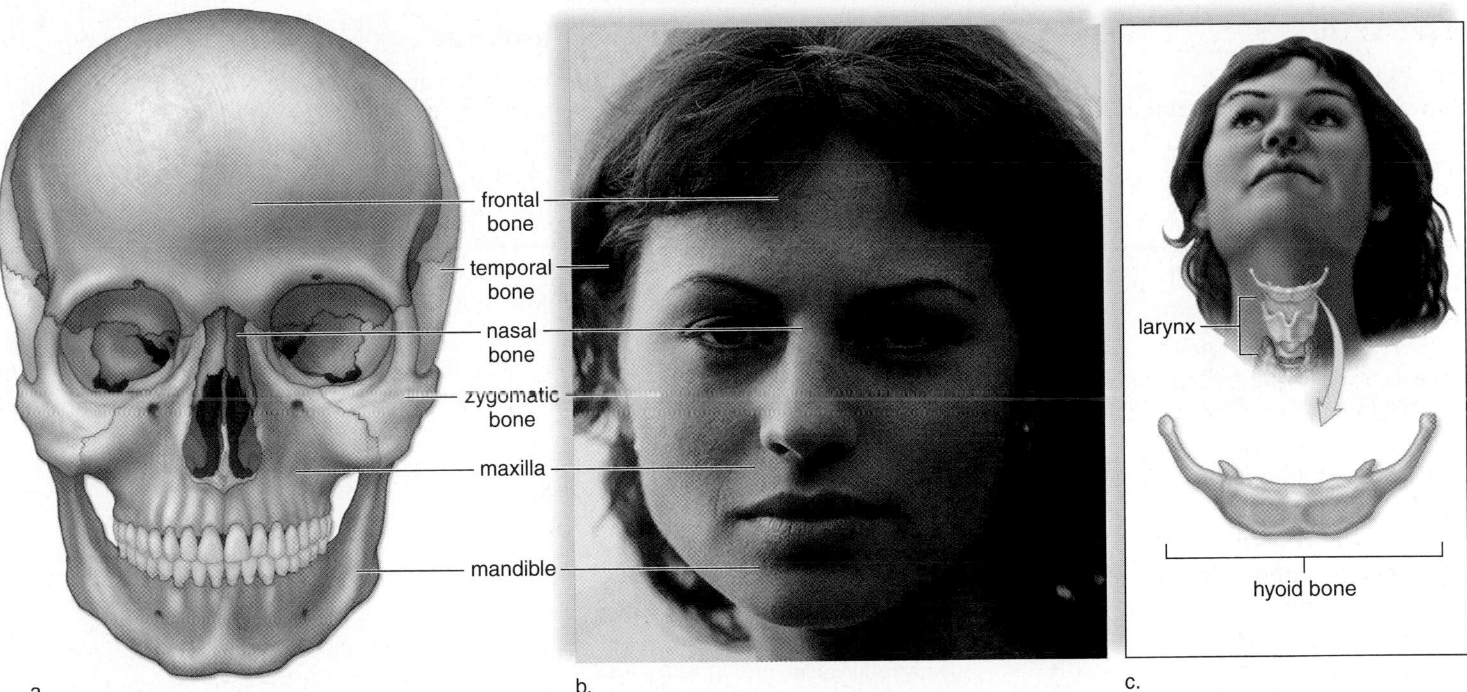

Figure 11.4 The bones of the face and the location of the hyoid bone.
a. The frontal bone forms the forehead and eyebrow ridges; the zygomatic bones form the cheekbones; and the maxillae have numerous functions. They assist in the formation of the eye sockets and the nasal cavity. They form the upper jaw and contain sockets for the upper teeth. The mandible is the lower jaw with sockets for the lower teeth. The mandible has a projection we call the chin. **b.** The maxillae, frontal, and nasal bones help form the external nose. **c.** The hyoid bone is located as shown.

the interior of the nose into two nasal cavities. The lacrimal bone (see Fig. 11.3*a*) contains the opening for the nasolacrimal canal, which drains tears from the eyes to the nose.

Certain cranial bones contribute to the face. The temporal bone and the wings of the sphenoid bone account for the flattened areas we call the temples. The frontal bone forms the forehead and has supraorbital ridges, where the eyebrows are located. Glasses sit where the frontal bone joins the nasal bones.

The exterior portions of ears are formed only by cartilage and not by bone. The nose is a mixture of bones, cartilages, and connective tissues. The cartilages complete the tip of the nose, and fibrous connective tissue forms the flared sides of the nose.

MP3 The Skull

The Hyoid Bone

The *hyoid bone* is not part of the skull but is mentioned here because it is a part of the axial skeleton. It is the only bone in the body that does not articulate with another bone (Fig. 11.4*c*). It is attached to the temporal bones by muscles and ligaments and to the larynx by a membrane. The larynx is the voice box at the top of the trachea in the neck region. The hyoid bone anchors the tongue and serves as the site for the attachment of muscles associated with swallowing. Due to its position, the hyoid bone is not usually easy to fracture in most situations. In cases of suspicious death, however, a fractured hyoid is a strong sign of manual strangulation.

The Vertebral Column

The **vertebral column** consists of 33 vertebrae (Fig. 11.5). Normally, the vertebral column has four curvatures that provide more resilience and strength for an upright posture than a straight column could provide. **Scoliosis** is an abnormal lateral (sideways) curvature of the spine. There are two other well-known abnormal curvatures. Kyphosis is an abnormal posterior curvature that often results in a "hunchback." An abnormal anterior curvature results in lordosis, or "swayback."

As the individual vertebrae are layered on top of one another, they form the vertebral column. The vertebral canal is in the center of the column, and the spinal cord passes through this canal. The intervertebral foramina (sing., foramen, a hole or opening) are found on either side of the column. Spinal nerves branch from the spinal cord and travel through the intervertebral foramina to locations throughout the body. Spinal nerves function to control skeletal muscle contraction, among other functions. If a vertebra is compressed, or slips out of position, the spinal cord and/or spinal nerves might be injured. The result can be paralysis or even death.

The spinous processes of the vertebrae can be felt as bony projections along the midline of the back. The transverse processes extend laterally. Both spinous and transverse processes serve as attachment sites for the muscles that move the vertebral column.

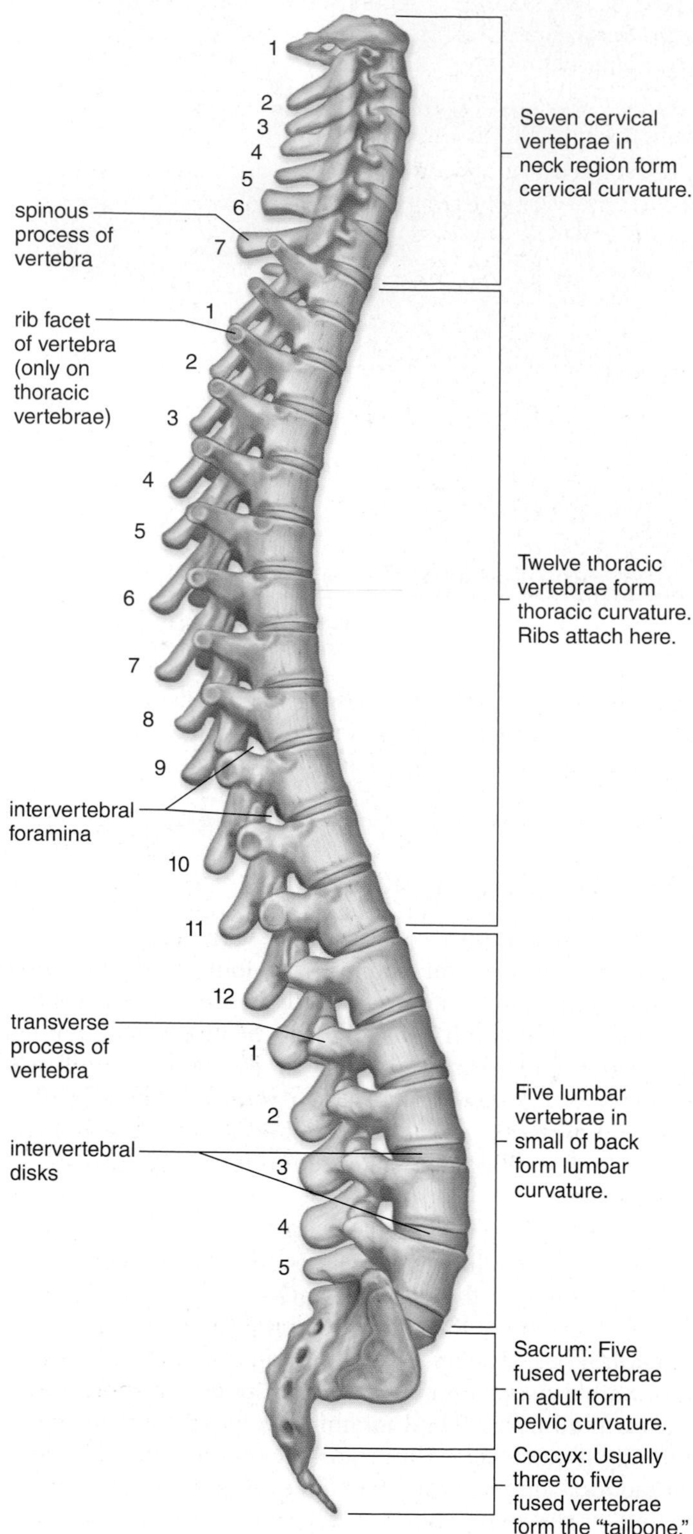

Figure 11.5 The vertebral column.
The vertebral column is made up of 33 vertebrae separated by intervertebral disks. The intervertebral disks make the column flexible. The vertebrae are named for their location in the vertebral column. For example, the thoracic vertebrae are located in the thorax. Humans have a coccyx, which is also called a tailbone.

Types of Vertebrae

The various vertebrae are named according to their location in the vertebral column. The cervical vertebrae are located in the neck. The first cervical vertebra, called the *atlas,* holds up the head. It is so named because Atlas, of Greek mythology, held up the world. Movement of the atlas permits the "yes" motion of the head. It also allows the head to tilt from side to side. The second cervical vertebra is called the *axis* because it allows a degree of rotation, as when we shake the head "no." The thoracic vertebrae have long, thin, spinous processes and articular facets for the attachment of the ribs (Fig. 11.6*a*). Lumbar vertebrae have large bodies and thick processes. The five sacral vertebrae are fused together in the sacrum. The coccyx, or tailbone, is usually composed of four fused vertebrae.

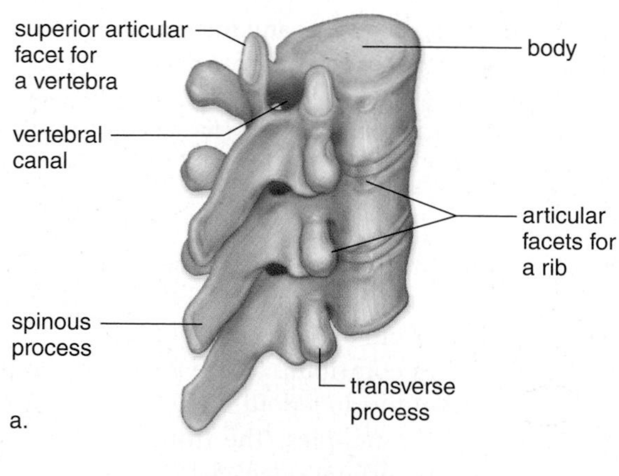

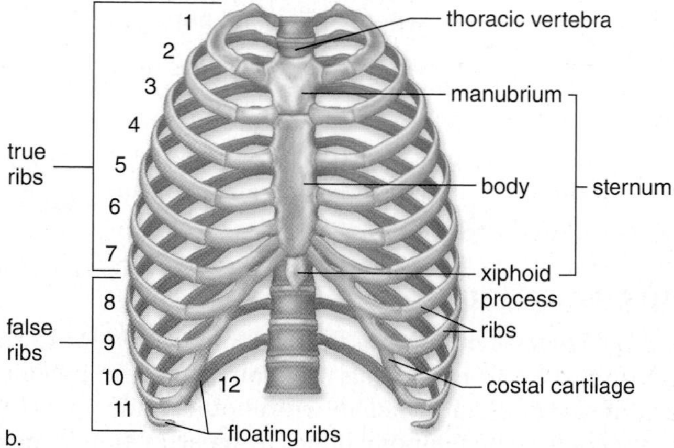

Figure 11.6 The thoracic vertebrae, ribs, and sternum.
a. The thoracic vertebrae articulate with one another and also with the ribs at articular facets. A thoracic vertebra has two facets for articulation with a rib; one is on the body, and the other is on the transverse process. **b.** The rib cage consists of the 12 thoracic vertebrae, the 12 pairs of ribs, the costal cartilages, and the sternum. The rib cage protects the lungs and the heart.

Intervertebral Disks

Between the vertebrae are **intervertebral disks** composed of fibrocartilage that act as padding. The disks prevent the vertebrae from grinding against one another. They also absorb shock caused by movements such as running, jumping, and even walking. The presence of the disks allows the vertebrae to move as we bend forward, backward, and from side to side. Unfortunately, these disks become weakened with age and can herniate and rupture. Pain results if a disk presses against the spinal cord and/or spinal nerves. If that occurs, surgical removal of the disk may relieve the pain.

The Rib Cage

The rib cage, also called the thoracic cage, is composed of the thoracic vertebrae, the ribs and their associated cartilages, and the sternum (Fig. 11.6*b*). The rib cage is part of the axial skeleton.

The rib cage demonstrates how the skeleton is protective but also flexible. The rib cage protects the heart and lungs, yet it swings outward and upward upon inspiration and then downward and inward upon expiration.

MP3 The Vertebral Column and Thoracic Cage

The Ribs

A rib is a flattened bone that originates at the thoracic vertebrae and proceeds toward the anterior thoracic wall. There are 12 pairs of ribs. All 12 pairs connect directly to the thoracic vertebrae in the back. A rib articulates with the body and transverse process of its corresponding thoracic vertebra. Each rib curves outward and then forward and downward.

The upper seven pairs of ribs (ribs 1 through 7; Fig.11.6*b*) connect directly to the sternum by means of a long strip of hyaline cartilage called costal cartilage. These are called "true ribs." Ribs 8 through 12 are called "false ribs,"because their costal cartilage at the end of the ribs does not connect directly to the sternum. Ribs 11 and 12 are also called "floating ribs" because they have no connection with the sternum.

The Sternum

The *sternum* lies in the midline of the body. Along with the ribs, it helps protect the heart and lungs. The sternum, or breastbone, is a flat bone that has the shape of a knife.

The sternum is composed of three bones. These bones are the manubrium (the handle), the body (the blade), and the xiphoid process (the point of the blade). The manubrium articulates with the clavicles of the appendicular skeleton. Costal cartilages from the first pair of ribs also join to the manubrium. The manubrium joins with the body of the sternum at an angle. This is an important anatomical landmark because it occurs at the level of the second rib and therefore allows the ribs to be counted. Counting the ribs is sometimes done to determine where the apex of the heart is located—usually between the fifth and sixth ribs.

The xiphoid process is the third part of the sternum. The variably shaped xiphoid process serves as an attachment site for the diaphragm, which separates the thoracic cavity from the abdominal cavity.

CHECK YOUR PROGRESS 11.2

1. List the bones of the axial skeleton.
2. Identify the bones of the cranium and face, and describe how they contribute to facial features.
3. Describe the various types of vertebrae and state their function.

CONNECTING THE CONCEPTS

For more information on the interaction of the axial skeleton with other organ systems in the body, refer to the following discussions:

Section 9.4 examines how the rib cage is involved in respiration.

Section 13.2 details how the vertebral column and skull protect components of the central nervous system.

11.3 Bones of the Appendicular Skeleton

LEARNING OUTCOMES

Upon completion of this section, you should be able to

1. Identify the bones of the pelvic and pectoral girdles.
2. Identify the bones of the upper and lower limbs.

The **appendicular skeleton** consists of the bones within the pectoral and pelvic girdles and their attached limbs. A pectoral (shoulder) girdle and upper limb are specialized for flexibility. The pelvic (hip) girdle and lower limbs are specialized for strength.

MP3 The Appendicular Skeleton

APPLICATIONS AND MISCONCEPTIONS

Why are human toes shorter than the fingers?

Though some scientists believe that toes are an example of a vestigial organ, there is growing evidence that short toes play an important role in our evolutionary history. According to some scientists, the shortness of our toes contributes to our ability to run long distances, a rare trait in the animal kingdom. Studies have shown that long toes require more energy to start and stop running. In most animals with toes, the fingers and toes are approximately the same length. As toe length shortens, the ability of the animal to run long distances increases. The shortness of our toes may have contributed to the success of our species in hunting large prey on the open plains of Africa.

The Pectoral Girdle and Upper Limb

The body has left and right **pectoral girdles.** Each consists of a scapula (shoulder blade) and a clavicle (collarbone) (Fig. 11.7). The *clavicle* extends across the top of the thorax. It articulates with (joins with) the sternum and the acromion process of the *scapula,* a visible bone in the back. The muscles of the arm and chest attach to the coracoid process of the scapula. The *glenoid cavity* of the scapula articulates with and is much smaller than the head of the humerus. This allows the arm to move in almost any direction but reduces stability. This is the joint most apt to dislocate. Ligaments and also tendons stabilize this joint. Tendons that extend to the humerus from four small muscles originating on the scapula form the *rotator cuff.* Vigorous circular movements of the arm can lead to rotator cuff injuries.

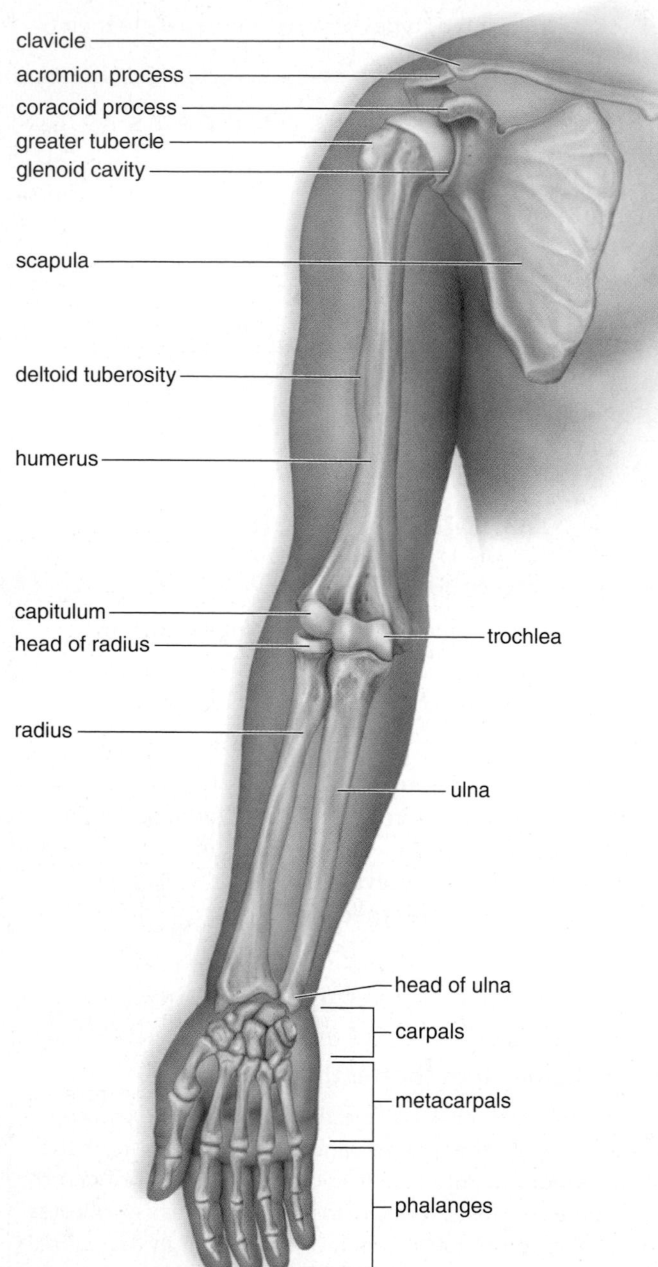

Figure 11.7 The bones of the pectoral girdle and upper limb. The pectoral girdle consists of the clavicle (collarbone) and the scapula (shoulder blade). The humerus is the single bone of the arm. The forearm is formed by the radius and ulna. A hand contains carpals, metacarpals, and phalanges.

The components of a pectoral girdle freely follow the movements of the upper limb, which consists of the humerus within the arm and the radius and ulna within the forearm. The *humerus,* the single long bone in the arm, has a smoothly rounded head that fits into the glenoid cavity of the scapula as mentioned. The shaft of the humerus has a tuberosity (protuberance) where the deltoid, a shoulder muscle, attaches. You can determine, even after death, whether the person did a lot of heavy lifting during his or her lifetime by the size of the deltoid tuberosity.

The far end of the humerus has two protuberances, called the capitulum and the trochlea, which articulate respectively with the *radius* and the *ulna* at the elbow. The bump at the back of the elbow is the olecranon process of the ulna.

When the upper limb is held so that the palm is turned forward, the radius and ulna are about parallel to each other. When the upper limb is turned so that the palm is turned backward, the radius crosses in front of the ulna, a feature that contributes to the easy twisting motion of the forearm.

The hand has many bones, and this increases its flexibility. The wrist has eight *carpal* bones, which look like small pebbles. From these, five *metacarpal* bones fan out to form a framework for the palm. The metacarpal bone that leads to the thumb is opposable to the other digits. An opposable thumb can touch each finger separately or cross the palm to grasp an object. (*Digits* is a term that refers to either fingers or toes.) The knuckles are the enlarged distal ends of the metacarpals. Beyond the metacarpals are the *phalanges,* the bones of the fingers and the thumb. The phalanges of the hand are long, slender, and lightweight.

Pectoral girdle
- scapula and clavicle

Upper limb
- arm: humerus
- forearm: radius and ulna
- hand: carpals, metacarpals, and phalanges

The Pelvic Girdle and Lower Limb

Pelvic girdle
- coxal bones

Lower limb
- thigh: femur
- leg: tibia and fibula
- foot: tarsals, metatarsals, and phalanges

Figure 11.8 shows how the lower limb is attached to the pelvic girdle. The **pelvic girdle** (hip girdle) consists of two heavy, large coxal bones (hip bones). The **pelvis** is a basin composed of the pelvic girdle, sacrum, and coccyx. The pelvis bears the weight of the body, protects the organs within the pelvic cavity, and serves as the place of attachment for the legs.

Each *coxal bone* has three parts: the ilium, the ischium, and the pubis, which are fused in the adult (Fig. 11.8). The hip socket, called the acetabulum, occurs where these three bones meet. The ilium is the largest part of the coxal bones, and our hips form where it flares out. We sit on the ischium, which has a posterior spine called the ischial spine for muscle attachment. The pubis, from which the term *pubic hair* is derived, is the anterior part of a coxal bone. The two pubic bones are joined by a fibrocartilaginous joint, the *pubic symphysis.*

The male and female pelves differ from each other. In the female, the iliac bones are more flared and the pelvic cavity is more shallow, but the outlet is wider. These adaptations facilitate the birthing process during vaginal delivery.

The *femur* (thighbone) is the longest and strongest bone in the body. The head of the femur articulates with the coxal bones at the acetabulum, and the short neck better positions the legs for walking. The femur has two large processes, the greater and lesser trochanters, which are places of attachment for thigh muscles, buttock muscles, and hip flexors. At its distal end, the femur has medial and lateral condyles that articulate with the *tibia* of the leg. This is the region of the knee and the *patella,* or kneecap. The patella is held in place by the quadriceps tendon, which continues as a ligament that attaches to the tibial tuberosity. At the distal end, the medial malleolus of the tibia causes the inner bulge of the ankle. The *fibula* is the more slender bone in the leg. The fibula has a head that articulates with the tibia and a distal lateral malleolus that forms the outer bulge of the ankle.

Each foot has an ankle, an instep, and five toes. The many bones of the foot give it considerable flexibility, especially on rough surfaces. The ankle contains seven *tarsal* bones, one of which (the talus) can move freely where it joins the tibia and fibula. The calcaneus, or heel bone, is also considered part of the ankle. The talus and calcaneus support the weight of the body.

The instep has five elongated *metatarsal* bones. The distal ends of the metatarsals form the ball of the foot. If the ligaments that bind the metatarsals together become weakened, flat feet are apt to result. The bones of the toes are called *phalanges,* just like those of the fingers. In the foot, the phalanges are stout and extremely sturdy.

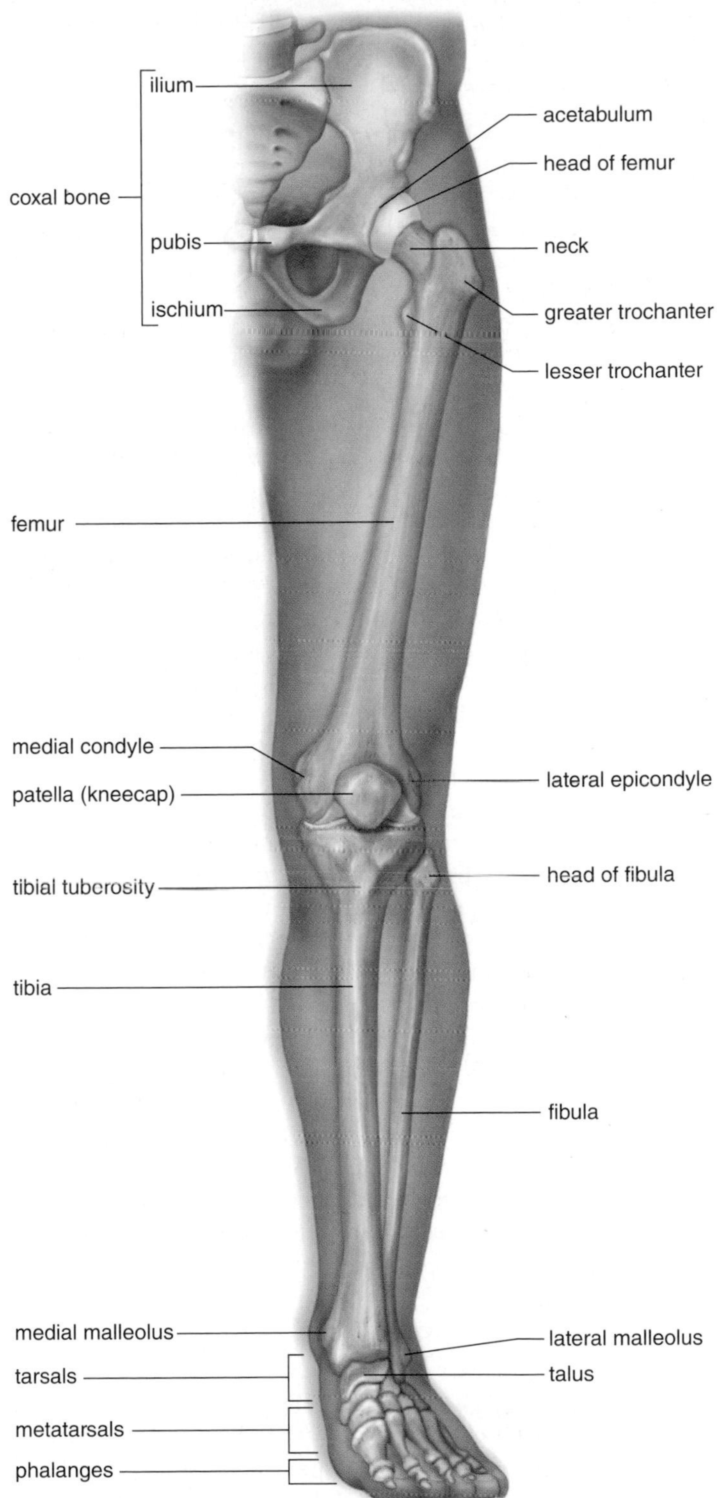

Figure 11.8 The coxal bones and bones of the pelvis and lower limb.
The ilium, ischium, and pubis join at the acetabulum (hip socket) to form a coxal bone. The pelvis is completed by the addition of the sacrum and coccyx. The femur (thigh bone) and tibia and fibula (shin bones) form the leg. Tarsals, metatarsals, and phalanges construct the foot.

BIOLOGY MATTERS  **Science**

Identifying Skeletal Remains

Regardless of how, when, and where human bones are found unexpectedly, many questions must be answered. How old was this person at the time of death? Are these the bones of a male or female? What was the ethnicity? Are there any signs that this person was murdered?

Clues about the identity and history of the deceased person are available throughout the skeleton (Fig. 11A). Age is approximated by *dentition,* or the structure of the teeth in the upper jaw (maxilla) and lower jaw (mandible). For example, infants between 0 and 4 months of age will have no teeth present; children approximately 6 to 10 years of age will have missing deciduous, or "baby teeth"; young adults acquire their last molars, or "wisdom teeth," around age 20. The age of older adults may be approximated by the number and location of missing or broken teeth. Studying areas of bone ossification also gives clues to the age of the deceased at the time of death. In older adults, signs of joint breakdown provide additional information about age. Hyaline cartilage becomes worn, yellowed, and brittle with age, and the hyaline cartilages covering bone ends wear down over time. The amount of yellowed, brittle, or missing cartilage helps scientists guess the person's age.

If skeletal remains include the individual's pelvic bones, these provide the best method for determining an adult's gender. The pelvis is more shallow and wider in the female than in the male. The long bones, particularly the humerus and femurs, give information about gender as well. Long bones will be thicker and denser in males, and points of muscle attachment will be bigger and more prominent. The skull of a male will have a square chin and more prominent ridges above the eye sockets or orbits.

Determining ethnic origin of skeletal remains can be difficult, because so many people have a mixed racial heritage. Forensic anatomists rely on observed racial characteristics of the skull. In general, individuals of African or African-American descent have a greater distance between the eyes, eye sockets that are roughly rectangular, and a jaw that is large and prominent. Skulls of Native Americans typically have round eye sockets, prominent cheek (zygomatic) bones, and a rounded palate. Caucasian skulls usually have a U-shaped palate, and a suture line between the frontal bones is often visible. Additionally, the external ear canals in Caucasians are long and straight, so that the auditory ossicles can be seen.

Once the identity of the individual has been determined, the skeletal remains can be returned to the family for proper burial.

Figure 11A Forensic investigators uncover a skeleton. A knowledge of bone structure and how bones age will help identify these remains.

Questions to Consider

1. Why do you think that compact bones are most often found by forensic investigators?
2. How does an examination of bone ossification provide an indication of age?

CHECK YOUR PROGRESS 11.3

1. List the bones of the pectoral girdle and the upper limb.
2. List the bones of the pelvic girdle and the lower limb.
3. Describe how you can tell the difference in gender from looking at the bones of the pelvic girdle.

CONNECTING THE CONCEPTS

For more information on the limbs of the body, refer to the following discussions:

Figure 5.11 diagrams the major blood vessels of the arms and legs.

Figure 12.5 illustrates the major muscles of the arms and legs.

Section 13.2 details the areas of the brain that are responsible for the movement of the arms and legs.

11.4 Articulations

LEARNING OUTCOMES

Upon completion of this section, you should be able to

1. List the three types of joints that connect bones.
2. Describe the structure and operation of a synovial joint.
3. Summarize the types of movement that are made possible by a synovial joint.

Bones are joined at the joints, classified as fibrous, cartilaginous, or synovial. Many fibrous joints, such as the **sutures** between the cranial bones, are immovable. Cartilaginous joints may be connected by hyaline cartilage, as in the costal cartilages that join the ribs to the sternum. Other cartilaginous joints are formed by fibrocartilage, as in the intervertebral disks. Cartilaginous joints tend to be slightly movable. **Synovial joints** are freely movable (Fig. 11.9).

There are several general classes of synovial joints, two of which are shown in Figure 11.9. Figure 11.9*b* illustrates the general anatomy of a freely movable synovial joint. Ligaments connect bone to bone and support or strengthen the joint. A fibrous joint capsule formed by ligaments surrounds the bones at the joint. This capsule is not shown in Figure 11.9 so that the inner structure of the joint may be revealed. The joint capsule is lined with synovial membrane that secretes a small amount of synovial fluid to lubricate the joint. Fluid-filled sacs called bursae (sing., bursa) ease friction between bare areas of bone and overlapping muscles or between skin and tendons. The full joint contains menisci (sing., meniscus), which are C-shaped pieces of hyaline cartilage between the bones. These give added stability and act as shock absorbers.

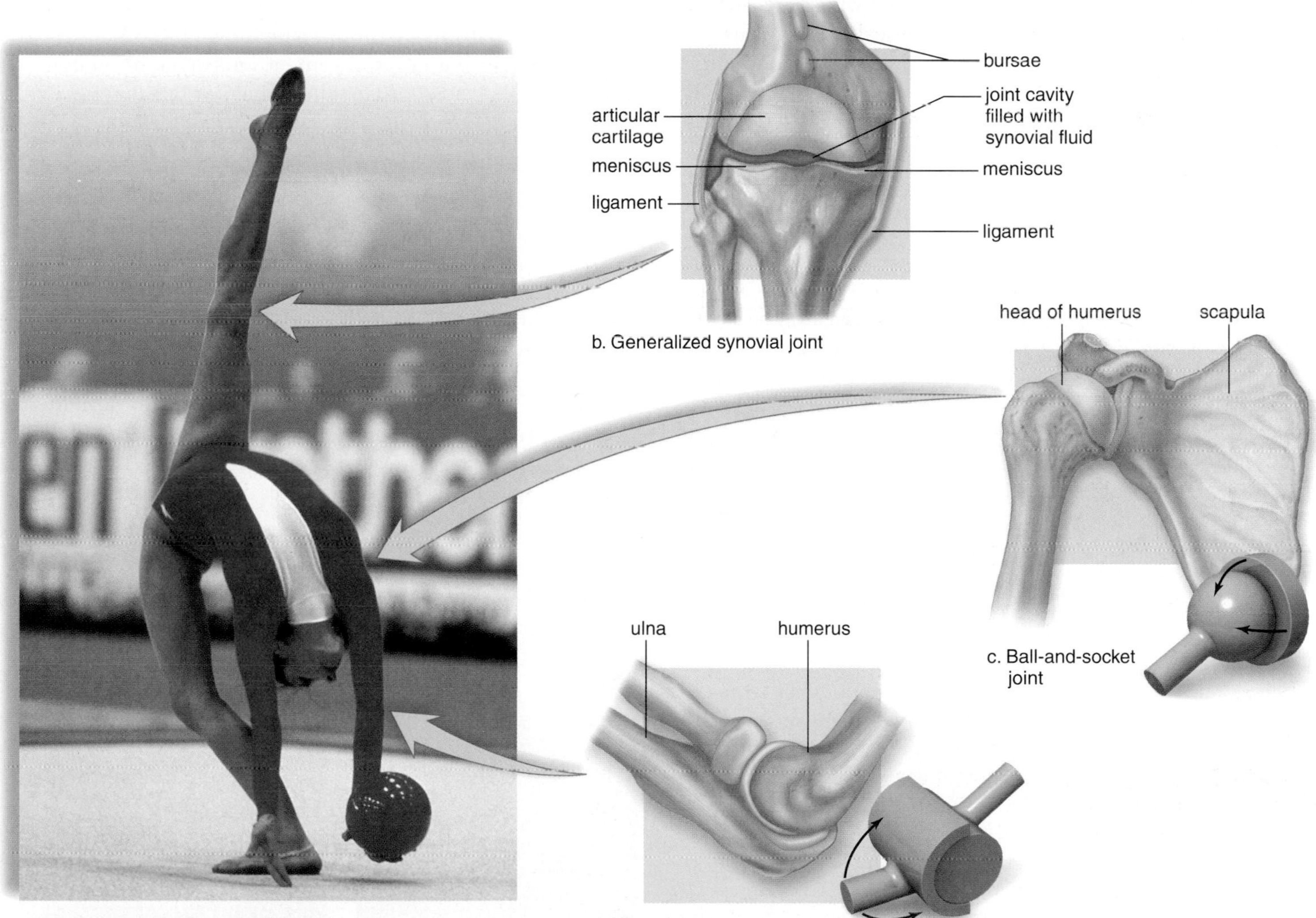

Figure 11.9 **The structure of a synovial joint.**
a. Synovial joints are movable and therefore flexible. **b.** The bones of joints are joined by ligaments that form a capsule. The capsule is lined with synovial membrane, which gives off synovial fluid as a lubricant. Bursae are fluid-filled sacs that reduce friction. Menisci (sing., meniscus), formed of cartilage, stabilize a joint, and articular cartilage caps the bones. **c.** Ball-and-socket joints form the hip and shoulder. **d.** Hinge joints construct the knee and elbow.

BIOLOGY MATTERS  Science

Osteoarthritis and Joint-Replacement Surgery

Osteoarthritis is a condition that afflicts nearly everyone, to a greater or lesser degree, as each person ages. The bones that unite to form joints, or articulations, are covered with a slippery cartilage. This articular cartilage wears down over time, as friction in the joint wears it away (Fig. 11B). By age 80, people typically have osteoarthritis in one or more joints. By contrast, rheumatoid arthritis is an autoimmune disorder (see section 7.6) that causes inflammation within the joint. Unlike osteoarthritis, which typically affects older people, rheumatoid arthritis can afflict a person of any age, even young children. Both forms of arthritis cause a loss of the joint's natural smoothness. This is what causes the pain and stiffness associated with arthritis. Arthritis is first treated with medications for joint inflammation and pain and physical therapy to maintain and strengthen the joint. However, if these treatments fail, a total joint replacement is often performed. Successful replacement surgeries are now routine, thanks to the hard work and dedication of the British orthopedic surgeon, Dr. John Charnley.

Early experimental surgeries by Charnley and others had been very disappointing. Fused joints were immobile, and fusion didn't always relieve the patient's pain. Postsurgical infection was common. The bones attached to the artificial joint eroded, and the supporting muscles wasted away because the joint wasn't useful. Charnley wanted to design a successful prosthetic hip, with the goal of replacing both parts of the diseased hip joint: the acetabulum, or "socket," as well as the ball-shaped head of the femur. Charnley soon determined that surgical experimentation alone wasn't enough. He studied bone repair, persuading a colleague to operate on his own tibia, or shinbone, to see how repair occurred. He studied the mechanics of the hip joint, testing different types of synthetic materials. He achieved his first success using a hip socket lined with Teflon but soon discovered that the surrounding tissues became inflamed. After multiple attempts, his perfected hip consisted of a socket of durable polyethylene, still used as the joint's plastic component. The head of his prosthetic femur was a small, highly polished metal ball. Stainless steel, cobalt, and titanium, as well as chrome alloys, form the metal component today. Various techniques for cementing the polyethylene socket onto the pelvic bone had failed when bone pulled away from the cemented surface and refused to grow. Charnley's surgery used dental cement, slathered onto the bone surfaces. When the plastic components were attached, cement was squeezed into every pore of the bone, allowing the bone to regenerate and grow around the plastic. Finally, Charnley devised a specialized surgical tent and instrument tray to minimize infection.

Charnley's ideas were innovative and unorthodox and he was reassigned to a former tuberculosis hospital, which he converted into a center for innovation in orthopedic surgery. His colleagues developed a prosthetic knee joint similar to the Charnley hip. In knee-replacement surgery (see the chapter opener), the damaged ends of bones are removed and replaced with artificial components that resemble the original bone ends. Hip and knee replacements remain the most common joint-replacement surgeries, but ankles, feet, shoulders, elbows, and fingers can also be replaced. Though many improvements on the procedure continue, the Charnley hip replacement remains the technique after which all others are modeled.

When a joint replacement is complete, the patient's hard work is vital to ensure the success of the procedure. Exercise and activity are critical to the recovery process. After surgery, the patient is encouraged to use the new joint as soon as possible. The extent of improvement and range of motion of the joint depends on its stiffness before the surgery, as well as the amount of patient effort during therapy following surgery. A complete recovery varies in time from patient to patient, but typically takes several months. Older patients can expect their replacements to last about ten years. However, younger patients may need a second replacement if they wear out their first prosthesis. Still, individuals who have joint-replacement surgery can expect an improved quality of life and a bright future with greater independence and healthier pain-free activity.

Video
Nano Bones

Questions to Consider

1. Compare each component of Charnley's artificial joint with that of a real synovial joint.
2. What specifically in a joint does rheumatoid arthritis target?

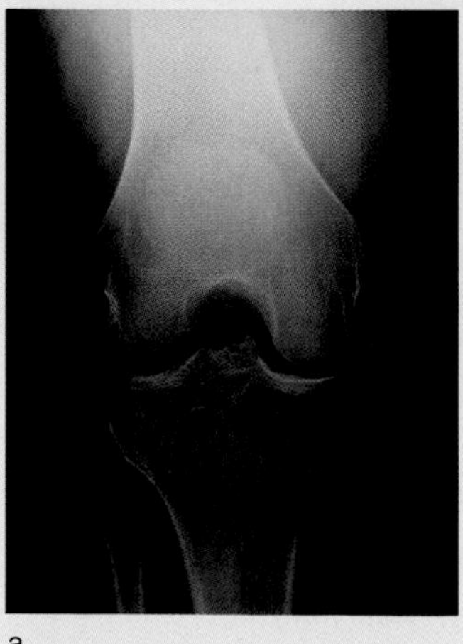

a.

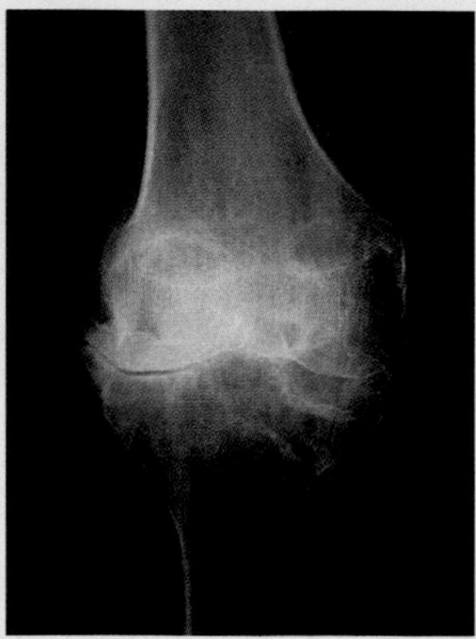

b.

Figure 11B **Osteoarthritis.**
A comparison of a normal knee (**a**) and a knee with osteoarthritis (**b**).

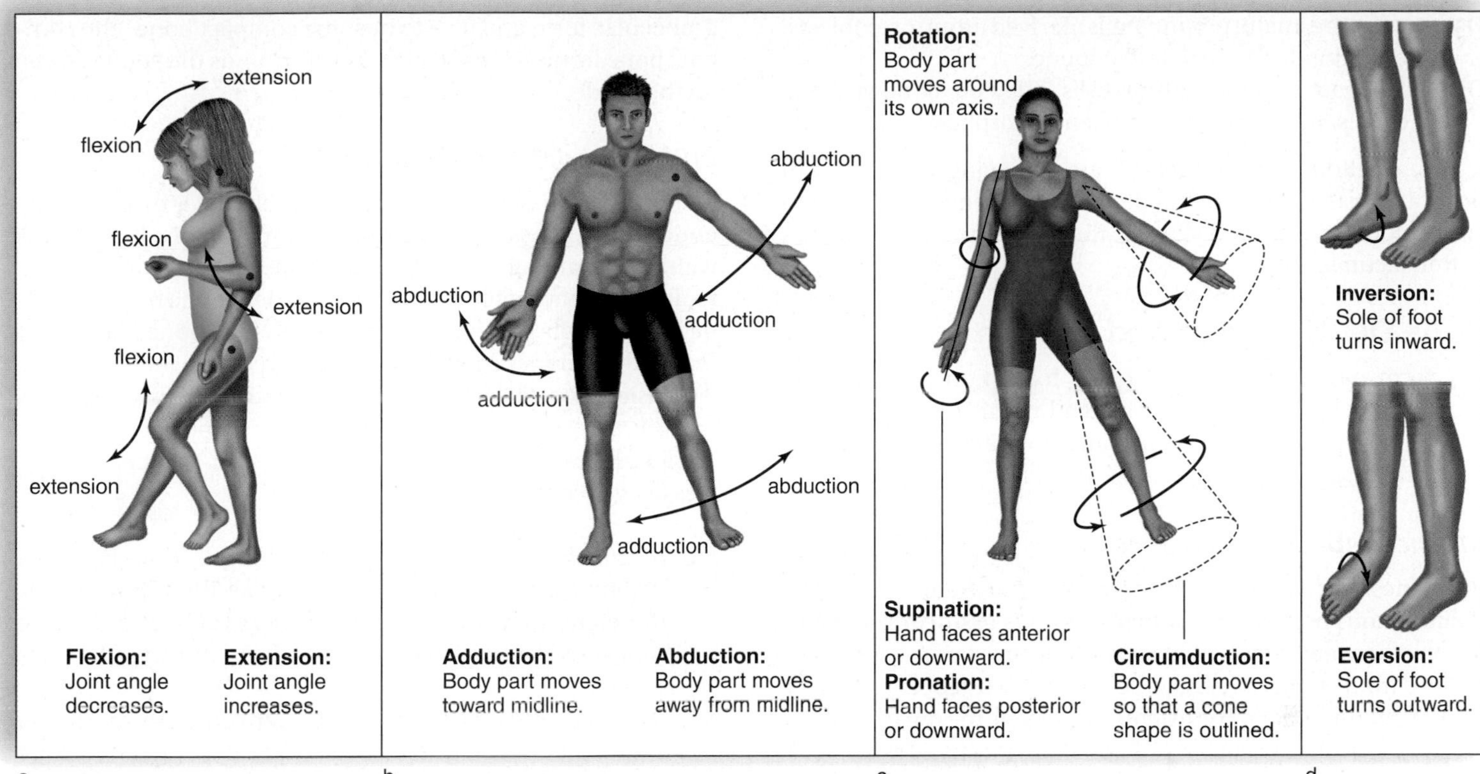

Figure 11.10 **Synovial joints allow for a variety of movement.**
a. Flexion and extension. **b.** Adduction and abduction. **c.** Rotation and circumduction. **d.** Inversion and eversion. Red dots indicate pivot points.

The ball-and-socket joints at the hips and shoulders allow movement in all planes, even rotational movement. The elbow and knee joints are synovial joints called hinge joints. Like a hinged door, they largely permit movement in one direction only.

Movements Permitted by Synovial Joints

Intact skeletal muscles are attached to bones by tendons that span joints. When a muscle contracts, one bone moves in relation to another bone. The more common types of movements are described in Figure 11.10.

Video Smart Robo Knee

CHECK YOUR PROGRESS 11.4

1. List the three major types of joints.
2. Describe the basic movements of cartilaginous and fibrous joints, and give an example of each in the body.
3. Describe all the different movements of synovial joints, and give an example of each in the body.

CONNECTING THE CONCEPTS

For more information on ligaments and tendons, refer to the following discussion:

Section 4.2 describes the connective tissue that is found in the tendons and ligaments.

11.5 Bone Growth and Homeostasis

LEARNING OUTCOMES

Upon completion of this section, you should be able to

1. Summarize the process of ossification and list the types of cells involved.
2. Describe the process of bone remodeling.
3. Explain the steps in the repair of bone.

The importance of the skeleton to the human form is evident by its early appearance during development. The skeleton starts forming at about six weeks, when the embryo is only about 12 mm (0.5 in.) long. Most bones grow in length and width through adolescence, but some continue enlarging until about age 25. In a sense, bones can grow throughout a lifetime because they are able to respond to stress by changing size, shape, and strength. This process is called remodeling. If a bone fractures, it can heal by bone repair.

Bones are living tissues, as shown by their ability to grow, remodel, and undergo repair. Several different types of cells are involved in bone growth, remodeling, and repair:

Osteoblasts are bone-forming cells. They secrete the organic matrix of bone and promote the deposition of calcium salts into the matrix.

Osteocytes are mature bone cells derived from osteoblasts. They maintain the structure of bone.

Osteoclasts are bone-absorbing cells. They break down bone and assist in returning calcium and phosphate to the blood.

Throughout life, osteoclasts are removing the matrix of bone and osteoblasts are building it up. When osteoblasts are surrounded by calcified matrix, they become the osteocytes within lacunae.

Bone Development and Growth

The term **ossification** refers to the formation of bone. The bones of the skeleton form during embryonic development in two distinctive ways: intramembranous ossification and endochondral ossification (Fig. 11.11).

Intramembranous Ossification

Flat bones, such as the bones of the skull, are examples of intramembranous bones. In **intramembranous ossification,** bones develop between sheets of fibrous connective tissue. Here, cells derived from connective tissue cells become osteoblasts located in ossification centers. The osteoblasts secrete the organic matrix of bone. This matrix consists of mucopolysaccharides and collagen fibrils. Calcification occurs when calcium salts are added to the organic matrix. The osteoblasts promote calcification, or ossification, of the matrix. Ossification results in the formation of soft sheets, or trabeculae, of spongy bone. Spongy bone remains inside a flat bone. The spongy bone of flat bones, such as those of the skull and clavicles (collarbones), contains red bone marrow.

A periosteum forms outside the spongy bone. Osteoblasts derived from the periosteum carry out further ossification. Trabeculae form and fuse to become compact bone. The compact bone forms a bone collar that surrounds the spongy bone on the inside.

Endochondral Ossification

Most of the bones of the human skeleton are formed by **endochondral ossification,** which means that the bone forms within the cartilage. During endochondral ossification, bone replaces the cartilaginous (hyaline) models of the bones. Gradually, the cartilage is replaced by the calcified bone matrix that makes these bones capable of bearing weight.

Inside, bone formation spreads from the center to the ends, and this accounts for the term used for this type of ossification. The long bones, such as the tibia, provide examples of endochondral ossification (Fig. 11.11).

Animation Bone Growth in Width

1. *The cartilage model.* In the embryo, chondrocytes lay down hyaline cartilage, which is shaped like the future bones. Therefore, they are called cartilage models of the future bones. As the cartilage models calcify, the chondrocytes die off.
2. *The bone collar.* Osteoblasts are derived from the newly formed periosteum. Osteoblasts secrete the organic bone matrix, and the matrix undergoes calcification. The result is a bone collar, which covers the diaphysis (Fig. 11.11). The bone collar is composed of compact bone. In time, the bone collar thickens.
3. *The primary ossification center.* Blood vessels bring osteoblasts to the interior, and they begin to lay down spongy bone. This region is called a primary ossification center because it is the first center for bone formation.

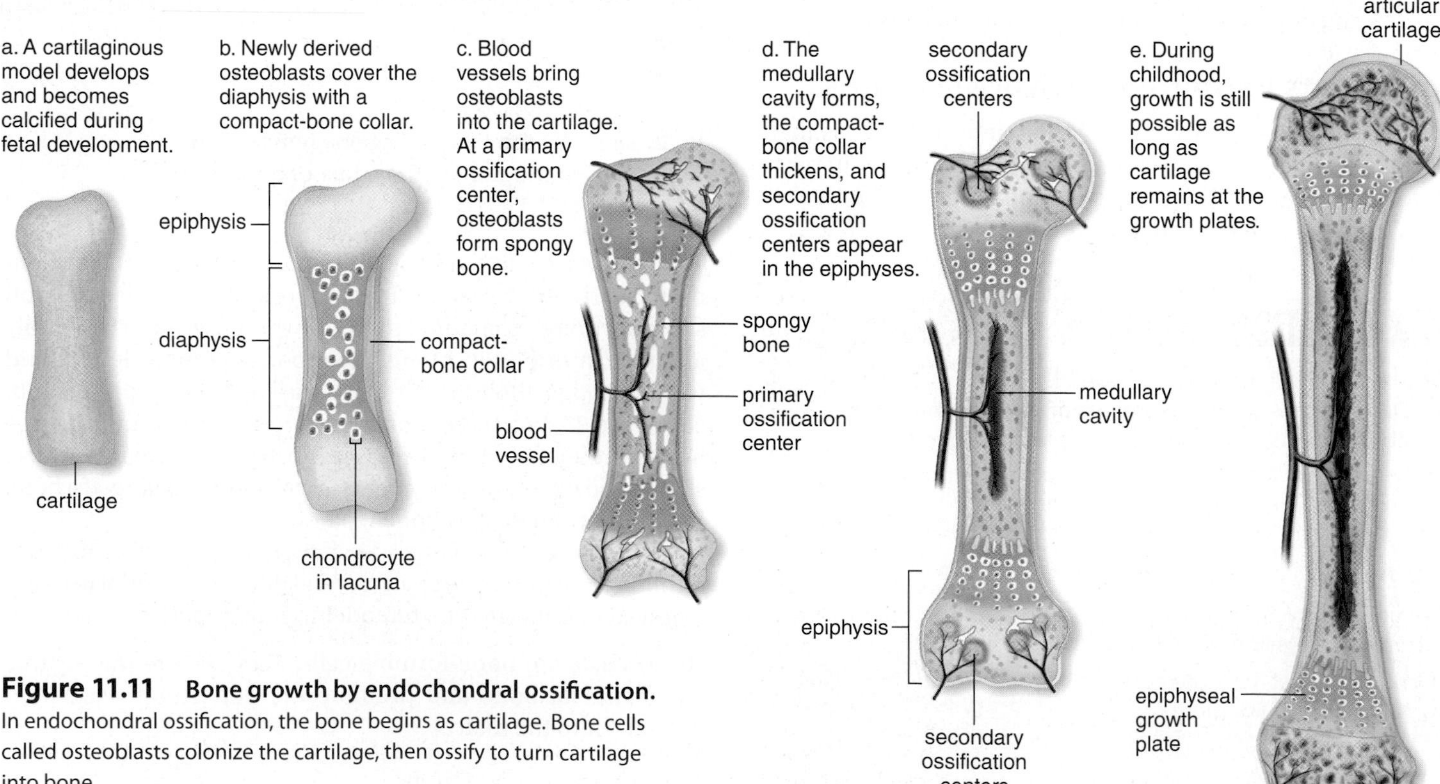
a. A cartilaginous model develops and becomes calcified during fetal development.
cartilage
b. Newly derived osteoblasts cover the diaphysis with a compact-bone collar.
epiphysis
diaphysis
compact-bone collar
chondrocyte in lacuna
c. Blood vessels bring osteoblasts into the cartilage. At a primary ossification center, osteoblasts form spongy bone.
spongy bone
primary ossification center
blood vessel
d. The medullary cavity forms, the compact-bone collar thickens, and secondary ossification centers appear in the epiphyses.
secondary ossification centers
medullary cavity
epiphysis
secondary ossification centers
articular cartilage
e. During childhood, growth is still possible as long as cartilage remains at the growth plates.
epiphyseal growth plate

Figure 11.11 Bone growth by endochondral ossification. In endochondral ossification, the bone begins as cartilage. Bone cells called osteoblasts colonize the cartilage, then ossify to turn cartilage into bone.

4. *The medullary cavity and secondary ossification sites.* The spongy bone of the diaphysis is absorbed by osteoclasts, and the cavity created becomes the medullary cavity. Shortly after birth, secondary ossification centers form in the epiphyses. Spongy bone persists in the epiphyses, and it persists in the red bone marrow for quite some time. Cartilage is present at two locations: the epiphyseal (growth) plate and articular cartilage, which covers the ends of long bones.
5. *The epiphyseal (growth) plate.* A band of cartilage called the **epiphyseal plate** (also called a growth plate) remains between the primary ossification center and each secondary center. The limbs keep increasing in length as long as the epiphyseal plates are still present.

Figure 11.12 shows that the epiphyseal plate contains four layers. The layer nearest the epiphysis is the resting zone, where cartilage remains. The next layer is the proliferating zone, in which chondrocytes are producing new cartilage cells. In the third layer, the degenerating zone, the cartilage cells are dying off; and in the fourth layer, the ossification zone, bone is forming. Bone formation here causes the length of the bone to increase. The inside layer of articular cartilage also undergoes ossification in the manner described.

The diameter of a bone enlarges as a bone lengthens. Osteoblasts derived from the periosteum are active in new bone deposition as osteoclasts enlarge the medullary cavity from inside.

Final Size of the Bones When the epiphyseal plates close, bone lengthening can no longer occur. The epiphyseal plates in the arms and legs of women typically close at about age 16 to 18, and they do not close in men until about age 20. Portions of other types of bones may continue to grow until age 25. Hormones are chemical messengers secreted by the endocrine glands and distributed about the body by the bloodstream. Hormones control the activity of the epiphyseal plate, as is discussed next.

Hormones Affect Bone Growth

The importance of bone growth is signified by the involvement of several hormones in bone growth. A **hormone,** a chemical messenger, is produced by one part of the body and acts on a different part of the body.

Vitamin D is formed in the skin when it is exposed to sunlight, but it can also be consumed in the diet. Milk, in particular, is often fortified with vitamin D. In the kidneys, vitamin D is converted to a hormone that acts on the intestinal tract. The chief function of vitamin D is intestinal absorption of calcium. In the absence of vitamin D, children can develop rickets, a condition marked by bone deformities including bowed long bones.

Growth hormone (GH) directly stimulates growth of the epiphyseal plate, as well as bone growth in general. However, growth hormone will be somewhat ineffective if the metabolic activity of cells is not promoted. Thyroid hormone, in particular, promotes the metabolic activity of cells. Too little growth hormone in childhood results in dwarfism. Too much growth hormone during childhood (prior to epiphyseal fusion) can produce excessive growth and even gigantism (see Fig. 15.8). Acromegaly results from excess GH in adults following epiphyseal fusion. This condition produces excessive growth of bones in the hands and face (see Fig. 15.9).

Adolescents usually experience a dramatic increase in height, called the growth spurt, due to an increased level of sex hormones. These hormones apparently stimulate osteoblast activity. Rapid growth causes epiphyseal plates to become "paved over" by the faster-growing bone tissue within one or two years of the onset of puberty.

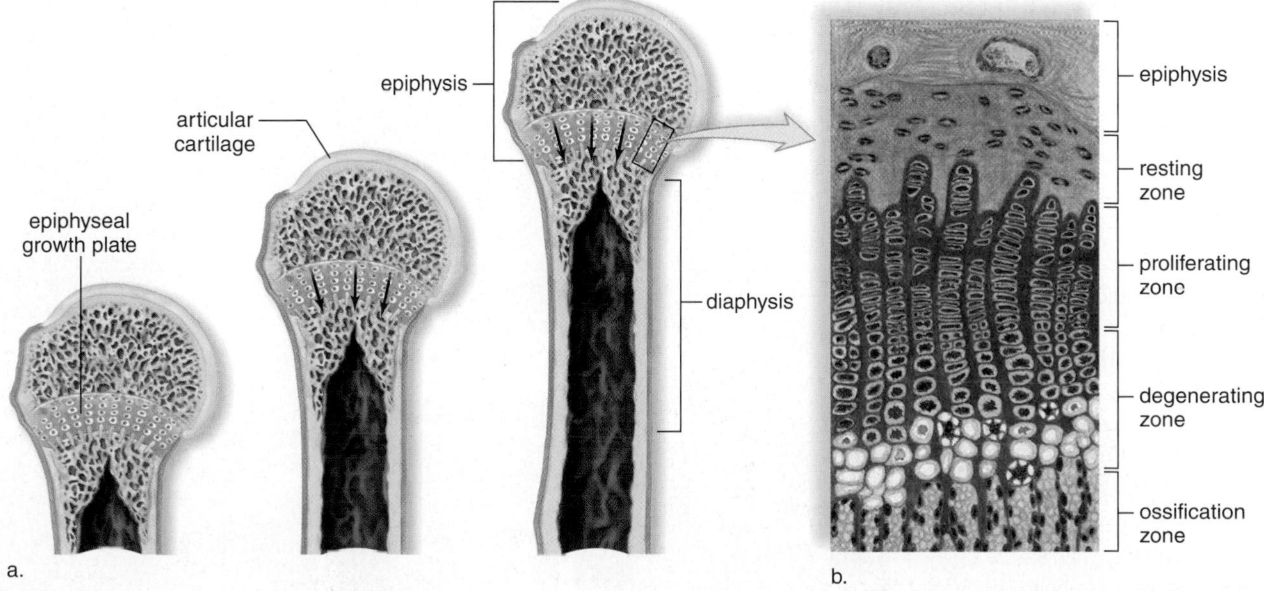

Figure 11.12 Increasing bone length.
a. Length of a bone increases when cartilage is replaced by bone at the growth plate. Arrows indicate the direction of ossification. **b.** Chondrocytes produce new cartilage in the proliferating zone, and cartilage becomes bone in the ossification zone closest to the diaphysis.

Bone Remodeling and Calcium Homeostasis

Bone is constantly being broken down by osteoclasts and re-formed by osteoblasts in the adult. As much as 18% of bone is recycled each year. This process of bone renewal, often called **bone remodeling,** normally keeps bones strong (Fig. 11.13). In Paget's disease, new bone is generated at a faster-than-normal rate. This rapid remodeling produces bone that's softer and weaker than normal bone and can cause bone pain, deformities, and fractures.

Bone recycling allows the body to regulate the amount of calcium in the blood. To illustrate that the blood calcium level is critical, recall that calcium is required for blood to clot (see section 6.4). Also, if the blood calcium concentration is too high, neurons and muscle cells no longer function. If calcium falls too low, nerve and muscle cells become so excited that convulsions occur. They are also necessary for the regulation of cellular metabolism by acting in cellular messenger systems. Thus, the skeleton acts as a reservoir for storage of this important mineral—if the blood calcium rises above normal, at least some of the excess is deposited in the bones. If the blood calcium dips too low, calcium is removed from the bones to bring it back up to the normal level.

Two hormones in particular are involved in regulating the blood calcium level. Parathyroid hormone (PTH) stimulates osteoclasts to dissolve the calcium matrix of bone. In addition, parathyroid hormone promotes calcium reabsorption in the small intestine and kidney, increasing blood calcium levels. Vitamin D is needed for the absorption of Ca^{2+} from the digestive tract, which is why vitamin D deficiency can result in weak bones. It is easy to get enough of this vitamin, because your skin produces it when exposed to sunlight, and the milk you buy at the grocery store is fortified with vitamin D.

Calcitonin is a hormone that acts opposite to PTH. The female sex hormone estrogen can actually increase the number of osteoblasts; the reduction of estrogen in older women is often given as reason for the development of weak bones, called osteoporosis. Osteoporosis is discussed in the Health feature, "You Can Avoid Osteoporosis." In the young adult, the activity of osteoclasts is matched by the activity of osteoblasts, and bone mass remains stable until about age 45 in women. After that age, bone mass starts to decrease.

Video Bear Bones

Bone remodeling also accounts for why bones can respond to stress. If you engage in an activity that calls upon the use of

BIOLOGY MATTERS  **Health**

You Can Avoid Osteoporosis

Osteoporosis is a condition in which the bones are weakened due to a decrease in the bone mass that makes up the skeleton. The skeletal mass continues to increase until ages 20 to 30. After that, there is an equal rate of formation and breakdown of bone mass until ages 40 to 50. Then, reabsorption begins to exceed formation, and the total bone mass slowly decreases (Fig. 11C).

Animation Osteoporosis

Over time, men are apt to lose 25% and women to lose 35% of their bone mass. But we have to consider that men—unless they have taken asthma medications that decrease bone formation—tend to have denser bones than women anyway. Whereas a man's testosterone (male sex hormone) level generally declines slowly after the age of 45, estrogen (female sex hormone) levels in women begin to decline significantly at about age 45. Sex hormones play an important role in maintaining bone strength, so this difference means that women are more likely than men to suffer a higher incidence of fractures, involving especially the hip, vertebrae, long bones, and pelvis. Although osteoporosis may at times be the result of various disease processes, it is essentially a disease that occurs as we age.

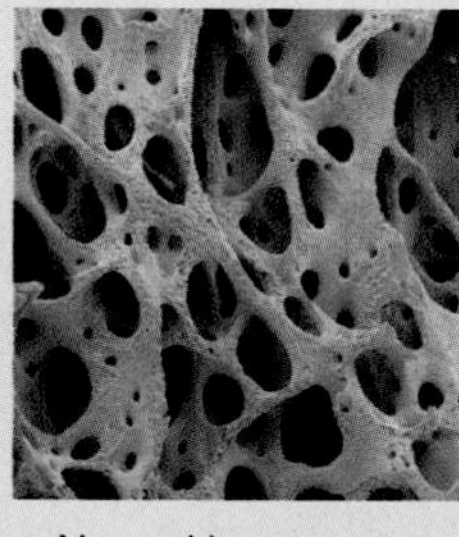

a. Normal bone

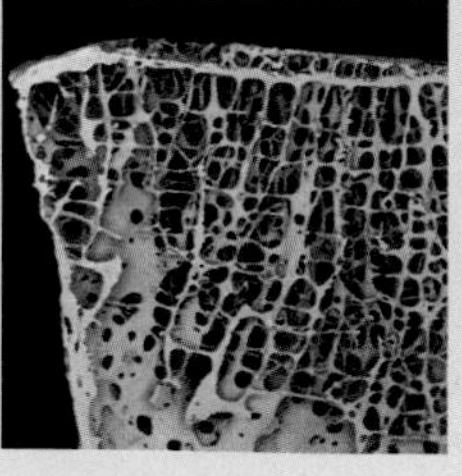

b. Osteoporosis

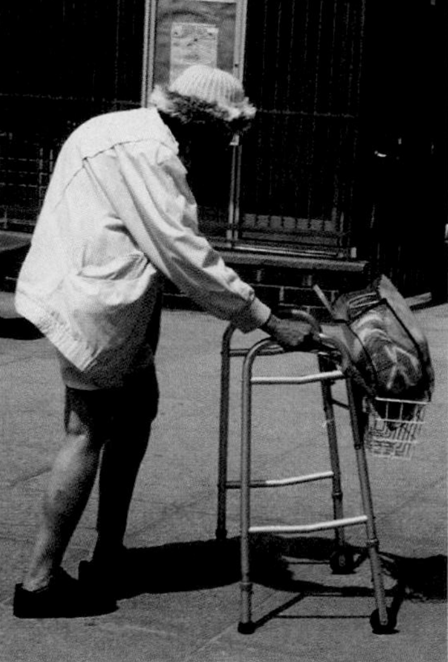

Figure 11C **Preventing osteoporosis.** Weight-bearing exercise, when we are young, can help prevent osteoporosis when we are older. **a.** Normal bone. **b.** Bone from a person with osteoporosis.

How to Avoid Osteoporosis

Everyone can take measures to avoid having osteoporosis when they get older. Adequate dietary calcium throughout life is an important protection against osteoporosis. The National Osteoporosis Foundation (www.nof.org) recommends that adults

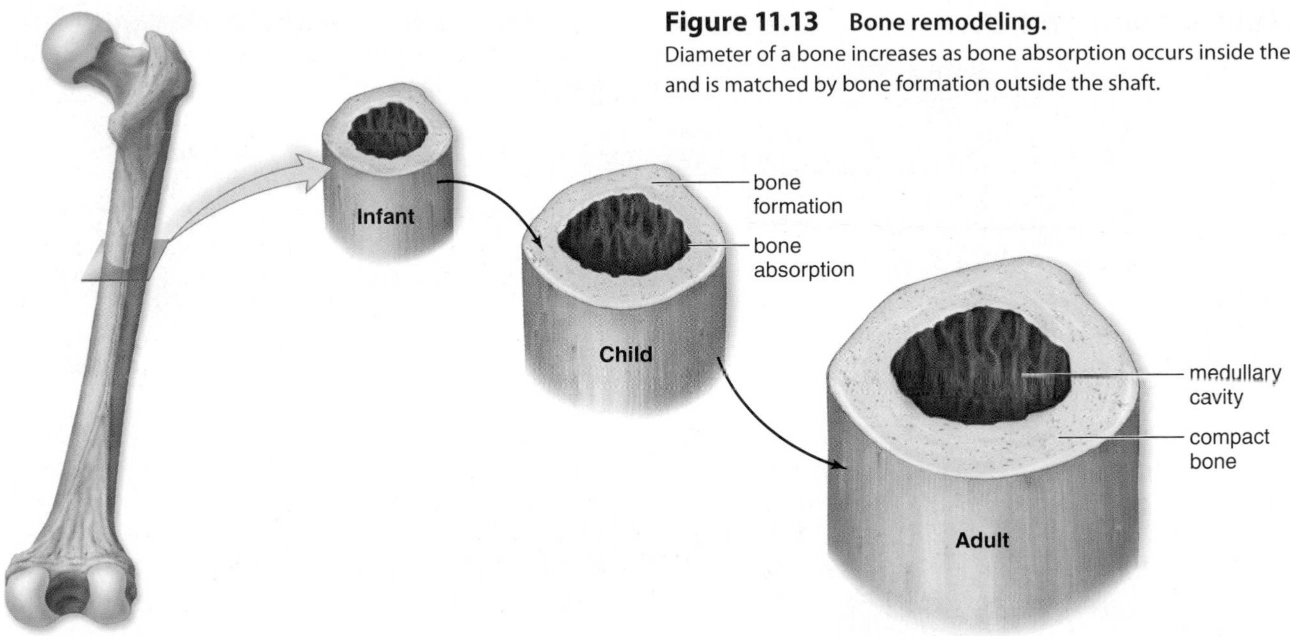

Figure 11.13 **Bone remodeling.**
Diameter of a bone increases as bone absorption occurs inside the shaft and is matched by bone formation outside the shaft.

under the age of 50 take in 1,000 mg of calcium per day. After the age of 50, the daily intake should exceed 1,200 mg per day.

A small daily amount of vitamin D is also necessary for the body to use calcium correctly. Exposure to sunlight is required to allow skin to synthesize a precursor to vitamin D. If you reside on or north of a "line" drawn from Boston to Milwaukee, to Minneapolis, to Boise, chances are you're not getting enough vitamin D during the winter months. Therefore, you should take advantage of the vitamin D present in fortified foods such as low-fat milk and cereal. If you are under age 50, you should be receiving 400–800 IU of vitamin D per day. After age 50, this amount should increase to 800–1,000 IU of vitamin D daily.

Very inactive people, such as those confined to bed, lose bone mass 25 times faster than people who are moderately active. On the other hand, moderate weight-bearing exercise, such as regular walking or jogging, is another good way to maintain bone strength (Fig. 11C).

Diagnosis and Treatment

Postmenopausal women with any of the following risk factors should have an evaluation of their bone density:

- white or Asian race
- thin body type
- family history of osteoporosis
- early menopause (before age 45)
- smoking
- a diet low in calcium, or excessive alcohol consumption and caffeine intake
- sedentary lifestyle

Bone density is measured by a method called dual-energy X-ray absorptiometry (DEXA). This test measures bone density based on the absorption of photons generated by an X-ray tube. Soon there may be blood and urine tests to detect the biochemical markers of bone loss. Then it will be made easier for physicians to screen older women and at-risk men for osteoporosis.

If the bones are thin, it is worthwhile to take all possible measures to gain bone density because even a slight increase can significantly reduce fracture risk. Although estrogen therapy does reduce the incidence of hip fractures, long-term estrogen therapy is rarely recommended for osteoporosis. Estrogen is known to increase the risk of breast cancer, heart disease, stroke, and blood clots. Other medications are available, however. Calcitonin, a thyroid hormone, has been shown to increase bone density and strength, while decreasing the rate of bone fractures. Also, the bisphosphonates are a family of nonhormonal drugs used to prevent and treat osteoporosis. To achieve optimal results with calcitonin or one of the bisphosphonates, patients should also receive adequate amounts of dietary calcium and vitamin D.

Questions to Consider

1. How may long-term digestive system problems promote the chances of developing osteoporosis?
2. Why are individuals at risk for osteoporosis encouraged to increase their exercise regimes, including load-bearing exercises?

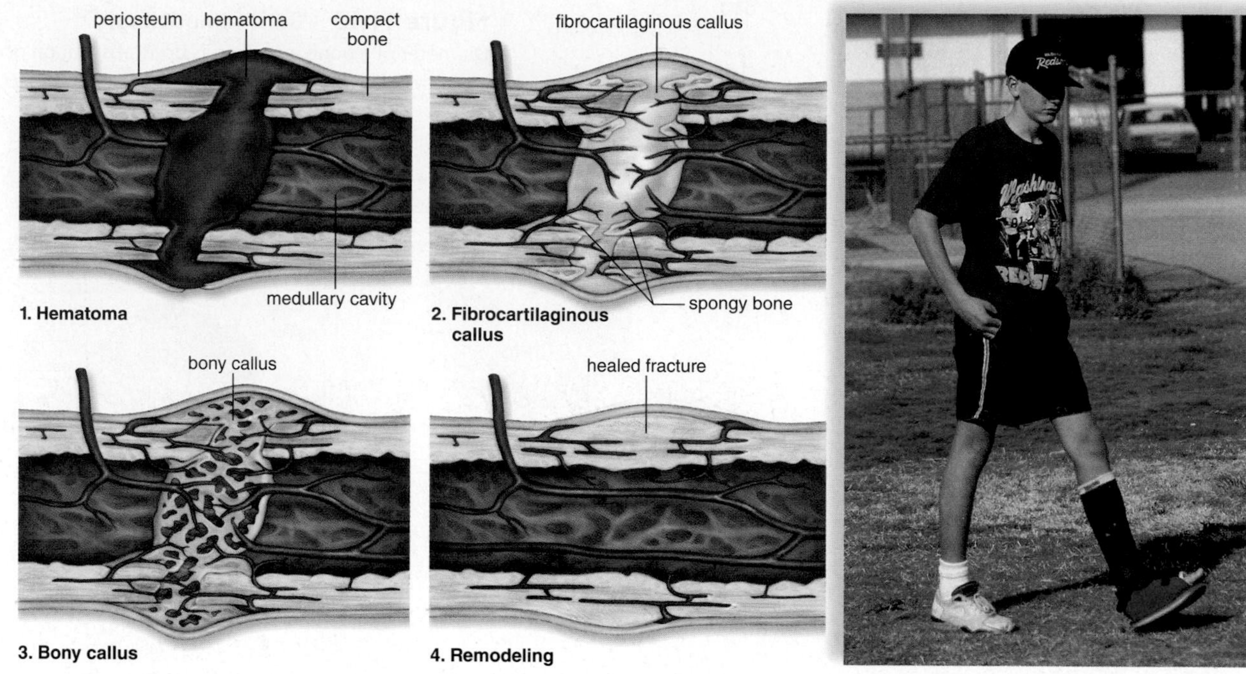

Figure 11.14 Bone repair following a fracture. **a.** Steps in the repair of a fracture. **b.** A cast helps stabilize the bones while repair takes place. Fiberglass casts are now replacing plaster of paris as the usual material for a cast.

a particular bone, the bone enlarges in diameter at the region most affected by the activity. During this process, osteoblasts in the periosteum form compact bone around the external bone surface and osteoclasts break down bone on the internal bone surface around the medullary cavity. Increasing the size of the medullary cavity prevents the bones from getting too heavy and thick. Today, exercises such as walking, jogging, and weight lifting are recommended. These exercises strengthen bone because they stimulate the work of osteoblasts instead of osteoclasts.

Bone Repair

Repair of a bone is required after it breaks or fractures. Fracture repair takes place over a span of several months in a series of four steps, shown in Figure 11.14 and listed here:

1. *Hematoma.* After a fracture, blood escapes from ruptured blood vessels and forms a hematoma (mass of clotted blood) in the space between the broken bones. The hematoma forms within 6 to 8 hours.
2. *Fibrocartilaginous callus.* Tissue repair begins, and a fibrocartilaginous callus fills the space between the ends of the broken bone for about three weeks.
3. *Bony callus.* Osteoblasts produce trabeculae of spongy bone and convert the fibrocartilage callus to a bony callus that joins the broken bones together. The bony callus lasts about three to four months.
4. *Remodeling.* Osteoblasts build new compact bone at the periphery. Osteoclasts absorb the spongy bone, creating a new medullary cavity.

In some ways, bone repair parallels the development of a bone except that the first step, hematoma, indicates that injury has occurred. Further, a fibrocartilaginous callus precedes the production of compact bone.

The naming of fractures tells you what type of break occurred. A fracture is complete if the bone is broken clear through and incomplete if the bone is not separated into two parts. A fracture is simple if it does not pierce the skin and compound if it does pierce the skin. Impacted means that the broken ends are wedged into each other. A spiral fracture occurs when the break is ragged due to twisting of a bone.

MP3 Remodeling and Repair

Blood Cells Are Produced in Bones

The bones of your skeleton contain two types of marrow: yellow and red. Fat is stored in yellow bone marrow, thus making it part of the body's energy reserves.

Red bone marrow is the site of blood cell production. The red blood cells are the carriers of oxygen in the blood. Oxygen is necessary for the production of ATP by aerobic cellular respiration. White blood cells also originate in the red bone marrow. The white cells are involved in defending your body against pathogens and cancerous cells; without them, you would soon succumb to disease and die.

CHECK YOUR PROGRESS 11.5

1. Describe how bone growth occurs during development.
2. Summarize the stages in the repair of bone.
3. Explain how the skeletal system is involved in calcium homeostasis.

CONNECTING THE CONCEPTS

For more on bone development and the hormones that influence bone growth, refer to the following discussions:

Section 8.6 provides additional information on inputs of vitamin D and calcium in the diet.

Section 15.2 examines the role of growth hormones in the body.

Section 15.3 describes the action of the hormones calcitonin and PTH.

CASE STUDY CONCLUSION

For the first painful month or so after having her knee replaced, Jackie wondered if she had made the right decision. Just walking down the hall or up stairs was excruciating at first. Within two months, however, she was walking and swimming. Her physical therapist attributed her rapid return to her previous habits of staying in shape. But Jackie knows that without twenty-first-century medicine, she might have had a difficult time walking by the time she was 60. Still, she has been reminded by her doctor that all bones, even those of adults, are dynamic structures. Whereas her bone could be replaced by bone remodeling, the plastic and ceramic parts of her knees would eventually wear out. So there was a very good chance that she would have to undergo a repeat replacement of her knee within 10 to 20 years. For Jackie, the ability to once again lead an active lifestyle was a worthwhile trade for a few months of discomfort.

MEDIA STUDY TOOLS

Enhance your study of this chapter with media! Visit **www.mhhe.com/maderhuman13e** and go to "Media Study Tools" for this chapter to access the following:

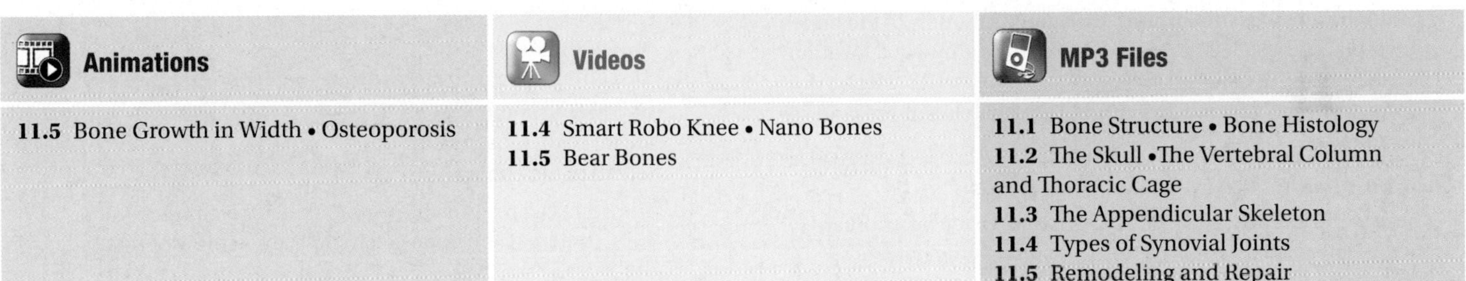

Animations	Videos	MP3 Files
11.5 Bone Growth in Width • Osteoporosis	**11.4** Smart Robo Knee • Nano Bones **11.5** Bear Bones	**11.1** Bone Structure • Bone Histology **11.2** The Skull • The Vertebral Column and Thoracic Cage **11.3** The Appendicular Skeleton **11.4** Types of Synovial Joints **11.5** Remodeling and Repair

SUMMARIZE

11.1 Overview of the Skeletal System

Functions of the **skeletal system:**

- It supports and protects the body.
- It produces blood cells.
- It stores mineral salts, particularly calcium phosphate. It also stores fat.
- Along with the muscles, it permits flexible body movement.

The bones of the skeleton are composed of bone tissues and cartilage. In a long bone:

- Cartilage covers the ends of a long bone. **Periosteum** (fibrous connective tissue) covers the rest of the bone.
- **Spongy bone,** containing **red bone marrow,** is in the epiphyses.
- Yellow bone marrow is in the **medullary cavity** of the diaphysis.
- **Compact bone** makes up the wall of the diaphysis.
- Bone cells called **osteocytes** reside within the **lacunae** of compact bone.

Cartilage is a connective tissue that is more flexible than bone. Cartilage is formed by cells called **chondrocytes.**

Ligaments composed of **fibrous connective tissue** connect bones at **joints.**

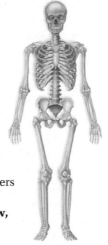

11.2 Bones of the Axial Skeleton

The **axial skeleton** consists of the skull, the hyoid bone, the vertebral column, and the rib cage.

- The **skull** is formed by the cranium, which protects the brain, and the facial bones. Membranous regions called **fontanels** connect the bones at birth. **Sinuses** are air spaces that reduce the weight of the skull. The **foramen magnum** is the opening through which the spinal cord passes.
- The hyoid bone anchors the tongue and is the site of attachment of muscles involved with swallowing.
- The **vertebral column** is composed of vertebrae separated by shock-absorbing disks, which make the column flexible. **Intervertebral disks** separate and pad the vertebrae. It supports the head and trunk, protects the spinal cord, and is a site for muscle attachment. Curving of the spinal column is called **scoliosis.**
- The rib cage is composed of the thoracic vertebrae, ribs, costal cartilages, and sternum. It protects the heart and lungs.

11.3 Bones of the Appendicular Skeleton

The **appendicular skeleton** consists of the bones of the pectoral girdles, upper limbs, pelvic girdle, and lower limbs.

- The **pectoral girdles** and upper limbs are adapted for flexibility.
- The **pelvic girdle** and the lower limbs are adapted for supporting weight. The **pelvis** consists of the pelvic girdle, sacrum, and coccyx. The femur is the longest and strongest bone in the body.

11.4 Articulations

Bones are joined at joints, of which there are three types:

- Fibrous joints (such as the **sutures** of the cranium) are immovable.
- Cartilaginous joints (such as those between the ribs and sternum and the pubic symphysis) are slightly movable.
- **Synovial joints** (which have a synovial membrane) are freely movable.

11.5 Bone Growth and Homeostasis

Cells involved in growth, remodeling, and repair of bone are

- **osteoblasts,** bone-forming cells;
- **osteocytes,** mature bone cells derived from osteoblasts; and
- **osteoclasts,** which break down and absorb bone.

Bone Development and Growth

- **Ossification** refers to the formation of bone.
- **Intramembranous ossification:** Bones develop between sheets of fibrous connective tissue. Examples are flat bones such as bones of the skull.
- **Endochondral ossification:** Cartilaginous models of the bones are replaced by calcified bone matrix. Bone grows at a location called the **epiphyseal plate.**
- Bone growth is affected by vitamin D, growth hormone, and sex hormones.

Bone Remodeling and Its Role in Homeostasis

- **Bone remodeling** is the renewal of bone. Osteoclasts break down bone and osteoblasts re-form bone. Some bone is recycled each year.
- Bone recycling allows the body to regulate blood calcium.
- Two hormones, parathyroid hormone and calcitonin, direct bone remodeling and control blood calcium.

Bone Repair

Repair of a fracture requires the following four steps:

- hematoma formation,
- fibrocartilaginous callus,
- bony callus, and
- remodeling.

ASSESS

Testing Your Knowledge of the Concepts

1. What are the five functions of the skeletal system? (page 239)
2. Where would you expect to find spongy bone in a long bone? What are its functions? (pages 239–240)
3. Differentiate between compact and spongy bone. (page 239)
4. Describe the structure of hyaline cartilage. Where can it be found in the body? (page 239)
5. Explain the terms *axial skeleton* and *appendicular skeleton.* (pages 241, 245)
6. List the bones that make up the cranium and the face. (pages 242–243)
7. What are the types of vertebrae, and how many of each are there? (page 244)
8. What bones make up the rib cage? What are the functions of the rib cage? (page 245)
9. Name the bones of the pectoral girdle and upper limbs. (page 246)
10. Name the bones of the pelvic girdle and lower limbs. (page 247)
11. What are the two types of cartilaginous joints, and what type of movement do cartilaginous joints have? (page 249)
12. Describe the different types of movements made by synovial joints. (pages 249–251)
13. What is the function of osteoclasts, and why are they needed? (pages 251–252)
14. Explain how bone is formed by intramembranous ossification. (page 252)
15. Describe the process of endochondral ossification. (pages 252–253)
16. Describe the actions of the hormones involved in bone growth and bone remodeling. (pages 254–256)
17. Provide the rationale for the four steps of bone repair. (page 256)
18. Spongy bone
    - **a.** contains osteons.
    - **b.** contains red bone marrow, where blood cells are formed.
    - **c.** weakens bones.
    - **d.** takes up most of a leg bone.
    - **e.** All of the choices are correct.
19. The bone cell responsible for breaking down bone tissue is the _____, whereas the bone cell that produces new bone tissue is the _____.
    - **a.** osteoclast, osteoblast
    - **b.** osteocyte, osteoclast
    - **c.** osteoblast, osteocyte
    - **d.** osteocyte, osteoblast
    - **e.** osteoclast, osteocyte
20. All blood cells—red, white, and platelets—are produced by which of the following?
    - **a.** yellow bone marrow
    - **b.** red bone marrow
    - **c.** periosteum
    - **d.** medullary cavity
21. This bone is the only movable bone of the skull.
    - **a.** sphenoid
    - **b.** frontal
    - **c.** mandible
    - **d.** maxilla
    - **e.** temporal
22. Which of the following is not a function of the skeletal system?
    - **a.** production of blood cells
    - **b.** storage of minerals
    - **c.** involved in movement
    - **d.** storage of fat
    - **e.** production of body heat
23. Which of the following is not a bone of the appendicular skeleton?
    - **a.** the scapula
    - **b.** a rib
    - **c.** a metatarsal bone
    - **d.** the patella
24. Which of the following statements is incorrect?
    - **a.** A growth plate occurs between the primary ossification center and a secondary center.
    - **b.** Each temporal bone has an opening that leads to the middle ear.
    - **c.** Intervertebral disks are composed of fibrocartilage.
    - **d.** Bone cells are rigid and hard because they are dead.
    - **e.** The sternum is composed of three bones that fuse during fetal development.

In questions 25–29, indicate whether the statement is true (T) or false (F).

25. The pectoral girdle is specialized for weight bearing, and the pelvic girdle is specialized for flexibility of movement. ________

26. The term *phalanges* refers to the bones in both the fingers and the toes. ________

27. Bones synthesize vitamin D for the body. ________

28. Bones store minerals and fat. ________

29. Most bones develop through endochondral ossification.

____

30. Label this diagram of a skeleton.

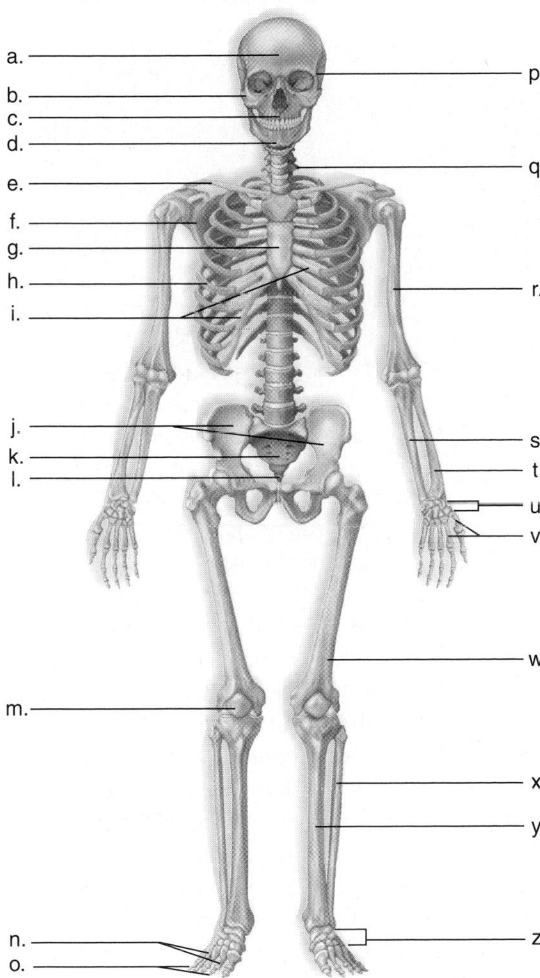

ENGAGE

Thinking Critically About the Concepts

The ligaments that connect children's bones are flexible, yet weak. Dislocations of bones forming a synovial joint are common in children and are usually not serious. Historically, this type of injury was termed "nursemaid's elbow," indicating that the injury resulted from careless handling of a child by a caregiver. Likewise, bone fractures are common to children. The most common fracture in a child is called a "greenstick" fracture, in which the bone splinters but does not break clean through. A greenstick fracture is most common in the radius and ulna from children falling on their arms. Most strains, sprains, dislocations, and fractures result from a child's active lifestyle. Injuries like these are treated with pain management, maneuvers to return a bone/joint to normal, immobilization in a cast or splint, and surgery if necessary. The vast majority of these injuries heal quickly, with no lingering effects. The child is encouraged to be as active as possible during the healing process, because inactivity causes increased bone depletion. Balancing safety with the need for children to exercise and play is critical for normal growth.

1. What nutritional and personal habits would contribute to rapid bone repair?

2. Individuals who spend the majority of their day indoors (e.g., residents in nursing homes) are more susceptible to fracture. Why?

3. Why do the broken bones of older people take much longer to mend than the broken bones of children and young adults?

4. Pediatricians are becoming increasingly concerned by the increased incidence of rickets in children eating a typical fast-food diet. What nutrients are missing from the diet?

5. Two athletes show up in the emergency room following a college football game. One has a fracture of the fibula. The second has a severe ankle sprain. Which player is most likely to return to play first?

6. a. What stimulates the release of parathyroid hormone?

b. What disease or diseases is (are) likely to result from hyperparathyroidism?

7. Why is the spinal cord almost always damaged when someone breaks one or more cervical vertebrae?

CHAPTER

12

Muscular System

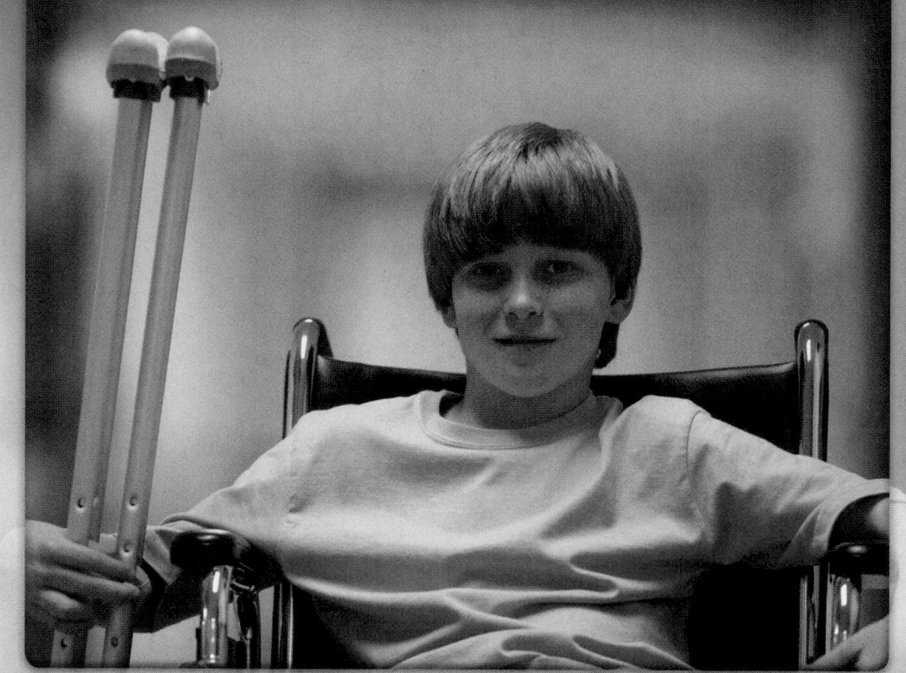

CHAPTER CONCEPTS

12.1 Overview of the Muscular System

The muscular system includes three types of muscles—skeletal, cardiac, and smooth. This chapter concentrates on skeletal muscle, which is attached to the bones of the skeleton. The skeletal muscles are named for their size, shape, location, and other attributes.

12.2 Skeletal Muscle Fiber Contraction

A skeletal muscle contains muscle fibers that are highly organized to carry out contraction. Muscle contraction occurs when the protein myosin interacts with the protein actin.

12.3 Whole Muscle Contraction

The degree of whole muscle contraction depends on how many motor units are maximally contracted. The energy for whole muscle contraction comes from carbohydrates and fats. Submaximal contraction for an extended period offers the best opportunity to burn fat.

12.4 Muscular Disorders

Muscular disorders include spasms and injuries, as well as such diseases as fibromyalgia, muscular dystrophy, and myasthenia gravis.

12.5 Homeostasis

The muscular system works closely with the skeletal system to help maintain homeostasis. Together, these systems move and protect body parts. The muscular system also helps maintain body temperature.

BEFORE YOU BEGIN

Before beginning this chapter, take a few moments to review the following discussions:

Section 3.6 How do fermentation and cellular respiration produce ATP?

Figure 3.20 What are the stages of the ATP cycle?

Section 4.8 How do feedback mechanisms contribute to homeostasis?

CASE STUDY MUSCULAR DYSTROPHY

Shortly after his 14th birthday, Kyle began to notice that he was becoming very clumsy. At the dinner table and at school, he constantly knocked over glasses and dropped his books. He was also beginning to have other problems; he frequently fell and had difficulty keeping his balance. At first, he just thought that this was the normal result of a teenage growth spurt. However, over the next several months, the episodes of clumsiness and falling occurred much more frequently. In addition, Kyle noticed that his muscles tired easily. Tasks that he normally had completed without any difficulties now seemed to take much more of an effort.

Concerned over the changes in Kyle, his parents scheduled an appointment with his longtime physician. His doctor asked that he stop by the local clinic and have blood drawn to see if Kyle was fighting any long-term infections. At the appointment, Kyle's doctor tested his reflexes and performed a complete examination. Because neither the blood work nor the examination indicated any of the usual causes of Kyle's symptoms, his doctor scheduled him for an electromyography (EMG) procedure the next week. An EMG records the electrical signals that control the skeletal muscles of the body. During an EMG, small pins are placed in the muscles of the arms or legs, and the patent is then asked to move the muscle while a technician records the electrical impulses. In Kyle's case, the results were abnormal—his muscles did not respond to the electrical signals. A biopsy of Kyle's muscle tissue indicated that he had a mutation in the protein dystrophin, which is responsible for holding the muscle fibers together. Defects in this protein are a leading cause of muscular dystrophy, a disease characterized by the wasting away of muscle tissue.

As you read through the chapter, think about the following questions:

1. Why would the physician order an electromyography procedure to test for muscle function?
2. Why is it important for the individual muscle fibers to be held together in a muscle?
3. What would be the long-term effects of not being able to control muscle contraction?

12.1 Overview of the Muscular System

LEARNING OUTCOMES

Upon completion of this section, you should be able to

1. List the three types of muscles and provide a function for each.
2. Describe the general structure of a skeletal muscle.
3. Identify the major skeletal muscles of the human body.

The **muscular system** is involved with movement. This may be the movement of the entire organism (walking or running), or the movement of materials (blood, food) within the organism. The muscular system is made up of muscles. All muscles, regardless of type, contract or shorten. It is this activity that causes movement.

MP3 Muscular System

Types of Muscles

Humans have three types of muscle tissue: smooth, cardiac, and skeletal (Fig. 12.1). The cells of these tissues are called **muscle fibers.**

Smooth muscle fibers are shaped like narrow cylinders with pointed ends. Each has a single nucleus (uninucleated). The cells are usually arranged in parallel lines, forming sheets. Striations (bands of light and dark) are seen in cardiac and skeletal muscle but not in smooth muscle. Smooth muscle is located in the walls of hollow internal organs and blood vessels and causes these walls to contract (see section 8.1). Contraction of smooth muscle is involuntary, occurring without conscious control. Although smooth muscle is slower to contract than skeletal muscle, it can sustain prolonged contractions and does not fatigue easily.

Cardiac muscle forms the heart wall (see section 5.3). Its fibers are generally uninucleated, striated, and tubular. Branching allows the fibers to interlock at **intercalated disks.**

APPLICATIONS AND MISCONCEPTIONS

How many muscles are there in the human body?

Most experts agree that there are over 600 muscles in the human body. The exact number varies, because some experts lump muscles together under one name and others split them apart. The smallest of these, the stapedius, is a 1.27-mm-long muscle located in the middle ear. The longest muscle is the sartorius, which starts at the hip and extends to the knee. The biggest muscle (in terms of mass) is the gluteus maximus, the muscle that makes up the majority of the buttocks.

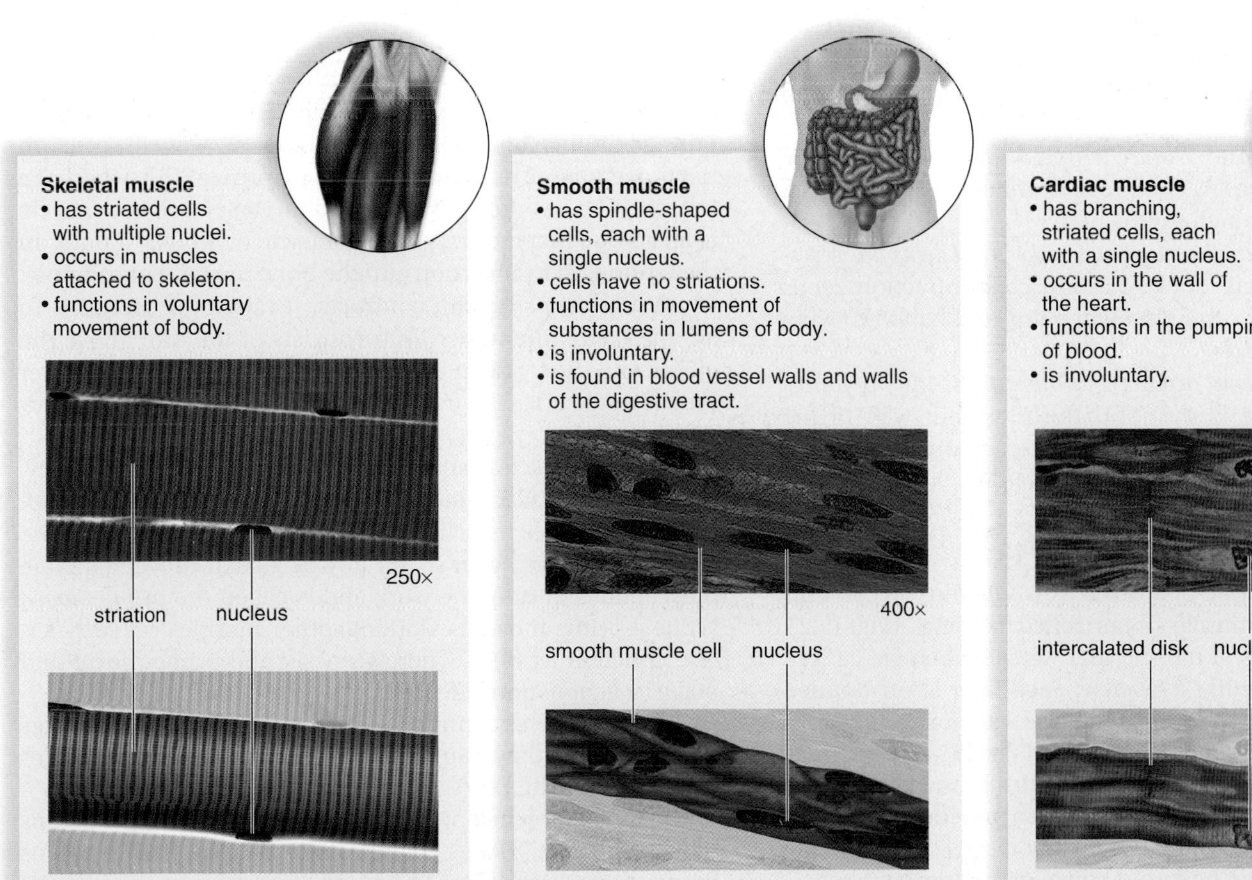
Skeletal muscle
• has striated cells with multiple nuclei.
• occurs in muscles attached to skeleton.
• functions in voluntary movement of body.
250×
striation
nucleus
Smooth muscle
• has spindle-shaped cells, each with a single nucleus.
• cells have no striations.
• functions in movement of substances in lumens of body.
• is involuntary.
• is found in blood vessel walls and walls of the digestive tract.
400×
smooth muscle cell
nucleus
Cardiac muscle
• has branching, striated cells, each with a single nucleus.
• occurs in the wall of the heart.
• functions in the pumping of blood.
• is involuntary.
250×
intercalated disk
nucleus
a.
b.
c.

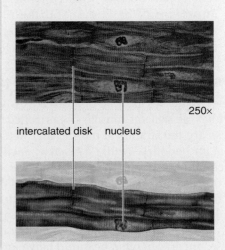

Figure 12.1 **The three classes of muscles in humans.**
Human muscles are of three types: (**a**) skeletal, (**b**) smooth, and (**c**) cardiac. Each muscle type has different characteristics.

The plasma membranes at intercalated disks contain gap junctions (see section 3.5) that permit contractions to spread quickly throughout the heart wall. Cardiac fibers relax completely between contractions, which prevents fatigue. Contraction of cardiac muscle is rhythmic. It occurs without outside nervous stimulation and without conscious control. Thus, cardiac muscle contraction is involuntary. **Skeletal muscle** fibers are tubular, multinucleated, and striated and make up the skeletal muscles attached to the skeleton. Fibers run the length of the muscle and can be quite long. Skeletal muscle is voluntary because we can decide to move a particular part of the body, such as the arms and legs.

Skeletal Muscles of the Body

Humans belong to a class of animals called the *vertebrates.* Vertebrate animals possess an internal vertebral column and a skeleton with jointed appendages. Our skeletal muscles are attached to the skeleton, and their contraction causes the movement of bones at a joint.

Functions of Skeletal Muscles

The skeletal muscles of the body have a wide variety of functions, including the following:

Support. Skeletal muscle contraction opposes the force of gravity and allows us to remain upright.

Movements of bones and other body structures. Muscle contraction accounts not only for the movement of arms and legs but also for movements of the eyes, facial expressions, and breathing.

Maintenance of a constant body temperature. Skeletal muscle contraction causes ATP to break down, releasing heat that is distributed throughout the body.

Movement of fluids in the cardiovascular and lymphatic systems. The pressure of skeletal muscle contraction keeps blood moving in cardiovascular veins and lymph moving in lymphatic vessels.

Protection of the internal organs and the stabilization of joints. Muscles pad the bones, and the muscular wall in the abdominal region protects the internal organs. Muscle tendons help hold bones together at joints.

Basic Structure of Skeletal Muscles

Skeletal muscles are well organized. A whole muscle contains bundles of skeletal muscle fibers called fascicles (Fig. 12.2). These are the strands of muscle that we see when we cut red meat and poultry. Within a fascicle, each fiber is surrounded by connective tissue; the fascicle is also surrounded by connective tissue. Muscles are covered with fascia, a type of connective tissue that extends beyond the muscle and becomes its **tendon.** Tendons quite often extend past a joint before anchoring a muscle to a bone. Small, fluid-filled sacs called **bursae** (sing., bursa) can often be found between tendons and bones. The bursae act as "cushions," allowing ease of movement.

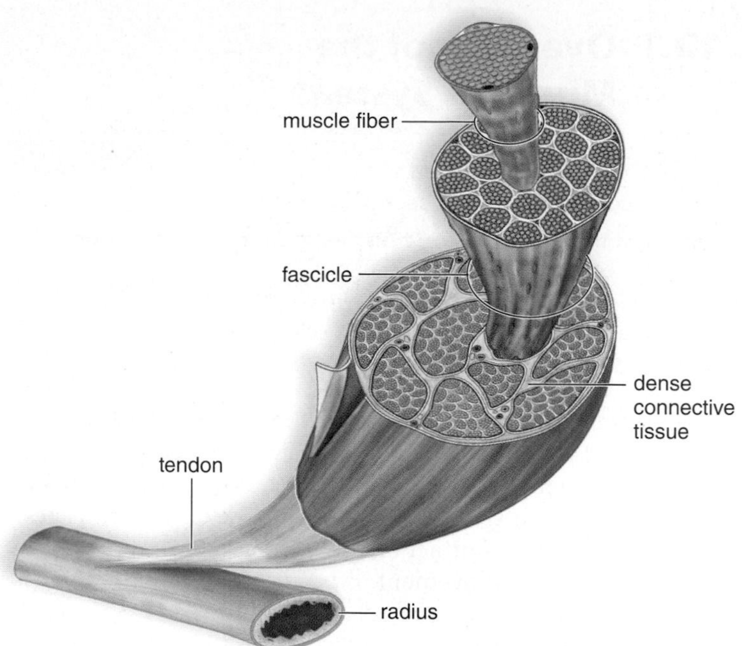

Figure 12.2 **Connecting muscle to bone.**
Connective tissue separates bundles of muscle fibers that make up a skeletal muscle. A layer of connective tissue covering the muscle contributes to the tendon, which attaches muscle to bone.

Skeletal Muscles Work in Pairs

In general, each muscle is concerned with the movement of only one bone. To simplify the discussion, we will focus on the movement of a single bone and no others. The **origin** of a muscle is on a stationary bone, and the **insertion** of a muscle is on a bone that moves. When a muscle contracts, it pulls on the tendons at its insertion and the bone moves. For example, when the biceps brachii contracts, it raises the forearm. To an anatomist, the upper limb consists of the arm above the elbow and the forearm below the elbow. Likewise, the lower limb is made up of the thigh above the knee and the leg below the knee.

Skeletal muscles usually function in groups. Consequently, to make a particular movement, your nervous system does not stimulate a single muscle. Rather, it stimulates an appropriate group of muscles. Even so, for any particular movement, one muscle does most of the work and is called the *prime mover.* While a prime mover is working, other muscles called *synergists* function as well. Synergists assist the prime mover and make its action more effective.

When muscles contract, they shorten. Therefore, muscles can only pull; they cannot push. This means that muscles work in opposite pairs. The muscle that acts opposite to a prime mover is called an *antagonist.* For example, the biceps brachii and the triceps brachii are antagonists. The biceps flexes the forearm (Fig. 12.3*a*), and the triceps extends the forearm (Fig. 12.3*b*). If both of these muscles contracted at

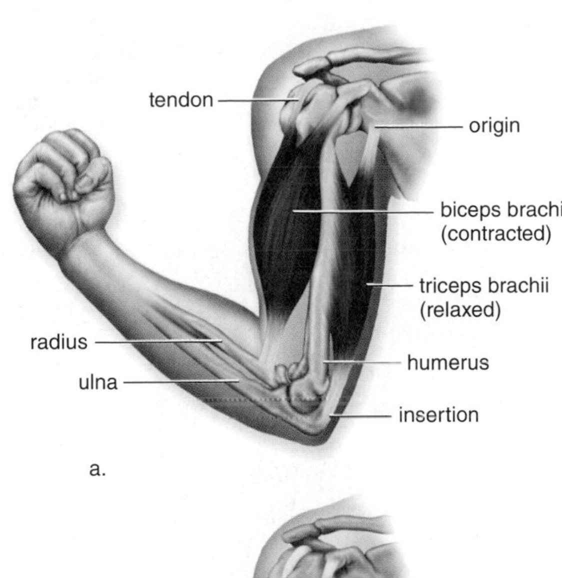

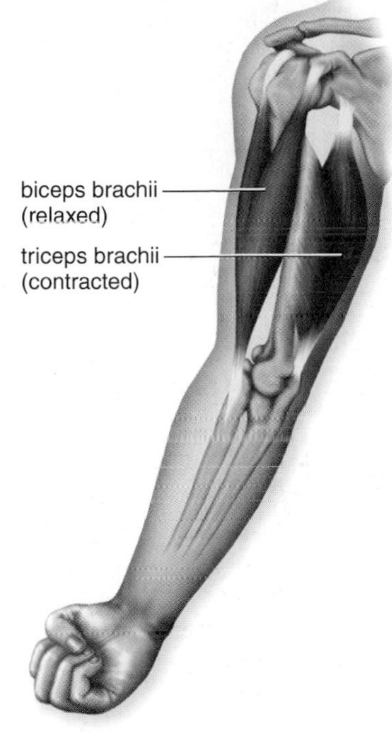

Figure 12.3 Skeletal muscles often work in pairs.
a. When the biceps brachii contracts, the forearm flexes. **b.** When the triceps brachii contracts, the forearm extends. Therefore, these two muscles are antagonistic. The origin of a skeletal muscle is on a bone that remains stationary, and the insertion of a muscle is on a bone that moves when the muscle contracts.

once, the forearm would remain rigid. Smooth body movements depend on an antagonist relaxing when a prime mover is acting.

Not all skeletal muscles are involved in the movement of limbs. For example, the facial muscles (Fig. 12.4) produce the facial expressions that tell us about the emotions and mood of a person and therefore play an important role in our interactions with other people.

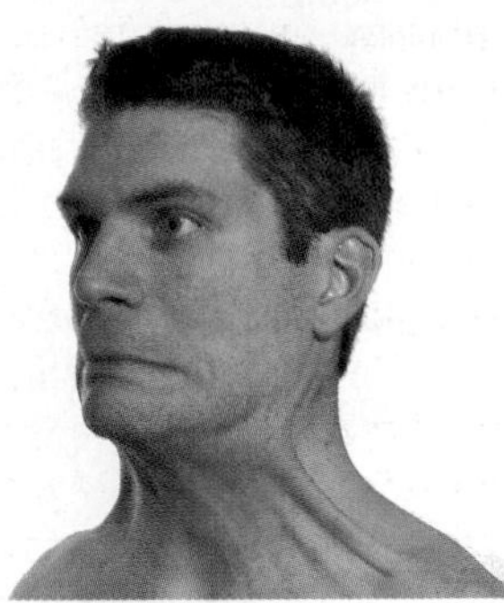

Figure 12.4 Facial expressions.
Our many facial expressions are due to muscle contractions.

Names and Actions of Skeletal Muscles

Figure 12.5, *a* and *b*, illustrates the location of some of the major skeletal muscles and gives their actions. (Not all the muscles mentioned are featured in Figure 12.5, but most are.)

When learning the names of muscles, considering what the name means will help you remember it. The names of the various skeletal muscles are often combinations of the following terms used to characterize muscles:

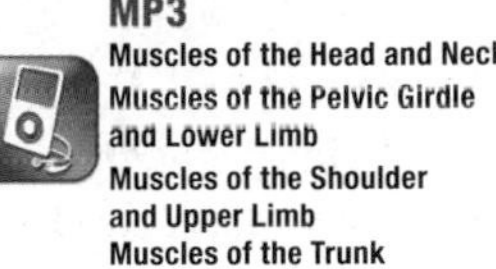

MP3
Muscles of the Head and Neck
Muscles of the Pelvic Girdle and Lower Limb
Muscles of the Shoulder and Upper Limb
Muscles of the Trunk

1. **Size.** The *gluteus maximus* is the largest muscle that makes up the buttocks. The *gluteus minimus* is the smallest of the gluteal muscles. Other terms used to indicate size are *vastus* (huge), *longus* (long), and *brevis* (short).
2. **Shape.** The *deltoid* is shaped like a triangle. (The Greek letter *delta* has this appearance: Δ.) The *trapezius* is shaped like a trapezoid. Other terms used to indicate shape are *latissimus* (wide) and *teres* (round).
3. **Location.** The *external oblique* muscles are located outside the *internal obliques.* The *frontalis* muscle overlies the frontal bone. Other terms used to indicate location are *pectoralis* (chest), *gluteus* (buttock), *brachii* (arm), and *sub* (beneath).
4. **Direction of muscle fibers.** The *rectus abdominis* is a longitudinal muscle of the abdomen (*rectus* means "straight"). The *orbicularis oculi* is a circular muscle around the eye. Other terms used to indicate direction are *transverse* (across) and *oblique* (diagonal).
5. **Attachment.** The *sternocleidomastoid* is attached to the sternum, clavicle, and mastoid process. The mastoid process is located on the temporal bone of the skull. The *brachioradialis* is attached to the brachium (arm) and the radius (forearm).
6. **Number of attachments.** The *biceps brachii* has two attachments, or origins, and is located on the arm. The *quadriceps femoris* has four origins and is located on the femur.
7. **Action.** The *extensor digitorum* extends the fingers or digits. The *adductor longus* is a large muscle that adducts the thigh. Adduction is the movement of a body part toward the midline. Other terms used to indicate action are *flexor* (to flex or bend), *masseter* (to chew), and *levator* (to lift).

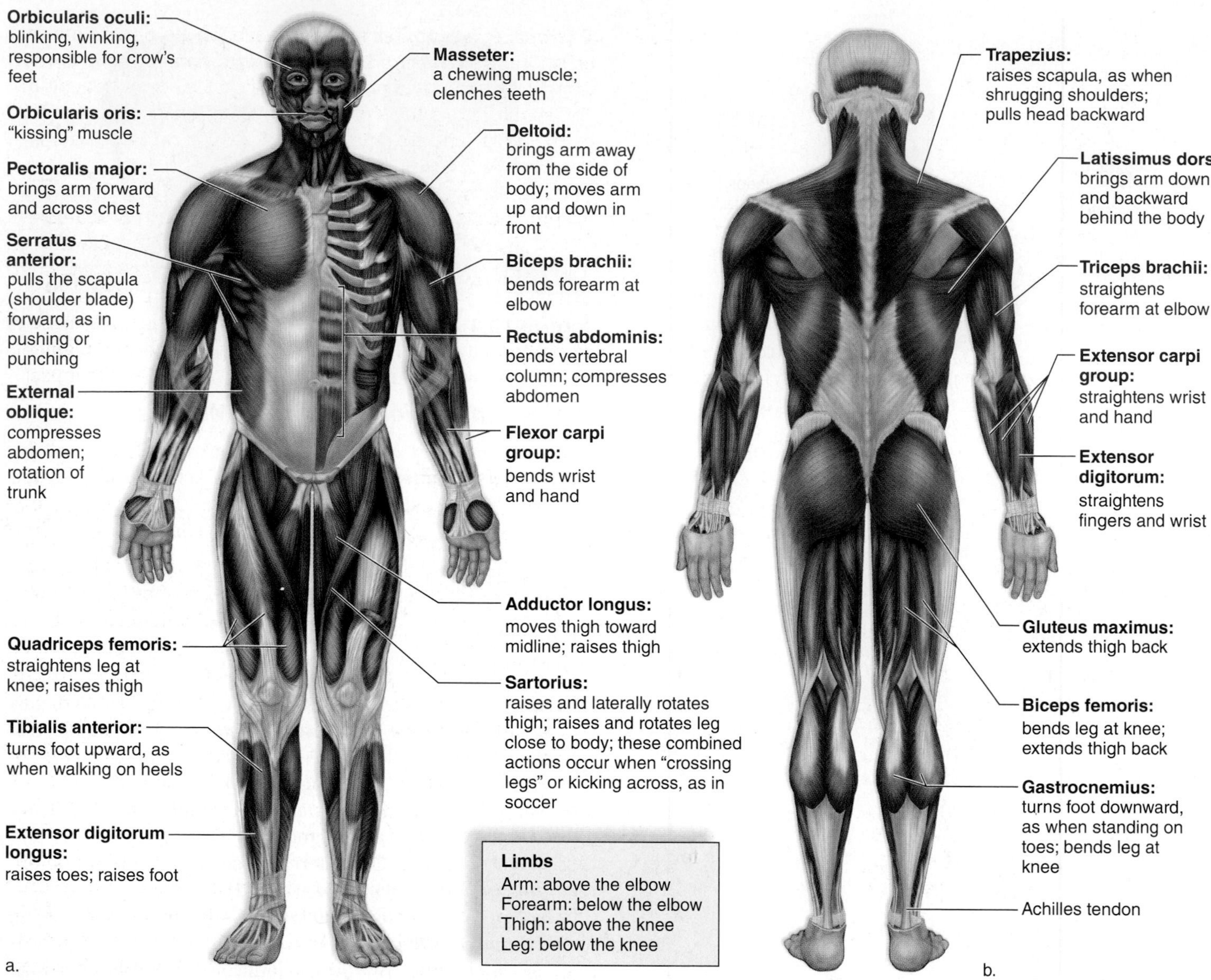
Orbicularis oculi: blinking, winking, responsible for crow's feet
Masseter: a chewing muscle; clenches teeth
Orbicularis oris: "kissing" muscle
Deltoid: brings arm away from the side of body; moves arm up and down in front
Pectoralis major: brings arm forward and across chest
Serratus anterior: pulls the scapula (shoulder blade) forward, as in pushing or punching
Biceps brachii: bends forearm at elbow
Rectus abdominis: bends vertebral column; compresses abdomen
External oblique: compresses abdomen; rotation of trunk
Flexor carpi group: bends wrist and hand
Adductor longus: moves thigh toward midline; raises thigh
Quadriceps femoris: straightens leg at knee; raises thigh
Sartorius: raises and laterally rotates thigh; raises and rotates leg close to body; these combined actions occur when "crossing legs" or kicking across, as in soccer
Tibialis anterior: turns foot upward, as when walking on heels
Extensor digitorum longus: raises toes; raises foot
Limbs
Arm: above the elbow
Forearm: below the elbow
Thigh: above the knee
Leg: below the knee
a.
Trapezius: raises scapula, as when shrugging shoulders; pulls head backward
Latissimus dorsi: brings arm down and backward behind the body
Triceps brachii: straightens forearm at elbow
Extensor carpi group: straightens wrist and hand
Extensor digitorum: straightens fingers and wrist
Gluteus maximus: extends thigh back
Biceps femoris: bends leg at knee; extends thigh back
Gastrocnemius: turns foot downward, as when standing on toes; bends leg at knee
Achilles tendon
b.

Figure 12.5 The major skeletal muscles of the human body.
a. Anterior view. **b.** Posterior view

APPLICATIONS AND MISCONCEPTIONS

Which muscles are best to use for intramuscular injections?

Healthcare providers have to choose muscles that are sufficiently large and well developed to tolerate intramuscular injections. At the same time, muscles containing large blood vessels or nerves must be avoided. An injection in these muscles could pierce a blood vessel or damage a nerve. Typically, one of three preferred injection sites is chosen. The deltoid muscle on the upper arm is usually well developed in older children and adults. The vastus lateralis on the side of the thigh (part of the quadriceps group) is the best site for infants and young children. The gluteus medius is on the lower back, above the buttock. However, a clinician injecting into the gluteus medius must be careful to avoid the gluteus maximus (buttock) muscle. The body's largest nerve, the sciatic nerve, lies underneath and within the gluteus maximus.

CHECK YOUR PROGRESS 12.1

1. State the three types of muscles in the human body.
2. Summarize the functions of skeletal muscles.
3. Identify the major skeletal muscles of the body.
4. Explain how skeletal muscles work together to cause bones to move.

CONNECTING THE CONCEPTS

For more information on the three types of muscles, refer to the following discussions:

Section 4.3 describes the general structure of cardiac, smooth, and skeletal muscle.

Section 5.3 examines the function of cardiac muscle in the heart.

Section 8.1 illustrates how smooth muscle lines the wall of the digestive tract.

12.2 Skeletal Muscle Fiber Contraction

LEARNING OUTCOMES

Upon completion of this section, you should be able to

1. Identify the structures of a muscle fiber.
2. Explain how the sliding filament model is responsible for muscle contraction.
3. Summarize how activities within the neuromuscular junction control muscle fiber contraction.

We have already examined the structure of skeletal muscle, as seen with the light microscope (see Fig. 12.1). Skeletal muscle tissue has alternating light and dark bands, giving it a striated appearance. These bands are due to the arrangement of myofilaments in a muscle fiber.

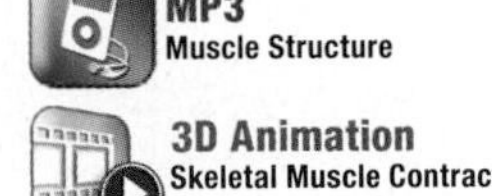

Muscle Fibers and How They Slide

A muscle fiber is a cell containing the usual cellular components, but special names have been assigned to some of these components (Table 12.1 and Fig. 12.6). For example, the plasma membrane is called the **sarcolemma** (*sarco* means "muscle"); the cytoplasm is the sarcoplasm; and the endoplasmic reticulum is the **sarcoplasmic reticulum.** A muscle fiber also has some unique anatomical characteristics. One feature is its T (for transverse) system. The sarcolemma forms **T (transverse) tubules** that penetrate, or dip down, into the cells. The transverse tubules come into contact—but do not fuse—with expanded portions of the sarcoplasmic reticulum. The expanded portions of the sarcoplasmic reticulum are calcium storage sites. Calcium ions (Ca^{2+}), as we shall see, are essential for muscle contraction.

The sarcolemma encases hundreds and sometimes even thousands of **myofibrils,** each about 1 μm in diameter. Myofibrils are the contractile portions of the muscle fibers. Any other organelles, such as mitochondria, are located in the sarcoplasm between the myofibrils. The sarcoplasm also contains glycogen, which provides stored energy for muscle contraction. In addition, sarcoplasm includes the red pigment myoglobin, which binds oxygen until it is needed for muscle contraction.

Table 12.1 Microscopic Anatomy of a Muscle Fiber

Name	Function
Sarcolemma	Plasma membrane of a muscle fiber that forms T tubules
Sarcoplasm	Cytoplasm of a muscle fiber that contains the organelles, including myofibrils
Glycogen	A polysaccharide that stores energy for muscle contraction
Myoglobin	A red pigment that stores oxygen for muscle contraction
T tubule	Extension of the sarcolemma that extends into the muscle fiber and conveys impulses that cause Ca^{2+} to be released from the sarcoplasmic reticulum
Sarcoplasmic reticulum	The smooth endoplasmic reticulum (ER) of a muscle fiber that stores Ca^{2+}
Myofibril	A bundle of myofilaments that contracts
Myofilament	Actin filaments or myosin filaments, whose structure and functions account for muscle striations and contractions

Myofibrils and Sarcomeres

As you see in Figure 12.6, a muscle cell or fiber is roughly cylindrical in shape. Grouped inside this larger cylinder are smaller cylinders called myofibrils. Myofibrils run the entire length of the muscle fiber. Myofibrils are composed of even smaller cylinders called myofilaments. Thus, the muscle cell is a set of small cylinders (myofilaments) assembled into larger cylinders (myofibrils) clustered within the largest cylinder (the muscle fiber). The light microscope shows that skeletal muscle fibers have light and dark bands called striations. At the higher magnification provided by an electron microscope, one can see that the striations of skeletal muscle fibers are formed by the placement of myofilaments within myofibrils. There are two types of myofilaments. Thick myofilaments are made up of a protein called **myosin,** and thin myofilaments are composed of a second protein termed **actin.** Myofibrils are further divided vertically into **sarcomeres**. A sarcomere extends between two dark vertical lines called the Z lines. The I bands on either side of the Z line are light colored because each contains only the thin actin myofilaments. The dark central A band within the sarcomere is composed of overlapping actin and myosin myofilaments. Centered within the A band is a vertical H band. In an uncontracted sarcomere, the H band lacks thin actin myofilaments and contains only thick myosin myofilaments.

Myofilaments

The thick and thin filaments differ in the following ways:

Thick filaments. A thick filament is composed of several hundred molecules of the protein myosin. Each myosin molecule is shaped like a golf club, with the straight portion of the molecule ending in a globular head, or cross-bridge. The cross-bridges occur on each side of a sarcomere but not in the middle (Fig. 12.6).

Thin filaments. Primarily, a thin filament consists of two intertwining strands of the protein actin. Two other proteins, called tropomyosin and troponin, also play a role, as we will discuss later in this section.

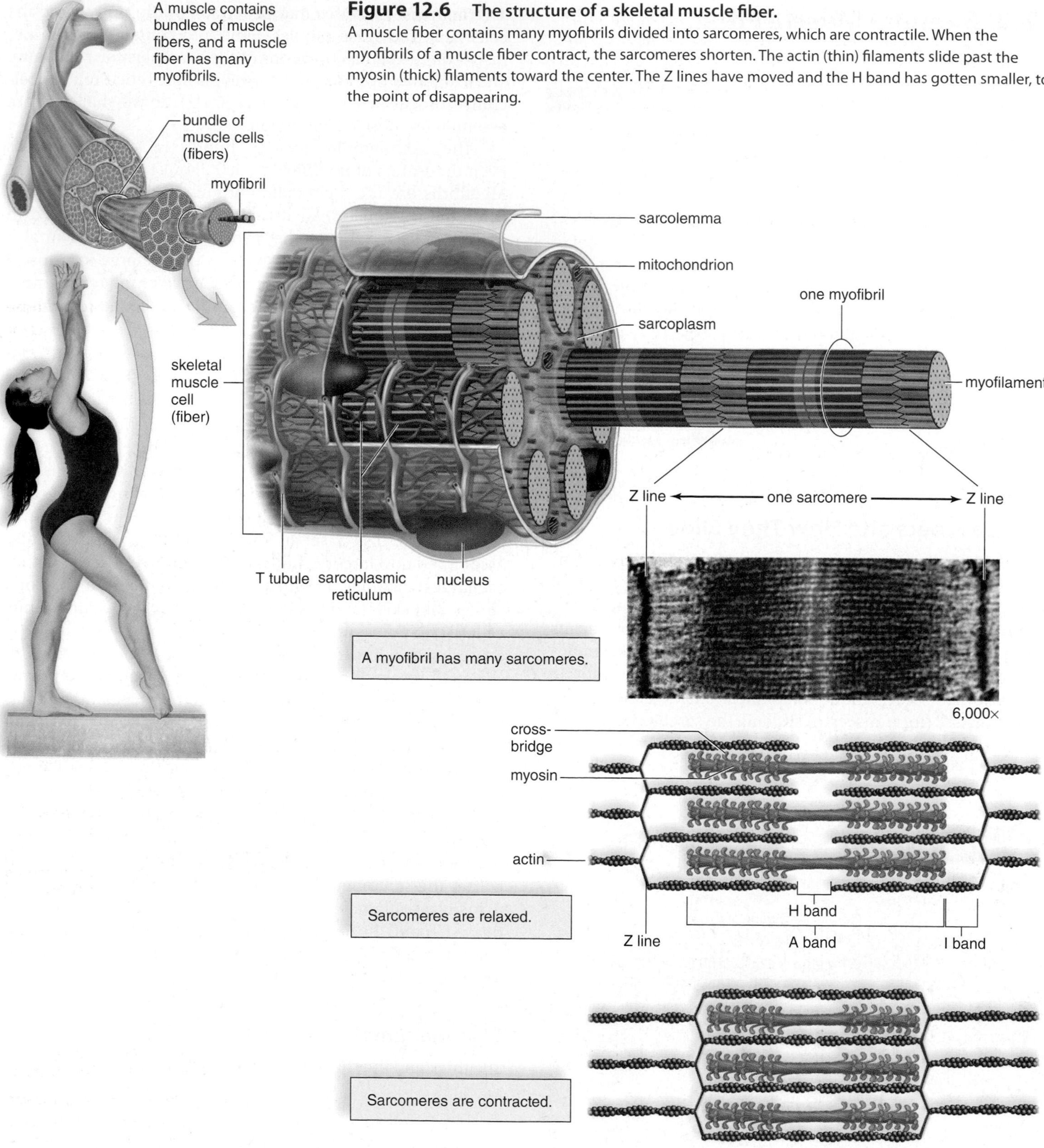

Figure 12.6 **The structure of a skeletal muscle fiber.**
A muscle fiber contains many myofibrils divided into sarcomeres, which are contractile. When the myofibrils of a muscle fiber contract, the sarcomeres shorten. The actin (thin) filaments slide past the myosin (thick) filaments toward the center. The Z lines have moved and the H band has gotten smaller, to the point of disappearing.

Sliding filaments. We will also see that when muscles are stimulated, electrical signals travel across the sarcolemma and then down a T tubule. In turn, this signals calcium to be released from the sarcoplasmic reticulum. Now the muscle fiber contracts as the sarcomeres within the myofibrils shorten. As you compare the relaxed sarcomere (Fig. 12.6) with the contracted sarcomere (Fig. 12.6), note that the filaments themselves remain the same length. When a sarcomere shortens, the actin (thin) filaments approach one another as they slide past the myosin (thick) filaments. This causes the I band to shorten, the Z line to move inward, and the H band to almost or completely disappear (Fig. 12.6). The sarcomere changes from a rectangular shape to a square as it shortens. The movement of actin filaments in relation to myosin filaments is called the **sliding filament model** of muscle contraction. ATP supplies the energy for muscle contraction. Although the actin filaments slide past the myosin filaments, it is the myosin filaments that do the work. Myosin filaments break down ATP, and their cross-bridges pull the actin filament toward the center of the sarcomere.

As an analogy, think of yourself and a group of friends as myosin. Collectively, your hands are the cross-bridges, and you are pulling on a rope (actin) to get an object tied to the end of the rope (the Z line). As you pull the rope, you grab, pull, release, and then grab farther along on the rope.

MP3
Sliding Filament Theory

Control of Muscle Fiber Contraction

Muscle fibers are stimulated to contract by motor neurons whose axons are grouped together to form nerves. The axon of one motor neuron can stimulate from a few to several muscle fibers of a muscle because each axon has several branches (Fig. 12.7*a*). Each branch of an axon ends in an axon terminal that lies in close

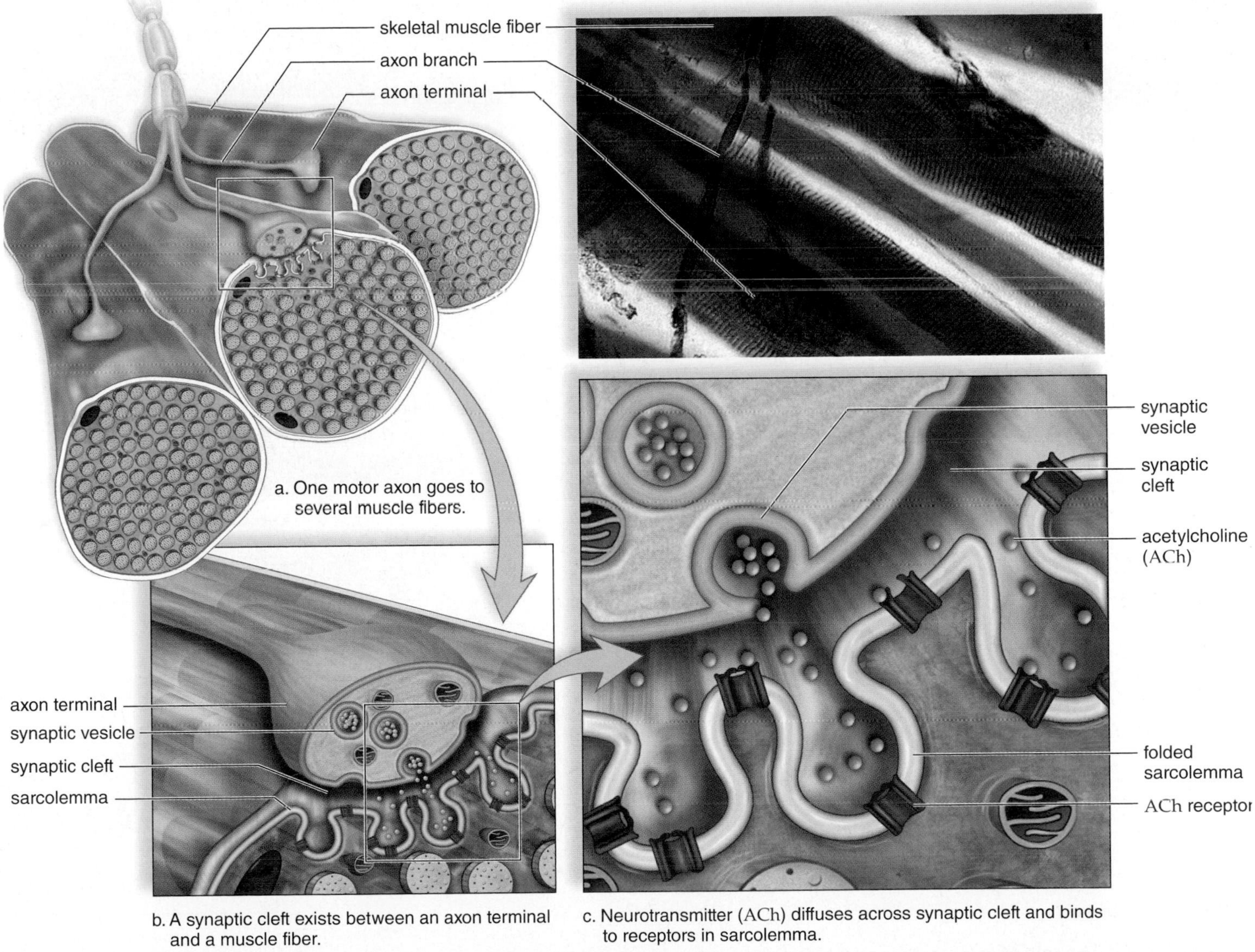

a. One motor axon goes to several muscle fibers.

b. A synaptic cleft exists between an axon terminal and a muscle fiber.

c. Neurotransmitter (ACh) diffuses across synaptic cleft and binds to receptors in sarcolemma.

Figure 12.7 Motor neurons and skeletal muscle fibers join neuromuscular junctions.
a. The branch of a motor nerve fiber terminates in an axon terminal. **b.** A synaptic cleft separates the axon terminal from the sarcolemma of the muscle fiber. **c.** Nerve impulses traveling down a motor fiber cause synaptic vesicles to discharge acetylcholine, which diffuses across the synaptic cleft and binds to ACh receptors. Impulses travel down the T tubules of a muscle fiber, and the muscle fiber contracts.

proximity to the sarcolemma of a muscle fiber. A small gap, called a synaptic cleft, separates the axon terminal from the sarcolemma (Fig. 12.7*b*). This entire region is called a **neuromuscular junction.**

MP3 Neuromuscular Junction

Axon terminals contain synaptic vesicles filled with the neurotransmitter acetylcholine (ACh). Nerve signals travel down the axons of motor neurons and arrive at an axon terminal. The signals trigger the synaptic vesicles to release ACh into the synaptic cleft (Fig. 12.7*c*). When ACh is released, it quickly diffuses across the cleft and binds to receptors in the sarcolemma. Now, the sarcolemma generates electrical signals that spread across the sarcolemma and down the T tubules. Recall that the T tubules lie adjacent to the sarcoplasmic reticulum, but the two structures are not connected. Nonetheless, signaling from the T tubules causes the release of Ca^{2+} from the sarcoplasmic reticulum, which leads to sarcomere contraction, as explained in Figure 12.8.

Two other proteins are associated with an actin filament. Threads of **tropomyosin** wind about an actin filament, covering binding sites for myosin located on each actin molecule. **Troponin** occurs at intervals along the threads. When calcium ions (Ca^{2+}) are released from the sarcoplasmic reticulum, they combine with troponin. This causes the tropomyosin threads to shift their position, exposing myosin-binding sites. In other words, myosin can now bind to actin (Fig.12.8*a*).

To fully understand muscle contraction, study Figure 12.8*b*. (1) The heads of a myosin filament have ATP-binding sites. At this site, ATP is hydrolyzed, or split, to form ADP and Ⓟ. (2) The ADP and Ⓟ remain on the myosin heads, and the heads attach to an actin-binding site. Joining myosin to actin forms temporary bonds called cross-bridges. (3) Now, ADP and Ⓟ are released and the cross-bridges bend sharply. This is the power stroke that pulls the actin filament toward the center of the sarcomere. (4) When ATP molecules again bind to the myosin heads, the cross-bridges

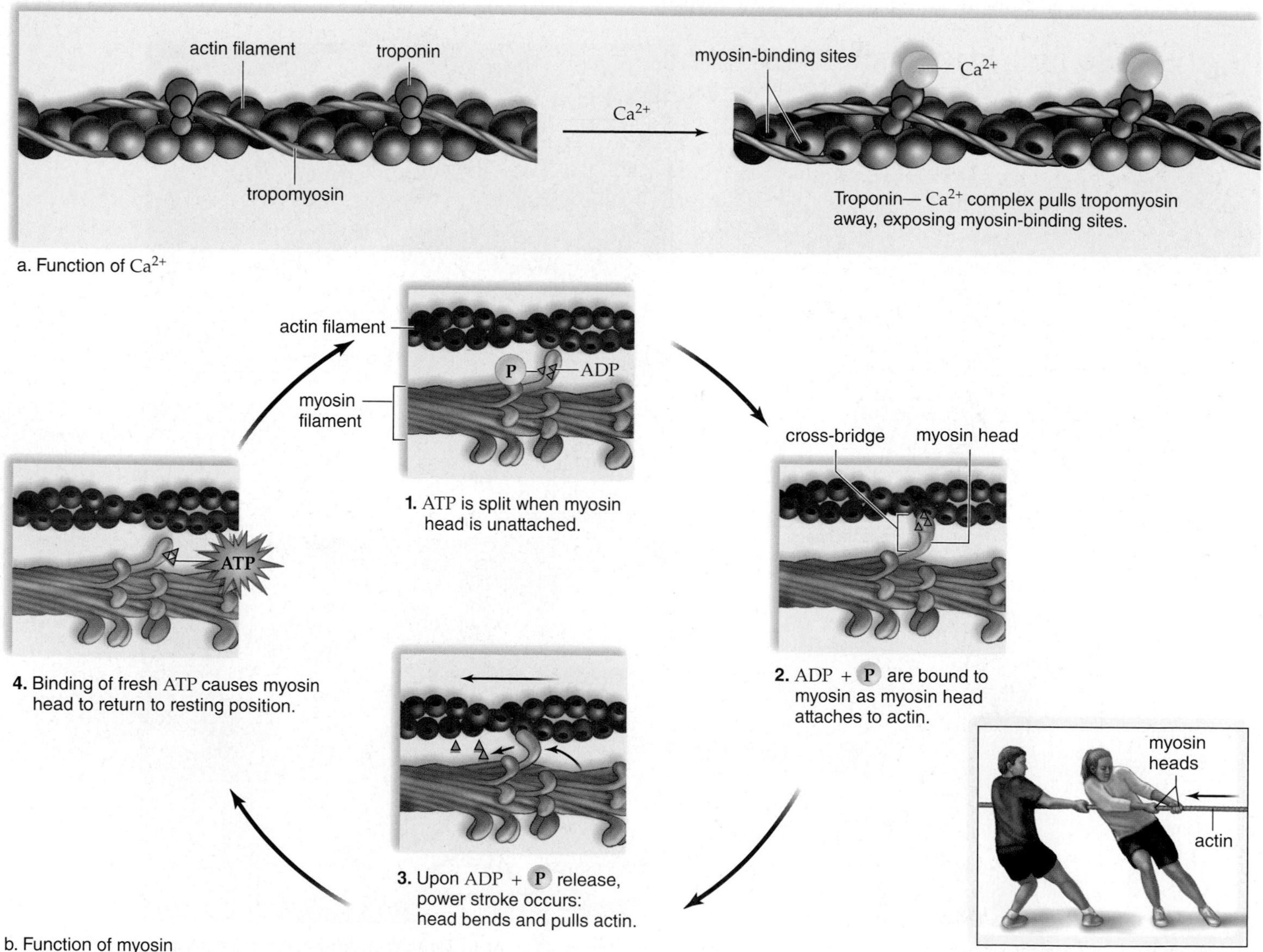
actin filament
troponin
tropomyosin
Ca^{2+}
myosin-binding sites
Ca^{2+}
Troponin— Ca^{2+} complex pulls tropomyosin away, exposing myosin-binding sites.
a. Function of Ca^{2+}
actin filament
myosin filament
P
ADP
1. ATP is split when myosin head is unattached.
cross-bridge
myosin head
2. ADP + P are bound to myosin as myosin head attaches to actin.
3. Upon ADP + P release, power stroke occurs: head bends and pulls actin.
ATP
4. Binding of fresh ATP causes myosin head to return to resting position.
myosin heads
actin
b. Function of myosin

Figure 12.8 The role of calcium ions and ATP during muscular contraction.
a. Calcium ions (Ca^{2+}) bind to troponin, exposing myosin-binding sites. **b.** Follow steps 1 through 4 to see how myosin uses ATP and does the work of pulling actin toward the center of the sarcomere, much as a group of people pulling a rope (*right*).

BIOLOGY MATTERS  Science

Botox and Wrinkles

Several of the most important bacterial pathogens that cause human diseases—including cholera, diphtheria, tetanus, and botulism—do so by secreting potent toxins capable of sickening or killing their victims. The botulinum toxin, produced by the bacterium *Clostridium botulinum,* is one of the most lethal substances known. Less than a microgram (μg) of the purified toxin can kill an average size person, and 4 kilograms (kg) (8.8 pounds [lb]) would be enough to kill all humans on Earth! Given this scary fact, it seems that the scientists who discovered the lethal activity of this bacterial toxin nearly 200 years ago could never have anticipated that the intentional injection of a very dilute form of botulinum toxin (now known as Botox) would become the most common nonsurgical cosmetic procedure performed by many physicians. As with many breakthroughs in science and medicine, the pathway from thinking about botulism as a deadly disease to using botulinum toxin as a beneficial treatment involved the hard work of many scientists, mixed with a considerable amount of luck.

In the 1820s, a German scientist, Justinus Kerner, was able to prove that the deaths of several people had been caused by their consumption of spoiled sausage (in fact, botulism is named for the Latin word for sausage, botulus). A few decades later, a Belgian researcher named Emile Pierre van Ermengem identified the specific bacterium responsible for producing the botulinum toxin, which could cause symptoms ranging from droopy eyelids to paralysis and respiratory failure.

By the 1920s, medical scientists at the University of California had obtained the toxin in pure form, which allowed them to determine that it acted by preventing nerves from communicating with muscles, specifically by interfering with the release of acetylcholine from the axon terminals of motor nerves.

Scientists soon began testing very dilute concentrations of the toxin as a treatment for conditions in which the muscles contract too much, such as crossed eyes or spasms of the facial muscles or vocal cords. In 1989, the FDA first approved diluted botulinum toxin (Botox) for treating specific eye conditions called blepharospasm (eyelid spasms) and strabismus (crossing of the eyes).

Right around this time, a lucky break occurred that eventually would open the medical community's eyes to the greater potential of the diluted toxin. A Canadian ophthalmologist, Jean Carruthers, had been using it to treat her patients' eye conditions when she noticed that some of their wrinkles had also subsided. One night at a family dinner, Dr. Carruthers shared this information with her husband, a dermatologist, who decided to investigate whether he could reduce the deep wrinkles of some of his patients by injecting the dilute toxin into their skin. The treatment worked well, and after trying it on several more patients, (as well as on themselves!), the Canadian doctors spent several years presenting their findings at scientific meetings and in research journals. Although they were initially considered "crazy," the Carrutheres eventually were able to convince the scientific community that diluted botulinum toxin was effective in treating wrinkles; however, they never patented it for that use, so they missed out on much of the $1.3 billion in annual sales the drug now earns for the company that did patent it.

The uses of diluted botulinum toxin seem to be growing since it was FDA-approved for the treatment of frown lines in 2002 (Fig. 12A). In March 2010, it was approved for the treatment of muscle stiffness in people with upper limb spasticity, and the company is currently seeking approval for as many as 90 uses of diluted botulinum toxin, including treatment of migraine headaches.

Questions to Consider

1. Considering that botulism is caused by a preformed toxin, how do you suppose it can be treated?
2. Do you think companies should be allowed to patent a naturally occurring molecule like botulinum toxin? Why or why not?

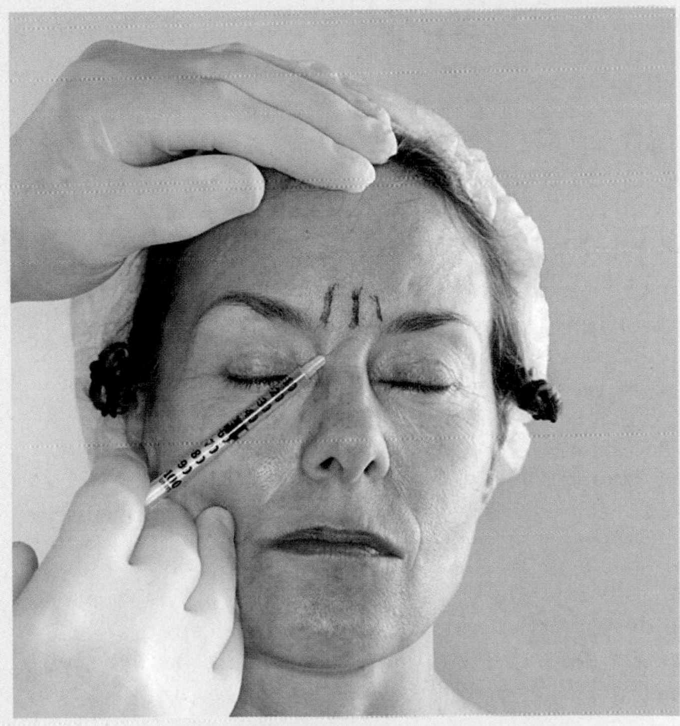

Figure 12A Treating wrinkles with diluted botulinum toxin.

are broken. Myosin heads detach from the actin filament. This is the step that does not happen during rigor mortis. Relaxing the muscle is impossible, because ATP is needed to break the bond between an actin-binding site and the myosin cross-bridge.

In living muscle, the cycle begins again and myosin reattaches farther along the actin filament. The cycle recurs until calcium ions are actively returned to the calcium storage sites. This step also requires ATP.

3D Animation
Skeletal Muscle Contraction: Regulation by Calcium Atoms

CHECK YOUR PROGRESS 12.2

1. Identify the myofibril, myofilament, and sarcomere in a muscle fiber.
2. Explain how the thin and thick filaments interact in the sliding filament model.
3. Describe the role of both ATP and calcium ions in muscle contraction.

CONNECTING THE CONCEPTS

For more information on ATP and how the nervous system controls the contraction of skeletal muscle, refer to the following discussions:

Figure 3.20 illustrates the ATP–ADP cycle.

Section 13.2 explains the role of neurotransmitters, such as acetylcholine, in the nervous system.

Figure 13.4 illustrates the action of neurotransmitters in the synaptic cleft.

12.3 Whole Muscle Contraction

LEARNING OUTCOMES

Upon completion of this section, you should be able to

1. List the stages of a muscle twitch and explain what is occurring in each stage.
2. Explain how summation and tetanus increase the strength of whole muscle contraction.
3. Summarize how muscle cells produce ATP for muscle contraction.
4. Distinguish between fast-twitch and slow-twitch muscle fibers.

In order for a whole muscle, such as the biceps or triceps, to contract, the individual muscle fibers must be activated by signals from the nervous system.

Muscles Have Motor Units

As already mentioned, each axon within a nerve stimulates a number of muscle fibers. A nerve fiber with all of the muscle fibers it innervates is called a **motor unit.** A motor unit obeys a principle called the *all-or-none law.* Why? Because all the muscle fibers in a motor unit are stimulated at once. They all either contract or do not contract. A variable of interest is the number of muscle fibers within a motor unit. For example, in the ocular muscles that move the eyes, the innervation ratio is one motor axon per 23 muscle fibers. By contrast, in the gastrocnemius muscle of the leg, the ratio is about one motor axon per 1,000 muscle fibers. Thus, moving the eyes requires finer control than moving the legs.

When a motor unit is stimulated by infrequent electrical impulses, a single contraction occurs. This response is called a **muscle twitch** and lasts only a fraction of a second. A muscle twitch is customarily divided into three stages. We can use our knowledge of muscle fiber contraction to understand these events. The latent period is the time between stimulation and initiation of contraction (Fig. 12.9*a*). During this time, we can imagine that the events that begin muscle contraction are occurring. The neurotransmitter ACh diffuses across the synaptic cleft, causing an electrical signal to spread across the sarcolemma and down the T tubules. The contraction period follows as calcium leaves the sarcoplasmic reticulum and myosin-actin cross-bridges form. As you know, the muscle shortens as it contracts. On the graph, the force increases as

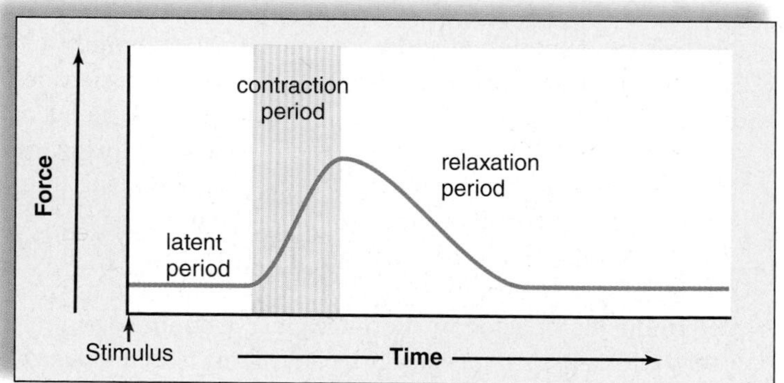

a.

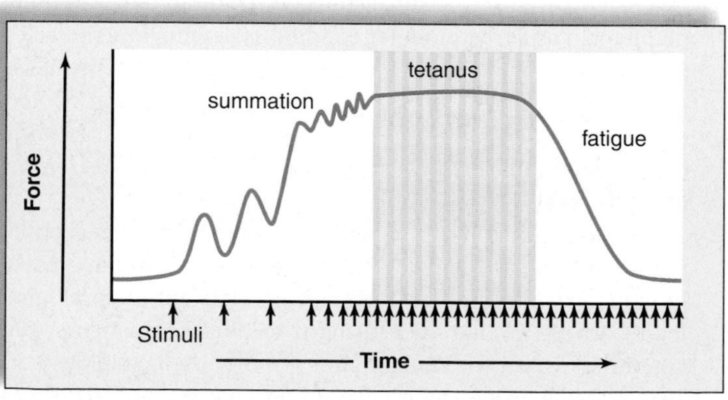

b.

Figure 12.9 The three phases of a single muscle twitch and how summation and tetanus increase the force of contraction. **a.** Stimulation of a muscle by a single electrical signal results in a simple muscle twitch: first, a latent period, followed by contraction and relaxation. **b.** Repeated stimulation results in summation and tetanus, which creates greater force because the motor unit cannot relax between stimuli.

the muscle contracts. Finally, the relaxation period completes the muscle twitch. Myosin–actin cross-bridges are broken, and calcium returns to the sarcoplasmic reticulum. Force diminishes as the muscle returns to its former length.

If a motor unit is given a rapid series of stimuli, it can respond to the next stimulus without relaxing completely. **Summation** is increased muscle contraction until maximal sustained contraction, called **tetanus,** is achieved (Fig. 12.9*b*). Tetanus continues until the muscle fatigues due to depletion of energy reserves. Fatigue is apparent when a muscle relaxes, even though stimulation continues. The *tetanus* of muscle cells is not the same as the infection called tetanus. The infection called tetanus is caused by the bacterium *Clostridium tetani.* Death occurs because the muscles, including the respiratory muscles, become fully contracted and do not relax.

A whole muscle typically contains many motor units. As the intensity of nervous stimulation increases, more motor units in a muscle are activated. This phenomenon is known as *recruitment.* Maximum contraction of a muscle would require that all motor units be undergoing tetanic contraction. This rarely happens because they could all fatigue at the same time. Instead, some motor units are contracting maximally while others are resting, allowing sustained contractions to occur.

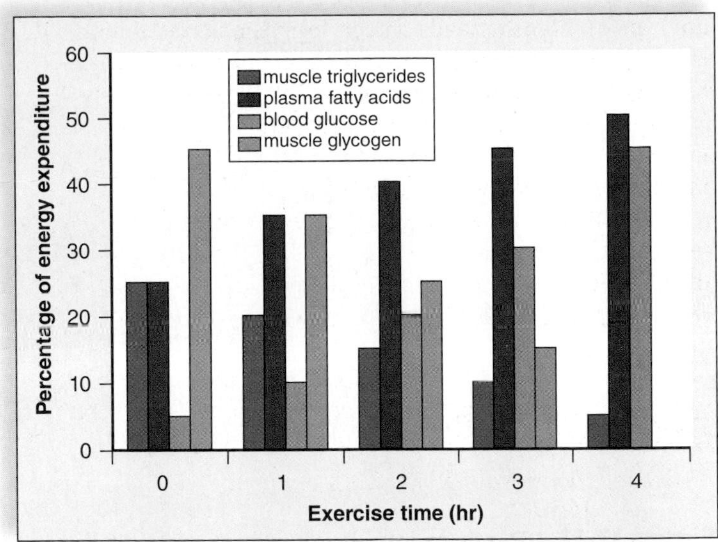

Figure 12.10 The sources of energy for muscle contraction. The percentage of energy derived from each of the four major fuel sources during submaximal exercise (65–75% of effort) is illustrated. The amount from plasma fatty acids increases during the time span shown.

APPLICATIONS AND MISCONCEPTIONS

What is the difference between an eyelid tic and an eyelid twitch?

The answer is based on whether the action may be controlled or not. If the individual is able to control the movement, even temporarily, it is called a tic. If the movement cannot be controlled, it is considered to be a twitch. No one is precisely sure what causes a tic, although the connection with medications for attention deficit hyperactivity disorder (ADHD) has been shown not to be the case. Twitches are believed to be the result of signaling problems in a specific area of the brain. The severity of both tics and twitches may be influenced by stress and sleep problems.

Muscle Tone

One desirable effect of exercise is to have good **muscle tone.** Muscles that have good tone are firm and solid, as opposed to being soft and flabby. The amount of muscle tone is dependent on muscle contraction. Some motor units are always contracted—but not enough to cause movement.

Energy for Muscle Contraction

Muscles can use various fuel sources for energy, and they have various ways of producing ATP needed for muscle contraction.

Fuel Sources for Exercise

A muscle has four possible energy sources (Fig. 12.10). Two of these are stored in muscle (glycogen and triglycerides), and two are acquired from blood (glucose and fatty acid). The amount that each of these is used depends on exercise intensity and duration. Figure 12.10 shows the percentage of energy derived from these sources due to submaximal exercise (65–75% of effort that an individual is capable of) over time. Notice that as the length of the exercise period is increased, use of muscle glycogen decreases and use of muscle triglyceride (fat) increases.

Muscles also make use of blood glucose and plasma fatty acids as energy sources. Both of these are delivered to muscles by circulating blood. Some of us specifically exercise to lose weight, and we are particularly interested in the use of plasma fatty acids by exercising muscles. Adipose tissue is the source of plasma fatty acids for muscle contraction, but it also tends to make us look fat. Figure 12.10 shows that the amount of fat burned increases when more time is spent in exercise. Therefore a diet that restricts the amount of fat eaten, when combined with exercise, will decrease body fat. Submaximal exercise burns fat better than maximal exercise, for reasons we will now explore.

Sources of ATP for Muscle Contraction

Muscle cells store limited amounts of ATP. Once stored ATP is used up, the cells have three ways to produce more ATP (Fig. 12.11). The three ways include (1) formation of ATP by the creatine phosphate (CP) pathway; (2) formation of ATP by fermentation; and (3) formation of ATP by cellular respiration, which involves the use of oxygen by mitochondria. Aerobic exercising depends on cellular respiration to supply ATP. Neither formation of ATP by the CP pathway nor by fermentation involves the need for oxygen. Both are anaerobic processes.

MP3
Energy Source for Muscle Contraction

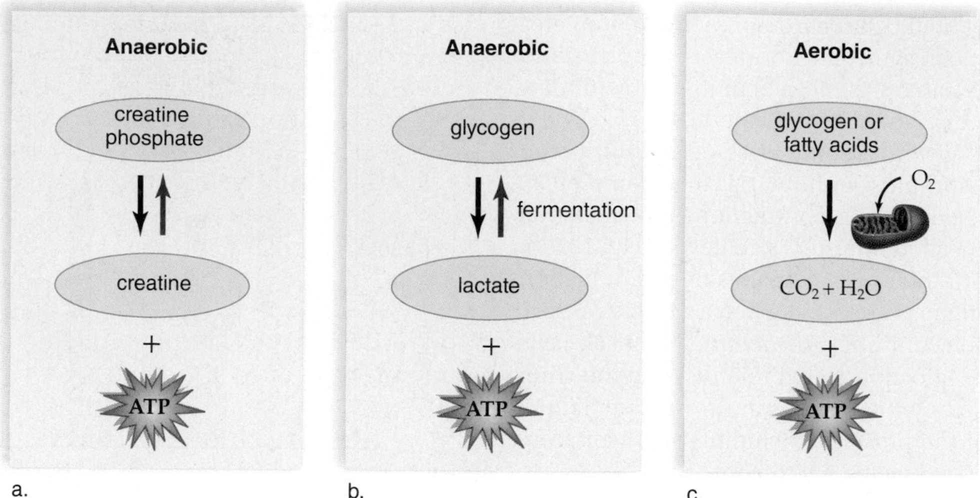

Figure 12.11 **The three pathways by which muscle cells produce the ATP energy needed for contraction.** **a.** When contraction begins, muscle cells break down creatine phosphate to produce ATP. When resting, muscle cells rebuild their supply of creatine phosphate (red arrow). **b.** Muscle cells also use fermentation to produce ATP quickly. When resting, muscle cells metabolize lactate, re-forming as much glucose and then glycogen as possible (red arrow). **c.** For the long term, muscle cells switch to cellular respiration to produce ATP aerobically.

The CP Pathway The simplest and most rapid way for muscle to produce ATP is to use the CP pathway because it only consists of one reaction (Fig. 12.11*a*), as shown in the following graphic:

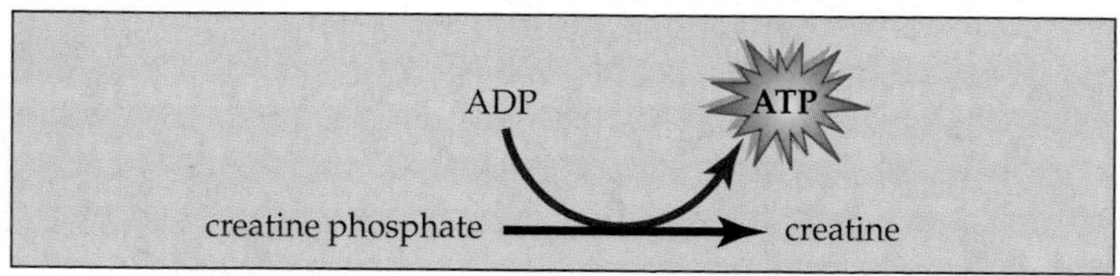

This reaction occurs in the midst of sliding filaments; therefore, this method of supplying ATP is the speediest energy source available to muscles. Creatine phosphate is formed only when a muscle cell is resting, and only a limited amount is stored. The CP pathway is used at the beginning of submaximal exercise and during short-term, high-intensity exercise that lasts less than 5 seconds. The energy to complete a single play in a football game comes principally from the CP system. Intense activities lasting longer than 5 seconds also make use of fermentation.

Fermentation Fermentation (see section 3.6) produces two ATP molecules from the anaerobic breakdown of glucose to lactate. This pathway is the one most likely to begin with glycogen. Hormones provide the signal to muscle cells to break down glycogen, making glucose available as an energy source.

Fermentation, like the CP pathway, is fast-acting, but it results in the buildup of lactate (Fig. 12.11*b*). Formation of lactate is noticeable because it produces short-term muscle aches and fatigue upon exercising. We have all had the experience of needing to continue breathing following strenuous exercise. This continued intake of oxygen, called **oxygen debt,** is required, in part, to complete the metabolism of lactate and restore cells to their original energy state. The lactate is transported to the liver, where 20% of it is completely broken down to carbon dioxide and water. The ATP gained by this respiration is then used to reconvert 80% of the lactate to glucose and then glycogen. In persons who train, the number of mitochondria in individual muscles increases. There is a greater reliance on these additional mitochondria to produce ATP. Muscles rely less on fermentation as a result.

Cellular Respiration Cellular respiration is the slowest of all three mechanisms used to produce ATP. However, it is also the most efficient, typically producing several dozen molecules of ATP from each food molecule. As was discussed in Chapter 3, cellular respiration occurs in the mitochondria. Thus, the process is aerobic and oxygen is supplied by the respiratory system. In addition, a protein called **myoglobin** found within muscle cells delivers oxygen directly to the mitochondria. Cellular respiration can make use of glucose from the breakdown of stored muscle glycogen, glucose taken up from blood, and/or fatty acids from fat digestion (Fig. 12.11*c*). Also, cellular respiration is more likely to supply ATP when exercise is submaximal in intensity. According to Figure 12.9, if you are interested in exercising to lose weight, you should do so at a lower intensity and for a generous amount of time. This means that your aerobic exercise class at the local gym burns triglyceride, or fat, by making ATP using cellular respiration.

Fast-Twitch and Slow-Twitch Muscle Fibers

We have seen that all muscle fibers metabolize aerobically and anaerobically. However, some muscle fibers use one method more than the other to provide myofibrils with ATP. Fast-twitch fibers tend to rely on the creatine phosphate pathway and fermentation, anaerobic means of supplying ATP to muscles. Slow-twitch fibers tend to prefer cellular respiration, which is aerobic.

Fast-Twitch Fibers

Fast-twitch fibers are usually anaerobic and seem to be designed for strength because their motor units contain many fibers (Fig. 12.12). They provide explosions of energy and are most helpful in sports activities such as sprinting, weight lifting, swinging a golf club, or throwing a shot. Fast-twitch fibers are light in color because they have fewer mitochondria, little or no myoglobin, and fewer blood vessels than slow-twitch fibers do. Fast-twitch fibers can develop maximum tension more rapidly than slow-twitch fibers can. In addition, their maximum tension is greater. However, their dependence on anaerobic energy leaves them vulnerable to an accumulation of lactate, which causes them to fatigue quickly.

Slow-Twitch Fibers

Despite having motor units with smaller numbers of muscle fibers, slow-twitch fibers have more stamina and a steadier "tug." These muscle fibers are most helpful in endurance sports, such as long distance running, biking, jogging, and swimming. They produce most of their energy aerobically, so they tire only when their fuel supply is gone. Slow-twitch fibers have many mitochondria and are dark in color because they contain myoglobin, the respiratory pigment found in muscles (Fig. 12.12). They are also surrounded by dense capillary beds and draw more blood and oxygen than fast-twitch fibers. Slow-twitch fibers have a low maximum tension, which develops slowly, but the muscle fibers are highly resistant to fatigue. Slow twitch fibers have a substantial reserve of glycogen and fat, so their abundant mitochondria can maintain a steady, prolonged production of ATP when oxygen is available.

> **APPLICATIONS AND MISCONCEPTIONS**
>
> **Why don't muscle cells use hemoglobin?**
>
> Hemoglobin is the primary transport pigment of oxygen in the bloodstream. However, as the temperature of the tissue rises, and the pH becomes more acidic, hemoglobin loses its ability to bind the oxygen molecules. This is where myoglobin comes in. Like hemoglobin, myoglobin binds oxygen molecules. However, myoglobin has a higher affinity (or attraction) for oxygen than hemoglobin. So when you exercise, or use a muscle group, the increase in temperature causes the oxygen in the hemoglobin to be transferred to the myoglobin of the muscle cells. This allows for an efficient transfer of oxygen to those muscles that are actively contracting.

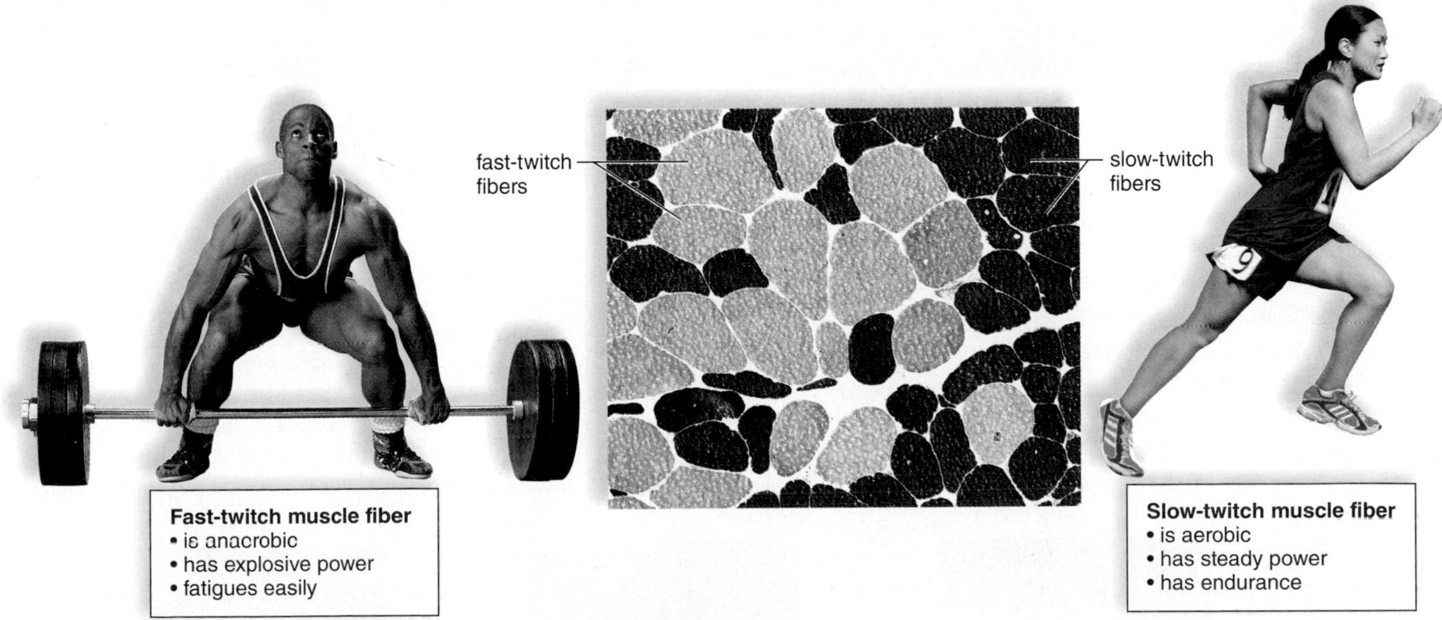

Figure 12.12 **Fast-twitch and slow-twitch muscle fibers differ in structure.**
If your muscles contain many fast-twitch fibers (light color), you would probably do better at a sport like weight lifting. If your muscles contain many slow-twitch fibers (dark color), you would probably do better at a sport like cross-country running.

BIOLOGY MATTERS  Health

Exercise, Exercise, Exercise

Exercise programs improve muscular strength, muscular endurance, and flexibility. Muscular strength is the force a muscle group (or muscle) can exert against a resistance in one maximal effort. Muscular endurance is judged by the ability of a muscle to contract repeatedly or to sustain a contraction for an extended period. Flexibility is tested by observing the range of motion about a joint.

Exercise also improves cardiorespiratory endurance. The heart rate and capacity increase, and the air passages dilate so that the heart and lungs are able to support prolonged muscular activity. The blood level of high-density lipoprotein (HDL) increases. HDL is the molecule that slows the development of artherosclerotic plaques in blood vessels (see Chapter 5). Also, body composition—the proportion of protein to fat—changes favorably when you exercise.

Exercise also seems to help prevent certain types of cancer. Cancer prevention involves eating properly, not smoking, avoiding cancer-causing chemicals and radiation, undergoing appropriate medical screening tests, and knowing the early warning signs of cancer. However, studies show that people who exercise are less likely to develop colon, breast, cervical, uterine, and ovarian cancers.

Physical training with weights can improve the density and strength of bones and the strength and endurance of muscles in all adults, regardless of age. Even men and women in their eighties and nineties can make substantial gains in bone and muscle strength that help them lead more independent lives. Exercise helps prevent osteoporosis, a condition in which the bones are weak and tend to break (see Chapter 11). Exercise promotes the activity of osteoblasts in young as well as older people. The stronger the bones when a person is young, the less chance of osteoporosis as that person ages. Exercise helps prevent weight gain, not only because the level of activity increases but also because muscles metabolize faster than other tissues. As a person becomes more muscular, the body is less likely to accumulate fat.

Exercise relieves depression and enhances the mood. Some people report that exercise actually makes them feel more energetic. Further, after exercise, particularly in the late afternoon, people sleep better that night. Self-esteem rises because of improved appearance, as well as other factors that are not well understood. For example, vigorous exercise releases endorphins, hormonelike chemicals known to alleviate pain and provide a feeling of tranquility.

A sensible exercise program is one that provides all of these benefits without the detriments of a too-strenuous program. Overexertion can be harmful to the body and may result in sports injuries, such as lower back strains or torn ligaments of the knees. The beneficial programs suggested in Table 12A are tailored according to age.

Dr. Arthur Leon at the University of Minnesota performed a study involving 12,000 men, and the results showed that only moderate exercise is needed to lower the risk of a heart attack by one-third. In another study conducted by the Institute for Aerobics Research in Dallas, Texas, which included 10,000 men and more than 3,000 women, even a little exercise was found to lower the risk of death from cardiovascular diseases and cancer. Increasing daily activity by walking to the corner store instead of driving and by taking the stairs instead of the elevator can improve your health.

Questions to Consider

1. At the level of the muscle fiber, how does exercise increase muscle strength?
2. How might an overly ambitious workout damage muscle fibers?

Table 12A Staying Fit

Exercise	Children, 7–12	Teenagers, 13–18	Adults, 19–55
Amount	Vigorous activity 1–2 hr daily	Vigorous activity 1 hr, 3–5 days a week; otherwise, ½ hr daily moderate activity	Vigorous activity 1 hr, 3 days a week; otherwise, ½ hr daily moderate activity
Purpose	Free play	Build muscle with calisthenics	Exercise to prevent lower back pain: aerobics, stretching, or yoga
Organized	Build motor skills through team sports, dancing, or swimming	Continue team sports, dancing, hiking, or swimming	Do aerobic exercise to control buildup of fat cells
Group	Enjoy more exercise outside of physical education classes	Pursue sports that can be enjoyed for a lifetime: tennis, swimming, or horseback riding	Find exercise partners: join a running club, bicycle club, or outing group
Family	Participate in family outings: bowling, boating, camping, or hiking	Take active vacations: hike, bicycle, or cross-country ski	Initiate family outings: bowling, boating, camping, or hiking

BIOLOGY MATTERS  Science

Rigor Mortis

When a person dies, the physiological events that accompany death occur in an orderly progression. Respiration ceases, the heart ultimately stops beating, and tissue cells begin to die. The first tissues to die are those with the highest oxygen requirement. Brain and nervous tissues have an extremely high requirement for oxygen. Deprived of oxygen, these cells typically die after only 6 minutes because of a lack of ATP. However, tissues that can produce ATP by fermentation (which does not require oxygen) can "live" for an hour or more before ATP is completely depleted. Muscle is capable of generating ATP by fermentation. Therefore, muscle cells can survive for a time after clinical death occurs. Muscle death is signaled by a process termed *rigor mortis,* the "stiffness of death" (Fig. 12B).

Stiffness occurs because—for biochemical reasons discussed in section 12.3 —muscles cannot relax unless they have a supply of ATP. Without ATP, the muscles remain fixed in their last state of contraction. If, for example, a murder victim dies while sitting at a desk, the body in rigor mortis will be frozen in the sitting position. Rigor mortis resolves approximately 24 to 36 hours after death. Muscles lose their stiffness because lysosomes rupture. The lysosomes release enzymes that break the bonds between the muscle proteins actin and myosin.

Body temperature and the presence or absence of rigor mortis allow the time of death to be estimated. For example, the body of someone dead for 3 hours or less is still warm (close to body temperature, 98.6°F [37°C]) and rigor mortis is absent. After approximately 3 hours, the body is significantly cooler than normal and rigor mortis begins to develop. The corpse of an individual dead at least 8 hours is in full rigor mortis, and the temperature of the body is the same as the surroundings. Forensic pathologists know that a person has been dead for more than 24 hours if the body temperature is the same as the environment and there is no longer a trace of rigor mortis.

Figure 12B **Rigor mortis can help estimate the time of death.**

Questions to Consider

1. At the level of the sarcomere, explain what is happening during the onset of rigor mortis.
2. Which muscle types would obtain rigor mortis first, cardiac, smooth, or skeletal?

APPLICATIONS AND MISCONCEPTIONS

What causes muscles to be sore a few days after exercising?

Many of us have experienced delayed onset muscle soreness (DOMS), which generally appears some 24 to 48 hours after strenuous exercise. It is thought that DOMS is due to tissue injury that takes several days to heal. Any movement you aren't used to can lead to DOMS, but it is especially associated with any activity that causes muscles to contract while they are lengthening. Examples include walking down stairs, running downhill, lowering weights, and the downward motion of squats and push-ups. To prevent DOMS, try warming up thoroughly and cooling down completely. Stretch after exercising. When beginning a new activity, start gradually and build up your endurance gradually. Avoid making sudden major changes in your exercise routine.

CHECK YOUR PROGRESS 12.3

1. List the stages of a muscle twitch.
2. Contrast the activities of a single muscle twitch with the action of summation and tetanus.
3. Summarize how the CP pathway, fermentation, and aerobic respiration produce ATP for muscle contraction.
4. Explain why weight lifters are not well adapted for distance running.

CONNECTING THE CONCEPTS

For more information on energy sources for muscle contraction and the pathways for generating ATP, refer to the following discussions:

Sections 2.4 and 2.5 describe the structure of carbohydrates and lipids and examine their function as energy nutrients.

Section 3.6 explores how ATP is generated by the cellular respiration and fermentation pathways.

Figure 3.20 illustrates the ATP–ADP cycle.

BIOLOGY MATTERS  **Bioethical**

Anabolic Steroid Use

They're called "performance-enhancing steroids," and their use is alleged to be widespread by athletes, both amateur and professional. Whether the sport is baseball, football, professional cycling, or track and field events—no activity seems to be safe from drug abuse. Steroid abuse admitted by Marion Jones forever changed Olympic history. Jones was the first female athlete to win five medals for track and field events during the 2000 Sydney Olympics. In 2008, she was stripped of all medals she had earned, as well as disqualified from a fifth-place finish in the 2004 Athens games. Future Olympic record books will not include her name. The records of her teammates in the relay events have also been tainted.

Baseball records will also likely require revisions. The exciting slugfest between Mark McGwire and Sammy Sosa in the summer of 1998 was largely credited with reviving national interest in baseball. However, the great home run competition drew unwanted attention to the darker side of professional sports when it was alleged that McGwire and Sosa were using anabolic steroids at the time. Since then, players such as Jose Canseco (Fig. 12C) and McGwire have admitted using anabolic steroids to recover from baseball-related injuries. However, the controversy continues. Similar charges of drug abuse may prevent baseball great Roger Clemens from entering the Baseball Hall of Fame, despite holding the record for Cy Young awards. Likewise, because controversy continues to surround baseball legend Barry Bonds, this talented athlete may never achieve Hall of Fame status. Though he scored a record 715 home runs and won more Most Valuable Player awards than anyone in history, Bonds remains accused of steroid abuse.

Use of anabolic steroids in professional sports continues to be denied by most athletes and officials. However, many people from both inside and outside the industry maintain that such abuse has been going on for many years—and that it continues despite the negative publicity. Congress continues to investigate the controversy, yet the finger pointing and accusations steadily increase.

What Are Anabolic Steroids?

Steroids encompass a large category of substances, both beneficial and harmful. Anabolic steroids are a class of steroids that generally cause tissue growth by promoting protein production. They are naturally occurring hormones created by the body and commonly used to regulate many physiological processes, from growth to sexual function. Most anabolic steroids are closely related to male sex hormones, such as testosterone.

These metabolically potent drugs are controlled substances available only by prescription under the close supervision of a physician because of their many side effects and vast potential for abuse. However, a few anabolic steroids are still legal due to loopholes in drug laws, despite being banned by most professional sports organizations.

Robust muscle growth is not possible from prescription doses of anabolic steroids, so large doses must be used to obtain that effect. Athletes may take dangerously large amounts of anabolic steroids to enhance athletic performance or to increase strength, often with serious consequences. Steroid abusers may vary the type and quantity of drug taken (called "stacking") or may take the drugs and then stop for a time, only to resume later (called "cycling"). Stacking and cycling are done to minimize serious side effects while maximizing the desired effects of the drugs. Even so, dangerous health consequences can occur.

Dangerous Health Consequences

The most common health consequences include high blood pressure, jaundice (yellowing of the skin), acne, and a greatly increased risk of cancer. In women, anabolic steroid abuse may cause masculinization, including a deepened voice, excessive facial and body hair, coarsening of the hair, menstrual cycle irregularities, and enlargement of the clitoris. Anabolic steroid abuse is even more dangerous during adolescence. When taken prior to or during the teenage growth spurt, steroids may result in permanently shortened height or early onset of puberty. Ironically, whereas proper use of anabolic steroids has been helpful in treating many cases of impotence in males, abuse of these drugs may cause impotence and even shrinking of the testicles.

Perhaps the most frightening aspects of anabolic steroid abuse are the reports of increased aggressive behavior and violent mood swings. Furthermore, many users have reported extremely severe withdrawal symptoms upon quitting. Also, many anabolic steroids have also been identified as "gateway drugs," leading abusers to escalate their drug habit to more dangerous drugs such as heroin and cocaine.

Questions to Consider

1. Should recognitions such as admission to the Hall of Fame be denied to athletes if steroid abuse is alleged but cannot be proven?
2. Do you believe the techniques athletes use to train and enhance performance should be regulated? If so, who can or should enforce regulation?

Figure 12C Jose Canseco wrote a book in 2006 about anabolic steroid abuse in major league baseball.

12.4 Muscular Disorders

LEARNING OUTCOMES

Upon completion of this section, you should be able to

1. Distinguish between common muscle conditions such as strains and sprains.
2. Summarize the causes of fibromyalgia, muscular dystrophy, myasthenia gravis, and muscle cancer.

Muscular disorders are common for most people. However, there are some disorders that can be life-threatening.

Common Muscular Conditions

Spasms are sudden and involuntary muscular contractions most often accompanied by pain. Spasms can occur in smooth and skeletal muscles. A spasm of the smooth muscle in the intestinal tract is a type of colic sometimes called a bellyache. Multiple spasms of skeletal muscles are called a seizure, or **convulsion. Cramps** are strong, painful spasms, especially of the leg and foot, usually due to strenuous activity. Cramps can even occur when sleeping after a strenuous workout. **Facial tics,** such as periodic eye blinking, head turning, or grimacing, are spasms that can be controlled voluntarily, but only with great effort.

A **strain** is caused by stretching or tearing of a muscle. A **sprain** is a twisting of a joint leading to swelling and injury, not only of muscles but also of ligaments, tendons, blood vessels, and nerves. The ankle and knee are often subject to sprains. When a tendon is inflamed by a sprain, **tendinitis** results. Tendinitis may irritate the bursa underlying the tendon, causing **bursitis.**

Muscular Diseases

These conditions are more serious and always require close medical care.

Myalgia and Fibromyalgia

Myalgia refers to achy muscles. The most common cause for myalgia is either overuse or overstretching of a muscle or group of muscles. Myalgia without a traumatic history is often due to viral infections. Myalgia may accompany myositis (inflammation of the muscles), either in response to viral infection or as an immune system disorder. **Fibromyalgia** is a chronic condition whose symptoms include achy pain, tenderness, and stiffness of muscles. Its precise cause is not known, but it may also be due to an underlying infection that is not obvious at first.

Muscular Dystrophy

Muscular dystrophy is a broad term applied to a group of disorders characterized by a progressive degeneration and weakening of muscles. As muscle fibers die, fat and connective tissue take their place. **Duchenne muscular dystrophy,** the most common type, is inherited through a flawed gene on the X chromosome. It is now known that the lack of a protein called dystrophin causes the condition. When dystrophin is absent, calcium leaks into the cell and activates an enzyme that dissolves muscle fibers. In an attempt to treat the condition, muscles have been injected with immature muscle cells that do produce dystrophin.

Myasthenia Gravis

Myasthenia gravis is an autoimmune disease characterized by weakness that especially affects the muscles of the eyelids, face, neck, and extremities. Muscle contraction is impaired because the immune system mistakenly produces antibodies that destroy acetylcholine (ACh) receptors. In many cases, the first sign of the disease is a drooping of the eyelids and double vision. Treatment includes drugs that inhibit the enzyme that digests acetylcholine so that ACh accumulates in neuromuscular junctions.

Muscle Cancer

Cancers that originate in muscle, or the connective tissue associated with muscle, belong to a group called the soft tissue **sarcomas.** In general, sarcomas may occur in a variety of tissues, including bone, adipose, and cartilage. Soft tissue sarcomas may occur in both smooth and skeletal muscles. One of the more common forms of smooth muscle cancer is leiomyoma, which occurs in the uterine wall. Rhabdomyosarcomas are a rare form of cancer that may originate in the skeletal muscle, or move into the muscle from another location in the body. Both of these types of sarcomas may be either benign or malignant.

CHECK YOUR PROGRESS 12.4

1. Distinguish between a strain and a sprain.
2. Compare and contrast the potential causes of myalgia and myasthenia gravis.
3. Discuss the potential symptoms of muscular dystrophy.

CONNECTING THE CONCEPTS

For background information on the physiology of these diseases, refer to the following discussions:

Section 7.3 describes the inflammatory response.

Section 7.6 examines how an autoimmune response causes diseases in humans.

Section 20.4 provides additional information on sex-linked inheritance.

12.5 Homeostasis

LEARNING OUTCOMES

Upon completion of this section, you should be able to

1. Summarize the role of the muscular and skeletal systems in movement.
2. Summarize the role of the muscular system in body temperature homeostasis.

In this section, our discussion centers on the contribution of the muscular system to homeostasis (Fig. 12.13). In many cases, the muscular system works closely with the skeletal system, for example, in the protection of body parts and movement.

Both Systems Produce Movement

Movement is essential to maintaining homeostasis. The skeletal and muscular systems work together to enable body movement. This is most evidently illustrated by what happens when skeletal muscles contract and pull on the bones to which they are attached, causing movement at joints. Body movement of this sort allows us to respond to certain types of changes in the environment. For instance, if you are sitting in the sun and start to feel hot, you can get up and move to a shady spot.

The muscular and skeletal systems work for other types of movements that are just as important for maintaining homeostasis. Contraction of skeletal muscles associated with the jaw and tongue allow you to grind food with the teeth. The rhythmic smooth muscle contractions of peristalsis move

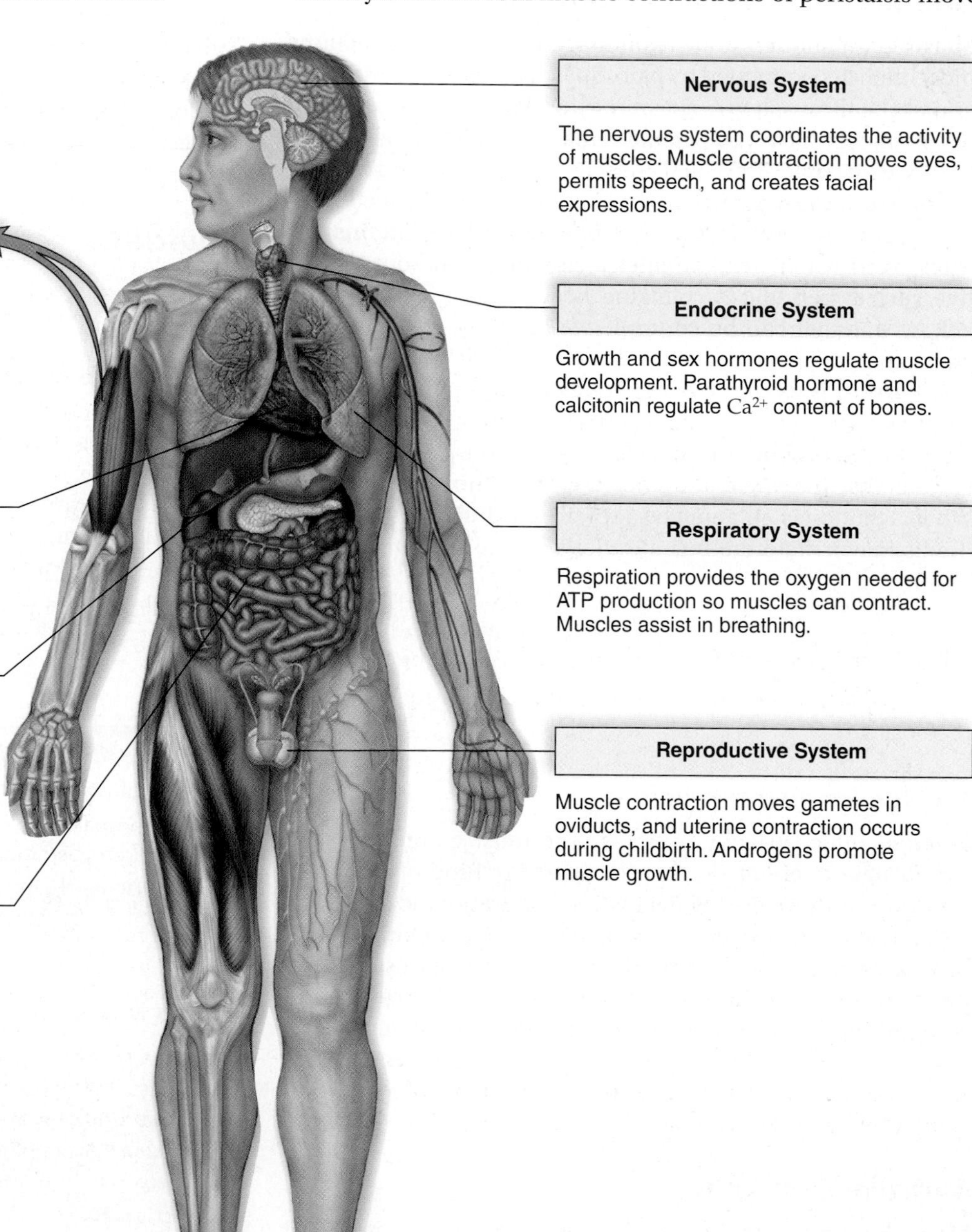

The muscular and skeletal systems work together to maintain homeostasis. The systems listed here in particular also work with these two systems.
Muscular System
The muscular system works with the skeletal system to allow movement and support and protection for internal organs. Muscle contraction provides heat to warm the body; bones play a role in Ca^{2+} balance. These systems specifically help the other systems as mentioned below.
Cardiovascular System
Muscle contraction keeps blood moving in the heart and blood vessels, particularly the veins.
Urinary System
Muscle contraction moves the fluid within ureters, bladder, and urethra. Kidneys activate vitamin D needed for Ca^{2+} absorption and help maintain the blood level of Ca^{2+} for muscle contraction.
Digestive System
Muscle contraction accounts for chewing of food and peristaltic movement. The digestive system absorbs ions needed for muscle contraction.
Nervous System
The nervous system coordinates the activity of muscles. Muscle contraction moves eyes, permits speech, and creates facial expressions.
Endocrine System
Growth and sex hormones regulate muscle development. Parathyroid hormone and calcitonin regulate Ca^{2+} content of bones.
Respiratory System
Respiration provides the oxygen needed for ATP production so muscles can contract. Muscles assist in breathing.
Reproductive System
Muscle contraction moves gametes in oviducts, and uterine contraction occurs during childbirth. Androgens promote muscle growth.

Figure 12.13 The muscular system's contributions to homeostasis.

ingested materials through the digestive tract. These processes are necessary for supplying the body's cells with nutrients. The ceaseless beating of your heart, which propels blood into the arterial system, is caused by the contraction of cardiac muscle. Contractions of skeletal muscles in the body, especially those associated with breathing and leg movements, aid in the process of venous return by pushing blood back toward the heart. This is why soldiers and members of marching bands are cautioned not to lock their knees when standing at attention. The reduction in venous return causes a drop in blood pressure that can result in fainting. The pressure exerted by skeletal muscle contraction also helps to squeeze tissue fluid into the lymphatic capillaries, where it is referred to as lymph.

Both Systems Protect Body Parts

The skeletal system plays an important role by protecting the soft internal organs of your body. The brain, heart, lungs, spinal cord, kidneys, liver, and most of the endocrine glands are shielded by the skeleton. In particular, the nervous and endocrine organs must be defended so they can carry out activities necessary for homeostasis.

The skeletal muscles pad and protect the bones, and the tendons and bursae associated with skeletal muscles reinforce and cushion the joints. Muscles of the abdominal wall offer additional protection to the soft internal organs. Examples of these muscles include the rectus abdominis and external oblique muscles illustrated in Figure 12.5.

Muscles Help Maintain Body Temperature

The muscular system helps to regulate body temperature. When you are very cold, smooth muscle constricts inside the blood vessels supplying the skin. Thus, the amount of blood close to the surface of the body is reduced. This helps to conserve heat in the body's core, where vital organs lie. If you are cold enough, you may start to shiver. Shivering is caused by involuntary skeletal muscle contractions. This is initiated by temperature-sensitive neurons in the hypothalamus of the brain. Skeletal muscle contraction requires ATP, and using ATP generates heat. You may also notice that you get goose bumps when you are cold. This is because arrector pili muscles contract. These tiny bundles of smooth muscle are attached to the hair follicles and cause the hairs to stand up. This is not very helpful in keeping humans warm, but it is quite effective in our furrier fellow mammals. Think of a cat or dog outside on a cold winter day. Its fur is a better insulator when standing up than when lying flat. Goose bumps can also be a sign of fear. Although a human with goose bumps may not look very impressive, a frightened or aggressive animal whose fur is standing on end looks bigger and (it is hoped) more intimidating to a predator or rival.

CHECK YOUR PROGRESS 12.5

1. Summarize the importance of movement in homeostasis.
2. Summarize how the muscular system works to maintain body temperature.
3. Explain how the muscular system interacts with the digestive system.

CONNECTING THE CONCEPTS

For more information on calcium and body temperature homeostasis, refer to the following discussions:

Section 4.8 explores how the body maintains homeostasis using feedback mechanisms.

Figure 4.18 examines how the hypothalamus is involved in body temperature regulation.

Section 15.3 describes how the thyroid and parathyroid glands are involved in calcium homeostasis.

CASE STUDY CONCLUSION

There are nine different classes of muscular dystrophy. In the most common types of muscular dystrophy, symptoms occur very early in life. In Kyle's case, the relatively late onset of the disease suggested that he had a rarer form called Becker muscular dystrophy. For Kyle, the good news was that this was a much slower-progressing form of the disease, with most patients living well into their thirties without being confined to a wheelchair. Furthermore, many of the symptoms of Becker muscular dystrophy can be controlled with medication. Becker muscular dystrophy is also known to cause heart problems later in life. However, researchers are actively studying whether it may be possible to use gene therapy (see Chapter 21) to replace the defective dystrophin gene. In the interim, patients of Becker muscular dystrophy, such as Kyle, are recommended to regularly exercise to slow the loss of muscle tissue over time.

MEDIA STUDY TOOLS

Enhance your study of this chapter with media! Visit **www.mhhe.com/maderhuman13e** and go to "Media Study Tools" for this chapter to access the following:

Animations

12.2 Sarcomere Contraction • Breakdown of ATP and Cross-Bridge Movement • Muscle Contraction
12.3 Energy Sources for Prolonged Exercise

MP3 Files

12.1 Muscular System • Muscles of the Head and Neck • Muscles of the Pelvic Girdle and Lower Limb • Muscles of the Shoulder and Upper Limb • Muscles of the Trunk
12.2 Muscle Structure • Sliding Filament Theory • Neuromuscular Junction
12.3 Energy Source for Muscle Contraction

3D Animation

McGraw-Hill's 3D animation, "Skeletal Muscle Contraction," provides a dynamic exploration of the key concepts of this chapter and is available through McGraw-Hill Connect®.

SUMMARIZE

12.1 Overview of the Muscular System

The **muscular system** is involved in movement, both externally and internally. Muscle tissue is composed of cells called **muscle fibers**. Muscle fibers may be found in three types of muscle tissue:

- **Smooth muscle** is involuntary and occurs in walls of internal organs.
- **Cardiac muscle** is involuntary and occurs in walls of the heart. Cardiac muscle has **intercalated disks** that permit rapid contraction.
- **Skeletal muscle** is voluntary, contains bundles of muscle fibers called fascicles, and is usually attached by tendons to the skeleton.

Skeletal muscle is involved in support, movement, and protection. Skeletal muscles are connected to bones by **tendons. Bursae** provide a cushion between the muscle and bone.

Skeletal Muscles of the Body

Skeletal muscles all possess both an **origin** (stationary point) and **insertion** (movement point). When achieving movement, some muscles are prime movers, some are synergists, and others are antagonists.

Names and Actions of Skeletal Muscles

Muscles are named for their size, shape, location, direction of fibers, number of attachments, and action.

12.2 Skeletal Muscle Fiber Contraction

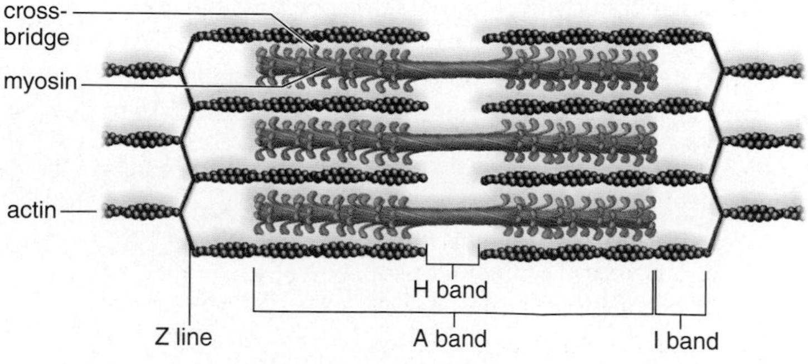

Muscle fibers contain **myofibrils,** and myofibrils contain **actin** and **myosin** filaments. Muscle contraction occurs when **sarcomeres** shorten and actin filaments slide past myosin filaments.

- Nerve impulses travel down motor neurons and stimulate muscle fibers at **neuromuscular junctions.**
- The **sarcolemma** of a muscle fiber forms **T (transverse) tubules** that almost touch the **sarcoplasmic reticulum,** which stores calcium ions.
- When calcium ions are released into muscle fibers, actin filaments slide past myosin filaments.
- At a neuromuscular junction, synaptic vesicles release acetylcholine (neurotransmitter), which diffuses across the synaptic cleft.
- When acetylcholine (ACh) is received by the sarcolemma, electrical signals begin and lead to the release of calcium.
- Calcium ions bind to **troponin,** causing **tropomyosin** proteins to shift, thus exposing myosin binding sites.
- Myosin filaments break down ATP and attach to actin filaments, forming cross-bridges.
- When ADP and Ⓟ are released, cross-bridges change their positions.
- This **sliding filament model** pulls actin filaments to the center of a sarcomere.

12.3 Whole Muscle Contraction

Muscles Have Motor Units

- A muscle contains **motor units:** several fibers under the control of a single motor axon.
- Motor unit contraction is described in terms of a **muscle twitch, summation,** and **tetanus.**
- The strength of muscle contraction varies according to recruitment of motor units.
- In the body, a continuous slight tension, called **muscle tone,** is maintained by muscle motor units that take turns contracting.

Energy for Muscle Contraction

A muscle fiber has three ways to acquire ATP for muscle contraction:

- Creatine phosphate (CP) transfers a phosphate to ADP, and ATP results. This CP pathway is the most rapid.
- Fermentation also produces ATP quickly. Fermentation is associated with an **oxygen debt** because oxygen is needed to metabolize the lactate that accumulates.

- Cellular respiration provides most of the muscle's ATP but takes longer because much of the glucose and oxygen must be transported in blood to mitochondria. The **myoglobin** in muscle cells delivers the oxygen to the mitochondria. Cellular respiration occurs during aerobic exercise and burns fatty acids in addition to glucose.

Fast-Twitch and Slow-Twitch Muscle Fibers

- Fast-twitch fibers, for sports like weight lifting, rely on an anaerobic means of acquiring ATP; have few mitochondria and myoglobin, but motor units contain more muscle fibers; and are known for explosive power but fatigue quickly.
- Slow-twitch fibers, for sports like running and swimming, rely on aerobic respiration to acquire ATP and have a plentiful supply of mitochondria and myoglobin, which gives them a dark color.

12.4 Muscular Disorders

- Muscular disorders include **spasms, convulsions, cramps,** and **facial tics.**
- Muscular system injuries include **strains, sprains, tendinitis,** and **bursitis.**
- Diseases of the muscular system include **myalgias (fibromyalgia)**; **muscular dystrophy (Duchenne muscular dystrophy)**; **myasthenia gravis;** and **cancer of the muscles (sarcomas)**.

12.5 Homeostasis

- The muscles and bones produce movement and protect body parts.
- The muscles produce the heat that gives us a constant body temperature.

ASSESS

Testing Your Knowledge of the Concepts

1. What are the characteristics of the three types of muscles in the human body? Where is each type found? (pages 261–262)
2. Give an example to show that skeletal muscles work in antagonistic pairs. Explain. (page 262)
3. What criteria are used to name muscles? Give an example of each one. (page 263)
4. What are the functions of a muscle fiber's components? (page 265)
5. Describe the sliding filament model of muscle contraction within a sarcomere. Begin with the nerve impulse and end with the relaxation of the muscle. (pages 265–270)
6. What are the four possible energy sources for a muscle and the three sources of ATP for muscle contraction? (pages 271–272)
7. Compare fast- and slow-twitch muscle fibers. (page 273)
8. What are some common muscular disorders and some more serious muscular diseases? (page 277)
9. How does the muscular system help maintain homeostasis? (pages 278–279)
10. Impulses that move down the T tubules of a muscle fiber most directly cause
    a. movement of tropomyosin.
    b. attachment of the cross-bridges to myosin.
    c. release of Ca^{2+} from the sarcoplasmic reticulum.
    d. splitting of ATP.
11. Which of the following statements about cross-bridges is false?
    a. They are composed of myosin.
    b. They bind to ATP after they detach from actin.
    c. They contain an ATPase.
    d. They split ATP before they attach to actin.
12. Which statement about sarcomere contraction is incorrect?
    a. The A bands shorten.
    b. The H bands shorten.
    c. The I bands shorten.
    d. The sarcomeres shorten.
13. The thick filaments of a muscle fiber are made up of
    a. actin.
    b. troponin.
    c. fascia.
    d. myosin.
14. As ADP and Ⓟ are released from a myosin head,
    a. actin filaments move toward the H band.
    b. myosin cross-bridges pull the thin filaments.
    c. a sarcomere shortens.
    d. Only a and c are correct.
    e. All of the choices are correct.
15. Which of these is a direct source of energy for muscle contraction?
    a. ATP
    b. creatine phosphate
    c. lactic acid
    d. glycogen
    e. Both a and b are correct.
16. When muscles contract,
    a. sarcomeres increase in length.
    b. actin breaks down ATP.
    c. myosin slides past actin.
    d. the H band disappears.
    e. calcium is taken up by the sarcoplasmic reticulum.
17. Nervous stimulation of muscles
    a. occurs at a neuromuscular junction.
    b. results in an impulse that travels down the T tubules.
    c. causes calcium to be released from expanded regions of the sarcoplasmic reticulum.
    d. All of the choices are correct.
18. Lack of calcium in muscles
    a. results in no contraction.
    b. causes weak contraction.
    c. causes strong contraction.
    d. has no effect.
    e. None of the choices are correct.
19. During muscle contraction,
    a. ATP is hydrolyzed when the myosin head is unattached.
    b. ADP and Ⓟ are released as the myosin head attaches to actin.
    c. ADP and Ⓟ release causes the head to change position and actin filaments to move.
    d. release of ATP causes the myosin head to return to resting position.
20. Proper functioning of a neuromuscular junction requires the
    a. presence of acetylcholine.
    b. presence of a synaptic cleft.
    c. presence of a motor terminal.
    d. sarcolemma of a muscle cell.
    e. All of the choices are correct.

21. Label this diagram of a muscle fiber, using these terms: *myofibril, T tubule, sarcomere, sarcolemma, sarcoplasmic reticulum, Z line.*

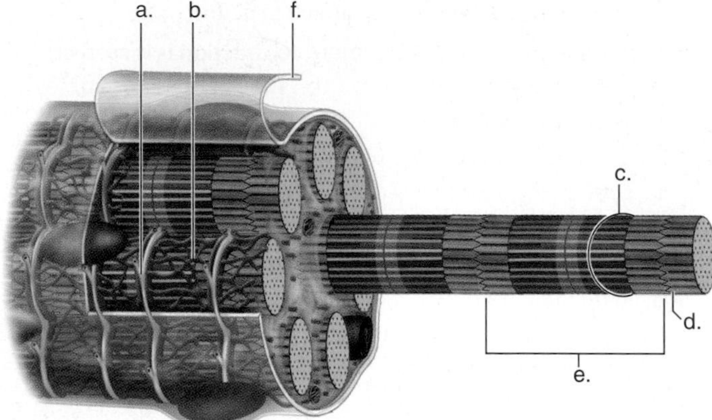

ENGAGE

The virtual lab "Muscle Stimulation" provides an interactive look at how skeletal muscles are stimulated by workload.

Thinking Critically About the Concepts

1. The dystrophin protein is located between the sarcolemma and the outer myofilaments of the muscle. It is responsible for conducting the force of the muscle contraction from the myofilaments to the connective tissue of the muscle. In Kyle's case, a mutation in this protein was causing the symptoms of muscular dystrophy.
   a. Why would a loss of dystrophin protein cause weakness and a loss of coordination?
   b. Muscular dystrophy is frequently referred to as a muscle-wasting disease, in which the muscles lose mass over time. How would a defect in dystrophin contribute to the wasting of muscle tissue?
2. You learned about rigor mortis in the Science feature, "Rigor Mortis," on page 275. Perhaps you're also a fan of crime scene shows as well. If so, you know that the onset of rigor mortis in a deceased person can be influenced by a number of factors. Consider the following:
   a. If a body was rapidly cooled after death, how would this affect the timing of rigor mortis?
   b. Discuss what factors, besides cooling, might delay or accelerate the onset of rigor mortis.
3. Rigor mortis is usually complete within one to two days after death (depending on environmental variables). Why would rigor mortis diminish after several days?
4. What causes a person with amyotrophic lateral sclerosis (Lou Gehrig's disease) to die?

CHAPTER 13

Nervous System

CASE STUDY MULTIPLE SCLEROSIS

On her way to work, Sarah noticed that the colors of the traffic lights didn't seem quite right; the red lights appeared to be more orange than red. At work, she noticed that she was having trouble reading her e-mail. By the end of the day, she had a splitting headache. She kept telling herself that she had just been working too hard. But even as she tried to remain calm, deep down she had a bad feeling. Within a few weeks, she was almost completely blind in one eye and the sensations in her feet felt muffled, like they were wrapped in gauze. Her doctor referred her to a neurologist, who immediately ordered a magnetic resonance imaging (MRI) scan of her brain and a series of somatosensory evoked potential (SSEP) tests to examine how her nervous system was processing electrical impulses.

The results indicated that Sarah had multiple sclerosis (MS), which is an inflammatory disease. This disease affects the myelin sheaths, which wrap parts of some nerve cells like insulation around an electrical cord. As these sheaths deteriorate, the nerves no longer conduct impulses normally. For unknown reasons, multiple sclerosis often attacks the optic nerves first before proceeding to other areas of the brain. Sarah's doctors were able to treat her MS symptoms using high doses of immunosuppressive medications. Unfortunately, there is no cure for MS, but most patients can control the symptoms with daily injections of medication.

As you read through the chapter, think about the following questions:

1. Why would a deterioration of the myelin sheaths cause a nerve cell to function incorrectly?
2. How would an MRI and SSEP test indicate that there was a problem with Sarah's neurological functions?
3. Why are many individuals who contract MS eventually confined to a wheelchair?

CHAPTER CONCEPTS

13.1 Overview of the Nervous System
In the nervous system, reception of stimuli is associated with sensory neurons; integration is associated with interneurons; and motor output is associated with motor neurons. All neurons use the same methods to transmit nerve impulses along neurons and across synapses.

13.2 The Central Nervous System
The central nervous system consists of the brain and the spinal cord. The brain is divided into portions, each with specific functions, and the spinal cord communicates with the brain. The spinal cord provides input to, and output from, the brain.

13.3 The Limbic System and Higher Mental Functions
The limbic system involves many parts of the brain. It gives emotional overtones to the activities of the brain, and is important in the processes of learning and memory.

13.4 The Peripheral Nervous System
The peripheral nervous system consists of nerves that project from the CNS. Cranial nerves project from the brain. The spinal cord gives rise to spinal nerves.

13.5 Drug Therapy and Drug Abuse
Although neurological drugs are quite varied, each type has been found to either promote, prevent, or replace the action of a particular neurotransmitter at a synapse.

BEFORE YOU BEGIN

Before beginning this chapter, take a few moments to review the following discussions:

Section 2.1 How does an ion differ from an atom of an element?

Section 3.3 How does the sodium–potassium pump move ions across the cell membrane?

Section 4.1 What is the function of nervous tissue in the body?

13.1 Overview of the Nervous System

LEARNING OUTCOMES

Upon completion of this section, you should be able to

1. Distinguish between the central nervous system and peripheral nervous system with regard to location and function.
2. List the three types of neurons and provide a function for each.
3. Summarize the activities that generate and propagate an action potential.
4. Explain the role of neurotransmitters and the process of synaptic integration.

The **nervous system** is responsible for the reception and processing of sensory information from both the external and internal environments. The nervous system has two major divisions (Fig. 13.1*a*). The **central nervous system** (**CNS**) consists of the brain and spinal cord. The brain is completely surrounded and protected by the skull. It connects directly to the spinal cord, similarly protected by the vertebral column. The **peripheral nervous system** (**PNS**) consists of nerves. Nerves lie outside the CNS. The division between the CNS and the PNS is arbitrary. The two systems work together and are connected to each other (Fig. 13.1*b*).

MP3 Organization of the Nervous System

The nervous system has three specific functions:

1. The nervous system receives sensory input. Sensory receptors in skin and other organs respond to external and internal stimuli by generating nerve signals that travel by way of the PNS to the CNS. For example, if you smell baking cookies, olfactory (smell) receptors in the nose use the PNS to transmit that information to the CNS.
2. The CNS performs information processing and integration, summing up the input it receives from all over the body. The CNS reviews the information, stores the information as memories, and creates the appropriate motor responses. The smell of those baking cookies evokes

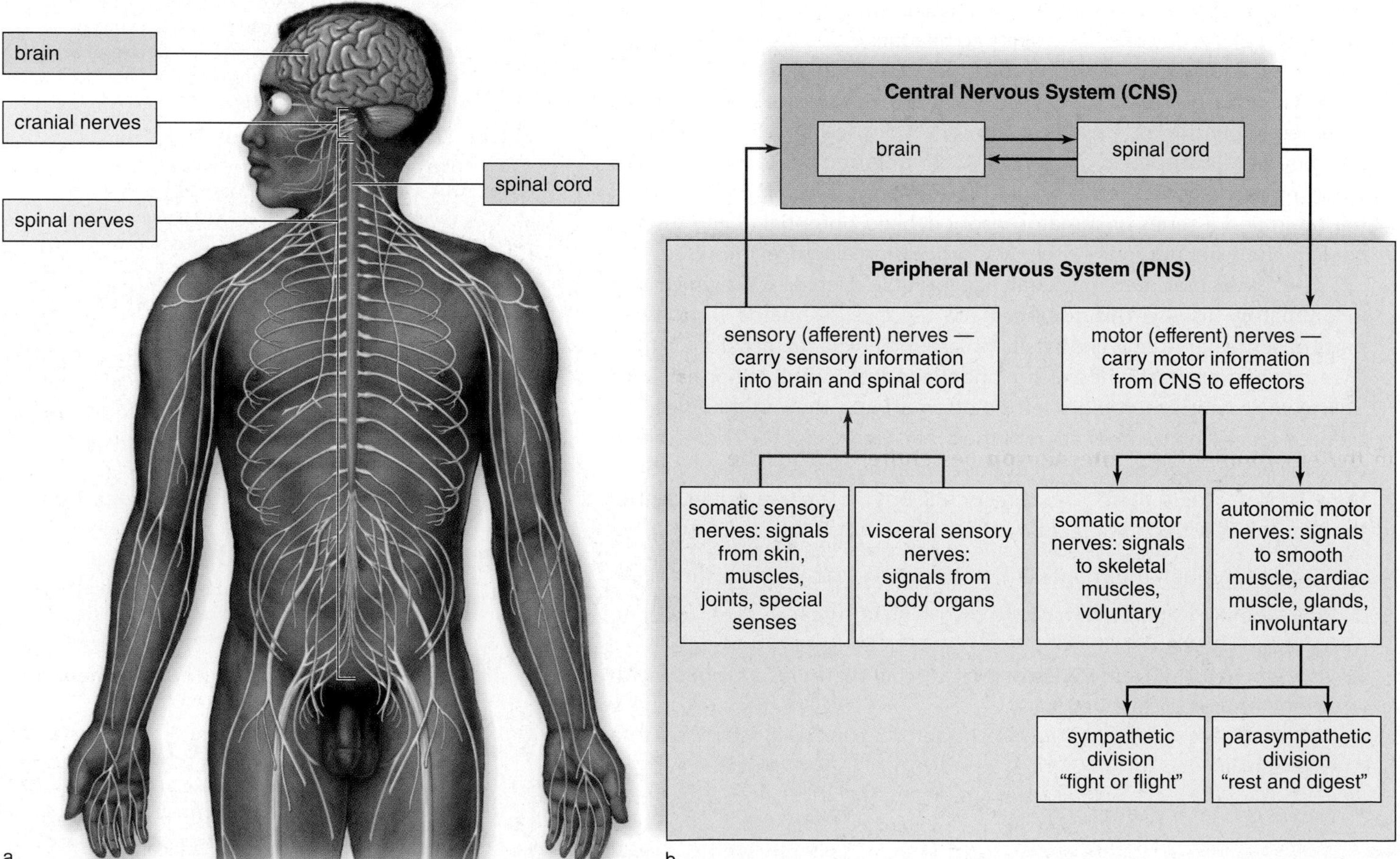

Figure 13.1 The two divisions of the nervous system.
a. The central nervous system (CNS) consists of the brain and spinal cord. The peripheral nervous system (PNS) consists of the nerves, which lie outside the CNS. **b.** The red arrows are the pathway by which the CNS receives sensory information. The black arrows are the pathway by which the CNS communicates with the somatic nervous system and the autonomic nervous system, two divisions of the PNS.

pleasant memories of their taste, and you quickly conceive the idea of getting some for yourself. Other stored memories from childhood lessons in proper behavior remind you that it's rude to grab those cookies without asking first!

3. The CNS generates motor output. Nerve signals from the CNS go by way of the PNS to the muscles, glands, and organs, all in response to the cookies. Signals to the salivary glands make you salivate. Your stomach generates the acid and enzymes needed to digest the cookies—even before you've had a bite. The CNS also coordinates the movement of your arms and hands as you reach for the cookies.

Nervous Tissue

Nervous tissue contains two types of cells: neurons and neuroglia (neuroglial cells). **Neurons** are the cells that transmit nerve impulses between parts of the nervous system; **neuroglia** support and nourish neurons. Neuroglia, or glial cells (see section 4.4), greatly outnumber neurons in the brain. There are several different types of neuroglia in the CNS, each with specific functions. *Microglia* are phagocytic cells that help remove bacteria and debris, whereas *astrocytes* provide metabolic and structural support directly to the neurons. The myelin sheath (Fig. 13.2), is formed from the membranes of tightly spiraled neuroglia. In the PNS, *Schwann cells* perform this function, leaving gaps called nodes of Ranvier. In the CNS, neuroglial cells called *oligodendrocytes* form the myelin sheath. We will focus our attention on the anatomy and physiology of neurons.

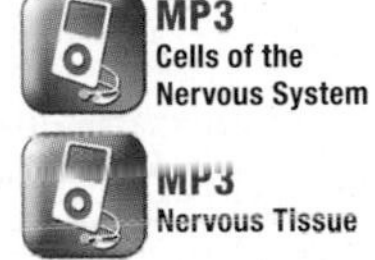

Anatomy of a Neuron

Classified according to function, the three types of neurons are sensory neurons, interneurons, and motor neurons (Fig. 13.2). Their functions are best described relative to the CNS. A **sensory neuron** takes nerve signals from a sensory receptor to the CNS. **Sensory receptors** are special structures that detect changes in the environment. An **interneuron** lies entirely within the CNS. Interneurons can receive input from sensory neurons and also from other interneurons in the CNS. Thereafter, they sum up all the information received from other neurons before they communicate with motor neurons. A **motor neuron** takes nerve impulses away from the CNS to an effector (muscle fiber, organ, or gland). **Effectors** carry out our responses to environmental changes, whether these are external or internal.

Neurons vary in appearance, but all of them have three distinct structures: a cell body, dendrites, and an axon. The **cell body** contains the nucleus, as well as other organelles. **Dendrites** are short extensions that receive signals from sensory receptors or other neurons. Incoming signals from dendrites can result in nerve signals that are then conducted by an axon. The **axon** is the portion of a neuron that conducts nerve impulses. An axon can be quite long. Individual axons are termed *nerve fibers*. Collectively, they form a **nerve**.

In sensory neurons, a very long axon carries nerve signals from the dendrites associated with a sensory receptor to the CNS, and this axon is interrupted by the cell body. In interneurons and motor neurons, on the other hand, multiple dendrites take signals to the cell body, and then an axon conducts nerve signals away from the cell body.

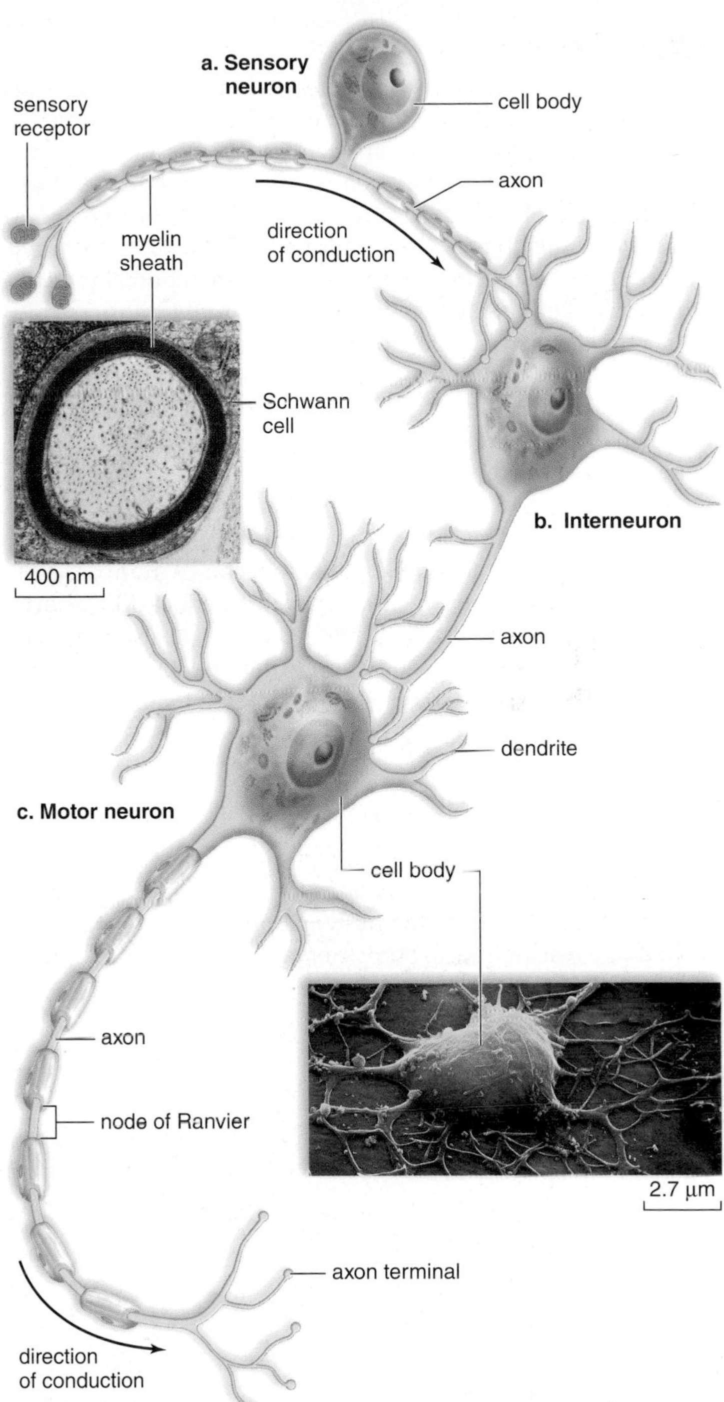

Figure 13.2 The structure of sensory neurons, interneurons, and motor neurons.
a. A sensory neuron has a long axon covered by a myelin sheath that takes nerve impulses all the way from dendrites to the CNS. **b.** In the CNS, some interneurons, such as this one, have a short axon that is not covered by a myelin sheath. **c.** A motor neuron has a long axon covered by a myelin sheath that takes nerve impulses from the CNS to an effector.

Myelin Sheath

Many axons are covered by a protective **myelin sheath.** The myelin sheath develops when Schwann cells (PNS) or oligodendrocytes (CNS) wrap their membranes around an axon many times. Each neuroglia cell covers only a portion of an axon, so the myelin sheath is interrupted. The gaps where there is no myelin sheath are called **nodes of Ranvier** (Fig. 13.2). As is presented in the section "Propagation of an Action Potential" later in this section, the myelin sheath plays an important role in the rate at which signals move through the neuron.

Long axons tend to have a myelin sheath, but short axons do not. The gray matter of the CNS is gray because it contains no myelinated axons; the white matter of the CNS is white because it does. In the PNS, myelin gives nerve fibers their white, glistening appearance and serves as an excellent insulator. When the myelin breaks down, as is the case in **multiple sclerosis (MS)** as discussed in the chapter opener, then it becomes more difficult for the neurons to transmit information. In effect, MS "short-circuits" the nervous system. The myelin sheath also plays an important role in nerve regeneration within the PNS. If an axon is accidentally severed, the myelin sheath remains and serves as a passageway for new fiber growth.

Physiology of a Neuron

Nerve signals convey information within the nervous system. In the past, nerve signals could be studied only in excised neurons (neurons taken from the body). Sophisticated techniques now enable researchers to study nerve signals in single, intact nerve cells.

Resting Potential

Think of all the devices, such as your cell phones and laptop, that are battery-powered. Every battery is an energy source manufactured by separating positively charged ions across a membrane from negative ions. The battery's *potential energy* can be used to perform work, for example, using your phone or lighting a flashlight. A resting neuron also has potential energy, much like a fully charged battery. This energy, called the **resting potential,** exists because the cell membrane is *polarized:* positively charged ions are stashed outside the cell, with negatively charged ions inside.

As Figure 13.3*a* shows, the outside of the cell is positive because positively charged sodium ions (Na^+) gather around the outside of the cell membrane. At rest, the neuron's cell membrane is permeable to potassium, but not to sodium. Thus, positively charged potassium ions (K^+) contribute to the positive charge by diffusing out of the cell to join the sodium ions. The inside of the cell is negative in relation to the exterior of the cell because of the presence of large, negatively charged proteins and other molecules that remain inside the cell because of their size.

Like a battery, the neuron's resting potential energy can be measured in volts. Whereas a D-size flashlight battery has 1.5 volts, a nerve cell typically has 0.070 volts, or 70 millivolts (mV), of stored energy (Fig. 13.3*a*). By convention, the voltage measurement is always a negative number. This is because scientists compare the inside of the cell—where negatively charged proteins and other large molecules are clustered—to the outside of the cell—where positively charged sodium and potassium ions are gathered.

Figure 13.3 Generation of an action potential.
a. Resting potential occurs when a neuron is not conducting a nerve impulse. During an action potential, (**b**) the stimulus causes the cell to reach its threshold. **c.** Depolarization is followed by (**d**) repolarization. **e.** Visualizing an action potential.

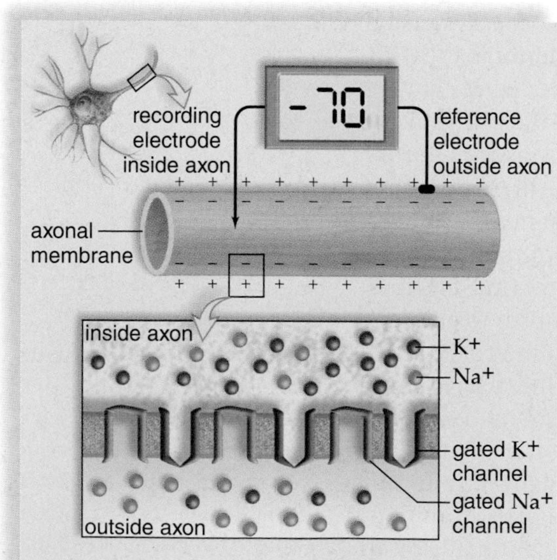

a. Resting potential: Na^+ outside the axon, K^+ and large anions inside the axon. Separation of charges polarizes the cell and causes the resting potential.

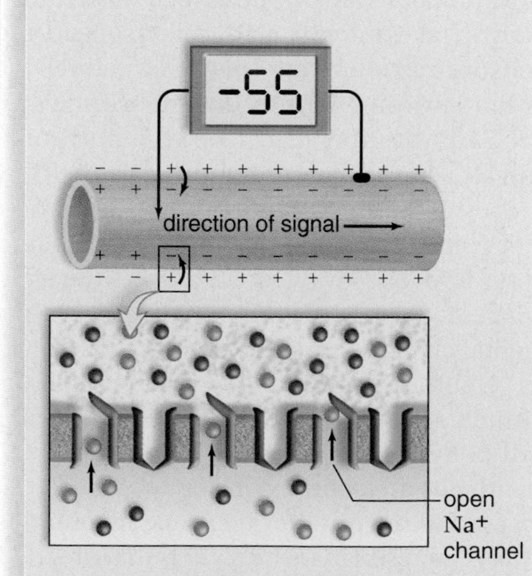

b. Stimulus causes the axon to reach its threshold; the axon potential increases from –70 to –55. The action potential has begun.

Just like rechargeable batteries, neurons must maintain their resting potential to be able to work. To do so, neurons actively transport sodium ions out of the cell and return potassium ions to the cytoplasm. A protein carrier in the membrane, called the **sodium-potassium pump,** pumps sodium ions (Na^+) out of the neuron and potassium ions (K^+) into the neuron (see section 3.3). This action effectively "recharges" the cell, so like a fresh battery, it can perform work.

Animation
How the Sodium–Potassium Pump Works

Action Potential

The resting potential energy of the neuron can be used to perform the work of the neuron: conduction of nerve signals. The process of conduction is termed an **action potential,** and it occurs in the axons of neurons. A **stimulus** activates the neuron and begins the action potential. For example, a stimulus for pain neurons in the skin would be the prick of a sharp pin. However, the stimulus must be strong enough to cause the cell to reach **threshold,** the voltage that will result in an action potential. In Figure 13.3*b*, the threshold voltage is −55 mV. An action potential is an all-or-nothing event. Once threshold is reached, the action potential happens automatically and completely. On the other hand, if the threshold voltage is never reached, the action potential does not occur. Increasing the strength of a stimulus (like pressing harder with the pin) does not change the strength of an action potential. However, it may cause more action potentials to occur in a given period; as a result, the person may perceive that pain has increased.

Sodium Gates Open Protein channels specific for sodium ions are located in the cell membrane of the axon. When an action potential begins in response to a threshold stimulus, these protein channels open, and sodium ions rush into the cell. Adding positively charged sodium ions causes the inside of the axon to become positive compared to the outside (Fig. 13.3*c*). This change is called **depolarization** because the charge (or polarity) inside the axon changes from negative to positive.

Potassium Gates Open Almost immediately after depolarization, the channels for sodium close and a separate set of potassium protein channels opens. Potassium flows rapidly from the cell. As positively charged potassium ions exit the cell, the inside of the cell becomes negative again because of the presence of large, negatively charged ions trapped inside the cell. This change in polarity is called **repolarization,** because the inside of the axon resumes a negative charge as potassium exits the axon (Fig. 13.3*d*). Finally, the sodium-potassium pump completes the action potential. Potassium ions are returned to the inside of the cell and sodium ions to the outside, and resting potential is restored.

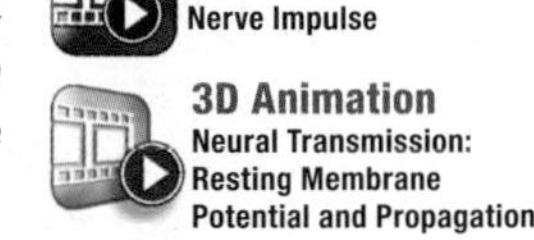

Visualizing an Action Potential

To visualize such rapid fluctuations in voltage across the axonal membrane, researchers generally find it useful to plot the voltage changes over time (Fig. 13.3*e*). During depolarization, the voltage increases from −70 mV to −55 mV to +35 mV, as sodium ions move to the inside of the axon. In repolarization, the opposite change occurs when potassium ions leave the axon. The entire process is very rapid, requiring only 3 to 4 milliseconds (ms) to complete.

Animation
Voltage-Gated Channels and the Action Potential

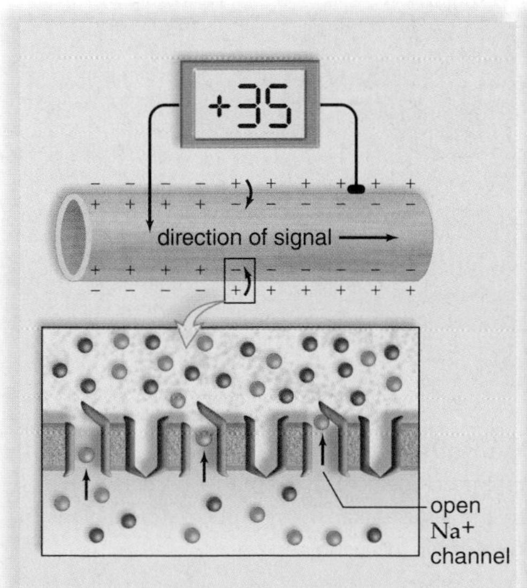

c. Depolarization continues as Na^+ gates open and Na^+ moves inside the axon.

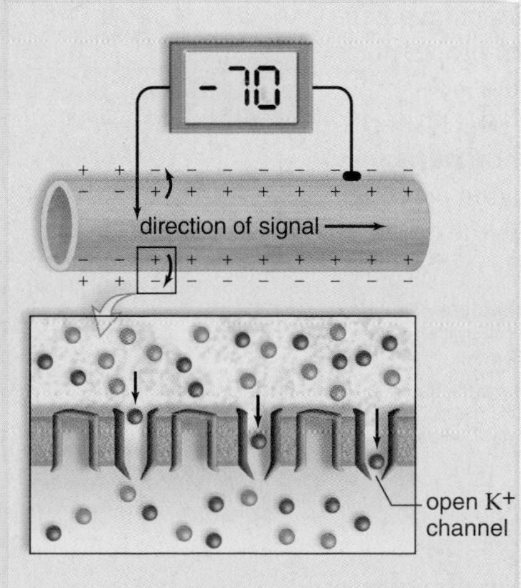

d. Action potential ends: repolarization occurs when K^+ gates open and K^+ moves to outside the axon. The sodium-potassium pump returns the ions to their resting positions.

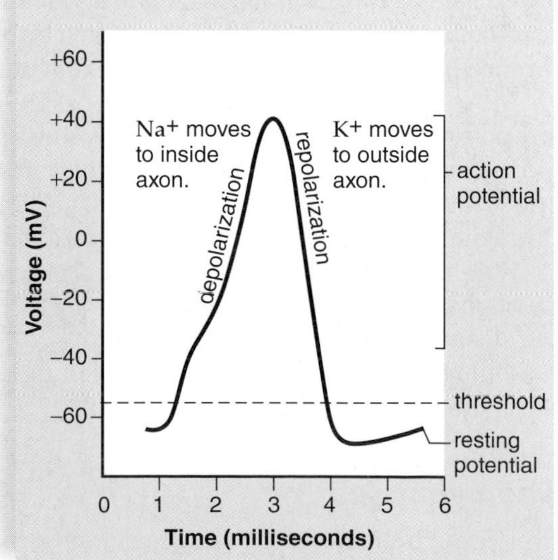

e. An action potential can be visualized if voltage changes are graphed over time.

Propagation of an Action Potential

If an axon is unmyelinated, an action potential at one locale stimulates an adjacent part of the axon membrane to produce an action potential. Conduction along the entire axon in this fashion can be rather slow—approximately 1 meter/second (1 m/s) in thin axons—because each section of the axon must be stimulated.

In myelinated fibers, an action potential at one node of Ranvier causes an action potential at the next node, jumping over the entire myelin-coated portion of the axon. This type of conduction is called **saltatory conduction** (*saltatio* is a Latin word that means "to jump") and is much faster. In thick, myelinated fibers, the rate of transmission is more than 100 m/s. Regardless of whether an axon is myelinated or not, its action potentials are self-propagating. Each action potential generates another, along the entire length of the axon.

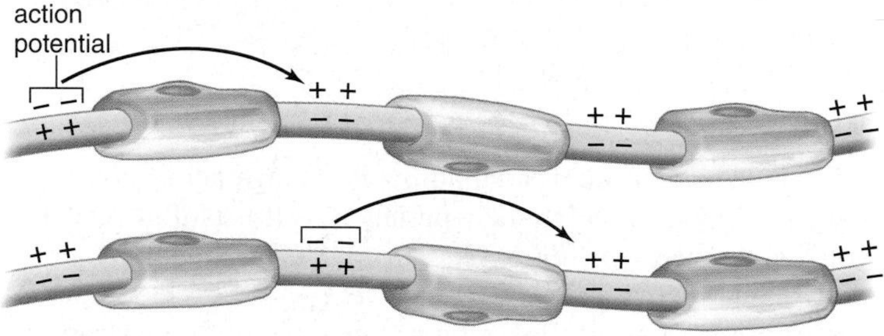
action potential

Like the action potential itself, conduction of an action potential is an all-or-none event—either an axon conducts its action potential or it does not. The intensity of a message is determined by how many action potentials are generated within a given time. An axon can conduct a volley of action potentials very quickly, because only a small number of ions are exchanged with each action potential. Once the action potential is complete, the ions are rapidly restored to their proper place through the action of the sodium–potassium pump.

As soon as the action potential has passed by each successive portion of an axon, that portion undergoes a short **refractory period** during which it is unable to conduct an action potential. This ensures the one-way direction of a signal from the cell body down the length of the axon to the axon terminal.

It is interesting to observe that all functions of the nervous system, from our deepest emotions to our highest reasoning abilities, are dependent on the conduction of nerve signals.

Animation
Action Potential Propagation
3D Animation
Neural Transmission: Action Potential Propagation

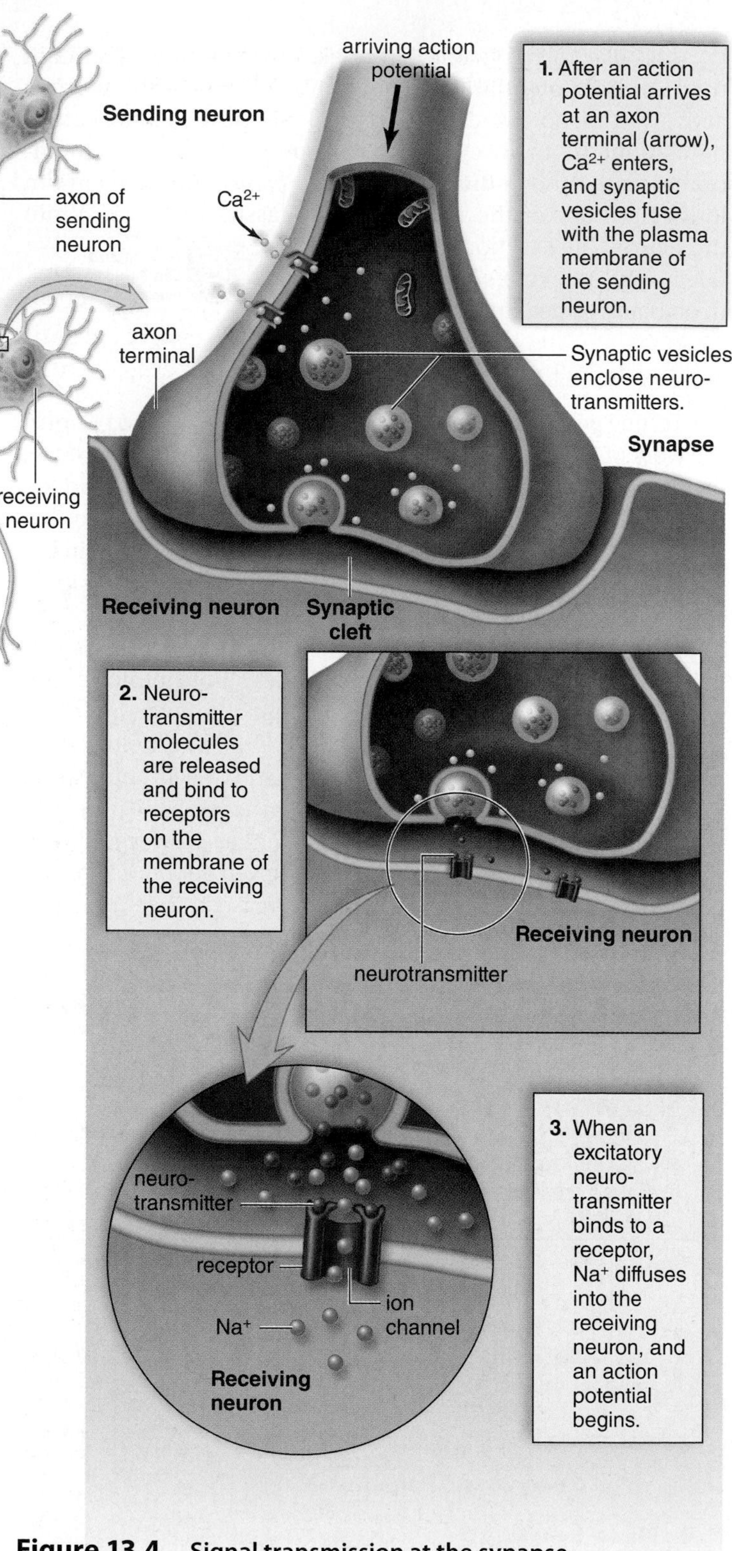
Sending neuron
axon of sending neuron
axon terminal
receiving neuron
arriving action potential
Ca2+
1. After an action potential arrives at an axon terminal (arrow), Ca2+ enters, and synaptic vesicles fuse with the plasma membrane of the sending neuron.
Synaptic vesicles enclose neuro-transmitters.
Synapse
Receiving neuron
Synaptic cleft
2. Neuro-transmitter molecules are released and bind to receptors on the membrane of the receiving neuron.
Receiving neuron
neurotransmitter
neuro-transmitter
receptor
Na+
ion channel
Receiving neuron
3. When an excitatory neuro-transmitter binds to a receptor, Na+ diffuses into the receiving neuron, and an action potential begins.

Figure 13.4 Signal transmission at the synapse. Transmission across a synapse from one neuron to another occurs when a neurotransmitter is released and diffuses across a synaptic cleft and binds to a receptor in the membrane of the receiving neuron.

The Synapse

Every axon branches into many fine endings, each tipped by a small swelling called an **axon terminal.** Each terminal lies very close to either the dendrite or the cell body of another neuron. This region of close proximity is called a **synapse** (Fig. 13.4). At a synapse, a small gap called the **synaptic cleft** separates the sending neuron from the receiving neuron. The nerve signal is unable to jump the cleft. Therefore, another means is needed to pass the nerve signal from the sending neuron to the receiving neuron.

MP3 Synapses

Transmission across a synapse is carried out by molecules called **neurotransmitters,** stored in synaptic vesicles in the axon terminals. (See section 3.2 for a review of vesicle function.) The events (Fig. 13.4) at a synapse are (1) nerve signals traveling along an axon to reach an axon terminal; (2) calcium ions entering the terminal and stimulating synaptic vesicles to merge with the sending membrane; and (3) neurotransmitter molecules releasing into the synaptic cleft and diffusing across the cleft to the receiving membrane; there, neurotransmitter molecules bind with specific receptor proteins. Depending on the type of neurotransmitter, the response of the receiving neuron can be toward excitation or toward inhibition. In Figure 13.5, excitation occurs because the neurotransmitter, such as acetylcholine (ACh), has caused the sodium gate to open. Sodium ions diffuse into the receiving neuron. Inhibition would occur if a neurotransmitter caused potassium ions to exit the receiving neuron.

Animation Chemical Synapses

Once a neurotransmitter has been released into a synaptic cleft and has initiated a response, it is removed from the cleft. In some synapses, the receiving membrane contains enzymes that rapidly inactivate the neurotransmitter. For example, the enzyme **acetylcholinesterase** (**AChE**) breaks down acetylcholine. In other synapses, the sending membrane rapidly reabsorbs the neurotransmitter, possibly for repackaging in synaptic vesicles or for molecular breakdown.

The short existence of neurotransmitters at a synapse prevents continuous stimulation (or inhibition) of receiving membranes. The receiving cell needs to be able to respond quickly to changing conditions. If the neurotransmitter were to linger in the cleft, the receiving cell would be unable to respond to a new signal from a sending cell.

3D Animation Neural Transmission: Synapse

Neurotransmitter Molecules

More than 100 substances are known or suspected to be neurotransmitters. Some of the more common ones in humans are acetylcholine, norepinephrine, dopamine, serotonin, glutamate, and GABA (gamma aminobutyric acid). Neurotransmitters transmit signals between nerves. Nerve-muscle, nerve-organ, and nerve-gland synapses also communicate using neurotransmitters.

Acetylcholine (ACh) and **norepinephrine** are active in both the CNS and PNS. In the PNS, these neurotransmitters act at synapses called neuromuscular junctions. Neuromuscular junctions are discussed in Chapter 12.

In the PNS, ACh excites skeletal muscle but inhibits cardiac muscle. It has either an excitatory or inhibitory effect on smooth muscle or glands, depending on their location.

Norepinephrine generally excites smooth muscle. In the CNS, norepinephrine is important to dreaming, waking, and mood. **Serotonin** is involved in thermoregulation, sleeping, emotions, and perception. Many drugs that affect the nervous system act at the synapse. Some interfere with the actions of neurotransmitters, and other drugs prolong the effects of neurotransmitters (see section 13.5).

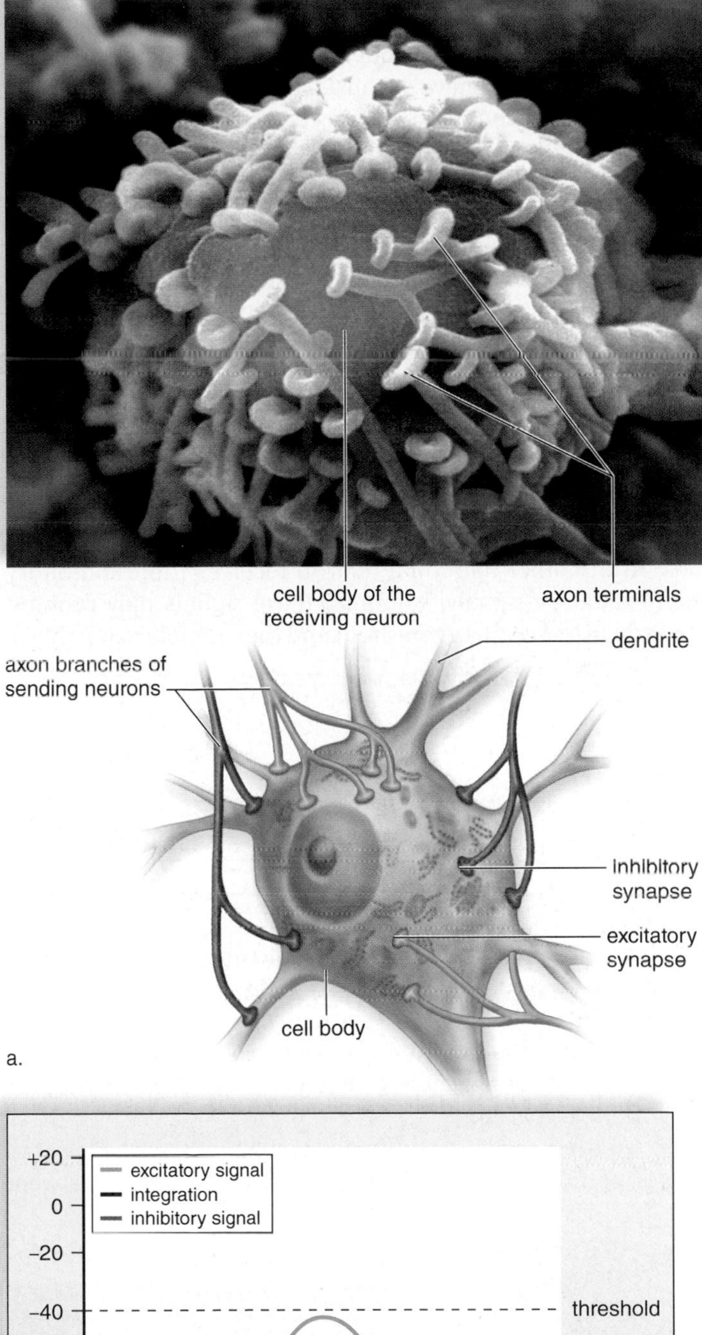

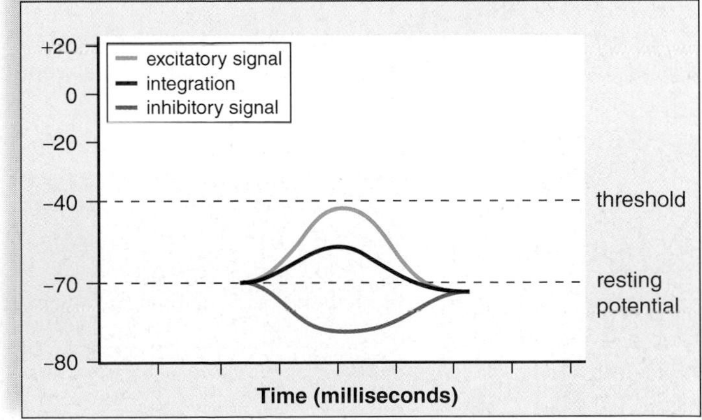

Figure 13.5 Integration of excitatory and inhibitory signals at the synapse.
a. Inhibitory signals and excitatory signals are summed up in the dendrite and cell body of the postsynaptic neuron. Only if the combined signals cause the membrane potential to rise above threshold does an action potential occur. **b.** In this example, threshold was not reached.

Synaptic Integration

A single neuron has a cell body and may have many dendrites (Fig. 13.5*a*). All can have synapses with many other neurons. Therefore, a neuron is on the receiving end of many signals, which can either be excitatory or inhibitory. Recall that an excitatory neurotransmitter produces an excitatory signal by opening sodium gates at a synapse. This drives the neuron closer to its threshold (illustrated by the green line in Fig. 13.5*b*). If threshold is reached, an action potential is inevitable. On the other hand, an inhibitory neurotransmitter drives the neuron farther from an action potential (red line in Fig. 13.5*b*) by opening the gates for potassium.

Neurons integrate these incoming signals. **Integration** is the summing up of excitatory and inhibitory signals. If a neuron receives enough excitatory signals (either from different synapses or at a rapid rate from a single synapse) to outweigh the inhibitory ones, chances are the axon will transmit a signal. On the other hand, if a neuron receives more inhibitory than excitatory signals, summing these signals may prohibit the axon from reaching threshold and then depolarizing (black solid line in Fig. 13.5*b*).

CHECK YOUR PROGRESS 13.1

1. Describe the three types of neurons, and list the three main parts of a neuron.
2. Describe how a nerve impulse is propagated.
3. Summarize how a nerve impulse is transmitted from one neuron to the next.

CONNECTING THE CONCEPTS

For more information on neurons and the nervous systems, refer to the following discussions:

Section 4.4 explores how stem cells may be used to regenerate nervous tissue.

Figure 12.7 illustrates the role of the synapse in the neuromuscular junction.

Section 14.1 explains how the peripheral nervous system sends information to and from the central nervous system.

BIOLOGY MATTERS  **Science**

Discovery of Neurons

In any discipline—humanities, music and the arts, history, mathematics, education, social sciences—innovation and discovery happen because of the hard work of dedicated people. The contributions of the passionate scientist Dr. Santiago Ramon y Cajal established a foundation for ongoing studies of the nervous system.

In the late-nineteenth century, the brain was believed to be a continuous network of "filaments," and scientists were not convinced that the filaments were even cells. Using a new technique, Cajal stained samples of brain tissue with a dye containing metallic silver. Careful microscopic studies showed Cajal that the brain was composed of individual cells. A later researcher named the nerve cells *neurons.* Cajal then discovered that the neurons were not directly connected to one another. This discovery allowed later scientists to research this gap—the synapse—as well as the neurotransmitters that allow nerve cells to communicate across a synapse. An artist as well as a researcher, Cajal illustrated his microscopic discoveries. His drawings were reproduced in textbooks for decades.

Cajal's theory of "dynamic polarization" described the idea of the resting and action potential. He proposed that neurons received signals at the cell body and dendrites and transmitted these signals via their axons to other neurons. Cajal described this basic principle of neuron function long before methods were devised to prove his theories.

Cajal was awarded the Nobel Prize in 1906 for his discoveries in the structure and function of the nervous system.

a.

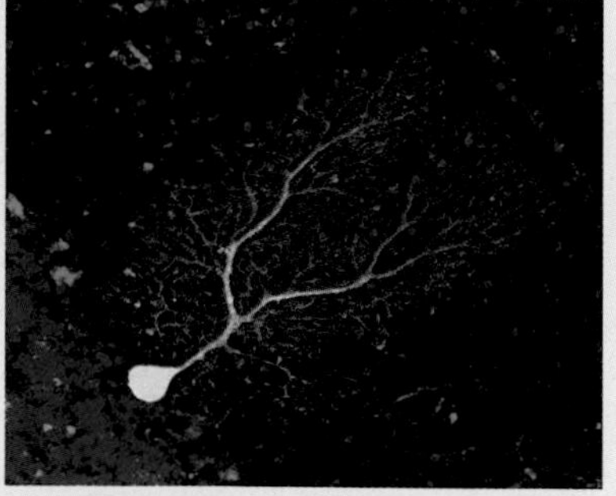

b.

Figure 13A Structure of a neuron.
a. Santiago Ramon y Cajal's sketch of neuron. **b.** Actual photomicrograph of the same neuron.

Question to Consider

1. Can you identify the structures of a neuron in Figure 13A*a* and *b*?

13.2 The Central Nervous System

LEARNING OUTCOMES

Upon completion of this section, you should be able to

1. Identify the structures of the spinal cord, and provide a function for each.
2. Identify the structures of the brain, and provide a function for each.
3. Identify the lobes and major areas of the human brain.
4. Distinguish between the functions of the primary motor and the primary somatosensory areas of the brain.

The spinal cord and the brain make up the CNS, where sensory information is received and motor control is initiated. As mentioned previously, both the spinal cord and the brain are protected by bone. The spinal cord is surrounded by vertebrae, and the brain is enclosed by the skull. Also, both the spinal cord and the brain are wrapped in protective membranes known as **meninges.** *Meningitis* is an infection of the meninges and may be caused by either bacterial or viral pathogens. The spaces between the meninges are filled with **cerebrospinal fluid,** which cushions and protects the CNS. In a spinal tap (lumbar puncture), a small amount of this fluid is withdrawn from around the spinal cord for laboratory testing. Cerebrospinal fluid is also contained within the ventricles of the brain and in the central canal of the spinal cord. The brain has four **ventricles,** interconnecting chambers that produce and serve as a reservoir for cerebrospinal fluid (Fig. 13.6). Normally, any excess cerebrospinal fluid drains away into the cardiovascular system. However, blockages can occur. In an infant, the brain can enlarge due to cerebrospinal fluid accumulation, resulting in a condition called hydrocephalus ("water on the brain"). If cerebrospinal fluid collects in an adult, the brain cannot enlarge. Instead, it is pushed against the skull. Such situations cause severe brain damage and can be fatal unless quickly corrected.

The CNS is composed of two types of nervous tissue—gray matter and white matter. **Gray matter** contains cell bodies and short, nonmyelinated fibers. **White matter** contains myelinated axons that run together in bundles called **tracts.**

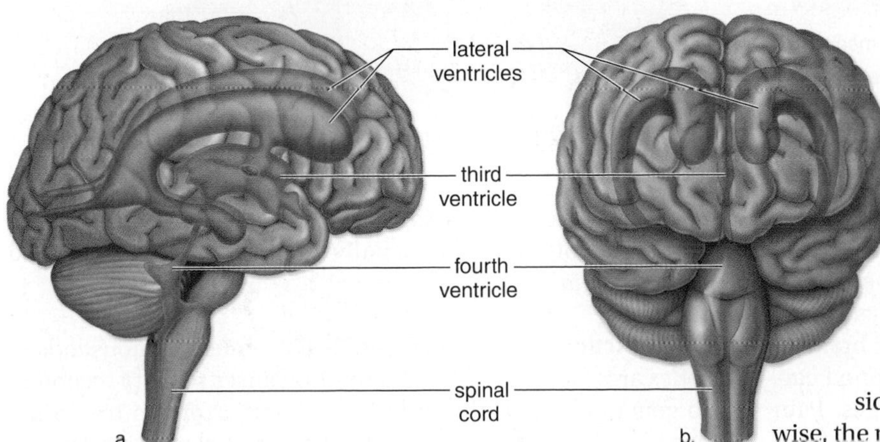

Figure 13.6 The ventricles of the brain.
The brain has four ventricles. A lateral ventricle is found on each side of the brain. They join at the third ventricle. The third ventricle connects with the fourth ventricle superiorly; the central canal of the spinal cord joins the fourth ventricle inferiorly. All structures are filled with cerebrospinal fluid. **a.** Lateral view of ventricles seen through a transparent brain. **b.** Anterior view of ventricles seen through a transparent brain.

The Spinal Cord

The **spinal cord** extends from the base of the brain through a large opening in the skull called the foramen magnum (see Fig. 11.3). From the foramen magnum, the spinal cord proceeds inferiorly in the vertebral canal.

Structure of the Spinal Cord

A cross section of the spinal cord shows a central canal, gray matter, and white matter (Fig. 13.7*a*). Figure 13.7*b* shows how an individual vertebra protects the spinal cord. The spinal nerves project from the cord through small openings called intervertebral foramina. Fibrocartilage intervertebral disks separate the vertebrae. If the disk ruptures or herniates, the vertebrae compress a spinal nerve, resulting in pain and a loss of mobility.

The central canal of the spinal cord contains cerebrospinal fluid, as do the meninges that protect the spinal cord. The gray matter is centrally located and shaped like the letter H (Fig. 13.7*a–c*). Portions of sensory neurons and motor neurons are found in gray matter, as are interneurons that communicate with these two types of neurons. The dorsal root of a spinal nerve contains sensory fibers entering the gray matter. The ventral root of a spinal nerve contains motor fibers exiting the gray matter. The dorsal and ventral roots join before the spinal nerve leaves the vertebral canal (Fig. 13.7*c, d*), forming a mixed nerve. Spinal nerves are a part of the PNS.

The white matter of the spinal cord occurs in areas around the gray matter. The white matter contains ascending tracts taking information to the brain (primarily located posteriorly) and descending tracts taking information from the brain (primarily located anteriorly). Many tracts cross just after they enter and exit the brain, so the left side of the brain controls the right side of the body. Likewise, the right side of the brain controls the left side of the body.

Functions of the Spinal Cord

The spinal cord provides a means of communication between the brain and the peripheral nerves that leave the cord. When someone touches your hand, sensory receptors generate nerve signals that pass through sensory fibers to the spinal cord and up ascending tracts to the brain (see Fig. 13.1*b*, red arrows).

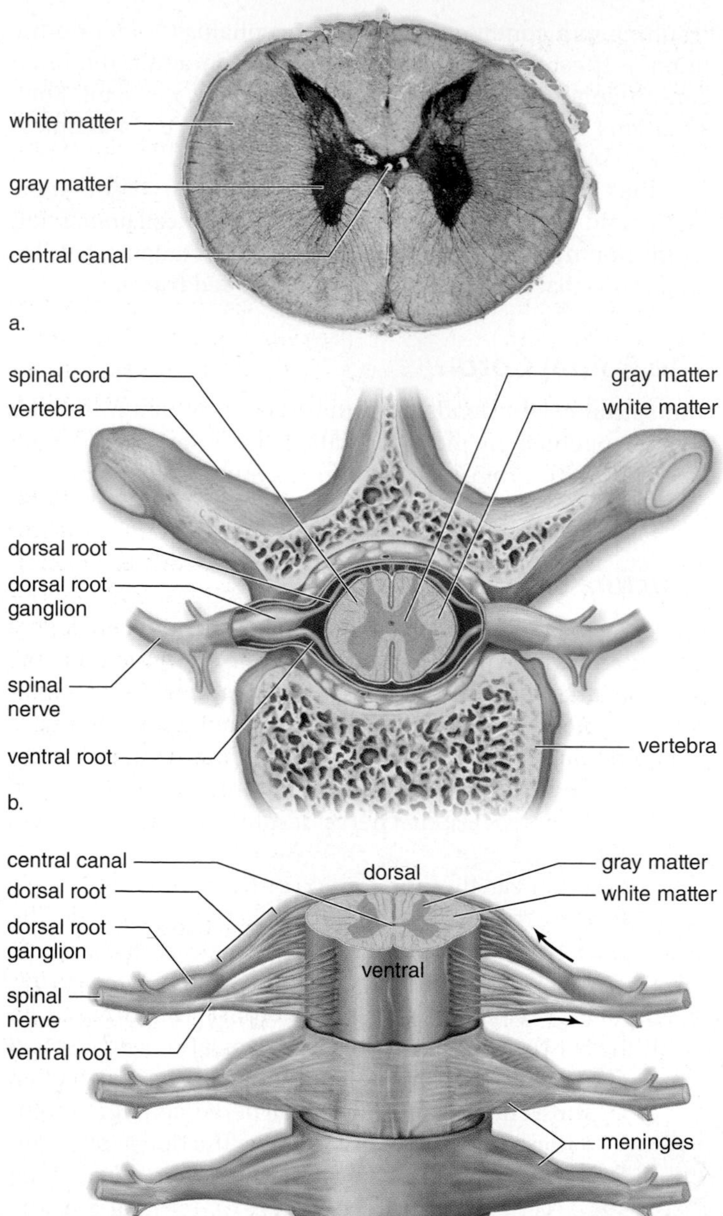

Figure 13.7 The organization of white and gray matter in the spinal cord and the spinal nerves.
a. Cross section of the spinal cord, showing arrangements of white and gray matter. **b.** Spinal nerves originating from the spinal cord. **c.** The spinal cord is protected by vertebrae. **d.** Spinal nerves emerging from the cord.

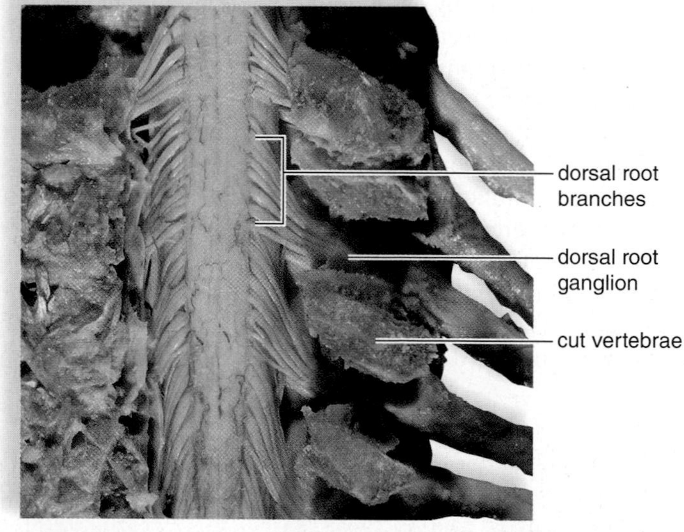

d. Dorsal view of spinal cord and dorsal roots of spinal nerves

The gate control theory of pain proposes that the tracts in the spinal cord have "gates" and that these "gates" control the flow of pain messages from the peripheral nerves to the brain. Depending on how the gates process a pain signal, the pain message can be allowed to pass directly to the brain or can be prevented from reaching the brain. As mentioned earlier, endorphins can temporarily block pain messages. Pain messages may also be blocked by other inputs, such as those received from touch receptors.

The brain coordinates the voluntary control of our limbs. Motor signals originating in the brain pass down descending tracts to the spinal cord and out to our muscles by way of motor fibers (see Fig. 13.1*b*, black arrows). Therefore, if the spinal cord is severed, we suffer a loss of sensation and a loss of voluntary control—paralysis. If the cut occurs in the thoracic region, the lower body and legs are paralyzed, a condition known as *paraplegia*. If the injury is in the neck region, all four limbs are usually affected, a condition called *quadriplegia*.

Reflex Actions The spinal cord is the center for thousands of reflex arcs (see Fig. 13.16). A stimulus causes sensory receptors to generate signals that travel in sensory axons to the spinal cord. Interneurons integrate the incoming data and relay signals to motor neurons. A response to the stimulus occurs when motor axons cause skeletal muscles to contract. Motor neurons in a reflex arc may also affect smooth muscle, organs, or glands. Each interneuron in the spinal cord synapses with numerous other neurons. From there, interneurons send signals to other interneurons and motor neurons.

Similarly, the spinal cord creates reflex arcs for the internal organs. For example, when blood pressure falls, internal receptors in the carotid arteries and aorta generate nerve signals that pass through sensory fibers to the cord and then up an ascending tract to a cardiovascular center in the brain. Thereafter, nerve signals pass down a descending tract to the spinal cord. Motor signals then cause blood vessels to constrict so that the blood pressure rises.

The Brain

The human **brain** has been called the last great frontier of biology. The goal of modern neuroscience is to understand the structure and function of the brain's various parts so well that it will be possible to prevent or correct the thousands of mental disorders that rob humans of a normal life. This section gives only a glimpse of what is known about the brain and the modern avenues of research.

Video Brain Bank

We discuss the parts of the brain with reference to the cerebrum, the diencephalon, the cerebellum, and the brain stem. The brain's four ventricles (see Fig. 13.6) are called, in turn, the two lateral ventricles, the third ventricle, and the fourth ventricle. It may be helpful for you to associate the cerebrum with the two lateral ventricles, the diencephalon with the third ventricle, and the brain stem and the cerebellum with the fourth ventricle (Fig. 13.8*a*).

MP3 The Brain

The Cerebrum

The **cerebrum,** also called the telencephalon, is the largest portion of the brain in humans. The cerebrum is the last center to receive sensory input and carry out integration before commanding voluntary motor responses. It communicates with and coordinates the activities of the other parts of the brain.

MP3 The Cerebrum

Cerebral Hemispheres Just as the human body has two halves, so does the cerebrum. These halves are called the left and right **cerebral hemispheres** (Fig. 13.8*b*). A deep groove called the longitudinal fissure divides the left and right cerebral hemispheres. The two cerebral hemispheres communicate via the **corpus callosum,** an extensive bridge of nerve tracts.

The characteristic appearance of the cerebrum is the result of thick folds, called *gyri* (sing., gyrus) separated by shallow grooves called *sulci* (sing., sulcus). The sulci divide each hemisphere into lobes (Fig. 13.9). The *frontal lobe* is the most anterior of the lobes (directly behind the forehead). The *parietal lobe* is posterior to the frontal lobe. The *occipital lobe* is posterior to the parietal lobe (at the rear of the head). The *temporal lobe* lies inferior to the frontal and parietal lobes (at the temple and the ear).

Each lobe is associated with particular functions, as indicated in Figure 13.9.

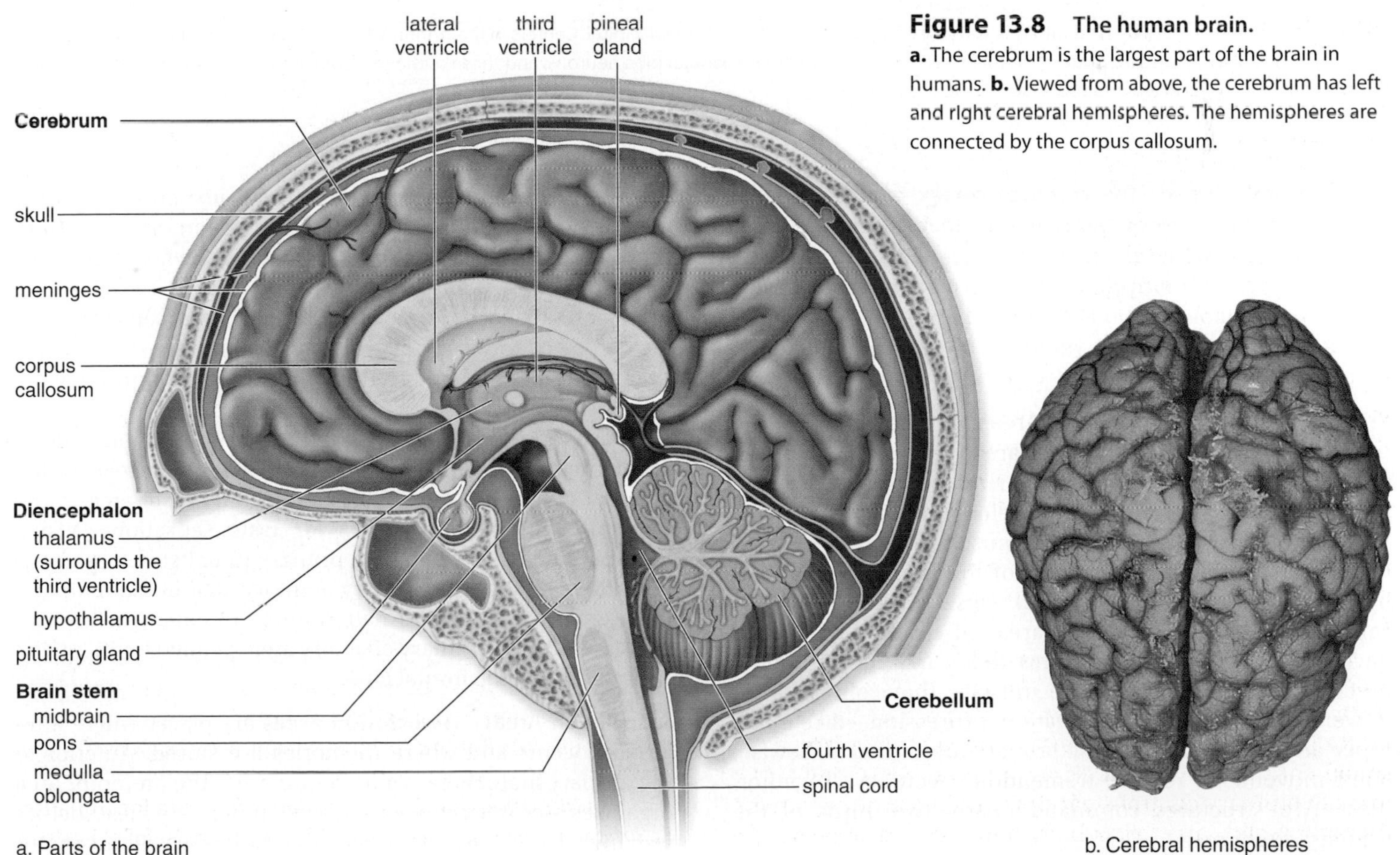

Figure 13.8 **The human brain.**
a. The cerebrum is the largest part of the brain in humans. **b.** Viewed from above, the cerebrum has left and right cerebral hemispheres. The hemispheres are connected by the corpus callosum.

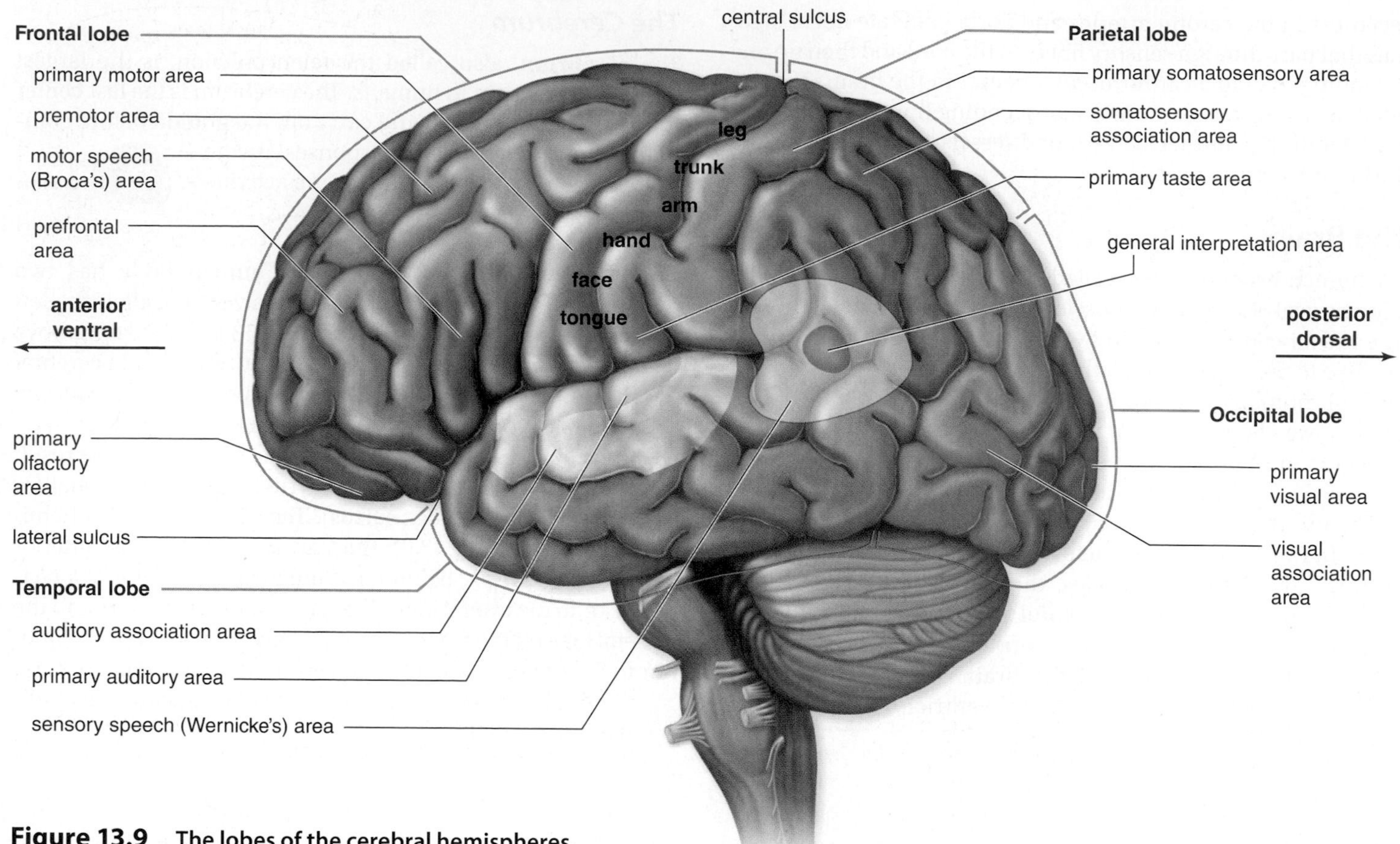

Figure 13.9 **The lobes of the cerebral hemispheres.**
Each cerebral hemisphere is divided into four lobes: frontal, parietal, temporal, and occipital. Centers in the frontal lobe control movement and higher reasoning, as well as the smell sensation. Somatic sensing is carried out by parietal lobe neurons, and those of the temporal lobe receive sound information. Visual information is received and processed in the occipital lobe.

The Cerebral Cortex The **cerebral cortex** is a thin, highly convoluted outer layer of gray matter that covers the cerebral hemispheres. (Recall that gray matter consists of neurons whose axons are unmyelinated.) The cerebral cortex contains over 1 billion cell bodies and is the region of the brain that accounts for sensation, voluntary movement, and all the thought processes we associate with consciousness.

Primary Motor and Sensory Areas of the Cortex The cerebral cortex contains motor areas and sensory areas, as well as association areas. The **primary motor area** is in the frontal lobe just anterior to (before) the central sulcus. Voluntary commands to skeletal muscles begin in the primary motor area, and each part of the body is controlled by a certain section (Fig. 13.10*a*). Observe the illustration carefully. You'll see that large areas of cerebral cortex are devoted to controlling structures that carry out very fine, precise movements. Thus, the muscles that control facial movements—swallowing, salivation, expression—take up an especially large portion of the primary motor area. Likewise, hand movements require tremendous accuracy. Together, these two structures command nearly two-thirds of the primary motor area.

The **primary somatosensory area** is just posterior to the central sulcus in the parietal lobe. Sensory information from the skin and skeletal muscles arrives here, where each part of the body is sequentially represented (Fig. 13.10*b*). Like the primary motor cortex, large areas of the primary sensory cortex are dedicated to those body areas with acute sensation. Once again, the face and hands require the largest proportion of the sensory cortex.

Reception areas for the other primary sensations—taste, vision, hearing, and smell—are located in other areas of the cerebral cortex (see Fig. 13.9). The primary taste area in the parietal lobe (pink) accounts for taste sensations. Visual information is received by the primary visual cortex (blue) in the occipital lobe. The primary auditory area in the temporal lobe (green) accepts information from our ears. Smell sensations travel to the primary olfactory area (yellow) found on the deep surface of the frontal lobe.

Association Areas **Association areas** are places where integration occurs and where memories are stored. Anterior to the primary motor area is a premotor area. The premotor area organizes motor functions for skilled motor activities, such as walking and talking at the same time. Next, the primary motor

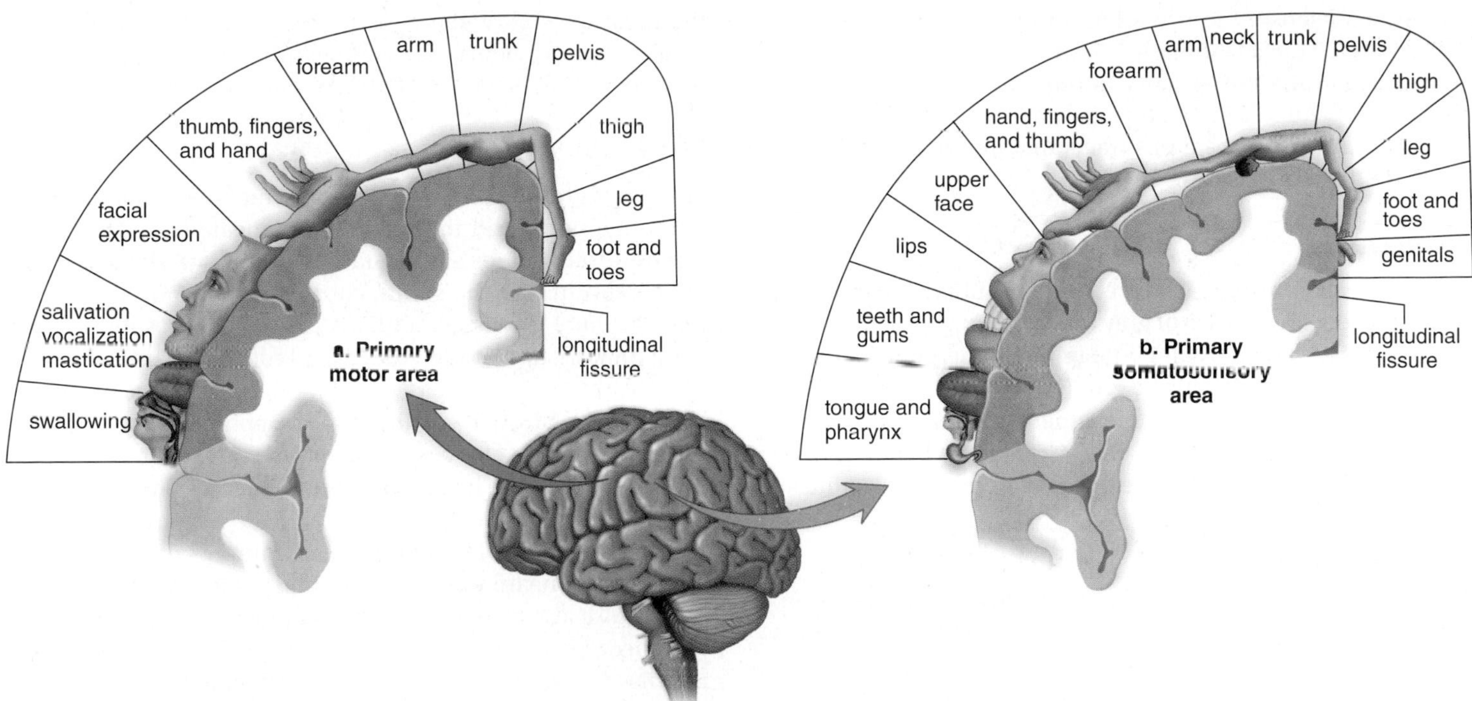

Figure 13.10 The primary motor and primary somatosensory areas of the brain.
a. The primary motor area (blue) is located in the frontal lobe, adjacent to (**b**) the primary somatosensory area in the parietal lobe. The primary taste area is colored pink. The size of each body region shown indicates the relative amount of cortex devoted to control of that body region.

area sends signals to the cerebellum, which integrates them. A momentary lack of oxygen during birth can damage the motor areas of the cerebral cortex. Cerebral palsy, a condition characterized by a spastic weakness of the arms and legs, results. The *somatosensory association area,* located just posterior to the primary somatosensory area, processes and analyzes sensory information from the skin and muscles. The *visual association area* in the occipital lobe associates new visual information with stored visual memories. It might "decide," for example, if we have seen a face, scene, or symbol before. The *auditory association area* in the temporal lobe performs the same functions with regard to sounds.

Processing Centers Processing centers of the cortex receive information from the other association areas and perform higher-level analytical functions. The **prefrontal area,** an association area in the frontal lobe, receives information from the other association areas and uses this information to reason and plan our actions. Integration in this area accounts for our most cherished human abilities. Reasoning, critical thinking, and formulating appropriate behaviors are possible because of integration carried out in the prefrontal area.

The unique ability of humans to speak is partially dependent upon two processing centers found only in the left cerebral cortex. **Wernicke's area** is located in the posterior part of the left temporal lobe. **Broca's area** is located in the left frontal lobe. Broca's area is located just anterior to the portion of the primary motor area for speech musculature (lips, tongue, larynx, and so forth) (see Fig. 13.9). Wernicke's area helps us understand both the written and spoken word and sends the information to Broca's area. Broca's area adds grammatical refinements and directs the primary motor area to stimulate the appropriate muscles for speaking and writing.

Central White Matter Much of the rest of the cerebrum is composed of white matter. Myelination occurs and white matter develops as a child grows. Progressive myelination enables the brain to grow in size and complexity. For example, as neurons become myelinated within tracts designed for language development, children become more capable of speech. Descending tracts from the primary motor area communicate with lower brain centers, and ascending tracts from lower brain

APPLICATIONS AND MISCONCEPTIONS

Why does a stroke on the right side of the brain cause weakness or paralysis, as well as decreased sensation, on the left side of the body?

Descending motor tracts (from the primary motor area) and ascending sensory tracts (from the primary somatosensory area) cross over in the spinal cord and medulla. Motor neurons in the right cerebral hemisphere control the left side of the body and vice versa because of crossing-over. Likewise, sensation from the left half of the body travels to the right cerebral hemisphere. Destruction of brain tissue by a stroke interferes with outgoing motor signals to the opposite side of the body, as well as incoming sensory information.

centers send sensory information up to the primary somatosensory area. Tracts within the cerebrum also take information between the different sensory, motor, and association areas pictured in Figure 13.9. As previously mentioned, the corpus callosum contains tracts that join the two cerebral hemispheres.

Video Brain Surgery

Basal Nuclei

Though the majority of each cerebral hemisphere is composed of tracts, there are masses of gray matter deep within the white matter. These **basal nuclei** integrate motor commands to ensure that the proper muscle groups are stimulated or inhibited. Integration ensures that movements are coordinated and smooth. **Parkinson disease** (see Chapter 17) is believed to be caused by degeneration of specific neurons in the basal nuclei.

The Diencephalon

The hypothalamus and the thalamus are in the **diencephalon,** a region that encircles the third ventricle. The **hypothalamus** forms the floor of the third ventricle. The hypothalamus is an integrating center that helps maintain homeostasis. It regulates hunger, sleep, thirst, body temperature, and water balance. The hypothalamus controls the pituitary gland and thereby serves as a link between the nervous and endocrine systems.

The **thalamus** consists of two masses of gray matter located in the sides and roof of the third ventricle. The thalamus is on the receiving end for all sensory input except the sense of smell. Visual, auditory, and somatosensory information arrives at the thalamus via the cranial nerves and tracts from the spinal cord. The thalamus integrates this information and sends it on to the appropriate portions of the cerebrum. The thalamus is involved in arousal of the cerebrum, and it also participates in higher mental functions such as memory and emotions.

The pineal gland, which secretes the hormone melatonin, is located in the diencephalon. Presently there is much popular interest in the role of melatonin in our daily rhythms. Some researchers believe it can help alleviate jet lag or insomnia. Scientists are also interested in the possibility that the hormone may regulate the onset of puberty.

The Cerebellum

The **cerebellum** lies under the occipital lobe of the cerebrum and is separated from the brain stem by the fourth ventricle. The cerebellum has two portions joined by a narrow median portion. Each portion is primarily composed of white matter. In a longitudinal section, the white matter has a treelike pattern called *arbor vitae.* Overlying the white matter is a thin layer of gray matter that forms a series of complex folds.

The cerebellum receives sensory input from the eyes, ears, joints, and muscles about the present position of body parts. It also receives motor output from the cerebral cortex about where these parts should be located. After integrating this information, the cerebellum sends motor signals by way of the brain stem to the skeletal muscles. In this way, the cerebellum maintains posture and balance. It also ensures that all of the muscles work together to produce smooth, coordinated, voluntary movements. The cerebellum assists the learning of new motor skills such as playing the piano or hitting a baseball.

The Brain Stem

The tracts cross in the **brain stem,** which contains the midbrain, the pons, and the medulla oblongata (see Fig. 13.8*a*). The **midbrain** acts as a relay station for tracts passing between the cerebrum and the spinal cord or cerebellum. It also has reflex centers for visual, auditory, and tactile responses. The word **pons** means "bridge" in Latin. True to its name, the pons contains bundles of axons traveling between the cerebellum and the rest of the CNS. In addition, the pons functions with the medulla oblongata to regulate breathing rate. Reflex centers in the pons coordinate head movements in response to visual and auditory stimuli.

The **medulla oblongata** contains a number of reflex centers for regulating heartbeat, breathing, and vasoconstriction (blood pressure). It also contains the reflex centers for vomiting, coughing, sneezing, hiccuping, and swallowing. The medulla oblongata lies just superior to the spinal cord, and it contains tracts that ascend or descend between the spinal cord and higher brain centers. Recall that tracts are groups of axons that travel together. Ascending tracts convey sensory information. Motor information is transmitted on descending tracts.

The Reticular Formation The **reticular formation** is a complex network of *nuclei,* which are masses of gray matter, and fibers that extend the length of the brain stem (Fig. 13.11). The reticular formation is a major component of the reticular

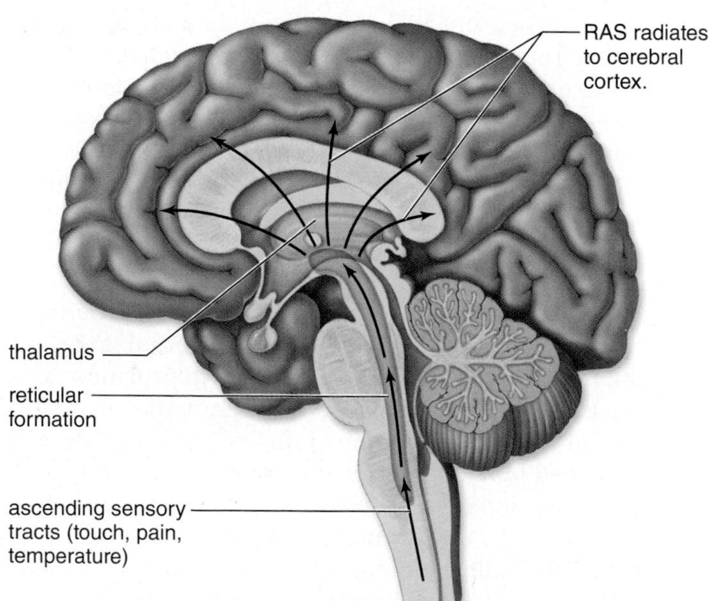

Figure 13.11 The reticular formation of the brain.
The reticular formation receives and sends on motor and sensory information to various parts of the CNS. One portion, the reticular activating system (RAS; see arrows), arouses the cerebrum and, in this way, controls alertness versus sleep.

activating system (RAS). The RAS receives sensory signals and sends them to higher centers. Motor signals received by the RAS are sent to the spinal cord.

The RAS arouses the cerebrum via the thalamus and causes a person to be alert. If you want to awaken the RAS, surprise it with sudden stimuli, such as an alarm clock ringing, bright lights, smelling salts, or splashing cold water on your face. The RAS can filter out unnecessary sensory stimuli, explaining why you can study with the TV on. Similarly, the RAS allows you to take a test without noticing the sounds of the people around you—unless the sounds are particularly distracting. To inactivate the RAS, remove visual or auditory stimuli, allowing yourself to become drowsy and drop off to sleep. General anesthetics function by artificially suppressing the RAS. A severe injury to the RAS can cause a person to be comatose, from which recovery may be impossible.

CHECK YOUR PROGRESS 13.2

1. List the functions of the spinal cord.
2. Summarize the major regions of the brain, and describe the general function of each.
3. Relate how the RAS aids in homeostasis.

CONNECTING THE CONCEPTS

For more information on the central nervous system, refer to the following discussions:

Section 9.5 examines how the central nervous system controls breathing.

Section 17.5 explores how aging and diseases such as Alzheimer disease influence the brain.

Sections 22.3 and 22.4 outline how the size of the brain has changed over the course of human evolution.

13.3 The Limbic System and Higher Mental Functions

LEARNING OUTCOMES

Upon completion of this section, you should be able to

1. Identify the structures of the limbic system.
2. Explain how the limbic system is involved in memory, language, and speech.
3. Summarize the types of memory associated with the limbic system.

The limbic system integrates our emotions (fear, joy, sadness) with our higher mental functions (reason, memory). Because of the limbic system, activities such as sexual behavior and eating seem pleasurable and mental stress can cause high blood pressure.

Limbic System

The **limbic system** is an evolutionary ancient group of linked structures deep within the cerebrum. It is a functional grouping rather than an anatomical one (Fig. 13.12). The limbic system blends primitive emotions and higher mental functions into a united whole. As already noted, it accounts for why activities such as sexual behavior and eating seem pleasurable. Conversely, unpleasant sensations or emotions (pain, frustration, hatred, despair) are translated by the limbic system into a stress response.

Two significant structures within the limbic system are the amygdala and the hippocampus. The **amygdala,** in particular, can cause experiences to have emotional overtones, and it creates the sensation of fear. This center can use past knowledge

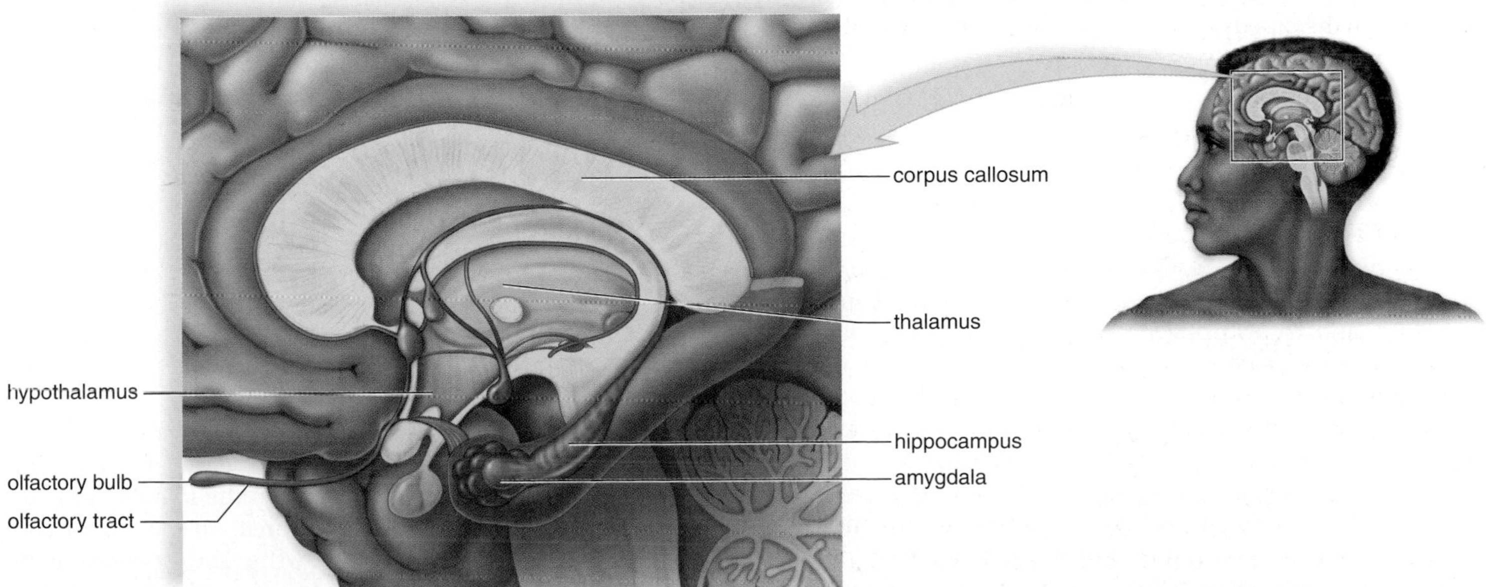

Figure 13.12 The regions of the brain associated with the limbic system.
In the limbic system (purple), structures deep within each cerebral hemisphere and surrounding the diencephalon join higher mental functions, such as reasoning, with more primitive feelings, such as fear and pleasure. Therefore, primitive feelings can influence our behavior, but reason can also keep them in check.

fed to it by association areas to assess a current situation. If necessary, the amygdala can trigger the fight-or-flight reaction. So if you are out late at night and you turn to see someone in a ski mask following you, the amygdala may immediately cause you to start running. The frontal cortex can override the limbic system and cause you to rethink the situation and prevent you from acting out strong reactions.

The **hippocampus** is believed to play a crucial role in learning and memory. The hippocampal region acts as an information gateway during the learning process. It determines what information about the world is to be sent to memory and how this information is to be encoded and stored by other regions in the brain. Most likely, the hippocampus can communicate with the frontal cortex because we know that memories are an important part of our decision-making processes.

It's hoped that research into the functioning of the hippocampus will help in understanding **Alzheimer disease,** a brain disorder characterized by gradual loss of memory (see Chapter 17).

Higher Mental Functions

As in other areas of biological research, brain research has progressed due to technological breakthroughs. Neuroscientists now have a wide range of techniques at their disposal for studying the human brain, including modern technologies that allow us to record its functioning.

Memory and Learning

Just as the connecting tracts of the corpus callosum are evidence that the two cerebral hemispheres work together, so the limbic system indicates that cortical areas may work with lower centers to produce learning and memory. **Memory** is the ability to hold a thought in mind or to recall events from the past, ranging from a word we learned only yesterday to an early emotional experience that has shaped our lives. **Learning** takes place when we retain and use past memories.

Types of Memory We have all tried to remember a seven-digit telephone number for a short time. If we say we are trying to keep it in the forefront of our brain, we are exactly correct. The prefrontal area, active during **short-term memory,** lies just posterior to our forehead! There are some telephone numbers that we have memorized. In other words, they have gone into **long-term memory.** Think of a telephone number you know by heart, and try to bring it to mind without also thinking about the place or person associated with that number. Most likely you cannot. Typically, long-term memory is a mixture of what is called **semantic memory** (numbers, words, etc.) and **episodic memory** (persons, events, etc.).

Skill memory is another type of memory that can exist independent of episodic memory. Skill memory is involved in performing motor activities such as riding a bike or playing ice hockey. When a person first learns a skill, more areas of the cerebral cortex are involved than after the skill is perfected. In other words, you have to think about what you are doing when you learn a skill, but later the actions become automatic. Skill memory involves all the motor areas of the cerebrum below the level of consciousness.

Long-Term Memory Storage and Retrieval Our long-term memories are apparently stored in bits and pieces throughout the sensory association areas of the cerebral cortex. Visions are stored in the vision association area, sounds are stored in the auditory association area, and so forth. As previously mentioned, the hippocampus serves as a bridge between the sensory association areas (where memories are stored) and the prefrontal area (where memories are used). The prefrontal area communicates with the hippocampus when memories are stored and when these memories are brought to mind. Some memories are emotionally charged because the amygdala seems to be responsible for fear conditioning and associating danger with sensory stimuli received from various parts of the brain.

Long-Term Potentiation Though it is helpful to know the memory functions of various portions of the brain, an important step toward curing mental disorders is understanding memory on the cellular level. After synapses have been used intensively for a short time, they release more neurotransmitters than before. This phenomenon, called *long-term potentiation,* may be involved in memory storage.

APPLICATIONS AND MISCONCEPTIONS

What is amnesia?

Amnesia results from disruption of the memory pathways and can be temporary or permanent. In anterograde amnesia, injury to the limbic system separates long-term memories of events that occurred prior to the injury from events that occur in the here and now. An affected person might carry on a conversation about past events (memories of a long-ago birthday) but cannot recall a breakfast menu from that morning. In retrograde amnesia, a blow to the head or similar injury abolishes all memories for a variable time before the injury. For example, a head injury occurring during a car accident may abolish all memories from hours to days prior to the accident.

Language and Speech

Language depends on semantic memory. Therefore, we would expect some of the same areas in the brain to be involved in both memory and language. Any disruption of these pathways could contribute to an inability to comprehend our environment and use speech correctly.

Seeing and hearing words depends on sensory centers in the occipital and temporal lobes, respectively. Damage to Wernicke's area, discussed earlier, results in the inability to comprehend speech. Damage to Broca's area, on the other hand, results in the inability to speak and write. The functions of the visual cortex, Wernicke's area, and Broca's area are shown in Figure 13.13.

One interesting aside pertaining to language and speech is the recognition that the left brain and the right brain may

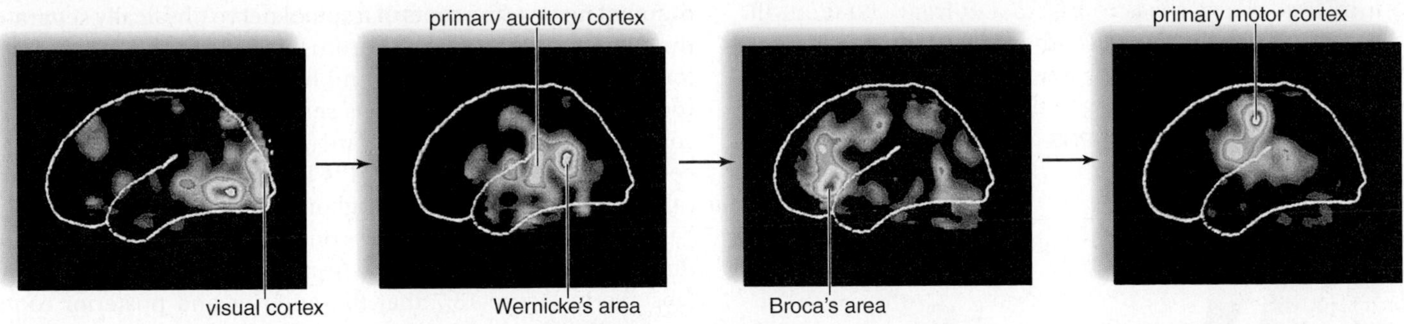
primary auditory cortex
primary motor cortex
visual cortex
Wernicke's area
Broca's area
1. The word is seen in the visual cortex.
2. Information concerning the word is interpreted in Wernicke's area.
3. Information from Wernicke's area is transferred to Broca's area.
4. Information is transferred from Broca's area to the primary motor area.

Figure 13.13 The areas of the brain involved in reading.
These functional images were captured by a high-speed computer during a PET (positron-emission tomography) scan of the brain. A radioactively labeled solution is injected into the subject, and then the subject is asked to perform certain activities. Cross-sectional images of the brain generated by the computer reveal where activity is occurring because the solution is preferentially taken up by active brain tissue and not by inactive brain tissue. These PET images show the cortical pathway for reading words and then speaking them. Red indicates the most active areas of the brain, and blue indicates the least active areas.

have different functions. Recall that the left hemisphere contains both Broca's area and Wernicke's area. As you might expect, it appears that the left hemisphere plays a role of great importance in language functions. The role of the isolated left hemisphere can be studied in patients after surgery to sever the corpus callosum. This procedure is used for seizure control in patients with epilepsy. After surgery, the patient is termed "split brain," because there is no longer direct communication between the two cerebral hemispheres. If a split-brain individual views an object with only the right eye, its image will be sent only to the right hemisphere. This person will be able to choose the proper object for a particular use—scissors to cut paper, for example—but will be unable to name that object.

Research on the split brain is ongoing. In a very general way, the left brain can be contrasted with the right brain.

Left Hemisphere	Right Hemisphere
Verbal	Nonverbal, visuospatial
Logical, analytical	Intuitive
Rational	Creative

Researchers now believe that the hemispheres process the same information differently. The left hemisphere is more global, whereas the right hemisphere is more specific in its approach.

CHECK YOUR PROGRESS 13.3

1. Define the function of the limbic system.
2. List what limbic system structures are involved in the fight-or-flight reaction, learning, and long-term memory.
3. Describe the relationship between the left and right sides of the brain and language and speech.

CONNECTING THE CONCEPTS

For more information on these topics, refer to the following discussion:

Section 17.5 examines the effects of aging on the body, including the nervous system.

13.4 The Peripheral Nervous System

LEARNING OUTCOMES

Upon completion of this section, you should be able to

1. Describe the series of events during a spinal reflex.
2. Distinguish between the somatic and autonomic divisions of the peripheral nervous system.
3. Distinguish between the sympathetic and parasympathetic divisions of the autonomic division.

The peripheral nervous system (PNS), which lies outside the central nervous system, contains the nerves. Nerves are designated as cranial nerves when they arise from the brain and spinal nerves when they arise from the spinal cord. In any case, all nerves carry signals to and from the CNS. So right now, your eyes are sending messages by way of a cranial nerve to the brain, allowing you to read a page. When you're finished, your brain will direct the muscles in your fingers to turn the page by way of the spinal cord and a spinal nerve.

Figure 13.14 illustrates the anatomy of a nerve. The cell body and the dendrites of neurons are in either the CNS or ganglia. **Ganglia** (sing., **ganglion**) are collections of nerve cell bodies outside the CNS. The axons of neurons project from the CNS and form the spinal cord. In other words, nerves, whether cranial or spinal, are composed of axons, the long part of neurons.

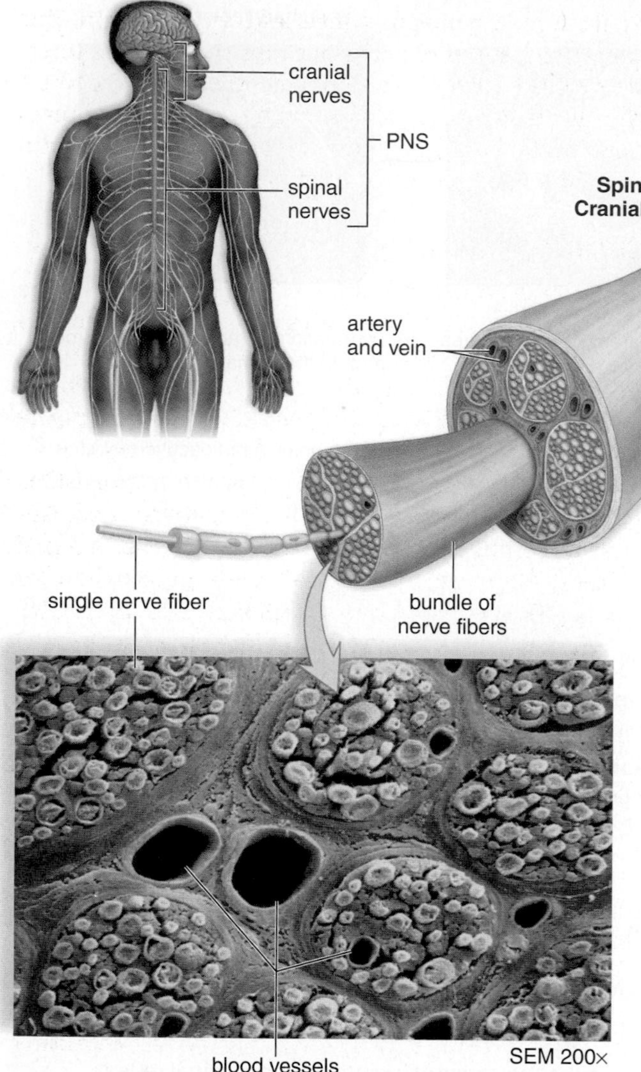

Figure 13.14 The structure of a nerve.
The peripheral nervous system consists of the cranial nerves and the spinal nerves. A nerve is composed of bundles of axons separated from one another by connective tissue.

Humans have 12 pairs of **cranial nerves** attached to the brain. By convention, the pairs of cranial nerves are referred to by Roman numerals (Fig. 13.15). Some cranial nerves are sensory nerves—they contain only sensory fibers; some are motor nerves that contain only motor fibers; and others are mixed nerves that contain both sensory and motor fibers. Cranial nerves are largely concerned with the head, neck, and facial regions of the body. However, the vagus nerve (X) has branches not only to the pharynx and larynx but also to most of the internal organs. From which part of the brain do you think the vagus arises? It arises from the brain stem, specifically, the medulla oblongata that communicates so well with the hypothalamus. These two parts of the brain control the internal organs.

As you know, the **spinal nerves** of humans emerge from either side of the spinal cord (see Fig. 13.7). There are 31 pairs of spinal nerves. The roots of a spinal nerve physically separate the axons of sensory neurons from the axons of motor neurons, forming an arrangement resembling a letter Y. The posterior root of a spinal nerve contains sensory fibers that direct sensory receptor information inward (toward the spinal cord). The cell body of a sensory neuron is in a posterior-root ganglion (also termed a dorsal-root ganglion). The anterior (also termed ventral) root of a spinal nerve contains motor fibers that conduct impulses outward (away from the cord) to the effectors. Observe in Figure 13.7 that the anterior and posterior roots rejoin to form a spinal nerve. All spinal nerves are called mixed nerves because they contain both sensory and motor fibers. Each spinal nerve serves the particular region of the body in which it is located. For example, the intercostal muscles of the rib cage are innervated by thoracic nerves.

Somatic System

The PNS has divisions, and right now we are going to consider the somatic system. The nerves in the **somatic system** serve the skin, skeletal muscles, and tendons (see Fig. 13.1). The somatic system sensory nerves take sensory information from external sensory receptors to the CNS. Motor commands leaving the CNS travel to skeletal muscles via somatic motor nerves.

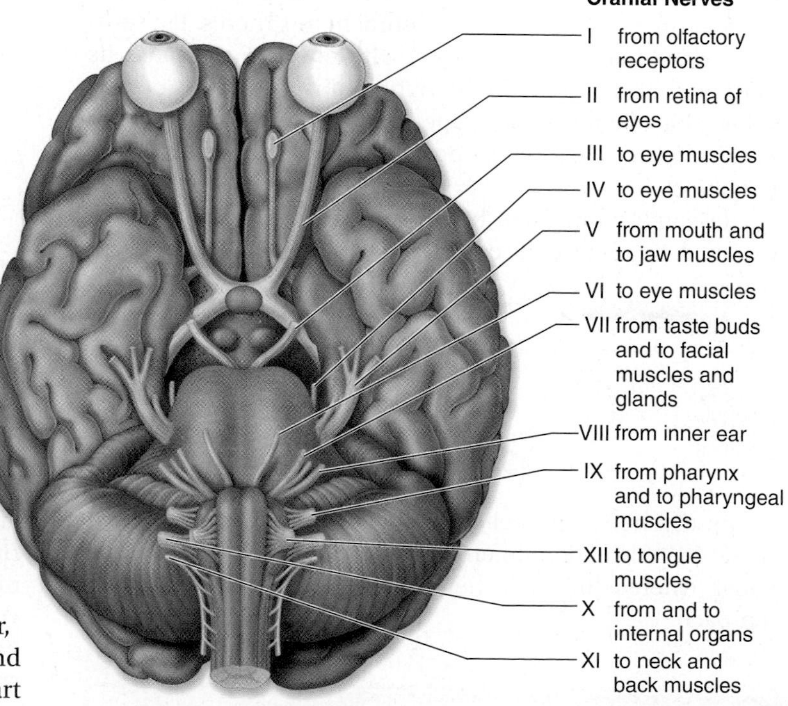

Figure 13.15 The cranial nerves.
Overall, cranial nerves receive sensory input from and send motor outputs to the head region. The spinal nerves receive sensory input from and send motor outputs to the rest of the body. Two important exceptions are the vagus nerve, X, which communicates with internal organs, and the spinal accessory nerve, XI, which controls neck and back muscles.

Not all somatic motor actions are voluntary. Some actions are automatic. Automatic responses to a stimulus in the somatic system are called **reflexes.** A reflex occurs quickly, without your even having to think about it. For example, a reflex may cause you to blink your eyes in response to a flash of light, without your willing it. We will study the path of a reflex because it allows us to study in detail the path of nerve signals to and from the CNS.

The Reflex Arc

Figure 13.16 illustrates the path of a reflex that involves only the spinal cord. If your hand touches a sharp pin, sensory receptors in the skin generate nerve signals that move along sensory fibers through the dorsal-root ganglia toward the spinal cord. Sensory neurons that enter the cord dorsally (posteriorly) pass signals on to many interneurons. Some of these interneurons synapse with motor neurons whose short dendrites and cell bodies are in the spinal cord. Nerve signals travel along these motor fibers to an effector, which brings about a response to the stimulus. In this case, the effector is a muscle, which contracts so that you withdraw your hand from the pin. Various other reactions are also possible—you will most likely look at the pin, wince, and cry out in pain. This whole series of responses occur because some of the interneurons involved carry nerve signals to the brain. The brain makes you aware of the stimulus and directs these other reactions to it. In other words, you don't feel pain until the brain receives the information and interprets it.

Autonomic System

The **autonomic system** is also in the PNS (see Fig. 13.1). The autonomic system regulates the activity of cardiac and smooth muscles, organs, and glands. The system is divided into the sympathetic and parasympathetic divisions (Fig. 13.17). Activation of these two systems generally causes opposite responses.

Although their functions are different, the two divisions share some features: (1) They function automatically and usually in an involuntary manner; (2) they innervate all internal organs; and (3) they use two neurons and one ganglion for each impulse. The first neuron has a cell body within the CNS and a preganglionic fiber that enters the ganglion. The second neuron has a cell body within a ganglion and a postganglionic fiber that leaves the ganglion.

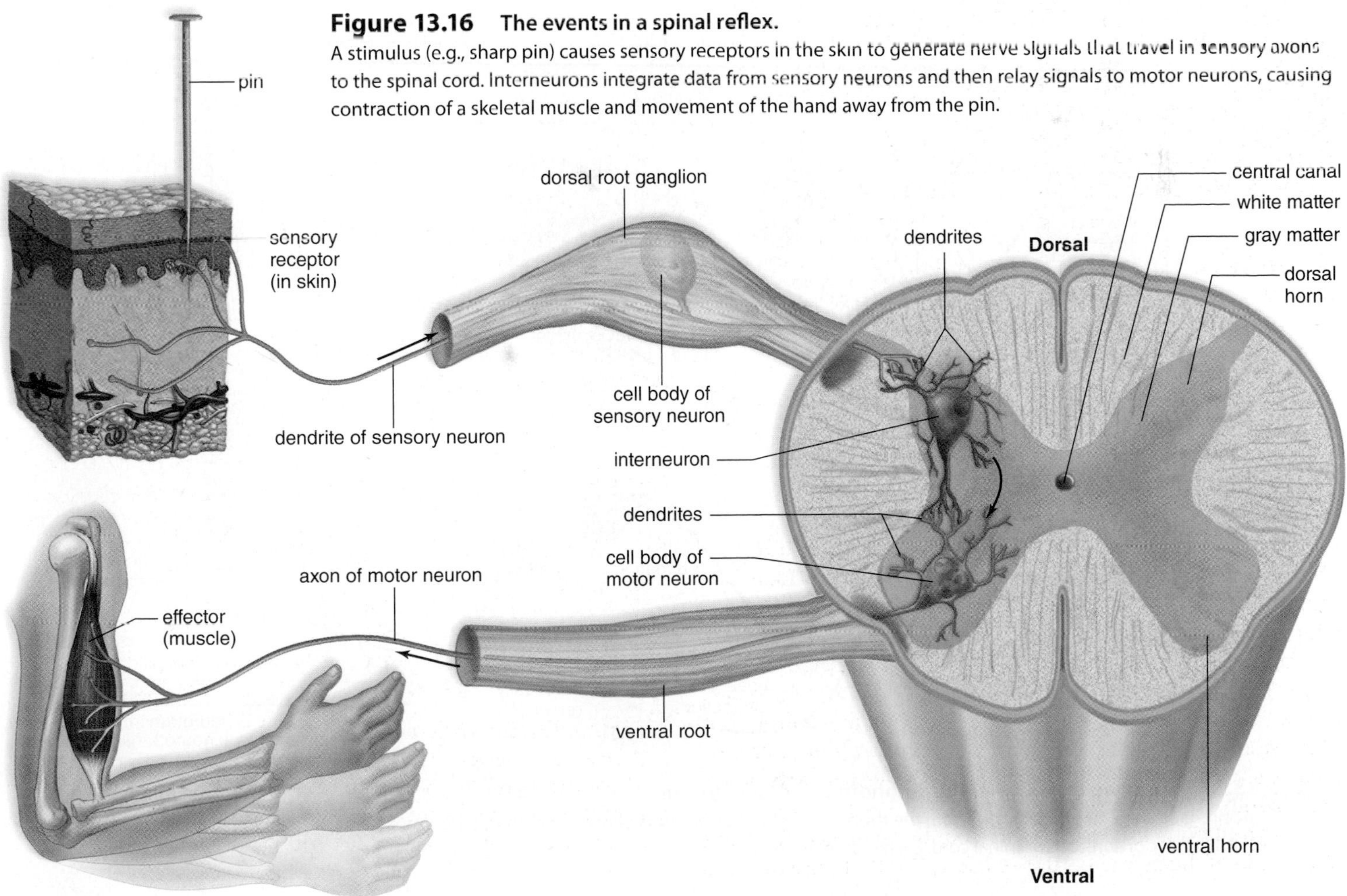

Figure 13.16 The events in a spinal reflex.
A stimulus (e.g., sharp pin) causes sensory receptors in the skin to generate nerve signals that travel in sensory axons to the spinal cord. Interneurons integrate data from sensory neurons and then relay signals to motor neurons, causing contraction of a skeletal muscle and movement of the hand away from the pin.

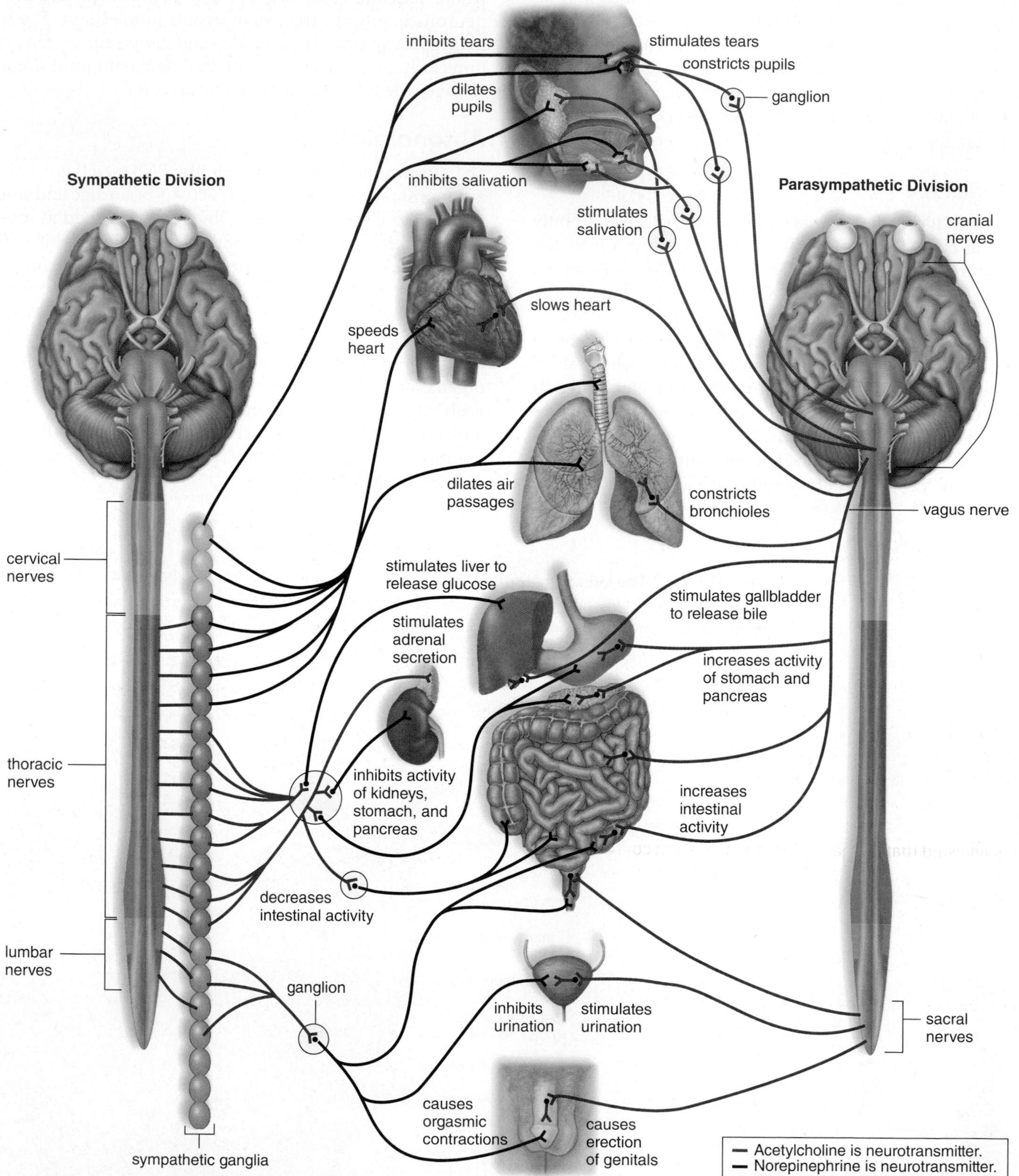

Figure 13.17 The two divisions of the autonomic nervous system.
Sympathetic preganglionic fibers (*left*) arise from the thoracic and lumbar portions of the spinal cord; parasympathetic preganglionic fibers (*right*) arise from the cranial and sacral portions of the spinal cord. Each system innervates the same organs but has contrary effects.

Reflex actions, such as those that regulate blood pressure and breathing rate, are especially important to the maintenance of homeostasis. These reflexes begin when the sensory neurons in contact with internal organs send messages to the CNS. They are completed by motor neurons within the autonomic system.

Sympathetic Division

Most preganglionic fibers of the **sympathetic division** arise from the middle, or thoracolumbar, portion of the spinal cord. They terminate almost immediately in ganglia that lie near the cord. Therefore, in this division, the preganglionic fiber is short, but the postganglionic fiber that contacts an organ is long.

The sympathetic division is especially important during emergency situations when you might be required to fight or take flight. It accelerates the heartbeat and dilates the bronchi—active muscles, after all, require a ready supply of glucose and oxygen. Sympathetic neurons inhibit the digestive organs, as well as the kidneys and urinary bladder. The activities of these organs—digestion, defecation, and urination—are not immediately necessary if you're under attack. The neurotransmitter released by the postganglionic axon is primarily norepinephrine (NE). The structure of NE is like that of epinephrine (adrenaline), an adrenal medulla hormone that usually increases heart rate and contractility.

Parasympathetic Division

The **parasympathetic division** includes a few cranial nerves (e.g., the vagus nerve) as well as fibers that arise from the sacral (bottom) portion of the spinal cord. Therefore, this division is often referred to as the craniosacral portion of the autonomic system. In the parasympathetic division, the preganglionic fiber is long, and the postganglionic fiber is short because the ganglia lie near or within the organ.

The parasympathetic division, sometimes called the housekeeper division, promotes all the internal responses we associate with a relaxed state. For example, it causes the pupil of the eye to contract, promotes digestion of food, and slows heart rate. It has been suggested that the parasympathetic system could be called the *rest-and-digest* system. The neurotransmitter used by the parasympathetic division is acetylcholine (ACh).

The Somatic Versus the Autonomic Systems

Recall that the PNS includes the somatic system and the autonomic system. Table 13.1 compares the features and functions of the somatic motor pathway with the motor pathways of the autonomic system.

APPLICATIONS AND MISCONCEPTIONS

How does aspirin work?

Aspirin is made of a chemical called acetylsalicylic acid (ASA). When tissue is damaged, it produces large amounts of a type of fatty acid called prostaglandin. Prostaglandin acts as a signal to the nervous system that tissue damage has occurred, which is interpreted by the brain as pain. Prostaglandins are manufactured in the cell by an enzyme called COX (cyclooxygenase). ASA reduces the capabilities of this enzyme, thus lowering the amount of prostaglandin produced and the perception of pain.

CHECK YOUR PROGRESS 13.4

1. Contrast cranial and spinal nerves.
2. Detail the fastest way for you to react to a stimulus.
3. Predict what might happen to homeostasis if the autonomic nervous system failed.

CONNECTING THE CONCEPTS

For more on the interaction of the PNS with the other systems of the body, refer to the following discussions:

Section 5.3 explores how the divisions of the autonomic system regulate the heart rate and help maintain homeostasis.

Section 9.5 examines how signals between the brain and the diaphragm control the rate of breathing.

Section 14.1 provides an overview of the types of sensory inputs processed by the peripheral nervous system.

Table 13.1 Comparison of Somatic Motor and Autonomic Motor Pathways

		Autonomic Motor Pathways	
	Somatic Motor Pathway	**Sympathetic**	**Parasympathetic**
Type of control	Voluntary/involuntary	Involuntary	Involuntary
Number of neurons per message	One	Two (preganglionic shorter than postganglionic)	Two (preganglionic longer than postganglionic)
Location of motor fiber	Most cranial nerves and all spinal nerves	Thoracolumbar spinal nerves	Cranial (e.g., vagus) and sacral spinal nerves
Neurotransmitter	Acetylcholine	Norepinephrine	Acetylcholine
Effectors	Skeletal muscles	Smooth and cardiac muscle, glands and organs	Smooth and cardiac muscle, glands and organs

13.5 Drug Therapy and Drug Abuse

LEARNING OUTCOMES

Upon completion of this section, you should be able to

1. Explain the ways that drugs interact with the nervous system.
2. Classify drugs as to whether they have a depressant, stimulant, or psychoactive effect on the nervous system.
3. List the long-term effects of drug use on the body.

As you are reading these words, synapses throughout your brain are organizing, integrating, and cataloging the information you take in. Neurotransmitters at these synapses control the firing of countless action potentials, thus creating a network of neural circuits. It is amazing to realize that all thoughts, feelings, and actions of a human are dependent on neurotransmitters in the CNS and PNS. By modifying or controlling synaptic transmission, a wide variety of drugs with neurological activity, both legal pharmaceuticals and illegal drugs of abuse, can alter mood, emotional state, behavior, and personality.

Drug Mode of Action

As mentioned previously (page 289), there are more than 100 known neurotransmitters. The most widely studied neurotransmitters to date are acetylcholine, norepinephrine, dopamine, serotonin, and gamma-aminobutyric acid (GABA). Acetylcholine is an essential CNS neurotransmitter for memory circuits in the limbic system. Norepinephrine is important to dreaming, waking, and mood. The neurotransmitter **dopamine** plays a central role in the brain's regulation of mood. Dopamine is also the basal nuclei neurotransmitter that helps to organize coordinated movements. Serotonin is involved in thermoregulation, sleeping, emotions, and perception. GABA is an abundant inhibitory neurotransmitter in the CNS.

Neuromodulators are naturally occurring molecules that block the release of a neurotransmitter or modify a neuron's response to a neurotransmitter. Two well-known neuromodulators are *substance P* and *endorphins.* Substance P is a neuropeptide that is released by sensory neurons when pain is present. Endorphins block the release of substance P and serve as natural painkillers. Endorphins are produced by the brain during times of physical and/or emotional stress. They are associated with the "runner's high" of joggers.

Both pharmaceuticals and illegal drugs have several basic modes of action:

- They promote the action of a neurotransmitter, usually by increasing the amount of neurotransmitter at a synapse. Examples include drugs such as alprazolam (Xanax) and diazepam (Valium), which increase GABA. These medications are used for panic attacks and anxiety. Reduced levels of norepinephrine and serotonin are linked to depression. Drugs such as fluoxetine (Prozac), paroxetine (Paxil), and duloxetine (Cymbalta) allow norepinephrine and/or serotonin to accumulate at the synapse, which explains their effectiveness as anti-depressants. Alzheimer disease causes a slow, progressive loss of memory (see Chapter 17). Drugs used for Alzheimer disease allow acetylcholine to accumulate at synapses in the limbic system.

  Video
  Apples for Alzheimers

- They interfere with or decrease the action of a neurotransmitter. For instance, antipsychotic drugs used for the treatment of schizophrenia decrease the activity of dopamine. The caffeine in coffee, chocolate, and tea keeps us awake by interfering with the effects of inhibitory neurotransmitters in the brain.
- They replace or mimic a neurotransmitter or neuromodulator. The opiates—namely, codeine, heroin, and morphine—bind to endorphin receptors and in this way reduce pain and produce a feeling of well-being.

Ongoing research into neurophysiology and neuropharmacology (the study of nervous system function and the way drugs work in the nervous system) continues to provide evidence that mental illnesses are caused by imbalances in neurotransmitters. These studies will undoubtedly improve treatments for mental illness, as well as provide insight into the problem of drug abuse.

Drug Abuse

Like mental illness, drug abuse is also linked to neurotransmitter levels. As mentioned previously, the neurotransmitter dopamine is essential for mood regulation. Dopamine plays a central role in the working of the brain's built-in *reward circuit.* The reward circuit is a collection of neurons that, under normal circumstances, promotes healthy, pleasurable activities, such as consuming food. It's possible to abuse behaviors such as eating, spending, or gambling because the behaviors stimulate the reward circuit and make us feel good. Drug abusers take drugs that artificially affect the reward circuit to the point that they neglect their basic physical needs in favor of continued drug use.

Drug abuse is apparent when a person takes a drug at a dose level and under circumstances that increase the potential for a harmful effect. Drug abusers are apt to display a psychological and/or physical dependence on the drug. Psychological dependence is apparent when a person craves the drug, spends time seeking the drug, and takes it regularly. With physical dependence, formerly called "addiction," the person has become tolerant to the drug. More is needed to get the same effect, and withdrawal symptoms occur when he or she stops taking the drug. This is true for not only teenagers and adults but also newborn babies of mothers who abuse and are addicted to drugs. Alcohol, drugs, and tobacco can all adversely affect the developing embryo, fetus, or newborn.

Alcohol

With the exception of caffeine, alcohol (ethanol) consumption is the most socially accepted form of drug use in the United States. According to a 2006 survey, nearly one-third of all U.S. high school students reported hazardous drinking (five or more drinks in one setting) during the 30 days preceding the survey. Notably, 80% of college-age young adults drink. According to a U.S. government study, drinking in college contributes to an estimated 1,400 student deaths, 500,000 injuries, and 70,000 cases of sexual assault or date rape each year.

Alcohol acts as a *depressant* on many parts of the brain (Table 13.2) by increasing the action of GABA, an inhibitory neurotransmitter. Depending on the amount consumed, the effects of alcohol on the brain can lead to a feeling of relaxation, lowered inhibitions, impaired concentration and coordination, slurred speech, and vomiting. If the blood level of alcohol becomes too high, coma or death can occur.

Beginning in about 2005, several manufacturers began selling alcoholic energy drinks. With names like Four Loko, JOOSE, and Sparks, these drinks combine fairly high levels of alcohol with caffeine and other ingredients. Although interactions between drugs can be complex, the stimulant effects of caffeine can counteract some of the depressant effects of alcohol, so that users feel able to drink more. Because caffeine does not reduce the intoxicating effects of alcohol, many state legislatures are banning these products, and in November 2010, the U.S. Food and Drug Administration warned several manufacturers that they would no longer be allowed to mix caffeine with alcohol in their products.

Nicotine

About 23% of U.S. high school students, and 8% of middle school students, reported smoking cigarettes in 2006. Young adults between age 18 and 25 reported the highest tobacco usage of any age group, at 45%. When tobacco is smoked or chewed, nicotine is rapidly delivered throughout the body. It causes a release of epinephrine from the adrenal glands, increasing blood sugar and causing the initial feeling of stimulation. As blood sugar falls, depression and fatigue set in, causing the user to seek more nicotine. In the CNS, nicotine stimulates neurons to release dopamine, a neurotransmitter that promotes a temporary sense of pleasure, and reinforces dependence on the drug. About 70% of people who try smoking become addicted.

As mentioned in earlier chapters, smoking is strongly associated with serious diseases of the cardiovascular and respiratory systems. Once addicted, however, only 10–20% of smokers are able to quit. Most medical approaches to quitting smoking involve the administration of nicotine in safer forms, like skin patches, gum, or a newly developed nicotine inhaler, so that withdrawal symptoms can be minimized while dependence is gradually reduced. As of 2010, an antinicotine vaccine called NicVAX was showing promise in early clinical trials. The vaccine stimulates the production of antibodies that prevent nicotine from entering the brain.

Table 13.2 Drug Influence on CNS and Route

Substance	Effect	Mode of Transmission
Alcohol	Depressant	Drink
Nicotine	Stimulant	Smoked or smokeless tobacco
Cocaine	Stimulant	Sniffed/snorted, injected, or smoked
Methamphetamine/ Ecstasy	Stimulant	Smoked or pill form
Heroin	Depressant	Sniffed/snorted, injected, or smoked
Marijuana/K2	Psychoactive	Smoked or consumed

Cocaine and Crack

Cocaine is an alkaloid derived from the shrub *Erythroxylon coca.* Approximately 35 million Americans have used cocaine by sniffing/snorting, injecting, or smoking. Cocaine is a powerful stimulant in the CNS that interferes with the reuptake of dopamine at synapses, increasing overall brain activity (Fig. 13.18). The result is a rush of well-being that lasts from 5 to 30 minutes.

"Crack" is the street name given to cocaine that is processed to a free base form for smoking. The term crack refers to the crackling sound heard when the drug is smoked. Smoking allows high doses of the drug to reach the brain rapidly,

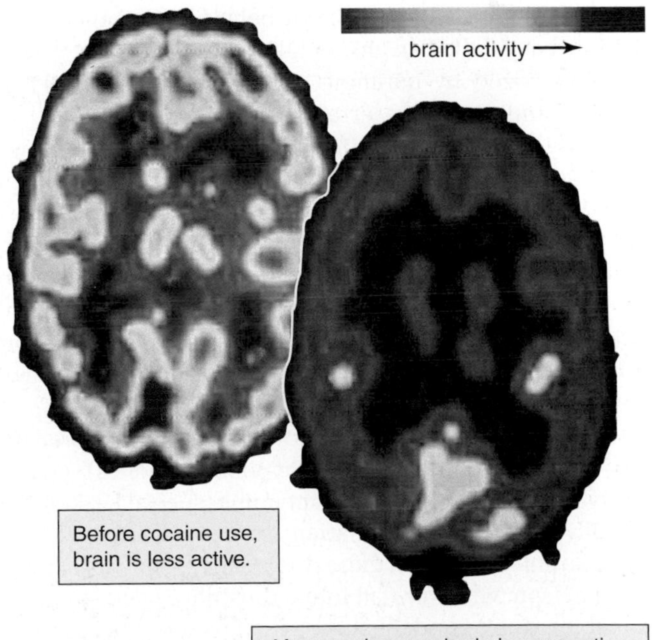
brain activity
Before cocaine use, brain is less active.
After cocaine use, brain is more active.

Figure 13.18 **Cocaine use.**
Brain activity before and after the use of cocaine.

providing an intense and immediate high, or "rush." Approximately 8 million Americans use crack.

A cocaine binge is a period in which a user takes the drug at ever-higher doses. The user is hyperactive, with little desire for food or sleep but an increased sex drive. This is followed by a crash period, characterized by fatigue, depression, irritability, and lack of interest in sex. In fact, men who use cocaine often become impotent.

Cocaine is highly addictive; with continued use, the brain makes less dopamine to compensate for a seemingly endless supply. The user experiences withdrawal symptoms and an intense craving for cocaine. Overdosing on cocaine can cause cardiac and/or respiratory arrest.

Methamphetamine and Ecstasy

Methamphetamine and Ecstasy are considered club or party drugs. Methamphetamine (commonly called meth or crank) is a powerful CNS stimulant. Meth is often produced in makeshift home laboratories, usually starting with ephedrine or pseudoephedrine, common ingredients in many cold and asthma medicines. As a result, many states have passed laws making these medications more difficult to purchase. The number of toxic chemicals used to prepare the drug makes a former meth lab site hazardous to humans and to the environment. Over 9 million people in the United States have used methamphetamine at least once in their lifetime. It is available as a powder that can be snorted, or as crystals (crystal meth or ice) that can be smoked.

The structure of methamphetamine is similar to that of dopamine, and the most immediate effect of taking meth is a rush of euphoria, energy, alertness, and elevated mood. However, this is typically followed by a state of agitation that, in some individuals, leads to violent behavior. Chronic use can result in what is called an amphetamine psychosis, characterized by paranoia, hallucinations, irritability, and aggressive, erratic behavior.

Video Meth and the Brain

Ecstasy is the street name for MDMA (methylenedioxymethamphetamine), which is chemically similar to methamphetamine. Many users say that "X," taken as a pill that looks like an aspirin or candy, increases their feelings of well-being and love for other people. However, it has many of the same side effects as other stimulants, plus it can interfere with temperature regulation, leading to hyperthermia, high blood pressure, and seizures. In June 2010, a 15-year-old girl died from Ecstasy-related causes after attending a large rave party in Los Angeles.

Drugs with sedative effects, known as date rape or predatory drugs, include flunitrazepam (Rohypnol, or roofies), gamma-hydroxybutyric acid (GHB), and ketamine (special K). Ketamine is actually a drug that veterinarians sometimes use to perform surgery on animals. Any of these drugs can be given to an unsuspecting person, who may fall into a dreamlike state where they are unable to move, and thus are vulnerable to sexual assault.

Heroin

Heroin is derived from the resin or sap of the opium poppy plant, which is widely grown in a region from Turkey to Southeast Asia, and in parts of Latin America. Drugs derived from opium are called opiates, a class that also includes morphine and codeine. After heroin is injected, snorted, or smoked, a feeling of euphoria, along with relief of any pain, occurs within a few minutes. It is estimated that 4 million Americans have used heroin some time in their lives, and over 300,000 people use heroin annually.

As with other drugs of abuse, addiction is common. Heroin binds to receptors meant for the endorphins, naturally occurring neurotransmitters that kill pain and produce feelings of tranquility. With repeated heroin use, the body's production of endorphins decreases. Tolerance develops so that the user needs to take more of the drug just to prevent withdrawal symptoms (tremors, restlessness, cramps, vomiting), and the original euphoria is no longer felt. Long-term users commonly acquire hepatitis, HIV/AIDS, and various bacterial infections due to the use of shared needles, and heavy users may experience convulsions and death by respiratory arrest.

Heroin addiction can be treated with synthetic opiate compounds, such as methadone or buprenorphine and naloxone (Suboxone), that decrease withdrawal symptoms and block heroin's effects. However, methadone itself can be addictive, and methadone-related deaths are on the rise.

Marijuana and K2

The dried flowering tops, leaves, and stems of the marijuana plant, *Cannabis sativa,* contain and are covered by a resin that is rich in THC (tetrahydrocannabinol). The names cannabis and marijuana apply to either the plant or THC. Marijuana can be ingested, but usually it is smoked in a cigarette called a "joint." Although the drug was banned in the United States in 1937, an estimated 22 million Americans use marijuana, making it the most commonly used illegal drug in the United States. Beginning with California in 1996, several states have legalized its use for medical purposes, such as treating cancer, AIDS, or glaucoma. In 2005, however, the Supreme Court ruled that patients prescribed medical marijuana can still be prosecuted by federal agencies.

Researchers have found that THC binds to a receptor for anandamide, a naturally occurring neurotransmitter that is important for short-term memory processing, and perhaps for feelings of contentment. The occasional marijuana user experiences mild euphoria, along with alterations in vision and judgment. Heavy use can cause hallucinations, anxiety, depression, paranoia, and psychotic symptoms. Some researchers believe that long-term marijuana use leads to brain impairment.

In recent years, awareness is increasing about a synthetic compound called K2, or Spice. Originally synthesized by an organic chemist at Clemson University, K2 is about ten times as potent as THC. The chemical is typically sprayed onto a mixture of other herbal products, and smoked. However, because there is no regulation of how it is produced, the amount of K2 itself, or contaminants, can vary greatly. This may account for the several reports of serious medical problems and even deaths in K2 users.

CHECK YOUR PROGRESS 13.5

1. Contrast drug therapy and drug abuse.
2. List how the abuse of drugs, including alcohol and nicotine, affects the nervous system.
3. Detail several different modes of action of pharmaceutical and illegal drugs.

CONNECTING THE CONCEPTS

For more on the long-term effects of drug use on the systems of the body, refer to the following discussions:

Section 5.7 explores the negative long-term effects of smoking on the cardiovascular system.

Section 10.4 provides information on how alcohol acts as a diuretic in the urinary system.

Section 19.2 examines the relationship between smoking and alcohol use and the increased risk of cancer.

CASE STUDY CONCLUSION

The cause of multiple sclerosis (MS) is still unknown, but most researchers agree that there are most likely many contributing factors, including environmental influences, genetics, and a faulty immune system. Many individuals with MS are able to control their symptoms by using immunosuppressive medications such as beta interferons. The fact that this treatment works suggests that in many cases, MS is caused by the immune system incorrectly identifying the myelin sheaths as foreign material. The breakdown of the myelin can be detected using both the MRI and SSEP tests (presented in the case study introduction). However, environmental conditions are also suspected to cause MS. Studies have shown that the risk of contracting MS is influenced in part by where in the world you live, although the specific environmental factor or pollutant has not yet been identified. Genetics is also believed to play a role in some cases. However, most researchers believe that a defect in a single gene is unlikely. Rather, it is more likely that a certain combination of genetic factors places an individual at a higher risk of contracting MS. Though there is no cure for MS, researchers have been very successful in developing disease-modifying drugs that reduce the symptoms and allow the individual to lead a normal life.

MEDIA STUDY TOOLS

Enhance your study of this chapter with media! Visit **www.mhhe.com/maderhuman13e** and go to "Media Study Tools" for this chapter to access the following:

 **Animations**	 **Videos**	 **MP3 Files**
13.1 How the Sodium-Potassium Pump Works • Nerve Impulse • Voltage-Gated Channels and the Action Potential • Action Potential Propagation • Chemical Synapses	**13.2** Brain Bank • Brain Surgery **13.5** Apples for Alzheimers • Meth and the Brain	**13.1** Organization of the Nervous System • Nervous Tissue • Cells of the Nervous System • Synapses **13.2** The Brain • The Cerebrum

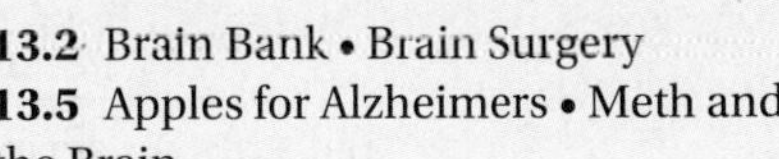

3D Animation

McGraw-Hill's 3D animation, "Neural Transmission," provides a dynamic exploration of the key concepts of this chapter and is available through McGraw-Hill Connect®.

SUMMARIZE

13.1 Overview of the Nervous System

The **nervous system**

- is divided into the **central nervous system** (**CNS**) and the **peripheral nervous system** (**PNS**).
- has three functions: (1) reception of input, (2) integration of data, and (3) generation of motor output.

Nervous Tissue

Nervous tissue contains the following two types of cells: **neurons** and **neuroglia:**

- neurons transmit nerve signals using **action potentials;** and
- neuroglia nourish and support neurons.

Anatomy of a Neuron

A **neuron** is composed of **dendrites,** a **cell body,** and an **axon.** The axons of neurons may be clustered into **nerves.** There are three types of neurons:

- **sensory neurons** take nerve signals from **sensory receptors** to the CNS;
- **interneurons** occur within the CNS; and
- **motor neurons** take nerve signals from the CNS to **effectors** (muscles or glands).

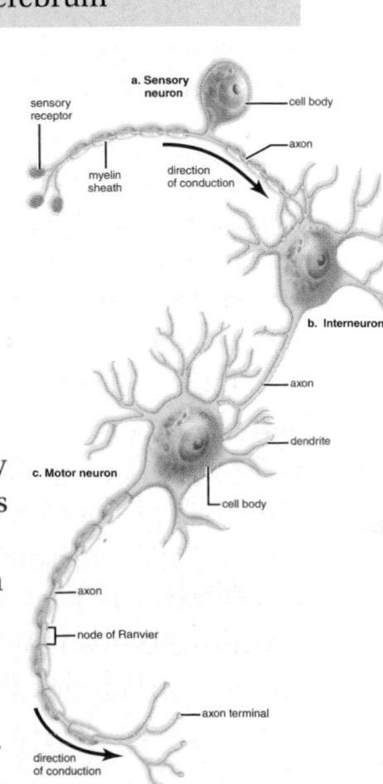

Myelin Sheath

- Long axons are covered by a **myelin sheath,** which is formed by the neuroglial cells. Gaps in the myelin sheath are called **nodes of Ranvier.**
- **Multiple sclerosis** (**MS**) occurs when the myelin sheath breaks down, causing a short-circuiting of nerve signals.

Physiology of a Neuron

Nerve signals move information within the nervous system. The generation of a nerve signal is based upon the polarity across the membrane of the neuron.

- *Resting Potential:* More Na^+ outside the axon and more K^+ inside the axon. The axon does not conduct a signal. The **resting potential** is maintained by active transport using the **sodium–potassium pump.**
- *Action Potential:* Upon receipt of a **stimulus** strong enough to overcome the **threshold,** a change in polarity across the axonal membrane as a nerve signal occurs: When Na^+ gates open, Na^+ moves to the inside of the axon, and a **depolarization** occurs. When K^+ gates open, K^+ moves to the outside of the axon, and a **repolarization** occurs.
- *Signal Propagation:* The presence of the myelin sheath speeds the movement of the nerve signal by **saltatory conduction.** After the action potential has passed, a **refractory period** occurs during which no additional action potentials may be processed.

The Synapse

- At the end of each axon is an **axon terminal,** which borders the **synapse** between another neuron or target cell.
- When a neurotransmitter is released into a **synaptic cleft,** transmission of a nerve signal occurs.
- Binding of the **neurotransmitter** to receptors in the receiving membrane causes excitation or inhibition.
- Enzymes, such as **acetylcholinesterase** (**AChE**), assist in removing the neurotransmitter from the synaptic cleft.

Neurotransmitters

- Neurotransmitters, such as **acetylcholine, norepinephrine,** and **serotonin,** are used to convey signals across the synapses.
- **Integration** is the summing of excitatory and inhibitory signals.

13.2 The Central Nervous System

The CNS receives and integrates sensory input and formulates motor output. The CNS consists of the spinal cord and brain. The CNS is protected by the **meninges,** which are filled with **cerebrospinal fluid.** This same fluid fills the four **ventricles** of the brain. In the CNS, **gray matter** contains cell bodies and nonmyelinated fibers. **White matter** contains myelinated axons that are organized as **tracts.**

The Spinal Cord

- The **spinal cord** is responsible for conduction of information to and from the **brain** and carries out reflex actions.

The Brain

The Cerebrum: The **cerebrum** has two **cerebral hemispheres** connected by the **corpus callosum.**

- Sensation, reasoning, learning and memory, and language and speech take place in the cerebrum.
- The **cerebral cortex** of each cerebral hemisphere has four lobes: frontal, parietal, occipital, and temporal.
- The **primary motor area** in the frontal lobe sends out motor commands to lower brain centers, which pass them on to motor neurons.
- The **primary somatosensory area** in the parietal lobe receives sensory information from lower brain centers in communication with sensory neurons.
- **Association areas** are located in all the lobes. The **prefrontal area** in the frontal lobe is involved in reasoning and planning of actions.
- **Wernicke's area** and **Broca's area** are two processing centers that are involved in speech.

Basal Nuclei: The **basal nuclei** integrate commands to the muscles to coordinate movement. **Parkinson disease** is associated with the degradation of neurons in this area.

The Diencephalon: The **diencephalon** contains both the hypothalamus and thalamus. The **hypothalamus** controls homeostasis. The **thalamus** sends sensory input on to the cerebrum.

The Cerebellum: The **cerebellum** coordinates skeletal muscle contractions.

The Brain Stem: The **brain stem** includes the midbrain, the pons, and the medulla oblongata.

- The **medulla oblongata** and **pons** have centers for breathing and the heartbeat.
- The **midbrain** serves as a relay station between the cerebrum and spinal cord or cerebellum.
- The **reticular formation** is part of the reticular activating system (RAS) that transfers sensory signals to higher processing centers in the brain.

13.3 The Limbic System and Higher Mental Functions

- The **limbic system,** located deep in the brain, is involved in determining emotions and higher mental functions, such as **learning.**
- The **amygdala** determines when a situation deserves the emotion we call "fear."
- The **hippocampus** is particularly involved in storing and retrieving memories. Alzheimer disease affects the cells of the hippocampus.
- A **memory** may be processed as either **short-term memory** or **long-term memory.** Long-term memory may be classified as either **semantic memory** or **episodic memory. Skill memory** is involved with processes like riding a bike.

13.4 The Peripheral Nervous System

- The PNS contains only nerves and **ganglia** (sing., **ganglion**).
- **Cranial nerves** take impulses to and from the brain.
- **Spinal nerves** take impulses to and from the spinal cord.
- The PNS is divided into the somatic system and the autonomic system.

Somatic System

The **somatic system** serves the skin, skeletal muscles, and tendons.

- Some actions are due to **reflexes,** which are automatic and involuntary.
- Other actions are voluntary and originate in the cerebral cortex.

Autonomic System

The **autonomic system** is further divided into the **sympathetic division** and the **parasympathetic division.**

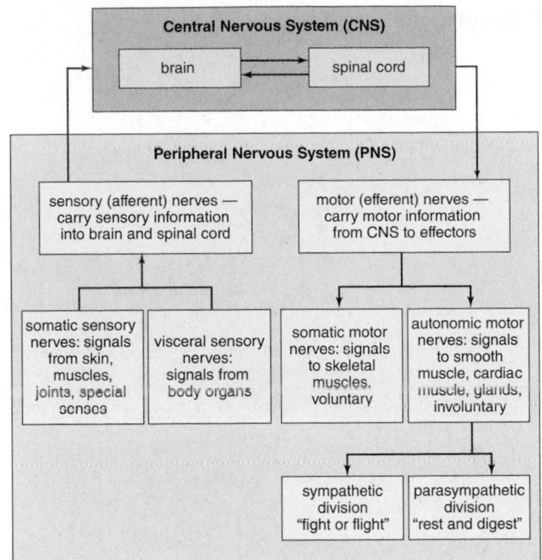

Sympathetic Division: Responses that occur during times of stress.

Parasympathetic Division: Responses that occur during times of relaxation.

- Actions in these divisions are involuntary and automatic.
- These divisions innervate internal organs.
- Two neurons and one ganglion are used for each impulse.

13.5 Drug Therapy and Drug Abuse

- Neurotransmitters, such as acetylcholine, norepinephrine, **dopamine,** and serotonin, play an important role in moving signals within the nervous system.
- Neuromodulators block the release of a neurotransmitter.
- Neurological drugs promote, prevent, or mimic the action of a particular neurotransmitter.
- Drugs, such as alcohol, nicotine, and marijuana, may have depressant, stimulant, or psychoactive effects.
- Dependency occurs when the body compensates for the presence of neurological drugs.

ASSESS

Testing Your Knowledge of the Concepts

1. What are the three functions of the nervous system? (pages 284–285)
2. What are the functions performed by the three types of neurons? Describe the structure and functions of the three parts of a neuron. (page 285)
3. What is the resting potential, and how is it created and maintained? (pages 286–287)
4. What is an action potential? Describe the two parts of the process. (page 287)
5. Explain transmission of a nerve signal across a synapse. (pages 288–289)
6. List and describe the functions of the major neurotransmitters. (page 289)
7. Describe the structure and functions of the spinal cord. (pages 291–292)
8. Name the major parts of the brain, and give a function for each part. (pages 293–297)
9. What are the lobes of the cerebrum? Give the location and function of the following areas: primary motor area, primary somatosensory area, primary visual area, primary auditory area, primary taste area, and primary olfactory area. (pages 293–296)
10. What are association areas, and what roles do they play? (pages 294–295)
11. What is the reticular formation, and what is the role of the RAS? (pages 296–297)
12. What is the limbic system? What are the two main parts and their functions? (pages 297–298)
13. Describe the various types of memory. (page 298)
14. What types of nerves make up the PNS? How many of each type are there, and what areas of the body does each serve? (pages 299–300)
15. Distinguish between the somatic and the autonomic nervous systems as to structure, effectors, neurotransmitters, and functions. (pages 300–303)
16. Trace the path of a reflex arc. (page 301)
17. Describe the physiological effects and mode of action of alcohol, nicotine, cocaine, methamphentamine and ectasy, heroin, and marijuana (K2). (pages 305–306)
18. What type of neuron lies completely in the CNS?
    a. motor neuron
    b. interneuron
    c. sensory neuron
19. Which of the following neuron parts receive(s) signals from sensory receptors of other neurons?
    a. cell body
    b. axon
    c. dendrites
    d. Both a and c are correct.
20. The neuroglial cells that form myelin sheaths in the PNS are
    a. oligodendrocytes.
    b. ganglionic cells.
    c. Schwann cells.
    d. astrocytes.
    e. microglia.
21. Which of these correctly describes the distribution of ions on either side of an axon when it is not conducting a nerve signal?
    a. more sodium ions (Na^+) outside and more potassium ions (K^+) inside
    b. more K^+ outside and less Na^+ inside
    c. charged protein outside; Na^+ and K^+ inside
    d. Na^+ and K^+ outside and water only inside
    e. chloride ions (Cl^-) outside and K^+ and Na^+ inside
22. When the action potential begins, sodium gates open, allowing Na^+ to cross the membrane. Now the polarity changes to
    a. negative outside and positive inside.
    b. positive outside and negative inside.
    c. neutral outside and positive inside.
    d. All of the choices are correct.
23. Repolarization of an axon during an action potential is produced by
    a. inward diffusion of Na^+.
    b. outward diffusion of K^+.
    c. inward active transport of Na^+.
    d. active extrusion of K^+.

24. Transmission of the nerve signal across a synapse is accomplished by the
    a. movement of Na^+ and K^+.
    b. release of a neurotransmitter by a dendrite.
    c. release of a neurotransmitter by an axon.
    d. release of a neurotransmitter by a cell body.
    e. All of the choices are correct.
25. Which of the following cerebral areas is not correctly matched with its function?
    a. occipital lobe—vision
    b. parietal lobe—somatosensory area
    c. temporal lobe—primary motor area
    d. frontal lobe—Broca's motor speech area
26. The hypothalamus does not
    a. control skeletal muscles.
    b. regulate thirst.
    c. control the pituitary gland.
    d. regulate body temperature.
27. Which of the following brain regions is not correctly matched to its function?
    a. medulla oblongata regulates heartbeat, breathing, and blood pressure
    b. cerebellum coordinates voluntary muscle movements
    c. thalamus secretes melatonin that regulates daily body rhythms
    d. midbrain acts as reflex center for visual, auditory, and tactile responses
28. Which of these are the first and last elements in a spinal reflex?
    a. axon and dendrite
    b. sensory receptor and muscle effector
    c. ventral horn and dorsal horn
    d. brain and skeletal muscle
    e. motor neuron and sensory neuron
29. Which of the following is incorrect regarding the autonomic nervous system?
    a. It includes both cranial and spinal nerves.
    b. It is divided into sympathetic and parasympathetic divisions.
    c. Most organs under ANS control are under dual innervation, that is, by both sympathetic and parasympathetic divisions.
    d. Major responsibilities are regulation and repair of cardiac muscle, smooth muscles, organs, and glands.
30. The sympathetic division of the autonomic system does not cause
    a. the liver to release glycogen.
    b. dilation of bronchioles.
    c. the gastrointestinal tract to digest food.
    d. an increase in the heart rate.
31. Label this diagram.

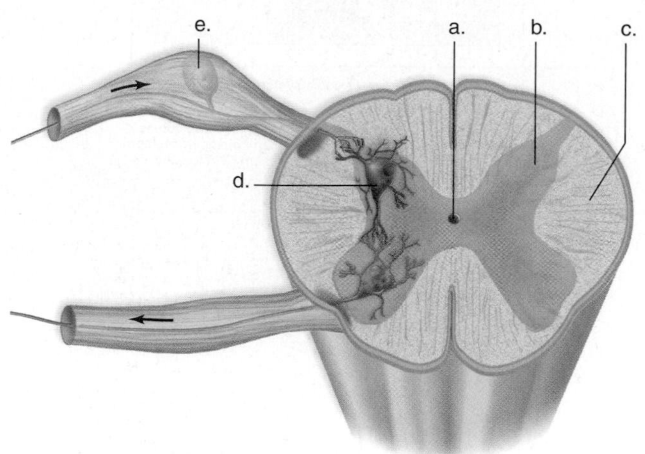

ENGAGE

Thinking Critically About the Concepts

Demyelinating disorders, like the multiple sclerosis discussed in the chapter opener, are the subject of numerous research projects. Many investigations focus on the cells that create myelin: the Schwann cells of the PNS and oligodendrocytes in the CNS. Other studies focus on immune system cells that attack this myelin sheath. The goal of this research is to determine how to restore lost myelin, which might help (or possibly cure) folks living with MS and other demyelinating diseases. Investigations into the role played by the sheath in nerve regeneration may offer hope to victims of spinal cord injury.

1. Why are impulses transmitted more quickly down a myelinated axon than an unmyelinated axon?
2. A buildup of very long chain saturated fatty acids is believed to be the cause of myelin loss in adrenoleukodystrophy. This rare disease is a demyelinating disorder like MS. It is the subject of the film *Lorenzo's Oil.* This real-life drama focuses on Lorenzo Odone, whose parents successfully developed a diet that helped their son.
    a. From your study of chemistry in Chapter 2, a fatty acid is a part of what type of molecule?
    b. What distinguishes a saturated fatty acid from an unsaturated fatty acid?
    c. From your study of nutrition in Chapter 8, what types of foods would contain saturated fatty acids?
3. Why would you expect the motor skills of a child to improve as myelination continues during early childhood development?

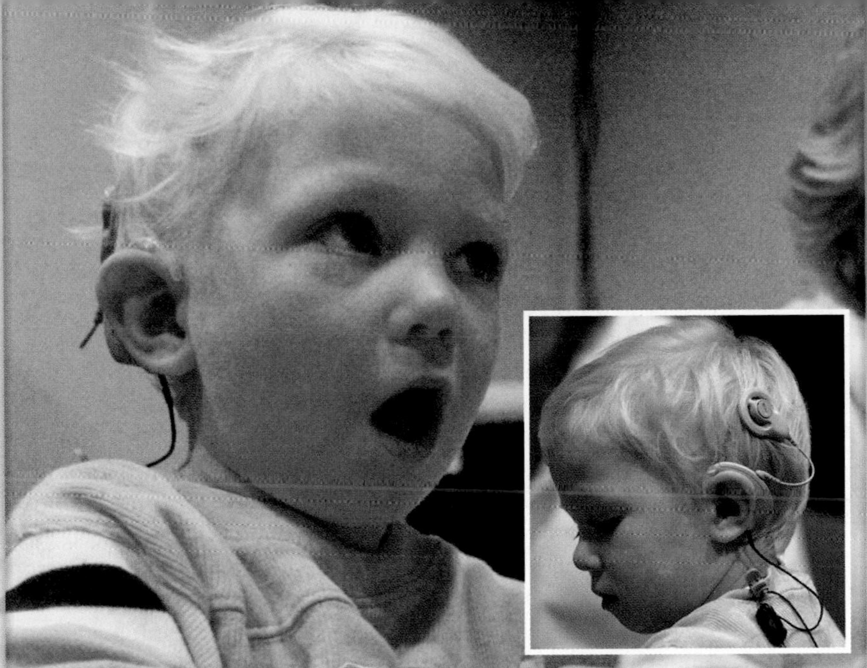

CHAPTER

14

Senses

CASE STUDY COCHLEAR IMPLANTS

Jacob and Marlene were married for almost a year when they learned that Marlene was pregnant with their first child. Almost immediately, they began discussing what they would do if their baby was born with hearing problems. Marlene had been born deaf because of a genetic condition called *nonsyndromic deafness.* The term *nonsyndromic* means that Marlene's deafness was not caused by another medical problem such as diabetes or a genetic syndrome such as Charcot–Marie–Tooth disease or Waardenburg syndrome. Researchers now know that many forms of nonsyndromic deafness are caused by a deletion in a gene on chromosome 13 called *GJB2*. *GJB2* produces a protein that is responsible for the formation of the cochlea, the key structure in hearing. However, when Marlene was a child, little was known as to the causes of deafness, so Marlene had compensated for her disability by learning sign language and had even become a teacher at a school for the deaf. Since then, advances in technology and medicine have made it possible to provide hearing for babies with many forms of deafness.

One option for treatment is cochlear implants. Cochlear implants are different from hearing aids, which simply amplify sounds. A cochlear implant consists of an external device that is surgically implanted under the skin. The external part picks up sounds from the environment and converts them to electrical impulses, which are sent directly to different regions of the auditory nerve and then to the brain, as we will see in this chapter. According to the National Institute on Deafness and other Communication Disorders (NIDCD), an estimated 188,000 people worldwide have received cochlear implants, including 25,500 children and 41,500 adults in the United States.

As you read through the chapter, think about the following questions:

1. Where is the cochlea located?
2. What are the roles of the cochlea and auditory nerve in hearing?
3. How might the technology associated with cochlear implants also be used to help people with vision problems?

CHAPTER CONCEPTS

14.1 Overview of Sensory Receptors and Sensations
Each type of sensory receptor detects a particular type of stimulus. When stimulation occurs, sensory receptors initiate nerve signals transmitted to the CNS. Sensation occurs when nerve impulses reach the cerebral cortex.

14.2 Somatic Senses
The somatic senses include proprioception, an awareness of the muscles and joints, as well as the sensations of touch, pressure, temperature, and pain.

14.3 Senses of Taste and Smell
Taste and smell involve the activity of sensory receptors in the mouth and nose, respectively.

14.4 Sense of Vision
Vision depends on the sensory receptors in the eyes, the optic nerves, and the visual cortex.

14.5 Sense of Hearing
Hearing depends on sensory receptors in the ears, the cochlear nerve, and the auditory cortex.

14.6 Sense of Equilibrium
The inner ear contains sensory receptors for our sense of equilibrium.

BEFORE YOU BEGIN

Before beginning this chapter, take a few moments to review the following discussions:

Figure 13.1 What are the roles of the central and peripheral nervous systems in the body?

Section 13.2 What is the role of the primary somatosensory area of the cerebral cortex?

Section 13.2 How does the cerebellum help maintain balance?

14.1 Overview of Sensory Receptors and Sensations

LEARNING OUTCOMES

Upon completion of this section, you should be able to

1. List the four categories of sensory receptors and describe what each responds to.
2. Distinguish between perception and sensation.
3. Explain the purpose of integration and sensory adaptation.

A **sensory receptor** is able to convert some type of event, or *stimulus* (sing., stimuli), occurring in the environment into a nerve impulse. This process is known as *sensory transduction.* Some sensory receptors are modified neurons, and others are specialized cells closely associated with neurons. Sensory receptors may detect stimuli originating from both the internal and external environments. **Exteroceptors** are sensory receptors that detect stimuli from outside the body, such as those that result in taste, smell, vision, hearing, and equilibrium (Table 14.1). **Interoceptors** receive stimuli from inside the body. Examples of interoceptors are the baroreceptors (also called pressoreceptors) that respond to changes in blood pressure, osmoreceptors that monitor the body's water-salt balance, and chemoreceptors that monitor the pH of the blood.

Interoceptors are directly involved in homeostasis and are regulated by a negative feedback mechanism (see Fig. 4.16). For example, when blood pressure rises, pressoreceptors signal a regulatory center in the brain. The brain responds by sending out nerve signals to the arterial walls, causing their smooth muscle to relax. The blood pressure then falls. Once blood pressure is returned to normal, the pressoreceptors are no longer stimulated.

Exteroceptors, such as those in the eyes, ears, and skin, continuously send messages to the central nervous system. In this way, they keep us informed regarding the conditions of the external environment.

Types of Sensory Receptors

Sensory receptors in humans can be classified into four categories: chemoreceptors, photoreceptors, mechanoreceptors, and thermoreceptors.

Chemoreceptors respond to chemical substances in the immediate vicinity. As Table 14.1 indicates, taste and smell, which detect external stimuli, use chemoreceptors. However, so do various other organs sensitive to internal stimuli. Chemoreceptors that monitor blood pH are located in the carotid arteries and aorta. If the pH lowers, the breathing rate increases. As more carbon dioxide is exhaled, the blood pH rises. **Nociceptors** (pain receptors) are a type of chemoreceptor. They are naked dendrites that respond to chemicals released by damaged tissues. Nociceptors are protective because they alert us to possible danger. For example, without the pain of appendicitis, we might never seek the medical help needed to avoid a ruptured appendix.

Photoreceptors respond to light energy. Our eyes contain photoreceptors that are sensitive to light rays and thereby provide us with a sense of vision. Stimulation of the photoreceptors known as rod cells results in black-and-white vision. Stimulation of the photoreceptors known as cone cells results in color vision.

Mechanoreceptors are stimulated by mechanical forces, which most often result in pressure of some sort. When we hear, airborne sound waves are converted to fluid-borne pressure waves that can be detected by mechanoreceptors in the inner ear. Mechanoreceptors are responding to fluid-borne pressure waves when we detect changes in gravity and motion, helping us keep our balance. These receptors are in the vestibule and semicircular canals of the inner ear.

The sense of touch depends on pressure receptors sensitive to either strong or slight pressures. Baroreceptors located in certain arteries detect changes in blood pressure, and stretch receptors in the lungs detect the degree of lung inflation. Proprioceptors respond to the stretching of muscle fibers, tendons, joints, and ligaments. Signals from proprioceptors make us aware of the position of our limbs.

Thermoreceptors located in the hypothalamus and skin are stimulated by changes in temperature. They respond to both heat and cold, and play a major role in the regulation of internal body temperature (see Fig. 4.18).

Table 14.1 Exteroceptors

Sensory Receptor	Stimulus	Category	Sense	Sensory Organ
Taste cells	Chemicals	Chemoreceptor	Taste	Taste bud
Olfactory cells	Chemicals	Chemoreceptor	Smell	Olfactory epithelium
Rod cells and cone cells in retina	Light rays	Photoreceptor	Vision	Eye
Hair cells in spiral organ of the inner ear	Sound waves	Mechanoreceptor	Hearing	Ear
Hair cells in semicircular canals of the inner ear	Motion	Mechanoreceptor	Rotational equilibrium	Ear
Hair cells in vestibule of the inner ear	Gravity	Mechanoreceptor	Gravitational equilibrium	Ear

How Sensation Occurs

Sensory receptors respond to environmental stimuli by generating nerve signals. When the nerve signals arrive at the cerebral cortex of the brain, **sensation,** the conscious perception of stimuli, occurs.

MP3 Sensations and Receptors

As discussed in Chapter 13, sensory receptors are the first element in a reflex arc. We are aware of a reflex action only when sensory information reaches the brain. At that time, the brain integrates this information with other information received from other sensory receptors. Consider what happens if you burn yourself and quickly remove your hand from a hot stove. The brain receives information not only from your skin but also from your eyes, nose, and all sorts of sensory receptors.

Some sensory receptors are free nerve endings or encapsulated nerve endings, and others are specialized cells closely associated with neurons. Often, the plasma membrane of a sensory receptor contains receptor proteins that react to the stimulus. For example, the receptor proteins in the plasma membrane of chemoreceptors bind to certain chemicals. When this happens, ion channels open, and ions flow across the plasma membrane. If the stimulus is sufficient, nerve signals begin and are carried by a sensory nerve fiber within the PNS to the CNS (Fig. 14.1). The stronger the stimulus, the greater the frequency of nerve signals. Nerve signals that reach the spinal cord first are conveyed to the brain by ascending tracts. If nerve signals finally reach the cerebral cortex, sensation and perception occur.

All sensory receptors initiate nerve signals. The sensation that results depends on the part of the brain receiving the nerve signals. Nerve signals that begin in the optic nerve eventually reach the visual areas of the cerebral cortex. Thereafter, we see objects. Nerve signals that begin in the auditory nerve eventually reach the auditory areas of the cerebral cortex. We hear sounds when the auditory cortex is stimulated. If it were possible to switch these nerves, stimulation of the eyes would result in hearing!

Before sensory receptors initiate nerve signals, they also carry out **integration,** the summing up of signals. One type of integration is called **sensory adaptation,** a decrease in response to a stimulus. We have all had the experience of smelling an odor when we first enter a room and then later not being aware of it. When sensory adaptation occurs, sensory receptors send fewer impulses to the brain. Without these impulses, the sensation of the stimuli is decreased. The functioning of our sensory receptors makes a significant contribution to homeostasis. Without sensory input, we would not receive information about our internal and external environment. This information leads to appropriate reflex and voluntary actions to keep the internal environment constant.

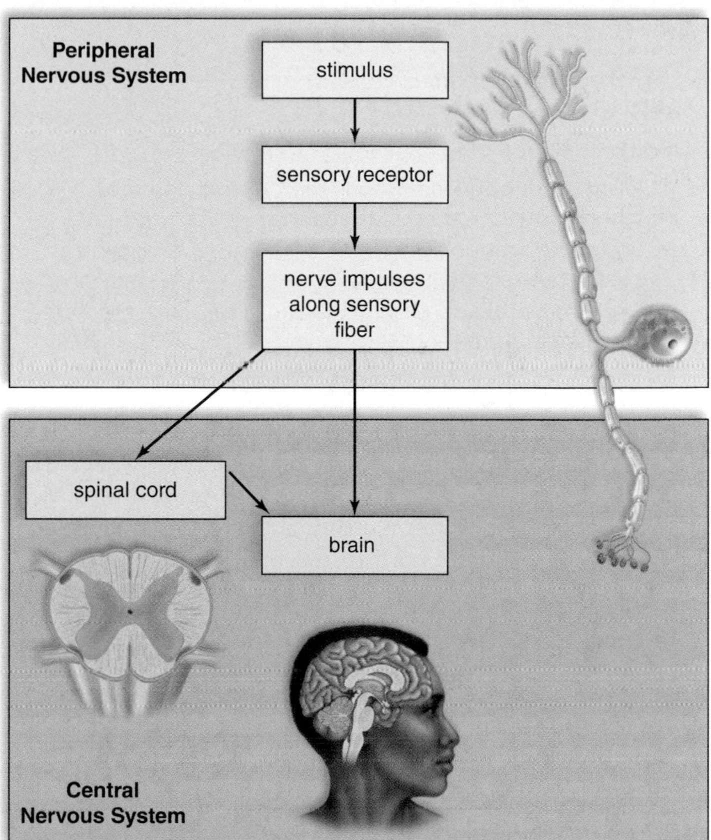

Figure 14.1 The role of the CNS and PNS in sensation and sensory perception.

After detecting a stimulus, sensory receptors initiate nerve signals within the peripheral nervous system (PNS). These signals give the central nervous system (CNS) information about the external and internal environment. The CNS integrates all incoming information, and then initiates a motor response to the stimulus.

CHECK YOUR PROGRESS 14.1

1. Describe the functions of the four different types of sensory receptors.
2. Explain how perception differs from sensation and give an example.
3. Summarize the importance of sensory receptors in the maintenance of homeostasis in the body.

CONNECTING THE CONCEPTS

For more information on the regions of the brain that are associated with sensation, refer to the following discussions:

Section 13.2 describes the location and function of the reticular activating system (RAS).

Figure 13.10 illustrates the somatosensory regions of the cerebral cortex.

Figure 13.16 illustrates the portions of the peripheral nervous system that are involved in a reflex arc.

14.2 Somatic Senses

LEARNING OUTCOMES

Upon completion of this section, you should be able to

1. Distinguish between proprioceptors and cutaneous receptors with regard to function.
2. State the location and general function of each type of cutaneous receptor.
3. Explain the role of nociceptors and summarize the type of sensory input that they detect.

Senses whose receptors are associated with the skin, muscles, joints, and viscera are termed the *somatic senses.* These receptors can be categorized into three types: proprioceptors, cutaneous receptors, and pain receptors. All of these send nerve impulses via the spinal cord to the primary somatosensory areas of the cerebral cortex (see Fig. 13.10).

Proprioceptors

Proprioceptors are mechanoreceptors involved in reflex actions that maintain muscle tone, and thereby the body's equilibrium and posture. For example, proprioceptors called muscle spindles are embedded in muscle fibers (Fig. 14.2). If a muscle relaxes too much, the muscle spindle stretches, generating nerve impulses that cause the muscle to contract slightly. Conversely, when muscles are stretched too much, proprioceptors called *Golgi tendon organs,* buried in the tendons that attach muscles to bones, generate nerve impulses that cause the muscles to relax. Both types of receptors act together to maintain a functional degree of muscle tone. The knee-jerk reflex, which involves muscle spindles, offers an opportunity for physicians to test a reflex action. The information sent by muscle spindles to the CNS is used to maintain the body's equilibrium and posture. Proper balance and body position are maintained, despite the force of gravity always acting upon the skeleton and muscles.

Cutaneous Receptors

The skin is composed of two layers: the epidermis and the dermis (see section 4.6). The dermis contains **cutaneous receptors** (Fig. 14.3), which make the skin sensitive to touch, pressure, pain, and temperature (warmth and cold). The dermis is a mosaic of these tiny receptors, as you can determine by slowly passing a metal probe over your skin. At certain points, you will feel touch or pressure; at others, you will feel heat or cold (depending on the probe's temperature).

Three types of cutaneous receptors are sensitive to fine touch. These receptors give a person specific information such as the location of the touch as well as its shape, size, and texture. *Meissner corpuscles* and *Krause end bulbs* are concentrated in the fingertips, the palms, the lips, the tongue, the nipples, the penis, and the clitoris. *Merkel disks* are found where the epidermis meets the dermis. A free nerve ending called a *root hair plexus* winds around the base of a hair follicle. This receptor responds if the hair is touched.

Two types of cutaneous receptors sensitive to pressure are Pacinian corpuscles and Ruffini endings. *Pacinian corpuscles* are onion-shaped sensory receptors that lie deep inside the dermis. *Ruffini endings* are encapsulated by sheaths of connective tissue and contain lacy networks of nerve fibers.

Temperature receptors are simply free nerve endings in the epidermis. Some free nerve endings are responsive to cold. Others respond to warmth. Cold receptors are far more numerous than warmth receptors, but the two types have no known structural differences.

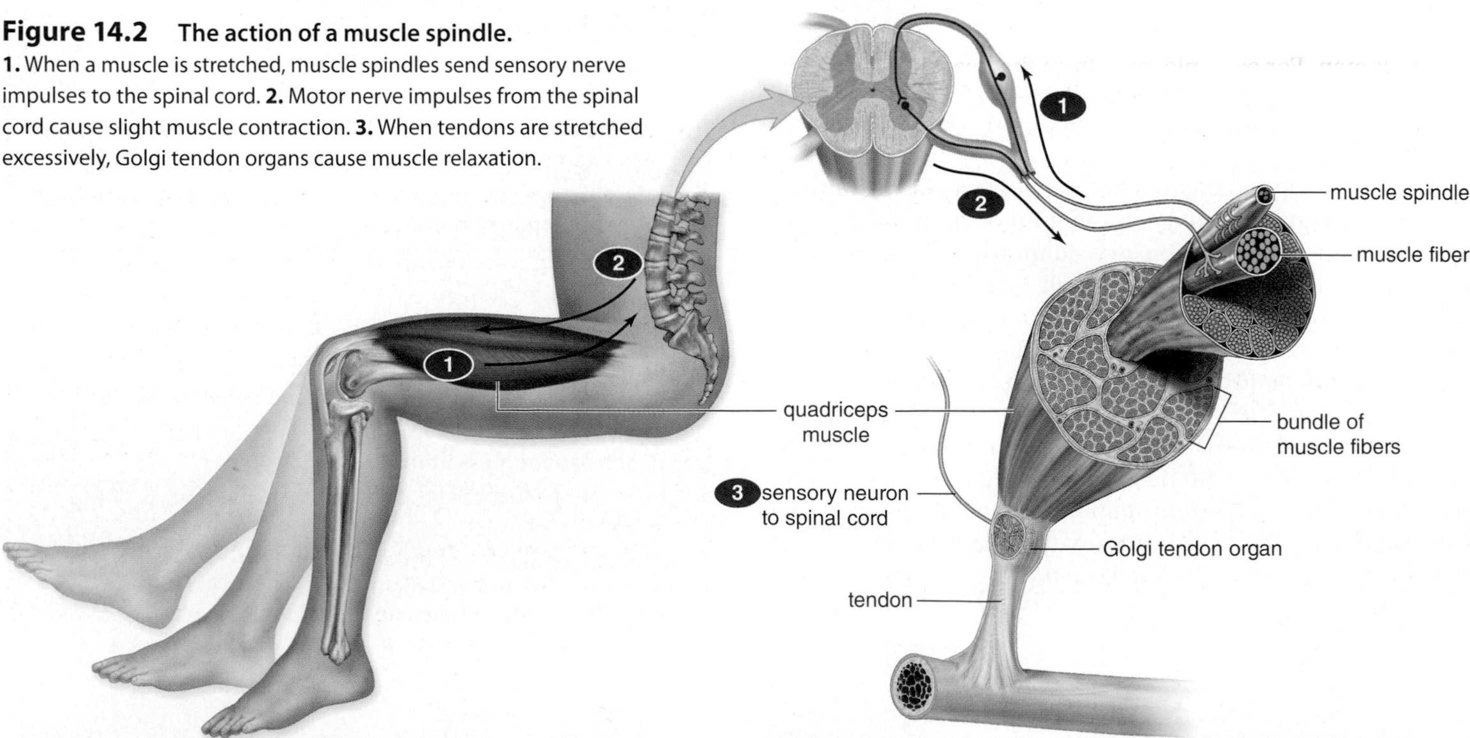

Figure 14.2 **The action of a muscle spindle.**
1. When a muscle is stretched, muscle spindles send sensory nerve impulses to the spinal cord. **2.** Motor nerve impulses from the spinal cord cause slight muscle contraction. **3.** When tendons are stretched excessively, Golgi tendon organs cause muscle relaxation.

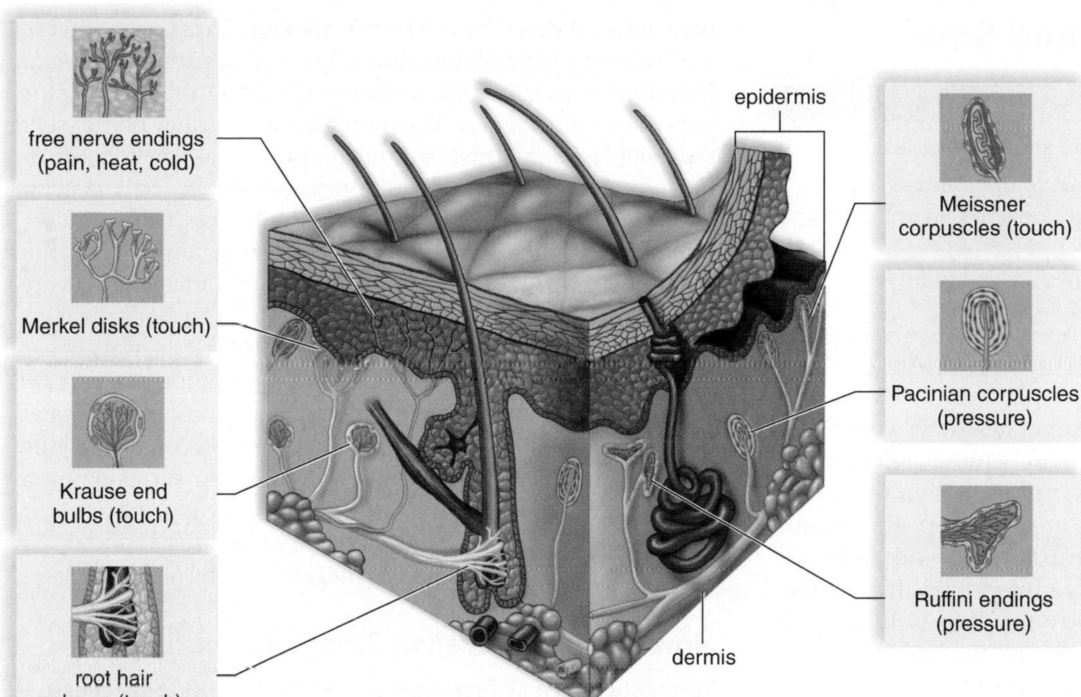

Figure 14.3 Sensory receptors of the skin. The classical view is that each sensory receptor has the main function shown here. However, investigators report that matters are not so clear-cut. For example, microscopic examination of the skin of the ear shows only free nerve endings (pain receptors); yet, the skin of the ear is sensitive to all sensations. Therefore, it appears that the receptors of the skin are somewhat—but not completely—specialized.

Pain Receptors

Like the skin, many internal organs have nociceptors, which respond to the stimuli of pain. These receptors are sensitive to chemicals released by damaged tissues. When inflammation occurs because of mechanical, thermal, or electrical stimuli or toxic substances, cells release chemicals that stimulate pain receptors. Aspirin and ibuprofen reduce pain by inhibiting the synthesis of one class of these chemicals.

Sometimes, stimulation of internal pain receptors is felt as pain from the skin, as well as the internal organs. This is called *referred pain.* Some internal organs have a referred pain relationship with areas located in the skin of the back, groin, and abdomen. For example, pain from the heart is often felt in the left shoulder and arm. This most likely happens when nerve impulses from the pain receptors of internal organs travel to the spinal cord and synapse with neurons also receiving impulses from the skin. Frequently, this type of referred pain is more common in men than in women. The nonspecific symptoms that women often experience during a heart attack may delay a diagnosis.

CHECK YOUR PROGRESS 14.2

1. Describe how proprioceptors are used by the body to indicate the position of the arms and legs.
2. Summarize the role of each type of cutaneous receptor and classify it as to whether it is located in the epidermis or the dermis.
3. Explain why the sensation of pain is important for the maintenance of homeostasis.

CONNECTING THE CONCEPTS

For more information on the material in this section, refer to the following discussions:

Figure 4.9 provides a more detailed look at the structure of human skin.

Section 12.2 provides an overview of muscle fiber contraction.

Section 13.2 presents the gate control theory of how the brain responds to input from pain receptors.

APPLICATIONS AND MISCONCEPTIONS

What are phantom sensation and phantom pain?

Suppose a person loses a foot and a leg due to an injury. In addition to dealing with loss of a limb, an amputee often must cope with the phenomenon of phantom sensation or phantom pain—or both. Phantom sensation is a painless awareness of the amputated limb. For example, a patient whose foot and lower leg have been removed may have an itchy or tingly sensation in the "foot," though the foot is no longer there. Similarly, pain can be sensed as originating from the absent body part. Researchers believe that any stimulus (such as a touch) to the stump will fool the brain into a perceived sensation, because the brain has received signals from the leg and foot for such a long time.

Phantom sensation may last for years but usually disappears without treatment. Phantom pain must be treated with a combination of medication, massage, and physical therapy.

14.3 Senses of Taste and Smell

LEARNING OUTCOMES

Upon completion of this section, you should be able to

1. Compare and contrast the senses of taste and smell.
2. Identify the structures of the tongue and olfactory areas of the nose.
3. Summarize how the brain receives taste and odor information.

Taste and smell are called chemical senses because their receptors are sensitive to molecules in the food we eat and the air we breathe.

Taste cells and olfactory cells bear chemoreceptors. Chemoreceptors are also in the carotid arteries and in the aorta, where they are primarily sensitive to the pH of the blood. These chemoreceptors are called carotid and aortic bodies. They communicate via sensory nerve fibers with the respiratory center in the medulla oblongata. When the pH drops, they signal this center. Immediately thereafter, the breathing rate increases. Exhaling CO_2 raises the pH of the blood.

Chemoreceptors are plasma membrane receptors that bind to particular molecules. They are divided into two types: those that respond to distant stimuli and those that respond to direct stimuli. Olfactory cells act from a distance and taste cells act directly. pH receptors also respond to direct stimuli.

Sense of Taste

In adult humans, approximately 4,000 **taste buds** are located primarily on the tongue (Fig. 14.4). Many taste buds lie along the walls of the papillae. These small elevations on the tongue are visible to the naked eye. Isolated taste buds are also present on the hard palate, the pharynx, and the epiglottis. In 2010, researchers at the University of Maryland found chemoreceptors in the human lung that are sensitive only to chemicals that normally taste bitter. These receptors are not clustered in buds, and they do not send taste signals to the brain. Stimulation of these receptors causes the airways to dilate, leading the researchers to speculate about implications for new medications to treat diseases like asthma.

Humans have five main types of taste receptors: sweet, sour, salty, bitter, and umami (from the Japanese, meaning "savory, delicious"). Foods rich in certain amino acids, such as the common seasoning monosodium glutamate (MSG), as well as certain flavors of cheese, beef broth, and some seafood, produce the taste of umami. Taste buds for each of these tastes are located throughout the tongue, although certain regions may be slightly more sensitive to particular tastes. A food can stimulate more than one of these types of taste buds. The brain appears to survey the overall pattern of incoming sensory impulses and to take a "weighted average" of their taste messages as the perceived taste.

How the Brain Receives Taste Information

Taste buds open at a taste pore. They have supporting cells and a number of elongated taste cells that end in microvilli. When molecules bind to receptor proteins of the microvilli, nerve signals are generated in sensory nerve fibers that go to the brain. Signals reach the gustatory (taste) cortex, located primarily in the parietal lobe, where they are interpreted as particular tastes.

Sense of Smell

Approximately 80–90% of what we perceive as "taste" actually is due to the sense of smell. This accounts for how dull food tastes when we have a head cold or a stuffed-up nose. Our sense of smell depends on between 10 and 20 million **olfactory cells** located within olfactory epithelia high in the roof of the

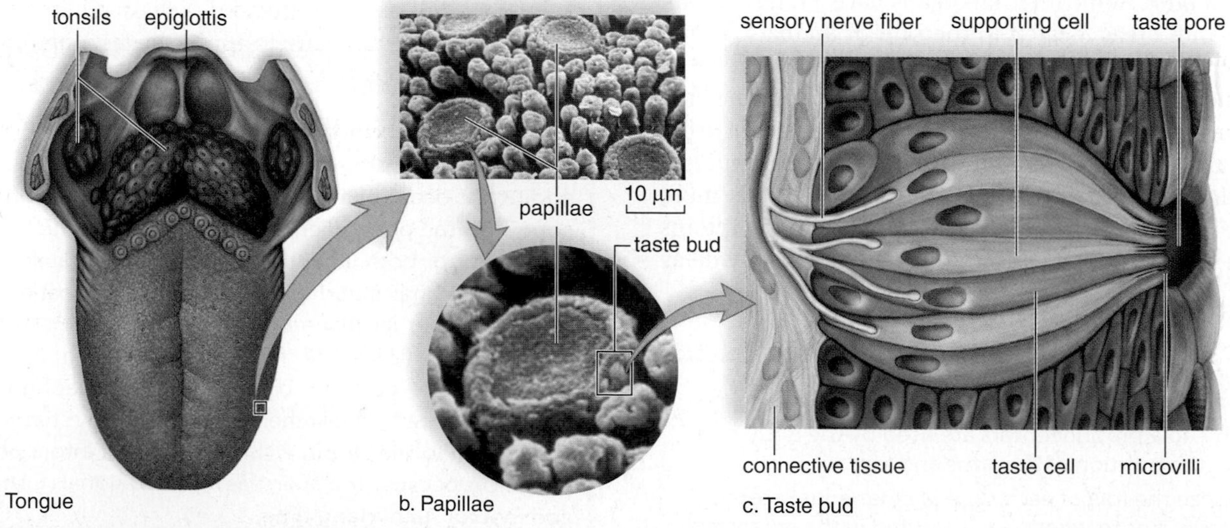

Figure 14.4 The tongue and the sense of taste.
a. Papillae on the tongue contain taste buds sensitive to sweet, sour, salty, bitter, and umami. **b.** Photomicrograph and enlargement of the papillae. **c.** Taste buds occur along the walls of the papillae. **d.** Taste cells in microvilli that bear receptor proteins for certain molecules. When molecules bind to the receptor proteins, nerve signals are generated and go to the brain, where the sensation of taste occurs.

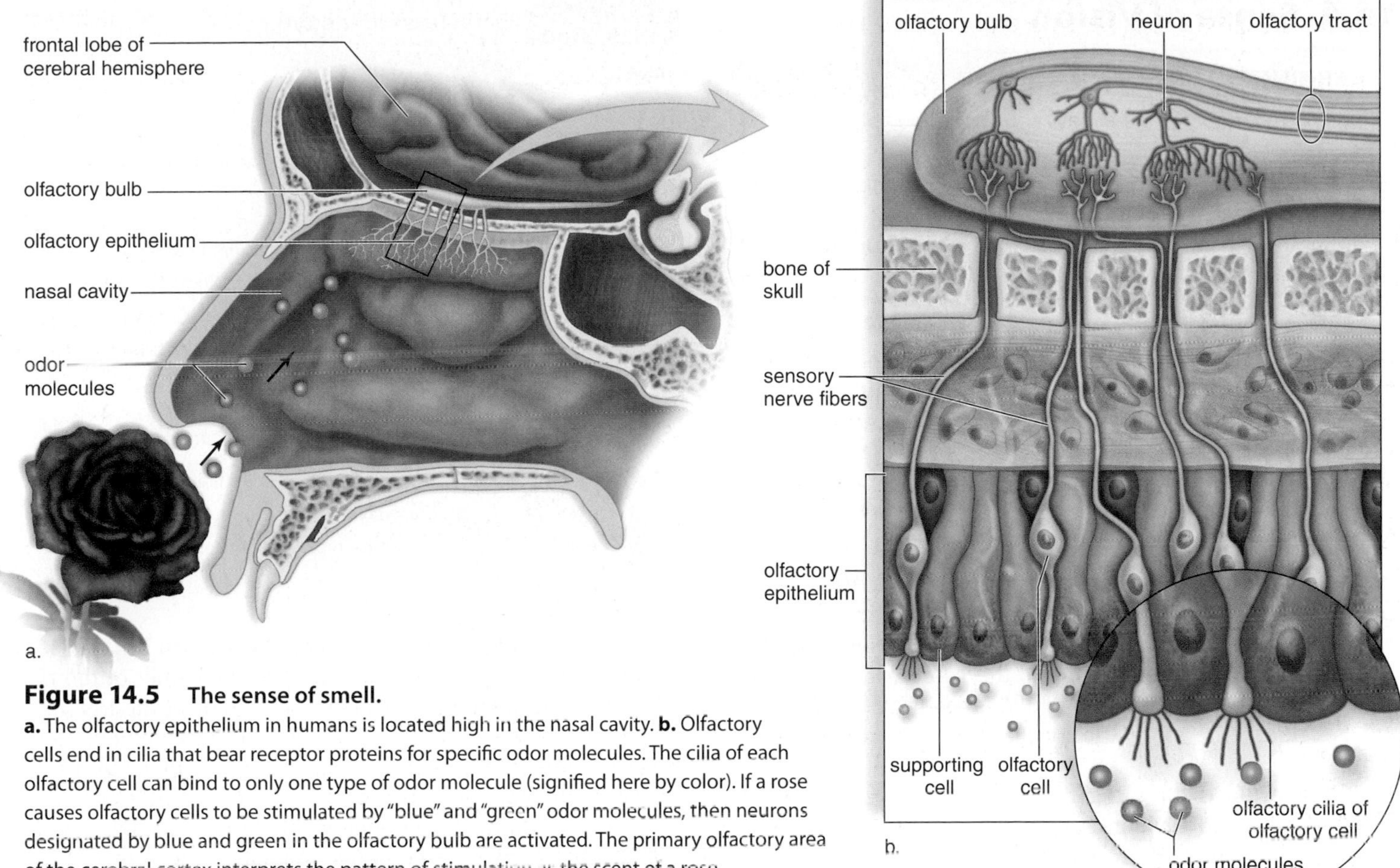

Figure 14.5 The sense of smell.
a. The olfactory epithelium in humans is located high in the nasal cavity. **b.** Olfactory cells end in cilia that bear receptor proteins for specific odor molecules. The cilia of each olfactory cell can bind to only one type of odor molecule (signified here by color). If a rose causes olfactory cells to be stimulated by "blue" and "green" odor molecules, then neurons designated by blue and green in the olfactory bulb are activated. The primary olfactory area of the cerebral cortex interprets the pattern of stimulation as the scent of a rose.

nasal cavity (Fig. 14.5). Olfactory cells are modified neurons. Each cell ends in a tuft of about five olfactory cilia, which bear receptor proteins for odor molecules.

How the Brain Receives Odor Information

Each olfactory cell has only 1 out of 1,000 different types of receptor proteins. Nerve fibers from similar olfactory cells lead to the same neuron in the olfactory bulb (an extension of the brain). An odor contains many odor molecules, which activate a characteristic combination of receptor proteins. For example, a rose may stimulate olfactory cells, designated by the blue and green dots in Figure 14.5, whereas a dandelion may stimulate a different combination. An odor's signature in the olfactory bulb is determined by which neurons are stimulated. When the neurons communicate this information via the olfactory tract to the olfactory areas of the cerebral cortex, we know we have smelled a rose or a carnation.

The olfactory cortex is located in the temporal lobe. Some areas of the olfactory cortex receive smell sensations, and other areas contain olfactory memories.

Have you ever noticed that a certain aroma vividly brings to mind a certain person or place and can re-create emotions you feel about that person or place? For example, a smell of a certain food may remind you of a favorite vacation. This is because the olfactory bulbs have direct connections with the limbic system and its centers for emotion and memory. One investigator showed that when subjects smelled an orange while viewing a painting, memories of the painting were more vividly recalled and the subjects also had many deep feelings about the painting.

The number of olfactory cells declines with age. This can be dangerous if an older person can't smell smoke or a gas leak. Older people also tend to apply excessive amounts of perfume or cologne before they can detect its smell.

CHECK YOUR PROGRESS 14.3

1. Identify the structures of the tongue and nose that are involved in the sense of taste and smell.
2. Compare and contrast the function of the chemoreceptors on the tongue and in the nose.
3. Summarize the pathway of sensory information on taste and smell from the receptors to the brain.

CONNECTING THE CONCEPTS

For more information on chemoreceptors, refer to the following discussions:

Section 9.5 describes the function of the respiratory center in the medulla oblongata.

Section 13.3 explains the role of the limbic system in maintaining memories such as smell and taste.

14.4 Sense of Vision

LEARNING OUTCOMES

Upon completion of this section, you should be able to

1. Identify the structures of the human eye.
2. Explain how the eye focuses on near and far objects.
3. Describe the role of photoreceptors in vision.
4. Summarize the abnormalities of the eye that produce vision problems.

Vision requires the work of the eyes and the brain. As we shall see, integration of stimuli occurs in the eyes before nerve signals are sent to the brain. Still, researchers estimate that at least a third of the cerebral cortex takes part in processing visual information.

Table 14.2 Structures of the Eye

Part	Function
Sclera	Protects and supports the eye
Cornea	Refracts light rays
Pupil	Admits light
Choroid	Absorbs stray light
Ciliary body	Holds lens in place, accommodation
Iris	Regulates light entrance
Retina	Contains sensory receptors for sight
Rod cells	Make black-and-white vision possible
Cone cells	Make color vision possible
Fovea centralis	Makes acute vision possible
Other	
Lens	Refracts and focuses light rays
Humors	Transmit light rays and support the eye
Optic nerve	Transmits impulses to brain

Anatomy and Physiology of the Eye

Table 14.2 lists the major structures of the eye and their function. The eye is an elongated sphere about 2.5 cm in diameter. It has three layers, or coats: the sclera, the choroid, and the retina (Fig. 14.6). The outer layer, the **sclera,** is white and fibrous except for the **cornea,** which is made of transparent collagen fibers. Therefore, the cornea is known as the window of the eye.

The **choroid** is the thin middle coat. It has an extensive blood supply, and its dark pigment absorbs stray light rays that photoreceptors have not absorbed. This helps visual acuity. Toward the front, the choroid becomes the doughnut-shaped **iris.** The iris regulates the size of the **pupil,** a hole in the center of the iris through which light enters the eye. The color of the iris (and therefore the color of the eyes) correlates with its pigmentation. Heavily pigmented eyes are brown, and lightly pigmented eyes are green or blue. Behind the iris, the choroid thickens and forms the circular ciliary body. The ciliary body contains the ciliary muscle, which controls the shape of the lens for near and far vision.

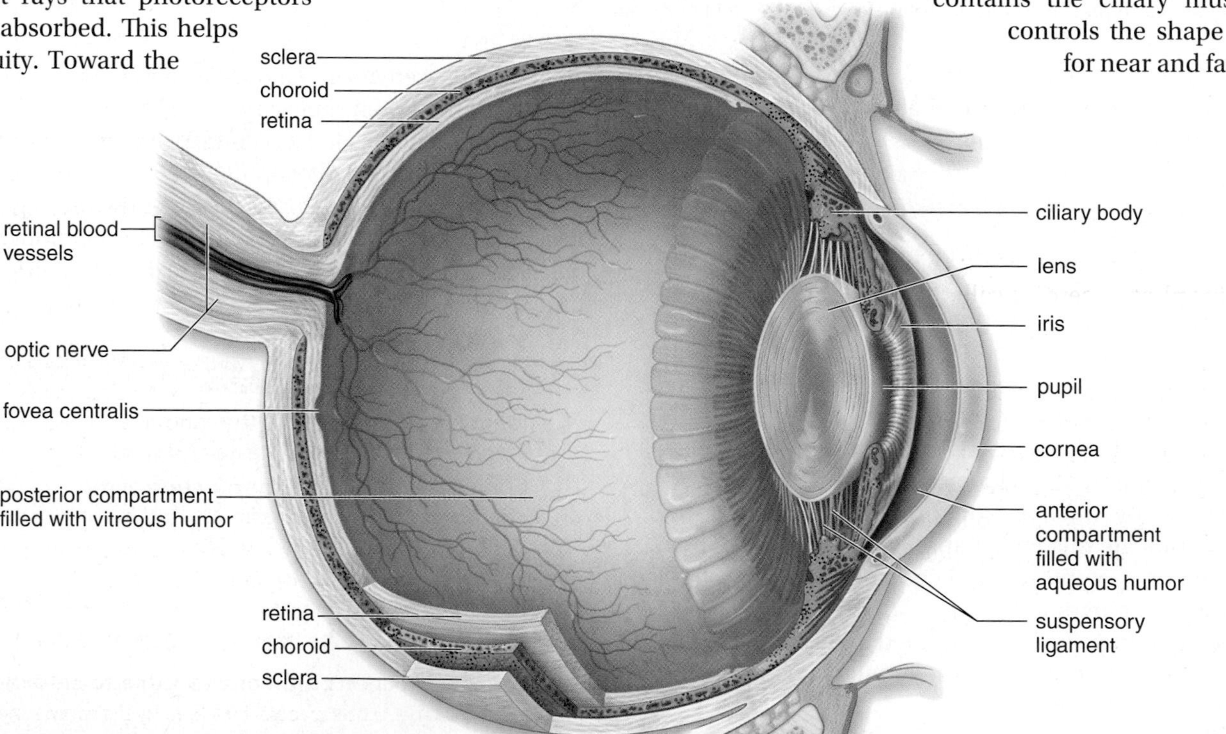

Figure 14.6 **The structures of the human eye.**
The sclera (the outer layer of the eye) becomes the cornea and the choroid (the middle layer) is continuous with the ciliary body and the iris. The retina (the inner layer) contains the photoreceptors for vision. The fovea centralis is the region where vision is most acute.

APPLICATIONS AND MISCONCEPTIONS

What is pinkeye?

At some point in their lives, most people have suffered from conjunctivitis, or pinkeye. Conjunctivitis is the inflammation of a mucous membrane called the conjunctiva, which covers the eye (except the cornea) and the inner part of the eyelid. The purpose of the conjunctiva is to lubricate the eye and keep it from drying out. In the case of viral conjunctivitis, the most common type, this membrane becomes inflamed as part of an immune response against the viral pathogens. Viral conjunctivitis is highly contagious; individuals with the condition must be careful not to spread the disease. However, not all conjunctivitis is contagious; allergies and other medical conditions may cause pinkeyelike symptoms. Treatment usually involves the use of eyedrops that help lubricate the eye and reduce inflammation.

The **lens** is attached to the ciliary body by suspensory ligaments and divides the eye into two compartments. The anterior compartment is in front of the lens, and the posterior compartment is behind it. The anterior compartment is filled with a clear, watery fluid called the **aqueous humor.** A small amount of aqueous humor is continually produced each day. Normally, it leaves the anterior compartment by way of tiny ducts. When a person has **glaucoma,** these drainage ducts are blocked and aqueous humor builds up. If glaucoma is not treated, the resulting pressure compresses the arteries that serve the nerve fibers of the retina, where photoreceptors are located. The nerve fibers begin to die because of lack of nutrients, and the person becomes partially blind. Eventually, total blindness can result.

The third layer of the eye, the **retina,** is located in the posterior compartment. This compartment is filled with a clear, gelatinous material called the **vitreous humor.** The vitreous humor holds the retina in place and supports the lens. The retina contains photoreceptors called rod cells and cone cells. The rods are very sensitive to light, but they do not detect color. Therefore, at night or in a darkened room, we see only shades of gray. The cones, which require bright light, are sensitive to different wavelengths of light. This sensitivity gives us the ability to distinguish colors. The retina has a very special region called the **fovea centralis** where cone cells are densely packed. Light is normally focused on the fovea when we look directly at an object. This is helpful because the sharpest images are produced by the fovea centralis. Sensory fibers from the retina form the **optic nerve,** which takes nerve signals to the visual cortex.

Function of the Lens

The cornea, assisted by the lens and the humors, focuses images on the retina. Focusing starts with the cornea and continues as the rays pass through the lens and the humors. The image produced is much smaller than the object because light rays are bent (refracted) when they are brought into **focus.** If the eye is too long or too short, the person may need corrective lenses to bring the image into focus. The image on the retina is inverted (upside down) and reversed from left to right.

Visual accommodation occurs for close vision. During visual accommodation, the lens changes its shape to bring the image into focus on the retina. The shape of the lens is controlled by the ciliary muscle, within the ciliary body. When we view a distant object, the ciliary muscle is relaxed, causing the suspensory ligaments attached to the ciliary body to be taut. The ligaments put tension on the lens and cause it to remain relatively flat (Fig. 14.7*a*). When we view a near object, the ciliary muscle contracts, releasing the tension on the suspensory ligaments. The lens becomes round and thick due to its natural elasticity (Fig. 14.7*b*). Thus, contraction or relaxation of the ciliary muscle allows the image to be focused on the retina. Close work requires contraction of the ciliary muscle, so it often causes muscle fatigue, known as eyestrain. Eyestrain is more common after the age of 40, because the lens loses some of its elasticity and is unable to accommodate. It is also common for those who work with computers because the intense focusing causes the person to blink less, allowing the eyes to dry out. Eyedrops, and/or corrective lenses, either eyeglasses or contact lenses, may be necessary to reduce eyestrain.

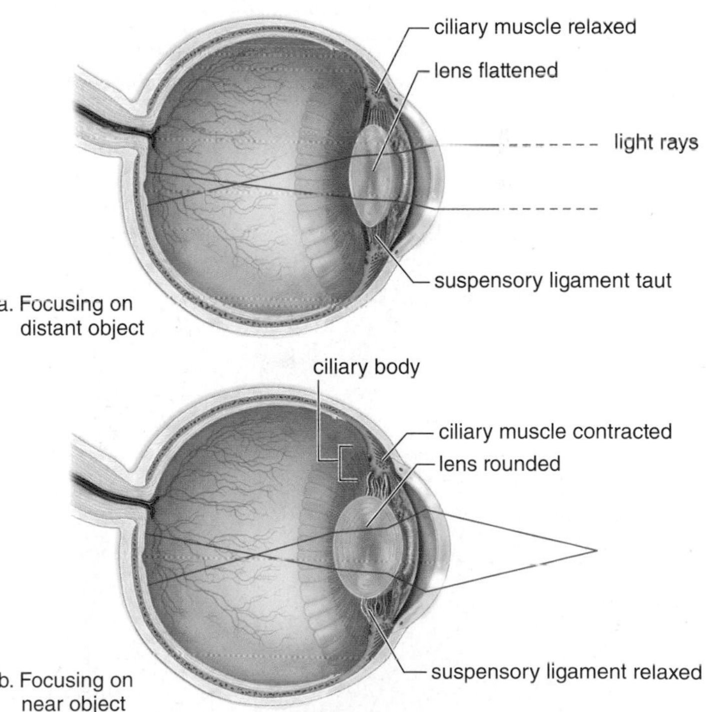

Figure 14.7 Focusing light on the retina of the eye. Light rays from each point on an object are bent by the cornea and the lens in such a way that an inverted and reversed image of the object forms on the retina. **a.** When focusing on a distant object, the lens is flat because the ciliary muscle is relaxed and the suspensory ligament is taut. **b.** When focusing on a near object, the lens accommodates—it becomes rounded because the ciliary muscle contracts, causing the suspensory ligament to relax.

APPLICATIONS AND MISCONCEPTIONS

Why does rubbing your closed eyes (as you might if you're tired or your eyes itch) produce a visual sensation?

The eye is a flexible container filled with fluid and a soft gelatinous material. Compressing the eyes by rubbing on them increases pressure in the eyes. In turn, the photoreceptors of the eye are stimulated by the increased eye pressure. When the nerve signals are conveyed to the brain, the brain senses "vision." We "see stars" because nerve signals from the eyes can only result in sight.

Visual Pathway to the Brain

The pathway for vision begins once light has been focused on the photoreceptors in the retina. Some integration occurs in the retina, where nerve signals begin before the optic nerve transmits them to the brain.

Function of Photoreceptors Figure 14.8*a* illustrates the structure of the photoreceptors called **rod cells** and **cone cells.** Both rods and cones have an outer segment joined to an inner segment by a short stalk. Pigment molecules are embedded in the membrane of the many disks present in the outer segment. Synaptic vesicles are located at the synaptic endings of the inner segment.

The visual pigment in rods is a deep purple pigment called rhodopsin (Fig. 14.8*b*). **Rhodopsin** is a complex molecule made up of the protein opsin and a light-absorbing molecule called **retinal,** a derivative of vitamin A. When a rod absorbs light, rhodopsin splits into opsin and retinal. This leads to a cascade of reactions and the closure of ion channels in the rod cell's plasma membrane. The release of inhibitory transmitter molecules from the rod's synaptic vesicles ceases. Thereafter, signals go to other neurons in the retina. Rods are very sensitive to light and, therefore, are suited to night vision. Carrots, and other orange and yellow vegetables, are rich in vitamin A, so it is true that eating carrots can improve your night vision. Rod cells are plentiful throughout the entire retina, except the fovea. Therefore, rods also provide us with peripheral vision and perception of motion.

The cones, on the other hand, are located primarily in the fovea and are activated by bright light. They allow us to detect the fine detail and the color of an object. Color vision depends on three different types of cones, which contain pigments called the B (blue), G (green), and R (red) pigments. Each pigment is made up of retinal and opsin, but there is a slight difference in the opsin structure of each. This accounts for their individual absorption patterns. Various combinations of cones are believed to be stimulated by in-between shades of color.

Function of the Retina The retina has three layers of neurons (Fig. 14.9). The layer closest to the choroid contains the rod cells and cone cells. A layer of bipolar cells covers the rods and cones. The innermost layer contains ganglion cells, whose sensory fibers become the optic nerve. Only the rod cells and the cone cells are sensitive to light; therefore, light must penetrate to the back of the retina before the rods and cones are stimulated.

The rod cells and the cone cells synapse with the bipolar cells. Next, signals from bipolar cells stimulate ganglion cells whose axons become the optic nerve. Notice in Figure 14.9 that there are many more rod cells and cone cells than ganglion cells. Although the precise number is not known, the retina has around

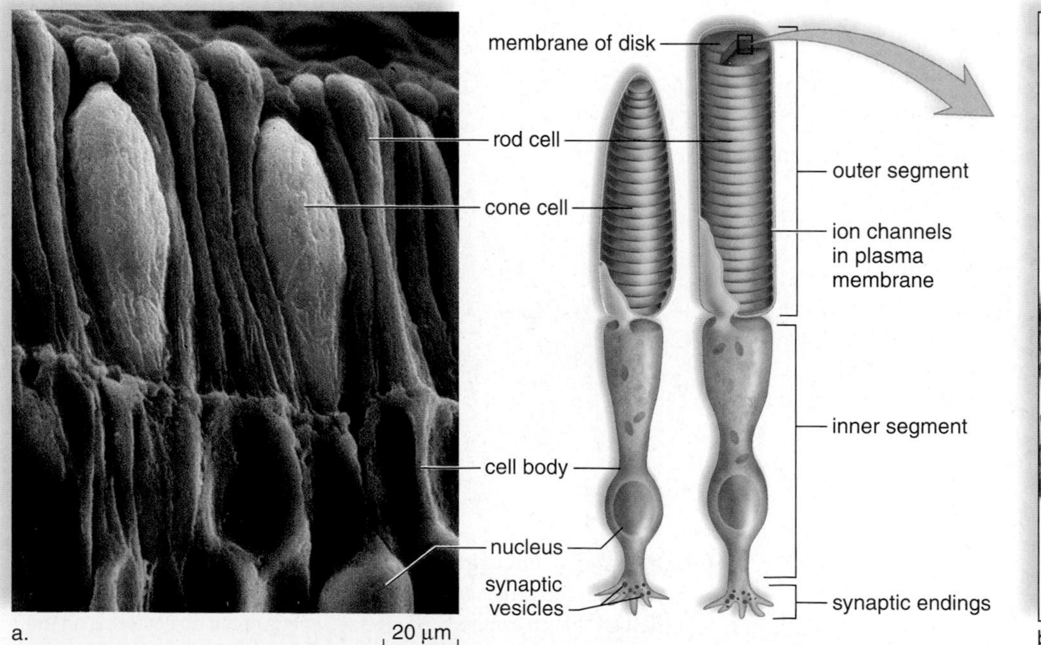

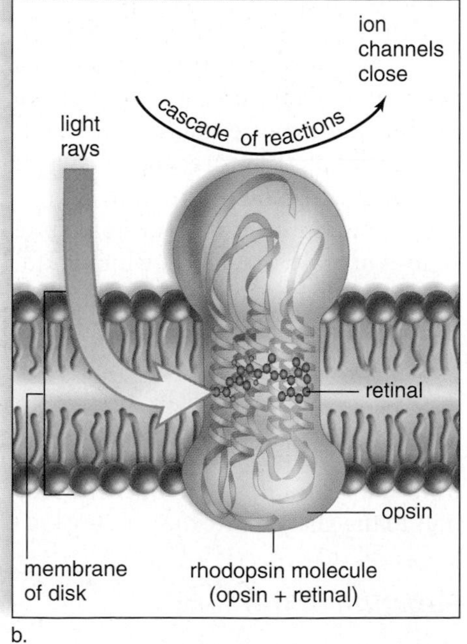

Figure 14.8 The two types of photoreceptors in the eye.
a. The outer segment of rods and cones contains stacks of membranous disks, which contain visual pigments. **b.** In rods, the membrane of each disk contains rhodopsin, a complex molecule containing the protein opsin and the pigment retinal. When rhodopsin absorbs light energy, it splits, releasing retinal, which sets in motion a cascade of reactions that causes ion channels in the plasma membrane to close. Thereafter, nerve signals go to the brain.

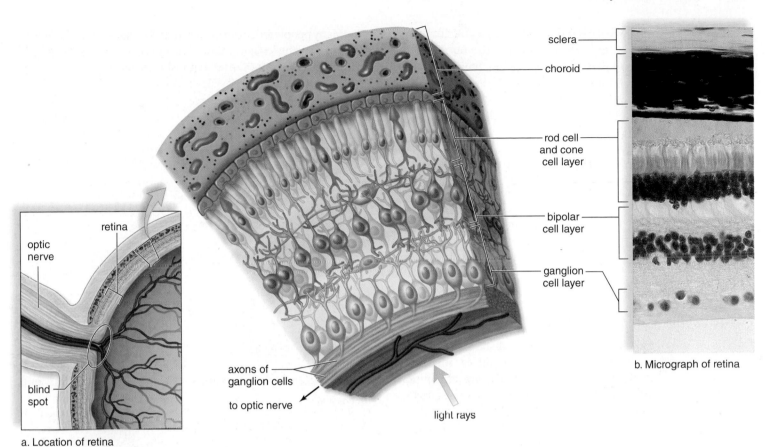

a. Location of retina

b. Micrograph of retina

Figure 14.9 The structure of the retina.
a. The retina is the inner layer of the eye. Rod and cone cells, located at the back of the retina nearest the choroid, synapse with bipolar cells, which synapse with ganglion cells. Integration of signals occurs at these synapses. Cone cells, in general, distinguish more detail than do rod cells. **b.** Micrograph shows that the sclera and choroid are relatively thin compared to the retina, which has several layers of cells.

150 million rod cells and 6.5 million cone cells, but only 1 million ganglion cells. The sensitivity of cones versus rods is mirrored by how directly they connect to ganglion cells. As many as 150 rods may activate the same ganglion cell. No wonder stimulation of rods results in vision that is blurred and indistinct. In contrast, some cone cells in the fovea centralis activate only one ganglion cell. This explains why cones, especially in the fovea centralis, provide us with a sharper, more detailed image of an object.

As signals pass to bipolar cells and ganglion cells, integration occurs. Therefore, considerable processing occurs in the retina before ganglion cells generate nerve signals. Ganglion cells converge to form the optic nerve, which transmits information to the visual cortex. Additional integration occurs in the visual cortex.

Blind Spot Figure 14.9 also provides an opportunity to point out that there are no rods and cones where the optic nerve exits the retina. Therefore, no vision is possible in this area. You can prove this to yourself by putting a dot to the right of center on a piece of paper. Use your right hand to move the paper slowly toward your right eye, and make sure you look straight ahead. The dot will disappear at one point—this is your right eye's **blind spot.** The two eyes together provide complete vision because the blind spot for the right eye is not the same as the blind spot for the left eye. The blind spot for the right eye is right of center and the blind spot for the left eye is left of center.

From the Retina to the Visual Cortex To reach the visual cortex, the optic nerves carry nerve impulses from the eyes to the optic chiasma (Fig. 14.10). The **optic chiasma** has an X shape, formed by a crossing-over of optic nerve fibers. After exiting the optic chiasma, the optic nerves continue as **optic tracts.**

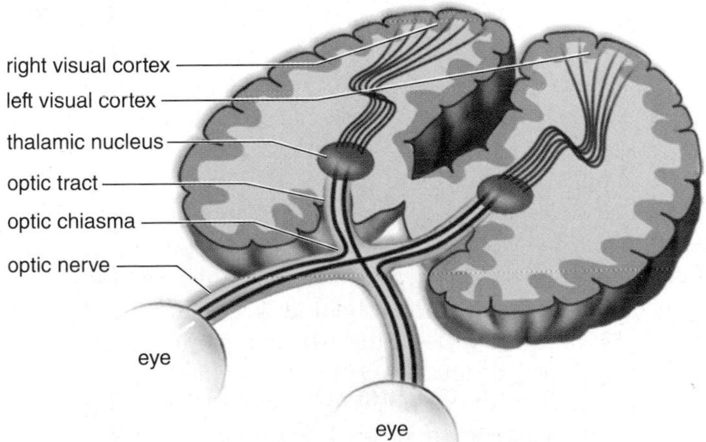

Figure 14.10 The function of the optic chiasma.
Because of the optic chiasma, data from the right half of each retina go to the right visual cortex, and data from the left half of the retina go to the left visual cortex. These data are then combined to allow us to see the entire visual field.

Fibers from the right half of each retina converge and continue on together in the right optic tract. Similarly, the nerve fibers from the left half of each retina join to form the left optic tract, traveling together to the brain.

The optic tracts sweep around the hypothalamus, and most fibers synapse with neurons in nuclei (masses of neuron cell bodies) within the thalamus. Axons from the thalamic nuclei form optic radiations that take nerve impulses to the *visual cortex* within the occipital lobe. The image is split in the visual cortex. This division of incoming information happens because the right visual cortex receives information from the right optic tract, and the left visual cortex receives information from the left optic tract. For good depth perception, the right and left visual cortices communicate with each other. Also, because the image is inverted and reversed, it must be righted in the brain for us to correctly perceive the visual field.

MP3
Sense of Vision

Animation
Vision

Abnormalities of the Eye

Color blindness and changes in the physical shape of the eye are two of the more common vision abnormalities. There are several forms of color blindness, all of which are attributed to a genetic mutation (see section 20.5). In most instances, only one type of cone is defective or deficient in number. The most common mutation is the inability to see the colors red and green. The gene for red-green color blindness is on the X chromosome, therefore males (who only possess one X chromosome) are more susceptible. This abnormality affects 5–8% of the male population. If the eye lacks cones that respond to red wavelengths, green colors are accentuated, and vice versa.

Distance Vision

If you can see from 20 ft what a person with normal vision can see from 20 ft, you are said to have 20/20 vision. Persons who can see close objects but can't see the letters on an optometrist's chart from 20 ft are said to be nearsighted. **Nearsighted** people can see close objects better than they can see objects at a distance. The shape of the eye in these individuals is elongated, and when they attempt to look at a distant object, the image is brought to focus in front of the retina (Fig. 14.11*a*). They can see close objects because their lens can compensate for the elongated shape of the eye. To see distant objects, these people can wear concave lenses, which spread the light rays so that the image focuses on the retina.

Persons who can easily see the optometrist's chart but cannot see close objects well are **farsighted.** These individuals can see distant objects better than they can see close objects. The shape of their eye is shortened, and when they try to see close objects, the image is focused behind the retina (Fig. 14.11*b*). When the object is distant, the lens can compensate for the shortened shape of the eye. When the object is close, these persons can wear convex lenses to increase the bending of light rays so that the image can be focused on the retina.

When the cornea or lens is uneven, the image is fuzzy. The light rays cannot be evenly focused on the retina. This condition, called **astigmatism,** can be corrected by an unevenly ground lens to compensate for the uneven cornea (Fig. 14.11*c*).

Many people today opt to have LASIK surgery instead of wearing lenses. LASIK surgery is discussed in the Health feature, "Correcting Vision Problems."

Video
Artificial Eye

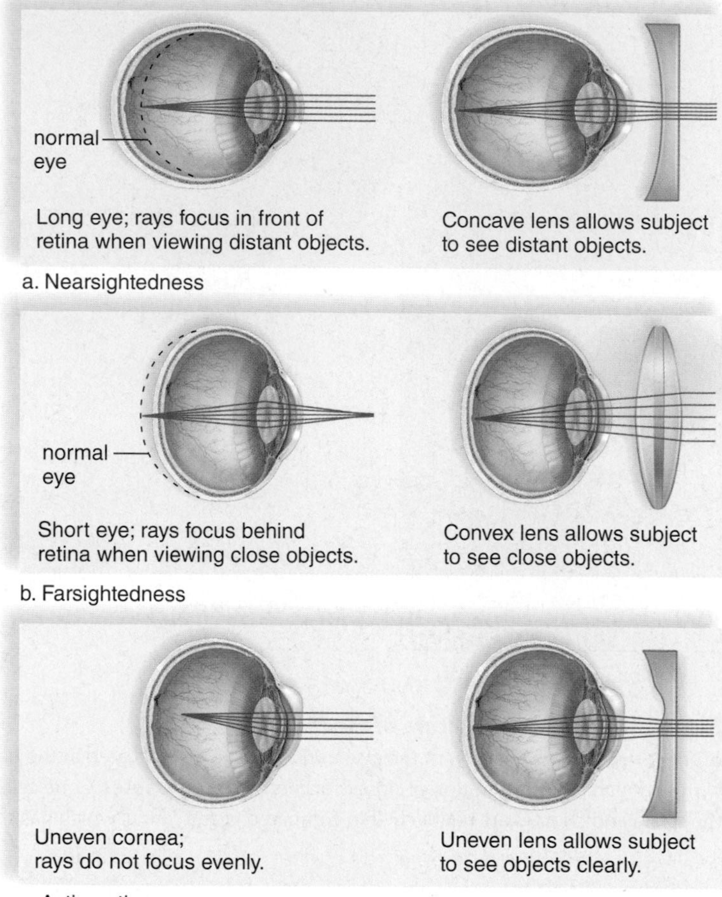
normal eye
Long eye; rays focus in front of retina when viewing distant objects.
Concave lens allows subject to see distant objects.
a. Nearsightedness
normal eye
Short eye; rays focus behind retina when viewing close objects.
Convex lens allows subject to see close objects.
b. Farsightedness
Uneven cornea; rays do not focus evenly.
Uneven lens allows subject to see objects clearly.
c. Astigmatism

Figure 14.11 **How corrective lenses correct vision problems.** **a.** A concave lens in nearsighted persons focuses light rays on the retina. **b.** A convex lens in farsighted persons focuses light rays on the retina. **c.** An uneven lens in persons with astigmatism focuses light rays on the retina.

CHECK YOUR PROGRESS 14.4

1. Identify the structures of the eye and provide a function of each.
2. Describe the two types of photoreceptors and state the function of each.
3. Summarize the movement of sensory information from the photoreceptors to the visual cortex.

CONNECTING THE CONCEPTS

For more information on the material presented in this section, refer to the following discussions:

Table 8.7 describes the function and dietary sources of vitamin A.

Section 13.2 describes the function of the visual association area in the cerebral cortex of the brain.

Section 20.4 explores the pattern of inheritance associated with color blindness.

BIOLOGY MATTERS **Health**

Correcting Vision Problems

Poor vision can be due to a number of problems, some more serious than others. For example, retinal detachment, cataracts, and glaucoma are three conditions that need medical attention.

Cataracts and Glaucoma

Cataracts develop when the lens of the eye becomes cloudy. Normally, the lens is clear and allows light to pass through easily. A cloudy lens decreases light levels that reach the retina and slowly causes vision loss. Fortunately, a doctor can surgically remove the cloudy lens and replace it with a clear plastic lens, which often restores the light level passing through the lens and improves the patient's vision.

Glaucoma is caused by a buildup of fluid pressure inside the eye and may lead to a decrease in vision. The condition may eventually cause blindness. Special eyedrops and oral medications are often prescribed to help reduce the interior pressure. If eyedrops and medications are not capable of controlling the pressure, surgery may be the only option. During glaucoma surgery, the doctor uses a laser to create tiny holes in the eye where the cornea and iris meet. This increases fluid drainage from the eye and decreases the pressure inside the eye.

The Benefits of LASIK Surgery

For many people, a sign of aging is the slow and steady decrease of their ability to see close-up. Difficulties in reading small print, or reading in low-light conditions, is usually the first sign of presbyopia. Many people who suffer from presbyopia experience headaches while reading. The condition usually begins in the late thirties and is common by age 55. Reading in low light situations becomes more difficult, and letters begin to look fuzzy when reading close-up. While historically people accommodated for the condition by using a magnifying lens, today, most people wear bifocal lenses. Bifocals are designed to correct vision at a distance of 12 to 18 inches. These lenses work well for most people while reading but pose a problem for people who use a computer. Computer monitors are usually 19 to 24 inches away. This forces bifocal lens wearers to constantly move their head up and down in an attempt to switch between the close and distant viewing sections of the bifocals. Another solution is to wear contact lenses with one eye corrected to see close objects and the other eye corrected to see distant objects. The same type of correction can be done with LASIK surgery (Fig. 14A). LASIK, which stands for *laser-assisted in situ keratomileusis,* is generally a safe and effective treatment option for a wide array of vision problems.

LASIK is a quick and painless procedure that involves the use of a laser to permanently change the shape of the cornea. For the majority of patients, LASIK improves their vision and reduces their dependency on corrective lenses. The ideal LASIK candidate is over 18 years of age and has had a stable contact or glasses prescription for at least two years. Patients need to have a cornea thick enough to allow the surgeon to safely create a clean corneal flap of appropriate depth. Typically, patients affected by common vision problems (nearsightedness, astigmatism, or farsightedness) respond well to LASIK. Anyone who suffers from any disease that decreases his or her ability to heal properly after surgery is not a candidate for LASIK. Candidates should thoroughly discuss the procedure with their eye-care professional before electing to have LASIK. People need to realize that the goal of LASIK is to reduce their dependency on glasses or contact lenses, not completely eliminate them.

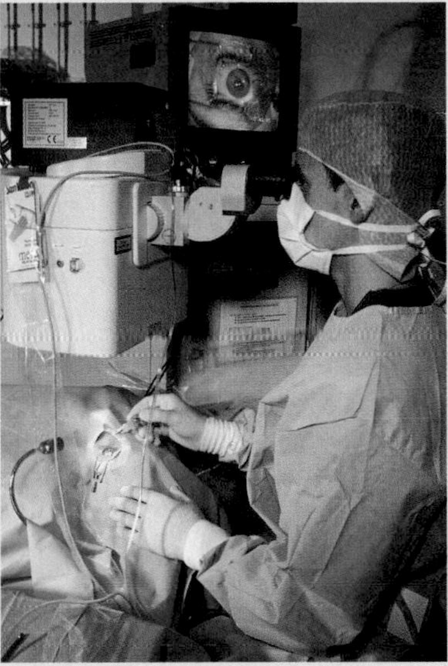

Figure 14A LASIK surgery.

Individuals who suffer from cataracts, advanced glaucoma, or corneal or other eye diseases are not considered for LASIK. Patients who expect LASIK to completely correct their visual problems and make them completely independent of their corrective lenses are not good candidates either.

The LASIK Procedure

During the LASIK procedure, a small flap of tissue (the conjunctiva) is cut away from the front of the eye. The flap is folded back exposing your cornea, allowing the surgeon to remove a defined amount of tissue from your cornea. Each pulse of the laser removes a small amount of corneal tissue, allowing the surgeon to flatten or increase the steepness of the curve of your cornea. After the procedure, the flap of tissue is put back into place and allowed to heal on its own. LASIK patients receive eyedrops or medications to help relieve the pain of the procedure. Improvements to your vision begin as early as the day after the surgery but typically take two to three months. Most patients will have vision close to 20/20, but your chances for improved vision are based in part on how good your eyes were before the surgery.

As with any surgery, complications are possible. Adverse effects include a sensation of having something in your eye or having blurred vision. You might also see halos around objects or be very sensitive to glare. In addition, dryness can cause eye irritation. Typically, these effects are temporary, and the rate of complications following surgery is very low. Always consult with your doctor before considering any type of surgery.

Questions to Consider

1. Using Figure 14.11 as a guide, explain how LASIK surgery corrects the flow of light into the eye.
2. What problems may be experienced for people who wear contacts that cause each eye to focus at different distances?

14.5 Sense of Hearing

LEARNING OUTCOMES

Upon completion of this section, you should be able to

1. Identify the structures of the ear that are involved in hearing.
2. Summarize how sound waves are converted into nerve signals.
3. Describe the pathway of sensory information from the ear to the brain.

The ear has two sensory functions: hearing and balance (equilibrium). The sensory receptors for both of these are located in the inner ear. Each consists of **hair cells** with **stereocilia** (long, stiff, microvilli) that are sensitive to mechanical stimulation. The stereocilia act as mechanoreceptors.

MP3 Sense of Hearing and Equilibrium

Anatomy and Physiology of the Ear

Figure 14.12 shows that the ear has three divisions: outer, middle, and inner. The **outer ear** consists of the **pinna** (external flap) and the **auditory canal.** The opening of the auditory canal is lined with fine hairs and sweat glands. Modified sweat glands are located in the upper wall of the canal. They secrete earwax, a substance that helps guard the ear against the entrance of foreign materials, such as air pollutants.

The **middle ear** begins at the **tympanic membrane** (eardrum) and ends at a bony wall containing two small openings covered by membranes. These openings are called the **oval window** and the **round window.** Three small bones are found between the tympanic membrane and the oval window. Collectively, they are called the **ossicles.** Individually, they are the **malleus** (hammer), the **incus** (anvil), and the **stapes** (stirrup) because their shapes resemble these objects. The malleus adheres to the tympanic membrane, and the stapes touches

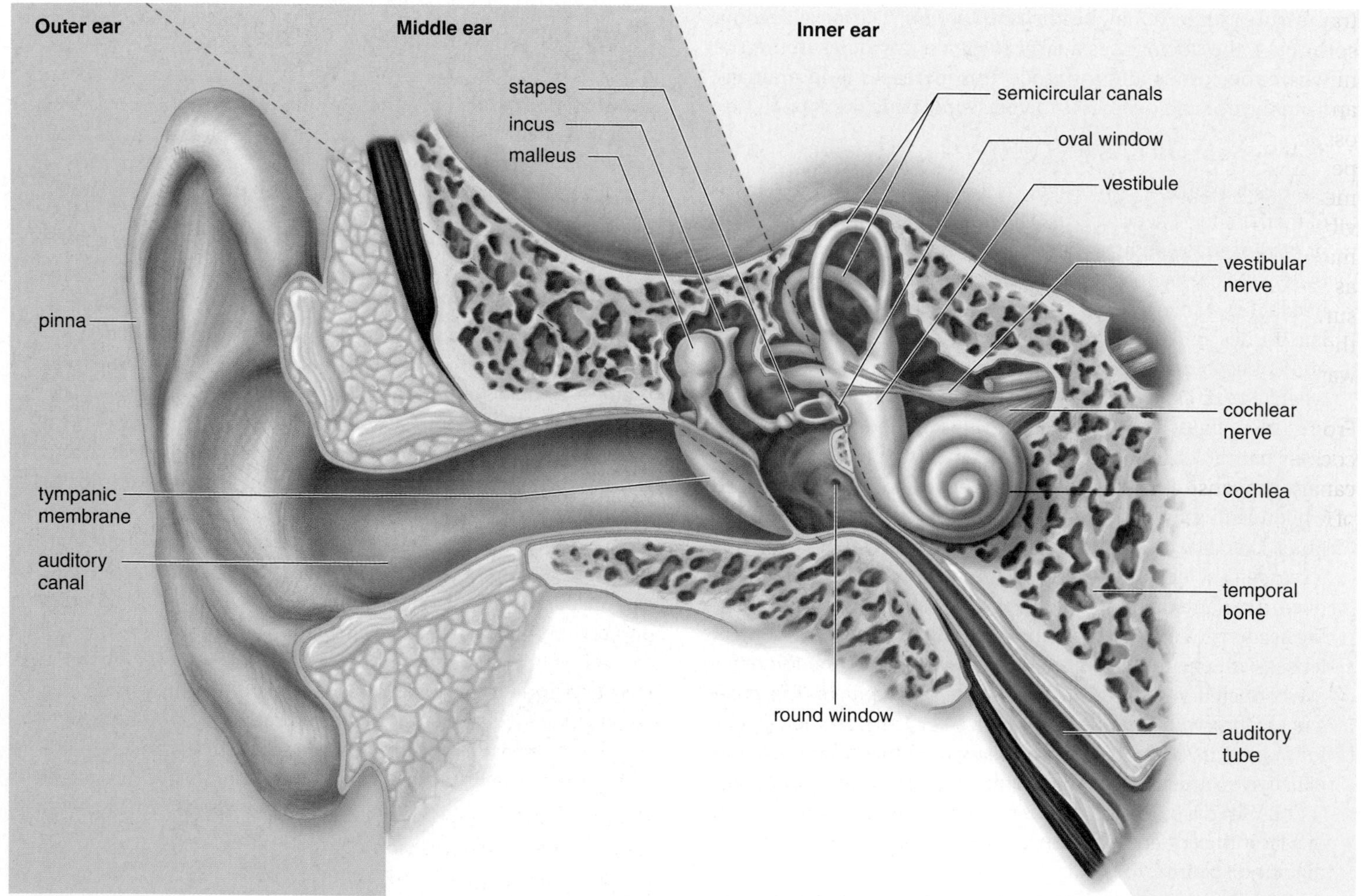

Figure 14.12 The three divisions of the human ear.
The external ear consists of the pinna (the structure commonly referred to as the "ear") and the auditory canal. The tympanic membrane separates the external ear from the middle ear. In the middle ear, the malleus (hammer), the incus (anvil), and the stapes (stirrup) amplify sound waves. In the inner ear, the mechanoreceptors for equilibrium are in the semicircular canals and the vestibule. The mechanoreceptors for hearing are in the cochlea.

the oval window. An **auditory tube,** also called the eustachian or pharyngotympanic tube, extends from the middle ear to the nasopharynx. Its purpose is to equalize air pressure across the tympanic membrane. When changing elevation, such as in an airplane, the act of chewing gum, yawning, or swallowing opens the auditory tubes wider. As this occurs, we often feel the ears "pop."

Whereas the outer ear and the middle ear contain air, the inner ear is filled with fluid. The **inner ear** has three areas: The **semicircular canals** and the **vestibule** are concerned with equilibrium; the **cochlea** is concerned with hearing. The cochlea resembles the shell of a snail because it spirals.

Auditory Pathway to the Brain

The auditory pathway begins with the auditory canal. Thereafter, hearing requires the other parts of the ear, the cochlear nerve, and the brain.

Through the Auditory Canal and Middle Ear The process of hearing begins when sound waves enter the auditory canal. Just as ripples travel across the surface of a pond, sound waves travel by the successive vibrations of molecules. Ordinarily, sound waves do not carry much energy. However, when a large number of waves strike the tympanic membrane, it moves back and forth (vibrates) ever so slightly. As you know, the auditory ossicles attach to one another: malleus to incus, incus to stapes. The malleus is attached to the inner wall of the tympanic membrane. Thus, vibrations of the tympanic membrane cause vibration of the malleus and, in turn, the incus and stapes. The magnitude of the original pressure wave increases significantly as the vibrations move along the auditory ossicles. The pressure is multiplied about 20 times. Finally, the stapes strikes the membrane of the oval window, causing it to vibrate. In this way, the pressure is passed to the fluid within the cochlea.

From the Cochlea to the Auditory Cortex By examining the cochlea in cross section (Fig. 14.13), you can see that it has three canals. The sense organ for hearing, called the **spiral organ** (organ of Corti), is located in the cochlear canal. The spiral organ consists of little hair cells and a gelatinous material called the *tectorial membrane.* The hair cells sit on the basilar membrane, and their stereocilia are embedded in the tectorial membrane.

Animation Effects of Sound Waves on Cochlear Structures

When the stapes strikes the membrane of the oval window, pressure waves move from the vestibular canal to the tympanic canal across the basilar membrane. The basilar membrane moves up and down, and the stereocilia of the hair cells embedded in the tectorial membrane bend. Then, nerve signals begin in the **cochlear nerve** and travel to the brain. When they reach the auditory cortex in the temporal lobe, they are interpreted as a sound.

Each part of the spiral organ is sensitive to different wave frequencies, or pitch. Near the tip, the spiral organ responds to low pitches, such as those of a tuba. Near the base (beginning), it responds to higher pitches, such as those of a bell or a whistle. The nerve fibers from each region along the length of the spiral

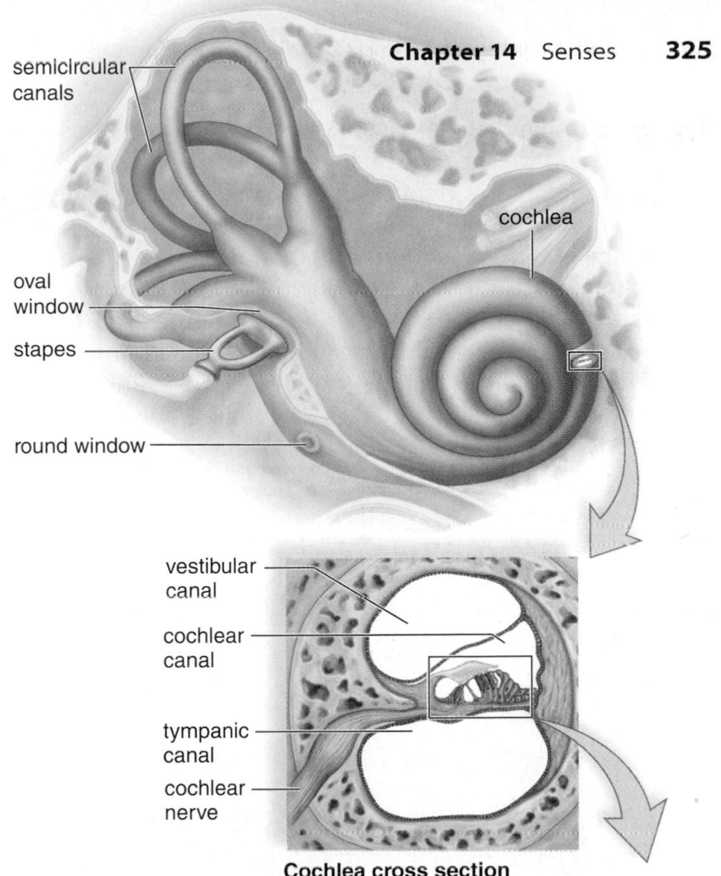

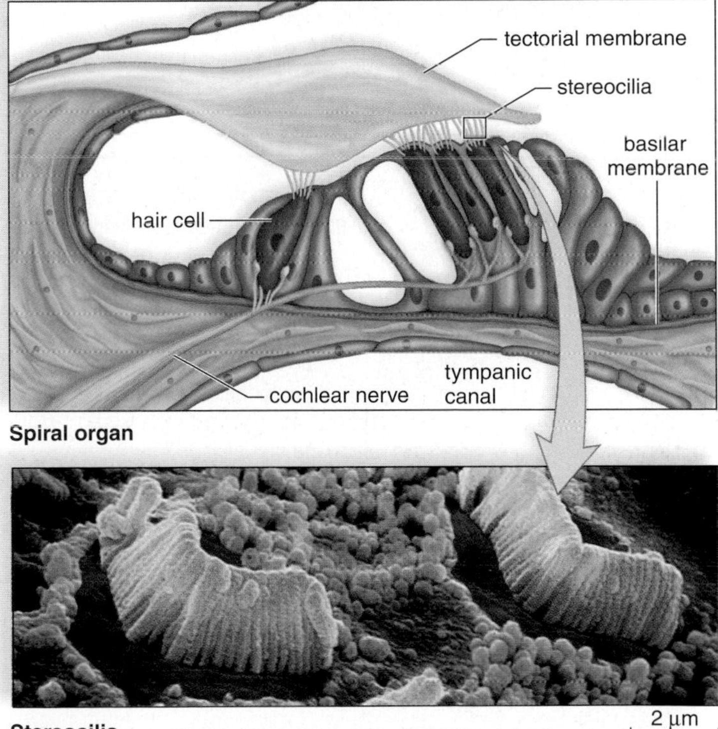

Figure 14.13 How the spiral organ (organ of Corti) translates sound waves into nerve signals.
The spiral organ (organ of Corti) is located within the cochlea. The spiral organ consists of hair cells resting on the basilar membrane, with the tectorial membrane above. Pressure waves moving through the canals cause the basilar membrane to vibrate. This causes the stereocilia embedded in the tectorial membrane to bend. Nerve impulses traveling in the cochlear nerve result in hearing.

organ lead to slightly different areas in the auditory cortex. The pitch sensation we experience depends upon which region of the basilar membrane vibrates and which area of the auditory cortex is stimulated.

Volume is a function of the amplitude (strength) of sound waves. Loud noises cause the fluid within the vestibular canal to exert more pressure and the basilar membrane to vibrate to a greater extent. The resulting increased stimulation is interpreted by the brain as volume. As discussed in the Health feature, "Noise Pollution," on page 328, noise levels above 85 decibels (Table 14.3) may cause permanent hearing loss.

Table 14.3 Noises That Affect Hearing

Type of Noise	Sound Level (Decibels)	Effect
"Boom car," jet engine, shotgun, rock concert	Over 125	Beyond threshold of pain; potential for hearing loss high
Nightclub, thunderclap	Over 120	Hearing loss likely
Earbuds in external ear canal	110–120	Hearing loss likely
Chain saw, pneumatic drill, jackhammer, symphony orchestra, snowmobile, garbage truck, cement mixer	100–200	Regular exposure of more than 1 min risks permanent hearing loss.
Farm tractor, newspaper press, subway, motorcycle	90–100	Fifteen minutes of unprotected exposure; potentially harmful
Lawn mower, food blender	85–90	Continuous daily exposure for more than 8 hr can cause hearing damage.
Diesel truck, average city traffic noise	80–85	Annoying; constant exposure may cause hearing damage.

CHECK YOUR PROGRESS 14.5

1. Identify the structures of the ear involved in hearing, and provide a function for each.
2. Describe the role of mechanoreceptors in the sense of hearing.
3. Summarize how the spiral organ translates sound waves to nerve impulses.

CONNECTING THE CONCEPTS

For more information on the material in this section, refer to the following discussions:

Section 13.2 describes the function of the cerebral cortex area of the brain in hearing.

Figure 13.14 illustrates the structure of a nerve.

APPLICATIONS AND MISCONCEPTIONS

Why does that annoying song you hear seem to replay itself in your head all day?

Scientists who study the brain and special senses call such phenomena *earworms.* These are most likely caused by songs with repetitive, silly, or catchy lyrics. Their cause is not fully understood. Scientists have documented that the auditory cortex—the site for storing memories of sounds you've heard—becomes very active when earworms run through your mind. Earworms may be a way to keep the brain "idling," much as a car idles when you're at a stoplight. When you're ready for active thinking and problem solving, your brain is ready to go.

The good news is, you can make an earworm go away. Distract yourself with a complex task, such as reading or solving a puzzle. Eat something very spicy—the taste sensation will stimulate other areas of the brain. If all else fails, just think of a song that is equally annoying and it will most likely replace the first!

14.6 Sense of Equilibrium

LEARNING OUTCOMES

Upon completion of this section, you should be able to

1. Explain how mechanoreceptors are involved in the sense of equilibrium.
2. Identify the structures of the ear involved in the sense of equilibrium.
3. Distinguish between rotational and gravitational equilibrium.

The vestibular nerve originates in the semicircular canals, saccule, and utricle. It takes nerve signals to the brain stem and cerebellum (Fig. 14.14). Through its communication with the brain, the vestibular nerve helps us achieve equilibrium, but other structures in the body are also involved. For example, we already mentioned that proprioceptors are necessary for maintaining our equilibrium. Vision, if available, usually provides extremely helpful input the brain can act upon. To explain, let's take a look at the two sets of mechanoreceptors for equilibrium.

MP3 Sense of Hearing and Equilibrium

Rotational Equilibrium Pathway

Mechanoreceptors in the **semicircular canals** detect rotational and/or angular movement of the head—**rotational equilibrium** (Fig. 14.14*a*). The three semicircular canals are arranged so that there is one in each dimension of space. The base, or *ampulla,* of each of the three canals is slightly enlarged. Little hair cells, whose stereocilia are embedded within a gelatinous material called a cupula, are found within the ampullae. Each ampulla responds to head rotation in a

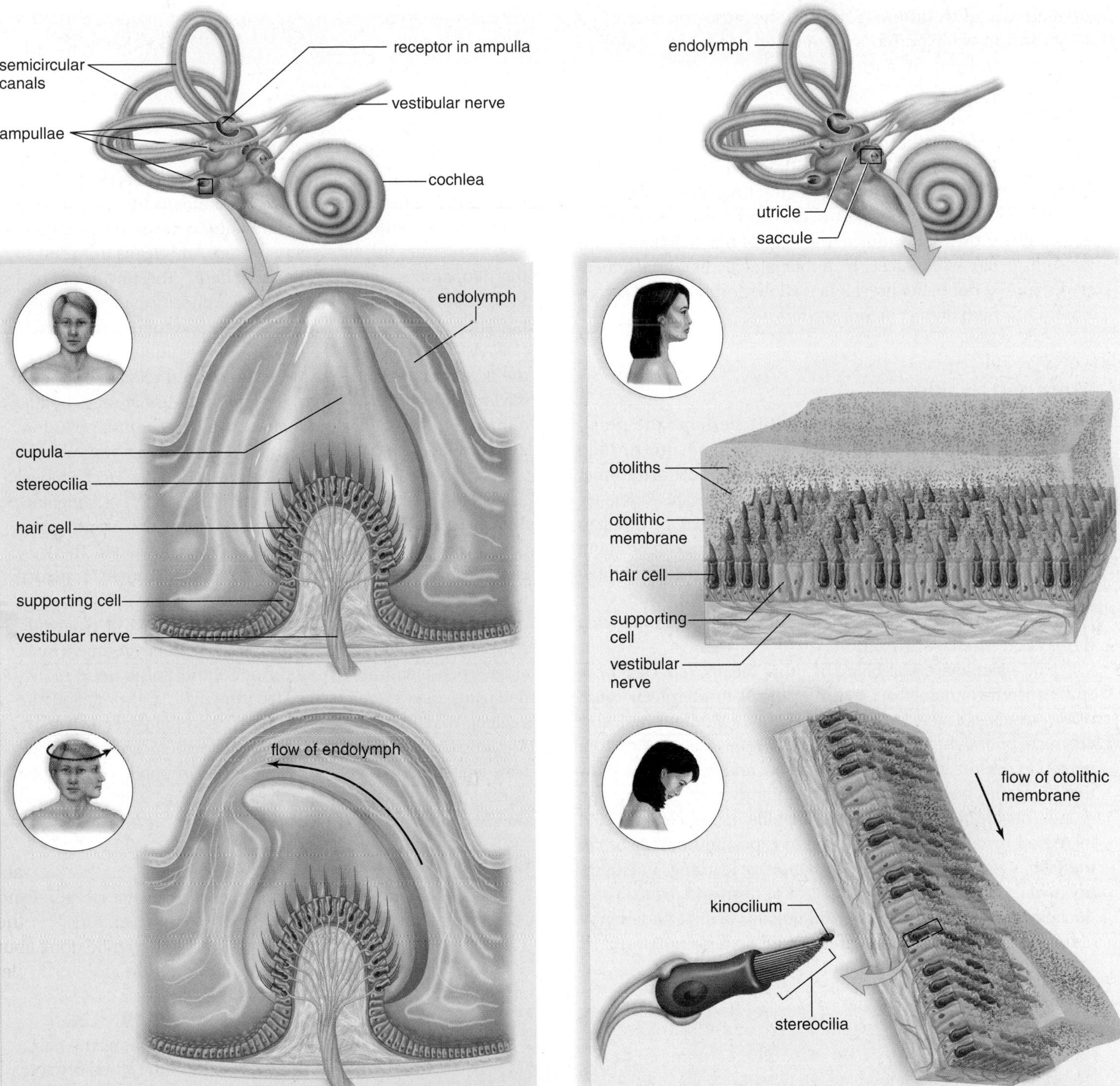

a. Rotational equilibrium: receptors in ampullae of semicircular canal

b. Gravitational equilibrium: receptors in utricle and saccule of vestibule

Figure 14.14 The mechanoreceptors of the inner ear and the sense of balance.
a. Rotational equilibrium is coordinated by receptors in the ampullae of the semicircular canals. **b.** Gravitational equilibrium is coordinated by receptors in the utricule and saccule located near the semicircular canals.

BIOLOGY MATTERS  Health

Noise Pollution

Though we can sometimes tune its presence out, unwanted noise is all around us. Noise pollution is noise from the environment that is annoying, distracting, and potentially harmful. It comes from airplanes, cars, lawn mowers, machinery, and our own loud music and that of our neighbors. It is present at our workplaces, in public spaces like amusement parks, and at home. Its prevalence allows loud noise to have a potentially high impact on our welfare.

Noise and Health

How does noise affect human health? Perhaps the greatest worry about noise pollution is that exposure to loud (over 85 decibels) or chronic noises can damage cells of the inner ear and cause hearing loss (Fig. 14B). When we are young, we often do not consider the damage that noise may be doing to our spiral organ. The stimulation of loud music is often sought by young people at rock concerts without regard to the possibility that their hearing may be diminished as a result. Over the years, loud noises can bring deafness and accompanying depression when we are seniors.

Noise can affect well-being by other means, too. Data from studies of environmental noise can be difficult to interpret because of the presence of other confounding factors, including physical or chemical pollution. The tolerance level for noise also varies from person to person. Nonetheless, laboratory and field studies show that noise may be detrimental in nonauditory ways. Its effects on mental health include annoyance, inability to concentrate, and increased irritability. Long-term noise exposure from air or car traffic may impair cognitive ability, language learning, and memory in children. Noise often causes loss of sleep and reduced productivity and can induce stress. Additionally, several studies have demonstrated a link between noise pollution and cardiovascular health, specifically hypertension.

Regulating Noise Pollution

Noise pollution has been a concern for several decades. In 1972, the Noise Control Act was passed as a means for coordinating federal noise control and research and to develop noise emission standards. The aim was to protect Americans from "noise that jeopardizes their health or welfare." The Environmental Protection Agency (EPA) had federal authority to regulate noise pollution, and their Office of Noise Abatement and Control (ONAC) worked on establishing noise guidelines. However, the activities of the ONAC were transferred to state and local governments in 1981. Today, there is no national noise policy, although the EPA does maintain a website on noise pollution, and its relationship to health, at www.epa.gov/air/noise.html.

Workplace noise exposure is controlled by the Occupational Safety and Health Administration (OSHA). OSHA has set guidelines for workplace noise. OSHA regulations require that protective gear be provided if sound levels exceed certain values. This may include noise-reducing earmuffs and other equipment for people who work around big equipment. However, OSHA guidelines don't cover things like telephone ringing and computer or typewriter noise that may be present in a nonindustrial environment such as an open-plan office. Aviation noise and traffic noise reduction plans are overseen by the Department of Transportation, the Federal Aviation Administration (FAA), and the Federal Highway Administration (FHWA), respectively. Local governments often have legislation that controls noise levels in public places, such as downtown areas and public parks. However, without national standards, the laws vary by location.

Questions to Consider

1. Given that noise pollution induces stress, what other body systems may be affected?
2. At a local level, do you think that more should be done to curb noise pollution in your neighborhood?

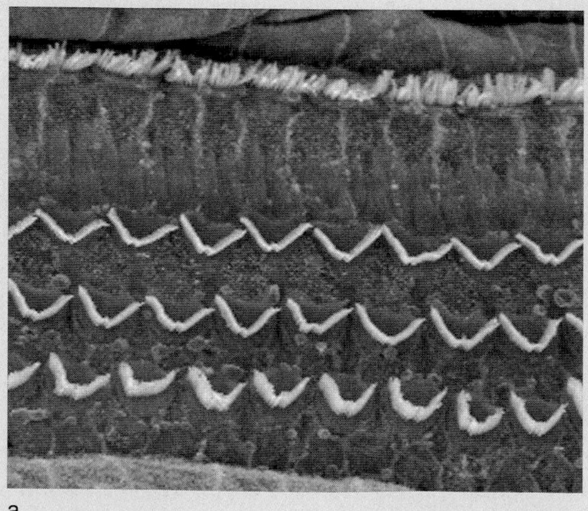

a.

b.

Figure 14B Loud noise damages the hair cells in the spiral organ. **a.** Normal hair cells in the spiral organ of a guinea pig. **b.** Damaged cells. This damage occurred after 24-hour exposure to a noise level equivalent to that at a rock concert (see Table 14.3). Hearing is permanently impaired because lost cells will not be replaced, and damaged cells may also die.

different plane of space because of the way the semicircular canals are arranged. As fluid within a semicircular canal flows over and displaces a cupula, the stereocilia of the hair cells bend. This causes a change in the pattern of signals carried by the vestibular nerve to the brain. The brain uses information from the hair cells within each ampulla of the semicircular canals to maintain equilibrium. Appropriate motor output to various skeletal muscles can correct our present position in space as needed.

Why does spinning around cause you to become dizzy? When we spin, the cupula slowly begins to move in the same direction we are spinning, and bending of the stereocilia causes hair cells to send messages to the brain. As time goes by, the cupula catches up to the rate we are spinning, and the hair cells no longer send messages to the brain. When we stop spinning, the slow-moving cupula continues to move in the direction of the spin and the stereocilia bend again, indicating that we are moving. Yet, the eyes know we have stopped. The mixed messages sent to the brain cause us to feel dizzy.

Gravitational Equilibrium Pathway

The mechanoreceptors in the utricle and saccule detect movement of the head in the vertical or horizontal planes, or **gravitational equilibrium.** The **utricle** and **saccule** are two membranous sacs located in the inner ear near the semicircular canals. Both of these sacs contain little hair cells whose stereocilia are embedded within a gelatinous material called an *otolithic membrane* (Fig. 14.14*b*). Calcium carbonate ($CaCO_3$) granules, or **otoliths,** rest on this membrane. The utricle is especially sensitive to horizontal (back-and-forth) movements and the bending of the head, and the saccule responds best to vertical (up-and-down) movements.

When the body is still, the otoliths in the utricle and the saccule rest on the otolithic membrane above the hair cells. When the head bends or the body moves in the horizontal and vertical planes, the otoliths are displaced. The otolithic membrane sags, bending the stereocilia of the hair cells beneath. If the stereocilia move toward the largest stereocilium, called the *kinocilium,* nerve impulses increase in the vestibular nerve. If the stereocilia move away from the kinocilium, nerve impulses decrease in the vestibular nerve. The frequency of nerve impulses in the vestibular nerve indicates whether you are moving up or down.

These data reach the cerebellum, which uses them to determine the direction of the movement of the head at that moment. Remember that the cerebellum (see Chapter 13) is vital to maintaining balance and gravitational equilibrium. The cerebellum processes information from the inner ear (the semicircular canals, utricle, and saccule) as well as visual and proprioceptive inputs. In addition, the motor cortex in the frontal lobe of the brain signals where the limbs should be located at any particular moment. After integrating all these nerve inputs, the cerebellum coordinates skeletal muscle contraction to correct our position in space if necessary.

Continuous stimulation of the stereocilia can contribute to motion sickness, especially when messages reaching the brain conflict with visual information from the eyes. Imagine that you are standing inside a ship that is tossing up and down on the waves. Your visual inputs signal that you are standing still, because you can see the wall in front of you and that wall isn't moving. However, the inputs from all three sensory areas of the inner ear tell your brain that you are moving up and down and from side to side. If you can match the two sets of information coming into the brain, you will begin to feel better. Thus, it makes sense to stand on deck if possible, so that visual signals and inner-ear signals both tell your brain that you're moving. Some antihistamine drugs, such as dimenhydrinate (Dramamine), reduce the excitability of the receptors in the inner ear, thus reducing the impulses received by the cerebellum, and alleviating motion sickness.

Animation Senses and Balance

CHECK YOUR PROGRESS 14.6

1. Explain the location and function of the structures that are involved in maintaining balance.
2. Describe how rotational equilibrium is achieved.
3. Contrast rotational and gravitational equilibrium and explain how the two work together to maintain balance.

CONNECTING THE CONCEPTS

For more information on the sense of equilibrium, refer to the following discussions:

Section 13.1 examines the structure of a neuron and the generation of a nerve impulse.

Section 13.2 explains the role of the cerebellum in the processing of sensory information regarding balance.

CASE STUDY CONCLUSION

The sounds that people hear with cochlear implants are different from what hearing people are familiar with. However, with training and experience, implant recipients can develop the ability to understand speech, and they function very much like a person with normal hearing. Several recent studies have shown that deaf babies that receive implants as young as six months of age develop language and speech skills very much like those of hearing children. By identifying the gene responsible for many forms of nonsyndromic deafness (*GJB2*), genetic counselors will soon be able to inform parents with a family history of deafness as to the chances of having a deaf child. Furthermore, by understanding the function of this gene, it may be possible to develop a greater understanding as to how the cochlea develops in the fetus and how environmental factors and genetics play a role in some forms of deafness.

MEDIA STUDY TOOLS

Enhance your study of this chapter with media! Visit **www.mhhe.com/maderhuman13e** and go to "Media Study Tools" for this chapter to access the following:

Animations	Videos	MP3 Files
14.1 Positive and Negative Feedback **14.4** Vision **14.5** Effects of Sound Waves on Cochlear Structures • Hearing **14.6** Senses and Balance	**14.3** Sex and the Senses **14.4** Artificial Eye	**14.1** Sensations and Receptors **14.3** Taste and Smell **14.4** Sense of Vision **14.5** Sense of Hearing and Equilibrium **14.6** Sense of Hearing and Equilibrium

SUMMARIZE

14.1 Overview of Sensory Receptors and Sensations

Signal transduction begins with the detection of stimuli by **sensory receptors.** These receptors may detect stimuli from within the body (**interoceptors**) or the external environment (**exteroceptors**). There are four classes of receptors:

- **Chemoreceptors** detect chemical stimuli. **Nociceptors** are a form of chemoreceptor that detects pain.
- **Photoreceptors** detect light stimuli.
- **Mechanoreceptors** detect stimuli generated by mechanical forces.
- **Thermoreceptors** detect stimuli caused by changes in temperature.

All of these classes function as follows:

- **Sensory receptors** perform **integration** of the incoming signals. They then initiate nerve signals to the spinal cord and/or brain. **Sensory adaptation** may occur if the stimuli are repeated continuously.
- **Sensation** occurs when nerve signals reach the cerebral cortex.
- Perception is an interpretation of sensations.

14.2 Somatic Senses

Somatic senses are associated with the skin, muscles, joints, and viscera. The sensory receptors associated with the somatic senses include the following:

- **Proprioceptors** (mechanoreceptors) are involved in reflex actions and help maintain equilibrium and posture.
- **Cutaneous receptors** in the skin sense touch, pressure, and temperature.
- Nociceptors detect pain by responding to chemical signals from damaged tissues.

14.3 Senses of Taste and Smell

Taste and smell are due to chemoreceptors stimulated by molecules in the environment.

Sense of Taste

Receptors for taste are found primarily on the **taste buds.** Microvilli of taste cells have receptor proteins for molecules that cause the brain to distinguish sweet, sour, salty, bitter, and umami.

Sense of Smell

The cilia of **olfactory cells** have receptor proteins for molecules that cause the brain to distinguish odors.

14.4 Sense of Vision

Vision depends on the eye, the optic nerves, and the visual areas of the cerebral cortex.

Anatomy and Physiology of the Eye

The eye has three layers:

- The **sclera** (outer layer) protects and supports the eye.
- The **choroid** (middle, pigmented layer) absorbs stray light rays.
- The **retina** (inner layer) contains the rod cells (sensory receptors for dim light) and cone cells (sensory receptors for bright light and color).
- *Function of the Lens:* Light enters the eye through the **pupil,** the size of which is regulated by the **iris.** The **lens,** with assistance from the **cornea, aqueous humor,** and **vitreous humor,** brings the light rays to **focus** on the retina, typically on the **fovea centralis** region of the retina. To see a close object, **visual accommodation** occurs as the lens becomes round and thick.
- *Photoreceptors:* Two types of photoreceptors are located on the retina: **rod cells** (black-white vision) and **cone cells** (color vision). Both contain **rhodopsin,** which includes retinal (vitamin A). An area called the **blind spot** lacks rods and cones.
- *Visual Pathway to the Brain:* The visual pathway begins when light strikes photoreceptors (rod cells and cone cells) in the retina. The **optic nerves** carry nerve impulses from the eyes to the **optic chiasma.** The nerve impulse leaves the optic chiasma along **optic tracts** to the thalamus before reaching the primary vision area in the occipital lobe of the brain.

Abnormalities of the Eye

Vision problems may be caused by a buildup of pressure in the eye (**glaucoma**), genetic factors (color blindness), or the shape of the eye (which can result in being **nearsighted, farsighted,** or having **astigmatism**).

14.5 Sense of Hearing

Hearing depends on the ear, the cochlear nerve, and the auditory areas of the cerebral cortex.

Anatomy and Physiology of the Ear

The ear has three parts:

- In the **outer ear,** the **pinna** and the **auditory canal** direct sound waves to the middle ear.
- In the **middle ear,** the **tympanic membrane** (including the **oval window** and **round window**), and the **ossicles** (**malleus, incus,** and **stapes**) amplify sound waves.
- In the **inner ear,** the **semicircular canals** and **vestibule** detect rotational equilibrium; the **utricle** and **saccule** detect gravitational equilibrium; and the **cochlea** houses the spiral organ, which contains mechanoreceptors, **hair cells** with **stereocilia,** for hearing.

The **auditory tube** (or eustachian tube) helps to equalize pressure across the tympanic membrane.

- *Auditory Pathway to the Brain:* The auditory pathway begins when the outer ear receives and the middle ear amplifies sound waves that then strike the oval window membrane.
- The mechanoreceptors for hearing are hair cells on the basilar membrane of the **spiral organ.**
- Nerve signals begin in the **cochlear nerve** and are carried to the primary auditory area in the temporal lobe of the cerebral cortex.

14.6 Sense of Equilibrium

The ear also contains mechanoreceptors for equilibrium.

Rotational Equilibrium Pathway

- **Rotational equilibrium** is due to mechanoreceptors (hair cells) in the semicircular canals that detect rotational and/or angular movement of the head.

Gravitational Equilibrium Pathway

- **Gravitational equilibrium** is due to mechanoreceptors (hair cells) in the utricle and saccule that detect head movement in the vertical or horizontal planes. Calcium carbonate granules called **otoliths** assist in this process.

ASSESS

Testing Your Knowledge of the Concepts

1. Contrast exteroceptors and interoceptors. (page 312)
2. What is sensory adaptation? (page 313)
3. What are proprioceptors, and what is the role of the Golgi tendon organs? (page 314)
4. List the cutaneous receptors and the type of stimulus to which each responds. (pages 314–315)
5. Describe the structure of a taste bud and explain how a taste cell functions. (pages 316–317)
6. Describe the structure and function of the olfactory epithelium. How does the sense of smell come about? (pages 316–317)
7. Describe the anatomy of the eye and the function of each part. (pages 318–319)
8. How does the eye respond when viewing an object far away? When viewing a close object? (pages 319–320)
9. Describe the sequence of events after rhodopsin absorbs light. (pages 320–321)
10. What are three abnormalities caused by a misshapen eye, and what type lens can be helpful? (page 322)
11. Describe the anatomy of the ear and the function of each part. (pages 324–325)
12. Explain the pathway of sound and how sound is produced. (pages 325–326)
13. Describe the structures that are involved in equilibrium and state their functions. (pages 326–327)
14. A sensory receptor
    a. is the first portion of a reflex arc.
    b. initiates nerve impulses.
    c. can be internal or external.
    d. All of the choices are correct.
15. Receptors sensitive to changes in blood pressure are
    a. interoceptors.
    b. exteroceptors.
    c. proprioceptors.
    d. nociceptors.
16. Conscious interpretation of changes in the internal and external environment is called
    a. responsiveness.
    b. perception.
    c. sensation.
    d. accommodation.
17. Pain perceived as coming from another location is known as
    a. intercepted pain.
    b. phantom pain.
    c. referred pain.
    d. parietal pain.
18. Which structure of the eye is incorrectly matched with its function?
    a. lens—focusing
    b. cones—color vision
    c. iris—regulation of amount of light
    d. choroid—location of cones
    e. sclera—protection
19. Which of the following gives the correct path for light rays entering the human eye?
    a. sclera, retina, choroid, lens, cornea
    b. fovea centralis, pupil, aqueous humor, lens
    c. cornea, pupil, lens, vitreous humor, retina
    d. cornea, fovea centralis, lens, choroid, rods
    e. optic nerve, sclera, choroid, retina, humors
20. Adjustment of the lens to focus on objects close to the viewer is called
    a. convergence.
    b. visual accommodation.
    c. focusing.
    d. constriction.
21. To focus on objects that are close to the viewer, the
    a. suspensory ligaments must be pulled tight.
    b. lens needs to become more rounded.
    c. ciliary muscle will be relaxed.
    d. image must focus on the area of the optic nerve.

22. Which abnormality of the eye is incorrectly matched with its cause?
- **a.** astigmatism—either the lens or cornea is not even
- **b.** farsightedness—eye is shorter than usual
- **c.** nearsightedness—image focuses behind the retina
- **d.** color blindness—genetic disorder in which certain types of cones may be missing

23. Which of the following is not involved in the sense of hearing?
- **a.** auditory canal
- **b.** tympanic membrane
- **c.** ossicles
- **d.** semicircular canals
- **e.** cochlea

24. The middle ear is separated from the inner ear by the
- **a.** oval window.
- **b.** tympanic membrane.
- **c.** round window.
- **d.** Both a and c are correct.

25. Which one of these correctly describes the location of the spiral organ?
- **a.** between the tympanic membrane and the oval window in the inner ear
- **b.** in the utricle and saccule within the vestibule
- **c.** between the tectorial membrane and the basilar membrane in the cochlear canal
- **d.** between the nasal cavities and the throat
- **e.** between the outer and inner ear within the semicircular canals

26. Which of the following structures would allow you to know that you were upside down, even if you were in total darkness?
- **a.** utricle and saccule
- **b.** cochlea
- **c.** semicircular canals
- **d.** tectorial membrane

27. Both olfactory receptors and sound receptors
- **a.** are chemoreceptors.
- **b.** are a part of the brain.
- **c.** are mechanoreceptors.
- **d.** initiate nerve impulses.
- **e.** All of the choices are correct.

28. Label this diagram of a human eye.

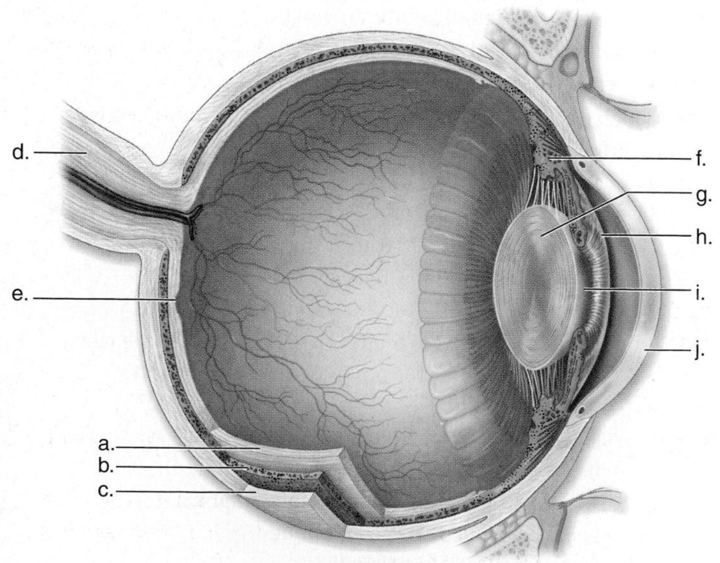

29. Label this diagram of the human ear.

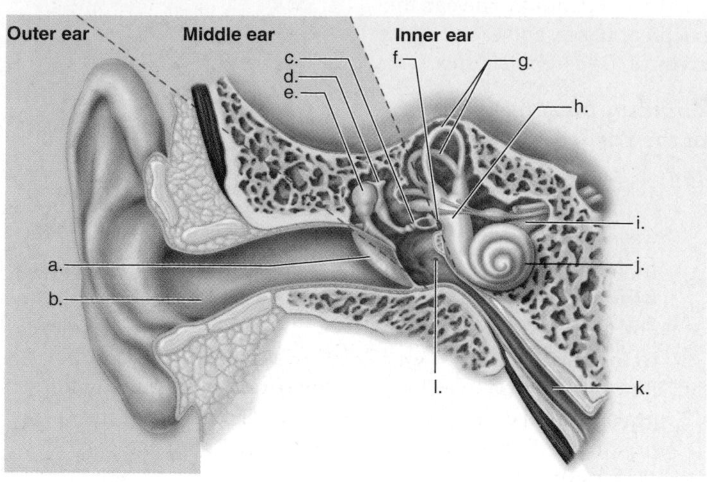

ENGAGE

Thinking Critically About the Concepts

Both exteroceptors and interoceptors are essential to provide information to the central nervous system. In the chapter opener, Marlene is deaf due to a birth defect. Loss of one of the special senses is immediately debilitating and can seriously affect homeostasis over time. Without vision, hearing, taste, or smell, we are vulnerable to dangerous conditions in our surroundings. Loss of cutaneous senses—touch, pressure, temperature, pain—can immediately impact survival. Imagine, for example, being unaware that you've been seriously cut because you can't feel pain. In that circumstance, it would be possible to bleed to death.

Keep in mind the hugely important role of interoceptors in maintaining homeostasis. Any unacceptable change in internal homeostatic conditions is reported to the brain by some type of interoceptor. Pressoreceptors of the cardiovascular system help to regulate blood pressure. Osmoreceptors in the hypothalamus help to control water-salt balance, and chemoreceptors monitor pH in the blood. Our lives depend on proper function of all components of the sensory system.

1. What receptors are activated when we enjoy supper in a pizza restaurant?
2. Besides the blood pH mentioned, what other homeostatic conditions are monitored by chemoreceptors?
3. If a person takes a blow to the back of the head, what sense is most likely to be affected?
4. Airport and construction workers are likely to be exposed to continuous, loud noises. What would you predict the long-term effect on their hearing to be? Why?
5. The acoustic and vestibular nerves travel together to the brain. If a tumor grows on this combined nerve, what sensations will be affected?
6. Explain how a brain tumor in the cerebral cortex might be diagnosed by a doctor.

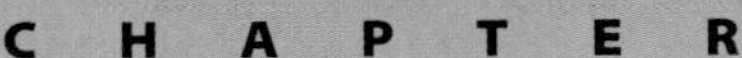

Endocrine System

CASE STUDY DIABETES

For some time, Hank had been feeling very sluggish and had been losing weight. At first, he attributed this to his very active lifestyle. Between school, work, and his social activities, Hank had very little time for sleep. However, he was beginning to notice that he was always thirsty and was urinating much more frequently than usual. Concerned about his health, Hank visited the local health clinic where he discussed his health history and symptoms with the physician. The doctor mentioned that his symptoms were consistent with many disorders, including both viral infections and diabetes. As a quick test, the doctor ordered a urinalysis test to see if there was any glucose in his urine, which would indicate that Hank's symptoms were caused by diabetes mellitus, a disease that affects over 25.8 million Americans. The results of the urinalysis indicated that there were small amounts of glucose in Hank's urine, a sign that Hank's body may not be adequately maintaining its blood glucose levels. The doctor scheduled Hank for a blood glucose test the following morning and instructed him to not eat or drink anything for 8 hours prior to the test.

During a blood glucose test, a small vial of blood is removed and the amount of glucose measured. Normally, after 8 hours of fasting, the blood glucose level should be between 70 and 100 mg per deciliter (mg/dl) of blood. Hank's value was slightly above this, but it was not high enough for the doctor to conclude that diabetes was the cause of Hank's symptoms. The next test was an oral glucose tolerance test (OGTT). In this test, Hank drank a solution containing 100 grams (g) of glucose. Then, over the next 3 hours, five additional vials of blood were drawn and tested for glucose levels. In a normal individual participating in this test, blood glucose levels rise rapidly and then fall to below 140 mg/dl within 2 hours. In Hank's case, the response was much slower, and his 2-hour blood glucose level was 150 mg/dl. The physician told Hank that the cause of his symptoms was most likely type 2 diabetes mellitus, a disease of the endocrine system, the organ system that is responsible for the long-term homeostasis of the body.

As you read through the chapter, think about the following questions:

1. What hormones control the level of glucose in the blood?
2. What is the difference between type 1 and type 2 diabetes?
3. How do feedback mechanisms help control blood glucose levels?

CHAPTER CONCEPTS

15.1 Endocrine Glands
Endocrine glands produce hormones that are secreted into the bloodstream and distributed to target cells where they alter cellular metabolism.

15.2 Hypothalamus and Pituitary Gland
The hypothalamus controls the secretions of the pituitary gland, which, in turn, controls the secretion of certain other glands.

15.3 Thyroid and Parathyroid Glands
The hormones of the thyroid and parathyroid glands stimulate cellular metabolism and help maintain blood calcium homeostasis.

15.4 Adrenal Glands
The adrenal glands release hormones to respond to both long-term and short-term stress.

15.5 Pancreas
The pancreas secretes insulin and glucagon, which together keep the blood glucose level fairly constant.

15.6 Other Endocrine Glands
Other endocrine glands include the testes and ovaries, the thymus, and the pineal gland.

15.7 Hormones and Homeostasis
The nervous and endocrine systems work together to control the other organ systems and maintain homeostasis throughout the body.

BEFORE YOU BEGIN

Before beginning this chapter, take a few moments to review the following discussions:

Section 2.5 What is the structure of a steroid?

Section 4.8 How are negative feedback mechanisms involved in homeostasis?

Section 13.2 What is the role of the hypothalamus in the nervous system?

15.1 Endocrine Glands

LEARNING OUTCOMES

Upon completion of this section, you should be able to

1. Distinguish between the mode of action of a neurotransmitter and that of a hormone.
2. Distinguish between endocrine and exocrine glands.
3. Identify the organs and glands of the endocrine system.
4. Compare the action of peptide and steroid hormones.

The nervous system and the endocrine system work together to regulate the activities of the other systems. Both systems use chemical signals when they respond to changes that might alter homeostasis. However, they have different means of delivering these signals (Fig. 15.1). As discussed in Chapter 13, the nervous system is composed of neurons. In this system, sensory receptors detect changes in the internal and external environment. The central nervous system (CNS) then integrates the information and responds by stimulating muscles and glands. Communication depends on nerve signals, conducted in axons, and neurotransmitters, which cross synapses. Axon conduction occurs rapidly and so does diffusion of a neurotransmitter across the short distance of a synapse. In other words, the nervous system is organized to respond rapidly to stimuli. This is particularly useful if the stimulus is an external event that endangers our safety—we can move quickly to avoid being hurt.

The endocrine system functions differently. The **endocrine system** is largely composed of glands (Fig. 15.2). These glands secrete **hormones,** carried by the bloodstream to target cells throughout the body. It takes time to deliver hormones, and it takes time for cells to respond. The effect initiated by the endocrine system is longer lasting. In other words, the endocrine system is organized for a slow but prolonged response. Table 15.1 summarizes the hormones of the endocrine system and provides the function and target of these hormones in the body.

MP3 Major Endocrine Glands

There is a difference in function between an endocrine gland and an exocrine gland. **Exocrine glands** have ducts and secrete their products into these ducts. The glands' products are carried to the lumens of other organs or outside the body. The accessory glands of the digestive system (see Chapter 8) are good examples of the exocrine glands. For example, the salivary glands send saliva into the mouth by way of the salivary

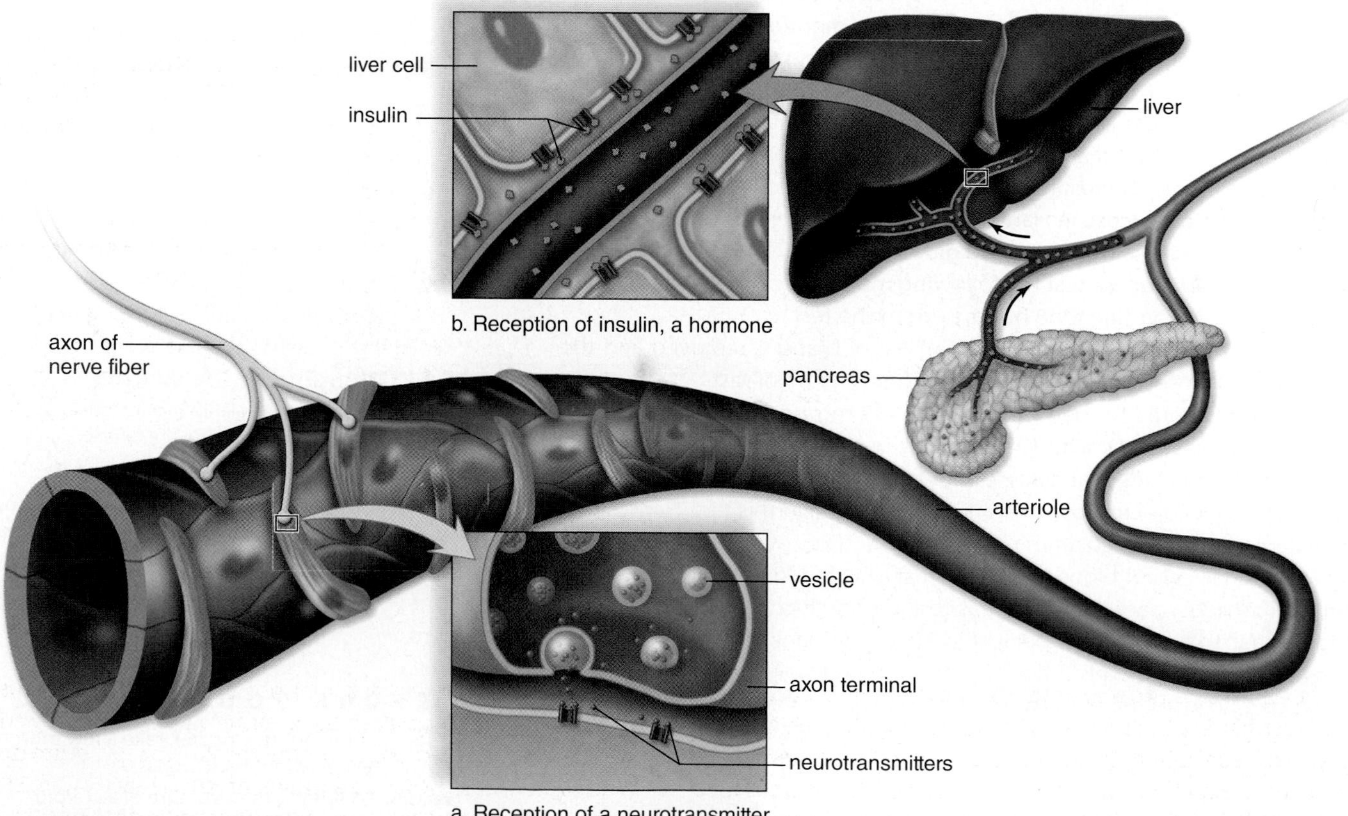

Figure 15.1 The action of a neurotransmitter differs from that of a hormone.
a. Nerve impulses passing along an axon cause the release of a neurotransmitter. The neurotransmitter, a chemical signal, causes the wall of an arteriole to constrict. **b.** The hormone insulin, a chemical signal, travels in the cardiovascular system from the pancreas to the liver, where it causes liver cells to store glucose as glycogen.

Hypothalamus

Pituitary Gland
Posterior pituitary
Anterior pituitary

Thyroid

Adrenal Gland
Adrenal cortex

Parathyroids

parathyroid glands
(posterior surface of thyroid)

Thymus

Pancreas

Gonads
Testes
Ovaries

testis
(male)

ovary (female)

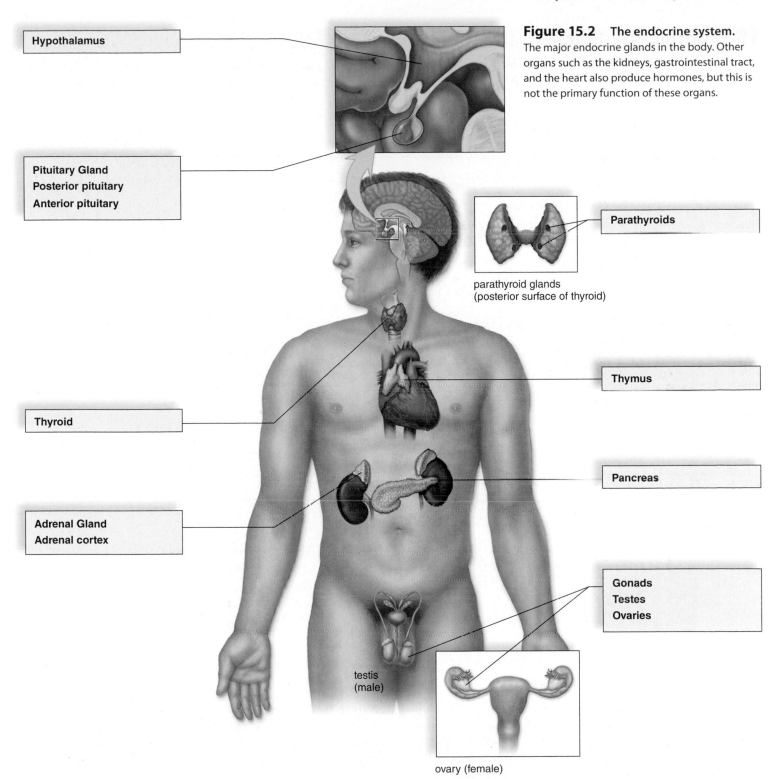

Figure 15.2 The endocrine system. The major endocrine glands in the body. Other organs such as the kidneys, gastrointestinal tract, and the heart also produce hormones, but this is not the primary function of these organs.

ducts. In contrast, **endocrine glands** secrete their products into the bloodstream, which delivers them throughout the body. It must be stressed that only certain cells, called target cells, can respond to a specific hormone. A target cell for a particular hormone will have a receptor protein for that hormone. The receptor protein and hormone bind together like a key that fits a lock. The target cell then responds to that hormone.

Both the nervous system and the endocrine system make use of negative feedback mechanisms. If the blood pressure falls, sensory receptors signal a control center in the brain. This center sends out nerve signals to the arterial walls so that they constrict, and blood pressure rises. Now the sensory receptors are no longer stimulated, and the feedback mechanism is inactivated. Similarly, a rise in blood glucose level causes the pancreas to release insulin. This, in turn, promotes glucose uptake by the liver, muscles, and other cells of the body (see Fig. 15.1). When the blood glucose level falls, the pancreas no longer secretes insulin.

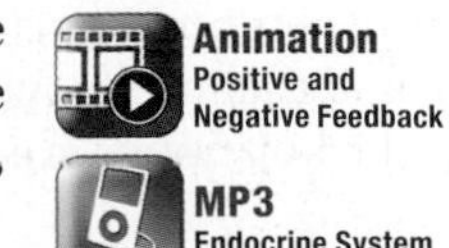

Table 15.1 Principal Endocrine Glands and the Hormones They Produce

Endocrine Gland	Hormone Released	Target Tissues/Organs	Chief Functions of Hormone
Hypothalamus	Hypothalamic-releasing	Anterior pituitary	Regulates anterior pituitary hormones and inhibiting hormones
Pituitary gland			
Posterior pituitary	Antidiuretic (ADH)	Kidneys	Stimulates water reabsorption by kidneys
	Oxytocin	Uterus, mammary glands	Stimulates uterine muscle contraction, release of milk by mammary glands
Anterior pituitary	Thyroid-stimulating (TSH)	Thyroid	Stimulates thyroid
	Adrenocorticotropic (ACTH)	Adrenal cortex	Stimulates adrenal cortex
	Gonadotropic (FSH, LH)	Gonads	Egg and sperm production; sex hormone production
	Prolactin (PRL)	Mammary glands	Milk production
	Growth (GH)	Soft tissues, bones	Cell division, protein synthesis, and bone growth
	Melanocyte-stimulating (MSH)	Melanocytes in skin	Unknown function in humans; regulates skin color in lower vertebrates
Thyroid	Thyroxine (T_4) and triiodothyronine (T_3)	All tissues	Increases metabolic rate; regulates growth and development
	Calcitonin	Bones, kidneys, intestine	Lowers blood calcium level
Parathyroids	Parathyroid (PTH)	Bones, kidneys, intestine	Raises blood calcium level
Adrenal gland			
Adrenal cortex	Glucocorticoids (cortisol)	All tissues	Raise blood glucose level; stimulate breakdown of protein
	Mineralocorticoids (aldosterone)	Kidneys	Reabsorb sodium and excrete potassium
	Sex hormones	Gonads, skin, muscles, bones	Stimulate reproductive organs and bring about sex characteristics
Adrenal medulla	Epinephrine and norepinephrine	Cardiac and other muscles	Released in emergency situations; raise blood glucose level
Pancreas	Insulin	Liver, muscles, adipose tissue	Lowers blood glucose level; promotes glycogen formation
	Glucagon	Liver, muscles, adipose tissue	Raises blood glucose level
Gonads			
Testes	Androgens (testosterone)	Gonads, skin, muscles, bones	Stimulate male sex characteristics
Ovaries	Estrogens and progesterone	Gonads, skin, muscles, bones	Stimulate female sex characteristics
Thymus	Thymosins	T lymphocytes	Stimulate production and maturation of T lymphocytes
Pineal gland	Melatonin	Brain	Controls circadian and circannual rhythms; possibly involved in maturation of sexual organs

Hormones Are Chemical Signals

Like other chemical signals, hormones are a means of communication between cells, between body parts, and even between individuals. They affect the metabolism of cells that have receptors to receive them (Fig. 15.3). In a condition called *androgen insensitivity,* an individual has X and Y sex chromosomes. The testes, which remain in the abdominal cavity, produce the sex hormone testosterone. However, the body cells lack receptors that are able to combine with testosterone. Therefore, the individual appears to be a normal female.

Like testosterone, most hormones act at a distance between body parts. They travel in the bloodstream from the gland that produced them to their target cells. Also considered to be hormones are the secretions produced by neurosecretory cells in the hypothalamus, a part of the brain. They travel in the capillary network that runs between the hypothalamus and the pituitary gland. Some of these secretions stimulate the pituitary to secrete its hormones, and others prevent it from doing so.

Not all hormones act between body parts. As we shall see, prostaglandins are a good example of *local hormones.* After prostaglandins are produced, they are not carried elsewhere in the bloodstream. Instead, they affect neighboring cells, sometimes promoting pain and inflammation. Also, growth factors are local hormones that promote cell division and mitosis.

Chemical signals that influence the behavior of other individuals are called **pheromones.** Other animals rely

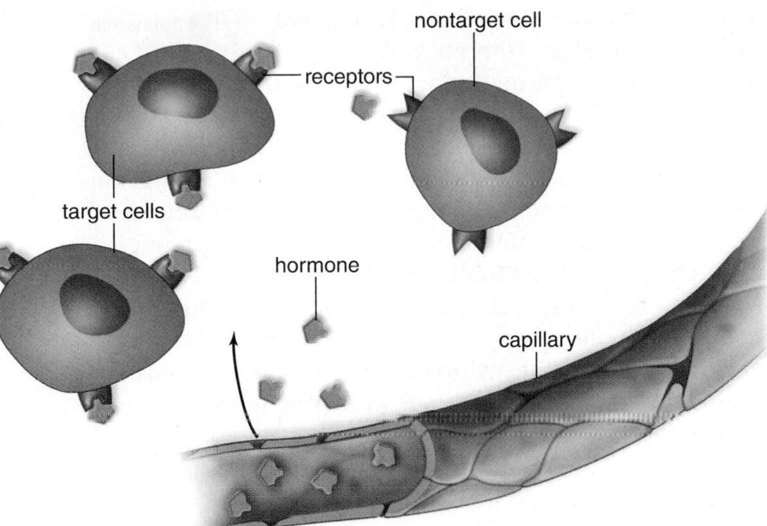

Figure 15.3 Hormones target specific cells.
Most hormones are distributed by the bloodstream to target cells. Target cells have receptors for the hormones, and a hormone combines with a receptor as a key fits a lock.

heavily on pheromones for communication. Pheromones are used to mark one's territory and to attract a mate. Humans produce pheromones too. A researcher has isolated one released by men that reduces premenstrual nervousness and tension in women. Women who live in the same household often have menstrual cycles in synchrony. This is likely caused by the armpit secretions of a woman who is menstruating affecting the menstrual cycle of other women in the household.

The Action of Hormones

Hormones have a wide range of effects on cells. Some of these effects induce a target cell to increase its uptake of particular substances (such as glucose) or ions (such as calcium). Other effects bring about an alteration of the target cell's structure in some way. A few hormones simply influence cell metabolism. Growth hormone is a peptide that influences cell metabolism leading to a change in the structure of bone. The term **peptide hormone** is used to include hormones that are peptides, proteins, glycoproteins, and modified amino acids. Growth hormone is a protein produced and secreted by the anterior pituitary. **Steroid hormones** have the same complex of four carbon rings because they are all derived from cholesterol (see Fig. 2.20).

The Action of Peptide Hormones Most endocrine glands secrete peptide hormones. The actions of peptide hormones can vary. We will concentrate on what happens in muscle cells after the hormone epinephrine binds to a receptor in the plasma membrane (Fig. 15.4). In muscle cells, the reception of epinephrine leads to the breakdown of glycogen to glucose, which provides energy for ATP production. The immediate result of binding is the formation of **cyclic adenosine monophosphate (cAMP)**. Cyclic AMP contains one phosphate group attached to adenosine at two locations. Therefore, the molecule is cyclic. Cyclic AMP activates a protein kinase enzyme in the cell. This enzyme, in turn, activates another enzyme, and so forth. The series of enzymatic reactions that follows cAMP formation is called an enzyme cascade. Each enzyme can be used over and over at every step of the cascade, so more enzymes are involved. Finally, many molecules of glycogen are broken down to glucose, which enters the bloodstream.

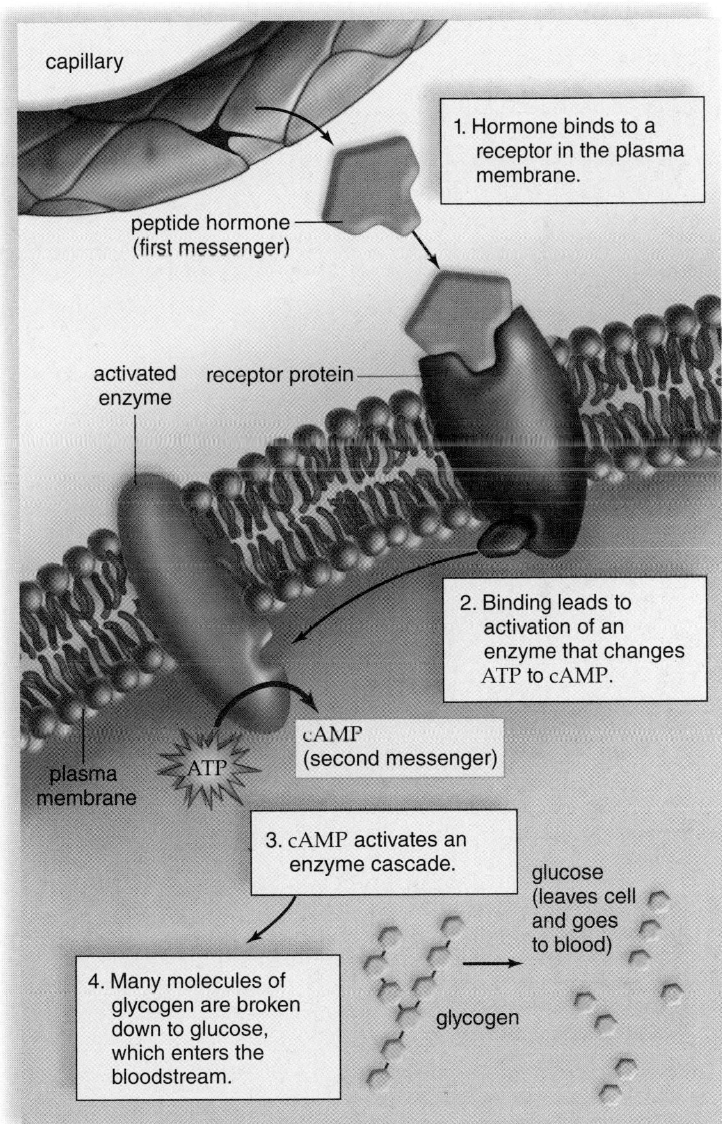

Figure 15.4 Action of a peptide hormone.
A peptide hormone (first messenger) binds to a receptor in the plasma membrane. Thereafter, cyclic AMP (second messenger) forms and activates an enzyme cascade.

Typical of a peptide hormone, epinephrine never enters the cell. Therefore, the hormone is called the first messenger; cAMP, which sets the metabolic machinery in motion, is called the **second messenger.** To explain this terminology, let's imagine that the adrenal medulla, which produces epinephrine, is like the home office that sends out a courier (i.e., the hormone epinephrine is the first messenger) to a factory (the cell). The courier doesn't have a pass to enter the factory, so when he arrives at the factory, he tells a supervisor through the screen door that the home office wants the factory to produce a particular product. The supervisor (i.e., cAMP, the second messenger) walks over and flips a switch that starts the machinery (the enzymatic pathway), and a product is made.

The Action of Steroid Hormones Only the adrenal cortex, the ovaries, and the testes produce steroid hormones. Thyroid hormones are amines and act similarly to steroid hormones, even though they have a different structure. Steroid hormones do not bind to plasma membrane receptors. Instead they are able to enter the cell because they are lipids (Fig. 15.5). Once inside, a steroid hormone binds to a receptor, usually in the nucleus but sometimes in the cytoplasm. Inside the nucleus, the hormone-receptor complex binds with DNA and activates certain genes. Messenger RNA (mRNA) moves to the ribosomes in the cytoplasm, and protein (e.g., enzyme) synthesis follows. To continue our analogy, a steroid hormone is like a courier that has a pass to enter the factory (the cell). Once inside, it makes contact with the plant manager (DNA), who sees to it that the factory (cell) is ready to produce a product.

Animation
Mechanism of Steroid Hormone Action

An example of a steroid hormone is aldosterone, which is produced by the adrenal glands. Aldosterone targets the kidneys where it helps to regulate the water-salt balance of the blood. In general, steroid hormones act more slowly than peptide hormones because it takes more time to synthesize new proteins than to activate enzymes already present in cells. Their action lasts longer, however.

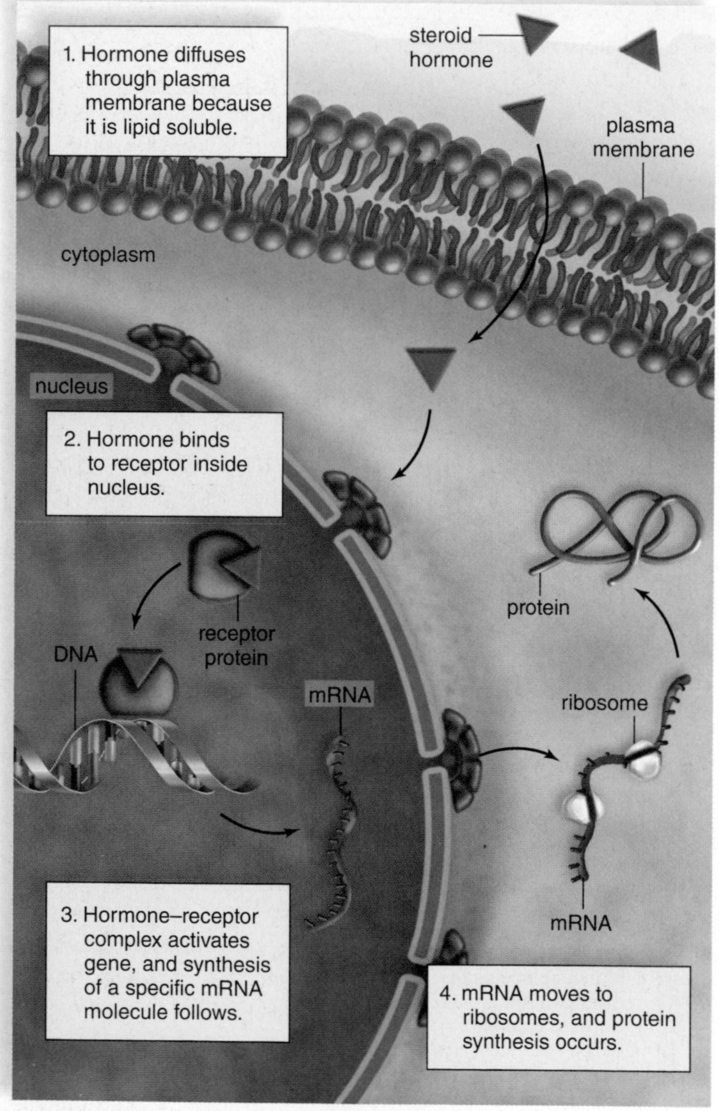

Figure 15.5 Action of a steroid hormone.
A steroid hormone passes directly through the target cell's plasma membrane before binding to a receptor in the nucleus or cytoplasm. The hormone–receptor complex binds to DNA, and gene expression follows.

CHECK YOUR PROGRESS 15.1

1. State the role of a hormone.
2. Compare and contrast the nervous and endocrine systems with regard to function and the types of signals used.
3. Summarize the differences between a peptide and steroid hormone.
4. Explain why second-messenger systems are needed for peptide hormones.

CONNECTING THE CONCEPTS

For more information on the interactions in this section, refer to the following discussions:

Sections 2.5 and 2.6 summarize the roles of steroids and proteins in the body.

Figure 3.6 illustrates the structure of the plasma membrane and the proteins that are associated with it.

Section 13.2 describes the location and function of the hypothalamus, which integrates the nervous and endocrine systems.

15.2 Hypothalamus and Pituitary Gland

LEARNING OUTCOMES

Upon completion of this section, you should be able to

1. Explain the role of the hypothalamus in the endocrine system.
2. List the hormones produced by the anterior and posterior pituitary glands, and provide a function for each.
3. Summarize the conditions produced by excessive and inadequate levels of growth hormone.

The **hypothalamus** acts as the link between the nervous and endocrine systems. It regulates the internal environment through communications with the autonomic nervous system. For example, it helps control body temperature and water-salt balance. The hypothalamus also controls the glandular secretions of the **pituitary gland.** The pituitary, a small gland about 1 cm in diameter, is connected to the hypothalamus by a stalklike structure. The pituitary has two portions: the posterior and the anterior pituitary.

MP3 Hormonal Secretion and Action

Posterior Pituitary

Neurons in the hypothalamus called neurosecretory cells produce the hormones **antidiuretic hormone** (ADH) and **oxytocin** (Fig. 15.6). These hormones pass through axons into the **posterior pituitary** where they are stored in axon endings. Certain neurons in the hypothalamus are sensitive to the water-salt balance of the blood. When these cells determine that the blood is too concentrated, ADH is released from the posterior pituitary. Upon reaching the kidneys, ADH causes more water to be reabsorbed into kidney capillaries. As the blood becomes dilute, ADH is no longer released. This is an example of control by negative feedback because the effect of the hormone (to dilute blood) acts to shut down the release of the hormone. Negative feedback maintains stable conditions and homeostasis.

Inability to produce ADH causes *diabetes insipidus.* A person with this type of diabetes produces copious amounts of urine. Excessive urination results in severe dehydration and loss of important ions from the blood. The condition can be corrected by the administration of ADH.

Oxytocin, the other hormone made in the hypothalamus, causes uterine contraction during childbirth and milk letdown when a baby is nursing. The more the uterus contracts during labor, the more nerve signals reach the hypothalamus, causing oxytocin to be released. Similarly, as a baby suckles while being breast fed, nerve signals from breast tissue reach the hypothalamus. As a result, oxytocin is produced by the hypothalamus and released from the posterior pituitary. The hormone causes the woman's breast milk to be released. The sound of a baby crying may also stimulate the release of oxytocin and milk letdown, much to the chagrin of women who are nursing. In both instances, the release of oxytocin from the posterior pituitary is controlled by positive feedback. The stimulus continues to bring about an effect that ever increases in intensity. Positive feedback terminates due to some external event. Therefore, positive feedback mechanisms are rarely used to maintain homeostasis; that role is typically associated with negative feedback mechanisms.

Animation Positive and Negative Feedback

APPLICATIONS AND MISCONCEPTIONS

How is labor induced if a woman's pregnancy extends past her due date?

After medication to prepare the birth canal for delivery, oxytocin (Pitocin) is used to induce labor. Pitocin is a synthetic version of the oxytocin released by the posterior pituitary. During labor, oxytocin may also be given to increase the strength of contractions. Stronger contractions speed the labor process if necessary (e.g., if the woman's uterus is contracting poorly or if the health of the mother or child is at risk during delivery). Oxytocin is routinely used following delivery to minimize postpartum bleeding by ensuring that strong uterine contractions continue.

Use of oxytocin must be monitored carefully, because it may cause excessive uterine contractions. Should this occur, the uterus could potentially tear itself. Further, reduced blood supply to the fetus caused by very strong contractions may be fatal to the baby. Though it reduces the actual duration of labor, inducing labor with oxytocin can be very painful for the mother. Whenever possible, physicians prefer gentler and more natural methods to induce labor and/or strengthen contractions.

Anterior Pituitary

A portal system, consisting of two capillary systems connected by a vein, lies between the hypothalamus and the **anterior pituitary.** The hypothalamus controls the anterior pituitary by producing hypothalamic-releasing and hypothalamic-inhibiting hormones, which pass from the hypothalamus to the anterior pituitary by way of the portal system. Examples are thyroid-releasing hormone (TRH) and thyroid-inhibiting hormone (TIH). The TRH stimulates the anterior pituitary to secrete thyroid-stimulating hormone, and the TIH inhibits the pituitary from secreting thyroid-stimulating hormone.

Four of the seven hormones produced by the anterior pituitary (Fig. 15.6, *right*) have an effect on other glands. **Thyroid-stimulating hormone** (**TSH**) stimulates the thyroid

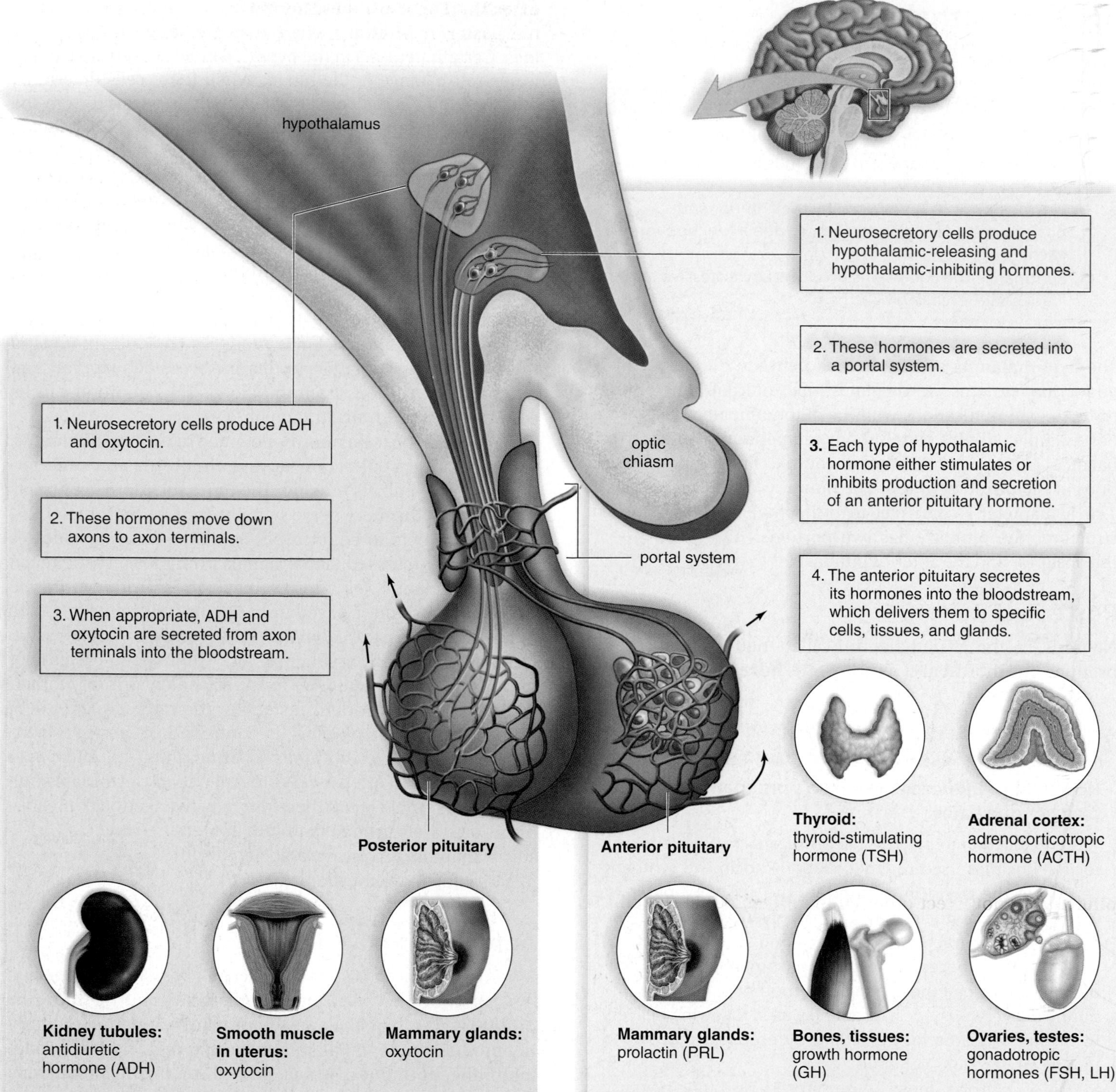
hypothalamus
1. Neurosecretory cells produce hypothalamic-releasing and hypothalamic-inhibiting hormones.
2. These hormones are secreted into a portal system.
3. Each type of hypothalamic hormone either stimulates or inhibits production and secretion of an anterior pituitary hormone.
4. The anterior pituitary secretes its hormones into the bloodstream, which delivers them to specific cells, tissues, and glands.
1. Neurosecretory cells produce ADH and oxytocin.
2. These hormones move down axons to axon terminals.
3. When appropriate, ADH and oxytocin are secreted from axon terminals into the bloodstream.
optic chiasm
portal system
Posterior pituitary
Anterior pituitary
Thyroid: thyroid-stimulating hormone (TSH)
Adrenal cortex: adrenocorticotropic hormone (ACTH)
Kidney tubules: antidiuretic hormone (ADH)
Smooth muscle in uterus: oxytocin
Mammary glands: oxytocin
Mammary glands: prolactin (PRL)
Bones, tissues: growth hormone (GH)
Ovaries, testes: gonadotropic hormones (FSH, LH)

Figure 15.6 Hormones produced by the hypothalamus and pituitary.
Left: The hypothalamus produces two hormones, ADH and oxytocin, stored and secreted by the posterior pituitary. *Right:* The hypothalamus controls the secretions of the anterior pituitary, and the anterior pituitary controls the secretions of the thyroid, adrenal cortex, and gonads, which are also endocrine glands.

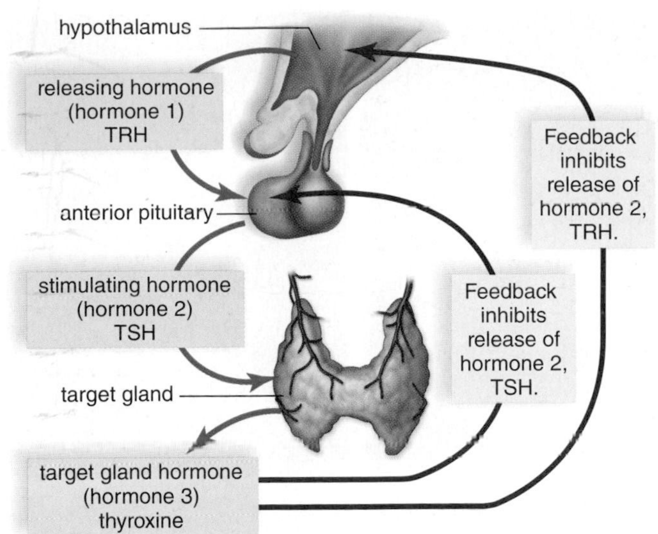

Figure 15.7 **Negative feedback mechanisms in the endocrine system.**
Feedback mechanisms (red arrows) provide a mechanism of controlling the amount of hormones produced (blue arrows) by the hypothalamus and pituitary glands.

to produce the thyroid hormones. **Adrenocorticotropic hormone** (**ACTH**) stimulates the adrenal cortex to produce cortisol. The **gonadotropic hormones, follicle-stimulating hormone** (**FSH**), and **luteinizing hormone** (**LH**), stimulate the gonads (the testes in males and the ovaries in females) to produce gametes and sex hormones. In each instance, the blood level of the last hormone in the sequence exerts negative feedback control over the secretion of the first two hormones (Fig. 15.7).

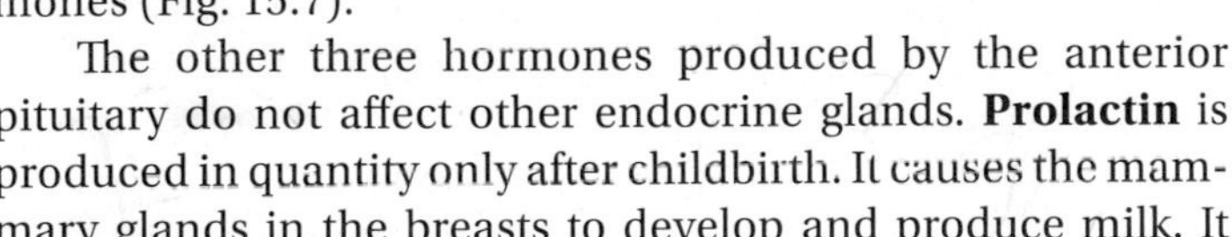

The other three hormones produced by the anterior pituitary do not affect other endocrine glands. **Prolactin** is produced in quantity only after childbirth. It causes the mammary glands in the breasts to develop and produce milk. It also plays a role in carbohydrate and fat metabolism.

Melanocyte-stimulating hormone causes skin-color changes in many fishes, amphibians, and reptiles having melanophores, special skin cells that produce color variations. The concentration of this hormone in humans is very low.

Growth hormone (**GH**), or somatotropic hormone, promotes skeletal and muscular growth. It stimulates the rate at which amino acids enter cells and protein synthesis occurs. It also promotes fat metabolism as opposed to glucose metabolism. The production of insulin-like growth factor 1 (IGF-1) by the liver is stimulated by growth hormone as well. IGF-1 is often measured as a means of determining GH level. Growth and development are also stimulated by IGF-1, and it may well be the means by which GH truly influences growth and development.

Effects of Growth Hormone

Growth hormone is produced by the anterior pituitary. The quantity is greatest during childhood and adolescence, when most body growth is occurring. If too little GH is produced during childhood, the individual has **pituitary dwarfism,** characterized by perfect proportions but small stature. If too much GH is secreted, a person can become a giant (Fig. 15.8). Giants usually have poor health, primarily because GH has a secondary effect on the blood sugar level, promoting an illness called *diabetes mellitus* (see pages 350–352). The Bioethical feature, "Growth Hormones and Pituitary Dwarfism," on page 342 discusses how a synthetic growth hormone may be used to treat some forms of dwarfism.

On occasion, GH is overproduced in the adult and a condition called **acromegaly** results. Long bone growth is no longer possible in adults, so only the feet, hands, and face (particularly the chin, nose, and eyebrow ridges) can respond, and these portions of the body become overly large (Fig. 15.9).

Figure 15.8 **Growth hormone influences height.**
a. The amount of growth hormone produced by the anterior pituitary during childhood affects the height of an individual. Plentiful growth hormone produces very tall basketball players. **b.** Too much growth hormone can lead to gigantism, whereas an insufficient amount results in limited stature and even pituitary dwarfism.

Age 9 Age 16 Age 33 Age 52

Figure 15.9 **Overproduction of growth hormone in adults leads to acromegaly.**
Acromegaly is caused by overproduction of GH in the adult. It is characterized by enlargement of the bones in the face, the fingers, and the toes as a person ages.

BIOLOGY MATTERS  **Bioethical**

Growth Hormones and Pituitary Dwarfism

Without treatment, children with a deficiency of growth hormone (GH) experience pituitary dwarfism: slow growth, short stature, and in some cases failure to begin puberty. Prior to the advent of biotechnology in the 1980s, treating these children was incredibly difficult and expensive. The GH needed to treat deficiencies had to be obtained from cadaver pituitaries. Whereas the treatment was generally very successful, the use of cadaveric GH caused Creutzfeldt-Jakob disease (a neurological disease similar to "mad cow" disease) in a small number of treated individuals.

Thanks to biotechnology, technologists are now able to synthesize human GH (HGH) using bacteria. These bacteria have had the gene for HGH inserted into their genetic information. The altered bacteria are then grown in laboratories and make unlimited amounts of GH. Children with insufficient GH can be treated more safely and inexpensively with this GH. Recombinant HGH can also be used to treat other disorders such as the chromosomal deficiency known as Turner syndrome (discussed in Chapter 18). It may even be possible to slow or reverse the aging process with HGH treatments.

There is some controversy surrounding treating short children without HGH deficiency for essentially cosmetic reasons. Unfortunately, Americans are obsessed with height. Shorter children are often bullied and teased by their peers. Some data suggest that shorter individuals are discriminated against at their jobs. Their salaries are often lower than those of their taller counterparts with equivalent education and experience. Many people of short stature report having greater self-esteem problems than individuals of average to above-average height. Treatment with HGH could be the solution to these problems.

Although the supply of HGH is seemingly unlimited, the cost of treatments is still quite high (though much cheaper than cadaveric GH), with annual treatments in the range of \$13,000 to \$30,000. In most cases, insurance companies will not cover these costs. Of greater concern, however, are the potential side effects of supplemental HGH therapy, which are not well understood. Moreover, it is not clear whether HGH treatment will result in a significant increase in the final height of short children.

Questions to Consider

1. Now that HGH is easier to obtain, what potential abuses would you predict?
2. Do you think insurance companies should be expected to pay for HGH treatment if a child shows no hormone deficiency and is simply short?
3. How may the increased availability of HGH be abused by some individuals?

CHECK YOUR PROGRESS 15.2

1. Explain how the endocrine system and nervous system communicate with one another.
2. List the hormones produced by the posterior pituitary and provide a function for each.
3. List the hormones produced by the anterior pituitary and provide a function for each.
4. Discuss what symptoms can occur if a person's anterior pituitary is inactivated by a tumor.

CONNECTING THE CONCEPTS

For more information on the hormones presented in this section, refer to the following discussions:

Section 11.2 examines the influence of growth hormone on bone growth.

Section 16.2 describes the role of pituitary hormones in the production of sperm cells in males.

Section 16.4 describes the role of pituitary hormones in the female ovarian cycle.

15.3 Thyroid and Parathyroid Glands

LEARNING OUTCOMES

Upon completion of this section, you should be able to

1. List the hormones produced by the thyroid and parathyroid glands and provide a function for each.
2. Describe the negative feedback mechanism that is involved in the maintenance of blood calcium homeostasis.
3. Summarize the diseases and conditions associated with the thyroid and parathyroid glands.

The **thyroid gland** is a large gland located in the neck, where it is attached to the trachea just below the larynx (see Fig. 15.2). The parathyroid glands are embedded in the posterior surface of the thyroid gland.

Thyroid Gland

The thyroid gland is composed of a large number of follicles. Each follicle is a small spherical structure made of thyroid cells filled with **triiodothyronine (T_3)**, which contains three iodine atoms, and **thyroxine (T_4)**, which contains four.

Effects of Thyroid Hormones

To produce triiodothyronine (T_3) and thyroxine (T_4), the thyroid gland actively requires iodine. The concentration of iodine in the thyroid gland can increase to as much as 25 times that of the blood. If iodine is lacking in the diet, the thyroid gland is unable to produce the thyroid hormones. In response to constant stimulation by TSH from the anterior pituitary, the thyroid enlarges, resulting in a **simple goiter** (Fig. 15.10*a*). In the 1920s, it was discovered that the use of iodized salt allows the thyroid to produce the thyroid hormones and, therefore, helps prevent simple goiter. However, iodine deficiencies are still common in many parts of the world, with an estimated 2 billion people still experiencing some degree of deficiency.

a. Simple goiter

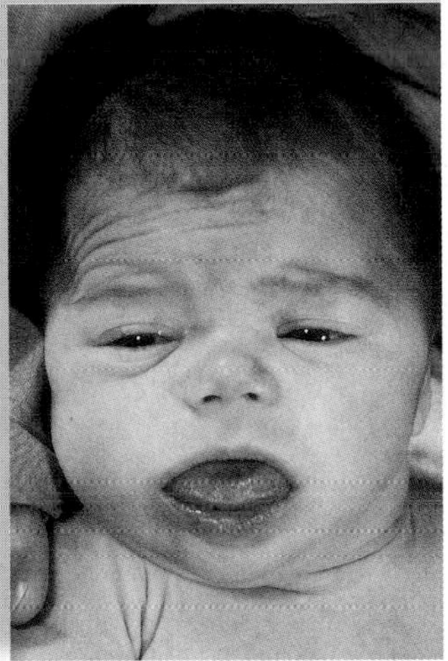

b. Congenital hypothyroidism

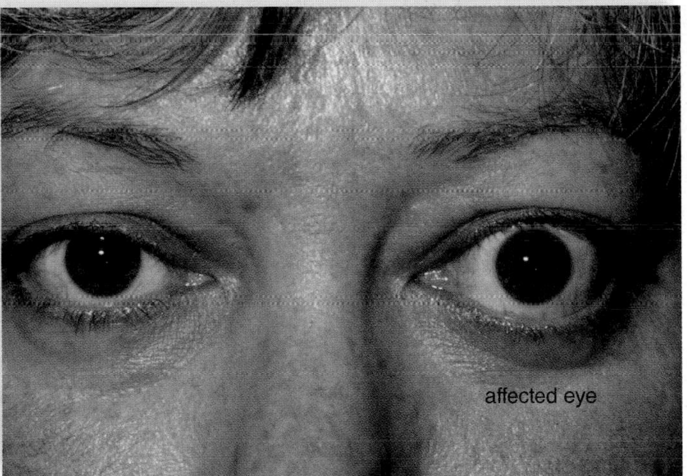

c. Exophthalmic goiter

Figure 15.10 Effects of insufficient dietary iodine, hypothyroidism, and hyperthyroidism.
a. An enlarged thyroid gland is often caused by a lack of iodine in the diet. Without iodine, the thyroid is unable to produce its hormones, and continued anterior pituitary stimulation causes the gland to enlarge.
b. Individuals who develop hypothyroidism during infancy or childhood do not grow and develop as others do. Unless medical treatment is begun, the body is short and stocky; intellectual disabilities are also likely.
c. In exophthalmic goiter, a goiter is due to an overactive thyroid and the eyes protrude because of edema in eye socket tissue.

While thyroid hormones increase the metabolic rate, they do not have a target organ. Instead, they stimulate all cells of the body to metabolize at a faster rate. More glucose is broken down, and more energy is used.

Animation Mechanism of Thyroxine Action

If the thyroid fails to develop properly, a condition called **congenital hypothyroidism** results (Fig. 15.10*b*). Individuals with this condition are short and stocky and have had extreme hypothyroidism (undersecretion of thyroid hormone) since infancy or childhood. Thyroid hormone therapy can initiate growth, but unless treatment is begun within the first two months of life, mental retardation results. The occurrence of hypothyroidism in adults produces the condition known as **myxedema.** Lethargy, weight gain, loss of hair, slower pulse rate, lowered body temperature, and thickness and puffiness of the skin are characteristics of myxedema. The administration of adequate doses of thyroid hormones restores normal function and appearance.

In the case of hyperthyroidism (oversecretion of thyroid hormone), the thyroid gland is overactive and a goiter forms. This type of goiter is called **exophthalmic goiter** (Fig. 15.10*c*). The eyes protrude because of edema in eye socket tissues and swelling of the muscles that move the eyes. The patient usually becomes hyperactive, nervous, and irritable and suffers from insomnia. Removal or destruction of a portion of the thyroid by means of radioactive iodine is sometimes effective in curing the condition. Hyperthyroidism can also be caused by a thyroid tumor, usually detected as a lump during physical examination. Again, the treatment is surgery in combination with administration of radioactive iodine. The prognosis for most patients is excellent.

Calcitonin

Calcium ions (Ca^{2+}) play a significant role in both nervous conduction and muscle contraction. They are also necessary for blood clotting. The blood calcium level is regulated in part by **calcitonin,** a hormone secreted by the thyroid gland when the blood calcium level rises (Fig. 15.11). The primary effect of calcitonin is to bring about the deposit of calcium ions in the bones (see section 11.5). It also temporarily reduces the activity and number of osteoclasts. When the blood calcium level lowers to normal, the release of calcitonin by the thyroid is inhibited.

Parathyroid Glands

Parathyroid hormone (**PTH**), produced by the **parathyroid glands,** causes the blood calcium level to increase. A low blood calcium level stimulates the release of PTH, which promotes the activity of osteoclasts and the release of calcium from the bones. PTH also activates vitamin D in the kidneys. Activated vitamin D, a hormone sometimes called *calcitriol,* then promotes calcium reabsorption by the kidneys. The absorption of calcium ions from the intestine is also stimulated by calcitriol. These effects bring the blood calcium level back to the normal range, and PTH secretion stops.

MP3 Calcium Homeostasis

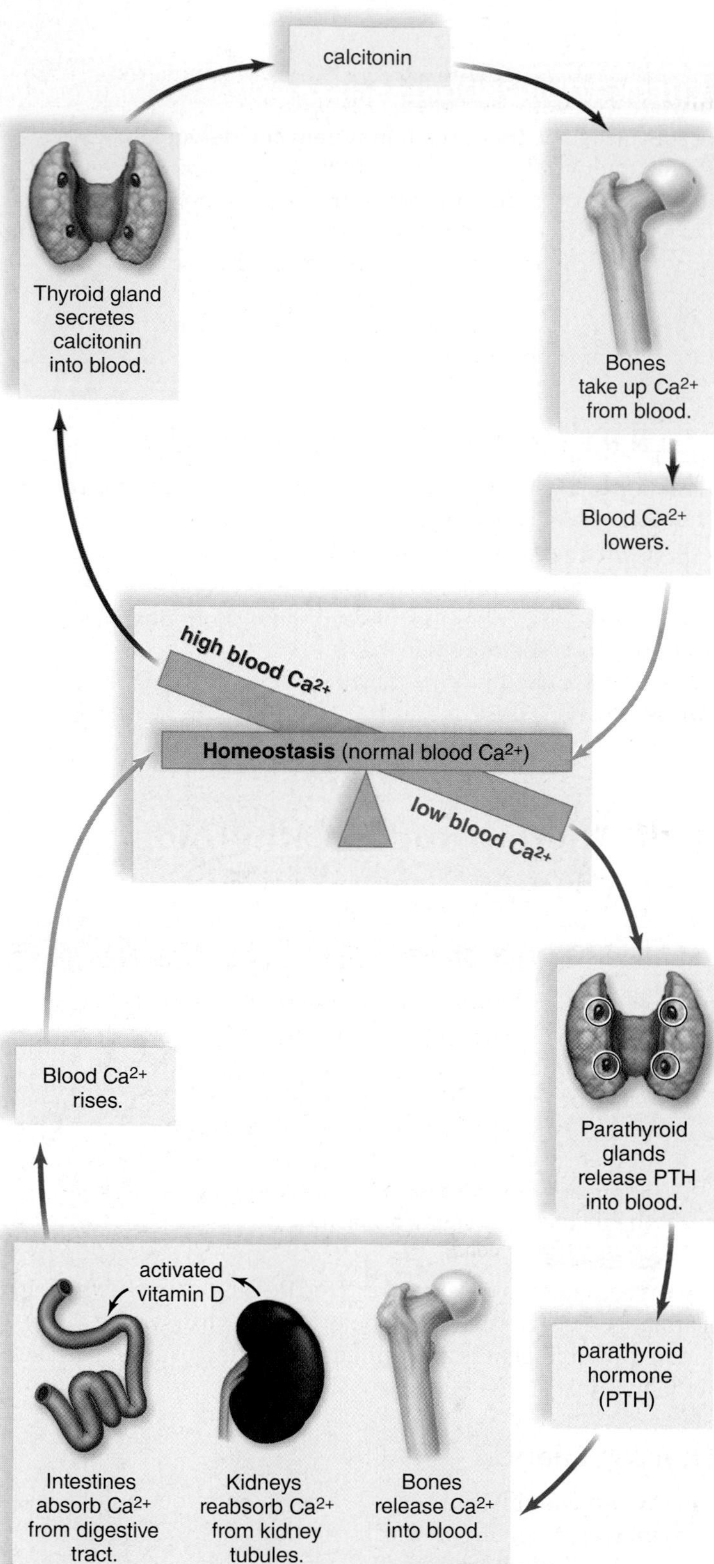

Figure 15.11 Blood calcium homeostasis.
Top: When the blood calcium level is high, the thyroid gland secretes calcitonin. Calcitonin promotes the uptake of calcium ions (Ca^{2+}) by the bones; therefore, the blood calcium level returns to normal. *Bottom:* When the blood calcium level is low, the parathyroid glands release parathyroid hormone (PTH). PTH causes the bones to release calcium ions (Ca^{2+}). It also causes the kidneys to reabsorb Ca^{2+} and activate vitamin D; thereafter, the intestines absorb Ca^{2+}. Therefore, the blood calcium level returns to normal.

Many years ago, the four parathyroid glands were sometimes mistakenly removed during thyroid surgery because of their size and location. Gland removal caused insufficient PTH production, which resulted in *hypoparathyroidism*. Hypoparathyroidism causes a dramatic drop in blood calcium, followed by excessive nerve excitability. Nerve signals happen spontaneously and without rest, causing a phenomenon called tetany. In **tetany,** the body shakes from continuous muscle contraction. Without treatment, severe hypoparathyroidism causes seizures, heart failure, and death.

Untreated *hyperparathyroidism* (oversecretion of PTH) can result in osteoporosis because of continuous calcium release from the bones. Hyperparathyroidism may also cause formation of calcium kidney stones.

When a bone is broken, homeostasis is disrupted. For the fracture to heal, osteoclasts will have to destroy old bone, and osteoblasts will have to lay down new bone. Many factors influence the formation of new bone, including parathyroid hormone, calcitonin, and vitamin D. The calcium needed to repair the fracture is made readily available as new blood capillaries penetrate the fractured area.

CHECK YOUR PROGRESS 15.3

1. Explain how the hormones of the thyroid gland influence the metabolic rate.
2. Describe how calcitonin and parathyroid hormones interact to regulate blood calcium levels.
3. Distinguish between hyperthyroidism and hyperparathyroidism with regard to the effects on the body.

CONNECTING THE CONCEPTS

For more information on the importance of calcium, refer to the following discussions:

Section 11.5 explains the role of the bones in maintaining calcium homeostasis.

Section 12.2 examines how calcium ions are involved in muscle contraction.

Section 13.1 explores how calcium ions are involved in the activity of a neural synapse.

15.4 Adrenal Glands

LEARNING OUTCOMES

Upon completion of this section, you should be able to

1. List the hormones produced by the adrenal medulla and adrenal cortex and provide a function for each.
2. Explain how the adrenal cortex is involved in the stress response.
3. Distinguish between mineralocorticoid and glucocorticoid hormones.

The **adrenal glands** sit atop the kidneys (see Fig. 15.2). Each adrenal gland consists of an inner portion called the **adrenal medulla** and an outer portion called the **adrenal cortex** (Fig. 15.12). These portions, like the anterior and the posterior pituitary, are two functionally distinct endocrine glands. The adrenal medulla is under nervous control. Portions of the adrenal cortex are under the control of corticotropin-releasing hormone (CRH) from the hypothalamus and ACTH, an anterior pituitary hormone. Stress of all types, including emotional and physical trauma, prompts the hypothalamus to stimulate a portion of the adrenal glands.

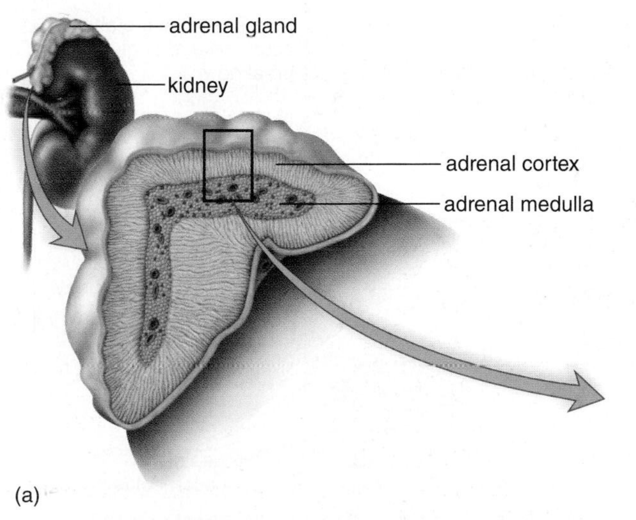

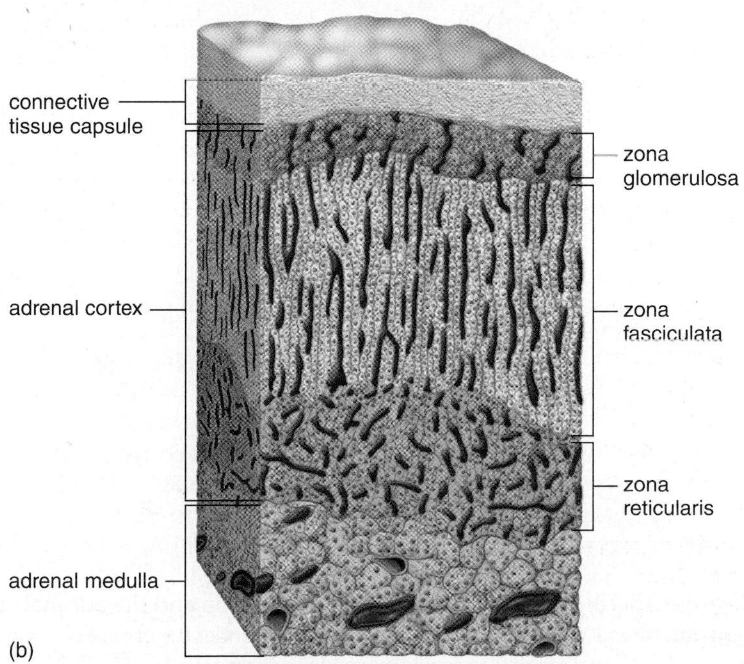

Figure 15.12 **The adrenal glands.**
a. The location of the adrenal glands relative to the kidney. **b.** The anatomy of the tissue layers within the adrenal glands.

Adrenal Medulla

The hypothalamus initiates nerve signals that travel by way of the brain stem, spinal cord, and preganglionic sympathetic nerve fibers to the adrenal medulla. These signals stimulate the adrenal medulla to secrete its hormones. The cells of the adrenal medulla are thought to be modified postganglionic neurons.

Epinephrine (adrenaline) and **norepinephrine** (noradrenaline) are the hormones produced by the adrenal medulla. These hormones rapidly bring about all the body changes that occur when an individual reacts to an emergency situation in a fight-or-flight manner. These hormones provide a short-term response to stress (Fig. 15.13, *left*).

Adrenal Cortex

The adrenal cortex is divided into three regions (see Fig. 15.12*b*). These are the zona glomerulosa, zona fasciculata, and the zona reticularis. In contrast to the adrenal medulla, the hormones produced by the adrenal cortex provide a long-term response to stress (Fig. 15.13, *right*). The two major types of hormones produced by the adrenal cortex are the glucocorticoids and the mineralocorticoids. The adrenal cortex also secretes a small amount of sex hormones in both males and females.

Glucocorticoids

The **glucocorticoids,** whose secretion is controlled by ACTH, regulate carbohydrate, protein, and fat metabolism. Glucocorticoids are produced in both the zona fasciculata and zona reticularis of the adrenal cortex. **Cortisol** is a glucocorticoid that is active in the stress response and the repair of damaged tissues in the body. Glucocorticoids raise the blood glucose level in at least two ways. (1) They promote the breakdown of muscle proteins to amino acids, taken up by the liver from the bloodstream. The liver then converts these excess amino acids to glucose, which enters the blood. (2) They promote the metabolism of fatty acids rather than carbohydrates, and this spares glucose.

The glucocorticoids also counteract the inflammatory response that leads to pain and swelling. Very high levels of glucocorticoids in the blood can suppress the body's defense

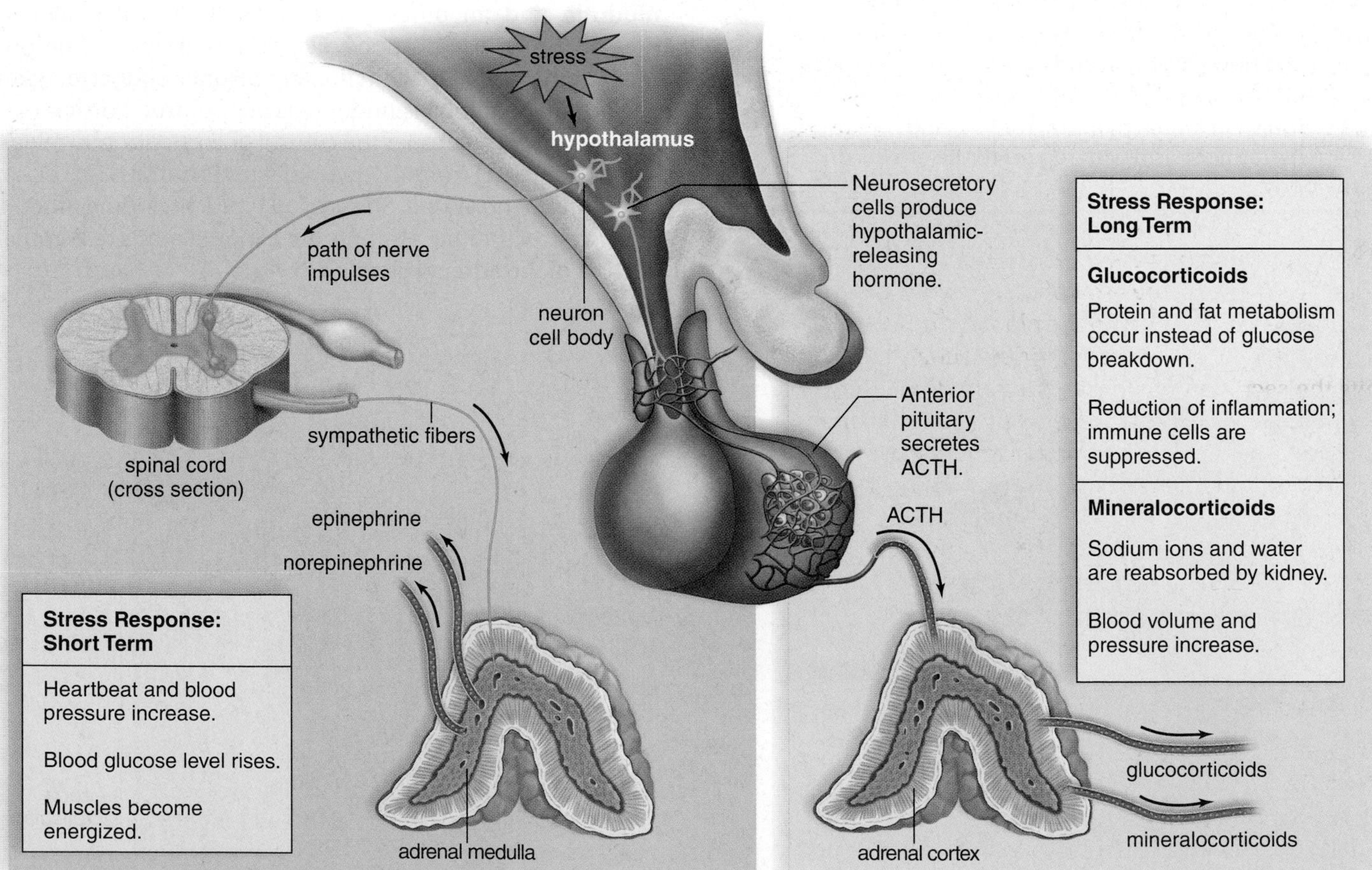

Figure 15.13 **Response of the adrenal medulla and the adrenal cortex to stress.**
Both the adrenal medulla and the adrenal cortex are under the control of the hypothalamus when they help us respond to stress. *Left:* Nervous stimulation causes the adrenal medulla to provide a rapid but short-term stress response. *Right:* The adrenal cortex provides a slower but long-term stress response. ACTH causes the adrenal cortex to release glucocorticoids. Independently, the adrenal cortex releases mineralocorticoids.

system, including the inflammatory response that occurs at infection sites. Cortisone and other glucocorticoids can relieve swelling and pain from inflammation. However, by suppressing pain and immunity, they can also make a person highly susceptible to injury and infection.

Animation Glucocorticoid Hormones

Mineralocorticoids

The **mineralocorticoids** regulate ion (electrolyte) balances in the body and are primarily produced by the zona glomerulosa in the adrenal cortex. **Aldosterone** is the most important of the mineralocorticoids. Aldosterone primarily targets the kidney, where it promotes renal absorption of sodium ions (Na^+) and renal excretion of potassium ions (K^+).

The secretion of mineralocorticoids is not controlled by the anterior pituitary. When the blood sodium level and pressure are low, the kidneys secrete **renin** (Fig. 15.14). Renin is an enzyme that converts the plasma protein angiotensinogen to angiotensin I. Angiotensin I is changed to angiotensin II by a converting enzyme found in lung capillaries. Angiotensin II stimulates the adrenal cortex to release aldosterone. The effect of this system, called the renin-angiotensin-aldosterone system, is to raise blood pressure in two ways. Angiotensin II constricts the arterioles, and aldosterone causes the kidneys to reabsorb sodium ions (Na^+). When the blood sodium level rises, water is reabsorbed, in part, because the hypothalamus secretes ADH (see page 339). Reabsorption means that water enters kidney capillaries and, thus, the blood. Then blood pressure increases to normal.

Recall that we studied the role of the kidneys in maintaining blood pressure (see Chapter 10). At that time, we mentioned that if the blood pressure rises due to the reabsorption of sodium ions (Na^+), the atria of the heart are apt to stretch. Due to a great increase in blood volume, cardiac cells release a chemical called atrial natriuretic hormone (ANH), which inhibits the secretion of aldosterone from the adrenal cortex. In other words, the heart is among various organs in the body that release a hormone but obviously not as the major function. Therefore, the heart is not included as an endocrine gland in Figure 15.2. The effect of this ANH is to cause *natriuresis,* the excretion of sodium ions (Na^+). When sodium ions are excreted, so is water; therefore, blood pressure lowers to normal.

Sex Steroids

In addition to glucocorticoids, the zona fasciculata and zona reticularis of the adrenal cortex secrete small amounts of sex steroids. These are both male sex hormones (**androgens**) and female sex hormones (estrogen). The primary androgen hormone is **dehydroepiandrosterone (DHEA),** which is a precursor for **testosterone,** the male sex hormone. While primarily active in males following puberty, androgens do play a role in the sexual development of both males and females. In addition, these regions also produce small amounts of **estradiol,**

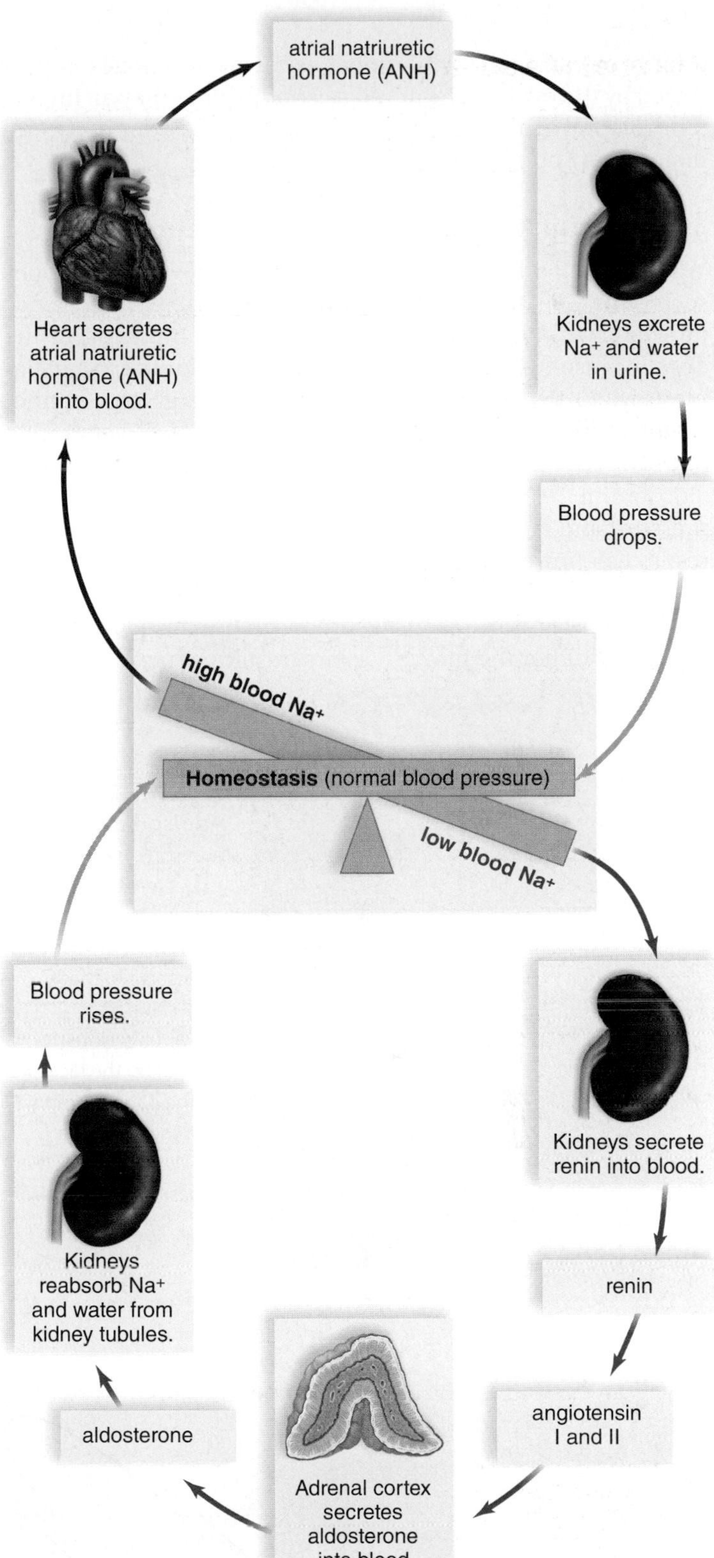

Figure 15.14 Regulation of blood pressure is under hormonal control.
Bottom: When the blood sodium level is low, a low blood pressure causes the kidneys to secrete renin. Renin leads to the secretion of aldosterone from the adrenal cortex. Aldosterone causes the kidneys to reabsorb sodium ions (Na^+), and water follows, so that blood volume and pressure return to normal. *Top:* When a high blood sodium level accompanies a high blood volume, the heart secretes atrial natriuretic hormone (ANH). ANH causes the kidneys to excrete sodium ions (Na^+), and water follows. The blood volume and pressure return to normal.

a form of **estrogen.** While most estrogen in females is produced by the ovaries, the adrenal estradiol does play an important role in regulating growth of the skeleton in puberty and maintaining bone mass.

Malfunction of the Adrenal Cortex

When the blood level of glucocorticoids is low due to hyposecretion, a person develops **Addison disease.** The presence of excessive but ineffective ACTH causes a bronzing of the skin because ACTH, like MSH, can lead to a buildup of melanin (Fig. 15.15). Without the glucocorticoids, glucose cannot be replenished when a stressful situation arises. Even a mild infection can lead to death. In some cases, hyposecretion of aldosterone results in a loss of sodium and water. Low blood pressure and, possibly, severe dehydration can develop as a result. Left untreated, Addison disease can be fatal.

Excessive levels of glucocorticoids result in **Cushing syndrome** (Fig. 15.16). This disorder can be caused by tumors that affect either the pituitary gland, resulting in excess ACTH secretion, or the adrenal cortex itself. The most common cause, however, is the administration of glucocorticoids to treat other conditions (e.g., to suppress chronic inflammation). Regardless of the source, excess glucocorticoids cause muscle protein to be metabolized and subcutaneous fat to be deposited in the midsection. Excess production of adrenal male sex hormones in women may result in masculinization, including an increase in body hair, deepening of the voice, and beard growth. Depending on the cause, treatment of Cushing syndrome may involve a careful reduction in the amount of cortisone being taken, the use of cortisol-inhibiting drugs, or surgery to remove any existing pituitary or adrenal tumor.

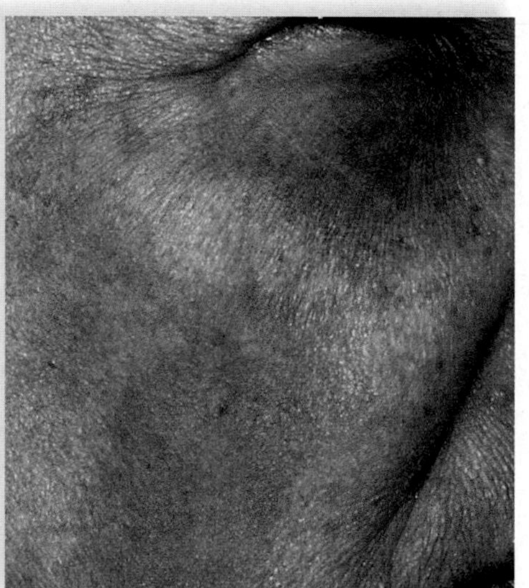

a.

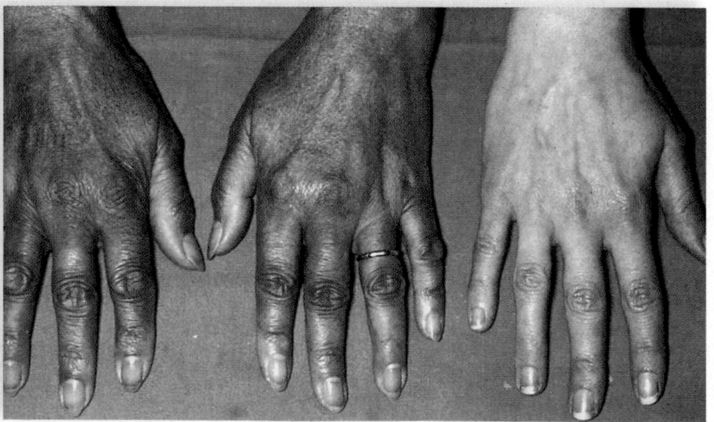

b.

Figure 15.15 Addison disease. Addison disease is characterized by a peculiar bronzing of the skin, particularly noticeable in these light-skinned individuals. Note the color of (**a**) the face and (**b**) the hands compared with the hand of an individual without the disease.

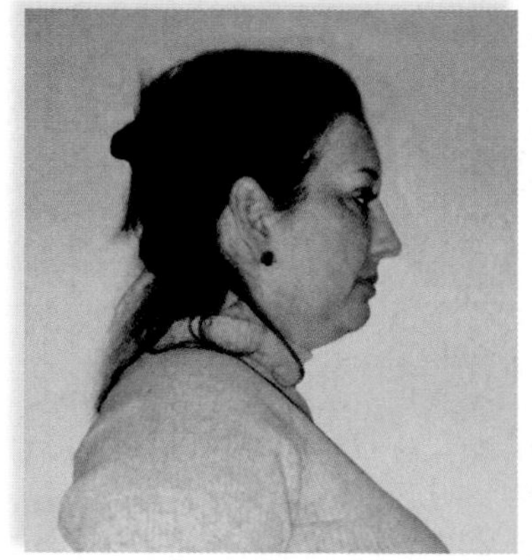

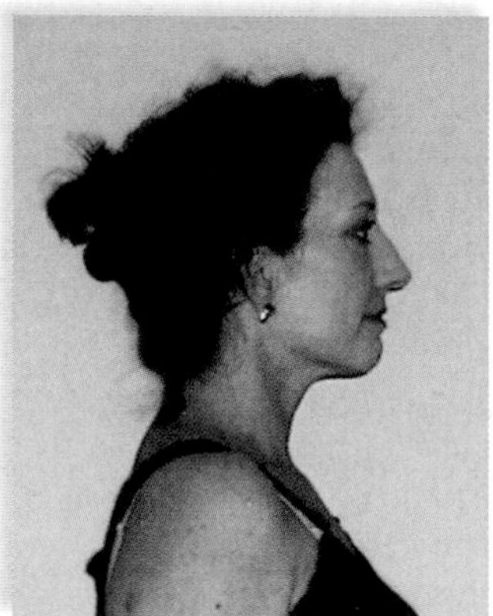

Figure 15.16 Cushing syndrome.
This 40-year-old woman was diagnosed with a small tumor in her pituitary gland, which secreted large amounts of ACTH. The high ACTH levels stimulated the adrenal glands to produce excessive amounts of cortisol.
Left: Patient at the time of surgery to remove her pituitary tumor.
Right: Patient's appearance one year later.

CHECK YOUR PROGRESS 15.4

1. List the hormones produced by the adrenal glands and indicate whether they are produced by the adrenal cortex or the adrenal medulla.
2. Summarize the involvement of the adrenal glands during a stress response.
3. Explain why maintaining the correct balance of water and salt is important in homeostasis.

CONNECTING THE CONCEPTS

For more information on the hormones produced by the adrenal glands, refer to the following discussions:

Section 5.3 describes how epinephrine and norepinephrine influence the heart rate.

Section 10.4 examines how aldosterone is involved in maintaining the water–salt balance of the body fluids.

15.5 Pancreas

LEARNING OUTCOMES

Upon completion of this section, you should be able to

1. Explain why the pancreas is both an endocrine and exocrine gland.
2. List the hormones produced by the pancreas and provide a function for each.
3. Describe how the pancreatic hormones help maintain blood glucose homeostasis.
4. Distinguish between type 1 and type 2 diabetes mellitus.

The **pancreas** is a fish-shaped organ that stretches across the abdomen behind the stomach and near the duodenum of the small intestine. It is composed of two types of tissue. Exocrine tissue produces and secretes digestive juices that go by way of ducts to the small intestine (see section 8.4). The endocrine tissue is called the **pancreatic islets** (islets of Langerhans). As Figure 15.17 illustrates, each pancreatic islet is surrounded by exocrine tissue. Within each islet are a variety of cell types, several of which play an important role in the endocrine functions of this organ. A cells are responsible for the secretion of the hormone glucagon, whereas B cells (not to be confused with the B cells of the immune system) secrete insulin. A third cell type, D cells, releases somatostatin, a hormone that is released at the same time as insulin to regulate the digestive processes. The pancreas is not under pituitary control. **Insulin** is secreted by the B cells when the blood glucose level is high, which usually occurs just after eating. Insulin stimulates the uptake of glucose by cells, especially liver cells, muscle cells, and adipose tissue cells. In liver and muscle cells, glucose is then stored as glycogen. In muscle cells, the glucose supplies energy for muscle contraction. Glucose enters the metabolic pool in fat cells and thereby supplies glycerol for the formation of fat. In these various ways, insulin lowers the blood glucose level (Fig. 15.18, *top*).

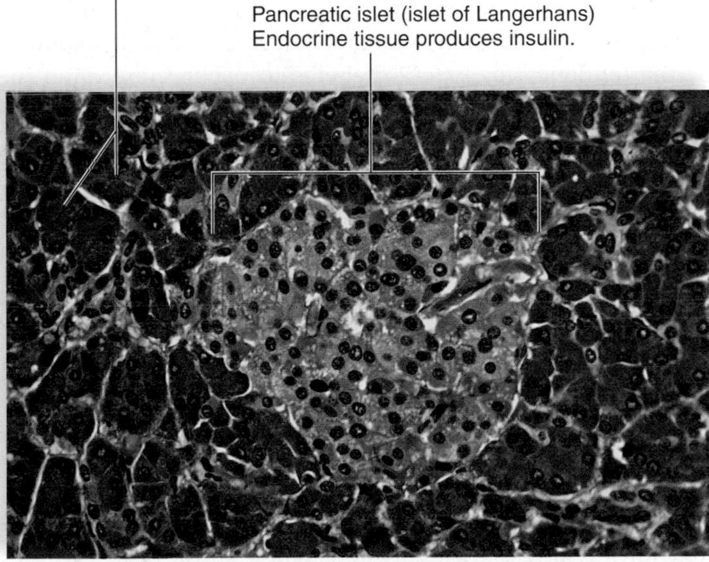

Figure 15.17 The pancreas is both an endocrine and exocrine gland.
This light micrograph shows that the pancreas has two types of cells. The exocrine tissue produces a digestive juice, and the endocrine tissue produces the hormones insulin and glucagon.

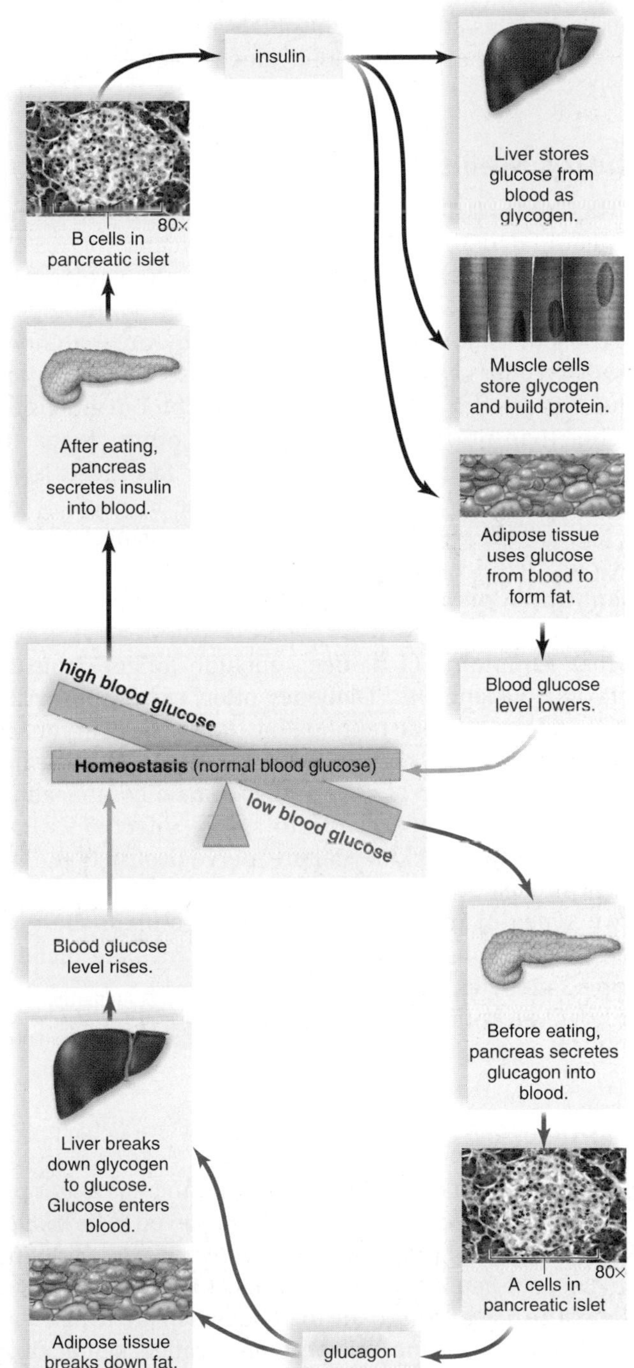

Figure 15.18 Blood glucose homeostasis.
Top: When the blood glucose level is high, the pancreas secretes insulin. Insulin promotes the storage of glucose as glycogen and the synthesis of proteins and fats. Therefore, insulin lowers the blood glucose level. *Bottom:* When the blood glucose level is low, the pancreas secretes glucagon. Glucagon acts opposite to insulin; therefore, glucagon raises the blood glucose level to normal.

Glucagon is secreted by the A cells of the pancreas, usually between eating, when the blood glucose level is low. The major target tissues of glucagon are the liver and adipose tissue. Glucagon stimulates the liver to break down glycogen to glucose. It also promotes the use of fat and protein in preference to glucose as energy sources. Adipose tissue cells break down fat to glycerol and fatty acids. The liver takes these up and uses them as substrates for glucose formation. In these ways, glucagon raises the blood glucose level (Fig. 15.18, *bottom*).

Diabetes Mellitus

In 2011, the National Diabetes Information Clearinghouse estimated that 25.8 million Americans (or 8.3% of the population) have **diabetes mellitus,** often referred to simply as diabetes. Of these, an estimated 7 million are undiagnosed. Diabetes is characterized by an inability of the body's cells, especially liver and muscle cells, to take up glucose as they should. This causes blood glucose to be higher than normal, and cells rely on other "fuels" like fatty acids for energy. Therefore, cellular famine exists in the midst of plenty. As the blood glucose level rises, glucose, along with water, is excreted in the urine (mellitus, from Greek, refers to "honey" or "sweetness"). This results in frequent urination and causes the diabetic to be extremely thirsty.

Other symptoms of diabetes include fatigue, constant hunger, and weight loss. Diabetics often experience vision problems due to diabetic retinopathy and swelling in the lens of the eye due to the high blood sugar levels. If untreated, diabetics often develop serious and even fatal complications. Sores that don't heal develop into severe infections. Blood vessel damage causes kidney failure, nerve destruction, heart attack, or stroke.

Two types of diabetes have been identified; these are termed type 1 diabetes, sometimes called juvenile diabetes, and type 2 diabetes, or adult-onset diabetes; however, both diseases may occur in children or adults.

Animation Blood Sugar Regulation in Diabetics

Types of Diabetes

Type 1 Diabetes In type 1 diabetes, the pancreas is not producing enough insulin. This condition is believed to be brought on by exposure to an environmental agent, most likely a virus, whose presence causes cytotoxic T cells to destroy the pancreatic islets. The body turns to the metabolism of fat, which leads to the buildup of ketones in the blood, called ketoacidosis, which increases acidity of the blood and can lead to coma and death.

Individuals with type 1 diabetes must have daily insulin injections. These injections control the diabetic symptoms but still can cause inconveniences because the blood sugar level may swing between hypoglycemia (low blood glucose) and hyperglycemia (high blood glucose). Without testing the blood glucose level, it is difficult to be certain which of these is present because the symptoms can be similar. These symptoms include perspiration, pale skin, shallow breathing, and anxiety. Whenever these symptoms appear, immediate attention is required to bring the blood glucose back within the normal range. If the problem is hypoglycemia, the treatment is one or two glucose tablets, hard candy, or orange juice. If the problem is hyperglycemia, the treatment is insulin. Better control of blood glucose levels can often be achieved with an insulin pump, a small device worn outside the body that is connected to a plastic catheter inserted under the skin.

Because diabetes is such a common problem, many researchers are working to develop more effective methods for treating diabetes. The most desirable would be an artificial pancreas, defined as an automated system that would provide insulin based on real-time changes in blood sugar levels. It is also possible to transplant a working pancreas, or even fetal pancreatic islet cells, into patients with type 1 diabetes. Another possibility is xenotransplantation, in which insulin-producing islet cells of another species, such as pigs, are placed inside a capsule that allows insulin to exit but prevents the immune system from attacking the foreign cells. Finally, researchers are close to testing a vaccine that could block the immune system's attack on the islet cells, perhaps by inducing T cells capable of suppressing these responses.

Type 2 Diabetes Most adult diabetics have type 2 diabetes. Often, the patient is overweight or obese, and adipose tissue produces a substance that impairs insulin receptor function. However, complex genetic factors can be involved, as shown by the tendency for type 2 diabetes to occur more often in certain families, or even ethnic groups. For example, the condition is 77% more common in African Americans than in non-Hispanic whites.

Normally, the binding of insulin to its plasma membrane receptor causes the number of protein carriers for glucose to increase, causing more glucose to enter the cell. In the type 2 diabetic, insulin still binds to its receptor, but the number of glucose carriers does not increase. Therefore, the cell is said to be insulin-resistant.

It is possible to prevent or at least control type 2 diabetes by adhering to a low-fat, low-sugar diet and exercising regularly. If this fails, oral drugs are available that stimulate the pancreas to secrete more insulin and enhance the metabolism of glucose in the liver and muscle cells. Millions of Americans may have type 2 diabetes without being aware of it; yet, the effects of untreated type 2 diabetes are as serious as those of type 1 diabetes.

Testing for Diabetes

The oral glucose tolerance test assists in the diagnosis of diabetes mellitus. After the patient is given 100 g of glucose, the blood glucose concentration is measured at intervals. In a diabetic, the blood glucose level rises greatly and remains elevated for several hours (Fig. 15.19). In the meantime, glucose appears in the urine. In a nondiabetic person, the blood glucose level rises somewhat and then returns to normal after about 2 hours.

BIOLOGY MATTERS  Science

Identifying Insulin as a Chemical Messenger

The pancreas is both an exocrine gland and an endocrine gland. It sends digestive juices to the duodenum by way of the pancreatic duct, and it secretes the hormones insulin and glucagon into the bloodstream.

In 1920, physician Frederick Banting decided to try to isolate insulin in order to identify it as a chemical messenger. Previous investigators had been unable to do this because the enzymes in the digestive juices destroyed insulin (a protein) during the isolation procedure. Banting hit upon the idea of ligating (tying off) the pancreatic duct, which he knew from previous research would lead to the degeneration only of the cells that produce digestive juices and not of the pancreatic islets (of Langerhans), where insulin is made. His professor, J. J. Macleod, made a laboratory available to him at the University of Toronto and also assigned a graduate student, Charles Best, to assist him.

Banting and Best had limited funds and spent that summer working, sleeping, and eating in the lab. By the end of the summer, they had obtained pancreatic extracts that did lower the blood glucose level in diabetic dogs. Macleod then brought in biochemists, who purified the extract. Insulin therapy for the first human patient began in 1922, and large-scale production of purified insulin from pigs and cattle followed. The basic experimental system used by Banting and Best is shown here:

Steps to Identify a Chemical Messenger (as used by Banting and Best)

1. Identify the source of the chemical	Pancreatic islets are source
2. Identify the effect to be studied	Presense of pancrease in body lowers blood glucose
3. Isolate the chemical	Insulin isolated from pancreatic secretions
4. Show that the chemical has the desired effect	Insulin lowers blood glucose

For their discovery, Banting and Macleod received the Nobel Prize in Medicine in 1923. The amino acid sequence of insulin was determined in 1953. Insulin is now synthesized using recombinant DNA technology, using the bacterium *Escherichia coli* to produce the hormone. The availability of recombinant insulin (Fig. 15A), sometimes also called synthetic insulin, has played a major role in improving the health of diabetics around the world.

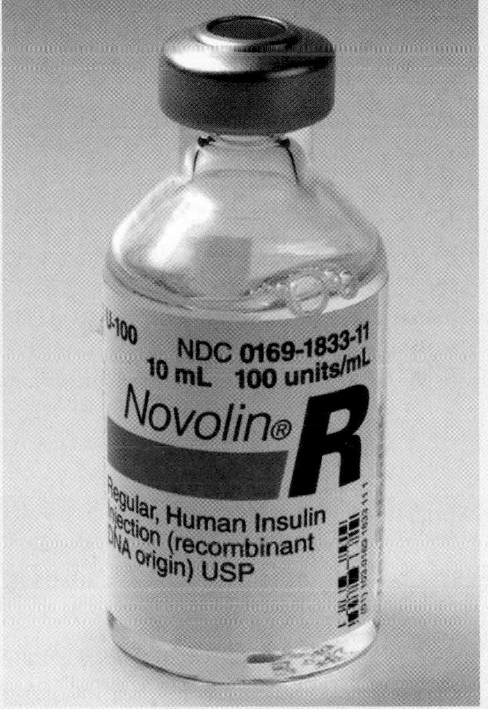

Figure 15A **Synthetic insulin.** Insulin is now a product of recombinant engineering and biotechnology.

Questions to Consider

1. What are some advantages, and potential disadvantages, of producing a medicine destined to be injected in humans (such as insulin) in a bacterium like *E. coli?*
2. Some people oppose the use of animals for medical research. Do you think that insulin would have eventually been discovered without animal experimentation? Why or why not?

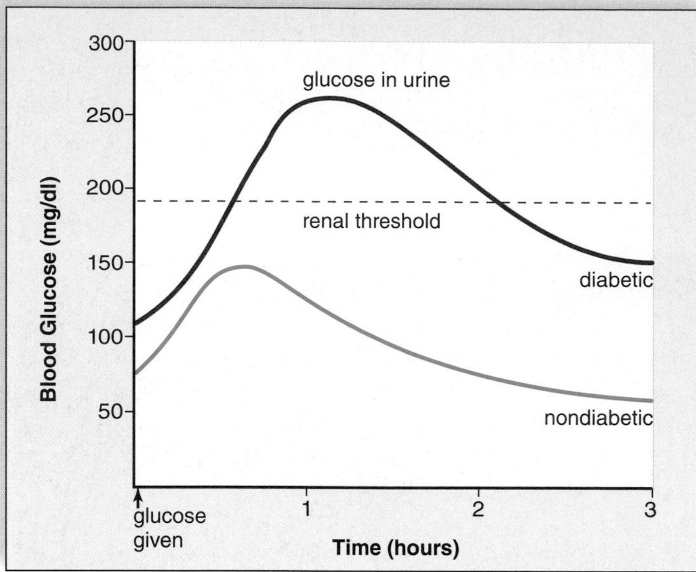

Figure 15.19 The results of a glucose tolerance test for diabetes.
Following the administration of 100 g of glucose, the blood glucose level rises dramatically in the diabetic, and glucose appears in the urine. Also, the blood glucose level at 2 hours is equal to more than 200 mg/dl.

APPLICATIONS AND MISCONCEPTIONS

What is gestational diabetes, and what causes it?

Women who were not diabetic prior to pregnancy but have high blood glucose during pregnancy have *gestational* diabetes. Gestational diabetes affects a small percentage of pregnant women. This form of diabetes is caused by insulin resistance—body insulin concentration is normal, but cells fail to respond normally. Gestational diabetes and insulin resistance generally develop later in the pregnancy. Carefully planned meals and exercise often control this form of diabetes, but insulin injections may be necessary.

If the woman is not treated, additional glucose crosses the placenta, causing high blood glucose in the fetus. The extra energy in the fetus is stored as fat, resulting in macrosomia or a "fat" baby. Delivery of a very large baby can be dangerous for both the infant and the mother. Cesarean section is often necessary. Complications after birth are common for these babies. Further, there is a greater risk that the child will become obese and develop type 2 diabetes mellitus later in life.

Gestational diabetes usually goes away after the birth of the child. However, once a woman has experienced gestational diabetes, she has a greater chance of developing it again during future pregnancies. These women also tend to develop type 2 diabetes later in life.

CHECK YOUR PROGRESS 15.5

1. Distinguish between the exocrine and endocrine functions of the pancreas.
2. Describe how the pancreatic hormones interact to regulate blood glucose levels.
3. Explain the difference in the function of the pancreas in type 1 and type 2 diabetes.

CONNECTING THE CONCEPTS

For additional information on the various forms of diabetes, refer to the following discussions:

Section 4.8 examines how feedback mechanisms are involved in maintaining blood glucose homeostasis.

Section 8.6 explains the body mass index (BMI), an indicator that is used to determine obesity and the subsequent risk of diabetes.

Section 10.3 examines the influence of diabetes on the urinary system.

15.6 Other Endocrine Glands

LEARNING OUTCOMES

Upon completion of this section, you should be able to

1. List the hormones produced by the sex organs, thymus, and pineal gland, and provide a function for each.
2. List the hormones that are produced by glands and organs outside of the endocrine system.

The **gonads** are the testes in males and the ovaries in females. The gonads are endocrine glands. Other lesser-known glands and some tissues also produce hormones.

Testes and Ovaries

The activity of the testes and ovaries is controlled by the hypothalamus and pituitary. The **testes** are located in the scrotum, and the **ovaries** are located in the pelvic cavity. The testes produce androgens (male sex hormones), such as testosterone. The ovaries produce estrogens and **progesterone,** the female sex hormones. These hormones feed back to control the hypothalamic secretion of gonadotropin-releasing hormone (GnRH). The pituitary gland secretion of follicle-stimulating hormone (FSH) and luteinizing hormone (LH), the gonadotropic hormones (Fig. 15.20), is controlled by feedback from the sex hormones, too. The activities of FSH and LH are discussed in more detail in Chapter 16.

Under the influence of the gonadotropic hormones, the testes begin to release increased amounts of testosterone at the time of puberty. Testosterone stimulates the growth of the penis and the testes. Testosterone also brings about and maintains

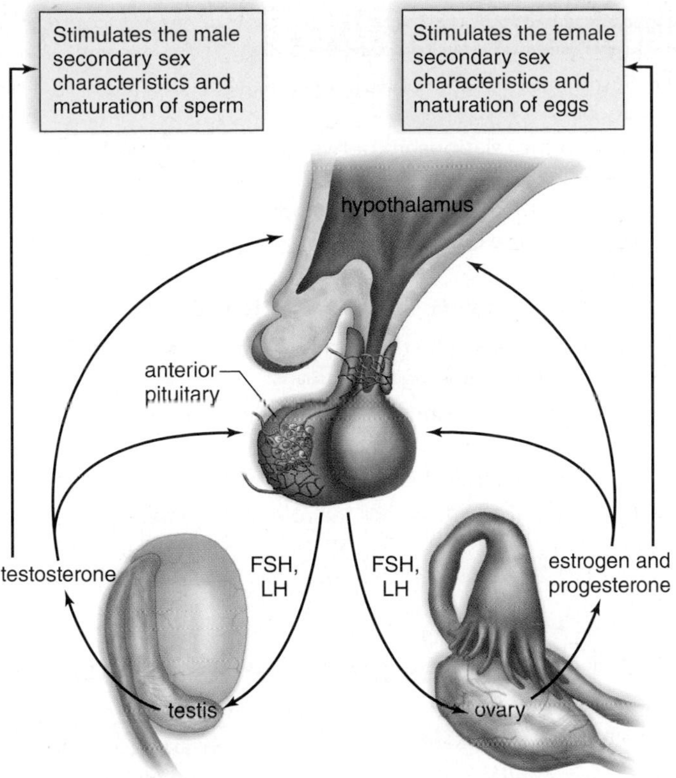

Figure 15.20 The hormones produced by the testes and the ovaries.
The testes and ovaries secrete the sex hormones. The testes secrete testosterone, and the ovaries secrete estrogens and progesterone. In each sex, secretion of GnRH from the hypothalamus and secretion of FSH and LH from the pituitary are controlled by their respective hormones.

the male secondary sex characteristics that develop during puberty. These include the growth of facial, axillary (underarm), and pubic hair. It prompts the larynx and the vocal cords to enlarge, causing the voice to lower. Testosterone also stimulates oil and sweat glands in the skin. It is largely responsible for acne and body odor. Another side effect of testosterone is baldness. Although females, like males, do inherit genes for baldness, baldness is seen more often in males because of the presence of testosterone. Testosterone is partially responsible for the muscular strength of males, and this is why some athletes take supplemental amounts of **anabolic steroids,** which are either testosterone or related chemicals. The Bioethical feature, "Anabolic Steroid Use," in Chapter 12 discusses the detrimental effect anabolic steroids can have on the body.

The female sex hormones, estrogens (often referred to in the singular) and progesterone, have many effects on the body. In particular, estrogen secreted at the time of puberty stimulates the growth of the uterus and the vagina. Estrogen is necessary for egg maturation and is largely responsible for the secondary sex characteristics in females. These include female body hair and fat distribution. In general, females have a more rounded appearance than males because of a greater accumulation of fat beneath the skin. Also, the pelvic girdle is wider in females than in males, resulting in a larger pelvic cavity. Both estrogen and progesterone are required for breast development and for regulation of the uterine cycle. This includes monthly menstruation (discharge of blood and mucosal tissues from the uterus).

Thymus

The lobular **thymus** lies just beneath the sternum (see Fig. 15.2). This organ reaches its largest size and is most active during childhood. With aging, the organ gets smaller and becomes fatty. Lymphocytes that originate in the bone marrow and then pass through the thymus are transformed into T lymphocytes. The lobules of the thymus are lined by epithelial cells that secrete hormones called **thymosins.** These hormones aid in the differentiation of lymphocytes packed inside the lobules. Although thymosins ordinarily work in the thymus, there is hope that these hormones could be injected into AIDS or cancer patients, where they would enhance T-lymphocyte function.

Video Lymphocytes

Pineal Gland

The **pineal gland,** located in the brain (see Fig. 15.2), produces the hormone **melatonin,** primarily at night. Melatonin is involved in our daily sleep-wake cycle. Normally, we grow sleepy at night when melatonin levels increase and awaken once daylight returns and melatonin levels are low (Fig. 15.21). Daily 24-hour cycles such as this are called **circadian rhythms.** These rhythms are controlled by a biological clock located in the hypothalamus.

Video Winter Mood

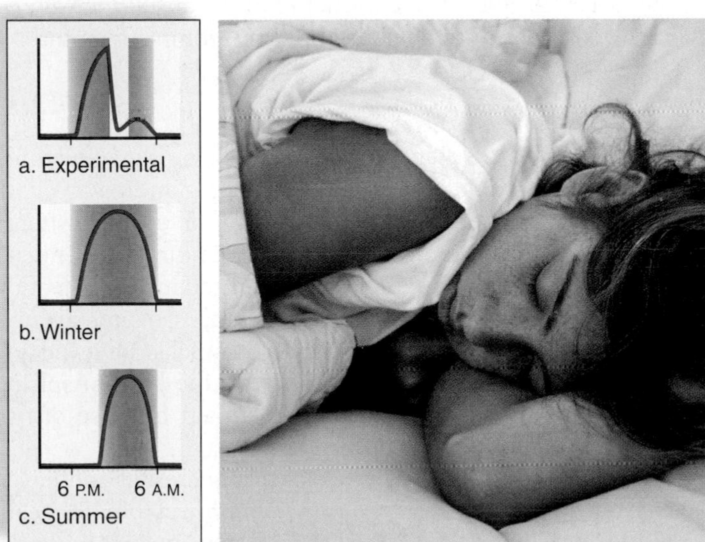

Figure 15.21 Melatonin production changes by season.
Melatonin production is greatest at night when we are sleeping. Light suppresses melatonin production (**a**), so it is secreted for a longer time in the winter (**b**) than in the summer (**c**).

Animal research suggests that melatonin also regulates sexual development. In keeping with these findings, it has been noted that children whose pineal glands have been destroyed due to brain tumors experience early puberty.

Hormones from Other Organs or Tissues

Some organs not usually considered endocrine glands do secrete hormones. We have already mentioned that the kidneys secrete renin and that the heart produces atrial natriuretic hormone (see page 347); recall also that the stomach and the small intestine produce peptide hormones that regulate digestive secretions. A number of other types of tissues produce hormones.

Erythropoietin

In response to a low oxygen blood level, the kidneys secrete erythropoietin (EPO). Erythropoietin stimulates red blood cell formation in the red bone marrow. A greater number of red blood cells results in increased blood oxygen. A number of different types of organs and cells also produce peptide growth factors, which stimulate cell division and mitosis. Growth factors can be considered hormones because they act on cell types with specific receptors to receive them. Some are released into the blood; others diffuse to nearby cells.

Leptin

Leptin is a protein hormone produced by adipose tissue. Leptin acts on the hypothalamus, where it signals satiety or fullness. Strange to say, the blood of obese individuals may be rich in leptin. It is possible that the leptin they produce is ineffective because of a genetic mutation or because their hypothalamic cells lack a suitable number of receptors for leptin.

Prostaglandins

Prostaglandins are potent chemical signals produced within cells from arachidonate, a fatty acid. Prostaglandins are not distributed in the blood. They act locally, quite close to where they were produced. In the uterus, prostaglandins cause muscles to contract. Therefore, they are implicated in the pain and discomfort of menstruation in some women. Also, prostaglandins mediate the effects of pyrogens, chemicals believed to reset the temperature regulatory center in the brain. Aspirin reduces body temperature and controls pain because of its effect on prostaglandins.

Certain prostaglandins reduce gastric secretion and have been used to treat gastric reflux. Others lower blood pressure and have been used to treat hypertension. Still others inhibit platelet aggregation and have been used to prevent thrombosis. However, different prostaglandins have contrary effects, and it has been very difficult to successfully standardize their use. Therefore, prostaglandin therapy is still considered experimental.

CHECK YOUR PROGRESS 15.6

1. Summarize the role of testosterone and estrogen in the body.
2. Explain the relationship between melatonin and the sleep–wake cycle.
3. Describe the response of the body to low levels of oxygen in the blood.

CONNECTING THE CONCEPTS

For more information on the hormones presented in this section, refer to the following discussions:

Figure 6.4 illustrates the role of erythropoietin in the manufacture of new red blood cells.

Section 8.4 examines the role of the digestive hormones.

Sections 16.2 to 16.4 explain the role of the male and female sex hormones.

15.7 Hormones and Homeostasis

LEARNING OUTCOMES

Upon completion of this section, you should be able to

1. Summarize how the endocrine and nervous systems respond to external changes in the body.
2. Summarize how the endocrine and nervous systems respond to internal changes in the body.

The nervous and endocrine systems exert control over the other systems and thereby maintain homeostasis (Fig. 15.22).

Responding to External Changes

The nervous system is particularly able to respond to changes in the external environment. Some responses are automatic as you can verify by trying this: Take a piece of clear plastic and hold it just in front of your face. Get someone to gently toss a soft object, such as a wadded-up piece of paper, at the plastic. Can you prevent yourself from blinking? This reflex protects your eyes.

The eyes and other organs that have sensory receptors provide us with valuable information about the external environment. The central nervous system, on the receiving end of millions of bits of information, integrates information, compares it with previously stored memories, and "decides" on the proper course of action. The nervous system often responds to changes in the external environment through body movement. It gives us the ability to stay in as moderate an environment as possible. Otherwise, we test the ability of the nervous system to maintain homeostasis despite extreme conditions.

Responding to Internal Changes

The governance of internal organs usually requires that the nervous and endocrine systems work together. This usually occurs below the level of consciousness. Subconscious control

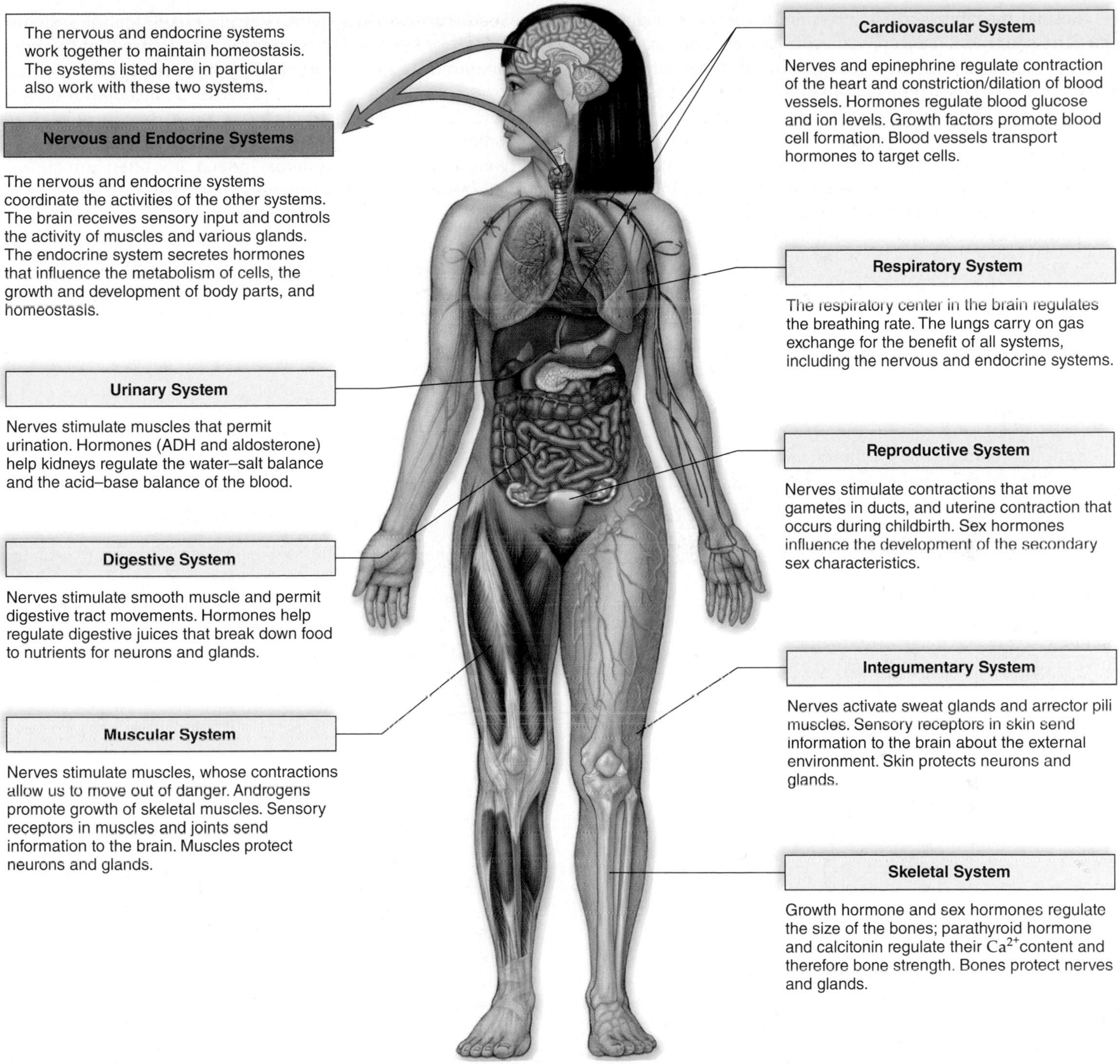

The nervous and endocrine systems work together to maintain homeostasis. The systems listed here in particular also work with these two systems.

Nervous and Endocrine Systems

The nervous and endocrine systems coordinate the activities of the other systems. The brain receives sensory input and controls the activity of muscles and various glands. The endocrine system secretes hormones that influence the metabolism of cells, the growth and development of body parts, and homeostasis.

Urinary System

Nerves stimulate muscles that permit urination. Hormones (ADH and aldosterone) help kidneys regulate the water–salt balance and the acid–base balance of the blood.

Digestive System

Nerves stimulate smooth muscle and permit digestive tract movements. Hormones help regulate digestive juices that break down food to nutrients for neurons and glands.

Muscular System

Nerves stimulate muscles, whose contractions allow us to move out of danger. Androgens promote growth of skeletal muscles. Sensory receptors in muscles and joints send information to the brain. Muscles protect neurons and glands.

Cardiovascular System

Nerves and epinephrine regulate contraction of the heart and constriction/dilation of blood vessels. Hormones regulate blood glucose and ion levels. Growth factors promote blood cell formation. Blood vessels transport hormones to target cells.

Respiratory System

The respiratory center in the brain regulates the breathing rate. The lungs carry on gas exchange for the benefit of all systems, including the nervous and endocrine systems.

Reproductive System

Nerves stimulate contractions that move gametes in ducts, and uterine contraction that occurs during childbirth. Sex hormones influence the development of the secondary sex characteristics.

Integumentary System

Nerves activate sweat glands and arrector pili muscles. Sensory receptors in skin send information to the brain about the external environment. Skin protects neurons and glands.

Skeletal System

Growth hormone and sex hormones regulate the size of the bones; parathyroid hormone and calcitonin regulate their Ca^{2+} content and therefore bone strength. Bones protect nerves and glands.

Figure 15.22 The nervous system and endocrine system interact to control homeostasis.
The nervous and endocrine systems work together to regulate and control the other systems.

often depends on reflex actions that involve the hypothalamus and the medulla oblongata. Let's take blood pressure as an example. After a 3-mile run, you decide to sit down under a tree to rest a bit. When you stand up to push off again, you feel faint. The feeling quickly passes because the medulla oblongata responds to input from the baroreceptors in the aortic arch and carotid arteries. The sympathetic system immediately acts to increase heart rate and constrict the blood vessels so that your blood pressure rises. Sweating may have upset the water–salt balance of your blood. If so, the hormone aldosterone from the adrenal cortex will act on the kidney tubules to conserve sodium ions (Na^+), and water reabsorption will follow. The hypothalamus can also help by sending antidiuretic hormone (ADH) to the posterior pituitary gland, which releases it into the blood. ADH actively promotes water reabsorption by the kidney tubules.

Recall from Chapter 13 that certain drugs, such as alcohol, can affect ADH secretion. When you consume alcohol, it is quickly absorbed across the stomach lining into the bloodstream, where it travels to the hypothalamus and inhibits ADH secretion. When ADH levels fall, the kidney tubules absorb less water. The result is increased production of dilute urine. Excessive water loss, or dehydration, is a disturbance of homeostasis. This is why drinking alcohol when you are exercising or perspiring heavily on a hot day is not a good idea. Instead of keeping you hydrated, an alcoholic beverage, such as beer, has the opposite effect.

Controlling the Reproductive System

Few systems intrigue us more than the reproductive system, which couldn't function without nervous and endocrine control. The hypothalamus controls the anterior pituitary, which controls the release of hormones from the testes and the ovaries and the production of their gametes. The nervous system directly controls the muscular contractions of the ducts that propel the sperm. Contractions of the uterine tubes, which move a developing embryo to the uterus where development continues, are stimulated by the nervous system, too. Without the positive feedback cycle involving oxytocin produced by the hypothalamus, birth might not occur.

The Neuroendocrine System

The nervous and endocrine systems work so closely together that they form what is sometimes called the neuroendocrine system. As we have seen, the hypothalamus certainly bridges the regulatory activities of the nervous and endocrine systems. In addition to producing the hormones released by the posterior pituitary, the hypothalamus produces hormones that control the anterior pituitary. The nerves of the autonomic system, which control other organs, are acted upon directly by the hypothalamus. The hypothalamus truly belongs to both the nervous and endocrine systems. Indeed, it is often and appropriately referred to as a neuroendocrine organ.

CHECK YOUR PROGRESS 15.7

1. Summarize the role of the nervous system in the monitoring of the internal and external environments.
2. Explain how the body restores its water–salt balance after it has lost water and salt through sweating.
3. Explain why the nervous and endocrine systems are integrated with one another.

CONNECTING THE CONCEPTS

For more information on the organ systems presented in this section, refer to the following discussions:

Section 5.3 examines the factors that regulate heart rate.

Section 10.4 explains the role of aldosterone and ADH on the function of the kidney.

Section 13.2 explores the roles of the hypothalamus and medulla oblongata in the CNS.

CASE STUDY CONCLUSION

For diabetics, the prospects of controlling their blood glucose levels and living a healthy life are better today than in the past. Prior to the development of recombinant DNA technology, which now allows human insulin to be produced in large quantities, insulin was derived from the pancreases of pigs or cows. This required laborious purification, and because the animal insulins were not identical to the human form, sometimes immunologic reactions occurred. Increasingly, insulin pumps are being used to treat diabetes. An insulin pump is a device a little bigger than a cell phone, which can deliver precise amounts of insulin under the skin using a small plastic catheter. The insulin pump more accurately mimics the pancreas's natural release of the correct amount of insulin needed by the body. Studies have shown that insulin pumps are more effective than traditional injections of insulin in controlling blood sugar levels. In the near future, it may be possible to implant a device—sometimes called an "artificial pancreas"—into patients with diabetes that will not only monitor the blood sugar level but will also provide the appropriate doses of insulin.

MEDIA STUDY TOOLS

Enhance your study of this chapter with media! Visit **www.mhhe.com/maderhuman13e** and go to "Media Study Tools" for this chapter to access the following:

Animations	Videos	MP3 Files
15.1 Positive and Negative Feedback • Second Messengers: cAMP • Peptide Hormone Action • Mechanism of Steroid Hormone Action **15.2** Hormonal Communication • Positive and Negative Feedback **15.3** Mechanism of Thyroxine Action **15.4** Glucocorticoid Hormones **15.5** Blood Sugar Regulation in Diabetics	**15.6** Lymphocytes • Winter Mood	**15.1** Major Endocrine Glands • Endocrine System **15.2** Hormonal Secretion and Action **15.3** Calcium Homeostasis

SUMMARIZE

15.1 Endocrine Glands

The **endocrine system** works with the nervous system to regulate the activities of the other body systems. **Endocrine glands** secrete hormones into the bloodstream; from there, they are distributed to target organs or tissues. This differs from **exocrine glands,** which secrete products into ducts.

- **Hormones,** a type of chemical signal, usually act at a distance between body parts. Hormones are either peptides or steroids.
- **Pheromones** are chemical signals that influence the behavior of another individual.
- Reception of a **peptide hormone** at the plasma membrane activates an enzyme cascade inside the cell. They typically use **second messenger** systems, such as **cyclic adenosine monophosphate** (**cAMP**).
- **Steroid hormones** combine with a receptor, and the complex attaches to and activates DNA. Protein synthesis follows.

15.2 Hypothalamus and Pituitary Gland

The endocrine system is controlled by the **hypothalamus,** which regulates the secretions of the **pituitary gland.** Neurosecretory cells in the hypothalamus produce **antidiuretic hormone** (**ADH**) and **oxytocin,** which are stored in axon endings in the **posterior pituitary** until released.

- The hypothalamus produces hypothalamic-releasing and hypothalamic-inhibiting hormones, which pass to the anterior pituitary by way of a portal system.
- The **anterior pituitary** produces several types of hormones, including **thyroid-stimulating hormone** (**TSH**), **adrenocorticotropic hormone** (**ACTH**), **gonadotropic hormones, follicle-stimulating hormone** (**FSH**), **luteinizing hormone** (**LH**), **prolactin, melanocyte-stimulating hormone,** and **growth hormone** (**GH**). Some of these stimulate other hormonal glands to secrete hormones.
- Endocrine disorders associated with growth hormones include **pituitary dwarfism** and **acromegaly.**

15.3 Thyroid and Parathyroid Glands

The **thyroid gland** requires iodine to produce **triiodothyronine** (**T_3**) and **thyroxine** (**T_4**), which increase the metabolic rate.

- If iodine is available in limited quantities, a **simple goiter** develops. **Congenital hypothyroidism** occurs if the thyroid does not develop correctly.
- In adults, an underactive thyroid leads to **myxedema,** while an overactive thyroid results in an **exophthalmic goiter.**
- The thyroid gland produces **calcitonin,** which helps lower the blood calcium level.
- The **parathyroid glands** secrete **parathyroid hormone** (**PTH**), which raises the blood calcium level. Hypoparathyroidism may result in **tetany.**

15.4 Adrenal Glands

The **adrenal glands** respond to stress.

Adrenal Medulla

The **adrenal medulla** immediately secretes **epinephrine** and **norepinephrine** (**NE**). Heartbeat and blood pressure increase, blood glucose level rises, and muscles become energized.

Adrenal Cortex

The **adrenal cortex** produces the **glucocorticoids** (**cortisol**), the **mineralocorticoids** (**aldosterone**), and sex steroids (**dehydroepiandrosterone** (**DHEA**), **testosterone,** and **estradiol** (**estrogen**). The glucocorticoids regulate carbohydrate, protein, and fat metabolism and also suppress the inflammatory response. Mineralocorticoids are influenced by **renin** from the kidneys, and regulate water and salt balance, leading to increases in blood volume and blood pressure. The male (**androgen**) and female sex steroids are used in both sexes.

- Problems with the adrenal cortex may result in **Addison disease** or **Cushing syndrome.**

15.5 Pancreas

The **pancreas** contains both endocrine and exocrine cells. The **pancreatic islets** secrete the hormones insulin and glucagon.

- **Insulin** lowers the blood glucose level.

- **Glucagon** raises the blood glucose level.
- **Diabetes mellitus** is due to the failure of the pancreas to produce insulin or the failure of the cells to take it up.

15.6 Other Endocrine Glands

Other endocrine glands produce hormones.

- The **testes** and **ovaries** (**gonads**) produce the sex hormones. Male sex hormones are the androgens (testosterone); female sex hormones are the estrogens and **progesterone. Anabolic steroids** mimic the action of testosterone.
- The **thymus** secretes **thymosins,** which stimulate T-lymphocyte production and maturation.
- The **pineal gland** produces **melatonin,** which may be involved in **circadian rhythms** and the development of the reproductive organs.

Some organs and tissues also produce hormones.

- Kidneys produce erythropoietin.
- Adipose tissue produces **leptin,** which acts on the hypothalamus.
- **Prostaglandins** are produced within cells and act locally.

15.7 Hormones and Homeostasis

The nervous and endocrine systems exert control over the other systems and thereby maintain homeostasis.

- The nervous system is able to respond to the external environment after receiving data from the sensory receptors. Sensory receptors are present in such organs as the eyes and ears.
- The nervous and endocrine systems work together to govern the subconscious control of internal organs. This control often depends on reflex actions involving the hypothalamus and medulla oblongata.
- The nervous and endocrine systems work so closely together that they form what is sometimes called the neuroendocrine system.

ASSESS

Testing Your Knowledge of the Concepts

1. Compare and contrast the nervous and endocrine systems. (pages 334–335)
2. How does the action of a peptide hormone differ from that of a steroid hormone? (pages 337–338)
3. Explain the relationship between the hypothalamus, the posterior pituitary gland, and the anterior pituitary gland. List the hormones secreted by the anterior pituitary and the posterior pituitary glands and their actions. (pages 339–340)
4. Give an example of the negative feedback relationship among the hypothalamus, the anterior pituitary, and other endocrine glands. (pages 340–341)
5. Discuss the action of growth hormone on the body. What occurs if there is too much or too little GH during the growing years? What occurs if there is too much GH in an adult? (pages 341–342)
6. What types of conditions are associated with a malfunctioning thyroid gland? Explain each type. (pages 343–344)
7. Explain how the thyroid and parathyroid glands work together to maintain blood calcium homeostasis. (page 344–345)
8. What hormones are secreted by the adrenal cortex and adrenal medulla, and what are their actions? (pages 345–347)
9. What are the causes and symptoms of Addison disease and Cushing syndrome? (page 348)
10. Explain how insulin and glucagon maintain blood glucose homeostasis. What are the two types of diabetes mellitus, and what are the major symptoms? (pages 349–350)
11. How does the neuroendocrine system work with other systems to maintain homeostasis? (pages 354–355)
12. Which type of glands are ductless?
    a. exocrine
    b. endocrine
    c. Both a and b are correct.
    d. Neither a nor b is correct.
13. Which hormones can cross cell membranes?
    a. peptide hormones
    b. steroid hormones
    c. Both a and b are correct.
    d. Neither a nor b is correct.
14. The anterior pituitary controls the secretions of both
    a. the adrenal medulla and the adrenal cortex.
    b. the thyroid and adrenal cortex.
    c. the ovaries and testes.
    d. Both b and c are correct.
15. Growth hormone is produced by the
    a. posterior adrenal gland.
    b. posterior pituitary.
    c. anterior pituitary.
    d. kidneys.
    e. None of the choices are correct.
16. __________ is released through positive feedback and causes __________ contractions.
    a. Insulin, stomach
    b. Oxytocin, stomach
    c. Oxytocin, uterine
    d. None of the choices are correct.
17. PTH causes the blood level of calcium to __________, and calcitonin causes it to __________.
    a. increase, increase
    b. increase, decrease
    c. decrease, increase
    d. decrease, decrease
18. Bodily response to stress includes
    a. water reabsorption by the kidneys.
    b. blood pressure increase.
    c. increase in blood glucose levels.
    d. heart rate increase.
    e. All of the choices are correct.
19. Lack of aldosterone will cause a blood imbalance of
    a. sodium.
    b. potassium.
    c. water.
    d. All of the choices are correct.
    e. None of the choices are correct.

20. Long-term complications of diabetes include
    a. blindness.
    b. kidney disease.
    c. circulatory disorders.
    d. All of the choices are correct.
    e. None of the choices are correct.
21. Diabetes mellitus is associated with
    a. too much insulin in the blood.
    b. too high a blood glucose level.
    c. blood that is too dilute.
    d. All of the choices are correct.
22. Which of these is not a pair of antagonistic hormones?
    a. insulin—glucagon
    b. calcitonin—parathyroid hormone
    c. cortisol—epinephrine
    d. aldosterone—atrial natriuretic hormone (ANH)
23. Which hormone and condition are mismatched?
    a. growth hormone—acromegaly
    b. thyroxine—goiter
    c. parathyroid hormone—tetany
    d. cortisol—myxedema
    e. insulin—diabetes

In questions 24–28, match the hormones to the correct gland in the key.

Key:

a. glucagon
b. prostaglandin
c. melatonin
d. insulin
e. leptin

24. Raises blood glucose levels
25. Conversion of glucose to glycogen
26. Hunger control
27. Controls circadian rhythms
28. Causes uterine contractions

ENGAGE

Thinking Critically About the Concepts

Blood tests are a way to diagnose any number of endocrine disorders because hormones are transported by the circulatory system. GH and IGF-1 can be checked to determine if deficiencies are the reason for a child's slow growth. Blood levels of TSH, T_3, and T_4 provide information about thyroid function. Some tests, such as the glucose tolerance test from the chapter opener, do not directly measure the level of the glucose-regulating hormones (in this case, insulin) but rather indirectly monitor whether an endocrine gland is performing correctly by measuring specific compounds in the blood.

1. How is follicle-stimulating hormone similar to growth hormone with regard to how their target cells respond to their signals?
2. It is possible to diagnose hypothyroidism by high levels of TSH in the blood. Explain what would cause a high TSH level. (*Hint:* You may want to consider what happens to TSH when the activity of the thyroid is normal.)
3. Why would a diabetic urinate frequently and always be thirsty?
4. Many diets advertise that they are specifically designed for diabetics. How would these diets be different from a "normal" diet?

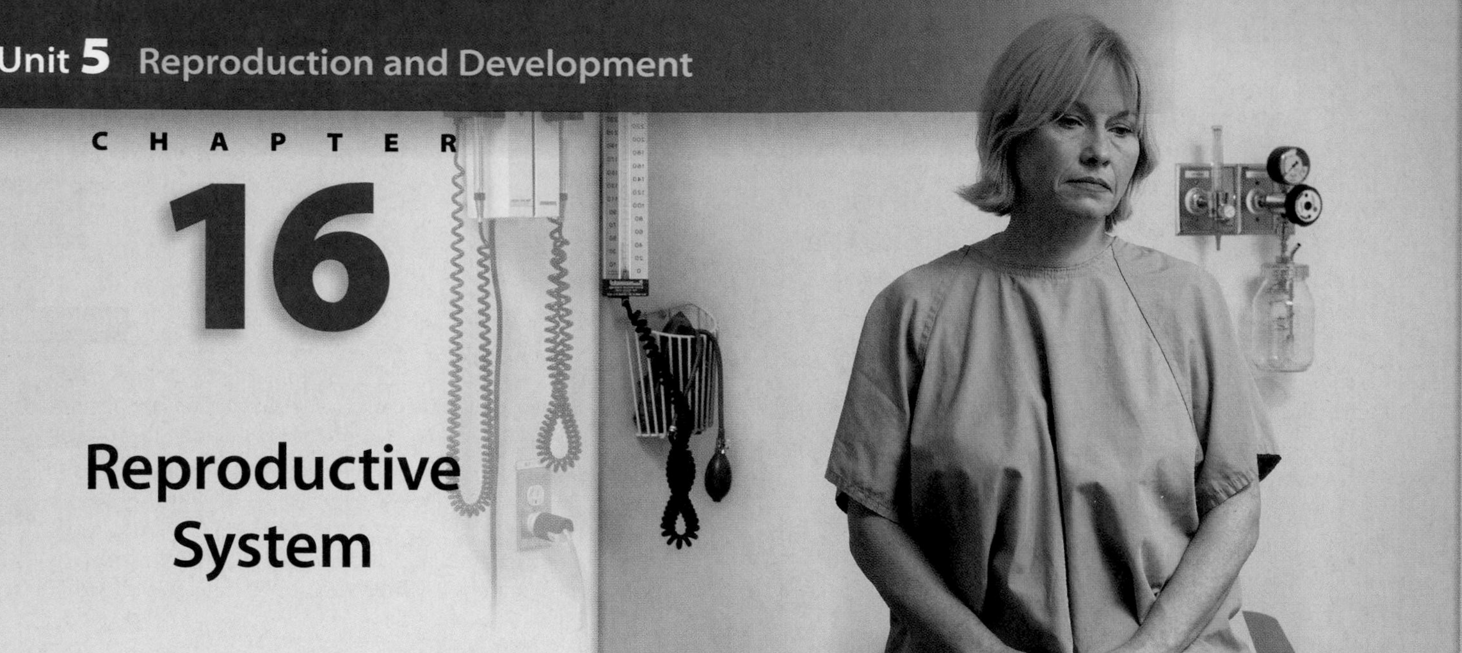

CHAPTER 16
Reproductive System

CHAPTER CONCEPTS

16.1 Human Life Cycle
The male reproductive system produces sperm, and the female reproductive system produces eggs, each of which have only 23 chromosomes due to meiosis.

16.2 Male Reproductive System
The male reproductive system produces sperm and the male sex hormones.

16.3 Female Reproductive System
In the female, the ovaries produce eggs and the female sex hormones.

16.4 The Ovarian Cycle
The female sex hormones fluctuate in monthly cycles, resulting in ovulation once a month followed by menstruation if pregnancy does not occur.

16.5 Control of Reproduction
Numerous birth control methods are available for those who wish to prevent pregnancy. Infertile couples may use assisted reproductive technologies to have a child.

16.6 Sexually Transmitted Diseases
Medications have been developed to control AIDS and genital herpes, but these STDs are not curable. STDs caused by bacteria are curable with antibiotic therapy, but resistance is making this more and more difficult.

BEFORE YOU BEGIN

Before beginning this chapter, take a few moments to review the following discussions:

Section S.1 What factors contribute to an increased risk of an HIV infection?

Section 15.6 Where is testosterone produced and what is its function in the male body?

Section 15.6 Where are estrogen and progesterone produced and what are their functions in the female body?

CASE STUDY CERVICAL CANCER

Ann always dreaded her visits to the gynecologist. At each visit, her doctor had warned her to quit smoking. However, for the past 20 years, she had been a regular smoker, and even though she had cut back considerably in the past few years, she was still smoking around a pack of cigarettes a day. Ann's annual Pap smears had always been normal, so Ann was beginning to view her annual trip to the gynecologist as just a formality. Since she had turned 40, her visits had become more sporadic; and over the past few years, she had stopped the visits completely.

Until recently, Ann had felt fine. However, starting just a few months ago, she had started to experience some abnormal vaginal bleeding, usually shortly after sexual intercourse with her partner. This was often accompanied by small amounts of pain. Concerned about these recent changes, Ann scheduled an appointment with her doctor.

At the appointment, the doctor performed a complete physical exam, which included a Pap smear. As the doctor expected, the results of her test were abnormal. Her doctor sent the results to an oncologist, a cancer specialist, who confirmed that Ann's symptoms were being caused by an early stage of cervical cancer. To check the extent of the cancer and to see if it had spread to any additional organs, the oncologist ordered a computed axial tomography (CAT) scan of Ann's pelvis and abdomen, as well as a series of blood tests to look for evidence of cancer. Both the CAT scan and the blood tests indicated that Ann was lucky—they had caught the cancer at an early stage of development. Her oncologist was convinced that a hysterectomy might be avoided, but Ann would have to immediately begin both chemotherapy and radiation treatment to stop the spread of the cancer.

As you read through the chapter, think about the following questions:

1. What is the role of the cervix in the female reproductive system?
2. What is a Pap smear test used to detect?
3. What is a hysterectomy?

16.1 Human Life Cycle

LEARNING OUTCOMES

Upon completion of this section, you should be able to

1. List the functions of the reproductive system in humans.
2. Describe the human life cycle and explain the role of mitosis and meiosis in this cycle.

Unlike the other systems of the body, the **reproductive system** is quite different in males and females. In both males and females, the reproductive system is responsible for the production of gametes, the cells that combine to form a new individual of the species. In females, the reproductive system has the added function of protecting and nourishing the developing fetus until birth. The reproductive organs, or genitals, have the following functions:

1. Males produce sperm within testes, and females produce eggs within ovaries.
2. Males nurture and transport the sperm in ducts until they exit the penis. Females transport the eggs in uterine tubes to the uterus.
3. The male penis functions to deliver sperm to the female vagina, which functions to receive the sperm. The vagina also transports menstrual fluid to the exterior and acts as the birth canal.
4. The uterus of the female allows the fertilized egg to develop within her body. After birth, the female breast provides nourishment in the form of milk.
5. The testes and ovaries produce the sex hormones. The sex hormones have a profound effect on the body because they bring about masculinization or feminization of various features. In females, the sex hormones also allow a pregnancy to continue.

Unlike many other animals, humans are not reproductively capable at birth. Instead, they undergo a sequence of events called *puberty,* during which a child becomes a sexually competent young adult. The reproductive system does not begin to fully function until puberty is complete. Sexual maturity typically occurs between the ages of 10 and 14 in girls and 12 and 16 in boys. At the completion of puberty, the individual is capable of producing children.

Mitosis and Meiosis

There are two different forms of cell division that occur during the human life cycle. First, it is important to recognize that our genetic instructions, or DNA, is distributed among 46 chromosomes within the nucleus. These 46 chromosomes exist in 23 pairs, with each pair containing a contribution from both the male and female parent. The majority of the cell types in the body have 46 chromosomes. During the majority of our life cycle, our cells divide by a process called **mitosis** (see section 18.3). Mitosis is *duplication division,* meaning that each of the cells that exit mitosis have the same complement of 46 chromosomes. In other words, when a cell divides, it produces exact copies of itself by mitosis, much like a copier machine does with a page of notes. In the life cycle of a human, mitosis is the type of cell division that plays an important role during growth and repair of tissues (Fig. 16.1).

For the purposes of reproduction, special cells in the body undergo a type of cell division called **meiosis.** Meiosis takes place only in the testes of males during the production of sperm and in the ovaries of females during the production of eggs.

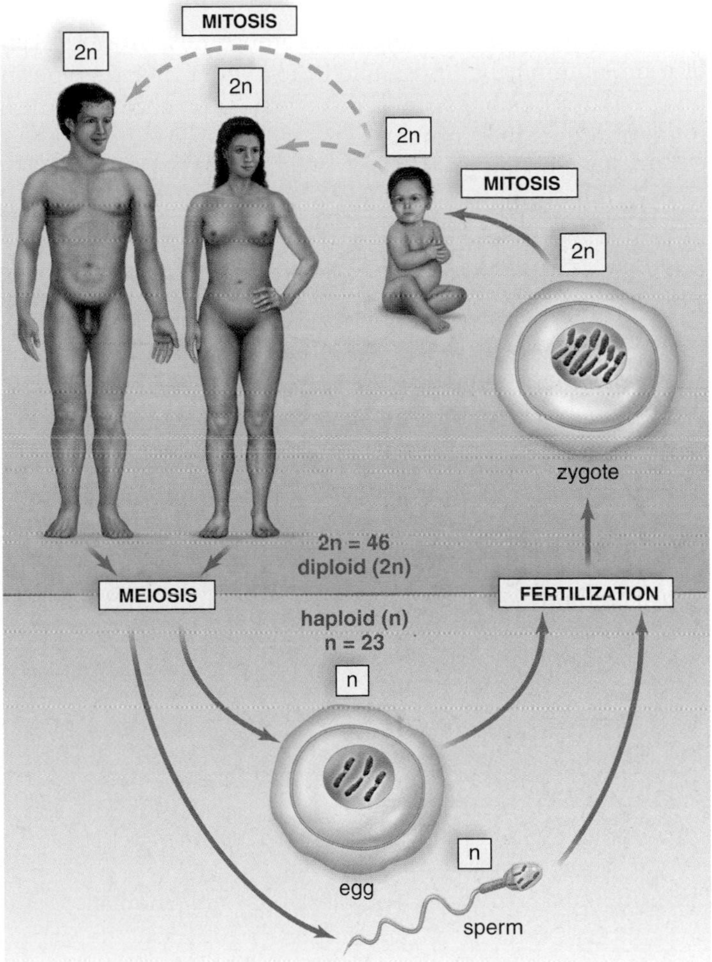

Figure 16.1 The human life cycle.
The human life cycle has two types of cell divisions: mitosis, in which the chromosome number stays constant, and meiosis, in which the the chromosome number is reduced. During growth or celsl repair, mitosis ensures that each new cell has 46 chromosomes. During production of sex cells, the chromosome number is reduced from 46 to 23. Therefore, an egg and a sperm each have 23 chromosomes so that when the sperm fertilizes the egg, the new cell, called a zygote, has 46 chromosomes.

Meiosis has two functions (see section 18.4), the first of which is called *reduction division.* During meiosis, the chromosome number is reduced from the normal 46 chromosomes, called the diploid or 2n number, down to 23 chromosomes, called the haploid or n number of chromosomes. This process requires two successive divisions, called meiosis I and meiosis II, and is involved in the formation of **gametes,** or sex cells. As is explained in section 18.4, meiosis also introduces genetic variation, thus ensuring that the new individual is not an exact copy of either parent.

Following meiosis, the haploid cells develop into either sperm (males) or eggs (female). The flagellated sperm is small compared to the egg. It is specialized to carry only chromosomes as it swims to the egg. The egg is specialized to await the arrival of a sperm and to provide the new individual with cytoplasm in addition to chromosomes. The first cell of a new human being is called the **zygote.** Since a sperm has 23 chromosomes and the egg has 23 chromosomes, the zygote has 46 chromosomes altogether. Without meiosis, the chromosome number in each generation of humans would double, and the cells would no longer be able to function.

APPLICATIONS AND MISCONCEPTIONS

What types of cells do not have 46 chromosomes?

In humans, the cell types that do not have the standard 23 pairs of chromosomes are the red blood cells and the cells of the liver. Recall from section 6.2 that red blood cells lack a nucleus; therefore, they do not have any chromosomes. Cells in the liver, called hepatocytes, typically have more than 3 copies of each chromosome (giving them 69 or more chromosomes). This condition is called polyploidy, and it is believed to provide the liver with its ability to degrade toxic compounds.

CHECK YOUR PROGRESS 16.1

1. Compare the functions of the reproductive system in males and females.
2. Contrast the two types of cell division in the human life cycle.
3. Explain the location of meiosis in males and females.

CONNECTING THE CONCEPTS

For more information on the topics presented in this section, refer to the following discussions:

Section 15.6 provides an introduction to the sex hormones.

Sections 18.3 and 18.4 examine the stages of mitosis and meiosis.

Section 18.5 compares the processes of mitosis and meiosis.

16.2 Male Reproductive System

LEARNING OUTCOMES

Upon completion of this section, you should be able to

1. Identify the structures of the male reproductive system and provide a function for each.
2. Describe the location and stages of spermatogenesis.
3. Summarize how hormones regulate the male reproductive system.

The male reproductive system includes the organs listed in Table 16.1 and shown in Figure 16.2. The male gonads, or primary sex organs, are paired **testes** (sing., testis), suspended within the sacs of the **scrotum.**

MP3
Male Reproductive Anatomy and Physiology

Sperm produced by the testes mature within the **epididymis** (pl., epididymides), a tightly coiled duct lying just outside each testis. Maturation seems to be required for sperm to swim to the egg. When sperm leave an epididymis, they enter a **vas deferens** (pl., vasa deferentia), also called the ductus deferens. The sperm may be stored for a time in the vas deferens. Each vas deferens passes into the abdominal cavity, where it curves around the bladder and empties into an ejaculatory duct. The ejaculatory ducts enter the **urethra.**

At the time of ejaculation, sperm leave the penis in a fluid called **semen.** The seminal vesicles, the prostate gland, and the bulbourethral glands add secretions to seminal fluid. A pair of **seminal vesicles** lie at the base of the bladder, and each has a duct that joins with a vas deferens. The **prostate gland** is a single, doughnut-shaped gland that surrounds the upper portion of the urethra just below the bladder. In older men, the prostate can enlarge and squeeze off the urethra, making urination painful and difficult. This condition is discussed in more detail in the Health feature, "Urinary Difficulties Due to an Enlarged Prostate," in Chapter 10. **Bulbourethral glands** (also called Cowper glands) are pea-sized organs that lie posterior to the prostate on either side of the urethra. Their secretion makes the seminal fluid gelatinous.

Each component of seminal fluid has a particular function. Sperm are more viable in a basic solution; seminal fluid, milky

Table 16.1 Male Reproductive Organs

Organ	Function
Testes	Produce sperm and sex hormones
Epididymides	Ducts where sperm mature and some sperm are stored
Vasa deferentia	Conduct and store sperm
Seminal vesicles	Contribute nutrients and fluid to semen
Prostate gland	Contributes fluid to semen
Urethra	Conducts sperm
Bulbourethral glands	Contribute mucus-containing fluid to semen
Penis	Organ of sexual intercourse

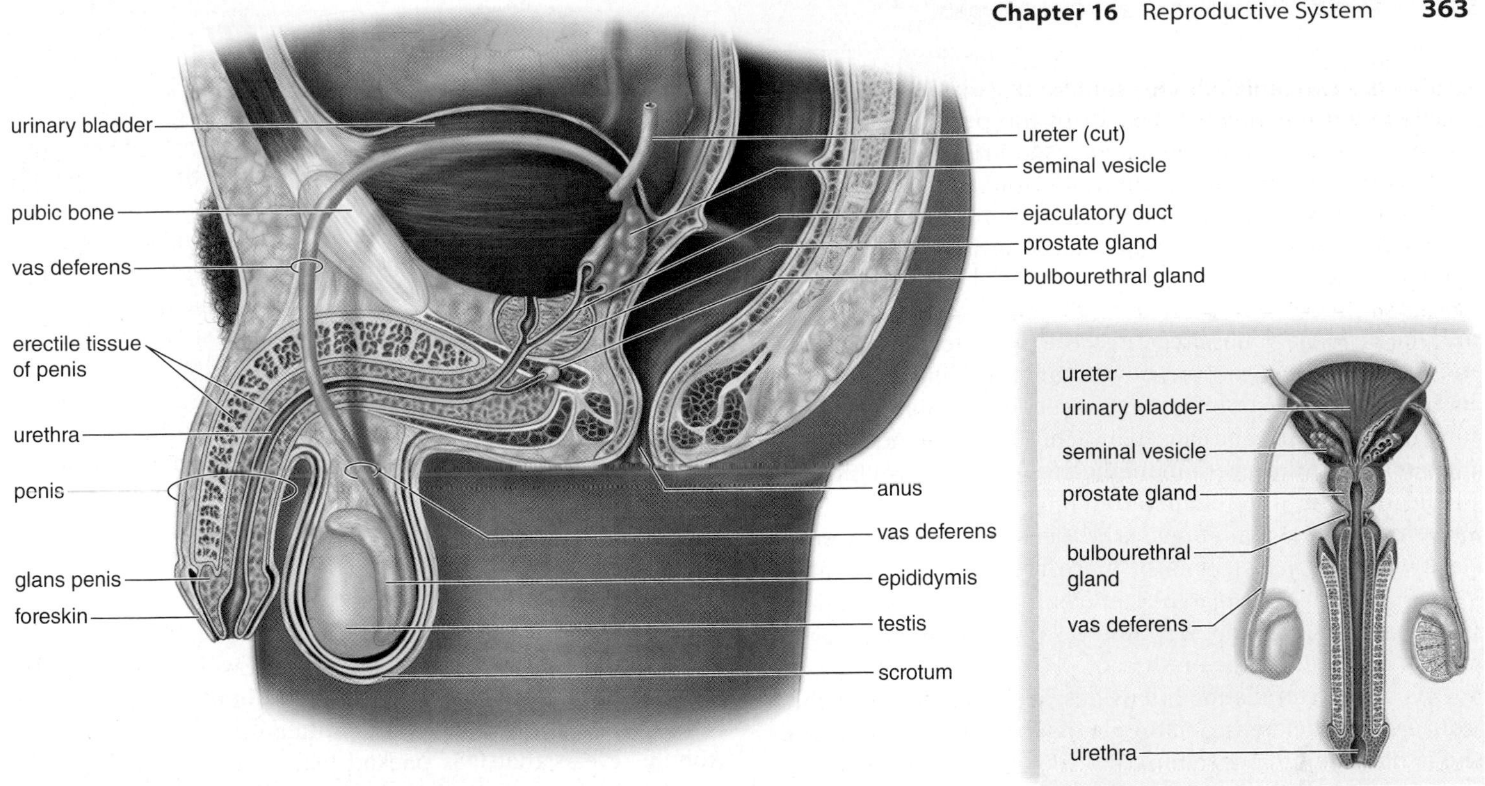

Figure 16.2 The male reproductive system.
The testes produce sperm. The seminal vesicles, the prostate gland, and the bulbourethral glands provide a fluid medium for the sperm, which move from the vas deferens through the ejaculatory duct to the urethra in the penis. The foreskin (prepuce) is removed when a penis is circumcised.

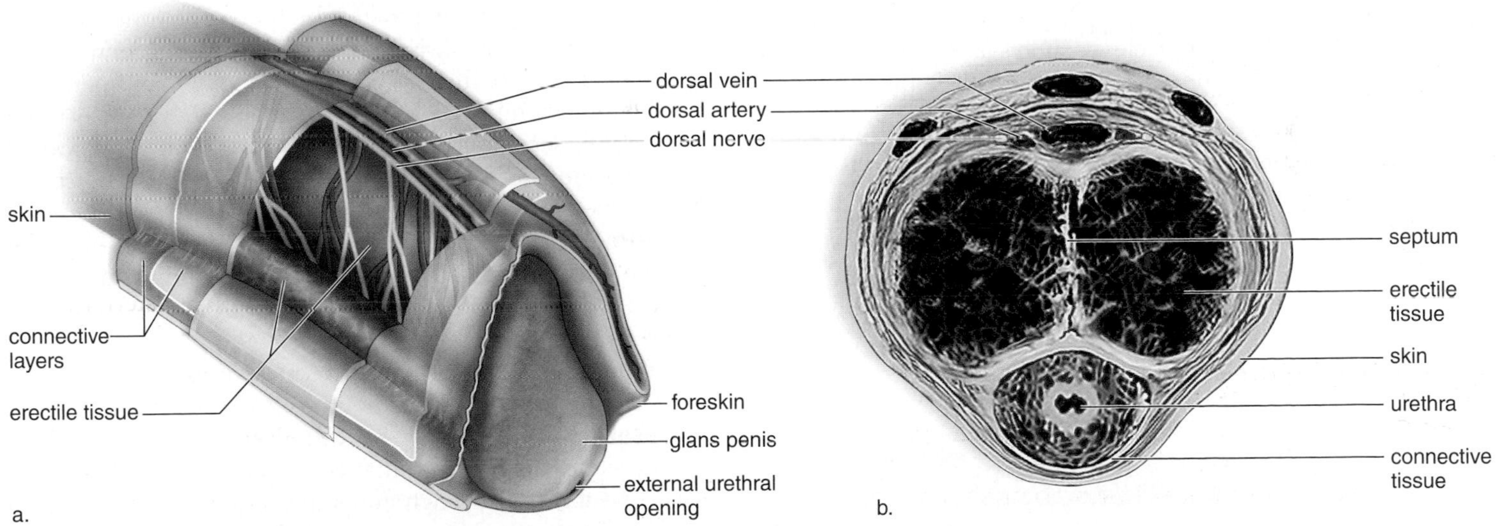

Figure 16.3 The structure of the penis.
a. The shaft of the penis ends in an enlarged tip called the glans penis. In uncircumsized males, this is partially covered by a foreskin (prepuce). **b.** Micrograph of shaft in cross section showing location of erectile tissue.

in appearance, has a slightly basic pH (about 7.5). Swimming sperm require energy; and seminal fluid contains the sugar fructose, which presumably serves as an energy source. Semen also contains prostaglandins, chemicals that cause the uterus to contract. Some investigators believe that uterine contractions help propel the sperm toward the egg.

The Penis and Male Orgasm

The **penis** (Fig. 16.3) is the male organ of sexual intercourse. It also contains the urethra of the urinary system. The penis has a long shaft and an enlarged tip called the glans penis. The layer of skin covering the glans penis, called the foreskin, may be removed surgically by **circumcision** shortly after birth.

Spongy, erectile tissue containing distensible blood spaces extends through the shaft of the penis. During sexual arousal, autonomic nerves release nitric oxide, NO. This stimulus leads to the production of cGMP (cyclic guanosine monophosphate), a high-energy compound similar to ATP. The cGMP causes the smooth muscle of incoming arterial walls to relax and the erectile tissue to fill with blood. The veins that take blood away from the penis are compressed, and the penis becomes erect. **Erectile dysfunction** (**ED**) (formerly called *impotency* but now commonly called *ED*) is the inability to achieve or maintain an erection suitable for sexual intercourse. ED may be caused by a number of factors, including poor blood flow, certain medications, and many illnesses. Medications for the treatment of erectile dysfunction inhibit the enzyme that breaks down cGMP, ensuring that a full erection will take place. Some of these medications can cause vision problems because the same enzyme occurs in the retina. During an erection, a sphincter closes off the bladder so that no urine enters the urethra.

As sexual stimulation intensifies, sperm enter the urethra from each vas deferens, and the glands contribute secretions to the seminal fluid. Once seminal fluid is in the urethra, rhythmic muscle contractions cause it to be expelled from the penis in spurts (ejaculation).

The contractions that expel seminal fluid from the penis are a part of male orgasm, the physiological and psychological sensations that occur at the climax of sexual stimulation. The psychological sensation of pleasure is centered in the brain. However, the physiological reactions involve the reproductive organs and associated muscles, as well as the entire body. Marked muscular tension is followed by contraction and relaxation. Following ejaculation and/or loss of sexual arousal, the penis returns to its normal flaccid state. Usually a period of time, called the refractory period, follows during which stimulation does not bring about an erection. The length of the refractory period increases with age.

There may be in excess of 400 million sperm in the 3.5 ml of semen expelled during ejaculation. The sperm count can be much lower than this, however, and fertilization of the egg by a sperm can still take place.

APPLICATIONS AND MISCONCEPTIONS

Boxers or briefs?

Most people seem to know that the scrotum's role is to keep the temperature of the testes lower than body temperature. The lower temperature is necessary for normal sperm production. It might follow that the man's type of underwear might change that temperature, affecting sperm production. However, research has not supported this assumption. The style of underwear worn by a man, loose or close fitting, has not been shown to affect sperm count or fertility significantly. So, boxers or briefs? It's up to you!

Male Gonads: The Testes

The testes, which produce sperm and also the male sex hormones, lie outside the abdominal cavity of the male, within the scrotum. The testes begin their development inside the abdominal cavity. They descend into the scrotal sacs through the inguinal canal during the last two months of fetal development. If the testes do not descend and the male is not treated or operated on to place the testes in the scrotum, sterility, or the inability to produce offspring, usually follows. This is because the internal temperature of the body is too high to produce viable sperm. The scrotum helps regulate the temperature of the testes by holding them closer to or farther away from the body.

Seminiferous Tubules

A longitudinal section of a testis shows that it is composed of compartments called lobules, each of which contains one to three tightly coiled **seminiferous tubules** (Fig. 16.4*a*). A microscopic cross section of a seminiferous tubule reveals that it is packed with cells undergoing **spermatogenesis** (Fig. 16.4*b*), the production of sperm.

Animation Spermatogenesis

During the production of sperm, spermatogonia divide to produce primary spermatocytes (2n). One of these cells does not proceed with the remainder of spermatogenesis, and instead serves as a stem cell, allowing spermatogenesis to continue throughout the lifetime of the male. The other primary spermatocyte moves away from the outer wall, increases in size, and undergoes meiosis I to produce secondary spermatocytes. Each secondary spermatocyte has only 23 chromosomes (Fig. 16.4*c*). Secondary spermatocytes (n) undergo meiosis II to produce four spermatids, each of which also has 23 chromosomes. Spermatids then develop into sperm. Note the presence of **Sertoli cells** (purple), which support, nourish, and regulate the process of spermatogenesis. It takes approximately 74 days for sperm to undergo development from spermatogonia to sperm.

Mature **sperm,** or spermatozoa, have three distinct parts: a head, a middle piece, and a tail (Fig. 16.4*d*). Mitochondria in the middle piece provide energy for the movement of the tail, which is a flagellum. The head contains a nucleus covered by a cap called the **acrosome,** which stores enzymes needed to penetrate the egg. The ejaculated semen of a normal human male contains several hundred million sperm, but only one sperm normally enters an egg. Sperm usually do not live more than 48 hours in the female genital tract.

Video Human Sperm

Interstitial Cells

The male sex hormones, the androgens, are secreted by cells that lie between the seminiferous tubules. These cells are called **interstitial cells.** The most important of the androgens is testosterone, whose functions are discussed next.

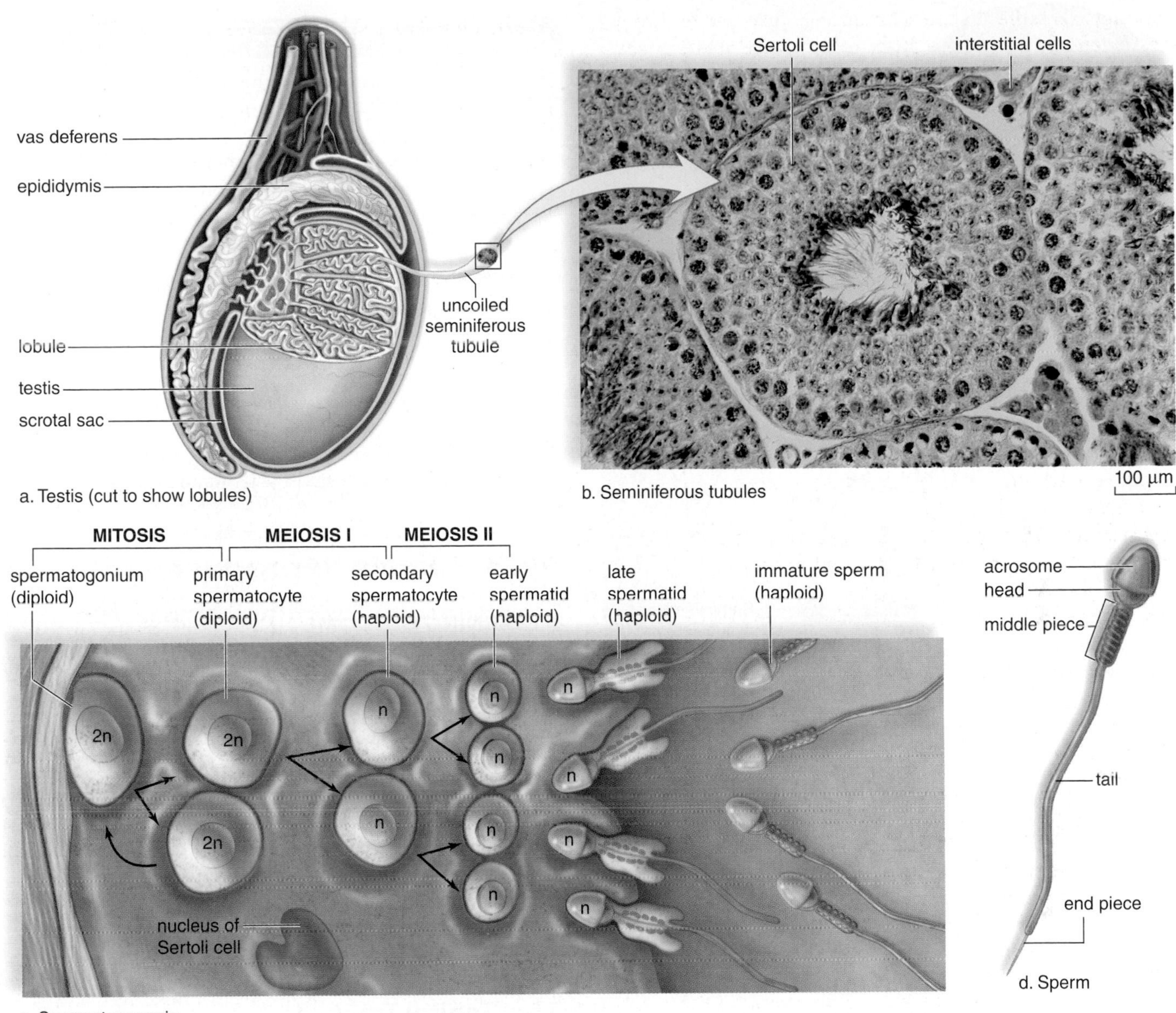

Figure 16.4 Spermatogenesis produces sperm cells.
a. The lobules of a testis contain seminiferous tubules. **b.** Electron micrograph of a cross section of the seminiferous tubules, where spermatogenesis occurs. Note the location of interstitial cells in clumps among the seminiferous tubules. **c.** Diagrammatic representation of spermatogenesis, which occurs in wall of tubules. **d.** A sperm has a head, a middle piece, and a tail. The nucleus is in the head, capped by the enzyme-containing acrosome.

Hormonal Regulation in Males

The hypothalamus has ultimate control of the testes' sexual function because it secretes a hormone called gonadotropin-releasing hormone (GnRH)(see Chapter 15). GnRH stimulates the anterior pituitary to secrete the gonadotropic hormones. There are two gonadotropic hormones, **follicle-stimulating hormone (FSH)** and **luteinizing hormone (LH)**, which are present in both males and females. In males, FSH promotes the production of sperm in the seminiferous tubules. LH in males controls the production of testosterone by the interstitial cells.

All these hormones are involved in a negative feedback relationship that maintains the fairly constant production of sperm and testosterone (Fig. 16.5). When the amount of testosterone in the blood rises to a certain level, it causes the hypothalamus and anterior pituitary to decrease their respective secretion of GnRH and LH. As the level of testosterone begins to fall, the hypothalamus increases its secretion of GnRH and the anterior pituitary increases its secretion of LH. These stimulate the interstitial cells to produce testosterone. A similar feedback mechanism maintains the continuous production of sperm. The Sertoli cells in the wall of the seminiferous tubules

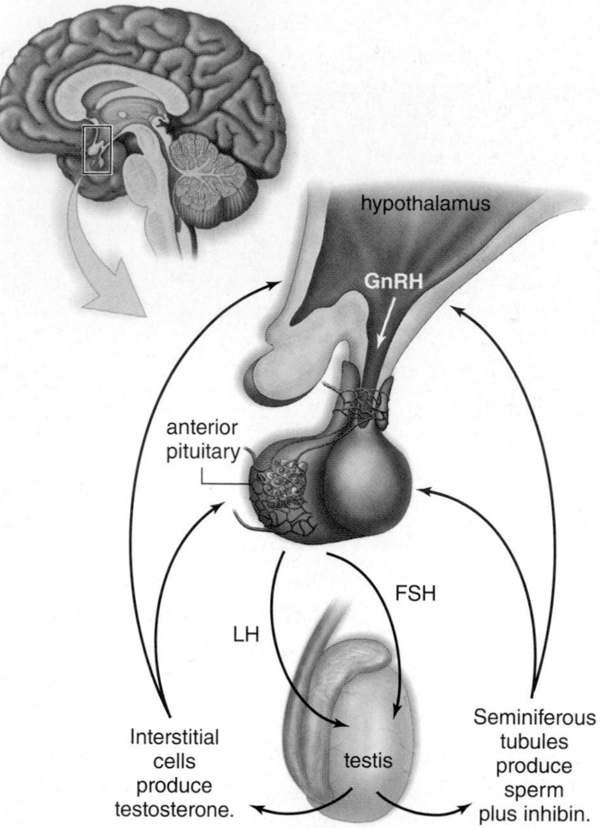

Figure 16.5 The hormones that control the production of sperm and testosterone by the testes.
Gonadotropin-releasing hormone (GnRH) stimulates the anterior pituitary to secrete the gonadotropic hormones: Follicle-stimulating hormone (FSH) stimulates the production of sperm, and luteinizing hormone (LH) stimulates the production of testosterone. Testosterone and inhibin exert negative feedback control over the hypothalamus and the anterior pituitary, and this regulates the level of testosterone in the blood and the production of sperm by the testes.

produce a hormone called *inhibin* that blocks GnRH and FSH secretion when appropriate (Fig. 16.5).

Testosterone, the main sex hormone in males, is essential for the normal development and functioning of the organs listed in Table 16.1. Testosterone also brings about and maintains the male secondary sex characteristics that develop at the time of puberty. Males are generally taller than females and have broader shoulders and longer legs relative to trunk length. The deeper voices of males compared with those of females are due to a larger larynx with longer vocal cords. The Adam's apple, part of the larynx, is usually more prominent in males than in females. Testosterone causes males to develop noticeable hair on the face, chest, and occasionally other regions of the body, such as the back. A related chemical also leads to the receding hairline and male-pattern baldness that occur in males.

Testosterone is responsible for the greater muscular development in males. Knowing this, both males and females sometimes take anabolic steroids, either testosterone or related steroid hormones resembling testosterone. Health problems involving the kidneys, the cardiovascular system, and hormonal imbalances can arise from such use.

CHECK YOUR PROGRESS 16.2

1. Identify the structures of the male reproductive system, and then trace the movement of sperm through the system.
2. Describe the process of spermatogenesis.
3. Explain the importance of testoterone to the male reproductive system.

CONNECTING THE CONCEPTS

For more information on the topics presented in this section, refer to the following discussions:

Figure 2.20 provides the chemical structure of testosterone.

Section 15.6 provides additional information on the male sex hormones.

Section 18.3 examines how meiosis reduces the chromosome number during spermatogenesis.

16.3 Female Reproductive System

LEARNING OUTCOME

Upon completion of this section, you should be able to

1. Identify the structures of the female reproductive system and provide a function for each.

The female reproductive system includes the organs listed in Table 16.2 and shown in Figure 16.6. The female gonads are paired **ovaries** that lie in shallow depressions, one on each side of the upper pelvic cavity. The ovaries produce **eggs,** also called the *ova* (sing., ovum), and the female sex hormones estrogen and progesterone.

MP3 Female Reproductive Anatomy and Physiology

The Genital Tract

The uterine tubes, also called the *oviducts* or *fallopian tubes,* extend from the uterus to the ovaries. However, the uterine tubes are not attached to the ovaries. Instead, they have fingerlike projections called **fimbriae** (sing., fimbria) that sweep over the ovaries. When an egg (ovum) bursts from an ovary during ovulation, it usually is swept into a uterine tube by the combined action of the fimbriae and the beating of cilia that line the uterine tube.

Table 16.2 Female Reproductive Organs

Organ	Function
Ovaries	Produce eggs and sex hormones
Uterine tubes (oviducts)	Conduct eggs; location of fertilization
Uterus (womb)	Houses developing fetus
Cervix	Contains opening to uterus
Vagina	Receives penis during sexual intercourse; serves as birth canal and as an exit for menstrual flow

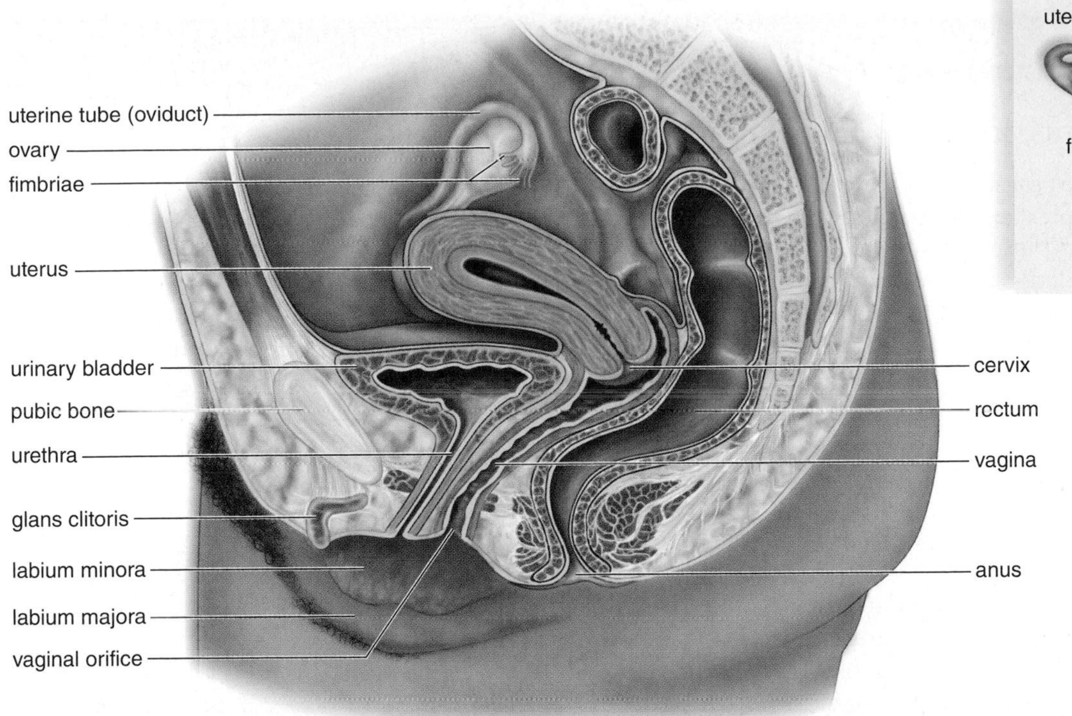

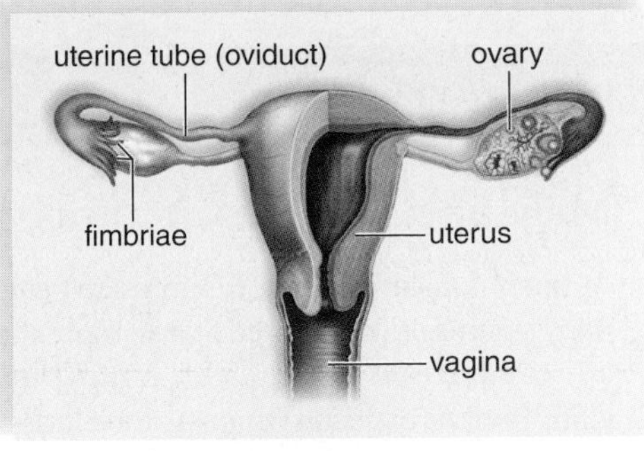

Figure 16.6 The female reproductive system.
The ovaries usually release one egg a month; fertilization occurs in the uterine tubes, and development occurs in the uterus. The vagina is the birth canal, as well as the organ of sexual intercourse and outlet for menstrual flow.

Once in the uterine tube, the egg is propelled slowly by ciliary movement and tubular muscle contraction toward the uterus. An egg lives approximately 6 to 24 hours, unless fertilization occurs. Fertilization, and therefore zygote formation, usually takes place in the uterine tube. A developing embryo normally arrives at the uterus after several days, and then **implantation** occurs. During implantation, the embryo embeds in the uterine lining, which has been prepared to receive it.

The **uterus** is a thick-walled, muscular organ about the size and shape of an inverted pear (Fig. 16.6). Normally, it lies above and is tipped over the urinary bladder. The uterine tubes join the uterus at its upper end; at its lower end, the **cervix** enters the vagina nearly at a right angle.

Cancer of the cervix is a common form of cancer in women (see chapter opener). Early detection is possible by means of a **Pap test,** which requires the removal of a few cells from the region of the cervix for microscopic examination. If the cells are cancerous, a physician may recommend a hysterectomy. A hysterectomy is the removal of the uterus, including the cervix. Removal of the ovaries in addition to the uterus is technically termed an *ovariohysterectomy* (radical hysterectomy). The vagina remains, so the woman can still engage in sexual intercourse.

Development of the embryo and fetus normally takes place in the uterus. This organ, sometimes called the womb, is approximately 5 cm wide in its usual state. It is capable of stretching to over 30 cm wide to accommodate a growing fetus. The lining of the uterus, called the **endometrium,** participates in the formation of the placenta (see page 372). The endometrium supplies nutrients needed for embryonic and fetal development. The endometrium has two layers: a functional layer that is shed during each menstrual period (see section 16.4) and a basal layer of reproducing cells. In the nonpregnant female, the functional layer of the endometrium varies in thickness according to a monthly reproductive cycle called the uterine cycle.

A small opening in the cervix leads to the vaginal canal. The **vagina** is a tube that lies at a 45° angle to the small of the back. The mucosal lining of the vagina lies in folds and can extend. This is especially important when the vagina serves as the birth canal, and it facilitates sexual intercourse when the vagina receives the penis. The vagina also acts as an exit for menstrual flow. Several different types of bacteria normally reside in the vagina and create an acidic environment. While this environment is protective against the possible growth of pathogenic bacteria, sperm prefer the basic environment provided by seminal fluid.

External Genitals

The external genital organs of the female are known collectively as the **vulva** (Fig. 16.7). The vulva includes two large, hair-covered folds of skin called the labia majora. The labia

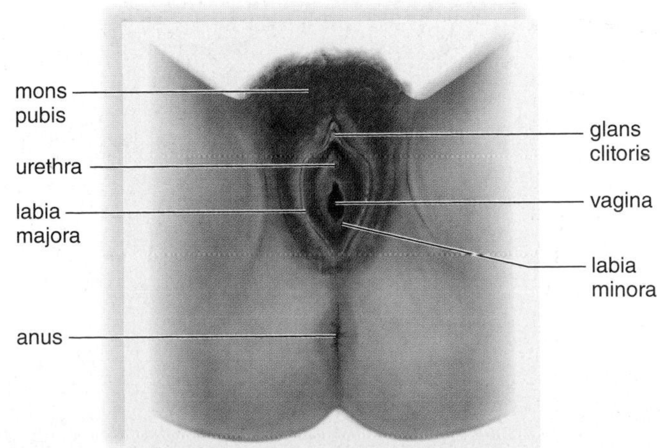

Figure 16.7 The external genitals of a female.
The external genitals of the female include the labia majora, labia minora, and glans clitoris. These organs are also referred to as the vulva.

BIOLOGY MATTERS  **Bioethical**

Male and Female Circumcision

At birth, a layer of skin (the foreskin) covers the end of a male baby's penis. In the United States, more than 50% of infant males are circumcised shortly after birth. During circumcision, the glans penis is exposed when the foreskin is removed during a surgical procedure. This procedure is done before babies go home from the hospital or during religious ceremonies in the home.

The decision to circumcise a baby boy is made by the parents and is often based on the religious or cultural beliefs of the parents. Circumcision may be done so that male children will resemble their father. Some choose circumcision because of concerns about cleanliness. Claims that circumcision increases or decreases sexual pleasure later in life have not been supported by research.

There is some evidence that suggests urinary tract infections are less common in circumcised infants. Research also shows that circumcision reduces the spread of HIV during heterosexual contact. In areas of the world where HIV infections are prevalent, circumcision may become an important means of limiting the spread of AIDS.

As with any type of surgery, there are risks associated with circumcision. The most common complications are minor bleeding and localized infections that can be treated easily. One of the biggest concerns is about the pain experienced by the baby during circumcision. The American Academy of Pediatrics (AAP) now recommends using a form of local anesthesia during the procedure. The AAP does not recommend nor argue against circumcision of male babies.

However, the circumcision of females is a highly controversial topic. Female circumcision (also referred to as female genital cutting [FGC], or female genital mutilation) is done strictly for cultural or religious reasons, though no religion specifically calls for its practice. The procedure involves partially or totally cutting away the external genitalia of a female. Cultures that practice FGC believe it to be a necessary rite of passage for girls. In the views of these cultures, FGC must be done to preserve the virginity of females and to prevent promiscuity. It is also done for aesthetic reasons, because the clitoris is thought to be an unhealthy and unattractive organ. Moreover, FGC is seen as an essential prerequisite for marriage. Females with an intact clitoris are believed to be unclean. Such women are considered to be potentially harmful to a man during intercourse or to a baby during childbirth if either is touched by the clitoris. Many believe that FGC enhances a husband's sexual pleasure and a woman's fertility.

Many girls die from infection after FGC. FGC also causes life-long urinary and reproductive tract infections, infertility, and pelvic pain. Victims report an absent or greatly diminished pleasurable response to sexual intercourse.

FGC is most commonly performed on girls between the ages of 4 and 12 and in countries in central Africa. It is also performed in some Middle Eastern countries and among Muslim groups in various other locations. With increasing immigration from these countries, there are also greater numbers of women who have been subjected to FGC. Likewise, there are more girls in the United States who are at risk for FGC.

Figure 16A Waris Dirie, Somalian-born supermodel, victim of FGC, advocates against female genital circumcision in her book *Desert Flower.*

Thanks to the efforts of mutilation victim Waris Dirie (Fig. 16A) and others like her, the need to eliminate FGC is now discussed openly, and action is being taken in many countries to outlaw the practice. FGC is considered to be a violation of human rights by the United Nations, UNICEF, and the World Health Organization. It is illegal to perform FGC in many African and Middle Eastern countries, but the practice continues because the laws are not enforced. In the United States, FGC is a criminal practice. In 1996, the United States granted asylum to a woman from Togo, who was trying to escape an arranged marriage and the FGC that would accompany it. Unfortunately, many immigrants to the United States continue the practice of FGC by sending their daughters abroad for the procedure or by importing someone to perform it. A number of educational approaches to eliminate FGC have been tried. These include community education that teaches about the harm done by FGC and substituting alternative rituals for the rite of passage to womanhood. Education may do even more to halt FGC, because more highly educated women are less likely to support having their daughters mutilated in this fashion.

Questions to Consider

1. In your view, is male circumcision unjustifiable? Why or why not?
2. Should families who accept the idea of FGC be allowed to immigrate?
3. How should the United States prosecute parents who have subjected their daughters to FGC?

majora extend backward from the mons pubis, a fatty prominence underlying the pubic hair. The labia minora are two small folds lying just inside the labia majora. They extend forward from the vaginal opening to encircle and form a foreskin for the glans clitoris. The glans clitoris is the organ of sexual arousal in females and, like the penis, contains a shaft of erectile tissue that becomes engorged with blood during sexual stimulation.

The cleft between the labia minora contains the openings of the urethra and the vagina. The vagina may be partially closed by a ring of tissue called the hymen. The hymen is ordinarily ruptured by sexual intercourse or by other types of physical activities. If remnants of the hymen persist after sexual intercourse, they can be surgically removed.

The urinary and reproductive systems in the female are entirely separate. For example, the urethra carries only urine, and the vagina serves only as the birth canal and the organ for sexual intercourse.

Orgasm in Females

Upon sexual stimulation, the labia minora, the vaginal wall, and the clitoris become engorged with blood. The breasts also swell, and the nipples become erect. The labia majora enlarge, redden, and spread away from the vaginal opening.

The vagina expands and elongates. Blood vessels in the vaginal wall release small droplets of fluid that seep into the vagina and lubricate it. Mucus-secreting glands beneath the labia minora on either side of the vagina also provide lubrication for entry of the penis into the vagina. Although the vagina is the organ of sexual intercourse in females, the clitoris plays a significant role in the female sexual response. The extremely sensitive clitoris can swell to two or three times its usual size. The thrusting of the penis and the pressure of the pubic symphyses of the partners act to stimulate the clitoris.

Orgasm occurs at the height of the sexual response. Blood pressure and pulse rate rise, breathing quickens, and the walls of the uterus and uterine tubes contract rhythmically. A sensation of intense pleasure is followed by relaxation when organs return to their normal size. Females have no refractory period, and multiple orgasms can occur during a single sexual experience.

CHECK YOUR PROGRESS 16.3

1. Distinguish the structures of the female reproductive system that (a) produce the egg, (b) transport the egg, (c) house a developing embryo, and (d) serve as the birth canal.
2. Explain the purpose of the vagina and uterus in the female reproductive system.
3. Discuss why the urinary and reproductive systems are separate in a female.

CONNECTING THE CONCEPTS

For more information on the topics presented in this section, refer to the following discussions:

Section 17.1 outlines the steps in the fertilization of an egg by a sperm cell.

Section 17.2 examines the stages of fetal development in the uterus.

Sections 19.1 and 19.2 examine the characteristics of cancer cells and the causes of cancer.

16.4 The Ovarian Cycle

LEARNING OUTCOMES

Upon completion of this section, you should be able to

1. List the stages of the ovarian cycle and explain what is occurring in each stage.
2. Describe the process of oogenesis.
3. Summarize how estrogen and progesterone influence the ovarian cycle.

Hormone levels cycle in the female on a monthly basis, and the ovarian cycle drives the uterine cycle, as discussed in this section.

Ovarian Cycle: Nonpregnant

An ovary contains many **follicles,** and each one contains an immature egg called an oocyte. A female is born with as many as 2 million follicles, but the number is reduced to 300,000 to 400,000 by the time of puberty. Only a small number of follicles (about 400) ever mature because a female usually produces only one egg per month during her reproductive years. As the follicle matures during the **ovarian cycle,** it changes from a primary to a secondary to a vesicular (Graafian) follicle (Fig. 16.8). Epithelial cells of a primary follicle surround a primary oocyte. Pools of follicular fluid bathe the oocyte in a secondary follicle. In a vesicular follicle, the fluid-filled cavity increases to the point that the follicle wall balloons out on the surface of the ovary.

APPLICATIONS AND MISCONCEPTIONS

Do women make testosterone?

The adrenal glands and ovaries of women make small amounts of testosterone. Women's small testosterone levels may affect libido, or sex drive. The use of supplemental testosterone to restore a woman's libido has not been well researched.

By the way, men make estrogen, too. Some estrogen is produced by the adrenal glands. Androgens are also converted to estrogen by enzymes in the gonads and peripheral tissues. Estrogen may prevent osteoporosis in males.

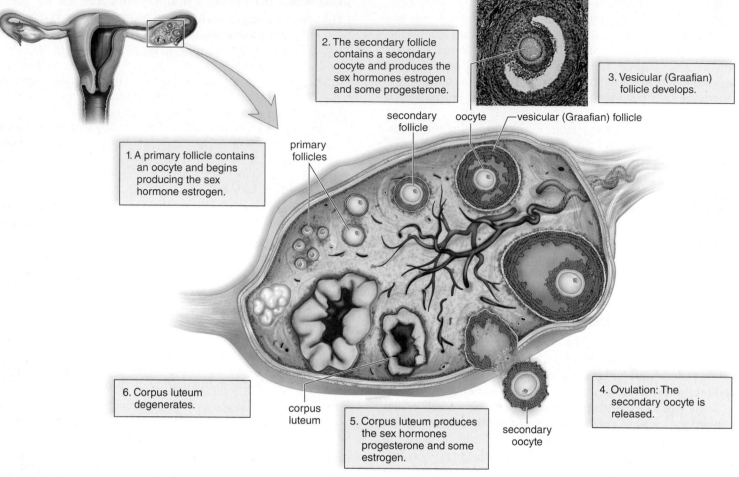

2. The secondary follicle contains a secondary oocyte and produces the sex hormones estrogen and some progesterone.

3. Vesicular (Graafian) follicle develops.

secondary follicle

oocyte

vesicular (Graafian) follicle

primary follicles

1. A primary follicle contains an oocyte and begins producing the sex hormone estrogen.

6. Corpus luteum degenerates.

corpus luteum

5. Corpus luteum produces the sex hormones progesterone and some estrogen.

secondary oocyte

4. Ovulation: The secondary oocyte is released.

Figure 16.8 **The ovarian cycle.**
A single follicle goes through all stages (1–6) in one place within the ovary. As the follicle matures, layers of follicle cells surround a secondary oocyte. Eventually, the mature follicle ruptures, and the secondary oocyte is released. The follicle then becomes the corpus luteum, which eventually disintegrates.

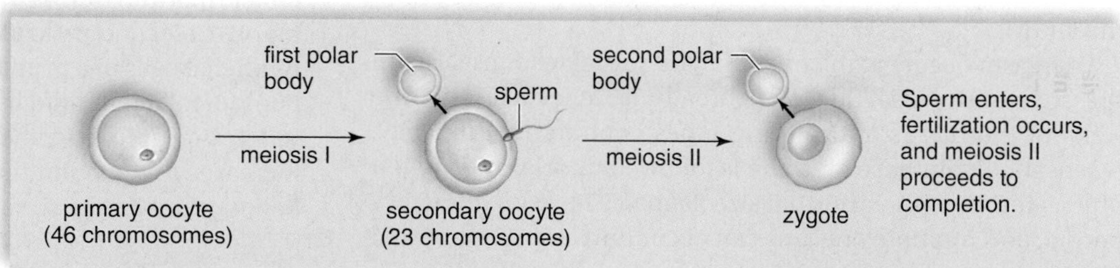

Figure 16.9 **Oogenesis produces egg cells.**
During oogenesis, the chromosome number is reduced from 46 to 23. Oogenesis produces a functional egg cell and nonfunctional polar bodies.

Figure 16.9 traces the steps of **oogenesis.** A primary oocyte undergoes meiosis I, and the resulting cells are haploid with 23 chromosomes each. One of these cells is called a **polar body.** A polar body is a sort of cellular "trash can" because its function is simply to hold discarded chromosomes. The secondary oocyte undergoes meiosis II but only if it is first fertilized by a sperm cell. If the secondary oocyte remains unfertilized, it never completes meiosis and will die shortly after being released from the ovary.

When appropriate, the vesicular follicle bursts, releasing the oocyte (often called an egg) surrounded by a clear membrane. This process is referred to as **ovulation.** Once a vesicular follicle has lost the oocyte, it develops into a **corpus luteum,** a glandlike structure. If the egg is not fertilized, the corpus luteum disintegrates.

As mentioned previously, the ovaries produce eggs and also the female sex hormones estrogen and progesterone. A primary follicle produces estrogen, and a secondary follicle produces estrogen and some progesterone. The corpus luteum produces progesterone and some estrogen.

Phases of the Ovarian Cycle

Similar to the testes, the hypothalamus has ultimate control of the ovaries' sexual function because it secretes gonadotropin-releasing hormone, or GnRH. GnRH stimulates the anterior pituitary to produce FSH and LH, and these hormones control

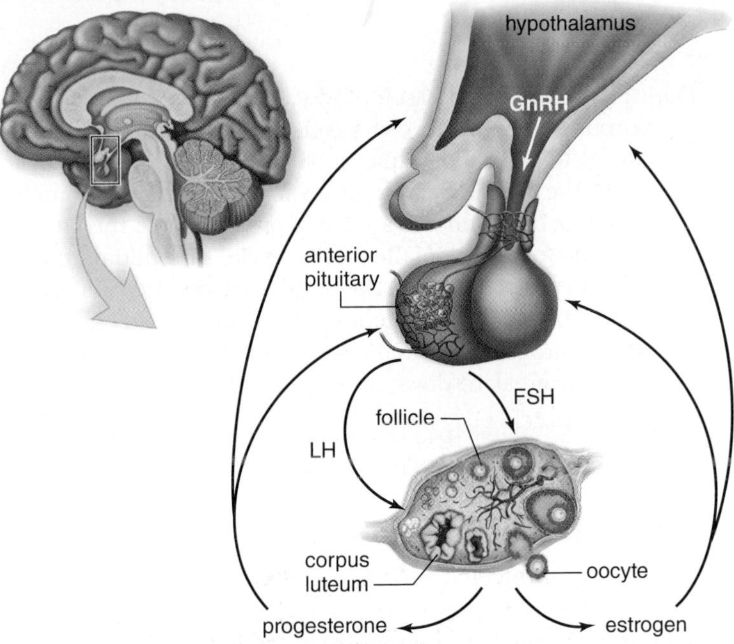

Figure 16.10 The hormones that control the production of estrogen and progesterone by the ovaries.
The hypothalamus produces gonadotropin-releasing hormone (GnRH). GnRH stimulates the anterior pituitary to produce follicle-stimulating hormone (FSH) and luteinizing hormone (LH). FSH stimulates the follicle to produce primarily estrogen, and LH stimulates the corpus luteum to produce primarily progesterone. Estrogen and progesterone maintain the sexual organs (e.g., uterus) and the secondary sex characteristics, and they exert feedback control over the hypothalamus and the anterior pituitary. Feedback control regulates the relative amounts of estrogen and progesterone in the blood.

the ovarian cycle. The gonadotropic hormones are not present in constant amounts. Instead they are secreted at different rates during the cycle. For simplicity's sake, it is convenient to emphasize that during the first half, or *follicular phase*, FSH promotes the development of follicles that primarily secrete estrogen (Fig. 16.10). As the estrogen level in the blood rises, it exerts negative feedback control over the anterior pituitary secretion of FSH. The follicular phase now comes to an end.

The estrogen spike at the end of the follicular phase has a positive feedback effect on the hypothalamus and pituitary gland. As a result, GnRH from the hypothalamus increases. There is a corresponding surge of LH released from the anterior pituitary. The LH surge triggers ovulation at about day 14 of a 28-day cycle.

Next, the *luteal phase* begins. During the luteal phase of the ovarian cycle, LH promotes the development of the corpus luteum. The corpus luteum secretes high levels of progesterone and some estrogen. When pregnancy does not occur, the corpus luteum regresses and a new cycle begins with menstruation (Fig. 16.11).

Animation Female Reproductive System

Estrogen and Progesterone

Estrogen and progesterone affect not only the uterus but other parts of the body as well. Estrogen is largely responsible for the secondary sex characteristics in females, including body hair and fat distribution. In general, females have a more rounded appearance than males because of a greater accumulation of fat beneath the skin. Like males, females develop axillary and

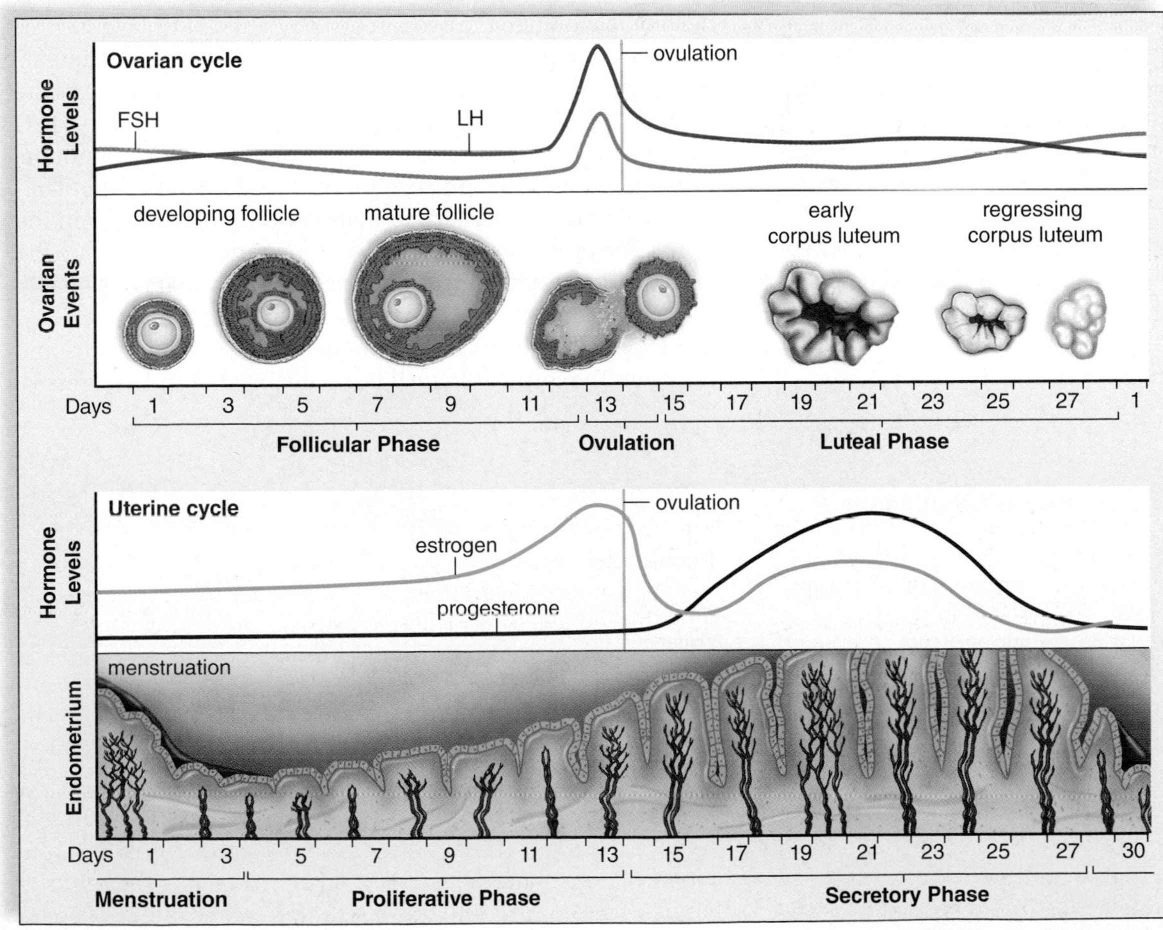

Figure 16.11 The effects of estrogen and progesterone on the endometrium during the uterine cycle.
During the follicular phase, FSH released by the anterior pituitary promotes the maturation of a follicle in the ovary. The ovarian follicle produces increasing levels of estrogen, which causes the endometrium to thicken during the proliferative phase of the uterine cycle. After ovulation and during the luteal phase of the ovarian cycle, LH promotes the development of the corpus luteum. Progesterone in particular causes the endometrial lining to become secretory. Menses, due to the breakdown of the endometrium, begins when progesterone production declines to a low level.

pubic hair during puberty. In females, the upper border of pubic hair is horizontal, but in males, it tapers toward the navel. Both estrogen and progesterone are also required for breast development. Other hormones are involved in milk production (prolactin) following pregnancy and milk letdown (oxytocin) when a baby begins to nurse.

The pelvic girdle is wider and deeper in females, so the pelvic cavity usually has a larger relative size compared with that of males. This means that females have wider hips than males and their thighs converge at a greater angle toward the knees. The female pelvis tilts forward, so females tend to have more of a lower back curve than males, an abdominal bulge, and protruding buttocks.

Menopause, the period in a woman's life during which the ovarian cycle ceases, is likely to occur between ages 45 and 55. The ovaries are no longer responsive to the gonadotropic hormones produced by the anterior pituitary, and the ovaries no longer secrete estrogen or progesterone. At the onset of menopause, menstruation becomes irregular, but as long as it occurs, it is still possible for a woman to conceive. Therefore, a woman is usually not considered to have completed menopause until menstruation is absent for a year.

Uterine Cycle: Nonpregnant

The female sex hormones, **estrogen** and **progesterone,** have numerous functions. One function of these hormones affects the endometrium, causing the uterus to undergo a cyclical series of events known as the **uterine cycle** (Fig. 16.11). Twenty-eight-day cycles are divided as follows:

During *days 1-5,* a low level of estrogen and progesterone in the body causes the endometrium to disintegrate and its blood vessels to rupture. On day 1 of the cycle, a flow of blood and tissues, known as the menses, passes out of the vagina during **menstruation,** also called the menstrual period.

During *days 6–13,* increased production of estrogen by a new ovarian follicle in the ovary causes the endometrium to thicken and become vascular and glandular. This is called the proliferative phase of the uterine cycle.

On *day 14* of a 28-day cycle, ovulation usually occurs.

During *days 15–28,* increased production of progesterone by the corpus luteum in the ovary causes the endometrium of the uterus to double or triple in thickness (from 1 mm to 2–3 mm). The uterine glands mature and produce a thick mucoid secretion in response to increased progesterone. This is called the secretory phase of the uterine cycle. The endometrium is now prepared to receive the developing embryo. If this does not occur, the corpus luteum in the ovary regresses. The low level of progesterone in the female body results in the endometrium breaking down during menstruation.

Table 16.3 compares the stages of the uterine cycle with those of the ovarian cycle when pregnancy does not occur.

Fertilization and Pregnancy

Following unprotected sexual intercourse, many sperm make their way into the uterine tubes, where the egg is located following ovulation. Only one sperm is needed to fertilize the egg, which is then called a zygote. Development begins even as the zygote travels down the uterine tube to the uterus. The endometrium is now prepared to receive the developing embryo. The embryo implants in the endometrial lining several days following fertilization. Implantation signals the beginning of a pregnancy. An abortion may be spontaneous (referred to as a miscarriage) or induced. Each type of abortion ends with the loss of the embryo or fetus.

The **placenta,** which sustains the developing embryo and later the fetus, originates from both maternal and fetal tissues. It is the region of exchange of molecules between fetal and maternal blood, although the two rarely mix. At first, the placenta produces **human chorionic gonadotropin** (**HCG**), which maintains the corpus luteum in the ovary. (A pregnancy test detects the presence of HCG in the blood or urine.) Rising amounts of HCG stimulate the corpus luteum to produce increasing amounts of progesterone. This progesterone shuts down the hypothalamus and anterior pituitary so that no new follicles begin in the ovary. The progesterone maintains the uterine lining where the embryo now resides. The absence of menstruation is a signal to the woman that she may be pregnant (Fig. 16.12).

Eventually, the placenta produces progesterone and some estrogen. The corpus luteum is no longer needed and it regresses.

Table 16.3 Ovarian and Uterine Cycles: Nonpregnant

Ovarian Cycle	Events	Uterine Cycle	Events
Follicular phase—days 1–13	FSH secretion begins.	Menstruation—days 1–5	Endometrium breaks down.
	Follicle maturation occurs.	Proliferative phase—days 6–13	Endometrium rebuilds.
	Estrogen secretion is prominent.		
Ovulation—day 14*	LH spike occurs.		
Luteal phase—days 15–28	LH secretion continues.	Secretory phase—days 15–28	Endometrium thickens, and glands are secretory.
	Corpus luteum forms.		
	Progesterone secretion is prominent.		

*Assuming a 28-day cycle.

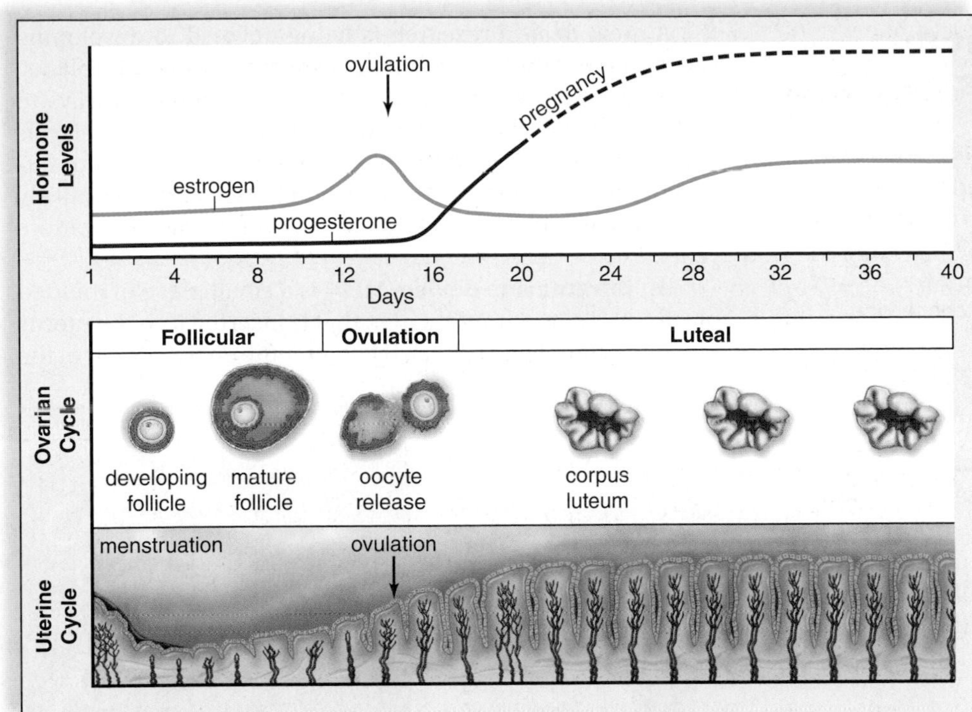

Figure 16.12 **The effect of pregnancy on the corpus luteum and endometrium.** If pregnancy occurs, the corpus luteum does not regress. Instead, the corpus luteum is maintained and secretes increasing amounts of progesterone. Therefore, menstruation does not occur and the uterine lining, where the embryo resides, is maintained.

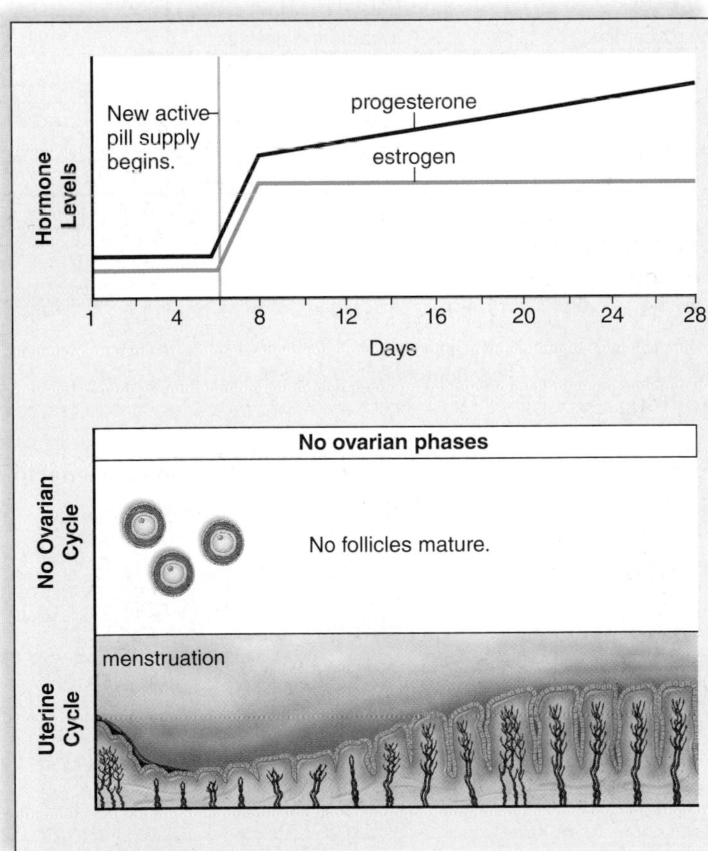

Figure 16.13 **The effect of birth control pills on the ovarian cycle.**
Active pills cause the uterine lining to build up, and this lining is shed when inactive pills are taken. Feedback inhibition of the hypothalamus and anterior pituitary means that the ovarian cycle does not occur.

Many women use birth control pills to prevent pregnancy (see section 16.5). The most commonly used pills include active pills, containing a synthetic estrogen and progesterone, taken for 21 days, followed by 7 days of taking inactive pills that do not contain these hormones (Fig. 16.13). The uterine lining builds up to some degree while the active pills are being taken. Progesterone decreases when the last of the active pills are taken, causing a minimenstruation to occur. Some women skip taking the inactive pills and start taking a new pack of active pills right away to skip menstruation (a period). Birth control pills are available that consist of three months of active pills. Women taking them have only four menstrual periods a year.

CHECK YOUR PROGRESS 16.4

1. Summarize the roles of estrogen and progesterone in the ovarian and uterine cycles.
2. Describe the changes that occur in the ovarian and uterine cycles during pregnancy.
3. Describe the changes that occur in the ovarian and uterine cycles when birth control pills are used.

CONNECTING THE CONCEPTS

For more information on the topics presented in this section, refer to the following discussions:

Section 15.6 provides additional information on the male sex hormones.

Section 17.1 outlines the steps in the fertilization of an egg by a sperm cell.

Section 18.3 examines how meiosis reduces the chromosome number during oogenesis.

16.5 Control of Reproduction

LEARNING OUTCOMES

Upon completion of this section, you should be able to

1. List the forms of birth control and summarize how each reduces the chances of fertilization of an egg by a sperm cell.
2. Explain the causes of infertility.
3. Describe how the use of assisted reproductive technologies can increase the chances of conceiving a child.

Several means are available to reduce or enhance our reproductive potential. **Birth control methods** are used to regulate the number of children an individual or couple has. For individuals who are experiencing infertility, or an inability to achieve pregnancy, a number of assisted reproductive technologies may be used to increase the chances of conceiving a child.

Birth Control Methods

The most reliable method of birth control is abstinence—not engaging in sexual intercourse. This form of birth control has the added advantage of preventing transmission of sexually transmitted diseases. Table 16.4 lists other means of birth control used in the United States and rates their effectiveness. For example, with the birth control pill, we expect 98% effectiveness, which means that, in a given year, 2% of sexually active women using this form of birth control may get pregnant. In comparison, the withdrawal method is 75% effective. Thus, 25% of women using this form of birth control can expect to get pregnant. That makes the withdrawal method one of the least effective methods of contraception.

Contraceptives are medications and devices that reduce the chance of pregnancy. Oral contraception, known as **birth control pills,** are the most effective form of contraception. These pills contain a combination of estrogen and progesterone. Most birth control pills are taken daily. For the first 21 days the pills contain active hormones. This is followed by seven days of inactive pills. The estrogen and progesterone in the birth control pill or a patch applied to the skin effectively shuts down the pituitary production of both FSH and LH. Follicle development in the ovary is prevented. Because ovulation does not occur, pregnancy cannot take place. Women taking birth control pills or using a patch should see a physician regularly because of possible side effects.

A great deal of research is being devoted to developing safe and effective hormonal birth control for men. Implants, pills, patches, and injections are being explored as ways to deliver testosterone and/or progesterone at adequate levels to suppress sperm production. Even the most successful formulations are still in the experimental stage and are unlikely to be available outside of clinical trials for at least a few more years.

An **intrauterine device (IUD)** is a small piece of molded plastic, and sometimes copper, that is inserted into the uterus by a physician (Fig. 16.14*a*). IUDs alter the environment of the

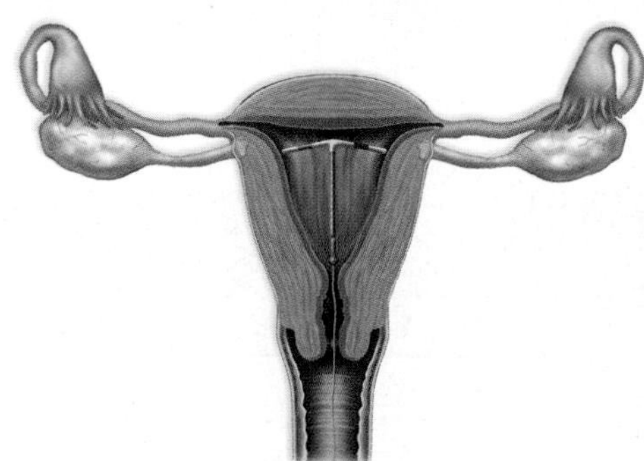

a. Intrauterine device placement

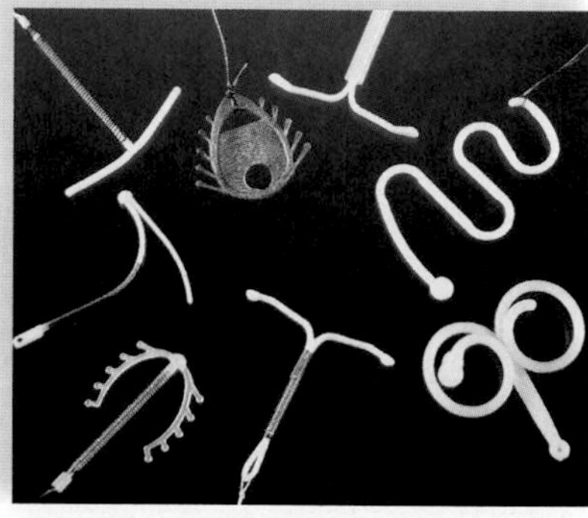

Intrauterine devices

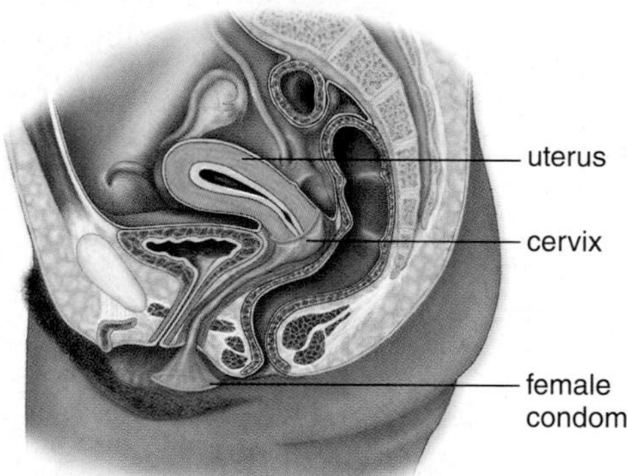

b. Female condom placement

Female condom

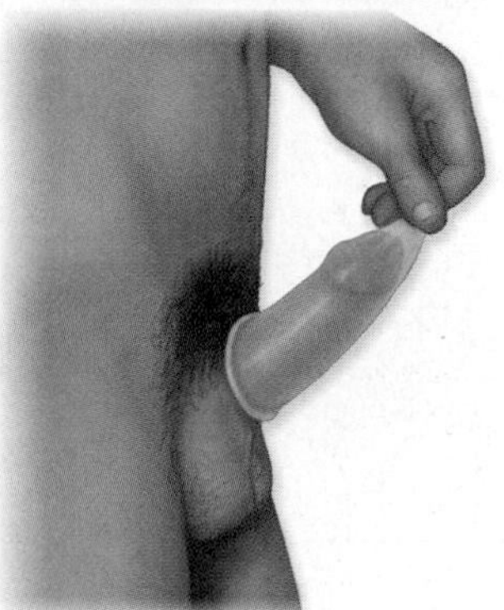

c. Male condom placement

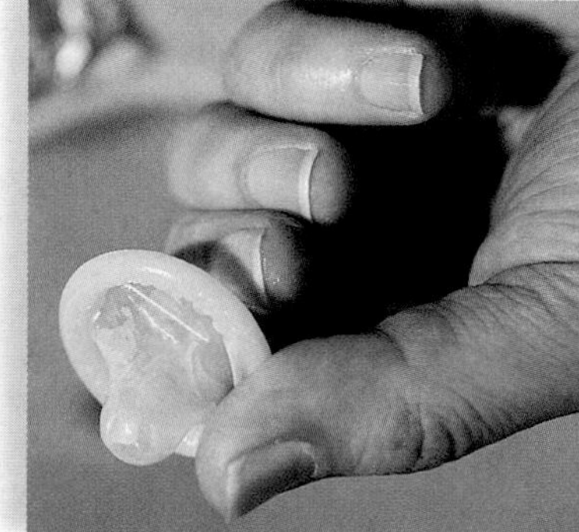

Male condom

Figure 16.14 Placement of birth control devices.
a. Intrauterine devices mechanically prevent implantation and can contain progesterone to prevent ovulation, prevent implantation, and thicken cervical mucus. **b.** Female condom that is fitted inside the vagina prevents sperm entry and protects against STDs. **c.** Male condom that fits over the penis prevents sperm from entering the vagina and protects against STDs.

Table 16.4 Common Methods of Contraception

Name	Procedure	How Does It Work?	Effectiveness?	Health Risk
Abstinence	Refrain from sexual intercourse	No sperm in vagina	100%	None; also protects against STDs
Natural family planning	Determine day of ovulation by keeping records	Intercourse avoided during the time that ovum is viable	80%	None
Withdrawal method	Penis withdrawn from vagina just before ejaculation	Ejaculation outside the woman's body; no sperm in vagina	75%	None
Douching	Vagina cleansed after intercourse	Washes sperm out of vagina	≥ 70%	May cause inflammation
Male condom	Sheath of latex, polyurethane, or natural material fitted over erect penis	Prevents entry of sperm into vagina; latex and polyurethane forms protect against STDs	89%	Latex allergy with latex forms, no protection against STDs with natural-material condoms
Female condom	Polyurethane liner fitted inside vagina	Prevents entry of sperm into vagina; some protection against STDs	79%	Possible allergy or irritation, urinary tract infection
Spermicide: jellies, foams, creams	Spermicidal products inserted into vagina before intercourse	Spermicide nonoxynol-9 kills large numbers of sperm cells.	50–80%	Irritation, allergic reaction, urinary tract infection
Contraceptive sponges	Sponge containing spermicide inserted into vagina and placed against cervix	Spermicide nonoxynol-9 kills large numbers of sperm cells.	72–86%	Irritation, allergic reaction, urinary tract infection, toxic shock syndrome
Combined hormone vaginal ring	Flexible plastic ring inserted into vagina; releases hormones absorbed into the bloodstream		98%	
Combined hormone pill	Pills are swallowed daily; chewable form also available		98%	
Combined hormone 91-day regimen	Pills are swallowed daily; user has three to four menstrual periods a year	Combined hormonal methods suppress ovulation by the combined actions of the hormones estrogen and progestin.	98%	Combined hormonal methods can cause dizziness; nausea; changes in menstruation, mood, and weight; rarely, cardiovascular disease, including high blood pressure, blood clots, heart attack, and strokes.
Combined hormone injection (Lunelle)	Injection of long-acting hormone given once a month		99%	
Combined hormone patch	Patch is applied to skin and left in place for 1 week; new patch applied		98%	
Progestin-only minipill	Pills are swallowed daily.	Thickens cervical mucus, preventing sperm from contacting egg	98%	Irregular bleeding, weight gain, breast tenderness
Progesterone-only injection (Depo-Provera)	Injection of progestin once every three months	Inhibits ovulation; prevents sperm from reaching the egg; prevents implantation	99%	Irregular bleeding, weight gain, breast tenderness, osteoporosis possible
Emergency contraception	Should be taken shortly after unprotected intercourse	Suppresses ovulation by the combined actions of the hormones estrogen and progestin; prevents implantation	80%	Nausea, vomiting, abdominal pain, fatigue, headache
Diaphragm	Latex cup, placed into vagina to cover cervix before intercourse	Blocks entrance of sperm into uterus, spermicide kills sperm	90% with spermicide	Irritation, allergic reaction, urinary tract infection, toxic shock syndrome
Cervical cap	Latex cap held over cervix	Blocks entrance of sperm into uterus, spermicide kills sperm	90% with spermicide	Irritation, allergic reaction, toxic shock syndrome, abnormal Pap smear
Cervical shield	Latex cap in upper vagina, held in place by suction	Blocks entrance of sperm into uterus, spermicide kills sperm	90% with spermicide	Irritation, allergic reaction, urinary tract infection, toxic shock syndrome
Intrauterine device Copper T	Placed in uterus	Causes cervical mucus to thicken; fertilized embryo cannot implant	99%	Cramps, bleeding, infertility, perforation of uterus
Intrauterine device progesterone-releasing type	Placed in uterus	Prevents ovulation; causes cervical mucus to thicken; fertilized embryo cannot implant	99%	Cramps, bleeding, infertility, perforation of uterus

uterus and uterine tubes to reduce the possibility of fertilization. If fertilization should occur, implantation cannot take place.

The **diaphragm** is a soft latex cup with a flexible rim that lodges behind the pubic bone and fits over the cervix. Each woman must be properly fitted by a physician, and the diaphragm can be inserted into the vagina no more than 2 hours before sexual relations. Also, it must be used with spermicidal jelly or cream and should be left in place at least 6 hours after sexual relations. The cervical cap is a minidiaphragm.

The male and female **condoms** also offer some protection against sexually transmitted diseases in addition to helping prevent pregnancy. Female condoms consist of a large polyurethane tube with a flexible ring that fits onto the cervix (Fig. 16.14*b*). The open end of the tube has a ring that covers the external genitals. A male condom is most often a latex sheath that fits over the erect penis (Fig. 16.14*c*). The ejaculate is trapped inside the sheath and thus does not enter the vagina. When used in conjunction with a spermicide, the protection is better than with the condom alone.

Contraceptive Injections and Vaccines

Contraceptive vaccines are in development. For example, a vaccine intended to immunize women against HCG, the hormone so necessary to maintaining the implantation of the embryo, was successful in a limited clinical trial. Because HCG is not normally present in the body, no autoimmune reaction is expected, but the immunization does wear off with time. Others believe that it would also be possible to develop a safe antisperm vaccine that could be used in women.

Contraceptive implants use a synthetic progesterone to prevent ovulation by disrupting the ovarian cycle. Most versions consist of a single capsule that remains effective for about three years. Contraceptive injections are available as progesterone only or a combination of estrogen and progesterone. The length of time between injections can vary from one to several months.

Emergency Contraception

Emergency contraception, or "morning-after pills," refers to medications that can prevent pregnancy after unprotected intercourse. The expression "morning-after" is a misnomer, in that some treatments can be started up to five days after unprotected intercourse.

The first FDA-approved medication produced for emergency contraception was a kit called Preven. Preven includes four synthetic progesterone pills; two are taken up to 72 hours after unprotected intercourse, and two more are taken 12 hours later. The hormone upsets the normal uterine cycle, making it difficult for an embryo to implant in the endometrium. One study estimated that Preven was 85% effective in preventing unintended pregnancies. The Preven kit also includes a pregnancy test; women are instructed to take the test first before using the hormone, because the medication is not effective on an established pregnancy.

In 2006, the FDA approved another drug called Plan B One-Step, which is up to 89% effective in preventing pregnancy if taken within 72 hours after unprotected sex. It is available without a prescription to women age 17 and older. In August 2010, ulipristal acetate (also known as ella) was also approved for emergency contraception. It can be taken up to five days after unprotected sex, and studies indicate it is somewhat more effective than Plan B One-Step. Unlike Plan B One-Step, however, a prescription is required.

Mifepristone, also known as RU-486 or the "abortion pill," can cause the loss of an implanted embryo by blocking the progesterone receptors of endometrial cells. This causes the endometrium to slough off, carrying the embryo with it. When taken in conjunction with a prostaglandin to induce uterine contractions, RU-486 is 95% effective at inducing an abortion up to the 49th day of gestation. Because of its mechanism of action, the use of RU-486 is more controversial compared to other medications, and while it is currently available in the United States for early medical abortion, it is not approved for emergency contraception.

Surgical Methods

Vasectomy and tubal ligation are two methods used to bring about sterility, the inability to reproduce (Fig. 16.15). **Vasectomy** consists of cutting and sealing the vas deferens from each testis so that the sperm are unable to reach the seminal fluid ejected at the time of orgasm. The sperm are then largely reabsorbed. Following this operation, which can be done in a doctor's office, the amount of ejaculate remains normal because sperm account for only about 1% of the volume of semen. Also, there is no effect on the secondary sex characteristics because testosterone continues to be produced by the testes.

Tubal ligation consists of cutting and sealing the uterine tubes. Pregnancy rarely occurs because the passage of the egg through the uterine tubes has been blocked. Using a method called laparoscopy, which requires only two small incisions, the surgeon inserts a small, lighted telescope to view the uterine tubes and a small surgical blade to sever them.

APPLICATIONS AND MISCONCEPTIONS

Are vasectomies 100% effective?

Vasectomies and tubal ligations are considered to be permanent forms of birth control. However, many men do not realize that it is still possible to father a child for several months following a vasectomy. This is because following the procedure some sperm remain in the vas deferens. Males who have had a vasectomy need to use alternate forms of birth control for 1 to 2 months or until their physician has performed a follow-up sperm count and verified that sperm are no longer present in the ejaculate. In very rare situations, the vas deferens may reconnect, allowing sperm to once again be ejaculated. While the only 100% effective form of contraception is abstinence, vasectomies are considered to be over 99.8% effective.

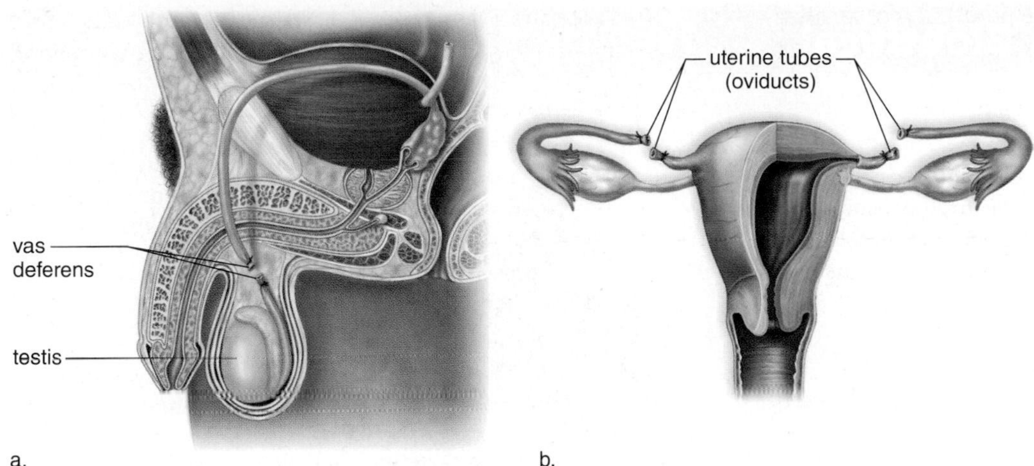

Figure 16.15 Vasectomies and tubal ligations.
a. Vasectomy involves making two small cuts in the skin of the scrotum. Each vas deferens is lifted out and cut. The cut ends are tied or sealed with an electrical current. The openings in the scrotum are closed with stitches. **b.** During tubal ligation, one or two small incisions are made in the abdomen. Using instruments inserted through the incisions, the uterine tubes are coagulated (burned), sealed shut with cautery, or cut and tied. The skin incision is then stitched closed.

It is best to view a vasectomy or tubal ligation as permanent. Even following successful reconnection, fertility is usually reduced by about 50%.

Infertility

Infertility is the failure of a couple to achieve pregnancy after one year of regular, unprotected intercourse. Estimates of the prevalence of infertility vary, but most professional organizations predict that around 15% of all couples are infertile. The cause of infertility can be evenly attributed to the male and female partners.

Causes of Infertility

The most frequent cause of infertility in males is low sperm count and/or a large proportion of abnormal sperm, which can be due to environmental influences. It appears that a sedentary lifestyle coupled with smoking and alcohol consumption is most often the cause of male infertility. When males spend most of the day sitting in front of a computer or the TV or driving, the testes temperature remains too high for adequate sperm production.

Body weight appears to be the most significant factor in causing female infertility. In women of normal weight, fat cells produce a hormone called leptin that stimulates the hypothalamus to release GnRH. FSH release and normal follicle development follow. In overweight women, leptin levels are higher, which impacts GnRH and FSH. The ovaries of overweight women often contain many small follicles that fail to ovulate. Other causes of infertility in females are blocked uterine tubes due to pelvic inflammatory disease (see pages 381–383) and endometriosis. **Endometriosis** is the presence of uterine tissue outside the uterus, particularly in the uterine tubes and on the abdominal organs. Backward flow of menstrual fluid allows living uterine cells to establish themselves in the abdominal cavity. The cells go through the usual uterine cycle, causing pain and structural abnormalities that make it more difficult for a woman to conceive.

Sometimes the causes of infertility can be corrected by medical intervention so that couples can have children. If no obstruction is apparent and body weight is normal, it is possible to give females fertility drugs. These drugs are gonadotropic hormones that stimulate the ovaries and bring about ovulation. As discussed in the Bioethical feature, "Should Infertility Be Treated," on page 378, these hormone treatments may cause multiple ovulations and multiple births.

When reproduction does not occur in the usual manner, many couples adopt a child. Others sometimes try one of the assisted reproductive technologies discussed in the following paragraphs.

Assisted Reproductive Technologies

Assisted reproductive technologies consist of techniques used to increase the chances of pregnancy. Often, sperm and/or eggs are retrieved from the testes and ovaries, and fertilization takes place in a clinical or laboratory setting.

Artificial Insemination by Donor (AID) During artificial insemination, sperm are placed in the vagina by a physician. Sometimes a woman is artificially inseminated by her partner's sperm. This is especially helpful if the partner has a low sperm count, because the sperm can be collected over time and concentrated so that the sperm count is sufficient to result in fertilization. Often, however, a woman is inseminated by sperm acquired from a donor who is a complete stranger to her. At times, a combination of partner and donor sperm is used.

A variation of AID is *intrauterine insemination* (*IUI*). In IUI, fertility drugs are given to stimulate the ovaries. Then the donor's sperm are placed in the uterus rather than in the vagina.

If the prospective parents wish, sperm can be sorted into those believed to be X-bearing or Y-bearing to increase the chances of having a child of the desired sex. Fertilization of an egg with an X-bearing sperm results in a female child. Fertilization by a Y-bearing sperm yields a male child.

BIOLOGY MATTERS  **Bioethical**

Should Infertility Be Treated?

Every day, couples make plans to start or expand their families. Yet, for many, their dreams might not be realized because conception is difficult or impossible. Before seeking medical treatment for infertility, a couple might want to decide how far they are willing to go to have a child. Here are some of the possible risks.

Some of the Procedures Used

If a man has low sperm count or motility, artificial or intrauterine insemination of his partner with a large number of specially selected sperm may be done to stimulate pregnancy. Although there are dangers in all medical procedures, artificial insemination is generally safe.

If a woman is infertile because of physical abnormalities in her reproductive system, she may be treated surgically. Whereas surgeries are now very sophisticated, they nonetheless have risks, including bleeding, infection, organ damage, and adverse reactions to anesthesia. Similar risks are associated with collecting eggs for in vitro fertilization (IVF). To ensure the collection of several eggs, a woman may be placed on hormone-based medications that stimulate egg production. Such medications may cause ovarian hyperstimulation syndrome—enlarged ovaries and abdominal fluid accumulation. In mild cases, the only symptom is discomfort; but in severe cases (though rare), a woman's life may be endangered. In any case, the fluid has to be drained.

Usually, IVF involves the creation of many embryos; the healthiest-looking ones are transferred into the woman's body. Others may be frozen for future attempts at establishing pregnancy, given to other infertile couples, donated for research, or destroyed. Of those that are transferred, none, one, or all might develop into fetuses. The significant increase of multifetal pregnancies in the United States in the last 15 years has been largely attributed to fertility treatment (Fig. 16B). Though the number of triplet and higher-number multiple pregnancies started to level off in 1999, twin pregnancies continue to climb.

They may seem like a dream come true, but multifetal pregnancies are difficult. The mother is more likely to develop complications such as gestational diabetes and high blood pressure than are women carrying single babies. Positioning of the babies in the uterus may make vaginal delivery less likely, and there is likely a chance of preterm labor. Babies born prematurely face numerous hardships. Infant death and long-term disabilities are also more common with multiple births. This is true even of twins. Even if all babies are healthy, parenting "multiples" poses unique challenges.

What Happens to Frozen Embryos?

Despite potential trials, thousands of people undergo fertility treatment every year. Its popularity has brought a number of ethical issues to light. For example, the estimated high numbers of stored frozen embryos (a few hundred thousand in the United States) has generated debate about their fate, complicated by the fact that the long-term viability of frozen embryos is not well understood. Scientists may worry that embryos donated to other couples are ones screened out from one implantation and not likely to survive. Some religious groups strongly oppose destruction of these embryos or their use in research. Patients for whom the embryos were created generally feel that they should have sole rights to make decisions about their fate. However, a fertility clinic may no longer be receiving monetary compensation for the storage of frozen embryos and may be unable to contact the couples for whom they were produced. The question then becomes whether the clinic now has the right to determine their fate.

Figure 16B Reproductive technologies may lead to mutliple births.

Who Should Be Treated?

Additionally, because fertility treatment is voluntary, is it ever acceptable to turn some people away? What if the prospect of a satisfactory outcome is very slim or almost nonexistent? This may happen when one of the partners is ill or the woman is at an advanced age. Should a physician go ahead with treatment even if it might endanger a woman's (or baby's) health? Those in favor of limiting treatment argue that a physician has a responsibility to prevent potential harm to a patient. On the other hand, there is concern that if certain people are denied fertility for medical reasons, might they be denied for other reasons also, such as race, religion, sexuality, or income?

Questions to Consider

1. Should couples go to all lengths to have children even if it could endanger the life of one or both spouses?
2. Should couples with "multiples" due to infertility treatment receive assistance from private and public services?
3. To what lengths should society go to protect frozen embryos?
4. Do you think that anyone should be denied fertility treatment? If so, what factors do you think a doctor should take into consideration when deciding whether to provide someone fertility treatment?

APPLICATIONS AND MISCONCEPTIONS

How many babies are born annually in the United States using ART?

The very first IVF baby born in the United States was Elizabeth Carr on December 28, 1981. Since that time, assisted reproductive technologies (ART) have improved along with their success rates. In 2010, more than 61,000 babies were born in the United States as a result of ART. This represents almost 1% of the total conceptions in the United States.

In Vitro Fertilization (IVF) During **in vitro fertilization (IVF)**, conception occurs in laboratory glassware. Ultrasound machines can now spot follicles in the ovaries that hold immature eggs; therefore, the latest method is to forgo the administration of fertility drugs and retrieve immature eggs by using a needle. The immature eggs are then brought to maturity in glassware, and then concentrated sperm are added. After about two to four days, the embryos are ready to be transferred to the uterus of the woman, who is now in the secretory phase of her uterine cycle. If desired, the embryos can be tested for a genetic disease, and only those found to be free of disease will be used. If implantation is successful, development is normal and continues to term.

Video In Vitro Fertilization

Video One Healthy Baby

Gamete Intrafallopian Transfer (GIFT) The term gamete refers to a sex cell, either a sperm or an egg. Gamete intrafallopian transfer (GIFT) was devised to overcome the low success rate (15–20%) of in vitro fertilization. The method is exactly the same as in vitro fertilization, except the eggs and the sperm are placed in the uterine tubes immediately after they have been brought together. GIFT has the advantage of being a one-step procedure for the woman—the eggs are removed and reintroduced all at the same time. A variation on this procedure is to fertilize the eggs in the laboratory and then place the zygotes in the uterine tubes.

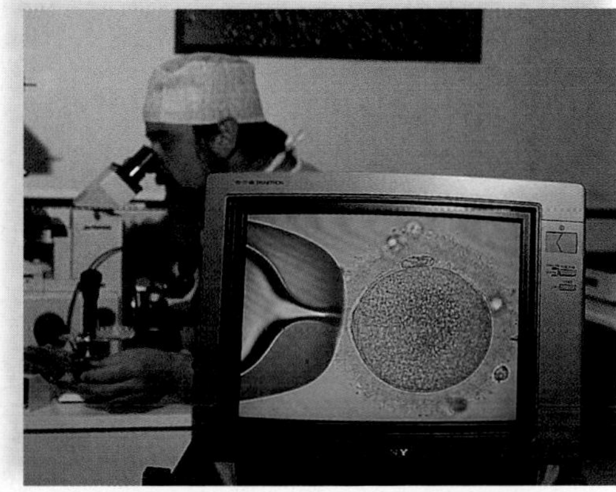

Figure 16.16 Intracytoplasmic sperm injection.
A microscope connected to a television screen is used to carry out intracytoplasmic sperm injection. A pipette holds the egg steady while a needle (not visible) introduces the sperm into the egg.

Surrogate Mothers In some instances, women are contracted and paid to have babies. These women are called surrogate mothers. The sperm and even the egg can be contributed by the contracting parents.

Intracytoplasmic Sperm Injection In this highly sophisticated procedure, a single sperm is injected into an egg (Fig. 16.16). It is used effectively when a man has severe infertility problems.

CHECK YOUR PROGRESS 16.5

1. List the major forms of birth control in order of effectiveness.
2. Explain why vasectomies and tubal ligations represent a permanent form of birth control.
3. Distinguish between an IVF and a GIFT procedure to compensate for infertility.

CONNECTING THE CONCEPTS

For more information on the hormones presented in this section, refer to the following discussions:

Section 15.2 explains the role of the gonadotropic hormones.

Section 15.6 describes the hormones produced by the female reproductive system.

16.6 Sexually Transmitted Diseases

LEARNING OUTCOMES

Upon completion of this section, you should be able to

1. Distinguish between sexually transmitted diseases (STDs) caused by viruses and those caused by bacteria.
2. Describe the causes and treatments of selected STDs.

Sexually transmitted diseases (STDs), sometimes also referred to as sexually transmitted infections (STIs), are caused by viruses, bacteria, fungi, and parasites.

STDs Caused by Viruses

Among those STDs caused by viruses, effective treatment is available for AIDS (acquired immunodeficiency syndrome) and genital herpes. However, treatment for HIV/AIDS and genital herpes cannot presently eliminate the virus from the person's body. Drugs used for treatment can merely slow replication of the viruses. Thus, neither viral disease is presently curable. Further, antiviral drugs have serious, debilitating side effects on the body.

HIV Infections

The Infectious Diseases Supplement, which begins on page 153, discusses HIV infections at greater length than this brief summary. At present, there is no vaccine to prevent an

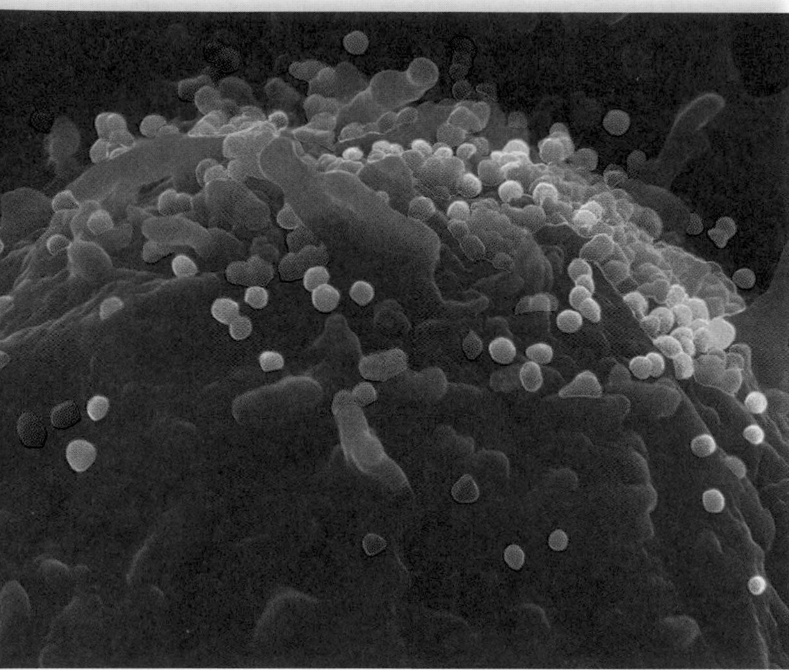

Figure 16.17 Cells infected by the HIV virus.
HIV viruses (yellow) can infect helper T cells (blue) and also macrophages, which work with helper T cells to stem the infection.

HIV infection (although several are in trials) nor is there a cure for AIDS. The best course of action is to follow the guidelines for preventing transmission of STDs outlined in the Health feature, "Preventing Transmission of STDs," on page 382.

The primary host for HIV is a helper T lymphocyte, or helper T cell (Fig. 16.17), although macrophages are also infected by the virus. The helper T cells are the very cells that stimulate an immune response (see section 7.4), so loss of these cells causes the immune system to become severely impaired in persons with AIDS. During the first stage of an HIV infection, symptoms are few, but the individual is highly contagious. Several months to several years after infection, the helper T-lymphocyte count falls. Following this decrease, infections, such as other sexually transmitted diseases, begin to appear. In the last stage of infection, called AIDS, the helper T-cell count falls way below normal. At least one opportunistic infection is present. Such diseases have the opportunity to occur only because the immune system is severely weakened. Persons with AIDS typically die from an opportunistic disease, such as *Pneumocystis* pneumonia.

Animation
How the HIV Infection Cycle Works

There is no cure for AIDS. A treatment called highly active antiretroviral therapy (HAART) is usually able to stop HIV reproduction to the extent that the virus becomes undetectable in the blood. The medications must be continued indefinitely because as soon as HAART is discontinued, the virus rebounds.

Genital Warts

Genital warts are caused by the human papillomaviruses (HPVs). Many times, carriers either do not have any sign of warts or merely have flat lesions. When present, the warts commonly are seen on the penis and foreskin of men and near the vaginal opening in women. A newborn can become infected while passing through the birth canal.

Individuals currently infected with visible growth may have those growths removed by surgery, freezing, or burning with lasers or acids. However, visible warts that are removed may recur. A vaccine has been released for the human papillomaviruses that most commonly cause genital warts. This development is an extremely important step in the prevention of cancer, as well as in the prevention of warts themselves. Genital warts are associated with cancer of the cervix (see chapter opener), as well as tumors of the vulva, vagina, anus, and penis. Researchers believe that these viruses may be involved in up to 90% of all cases of cancer of the cervix. HPV vaccinations, which are recommended for both women and men before the age of 26, might make such cancers a thing of the past.

Genital Herpes

Genital herpes is caused by herpes simplex virus. Type 1 usually causes cold sores and fever blisters, while type 2 more often causes genital herpes (Fig. 16.18).

Persons usually get infected with herpes simplex virus 2 when they are adults. Some people exhibit no symptoms. Others may experience a tingling or itching sensation before blisters appear on the genitals. Once the blisters rupture, they leave painful ulcers that may take between five days and three weeks to heal. The blisters may be accompanied by fever; pain on urination; swollen lymph nodes in the groin; and, in women, a copious discharge. At this time, the individual has an increased risk of acquiring an HIV infection.

After the ulcers heal, the disease is only latent, and blisters can recur, although usually at less frequent intervals and with milder symptoms. Fever, stress, sunlight, and menstruation are associated with recurrence of symptoms. Exposure to herpes in the birth canal can cause an infection in the newborn, which leads to neurological disorders and even death. Birth by cesarean section prevents this possibility. There are antiviral drugs available that reduce the number and length of outbreaks. However, these drugs are not a cure for genital herpes. Latex or polyurethane condoms are recommended by the FDA to prevent the transmission of the virus to sexual partners.

Hepatitis

Hepatitis infects the liver and can lead to liver failure, liver cancer, and death. There are six known viruses that cause hepatitis, designated A, B, C, D, E, and G. Hepatitis A is usually acquired from sewage-contaminated drinking water, but this infection can also be sexually transmitted through oral–anal contact. Hepatitis B is spread through sexual contact and by bloodborne transmission (accidental needlestick on the job, receiving a contaminated blood transfusion, a drug abuser sharing infected needles while injecting drugs, from mother to fetus, etc.). Simultaneous infection with hepatitis B and HIV is common, because both share the same routes of transmission.

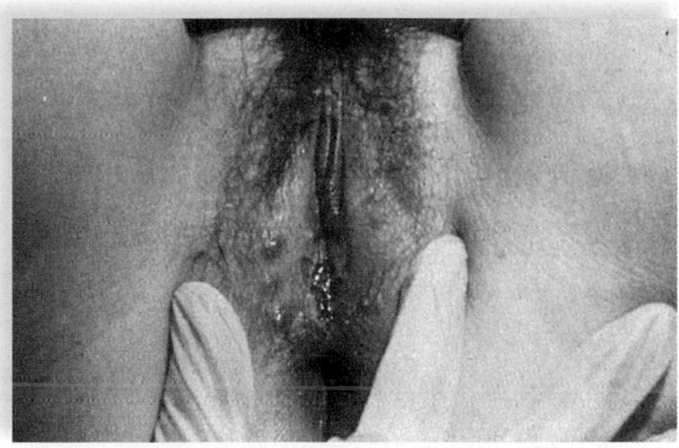

a.

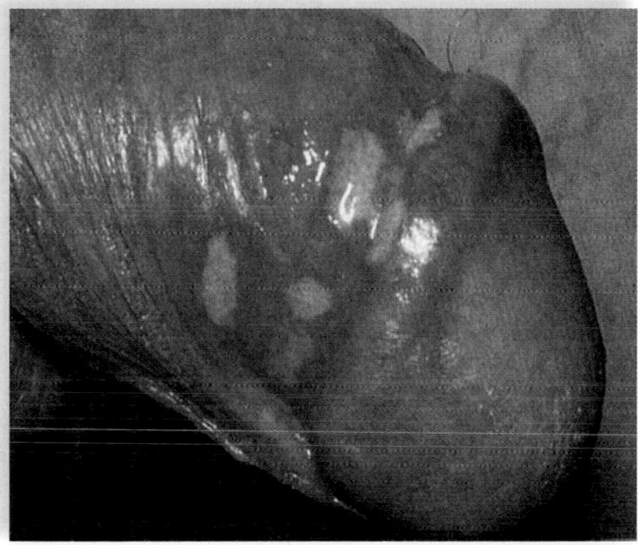

b.

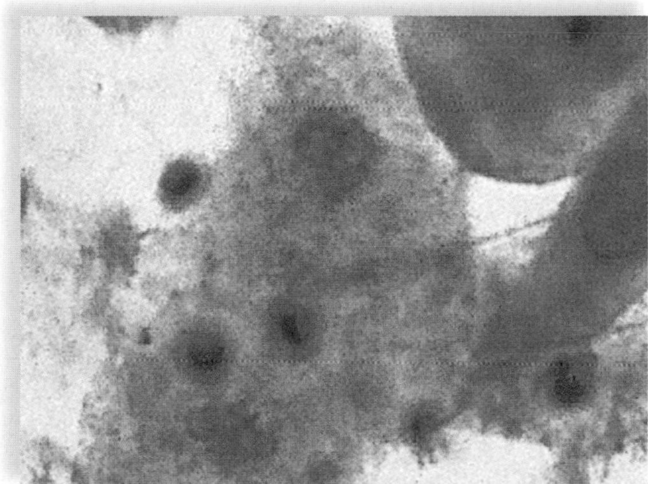

c.

Figure 16.18 Herpes simplex virus 2 and genital herpes. There are several types of viruses associated with herpes. Genital herpes is usually caused by herpes simplex virus 2. Symptoms of genital herpes include an outbreak of blisters, which can be present on the labia of females (**a**) or on the penis of males (**b**). **c.** A photomicrograph of cells infected with the herpes simplex virus.

Fortunately, a combined vaccine is available for hepatitis A and B. It is recommended that all children receive the vaccine to prevent infection (see Chapter 7). Hepatitis C (also called *non-A, non-B hepatitis*) causes most cases of posttransfusion hepatitis. Hepatitis D and G are sexually transmitted, and hepatitis E is acquired from contaminated water. Screening of blood and blood products can prevent transmission of hepatitis viruses during a transfusion. Proper water-treatment techniques can prevent contamination of drinking water.

Video Halting Hepatitis

STDs Caused by Bacteria

Only STDs caused by bacteria are curable with antibiotics. Antibiotic resistance acquired by these bacteria may require treatment with extremely strong drugs for an extended period to achieve a cure.

Chlamydia

Chlamydia is named for the tiny bacterium that causes it, *Chlamydia trachomatis* (Fig. 16.19). The incidence of new chlamydia infections has steadily increased since 1984.

Chlamydia infections of the lower reproductive tract are usually mild or asymptomatic, especially in women. About 18 to 21 days after infection, men may experience a mild burning sensation on urination and a mucoid discharge. Women may have a vaginal discharge along with the symptoms of a urinary tract infection. Chlamydia also causes cervical ulcerations, which increase the risk of acquiring HIV.

If the infection is misdiagnosed or if a woman does not seek medical help, there is a particular risk of the infection spreading from the cervix to the uterine tubes so that

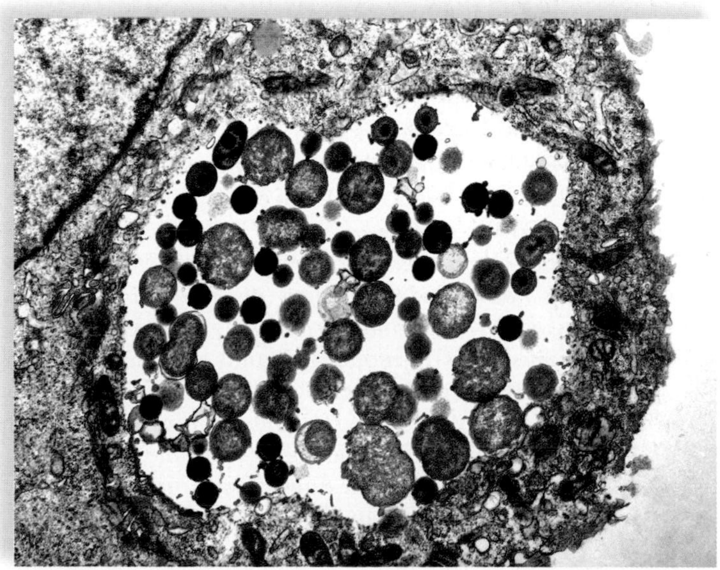

5000×

Figure 16.19 Chlamydial infection. The different stages of a *Chlamydia trachomatis* infection inside a cell are stained red, brown, and black.

BIOLOGY MATTERS  Health

Preventing Transmission of STDs

Sexual Activities Transmit STDs

Abstain from sexual intercourse or develop a long-term monogamous (always the same partner) sexual relationship with a partner who is free of STDs (Fig. 16C).

Refrain from having multiple sex partners or having relations with someone who has multiple sex partners. If you have sex with two other people and each of these has sex with two people, and so forth, the number of people who are relating is quite large.

Be aware that having relations with an intravenous drug user is risky because the behavior of this group risks AIDS and hepatitis B. Be aware that anyone who already has another sexually transmitted disease is more susceptible to an HIV infection.

Avoid anal intercourse (in which the penis is inserted into the rectum) because this behavior increases the risk of an HIV infection. The lining of the rectum is thin, and infected CD4 T cells can easily enter the body there. Also, the rectum is supplied with many blood vessels, and insertion of the penis into the rectum is likely to cause tearing and bleeding that facilitate the entrance of HIV. The vaginal lining is thick and difficult to penetrate, but the lining of the uterus is only one cell thick at certain times of the month and does allow CD4 T cells to enter.

Uncircumcised males are more likely to become infected than circumcised males because vaginal secretions can remain under the foreskin for a long time.

Practice Safer Sex

Always use a latex condom during sexual intercourse if you are not in a monogamous relationship. Be sure to follow the directions supplied by the manufacturer for the use of a condom. At one time, condom users were advised to use nonoxynol-9 in conjunction with a condom, but testing shows that this spermicide has no effect on viruses, including HIV.

Avoid fellatio (kissing and insertion of the penis into a partner's mouth) and cunnilingus (kissing and insertion of the tongue into the vagina) because they may be a means of transmission. The mouth and gums often have cuts and sores that facilitate catching an STD.

Practice penile, vaginal, oral, and hand cleanliness. Be aware that hormonal contraceptives make the female genital tract receptive to the transmission of sexually transmitted diseases, including HIV.

Figure 16C Sexual activities transmit STDs.

Figure 16D Sharing needles transmits STDs.

Be cautious about using alcohol or any drug that may prevent you from being able to control your behavior.

Drug Use Transmits HIV

Stop, if necessary, or do not start the habit of injecting drugs into your veins. Be aware that HIV and hepatitis B can be spread by blood-to-blood contact.

Always use a new sterile needle for injection or one that has been cleaned in bleach if you are a drug user and cannot stop your behavior (Fig. 16D).

Questions to Consider

1. Why might the use of female contraceptives actually increase the chances of contracting HIV?
2. What cells in the blood act as a host for HIV?

pelvic inflammatory disease (PID) results. This very painful condition can result in blockage of the uterine tubes with the possibility of sterility and infertility. If a baby comes in contact with chlamydia during birth, inflammation of the eyes or pneumonia can result.

Gonorrhea

Gonorrhea is caused by the bacterium *Neisseria gonorrhoeae.* Diagnosis in the male is not difficult, because typical symptoms are pain upon urination and a thick, greenish-yellow urethral discharge. In females, a latent infection leads to pelvic inflammatory disease (PID), which may result in damage to the uterus, ovaries, and other reproductive structures. If a baby is exposed during birth, an eye infection leading to blindness can result. All newborns are given eyedrops to prevent this possibility.

Gonorrhea proctitis, an infection of the anus characterized by anal pain and blood or pus in the feces, also occurs in patients. Oral–genital contact can cause infection of the mouth, throat, and tonsils. Gonorrhea can spread to internal parts of the body, causing heart damage or arthritis. If, by chance, the person touches infected genitals and then touches his or her eyes, a severe eye infection can result. Up to now, gonorrhea was curable by antibiotic therapy. However, resistance to antibiotics is becoming more and more common, and *Neisseria gonorrhoeae* is now classified as a "super-bug," meaning that it has developed resistance to a variety of antibiotics.

Syphilis

Syphilis is caused by a bacterium called *Treponema pallidum* (Fig. 16.20). As with many other bacterial diseases, penicillin is an effective antibiotic. Syphilis has three stages, often separated by latent periods, during which the bacteria are resting before multiplying again. During the primary stage, a hard chancre (ulcerated sore with hard edges) indicates the site of infection. The chancre usually heals spontaneously, leaving little scarring. During the secondary stage, the victim breaks out in a rash that does not itch and is seen even on the palms of the hands and the soles of the feet. Hair loss and infectious gray patches on the mucous membranes may also occur. These symptoms disappear of their own accord.

The tertiary stage lasts until the patient dies. During this stage, syphilis may affect the cardiovascular system by causing aneurysms, particularly in the aorta. In other instances, the disease may affect the nervous system, resulting in psychological disturbances. Also, gummas, large destructive ulcers, may develop on the skin or within the internal organs.

Congenital syphilis is caused by syphilitic bacteria crossing the placenta. The child is born blind and/or with numerous anatomical malformations. Control of syphilis depends on prompt and adequate treatment of all new cases. Therefore, it is crucial for all sexual contacts to be traced so they can be treated. Diagnosis of syphilis can be made by blood tests or by microscopic examination of fluids from lesions.

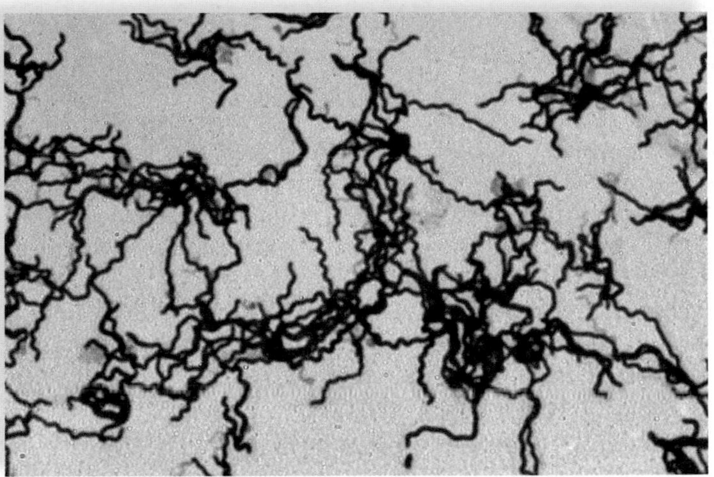

Figure 16.20 **Syphilis.**
Treponema pallidum, the cause of syphilis.

APPLICATIONS AND MISCONCEPTIONS

Can you catch an STD from a toilet seat?

When HIV/AIDS was first identified in the mid-1980s, many people were concerned about being infected by the virus on toilet seats. Toilet seats are plastic and inert, so they're not very hospitable to disease-causing organisms. So if you're deciding whether to hover or sit, remember sitting on a toilet seat will not give you an STD.

Vaginal Infections

The term *vaginitis* is used to describe any vaginal infection or inflammation. It is the most commonly diagnosed gynecologic condition. Bacterial vaginosis (BV) is believed to cause 40–50% of the cases of vaginitis in the United States. Overgrowth of certain bacteria inhabiting the vagina causes vaginosis. A common culprit is the bacterium *Gardnerella vaginosis.* Overgrowth of this organism and subsequent symptoms can occur for nonsexual reasons. However, symptomless males may pass on the bacterium to women, who do experience symptoms.

The symptoms of BV are vaginal discharge that has a strong odor, a burning sensation during urination, and/or itching or pain in the vulva. Some women with BV have no signs of the infection. How women acquire these infections is not well understood. Having a new sex partner or multiple sex partners seems to increase the risk of getting BV, but females who are not sexually active get BV as well. Douching also appears to increase the incidence of BV. Women with BV are more susceptible to infection by other STDs, including HIV, herpes, chlamydia, and gonorrhea. Pregnant women with BV are at greater risk of premature delivery.

The yeast *Candida albicans,* and a protozoan, *Trichomonas vaginalis,* are two other causes of vaginitis. *Candida albicans* is normally found living in the vagina. Under certain circumstances, its growth increases above normal, causing vaginitis. For example, women taking birth control pills or antibiotics may be prone to yeast infections. Both can alter the normal balance of vaginal organisms, causing a yeast infection. A yeast infection causes a thick, white, curdlike vaginal discharge and is accompanied by itching of the vulva and/or vagina. Antifungal medications inserted into the vagina are used to treat yeast infections. Trichomoniasis, caused by *Trichomonas vaginalis,* affects both males and females. The urethra is usually the site of infection in males. Infected males are often asymptomatic and pass the parasite to their partner during sexual intercourse. Symptoms of trichomoniasis in females are a foul-smelling, yellow-green frothy discharge and itching of the vulva/vagina. Having trichomoniasis greatly increases the risk of infection by HIV. Prescription drugs are used to treat trichomoniasis, but if one partner remains infected, reinfection will occur. It is recommended that both partners in a sexual relationship be treated and abstain from having sex until the treatment is completed.

CHECK YOUR PROGRESS 16.6

1. Explain what condition can occur due to both a chlamydial and gonorrheal infection.
2. Describe the medical conditions in women that are associated with genital warts.
3. Discuss the cause of most STDs.

CONNECTING THE CONCEPTS

For more information on the topics presented in this section, refer to the following discussions:

Section 7.1 explores the structure of both viruses and bacteria.

Section S.1 provides a detailed examination of the HIV virus, its replication, and the disease AIDS.

Section S.3 examines how antibiotic resistance occurs and its consequences in the treatment of disease.

CASE STUDY CONCLUSION

The good news for Ann is that early detection of cervical cancer is critical to successful treatment of the disease. For individuals with cervical cancer, the survival rate for those who have early detection is almost 100%, versus a less than 5% survival rate for those in whom the cancer has begun to spread, or metastasize, to other organs. In Ann's case, the years of smoking cigarettes probably were a major factor in her development of cervical cancer. However, for many cases of cervical cancer, the cause is the human papillomavirus, or HPV. There are over 15 forms of HPV that have been linked to cervical cancers. In 2006, the Food and Drug Administration (FDA) approved an HPV vaccine for women. The vaccine is designed to be administered as three doses starting between the ages of 11 and 12 years. Recently, the vaccine was also approved for men ages 9 to 26 to prevent genital warts and to reduce the chances that men will transmit HPV to their sexual partners. With the development of the HPV vaccine, it is hoped that the rates of cervical cancer in women will drop drastically over the next few decades.

MEDIA STUDY TOOLS

Enhance your study of this chapter with media! Visit **www.mhhe.com/maderhuman13e** and go to "Media Study Tools" for this chapter to access the following:

Animations

16.1 Comparison of Meiosis and Mitosis
16.2 Spermatogenesis
16.4 Maturation of the Follicle and Oocyte • Female Reproductive System
16.6 How the HIV Infection Cycle Works

Videos

16.2 Human Sperm
16.5 One Healthy Baby • In Vitro Fertilization
16.6 Halting Hepatitis

MP3 Files

16.2 Male Reproductive Anatomy and Physiology
16.3 Female Reproductive Anatomy and Physiology

SUMMARIZE

16.1 Human Life Cycle

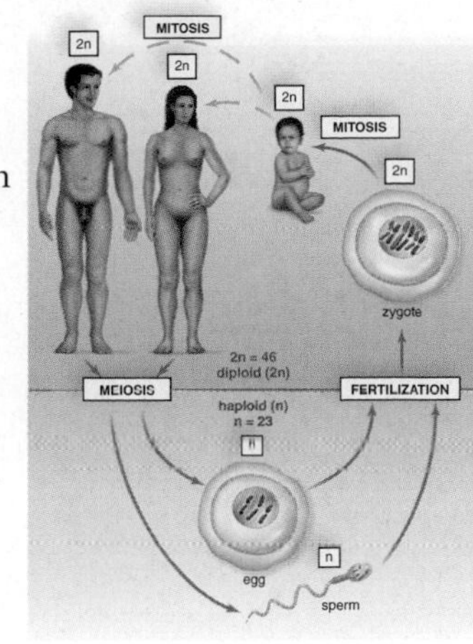

The **reproductive system** is responsible for the production of **gametes** in both sexes. In females, the system also provides the location for fertilization. Following **zygote** formation, the female reproductive system protects and nourishes the developing fetus.

The life cycle of higher organisms requires two types of cell division:

- **mitosis,** the growth and repair of tissues; and
- **meiosis,** the production of gametes.

16.2 Male Reproductive System

The external genitals of males are

- the **penis** (organ of sexual intercourse); and
- the **scrotum,** which contains the **testes.**

Spermatogenesis, occurring in **seminiferous tubules** of the testes, produces **sperm.**

- **Sertoli cells** regulate the process of spermatogenesis.
- The **epididymis** stores mature sperm cells. A mature sperm consists of a head containing the **acrosome,** a middle section containing mitochondria, and a tail.
- Sperm pass from the **vas deferens** to the **urethra.**
- The **seminal vesicles, prostate gland,** and **bulbourethral glands** secrete fluids that aid the sperm.
- Sperm and secretions are called **semen** or seminal fluid.

Orgasm in males results in ejaculation of semen from the penis.

Circumcision is the surgical removal of the foreskin. **Erectile dysfunction** (**ED**) is a condition in which the male is unable to achieve or sustain an erection.

Hormonal Regulation in Males

- Hormonal regulation, involving secretions from the hypothalamus, the anterior pituitary, and the testes, maintains a fairly constant level of testosterone.
- **Follicle-stimulating hormone** (**FSH**) from the anterior pituitary promotes spermatogenesis.
- **Luteinizing hormone** (**LH**) from the anterior pituitary promotes **testosterone** production by **interstitial cells.**

16.3 Female Reproductive System

Oogenesis occurring within the **ovaries** typically produces one mature follicle each month.

- This follicle balloons out of the ovary and bursts, releasing an **egg** (or ovum) that is moved by the **fimbriae** into the **uterine tubes.**
- The uterine tubes lead to the **uterus,** where **implantation** of the zygote and development occur. The **endometrium** is the lining of the uterus that participates in the formation of the placenta. **Endometriosis** occurs when uterine tissue is present outside of the uterus.
- The **cervix** is the term for the end of the uterus that links to the vagina. A **Pap test** may be used to screen the cervix for cancer.
- The female external genital area is called the vulva. It includes the vaginal opening, the clitoris, the labia minora, and the labia majora.
- The **vagina** is the organ of sexual intercourse and the birth canal in females.

Orgasm in females culminates in contractions of the uterus and uterine tubes.

16.4 The Ovarian Cycle

Ovarian Cycle: Nonpregnant

- **Oogenesis** in females produces egg cells and **polar bodies.**
- The **ovarian cycle** is under the hormonal control of the hypothalamus and the anterior pituitary.
- During the cycle's first half, FSH from the anterior pituitary causes maturation of a **follicle** that secretes estrogen and some progesterone.
- After **ovulation** and during the cycle's second half, LH from the anterior pituitary converts the follicle into the **corpus luteum.**
- The corpus luteum secretes progesterone and some estrogen.
- **Menopause** represents that stage of a woman's life when the ovarian cycle ceases.

Uterine Cycle: Nonpregnant

Estrogen and progesterone regulate the **uterine cycle.**

- **Estrogen** causes the endometrium to rebuild.
- Ovulation usually occurs on day 14 of a 28-day cycle.
- **Progesterone** produced by the corpus luteum causes the endometrium to thicken and become secretory.
- A low level of hormones causes the endometrium to break down as **menstruation** occurs.

Fertilization and Pregnancy

If fertilization takes place, the embryo implants in the thickened endometrium.

- The corpus luteum is maintained because of **human chorionic gonadotropin** (**HCG**) production by the **placenta;** therefore, progesterone production does not cease.
- Menstruation usually does not occur during pregnancy.

16.5 Control of Reproduction

Contraceptives are **birth control methods** that reduce the chance of pregnancy.

- A few of these are the **birth control pill, intrauterine device** (**IUD**), **diaphragm,** and **condom.**
- Contraceptive vaccines, implants, and injections are becoming increasingly available.
- Surgical procedures, such as a **vasectomy** or **tubal ligation,** make the individual sterile.

Assisted reproductive technologies may help couples who are experiencing **infertility.** Some of these technologies are

- artificial insemination by donor (AID),
- **in vitro fertilization** (**IVF**),
- gamete intrafallopian transfer (GIFT), and
- intracytoplasmic sperm injection.

16.6 Sexually Transmitted Diseases

STDs are caused by viruses, bacteria, fungi, and parasites.

STDs Caused by Viruses

- AIDS is caused by HIV (human immunodeficiency virus).
- Genital warts are caused by human papillomaviruses; these viruses cause warts or lesions on genitals and are associated with certain cancers.
- Genital herpes is caused by herpes simplex virus 2; causes blisters on genitals.
- Hepatitis is caused by hepatitis viruses A, B, C, D, E, and G. A and E are usually acquired from contaminated water; hepatitis B and C are from bloodborne transmission; and B, D, and G are sexually transmitted.

STDs Caused by Bacteria

- Chlamydia is caused by *Chlamydia trachomatis.*
- Gonorrhea is caused by *Neisseria gonorrhoeae.*
- Syphilis is caused by *Treponema pallidum.* It has three stages, with the third stage resulting in death.

Vaginal Infections

- Bacterial vaginosis commonly results from bacterial overgrowth. *Gardnerella vaginosis* often causes such infections.
- Infection with the yeast *Candida albicans* also occurs because of overgrowth, and antibiotics or hormonal contraceptives trigger this condition.
- The parasite *Trichomonas vaginalis* also causes vaginosis. This type affects both men and women, though men are often asymptomatic.

ASSESS

Testing Your Knowledge of the Concepts

1. What type of cell division produces the gametes? Why is this type of cell division necessary? (pages 361–362)
2. Trace the path of sperm. What glands contribute fluids to semen? (pages 362–364)
3. Where are sperm produced in the testes? What is the process called? Where is testosterone produced in the testes? (pages 363–364)
4. Name the hormones involved in maintaining the sex characteristics of the male, and tell what each does. (pages 365–366)
5. What are the organs of the female reproductive tract and their functions? (pages 366–367)
6. Describe the ovarian cycle in a nonpregnant female and the hormones involved. (pages 369–372)
7. Describe the uterine cycle in a nonpregnant female, and relate it to the ovarian cycle. (pages 372–373)
8. Describe the hormonal role of the placenta. (page 372)
9. Briefly describe various birth control methods, along with their effectiveness. (pages 374–377)
10. Briefly describe various types of assisted reproductive technologies. (pages 377–379)
11. What STDs are caused by viruses? List the causative agent, symptoms, and treatments. What STDs are caused by bacteria? List the causative agent, symptoms, and treatments. (pages 379–384)
12. Label this diagram of the male reproductive system.

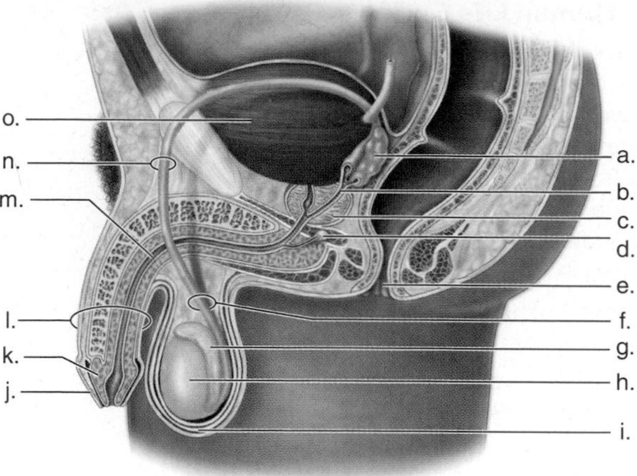

13. Follicle-stimulating hormone (FSH)
    a. is secreted by females but not by males.
    b. stimulates the seminiferous tubules to produce sperm.
    c. secretion is controlled by gonadotropin-releasing hormone (GnRH).
    d. Both b and c are correct.
14. Semen does not contain
    a. prostate fluid.
    b. urine.
    c. fructose.
    d. prostaglandins.
    e. Both b and d are correct.
15. Label this diagram of the female reproductive system.

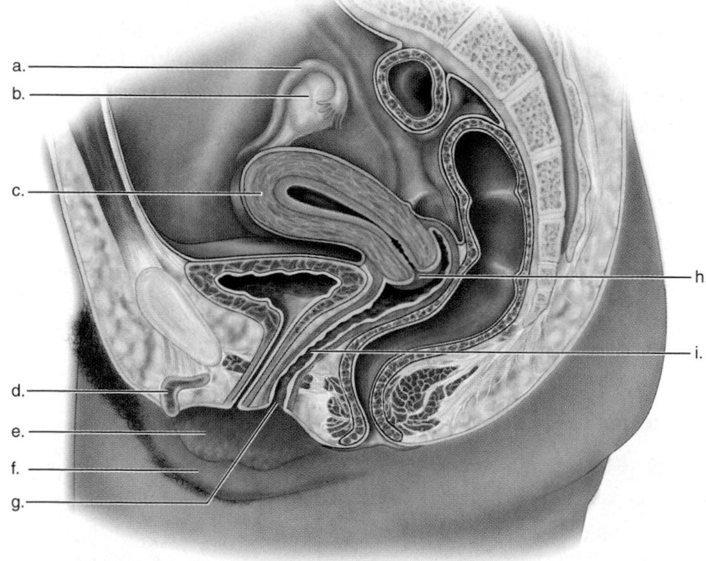

16. Testosterone is produced and secreted by
    a. spermatogonia.
    b. sustentacular cells.
    c. seminiferous tubules.
    d. interstitial cells.
17. The release of the oocyte from the follicle is caused by
    a. a decreasing level of estrogen.
    b. a surge in the level of follicle-stimulating hormone.
    c. a surge in the level of luteinizing hormone.
    d. progesterone released from the corpus luteum.

18. Which of the following is not an event of the ovarian cycle?
  a. FSH promotes the development of a follicle.
  b. The endometrium thickens.
  c. The corpus luteum secretes progesterone.
  d. Ovulation of an egg occurs.

19. An oocyte is fertilized in the
  a. vagina.
  b. uterus.
  c. uterine tubes.
  d. ovary.

20. During pregnancy,
  a. the ovarian and uterine cycles occur more quickly than before.
  b. GnRH is produced at a higher level than before.
  c. the ovarian and uterine cycles do not occur.
  d. the female secondary sex characteristics are not maintained.

21. Female oral contraceptives prevent pregnancy because
  a. the pill inhibits the release of luteinizing hormone.
  b. oral contraceptives prevent the release of an egg.
  c. follicle-stimulating hormone is not released.
  d. All of the choices are correct.

In questions 22–24, match each method of protection with a means of birth control in the key.

Key:

  a. vasectomy
  b. oral contraception
  c. intrauterine device (IUD)
  d. diaphragm
  e. male condom

22. Blocks entrance of sperm to uterus

23. Traps sperm and also prevents STDs

24. Prevents implantation of an embryo

ENGAGE

Thinking Critically About the Concepts

1. Female athletes who train intensively often stop menstruating. The important factor appears to be the reduction of body fat below a certain level. Give a possible evolutionary explanation for a relationship between body fat in females and reproductive cycles.
2. The average sperm count in males is now lower than it was several decades ago. The reasons for the lower sperm count usually seen today are not known. What data might be helpful in order to formulate a testable hypothesis?
3. Women who use birth control pills appear to have a lower risk of developing ovarian cancer. Women who use fertility enhancing drugs (which increase the number of follicles that develop) may increase their risk of developing ovarian cancer. Speculate about how these therapies may affect a woman's risk of developing ovarian cancer.

CHAPTER

17

Human Development and Aging

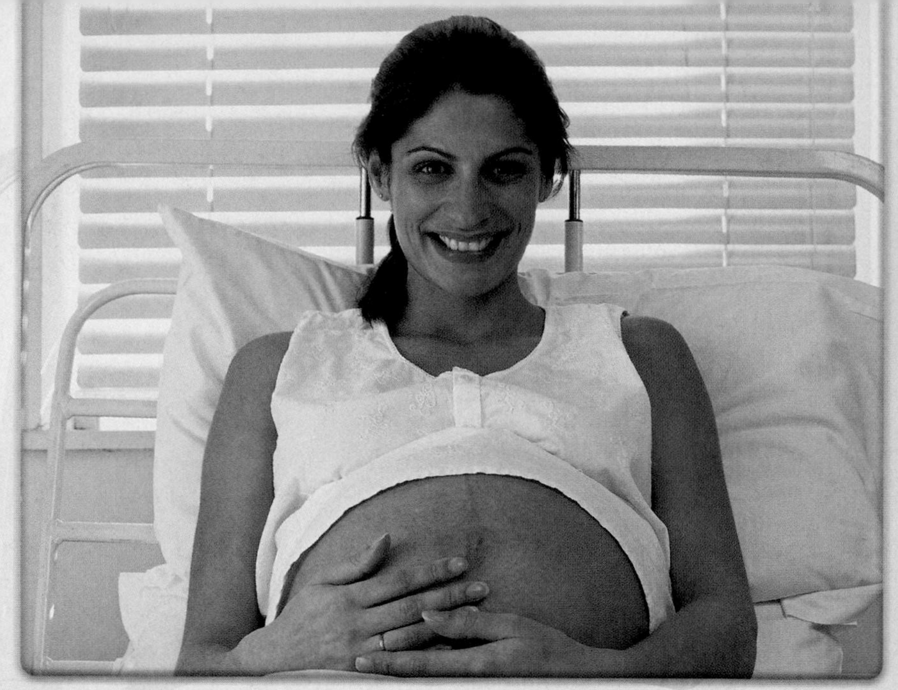

CHAPTER CONCEPTS

17.1 Fertilization
During fertilization, a sperm nucleus fuses with the egg nucleus. Once one sperm penetrates the plasma membrane, the egg undergoes changes that prevent any more sperm from entering.

17.2 Pre-Embryonic and Embryonic Development
Pre-embryonic development occurs between the time of fertilization and implantation in the uterine lining. By the end of embryonic development, all organ systems are established and there is a mature and functioning placenta.

17.3 Fetal Development
During fetal development, the gender becomes obvious, the skeleton continues to ossify, fetal movement begins, and the fetus gains weight.

17.4 Pregnancy and Birth
During pregnancy, the mother gains weight as the uterus comes to occupy most of the abdominal cavity. A positive feedback mechanism that involves uterine contractions and oxytocin explains the onset and continuation of labor so that the child is born.

17.5 Aging
Development after birth consists of infancy, childhood, adolescence, and adulthood. Aging is influenced by both our genes and external factors.

BEFORE YOU BEGIN

Before beginning this chapter, take a few moments to review the following discussions:

Figure 16.4 How does spermatogenesis produce sperm cells?

Figure 16.9 What are the differences between oogenesis and spermatogenesis?

Section 16.4 What are the roles of estrogen and progesterone in the female reproductive system?

CASE STUDY PREGNANCY TESTING

For several months, Amber and Kent had been trying to conceive a child. They had put off having children for several years while they pursued their careers. Now, as they both approached the age of 40, they were beginning to feel the pressures of time. As a precautionary method, Amber had begun taking prenatal vitamins; additionally, she was much more aware of the content of her diet. Although neither of them was ever really into physical exercise, both began walking several times a week in preparation for what they hoped would be news that Amber was pregnant. Finally, after two months, Amber proudly announced that the home pregnancy test was positive! They immediately scheduled an appointment with their regular physician to prepare for the next stage of their lives.

Both Amber and Kent were very satisfied with their choice of a doctor. At the first visit following the positive results of the home pregnancy test, their physician had performed a complete physical of Amber, as well as a blood test to confirm pregnancy. The physician informed the new parents that a blood test was much more accurate in detecting levels of the pregnancy hormone, human chorionic gonadotropin (HCG), than over-the-counter (OTC) urine tests. The results of the blood test confirmed what Amber and Kent suspected, that in a period of just 40 weeks, Amber and Kent would be parents.

Their physician promptly gave the parents-to-be a list of items to avoid. Amber was told to increase her level of exercise and watch her diet more closely. She needed to drink eight to ten glasses of water per day, as well as eat plenty of fruits and vegetables. The doctor informed them that it was crucial for Amber to inform her physician of any over-the-counter drugs she may want to take, especially in the first trimester. Her doctor told Amber that the first trimester was a period when critical organ systems developed in her baby and that most OTC medications and alcohol were now forbidden. Both Amber and Kent were up to the challenge and excited about the prospects of finally being parents.

As you read through the chapter, think about the following questions:

1. What is the role of the HCG hormone in pregnancy?
2. Why would the doctor request that Amber check before taking over-the-counter drugs?
3. What physiological changes should Amber expect over the course of her pregnancy?

17.1 Fertilization

LEARNING OUTCOMES

Upon completion of this section, you should be able to

1. Identify the structures of an egg and sperm cell and provide a function for each.
2. Describe the steps in the fertilization of an egg cell by a sperm.

In the previous chapter, we examined the structure and function of the reproductive system in males and females. One of the functions of a reproductive system is to produce gametes (egg and sperm cells) for the production of a new individual. **Fertilization** is the union of a sperm and egg to form a **zygote,** the first cell of that new individual (Fig. 17.1).

Steps of Fertilization

The tail of a sperm is a flagellum, which allows it to swim toward the egg. The middle piece contains energy-producing mitochondria. The head contains a nucleus capped by a membrane-bound acrosome. The acrosome is an organelle containing digestive enzymes. Only the nucleus from the sperm head fuses with the egg nucleus. Therefore, the zygote receives cytoplasm and organelles only from the mother.

Video Human Sperm

APPLICATIONS AND MISCONCEPTIONS

How many sperm compete to fertilize the egg?

Studies indicate that there are between 150 and 200 million sperm in a man's ejaculation. Of that number, only 20 to 30 or so make it near the egg. Out of those, only a single sperm fertilizes the egg.

The plasma membrane of the egg is surrounded by an extracellular matrix termed the *zona pellucida.* In turn, the zona pellucida is surrounded by a few layers of adhering follicular cells, collectively called the *corona radiata.* These cells nourished the egg when it was in a follicle of the ovary.

During fertilization, several sperm penetrate the corona radiata. Several sperm attempt to penetrate the zona pellucida, but only one sperm enters the egg. The acrosome plays a role in allowing sperm to penetrate the zona pellucida. After a sperm head binds tightly to the zona pellucida, the acrosome releases digestive enzymes that forge a pathway for the sperm through the zona pellucida. When a sperm binds to the egg, their plasma membranes fuse. This sperm (the head, the middle piece, and usually the tail) enters the egg. Fusion of the sperm nucleus and the egg nucleus follows.

To ensure proper development, only one sperm should enter an egg. Prevention of polyspermy (entrance of more than one sperm) depends on changes in the egg's plasma membrane and in the zona pellucida. As soon as a sperm touches an

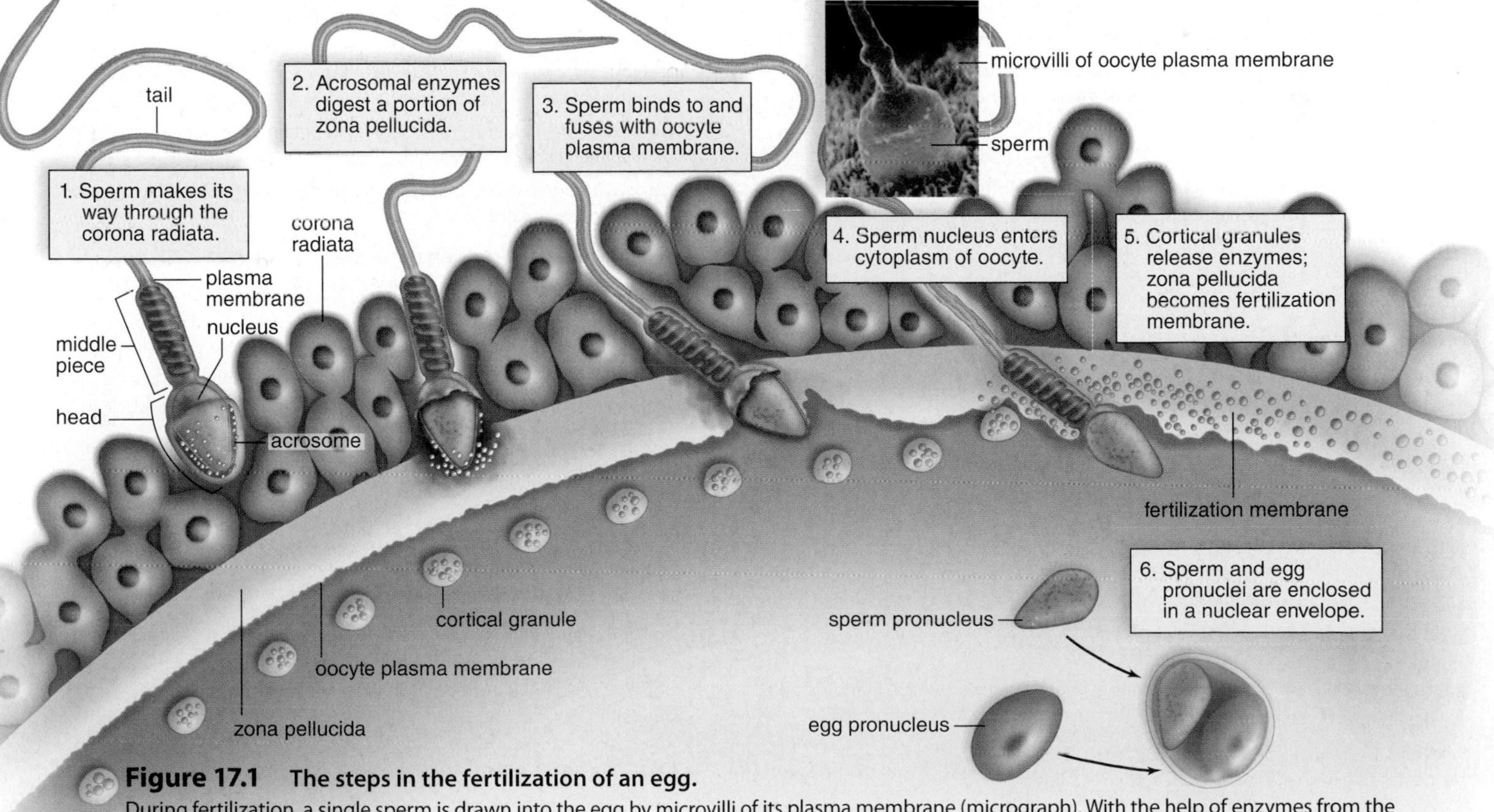

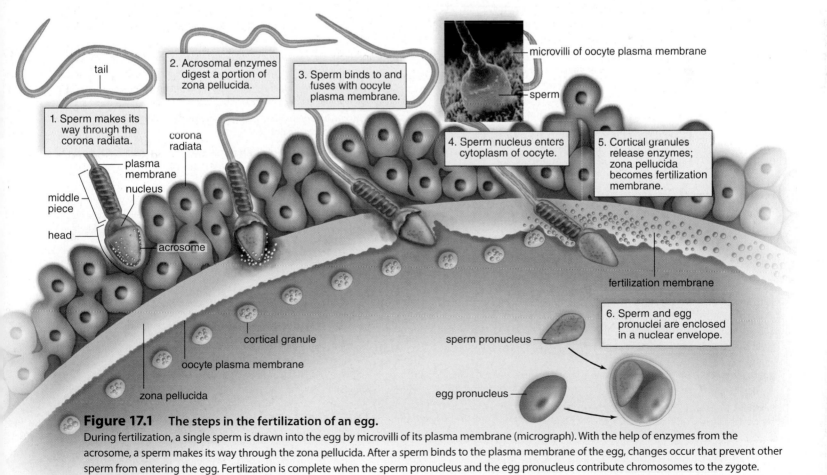

Figure 17.1 The steps in the fertilization of an egg.
During fertilization, a single sperm is drawn into the egg by microvilli of its plasma membrane (micrograph). With the help of enzymes from the acrosome, a sperm makes its way through the zona pellucida. After a sperm binds to the plasma membrane of the egg, changes occur that prevent other sperm from entering the egg. Fertilization is complete when the sperm pronucleus and the egg pronucleus contribute chromosomes to the zygote.

egg, the egg's plasma membrane depolarizes (from 265 mV to 10 mV), and this prevents the binding of any other sperm. Then, vesicles called cortical granules release enzymes that cause the zona pellucida to become an impenetrable fertilization membrane. Now sperm cannot bind to the zona pellucida either.

CHECK YOUR PROGRESS 17.1

1. Describe the steps in fertilization.
2. Distinguish between the function of the corona radiata and the zona pellucida.
3. Explain what prevents multiple sperm from fertilizing the same egg.

CONNECTING THE CONCEPTS

For more information on egg and sperm cells, refer to the following discussions:

Section 16.2 explains how sperm are produced by spermatogenesis.

Section 16.4 explains how egg cells are produced by oogenesis.

17.2 Pre-Embryonic and Embryonic Development

LEARNING OUTCOMES

Upon completion of this section, you should be able to

1. Recognize how cleavage, growth, morphogenesis, and differentiation all play a role in development.
2. Identify the extraembryonic membranes and provide a function for each.
3. Identify the organ systems that are formed from each of the primary germ layers.
4. Summarize the key events that occur at each stage of pre-embryonic and embryonic development.

Human development proceeds from pre-embryonic to embryonic development and then through fetal development. Table 17.1 outlines the major events during development.

Table 17.1 Human Development

Time	Events for Mother	Events for Baby
Pre-Embryonic Development		
First week	Ovulation occurs.	Fertilization occurs. Cell division begins and continues. Chorion appears.
Embryonic Development		
Second week	Symptoms of early pregnancy (nausea, breast swelling, and fatigue) are present. Blood pregnancy test is positive.	Implantation occurs. Amnion and yolk sac appear. Embryo has tissues. Placenta begins to form.
Third week	First menstruation is missed. Urine pregnancy test is positive. Symptoms of early pregnancy continue.	Nervous system begins to develop. Allantois and blood vessels are present. Placenta is well formed.
Fourth week		Limb buds form. Heart is noticeable and beating. Nervous system is prominent. Embryo has tail. Other systems form.
Fifth week	Uterus is the size of a hen's egg. Mother feels frequent need to urinate due to pressure of growing uterus on bladder.	Embryo is curved. Head is large. Limb buds show divisions. Nose, eyes, and ears are noticeable.
Sixth week	Uterus is the size of an orange.	Fingers and toes are present. Skeleton is cartilaginous.
Second month	Uterus can be felt above the pubic bone.	All systems are developing. Bone is replacing cartilage. Facial features are becoming refined. Embryo is about 38 mm (1.5 in.) long.
Fetal Development		
Third month	Uterus is the size of a grapefruit.	Gender can be distinguished by ultrasound. Fingernails appear.
Fourth month	Fetal movement is felt by a mother who has previously been pregnant.	Skeleton is visible. Hair begins to appear. Fetus is about 150 mm (6 in.) long and weighs about 170 grams (6 oz).
Fifth month	Fetal movement is felt by a mother who has not previously been pregnant. Uterus reaches up to level of umbilicus, and pregnancy is obvious.	Protective cheesy coating called vernix caseosa begins to be deposited. Heartbeat can be heard.
Sixth month	Doctor can tell where baby's head, back, and limbs are. Breasts have enlarged, nipples and areolae are darkly pigmented, and colostrum is produced.	Body is covered with fine hair called lanugo. Skin is wrinkled and reddish.
Seventh month	Uterus reaches halfway between umbilicus and rib cage.	Testes descend into scrotum. Eyes are open. Fetus is about 300 mm (12 in.) long and weighs about 1,350 grams (3 lb).
Eighth month	Weight gain is averaging about a pound a week. Standing and walking are difficult for the mother because her center of gravity is thrown forward.	Body hair begins to disappear. Subcutaneous fat begins to be deposited.
Ninth month	Uterus is up to rib cage, causing shortness of breath and heartburn. Sleeping becomes difficult.	Fetus is ready for birth. It is about 530 mm (20.5 in.) long and weighs about 3,400 grams (7.5 lb).

Processes of Development

As a human develops, these processes occur:

Cleavage. Immediately after fertilization, the zygote begins to divide so that there are first 2; then 4, 8, 16, and 32 cells; and so forth. Increase in size does not accompany these divisions (Fig. 17.2). Cell division during cleavage is mitotic, and each cell receives a full complement of chromosomes and genes.

Growth. During embryonic development, cell division is accompanied by an increase in size of the daughter cells.

Morphogenesis. Morphogenesis refers to the shaping of the embryo and is first evident when certain cells are seen to move, or migrate, in relation to other cells. By these movements, the embryo begins to assume various shapes.

Differentiation. When cells take on a specific structure and function, differentiation occurs. The first system to become visibly differentiated is the nervous system.

Stages of Development

Pre-embryonic development encompasses the events of the first week; **embryonic development** begins with the second week and lasts until the end of the second month.

Pre-Embryonic Development

The events of the first week of development are shown in Figure 17.2.

Immediately after fertilization, the zygote divides repeatedly as it passes down the uterine tube to the uterus. A **morula** is a compact ball of embryonic cells that becomes a **blastocyst.** The many cells of the blastocyst arrange themselves so that there is an **inner cell mass** surrounded by an outer layer of cells. The inner cell mass will become the **embryo,** and the layer of cells will become the chorion. The early appearance of the chorion emphasizes the complete dependence of the developing embryo on this extraembryonic membrane.

Video Blastocyst Formation

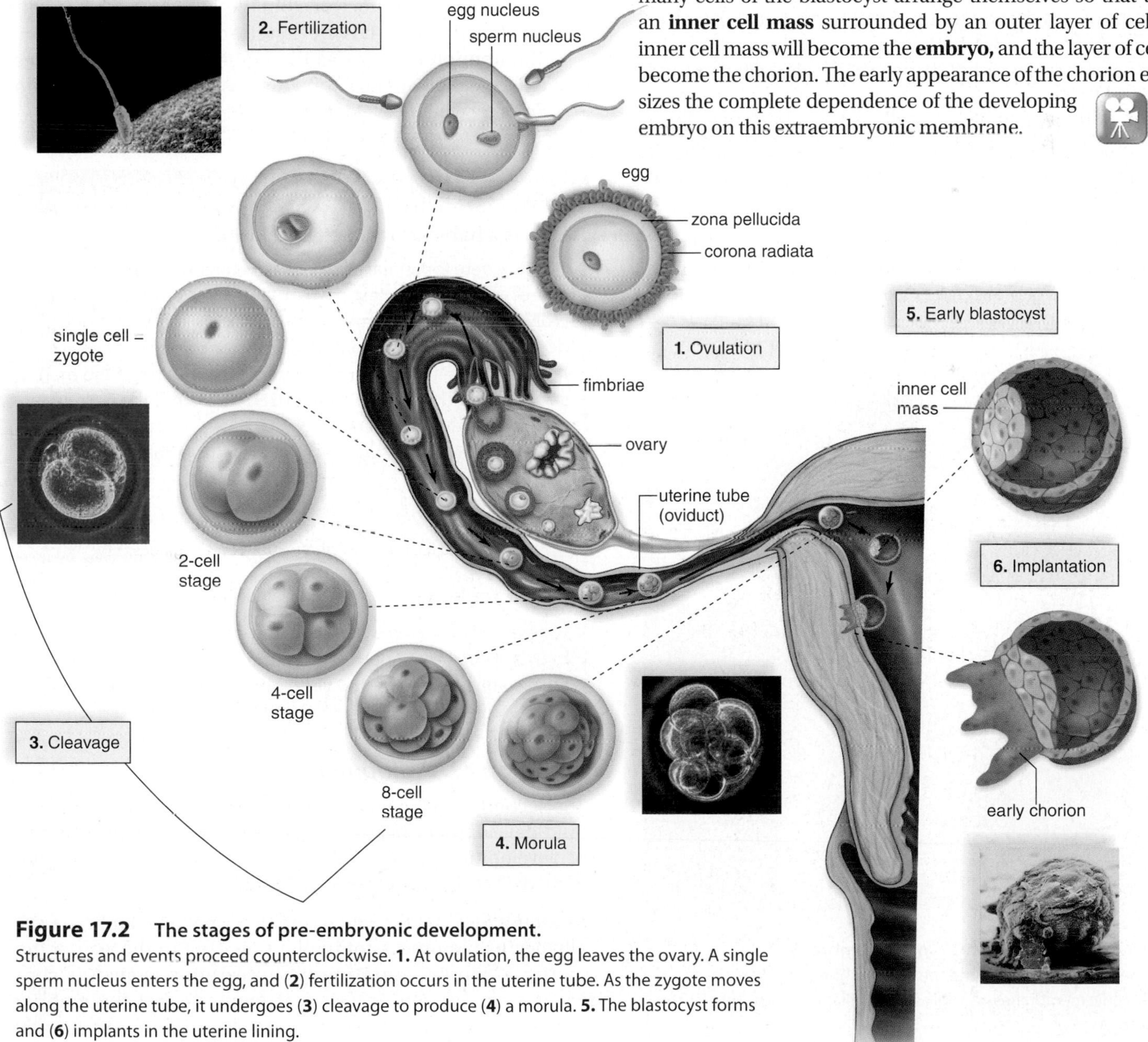

Figure 17.2 The stages of pre-embryonic development.
Structures and events proceed counterclockwise. **1.** At ovulation, the egg leaves the ovary. A single sperm nucleus enters the egg, and (**2**) fertilization occurs in the uterine tube. As the zygote moves along the uterine tube, it undergoes (**3**) cleavage to produce (**4**) a morula. **5.** The blastocyst forms and (**6**) implants in the uterine lining.

Each cell within the inner cell mass has the genetic capability of becoming any type of tissue. Sometimes, during development, the cells of the morula separate or the inner cell mass splits and two pre-embryos are present rather than one. If all goes well, these two pre-embryos will be identical twins because they have inherited exactly the same chromosomes. Fraternal twins arise when two different eggs are fertilized by two different sperm. They do not have identical chromosomes.

Extraembryonic Membranes

The **extraembryonic membranes** are not part of the embryo and fetus. Instead, as implied by their name, they are outside the embryo (Fig. 17.3). The names of the extraembryonic membranes in humans are strange to us because they are named for their function in animals, such as birds, that produce eggs with shells. In these animals, the chorion lies next to the shell and carries on gas exchange. The amnion contains the protective amniotic fluid, which bathes the developing embryo. The allantois collects nitrogenous wastes. The yolk sac surrounds the yolk, which provides nourishment.

The functions of the extraembryonic membranes are different in humans because humans develop inside the uterus. The extraembryonic membranes and their functions in humans follow.

1. The **chorion** develops into the fetal half of the **placenta,** the organ that provides the embryo/fetus with nourishment and oxygen and takes away its waste. The chorionic villi are fingerlike projections of the chorion that increase the absorptive area of the chorion. Blood vessels within the chorionic villi are continuous with the umbilical blood vessels.
2. The **allantois,** like the yolk sac, extends away from the embryo. It accumulates the small amount of urine produced by the fetal kidneys and later gives rise to the urinary bladder. For now, its blood vessels become the umbilical blood vessels, which take blood to and from the fetus. The umbilical arteries carry oxygen-poor blood from the fetus to the placenta, and the umbilical veins carry oxygen-rich blood from the placenta to the fetus.
3. The **yolk sac** is the first embryonic membrane to appear. In shelled animals such as birds, the yolk sac contains yolk, food for the developing embryo. In mammals such as humans, this function is taken over by the placenta and the yolk sac contains little yolk. But the yolk sac contains plentiful blood vessels. It is the first site of blood cell formation.
4. The **amnion** enlarges as the embryo and then the fetus enlarges. It contains fluid to cushion and protect the embryo, which develops into a fetus.

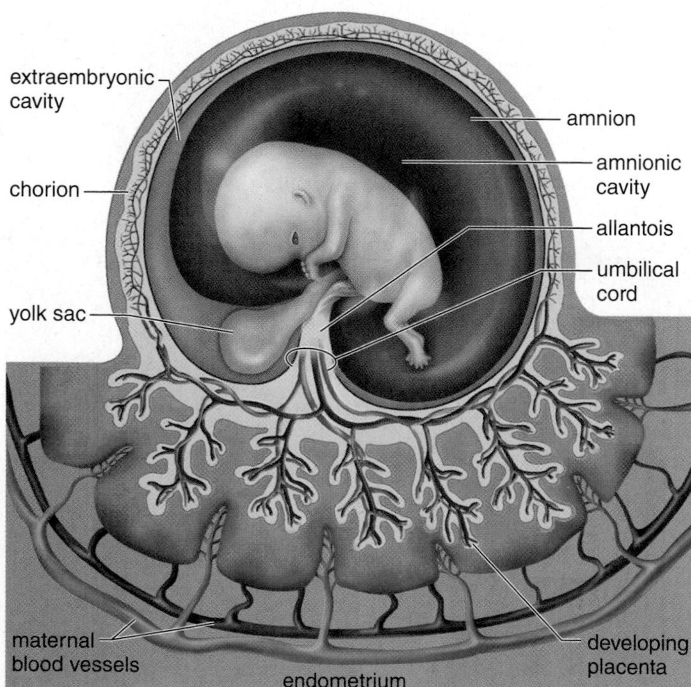

Figure 17.3 The extraembryonic membranes.
The chorion and amnion surround the embryo. The two other extraembryonic membranes, the yolk sac and allantois, contribute to the umbilical cord.

APPLICATIONS AND MISCONCEPTIONS

How is a baby's due date calculated?

The due date for arrival of the baby is calculated from the first day of the woman's last menstrual cycle before pregnancy. From conception to birth is approximately 266 days. Conception occurs approximately 14 days after the menstrual cycle begins (assuming ovulation occurs in the middle of the menstrual cycle). This gives a total of 280 days until the due date, or approximately 40 weeks.

To estimate the actual date, a calculation called Naegele's rule is often used:

1. Use the first day of the last period as a starting point.
2. Subtract three months from the month in which the period occurred.
3. Add seven days to the first day of the last period.

For example, if a woman's last period started on January 1, her baby's approximate due date is October 8. Ultrasound exams are also frequently used to verify the baby's due date.

Embryonic Development

The second week begins the process of implantation. Embryonic development lasts until the end of the second month of development. At the end of embryonic development, the embryo is easily recognized as human.

Second Week At the end of the first week, the embryo usually begins the process of implanting itself in the wall of the uterus. When **implantation** is successful, a woman is clinically pregnant. On occasion, it happens that the embryo implants itself in a location other than the uterus, usually one of the uterine tubes (oviducts). This is called an *ectopic pregnancy.* Because the uterine tubes are unable to expand to adjust for the growing embryo, this form of pregnancy is not successful and can pose health risks for the mother. During implantation, the chorion secretes enzymes to digest away some of the tissue and blood vessels of the endometrium of the uterus. The chorion also begins to secrete **human chorionic gonadotropin** (**HCG**), the hormone that is the basis for the pregnancy test. HCG acts like luteinizing hormone (LH) in that it serves to maintain the corpus luteum past the time it normally disintegrates. It is being stimulated, so the corpus luteum secretes progesterone, which maintains the endometrial wall. The endometrium is maintained, so the expected menstruation does not occur.

The embryo is now about the size of the period at the end of this sentence. As the week progresses, the inner cell mass becomes the *embryonic disk,* and two more extraembryonic membranes form (Fig. 17.4*a, b*). The yolk sac is the first site of blood cell formation. The amniotic cavity surrounds the embryo (and then the fetus) as it develops. In humans, amniotic fluid acts as an insulator against cold and heat. It also absorbs shock, such as that caused by the mother exercising.

The major event, called **gastrulation,** turns the inner cell mass into the embryonic disk. Gastrulation is an example of morphogenesis, during which cells move or migrate. In this case, cells migrate to become tissue layers called the **primary germ layers.** By the time gastrulation is complete, the embryonic disk has become an embryo with three primary germ layers: ectoderm, mesoderm, and endoderm. Figure 17.5 shows the significance of the primary germ layers. All the organs of an individual can be traced back to one of the primary germ layers.

Third Week Two important organ systems make their appearance during the third week. The nervous system is the first organ system to be visually evident. At first, a thickening appears along the entire posterior length of the embryo. Then, the center begins to fold inward, forming a pocket. The edges are called neural folds. When the neural folds meet at the

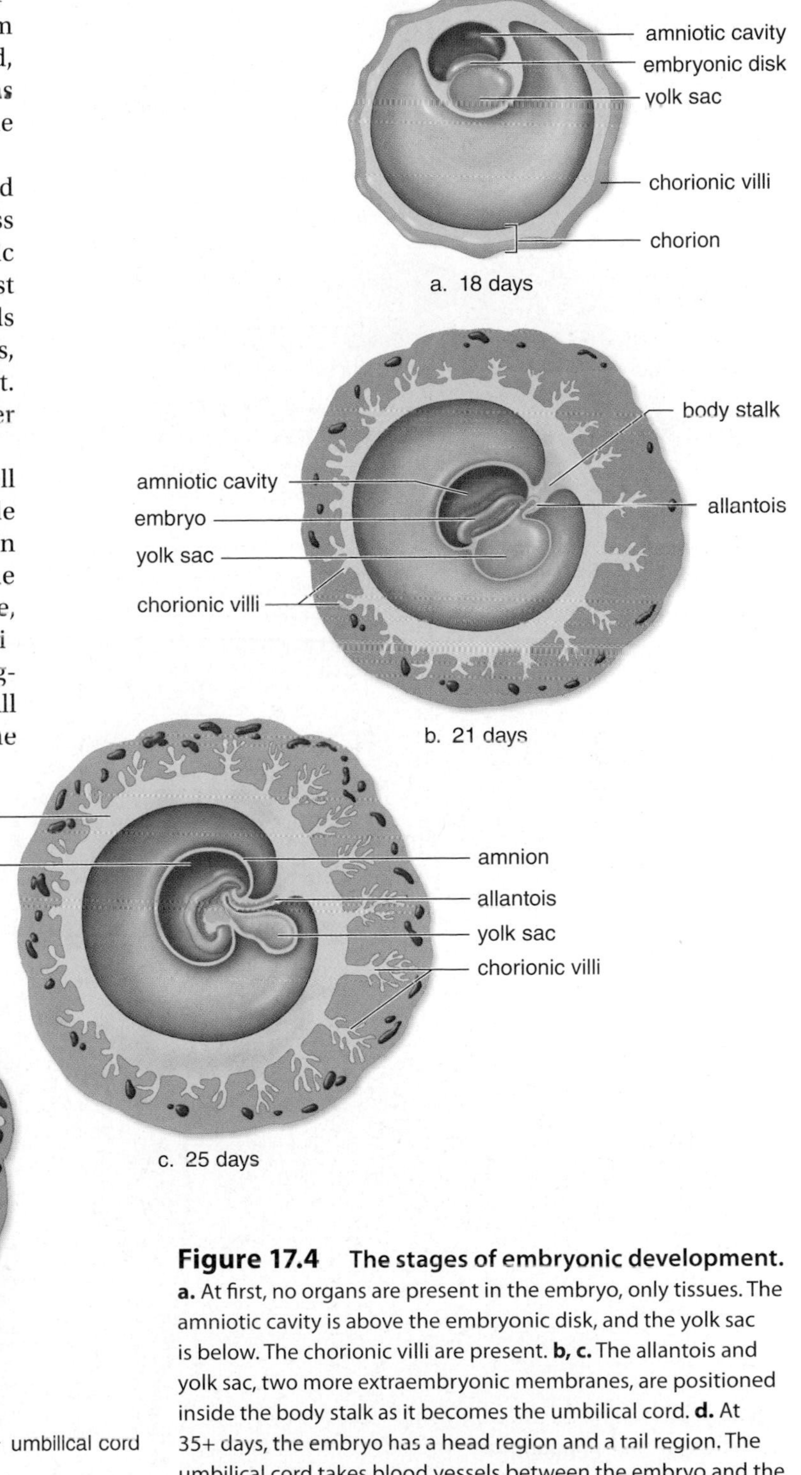

Figure 17.4 The stages of embryonic development. **a.** At first, no organs are present in the embryo, only tissues. The amniotic cavity is above the embryonic disk, and the yolk sac is below. The chorionic villi are present. **b, c.** The allantois and yolk sac, two more extraembryonic membranes, are positioned inside the body stalk as it becomes the umbilical cord. **d.** At 35+ days, the embryo has a head region and a tail region. The umbilical cord takes blood vessels between the embryo and the chorion (placenta).

midline, the pocket becomes a tube called the neural tube. The neural tube later develops into the brain and the spinal cord.

Development of the heart begins in the third week and continues into the fourth week. At first, cells from both sides of the body migrate to form the heart. When these fuse to form a continuous tube, the heart begins pumping blood, even though the chambers of the heart are not fully formed. The veins enter posteriorly, and the arteries exit anteriorly from this largely tubular heart. Later, the heart twists so that all major blood vessels are located anteriorly.

Fourth and Fifth Weeks At four weeks, the embryo is barely larger than the height of this print. A body stalk (future umbilical cord) connects the embryo to the chorion, which has tree-like projections called **chorionic villi** (see Fig. 17.4*c, d*). The fourth extraembryonic membrane, the allantois, lies within the body stalk. Its blood vessels become the umbilical blood vessels. The head and the tail then lift up, and the body stalk moves anteriorly by constriction. Once this process is complete, the **umbilical cord** is fully formed (see Fig. 17.4*d*). The umbilical cord connects the developing embryo to the placenta.

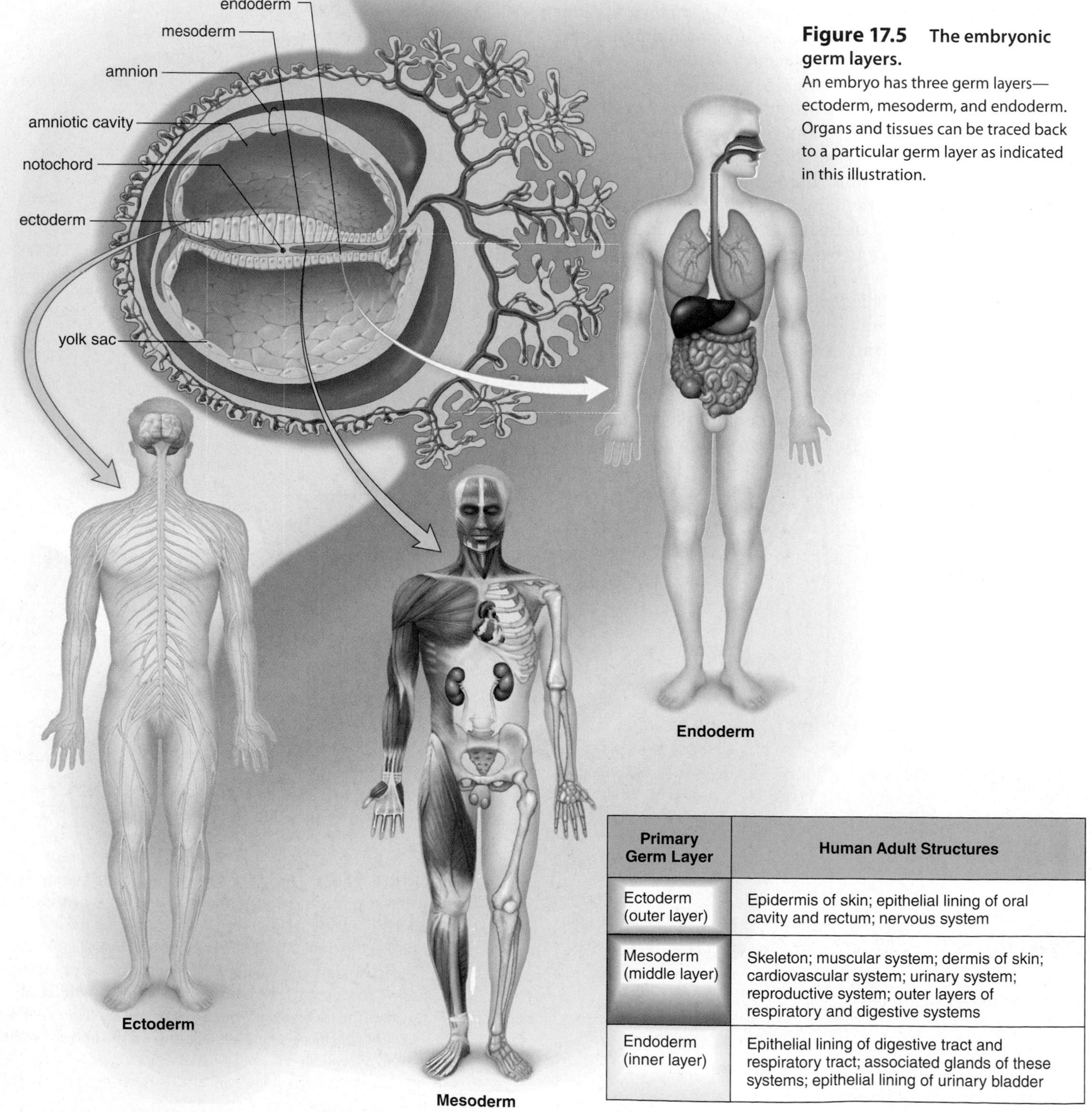

Figure 17.5 **The embryonic germ layers.**
An embryo has three germ layers—ectoderm, mesoderm, and endoderm. Organs and tissues can be traced back to a particular germ layer as indicated in this illustration.

Primary Germ Layer	Human Adult Structures
Ectoderm (outer layer)	Epidermis of skin; epithelial lining of oral cavity and rectum; nervous system
Mesoderm (middle layer)	Skeleton; muscular system; dermis of skin; cardiovascular system; urinary system; reproductive system; outer layers of respiratory and digestive systems
Endoderm (inner layer)	Epithelial lining of digestive tract and respiratory tract; associated glands of these systems; epithelial lining of urinary bladder

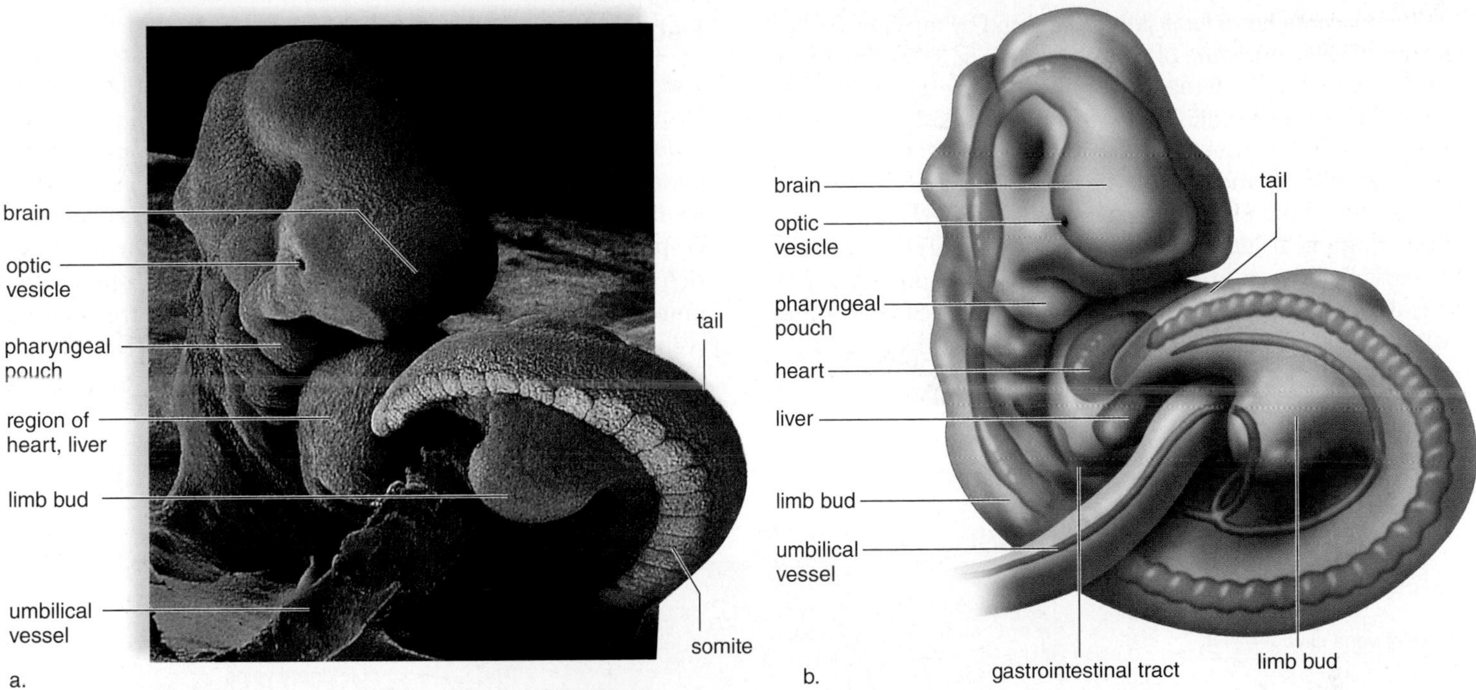

Figure 17.6 The human embryo after five weeks of development.
a. Scanning electron micrograph. **b.** The embryo is curled so that the head touches the heart and liver, the two organs whose development is farther along than the rest of the body. The organs of the gastrointestinal tract are forming, and the arms and the legs develop from the bulges called limb buds. The tail is an evolutionary remnant; its bones regress and become those of the coccyx (tailbone). The pharyngeal pouches become functioning gills in fishes and amphibian larvae; in humans, the first pair of pharyngeal pouches becomes the auditory tubes. The second pair becomes the tonsils, and the third and fourth become the thymus gland and the parathyroid glands.

Soon, limb buds appear (Fig. 17.6). Later, the arms and legs develop from the limb buds, and even the hands and feet become apparent. At the same time, during the fifth week, the head enlarges and the sense organs become more prominent. It is possible to make out the developing eyes and ears and even the nose.

Sixth Through Eighth Weeks During the sixth through eighth weeks of development, the embryo changes to a form easily recognized as a human being. Concurrent with brain development, the head achieves its normal relationship with the body as a neck region develops. The nervous system is developed well enough to permit reflex actions, such as a startle response to touch. At the end of this period, the embryo is about 38 mm (1.5 in.) long and weighs no more than an aspirin tablet, even though all organ systems have been established.

CHECK YOUR PROGRESS 17.2

1. Distinguish between the processes of morphogenesis and differentiation.
2. Describe the function of the extraembryonic membranes.
3. Summarize what occurs at each stage of pre-embryonic development and embryonic development.

CONNECTING THE CONCEPTS

For more information on the topics presented in this section, refer to the following discussions:

Section 4.7 summarizes the function of each of the organ systems in the body.

Section 16.4 examines the roles of HCG and progesterone in pregnancy.

17.3 Fetal Development

LEARNING OUTCOMES

Upon completion of this section you should be able to

1. State the roles of progesterone and estrogen in fetal development.
2. Describe the flow of blood in a fetus and explain the role of the placenta.
3. Summarize the major events in the development of the fetus from three to nine months.

The **placenta** is the source of progesterone and estrogen during pregnancy. These hormones have two functions: (1) Because of negative feedback on the hypothalamus and anterior pituitary, they prevent any new follicles from maturing; and (2) they maintain the endometrium. Menstruation does not usually occur during pregnancy.

The placenta has a fetal side contributed by the chorion and a maternal side consisting of uterine tissue (Fig. 17.7). The blood of the mother and the fetus never mix because exchange always takes place across the villi. Exchange occurs via diffusion. Carbon dioxide and other wastes move from the fetal side to the maternal side. Nutrients and oxygen move from the maternal side to the fetal side of the placenta. As discussed in the Health feature, "Preventing and Testing for Birth Defects," harmful chemicals can also cross the placenta, and this is of particular concern during the embryonic period, when various structures are first forming. Each organ or part seems to have a sensitive period during which a substance can alter its normal function.

Fetal Circulation

The umbilical cord stretches between the placenta and the fetus. It is the lifeline of the fetus because it contains the umbilical arteries and vein (Fig. 17.7). Blood within the fetal aorta travels to its various branches, including the iliac arteries. The iliac arteries connect to the *umbilical arteries* carrying O_2-poor blood to the placenta. The *umbilical vein* carries blood rich in nutrients and O_2 from the placenta to the fetus. The umbilical vein enters the liver and then joins the *ductus venosus* (venous duct). This merges with the inferior vena cava, a vessel that returns blood to the right atrium. This mixed blood enters the heart, and most of it is shunted to the left atrium through

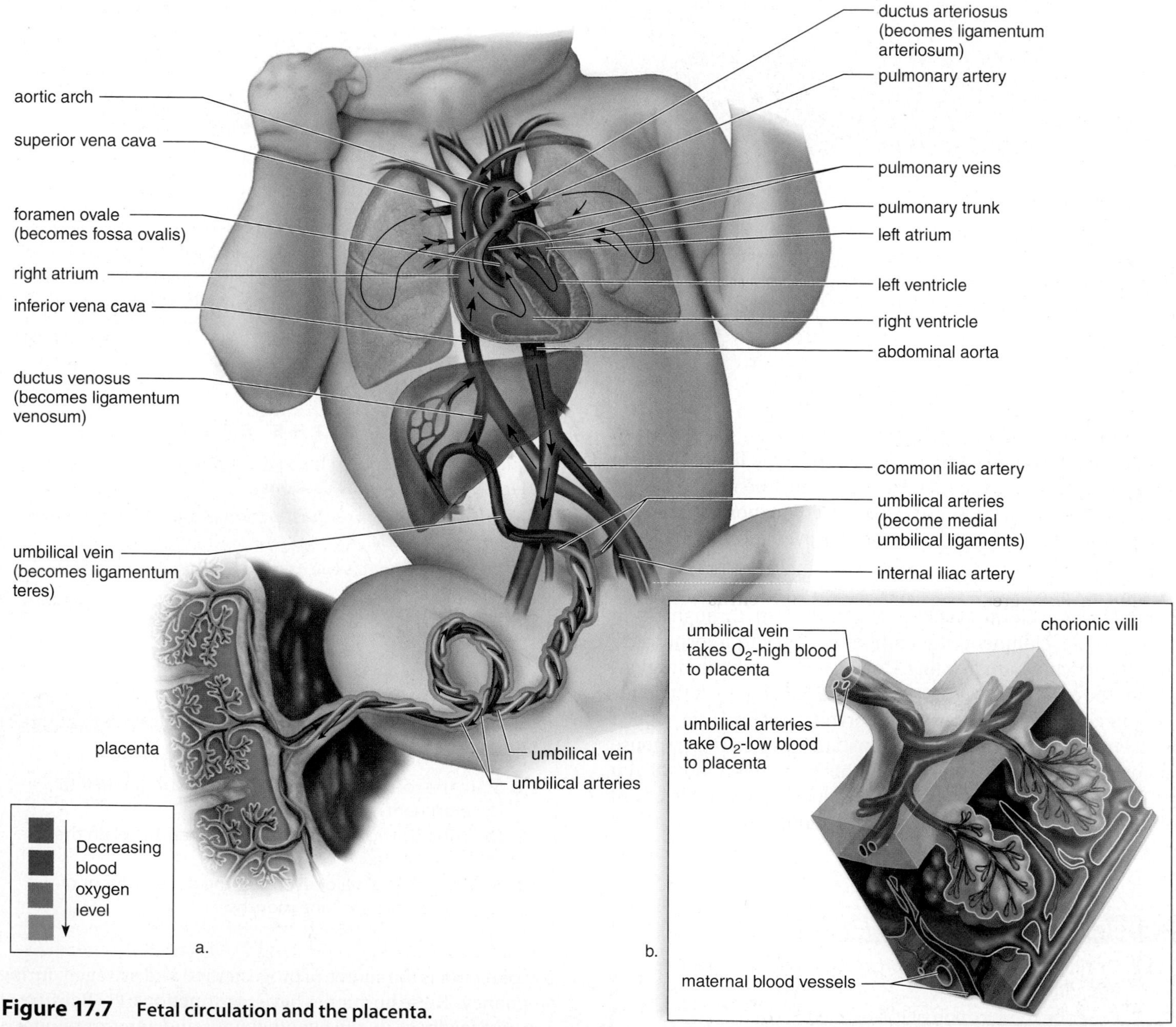

Figure 17.7 Fetal circulation and the placenta.
a. Trace the path of blood by following the arrows. **b.** At the placenta, an exchange of molecules between fetal and maternal blood takes place across the walls of the chorionic villi.

BIOLOGY MATTERS **Health**

Preventing and Testing for Birth Defects

Birth defects, or congenital disorders, are abnormal conditions that are present at birth. According to the Centers for Disease Control and Prevention (CDC), about 1 in 33 babies born in the United States has a birth defect. Genetic birth defects can sometimes be detected before birth by a variety of methods (Fig. 17A). However, these methods are not without risk.

Some birth defects are not serious, and not all can be prevented. But women can take steps to increase their chances of delivering a healthy baby.

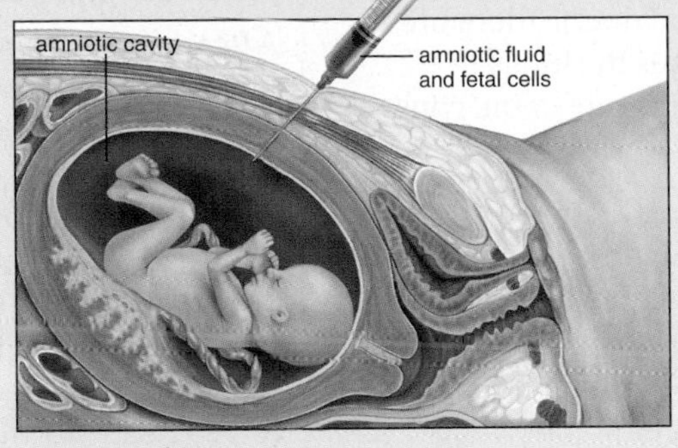

a. Amniocentesis

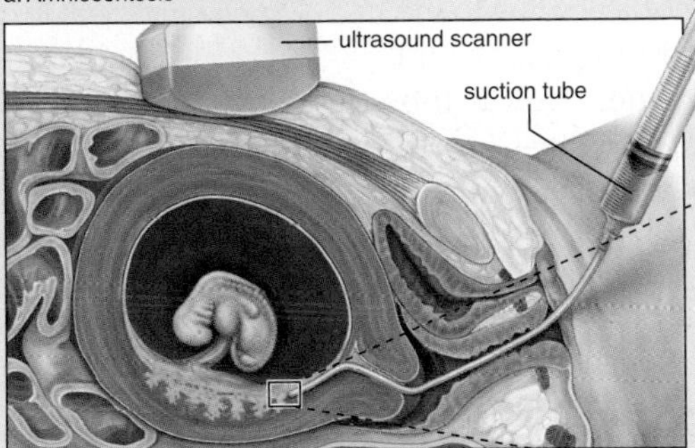

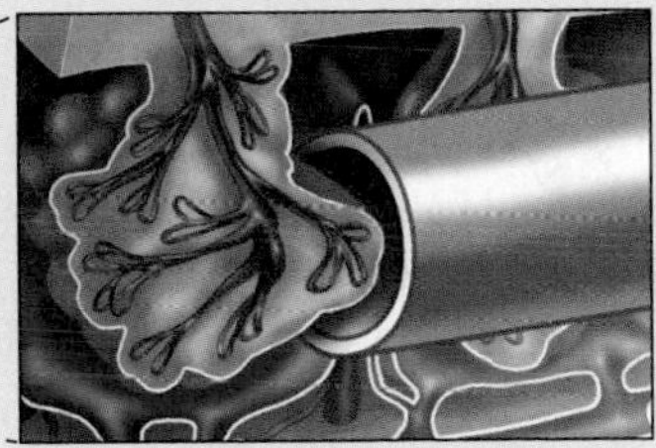

b. Chorionic villus sampling

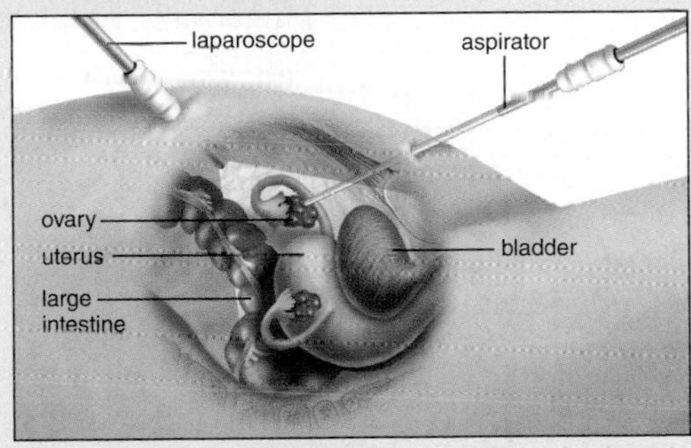

c. Preimplantation genetic diagnosis

Figure 17A Methods for genetic defect testing before birth. About 20% of all birth defects are due to genetic or chromosomal abnormalities, which may be detected before birth. **a.** Amniocentesis is usually performed from the fifteenth to the seventeenth week of pregnancy. **b.** Chorionic villus sampling is usually performed from the eighth to the twelfth week of pregnancy. **c.** Preimplantation genetic diagnosis can be performed prior to in vitro fertilization, either on oocytes that have been collected from the woman, or on the early embryo.

Eat a Healthy Diet

Certain birth defects may occur because of nutritional deficiencies. For example, women of childbearing age are urged to make sure they consume adequate amounts folic acid (a B vitamin) in order to prevent neural tube defects such as spina bifida and anencephaly. In spina bifida, part of the vertebral column fails to develop properly and cannot adequately protect the spinal cord. With anencephaly, most of the fetal brain fails to develop. Anencephalic infants are stillborn or survive for only a few days after birth.

Fortunately, folic acid is plentiful in leafy green vegetables, nuts, and citrus fruits. The CDC recommends that all women of childbearing age get at least 400 micrograms (μg) of folic acid every day through supplements, in addition to eating a healthy, folic acid-rich diet. Unfortunately, neural tube birth defects can occur just a few weeks after conception, when many women are still unaware that they are pregnant—especially if the pregnancy is unplanned.

Avoid Alcohol, Smoking, and Drug Abuse

Alcohol consumption during pregnancy is a leading cause of birth defects. In severe instances, the baby is born with fetal alcohol syndrome (FAS), estimated to occur in 0.2 to 1.5 of every 1,000 live births in the United States. Children with FAS are frequently underweight, have an abnormally small head, abnormal facial development, and intellectual disabilities. As they mature, children with FAS often exhibit short attention span, impulsiveness, and poor judgment, as well as problems with learning and memory. Heavy alcohol use also reduces a woman's folic acid level, increasing the risk of the neural tube defects.

Cigarette smoking causes many birth defects. Babies born to smoking mothers typically have low birth weight, and are more likely to have defects of the face, heart, and brain than children of nonsmokers.

Illegal drugs should also be avoided. For example, cocaine causes blood pressure fluctuations that deprive the fetus of oxygen. Cocaine-exposed babies may have problems with vision and coordination and may be intellectually disabled.

Video Inherited Pollution

Alert Medical Personnel If You Are or May Be Pregnant

Several medications that are safe for healthy adults may pose a risk to a developing fetus. Pregnant women who require immunizations should consult with their physicians.

Because the rapidly dividing cells of a developing embryo or fetus are very susceptible to damage from radiation, pregnant women should avoid unnecessary X-rays. If unavoidable, the woman should notify the X-ray technician so that her fetus can be protected as much as possible.

Avoid Infections That Cause Birth Defects

Certain pathogens, such as rubella, toxoplasmosis, herpes simplex, and cytomegalovirus, may cause birth defects.

Question to Consider

1. Besides tobacco, alcohol, and illegal drugs, what other potential risks should a pregnant woman avoid? Why?

the *foramen ovale* (oval opening). The left ventricle pumps this blood into the aorta. Oxygen-poor blood that enters the right atrium is pumped into the pulmonary trunk. It then joins the aorta by way of the *ductus arteriosus* (arterial duct). Therefore, most blood entering the right atrium bypasses the lungs.

Various circulatory changes occur at birth due to the tying of the cord and the expansion of the lungs:

1. Inflation of the lungs. This reduces the resistance to blood flow through the lungs. This allows an increased amount of blood flow from the right atrium to the right ventricle and into the pulmonary arteries. Now gas exchange occurs in the lungs, not at the placenta. Oxygen-rich blood returns to the left side of the heart through the pulmonary veins.
2. An increase in blood flow from the pulmonary veins to the left atrium. This increases the pressure in the left atrium causing a flap to cover the foramen ovale. Even if this mechanism fails, passage of blood from the right atrium to the left atrium rarely occurs because either the opening is small or it closes when the atria contract.
3. The ductus arteriosus closes at birth because endothelial cells divide and block off the duct.
4. Remains of the ductus arteriosus and parts of the umbilical arteries and vein later are transformed into connective tissue.

Events of Fetal Development

Fetal development includes the third through the ninth months of development. At this time, the fetus is recognizably human (Fig. 17.8), but many refinements still need to be added. The fetus usually increases in size and gains the weight that will be needed to allow it to live as an independent individual.

Third and Fourth Months

At the beginning of the third month, the fetal head is still very large relative to the rest of the body. The nose is flat, the eyes are far apart, and the ears are well formed. Head growth now begins to slow down as the rest of the body increases in length. Fingernails, nipples, eyelashes, eyebrows, and hair on the head appear.

Cartilage begins to be replaced by bone as ossification centers appear in most of the bones. Cartilage remains at the ends of the long bones, and ossification is not complete until age 18 or 20 years. The skull has six large membranous areas called **fontanels.** These permit a certain amount of flexibility as the head passes through the birth canal, and they allow rapid growth of the brain during infancy. Progressive fusion of the skull bones causes the fontanels to close, usually by 2 years of age.

Sometime during the third month, it is possible to distinguish males from females. As discussed on page 399, the presence of an *SRY* gene, usually on the Y chromosome, leads to the development of testes and male genitals. Otherwise, ovaries and female genitals develop. At this time, either testes or ovaries are located within the abdominal cavity. Later, in the last trimester of fetal development, the testes descend into the scrotal sacs (scrotum). Sometimes, the testes fail to descend. In that case, an operation may be done later to place them in their proper location.

During the fourth month, the fetal heartbeat is loud enough to be heard when a physician applies a stethoscope to the mother's abdomen. By the end of this month, the fetus is about 152 mm (6 in.) in length and weighs about 171 g (6 oz).

Fifth Through Seventh Months

During the fifth through seventh months (Fig. 17.8), the mother begins to feel movement. At first, there is only a fluttering sensation; but as the fetal legs grow and develop, kicks and jabs are felt. The fetus, though, is in the fetal position, with the head bent down and in contact with the flexed knees.

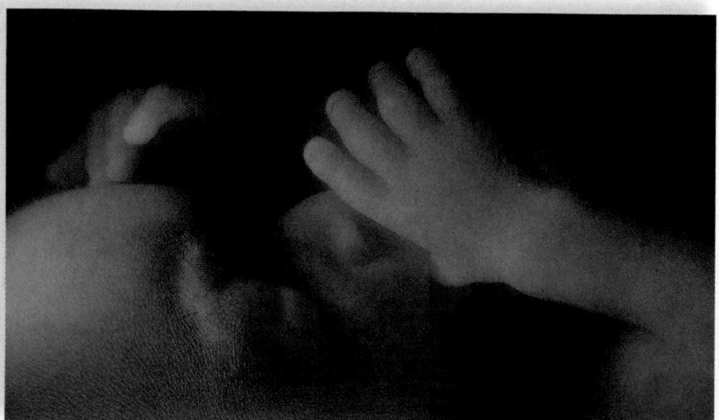

Figure 17.8 **A five- to seven-month-old fetus.** Wrinkled skin is covered by fine hair.

APPLICATIONS AND MISCONCEPTIONS

When can the heartbeat of a fetus first be detected?

A fetal heart will begin forming from tissues in the chest of the fetus first by forming a tube structure that will later develop into the chambers of the heart. By the fifth week of fetal development, the heart is formed; it is too small to hear but it can be seen on an ultrasound. By week 10, the heart is fully developed and has a heartbeat rate of 150 to 195 beats per minute. The sounds of the heart at this point are often referred to as fetal heart tones (FHTs) and are very rapid due to the heart's small size; but eventually they will settle into a rate of 120 to 160 beats per minute (usually after the twelfth week) and may even beat in-synch with the mother's rate. The fetal heart can be heard through the end of the first trimester and into the beginning of the second using an amplification machine called a Doppler instrument, which bounces harmless sound waves off the heart. After that point, depending on the position of the fetus, a stethoscope can detect the heartbeat as the fetus and its heart get larger.

APPLICATIONS AND MISCONCEPTIONS

Can a baby hear in the womb?

A baby's ears begin to develop around eight weeks but are not fully formed. At 18 weeks, the bones of the skull and middle ear, as well as their nerve connections to the brain, have developed so that a baby can hear. At this time, sounds such as the mother's heartbeat are heard. At 25 weeks, the baby listens to voices. The mother's voice, in particular, is recognized. The baby's heart rate will slow slightly—a sign that her voice comforts her baby.

For the rest of pregnancy, the baby's sense of hearing transmits information from the outside world. Soft music lulls him to sleep. Loud sounds startle the baby and wake him. However, all sounds reaching the baby in the womb are muffled. The baby's ears are filled with amniotic fluid, and the outer ear is covered with the waxy coating that protects the baby's skin.

The wrinkled, translucent skin is covered by a fine down called **lanugo.** This, in turn, is coated with a white, greasy, cheeselike substance called **vernix caseosa,** which probably protects the delicate skin from the amniotic fluid. The eyelids are now fully open.

At the end of this period, the fetus's length has increased to about 300 mm (12 in.), and it weighs about 1,380 g (3 lb). It is possible that, if born now, the baby will survive.

Eighth Through Ninth Months

At the end of nine months, the fetus is about 530 mm (20.5 in.) long and weighs about 3,400 g (7.5 lb). Weight gain is due largely to an accumulation of fat beneath the skin. Full-term babies have the best chance of survival. Premature babies are subject to various challenges, such as respiratory distress syndrome, because their lungs are underdeveloped (see Chapter 9), jaundice (see Chapter 8), and infections.

As the end of development approaches, the fetus usually rotates so that the head is pointed toward the cervix. However, if the fetus does not turn, a breech birth (rump first) is likely. It is very difficult for the cervix to expand enough to accommodate this form of birth, and asphyxiation of the baby is more likely to occur. Thus, a *cesarean section* (incision through the abdominal and uterine walls) may be prescribed for delivery of the fetus.

Development of Male and Female Sex Organs

The sex of an individual is determined at the moment of fertilization. Males have a pair of chromosomes designated as X and Y, and females have two X chromosomes. On the Y chromosome, a gene called *SRY* (*s*ex-determining *r*egion of the *Y*) determines whether the gonadal tissue in the embryo will develop into the male or female sex organs. The protein encoded by the *SRY* gene acts as a regulatory mechanism that controls the expression, or function, of other developmental genes in the body (see section 21.2).

Normal Development of the Sex Organs

Development of the internal and external sex organs is dependent upon the presence or absence of the *SRY* gene.

Internal Sex Organs During the first several weeks of development, it is impossible to tell by external inspection whether the unborn child is a boy or a girl. Gonads don't start developing until the seventh week of development. The tissue that gives rise to the gonads is called *indifferent* because it can become testes or ovaries, depending on the action of hormones.

In Figure 17.9, notice that ① at six weeks, both males and females have the same types of ducts. During this indifferent stage, an embryo has the potential to develop into a male or a female. If the *SRY* gene is present, a protein called

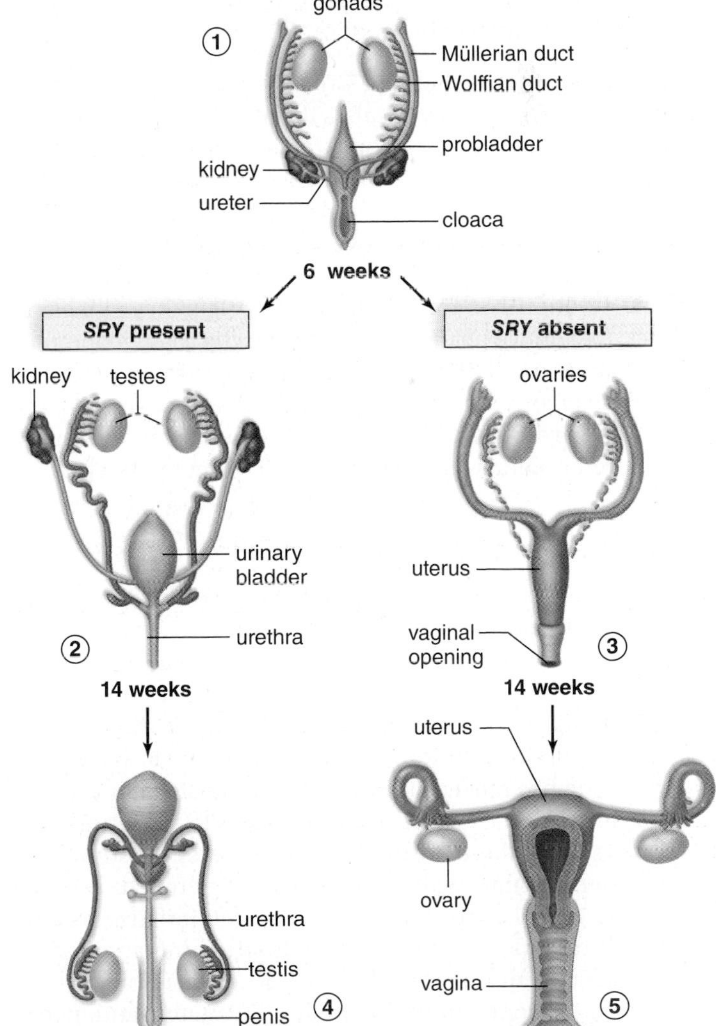

Figure 17.9 Development of the internal sex organs. The formation of the internal male and female sex organs is largely determined by the presence or absence of the *SRY* gene on the Y chromosome.

testis-determining factor is produced that regulates the initial development of the testes. The **testosterone** produced by the testes stimulates the Wolffian ducts to become male genital ducts. ② The Wolffian ducts enter the urethra, which belongs to both the urinary and reproductive systems in males. An anti-Müllerian hormone causes the Müllerian ducts to regress. In the absence of an *SRY* gene, ovaries develop instead of testes from the same indifferent tissue. ③ Now the Wolffian ducts regress, and because of an absence of testosterone, the Müllerian ducts develop into the uterus and uterine tubes. Estrogen has no effect on the Wolffian duct, which degenerates in females. A developing vagina also extends from the uterus. There is no connection between the urinary and genital systems in females.

At 14 weeks, both the primitive testes and ovaries are located deep inside the abdominal cavity. An inspection of the interior of the ovaries would indicate that they already contain large numbers of tiny follicles, each having an ovum. ④ Toward the end of development, the testes descend into the scrotal sac; ⑤ the ovaries remain in the abdominal cavity.

External Sex Organs Figure 17.10 shows the development of the external sex organs (genitals). These tissues are also indifferent at first—they can develop into either male or female sex organs. ① At six weeks, a small bud appears between the legs; this can develop into the male penis or the female clitoris. ② At nine weeks, a urogenital groove bordered by two swellings appears. ③ By 14 weeks, this groove has disappeared in males, and the scrotum has formed from the original swellings. In females, the groove persists and becomes the vaginal opening. Labia majora and labia minora are present instead of a scrotum. These changes are due to the presence or absence of the hormone dihydrotestosterone (DHT), which is manufactured in the adrenal glands and prostate glands from testosterone.

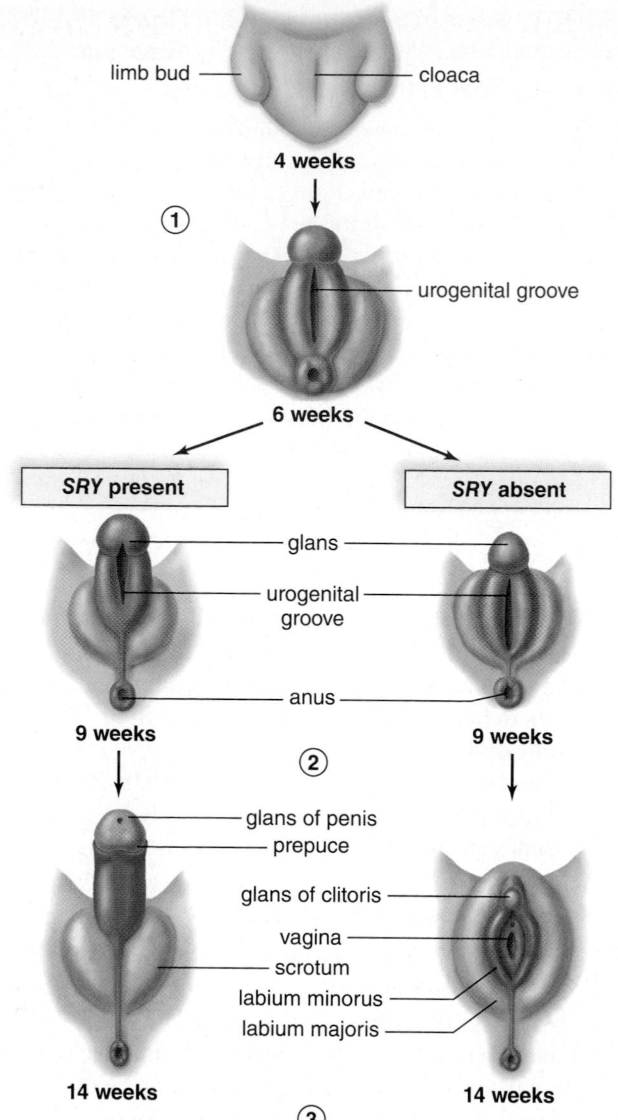

Figure 17.10 Development of the external sex organs. The presence or absence of the *SRY* gene on the Y chromosome plays an important role in the development of the external sex organs (or genitals).

Abnormal Development of the Sex Organs

It's not correct to say that all XY individuals develop into males. Some XY individuals become females (XY female syndrome). Similarly, some XX individuals develop into males (XX male syndrome). In individuals with XY female syndrome, a piece of the Y chromosome containing the *SRY* gene is missing (this is called a deletion). In individuals with XX male syndrome, the *SRY* gene has moved (this is called a translocation) to the X chromosome. The *SRY* gene causes testes to form, and then the testes secrete these hormones: (1) Testosterone stimulates development of the epididymides, vasa deferentia, seminal vesicles, and ejaculatory duct. (2) Anti-Müllerian hormone prevents further development of female structures and instead causes them to degenerate. (3) Dihydrotestosterone (DHT) directs the development of the urethra, prostate gland, penis, and scrotum.

Ambiguous Sex Determination The absence of any one or more of these hormones results in ambiguous sex determination. The individual has the external appearance of a female, although the gonads of a female are absent.

APPLICATIONS AND MISCONCEPTIONS

Why is the female gender sometimes referred to as the "default sex"?

The term "default sex" has to do with the presence or absence of the Y chromosome in the fetus. On the Y chromosome, there is a gene called *SRY* (sex determining region of the Y) that produces a protein that will cause Sertoli cells in the testes to produce Müllerian-inhibiting substance (MIS). This causes Leydig cells in the testes to produce testosterone, which signals the development of the male sex organs (vasa deferentia, epididymides, penis, etc.). Without the *SRY* gene and this hormone cascade effect, female structures (uterus, uterine tubes, ovaries, etc.) will begin to form.

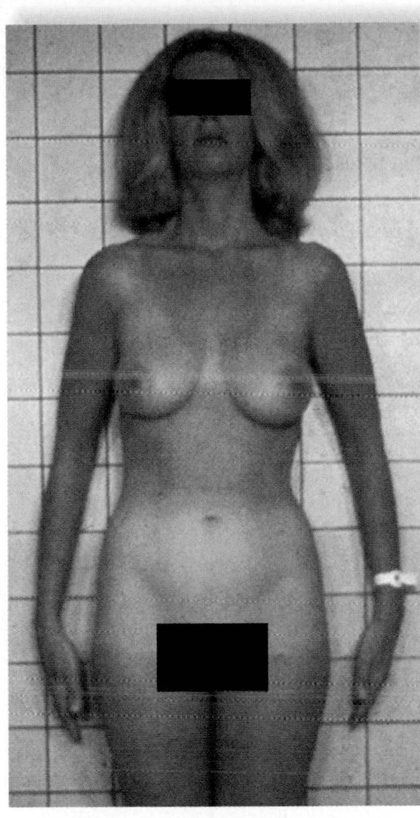

Figure 17.11 Androgen insensitivity affects sexual development. This individual has a female appearance but the XY chromosomes of a male. She developed as a female because her receptors for testosterone are ineffective. Underdeveloped testes are in the abdominal cavity, instead of a uterus and ovaries.

In *androgen insensitivity syndrome,* these three types of hormones are produced by testes during development, but the individual develops as a female because the receptors for testosterone are ineffective (Fig. 17.11). The external genitalia develop as female, and the Wolffian duct degenerates internally. The individual does not develop a scrotum, so the testes fail to descend and instead remain deep within the body. The individual develops the secondary sex characteristics of a female, and no abnormality is suspected until the individual fails to menstruate.

CHECK YOUR PROGRESS 17.3

1. Describe the path of blood flow in the fetus starting with the placenta, and name the structures unique to fetal circulation.
2. Summarize major events by month during fetal development.
3. Explain how the development of the genitals differs in males and females.

CONNECTING THE CONCEPTS

For more information on the topics presented in this section, refer to the following discussions:

Figures 16.2 and 16.3 illustrate the structures of the male reproductive system.

Figures 16.6 and 16.7 illustrate the structures of the female reproductive system.

17.4 Pregnancy and Birth

LEARNING OUTCOMES

Upon completion of this section, you should be able to

1. Explain the influence of progesterone and estrogen on female physiology during pregnancy.
2. Summarize the events that occur during each stage of birth.

Pregnancy

Major changes that take place in the mother's body during pregnancy (see Table 17.1) are due largely to the effects of the hormones progesterone and estrogen (Table 17.2).

Digestive System and Nutrition

When first pregnant, the mother may experience nausea and vomiting, loss of appetite, heartburn, constipation, and fatigue. These symptoms subside, and some mothers report increased energy levels and a general sense of well-being despite an increase in weight. During pregnancy, the mother gains weight due to breast and uterine enlargement; weight of the fetus; amount of amniotic fluid; size of the placenta; her own increase in total body fluid; and an increase in storage of proteins, fats, and minerals. The increased weight can lead to lordosis (swayback) and lower back pain.

A pregnant woman's metabolic rate may rise by up to 10–15% during pregnancy. While this may encourage overeating, the daily diet of a pregnant woman only needs to increase by about 300 kcal to meet the energy needs of the developing fetus. Of far greater importance is the quality of foods in the diet, because there is a greater demand on the woman's body for nutrients such as iron, calcium, and proteins.

The Circulatory System

Aside from an increase in weight, many of the physiological changes in the mother are due to the presence of the placental hormones that support fetal development. Progesterone decreases uterine motility by relaxing smooth muscle, including the smooth muscle in the walls of arteries. The arteries expand, and this leads to a low blood pressure that sets in motion the

Table 17.2 Effects of Placental Hormones on Mother

Hormone	Chief Effects
Progesterone	Relaxation of smooth muscle; reduced uterine motility; reduced maternal immune response to fetus
Estrogen	Increased uterine blood flow; increased renin–angiotensin–aldosterone activity; increased protein biosynthesis by the liver
Peptide hormones	Increased insulin resistance

renin–angiotensin–aldosterone mechanism, promoted by estrogen. Aldosterone activity promotes sodium and water retention, and blood volume increases until it reaches its peak sometime during weeks 28 to 32 of pregnancy. Altogether, blood volume increases from 5 l to 7 l—a 40% rise. An increase in the number of red blood cells follows. With the rise in blood volume, cardiac output increases by 20–30%. Blood flow to the kidneys, placenta, skin, and breasts rises significantly. Smooth muscle relaxation also explains the common gastrointestinal effects of pregnancy. The heartburn experienced by many is due to relaxation of the esophageal sphincter and reflux of stomach contents into the esophagus. Constipation is caused by a decrease in intestinal tract motility.

The Respiratory System

Of interest is the increase in pulmonary values in a pregnant woman. The bronchial tubes relax, but this alone cannot explain the typical 40% increase in vital capacity and tidal volume. The increasing size of the uterus from a nonpregnant weight of 60–80 g to 900–1,200 g contributes to an improvement in respiratory functions. The uterus comes to occupy most of the abdominal cavity, reaching nearly to the xiphoid process of the sternum. This increase in size not only pushes the intestines, liver, stomach, and diaphragm superiorly but it also widens the thoracic cavity. Compared with nonpregnant values, the maternal oxygen level changes little. Blood carbon dioxide levels fall by 20%, creating a concentration gradient favorable to the flow of carbon dioxide from fetal blood to maternal blood at the placenta.

Still Other Effects

The enlargement of the uterus may create problems. In the pelvic cavity, compression of the ureters and urinary bladder (Fig. 17.12) can result in stress **incontinence,** or the involuntary leakage of urine from the urinary tract. Compression of the inferior vena cava, especially when lying down, decreases venous return, and the result is edema and varicose veins.

Aside from the steroid hormones progesterone and estrogen, the placenta also produces some peptide hormones. One of these makes cells resistant to insulin, and the result can be gestational diabetes. Some of the integumentary changes observed during pregnancy are also due to placental hormones. "Stretch marks" typically form over the abdomen and lower breasts in response to increased steroid hormone levels rather than stretching of the skin. Melanocyte activity also increases during pregnancy. As a result, many women develop a dark line, the linea nigra, that extends from the pubic region to the umbilical region (belly button). In addition, darkening of certain areas of the skin, including the face, neck, and breast areolae, is common.

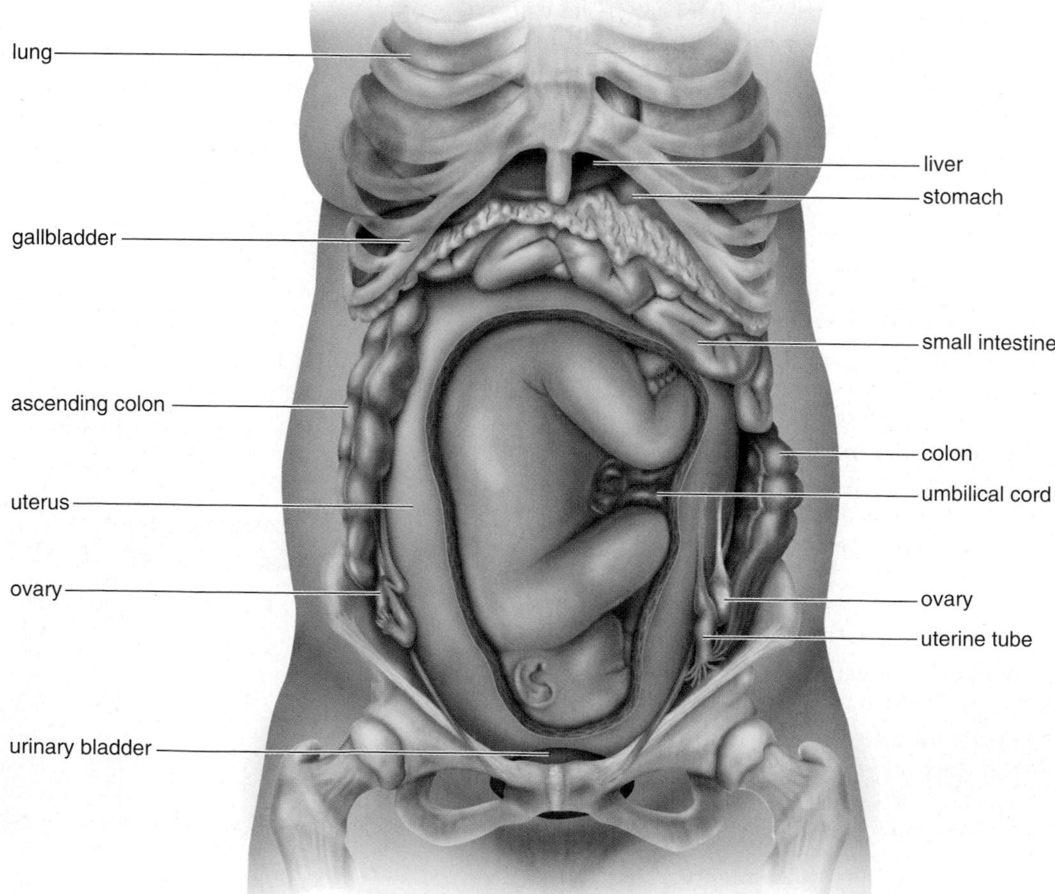

Figure 17.12 Changes to the internal anatomy of a pregnant woman.
The developing fetus compresses many of the organs within the body of the pregnant woman.

> **APPLICATIONS AND MISCONCEPTIONS**
>
> **Does the length or width of the linea nigra indicate the sex of a baby?**
>
> The linea nigra occurs due to the secretions of melanocytes in a pregnant woman. The line may be darker in fair-skinned women, and the length and width of the line may vary even between pregnancies. However, there is no correlation between the existence of the linea nigra and the sex, or any characteristics, of the developing fetus.

Birth

The uterus has contractions throughout pregnancy. At first, these are light, lasting about 20 to 30 seconds and occurring every 15 to 20 minutes. Near the end of pregnancy, the contractions may become stronger and more frequent so that a woman thinks she is in labor. "False labor" contractions are called **Braxton Hicks contractions.** However, the onset of true labor is marked by uterine contractions that occur regularly every 15 to 20 minutes and last for 40 seconds or longer.

Video
Triggering Birth

A positive feedback mechanism can explain the onset and continuation of labor. Uterine contractions are induced by a stretching of the cervix, which also brings about the release of oxytocin from the posterior pituitary gland. Oxytocin stimulates the uterine muscles, both directly and through the action of prostaglandins. Uterine contractions push the fetus downward, and the cervix stretches even more. This cycle keeps repeating itself until birth occurs.

Prior to or at the first stage of **parturition,** the process of giving birth to an offspring, there can be a "bloody show" caused by expulsion of a mucous plug from the cervical canal. This plug prevented bacteria and sperm from entering the uterus during pregnancy.

Stage 1

During the first stage of labor, the uterine contractions of labor occur in such a way that the cervical canal slowly disappears as the lower part of the uterus is pulled upward toward the baby's head. This process is called **effacement,** or "taking up the cervix." With further contractions, the baby's head acts as a wedge to assist cervical dilation (Fig. 17.13*a*). If the amniotic

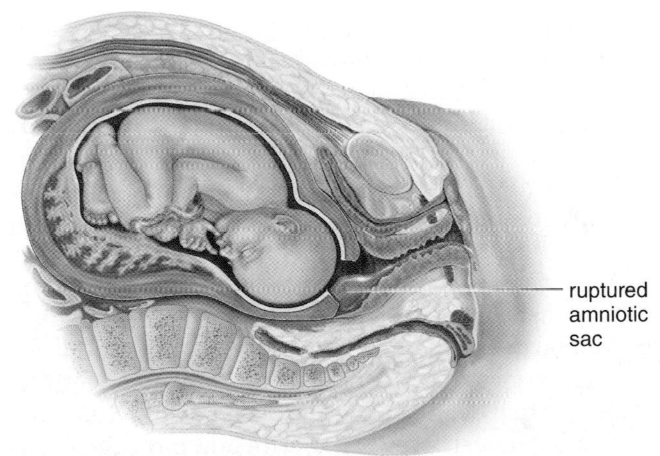

a. First stage of birth: Cervix dilates.

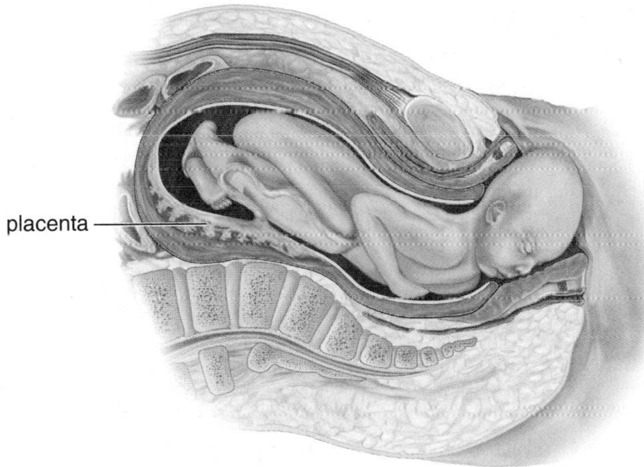

b. Second stage of birth: Baby emerges.

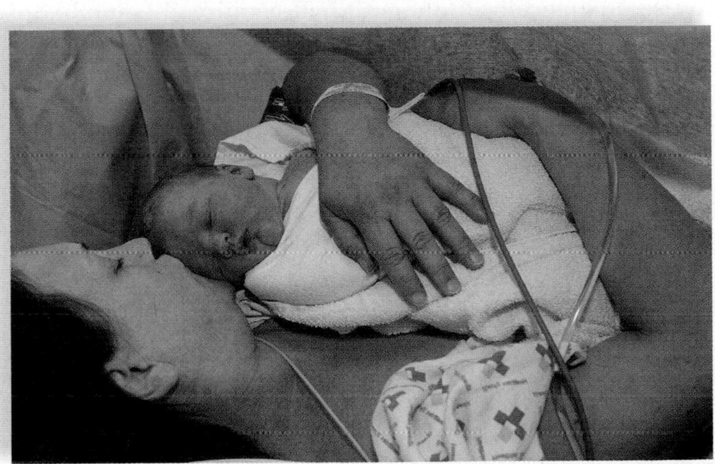

c. Baby has arrived.

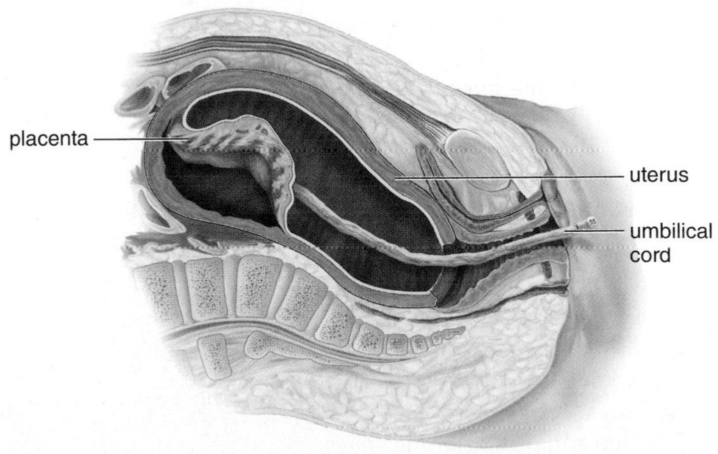

d. Third stage of birth: Afterbirth is expelled.

Figure 17.13 The stages of birth.
Birth proceeds in three stages. The baby emerges in stage two.

membrane has not already ruptured, it is apt to do so during this stage, releasing the amniotic fluid, which leaks out of the vagina (an event sometimes referred to as "breaking water"). The first stage of parturition ends once the cervix is dilated completely.

Stage 2

During the second stage of parturition, the uterine contractions occur every 1 to 2 minutes and last about 1 minute each. They are accompanied by a desire to push, or bear down. As the baby's head gradually descends into the vagina, the desire to push becomes greater. When the baby's head reaches the exterior, it turns so that the back of the head is uppermost (Fig. 17.13*b*). To enlarge the vaginal orifice, an **episiotomy** is often performed. This incision, which enlarges the opening, is sewn together later. As soon as the head is delivered, the physician may hold the head and guide it downward, while one shoulder and then the other emerges. The rest of the baby follows easily (Fig. 17.13*c*).

Once the baby is breathing normally, the umbilical cord is cut and tied, severing the child from the placenta. The stump of the cord shrivels and leaves a scar, the umbilicus (belly button).

Stage 3

The placenta, or **afterbirth,** is delivered during the third stage of parturition (Fig. 17.13*d*). About 15 minutes after delivery of the baby, uterine muscular contractions shrink the uterus and dislodge the placenta. The placenta then is expelled into the vagina. As soon as the placenta and its membranes are delivered, the third stage of parturition is complete.

CHECK YOUR PROGRESS 17.4

1. Identify the hormonal changes that occur in a female during pregnancy.
2. Explain the following: Maternal blood carbon dioxide levels fall by 20% during pregnancy. How does this benefit the fetus?
3. Describe the three stages of labor.

CONNECTING THE CONCEPTS

For more information on the content of this section, refer to the following discussions:

Section 4.9 explains positive feedback mechanisms.

Section 10.4 examines how aldosterone regulates urine formation in the kidneys.

Section 16.4 examines the role of the female sex hormones estrogen and progesterone.

17.5 Aging

LEARNING OUTCOMES

Upon completion of this section, you should be able to

1. Summarize the hypotheses on why humans age.
2. Summarize the effects of aging on the organ systems of the body.

Development does not cease once birth has occurred but continues throughout the stages of life: infancy, childhood, adolescence, and adulthood. Infancy, the toddler years, and preschool years are times of remarkable growth. During the birth to 5-year-old stage, humans acquire gross motor and fine motor skills. These include the ability to sit up and then to walk, as well as being able to hold a spoon and manipulate small objects. Language usage begins during this time and will become increasingly sophisticated throughout childhood. As infants and toddlers explore their environment, their senses—vision, taste, hearing, smell, and touch—mature dramatically. Socialization is very important as a child forms emotional ties with its caregivers and learns to separate self from others. Babies do not all develop at the same rate, and there is a large variation in what is considered normal.

The preadolescent years, from 6 to 12 years of age, are a time of continued rapid growth and learning. Preadolescents form identities apart from parents, and peer approval becomes very important. Adolescence begins with the onset of puberty as the young person achieves sexual maturity. For girls, puberty begins between 10 and 14 years of age, whereas for boys it generally occurs between ages 12 and 16. During this time, the sex-specific hormones (see Chapter 15) cause the secondary sexual characteristics to appear. Profound social and psychological changes are also associated with the transition from childhood to adulthood.

Aging encompasses the progressive changes from infancy until eventual death. Today, **gerontology,** the study of aging, is of great interest because there are now more older individuals in our society than ever before. The number is expected to rise dramatically; in the next half-century, the number of people over age 65 will increase 147%. The human life span is judged to be a maximum of 120 to 125 years. The present goal of gerontology is not necessarily to increase the life span but to increase the health span, the number of years that an individual enjoys the full functions of all body parts and processes (Fig. 17.14).

Cellular Aging

Aging is a complex process, and multiple factors can affect the aging process. Most scientists who study gerontology, believe that aging is partly genetically preprogrammed. This idea is supported by the observation that longevity runs in families, that is, the children of long-lived parents tend to live longer than those of short-lived parents. As would also be expected,

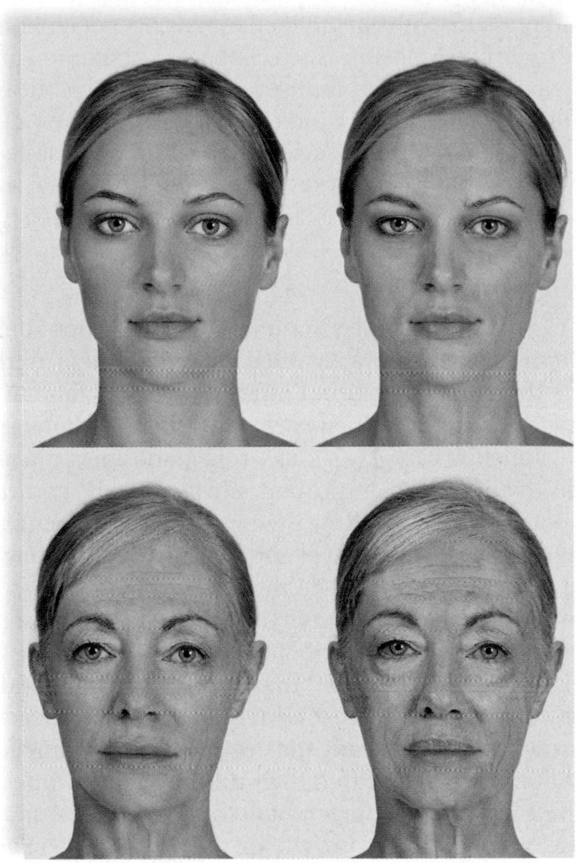

Figure 17.14 The effects of aging.
Aging is a slow process during which the body undergoes changes that eventually bring about death, even if no marked disease or disorder is present. Medical science is trying to extend the human life span and the health span, the length of time the body functions normally.

studies show that identical twins have a more similar life span than nonidentical twins.

Hormones

Many laboratory studies of aging have been performed in the nematode *C. elegans,* in which single-gene mutations have been shown to influence the life span. For example, mutations that decrease the activity of a hormone receptor similar to the insulin receptor more than double the life span of the worms, which also behave and look like much younger worms throughout their prolonged lives. Interestingly, small breed dogs like poodles or terriers, which may live 15 to 20 years, have lower levels of an analogous receptor, compared to large dog breeds that live 6 to 8 years.

Telomeres

Studies of the behavior of cells grown in the lab also suggest a genetic influence on aging. Most types of differentiated cells can only divide a limited number of times. One factor that may control the number of cell divisions is the length of the *telomeres,* sequences of DNA at the ends of chromosomes. Telomeres protect the ends of chromosomes from deteriorating or fusing with other chromosomes. Each time a cell divides, the telomeres normally shorten, and cells with shorter telomeres tend to undergo fewer divisions (see section 18.2). Some cells, such as stem cells, possess an enzyme called telomerase, which replenishes the length of the telomeres, effectively making stem cells immortal. Cancer cells, which behave in a similar manner to stem cells, frequently possess an active telomerase enzyme, which allows them to replicate continuously (see section 19.1). Studies using stem cells and cancer cells have begun to close in on the genetic factors that cause cellular aging. For example, in 2012 researchers used *gene therapy* (see section 21.4) to introduce an active telomerase enzyme into mice, thus slowing the aging process.

Mitochondria and Diet

The mitochondria are the powerhouses of the cell (see section 3.6). As the mitochondria harvest the energy contained within carbohydrates, fats, and proteins, they generate free radicals. Free radicals are unstable molecules that carry an extra electron. To become stable, free radicals donate an electron to another molecule, such as DNA or proteins (e.g., enzymes) or lipids, found in plasma membranes. Eventually, these molecules are unable to function, and the cell loses internal functions. This may lead to cell death. Scientists have determined that high-calorie diets increase the levels of free radicals, thus accelerating cellular aging. Multiple studies on model organisms, including mice, have supported that a low-calorie diet can expand the life span. It is also possible to reduce the negative effects of free radicals by increasing one's consumption of natural antioxidants, such as those present in brightly colored and dark-green vegetables and fruits. Chemicals in nuts, fish, shellfish, and red wine have also been shown to reduce our exposure to free radicals, and slow the aging process.

However, if aging were mainly a function of genes, we would expect much less variation in life span to be observed among individuals of a given species than is actually seen. For this reason, experts have estimated that in most cases, genes account only for about 25% of what determines the length of life.

Damage Accumulation

Another set of hypotheses propose that aging involves the accumulation of damage over time. In 1900, the average human life expectancy was around 45 years. A baby born in the United States in 2012 is expected to live an average of about 78 years. Because human genes have presumably not changed much in such a short time, most of this gain in life span is due to better medical care, along with the application of scientific knowledge about how to prolong our lives.

There are two basic types of cellular damage that can accumulate over time. The first type can be thought of as agents that are unavoidable; for example, the accumulation of harmful

DNA mutations, or the buildup of harmful metabolites. For some time medical researchers have known that proteins—such as the collagen fibers present in many support tissues—become increasingly cross-linked as people age. This cross-linking may account for the inability of such organs as the blood vessels, the heart, and the lungs to function as they once did. Some researchers have now found that glucose has the tendency to attach to any type of protein, which is the first step in a cross-linking process. They are currently experimenting with drugs that can prevent cross-linking. However, other sources of cellular damage may be avoidable, such as a poor diet or exposure to the sun.

Effect of Age on Body Systems

In the preceding chapters, we have examined the structure and function of the major body systems. Aging reduces the ability of the organ system to perform these functions, and in many cases impacts the ability of the system to contribute to homeostasis. Therefore, it seems appropriate to first consider the effects of aging on the various body systems.

Integumentary System

As aging occurs, the skin becomes thinner and less elastic because the number of elastic fibers decreases and the collagen fibers become increasingly cross-linked to each other, reducing their flexibility. There is also less adipose tissue in the subcutaneous layer; therefore, older people are more likely to feel cold. Together these changes typically result in sagging and wrinkling of the skin.

As people age, the sweat glands also become less active, resulting in decreased tolerance to high temperatures. There are fewer hair follicles, so the hair on the scalp and the limbs thins out. Older people also experience a decrease in the number of melanocytes, making their hair gray and their skin pale. In contrast, some of the remaining pigment cells are larger, and pigmented blotches ("age spots") appear on the skin.

Cardiovascular System

Common problems with cardiovascular function are usually related to diseases, especially atherosclerosis. However, even with normal aging, the heart muscle does weaken somewhat, and may increase slightly in size as it compensates for its decreasing strength. The maximum heart rate decreases even among the most fit older athletes, and it takes longer for the heart rate and blood pressure to return to normal resting levels following stress. Some part of this decrease in heart function may also be due to blood leaking back through heart valves that have become less flexible.

Aging also affects the blood vessels. The middle layer of arteries contains elastic fibers, which, like collagen fibers in the skin, become more cross-linked and rigid with time. These changes, plus a frequent decrease in the internal diameter of arteries due to atherosclerosis, contribute to a gradual increase in blood pressure with age. Indeed, nearly 50% of older adults have chronic hypertension. Such changes are common in individuals living in Western industrialized countries but not in agricultural societies. This indicates that a diet low in cholesterol and saturated fatty acids, along with a sensible exercise program, may help to prevent age-related cardiovascular disease.

Immune System

As people age, many of their immune system functions become compromised. Because a healthy immune system normally protects the entire body from infections, toxins, and at least some types of cancer, some investigators believe that losses in immune function can play a major role in the aging process.

The thymus is an important site for T-cell maturation. Beginning in adolescence, the thymus begins to involute, gradually decreasing in size, and eventually becoming replaced by fat and connective tissue. The thymus of a 60-year-old adult is about 5% of the size of the thymus of a newborn, resulting in a decrease in the ability of older people to generate T-cell responses to new antigens. The evolutionary rationale for this may be that the thymus is energetically expensive for an organism to maintain, and that compared to younger animals that must respond to a high number of new infections and other antigens, older animals have already responded to most of the antigens to which they will be exposed in their life.

Aging also affects other immune functions. Because most B-cell responses are dependent on T cells, antibody responses also begin to decline. This, in turn, may explain why the elderly do not respond as well to vaccinations as young people do. This presents challenges in protecting older people against diseases like influenza and pneumonia, which can otherwise be prevented by annual vaccination.

Not all immune functions decrease with age. The activity of natural killer cells, which are a part of the innate immune system, seems to change very little with age. Perhaps by investigating how these cells remain active throughout a normal human life span, researchers can learn to preserve other aspects of immunity in the elderly.

Digestive System

The digestive system is perhaps less affected by the aging process than other systems. Because secretion of saliva decreases, more bacteria tend to adhere to the teeth, causing more decay and periodontal disease. Blood flow to the liver is reduced, resulting in less efficient metabolism of drugs or toxins. This means that, as a person gets older, less medication is needed to maintain the same level in the bloodstream.

Respiratory System

Cardiovascular problems are often accompanied by respiratory disorders, and vice versa. Decreasing elasticity of lung tissues means that ventilation is reduced. Because we rarely use the entire vital capacity, these effects may not be noticed unless the demand for oxygen increases, such as during exercise.

Excretory System

Blood supply to the kidneys is also reduced. The kidneys become smaller and less efficient at filtering wastes. Salt and water balance are difficult to maintain, and the elderly dehydrate faster than young people. Urinary incontinence (lack of bladder control) increases with age, especially in women. In men, an enlarged prostate gland may reduce the diameter of the urethra, causing frequent or difficult urination.

Nervous System

Between the ages of 20 and 90, the brain loses about 20% of its weight and volume. Neurons are extremely sensitive to oxygen deficiency, and neuron death may be due not to aging itself but to reduced blood flow in narrowed blood vessels. However, contrary to previous scientific opinion, recent studies using advanced imaging techniques show that most age-related loss in brain function is not due to whatever loss of neurons is occurring. Instead, decreased function may occur due to alterations in complex chemical reactions, or increased inflammation in the brain. For example, an age-associated decline in levels of dopamine can affect brain regions involved in complex thinking.

Perhaps more important than the molecular details, recent studies have confirmed that lifestyle factors can affect the aging brain. For example, animals on restricted calorie diets developed fewer Alzheimer-like changes in their brains. For more on Alzheimer, see the Health feature, "Alzheimer Disease," on page 408. Other positive factors that may help maintain a healthy brain include attending college (the "use it or lose it" idea), regular exercise, and sufficient sleep.

Sensory Systems

In general, with aging more stimulation is needed for taste, smell, and hearing receptors to function as before. A majority of people over age 80 have a significant decline in their sense of smell, and about 15% suffer from anosmia, or a total inability to smell. The latter condition can be a serious health hazard, due to the inability to detect smoke, gas leaks, or spoiled food. After age 50, most people gradually begin to lose the ability to hear tones at higher frequencies, and this can make it difficult to identify individual voices and to understand conversation in a group.

Starting at about age 40, the lens of the eye does not accommodate as well, resulting in presbyopia, or difficulty focusing on near objects, which causes many people to require reading glasses as they reach middle age. Finally, cataracts and other eye disorders become much more common in the aged.

Musculoskeletal System

For the average person, muscle mass peaks between the ages of 16 and 19 for females and between 18 and 24 for males. Beginning in the twenties or thirties, but accelerating with increasing age, muscle mass generally decreases, due to decreases in both the size and number of muscle fibers. Most people who reach age 90 have 50% less muscle mass compared to when they were 20. Although some of this loss may be inevitable, regular exercise can slow this decline.

Like muscles, bones tend to shrink in size and density with age. Due to compression of the vertebrae, along with changes in posture, most of us lose height as we age. Those who reach age 80 will be about 2 in. shorter than they were in their twenties. Women lose bone mass more rapidly than men do, especially after menopause. Osteoporosis is a common disease in the elderly. Although some decline in bone mass is a normal result of aging, certain extrinsic factors are also important. A proper diet and moderate exercise program has been found to slow the progressive loss of bone mass.

Animation Osteoporosis

Endocrine System

As with the immune system, aging of the hormonal system can affect many different organs of the body. These changes are complex, however, with some hormone levels tending to decrease with age, while others increase. The activity of the thyroid gland generally declines, resulting in a lower basal metabolic rate. The production of insulin by the pancreas may remain stable, but cells become less sensitive to its effects, resulting in a rise in fasting glucose levels of about 10 mg/dl each decade after age 50.

Human growth hormone (HGH) levels also decline with age, but it is very unlikely that taking HGH injections will "cure" aging, despite Internet claims. In fact, one study found that people with lower levels of HGH actually lived longer than those with higher levels.

Reproductive System

Testosterone levels are highest in men in their twenties. After age 30, testosterone levels decrease by about 1% per year. Extremely low testosterone levels have been linked to a decreased sex drive, excessive weight gain, loss of muscle mass, osteoporosis, general fatigue, and depression. However, the levels below which testosterone treatment should be initiated remain controversial. Testosterone replacement therapy, whether through injection, patches, or gels, is associated with side effects like enlargement of the prostate, acne or other skin reactions, and the production of too many red blood cells.

BIOLOGY MATTERS

Health

Alzheimer Disease

In 1900, the average life span in the United States was 47 years of age. Today, it is 78 years of age. Normal aging does involve some changes in mental faculties, but many of the changes we associate with old age are related to disease, not aging. Two of the more common diseases are Alzheimer disease and Parkinson disease.

Alzheimer is characterized by the presence of abnormally structured neurons and a reduced amount of acetylcholine (ACh). A neuron that has been affected by Alzheimer has two characteristic features. Bundles of fibrous protein, called neurofibrillary tangles, extend from the axon to surround the nucleus of the neuron. Tangles form when the supporting protein, called tau, becomes malformed (Fig. 17B) and twists the neurofibrils, which are normally straight. In addition, protein-rich accumulations called amyloid plaques envelop branches of the axon. The plaques grow so dense that they trigger an inflammatory reaction that causes neuron death.

Video Prion Disease

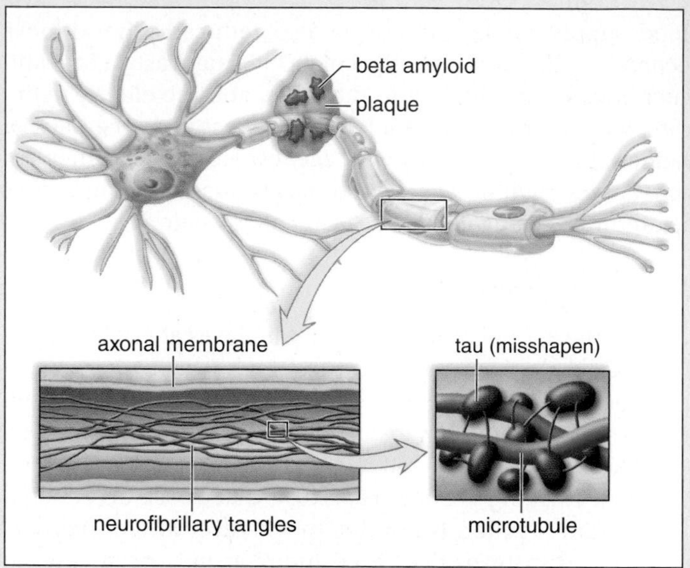

Figure 17B **Tau protein and Alzheimer disease.** Some of the neurons of Alzheimer disease (AD) patients have beta amyloid plaques and neurofibrillary tangles. AD neurons are present throughout the brain but concentrated in the hippocampus and amygdala.

Treatment for Alzheimer Disease

Treatment for Alzheimer involves using one of two categories of drugs. Cholinesterase inhibitors work at neuron synapses in the brain, slowing the activity of acetylcholinesterase, the enzyme that breaks down ACh. Allowing ACh to accumulate in synapses keeps memory pathways in the brain functional for a longer period. A second drug, memantine, blocks *excitotoxicity,* the tendency of diseased neurons to self-destruct. This recently approved medication is used only in moderately to severely affected patients. Using the drug allows neurons involved in memory pathways to survive longer in affected patients. Successes with these medications indicates that treatment for Alzheimer patients should begin as soon as possible after diagnosis and continue indefinitely. However, neither type of medication cures Alzheimer; both merely slow the progress of disease symptoms, allowing the patient to function independently for a longer time. Additional research is currently underway to test the effectiveness of anticholesterol *statin* drugs, as well as anti-inflammatory medications, in slowing the progress of the disease.

Much of current research on Alzheimer focuses on the prevention and cure of the disease. Scientists believe that curing Alzheimer will require an early diagnosis, because it is thought that the disease may begin in the brain 15 to 20 years before symptoms ever develop. Currently, diagnosis can't be made with absolute certainty until the brain is examined at autopsy. A new test on the cerebrospinal fluid may allow early detection of amyloid proteins and a much earlier diagnosis of the disease. Researchers are also testing vaccines for Alzheimer, which would target the patient's immune system to destroy amyloid plaques.

Early findings have shown that risk factors for cardiovascular disease also contribute to an increased incidence of Alzheimer. Risk factors for cardiovascular disease include elevated blood cholesterol and blood pressure, smoking, obesity, sedentary lifestyle, and diabetes mellitus. Thus, evidence suggests that a lifestyle tailored for good cardiovascular health may also prevent Alzheimer.

Video Apples for Alzheimers

Questions to Consider

1. What part of the brain does the accumulation of the tau protein directly affect?
2. How might anti-inflammatory drugs slow the progression of Alzheimer disease?

> **APPLICATIONS AND MISCONCEPTIONS**
>
> **What genes may be associated with longevity?**
>
> In the past several years, researchers have begun to use new molecular techniques to unravel some of the mysteries as to why some individuals live past the age of 100—the centenarians. A variation of one gene, *FOX03A,* has been identified to be more prevalent in centenarians than the general population. This gene regulates the insulin pathways of the body and also appears to control the genetic mechanisms that protect cells against free radicals. Both insulin regulation and protection against free radicals have been known for some time to enhance longevity in a variety of organisms. Interestingly, *FOX03A* is also believed to be involved in the process of apoptosis and may help protect the body against cancer. Although researchers do not think that *FOX03A* is the sole gene for longevity, it is providing a starting point for larger studies on human aging.

Menopause, the period in a woman's life during which the ovarian and uterine cycles cease, usually occurs between ages 45 and 55. The ovaries become unresponsive to the gonadotropic hormones produced by the anterior pituitary, and they no longer secrete estrogen or progesterone. At the onset of menopause, the uterine cycle becomes irregular, but as long as menstruation occurs, it is still possible for a woman to conceive. Therefore, a woman is usually not considered to have completed menopause (and thus be infertile) until menstruation has been absent for a year.

The hormonal changes during menopause often produce physical symptoms such as "hot flashes" (caused by circulatory irregularities), dizziness, headaches, insomnia, sleepiness, and depression. To ease these symptoms, female hormone replacement therapy (HRT) was routinely prescribed until 2002, when a large clinical study showed that in the long term, HRT caused more health problems than it prevented. For this reason, most doctors no longer recommend long-term HRT for the prevention of postmenopausal conditions.

It is also of interest that, as a group, females live longer than males. It is likely that estrogen offers women some protection against cardiovascular disorders when they are younger. Males suffer a marked increase in heart disease in their forties, but an increase is not noted in females until after menopause, when women lead men in the incidence of stroke. Men remain more likely than women to have a heart attack at any age, however.

Conclusion

We have listed many adverse effects of aging; however, though such effects are seen, they are not inevitable (Fig. 17.15). Diet and exercise are factors that are under our personal control. Just as it is wise to make the proper preparations to remain financially independent when older, it is also wise to realize that, biologically, successful old age begins with the health habits developed when we are younger.

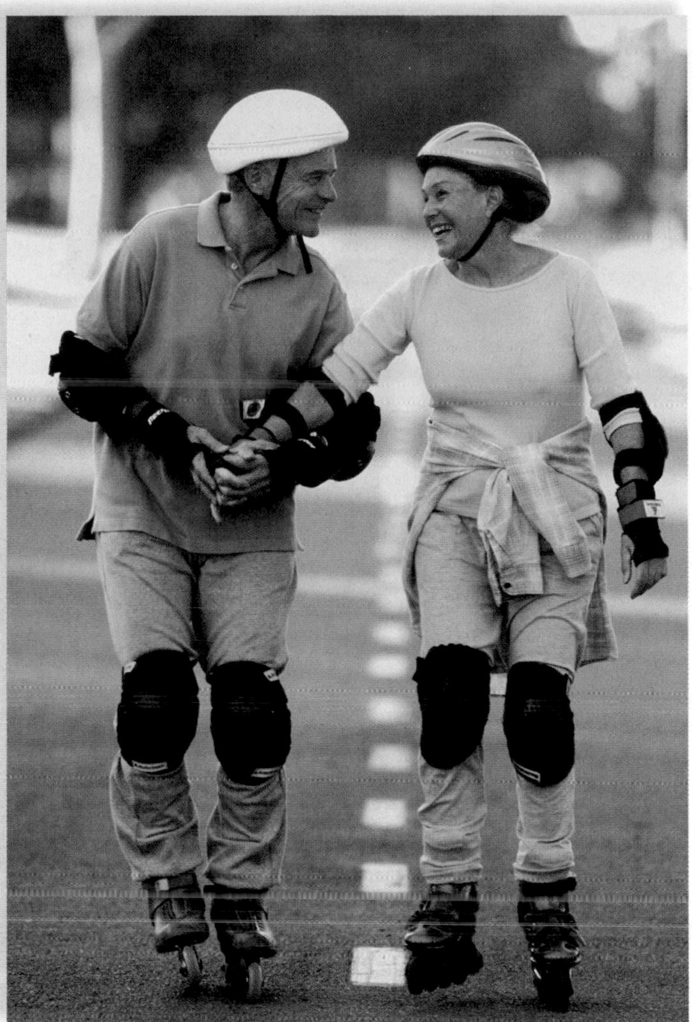

Figure 17.15 What steps can an individual take to increase health span?
Gerontology research has shown that regular physical exercise, as well as staying engaged both mentally and socially, can slow the progress of aging and lengthen the health span.

> **CHECK YOUR PROGRESS 17.5**
>
> 1. Distinguish between the two different hypotheses of aging presented in this section.
> 2. Summarize the effect of aging on the various body systems.
> 3. Discuss the best way to keep healthy, even though aging occurs.

CONNECTING THE CONCEPTS

For more information on the organ systems presented in this section, refer to the following discussions:

Section 5.1 provides an overview of the cardiovascular system.
Section 10.1 provides an overview of the urinary system.
Section 13.1 provides an overview of the nervous system.

BIOLOGY MATTERS  Bioethical

Reproductive and Therapeutic Cloning

In March 1997, Scottish investigators announced that they had cloned a sheep called Dolly, and their procedure is now routinely used (Fig. 17C). A donor 2n nucleus is substituted for the n nucleus of an egg. A stimulus is applied that triggers cell division, and the resulting embryo is implanted into a surrogate mother where it develops to term. Using the procedure developed by the Scottish researchers, it is now common practice to clone all sorts of farm animals (horses, cows, sheep, goats, pigs) and also cats and monkeys.

Success of Cloning

Even so, cloning of animals is still in its infancy, and many problems still exist. (1) The vast majority of pregnancies involving clones are not successful. To clone Dolly the sheep, it took 247 tries before one was successful. In many cases, the clone grows abnormally large and the uterus enlarges with fluid to the point at which it can rip apart. Almost all clone pregnancies spontaneously abort. (2) Of the small number of animal clones born, most have severe abnormalities; malfunctioning livers, abnormal blood vessels and heart problems, underdeveloped lungs, diabetes, and immune system deficiencies are seen in newborns. Several cow clones had head deformities—none survived very long. (3) Even if the newborn clone appears healthy, it usually soon develops diseases seen in older animals. Dolly was euthanized in 2003 because she was suffering from lung cancer and crippling arthritis. She had lived only half the normal life span for a Dorset sheep.

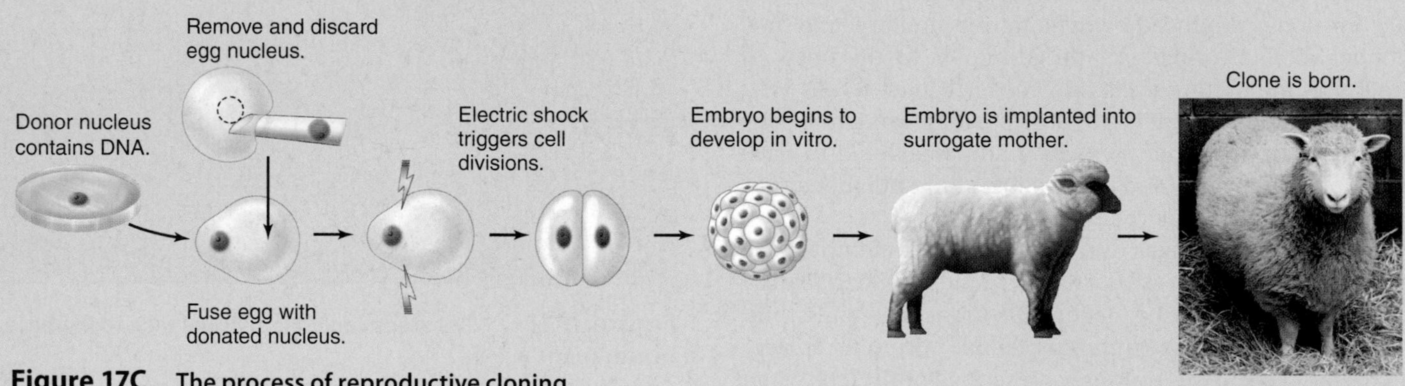

Figure 17C **The process of reproductive cloning.**

CASE STUDY CONCLUSION

Over the course of the next few office visits, Amber's doctor performed a variety of tests. Because Amber was over 35 years old when she conceived her first child, the doctor recommended that a chorionic villus sampling test be done to test for birth defects, such as Down syndrome. In addition, the doctor performed a glucose tolerance test to determine if Amber was experiencing gestational diabetes. An additional blood test, called an alpha-fetoprotein (AFP) test, screened for neural tube defects and additional evidence of Down syndrome. Much to the relief of Amber and Kent, all of these tests came back normal. Then, around week 20 of Amber's pregnancy, the doctor made an appointment for Amber to get an ultrasound exam, which is frequently used to confirm the due date of the baby but can also be used to look for birth defects such as cleft palate. Everything once again appeared normal. Finally, the technician asked the parents the question that they had been waiting for: were they interested in knowing the sex of their baby? Within a few minutes, Amber and Kent found out that, in just a few short months, they would be the parents of a baby girl.

Reproductive Cloning Versus Therapeutic Cloning

Animal cloning is a form of reproductive cloning. In *reproductive cloning,* the desired end is to create an individual. Even if such a feat were accomplished, a clone would not be identical to the person being cloned. Recall that mitochondria have genes, and these genes are contributed by the egg even if the egg nucleus is removed. The clone could be identical to its genetic parent only if the donor nucleus and the donor egg came from the same female. Further, the clone would be subject to different environmental factors and a different upbringing from his/her genetic parent.

In *therapeutic cloning,* the desired end is *not* an individual. Rather, it is the embryonic stem cells that could possibly be "coaxed" into becoming other types of cells. The goal is twofold. Stem-cell research will undoubtedly yield useful information about how cells develop and become specialized. However, the ultimate goal of therapeutic cloning is to provide cells and tissues that could treat human illnesses: insulin-secreting cells for diabetics, nerve cells for stroke patients or those with Parkinson disease, cardiac cells for those with heart disease, and so forth. Yet, ethical concerns about therapeutic cloning remain—after all, any pre-embryo is potentially a living, breathing human.

Video Heart Stem Cells

Anticipating intense interest in therapeutic cloning, the U.S. National Academy of Sciences proposed strict new guidelines for federally funded research in 2005. As a result of the Academy's recommendations, Embryonic Stem Cell Research Oversight (ESCRO) committees must approve embryonic stem-cell research before it is begun. Thus, this type of research is subject to two reviews: (1) an ESCRO committee analysis and (2) reviews already required by an Institutional Review Board. These committees work to ensure that the research procedures and use of the information collected from the study follow ethical guidelines.

ESCRO committees consist of bioethicists and legal experts, as well as members of the general public. All research requires informed consent from the donors of ova or sperm prior to beginning the research. Stem cells created for therapeutic cloning can never be used for reproductive purposes under the proposed guidelines. The guidelines also require that the embryos created cannot be grown in culture for longer than 14 days. At that point, the primitive streak of the developing nervous system begins to form.

Questions to Consider

1. Do you approve of the restrictions currently in place for therapeutic cloning? If not, how would you see these restrictions changed?
2. Scientific researchers have made great strides in stem-cell research, successfully converting both adult cells and umbilical cord cells back to stem-cell forms. In light of these successes, should therapeutic cloning be funded at all?
3. Several companies now provide cloning services for pets; should this be permitted when there are so many unwanted pets in the United States?
4. Should scientists be allowed to bring extinct species back to life (shades of *Jurassic Park*) using frozen tissue samples and reproductive cloning?

MEDIA STUDY TOOLS

Enhance your study of this chapter with media! Visit **www.mhhe.com/maderhuman13e** and go to "Media Study Tools" for this chapter to access the following:

 **Animations**

17.3 Fetal Development and Risk
17.5 Osteoporosis

 **Videos**

17.1 Human Sperm
17.2 Blastocyst Formation
17.3 Inherited Pollution
17.4 Triggering Birth
17.5 Prion Disease • Apples for Alzheimers • Heart Stem Cells

SUMMARIZE

17.1 Fertilization

Fertilization is the combination of an egg and sperm cell to produce a new individual, or **zygote.** During fertilization the acrosome of a sperm releases enzymes that digest a pathway for the sperm through the zona pellucida. The sperm nucleus then enters the egg and fuses with the egg nucleus.

17.2 Pre-Embryonic and Embryonic Development

Cleavage, growth, morphogenesis, and **differentiation** are the processes of development. Development of the fetus may be divided into **pre-embryonic development** events (first week) and **embryonic development** events (second week and later).

Pre-Embryonic Development

- Immediately after fertilization, the zygote begins to divide, first forming a **morula,** then a **blastocyst.** Within the blastocyst, the **inner cell mass** forms. This will become the **embryo.**
- The **extraembryonic membranes** (**chorion, allantois, yolk sac,** and **amnion**) begin to form.

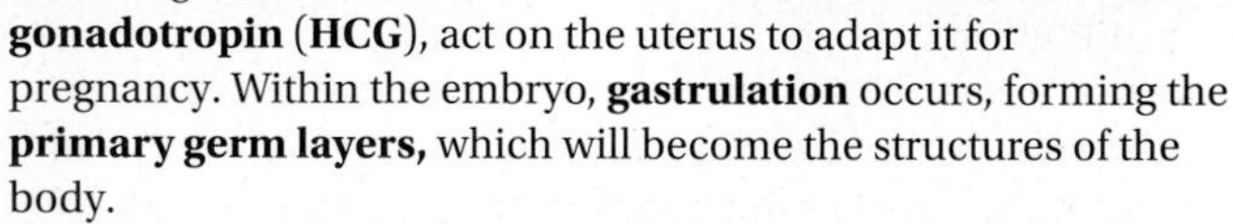

Embryonic Development

- During the second week, **implantation** of the embryo occurs. Several hormones, including **human chorionic gonadotropin** (**HCG**), act on the uterus to adapt it for pregnancy. Within the embryo, **gastrulation** occurs, forming the **primary germ layers,** which will become the structures of the body.
- Starting in the third week, the development of organ systems begins, including the nervous system and circulatory system (heart).
- In the fourth week, the embryo connects to the chorion using **chorionic villi.** The formation of the **umbilical cord** establishes a connection to the placenta. The limb buds appear and by the fifth week the sense organs develop in the head.

17.3 Fetal Development

At the end of the embryonic period, all organ systems are established and there is a mature and functioning **placenta.** The umbilical arteries and umbilical vein take blood to and from the placenta, where exchanges take place. Fetal development occurs during the third to ninth months.

Fetal Circulation

- Fetal circulation supplies the fetus with oxygen and nutrients and rids the fetus of carbon dioxide and wastes.
- The venous duct joins the umbilical vein to the inferior vena cava. The oval duct and arterial duct allow the blood to pass through the heart without going to the lungs. **Fetal development** extends from the third through the ninth months.

Third and Fourth Months

- During the third and fourth months, the skeleton is becoming ossified. In the skull, the **fontanels** will allow the head to pass through the birth canal.
- The sex of the fetus becomes distinguishable. If an *SRY* gene is present, **testosterone** directs the development of the testes and male sex organs. Otherwise, ovaries and female sex organs develop.

Fifth to Ninth Months

- During the fifth through the ninth months, the fetus continues to grow and to gain weight. At this point, the skin is covered by a fine down (**lanugo**) and a protective layer (**vernix caseosa**).

17.4 Pregnancy and Birth

Pregnancy

Major changes take place in the mother's body during pregnancy.

- Weight gain occurs as the uterus occupies most of the abdominal cavity.
- Many complaints, such as constipation, **incontinence,** heartburn, darkening of certain skin areas, and pregnancy-induced diabetes, are due to the presence of placental hormones.

Birth

A positive feedback mechanism that involves uterine contractions (including **Braxton Hicks contractions**) and oxytocin explains the onset and continuation of labor.

- During stage 1 of **parturition** (birth), **effacement** assists in the dilation of the cervix.
- During stage 2, the child is born. An **episiotomy** may assist in the widening of the vaginal orifice.
- During stage 3, the **afterbirth** is expelled.

17.5 Aging

Development after birth consists of infancy, childhood, adolescence, and adulthood. The science of **gerontology** examines the progressive changes that occur during **aging.** As we age, these changes contribute to an increased risk of infirmity, disease, and death.

Cellular Aging

There are several factors that contribute to aging at the cellular level.

- Telomeres contribute to the life span of cells.
- Receptors for certain hormones may not work efficiently
- Free radicals and other metabolites cause damage to cellular components.

Effect of Age on Body Systems

- Deterioration of organ systems can possibly be prevented or reduced in part by using good health habits.

ASSESS

Testing Your Knowledge of the Concepts

1. Describe the events that occur during fertilization. (pages 389–390)
2. Explain why development is divided into pre-embryonic, embryonic, and fetal development phases. (pages 391–392)
3. Name the four embryonic membranes and give a human function for each one. (page 392)
4. What are the three primary germ layers, and what body structures come from each germ layer? (pages 393–394)
5. Briefly summarize the weekly events of embryonic development. (pages 392–395)

6. Briefly summarize the monthly events of fetal development. (pages 395–401)
7. Explain how blood circulates to and from the placenta and the fetus. How is blood shunted away from the lungs? (pages 396–398)
8. List the hormones involved in the development of the male and female internal and external sex organs and state their functions. (pages 399–400)
9. Describe some of the changes that occur in the mother during pregnancy. (pages 401–402)
10. What event marks the end of each stage of birth? (pages 403–404)
11. Discuss how aging occurs at both the cellular and organismal (body) level. (pages 404–409)
12. Only one sperm enters an egg because
    a. sperm have an acrosome.
    b. the corona radiata gets larger.
    c. changes occur in the zona pellucida.
    d. the cytoplasm hardens.
    e. All of the choices are correct.
13. When all three germ layers are present (ectoderm, endoderm, and mesoderm), what event has occurred?
    a. blastulation
    b. limb formation
    c. gastrulation
    d. morulation
14. Which of these is not a process of development?
    a. cleavage
    b. parturition
    c. growth
    d. morphogenesis
    e. differentiation
15. Which of these is mismatched?
    a. chorion—sense perception
    b. yolk sac—first site of blood cell formation
    c. allantois—umbilical blood vessels
    d. amnion—contains fluid that protects embryo
16. In human development, which part of the blastocyst will develop into a embryo?
    a. trophoblast
    b. inner cell mass
    c. chorion
    d. yolk sac
17. Which primary germ layer is not correctly matched to an organ system or organ that develops from it?
    a. ectoderm—the nervous system
    b. endoderm—lining of the digestive tract
    c. mesoderm—skeletal system
    d. endoderm—cardiovascular system
18. Human chorionic gonadotropin is a
    a. hormone.
    b. basis of pregnancy tests.
    c. cause of ectopic pregnancy.
    d. Both a and b are correct.
19. Which is a correct sequence that ends with the stage that implants?
    a. morula, blastocyst, embryonic disk, gastrula
    b. ovulation, fertilization, cleavage, morula, early blastocyst
    c. embryonic disk, gastrula, primitive streak, neurula
    d. primitive streak, neurula, extraembryonic membranes, chorion
20. Differentiation is equivalent to which term?
    a. morphogenesis
    b. growth
    c. specialization
    d. gastrulation
21. Which process refers to the shaping of the embryo and involves cell migration?
    a. cleavage
    b. differentiation
    c. growth
    d. morphogenesis
22. Which association is not correct?
    a. third and fourth months—fetal heart has formed, but it does not beat
    b. fifth through seventh months—mother feels movement
    c. eighth through ninth months—usually head is now pointed toward the cervix
    d. All of the choices are correct.
23. At three months, the embryo has
    a. become a fetus.
    b. body systems already.
    c. a head, arms, and legs.
    d. ears and eyes, which don't function.
    e. All but b are correct.
24. Which of these structures is not a circulatory feature unique to the fetus?
    a. arterial duct
    b. oval opening
    c. umbilical vein
    d. pulmonary trunk
25. Which of these statements is correct?
    a. Fetal circulation, like adult circulation, takes blood equally to a pulmonary circuit and a systemic circuit.
    b. Fetal circulation shunts blood away from the lungs but makes full use of the systemic circuit.
    c. Fetal circulation includes exchange of substances between fetal blood and maternal blood at the placenta.
    d. Unlike adult circulation, fetal blood always carries oxygen-rich blood and therefore has no need for the pulmonary circuit.
    e. Both b and c are correct.
26. Which of these is a hormone involved in development of male and female sex organs?
    a. estrogen
    b. anti-Müllerian hormone
    c. dihydrotestosterone
    d. testosterone
    e. All of these hormones are involved.

27. After each cell division, these chromosomal structures shorten, thus regulating the life span of a cell.
    a. mitochondria
    b. free radicals
    c. telomeres
    d. plasma membranes
28. Label this diagram illustrating the placement of the extraembryonic membranes.

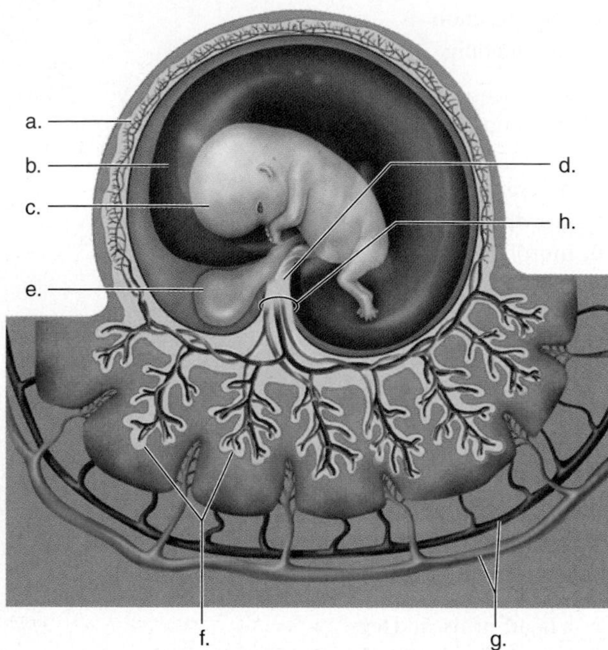

ENGAGE

Thinking Critically About the Concepts

Amber and Kent used a home pregnancy test to determine if she was pregnant. These tests detect the level of HCG (human chorionic gonadotropin; see page 393) in the urine. This hormone is released following implantation of the embryo into the uterus, usually around six days after fertilization. Some tests claim that they are sensitive enough to detect HCG on the date that menstruation is expected to begin. However, doctors recommend waiting until menstruation is one week late. If pregnant, a woman's level of HCG rises with each passing day, and testing is more likely to be accurate. However, even with a negative test result, the woman may still be pregnant if HCG levels are too low to be detected at the time of the first test. The test should be repeated later if menstruation doesn't begin. The home pregnancy tests contain a positive control. This is a visual sign (usually a line or a +) that appears if the test is working correctly. If this line does not appear, the test is not valid and must be repeated.

1. At home pregnancy tests check for the presence of HCG in a female's urine. Where does HCG come from? Why is HCG found in a pregnant woman's urine?
2. A blood test at a doctor's office can also check for the presence of HCG in a female's blood.
   a. Why would you expect to find HCG circulating in a pregnant female's blood?
   b. HCG is a protein; so how does HCG affect its target cells?
3. a. What pituitary hormone is checked with a blood test to diagnose menopause?
   b. Will levels of this hormone be increased or decreased if the female is in menopause?
   c. How does the changed (increased or decreased) level of this pituitary hormone cause the onset of menopause (cessation of menses)?

CHAPTER 18

Patterns of Chromosome Inheritance

CASE STUDY CELL CYCLE CONTROL

In August 2008, actress Christina Applegate, then 36 years old, revealed to shocked fans that she had been diagnosed with breast cancer. Every year, over 226,000 American women are diagnosed with breast cancer. Christina's mother is a breast cancer survivor, and Christina tested positive for a *BRCA1* (*br*east *ca*ncer susceptibility gene *1*) gene mutation, which is linked to both breast and ovarian cancer. Cancer results from a failure to control the cell cycle, a series of steps that all cells go through prior to initiating cell division. *BRCA1* is an important component of that control mechanism. *BRCA1* is a tumor suppressor gene, meaning that at specific points in the cell cycle, called checkpoints, *BRCA1* will slow down the cell cycle if it detects damage in the DNA. However, in Christina's case, the version of *BRCA1* that she inherited from her parents was not acting as a gatekeeper. Instead of preventing damaged cells from progressing into cell division, Christina's genes were allowing them to multiply and divide. The result of her body's failure to control the cell was a tumor in one of her breasts. It is estimated that 1 in 833 people possesses the mutant *BRCA1* allele. Although it is not the only genetic contribution to breast and ovarian cancers, it does play a major role. Several medical approaches, including increased surveillance and surgery, are available to those who test positive for a mutated *BRCA1* gene.

As you read through the chapter, think about the following questions:

1. What are the roles of checkpoints in the cell cycle?
2. At what checkpoint in the cell cycle would you think that *BRCA1* would normally be active?

CHAPTER CONCEPTS

18.1 Chromosomes
The genetic material in the cell is organized into chromosomes. A karyotype is a picture of chromosomes about to divide.

18.2 The Cell Cycle
The cell cycle consists of interphase and mitosis and is regulated by a series of checkpoints.

18.3 Mitosis
Mitosis is duplication division in which the daughter cells have the same number and types of chromosomes as the mother cell.

18.4 Meiosis
Meiosis is reduction division in which the daughter cells have half the number of chromosomes, and different combinations of genes, as the parent cell.

18.5 Comparison of Meiosis and Mitosis
Meiosis I uniquely pairs and separates the paired chromosomes so that the daughter cells have half the number of chromosomes. Meiosis II is exactly like mitosis, except the cells have half the number of chromosomes.

18.6 Chromosome Inheritance
Abnormalities in chromosome inheritance occur due to changes in chromosome number and changes in chromosome structure.

BEFORE YOU BEGIN

Before beginning this chapter, take a few moments to review the following discussions:

Section 2.7 What is the structure of a DNA molecule? How does this structure differ from that of RNA?

Section 3.4 What is chromatin?

Section 3.5 What is the function of microtubules and actin filaments in a cell?

18.1 Chromosomes

LEARNING OUTCOMES

Upon completion of this section, you should be able to

1. Distinguish between a chromosome and chromatin.
2. Explain the purpose of a karyotype.
3. Describe the purpose of the centromere in relation to the sister chromatids.

While a human nucleus is only about 5–8 μm long, it holds all the genetic material that is necessary to direct all of the functions in the body. The genetic material is arranged into **chromosomes,** structures that assist in the transmission of genetic information from one generation to the next. The instructions within each chromosome are contained within genes, which in turn are composed of DNA (see Chapter 21). Chromosomes also contain proteins that assist in the organizational structure of the chromosome. Collectively, the DNA and proteins are called **chromatin.** Humans have 46 chromosomes that occur in 23 pairs. Twenty-two of these pairs are called *autosomes.* All of these chromosomes are found in both males and females. One pair of chromosomes is called the *sex chromosomes* because this pair contains the genes that control gender. Males have the sex chromosomes X and Y, and females have two X chromosomes. Recall from Chapter 17 that a Y chromosome contains the *SRY* gene that causes testes to develop.

A Karyotype

Physicians and prospective parents sometimes want to view an unborn child's chromosomes to determine whether a chromosomal abnormality exists. Any cell in the body except red blood cells, which lack a nucleus, can be a source of chromosomes for examination. In adults, it is easiest to obtain and use white blood cells separated from a blood sample for the purpose of looking at the chromosomes. For an examination of fetal chromosomes, a physician may recommend procedures such as chorionic villus sampling or amniocentesis. These processes are described more fully in the Health feature, "Preventing and Testing for Birth Defects," in Chapter 17.

After a cell sample has been obtained, the cells are stimulated to divide in a culture medium. When a cell divides, chromatin condenses to form chromosomes. The nuclear envelope fragments, liberating the chromosomes. Next, a chemical is used to stop the division process when the chromosomes are most compacted and visible microscopically. Stains are applied to the slides, and the cells can be photographed with a camera attached to a microscope. Staining causes the chromosomes to have dark and light cross-bands of varying widths, and these can be used—in addition to size and shape—to help pair up the chromosomes. A computer is used to arrange the chromosomes in pairs (Fig. 18.1). The display is called a *karyotype.*

The karyotype in Figure 18.1 is that of a normal male. A normal karyotype tells us a lot about a body cell. First, we should notice that a normal body cell is diploid, meaning it has the full complement of 46 chromosomes. How does it happen

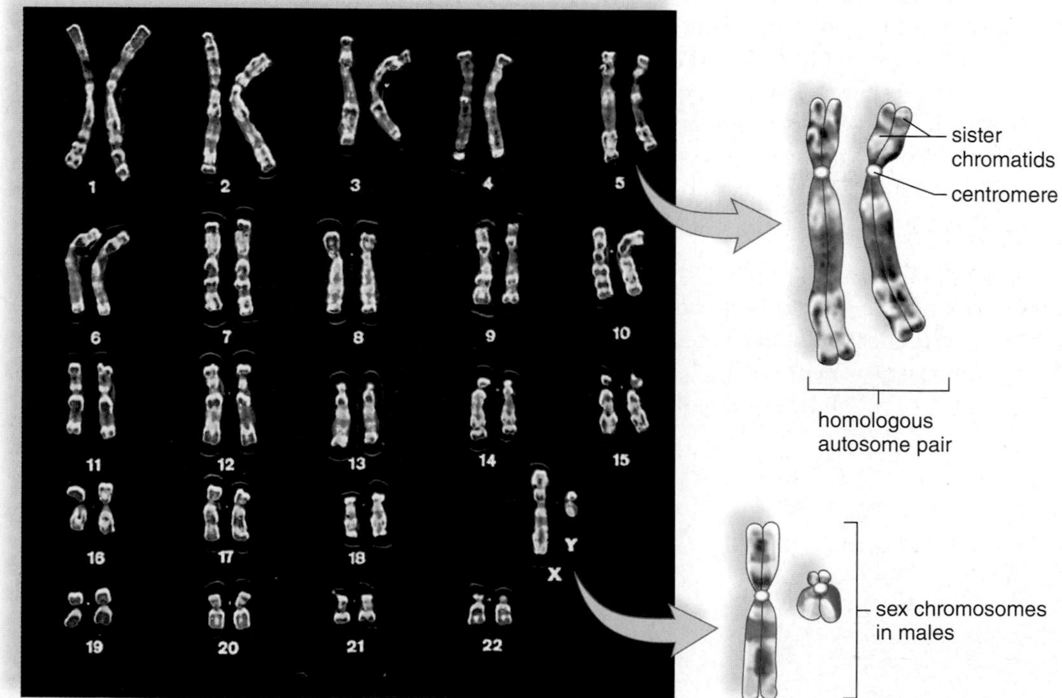

Figure 18.1 A karyotype of human chromosomes. In body cells, the chromosomes occur in pairs. In a karyotype, the pairs have been numbered and arranged by size from largest to smallest. These chromosomes are duplicated, and each one is composed of two sister chromatids.

that almost all of the cells in your body (red blood cells and liver cells are exceptions) have 46 chromosomes? Mitosis, or duplication division (see section 18.3), begins when the fertilized egg starts dividing, and ensures that every cell has 46 chromosomes.

The enlargement of a pair of chromosomes shows that in dividing cells each chromosome is composed of two identical parts, called **sister chromatids** (Fig. 18.1). These chromosomes are said to be *replicated* or *duplicated* chromosomes because the two sister chromatids contain the same genes. Genes are the units of heredity that control the cell. Replication of the chromosomes is possible only because each chromatid contains a DNA double helix.

The chromatids are held together at a region called the centromere. A **centromere** has the function of holding the chromatids together until a certain phase of mitosis when the centromere splits. Once separated, each sister chromatid is a chromosome. In this way, a duplicated chromosome gives rise to two individual daughter chromosomes. When daughter chromosomes separate, the new cell gets one of each type and, therefore, a full complement of chromosomes.

APPLICATIONS AND MISCONCEPTIONS

Does the number of chromosomes relate to the overall complexity of an organism?

In eukaryotes, like humans, the number of chromosomes varies considerably. A fruit fly has 8 chromosomes, and yeasts have 32. Humans have 46, and horses have 64. The largest number of chromosomes appears to be found in a particular type of fern. It has 1,252 chromosomes. The number of chromosomes doesn't seem to determine an organism's complexity.

CHECK YOUR PROGRESS 18.1

1. Explain the purpose of chromosomes in a cell.
2. Describe how a karyotype can be used to determine the number of chromosomes in a cell.
3. Explain why sister chromatids are genetically the same.

CONNECTING THE CONCEPTS

For more information on the topics in this section, refer to the following discussions:

Section 3.4 examines the role of the nucleus in a cell.

Section 3.4 explains the relationship of chromatin to the chromosomes.

Section 17.3 describes the stages of fetal development.

18.2 The Cell Cycle

LEARNING OUTCOMES

Upon completion of this section, you should be able to

1. List the stages of the cell cycle and state the purpose of each.
2. Describe the purpose of the checkpoints in the cell cycle.
3. Distinguish between mitosis and cytokinesis.

The **cell cycle** is an orderly process that has two parts: interphase and cell division. To understand the cell cycle, it is necessary to understand the structure of a cell (see Fig. 3.4). A human cell has a plasma membrane, which encloses the cytoplasm, the content of the cell outside the nucleus. In the cytoplasm are various organelles that carry on various functions necessary to the life of the cell. When a cell is not undergoing division, the chromatin (DNA and associated proteins) within a nucleus is a tangled mass of thin threads.

Interphase

As Figure 18.2 shows, most of the cell cycle is spent in **interphase.** This is the time when the organelles within the cell carry on their usual functions. As the cell continues through interphase, it gets ready to divide. The cell grows larger, the number of organelles doubles, and the amount of chromatin doubles as DNA replication occurs.

Interphase is divided into three stages: the G_1 stage occurs before DNA synthesis; the S stage includes DNA replication; and the G_2 stage occurs after DNA replication. Originally, *G* stood for

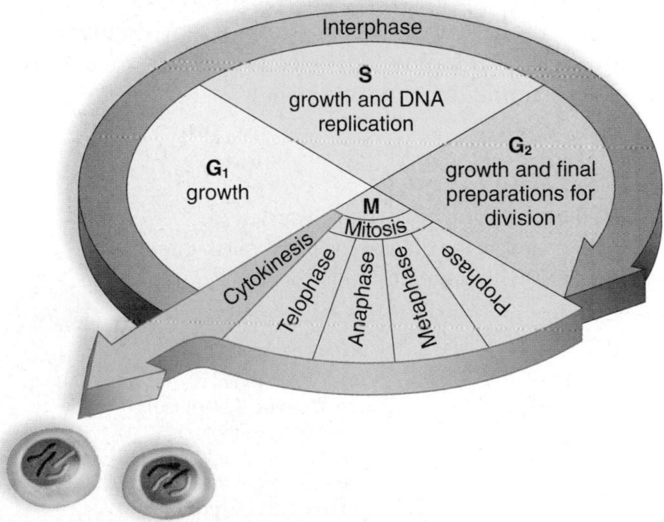

Figure 18.2 **Stages of the cell cycle.**
The cell cycle has four stages. During interphase, which consists of G_1, S, and G_2, the cell gets ready to divide. During the mitotic stage, nuclear division and cytokinesis (cytoplasmic division) occur.

the "gaps"—those times during interphase when DNA synthesis was not occurring. But now that we know growth happens during these stages, the *G* can be thought of as standing for "growth." Let us see what specifically happens during these stages.

G_1 stage. The cell returns to normal size and resumes its function within the body. A cell doubles its organelles (e.g., mitochondria and ribosomes), and it accumulates the materials needed for DNA synthesis.

S stage. A copy is made of all the DNA in the cell. DNA replication occurs, so each chromosome consists of two identical DNA double-helix molecules. These molecules occur in the strands called *sister chromatids.*

G_2 stage. The cell synthesizes the proteins needed for cell division, such as the protein found in microtubules. The role of microtubules in cell division is described in a later section.

The amount of time the cell takes for interphase varies widely. Some cells, such as nerve and muscle cells, typically do not complete the cell cycle and are permanently arrested in G_1. These cells are said to have entered a G_0 stage. Embryonic cells spend very little time in G_1 and complete the cell cycle in a few hours.

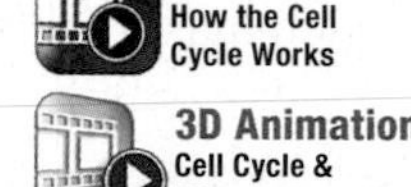

Mitosis and Cytokinesis

Following interphase, the cell enters the cell division part of the cell cycle. Cell division has two stages: M (for "mitotic") stage and cytokinesis. **Mitosis** is a type of nuclear division. Mitosis is also referred to as *duplication division* because each new nucleus contains the same number and type of chromosomes as the former cell. **Cytokinesis** is division of the cytoplasm.

During mitosis, the sister chromatids of each chromosome separate, becoming chromosomes distributed to two daughter nuclei. When cytokinesis is complete, two daughter cells are now present. Mammalian cells usually require only about 4 hours to complete the mitotic stage.

The cell cycle, including interphase and cell division, occurs continuously in certain tissues. Right now, your body is producing thousands of new red blood cells, skin cells, and cells that line your respiratory and digestive tracts. Mitosis is balanced by the process of **apoptosis,** or programmed cell death. Apoptosis occurs when cells are no longer needed, or have become excessively damaged.

Cell Cycle Control

In order for a cell to reproduce successfully, the cell cycle must be controlled. The cell cycle is controlled by **checkpoints** that can delay the cell cycle until certain conditions are met. The cell cycle has many checkpoints, but we will consider only three: G_1, G_2, and the mitotic checkpoint (Fig. 18.3). In addition, the cell cycle may be controlled by external factors, such as hormones and growth factors. Failure of the cell cycle control mechanisms may result in unrestricted cell growth, or cancer.

Checkpoints The *G_1 checkpoint* is especially significant because if the cell cycle passes this checkpoint, the cell is committed to divide. If the cell does not pass this checkpoint, it can enter a holding phase called G_0, during which it performs its normal functions but does not divide. The proper growth signals, such as certain growth factors, must be present for a cell to pass the G_1 checkpoint. Additionally, the integrity of the cell's DNA is also checked. If the DNA is damaged, the protein p53 can stop the cycle at this checkpoint and place the cell in G_0 phase. G_0 phase acts like a holding phase, if the DNA can be repaired, the cell may reenter the cell cycle. If not, then internal mechanisms cause the cell to undergo apoptosis. The cell cycle halts momentarily at the *G_2 checkpoint* until the cell verifies that DNA has replicated. This prevents the initiation of the M stage unless the chromosomes are duplicated. Also, if DNA is damaged, as from exposure to solar radiation or X-rays, arresting the cell cycle at this checkpoint allows time for the damage to be repaired so that it is not passed on to daughter cells.

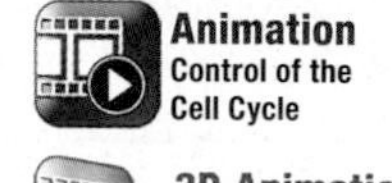

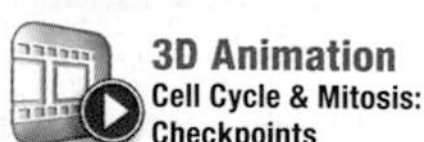

Figure 18.3 Control of the cell cycle.
Checkpoints regulate the progression of the cell through the cell cycle. This diagram shows three of the major checkpoints.

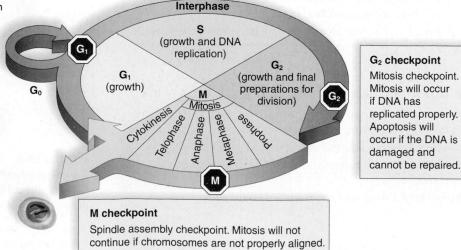

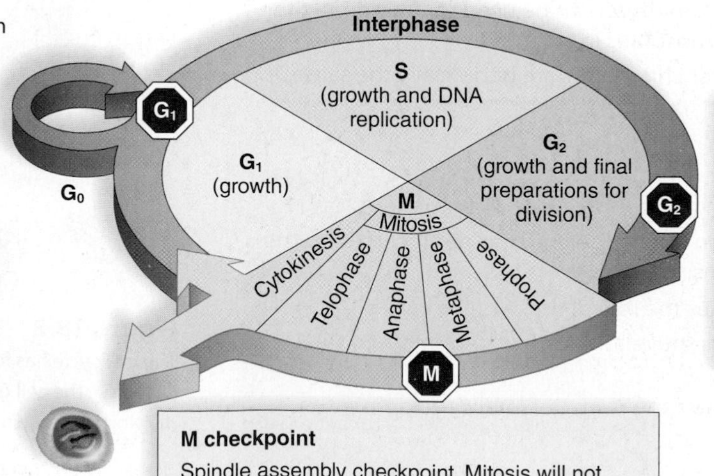

G_1 checkpoint
Cell cycle main checkpoint. If DNA is damaged, apoptosis will occur. Otherwise, the cell is committed to divide when growth signals are present and nutrients are available.

G_2 checkpoint
Mitosis checkpoint. Mitosis will occur if DNA has replicated properly. Apoptosis will occur if the DNA is damaged and cannot be repaired.

M checkpoint
Spindle assembly checkpoint. Mitosis will not continue if chromosomes are not properly aligned.

Another cell cycle checkpoint occurs during the mitotic stage. The cycle hesitates between metaphase and anaphase to make sure the chromosomes are properly attached to the spindle and will be distributed accurately to the daughter cells. The cell cycle does not continue until every chromosome is ready for the nuclear division process.

External Control The cell cycle control system extends from the plasma membrane to particular genes in the nucleus. Some external signals such as hormones and growth factors can stimulate a cell to go through the cell cycle. At a certain time in the menstrual cycle of females, the hormone progesterone stimulates cells lining the uterus to prepare the lining for implantation of a fertilized egg. Epidermal growth factor stimulates skin in the vicinity of an injury to finish the cell cycle, thereby repairing damage.

As shown in Figure 18.4, ①, during reception, an external signal delivers a message to a specific receptor embedded in the plasma membrane of a receiving cell. ② The receptor relays the signal to proteins inside the cell's cytoplasm. The proteins form a pathway called the signal transduction pathway because they pass the signal from one to the other. ③ The last signal activates genes whose protein product ④ stimulates or inhibits the cell cycle. Genes called proto-oncogenes stimulate the cell cycle, and genes called tumor suppressor genes inhibit the cell cycle. These genes are discussed further in Chapter 19, which is about cancer.

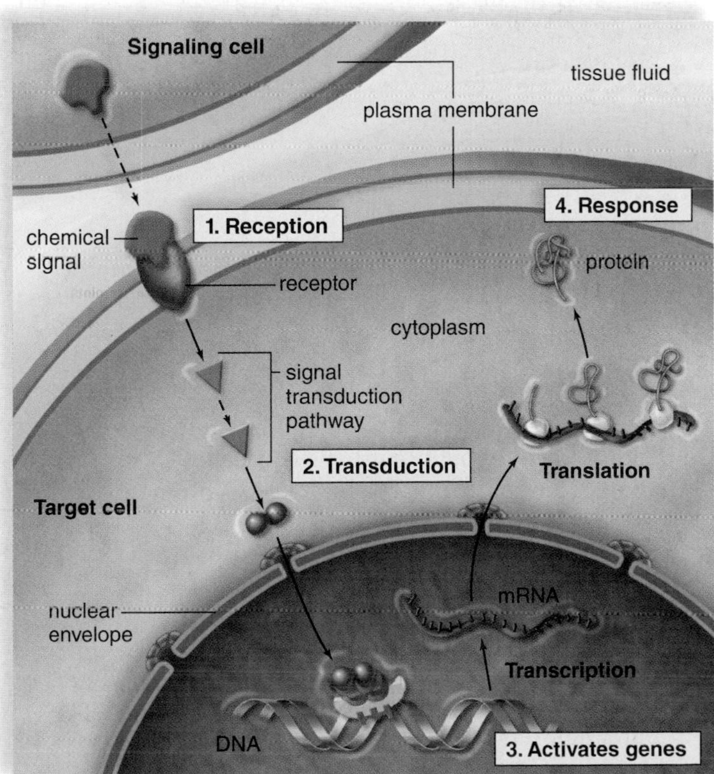

Figure 18.4 External controls of the cell cycle.
Growth factors stimulate a cell signaling pathway that stretches from the plasma membrane to the genes that regulate the occurrence of the cell cycle.

CHECK YOUR PROGRESS 18.2

1. Describe the cell cycle and list the locations of each phase and checkpoint.
2. Explain the purpose of the S phase in the cell cycle.
3. Explain how checkpoints help protect the cell against unregulated cell growth.
4. Summarize why external controls may be necessary to regulate the cell cycle.

CONNECTING THE CONCEPTS

For more information on the material presented in this section, refer to the following discussions:

Figure 15.5 illustrates how steroid hormones, such as progesterone, influence the internal activities of a cell.

Section 19.1 describes the differences between proto-oncogenes and tumor suppressor genes.

Section 19.2 explores how environmental factors, such as radiation, may cause cancer.

18.3 Mitosis

LEARNING OUTCOMES

Upon completion of this section, you should be able to

1. Explain the purpose of mitosis.
2. Explain the events that occur in each stage of mitosis.
3. State the purpose of cytokinesis.

The cell cycle, which includes mitosis, is very important to the well-being of humans. Mitosis is responsible for new cells in the developing embryo, fetus, and child. It is also responsible for replacement cells in an adult (Fig. 18.5). As mentioned, mitosis is *duplication division.* During mitosis, the cell that

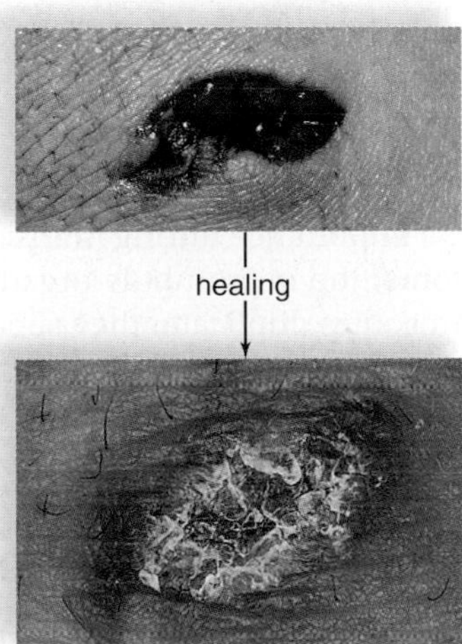

Figure 18.5 The importance of mitosis.
The cell cycle, including mitosis, occurs when humans grow and when tissues undergo repair.

divides is called the **parent cell,** and the new cells are called the **daughter cells.** At the conclusion of mitosis, the nuclei of the two new daughter cells have the same number and types of chromosomes as the parent cell, so they are genetically identical.

Overview of Mitosis

During S phase of the cell cycle, replication of the DNA occurs (see section 21.1), thus duplicating the chromosomes. Each chromosome now contains two identical parts, called sister chromatids, held together at a centromere. They are called *sister chromatids* because they contain the same genes:

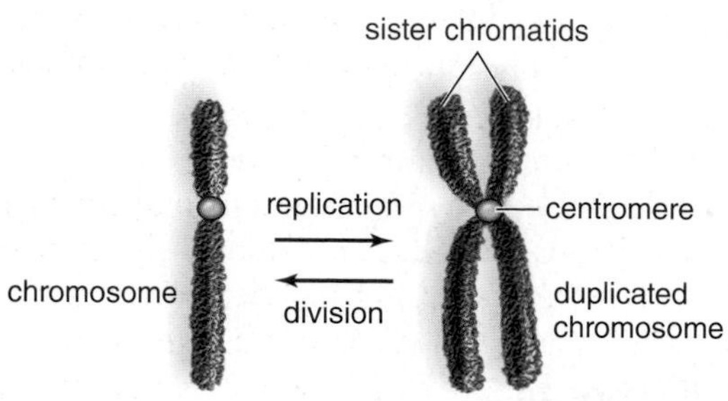

As mitosis begins, the proteins in the chromatin assist in causing the chromosomes to become highly condensed. Once this occurs, the chromosomes become visible under a light microscope. Figure 18.6 gives an overview of mitosis. For simplicity, only four chromosomes are depicted. (In determining the number of chromosomes, it is necessary to count only the number of independent centromeres.) As you know, the complete number of chromosomes is called the **diploid (2n)** number.

During mitosis, the centromeres divide and the sister chromatids separate. Following separation during mitosis, each chromatid is called a *chromosome.* Each daughter cell gets a complete set of chromosomes and is diploid. Therefore, each daughter cell receives the same number and types of chromosomes as the parent cell. Each daughter cell is genetically identical to the other and to the parent cell.

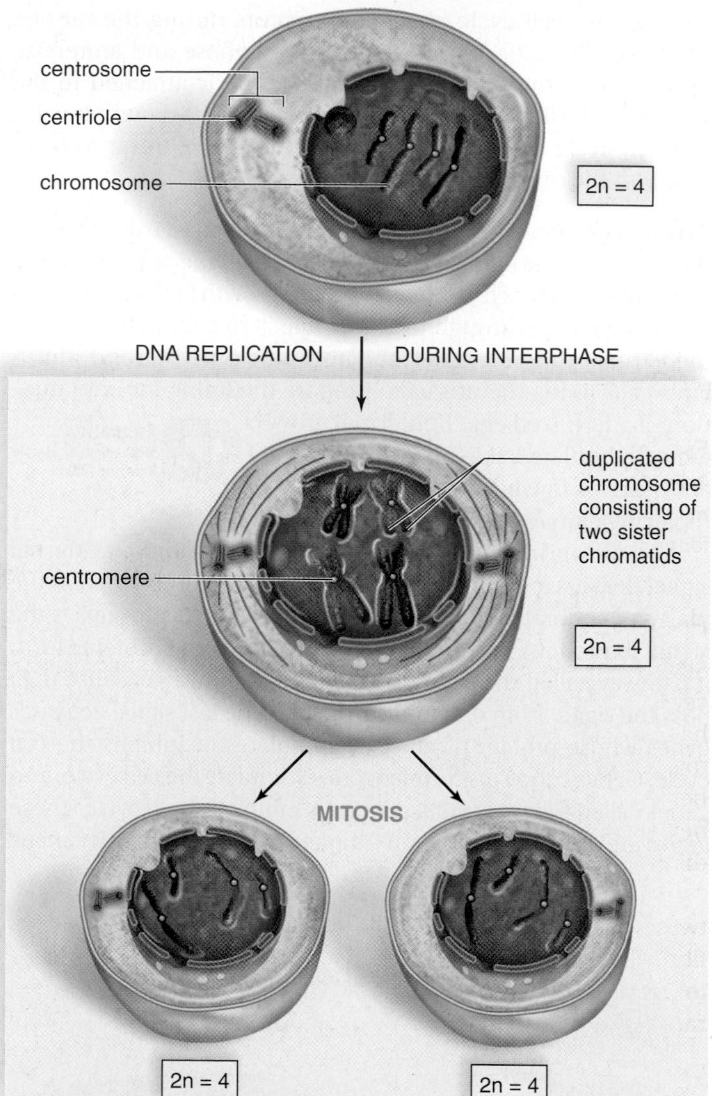

Figure 18.6 An overview of mitosis.
Following DNA replication, each chromosome is duplicated. When the centromeres split, the sister chromatids, now called chromosomes, move into daughter nuclei. (The blue and red chromosomes were inherited from different parents.)

The Mitotic Spindle

Another event of importance during mitosis is the duplication of the **centrosome,** the microtubule organizing center of the cell. After centrosomes duplicate, they separate and form the poles of the **mitotic spindle,** where they assemble the microtubules that make up the spindle fibers. The chromosomes are attached to the spindle fibers at their centromeres (Fig. 18.7). An array of microtubules called an *aster* (because it looks like a star) is also at the poles. Each centrosome contains a pair of **centrioles,** which consist of short cylinders of microtubules. The centrioles lie at right angles to one another. Centrioles are absent in plant cells, and their function in animals cells is not well understood, although it is believed that they assist in the formation of the spindle that separates the chromatids during mitosis.

Phases of Mitosis

As an aid in describing the events of mitosis, the process is divided into four phases: prophase, metaphase, anaphase, and telophase (Fig. 18.8). Although the stages of mitosis are depicted as if they were separate, they are continuous. One stage flows from the other with no noticeable interruption.

Animation
Mitosis

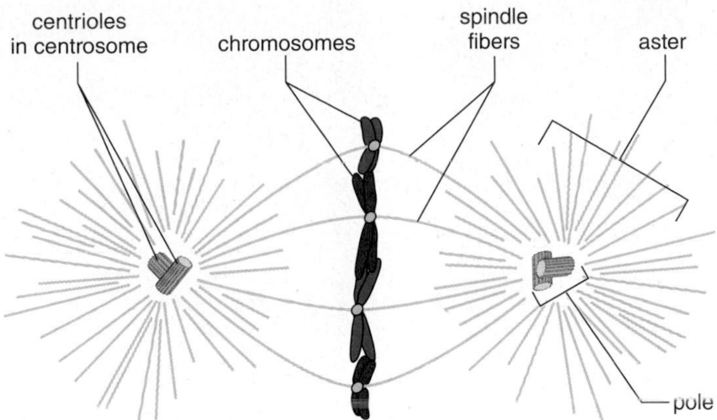

Figure 18.7 The mitotic spindle.
Before mitosis begins, the centrosome and the centrioles duplicate. During mitosis, they separate, and the mitotic spindle, composed of microtubules, forms between them.

Prophase

Several events occur during **prophase** that visibly indicate the cell is preparing to divide. The centrosomes outside the nucleus have duplicated, and they begin moving away from one another toward opposite ends of the nucleus. Spindle fibers appear between the separating centrosomes. The nuclear envelope begins to fragment. The nucleolus, a special region of DNA, disappears as the chromosomes coil and become condensed.

The chromosomes are now visible. Each is composed of two sister chromatids held together at a centromere. Spindle fibers attach to the centromeres as the chromosomes continue to shorten and to thicken. During prophase, chromosomes are randomly placed in the nucleus (Fig. 18.8).

Metaphase

During **metaphase,** the nuclear envelope is fragmented and the spindle occupies the region formerly occupied by the nucleus. The chromosomes are now at the equator (center) of the spindle. Metaphase is characterized by a fully formed spindle. The chromosomes, each with two sister chromatids, are aligned at the equator (Fig. 18.8).

Anaphase

At the start of **anaphase,** the centromeres uniting the sister chromatids divide. Then the sister chromatids separate, becoming chromosomes that move toward opposite poles of the spindle. Separation of the sister chromatids ensures that each cell receives a copy of each type of chromosome and thereby has a full complement of genes. Anaphase is characterized by the 2n (diploid) number of chromosomes moving toward each pole.

Remember that the number of centromeres indicates the number of chromosomes. Therefore, in Figure 18.8, each pole receives four chromosomes: two are red and two are blue.

Spindle Function in Anaphase The spindle brings about chromosomal movement. Two types of spindle fibers are involved in the movement of chromosomes during anaphase. One type extends from the poles to the equator of the spindle. There, they overlap. As mitosis proceeds, these fibers increase in length, which helps push the chromosomes apart. The chromosomes themselves are attached to other spindle fibers that extend from their centromeres to the poles. These fibers (composed of microtubules that can disassemble) become shorter as the chromosomes move toward the poles. Therefore, they pull the chromosomes apart.

Spindle fibers, as stated earlier, are composed of microtubules. Microtubules can assemble and disassemble by the addition or subtraction of tubulin (protein) subunits. This is what enables spindle fibers to lengthen and shorten and what ultimately causes the movement of the chromosomes.

Telophase

Telophase begins when the chromosomes arrive at the poles. During telophase, the chromosomes become indistinct chromatin again. The spindle disappears as the nuclear envelope components reassemble in each cell. Each nucleus has a nucleolus because each has a region of the DNA where ribosomal subunits are produced. Telophase is characterized by the presence of two daughter nuclei.

3D Animation Cell Cycle & Mitosis: Mitosis

Cytokinesis

Cytokinesis is the division of the cytoplasm and organelles. In human cells, a slight indentation called a **cleavage furrow** passes around the circumference of the cell. Actin filaments form a contractile ring; as the ring becomes smaller, the cleavage furrow pinches the cell in half. As a result, each cell becomes enclosed by its own plasma membrane.

Animation Mitosis and Cytokinesis

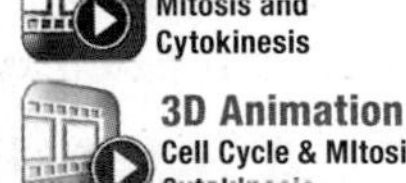

CHECK YOUR PROGRESS 18.3

1. Explain how the chromosome number of the daughter cell compares with the chromosome number of the parent cell following mitosis.
2. List the phases of mitosis, and explain what happens during each phase.
3. Describe how the cytoplasm is divided between the daughter cells following mitosis.

CONNECTING THE CONCEPTS

For more information on mitosis and cytokinesis, refer to the following discussions:

Section 3.4 examines the structure of the nucleus.

Section 3.5 describes the function of microtubules and actin filaments in a nondividing cell.

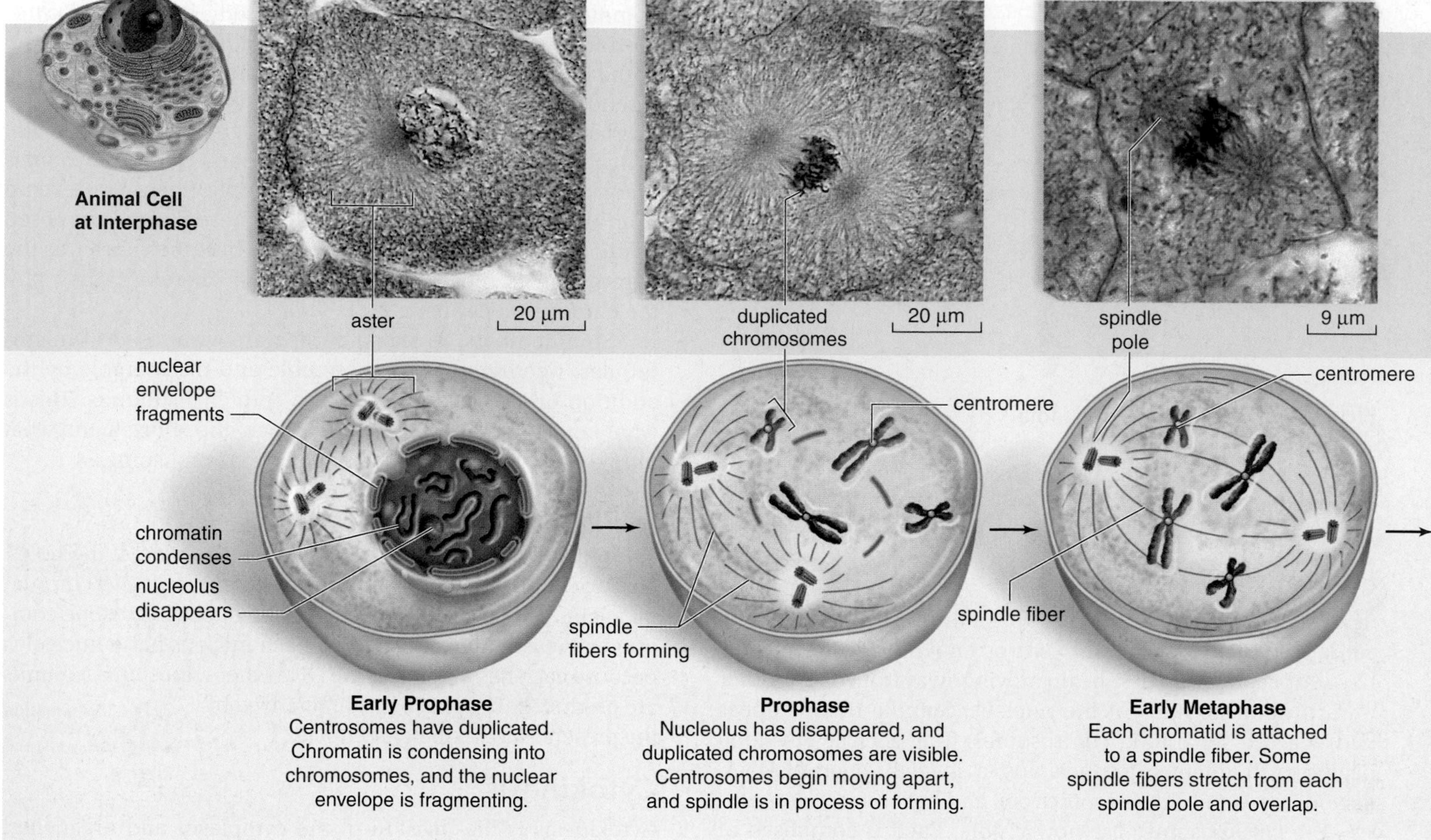
Animal Cell
at Interphase
aster
20 µm
duplicated
chromosomes
20 µm
spindle
pole
9 µm
nuclear
envelope
fragments
centromere
centromere
chromatin
condenses
nucleolus
disappears
spindle
fibers forming
spindle fiber
Early Prophase
Centrosomes have duplicated.
Chromatin is condensing into
chromosomes, and the nuclear
envelope is fragmenting.
Prophase
Nucleolus has disappeared, and
duplicated chromosomes are visible.
Centrosomes begin moving apart,
and spindle is in process of forming.
Early Metaphase
Each chromatid is attached
to a spindle fiber. Some
spindle fibers stretch from each
spindle pole and overlap.

Figure 18.8 The phases of mitosis.

18.4 Meiosis

LEARNING OUTCOMES

Upon completion of this section, you should be able to

1. List the stages of meiosis, and describe what occurs in each stage.
2. Explain how meiosis increases genetic variation.
3. Differentiate between spermatogenesis and oogenesis with regard to occurrence and the number of functional gametes produced by each process.

Meiosis is reduction division, meaning that it reduces the chromosome number in the daughter cells. To do this, meiosis involves two consecutive cell divisions, without an intervening interphase. The end result is four daughter cells, each of which has one of each type of chromosome and, therefore, half as many chromosomes as the parent cell. The parent cell has the diploid (2n) number of chromosomes; the daughter cells have half this number, called the **haploid (n)** number of chromosomes. In addition, meiosis introduces genetic variation, which means that each of the resulting daughter cells is not a genetic replicate of the parent cell, but rather possesses new combinations of the genetic material. In animals, including humans, the daughter cells that result from meiosis may go on to become the gametes.

Overview of Meiosis

At the start of meiosis, the parent cell is 2n, or diploid, and the chromosomes occur in pairs. For simplicity's sake, Figure 18.9 has only two pairs of chromosomes. In this figure, the diploid (2n) number of chromosomes is four chromosomes. The short chromosomes are one pair, and the long chromosomes are another. The members of a pair are called **homologous chromosomes,** or *homologues,* because they look alike and carry genes for the same traits, such as type of hair or color of eyes. Notice that the parent cell (*top*) has the diploid (2n) number of chromosomes; the daughter cells (*bottom*) have the haploid (n) number of chromosomes, equal to two chromosomes.

MP3 Meiosis

Meiosis I

The two cell divisions of meiosis are called *meiosis I* and *meiosis II.* Prior to meiosis I, DNA replication has occurred and the chromosomes are duplicated. Each chromosome consists of two chromatids held together at a centromere. During

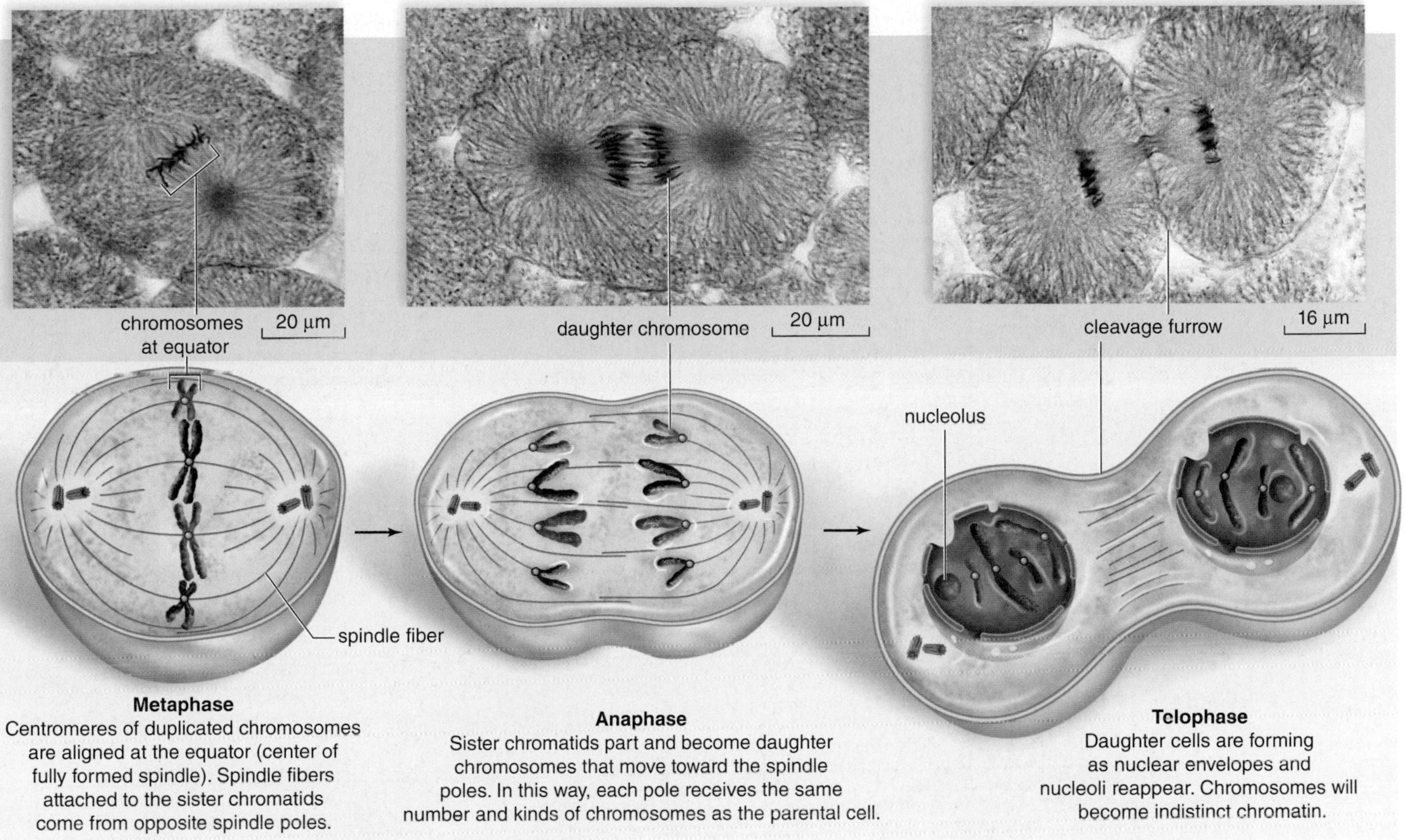

chromosomes at equator
20 μm
daughter chromosome
20 μm
cleavage furrow
16 μm
nucleolus
spindle fiber

Metaphase
Centromeres of duplicated chromosomes are aligned at the equator (center of fully formed spindle). Spindle fibers attached to the sister chromatids come from opposite spindle poles.

Anaphase
Sister chromatids part and become daughter chromosomes that move toward the spindle poles. In this way, each pole receives the same number and kinds of chromosomes as the parental cell.

Telophase
Daughter cells are forming as nuclear envelopes and nucleoli reappear. Chromosomes will become indistinct chromatin.

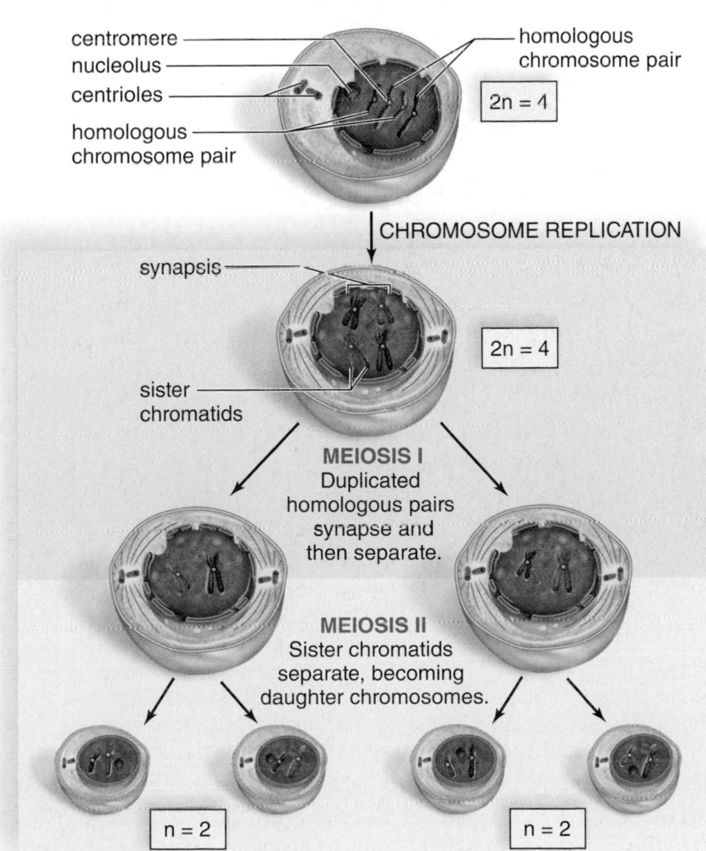

Figure 18.9 **The results of meiosis.**
DNA replication is followed by meiosis I when homologous chromosomes pair and then separate. During meiosis II, the sister chromatids become chromosomes that move into daughter nuclei.

meiosis I (Figure 18.10, *top*), the homologous chromosomes come together and line up side by side. This is called **synapsis,** and results in an association of four chromatids that stay in close proximity during the first two phases of meiosis I. Synapsis is significant because its occurrence leads to a reduction of the chromosome number.

There are pairs of homologous chromosomes at the equator during meiosis I because of synapsis. Only during meiosis I is it possible to observe paired chromosomes at the equator. When the members of these pairs separate, each daughter nucleus receives one member of each pair. Therefore, each daughter cell now has the haploid (n) number of chromosomes, as you can verify by counting its centromeres. Each chromosome, however, is still duplicated. No replication of DNA occurs between meiosis I and meiosis II. The time between meiosis I and meiosis II is called **interkinesis.**

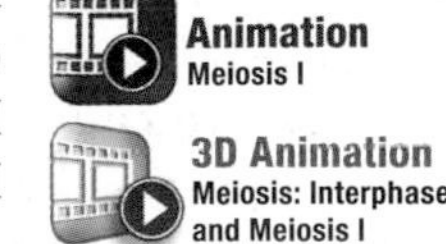

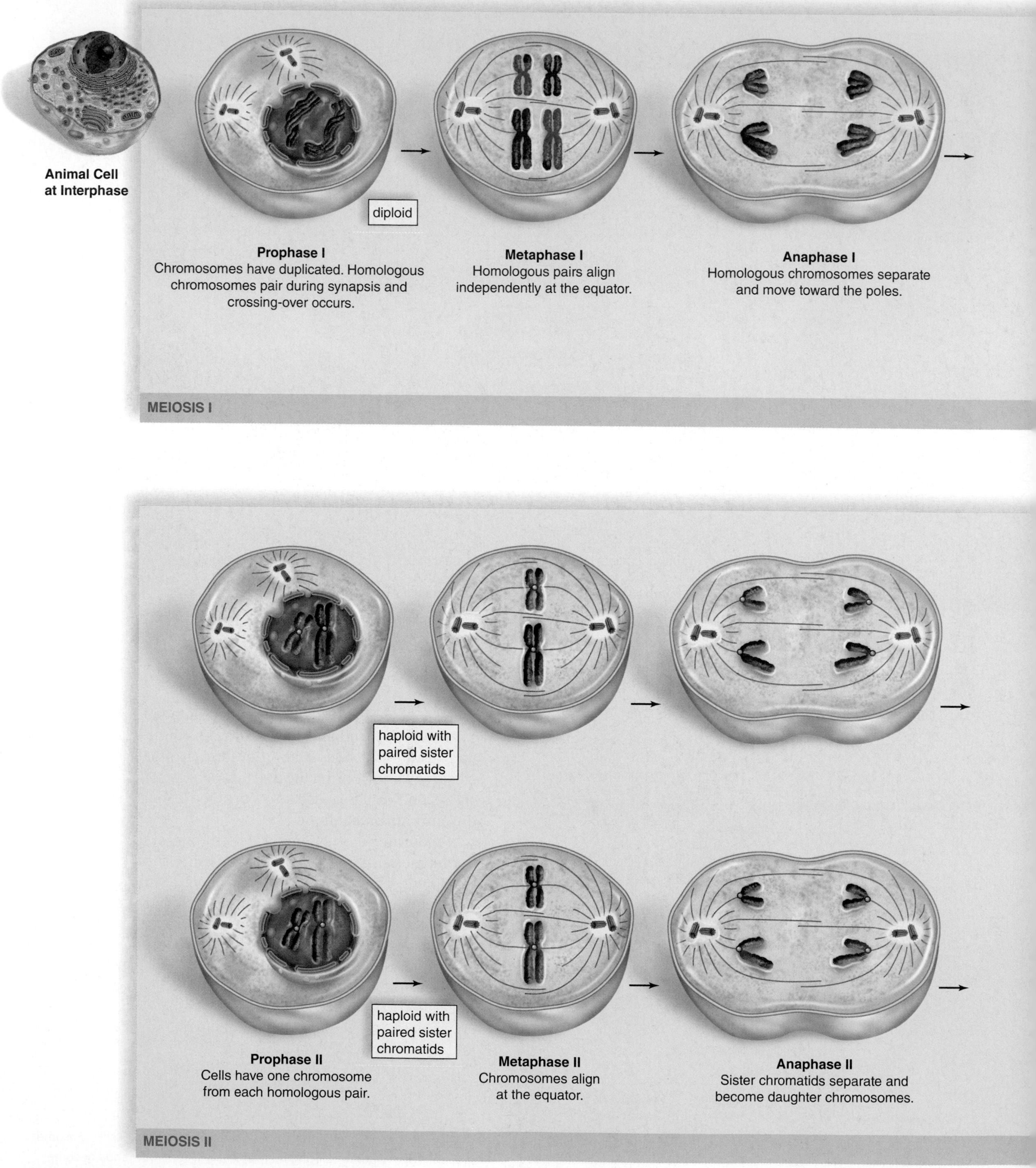

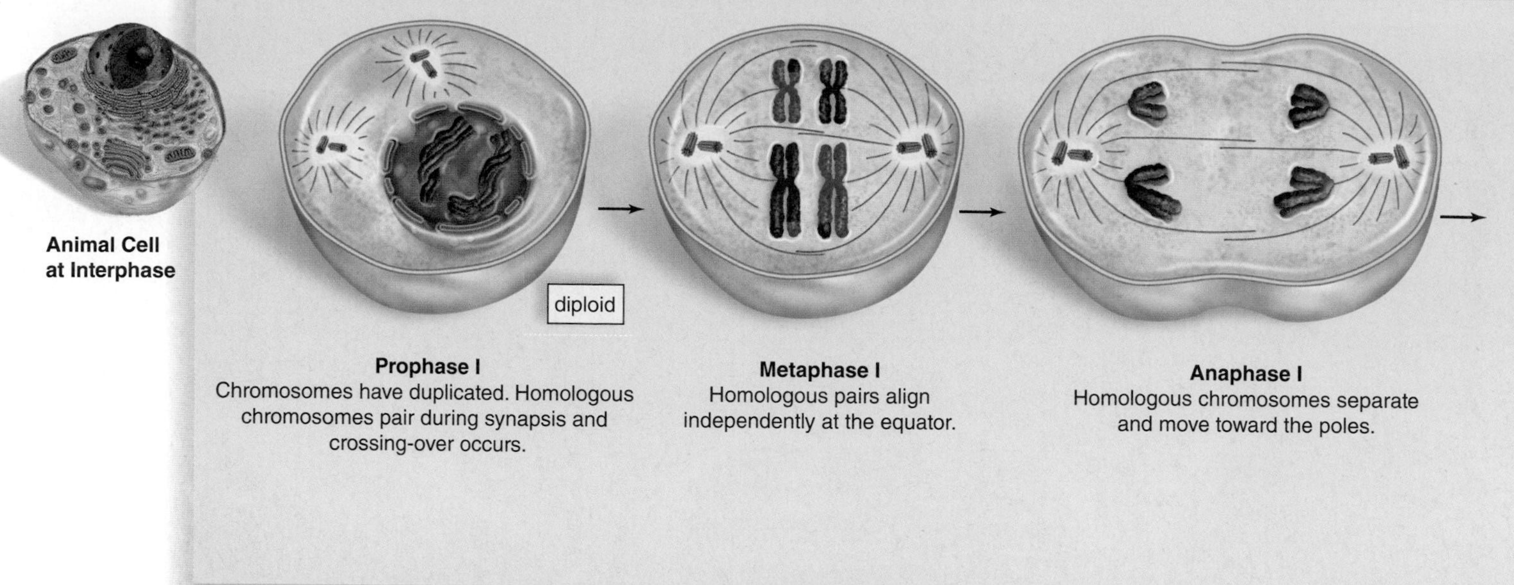
Animal Cell
at Interphase
diploid
Prophase I
Chromosomes have duplicated. Homologous chromosomes pair during synapsis and crossing-over occurs.
Metaphase I
Homologous pairs align independently at the equator.
Anaphase I
Homologous chromosomes separate and move toward the poles.
MEIOSIS I

haploid with paired sister chromatids

haploid with paired sister chromatids

Prophase II
Cells have one chromosome from each homologous pair.

Metaphase II
Chromosomes align at the equator.

Anaphase II
Sister chromatids separate and become daughter chromosomes.

MEIOSIS II

Figure 18.10 The phases of meiosis.
Homologous chromosomes pair and then separate during meiosis I. Crossing-over, which occurs during meiosis I, is discussed more fully on page 426. Chromatids separate, becoming daughter chromosomes during meiosis II. Following meiosis II, there are four haploid daughter cells.

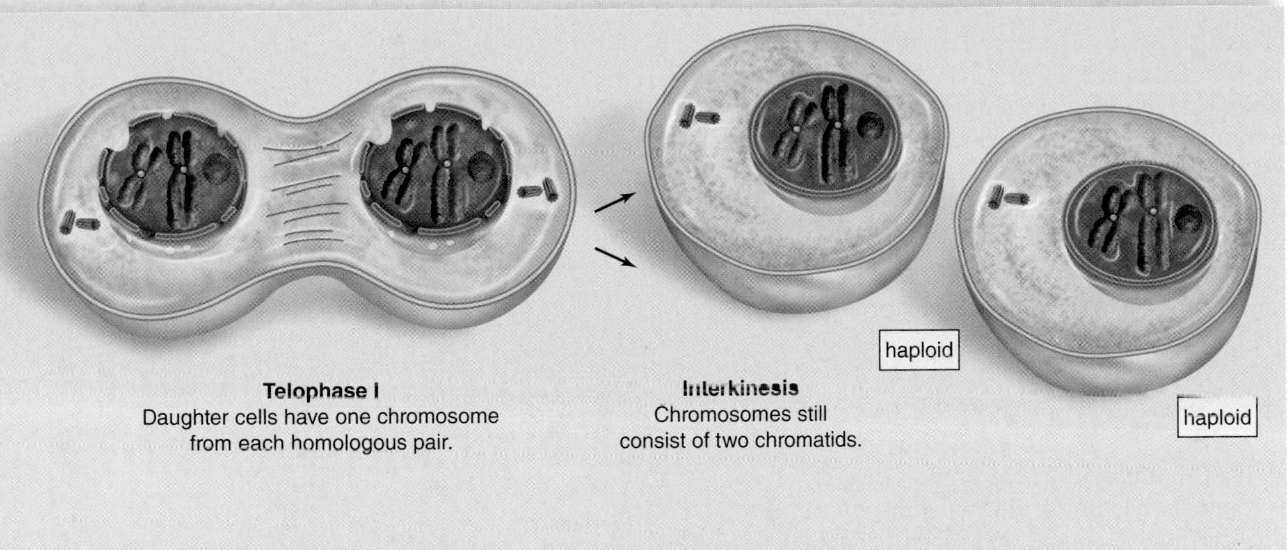

Telophase I
Daughter cells have one chromosome from each homologous pair.

Interkinesis
Chromosomes still consist of two chromatids.

MEIOSIS I cont'd

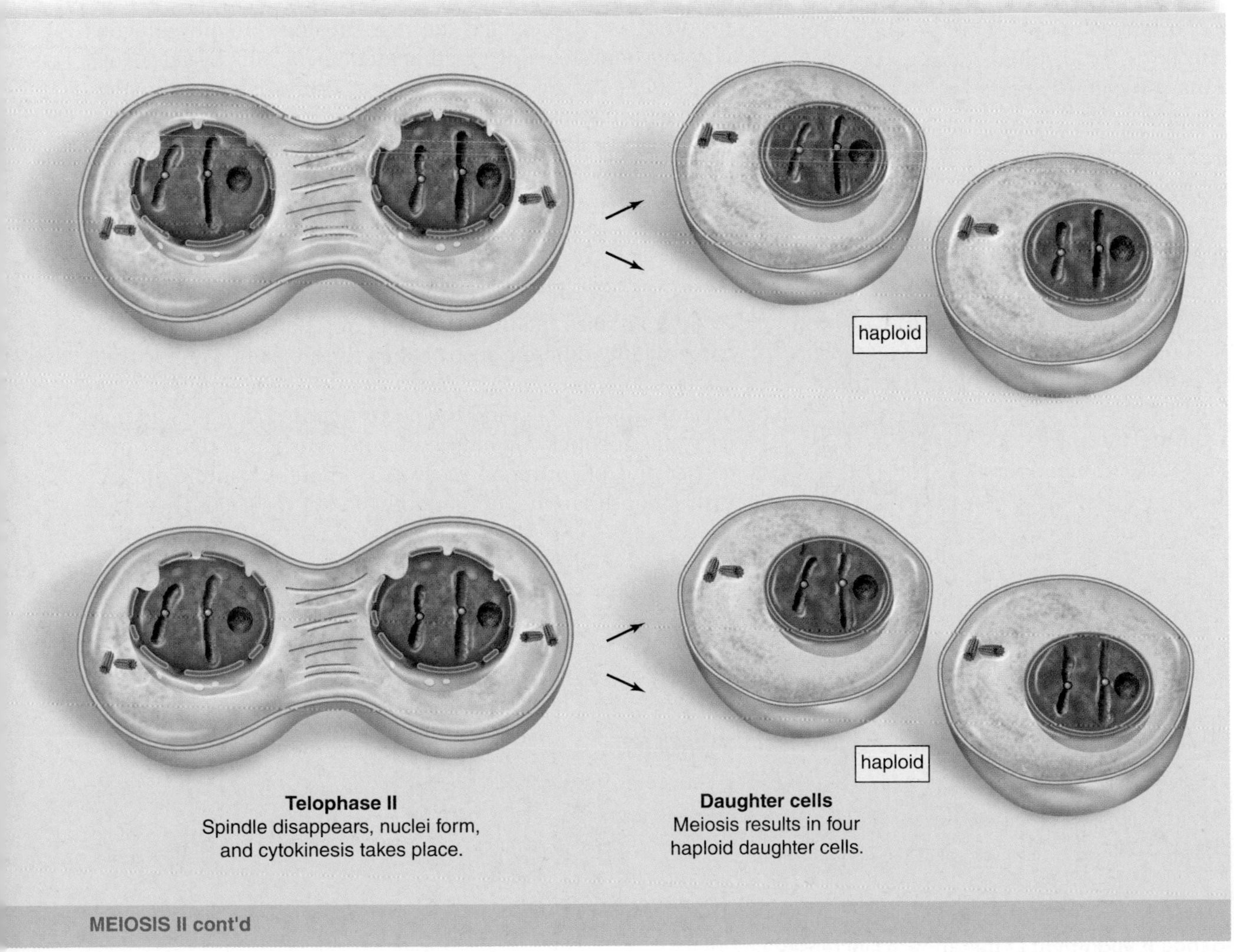

haploid

Telophase II
Spindle disappears, nuclei form, and cytokinesis takes place.

haploid

Daughter cells
Meiosis results in four haploid daughter cells.

MEIOSIS II cont'd

APPLICATIONS AND MISCONCEPTIONS

How common are mistakes made during meiosis that result in chromosomal abnormalities in the sperm or egg?

It is estimated that 8% of all clinically recognized pregnancies have some form of chromosomal aberration. In a spontaneous abortion (typically referred to as a *miscarriage*), the frequency of chromosomal abnormalities rises to approximately 50%. Most of these are the result of errors during meiosis in parents with normal karyotypes.

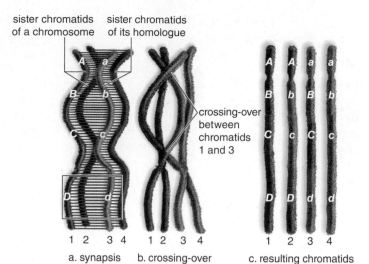

Meiosis II

During meiosis II (Fig. 18.10, *bottom*), the centromeres divide. The sister chromatids separate, becoming chromosomes that are distributed to daughter nuclei. In the end, each of four daughter cells has the n, or haploid, number of chromosomes. Each chromosome consists of one chromatid.

In humans, the daughter cells mature into gametes (sperm and egg) that fuse during fertilization. Fertilization restores the diploid number of chromosomes in the zygote, the first cell of the new individual. If the gametes carried the diploid instead of the haploid number of chromosomes, the chromosome number would double with each fertilization. After several generations, the zygote would be nothing but chromosomes.

Meiosis and Genetic Variation

Meiosis is a part of sexual reproduction. The process of meiosis ensures that the next generation of individuals will have the diploid number of chromosomes and a combination of genetic characteristics different from that of either parent. Though both meiosis I and meiosis II have the same four stages of nuclear division as did mitosis, as discussed previously, here we discuss only prophase I and metaphase I because special events occur during these phases that introduce new genetic combinations into the daughter cells.

sister chromatids of a chromosome
sister chromatids of its homologue
crossing-over between chromatids 1 and 3
1 2 3 4
a. synapsis
b. crossing-over
c. resulting chromatids

Figure 18.11 Synapsis and crossing-over increase variability.
a. During meiosis I, duplicated homologous chromosomes undergo synapsis and line up with each other. **b.** During crossing-over, nonsister chromatids break and then rejoin in the manner shown. **c.** Two of the resulting chromosomes have a different combination of genes than they had before.

Prophase I

In prophase I, synapsis occurs causing the homologous chromosomes to come together and line up side by side. Now, an exchange of genetic material may occur between the nonsister chromatids of the homologous pair (Fig. 18.11). This exchange is called **crossing-over.** Notice that in Figure 18.11, the crossing-over events (there may be more than one) have produced chromatids that are no longer identical. When the chromatids separate during meiosis II, the daughter cells receive chromosomes with recombined genetic material.

To appreciate the significance of crossing-over, it is necessary to realize that the members of a homologous pair can carry slightly different instructions for the same genetic trait. For example, one homologue may carry instructions for brown eyes and blond hair, and the corresponding homologue may carry instructions for blue eyes and red hair. Crossing-over causes the offspring to receive a different combination of instructions than the mother or the father received. Therefore,

Figure 18.12 Independent alignment at metaphase I increases variability.
When a parent cell has three pairs of homologous chromosomes, there are eight possible chromosome alignments at the equator due to independent assortment. Among the 16 daughter nuclei resulting from these alignments, there are eight different combinations of chromosomes.

offspring could receive brown eyes and red hair or blue eyes and blond hair.

Metaphase I

During metaphase I, the homologous pairs align independently at the equator. This means that the maternal or paternal member may be oriented toward either pole. Figure 18.12 shows the eight possible orientations for a cell that contains only three pairs of chromosomes. The first four orientations will result in gametes that have different combinations of maternal and paternal chromosomes. The next four will result in the same types of gametes as the first four. For example, the first cell and the last cell will both produce gametes with either three red or three blue chromosomes.

Once all possible orientations are considered, the result will be 2^3, or 8, possible combinations of maternal and paternal chromosomes in the resulting gametes from this cell. In humans, in whom there are 23 pairs of chromosomes, the number of possible chromosomal combinations in the gametes is a staggering 2^{23}, or 8,388,608—and this does not even consider the genetic variations introduced due to crossing-over.

The events of prophase I and metaphase I help ensure that gametes will not have the same combination of chromosomes and genes.

Animation Random Orientation of Chromosomes During Meiosis

Spermatogenesis and Oogenesis

Meiosis is a part of **spermatogenesis,** the production of sperm in males, and **oogenesis,** the production of eggs in females. Following meiosis, the daughter cells mature to become the gametes.

Spermatogenesis

After puberty, the time of life when the sex organs mature, spermatogenesis is continual in the testes of human males. As many as 300,000 sperm are produced per minute, or over 400 million per day.

Spermatogenesis is shown in Figure 18.13, *top*. The *primary spermatocytes,* which are diploid (2n), divide during meiosis I to form two *secondary spermatocytes,* which are haploid (n). Secondary spermatocytes divide during meiosis II to produce four *spermatids,* which are also haploid (n). What's the difference between the chromosomes in haploid secondary spermatocytes and those in haploid spermatids? The chromosomes in secondary spermatocytes are duplicated and consist of two chromatids, whereas those in spermatids consist of only one. Spermatids mature into sperm (spermatozoa). In human males, sperm have 23 chromosomes, the haploid number. The process of meiosis in males always results in four cells that become sperm. In other words, all four daughter cells—the spermatids—become sperm.

Oogenesis

As you know, the ovary of a female contains many immature follicles (see Fig. 16.8). Each of these follicles contains a primary oocyte arrested in prophase I. As shown in Figure 18.13, *bottom,* a primary oocyte, which is diploid (2n), divides during meiosis I into two cells, each of which is haploid. The chromosomes are duplicated. One of these cells, termed the *secondary oocyte,* receives almost all the cytoplasm. The other is the first polar body.

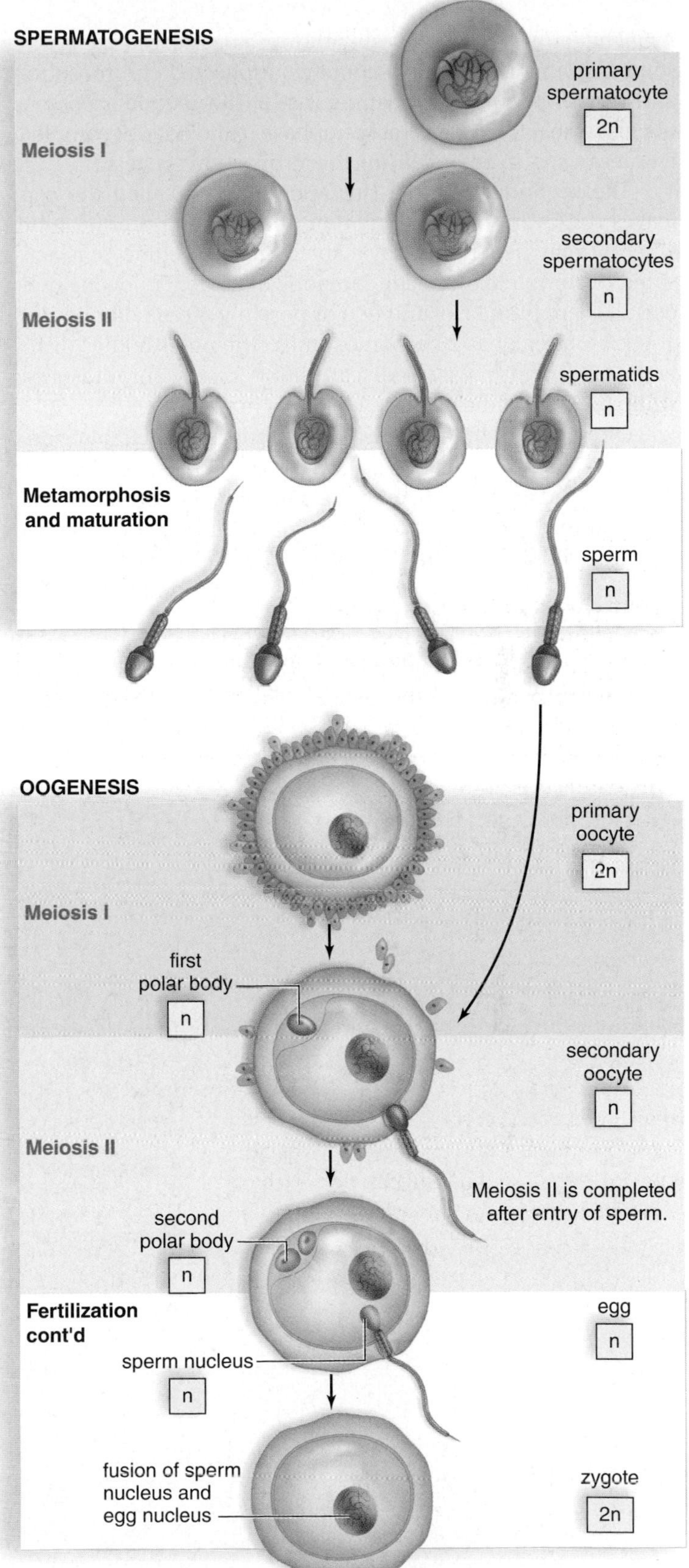

Figure 18.13 A comparison of spermatogenesis and oogenesis in mammals.

Spermatogenesis produces four viable sperm, whereas oogenesis produces one egg and at least two polar bodies. In humans, both sperm and egg have 23 chromosomes each; therefore, following fertilization, the zygote has 46 chromosomes.

A *polar body* acts like a trash can to hold discarded chromosomes. The first polar body contains duplicated chromosomes and occasionally completes meiosis II. The secondary oocyte begins meiosis II but stops at metaphase II and doesn't complete it unless a sperm enters during the fertilization process.

The secondary oocyte (for convenience, called the egg) leaves the ovary during ovulation and enters a uterine tube, where it may be fertilized by a sperm. If so, the oocyte is activated to complete the second meiotic division. Following meiosis II, there is one egg and two or possibly three polar bodies. The mature egg has 23 chromosomes. The polar bodies disintegrate, which is a way to discard unnecessary chromosomes while retaining much of the cytoplasm in the egg.

One egg can be the source of identical twins if, after one division of the fertilized egg during development, the cells separate and each one becomes a complete individual. On the other hand, the occurrence of fraternal twins requires that two eggs be ovulated and then fertilized separately.

Significance of Meiosis

In animals, meiosis is a part of gametogenesis, production of the sperm and egg. One function of meiosis is to keep the chromosome number constant from generation to generation. The gametes are haploid, so the zygote has only the diploid number of chromosomes.

An easier way to keep the chromosome number constant is to reproduce asexually. Unicellular organisms such as bacteria, protozoans, and yeasts (a fungi) reproduce by binary fission.

Video Binary Fission

Binary fission is a form of asexual reproduction because one parent produces identical offspring. Binary fission is a quick and easy way to asexually reproduce many organisms within a short time. A bacterium can increase to over 1 million cells in about 7 hours, for example. Then why do organisms expend the energy to reproduce sexually? It takes energy to find a mate, carry out a courtship, and produce eggs or sperm that may never be used for reproductive purposes. A human male produces over 400 million sperm per day, and very few of these will fertilize an egg.

Most likely, humans and other animals practice sexual reproduction that includes meiosis because it results in genetic recombination. Genetic recombination ensures that offspring will be genetically different compared to each other and to their parents. Genetic recombination occurs because of crossing-over and independent alignment of chromosomes. Also, at the time of fertilization, parents contribute genetically different chromosomes to the offspring.

All environments are subject to a change in conditions. Those individuals able to survive in a new environment are able to pass on their genes. Environments are subject to change, so

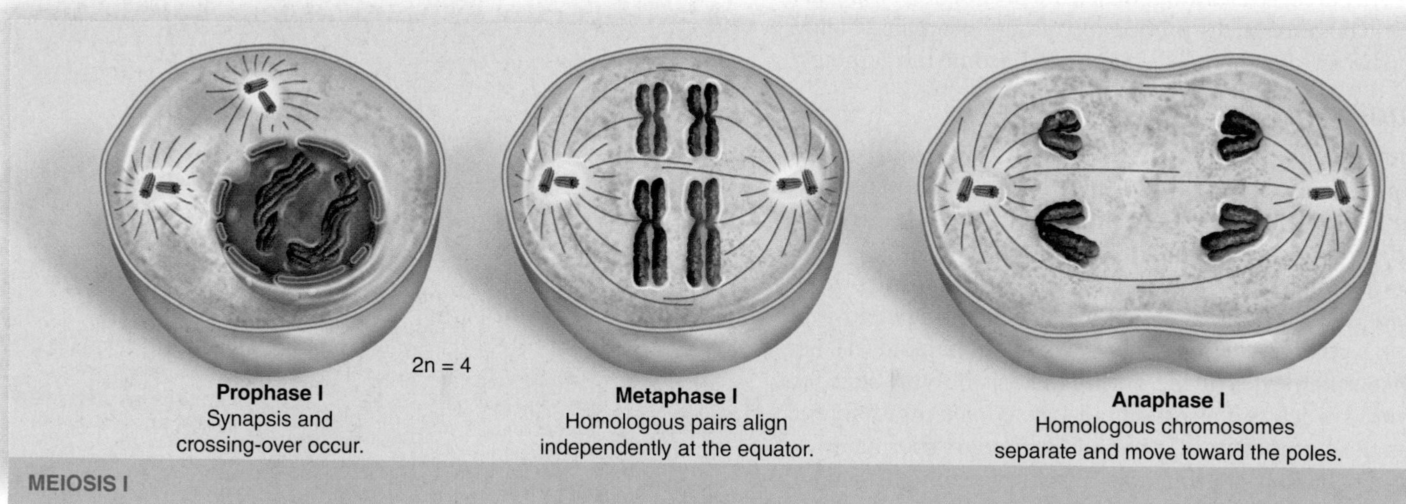

Prophase I Synapsis and crossing-over occur.

Metaphase I Homologous pairs align independently at the equator.

Anaphase I Homologous chromosomes separate and move toward the poles.

Prophase

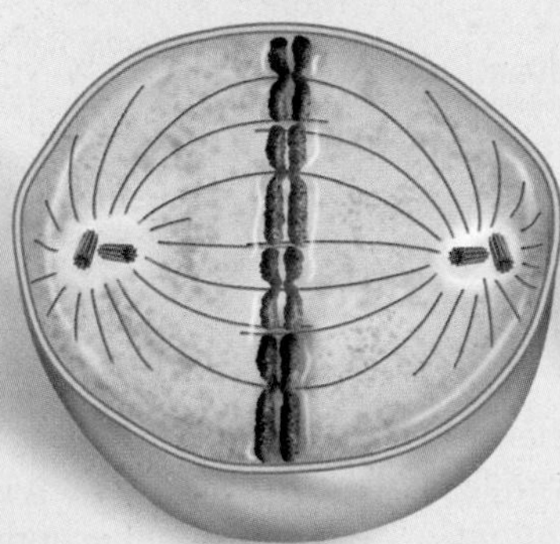

Metaphase Chromosomes align at the equator.

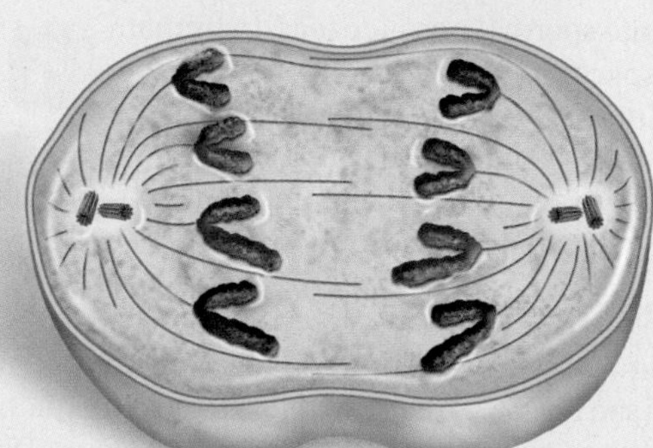

Anaphase Sister chromatids separate and become daughter chromosomes.

MITOSIS

sexual reproduction is advantageous. It generates the diversity needed so that at least a few will be suited to new and different environmental circumstances.

CHECK YOUR PROGRESS 18.4

1. Explain how, following meiosis, the chromosome number of the daughter cells compares to the chromosome number of the parent cell.
2. Explain how meiosis reduces the likelihood that gametes will have the same combination of chromosomes and genes.
3. Summarize the events during the two cell divisions of meiosis.
4. Compare and contrast the stages of oogenesis and spermatogenesis.

CONNECTING THE CONCEPTS

For more on the importance of meiosis, refer to the following discussions:

Figure 16.4 illustrates how meiosis relates to spermatogenesis.

Figure 16.9 demonstrates how meiosis produces eggs and polar bodies during oogenesis.

Figure 20.5 relates meiosis to the patterns of genetic inheritance.

18.5 Comparison of Meiosis and Mitosis

LEARNING OUTCOMES

Upon completion of this section, you should be able to

1. Distinguish between meiosis and mitosis with regard to the number of divisions and the number and chromosome content of the resulting cells.
2. Contrast the events of meiosis I and meiosis II with the events of mitosis.

Meiosis and mitosis are both nuclear divisions, but they differ in the number of cells produced and the genetic complement (haploid or diploid) of each of the daughter cells. Figure 18.14 provides a visual review of the similarities and differences between meiosis and mitosis.

Animation Comparison of Mitosis and Meiosis

General Comparison

DNA replication takes place only once prior to both meiosis and mitosis. Meiosis requires two nuclear divisions, but mitosis requires only one.

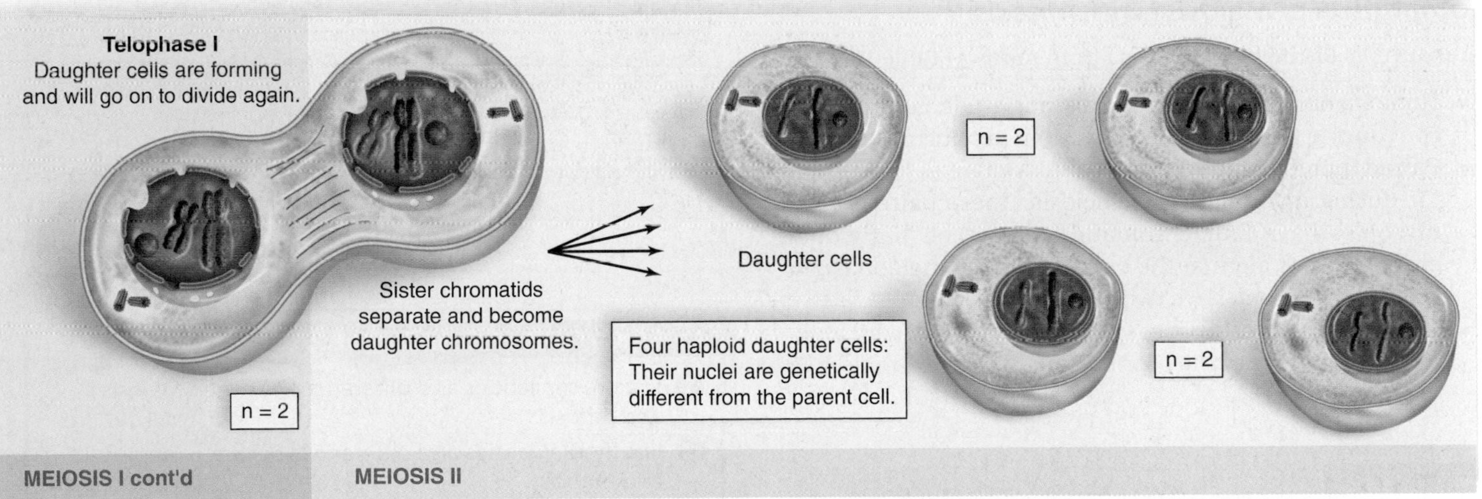
Telophase I
Daughter cells are forming and will go on to divide again.
Sister chromatids separate and become daughter chromosomes.
Daughter cells
n = 2
n = 2
n = 2
Four haploid daughter cells: Their nuclei are genetically different from the parent cell.
MEIOSIS I cont'd
MEIOSIS II

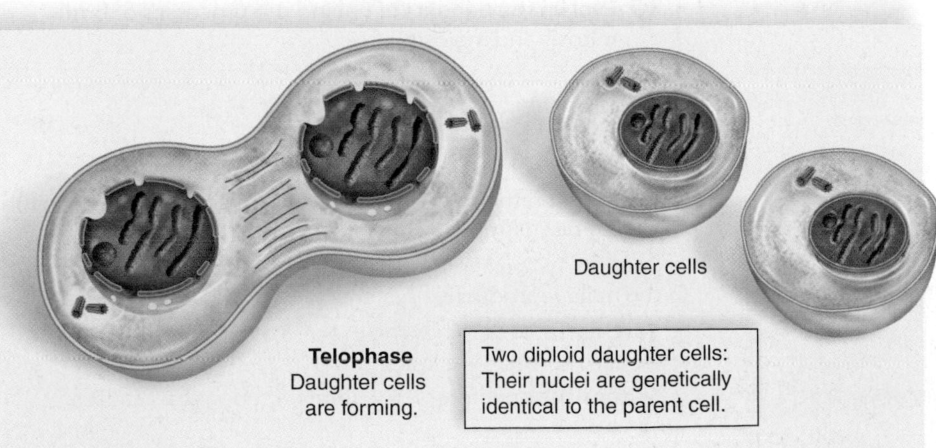
Daughter cells
Telophase
Daughter cells are forming.
Two diploid daughter cells: Their nuclei are genetically identical to the parent cell.
MITOSIS cont'd

Figure 18.14 **A comparison of meiosis and mitosis.**
Why does meiosis produce daughter cells with half the number and mitosis produces daughter cells with the same number of chromosomes as the parent cell? Compare metaphase I of meiosis to metaphase of mitosis. Only in metaphase I are the homologous chromosomes paired at the equator. Members of homologous chromosome pairs separate during anaphase I; therefore, the daughter cells are haploid. The blue chromosomes were inherited from the paternal parent, and the red chromosomes were inherited from the maternal parent. The exchange of color between nonsister chromatids represents the crossing-over that occurs during meiosis I.

- Four daughter nuclei are produced by meiosis; following cytokinesis, there are four daughter cells. Mitosis followed by cytokinesis results in two daughter cells.
- The four daughter cells following meiosis are haploid (n) and have half the chromosome number of the parent cell (2n). The daughter cells following mitosis have the same chromosome number as the parent cell—the 2n, or diploid, number.
- The daughter cells from meiosis are not genetically identical to each other or to the parent cell. The daughter cells from mitosis are genetically identical to each other and to the parent cell.

The specific differences between these nuclear divisions can be categorized according to occurrence and process.

Occurrence

Meiosis occurs only at certain times in the life cycle of sexually reproducing organisms. In humans, meiosis occurs only in the reproductive organs and produces the gametes. Mitosis is more common because it occurs in all tissues during growth and repair. Which type of cell division can lead to cancer? Mitosis can result in a proliferation of body cells. Abnormal mitosis can lead to cancer.

Process

Comparison of Meiosis I with Mitosis

These events distinguish meiosis I from mitosis (Table 18.1):

- Homologous chromosomes pair and undergo crossing-over during prophase I of meiosis but not during mitosis.
- Paired homologous chromosomes align at the equator during metaphase I in meiosis. These paired chromosomes have four chromatids altogether. Individual chromosomes align at the equator during metaphase in mitosis. They each have two chromatids.
- This difference makes it easy to tell whether you are looking at mitosis, meiosis I, or meiosis II. For example, if a cell has 16 chromosomes, then 16 chromosomes are at the equator during mitosis but only 8 chromosomes during meiosis II. Only meiosis I has paired duplicated chromosomes at the equator.
- Homologous chromosomes (with centromeres intact) separate and move to opposite poles during anaphase I of meiosis. Centromeres split, and sister chromatids, now called chromosomes, move to opposite poles during anaphase in mitosis.

Table 18.1 Comparison of Meiosis I with Mitosis

Meiosis I	Mitosis
Prophase I Pairing of homologous chromosomes	*Prophase* No pairing of chromosomes
Metaphase I Homologous duplicated chromosomes at equator	*Metaphase* Duplicated chromosomes at equator
Anaphase I Homologous chromosomes separate	*Anaphase* Sister chromatids separate, becoming daughter chromosomes that move to the poles
Telophase I Two haploid daughter cells	*Telophase* Two daughter cells, identical to the parent cell

Comparison of Meiosis II with Mitosis

The events of meiosis II are just like those of mitosis (Table 18.2) except that, in meiosis II, the nuclei contain the haploid number of chromosomes. If the parent cell has 16 chromosomes, then the cells undergoing meiosis II have 8 chromosomes, and the daughter cells have 8 chromosomes, for example.

Table 18.2 Comparison of Meiosis II with Mitosis

Meiosis II	Mitosis
Prophase II No pairing of chromosomes	*Prophase* No pairing of chromosomes
Metaphase II Haploid number of duplicated chromosomes at equator	*Metaphase* Duplicated chromosomes at equator
Anaphase II Sister chromatids separate, becoming daughter chromosomes that move to the poles	*Anaphase* Sister chromatids separate, becoming daughter chromosomes that move to the poles
Telophase II Four haploid daughter cells	*Telophase* Two daughter cells, identical to the parent cell

CHECK YOUR PROGRESS 18.5

1. List the similarities and differences between meiosis I and mitosis.
2. List the similarities and differences between meiosis II and mitosis.
3. Explain the benefits of having the processes of both mitosis and meiosis.

CONNECTING THE CONCEPTS

For more on the roles of mitosis and meiosis, refer to the following discussions:

Section 16.2 examines the process of spermatogenesis and the male reproductive system.

Section 16.4 examines the process of oogenesis and the female reproductive system.

Section 19.1 explores how unrestricted mitosis causes cancer.

18.6 Chromosome Inheritance

LEARNING OUTCOMES

Upon completion of this section, you should be able to

1. Explain how nondisjunction produces monosomy and trisomy chromosome conditions.
2. Describe the causes and consequences of trisomy 21.
3. List the major syndromes associated with changes in the number of sex chromosomes.
4. Describe the effects of deletions, duplications, inversions, and translocations on chromosome structure.

Normally, an individual receives 22 pairs of autosomes and two sex chromosomes. Each pair of autosomes carries alleles for particular traits. The alleles can be different, as when one contains instructions for freckles and one does not.

Changes in Chromosome Number

Sometimes individuals are born with either too many or too few autosomes or sex chromosomes, most likely due to an error, called nondisjunction, during meiosis. **Nondisjunction** is the failure of the homologous chromosomes, or daughter chromosomes, to separate correctly during meiosis I and meiosis II, respectively. Nondisjunction may occur during meiosis I, when both members of a homologous pair go into the same daughter cell. It can also occur during meiosis II, when the sister chromatids fail to separate and both daughter chromosomes go into the same gamete. Figure 18.15 assumes that nondisjunction has occurred during oogenesis. Some abnormal eggs have 24 chromosomes, whereas others have only 22 chromosomes. If an egg with 24 chromosomes is fertilized with a normal sperm, the result is a called a **trisomy** because one type of chromosome is present in three copies. If an egg with 22 chromosomes is fertilized with a normal sperm, the result is a called a **monosomy** because one type of chromosome is present in a single copy.

APPLICATIONS AND MISCONCEPTIONS

Down syndrome is a trisomy of chromosome 21. Are there trisomies of the other chromosomes?

There are other chromosomal trisomies. However, because most chromosomes are much larger than chromosome 21, the abnormalities associated with three copies of these other chromosomes are much more severe than those found in Down syndrome. The extra genetic material causes profound congenital defects, resulting in fatality. Trisomies of the X and Y chromosomes appear to be exceptions to this, as noted in the text.

Chromosome 8 trisomy occurs rarely. Affected fetuses generally do not survive to birth or die shortly after birth. There are also trisomies of chromosomes 13 (Patau syndrome) and 18 (Edwards syndrome). Again, these babies usually die within the first few days of life.

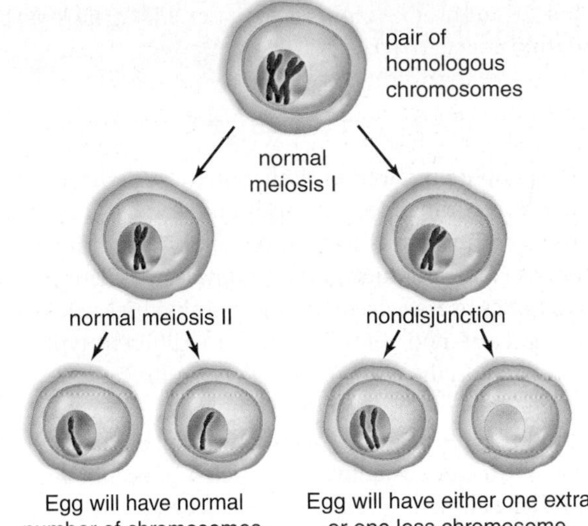
pair of homologous chromosomes
normal meiosis I
normal meiosis II
nondisjunction
a.
Egg will have normal number of chromosomes.
Egg will have either one extra or one less chromosome.

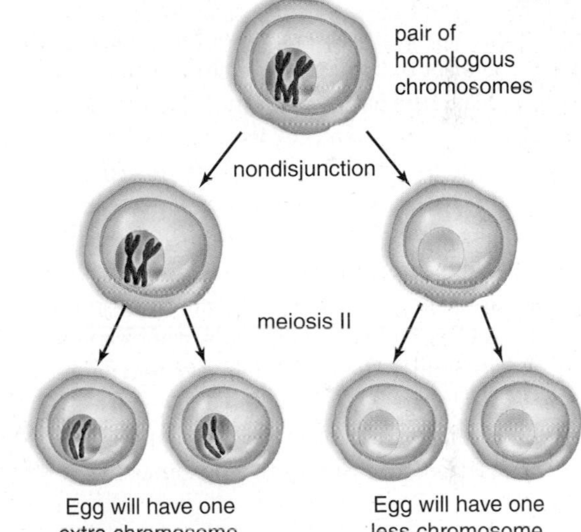
pair of homologous chromosomes
nondisjunction
meiosis II
b.
Egg will have one extra chromosome.
Egg will have one less chromosome.

Figure 18.15 The consequences of nondisjunction of chromosomes during oogenesis.
a. Nondisjunction can occur during meiosis II if the sister chromatids separate, but the resulting chromosomes go into the same daughter cell. Then the egg will have one more (24) or one less (22) than the usual number of chromosomes. **b.** Nondisjunction can also occur during meiosis I and result in abnormal eggs that also have one more or one less than the normal number of chromosomes.

Normal development depends on the presence of exactly two of each type of chromosome. An abnormal number of autosomes causes a developmental abnormality. Monosomy of all but the X chromosome is fatal. The affected infant rarely develops to full term. Trisomy is usually fatal, though there are some exceptions. Among autosomal trisomies, only trisomy 21 (Down syndrome) has a reasonable chance of survival after birth.

The chances of survival are greater when trisomy or monosomy involves the sex chromosomes. In normal XX females, one of the X chromosomes becomes a darkly staining mass of chromatin called a **Barr body** (named after the person who

BIOLOGY MATTERS  Science

Barr Bodies and Dosage Compensation

Most people are familiar with calico cats, whose fur contains patches of orange, black, and white. These cats are genetic mosaics. A mosaic is formed by combining different pieces to form a whole (a stained glass window is one example). Likewise, in genetics, a mosaic refers to an individual whose cells have at least two—and sometimes more—different types of genetic expression. In the case of the calico cat, the fur colors are due to expression of different genes. Some of the hair cells of these cats express the paternal copy of the gene. If an orange-haired father's copy of the gene is activated, a patch of orange hair develops. In other cells, the maternal gene is activated. A calico kitten with a black mother will grow black patches of hair scattered among the orange. Were you aware that human females are also mosaics?

The nucleus of human cells contains 46 chromosomes arranged into a set of 23 pairs. One chromosome from each pair is maternal, and the other is paternal. Each of the chromosomes in the first 22 pairs resembles its mate. Further, each member of a pair contains the same genes as the other member. Sex chromosomes that determine a person's gender are the last pair. Females have two X chromosomes, and males have one X and one Y chromosome. The Y chromosome is very small and contains far fewer genes than the X chromosome. Almost all of the genes on the X chromosome lack a corresponding gene on the Y chromosome. Thus, females have two copies of X genes, whereas males have only one. The body compensates for this extra dose of genetic material by inactivating one of the X chromosomes in each cell of the female embryo. Inactivation occurs early in development (at approximately the 100-cell stage). The inactivated X chromosome is called a Barr body, named after its discoverer. Barr bodies are highly condensed chromatin that appear as dark spots in the nucleus. Which X chromosome is inactivated in a given cell appears to be random. But every cell that develops from the original group of 100 cells will have the same inactivated X chromosome as its parent cell. Some of a woman's cells have inactivated the maternal X chromosome and other cells have inactivated the paternal X chromosome—she is a mosaic. Problems with inactivation of the X chromosome in humans could be linked to the development of cancer. For example, women who contain one defective copy of the breast cancer gene *BRCA1* have a greatly increased risk of developing breast and ovarian cancer. The BRCA1 protein produced from the gene is called a tumor suppressor. When the protein is functioning normally, it suppresses the development of cancer. This same protein is involved in X chromosome inactivation, although its exact role is uncertain. Presumably, increased cancer risk occurs because abnormal BRCA1 protein can neither inactivate the X chromosome nor function as a tumor suppressor.

Questions to Consider

1. Why would having an extra set of X chromosome genes be a problem for a female?
2. If X inactivation compensates for an extra X chromosome, why do Klinefelter males (XXY) have problems with development?

discovered it). A Barr body is an inactive X chromosome. The Science feature, "Barr Bodies and Dosage Compensation," provides additional information regarding Barr bodies.

We now know that the cells of females function with a single X chromosome just as those of males do. This is most likely the reason that a zygote with one X chromosome (Turner syndrome) can survive. Then, too, all extra X chromosomes beyond a single one become Barr bodies; this explains why poly-X females and XXY males are seen fairly frequently. An extra Y chromosome, called Jacobs syndrome, is tolerated in humans, most likely because the Y chromosome carries few genes. Jacobs syndrome (XYY) is due to nondisjunction during meiosis II of spermatogenesis. We know this because two Ys are present only during meiosis II in males.

Down Syndrome: An Autosomal Trisomy

The most common autosomal trisomy seen among humans is Down syndrome, also called trisomy 21. Persons with Down syndrome usually have three copies of chromosome 21 because the egg had two copies instead of one. However, about 20% of the time the sperm contributes the extra chromosome 21. The chances of a woman having a Down syndrome child increase rapidly with age, starting at about age 40. The reasons for this are still being investigated.

Although an older woman is more likely to have a Down syndrome child, most babies with Down syndrome are born to women younger than age 40 because this is the age group having the most babies. Karyotyping can detect a Down syndrome child. However, young women are not routinely encouraged to undergo the procedures necessary to get a sample of fetal cells (i.e., amniocentesis or chorionic villus sampling) because the risk of complications is greater than the risk of having a Down syndrome child. Fortunately, a test based on substances in maternal blood can help identify fetuses who may need to be karyotyped.

Down syndrome is easily recognized by these common characteristics: short stature; an eyelid fold; a flat face; stubby fingers; a wide gap between the first and second toes; a large, fissured tongue; a round head; and a palm crease, the so-called simian line. Unfortunately, intellectual disability, which can vary

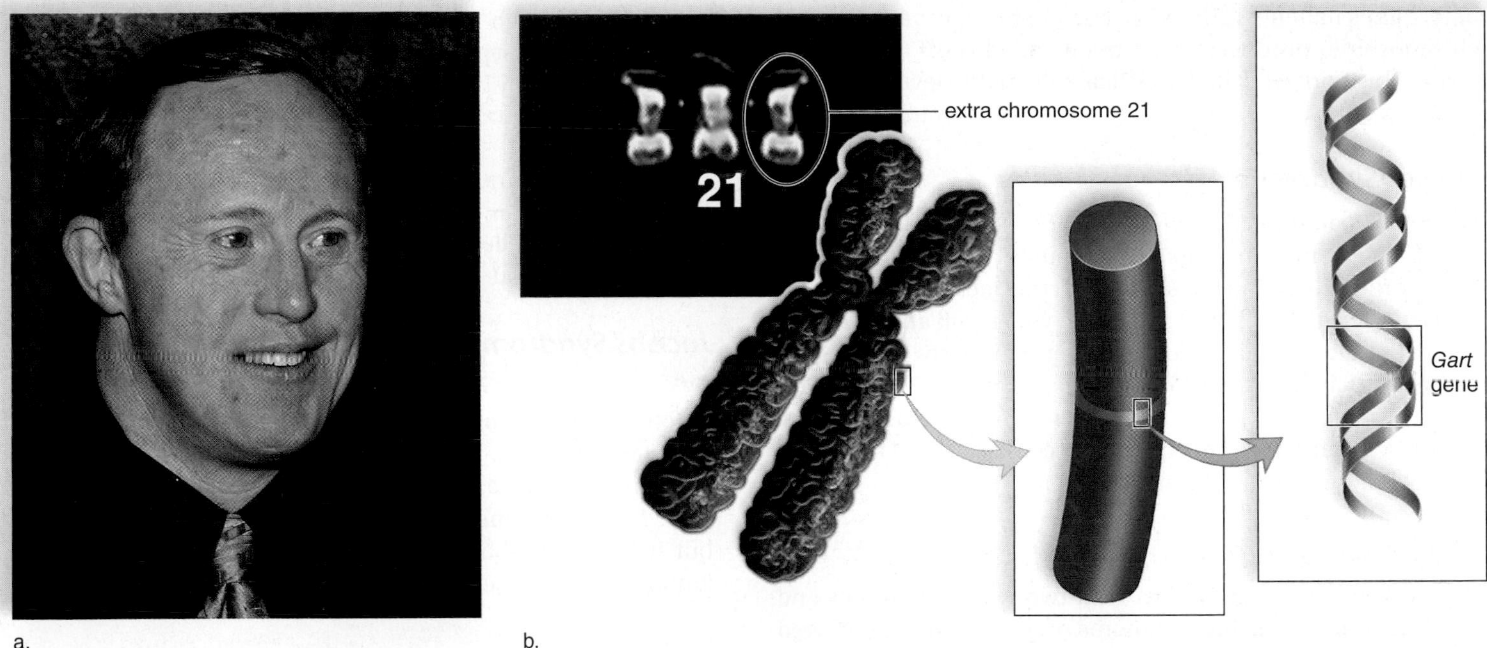

Figure 18.16 **Down syndrome.**
a. Chris Burke was born with Down syndrome. Common characteristics of the syndrome include a wide, rounded face, and a fold on the upper eyelids. Intellectual disability, along with an enlarged tongue, makes it difficult for a person with Down syndrome to speak distinctly. **b.** Karotype of an individual with Down syndrome shows an extra chromosome 21. More sophisticated technologies allow investigators to pinpoint the location of specific genes associated with the syndrome, such as the *Gart* gene.

in intensity, is also a characteristic. Chris Burke (Fig. 18.16*a*) was born with Down syndrome, and his parents were advised to put him in an institution. But Chris's parents didn't do that. They gave him the same loving care and attention they gave their other children, and it paid off. Chris is remarkably talented. He is a playwright, actor, and musician. He starred in *Life Goes On* (1989–1993), a TV series written just for him, and he is sometimes asked to be a guest star in other TV shows. His love of music and collaboration with other musicians have led to the release of several albums—like Chris, the songs are uplifting and inspirational. You can read more about this remarkable individual in his autobiography, *A Special Kind of Hero.*

APPLICATIONS AND MISCONCEPTIONS

Why is the age of a female a factor in Down syndrome?

One reason may be due to a difference in the timing of meiosis between males and females. Following puberty, males produce sperm continuously throughout their entire lives. In contrast, meiosis for females begins about five months after being conceived. However, the process is paused at prophase I of meiosis. Only after puberty are a selected few number of these cells allowed to continue meiosis as part of the female menstrual cycle. Because long periods of time may occur between the start and completion of meiosis, there is a greater chance that nondisjunction will occur; thus, as a female ages, there is a greater chance of producing a child with Down syndrome.

The genes that cause Down syndrome are located on the bottom third of chromosome 21 (Fig. 18.16*b*). Extensive investigative work has been directed toward discovering the specific genes responsible for the characteristics of the syndrome. Thus far, investigators have discovered several genes that may account for various conditions seen in persons with Down syndrome. For example, they have located genes most likely responsible for the increased tendency toward leukemia, cataracts, accelerated rate of aging, and intellectual disabilities. The gene associated with the intellectual disabilities, called the *Gart* gene, causes an increased level of purines in the blood, a finding associated with problems in intellectual development. One day, it may be possible to control the expression of the *Gart* gene even before birth so that at least this symptom of Down syndrome does not appear.

Changes in Sex Chromosome Number

An abnormal sex chromosome number is the result of inheriting too many or too few X or Y chromosomes. Figure 18.15 can be used to illustrate nondisjunction of the sex chromosomes during oogenesis if you assume that the chromosomes shown represent X chromosomes. Nondisjunction during oogenesis or spermatogenesis can result in gametes that have too few or too many X or Y chromosomes.

A person with Turner syndrome (XO) is a female, and a person with Klinefelter syndrome (XXY) is a male. The term **syndrome** indicates that there are a group of symptoms that always occur together. This shows that in humans, the presence of a Y chromosome, not the number of X chromosomes, determines maleness. The *SRY* gene, on the short arm of the Y chromosome, produces a hormone called *testis-determining factor.* This hormone plays a critical role in the development of male sex organs.

Turner Syndrome

From birth, an individual with Turner syndrome has only one sex chromosome, an X. As adults, Turner females are short, with a broad chest and folds of skin on the back of the neck. The ovaries, uterine tubes, and uterus are very small and underdeveloped. Turner females do not undergo puberty or menstruate, and their breasts do not develop. However, some have given birth following in vitro fertilization using donor eggs. They usually are of normal intelligence and can lead fairly normal lives if they receive hormone supplements.

Klinefelter Syndrome

One in 650 live males is born with two X chromosomes and one Y chromosome. The symptoms of this condition (referred to as "47, XXY") are often so subtle that only 25% are ever diagnosed, and those are usually not diagnosed until after age 15. Earlier diagnosis opens the possibility for educational accommodations and other interventions that can help mitigate common symptoms, which include speech and language delays. Those 47, XXY males who develop more severe symptoms as adults are referred to as having *Klinefelter syndrome.* All 47, XXY adults require assisted reproduction to father children. Affected individuals commonly receive testosterone supplementation beginning at puberty.

Poly-X Females

A poly-X female has more than two X chromosomes and extra Barr bodies in the nucleus. Females with three X chromosomes have no distinctive phenotype, aside from a tendency to be tall and thin. Although some have delayed motor and language development, most poly-X females do not have intellectual disabilities. Some may have menstrual difficulties, but many menstruate regularly and are fertile. Their children usually have a normal karyotype.

Females with more than three X chromosomes occur rarely. Unlike XXX females, XXXX females are more likely to possess problems with intellectual development. Various physical abnormalities are seen, but these females may menstruate normally.

Jacobs Syndrome

XYY males with Jacobs syndrome can only result from nondisjunction during spermatogenesis. Affected males are usually taller than average, suffer from persistent acne, and tend to have speech and reading problems. At one time, it was suggested that these men were likely to be criminally aggressive, but it has been shown that the incidence of such behavior among them may be no greater than among XY males.

Changes in Chromosome Structure

Another type of chromosomal mutation is described as "changes in chromosome structure." Various agents in the environment, such as radiation, certain organic chemicals, or even viruses, can cause chromosomes to break. Ordinarily, when breaks occur in chromosomes, the two broken ends reunite to give the same sequence of genes. Sometimes, however, the broken ends of one or more chromosomes do not rejoin in the same pattern as before. The result is various types of chromosomal mutation.

Changes in chromosome structure include deletions, duplications, inversions, and translocations of chromosome segments. A **deletion** occurs when an end of a chromosome breaks off or when two simultaneous breaks lead to the loss of an internal segment (Fig. 18.17*a*). Even when only one member

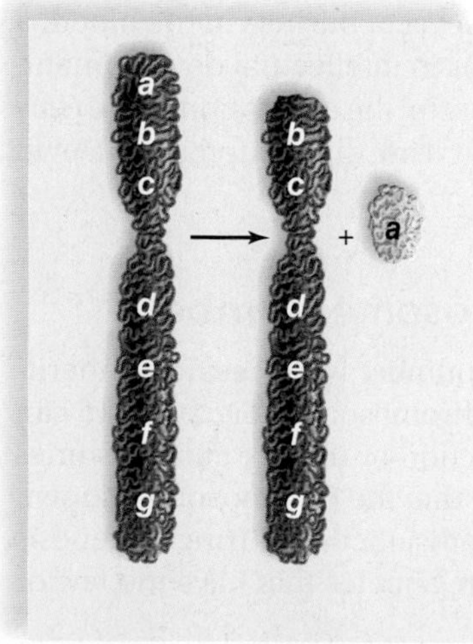

a. Deletion

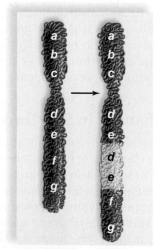

b. Duplication

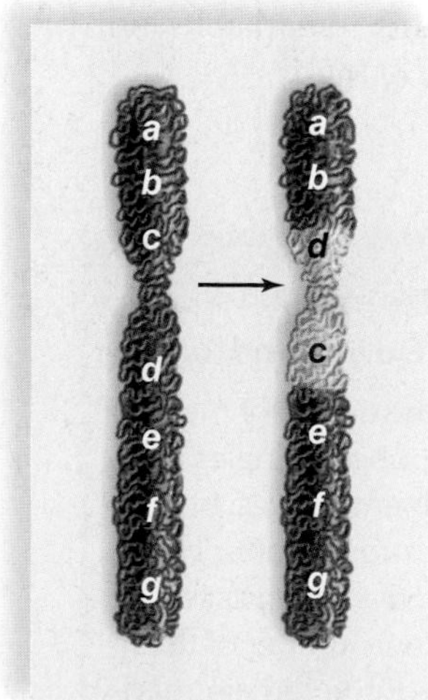

c. Inversion

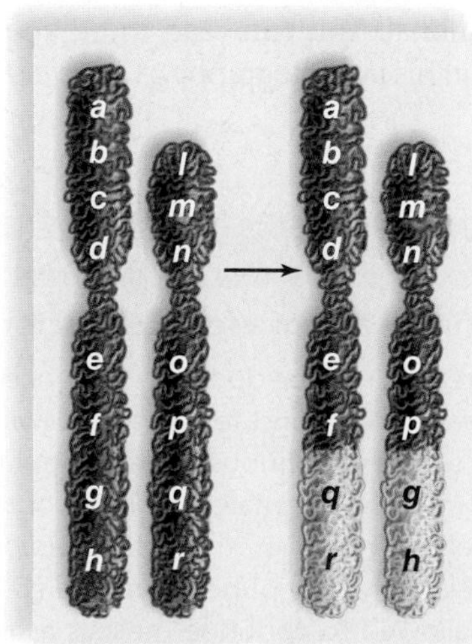

d. Translocation

Figure 18.17 The various types of chromosomal mutations.
a. Deletion is the loss of a chromosome piece. **b.** Duplication occurs when the same piece is repeated within the chromosome. **c.** Inversion occurs when a piece of chromosome breaks loose and then rejoins in the reversed direction. **d.** Translocation is the exchange of chromosome pieces between nonhomologous pairs.

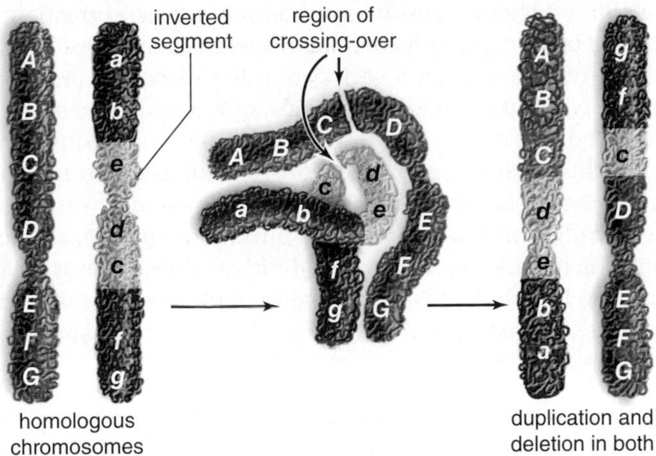

Figure 18.18 **A chromosomal inversion.**
Left: A segment of one homologue is inverted. In the shaded segment, *edc* occurs instead of *cde*. *Middle:* The two homologues can pair only when the inverted sequence forms an internal loop. After crossing-over, a duplication and a deletion can occur. *Right:* The homologue on the left has *AB* and *ab* sequences and neither *fg* nor *FG* genes. The homologue on the right has *gf* and *FG* sequences and neither *AB* nor *ab* genes.

of a pair of chromosomes is affected, a deletion often causes abnormalities.

A **duplication** is the presence of a chromosomal segment more than once in the same chromosome (Fig. 18.17*b*). An **inversion** has occurred when a segment of a chromosome is turned around 180° (Fig. 18.17*c*). While most inversions do not present problems for the individuals, because all of the genes are present, the reversed sequence of genes can lead to problems during prophase of meiosis when crossing-over occurs. Often, inversions may lead to the formation of deletions and duplications (Fig. 18.18).

Animation Consequence of Inversion

A **translocation** is the movement of a chromosome segment from one chromosome to another nonhomologous chromosome (see Fig. 18.17*d*). In 5% of cases, a translocation that occurred in a previous generation between chromosomes 21 and 14 is the cause of Down syndrome. In other words, because a portion of chromosome 21 is now attached to a portion of chromosome 14, the individual has three copies of the alleles that bring about Down syndrome when they are present in triplet copy. In these cases, Down syndrome is not related to the age of the mother but instead tends to run in the family of either the father or the mother.

Human Syndromes

Changes in chromosome structure occur in humans and lead to various syndromes, the genetics of which are just now being investigated.

Deletion Syndromes Williams syndrome occurs when chromosome 7 loses a tiny end piece (Fig. 18.19). Children who have this syndrome look like pixies, with turned-up noses, wide mouths, a small chin, and large ears. Although their academic skills are poor, they exhibit excellent verbal and musical abilities. The gene that governs the production of the protein elastin is missing. This affects the health of the cardiovascular system and causes their skin to age prematurely. Such individuals are very friendly but need an ordered life, perhaps because of the loss of a gene for a protein normally active in the brain.

Cri du chat (cat's cry) syndrome is seen when chromosome 5 is missing an end piece. The affected individual has a small head, is intellectually disabled, and has facial abnormalities. Abnormal development of the glottis and larynx results in the most characteristic symptom—the infant's cry resembles that of a cat.

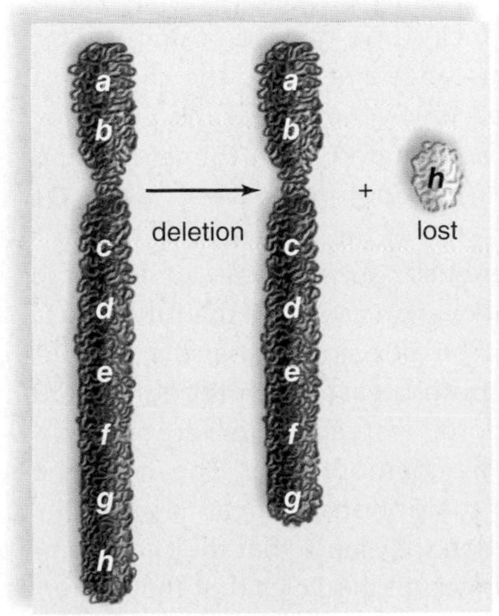

Figure 18.19 **A chromosomal deletion.**
a. When chromosome 7 loses an end piece, the result is Williams syndrome. **b.** These children, although unrelated, have the same appearance, health, and behavioral problems.

Translocation Syndromes A person who has both of the chromosomes involved in a translocation has the normal amount of genetic material and is healthy, unless the chromosome exchange broke an allele into two pieces. The person who inherits only one of the translocated chromosomes will no doubt have only one copy of certain alleles and three copies of certain other alleles. A genetics counselor begins to suspect a translocation has occurred when spontaneous abortions are commonplace and family members suffer from various syndromes.

Figure 18.20 shows the effect of an individual who has Alagille syndrome, a translocation between chromosomes 2 and 20. People with this syndrome ordinarily have a deletion on chromosome 20. The syndrome may also produce abnormalities of the eyes and internal organs. The symptoms of Alagille syndrome range from mild to severe, so some people may not be aware they have the syndrome. Translocations can also be responsible for a variety of other disorders including certain types of cancer. In the 1970s, new staining techniques identified that a translocation from a portion of chromosome 22 to chromosome 9 was responsible for chronic myelogenous leukemia. This translocated chromosome was called the Philadelphia chromosome. In Burkitt lymphoma, a cancer common in children in equatorial Africa, a large tumor develops from lymph glands in the region of the jaw. This disorder involves a translocation from a portion of chromosome 8 to chromosome 14.

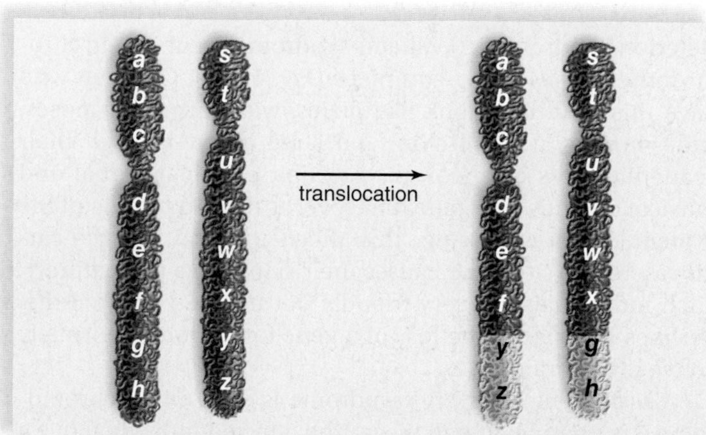

a.

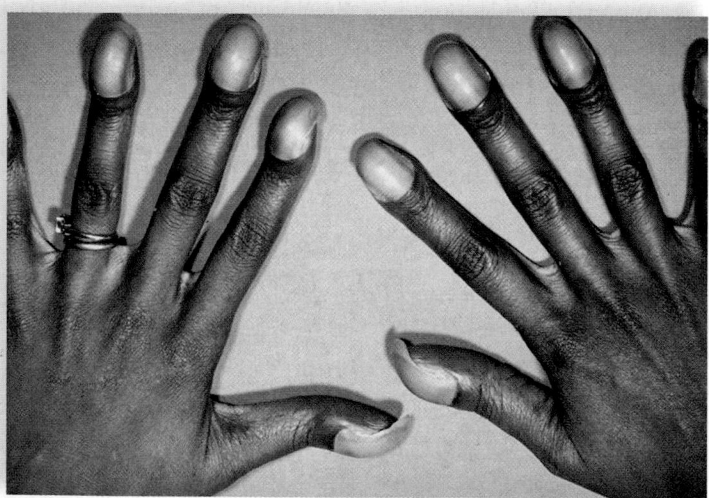

b.

Figure 18.20 A chromosomal translocation.
a. When chromosomes 2 and 20 exchange segments, (**b**) Alagille syndrome, with distinctive body features, sometimes results because the translocation disrupts an allele on chromosome 20.

CHECK YOUR PROGRESS 18.6

1. Explain what causes an individual to have an abnormal number of chromosomes.
2. Describe the specific chromosome abnormality of a person with Down syndrome.
3. List some syndromes that result from inheritance of an abnormal sex chromosome number.
4. Describe other types of chromosome mutations, aside from abnormal chromosome number.

CONNECTING THE CONCEPTS

For more information on the topics presented in this section, refer to the following discussions:

Section 17.3 explains how the *SRY* gene directs the formation of the male reproductive system.

Section 20.2 explores how chromosomes are involved in patterns of inheritance.

CASE STUDY CONCLUSION

Christina was lucky. She had been undergoing routine mammograms since age 30, so the cancer was caught at an early stage. However, most women may be unaware that they carry the mutated *BRCA1* gene. A blood test can easily screen for mutations in the *BRCA1* gene. Depending on the mutation, age of the patient, and health history of the patient, a genetics counselor can advise on the probability of breast or ovarian cancer in the future. In Christina's case, she made the decision to have a double mastectomy, removing both breasts even though the tumor was located in only one, instead of undergoing longer-term treatment such as chemotherapy. The next chapter will explore some of the options that cancer patients, such as Christina, face when they learn that they carry a potentially dangerous combination of alleles that make them susceptible to cancer.

MEDIA STUDY TOOLS

Enhance your study of this chapter with media! Visit **www.mhhe.com/maderhuman13e** and go to "Media Study Tools" for this chapter to access the following:

Animations	Videos	MP3 File
18.2 How the Cell Cycle Works • Control of the Cell Cycle • Mechanism of Steroid Hormone Action **18.3** Mitosis • Mitosis and Cytokinesis **18.4** Meiosis I • Genetic Diversity • Meiosis II • Meiosis Crossing-Over • Random Orientation of Chromosomes During Meiosis • Spermatogenesis **18.5** Comparison of Mitosis and Meiosis **18.6** X Inactivation • Changes in Chromosome Structure • Consequence of Inversion	**18.3** Mitosis **18.4** Binary Fission • Human Sperm	**18.3** Mitosis

3D Animations

3D Animation Cell Cycle & Mitosis

3D Animation Meiosis

McGraw-Hill's 3D animations, "Cell Cycle & Mitosis" and "Meiosis," provide a dynamic exploration of the key concepts of this chapter and are available through McGraw-Hill Connect®.

SUMMARIZE

18.1 Chromosomes

- The genetic material of the cell is organized as **chromosomes.** Chromosomes contain a combination of proteins and DNA called **chromatin.**
- Most human cells are diploid—therefore, chromosomes occur in pairs.
- Prior to mitosis, or duplication division, the chromosomes are replicated, forming **sister chromatids.** The sister chromatids are joined at the **centromere.**
- A karyotype is a visual display of an individual's chromosomes.

18.2 The Cell Cycle

The **cell cycle** occurs continuously and has four stages: G_1, S, G_2 (the **interphase** stages), and M (the mitotic stage), which includes **cytokinesis** and the stages of **mitosis.**

- In G_1, a cell doubles organelles and accumulates materials for DNA synthesis.
- In S, DNA replication occurs.
- In G_2, a cell synthesizes proteins needed for cell division.
- **Checkpoints** and external signals control the progression of the cell cycle. Cells that fail to pass checkpoints may undergo **apoptosis.**

18.3 Mitosis

Mitosis is duplication division that ensures that the **daughter cells** have the **diploid** (**2n**) number and the same types of chromosomes as the **parent cell.** The **mitotic spindle** plays an important role in the separation of the sister chromatids during mitosis. The mitotic spindle is organized by the **centrosomes** of the cell. Centrosomes contain clusters of microtubules called **centrioles.**

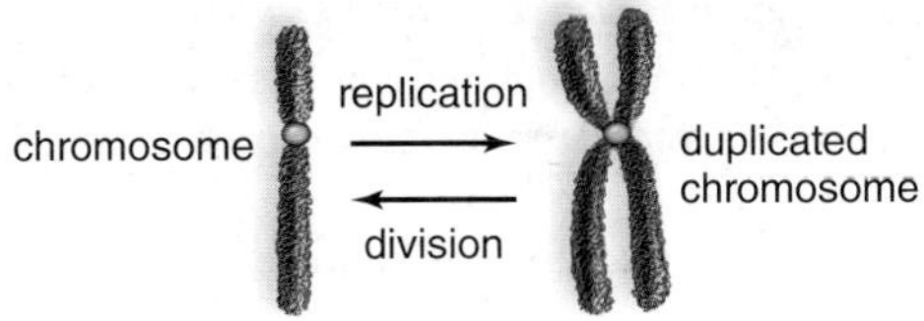

The phases of mitosis are prophase, metaphase, anaphase, and telophase:

- **Prophase** Chromosomes attach to spindle fibers.
- **Metaphase** Chromosomes align at the equator.
- **Anaphase** Chromatids separate, becoming chromosomes that move toward the poles.
- **Telophase** Nuclear envelopes form around chromosomes; cytokinesis begins.

Cytokinesis is the division of cytoplasm and organelles following mitosis.

- Cytokinesis in animal cells involves the formation of a **cleavage furrow** to separate the cytoplasm.

18.4 Meiosis

Meiosis is reduction division and serves to reduce the diploid (2n) chromosome number to a **haploid** (**n**) number. Meiosis involves two cell divisions—meiosis I and meiosis II.

Meiosis I

- **Homologous chromosomes** pair (**synapsis**) and then separate. **Interkinesis** follows meiosis I.

Meiosis II

- Sister chromatids separate, resulting in four cells with the haploid number of chromosomes that move into daughter nuclei.

Meiosis results in genetic recombination due to **crossing-over;** gametes have all possible combinations of chromosomes.
Upon fertilization, the zygote is restored to a diploid number of chromosomes.

Spermatogenesis and Oogenesis

- **Spermatogenesis** In males, produces four viable sperm.
- **Oogenesis** In females, produces one egg and several polar bodies. Oogenesis goes to completion if the sperm fertilizes the developing egg.

18.5 Comparison of Meiosis and Mitosis

- In prophase I, homologous chromosomes pair; there is no pairing in mitosis.
- In metaphase I, homologous duplicated chromosomes align at equator.
- In anaphase I, homologous chromosomes separate.

18.6 Chromosome Inheritance

Meiosis is a part of gametogenesis (spermatogenesis in males and oogenesis in females) and contributes to genetic diversity.

Changes in Chromosome Number

- **Nondisjunction** changes the chromosome number in gametes, resulting in **trisomy** or **monosomy.**
- Autosomal syndromes include trisomy and Down syndrome.

Changes in Sex Chromosome Number

- Nondisjunction during oogenesis or spermatogenesis can result in gametes that have too few or too many X or Y chromosomes.
- If more than one X chromosome is present in a cell, a **Barr body** may be formed.
- **Syndromes** include Turner, Klinefelter, poly-X, and Jacobs.

Changes in Chromosome Structure

- Chromosomal mutations can produce **deletions, duplications, inversions,** and **translocations.**
- These result in various syndromes such as Williams and cri du chat (deletion) and Alagille and certain cancers (translocation).

ASSESS

Testing Your Knowledge of the Concepts

1. Describe the two parts of the cell cycle. (pages 417–419)
2. Explain how the checkpoints regulate the cell cycle. (pages 418–419)
3. How do external signals, such as hormones, influence the cell cycle? (page 419)
4. What is the importance of mitosis? (pages 419–420)
5. What is the function of the mitotic spindle? (page 420)
6. How do the terms *diploid* (2n) and *haploid* (n) relate to meiosis? (page 422)
7. List and describe the events in meiosis I and meiosis II. (pages 422–426)
8. How does spermatogenesis differ from oogenesis? (page 427)
9. What is the significance of meiosis? (pages 428–429)
10. Contrast mitosis and meiosis I and meiosis II. (pages 429–430)
11. What is nondisjunction, when can it occur, and what are the results? (page 431)
12. What are some syndromes caused by changes in chromosome number, and what is the change for each? (pages 431–434)
13. Describe three changes that can occur in chromosome structure and list some syndromes caused by such changes. (pages 434–436)
14. The point of attachment for two sister chromatids is the
    a. centriole.
    b. centromere.
    c. chromosome.
    d. karyotype.
15. Label the drawing of the cell cycle; then list the main event of each stage.

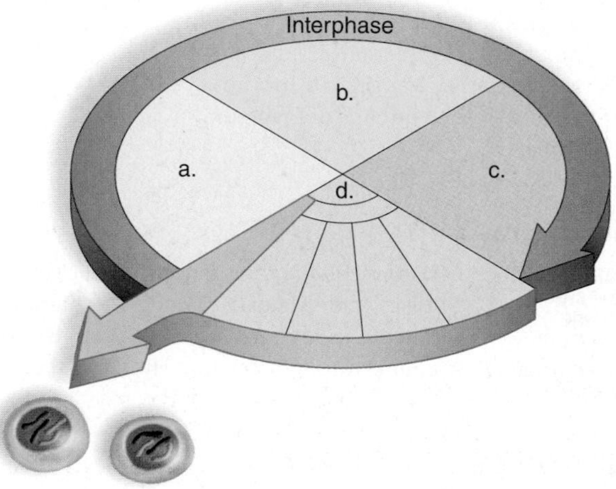

In questions 16–20, match the statement to interphase or the phase of mitosis in the key.

Key:

a. metaphase
b. interphase
c. telophase
d. prophase
e. anaphase

16. Spindle fibers begin to appear.
17. DNA replication occurs.
18. Chromosomes line up at the equator.
19. Duplicated chromosomes become visible.
20. Centromere splits and sister chromosomes move to opposite poles.
21. If a parent cell has a diploid number of 18 chromosomes before mitosis, how many chromosomes will the daughter cells have?
    a. 18
    b. 36
    c. 9
    d. 27
22. Crossing-over occurs between
    a. sister chromatids of the same chromosome.
    b. chromatids of nonhomologous chromosomes.
    c. nonsister chromatids of a homologous pair.
    d. Both b and c are correct.
23. The products of ________ are ________ cells.
    a. mitosis, diploid
    b. meiosis, haploid
    c. meiosis, diploid
    d. Both a and b are correct.

24. If a parent cell has 22 chromosomes, the daughter cells following meiosis II will have
    a. 22 chromosomes.
    b. 44 chromosomes.
    c. 11 chromosomes.
    d. Any of the choices could be correct.
25. Which of these helps to provide genetic diversity?
    a. independent alignment during metaphase I
    b. crossing-over during prophase I
    c. fusion of sperm and egg nuclei during fertilization
    d. All of the choices are correct.
26. Monosomy or trisomy occurs because of
    a. crossing-over.
    b. inversion.
    c. translocation.
    d. nondisjunction.
27. A person with Klinefelter syndrome is ________ and has ________ sex chromosomes.
    a. male, XYY
    b. male, XXY
    c. female, XXY
    d. female, XO

ENGAGE

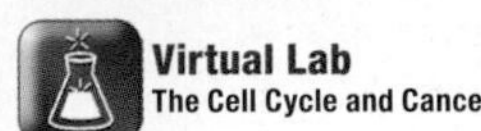

The virtual lab "The Cell Cycle and Cancer" provides a more detailed look at how a failure in the control system of the cell cycle can lead to cancer.

Thinking Critically About the Concepts

1. Benign and cancerous tumors occur when the cell cycle control mechanisms no longer operate correctly. What types of genes may be involved in these cell cycle control mechanisms?
2. Mitosis goes on continually in your body: in your blood cells, skin cells, and the cells that line the respiratory and digestive tracts. Why might problems with mitosis cause problems with wound healing?
3. Explain how the separation of homologous chromosomes during meiosis affects the appearance of siblings (that some resemble each other and others look very different from one another).
4. a. What would you conclude about the ability of nervous and muscle tissue to repair themselves if nerve and muscle cells are typically arrested in G_1 of interphase?
   b. What are the implications of this arrested state to someone who suffers a spinal cord injury or heart attack?

CHAPTER

19

Cancer

CHAPTER CONCEPTS

19.1 Cancer Cells
Cancer cells have a number of abnormal characteristics that prevent them from functioning in the same manner as normal cells. They divide repeatedly and form tumors in the place of origin and in other parts of the body.

19.2 Causes and Prevention of Cancer
Whether cancer develops is partially due to inherited genes, but exposure to carcinogens such as UV radiation, tobacco smoke, pollutants, industrial chemicals, and certain viruses also plays a significant role.

19.3 Diagnosis of Cancer
Cancer is usually diagnosed by certain screening procedures and by imaging the body and tissues using various techniques.

19.4 Treatment of Cancer
Surgery followed by radiation and/or chemotherapy has now become fairly routine. Immune therapy, bone marrow transplants, and other methods are under investigation.

CASE STUDY NEPHROBLASTOMA

Cody had all of the appearances of being a healthy three-year-old boy. He liked to play with his toys, fought with his sister, and chased the cat around the house. However, shortly after his last birthday, Cody had begun to complain of stomach aches and often said that he felt sick. At first, his parents paid little attention to his condition, until they noticed that his appetite was decreasing and that he kept running low fevers without any apparent cause. One morning, Cody yelled to his mom that his pee was pink. His mother rushed into the bathroom and discovered that Cody had blood in his urine. She immediately called their pediatrician and scheduled an appointment for that afternoon.

At the office, the doctor performed a complete physical exam and had Cody drink a couple of glasses of water to get a urine sample. During the exam, the doctor noticed a small lump in Cody's abdomen, just around the location of his kidney. The symptoms suggested that Cody may be having a kidney problem, so the doctor scheduled a magnetic resonance imaging (MRI) test the next morning for Cody.

The results of the MRI indicated that Cody had a mass (or tumor) on his kidney, which was a sign of nephroblastoma, or Wilms disease, a rare form of kidney cancer that affects only 500 children annually. The good news was that a follow-up computerized axial tomography (CAT) scan of Cody's abdomen did not suggest that the cancer had spread to any other organs. If this was the case, Cody's chances of survival were greater than 92%, but he was going to have to undergo surgery to remove the kidney, followed by several weeks of chemotherapy to make sure that no cancer cells were present outside the kidney. The doctor was very optimistic that they had caught Cody's cancer in time.

As you read through the chapter, think about the following questions:

1. What are the characteristics of cancer cells that distinguish them from normal cells?
2. Why do cancer cells form tumors?
3. Why would the doctor recommend both surgery and chemotherapy?

BEFORE YOU BEGIN

Before beginning this chapter, take a few moments to review the following discussions:

Section 8.6 What are antioxidants?

Section 18.1 What is the role of the checkpoints in the cell cycle?

Section 18.1 What external factors may regulate cell division?

19.1 Cancer Cells

LEARNING OUTCOMES

Upon completion of this section, you should be able to

1. Describe the characteristics of cancer cells that distinguish them from normal cells.
2. Distinguish between a proto-oncogene and a tumor suppressor gene with regard to their effects on the cell cycle.
3. Identify the common types of cancers and the body system with which each is associated.

Cancer is disease characterized by uncontrolled cell growth. Although there are many different types of cancer, and the causes vary widely, most cancers are a result of a cell accumulating mutations that ultimately cause a loss of control over the cell cycle.

Characteristics of Cancer Cells

While the effects of cancer are often noticed at the tissue, organ, or organismal levels, cancer is ultimately a cellular disease. Despite the large number of different cancer types, cancer cells share general traits that distinguish them from normal cells.

Cancer Cells Lack Differentiation

Differentiation is the process of cellular development by which a cell acquires a specific structure and function. Red blood cells are examples of differentiated cells in the circulatory system. In comparison, cancer cells are nonspecialized and do not contribute to the functioning of a body part. A cancer cell does not look like a differentiated epithelial, muscle, nervous, or connective tissue cell. Instead, it looks distinctly abnormal.

Cancer Cells Have Abnormal Nuclei

In Figure 19.1, you can compare the appearance of normal cervical cells (*a*) with that of precancerous (*b*) and cancerous (*c*) cervical cells. The nuclei of cancer cells are enlarged and may contain an abnormal number of chromosomes. The nuclei of the cervical cancer cells (Fig. 19.1*c*) have increased to the point that they take up most of the cell.

In addition to nuclear abnormalities, cancer cells have defective chromosomes. Some portions of the chromosomes may be duplicated, and/or some may be deleted. In addition, gene amplification (extra copies of specific genes) is seen much more frequently than in normal cells. Ordinarily, cells with damaged DNA undergo **apoptosis,** or programmed cell death. Cancer cells fail to undergo apoptosis, even though they are abnormal cells.

Tissues that divide frequently, such as those that line the respiratory and digestive tracts, are more likely to become cancerous. Cell division gives them the opportunity to undergo genetic mutations, each one making the cell more abnormal and giving it the ability to produce more of its own type.

Cancer Cells Have Unlimited Potential to Replicate

Ordinarily, cells divide about 60 to 70 times and then just stop dividing and eventually undergo apoptosis. Cancer cells are immortal and keep on dividing for an unlimited number of times.

Just as shoelaces are capped by small pieces of plastic, chromosomes in human cells end with special repetitive DNA sequences called **telomeres.** Specific proteins bind to telomeres in both normal and cancerous cells. These telomere proteins protect the ends of chromosomes from DNA repair enzymes. Though the enzymes effectively repair DNA in the center of the chromosome, they always tend to bind together the naked ends of chromosomes. In a normal cell, the telomeres get shorter after each cell cycle and protective telomere proteins gradually decrease. In turn, repair enzymes eventually cause the chromosomes' ends to bind together, causing the cell to undergo apoptosis and die. Telomerase is an enzyme that can rebuild telomere sequences and in that way prevent a cell from ever losing its potential to divide. The gene that codes for telomerase is constantly turned on in cancer cells, and telomeres are continuously rebuilt. The telomeres remain at a constant length, and the cell can keep dividing over and over.

Animation Telomerase Function

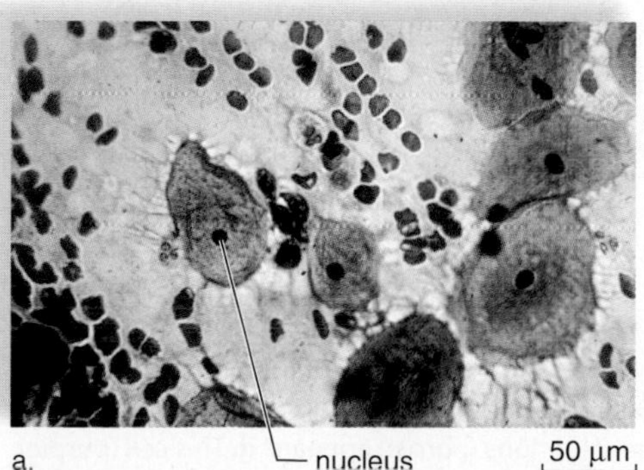

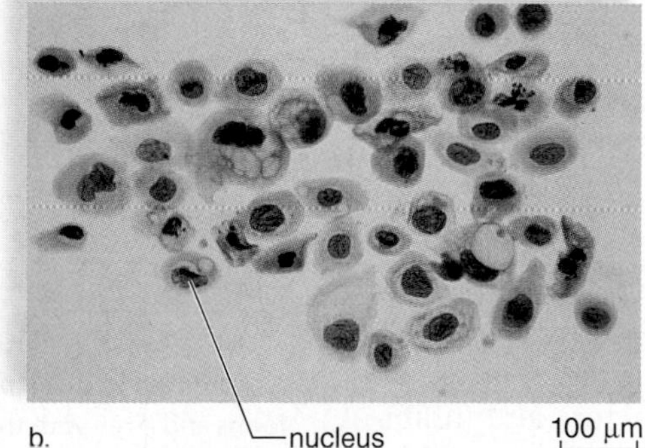

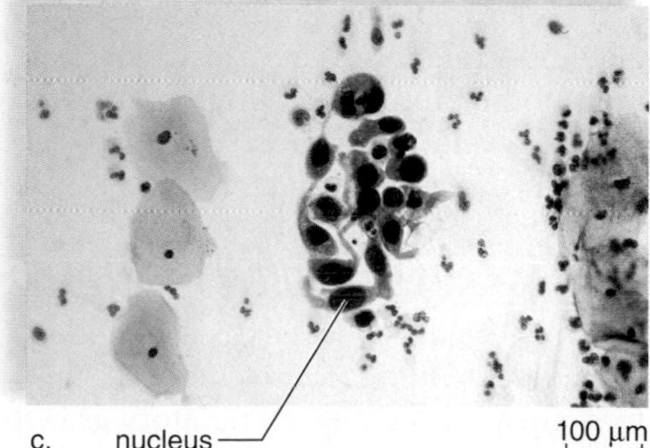

Figure 19.1 A comparison of normal tissue cells and cancer cells.
a. Normal cervical cells. **b.** Precancerous cervical cells. **c.** Cancerous cervical cells.

APPLICATIONS AND MISCONCEPTIONS

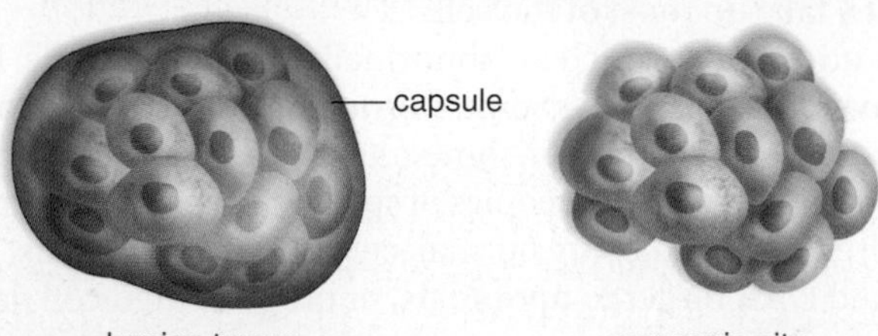
capsule
benign tumor
cancer in situ

What's the difference between a benign tumor and a malignant tumor?

A *benign* tumor is usually surrounded by a connective tissue capsule. A benign tumor does not invade adjacent tissue because of the capsule. The cells of a benign tumor resemble normal cells fairly closely. Chances are you have one of these tumors. If you have a mole, or nevus, on your body, you have a benign tumor of the skin melanocytes. Cells from a mole closely resemble other melanocytes.

However, despite the term benign, this type of tumor isn't necessarily harmless. If it presses on normal tissue or restricts the normal tissue's blood supply, a benign tumor can be fatal. For example, a benign neuroma (nerve cell tumor) growing near the brainstem eventually affects control centers for heartbeat and respiration.

On the other hand, a malignant tumor is able to invade surrounding tissues and its cells don't resemble normal cells. By traveling in blood or lymph vessels, the tumor can spread throughout the body. As a general rule, badly deformed cells are most likely to spread. Thus, they are the most malignant.

Cancer in situ (in place) is a malignant tumor found in its place of origin. As yet, the cancer has not spread beyond the basement membrane, the nonliving material that anchors tissues to one another. If recognized and treated early, cancer in situ is usually curable.

Cancer Cells Form Tumors

Normal cells anchor themselves to a substratum and/or adhere to their neighbors. They exhibit *contact inhibition,* meaning that when they come in contact with a neighbor, they stop dividing. Cancer cells have lost all restraint. They pile on top of one another and grow in multiple layers, forming a **tumor.** As cancer develops, the most aggressive cell becomes the dominant cell of the tumor (Fig. 19.2).

Cancer Cells Disregard Growth Factors

Chemical signals between cells tell them whether or not they should be dividing. These chemical signals, called growth factors, are of two types: stimulatory growth factors and inhibitory growth factors. Cancer cells keep on dividing even when stimulatory growth factors are absent, and they do not respond to inhibitory growth factors.

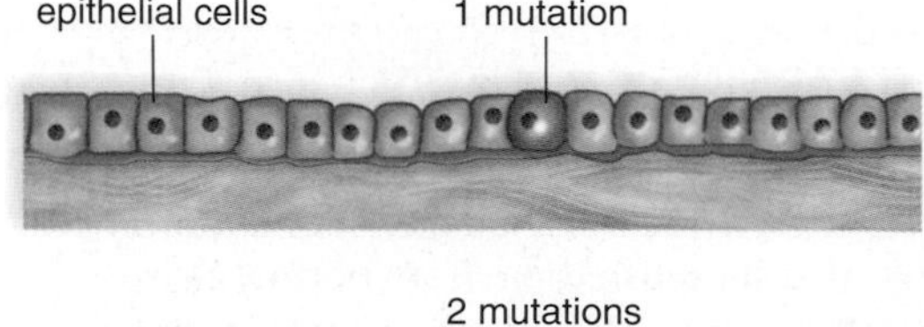
a. Cell (dark pink) acquires a mutation for repeated cell division.
epithelial cells
1 mutation

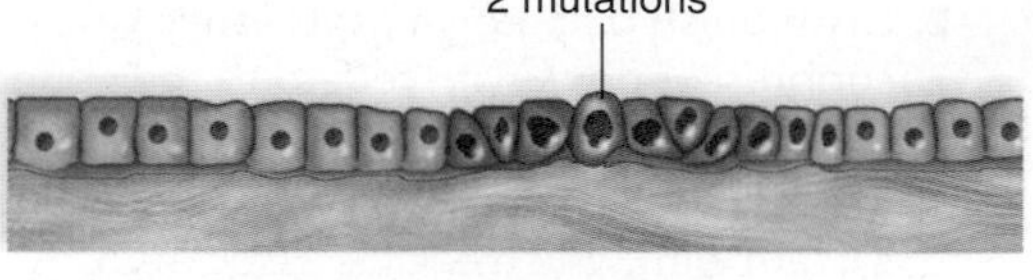
b. New mutations arise, and one cell (brown) has the ability to start a tumor.
2 mutations

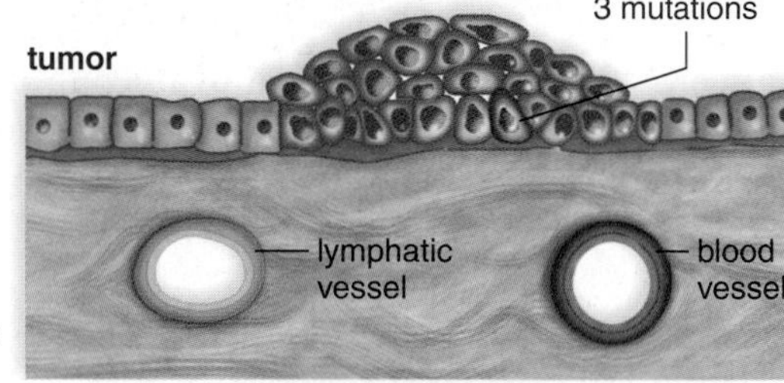
tumor
3 mutations
c. Cancer in situ. The tumor is at its place of origin. One cell (purple) mutates further.
lymphatic vessel
blood vessel

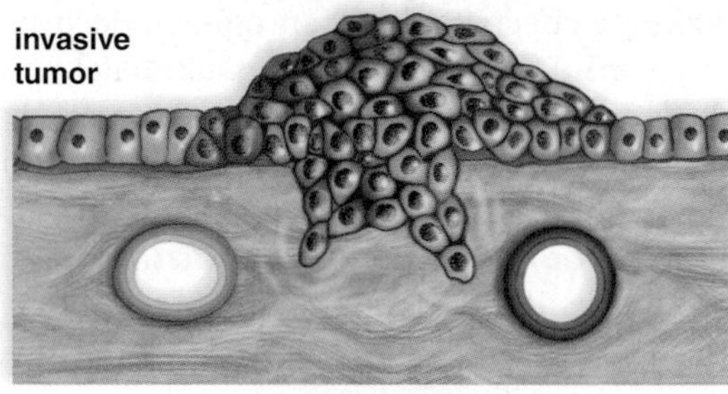
invasive tumor
d. Cells have gained the ability to invade underlying tissues by producing a proteinase enzyme.

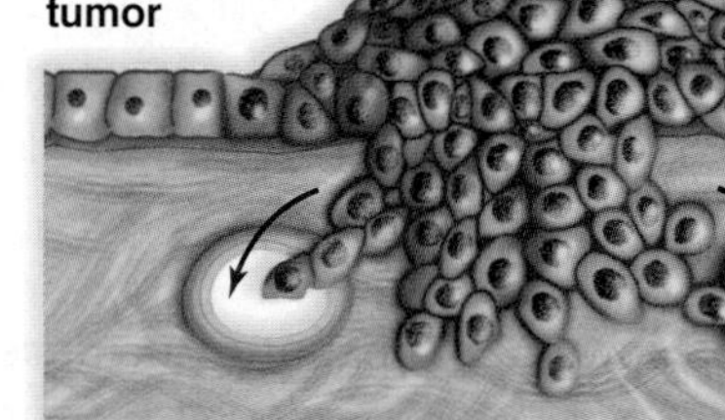
malignant tumor
e. Cancer cells now have the ability to invade lymphatic and blood vessels.

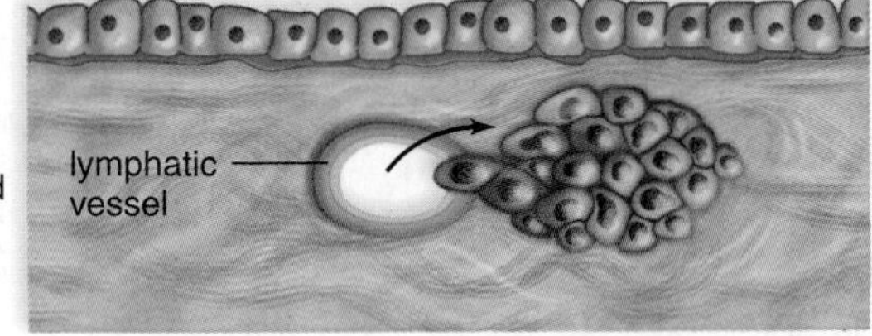
distant tumor
f. New metastatic tumors are found some distance from the original tumor.
lymphatic vessel

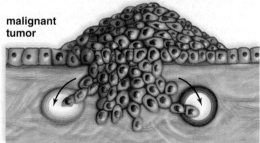

Figure 19.2 Progression from a single mutation to a tumor. **a.** One cell (dark pink) in a tissue mutates. **b.** This mutated cell divides repeatedly and a cell (brown) with two mutations appears. **c.** A tumor forms and a cell with three mutations (purple) appears. **d.** This cell (purple), which takes over the tumor, can invade underlying tissue. **e.** Tumor cells invade lymphatic and blood vessels. **f.** A new tumor forms at a distant location.

Cancer Cells Gradually Become Abnormal

Figure 19.2 illustrates that **carcinogenesis,** the development of cancer, is a multistage process that can be divided into these three phases:

Initiation: A single cell undergoes a mutation that causes it to begin to divide repeatedly (Fig. 19.2*a*).

Promotion: A tumor develops, and the tumor cells continue to divide. As they divide, they undergo mutations (Fig. 19.2*b, c*).

Progression: One cell undergoes a mutation that gives it a selective advantage over the other cells (Fig. 19.2*c*). This process is repeated several times; eventually, there is a cell that has the ability to invade surrounding tissues (Fig. 19.2*d, e*).

Cancer Cells Undergo Angiogenesis and Metastasis

To grow larger than about a billion cells (about the size of a pea), a tumor must have a well-developed capillary network to bring it nutrients and oxygen. **Angiogenesis** is the formation of new blood vessels. The low oxygen content in the middle of a tumor may turn on genes coding for angiogenic growth factors that diffuse into the nearby tissues and cause new vessels to form.

Due to mutations, cancer cells tend to be motile. They have a disorganized internal cytoskeleton and lack intact actin filament bundles. To metastasize, cancer cells must make their way across the basement membrane and invade a blood vessel or lymphatic vessel. Invasive cancer cells are abnormally-shaped (see Fig. 19.1*c*) and don't look at all like the normal cell seen nearby. Cancer cells produce proteinase enzymes that degrade the basement membrane and allow them to invade underlying tissues. Malignancy occurs when cancer cells are found in nearby lymph nodes. When these cells begin new tumors far from the primary tumor, **metastasis** has occurred (Fig. 19.2*f*). Not many cancer cells achieve this ability (maybe 1 in 10,000), but those that successfully metastasize to various parts of the body lower the prognosis (the predicted outcome of the disease) for recovery.

Cancer Results from Gene Mutation

Recall that the cell cycle consists of interphase, followed by mitosis. **Checkpoints** (see Fig. 18.3) in the cell cycle monitor the condition of the cell and regulate its ability to divide. Normally, a protein called **cyclin** directs the movement of a cell through the cell cycle. At each checkpoint specific factors, such as DNA damage, are assessed by checkpoint proteins. However, mutations in these checkpoint proteins cause the cell to lose control of the cell cycle, resulting in cancer. The two classes of checkpoint proteins are as follows:

Animation Control of the Cell Cycle

1. **Proto-oncogenes** code for proteins that promote the cell cycle and prevent apoptosis. They are often likened to the gas pedal of a car because they cause acceleration of the cell cycle.
2. **Tumor suppressor genes** code for proteins that inhibit the cell cycle and promote apoptosis. They are often likened to the brakes of a car because they inhibit acceleration.

Proto-Oncogenes Become Oncogenes

When proto-oncogenes mutate, they become cancer-causing genes called **oncogenes.** These mutations can be called "gain-of-function," or dominant, mutations because overexpression is the result (Fig. 19.3). Even though we possess two copies of each proto-oncogene, only one must be mutated to lose control of the cell cycle. In general, whatever a proto-oncogene does, an oncogene does it better.

A **growth factor** is a signal that activates a cell signaling pathway, resulting in cell division. Some proto-oncogenes code for a growth factor or for a receptor protein that receives a growth factor. When these proto-oncogenes become oncogenes, receptor proteins are easy to activate and may even be stimulated by a growth factor produced by the receiving cell. Several proto-oncogenes code for Ras proteins that promote mitosis by activating cyclin. *Ras* oncogenes are typically found in many different types of cancers. *Cyclin D* is a proto-oncogene that codes for cyclin directly. When this gene becomes an oncogene, cyclin is readily available all the time.

Tumor Suppressor Genes Become Inactive

When tumor suppressor genes mutate, their products no longer inhibit the cell cycle nor promote apoptosis. Therefore, these mutations can be called "loss-of-function," or recessive, mutations (Fig. 19.4). Unlike proto-oncogenes, both copies of the tumor suppressor gene in the cell must be mutated to lose cell cycle control.

Animation How Tumor Suppressor Genes Block Cell Division

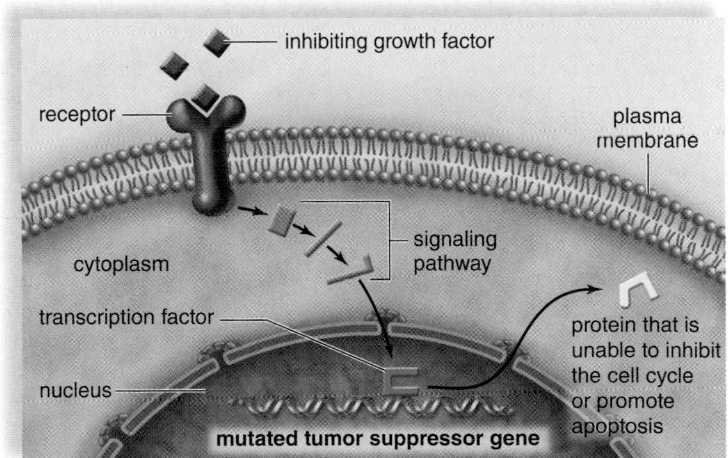

Figure 19.3 Mutations in proto-oncogenes produce oncogenes that stimulate the cell cycle.
Normally, proto-oncogenes stimulate cell division but are typically turned off in differentiated cells. A mutation may convert a proto-oncogene to an oncogene, which then produces a protein that overstimulates the cell cycle.

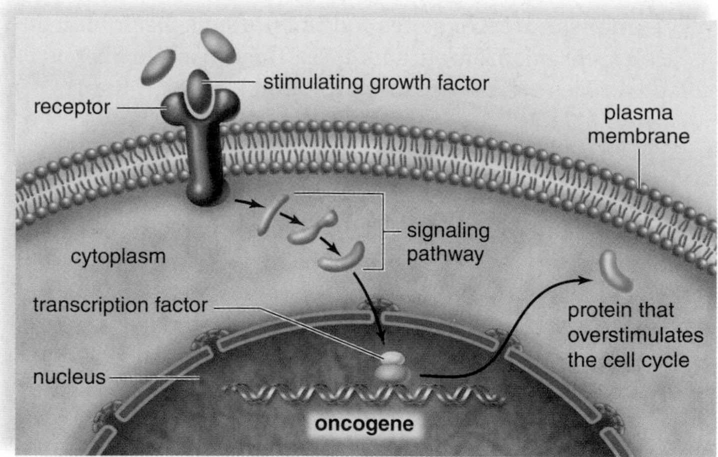

Figure 19.4 Mutations in tumor suppressor genes cause a loss of cell cycle control.
Normally, tumor suppressor genes code for a protein that inhibits the cell cycle. If mutations occur, the resulting protein loses the function of controlling the cell cycle.

Mutation of the tumor suppressor gene *Bax* is a good example. Its product, the protein Bax, promotes apoptosis. When *Bax* mutates, Bax protein is not present and apoptosis is less likely to occur. The *Bax* gene contains a line of eight consecutive G bases in its DNA (see Chapter 2). When the same base molecules are lined up in this fashion, the gene is more likely to be subject to mutation.

Another tumor suppressor protein, p53, activates DNA repair enzymes. At the same time, p53 turns on genes that stop the cell cycle from proceeding. If repair is impossible, the p53 protein promotes *apoptosis,* programmed cell death. Apoptosis is an important way for carcinogenesis to be prevented. Many tumors are lacking in p53 activity.

The *BRCA1* gene codes for another DNA repair enzyme, one that is responsible for fixing breaks in the DNA molecule. In fact, it works very closely with the p53 protein. *BRCA1* mutations prevent the body from recognizing DNA damage, allowing the cells to progress through the cell cycle unchecked. *BRCA1* mutations are associated with a number of cancers, including breast cancer (see the Chapter 18 opener).

Types of Cancer

Statistics indicate that one in three Americans will deal with cancer in their lifetime. Therefore, this is a topic of considerable importance to the health and well-being of every individual. **Oncology** is the study of cancer. A medical specialist in cancer is, therefore, known as an *oncologist.* The patient's prognosis (probable outcome) depends on (1) whether the tumor has invaded surrounding tissues and (2) whether there are metastatic tumors in distant parts of the body.

Tumors are classified according to their place of origin. **Carcinomas** are cancers of the epithelial tissues, and adenocarcinomas are cancers of glandular epithelial cells. Carcinomas

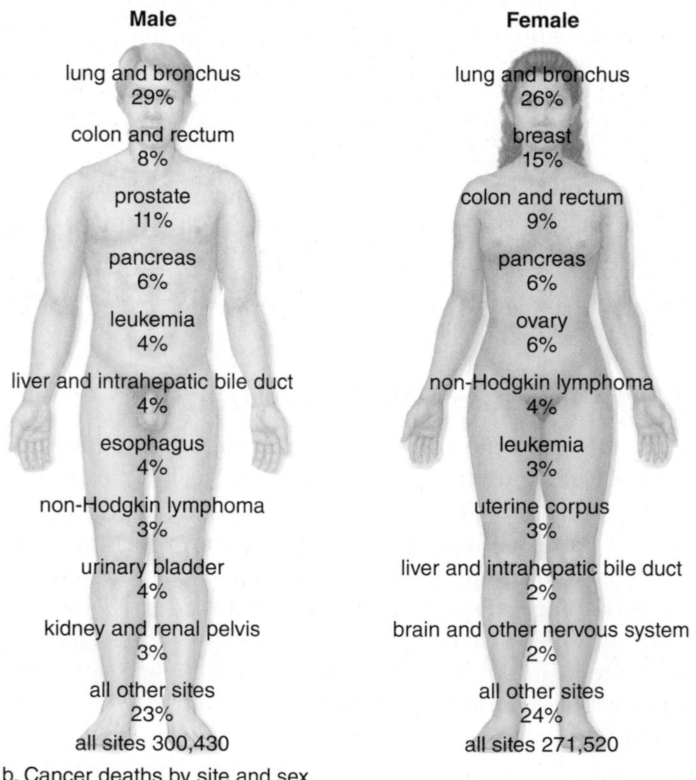

Figure 19.5 Estimated cases of cancer and cancer deaths in the United States.
a. Estimated new cancer cases by sex in 2011. **b.** Estimated deaths from cancer by sex in 2011.

include cancer of the skin, breast, liver, pancreas, intestines, lung, prostate, and thyroid. **Sarcomas** are cancers that arise in muscles and connective tissue, such as bone and fibrous connective tissue. **Leukemias** are cancers of the blood, and **lymphomas** are cancers of lymphatic tissue. A blastoma is a cancer composed of immature cells. Recall that the embryo is formed from three primary germ layers: ectoderm, mesoderm, and endoderm (see Chapter 17). Each blastoma cell resembles the cells in its original primary germ layer. For example, a nephroblastoma (see the chapter opener) has cells similar to mesoderm cells, because the kidney grows from mesoderm.

Common Cancers

Cancer occurs in all parts of the body, but some organs are more susceptible than others (Fig. 19.5). In the respiratory system, lung cancer is the most common type. Overall, this is one of the most common types of cancer, and smoking is known to increase a person's risk for this disease. Smoking also increases the risk for cancer in the oral cavity.

In the digestive system, colorectal cancer (colon/rectum) is another common tumor. Other cancers of the digestive system include those of the pancreas, stomach, esophagus, and other organs. In the cardiovascular system, cancers include leukemia and plasma cell tumors. In the lymphatic system, cancers are classified as either Hodgkin or non-Hodgkin lymphoma. Hodgkin lymphomas develop from mutated B cells. Non-Hodgkin lymphomas can arise from B cells or T cells. Thyroid cancer is the most common type of tumor in the endocrine system, whereas brain and spinal tumors are found in the central nervous system.

Breast cancer is one of the most common types of cancer. Although predominantly found in women, occasionally it is also found in men. Cancers of the cervix, ovaries, and other reproductive structures also occur in women. In males, prostate cancer is one of the most common cancers. Other cancers of the male reproductive system include cancer of the testis and the penis. Bladder and kidney cancers are associated with the urinary system. Skin cancers include melanoma, and basal and squamous cell carcinomas.

CHECK YOUR PROGRESS 19.1

1. List the characteristics of cancer cells that allow them to grow uncontrollably.
2. Describe how mutations in tumor suppressor genes and proto-oncogenes have opposite effects on the cell cycle.
3. Summarize some of the more common types of cancer.

CONNECTING THE CONCEPTS

For more information on the topics in this section, refer to the following discussions:

Section 18.2 describes the role of the checkpoints of the cell cycle.

Section 21.2 describes how the information in a gene is expressed to form a protein.

19.2 Causes and Prevention of Cancer

LEARNING OUTCOMES

Upon completion of this section, you should be able to

1. Explain how heredity and the environment may both contribute to cancer.
2. Identify the genetic mechanisms of select forms of cancer.
3. Summarize how protective behaviors and diet can help prevent cancer.

Our current understanding of the causes of cancer is incomplete. However, by studying patterns of cancer development in populations, scientists have determined that both heredity and environmental risk factors come into play. Obviously, one's genetic inheritance can't be changed; but avoiding risk factors and following dietary guidelines can help prevent cancer.

Heredity

The first gene associated with breast cancer was discovered in 1990. Scientists named this gene *BRCA1* (*br*east *ca*ncer *1*), early onset. The gene is now called *br*east *ca*ncer susceptibility gene *1*. Later, they found that breast cancer could also be due to another breast cancer gene they called *BRCA2*. These genes are tumor suppressor genes that follow a recessive pattern of inheritance (see section 20.1). We inherit two copies of every gene, one from each parent. If a mutated copy of *BRCA1* or *BRCA2* is inherited from either parent, a mutation in the other copy is required before predisposition to cancer is increased. Each cell in the body already has the single mutated copy, but cancer develops wherever the second mutation occurs. If the second mutation occurs in the breast, breast cancer may develop. Ovarian cancer may develop if a second cancer-causing mutation occurs in an ovary.

APPLICATIONS AND MISCONCEPTIONS

How does cancer cause death?

Often, cancer causes death by interfering with the ability of the body to maintain homeostasis. For example, bone cancer may interfere with calcium homeostasis or the ability of the body to produce red blood cells. The liver and pancreas play an important role in homeostasis, and thus cancers of these organs often cause death by altering the levels of important chemicals (hormones, nutrients) in the blood. Sometimes, a growing tumor may block an important pathway in the body, such as the lungs, digestive tract, or an artery or vein. In other cases, the tumor will produce chemicals (or metabolites) that are toxic to a certain group of tissues.

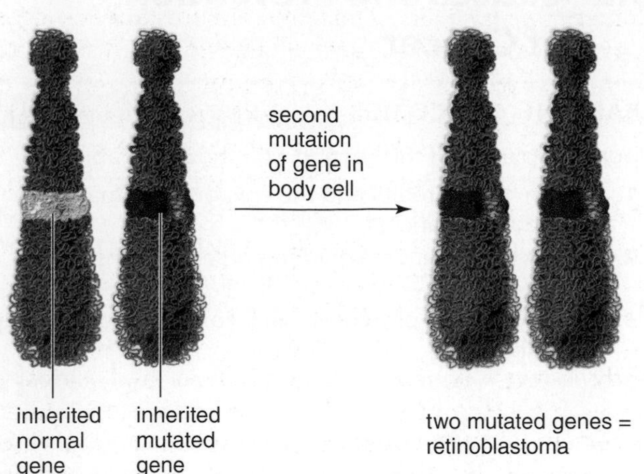

Figure 19.6 Inheritance of retinoblastoma.
A child is at risk for an eye tumor when a mutated copy of the *RB* gene is inherited, even though a second mutation in the normal copy is required before the tumor develops.

The *RB* gene is also a tumor suppressor gene. It takes its name from its association with retinoblastoma, a rare eye cancer that occurs almost exclusively in early childhood. In an affected child, a single mutated copy of the *RB* gene is inherited. A second mutation, this time to the other copy of the *RB* gene, causes the cancer to develop (Fig. 19.6). Retinoblastoma affecting a single eye is treated by removal of the eye, because total removal is more likely to result in a cure. Radiation, laser, and chemotherapy are options when both eyes are involved, but these treatments are less likely to produce a cure.

An abnormal *RET* gene, which predisposes an individual to thyroid cancer, can be passed from parent to child. *RET* is a proto-oncogene known to be inherited in an autosomal dominant manner (see section 20.1), meaning that only one mutation is needed to increase a predisposition to cancer. The remainder of the mutations necessary for a thyroid cancer to develop are acquired (not inherited).

Environmental Carcinogens

A **mutagen** is an agent that causes mutations. A simple laboratory test—an Ames test—is used to determine whether a substance is mutagenic. A **carcinogen** is a chemical that causes cancer by being mutagenic. Some carcinogens cause only initiation. Others cause initiation and promotion.

Heredity can predispose a person to cancer, but whether it develops or not depends on environmental mutagens, such as the ones we will be discussing.

Radiation

Ionizing radiation, such as in radon gas, nuclear fuel, and X-rays, is capable of affecting DNA and causing mutations. Though not a form of ionizing radiation, ultraviolet light also causes mutations. Ultraviolet radiation in sunlight and tanning lamps is most likely responsible for the dramatic increases seen in skin cancer in the past several years. Today, at least six cases of skin cancer occur for every one case of lung cancer. Nonmelanoma skin cancers are usually curable through surgery. However, melanoma skin cancer tends to metastasize and is responsible for 1–2% of cancer deaths annually in the United States.

Video Melanoma Marker

Another natural source of ionizing radiation is radon gas, which comes from the natural (radioactive) breakdown of uranium in soil, rock, and water. The Environmental Protection Agency recommends that every home be tested for radon because it is the second leading cause of lung cancer in the United States. The combination of inhaling radon gas and smoking cigarettes can be particularly dangerous. A vent pipe system and fan, which pull radon from beneath the house and vent it to the outside, are the most common tools to rid a house of radon after the house is constructed.

We are well aware of the damaging effects of a nuclear bomb explosion or accidental emissions from nuclear power plants. For example, cancer deaths are elevated in the vicinity of the Chernobyl Power Station (in Ukraine), which suffered a terrible accident in 1986. The 2011 incident at the Fukushima nuclear plant (in Japan) released radiation into the atmosphere. Medical professionals are still assessing the potential cancer risks from this event. Usually, however, diagnostic X-rays account for most of our exposure to artificial sources of radiation. The benefits of these procedures can far outweigh the possible risk, but it is still wise to avoid X-ray procedures that are not medically warranted. When X-rays are necessary for therapy, nearby noncancerous tissues should be shielded carefully.

Recently, there has been a great deal of public concern regarding the presumed danger of nonionizing radiation. This energy form is given off by cell phones, electrical lines, and appliances. However, extensive studies have not found any evidence that links nonionizing radiation to cancer.

Organic Chemicals

Certain chemicals, particularly synthetic organic chemicals, have been found to be risk factors for cancer. We will mention only two examples: the organic chemicals in tobacco and those pollutants in the environment. There are many other chemical carcinogens.

Tobacco Smoke Tobacco smoke contains a number of organic chemicals that are known mutagens, including *N*-nitrosonornicotine, vinyl chloride, and benzo[a]pyrene (a known suppressor of p53). The greater the number of cigarettes smoked per day and the earlier the habit starts, the more likely it is that cancer will develop. On the basis of data, such as those shown in Figure 19.7, scientists estimate that about 80% of all cancers, including oral cancer and cancers of the larynx, esophagus, pancreas, bladder, kidney, and cervix, are related to the use of tobacco products. When smoking is combined with drinking alcohol, the risk of these cancers increases.

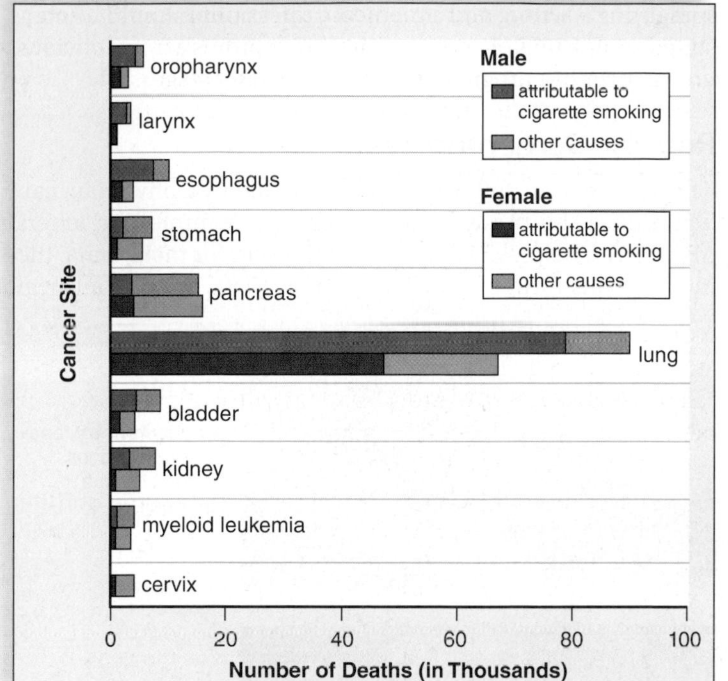

Figure 19.7 **Cancer deaths attributed to smoking.** These data from the Centers for Disease Control and Prevention (CDC) show that, overall, the majority of cancer deaths for men and women may be attributed to cigarette smoking.

Passive smoking, or inhalation of someone else's tobacco smoke, is also dangerous. Researchers continue to collect evidence that confirms the link between passive smoking and cancer.

Pollutants Being exposed to substances such as metals, dust, chemicals, or pesticides at work can increase the risk of cancer. Asbestos, nickel, cadmium, uranium, radon, vinyl chloride, benzidine, and benzene are well-known examples of carcinogens in the workplace. For example, inhaling asbestos fibers increases the risk of lung diseases, including cancer. The cancer risk is especially high for asbestos workers who smoke.

Data show the incidence in soft tissue sarcomas, malignant lymphomas, and non-Hodgkin lymphomas has increased in farmers living in Nebraska and Kansas. All used 2,4-D, a commonly used herbicidal agent, on crops and to clear weeds along railroad tracks.

Viruses

At least four types of DNA viruses—hepatitis B and C viruses, Epstein-Barr virus, and human papillomavirus (HPV)—are directly believed to cause human cancers.

In China, almost all the people have been infected with the hepatitis B virus. This correlates with the high incidence of liver cancer in that country. A combined vaccine is now available for hepatitis A and B. For a long time, circumstances suggested that cervical cancer was a sexually transmitted disease. Now, human papillomaviruses (HPVs) are routinely isolated from cervical cancers. Burkitt lymphoma occurs frequently in Africa, where virtually all children are infected with the Epstein–Barr virus. In China, the Epstein–Barr virus is isolated in nearly all nasopharyngeal cancer specimens.

RNA-containing retroviruses, in particular, are known to cause cancers in animals. In humans, the retrovirus HTLV-1 (human T-cell lymphotropic virus, type 1) has been shown to cause hairy cell leukemia. This disease occurs frequently in parts of Japan, the Caribbean, and Africa, particularly in regions where people are known to be infected with the virus. HIV, the virus that causes AIDS, and Kaposi's sarcoma-associated herpesvirus (KSHV) are responsible for the development of Kaposi's sarcoma and certain lymphomas. This occurs due to the suppression of proper immune system functions.

Animation Replication Cycle of a Retrovirus

Dietary Choices

Nutrition is emerging as a way to help prevent cancer. The incidence of breast and prostate cancer parallels a high-fat diet, as does obesity. The Health feature, "Prevention of Cancer," on page 449, discusses how to protect yourself from cancer. The American Cancer Society recommends consumption of fruits and vegetables, whole grains instead of processed (refined) grains, and limited consumption of red meats (especially high-fat and processed meats). Moderate to vigorous activity for 30 to 45 minutes a day, five or more days a week, is also recommended.

APPLICATIONS AND MISCONCEPTIONS

Can transposons cause cancer?

Transposons are small, mobile sequences of DNA that have the ability to move throughout the genome. From an evolutionary perspective, they are closely related to the retroviruses. Also known as "jumping genes," transposons are known to cause mutation as they move throughout the genome. Although the chances of a transposon disrupting the activity of a proto-oncogene or a tumor suppressor gene in a specific cell are very small (our genome has over 3.4 billion nucleotides), there have been cases when a transposon has caused a loss of cell cycle control and been a factor in the development of cancer. Transposon activity has also been associated with the development of other diseases, such as some forms of hemophilia and muscular dystrophy.

Animation Transposons: Shifting Segments of the Genome

CHECK YOUR PROGRESS 19.2

1. Summarize the evidence that there is a hereditary component to cancer.
2. List the environmental carcinogens that are known to play a role in the development of cancer.
3. Discuss the proactive steps that you can take to reduce your risk of cancer.

CONNECTING THE CONCEPTS

For more information on cancer and its prevention, refer to the following discussions:

Section 4.7 describes methods of protecting yourself against skin cancer.

Section 8.6 summarizes how the components of a healthy diet reduce the risk of cancer.

Section 9.7 examines the influence of smoking on lung cancer.

19.3 Diagnosis of Cancer

LEARNING OUTCOMES

Upon completion of this section, you should be able to

1. List the seven warning signs of cancer.
2. Describe the tests that may be used to diagnose cancer in an individual.

The earlier a cancer is detected, the more likely it can be effectively treated. At present, physicians have ways to detect several types of cancer before they become malignant. Researchers are always looking for new and better detection methods. A growing number of researchers now believe the future of early detection lies in testing for the "fingerprints"—molecular changes caused by cancer. Several teams of scientists are working on blood, saliva, and urine tests to catch cancerous gene and protein patterns in these bodily fluids before a tumor develops. In the meantime, cancer is usually diagnosed by the methods discussed here.

Seven Warning Signs

At present, diagnosis of cancer before metastasis is difficult, although treatment at this stage is usually more successful. The American Cancer Society publicizes seven warning signals, which spell out the word CAUTION and of which everyone should be aware:

C change in bowel or bladder habits
A a sore that does not heal
U unusual bleeding or discharge
T thickening or lump in breast or elsewhere
I indigestion or difficulty in swallowing
O obvious change in wart or mole
N nagging cough or hoarseness

Keep in mind that these signs do not necessarily mean that you have cancer. However, they are an indication that something is wrong and a medical professional should be consulted. Unfortunately, some of these symptoms are not obvious until cancer has progressed to one of its later stages.

Routine Screening Tests

Self-examination, followed by examination by a physician, can help detect the presence of cancer. For example, the letters ABCDE can help your self-exam of the skin for melanoma, the most serious form of skin cancer (Fig. 19.8). Breast cancer in

A = Asymmetry: one-half of mole does not look like the other half.

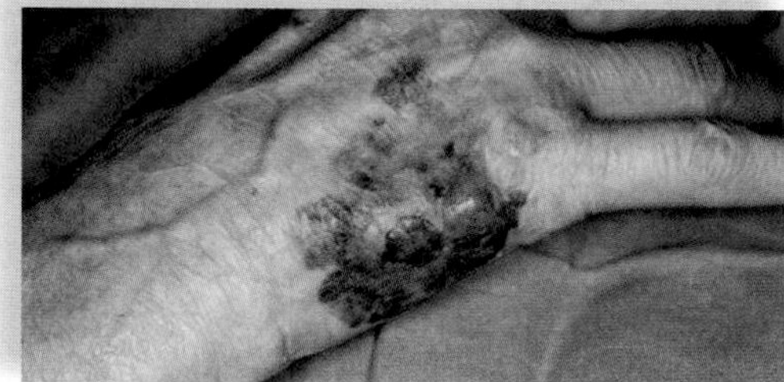

B = Border: irregular scalloped or poorly circumscribed border.

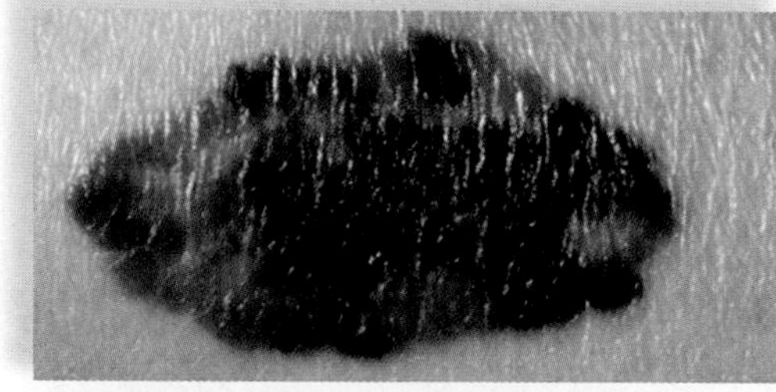

C = Color: varied from one area to another; shades of tan, brown, black, or sometimes white, red, or blue.

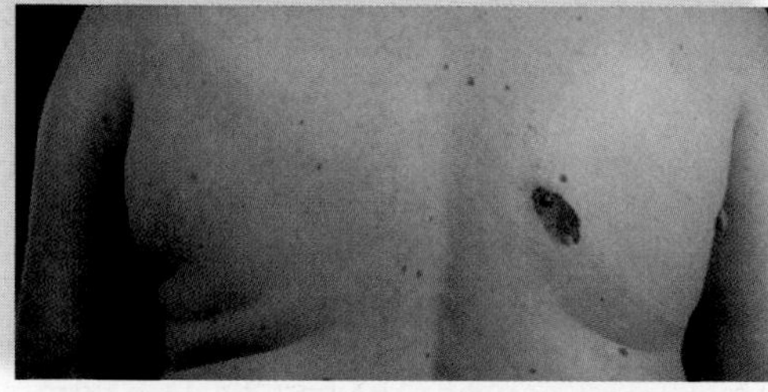

D = Diameter: larger than 6 mm (the diameter of a pencil eraser).

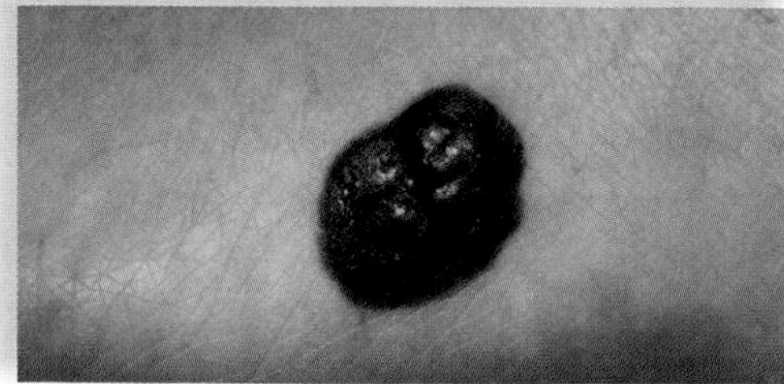

E = Elevated: above skin surface, and **evolving,** or changing over time

Figure 19.8 The ABCDE test for melanoma. Suspicion of melanoma can begin by discovering a mole that has one or more of these characteristics.

BIOLOGY MATTERS  Health

Prevention of Cancer

Protective Behaviors

These behaviors help prevent cancer:

Don't use tobacco. Cigarette smoking accounts for over 30% of cancer deaths. Smoking is responsible for 80% of lung cancer cases among men and 79% among women—about 87% on average. Smokers have lung cancer mortality rates 23 times greater than those of nonsmokers. Smokeless tobacco (chewing tobacco or snuff) increases the risk of cancers of the mouth, larynx, throat, and esophagus.

Don't sunbathe or use a tanning booth. Almost all cases of basal cell and squamous cell skin cancers are considered sun-related. Further, sun exposure is a major factor in the development of melanoma, and the incidence of this cancer increases for people living near the equator.

Avoid radiation. Excessive exposure to ionizing radiation can increase cancer risk. Even though most medical and dental X-rays are adjusted to deliver the lowest dose possible, unnecessary X-rays should be avoided. Excessive radon exposure in homes increases the risk of lung cancer, especially for cigarette smokers. It is best to test your home for radon and take the proper remedial actions.

Be tested for cancer. Do the shower check for breast cancer or testicular cancer. Have other exams done regularly by a physician. (See Table 19.1.)

Be aware of occupational hazards. Exposure to several different industrial agents (nickel, chromate, asbestos, vinyl chloride, etc.) and/or radiation increases the risk of various cancers. Risk from asbestos is greatly increased when combined with cigarette smoking.

Get vaccinated. Get vaccinated for HPV and hepatitis A and B. Consult with your physician or other health professional for advice regarding these vaccines and whether they are appropriate for you.

The Right Diet

Statistical studies have suggested that people who follow certain dietary guidelines are less likely to have cancer. The following dietary guidelines greatly reduce your risk of developing cancer:

Avoid obesity. Obesity increases the risk for many different types of cancer, especially those related to the reproductive systems in both men and women. Cancers of the colon, rectum, esophagus, breast, kidney, and prostate are all associated with being overweight.

To reduce the chances of obesity:

Eat plenty of high-fiber foods. Studies have indicated that a high-fiber diet (whole grain cereals, fruits, and vegetables) may protect against colon cancer, a frequent cause of cancer deaths. Foods high in fiber also tend to be low in fat!

Increase consumption of foods that are rich in vitamins A and C. Beta-carotene, a precursor of vitamin A, is found in dark-green, leafy vegetables; carrots; and various fruits. Vitamin C is present in citrus fruits. These vitamins are called antioxidants because in cells they prevent the formation of free radicals (organic ions having an unpaired electron) that can possibly damage DNA. Vitamin C also prevents the conversion of nitrates and nitrites into carcinogenic nitrosamines in the digestive tract.

Reduce consumption of salt-cured, smoked, or nitrite-cured foods. Salt-cured or pickled foods may increase the risk of stomach and esophageal cancers. Smoked foods, such as ham and sausage, contain chemical carcinogens similar to those in tobacco smoke. Nitrites are sometimes added to processed meats (e.g., hot dogs and cold cuts) and other foods to protect them from spoilage. As mentioned previously, nitrites are converted to nitrosamines in the digestive tract.

Include vegetables from the cabbage family in the diet. The cabbage family includes cabbage, broccoli, brussels sprouts, kohlrabi, and cauliflower. These vegetables may reduce the risk of gastrointestinal and respiratory tract cancers.

Drink alcohol in moderation. Risks of cancer development rise as the level of alcohol intake increases. The risk increases still further for those who smoke or chew tobacco while drinking. Heavy drinkers face a much greater risk for oral, pharyngeal, esophageal, and laryngeal cancer. Higher rates for cancers of the breast and liver are also linked to alcohol abuse. Men should limit daily alcohol consumption to two drinks or fewer. Women should have only one alcoholic drink daily. (A drink is defined as 12 oz. of beer, 5 oz. of wine, or 1.5 oz. of distilled spirits.)

BIOLOGY MATTERS  Health

Cancer Self-Examinations

The American Cancer Society urges women to do a breast self-exam and men to do a testicle self-exam every month. Breast cancer and testicular cancer are far more curable if found early, and we must all take on the responsibility of checking for one or the other.

Shower Self-Exam for Women

1. Check your breasts for any lumps, knots, or changes about one week after your period.
2. Place your right hand behind your head. Move your *left* hand over your *right* breast in a circle. Press firmly with the pads of your fingers (Fig. 19A). Also check the armpit.
3. Now place your left hand behind your head and check your *left* breast with your *right* hand in the same manner as before. Also check the armpit.
4. Check your breasts while standing in front of a mirror right after you do your shower check. First, put your hands on your hips and then raise your arms above your head (Fig. 19B). Look for any changes in the way your breasts look: dimpling of the skin, changes in the nipple, or redness or swelling.
5. If you find any changes during your shower or mirror check, see your doctor right away.

You should know that the best check for breast cancer is a mammogram. When your doctor checks your breasts, ask about getting a mammogram.

Shower Self-Exam for Men

1. Check your testicles once a month.
2. Roll each testicle between your thumb and finger as shown in Figure 19C. Feel for hard lumps or bumps.
3. If you notice a change or have aches or lumps, tell your doctor right away so he or she can recommend proper treatment.

Cancer of the testicles can be cured if you find it early. You should also know that prostate cancer is the most common cancer in men. Men over age 50 should have an annual health checkup that includes a prostate examination.

Information provided by the American Cancer Society. Used by permission.

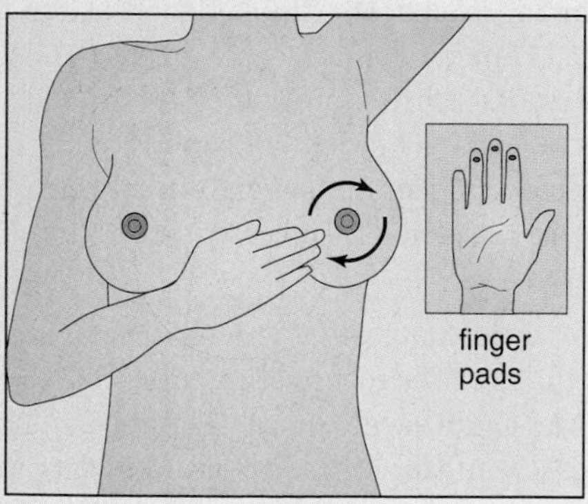

Figure 19A Shower check for breast cancer.

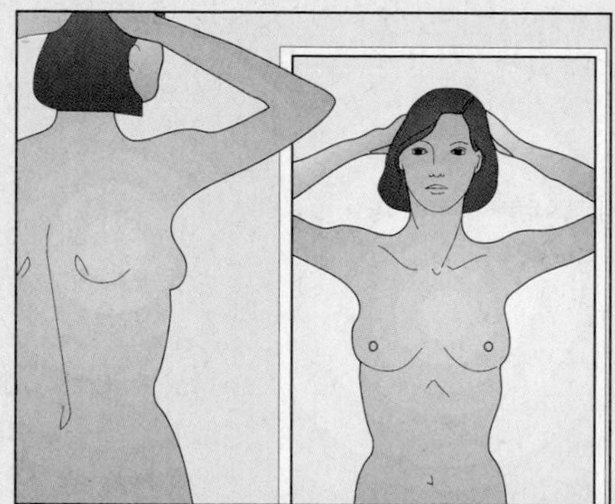

Figure 19B Mirror check for breast cancer.

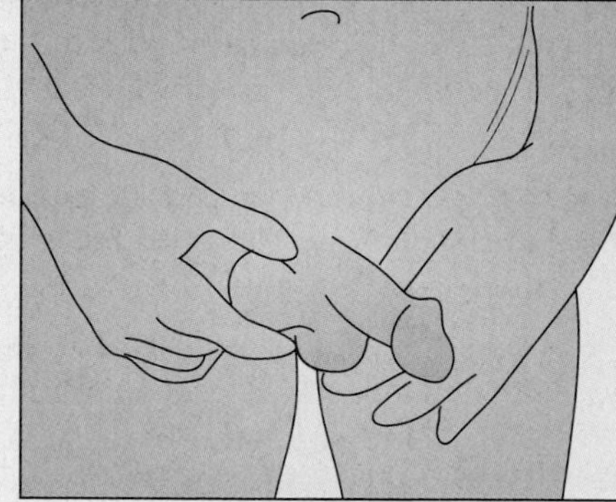

Figure 19C Shower check for testicular cancer.

women and testicular cancer in men can often be detected during a monthly shower check. The technique is discussed in the Health feature, "Cancer Self-Examinations."

The ideal tests for cancer are relatively easy to do, cost little, and are fairly accurate. The Pap test for cervical cancer fulfills these three requirements. A physician merely takes a sample of cells from the cervix, which are then examined microscopically for signs of abnormality (see Fig. 19.1). Any woman who chooses to receive the new HPV vaccine should still get regular Pap tests because (1) the vaccine will not protect against all types of HPV that cause cervical cancer, and (2) the vaccine does not protect against any HPV infections she may already have if she is sexually active. Regular Pap tests are credited with preventing over 90% of deaths from cervical cancer.

Screening for colon cancer also depends upon three types of testing. A digital rectal examination performed by a physician is of limited value because only a portion of the rectum can be reached. With flexible sigmoidoscopy, the second procedure, a much larger portion of the colon can be examined by using a thin, pliable, lighted tube. Finally, a stool blood test (fecal occult blood test) consists of examining a stool sample to detect any hidden blood. The sample is smeared on a slide, and a chemical is added that changes color in the presence of hemoglobin. This procedure is based on the supposition that a cancerous polyp bleeds, although some polyps do not bleed. Moreover, bleeding is not always because of a polyp. Therefore, the percentage of false negatives and false positives is high. All positive tests are followed up by a colonoscopy, an examination of the entire colon, or by X-ray after a barium enema. If the colonoscopy detects polyps, they can be destroyed by laser therapy. Blood tests are used to detect leukemia. Urinalysis aids in the diagnosis of bladder cancer.

Breast cancer is not as easily detected, but three procedures are recommended. First, every woman should do a monthly breast self-examination. But this is not a sufficient screen for breast cancer. Therefore, all women should have an annual physical examination, especially women over 40, when a physician will do the same procedure. Even then, this type of examination may not detect lumps before metastasis has already taken place. That is the goal of the third recommended procedure, *mammography,* an X-ray study of the breast (Fig. 19.9). However, mammograms do not show all cancers, and new tumors may develop in the interval between mammograms. The hope is that a mammogram will reveal a lump too small to be felt, at a time when the cancer is still highly curable. Table 19.1 outlines when routine screening tests should be done for various cancers, including breast cancer.

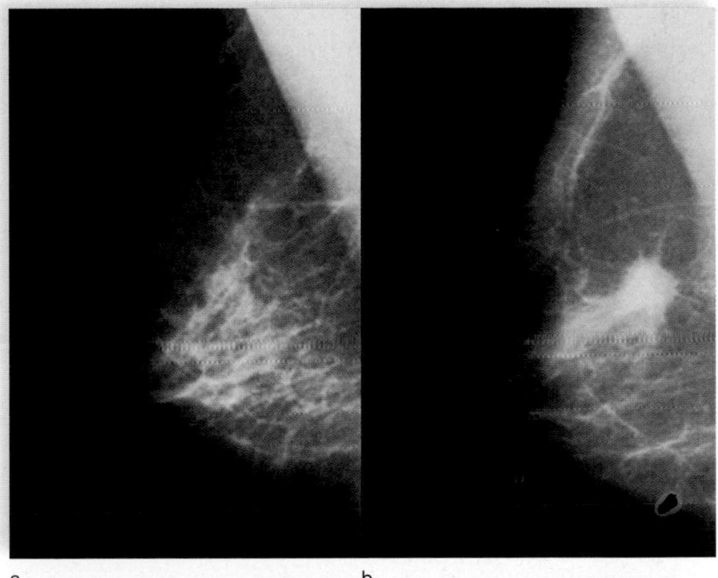

Figure 19.9 Mammograms can detect breast cancer. An X-ray image of the breast can find tumors too small to be felt. **a.** An image of a normal breast; (**b**) a breast with a tumor.

The diagnosis of cancer in other parts of the body may involve other types of imaging. A computerized axial tomography (CAT) scan uses computer analysis of scanning X-ray images to create cross-sectional pictures that portray a tumor's size and location (see Fig. 2.3). Magnetic resonance imaging (MRI) is another type of imaging technique that depends on computer analysis. MRI is particularly useful for analyzing tumors in tissues surrounded by bone, such as tumors of the brain or spinal cord. A radioactive scan obtained after a radioactive isotope is administered can reveal any abnormal isotope accumulation due to a tumor. During ultrasound,

Table 19.1 Recommendations for the Early Detection of Cancer in Average-Risk Asymptomatic People

Cancer	Population	Test or Procedure	Frequency
Breast	Women, age ≥20 years	Breast self-examination	Monthly, starting at age 20 years; see the Health feature, "Cancer Self-Examinations," page 450.
		Clinical breast examination and mammography	For women in 20s and 30s, every 3 years; aged 40 years and over, preferably annually
Colorectal*	Men and women, age ≥50 years	Fecal occult blood test (FOBT)	Annually, starting at age 50 years
		Fecal immunochemical test (FIT), flexible sigmoidoscopy, *or* double-contrast barium enema	Every 5 years, starting at age 50 years
		Colonoscopy	Every 10 years, starting at age 50 years
Prostate	Men, age ≥50 years	Digital rectal examination and prostate-specific antigen (PSA) test	Discussion with physician annually, starting at age 50 years, for men who have a life expectancy of at least 10 more years
Cervical	Women, age ≥21 years	Pap test	Every 3 years
	Women, age 30–65 years	Pap test + HPV DNA test	Every 5 years, with an alternate Pap test alone every 3 years. Women age >65 should stop screening unless she had a serious cervical pre-cancer or cancer in the last 20 years.
Endometrial†	Women, at menopause	Report any unexpected bleeding or spotting to a physician.	
Testicular	Men, age 20 years	Testicle, self-examination	Monthly, starting at age 20 years, see Health feature, "Cancer Self-Examinations," page 450.
Other cancers	Men and women, age 20 years	On the occasion of a periodic health examination, a cancer-related checkup should include examination for other cancers. See the Health feature, "Cancer Self-Examinations," page 450, for counseling about tobacco, sun exposure, diet and nutrition, risk factors, and environmental and occupational exposures.	

* follow one of the schedules. †average risk, increased, or high risk should contact physician. *Source:* American Cancer Society website: www.cancer.org.

echoes of high-frequency sound waves directed at a part of the body are used to reveal the size, shape, and location of tissue masses. Ultrasound can confirm tumors of the stomach, prostate, pancreas, kidney, uterus, and ovary.

Aside from various imaging procedures, a diagnosis of cancer can be confirmed without major surgery by performing a *biopsy* or viewing body parts. Needle biopsies allow removal of a few cells for examination. Sophisticated techniques, such as laparoscopy, permit viewing of body parts.

Tumor Marker Tests

Tumor marker tests are blood tests for antigens and/or antibodies. Blood tests are possible because tumors release substances that provoke an antibody response in the body. For example, if an individual has already had colon cancer, it is possible to use the presence of an antigen called *carcinoembryonic antigen* (*CEA*) to detect any relapses. When the CEA level rises, additional tumor growth has occurred.

There are also tumor marker tests that can be used for early cancer diagnosis. They are not reliable enough to count on solely, but in conjunction with physical examination and ultrasound, they are considered useful. For example, there is a *prostate-specific antigen* (*PSA*) *test* for prostate cancer, a *CA-125 test* for ovarian cancer, and an *alpha-fetoprotein* (*AFP*) *test* for liver tumors.

Video Melanoma Marker

Genetic Tests

Tests for genetic mutations in proto-oncogenes and tumor suppressor genes are making it possible to detect the likelihood of cancer before the development of a tumor. Tests are available that signal the possibility of colon, bladder, breast, and thyroid cancers, as well as melanoma. Physicians now believe that a mutated *RET* gene means that thyroid cancer is present or may occur in the future, and a mutated *p16* gene appears to be associated with melanoma. Genetic testing can also be used to determine if cancer cells still remain after a tumor has been removed.

A genetic test is also available for the presence of *BRCA1* (breast cancer gene 1). A woman who has inherited this gene can choose to either have prophylactic surgery or to be frequently examined for signs of breast cancer. Physicians can use microsatellites to detect chromosomal deletions that accompany bladder cancer. Microsatellites are small regions of DNA that always have two (di-), three (tri-), or four (tetra-) nucleotide repeats. They compare the number of nucleotide DNA repeats in a lymphocyte microsatellite with the number in a microsatellite of a cell found in urine. When the number of repeats is less in the cell from urine, a bladder tumor is suspected.

Telomerase, you will recall, is the enzyme that keeps telomeres a constant length in cells. The gene that codes for telomerase is turned off in normal cells but is active in cancer cells. Therefore, if the test for the presence of telomerase is positive, the cell is cancerous.

Video New Cancer Clue

CHECK YOUR PROGRESS 19.3

1. List the seven warning signs of cancer.
2. Discuss the routine screening tests that are available to detect and diagnose cancer.
3. Discuss how genetic tests and tumor marker tests may be used to prevent cancer.

CONNECTING THE CONCEPTS

For more information on the information listed in this section, refer to the following discussions:

Section 7.4 describes how the body generates antibodies against antigens.

Section 8.5 examines how polyps in the large intestine (colon) may lead to colon cancer.

Section 16.3 examines the structure of the female reproductive system and illustrates the location of the cervix.

19.4 Treatment of Cancer

LEARNING OUTCOMES

Upon completion of this section, you should be able to

1. Describe how radiation therapy, chemotherapy, and surgery may all be used to treat cancer.
2. Summarize some of the new advances in the treatment of cancer.

Certain therapies for cancer have been available for some time. Other methods of therapy are in clinical trials. If successful, these new therapies will become more generally available in the future.

Standard Therapies

Surgery, radiation therapy, and chemotherapy are the standard methods of cancer therapy.

Surgery

With the exception of cancers of the blood, surgery may be sufficient for cancer in situ. But because there is always the danger that some cancer cells have been left behind, surgery is often preceded by and/or followed by radiation therapy and chemotherapy.

Radiation Therapy

Ionizing radiation causes chromosomal breakage and cell cycle disruption. Therefore, rapidly dividing cells, such as cancer cells, are more susceptible to its effects than other cells. Powerful X-rays or gamma rays can be administered through an externally applied beam. In some instances, tiny radioactive sources can be implanted directly into the patient's body. Cancer of the cervix and larynx, early stages of prostate cancer, and Hodgkin disease are often treated with radiation therapy alone.

Although X-rays and gamma rays are the mainstays of radiation therapy, protons and neutrons also work well. Proton beams can be aimed at the tumor like a rifle bullet hitting the bull's-eye of a target.

Side effects of radiation therapy greatly depend upon which part of the body is being irradiated and how much radiation is used. Weakness and fatigue are very typical. Dry mouth, nausea, and diarrhea often affect the digestive tract. Dry, red, or irritated skin or even blistering burns may occur at the treatment site. Hair loss at the treatment site, which in some situations can be permanent, can be very depressing to the patient. Fortunately, most side effects of radiation treatment are temporary.

Chemotherapy

Radiation is localized therapy, whereas chemotherapy is a way to catch cancer cells that have spread (Figs. 19.10 and 19.11). Unlike radiation, which treats only the part of the body exposed to the radiation, chemotherapy treats the entire body. As a result, any cells that may have escaped from where the cancer originated are treated. Most chemotherapeutic drugs kill cells by damaging their DNA or interfering with DNA synthesis. The hope is that all cancer cells will be killed, while leaving enough normal cells untouched to allow the body to keep functioning. Combining drugs that have different actions at the cellular level may help destroy a greater number of cancer cells. Combinations may also reduce the risk of the cancer developing resistance to one particular drug. What chemicals are used is generally based on the type of cancer and the patient's age, general health, and perceived ability to tolerate potential side effects. Some of the types of chemotherapy medications commonly used to treat cancer include the following.

Alkylating agents. These medications interfere with the growth of cancer cells by blocking the replication of DNA.

Antimetabolites. These drugs block the enzymes needed by cancer cells to live and grow.

Antitumor antibiotics. These antibiotics—different from those used to treat bacterial infections—interfere with DNA, blocking certain enzymes and cell division and changing cell membranes.

Mitotic inhibitors. These drugs inhibit cell division or hinder certain enzymes necessary in the cell reproduction process.

Nitrosoureas. These medications impede the enzymes that help repair DNA.

Whenever possible, chemotherapy is specifically designed for the particular cancer. For example, in some cancers, a small portion of chromosome 9 is missing. Therefore, DNA metabolism differs in the cancerous cells compared with normal cells. Specific chemotherapy for the cancer can exploit this metabolic difference and destroy the cancerous cells.

One drug, paclitaxel (Taxol), extracted from the bark of the Pacific yew tree, was found to be particularly effective against advanced ovarian cancers, as well as breast, head, and neck tumors. Paclitaxel interferes with microtubules needed for cell division. Now, chemists have synthesized a family of related drugs, called taxoids, which may be more powerful and have fewer side effects than paclitaxel.

Certain types of cancer, such as leukemias, lymphomas, and testicular cancer, are now successfully treated by combination chemotherapy alone. The survival rate for children with childhood leukemia is 80%. Hodgkin disease, a lymphoma, once killed two out of three patients. Combination therapy, using four different drugs, can now wipe out the disease in a matter of months. Three out of four patients achieve a cure, even when the cancer is not diagnosed immediately. In other cancers—most notably, breast and colon cancer, chemotherapy can reduce the chance of recurrence after surgery has removed all detectable traces of the disease.

Figure 19.10 **Radiation treatment for cancer.** Most people who receive radiation therapy for cancer have external radiation on an outpatient basis. The patient's head is immobilized so that radiation can be delivered to a very precise area.

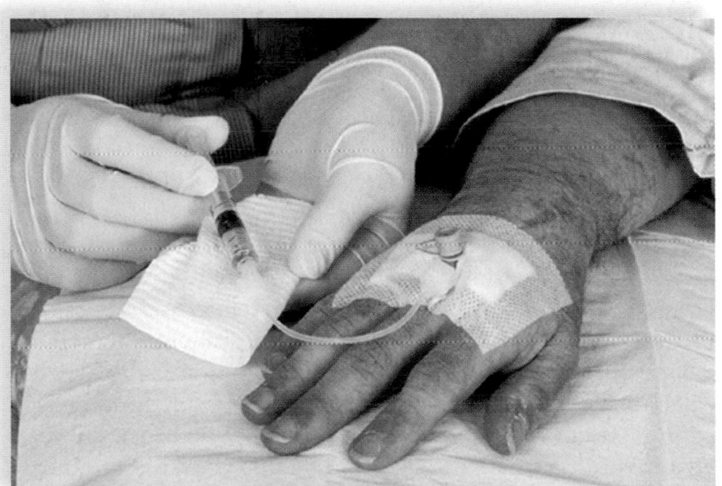

Figure 19.11 **Treating cancer with chemotherapy.** The intravenous route is the most common, allowing chemotherapy drugs to spread quickly throughout the entire body by way of the bloodstream.

APPLICATIONS AND MISCONCEPTIONS

What can be done to lessen the side effects of radiation and chemotherapy?

Victims of cancer will often describe the side effects of treatment as the worst part of the disease. Nausea, vomiting, diarrhea, weight loss and hair loss, anxiety and/or depression, and extreme fatigue are common symptoms that may be caused by chemotherapy or radiation therapy. However, strategies now exist to help sufferers deal with treatment side effects, and others continue to be devised. Antiemetic drugs alleviate nausea and vomiting. Genetically engineered erythropoietin will help to stimulate red blood cell production and reduce fatigue (see Chapter 6). Dronabinol (Marinol), a medication closely related to marijuana, stimulates appetite and helps with weight loss. Hair loss can be minimized by cryotherapy, which involves applying cold packs on the scalp during a treatment. Cold temperatures slow the metabolic rate of hair follicles, and more follicles survive treatment. Antidepressants and antianxiety medications will help the patient to deal with the psychological effects of the disease.

Increasingly, cancer patients are turning to alternative therapies as well for help in coping with the disease. Calming techniques, including yoga, meditation, and tai chi, help the sufferer to relax. Hypnosis can have the same restful effect. Aromatherapy, massage therapy, and music therapy seem to ease tension and stress, perhaps by stimulating pleasure centers in the brain. Exercise is known to produce endorphins, the brain's pain-relieving neurotransmitters (see Chapter 13).

Chemotherapy sometimes fails because cancer cells become resistant to one or several chemotherapeutic drugs, a phenomenon called multidrug resistance. This occurs because a plasma membrane carrier pumps the drug (or drugs) out of the cancer cell before it can be harmed. Researchers are testing drugs known to poison the pump in an effort to restore the effectiveness of the drugs. Another possibility is to use combinations of drugs with different toxic activities, because cancer cells can't become resistant to many different types at once.

Bone marrow transplants are sometimes done in conjunction with chemotherapy. The red bone marrow contains large populations of dividing cells. Therefore, red bone marrow is particularly prone to destruction by chemotherapeutic drugs. In bone marrow autotransplantation, a patient's stem cells are harvested and stored before chemotherapy begins. High doses of radiation or chemotherapeutic drugs are then given within a relatively short time. This prevents multidrug resistance from occurring, and the treatment is more likely to catch every cancer cell. The stored stem cells can then be returned to the patient by injection. They automatically make their way to bony cavities, where they resume blood cell formation.

Newer Therapies

Several therapies are now in clinical trials and are expected to be increasingly used to treat cancer.

Immunotherapy

We now know that the immune system has the ability to keep some cancers under control, meaning that it will slow tumor growth. However, in many cases the cancer cells are able to avoid the immune system even though they bear antigens that make them different from the body's normal cells. It has been possible for some time to vaccinate the body against viruses that are associated with cancer—for example, HPV—and scientists are beginning to develop vaccines against cancer cells directly. Several vaccines are in development, but currently none are available in the United States. Another idea is to use immune cells, genetically engineered to bear the tumor's antigens (Fig. 19.12). When these cells are returned to the body, they produce cytokines. Recall that cytokines are chemical messengers for immune cells (see Chapter 7). Cytokines stimulate the body's immune cells to attack the tumor. Further, the altered immune cells present the tumor antigen to cytotoxic T cells, which then go forth and destroy tumor cells in the body.

Passive immunotherapy is also possible. Monoclonal antibodies have the same structure because they are produced by the same plasma cell (see Fig. 7.17). Some monoclonal antibodies are designed to zero in on the receptor proteins of cancer cells. To increase the killing power of monoclonal antibodies, they are linked to radioactive isotopes or chemotherapeutic drugs. It is expected that soon they will be used as initial therapies, in addition to chemotherapy.

p53 Gene Therapy

Researchers believe that *p53* gene expression is needed for only 19 hours to trigger apoptosis, programmed cell death. And the *p53* gene seems to trigger cell death only in cancer cells—elevating the *p53* level in a normal cell doesn't do any harm, possibly because apoptosis requires extensive DNA damage.

Ordinarily, when adenoviruses infect a cell, they first produce a protein that inactivates *p53.* In a cleverly designed procedure, investigators genetically engineered an adenovirus that lacks the gene for this protein. Now, the adenovirus can infect and kill only cells that lack a *p53* gene. Which cells are those? Tumor cells, of course. Another plus to this procedure is that the injected adenovirus spreads through the cancer, killing tumor cells as it goes. This genetically engineered virus is now in clinical trials.

Other Therapies

Many other therapies are now being investigated. Among them, drugs that inhibit angiogenesis are a proposed therapy under investigation. Antiangiogenic drugs confine and reduce tumors by breaking up the network of new capillaries in the vicinity

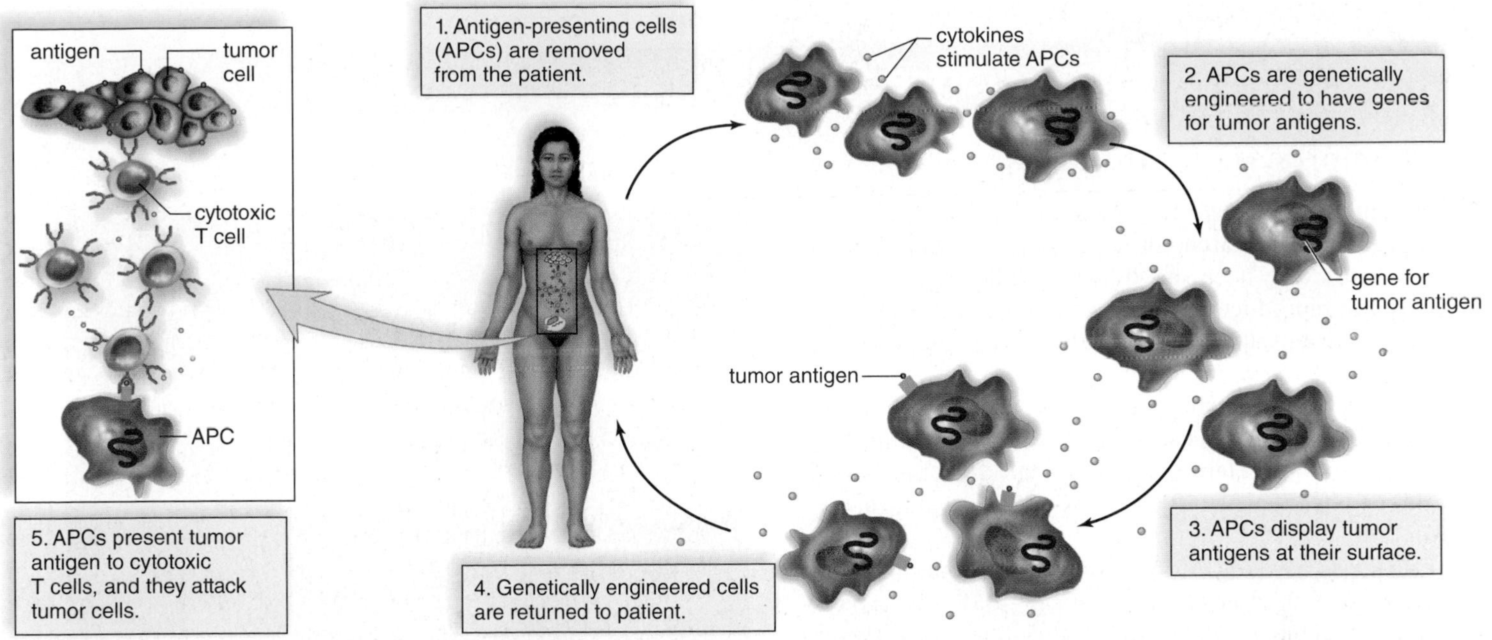

antigen
tumor cell
cytotoxic T cell
APC
5. APCs present tumor antigen to cytotoxic T cells, and they attack tumor cells.
1. Antigen-presenting cells (APCs) are removed from the patient.
cytokines stimulate APCs
2. APCs are genetically engineered to have genes for tumor antigens.
gene for tumor antigen
tumor antigen
3. APCs display tumor antigens at their surface.
4. Genetically engineered cells are returned to patient.

Figure 19.12 Use of immunotherapy to treat cancer.
1. Antigen-presenting cells (APCs) are removed and (**2**) are genetically engineered to (**3**) display tumor antigens. **4.** After these cells are returned to the patient, (**5**) they present the antigen to cytotoxic T cells, which then kill tumor cells.

of a tumor. A number of antiangiogenic compounds are currently being tested in clinical trials. Two highly effective drugs, called angiostatin and endostatin, have been shown to inhibit angiogenesis in laboratory animals and are expected to do the same in humans.

CHECK YOUR PROGRESS 19.4

1. Describe the three types of therapy that are presently the standard ways to treat cancer.
2. Discuss why cancer treatment may involve a combination of chemotherapy, radiation therapy, and surgery.
3. Discuss why a cancer treatment that targets *p53* may be successful.
4. Explain the rationale for antiangiogenesis therapy.

CONNECTING THE CONCEPTS

For more information on the topics presented in this section, refer to the following discussions:

Section 7.3 examines the role of chemical signals, such as cytokines, in protecting the body from infection.

Section 7.4 explores the function of cytotoxic T cells in the immune response.

Section 21.3 provides an overview on how genetic engineering can alter the genes within a virus.

CASE STUDY CONCLUSION

In Cody's case, his treatment involved a process called a simple nephrectomy. During this procedure, the surgeon removes the entire kidney. Typically, nephroblastoma does not occur in both kidneys, and because the MRI and CAT scan did not indicate any problems with Cody's other kidney, it most likely did not have the nephroblastoma. Because a single kidney is all that is needed to maintain blood homeostasis and water–salt balance in the body, Cody did not need a transplant. Initially, the chemotherapy treatment caused Cody to lose his hair, and his appetite decreased, but the doctors informed the concerned parents that this was normal, and very soon Cody would be back to his normal self.

Although Wilms disease is a rare form of cancer, researchers have identified a number of genes that appear to be associated with the condition. One of these is *WT1*, a gene that plays a major role in the differentiation of renal (kidney) tissue and the development of the kidneys. In Cody's case, he probably inherited one copy of a defective gene from one of his parents. Sometime during development, the other gene in one of his kidney cells acquired a mutation that initiated the formation of the tumor. This explained why Cody's other kidney was unaffected. Because *WT1* is active early in development, it was unlikely that Cody would have any additional complications from his cancer.

BIOLOGY MATTERS  Science

The Immortal Henrietta Lacks

The availability of human tissue cells is important for many areas of medical research. Unfortunately, growing human cells has historically been very difficult. While these cells can be coaxed to reproduce using a technique called tissue culture, they typically all die very rapidly, usually after dividing only a few times. However, metastatic cells (a form of cancer cells) can survive and thrive in tissue culture. The techniques used to grow human cells successfully in tissue culture are part of the legacy of Henrietta Lacks (Fig. 19D). The cancer that claimed her life ensured that a part of Henrietta will live forever.

Lacks was a young African-American woman and the mother of five children. In February of 1951, she experienced unexplained vaginal bleeding and sought help from a Baltimore hospital. Physicians found a quarter-sized tumor on Henrietta's cervix. Samples of the tumor were quickly obtained, then sent to the tissue culture laboratories at Johns Hopkins University. Though radiation treatment was attempted, the tumor ravaged Henrietta's body, appearing on all her major organs within months. After eight months of suffering, she died in October 1951. She was survived by her husband and five children, three of whom were still in diapers.

Figure 19D Henrietta Lacks, 1920–1951.

Henrietta's tumor cells ended up in the laboratory of George and Margaret Gey at Johns Hopkins. The couple had been trying to culture human cells, with little success, for more than two decades. The cells from Henrietta Lacks' tumor, now termed HeLa cells (Fig. 19E), not only lived but multiplied like wildfire. At last, here was a source of human cells that grew rapidly and seemed almost indestructible.

Using these cells, the Geys directed their research toward curing polio. Infantile paralysis, or polio, occurred in epidemics that damaged the brain and spinal cord in a small percentage of patients. Paralysis of major muscle groups resulted. If the diaphragm and other respiratory muscles were affected, the disease could be fatal. For the first time, the virus could be grown in human cells—HeLa cells—so that its characteristics could be studied. The results of these studies enabled Dr. Jonas Salk to develop a vaccine for polio. Today, polio is almost unheard of in the Western world.

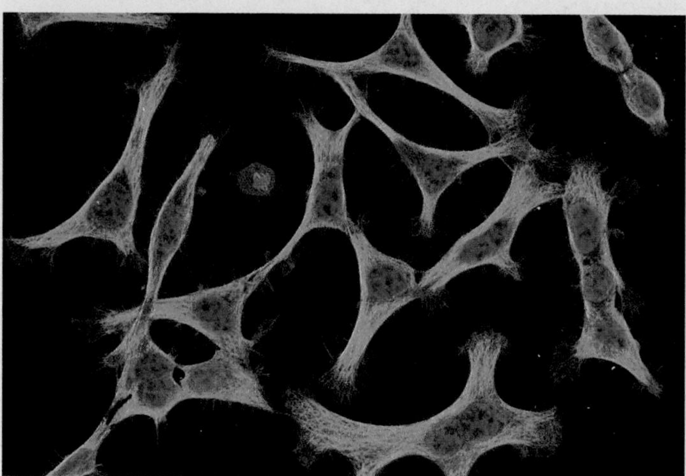

Figure 19E SEM of HeLa cells.

Research with HeLa cells did not stop with the Geys. For over 50 years, these durable cells have been used worldwide to study many types of viruses, as well as leukemia and other cancers. The cells are human in origin, so they have also been the test subjects to determine the harmful effects caused by drugs or radiation. Effective testing was developed for chromosome abnormalities and hereditary diseases using HeLa. Samples of HeLa cells have even been launched in the space shuttle for experiments involving a zero-gravity environment. HeLa cell colonies can be found around the world. The cells can even be purchased from biological supply companies.

HeLa cells are almost too sturdy. In 1974, researchers discovered that HeLa cells, like bacteria, could be transmitted through the air, on a researcher's glove, or on contaminated glassware. This discovery led to better sterile techniques not only in the laboratory but also in operating rooms where cancerous tumors were removed.

In 2011 Henrietta's story was presented in a book, *The Immortal Life of Henrietta Lacks,* by Rebecca Sloot. The book has received multiple awards for its ability to portray the importance of Henrietta's cells to science.

Questions to Consider

1. What characteristics of cancer cells would make HeLa cells immortal?
2. What other uses do you think are possible for these cells?

MEDIA STUDY TOOLS

Enhance your study of this chapter with media! Visit **www.mhhe.com/maderhuman13e** and go to "Media Study Tools" for this chapter to access the following:

Animations	Videos
19.1 Telomerase Function • Control of the Cell Cycle • How Tumor Suppressor Genes Block Cell Division **19.2** Replication Cycle of a Retrovirus **19.3** Transposons: Shifting Segments of the Genome	**19.2** Melanoma Marker **19.3** Melanoma Marker **19.3** New Cancer Clue

SUMMARIZE

19.1 Cancer Cells

Characteristics of Cancer Cells

Cancer represents uncontrolled cell growth. Certain characteristics are common to cancer cells. Cancer cells:

- lack **differentiation** and do not contribute to function;
- do not undergo **apoptosis**—they enter the cell cycle an unlimited number of times. This is often due to the presence of telomerase enzyme, which regulates the length of the **telomere;**
- form **tumors** and do not need growth factors to signal them to divide;
- gradually become abnormal—**carcinogenesis** is comprised of **initiation, promotion,** and **progression;** and
- undergo **angiogenesis** (the growth of blood vessels to support them) and can spread throughout the body (**metastasis**).

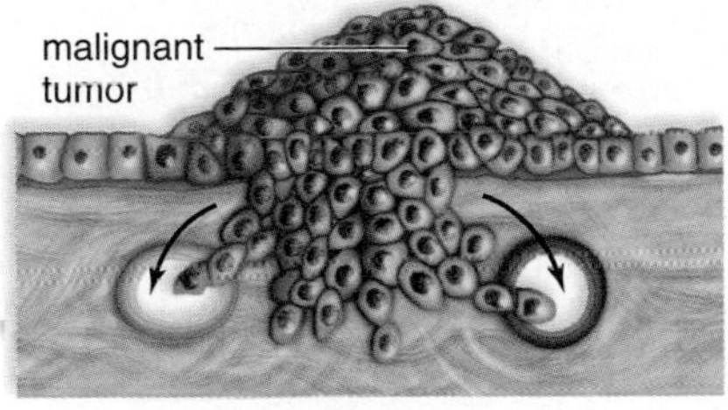

Cancer Results from Gene Mutation

The protein **cyclin** moves the cell through the stages of the cell cycle. The **cell cycle** contains **checkpoints** that are regulated by proteins. Cells become increasingly abnormal due to mutations in proto-oncogenes and tumor suppressor genes.

In normal cells, the cell cycle functions normally.

- **Proto-oncogenes** promote cell cycle activity, often using **growth factors,** and restrain apoptosis. Proto-oncogenes can mutate into **oncogenes.**
- **Tumor suppressor genes** restrain the cell cycle and promote apoptosis.

In cancer cells, the cell cycle is accelerated and occurs repeatedly.

- **Oncogenes** cause an unrestrained cell cycle and prevent apoptosis.
- **Mutated tumor suppressor genes** cause an unrestrained cell cycle and prevent apoptosis.

Types of Cancer

Oncology is the study of cancer. Cancers are classified according to their places of origin.

- **Carcinomas** originate in epithelial tissues.
- **Sarcomas** originate in muscle and connective tissues.
- **Leukemias** originate in blood.
- **Lymphomas** originate in lymphatic tissue.

Certain body organs are more susceptible to cancer than others.

19.2 Causes and Prevention of Cancer

Development of cancer is determined by a person's genetic profile, plus exposure to environmental **mutagens** and **carcinogens.**

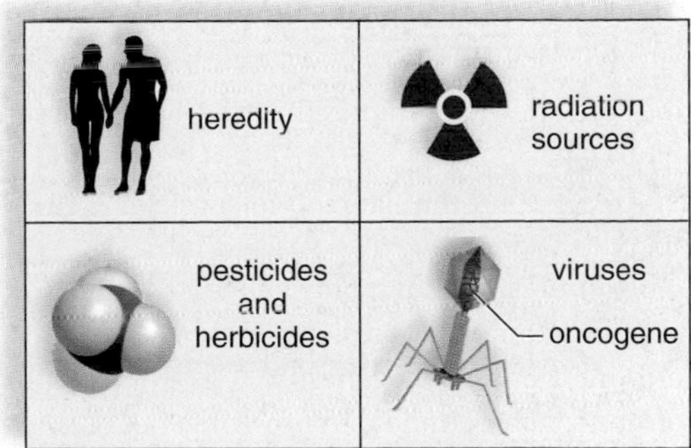

some sources contributing to development of cancer

- Cancers that run in families are most likely due to the inheritance of mutated genes (e.g., breast cancer, retinoblastoma tumor).
- Certain environmental factors are carcinogens (e.g., UV radiation, tobacco smoke, pollutants).
- Industrial chemicals (e.g., pesticides and herbicides) are carcinogenic.
- Certain viruses (e.g., hepatitis B and C, human papillomavirus, and Epstein-Barr virus) cause specific cancers.

19.3 Diagnosis of Cancer

The earlier a cancer is diagnosed, the more likely it can be effectively treated. Tests for cancer include the following:

- Pap test for cervical cancer;
- mammogram for breast cancer;
- tumor marker tests, which are blood tests that detect tumor antigens/antibodies;
- tests for genetic mutations of oncogenes and tumor suppressor genes; and
- biopsy and imaging—used to confirm the diagnosis of cancer.

19.4 Treatment of Cancer

Surgery, radiation, and chemotherapy are traditional methods of treating cancer. Other methods include the following:

- chemotherapy involving **bone marrow transplants**
- immunotherapy;
- *p53* gene therapy, which causes cancer cells to undergo apoptosis; and
- other therapies, such as inhibitory drugs for angiogenesis and metastasis.

ASSESS

Testing Your Knowledge of the Concepts

1. List and briefly discuss the seven characteristics of cancer cells that cause them to be abnormal. (pages 441–443)
2. What are the roles of the two genes that mutate, causing cancer to develop? (pages 443–444)
3. What is the difference between a carcinoma, sarcoma, leukemia, and lymphoma? (pages 444–445)
4. What is the role of heredity and the environment in causing cancer? (pages 445–447)
5. What are the seven warning signals of cancer? (page 448)
6. List and describe three tests designed to detect cancer. (pages 450–452)
7. Describe three standard therapies for cancer treatment. (pages 452–453)
8. Describe some newer therapies that may be successful in cancer treatment. (pages 454–455)
9. Whereas ________ stimulate the cell cycle, ________ inhibit the cell cycle.
   a. tumor suppressor genes, oncogenes
   b. oncogenes, tumor suppressor genes
   c. proto-oncogenes, oncogenes
   d. proto-oncogenes, tumor suppressor genes
10. Growth factors lead to
    a. increased cell division.
    b. the functioning of cyclin proteins.
    c. progression through the cell cycle.
    d. All of the choices are correct.
11. Which of these is not true of the gene *p53?*
    a. Mutations of both proto-oncogenes and tumor suppressor genes lead to inactivity of *p53.*
    b. Normally, *p53* functions to stop the cell cycle and initiate repair enzymes, when necessary.
    c. Normally, *p53* restores the length of telomeres.
    d. Cancer cells shut down the activity of *p53,* even though it may be present.
12. Following each cell cycle, telomeres
    a. get longer.
    b. get shorter.
    c. return to the same length.
    d. bind to cyclin proteins.
13. Angiogenesis
    a. stimulates the development of new blood vessels.
    b. activates tumor suppressor genes.
    c. changes proto-oncogenes into oncogenes.
    d. promotes metastasis.
14. A tumor with cells that spread to secondary locations is referred to as
    a. benign.
    b. cancer in situ.
    c. metastatic.
    d. encapsulated.
15. Which of the following is not a type of carcinogen?
    a. tobacco smoke
    b. radiation
    c. pollutants
    d. viruses
    e. All of these are carcinogens.
16. What type of cancer is associated with human papilloma viruses?
    a. breast
    b. cervical
    c. kidney
    d. lymphatic
17. Why is cancer called a genetic disease?
    a. Cancer is always inherited.
    b. Carcinogenesis is accompanied by mutations.
    c. Cancer causes mutations that are passed on to offspring.
    d. All of the choices are correct.
18. Leukemia is a form of cancer that affects
    a. lymphatic tissue.
    b. bone tissue.
    c. blood-forming cells.
    d. nervous system structures.
19. Which is the name of the tumor suppressor gene that causes retinoblastoma?
    a. *p21*
    b. *RB*
    c. *ras*
    d. *TGF-b*
20. Concerning the causes of cancer, which one is incorrect?
    a. Genetic mutations cause cancer.
    b. Genetic mutations can be caused by environmental influences, such as radiation, organic chemicals, and viruses.
    c. An active immune system, diet, and exercise can help prevent cancer.
    d. Heredity cannot be a cause of cancer.
21. Which of the following is not a warning signal for cancer?
    a. a sore that does not heal
    b. change in bowel or bladder habits
    c. nagging cough or hoarseness
    d. shortness of breath or fatigue
22. Following a biopsy, what does a doctor look for to diagnose cancer?
    a. abnormal-appearing cells
    b. whether the cells can divide
    c. whether the cells will respond to growth factors
    d. whether the chromosomes have telomeres

23. Most chemotherapeutic drugs kill cells by
    a. producing pores in plasma membranes.
    b. interfering with protein synthesis.
    c. interfering with cellular respiration.
    d. interfering with DNA and/or enzymes.
24. *p53* gene therapy
    a. triggers cytotoxic T cells to destroy tumor cells.
    b. triggers apoptosis in cancer cells.
    c. produces monoclonal antibodies against the tumor cells.
    d. reduces tumors by breaking up their blood vessels.

ENGAGE

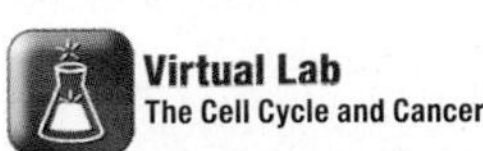

The virtual lab "The Cell Cycle and Cancer" provides a more detailed exploration of how failures in the cell cycle control mechanisms may lead to cancer.

Thinking Critically About the Concepts

Today, cancer is often treated with a combination of chemotherapy, radiation treatment, and surgery. In Cody's case, his cancer required surgery to remove the cancerous kidney, followed by chemotherapy treatment to ensure that no cancer cells had metastasized to other locations in his body. From what you have learned about cancer in this chapter, answer the following questions.

1. After his initial treatments in the hospital, Cody had to avoid public exposure. Based on what you know about the immune system, explain.
2. Cody's cancer was called a nephroblastoma. The suffix -blast refers to an immature cell. Would this type of cancer be more or less likely to spread throughout the body?
3. Cody's nephroblastoma caused blood to be found in his urine. What large blood vessels supplying the kidney might have been damaged by cancer? (See Chapter 5.)
4. In addition to blood in his urine, what other abnormal symptoms did Cody show?

The vaccine for HPV is recommended for both females and males between the ages 11–26. There is some controversy over this vaccine. Some believe vaccination for sexually transmitted disease might lull these young women into a false sense of security regarding safety during sexual intercourse. Others fear that vaccination will encourage girls to engage in sex.

5. Why would the vaccine be recommended for young girls who may not be sexually active yet?
6. What type of immunity would women develop from getting vaccinated for HPV? You may need to revisit Chapter 7 on immunity to answer this question.
7. If women are not vaccinated against HPV, how else could they protect themselves from infection?
8. Why are lymph nodes surrounding a breast with an invasive cancer often removed when a mastectomy is performed?
9. What are some healthy lifestyle choices you could make that might prevent cancer later in your life?
10. Many types of cancer (ovarian cancer, for example) show no symptoms until they are well advanced. What might be the consequence for the patient?

CHAPTER

20

Patterns of Genetic Inheritance

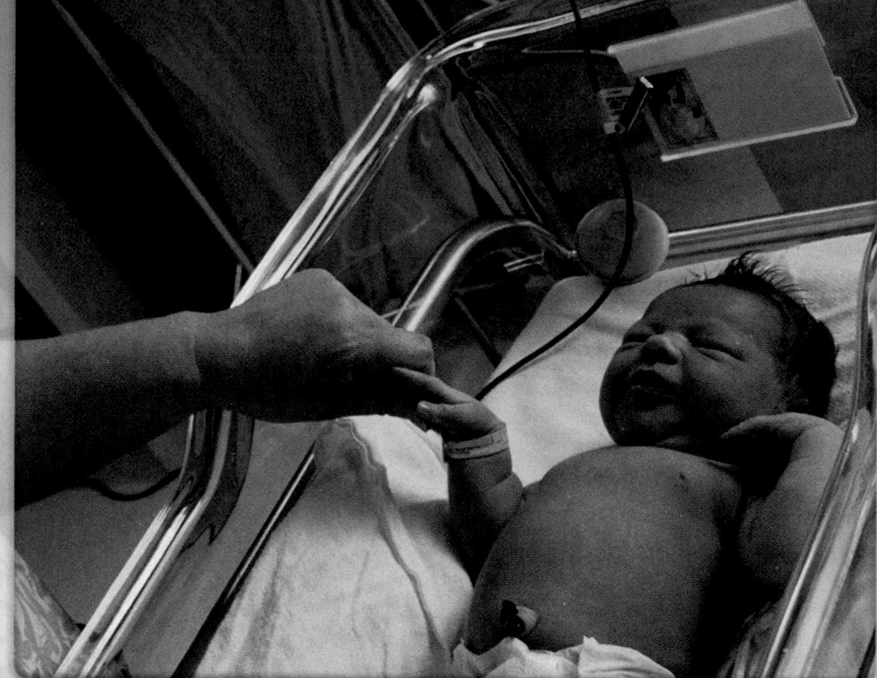

CHAPTER CONCEPTS

20.1 Genotype and Phenotype
The genotype refers to the genes for a particular trait. The phenotype refers to physical characteristics, such as hairline, blood type, color blindness, and the operation of cellular pathways.

20.2 One- and Two-Trait Inheritance
In humans, each trait is controlled by two alleles. In most cases, dominant alleles mask recessive alleles. It is possible to predict patterns of inheritance using Punnett squares and pedigrees. Many disorders, including some cellular disorders, are inherited in a dominant or recessive manner.

20.3 Inheritance of Genetic Disorders
Inheritance of traits can be traced throughout generations of a family using a pedigree chart.

20.4 Beyond Simple Inheritance Patterns
There are other inheritance patterns beyond simple dominant or recessive ones. The environment and other alleles may both influence the phenotype.

20.5 Sex-Linked Inheritance
Not all traits on the sex chromosomes are associated with sex. Because males only have one X chromosome, the alleles on that chromosome are always expressed. Therefore, males are more apt to have an X-linked disorder than are females.

BEFORE YOU BEGIN

Before beginning this chapter, take a few moments to review the following discussions:

Section 2.7 What is the role of DNA in a cell?

Figures 16.4 and 16.9 How are sperm and egg cells produced?

Section 18.4 How does meiosis produce new combinations of alleles?

CASE STUDY PHENYLKETONURIA

As part of the routine newborn screening test performed on most children born in the United States and other developed countries, a high-performance liquid chromatography (HPLC) test was performed by taking a small amount of blood from a heel prick of 12-hour-old Patrick. The test came back positive for phenylketonuria (PKU). Dr. Preston explained to Patrick's parents that PKU was a disorder in which the body has a deficiency in the hepatic enzyme phenylalanine hydroxylase (PAH). PAH is an important enzyme that the body needs to properly metabolize the amino acid phenylalanine into the amino acid tyrosine. When phenylalanine is not metabolized, it accumulates in the body and is converted into the compound phenylpyruvate. An accumulation of phenylpyruvate can lead to impaired brain development, intellectual disability, and seizure disorders. Patrick's parents wondered how he had developed this disorder. Dr. Preston explained that it was a genetic disorder; the cause of the disorder was found in the DNA Patrick had acquired from both his parents. The fact that neither of Patrick's parents showed any signs of PKU indicated that this disease is inherited in an autosomal recessive manner, and that both of the parents are carriers for the disease.

As you read through the chapter, think about the following questions:

1. What is an autosomal recessive disorder? What are the different types of genetic inheritance?
2. How can people pass on conditions if they do not show any signs?
3. How does the environment influence the expression of a trait?

20.1 Genotype and Phenotype

LEARNING OUTCOMES

Upon completion of this section, you should be able to

1. Distinguish between a genotype and a phenotype.
2. Define allele, gene, dominant, and recessive as they relate to patterns of inheritance.
3. Given the genotype of an individual, identify the phenotype.

Genotype

Genotype refers to the genes of the individual. Recall that genes are segments of DNA on a chromosome that code for a trait (characteristic). These genes are units of heredity existing in a specific position, or **locus** (pl., loci), on a chromosome. An **allele** can be easily described as an alternate form of a gene. For example, if the trait the gene coded for was eye color, an allele would be the code for blue eyes or the code for brown eyes. Alleles are classified as either dominant or recessive. In general, **dominant alleles** mask (or hide) the expression of **recessive alleles.** Therefore, if an allele is dominant, only one copy of that allele needs to be present in both chromosomes for that trait to appear. If an allele is recessive, both alleles need to code for the recessive trait for it to appear in the individual.

Alleles are often designated by an abbreviation. A dominant allele is assigned an uppercase letter, and the lowercase letter is used for recessive alleles. In humans, for example, unattached (free) earlobes are dominant over attached earlobes. A suitable key would be *E* for unattached earlobes and *e* for attached earlobes. In cystic fibrosis, the dominant allele is designated as *Cf*, whereas the recessive allele is *cf*.

In humans, we receive one chromosome from each parent; therefore, we inherit one allele from each parent, resulting in a pair of alleles for each trait. Figure 20.1 shows three possible fertilizations between different alleles for the earlobe trait and the resulting offspring of those fertilizations. In the first instance, the chromosomes of both the sperm and the egg carry the dominant trait, designated *E*, resulting in an individual with the alleles *EE*. This type of genotype is called **homozygous dominant.** In the second fertilization, the zygote has received two recessive alleles (*ee*), a genotype called **homozygous recessive,** and results in the appearance of the dominant phenotype—in this example, unattached earlobes. In the third fertilization, the resulting individual has the alleles *Ee*, called a **heterozygous** genotype.

Phenotype

The physical appearance of a trait is called the **phenotype.** In Figure 20.1, the phenotypes are attached (recessive) or unattached (dominant) earlobes. It is important to recognize that the genotype directs the phenotype of the individual. In most cases, the presence of a single dominant allele is all that is necessary to express the dominant phenotype (some exceptions are explained in section 20.4). Notice that this is the case in the heterozygous individual (*Ee*) in Figure 20.1. A heterozygote shows the dominant phenotype because the dominant allele only needs one copy to appear in the phenotype. Therefore, this individual has unattached earlobes. Homozygous recessive individuals (*ee*) have attached earlobes and homozygous dominant individuals (*EE*) have unattached earlobes.

From our example, you may get the impression that the phenotype has to be an easily observable trait. However, the phenotype can be any characteristic of the individual. This could include color blindness or a metabolic disorder such as the lack of an enzyme to metabolize the amino acid phenylalanine.

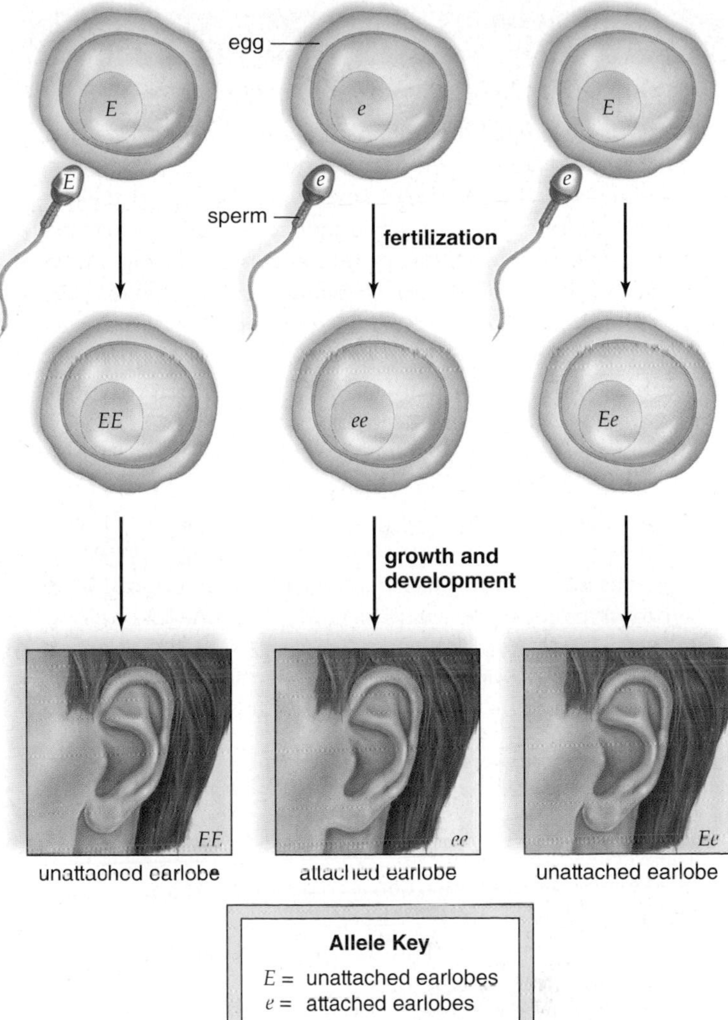

Figure 20.1 Genetic inheritance affects our characteristics. For each trait, such as the shape of the earlobes, we inherit one allele from each parent. Therefore, we possess two alleles for every trait. The inheritance of a single dominant allele (resulting in either *EE* or *Ee*) causes an individual to have unattached earlobes; two recessive alleles (*ee*) cause an individual to have attached earlobes.

APPLICATIONS AND MISCONCEPTIONS

Why are some alleles dominant to other alleles?

In a simple example, the dominant allele (*A*) codes for a particular protein. Let's assume this gene codes for an enzyme (protein) responsible for brown eye color. The recessive allele (*a*) is a mutated allele that no longer codes for the enzyme. At the molecular level, having the enzyme is dominant to the lack of the enzyme. In other words, having brown in the eye is dominant over not having dark pigment in the eye. Using this example, a homozygous dominant individual (*AA*) may have twice the amount of enzyme as a heterozygous individual (*Aa*). However, it may not be possible to detect a difference between homozygous and heterozygous, as long as there is enough enzyme to bring about the dominant phenotype. Half the amount of enzyme may still turn the eyes completely brown. So a person with the genotype *Aa* would have eyes just as brown as a person with the genotype *AA*. A homozygous recessive individual (*aa*) makes no enzyme, so the brown phenotype is absent. Gene expression is discussed further in Chapter 21.

CHECK YOUR PROGRESS 20.1

1. Define the following terms: *gene, allele, locus, chromosome, dominant,* and *recessive.*
2. Describe the difference between genotype and phenotype.
3. Summarize the three possible genotypes and their corresponding phenotypes.

CONNECTING THE CONCEPTS

The genotype of an individual is based on the information in the DNA of its cells. For more information on DNA, refer to the following discussions:

Sections 2.7 and 21.1 describe the basic structure of the DNA molecule.

Section 21.2 examines how the information in the DNA is expressed as a protein.

20.2 One- and Two-Trait Inheritance

LEARNING OUTCOMES

Upon completion of this section, you should be able to

1. Understand how probability is involved in solving one- and two-trait crosses.
2. Calculate the probability of a specific genotype or phenotype in an offspring of a genetic cross.

In one-trait crosses (one-characteristic crosses), the inheritance of only one set of alleles is being considered. In two-trait crosses (two-characteristic crosses), the inheritance of two different alleles is being considered. For both types of crosses, it will be necessary to determine the gametes of both individuals who are reproducing.

Forming the Gametes

During meiosis, the chromosome number of the cells that will form the gametes (egg and sperm) is divided in half (see section 18.4). This is accomplished by the separation of homologous chromosomes. Each individual has 46 chromosomes, arranged as 23 homologous pairs. For each pair, one of the homologous chromosomes was originally from the mother, and the other was donated by the father. During meiosis, the homologous chromosomes, and the alleles they contain, are separated. Therefore, the sperm or egg only has 23 chromosomes. If reduction of the chromosome number didn't happen, the new individual would have twice as many chromosomes after fertilization, which would not result in a viable embryo. For example, let's say that the gene for unattached or attached earlobes is on a particular chromosome. On the homologous chromosome originally from the mother, the allele is an *E*. On the father's paired chromosome, the allele is also an *E*. Therefore, *E* is the only option for alleles on either chromosome, so every gamete formed will carry an *E*. This occurs whether the gamete gets the mother's or the father's original chromosome. Similarly, if the homologous chromosome originally from the mother carries an *e* and the father's chromosome also bears an *e*, then all gametes will have the *e* allele. What if the mother's homologous chromosome has an *E* allele and the father's has an *e*? Then, half of the gametes formed will receive an *E* allele from the mother's homologue. The other half will get the father's homologue with the *e*.

Figure 20.2 shows the genotypes and phenotypes for other observable traits in humans. You can practice deciding what alleles the gametes would carry in order to produce these genotypes. Humans contain combinations of these traits (e.g., earlobe attachment and finger length), and these can be expressed as a single genotype. For example, if the genotype is *EeSs*, all combinations of any two different letters (*ES, eS, Es,* and *es*) can be present in the gametes.

One-Trait Crosses

Parents often like to know the chances of having a child with a certain genotype and, therefore, a certain phenotype. To illustrate, let us consider a cross involving freckles. What happens when two parents without freckles have children? Will the children of this couple have freckles? In solving the problem, we will (1) use *F* to indicate the dominant allele for freckles, and *f* to indicate the recessive allele of no freckles; (2) determine the possible gametes for each parent; (3) combine all possible gametes; and (4) finally, determine the genotypes and the phenotypes of all the offspring. If both parents do not have freckles, then their genotypes are both *ff*. The only gametes they can produce contain the *f* allele. All of the children

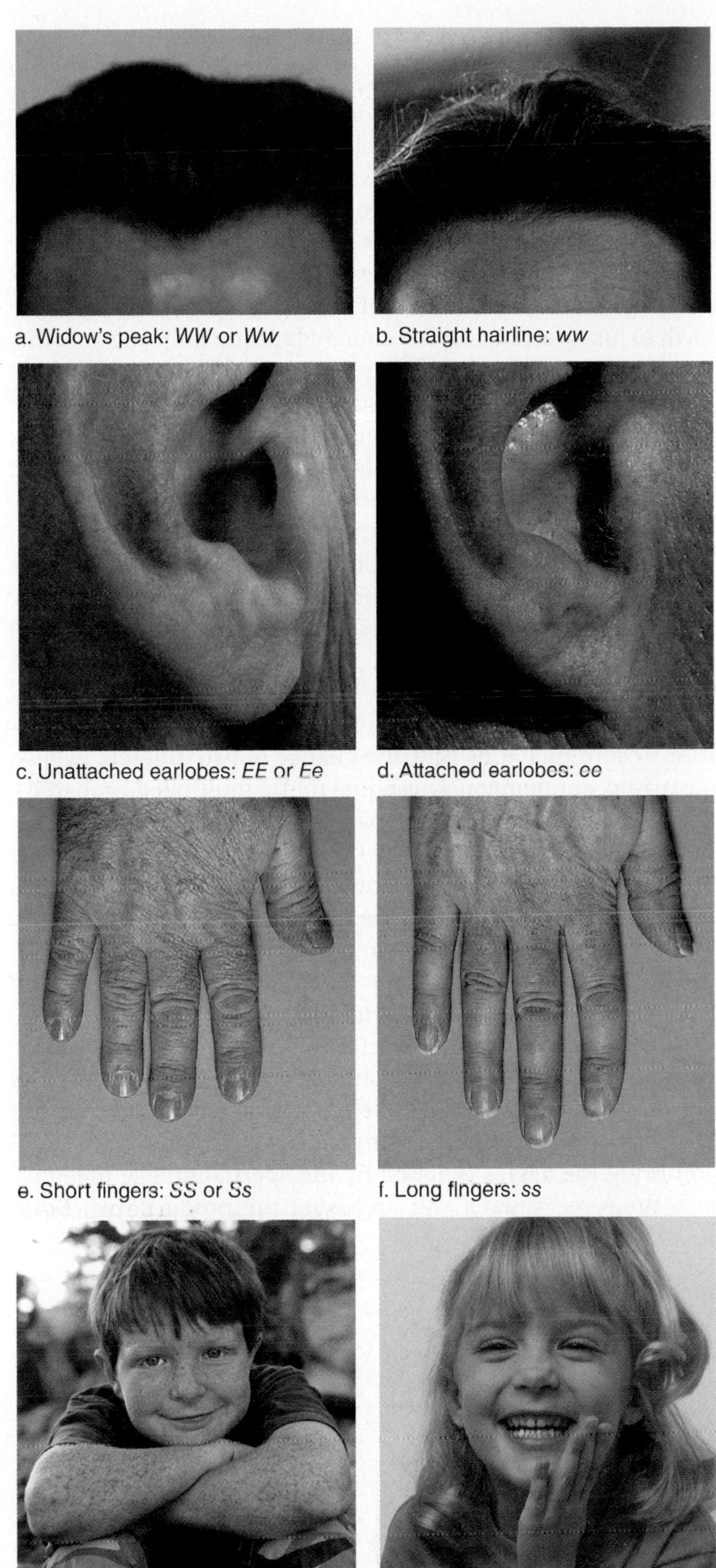

Figure 20.2 **Common inherited traits in humans.**
The allele keys indicate which traits are dominant and which are recessive.

will therefore be *ff* and will not have freckles. In the following diagram, the letters in the first row give the genotypes of the parents. Each parent has only one type of gamete with regard to freckles; therefore, all the children have a similar genotype and phenotype.

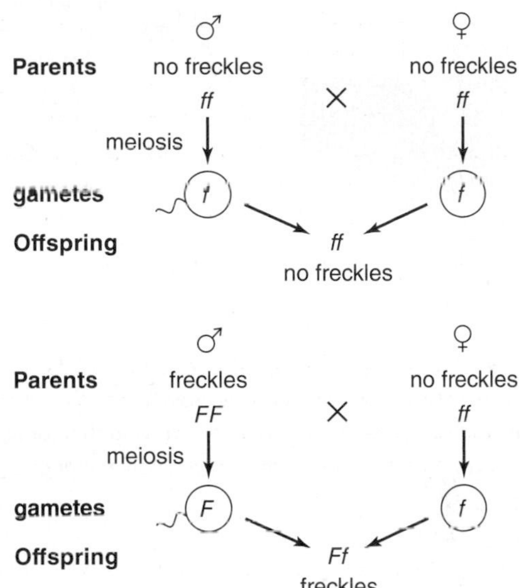

Let's also consider the results when a homozygous dominant man with freckles has children with a woman with no freckles. Will the children of this couple have freckles? The children are heterozygous (*Ff*) and have freckles. When writing a heterozygous genotype, always put the capital letter (for the dominant allele) first to avoid confusion.

A one-trait cross is often referred to as a **monohybrid cross,** because it involves only a single trait. Notice that in this example the children are heterozygous with regard to one pair of alleles. If these individuals reproduce with someone else of the same genotype, will their children have freckles? In this problem (*Ff* × *Ff*), each parent has two possible types of gametes (*F* or *f*), and we must ensure that all types of sperm have an equal chance to fertilize all possible types of eggs. One way to do this is to use a **Punnett square** (Fig. 20.3), in which all possible types of sperm are lined up vertically and all possible types of eggs are lined up horizontally (or vice versa). Every possible combination of gametes occurs within the squares.

After we determine the genotypes and the phenotypes of the offspring, we can determine both the genotypic and phenotypic ratios. The genotypic ratio is 1 *FF*: 2 *Ff*: 1 *ff* or simply 1:2:1, but the phenotypic ratio is 3:1. Why? Three individuals will have freckles (the *FF* and the two *Ff*) and one will not have freckles (the *ff*).

This 3:1 phenotypic ratio is always expected for a monohybrid cross when one allele is completely dominant over the other. The exact ratio is more likely to be observed if a large number of matings take place and if a large number of offspring result. Only then do all possible types of sperm have an equal chance of fertilizing all possible types of eggs. Naturally, we do not routinely observe hundreds of offspring

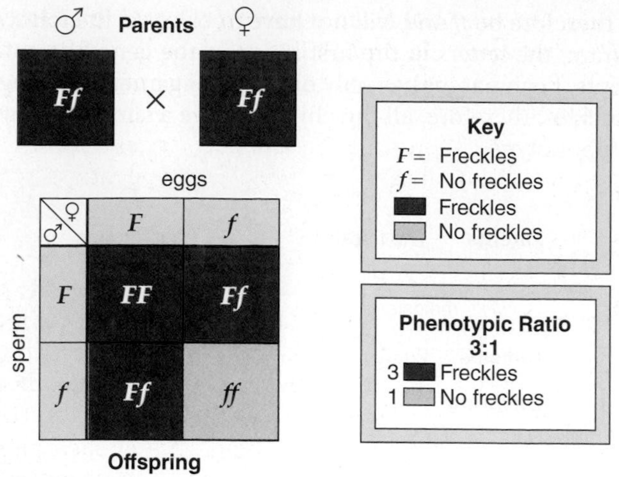

Figure 20.3 Expected results of a monohybrid cross.
A Punnett square diagrams the results of a cross. When the parents are heterozygous, each child has a 75% chance of having the dominant phenotype and a 25% chance of having the recessive phenotype.

from a single type of cross in humans. The best interpretation of Figure 20.3 in humans is to say that each child has three chances out of four to have freckles, or one chance out of four to not have freckles.

Every fertilization has the exact same chance for combinations as the previous fertilization. For example, if two heterozygous parents already have three children with freckles and are expecting a fourth child, this child still has a 75% chance of having freckles and a 25% chance of not having freckles, just like each of its siblings did. The chance of achieving a new phenotype not previously shown does not increase with each fertilization; it stays the exact same with every fertilization. Every new fertilization is not influenced by any previous fertilizations. Each one is considered an individual occurrence, each time subject to the probabilities of the gametes of the parents.

Other One-Trait Crosses

It is not possible to tell by inspection if a person expressing a dominant allele is homozygous dominant or heterozygous. However, it is sometimes possible to tell by the results of a cross. For example, Figure 20.4 shows two possible results when a man with a widow's peak reproduces with a woman who has a straight hairline. If the man is homozygous dominant, all his children will have a widow's peak. If the man is heterozygous, each child has a 50% chance of having a straight hairline. The birth of just one child with a straight hairline indicates that the man is heterozygous.

Consider, also, that a person's parentage sometimes tells you that he or she is heterozygous. In Figure 20.4, each of the offspring with a widow's peak has to be heterozygous. Why? One of the parents was homozygous recessive and therefore had to give each offspring a *w*. Be sure to do all the practice problems in the Testing Your Knowledge of the Concepts section at the end of the chapter to ensure that you can do one-trait genetics problems.

The Punnett Square and Probability

Two laws of probability apply to genetics. The first is the product rule. According to this rule, the chance of two different events occurring simultaneously is equal to the multiplied probabilities of each event occurring separately. For example, what is the probability that a coin toss will be "heads"? There are only two options, so the probability of "heads" is 1 out of 2 or 50% (0.50). If we wanted to know what the probability was of a first coin toss being "heads" *and* a second coin toss being "heads," we use the product rule. *The product rule is often applied to cases in which the word "and" is used.* This would be $\frac{1}{2} \times \frac{1}{2} = \frac{1}{4}$, or 25% (0.25). Probabilities range from 0.0 (0%, or an event that will not happen) to 1.0 (100%, or an event that will always happen). The Punnett square allows you to determine the probability that an offspring will have a particular genotype or phenotype. When you bring the alleles donated by the sperm and egg together into the same square, you are using the product rule. Both

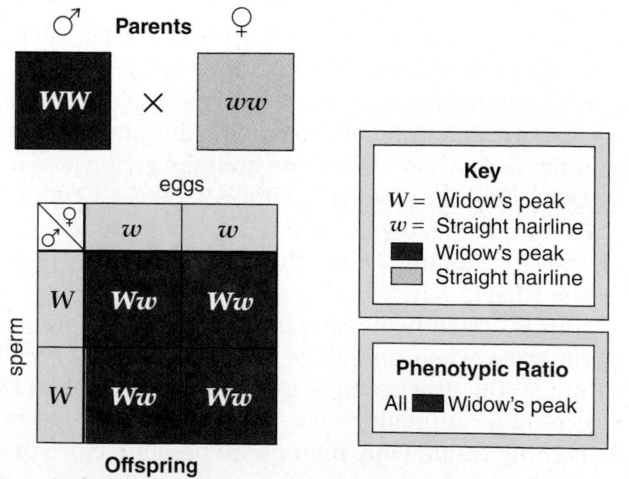

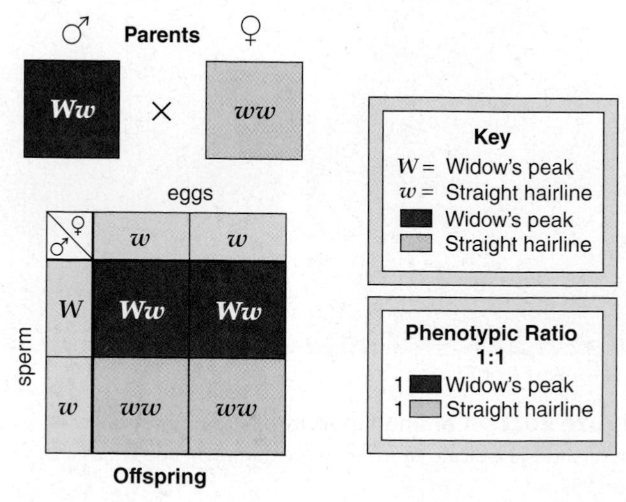

Figure 20.4 Determining if a dominant phenotype is homozygous or heterozygous.
This cross will determine if an individual with the dominant phenotype is homozygous or heterozygous. **a.** All offspring show the dominant characteristic, so the individual is most likely homozygous, as shown. **b.** The offspring show a 1:1 phenotypic ratio, so the individual is heterozygous, as shown.

father and mother each give a chance of having a particular allele, so the probability for that allele in an offspring is the multiple of these two separate chances.

The second law of probability is the sum rule. Using this rule, individual probabilities can be added to determine total probability for an event. If we toss a coin and we want to know the probability that it will be either heads *or* tails, we use the sum rule. *The sum rule is often applied to cases in which the word "or" is used.* The probability of heads is ½. The probability of tails is ½. Therefore, the probability of the coin toss giving heads or tails is ½ + ½ = 1 . In the Punnett square, you use the sum rule when you add up the results of each square to determine the final genotype and phenotype ratios.

APPLICATIONS AND MISCONCEPTIONS

Why is it called a Punnett square?

The Punnett square was first proposed by the English geneticist Reginald Punnett (1875–1967). Punnett first used the diagram as a teaching tool to explain basic patterns of inheritance to his introductory genetics classes in 1909. This teaching tool has remained basically unchanged for over 100 years!

Punnett squares are useful for visualizing crosses that involve one or two traits, but they quickly become cumbersome with complex traits involving many different factors. Most geneticists rely on statistical analysis to predict the outcomes of complex crosses.

Two-Trait Crosses

Figure 20.5 allows you to relate the events of meiosis to the formation of gametes when a cross involves two traits. In the example given, a cell has two pairs of homologues, recognized by length. One pair of homologues is short, and the other is long. The color in the figure signifies that we inherit chromosomes from our parents. One homologue of each pair is the "paternal" chromosome, and the other is the "maternal" chromosome.

The homologues separate during meiosis I, so each gamete receives one member from each pair of homologues. The homologues separate independently so it does not matter which member of a pair goes into which gamete. All possible combinations of alleles occur in the gametes. In the simplest of terms, a gamete in Figure 20.5 will *receive one*

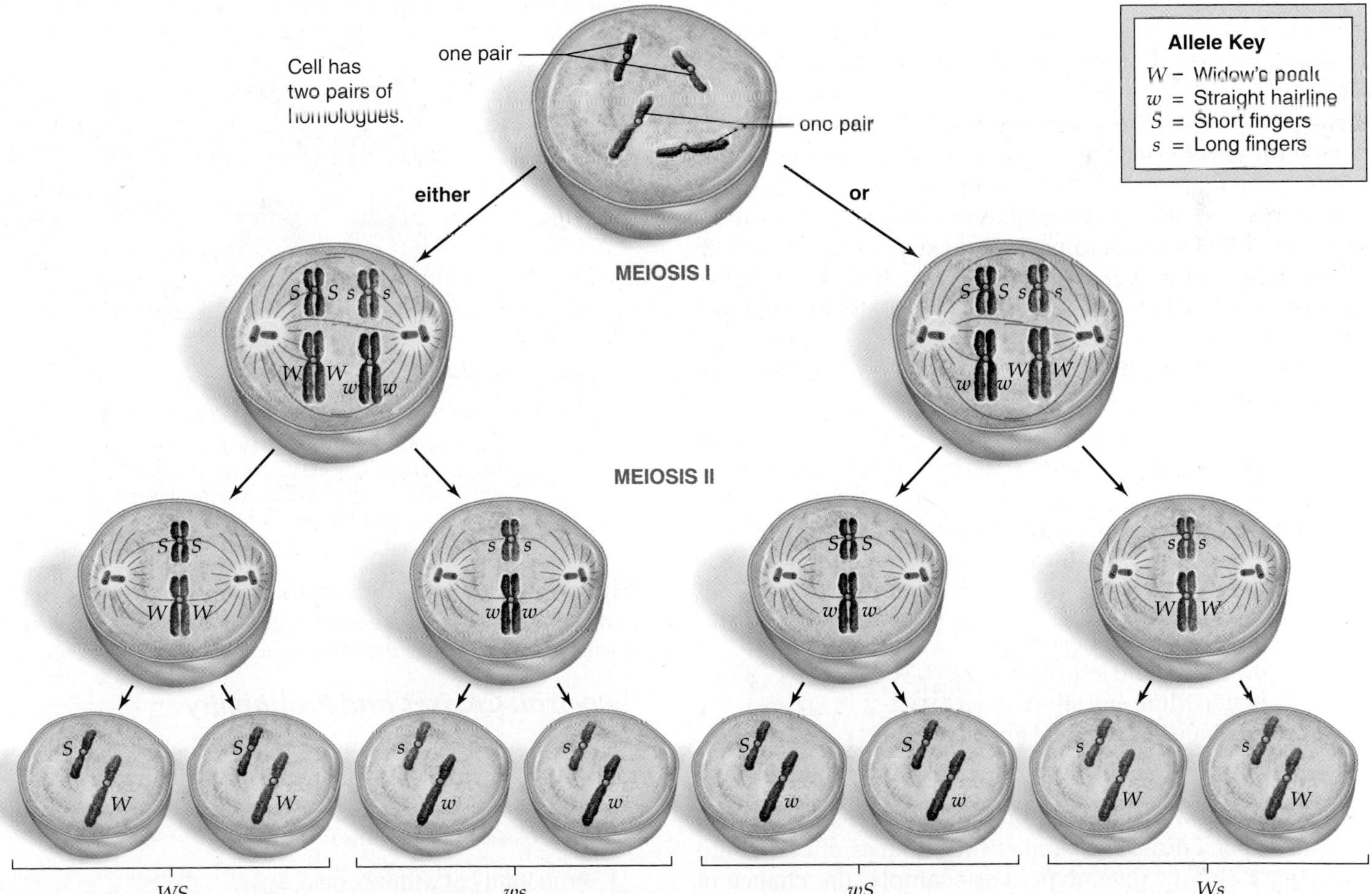

Figure 20.5 Meiosis results in genetic diversity of gametes.
A cell has two pairs of homologous chromosomes (homologues), recognized by length, not color. The long pair of homologues carries alleles for type of hairline and the short pair of homologues carries alleles for finger length. The homologues, and the alleles they carry, align independently during meiosis. Therefore, all possible combinations of chromosomes and alleles occur in the gametes as shown in the last row of cells.

short and one long chromosome of either color. Therefore, all possible combinations of chromosomes and alleles are in the gametes.

Specifically, assume that the alleles for two genes are on these homologues. The alleles *S* and *s* are on one pair of homologues, and the alleles *W* and *w* are on the other pair of homologues. The homologues separate, so a gamete will have either an *S* or an *s* and either a *W* or a *w*. They will never have two of the same letter of the alphabet. Also, because the homologues align independently at the equator, either the paternal or maternal chromosome of each pair can face either pole.

Therefore, there are no restrictions as to which homologue goes into which gamete. A gamete can receive either an *S* or an *s* and either a *W* or a *w* in any combination. In the end, the gametes will collectively have all possible combinations of alleles. You should be able to transfer this information to any cross that involves two traits. In other words, the process of meiosis explains why a person with the genotype *EeFf* would produce the gametes *EF, ef, Ef,* and *eF* in equal number.

The Dihybrid Cross

In the two-trait cross depicted in Figure 20.6, a person homozygous for widow's peak and short fingers (*WWSS*) reproduces with one who has a straight hairline and long fingers (*wwss*). Because this cross involves two traits, it is also referred to as a **dihybrid cross.** In this example, the gametes for the *WWSS* parent must be *WS* and the gametes for the *wwss* parent must be *ws*. Therefore, the offspring will all have the genotype *WwSs* and the same phenotype (widow's peak with short fingers). This genotype is called a dihybrid because the individual is heterozygous in two regards: hairline and fingers.

When a dihybrid *WwSs* has children with another dihybrid that is *WwSs,* what gametes are possible? Each gamete can have only one letter of each type in all possible combinations. Therefore, these are the gametes for both dihybrids: *WS, Ws, wS,* and *ws*.

A Punnett square makes sure that all possible sperm fertilize all possible eggs. If so, these are the expected phenotypic results:

9 widow's peak and short fingers
3 widow's peak and long fingers
3 straight hairline and short fingers
1 straight hairline and long fingers

This 9:3:3:1 phenotypic ratio is always expected for a dihybrid cross when simple dominance is present. We can use this expected ratio to predict the chances of each child receiving a certain phenotype. For example, the chance of getting the two dominant phenotypes together is 9 out of 16. The chance of getting the two recessive phenotypes together is 1 out of 16.

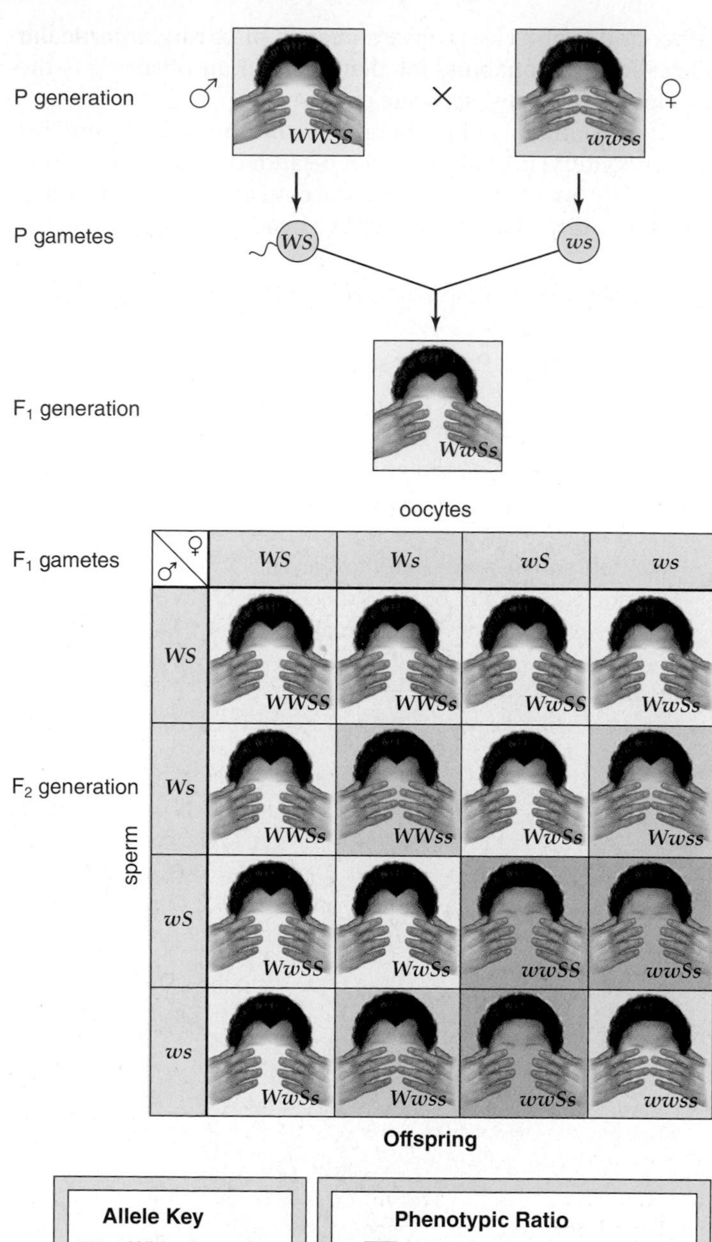

Figure 20.6 Expected results of a dihybrid cross.
Each dihybrid can form four possible types of gametes, so four different phenotypes occur among the offspring in the proportions shown.

Two-Trait Crosses and Probability

It is also possible to use the rules of probability we discussed on pages 464–465 to predict the results of a dihybrid cross. For example, we know the probable results for two separate monohybrid crosses are as follows:

1. Probability of widow's peak = ¾
   Probability of straight hairline = ¼
2. Probability of short fingers = ¾
   Probability of long fingers = ¼

Using the product rule, we can calculate the probable outcome of a dihybrid cross as follows:

Probability of widow's peak and short fingers =
$\frac{3}{4} \times \frac{3}{4} = \frac{9}{16}$

Probability of widow's peak and long fingers =
$\frac{3}{4} \times \frac{1}{4} = \frac{3}{16}$

Probability of straight hairline and short fingers =
$\frac{1}{4} \times \frac{3}{4} = \frac{3}{16}$

Probability of straight hairline and long fingers =
$\frac{1}{4} \times \frac{1}{4} = \frac{1}{16}$

In this way, the rules of probability tell us that the expected phenotypic ratio when all possible sperm fertilize all possible eggs is 9:3:3:1.

Other Two-Trait Crosses

It is not possible to tell by inspection whether an individual expressing the dominant alleles for two traits is homozygous dominant or heterozygous. But, if the individual has children with a homozygous recessive individual, it may be possible to tell. For example, if a man homozygous dominant for widow's peak and short fingers reproduces with a female homozygous recessive for both traits, then all his children will have the dominant phenotypes. However, if a man is heterozygous for both traits, then each child has a 25% chance of showing either one or both recessive traits. A Punnett square (Fig. 20.7) shows that the expected ratio is 1 widow's peak with short fingers: 1 widow's peak with long fingers: 1 straight hairline with short fingers: 1 straight hairline with long fingers, or 1:1:1:1.

For practical purposes, if a parent with the dominant phenotype in either trait has an offspring with the recessive phenotype, the parent has to be heterozygous for that trait. Also, it is possible to tell if a person is heterozygous by knowing the parentage. In Figure 20.7, no offspring showing a dominant phenotype is homozygous dominant for either trait. Why? The mother is homozygous recessive for that trait.

Table 20.1 gives the phenotypic results for certain crosses we have been studying. These crosses always give these phenotypic results. Therefore, it is not necessary to do a Punnett square to arrive at the results. To facilitate doing crosses, study Table 20.1 so that you can understand why these are the results expected for these crosses.

Table 20.1 Phenotypic Ratios of Common Crosses

Genotypes	Phenotypes
Monohybrid *Gg* × monohybrid *Gg*	3:1 (dominant to recessive)
Monohybrid *Gg* × recessive *gg*	1:1 (dominant to recessive)
Dihybrid *GgRr* × dihybrid *GgRr*	9:3:3:1 (9 both dominant: 3 one dominant: 3 other dominant: 1 both recessive)
Dihybrid *GgRr* × recessive *ggrr*	1:1:1:1 (all possible combinations in equal number)

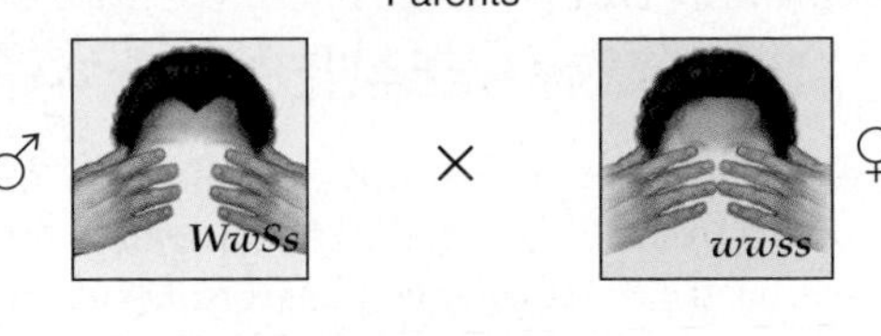

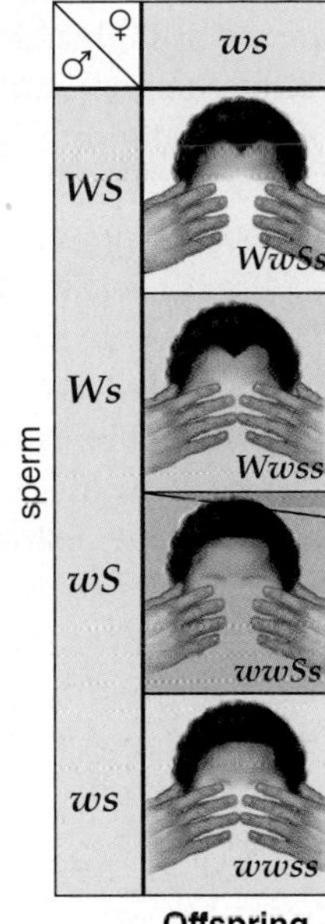

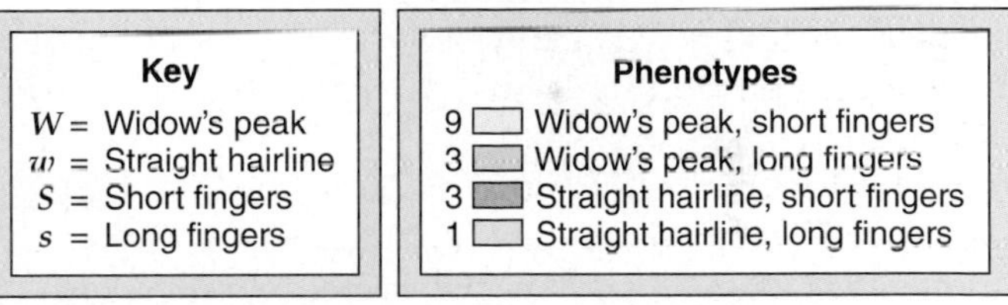

Figure 20.7 Two-trait cross.
The results of this cross indicate that the individual with the dominant phenotypes is heterozygous for both traits because some of the children are homozygous recessive for one or both traits. The chance of receiving any possible phenotype is 25%.

CHECK YOUR PROGRESS 20.2

1. Explain how the results of a dihybrid cross are related to the events of meiosis.
2. Predict what genotype the children will have if one parent is homozygous recessive for earlobes and homozygous dominant for hairline (*eeWW*) and the other parent is homozygous dominant for unattached earlobes and homozygous recessive for hairline (*EEww*).
3. Determine the genotypes of each of the following individuals: A child who does not have dimples or freckles is born to a man who has dimples and freckles (both dominant traits) and a woman who does not.

BIOLOGY MATTERS  **Bioethical**

Preimplantation Genetic Diagnosis

If prospective parents are heterozygous for one of the genetic disorders discussed on pages 469–472, they may want the assurance that their offspring will be free of the disorder. Determining the genotype of the embryo will provide this assurance. For example, if both parents are *Aa* for a recessive disorder, the embryo will develop normally if it has the genotype *AA* or *Aa*. On the other hand, if one of the parents is *Aa* for a dominant disorder, the embryo will develop normally only if it has the genotype *aa*.

Following in vitro fertilization (IVF), the zygote (fertilized egg) divides. When the embryo has eight cells (Fig. 20A*a*), removal of one of these cells for testing purposes has no effect on normal development. Only embryos that will not have the genetic disorders of interest are placed in the uterus to continue developing.

It is estimated that thousands of children have been born worldwide with normal genotypes following preimplantation embryo analysis for genetic disorders that run in their families. No American agency currently tracks these statistics, however. In the future, it's possible that embryos that test positive for a disorder could be treated by gene therapy so that they, too, would be allowed to continue to term.

Testing the egg is possible if the condition of concern is recessive. Recall that meiosis in females results in a single egg and at least two polar bodies (see Fig. 16.9). Polar bodies later disintegrate. They receive very little cytoplasm, but they do receive a haploid number of chromosomes. When a woman is heterozygous for a recessive genetic disorder, about half of the first polar bodies have received the mutated allele. In these instances, the egg received the normal allele. Therefore, if a polar body tests positive for a recessive mutated allele, the egg received the normal dominant allele. Only normal eggs are then used for IVF. Even if the sperm should happen to carry the mutation, the zygote will, at worst, be heterozygous. But the phenotype will appear normal.

Questions to Consider

1. Of the two diagnostic procedures described, does either seem more ethically responsible? Why?
2. Caring for an individual with a genetic condition can be very costly. Should society require preimplantation studies for the carriers of a genetic disease?

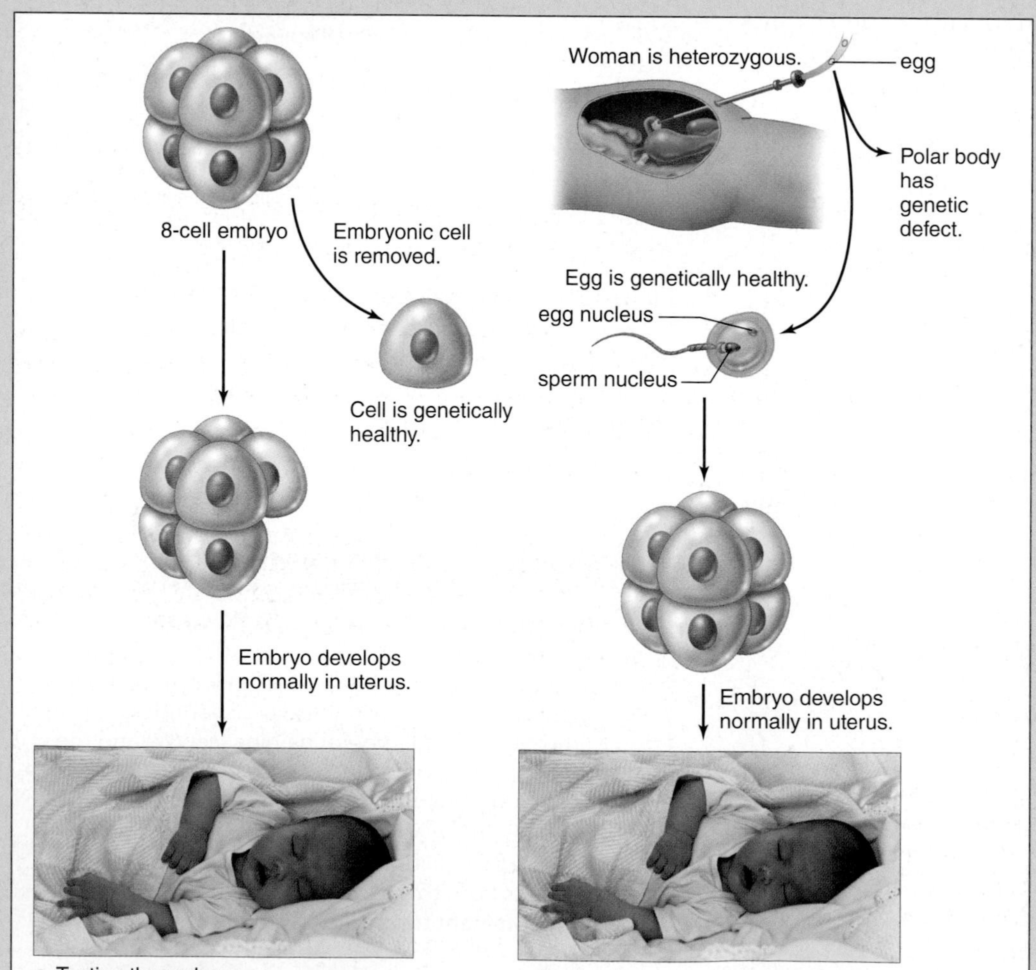

Figure 20A **The process of preimplantation genetic diagnosis.**
a. Following IVF and cleavage, genetic analysis is performed on one cell removed from an eight-cell embryo. If it is found to be free of the genetic defect of concern, the seven-cell embryo is implanted in the uterus and develops into a newborn with a normal phenotype. **b.** Chromosomal and genetic analysis is performed on a polar body attached to an egg. If the egg is free of a genetic defect, it is used for IVF and the embryo is implanted in the uterus for further development.

CONNECTING THE CONCEPTS

For more information on the relationship between meiosis and patterns of inheritance, refer to the following discussions:

Section 16.2 explains gamete production in males.

Section 16.3 explores the process of gamete formation in females.

Section 18.4 describes how meiosis introduces genetic variation.

20.3 Inheritance of Genetic Disorders

LEARNING OUTCOMES

Upon completion of this section, you should be able to

1. Interpret a human pedigree to identify the pattern of inheritance for a trait.
2. Understand the genetic basis of select human autosomal dominant and autosomal recessive genetic disorders.

We inherit many different traits from our parents—not only our hair and eye color but also traits for diseases and disorders. Many of these diseases occur as a result of changes, or mutations, in our parents' genetic codes. The abnormal gene could be present in each of your parents' cells and thus passed down in the sperm or egg. Your parent may or may not have been affected by this genetic mutation. Alternatively, the genetic mutation might have occurred only in the sperm or egg that became a part of you. Some genetic diseases require two damaged alleles for the disease to manifest itself. Others need only one. When a genetic disorder is autosomal dominant, an individual with the alleles *AA* or *Aa* will have the disorder. When a genetic disorder is autosomal recessive, only individuals with the alleles *aa* will have the disorder. Genetic counselors often construct pedigrees to determine whether a condition that runs in the family is dominant or recessive. A pedigree shows the pattern of inheritance for a particular condition. Consider these two possible patterns of inheritance:

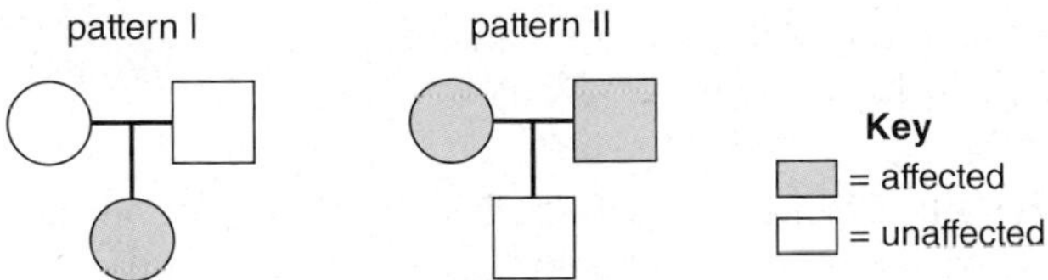

In both patterns, males are designated by squares and females by circles. Shaded circles and squares are affected individuals. A line between a square and a circle represents a mating. A vertical line going downward leads, in these patterns, to a single child. (If there are more children, they are placed off a horizontal line.) Which pattern of inheritance do you suppose represents an autosomal recessive characteristic? Which represents an autosomal dominant characteristic?

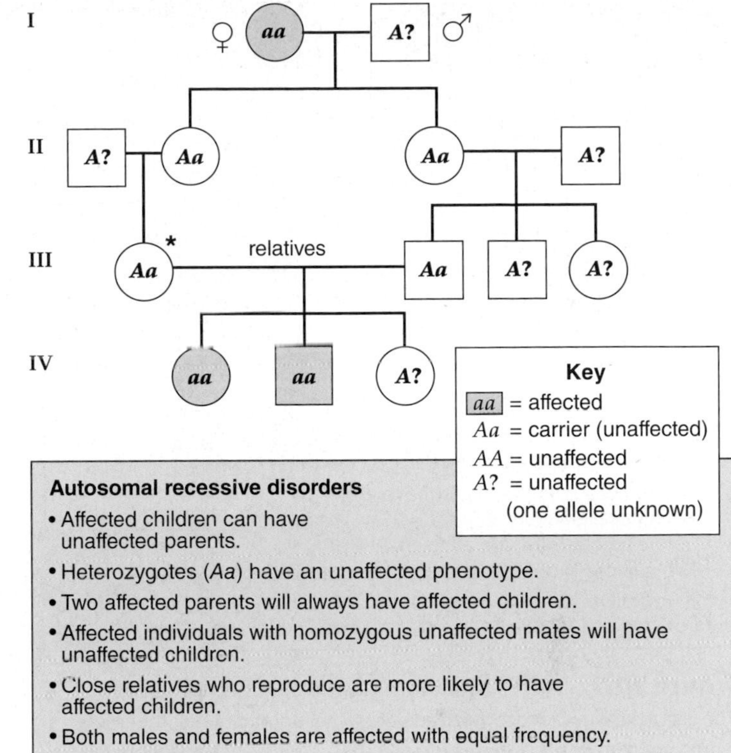

Autosomal recessive disorders

- Affected children can have unaffected parents.
- Heterozygotes (*Aa*) have an unaffected phenotype.
- Two affected parents will always have affected children.
- Affected individuals with homozygous unaffected mates will have unaffected children.
- Close relatives who reproduce are more likely to have affected children.
- Both males and females are affected with equal frequency.

Figure 20.8 **Autosomal recessive disorder pedigree.** The list gives ways to recognize an autosomal recessive disorder. How would you know that the individual at the asterisk is heterozygous? (Answer is provided in Appendix C.)

Autosomal recessive disorder: In pattern I, the child is affected but neither parent is. This can happen if the condition is recessive and the parents are *Aa*. The parents are **carriers** because they carry the recessive trait in their DNA but their phenotype is dominant. Figure 20.8 shows a typical pedigree chart for a recessive genetic disorder. Other ways to recognize an autosomal recessive pattern of inheritance are also listed in the figure. If both parents are affected, all the children are affected. Why? The parents can pass on only recessive alleles for this condition. All children will be homozygous recessive, just like the parents.

Autosomal dominant disorder: In pattern II, the child is unaffected but the parents are affected. Of the two patterns, this one shows a dominant pattern of inheritance. The condition is dominant, so the parents can be *Aa* (heterozygous). The child inherited a recessive allele from each parent and, therefore, is unaffected. Figure 20.9 shows a typical pedigree for a dominant disorder. Other ways to recognize an autosomal dominant pattern of inheritance are also listed. When a disorder is dominant, an affected child must have at least one affected parent.

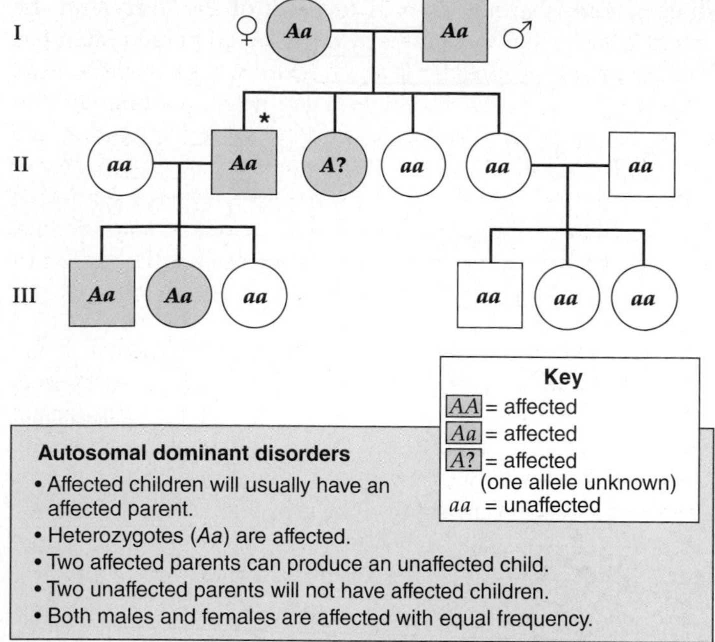

Figure 20.9 **Autosomal dominant disorder pedigree.** The list gives ways to recognize an autosomal dominant disorder. How would you know that the individual at the asterisk is heterozygous? (Answer is provided in Appendix C.)

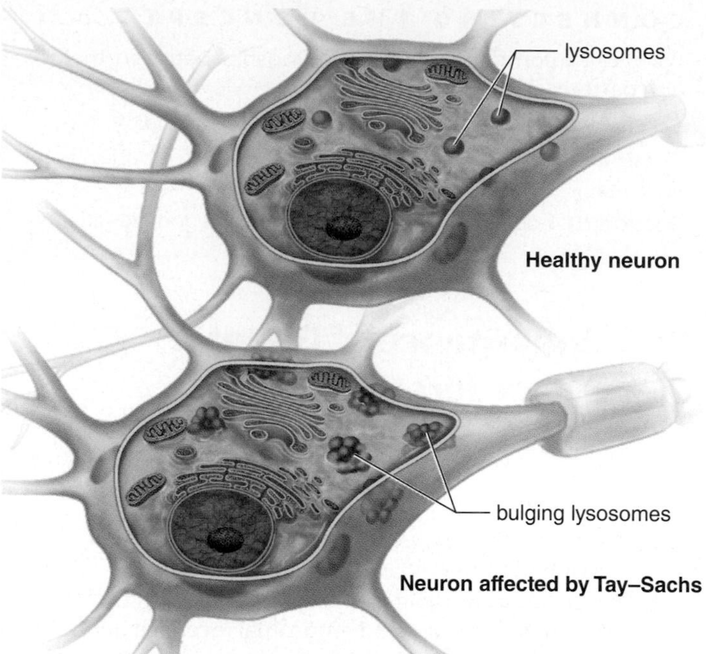

Figure 20.10 **Neuron affected by Tay–Sachs disease.** In Tay–Sachs disease, a lysosomal enzyme is missing. This causes the substrate of that enzyme to accumulate within the lysosomes.

Genetic Disorders of Interest

Medical genetics has traditionally focused on disorders caused by single gene mutations; they are well understood due to their straightforward patterns of inheritance. Presently, it is estimated that there are over 4,000 identified disorders caused by single gene mutations in humans. Here, we will focus on only a few.

MP3 Genetics and Inheritance

Autosomal Recessive Disorders

Inheritance of two recessive alleles is required for an autosomal recessive disorder to be the expressed phenotype.

Tay–Sachs Disease **Tay–Sachs disease** is a well-known autosomal recessive disorder that occurs usually among Ashkenazic Jewish people (those from Central and Eastern Europe) and their descendants. Tay-Sachs disease results from a lack of a lysosome enzyme, hex A, which clears out fatty acid proteins that build up in cells of the brain. Without this enzyme, the buildup will interfere with proper brain development and growth and cause malfunctions in vision, movement, hearing, and overall mental development. This impairment leads to blindness, seizures, and paralysis. Currently, there is no cure for Tay–Sachs disease; affected children normally die by the age of five (Fig. 20.10).

Cystic Fibrosis **Cystic fibrosis** (**CF**) is an autosomal recessive disorder that occurs among all ethnic groups but is most prevalent in Caucasians. Cystic fibrosis is caused by a defective chloride ion channel that is encoded by the *c*ystic *f*ibrosis conductance *t*ransmembrane *r*egulator (*CFTR*) allele on chromosome 7. It is estimated that 1 in 29 Caucasians in the United States carries this allele. Research has demonstrated that chloride ions (Cl^-) fail to pass through the defective version of the *CFTR* chloride ion channel (Fig. 20.11), which is located on the plasma membrane. Ordinarily, after chloride ions have passed through the channel to the other side of the membrane, sodium ions (Na^+) and water follow. It is believed that lack of water is the cause of the abnormally thick mucus in the bronchial tubes and pancreatic ducts.

In the past few years, new treatments have raised the average life expectancy for CF patients to as much as 35 years of age. It is hoped that other novel treatments, such as gene therapy, may be able to correct the defect by replacing a faulty copy of the gene with a normal one. Some scientists have suggested that the mutated *CFTR* allele has persisted in the human population as a means of surviving potentially fatal diseases, such as cholera.

Video Good Poison

Sickle-Cell Disease **Sickle-cell disease** is an autosomal recessive disorder in which the red blood cells are not biconcave disks like normal red blood cells. Many are sickle- or boomerang-shaped, and these red blood cells live for only about two weeks, unlike the average four-month life span of a normal red blood cell. The defect is caused by an abnormal hemoglobin that differs from normal hemoglobin by one amino acid in the protein globin. The single amino acid change causes hemoglobin molecules to stack up and form insoluble rods. This causes the red blood cells to become sickle-shaped.

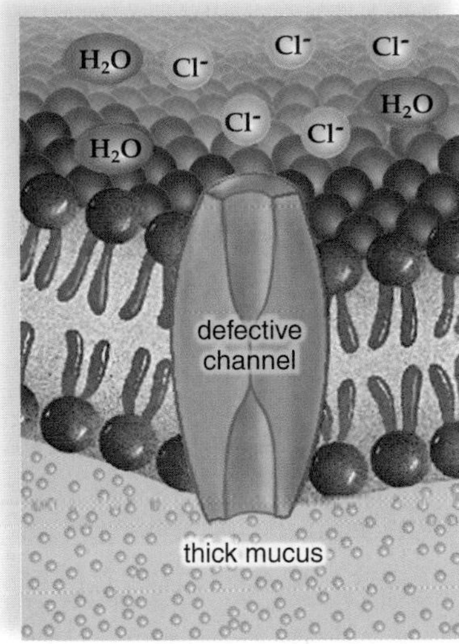

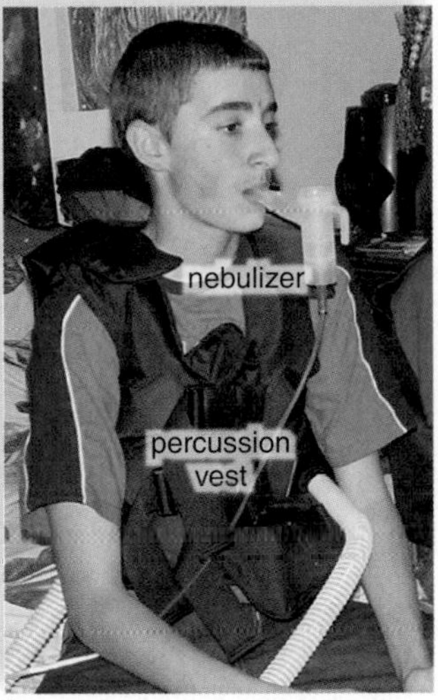

Figure 20.11 Cystic fibrosis disease.
a. Cystic fibrosis is due to a faulty protein that is supposed to regulate the flow of chloride ions into and out of cells through a channel protein. **b.** The nebulizer delivers drugs in aerosol form to the bronchial passages.

This single gene defect can affect any race but is prevalent among African Americans; it is estimated that 1 in every 625 African Americans is affected by sickle-cell disease.

Sickle-shaped cells can't pass along narrow capillary passageways as disk-shaped cells can, so they clog the vessels preventing adequate circulation. This results in anemia, tissue damage, jaundice, joint pain, and gallstones. Those affected by sickle-cell disease are also susceptible to many types of bacterial infections and have a higher incidence of stroke. Many treatment options, including blood transfusions and bone marrow transplants, are highly effective. The most common medicinal treatment, hydroxyurea, has been on the market for over a decade and is considered one of the most effective daily treatments for the reduction in sickle-cell-related anemia, joint pain, and tissue damage.

The sickle-cell gene is one of only a few autosomal recessive disorders in which the heterozygote can express variations of the recessive phenotype. Sickle-cell heterozygotes have sickle-cell traits in which the blood cells are normal unless they experience dehydration or mild oxygen deprivation. Intense exertion may cause sickling of some red blood cells for a short period of time. These patients can experience episodes and symptoms very similar to those patients with the autosomal recessive genotype.

Autosomal Dominant Disorders

Inheritance of only one dominant allele is necessary for an autosomal dominant genetic disorder to be displayed. Here, we discuss just a few of the known autosomal dominant disorders.

Marfan Syndrome The autosomal dominant disorder **Marfan syndrome** is caused by a defect in the production of an elastic connective tissue protein called fibrillin. This protein is normally abundant in the lens of the eye; the bones of limbs, fingers, and ribs; and also in the wall of the aorta and the blood vessels. This explains why the affected person often has a dislocated lens, long limbs and fingers, and a caved-in chest. The wall of the aorta is weak and can possibly burst without warning. Marfan syndrome is considered a "rare" disorder, currently affecting less than 200,000 people in the United States or 1 in every 2,000 Americans. Treatments for Marfan syndrome include beta blockers to control the cardiovascular symptoms, corrective lenses or eye surgery, and braces or orthopedic surgery for musculoskeletal symptoms.

Osteogenesis Imperfecta **Osteogenesis imperfecta** is an autosomal dominant genetic disorder that results in weakened, brittle bones. Although at least nine types of the disorder are known, most are linked to mutations in two genes necessary to the synthesis of a type I collagen—one of the most abundant proteins in the human body. Collagen has many roles, including providing strength and rigidity to bone and forming the framework for most of the body's tissues. Osteogenesis imperfecta leads to a defective collagen I that causes the bones to be brittle and weak. Because the mutant collagen can cause structural defects even when combined with normal collagen I, osteogenesis imperfecta is generally considered to be dominant.

Osteogenesis imperfecta, which has an incidence of approximately 1 in 5,000 live births, affects all racial groups similarly and has been documented as long as 300 years ago. Some historians think that the Viking chieftain Ivar Ragnarsson, who was known as Ivar the Boneless and was often carried into battle on a shield, had this condition. In most cases, the diagnosis is made in young children who visit the emergency room frequently due to broken bones. Some children with the disorder have an unusual blue tint in the sclera, the white portion of the eye; reduced skin elasticity; weakened teeth; and occasionally heart valve abnormalities. Currently, the disorder is treatable with a number of drugs that help to increase bone mass, but these drugs must be taken long term.

Huntington Disease An autosomal dominant neurological disorder that leads to progressive degeneration of brain cells (Fig. 20.12), **Huntington disease** is caused by a mutated copy

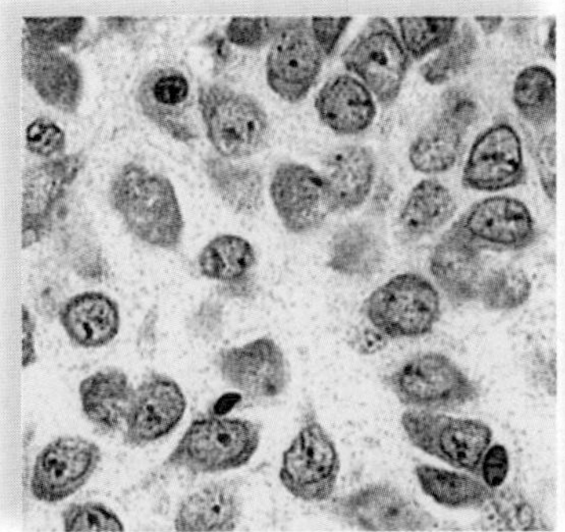

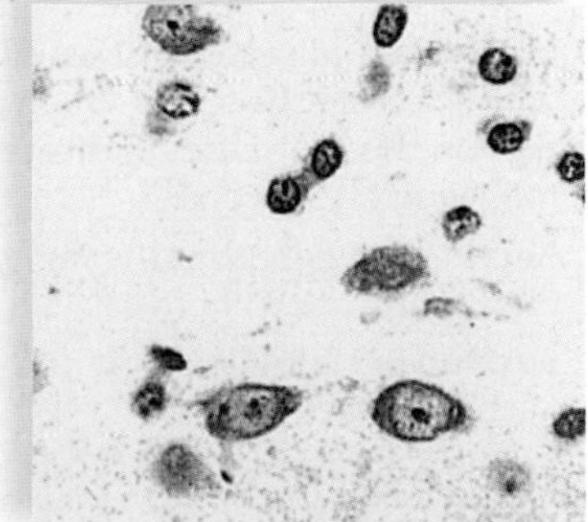

Figure 20.12 Huntington disease.
Huntington disease is caused by a loss of nerve cells.

of the gene for a protein called huntingtin. The defective gene contains segments of DNA in which the base sequence CAG repeats again and again. This type of structure, called a trinucleotide repeat, causes the huntingtin protein to have too many copies of the amino acid glutamine. The normal version of huntingtin has stretches of between 10 and 25 glutamic acid. If huntingtin has more than 36 glutamic acid, as is seen in the mutated Huntington gene, it changes shape and forms clumps inside neurons. These clumps attract other proteins to clump, rendering them inactive. One of these proteins that attaches itself to the clumps, called CBP, helps nerve cells survive.

The onset of symptoms for Huntington disease is normally not first seen until later in life (average age of onset is late thirties to late forties) although it is not unusual for an affected person to develop symptoms as early as their late teens. Huntington disease has a range of symptoms but is normally characterized by uncontrolled movements, unsteady gait, dementia, and speech impairment. On average, patients live 15 to 20 years after onset of symptoms, as the disease progresses rapidly. Current effective treatments include medications that slow the progression of the disease. Additionally, dopamine blockers have been found to be very effective in reducing uncontrolled movements and improving eye–hand coordination and steadiness during walking.

CHECK YOUR PROGRESS 20.3

1. Solve the following: In a pedigree, all the members of one family are affected. Based on this knowledge, list the genotypes of the parents (a) if the trait is recessive and (b) if the trait is dominant.
2. Predict the chances that homozygous normal parents for cystic fibrosis will have a child with cystic fibrosis.
3. Explain why some incidences of autosomal recessive disorders are higher in one race or culture.

CONNECTING THE CONCEPTS

Many of the diseases discussed in this section are associated with specific aspects of human physiology. For more information on these diseases (indicated in parentheses), refer to the following discussions:

Section 3.3 describes the function of proteins in the cell membrane (cystic fibrosis).

Section 3.4 contains descriptions of the lysosomes (Tay–Sachs disease).

Section 4.2 reviews the role of connective tissue in the body (Marfan syndrome).

Section 6.2 overviews the role of the red blood cells (sickle-cell disease).

20.4 Beyond Simple Inheritance Patterns

LEARNING OUTCOMES

Upon completion of this chapter, you should be able to

1. Summarize how polygenic inheritance, pleiotropy, codominance, and incomplete dominance differ from simple one-trait crosses.
2. Explain how a combination of genetics and the environment can influence a phenotype.
3. Predict a person's blood type based upon his or her genotype.

Certain traits, such as those studied in section 20.2, are controlled by one set of alleles that follows a simple dominant or recessive inheritance. We now know of many other types of inheritance patterns.

Polygenic Inheritance

Polygenic traits, such as skin color and height, are governed by several sets of alleles. The individual has a copy of all allelic pairs, possibly located on many different pairs of chromosomes. Each dominant allele codes for a product; therefore, the dominant alleles have a quantitative effect on the phenotype, that is, these effects are additive. The result is a *continuous variation* of phenotypes, resulting in a distribution of these phenotypes that resembles a bell-shaped curve. The more genes involved, the more continuous the variations and distribution of the phenotypes. Also, environmental effects cause many intervening phenotypes. In the case of height, differences in nutrition bring about a bell-shaped curve (Fig. 20.13).

Skin Color

Skin color is the result of pigmentation produced by cells called melanocytes in the skin (see section 4.6). We now know that there are over 100 different genes that influence skin color. For simplicity, skin color is an example of a polygenic trait that is likely controlled by many pairs of alleles.

APPLICATIONS AND MISCONCEPTIONS

Is skin color a good indication of a person's race?

The simple answer to this question is, No, an individual's skin color is not a true indicator of his or her genetic heritage. People of the same skin color are not necessarily genetically related. Because only a few genes control skin color, a person's skin color may not indicate his or her ancestors' origins. Scientists are actively investigating the genetic basis of race and have found some interesting relationships between a person's genetic heritage and medicine. This will be explored in greater detail in Chapter 22.

Figure 20.13 **Height is a polygenic trait in humans.**
When you record the heights of a large group of people chosen at random, the values follow a bell-shaped curve. Such a continuous distribution is because of control of a trait by several sets of alleles. Environmental effects are also involved.

Even so, we will use the simplest model and we will assume that skin has only three pairs of alleles (*Aa* and *Bb* and *Cc*) and that each capital letter contributes pigment to the skin. When a very dark person reproduces with a very light person, the children have medium-brown skin. When two people with the genotype *AaBbCc* reproduce with one another, individuals may range in skin color from very dark to very light. The distribution of these phenotypes typically follows a bell-shaped curve, meaning that few people have the extreme phenotypes and most people have the phenotype that lies in the middle. A bell-shaped curve is a common identifying characteristic of a polygenic trait (Fig. 20.14).

However, skin color is also influenced by the sunlight in the environment. Notice again that a range of phenotypes exists for each genotype. For example, individuals who are *AaBbCc* may vary in their skin color, even though they possess the same genotype, and several possible phenotypes fall between the two extremes. The interaction of the environment with polygenic traits is discussed in the next section.

Multifactorial Traits

Many human disorders, such as cleft lip and/or palate, clubfoot, schizophrenia, diabetes, phenylketonuria (see chapter opener), and even allergies and cancers, are most likely controlled by polygenes subject to environmental influences. These are commonly called **multifactorial traits.** The coats

APPLICATIONS AND MISCONCEPTIONS

Why do diet sodas carry the warning, "Phenylketonurics: Contains Phenylalanine"?

The sweetener used in diet sodas is aspartame. Aspartame is formed by the combination of two amino acids: aspartic acid and phenylalanine. When aspartame is broken down by the body, phenylalanine is released. Phenylalanine can be toxic for those with PKU and must be avoided. Other foods high in phenylalanine include eggs, meat, milk, and bananas.

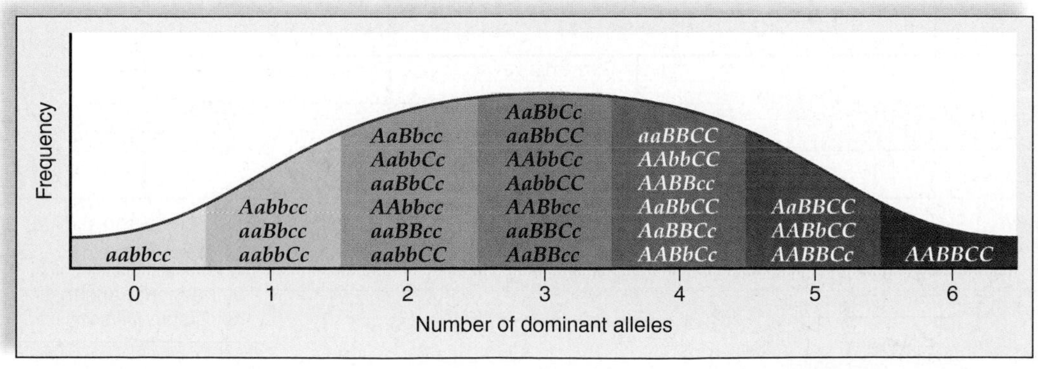

Figure 20.14 **Polygenic inheritance and skin color.**
Skin color is controlled by many pairs of alleles, which results in a range of phenotypes. The vast majority of people have skin colors in the middle range, whereas fewer people have skin colors in the extreme range.

Figure 20.15 Himalayan rabbit with temperature-susceptible coat color. The dark ears, nose, and feet of this rabbit are believed to be because of a lower body temperature in these areas.

of Siamese cats and Himalayan rabbits are darker in color at the ears, nose, paws, and tail (Fig. 20.15). Himalayan rabbits are homozygous for the allele *ch,* involved in the production of melanin. Experimental evidence suggests that the enzyme coded for by this gene is active only at a low temperature. Therefore, black fur occurs only at the extremities where body heat is lost to the environment.

Recent studies have reported that all sorts of behavioral traits, such as alcoholism, phobias, and even suicide, can be associated with particular genes. However, in almost all cases, the environment plays an important role in the severity of the phenotype. Therefore, they must be multifactorial traits. Current research focuses on determining what percentage of the trait is due to nature (inheritance) and what percentage is due to nurture (the environment). Some studies use identical and fraternal twins separated at birth and raised in different environments. The supposition is that, if identical twins in different environments share the same trait, that trait is most likely inherited. Identical twins are more similar in their intellectual talents, personality traits, and levels of lifelong happiness than are fraternal twins separated from birth. This substantiates the belief that behavioral traits are partly heritable. It also supports the belief that genes exert their effects by acting together in complex combinations susceptible to environmental influences.

Pleiotropy

Pleiotropy occurs when a single mutant gene affects two or more distinct and seemingly unrelated traits. For example, persons with Marfan syndrome have disproportionately long arms, legs, hands, and feet; a weakened aorta; poor eyesight; and other characteristics (Fig. 20.16). All of these characteristics are due to the production of abnormal connective tissue.

Marfan syndrome has been linked to a mutation in a gene (*FBN1*) on chromosome 15 that ordinarily specifies a functional protein called fibrillin. Fibrillin is essential for the formation of elastic fibers in connective tissue. Without the structural support of normal connective tissue, the aorta can burst, particularly if the person is engaged in a strenuous sport, such as volleyball or basketball. Flo Hyman may have been the best American woman volleyball player ever, but she fell to the floor and died at the age of only 31 because her aorta gave way during a game.

Incomplete Dominance and Codominance

Incomplete dominance occurs when the heterozygote is intermediate between the two homozygotes. For example, when a curly-haired individual has children with a straight-haired individual, their children have wavy hair. When two wavy-haired persons have children, the expected phenotypic ratio among the offspring is 1:2:1—one curly-haired child to two with wavy hair to one with straight hair. We can explain incomplete dominance by assuming that only one allele codes for a product and the single dose of the product gives the intermediate result.

Codominance occurs when alleles are equally expressed in a heterozygote. A familiar example is the human blood type AB, in which the red blood cells have the characteristics of both type A and type B blood. We can explain codominance by assuming that

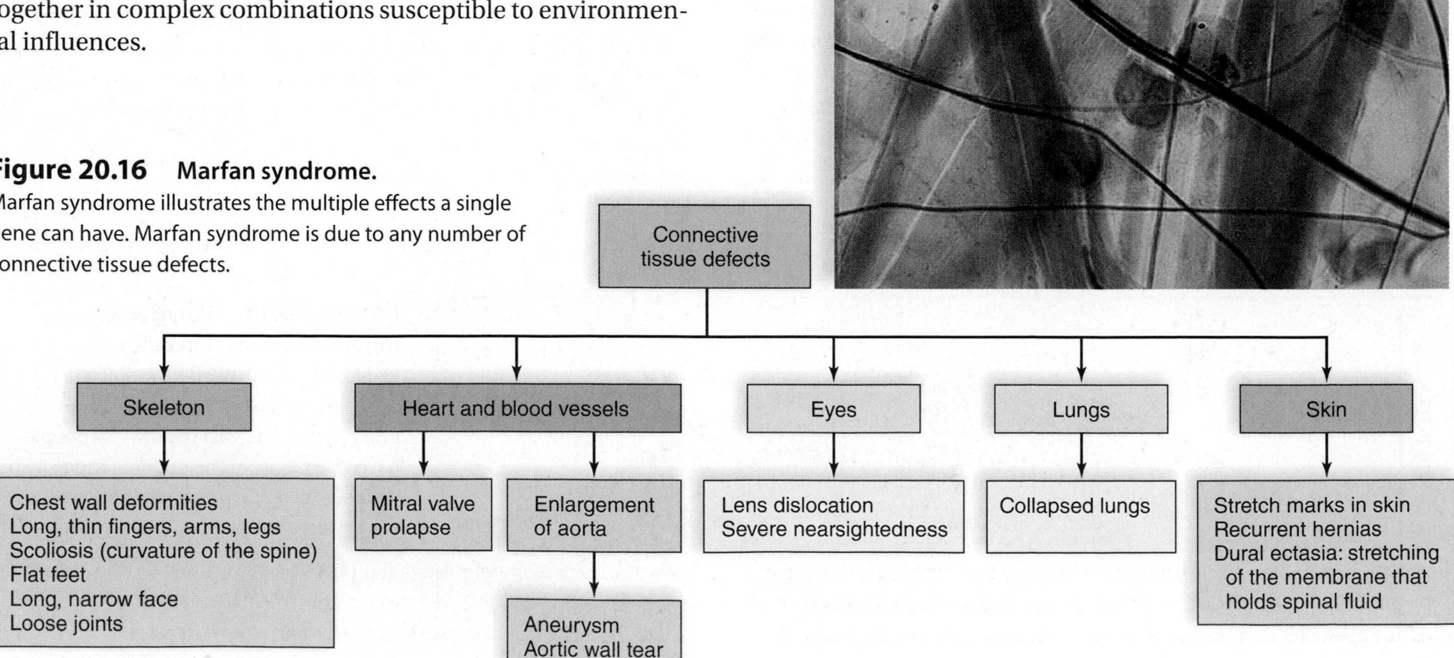

Figure 20.16 Marfan syndrome. Marfan syndrome illustrates the multiple effects a single gene can have. Marfan syndrome is due to any number of connective tissue defects.

both genes code for a product, and we observe the results of both products being present. Blood type inheritance is said to be an example of multiple alleles, described in the next section.

Incompletely Dominant Disorders

The prognosis in **familial hypercholesterolemia** parallels the number of LDL-cholesterol receptor proteins in the plasma membrane. A person with two mutated alleles lacks LDL-cholesterol receptors. A person with only one mutated allele has half the normal number of receptors, and a person with two normal alleles has the usual number of receptors. People with the full number of receptors do not have familial hypercholesterolemia. When receptors are completely absent, excessive cholesterol is deposited in various places in the body, including under the skin (Fig. 20.17).

The presence of excessive cholesterol in the blood causes cardiovascular disease. Therefore, those with no receptors die of cardiovascular disease as children. Individuals with half the number of receptors may die when young or after they have reached middle age.

Multiple-Allele Inheritance

When a trait is controlled by **multiple alleles,** the gene exists in several allelic forms. However, each person can only have two of the possible alleles.

ABO Blood Types

Three alleles for the same gene control the inheritance of ABO blood types. These alleles determine the presence or absence of antigens on red blood cells:

I^A = A antigen on red blood cells
I^B = B antigen on red blood cells
i = neither A nor B antigen on red blood cells

Each person has only two of the three possible alleles, and both I^A and I^B are dominant over i. Therefore, there are two possible genotypes for type A blood and two possible genotypes for type B blood. On the other hand, I^A and I^B are fully expressed in the presence of the other. Therefore, if a person inherits one of each of these alleles, that person will have type AB blood. Type O blood can result only from the inheritance of two i alleles.

The possible genotypes and phenotypes for blood type are as follows:

Phenotype	Genotype
A	I^AI^A, I^Ai
B	I^BI^B, I^Bi
AB	I^AI^B
O	ii

Blood typing was traditionally used in paternity suits. However, a blood test of a supposed father can only suggest that he *might* be the father, not that he definitely *is* the father. For example, it is possible, but not definite, that a man with type A blood (genotype I^Ai) is the father of a child with type O blood. On the other hand, a blood test sometimes can definitely prove that a man is not the father. For example, a man with type AB blood cannot possibly be the father of a child with type O blood. Therefore, blood tests can be used in legal cases only to try to exclude a man from possible paternity. Modern identification relies more heavily on analysis of patterns in the DNA, called DNA profiling (or DNA fingerprinting) (see section 21.4).

Figure 20.18 shows that matings between certain genotypes can have surprising results in terms of blood type. Parents with type A and type B blood can have offspring with all four possible blood types. Other blood types, such as Rh factor, are discussed in section 6.5.

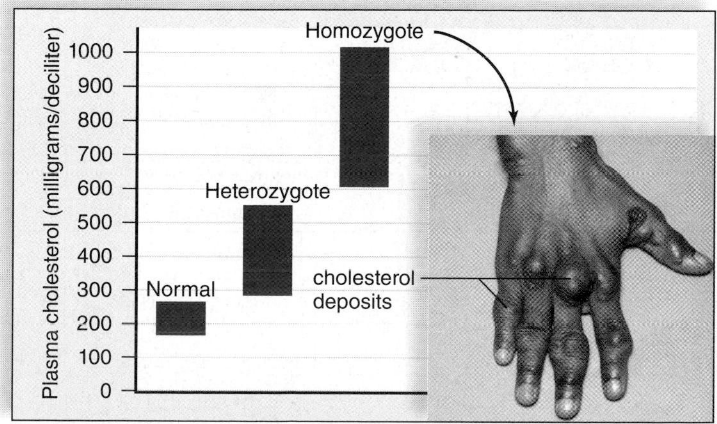

Figure 20.17 The inheritance of familial hypercholesterolemia. Familial hypercholesterolemia is incompletely dominant. Persons with one mutated allele have an abnormally high level of cholesterol in the blood, and those with two mutated alleles have a higher level still.

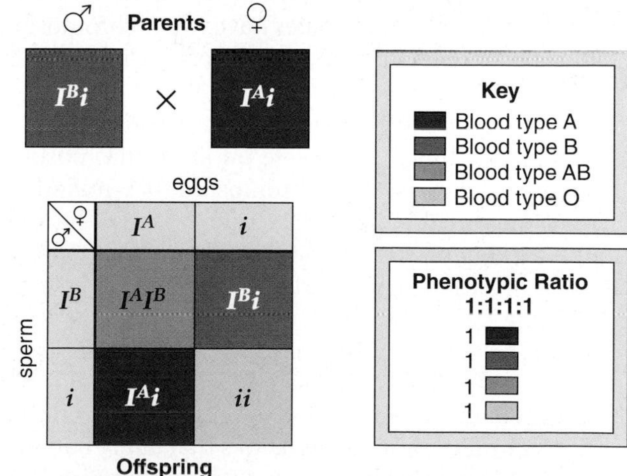

Figure 20.18 The inheritance of ABO blood types. Blood type exemplifies multiple-allele inheritance. The *I* gene has two codominant alleles, designated as I^A and I^B, and one recessive allele, designated by i. Therefore, a mating between individuals with type A blood and type B blood can result in any one of the four blood types. Why? The parents are I^Ai and I^Bi. If both parents were type AB blood, no child would have what blood type? See Appendix C for answers.

CHECK YOUR PROGRESS 20.4

1. Detail why polygenic inheritance follows a bell-shaped curve.
2. Describe a multifactorial trait that could have diet and nutrition as environmental influences.
3. Discuss the potential evolutionary advantages of having multiple alleles for a trait.

CONNECTING THE CONCEPTS

For more information on the topics presented in this section, refer to the following discussions:

Section 4.6 provides more information on the melanocyte cells that control skin color.

Section 6.2 describes the role of red blood cells in the human body.

Section 22.5 explores the evolution of humans and human races.

20.5 Sex-Linked Inheritance

LEARNING OUTCOMES

Upon completion of this section, you should be able to

1. Understand the differences between autosomal and sex-linked patterns of inheritance.
2. Interpret a human pedigree to determine sex-linked inheritance of a trait.

Normally, both males and females have 23 pairs of chromosomes; 22 pairs are called **autosomes,** and one pair is the sex chromosomes. These are called the **sex chromosomes** because they differ between the sexes. In humans, males have the sex chromosomes X and Y, and females have two X chromosomes. The Y chromosome contains the gene responsible for determining male gender.

Traits controlled by genes on the sex chromosomes are said to be **sex-linked.** An allele on an X chromosome is **X-linked,** and an allele on the Y chromosome is **Y-linked.** Most sex-linked genes are only on the X chromosomes. The Y chromosome is lacking these. Very few Y-linked alleles associated with human disorders have been found on the much smaller Y chromosome.

Many of the genes on the X chromosomes, such as those that determine normal as opposed to red-green color blindness, are unrelated to the gender of the individual. In other words, the X chromosome carries genes that affect both males and females. It would be logical to suppose that a sex-linked trait is passed from father to son or from mother to daughter, but this is not the case. A male always receives an X-linked allele from his mother, from whom he inherited an X chromosome. *The Y chromosome from the father does not carry an allele for the trait.* Usually, a sex-linked genetic disorder is recessive. Therefore, a female must receive two alleles, one from each parent, before she has the disorder.

X-Linked Alleles

When considering X-linked traits, the allele on the X chromosome is shown as a letter attached to the X chromosome. For example, this is the key for red-green color blindness, a well-known X-linked recessive disorder:

X^B = normal vision
X^b = color blindness

The possible genotypes and phenotypes in both males and females follow.

Genotypes	Phenotypes
X^BX^B	Female who has normal color vision
X^BX^b	Carrier female who has normal color vision
X^bX^b	Female who is color-blind
X^BY	Male who has normal vision
X^bY	Male who is color-blind

The second genotype is a carrier female. Although a female with this genotype appears normal, she is capable of passing on an allele for color blindness. Color-blind females are rare because they must receive the allele from both parents. Color-blind males are more common because they need only one recessive allele to be color-blind. The allele for color blindness must be inherited from their mother because it is on the X chromosome. Males only inherit the Y chromosome from their father (Fig. 20.19).

Now let us consider a mating between a man with normal vision and a heterozygous woman (Fig. 20.19). What is the chance that this couple will have a color-blind daughter? A color-blind son? All daughters will have normal color vision because they all receive an X^B from their father. The sons,

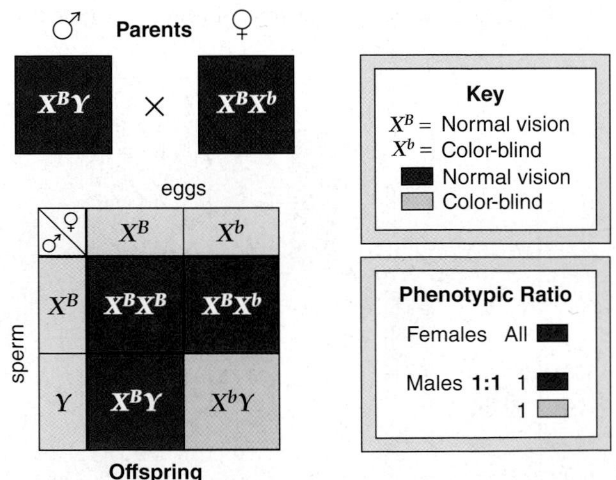

Figure 20.19 **Results of an X-linked cross.** The male parent is normal, but the female parent is a carrier—an allele for color blindness is located on one of her X chromosomes. Therefore, each son has a 50% chance of being color-blind. The daughters will appear normal, but each one has a 50% chance of being a carrier.

however, have a 50% chance of being color-blind, depending on whether they receive an X^B or an X^b from their mother. The inheritance of a Y chromosome from their father cannot offset the inheritance of an X^b from their mother. The Y chromosome doesn't have an allele for the trait, so it can't possibly prevent color blindness in a son. Notice in Figure 20.19 that the phenotypic results for sex-linked traits are given separately for males and females.

Pedigree for X-Linked Disorders

Like color blindness, most sex-linked disorders are usually carried on the X chromosome. Figure 20.19 gives a pedigree for an *X-linked recessive disorder.* More males than females have the disorder because recessive alleles on the X chromosome are always expressed in males. The Y chromosome lacks an allele for the disorder. X-linked recessive conditions often pass from grandfather to grandson because the daughters of a male with the disorder are carriers. Figure 20.20 lists various ways to recognize a recessive X-linked disorder.

Only a few known traits are *X-linked dominant.* If a disorder is X-linked dominant, affected males pass the trait *only* to daughters, who have a 100% chance of having the condition. Females can pass an X-linked dominant allele to both sons and daughters. If a female is heterozygous and her partner is normal, each child has a 50% chance of escaping an X-linked dominant disorder. This depends on the maternal X chromosome that is inherited.

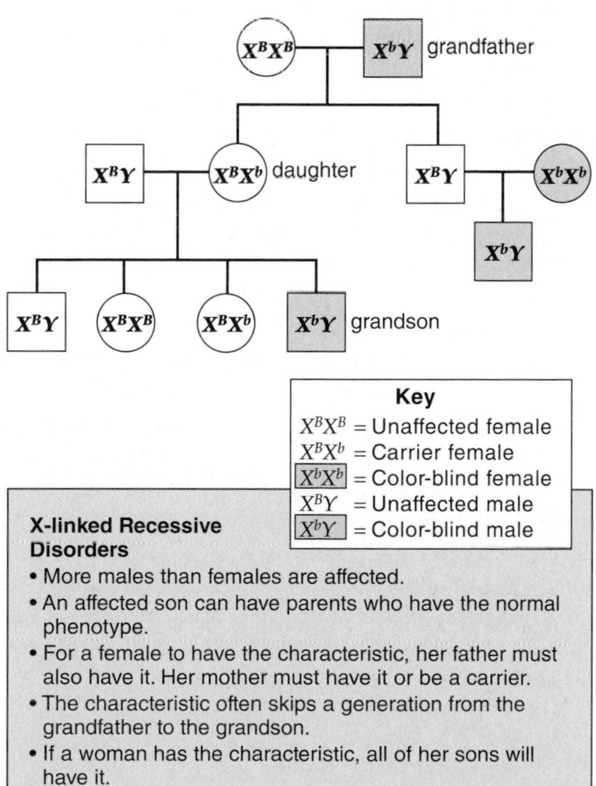
X^BX^B
X^bY grandfather
X^BY
X^BX^b daughter
X^BY
X^bX^b
X^bY
X^BY
X^BX^B
X^BX^b
X^bY grandson
Key
X^BX^B = Unaffected female
X^BX^b = Carrier female
X^bX^b = Color-blind female
X^BY = Unaffected male
X^bY = Color-blind male
X-linked Recessive Disorders
• More males than females are affected.
• An affected son can have parents who have the normal phenotype.
• For a female to have the characteristic, her father must also have it. Her mother must have it or be a carrier.
• The characteristic often skips a generation from the grandfather to the grandson.
• If a woman has the characteristic, all of her sons will have it.

Figure 20.20 **X-linked recessive disorder pedigree.** This pedigree for color blindness exemplifies the inheritance pattern of an X-linked recessive disorder. The list gives various ways of recognizing the X-linked recessive pattern of inheritance.

X-Linked Recessive Disorders of Interest

Color blindness, an X-linked recessive disorder, is the inability to differentiate certain color perceptions. There is no treatment for color blindness, but the disorder itself does not lead to a significant disability for the patient. Color blindness affects about 8–12% of Caucasian males and 0.5% of Caucasian females in the United States. Most people affected see brighter greens as tans, olive greens as browns, and reds as reddish-browns. A few cannot tell reds from greens at all. They see only yellows, blues, blacks, whites, and grays.

Duchenne muscular dystrophy is an X-linked recessive disorder characterized by a degeneration of the muscles. Symptoms, such as waddling gait, toe walking, frequent falls, and difficulty in rising, may appear as soon as the child starts to walk. Muscle weakness intensifies and respiratory and cardiovascular conditions progress until the individual is confined to a wheelchair, normally by ages 7 to 10. Death usually occurs by ages 20 to 25. Therefore, affected males are rarely fathers. The recessive allele remains in the population by passage from carrier mother to carrier daughter.

The absence of a protein, now called dystrophin, is the cause of Duchenne muscular dystrophy. Much investigative work determined that dystrophin is involved in the release of calcium from the sarcoplasmic reticulum in muscle fibers. The lack of dystrophin causes calcium to leak into the cell, which promotes the action of an enzyme that dissolves muscle fibers. When the body attempts to repair the tissue, fibrous tissue forms (Fig. 20.21). This cuts off the blood supply so that more and more cells die. Immature muscle cells can be injected into muscles, but it takes 100,000 cells for dystrophin production to increase by 30–40%.

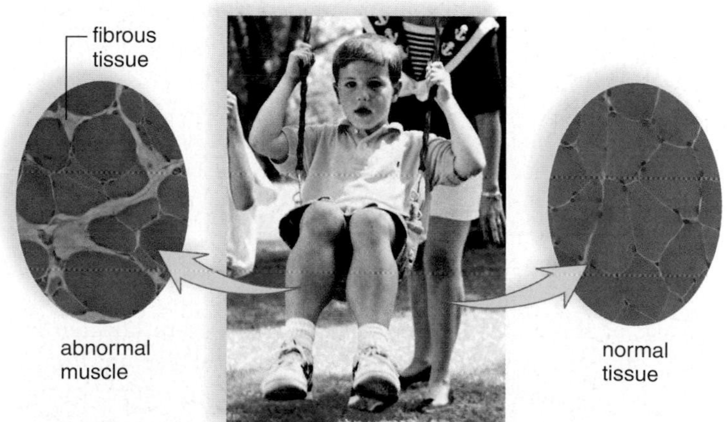
fibrous tissue
abnormal muscle
normal tissue

Figure 20.21 **Muscular dystrophy.** In muscular dystrophy, an X-linked recessive disorder, calves enlarge because fibrous tissue develops as muscles waste away, due to lack of the protein dystrophin.

Fragile X syndrome is the most common cause of inherited mental impairment. These impairments can range from mild learning disabilities to more severe intellectual disabilities. It is also the most common known cause of *autism,* a class of social, behavioral, and communication disorders. Fragile X syndrome affects 1 out of every 4,000 males and 1 out of every 8,000 females on average. Males with full symptoms of the condition have characteristic physical abnormalities—a long face, prominent jaw, large ears, joint laxity (excessively flexible joints), and genital abnormalities. Other common characteristics include tactile defensiveness (dislike of being touched), poor eye contact, repetitive speech patterns, hand flapping, and distractibility. Females with the condition present with variable symptoms. Most of the same traits seen in males with fragile X syndrome have been reported in females as well, but often the symptoms are milder in females and present with lower frequency.

A person with fragile X syndrome does not make the fragile X mental retardation protein (FMRP). Lack of this protein in the brain results in the various manifestations of the disease. As the name implies, a gene defect is found on the X chromosome. The genetic basis of fragile X syndrome is similar to that seen in Huntington disease (see page 471). In both cases, the gene

BIOLOGY MATTERS  **Science**

Hemophilia: The Royal Disease

The pedigree in Figure 20B shows why hemophilia is often referred to as "The Royal Disease." Queen Victoria of England, who reigned from 1837 to 1901, was the first of the royals to carry the gene. From her, the disease eventually spread to the Prussian, Spanish, and Russian royal families. In that era, monarchs arranged marriages between their children to consolidate political alliances. This practice allowed the gene for hemophilia to spread throughout the royal families. It is assumed that a spontaneous mutation arose either in Queen Victoria after her conception or in one of the gametes of her parents. However, in the book *Queen Victoria's Gene* by D. M. Potts, the author postulates that Edward Augustus, Duke of Kent, may not have been Queen Victoria's father. Potts suggests that Victoria may have instead been the illegitimate child of a hemophiliac male. Regardless of her parentage, had Victoria not been crowned, the fate of the various royal households may have been very different. Further, the history of Europe could have been changed dramatically as well.

However, Victoria did become queen. Queen Victoria and her husband Prince Albert had nine children. Fortunately, only one son, Leopold, suffered from hemophilia. He experienced severe hemorrhages and died in 1884 at the age of 31 as the result of a minor fall. He left behind a daughter Alice, a carrier for the disease. Her son Rupert also suffered from hemophilia and in 1928 died of a brain hemorrhage as a result of a car accident. Queen Victoria's eldest son Edward VII, the heir to the throne, did not have the disease; thus, the current British royal family is free of the disease.

Two of Queen Victoria's daughters, Alice and Beatrice, were carriers of the disease. Alice married Louis IV, the Grand Duke of Hesse. Of her six children, three were affected by hemophilia. Her son Frederick died of internal bleeding from a fall. Alice's daughter Irene married Prince Henry of Prussia, her first cousin. Two of their three sons suffered from hemophilia. One of these sons, Waldemar, died at age 56 due to the lack of blood-transfusion supplies during World War I. Henry, Alice's other son, bled to death at the age of 4.

Alice's daughter Alexandra married Nicholas II of Russia. Alexandra gave birth to four daughters before giving birth to Alexei, the heir to the Russian throne. It was obvious almost from birth that Alexei had hemophilia. Every fall caused bleeding into his joints, which led to crippling of his limbs and excruciating pain. The best medical doctors could not help Alexei. Desperate to relieve his suffering, his parents turned to the monk Rasputin. Rasputin was able to relieve some of Alexei's suffering by hypnotizing him and putting him to sleep. Alexandra and Nicholas, the czar and czarina, put unlimited trust in Rasputin. The illness of the only heir to the czar's throne, the strain Alexei's illness placed on the czar and czarina, and the power of Rasputin were all factors leading to the Russian Revolution of 1917. The czar and czarina, as well as their children, were all murdered during the revolution.

Queen Victoria's other carrier daughter, Beatrice, married Prince Henry of Battenberg. Her son Leopold was a hemophiliac, dying at 32 during a knee operation. Beatrice's daughter Victoria Eugenie married Alfonso XII of Spain. Queen Ena, as Victoria Eugenie came to be known, was not popular with the Spanish people. Her firstborn son Alfonso, the heir to the Spanish throne, did not stop bleeding upon his circumcision. When it became obvious that she had given her son hemophilia, it is alleged that her husband never forgave her. Like his cousin Rupert, Alfonso died in 1938 from internal bleeding after a car accident. Victoria's youngest son Gonzalo was also a hemophiliac whose life was claimed by a car accident in 1934.

Today, no members of any European royal family are known to have hemophilia. Individuals with the disease gene born in the late 1800s and early 1900s have all died, eliminating the gene from the current royal houses.

Questions to Consider

1. What was the probability that Alice's and Louis's sons would have hemophilia? Daughters? How does that relate to the actual occurrence in their children?
2. Assume that the mutation for hemophilia did not originate with Victoria. What does this tell you about the genotypes of her parents?

in question has too many repeated copies of a DNA sequence containing three nucleotides (called a trinucleotide repeat). In fragile X syndrome, the DNA sequence is CGG. (Recall that in Huntington disease, the repeated sequence is CAG.) Fewer than 59 copies of the repeated sequence is considered normal. Between 59 and 200 copies is considered "premutation." Generally, both males and females with the premutation genotype have normal intellect and appearance, although they can have subtle intellectual or behavioral symptoms. Persons whose DNA has over 200 copies of the repeat have "full mutation" and show the physical and behavioral traits of fragile X syndrome.

The trinucleotide-repeat disorders, like Huntington disease and fragile X syndrome, exhibit what is called *anticipation.* This means that the number of repeats in the gene can increase in each successive generation. For example, a female with 100 copies of the repeat is considered to have a premutation. When this female passes her X chromosome with the premutation to her offspring, the number of repeats may expand to over 200. This would result in a full mutation in her child. In a recent study, maternal premutations of between 90 and 200 repeats resulted in an expansion to full mutation in 80–100% of the offspring.

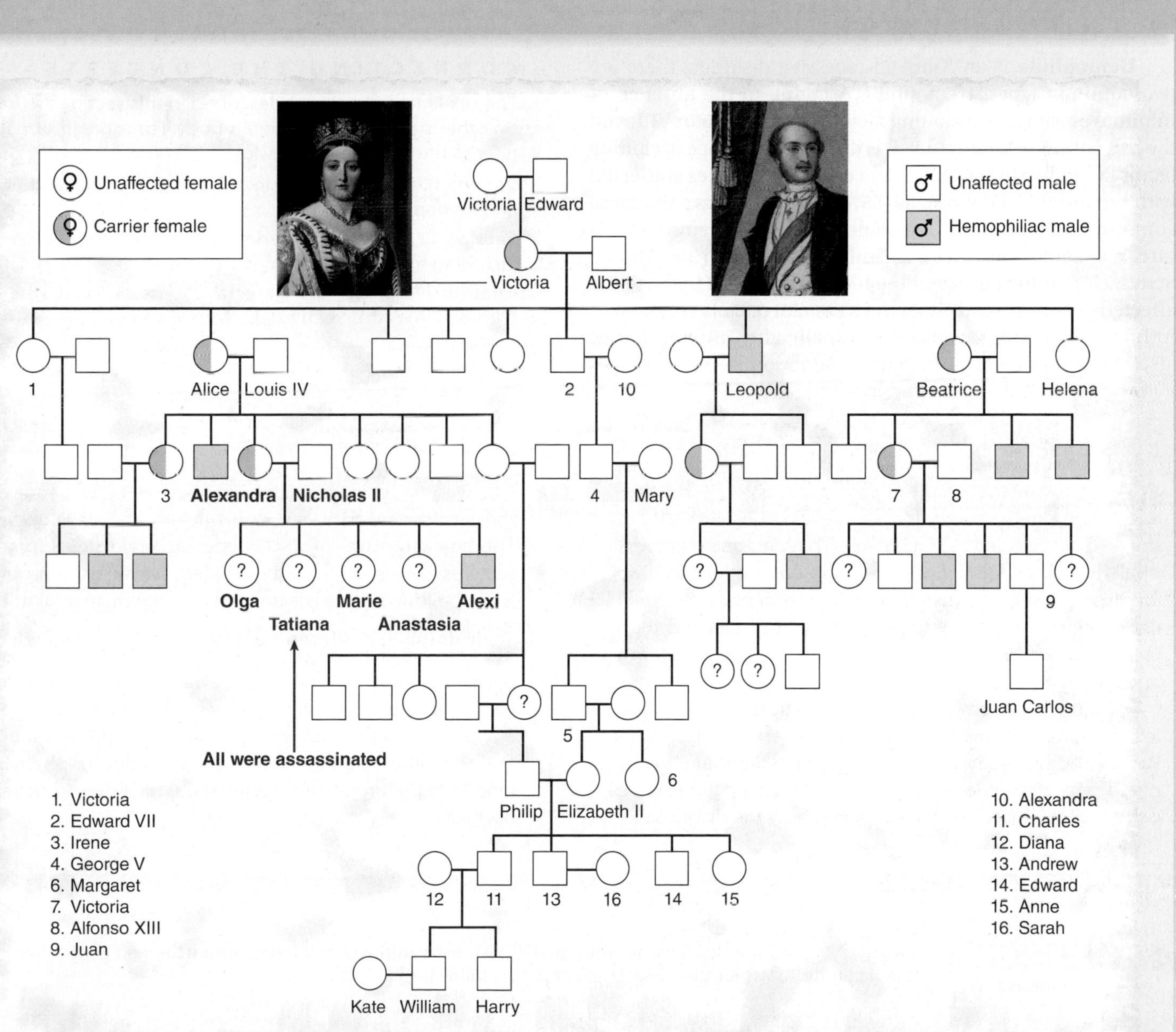

Figure 20B The royal families' X-linked pedigree.
Queen Victoria was a carrier, so each of her sons had a 50% chance of having the disorder, and each of her daughters had a 50% chance of being a carrier. This pedigree shows only the affected descendants. Many others are unaffected, including the members of the present British royal family.

APPLICATIONS AND MISCONCEPTIONS

Why is this disease called "fragile" X syndrome?

The name of fragile X syndrome may make you think that the disease has something to do with a breakable X chromosome. Actually, the name refers to the appearance of the chromosome under a microscope. In extreme cases of fragile X syndrome, in which there are many copies of the trinucleotide repeat, a portion of the X chromosome appears to be hanging from a thread. However, the appearance of the chromosome is not a cause of the disease. Fragile X syndrome is a result of a malfunction in the fragile X mental retardation protein (FMRP).

Hemophilia is an X-linked recessive disorder. There are two common types. Hemophilia A is due to the absence or minimal presence of a clotting factor known as factor VIII, and the less common hemophilia B is due to the absence of clotting factor IX. In the United States, 1 in every 5,000 males is affected with hemophilia. Of those, 80–85% specifically have the more common hemophilia A. A female affected by hemophilia is rare, averaging 1 in every 50 million women in the United States. Hemophilia is called the bleeder's disease because the affected person's blood either does not clot or clots very slowly. Although hemophiliacs bleed externally after an injury, they also bleed internally, particularly around joints. Hemorrhages can be stopped with transfusions of fresh blood (or plasma) or concentrates of the clotting protein. Also, factors VIII and IX are now available as biotechnology products.

CHECK YOUR PROGRESS 20.5

1. Solve the following: In a given family, a man and woman have two children, a boy and a girl, and both are colorblind. List the possible genotypes of the parents if both parents have normal vision.
2. Predict if a woman affected by an X-linked dominant disorder can have a child that is not affected. Why or why not?
3. Discuss why X-linked disorders are more common than Y-linked disorders.

CONNECTING THE CONCEPTS

The sex-linked conditions described in this section are based on other systems of the human body. For more information, refer to the following discussions:

Section 6.4 provides more information on blood clotting and hemophilia.

Section 12.4 describes a variety of muscular disorders, including muscular dystrophy.

Chapter 17 outlines how genetics counselors obtain chromosomes to screen for diseases such as fragile X syndrome.

CASE STUDY CONCLUSION

Dr. Preston referred Patrick's parents to Ms. Wolfe, the hospital's genetics counselor. A genetics counselor works in the medical field as a liaison between patients and members of all health-care disciplines. Their job descriptions include identifying couples at risk for passing on genetic disorders and working with patients affected by disorders and their families to better understand diagnoses, treatment options, and counseling.

Ms. Wolfe explained that PKU was autosomal recessive in inheritance. She drew out a pedigree chart depicting both Patrick and his parents. Patrick was affected and neither of his parents was, which meant that both of his parents were carriers of the disorder. She explained how the diagnosis of PKU was made on Patrick by use of the HPLC test and what the disorder might mean to his health and development. The characteristics of PKU—reduced brain development, seizures, and so on—can be avoided; because the problem is the buildup of phenylalanine levels, the main goal during developmental years is to eliminate or reduce the intake of phenylalanine. She explained that a diet restricting phenylalanine intake starting right away and continuing until Patrick finished puberty would result in little to no adverse effects. This diet, combined with prescribed protein supplements, would help Patrick avoid the buildup of phenylalanine in his neurons and would allow his brain to develop normally.

MEDIA STUDY TOOLS

Enhance your study of this chapter with media! Visit **www.mhhe.com/maderhuman13e** and go to "Media Study Tools" for this chapter to access the following:

 **Animation**	 **Video**	 **MP3 File**
20.3 Lysosomes	**20.3** Good Poison	**20.3** Genetics and Inheritance

SUMMARIZE

20.1 Genotype and Phenotype

An allele is a variation of a gene. Each allele exists at a specific locus on a chromosome. Genotype refers to the alleles of the individual, and phenotype refers to the physical characteristics associated with these alleles. Dominant alleles mask the expression of recessive alleles.

- **Homozygous dominant** individuals have the dominant phenotype (i.e., *EE* = unattached earlobes).
- **Homozygous recessive** individuals have the recessive phenotype (i.e.,*ee* = attached earlobes).
- **Heterozygous** individuals have the dominant phenotype (i.e., *Ee* = unattached earlobes).

20.2 One- and Two-Trait Inheritance

One-Trait Crosses

The first step in doing a problem with a one-trait, or **monohybrid cross,** is to determine the genotype and then the gametes.

- An individual has two alleles for every trait, but a gamete has one allele for every trait.

The next step is to combine all possible sperm with all possible eggs. If there is more than one possible sperm and/or egg, a **Punnett square** is helpful in determining the genotypic and phenotypic ratio among the offspring.

- For a monohybrid cross between two heterozygous individuals, a 3:1 ratio is expected among the offspring.
- For a monohybrid cross between a heterozygous and homozygous recessive individual, a 1:1 ratio is expected among the offspring.
- The expected ratio can be converted to the chance of a particular genotype or phenotype. For example, a 3:1 ratio = a 75% chance of the dominant phenotype and a 25% chance of the recessive phenotype.

Two-Trait Crosses

A problem consisting of two traits is often referred to as a **dihybrid cross.**

- If an individual is heterozygous for two traits, four gamete types are possible, as can be substantiated by knowledge of meiosis.
- For a cross between two heterozygous individuals (*AaBb* × *AaBb*), a 9:3:3:1 ratio is expected among the offspring.
- For a cross between a heterozygous and homozygous recessive individual (*AaBb* × *aabb*), a 1:1:1:1 ratio is expected among the offspring.

20.3 Inheritance of Genetic Disorders

A pedigree shows the pattern of inheritance for a trait from generation to generation of a family. This first pattern appears in a family pedigree for a recessive disorder—both parents are **carriers.** The second pattern appears in a family pedigree for a dominant disorder. Both parents are again heterozygous.

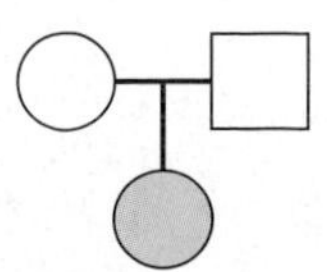

Trait is recessive.

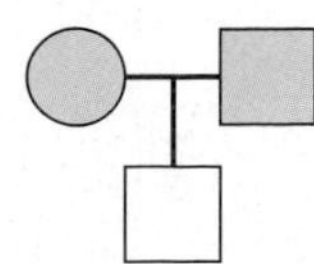

Trait is dominant.

Genetic Disorders of Interest

- **Tay–Sachs disease, cystic fibrosis** (CF), and **sickle-cell disease** are autosomal recessive disorders.
- **Marfan syndrome, osteogenesis imperfecta,** and **Huntington disease** are autosomal dominant disorders.

20.4 Beyond Simple Inheritance Patterns

In some patterns of inheritance, the alleles are not just dominant or recessive.

Polygenic Inheritance

Polygenic traits, such as skin color and height, are controlled by more than one set of alleles. The alleles have an additive effect on the phenotype. **Multifactorial traits** are usually polygenic with an environmental influence.

Incomplete Dominance and Codominance

In **incomplete dominance** (e.g., **familial hypercholesterolemia**), the heterozygote is intermediate between the two homozygotes. In **codominance** (e.g., type AB blood), both dominant alleles are expressed equally.

Multiple-Allele Inheritance

In humans, an example of a trait involving **multiple alleles** is the ABO blood types. Every individual has two out of three possible alleles: I^A, I^B, or *i*. Both I^A and I^B are expressed. Therefore, this is also a case of codominance.

20.5 Sex-Linked Inheritance

X-Linked Alleles

Humans contain 22 pairs of **autosomes,** and 1 pair of **sex chromosomes.** Traits on the sex chromosomes are said to be **sex-linked. X-linked** traits, such as those that determine normal vision as opposed to color blindness, are unrelated to the gender of the individual. Common X-linked genetic crosses are

- $X^BX^b \times X^BY$ All daughters will be normal, even though they have a 50% chance of being carriers, but sons have a 50% chance of being color-blind.
- $X^BX^B \times X^bY$ All children are normal (daughters will be carriers).

Pedigree for X-Linked Disorders

- A pedigree for an X-linked recessive disorder shows that the trait often passes from grandfather to grandson by way of a carrier daughter. Also, more males than females have the characteristic.
- Like most X-linked disorders, **color blindness, Duchenne muscular dystrophy, fragile X syndrome,** and **hemophilia** are recessive.

ASSESS

Testing Your Knowledge of the Concepts

1. Distinguish between the phenotype and genotype, and explain why the phenotype does not always give the complete genotype. Explain in terms of phenotype and genotype. (pages 461–462)
2. Explain why the gametes have only one allele for a trait. (pages 462–463)

3. Describe the genotypic and phenotypic ratios of a monohybrid cross between two heterozygotes. (pages 462–464)
4. Explain how the genetic combinations of the gametes in a dihybrid cross are explained by meiosis. (pages 465–466)
5. Describe the phenotypic ratio of the offspring of a dihybrid cross between two heterozygotes. (pages 466–467)
6. Using a pedigree, show an autosomal recessive disorder pattern and an autosomal dominant disorder pattern. (pages 469–470)
7. What is required for an autosomal recessive disorder to appear? Name four autosomal recessive disorders. What is required for an autosomal dominant disorder to appear? Name two autosomal dominant disorders. (pages 469–472)
8. Define and give an example of each of the following inheritance patterns: polygenic inheritance, multifactorial trait, incomplete dominance, codominance, pleiotropy, and multiple alleles. (pages 472–475)
9. What are the phenotypes and genotypes for the ABO blood groups? If both parents are type AB blood, what are the possible genotypes and phenotypes for their children, including the expected percentages? (page 475)
10. How is an X-linked trait different from an autosomal trait? Show a pedigree of an X-linked trait. (pages 476–477)
11. Which of these is a correct statement?
    - **a.** Each gamete contains two alleles for each trait.
    - **b.** Each individual has one allele for each trait.
    - **c.** Fertilization gives each new individual one allele for each trait.
    - **d.** All of the choices are correct.
    - **e.** None of the choices are correct.
12. What possible gametes can be produced by *AaBb?*
    - **a.** *Aa, Bb*
    - **b.** *A, a, B, b*
    - **c.** *AB, ab*
    - **d.** *AB, Ab, aB, ab*
13. A straight hairline is recessive. If two parents with a widow's peak have a child with a straight hairline, then what is the chance that their next child will have a straight hairline?
    a. no chance **b.** $\frac{1}{4}$ **c.** $\frac{3}{16}$ **d.** $\frac{1}{2}$ **e.** $\frac{1}{16}$
14. What is the chance that an *Aa* individual will be produced from an *Aa*–*Aa* cross?
    a. 50% **b.** 75% **c.** 0% **d.** 25% **e.** 100%
15. The genotype of an individual with the dominant phenotype can be determined best by reproduction with
    - **a.** the recessive genotype or phenotype.
    - **b.** a heterozygote.
    - **c.** the dominant phenotype.
    - **d.** the homozygous dominant.
    - **e.** Both a and b are correct.
16. The homologous chromosomes align independently at the equator during meiosis so
    - **a.** all possible combinations of alleles can occur in the gametes.
    - **b.** only the parental combinations of gametes can occur in the gametes.
    - **c.** only the nonparental combinations of gametes can occur in the gametes.
17. Which of the following is not a feature of multifactorial inheritance?
    - **a.** Effects of dominant alleles are additive.
    - **b.** Genes affecting the trait may be on multiple chromosomes.
    - **c.** Environment influences phenotype.
    - **d.** Recessive alleles are harmful.
18. The ABO blood system exhibits
    - **a.** codominance.
    - **b.** multiple alleles.
    - **c.** incomplete dominance.
    - **d.** Both a and b are correct.
19. If a child has type O blood and the mother is type A, then which of the following could be the blood type of the child's father?
    - **a.** A only
    - **b.** B only
    - **c.** O only
    - **d.** A or O
    - **e.** A, B, or O
20. In which of the following is a trait influenced by both polygenic and environmental factors?
    - **a.** multifactorial trait
    - **b.** sex-linked
    - **c.** incomplete dominance
    - **d.** multiple-allele inheritance

ENGAGE

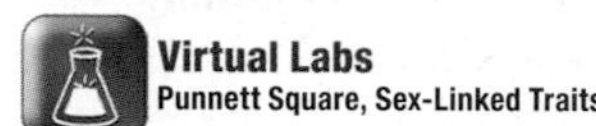

The virtual labs "Punnett Square" and "Sex-Linked Traits" allow you to test your knowledge of monohybrid, dihybrid, and sex-linked crosses.

Thinking Critically About the Concepts

In the case study, Patrick was affected with an autosomal recessive disorder, PKU. The chapter detailed many other autosomal recessive disorders, other types of genetic inheritance, and one- and two-trait inheritance patterns. Think about the basics learned from this chapter when answering the following questions:

1. In the pedigree for an autosomal recessive disorder on page 469, the chart depicts a skip in the generations of affected individuals. Is this always the case for autosomal recessive disorders? Why or why not?
2. In two-trait crosses, does one trait being dominant have an effect on the inheritance of the second trait?
3. Would you think an X-linked disorder or an autosomal disorder would appear more in a family pedigree? Explain your reasoning.

Assume for a moment that you are working in a genetics lab with *Drosophila melanogaster,* a model organism for genetic research. You want to determine whether a newly found *Drosophila* characteristic is dominant or recessive.

4. Explain how you would construct such a cross, and the expected outcomes that would indicate dominance.
5. Why would working with *Drosophila* be easier than working with humans?

CHAPTER

21

DNA Biology and Technology

CASE STUDY DIABETES

At eight years old, Kaya was diagnosed with type 1 diabetes mellitus. For months before her diagnosis, she had been fatigued. She had also had noticeable weight loss; an almost insatiable thirst; and dry, itchy skin that was slow to heal after a cut or injury. Over the next few days, Kaya's pediatrician, Dr Weber, ran a series of diagnostic tests, including a fasting plasma glucose (FPG) test and a random plasma glucose test. Kaya's doctor was administering these tests to detect her blood glucose levels at different times during the day and in response to fasting and nonfasting situations. Type 1 diabetes results from the body's failure to either produce insulin or produce enough insulin for the body's needs. Insulin is a hormone that helps move glucose from the bloodstream into the cells of the body where it can be used to make the fuel, ATP, for our cells. Kaya's pediatrician explained to her and her parents that this type of diabetes was a chronic disease that she would have to control throughout her life.

Left untreated, type 1 diabetes can lead to blindness, kidney failure, nerve damage, cardiovascular disease, and death. Dr. Weber explained that over 25 million people in the United States are afflicted with diabetes and, like Kaya, approximately 5% of them have type 1. In fact, diabetes is the third leading major cause of death in the United States behind heart disease and cancer. To treat this conditon, Kaya would need to monitor her blood glucose level several times per day and use insulin injections multiple times per day when needed. Kaya's parents asked where the insulin she would be injecting came from. Dr. Weber explained that at one time it was derived from the pancreas of cows and pigs, but now with the advances in biotechnology, human insulin is used to treat diabetes patients. The gene for human insulin can be inserted into a bacteria cell and the bacteria can produce the insulin Kaya will inject.

As you read through the chapter, think about the following questions:

1. What advances in biotechnology make the production of human insulin in bacterial cells possible?
2. What is gene expression? What processes are involved to express a gene in a cell?
3. How do scientists clone a gene? Why is this done?

CHAPTER CONCEPTS

21.1 DNA and RNA Structure and Function

DNA is a double helix composed of two polynucleotide strands. When DNA replicates, each strand serves as a template for a new strand. RNA is involved with processing the information contained within the DNA molecule.

21.2 Gene Expression

Gene expression results in an RNA or protein product. Each protein has a sequence of amino acids. The blueprint for building a protein is coded in the sequence of nucleotides in the DNA.

21.3 DNA Technology

DNA technology allows us to clone a portion of DNA for various purposes, including DNA fingerprinting and transfer of DNA to other organisms. Transgenic (genetically modified) organisms receive and express foreign DNA. During gene therapy, humans receive foreign DNA to cure some particular condition.

21.4 Genomics and Gene Therapy

Genomics is the study of genetic information in a particular cell or organism, including humans and other organisms. Advances in gene therapy are making it possible to treat human disease.

BEFORE YOU BEGIN

Before beginning this chapter, take a few moments to review the following discussions:

Section 2.7 What are the roles of nucleic acids in a cell?

Section 3.4 What is the structure of the nucleus?

Section 18.1 When are the chromosomes replicated in the cell cycle?

21.1 DNA and RNA Structure and Function

LEARNING OUTCOMES

Upon completion of this section, you should be able to

1. Describe the structure of a DNA molecule.
2. Explain the process of DNA replication and list the major enzymes involved.
3. Distinguish between the structures of DNA and RNA.
4. State the roles of RNA in a cell.

For life on Earth, **DNA (deoxyribonucleic acid)** is the genetic material. Within the genetic material, information is organized as **genes,** short segments of DNA that contain instructions for a specific trait. Genes are organized into structures called chromosomes—the mechanism by which genetic information is transmitted, at both the cellular and organismal level, from one generation to the next. In eukaryotic cells, including those of humans, the majority of the DNA is located in the nucleus. A small amount of DNA is also found in the mitochondria, but for this we will focus on the nuclear DNA.

As the genetic material, DNA must be able to do three things: (1) replicate so that it can be transmitted to the next generation, (2) store information, and (3) undergo change (mutation) to provide genetic variability. DNA meets all three of these criteria.

Structure of DNA

In the twentieth century, scientists discovered that the DNA molecule is a **double helix** (see the Science feature, "Discovering the Structure of DNA"). DNA is composed of two strands that spiral about each other (Fig. 21.1*a*). Each strand is a polynucleotide because it is composed of a series of nucleotides. A nucleotide is a molecule composed of three subunits—phosphoric acid (phosphate), a pentose sugar (deoxyribose), and a nitrogen-containing base (either adenine [A], cytosine [C], guanine [G], or thymine [T]; see Fig. 2.24). Recall that adenine and guanine are purines (two rings), and cytosine and thymine are pyrimidines (one ring) (Fig. 21.1*c*).

Looking at just one strand of DNA, notice that the phosphate and sugar molecules make up a backbone and the bases project to one side. Put the two strands together, and DNA resembles a ladder (Fig. 21.1*b*). The sugar-phosphate backbones make up the supports of the ladder. The rungs of the ladder are the paired bases, which are held together by hydrogen bonding: A pairs with T by forming two hydrogen bonds, and G pairs with C by forming three hydrogen bonds, or vice versa. This is called **complementary base pairing** (Fig. 21.1*c*).

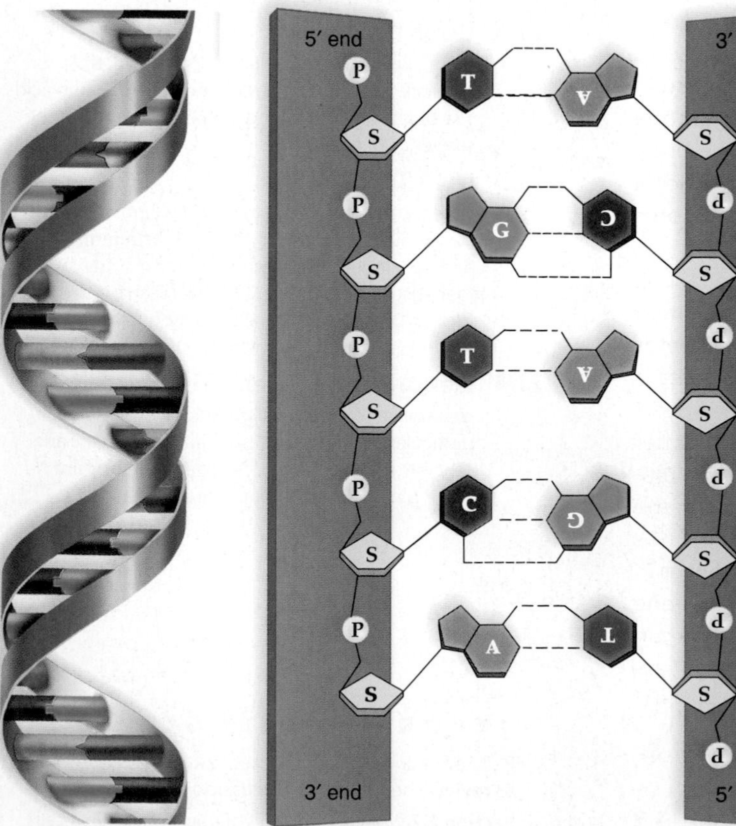

Figure 21.1 The structure of DNA.
a. DNA double helix. **b.** An unwound helix showing a ladder configuration. The supports are composed of sugar (S) and phosphate (P) molecules. The rungs are complementary bases. The DNA bases pair in such a way that the sugar-phosphate backbones are oriented in different directions. **c.** The DNA strands are antiparallel. Notice the numbering of the carbon atoms (5′–3′) in deoxyribose.

BIOLOGY MATTERS  **Science**

Discovering the Structure of DNA

In 1953, James Watson, an American biologist, began an internship at the University of Cambridge, England. There he met Francis Crick, a British physicist, who was interested in molecular structures. Together, they set out to determine the structure of DNA and to build a model that would explain how DNA, the genetic material, can vary from species to species and even from individual to individual. They also discovered how DNA replicates (makes a copy of itself) so that daughter cells receive an identical copy.

The bits and pieces of data available to Watson and Crick were like puzzle pieces they had to fit together. This is what they knew from the research of others:

1. DNA is a polymer of nucleotides, each one having a phosphate group, the sugar deoxyribose, and a nitrogen-containing base. There are four types of nucleotides because there are four different bases: adenine (A) and guanine (G) are purines, whereas cytosine (C) and thymine (T) are pyrimidines.
2. A chemist, Erwin Chargaff, had determined in the late 1940s that regardless of the species under consideration, the number of purines in DNA always equals the number of pyrimidines. Further, the amount of adenine equals the amount of thymine (A = T), and the amount of guanine equals the amount of cytosine (G = C). These findings came to be known as Chargaff's rules.
3. Rosalind Franklin (Fig. 21A*a*), working with Maurice Wilkins at King's College, London, had just prepared an X-ray diffraction photograph of DNA. It showed that DNA is a double helix of constant diameter and that the bases are regularly stacked on top of one another (Fig. 21A*b*).

Using these data, Watson and Crick deduced that DNA has a twisted, ladderlike structure. The sugar-phosphate molecules make up the sides of the ladder, and the bases make up the rungs. Further, they determined that if A is normally hydrogen-bonded with T, and G is normally hydrogen-bonded with C (in keeping with Chargaff's rules), then the rungs always have a constant width, consistent with the X-ray photograph.

Watson and Crick built an actual model of DNA out of wire and tin. This double-helix model does indeed allow for differences in DNA structure between species because the base pairs can be in any order. Also, the model suggests that complementary base pairing plays a role in the replication of DNA. As Watson and Crick pointed out in their original paper, "It has not escaped our notice that the specific pairing we have postulated immediately suggests a possible copying mechanism for the genetic material."

When the Nobel Prize for the discovery of the double helix was awarded in 1962, the honor went to James Watson, Francis Crick, and Maurice Wilkins. Rosalind Franklin was not listed as one of the recipients. Tragically, Franklin developed ovarian cancer in 1956, and she died in 1958 at the age of 37. At the time, the rules of the Nobel Prize committee prevented the award being given posthumously.

Video Dark Lady of DNA

Questions to Consider

1. Watson and Crick's discovery of DNA is clearly one of the most important biological discoveries in the last century. What advances in medicine and science can you think of that are built on knowing the structure of DNA?
2. Describe why the structure of DNA led Watson and Crick to point out "a possible copying mechanism for the genetic material."

a.

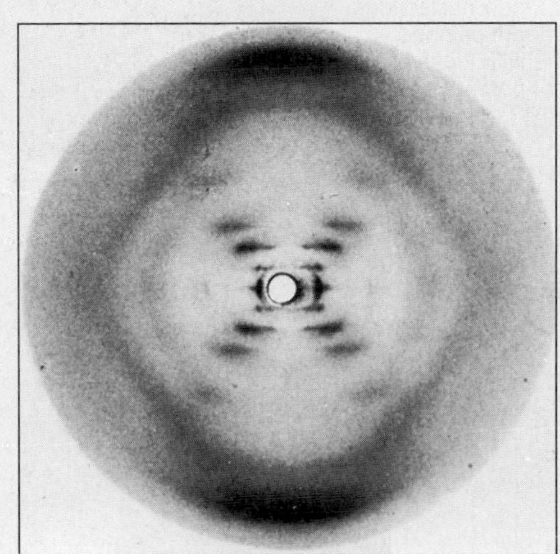

b.

Figure 21A X-ray diffraction of DNA.
a. Rosalind Franklin, 1920–1958.
b. The diffraction pattern of DNA produced by Rosalind Franklin. The crossed (X) pattern in the center told investigators that DNA is a helix, and the dark portions at the top and the bottom indicated that some feature is repeated over and over. Watson and Crick determined that this feature was the hydrogen-bonded bases.

Complementary base pairing is important to the functioning of DNA. The two strands of DNA are antiparallel, meaning they run in opposite directions. Notice in Figure 21.1*b* that in one strand, the sugar molecules appear right side up, and in the other, they appear upside down. This is due to the position of carbon molecules on the deoxyribose sugar molecules. When looking at a double helix, one side will have the 5′ carbon at one end, and the other side will have the 3′ carbon (Fig. 21.1*b*). This orientation becomes important when the DNA is replicated.

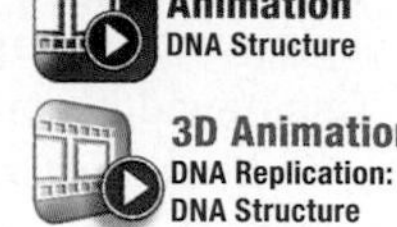

Replication of DNA

When cells divide, each new cell gets an exact copy of DNA. The process of copying a DNA helix is called **DNA replication.** During the S phase of interphase during mitosis when DNA is replicated, the double-stranded structure of DNA allows each original strand to serve as a **template** for the formation of a complementary new strand. DNA replication is termed *semiconservative* because each new double helix has one original strand and one new strand. In other words, one of the original strands is conserved, or present, in each new double helix. Each original strand has produced a new strand through complementary base pairing, so there are now two DNA helices identical to each other and to the original molecule (Fig. 21.2).

At the molecular level, several enzymes and proteins participate in the semiconservative replication of the new DNA strands. This process is summarized in Figure 21.3:

1. The enzyme D*NA helicase* unwinds and "unzips" double-stranded DNA by breaking the weak hydrogen bonds between the paired bases.

Figure 21.2 Semiconservative replication.
After the DNA double helix unwinds, each parental strand serves as a template for the formation of the new daughter strands. Complementary free nucleotides hydrogen bond to a matching base (e.g., A with T; G with C) in each parental strand, and are joined to form a complete daughter strand. Two helices, each with a daughter and parental strand, are produced following replication.

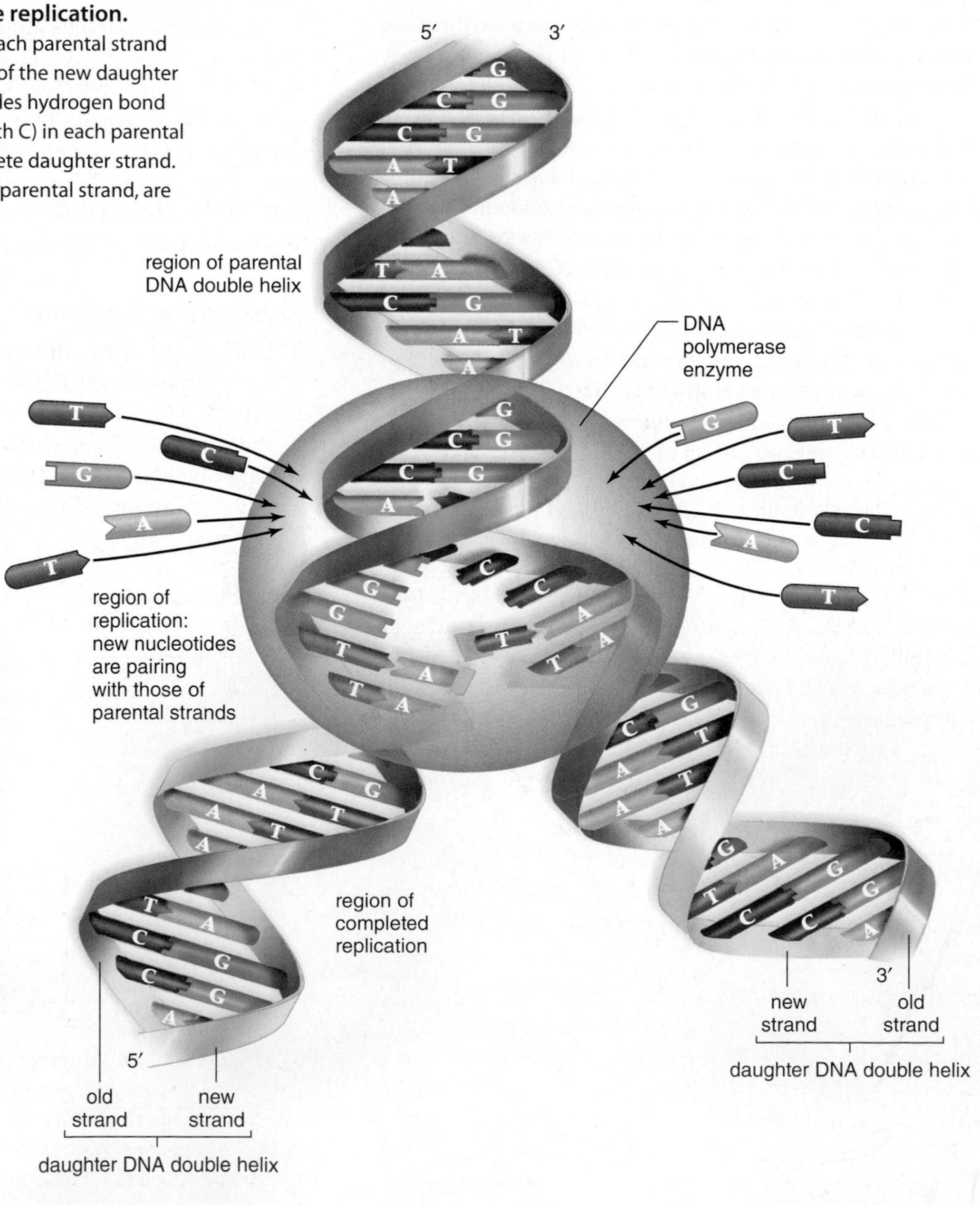

Figure 21.3 Enzymes of DNA replication.

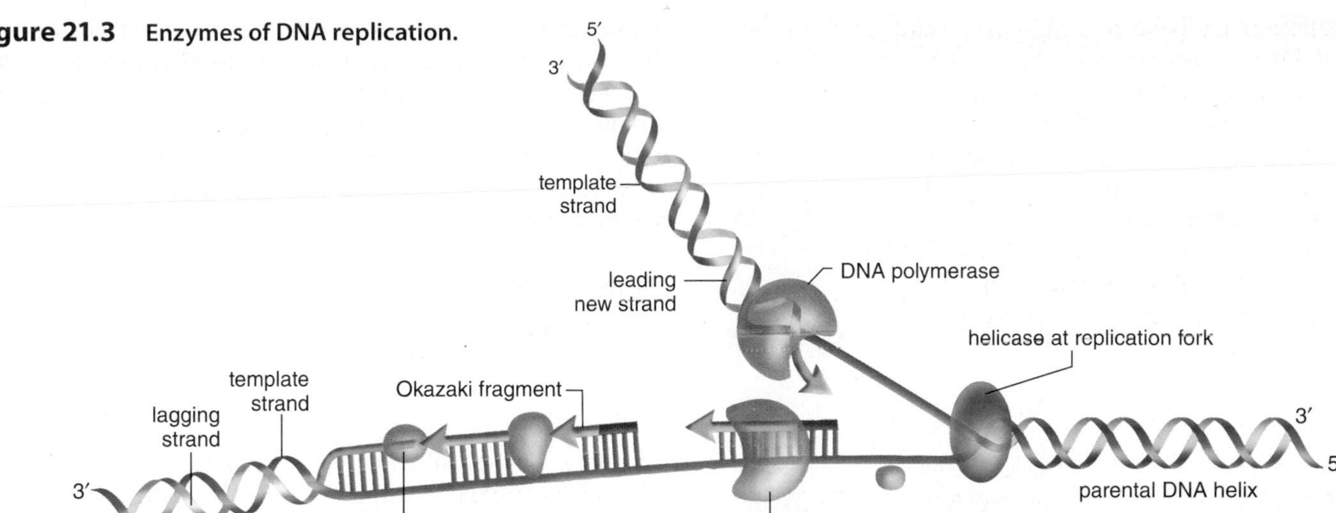

2. New complementary DNA nucleotides (composed of a sugar, phosphate, and one nitrogen-containing base), always present in the nucleus, fit into place by the process of complementary base pairing. These are positioned and joined by the enzyme *DNA polymerase.*
3. Because the strands of DNA are oriented in an antiparallel configuration, and the DNA polymerase may add new nucleotides onto one end of the chain, DNA synthesis occurs in opposite directions. The *leading strand* follows the helicase enzyme, while synthesis on the *lagging strand* results in the formation of short segments of DNA called *Okazaki fragments.*
4. To complete replication, the enzyme ligase seals any breaks in the sugar-phosphate backbone. The DNA returns to its coiled structure.

Animation DNA Replication

5. The two double-helix molecules are identical to each other and to the original DNA molecule.

DNA replication is a relatively fast process—the DNA polymerase is able to add new nucleotides at the rate of around 50 per second. Rarely, a replication error occurs (around 1 in every 100 million nucleotides), making the sequence of the bases in the new strand different from the parental strand. But, if an error does occur, the cell has repair enzymes that usually fix it. A replication error that persists is a **mutation,** a permanent change in the sequence of bases. A mutation is not always a bad thing; a mutation may introduce variability by producing a new allele that alters the phenotype. Such variabilities make you different from your neighbor and humans different from other animals.

MP3 DNA Replication and the Cell Cycle

APPLICATIONS AND MISCONCEPTIONS

How long does it take to copy the DNA in one human cell?

The enzyme DNA polymerase in humans can copy approximately 50 bases per second. If only one DNA polymerase were used to copy human DNA, it would take almost three weeks! However, multiple DNA polymerases copy the human genome by starting at many different places. The entire 3 billion base pairs can be copied in 8 hours in a rapidly dividing cell.

The Structure and Function of RNA

RNA (ribonucleic acid) is made up of nucleotides containing the sugar ribose. This sugar accounts for the scientific name of this polynucleotide. The four nucleotides that make up the RNA molecule have the following bases: adenine (A), uracil (U), cytosine (C), and guanine (G) (Fig. 21.4). One of the

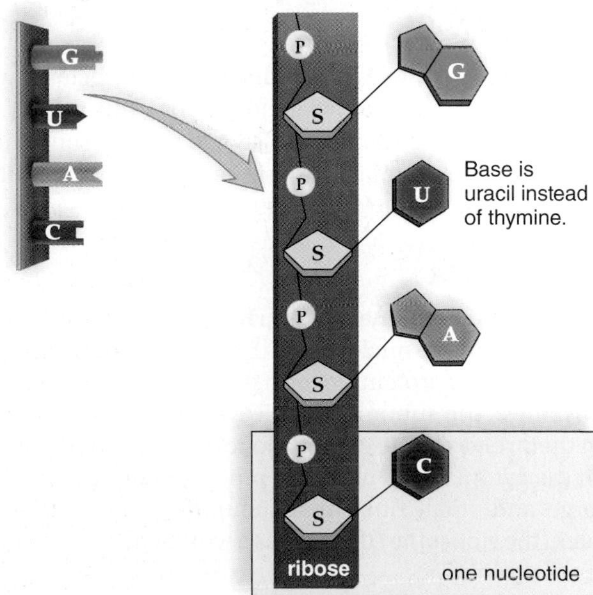

Figure 21.4 The structure of RNA.
Like DNA, RNA is a polymer of nucleotides. In an RNA nucleotide, the sugar ribose is attached to a phosphate molecule and to a base: G, U, A, or C. In RNA, the base uracil replaces thymine as one of the pyrimidine bases. RNA is single-stranded, whereas DNA is double-stranded.

Table 21.1 DNA–RNA Similarities and Differences

DNA–RNA Similarities	
Nucleic acids	
Composed of nucleotides	
Sugar–phosphate backbone	
Four different types of bases	
DNA–RNA Differences	
DNA	**RNA**
Found in nucleus and mitochondria	Found in nucleus and cytoplasm
The genetic material	Helper to DNA
Sugar is deoxyribose	Sugar is ribose
Bases are A, T, C, G	Bases are A, U, C, G
Double-stranded	Single-stranded
Is transcribed (to give RNA)	Can be translated (to give proteins)

differences between RNA and DNA is the replacement of thymine with uracil. As with DNA, complementary base pairing may occur in RNA; cytosine pairs with guanine and adenine pairs with uracil. However, unlike the double-helix structure of DNA, most RNA is single-stranded (Fig. 21.4). Table 21.1 lists the similarities and differences between DNA and RNA.

RNA is divided into coding and noncoding RNAs. The coding RNA is messenger RNA (mRNA), which is translated into protein. Noncoding RNAs are divided into ribosomal RNA (rRNA), transfer RNA (tRNA), and the small RNAs. The small RNAs are involved in the expression of the genes that code for mRNA and rRNA.

Messenger RNA

Messenger RNA (mRNA) is produced in the nucleus, where DNA serves as a template for its formation. This type of RNA carries genetic information from DNA to the ribosomes in the cytoplasm, where protein synthesis occurs. Messenger RNA is a linear molecule, as shown in Figure 21.4.

Ribosomal RNA

In eukaryotes, **ribosomal RNA (rRNA)** is produced using a DNA template in the nucleolus of the nucleus. Ribosomal RNA joins with specific proteins to form the large and small subunits of ribosomes. The subunits then leave the nucleus and either attach themselves to the endoplasmic reticulum or remain free within the cytoplasm. During the process of protein synthesis, the large and small ribosomal subunits combine to form a complex (the ribosome) that acts as a workbench for the manufacture of proteins.

Transfer RNA

Transfer RNA (tRNA) is produced in the nucleus, and a portion of DNA also serves as a template for its production. Appropriate to its name, tRNA transfers amino acids to the ribosomes. At the ribosomes, the amino acids are bonded together in the correct order to form a protein. There are 20 different types of amino acids used to make proteins. Each type of tRNA carries only one type of amino acid; therefore, at least 20 different tRNA molecules must be functioning in the cell to properly make a protein.

Small RNAs

Small RNAs are divided into several classes. Small nuclear RNAs (snRNAs) are involved in splicing the mRNA (see pages 490–491) before it is exported from the nucleus to the cytoplasm for translation. Small nucleolar RNAs (snoRNAs) modify the ribosomal RNAs within the nucleolus of the cell. MicroRNAs (miRNAs) attach to mRNAs in the cytoplasm. Messenger RNAs are thus prevented from being translated unnecessarily. Small interfering RNAs (siRNAs) also bind to mRNAs. Attachment of an siRNA prepares the mRNA for degradation.

Video Halting Hepatitis

Video Tiny Genes Big Role

Video Drug Discovery

APPLICATIONS AND MISCONCEPTIONS

What is a microRNA and how is it used?

A microRNA (miRNA) is a small, noncoding gene that plays a role in developmental biology specifically by acting to regulate the events of gene expression. Historically, miRNAs were commonly used in early detection of various forms of cancer. Researchers currently are using them to determine evolutionary relationships between organisms. They have discovered that once an miRNA is fixed in a genome, it is rarely lost, so organisms with similar miRNA sequences are closely related. This was discovered by studying the miRNA sequences in the annelids (segmented worms). Earthworms, leeches, and bristle worms from all over the globe had retained similar sequences despite differences in speciation over millions of years.

CHECK YOUR PROGRESS 21.1

1. Compare and contrast the structure and function of DNA and RNA.
2. Describe the function of the different types of RNA.
3. Summarize the process of DNA replication.

CONNECTING THE CONCEPTS

For more information on DNA and RNA, refer to the following discussions:

Section 2.7 describes the role of DNA and RNA as organic molecules.

Section 18.6 examines the role of chromosomes in inheritance, as well as the consequences of changes in chromosome number and structure.

Section 22.3 explores how mitochondrial DNA is being used to study human evolution.

21.2 Gene Expression

LEARNING OUTCOMES

Upon completion of this section, you should be able to

1. Summarize how the information contained within the DNA is expressed as a protein.
2. Interpret a sequence of DNA nucleotides.

As we shall see, DNA provides the cell with a blueprint for synthesizing proteins. In simplest terms, DNA acts as a template for making RNA, which in turn acts as a template for the manufacture of proteins. In gene expression, also known as *protein synthesis*, the process of transcription makes an RNA copy of DNA and the process of translation makes protein from the RNA.

MP3 Protein Synthesis

Before discussing the mechanics of gene expression, let's review the structure of proteins.

Structure and Function of Proteins

Proteins are composed of subunits called amino acids (see section 2.6). Twenty different amino acids are commonly found in proteins. Proteins differ because the number and order of their amino acids differ. The sequence of amino acids in a protein leads to its particular shape. (See Fig. 2.23 for more detail on the levels of protein structure.) Proteins have many different functions in the body as they determine the structure and function of various cells in the body. Proteins are used as structural and regulatory components of cells. They are used as enzymes to catalyze chemical reactions, neurotransmitters to aid in the function of the nervous system, antibodies for the immune system, and hormones to change activities of certain cells. Proteins have this great diversity of functions due to the arrangement of these 20 different amino acids in their individual structures.

Gene Expression: An Overview

The first step in gene expression is called *transcription* and the second step is called *translation* (Fig. 21.5). During **transcription,** a strand of mRNA forms that is complementary to the template strand of the DNA molecule. The mRNA molecule that forms (mRNA) is a *transcript* of a gene. *Transcription* means "to make a faithful copy." In this case, the sequence of nucleotides in DNA is copied to a sequence of nucleotides in mRNA.

APPLICATIONS AND MISCONCEPTIONS

Are all mutations bad?

A mutation simply means a change from the original structure of DNA during replication or division. In some cases, these changes can result in a faulty or detrimental product that can negatively affect the organism. In other cases, all the diversity in our traits can be considered mutations. Without the ability to change, we would not have been able to adapt to changing environments; thus, evolution would not have been possible.

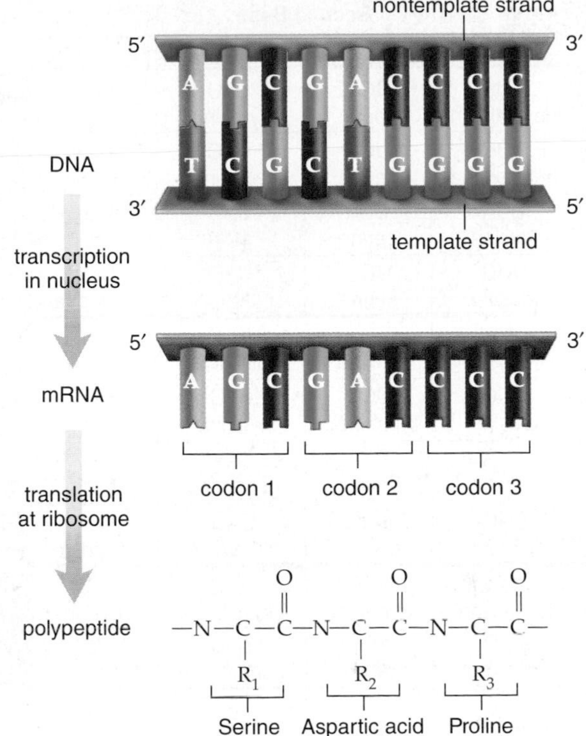

Figure 21.5 Summary of gene expression.
One strand of DNA acts as a template for mRNA synthesis, and the sequence of bases in mRNA determines the sequence of amino acids in a polypeptide.

Protein synthesis requires the process of **translation.** *Translation* means "to put information into a different language." In this case, a sequence of nucleotides (the mRNA) is translated into the sequence of amino acids (the protein). This is possible only if the bases in DNA and mRNA code for amino acids. This code is called the genetic code.

The Genetic Code

The genetic code (Fig. 21.6) corresponds to a three-base sequence in the mRNA molecule called a **codon.** Each codon represents a specific amino acid. The reason that each codon contains three bases instead of one or two is a matter of mathematics. There are 20 different amino acids that are used to build proteins. If a codon consisted of just a single base, then 16 amino acids could not be coded for. If each codon contained two bases, only 16 amino acids would be covered. However, the use of three bases allows for 64 possible codons, more than enough to code for the 20 amino acids. A closer examination of the genetic code (Fig. 21.6.) shows that 61 codons correspond to a particular amino acid. The remaining three are stop codons, which signal polypeptide termination. One of the codons (AUG) stands for the amino acid methionine, the amino acid that is used to signal the beginning of the polypeptide. Notice as well that most amino acids have more than one codon. For example, leucine, serine, and arginine each have six different codons. This redundancy in the genetic

First Base	Second Base: U	Second Base: C	Second Base: A	Second Base: G	Third Base
U	UUU phenylalanine	UCU serine	UAU tyrosine	UGU cysteine	U
U	UUC phenylalanine	UCC serine	UAC tyrosine	UGC cysteine	C
U	UUA leucine	UCA serine	UAA ***stop***	UGA ***stop***	A
U	UUG leucine	UCG serine	UAG ***stop***	UGG tryptophan	G
C	CUU leucine	CCU proline	CAU histidine	CGU arginine	U
C	CUC leucine	CCC proline	CAC histidine	CGC arginine	C
C	CUA leucine	CCA proline	CAA glutamine	CGA arginine	A
C	CUG leucine	CCG proline	CAG glutamine	CGG arginine	G
A	AUU isoleucine	ACU threonine	AAU asparagine	AGU serine	U
A	AUC isoleucine	ACC threonine	AAC asparagine	AGC serine	C
A	AUA isoleucine	ACA threonine	AAA lysine	AGA arginine	A
A	**AUG *(start)* methionine**	ACG threonine	AAG lysine	AGG arginine	G
G	GUU valine	GCU alanine	GAU aspartic acid	GGU glycine	U
G	GUC valine	GCC alanine	GAC aspartic acid	GGC glycine	C
G	GUA valine	GCA alanine	GAA glutamic acid	GGA glycine	A
G	GUG valine	GCG alanine	GAG glutamic acid	GGG glycine	G

Figure 21.6 The genetic code.
In this chart, each of the codons (white rectangles) is composed of three letters representing the first base, second base, and third base. For example, find the rectangle where C for the first base and A for the second base intersect. You will see that U, C, A, or G can be the third base. CAU and CAC are codons for histidine; CAA and CAG are codons for glutamine.

code offers some protection against possibly harmful mutations that change the sequence of the bases.

The genetic code is almost universal in living organisms, that is, it is the same for most living organisms. This suggests that the code dates back to the very first organisms on Earth and that all life is related. In other words, all organisms share an evolutionary heritage.

Transcription

During transcription, a segment of the DNA serves as a template for the production of an RNA molecule. Although all classes of RNA are formed by transcription, we will focus on transcription to form mRNA.

Forming mRNA

Transcription begins in the nucleus when the enzyme **RNA polymerase** opens up the DNA helix so that complementary base pairing can occur. Recall that RNA contains uracil instead of thymine, so base pairing between DNA and the mRNA strand will be A–U and C–G (Fig. 21.7). Then, RNA polymerase joins the RNA nucleotides, and an mRNA molecule with a sequence of bases complementary to the DNA segment results.

3D Animation
Molecular Biology of the Gene: Transcription

Processing mRNA

Before the transcribed mRNA leaves the nucleus for translation in the cytoplasm, it undergoes a series of processing steps.

The newly synthesized *primary* mRNA molecule becomes a *mature* mRNA molecule after processing (Fig. 21.8). Most genes

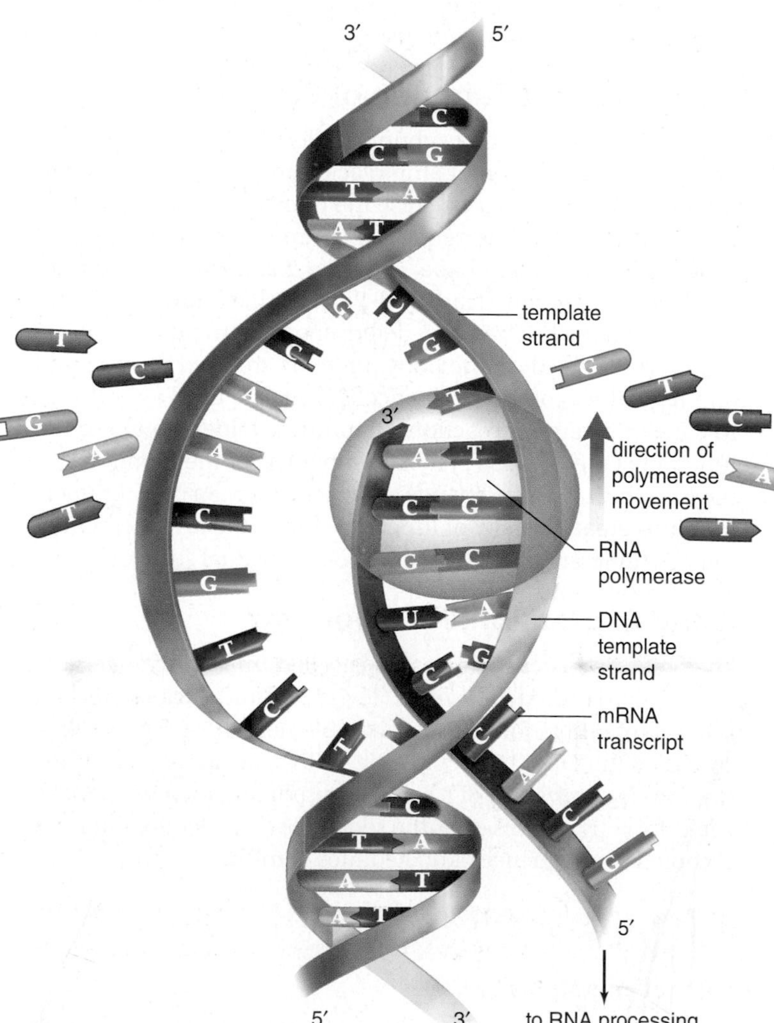

Figure 21.7
Transcription of DNA into mRNA.
During transcription, complementary RNA is made from a DNA template. A portion of DNA unwinds and unzips at the point of attachment of RNA polymerase. A strand of mRNA is produced when complementary bases join in the order dictated by the sequence of bases in template DNA.

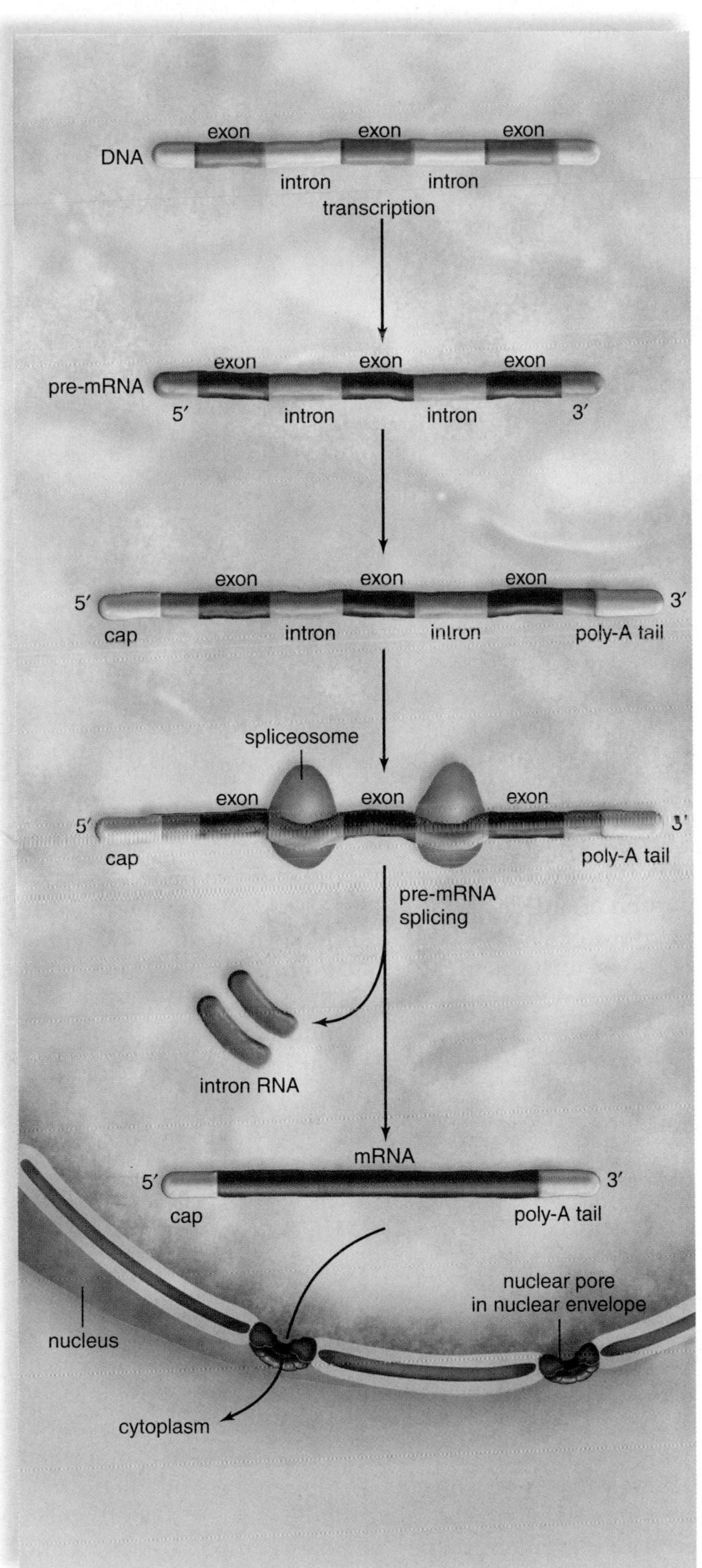

Figure 21.8 mRNA processing.
During processing, a cap and tail are added to mRNA, and the introns are removed so that only exons remain. Then the mRNA molecule is ready to leave the nucleus.

in humans are interrupted by segments of DNA that are not part of the gene. These portions are called *introns* because they are intragene segments and do not code for a functional protein. The other portions of the gene are called *exons* because they are ultimately expressed. Only exons result in a protein product.

Primary mRNA contains bases complementary to both exons and introns, but during processing, (1) one end of the mRNA is capped by the addition of an altered guanine nucleotide. The other end is given a tail, by the addition of multiple adenosine nucleotides. (2) The introns are removed, and the exons are joined to form a mature mRNA molecule consisting of continuous exons. This *splicing* of mRNA is done by a complex called a *spliceosome,* which is composed of both RNA and protein. Surprisingly, the RNA portion, not the protein, is functioning as the enzyme, and so it is called a *ribozyme.*

Animation
How Spliceosomes Process RNA

Ordinarily, processing brings together all the exons of a gene. In some instances, cells use only certain exons rather than all of them to form a mature RNA transcript. Alternate mRNA splicing is believed to account for the ability of a single gene to result in different proteins in a cell and the different complexities between all living organisms despite the universal genetic code. Increasingly, small RNA molecules have been found that regulate not only mRNA processing but also transcription and translation. DNA codes for proteins, but RNA orchestrates the outcome.

3D Animation
Molecular Biology of the Gene: mRNA Modifications

Translation

During translation, transfer RNA (tRNA) molecules bring amino acids to the ribosomes (Fig. 21.9), where polypeptide synthesis occurs. The ribosome consists of a small and a large subunit that will join together during translation and bind to the mRNA strand. This creates a translation complex (Fig. 21.9*a*). The ribosome contains special binding sites called the A and P sites where individual tRNA molecules can bind with the mRNA strand. A tRNA molecule has an almost cloverleaf shape with an area for binding onto an amino acid and a region called an **anticodon** (Fig. 21.9*c*). The anticodon is a three-base sequence of RNA that will complementary base pair with the codons of mRNA. This complementary base pairing between an anticodon and a codon is how the tRNA brings the correct amino acid into the correct order instructed by the mRNA strand. Each amino acid, coded by specific codons, has a tRNA molecule with a specific anticodon that carries it to the translation complex. The tRNA molecules attach to the translation complex at the A site where each anticodon pairs with the complementary codon (Fig. 21.9*b*).

The order in which tRNA molecules link to the ribosome is directed by the sequence of the mRNA codons. In this way, the order of codons in mRNA brings about a particular order of amino acids in a protein. The tRNA attached to the growing polypeptide moves from the A site to the P site of a ribosome, as shown in Figure 21.9*b*.

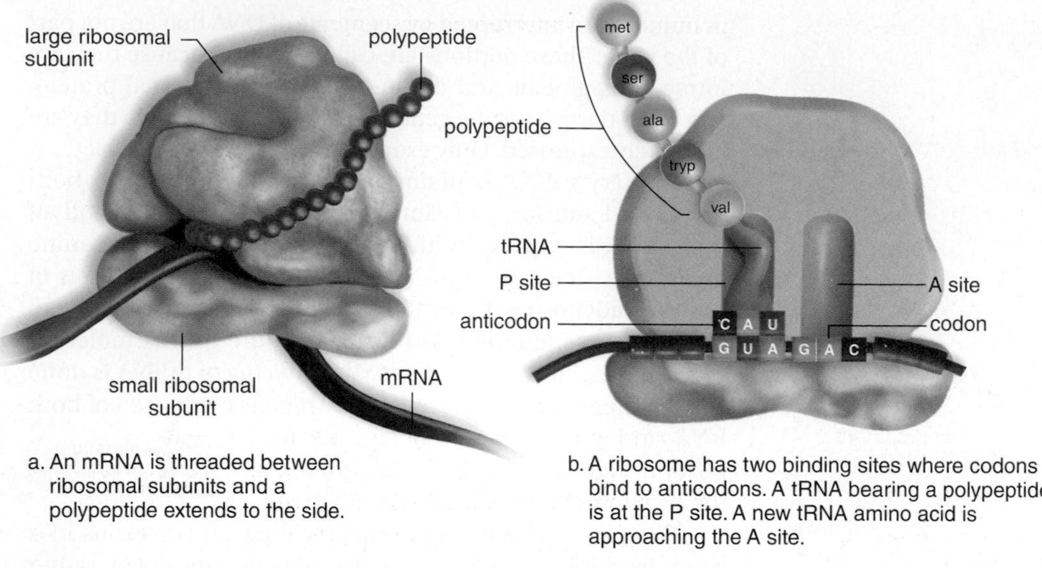

a. An mRNA is threaded between ribosomal subunits and a polypeptide extends to the side.

b. A ribosome has two binding sites where codons bind to anticodons. A tRNA bearing a polypeptide is at the P site. A new tRNA amino acid is approaching the A site.

c. A tRNA amino acid is coming to the ribosome. Upon arrival, its anticodon, CUG, will bind to its codon, GAC.

Figure 21.9 The roles of all three forms of RNA in translation.
Protein synthesis occurs at a ribosome. **a.** Side view of a ribosome showing mRNA and a growing polypeptide. **b.** The large ribosomal subunit contains two binding sites for tRNAs. **c.** tRNA structure and function.

If the codon sequence in a portion of the mRNA is ACC, GUA, and AAA, what will be the sequence of amino acids in a portion of the polypeptide? The genetic code chart in Figure 21.6 allows us to determine this:

Codon	Anticodon	Amino Acid
ACC	UGG	Threonine
GUA	CAU	Valine
AAA	UUU	Lysine

Polypeptide synthesis requires three steps: initiation, elongation, and termination (Fig. 21.10):

1. During *initiation,* mRNA binds to the smaller of the two ribosomal subunits. Then the larger ribosomal subunit associates with the smaller one, forming the translation complex.
2. During *elongation,* the polypeptide lengthens, one amino acid at a time, about five amino acids per second. An incoming tRNA arrives at the A site and then receives the growing peptide chain from the outgoing tRNA. The ribosome moves laterally down the mRNA strand one codon at a time so that again the P site is filled by a tRNA–peptide complex. The A site is now available to receive another incoming tRNA as the complex has moved down one codon. In this manner, the peptide grows, and the linear structure of a polypeptide is made. The particular shape of a polypeptide (see section 2.6) begins to form as the linear structure is established.
3. Then *termination* of synthesis occurs when one of the three stop codons is reached by the A site. The ribosome dissociates into its two subunits and falls off the mRNA molecule. The individual portions of the translation complex can then re-form at the beginning of the mRNA strand to repeat this process and make another polypeptide.

Additionally, many ribosomes are at work forming the same polypeptide at the same time. As soon as the initial portion of mRNA has been translated by one ribosome and the ribosome has begun to move down the mRNA, another ribosome attaches to the mRNA to begin translation. Therefore, several ribosomes, collectively called a *polyribosome,* can move along one mRNA at a time. Several polypeptides of the same type can be synthesized using one mRNA molecule (Fig. 21.11). This, in addition to the recycling of the translation complex, gives every cell the ability to make sufficient amounts of proteins.

3D Animation
Molecular Biology of the Gene: Translation

Review of Gene Expression

DNA in the nucleus contains genes that are transcribed into RNAs. Some of these RNAs are mRNAs that will then be translated into proteins. During transcription, a segment of a DNA strand (a gene) serves as a template for the formation of RNA. The bases in RNA are complementary to those in DNA. In mRNA, every three bases is a *codon* for a certain amino acid (Fig. 21.12 and Table 21.2). Messenger RNA is processed before it leaves the nucleus. During processing, the introns are removed and the ends are modified. The mRNA carries a sequence of codons to the *ribosomes,* composed of rRNA and proteins. A tRNA bonded to a particular amino acid has an *anticodon* that pairs with a codon in mRNA. During translation, tRNAs and their attached amino acids arrive at the ribosomes. The linear sequence of codons of mRNA determines the order in which amino acids become incorporated into a protein.

met
polypeptide bond
ser
thr
ala
tryp
UGG
val
asp
CAU
CUG
UA
GAC ACC

polypeptide
met
ser
amino acid
ala
tryp
val
asp
CAU CUG
GUA GAC

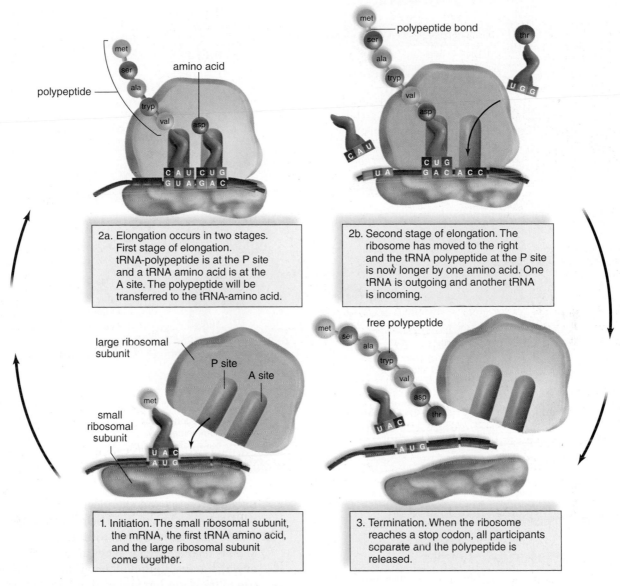

2a. Elongation occurs in two stages. First stage of elongation. tRNA-polypeptide is at the P site and a tRNA amino acid is at the A site. The polypeptide will be transferred to the tRNA-amino acid.

2b. Second stage of elongation. The ribosome has moved to the right and the tRNA polypeptide at the P site is now longer by one amino acid. One tRNA is outgoing and another tRNA is incoming.

large ribosomal subunit
P site
A site
met
small ribosomal subunit
UAC
AUG

free polypeptide
met
ser
ala
tryp
val
asp
thr
UAC
AUG

1. Initiation. The small ribosomal subunit, the mRNA, the first tRNA amino acid, and the large ribosomal subunit come together.

3. Termination. When the ribosome reaches a stop codon, all participants separate and the polypeptide is released.

Figure 21.10 **Formation of the polypeptide during translation.**
Polypeptide synthesis takes place at a ribosome and has three steps: (**1**) initiation, (**2**) elongation, and (**3**) termination.

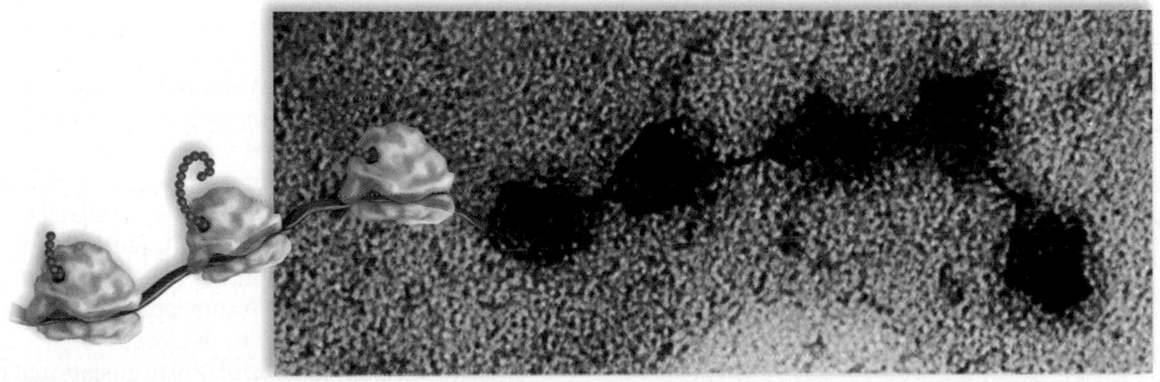

Figure 21.11 **Structure and function of a polyribosome.**
Several ribosomes, collectively called a polyribosome, move along an mRNA molecule at one time. They function independently of one another; therefore, several polypeptides can be made simultaneously.

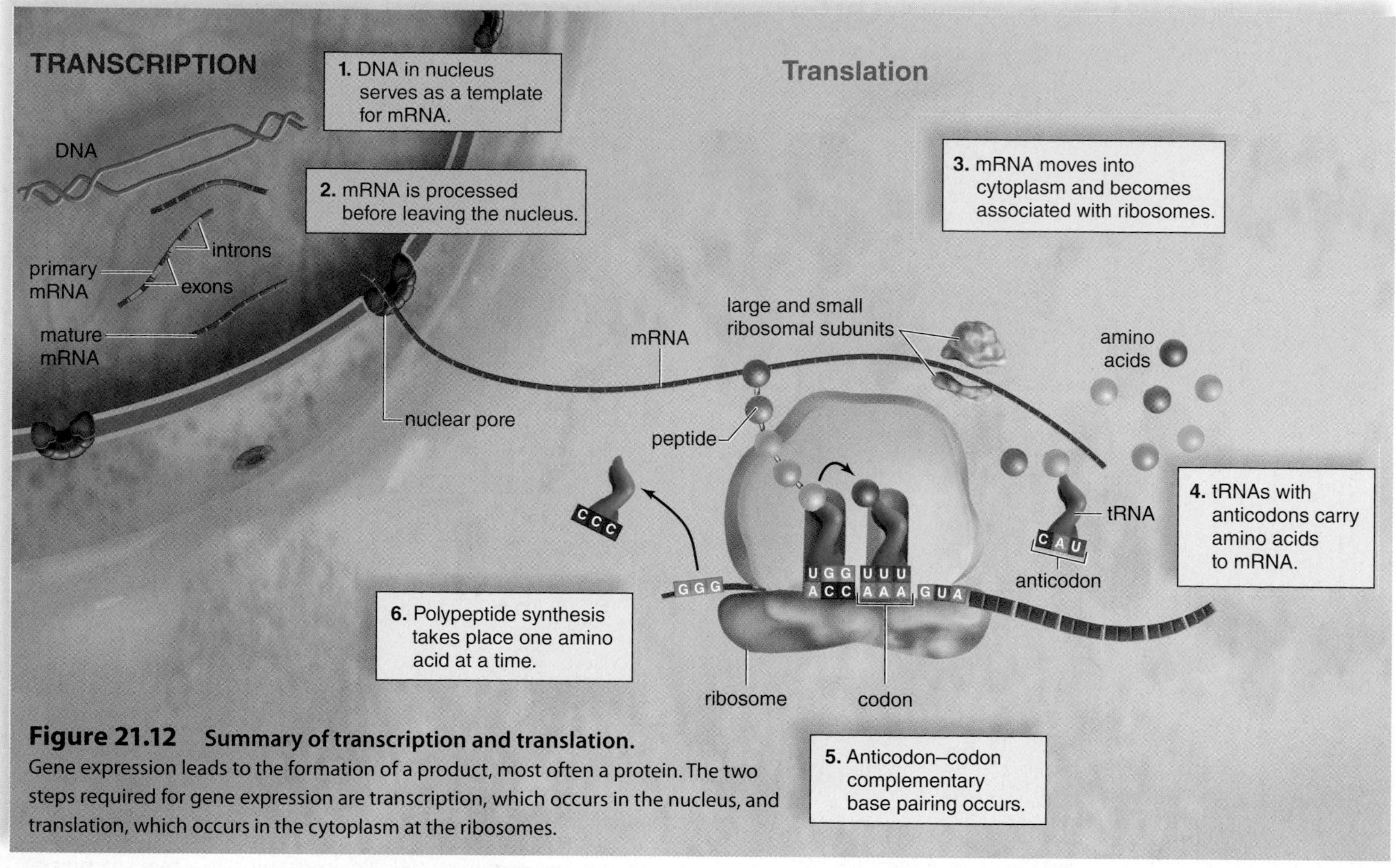

Figure 21.12 Summary of transcription and translation. Gene expression leads to the formation of a product, most often a protein. The two steps required for gene expression are transcription, which occurs in the nucleus, and translation, which occurs in the cytoplasm at the ribosomes.

Table 21.2 Participants in Gene Expression

Name of Molecule	Special Significance	Definition
DNA	Genetic information	Sequence of DNA bases
mRNA	Codons	Sequence of three RNA bases complementary to DNA
tRNA	Anticodon	Sequence of three RNA bases complementary to codon
rRNA	Ribosome	Site of protein synthesis
Amino acid	Building block for protein	Transported to ribosome by tRNA
Protein	Enzyme, structural protein, or secretory product	Amino acids joined in a predetermined order

The Regulation of Gene Expression

All cells receive a copy of all genes. However, cells differ as to which genes are actively expressed. Muscle cells, for example, have a different set of genes that are turned on in the nucleus and different proteins that are active in the cytoplasm than do nerve cells. A variety of mechanisms regulate gene expression, from pretranscriptional access to the DNA to posttranslational control of protein activity, in our cells (Fig. 21.13):

Animation Control of Gene Expression in Eukaryotes

1. *Pretranscriptional control:* In the nucleus, the DNA must be available to the enzymes necessary for transcription. The chromosome in the region must decondense, or uncoil. Proteins and chemical modifications that protect the DNA must be removed before transcription can begin.
2. *Transcriptional control:* In the nucleus, a number of mechanisms regulate which genes are transcribed and the rate at which transcription of genes occurs. These include the use of transcription factors that initiate transcription, the first step in gene expression.
3. *Posttranscriptional control:* Posttranscriptional control occurs in the nucleus after DNA is transcribed and mRNA is formed. How mRNA is processed before it leaves the nucleus and also how fast mature mRNA leaves the nucleus can affect the amount of gene expression.
4. *Translational control:* Translational control occurs in the cytoplasm after mRNA leaves the nucleus but before there is a protein product. Some mRNAs require additional changes before translation, and this affects not only life expectancy of mRNA molecules in the cytoplasm, but also their ability to bind to ribosomes. Two of the small RNA classes, microRNAs and small interfering RNAs (siRNA), are involved at this level of gene expression. MicroRNAs bind to mRNAs, inhibiting translation. siRNAs prevent translation by marking the mRNA for destruction by nucleases.
5. *Posttranslational control:* Posttranslational control, which also occurs in the cytoplasm, occurs after protein synthesis. The polypeptide product may have to undergo additional changes before it is biologically functional. Also, a functional enzyme is subject to feedback control—the binding of an enzyme's product can change its shape so that it is no longer able to carry out its reaction.

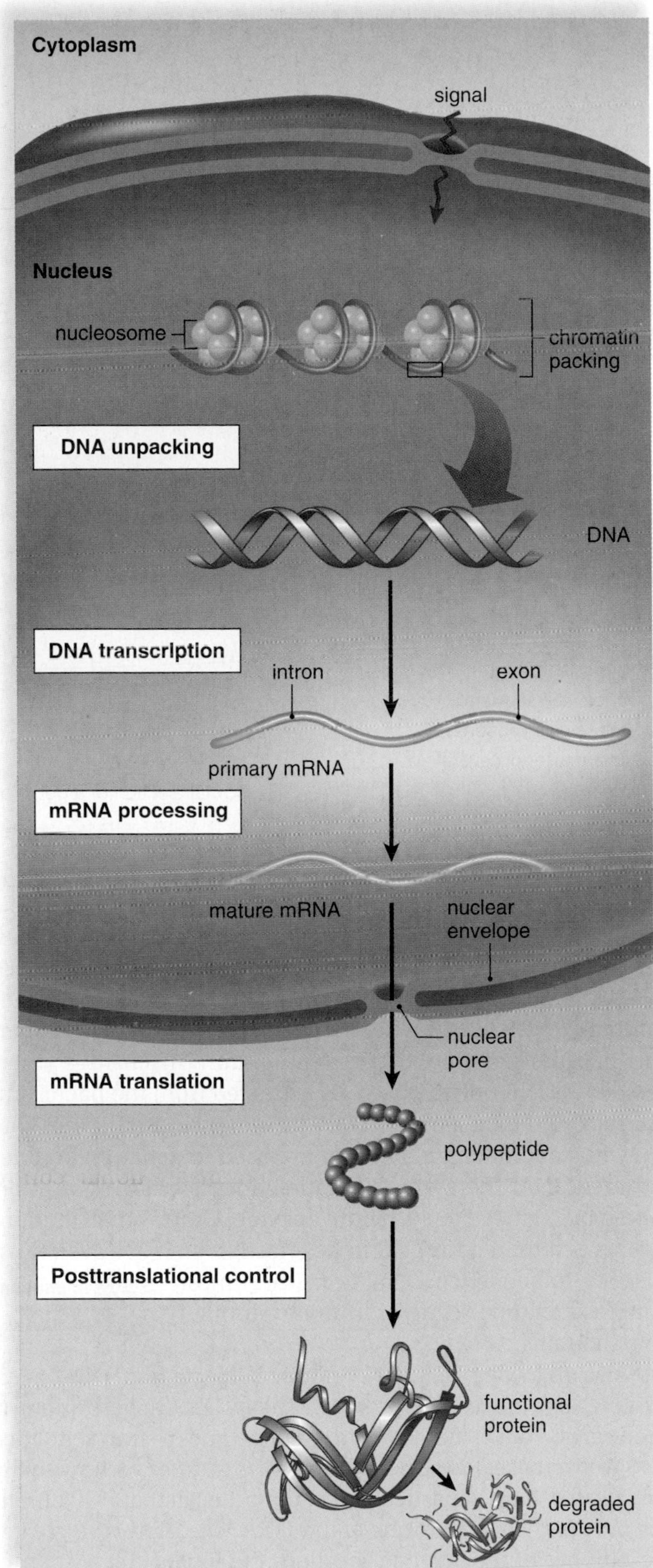

Figure 21.13 Control of gene expression in eukaryotic cells. Gene expression is controlled at various levels in eukaryotic cells. There are three mechanisms pertaining to the nucleus and two mechanisms that pertain to the cytoplasm. An external signal (red) may also alter gene expression.

Transcription Factors

In human cells, **transcription factors** are DNA-binding proteins. At one time, scientists thought that transcription factors simply acted as on/off switches for genes. We now know that this role of transcription factors is much more complex. A better analogy is that the level of gene expression is controlled like a variable light switch, with a variety of transcription factors fine-tuning the transcription process. Every cell contains many different types of transcription factors. After the right combination of transcription factors and other associated regulatory proteins bind to DNA, an RNA polymerase attaches to the DNA and begins the process of transcription.

Animation Transcription Factors

As cells mature, they differentiate and become specialized. Specialization is determined by which genes are active and, therefore, perhaps, by which transcription factors are active in that cell. Signals received from inside and outside the cell could turn on or off genes that code for certain transcription factors. For example, in an embryo, the hand is flattened and paddle-shaped. The five fingers are joined by webs of skin and connective tissue. The gene for apoptosis (programmed cell death) is turned on in the embryo's hand. Separated fingers are formed when the webs of tissue die off.

APPLICATIONS AND MISCONCEPTIONS

Why is the genetic code being universal important?

With a few exceptions, the genetic code is universal, meaning that the amino acid a particular codon codes for in a human is the same amino acid it will code for in a monkey, a fern, or a flea. A notable exception is the codon UGA, which in some protists and in the mitochondria of the cell, codes for the amino acid tryptophan, instead of a stop codon. This universality among living organisms suggests a common evolutionary heritage. The genetic code supports the underlying concept in biology that all living organisms are related.

CHECK YOUR PROGRESS 21.2

1. Describe the processes of transcription and translation.
2. Discuss the genetic code and explain how it works with different types of RNA to make a protein.
3. Explain the significance of the genetic code being universal.
4. Identify the various means of gene regulation and tell why they are important to homeostasis.

CONNECTING THE CONCEPTS

For more information on DNA, RNA, and proteins, refer to the following discussions:

Section 2.6 examines the structure of proteins and their role in the body.

Section 2.7 describes the role of DNA and RNA as organic molecules.

21.3 DNA Technology

LEARNING OUTCOMES

Upon completion of this section, you should be able to

1. Recognize the importance of DNA sequencing to the study of biology.
2. State the purpose of the polymerase chain reaction and DNA cloning.
3. Summarize some of the products that have been produced using biotechnology.

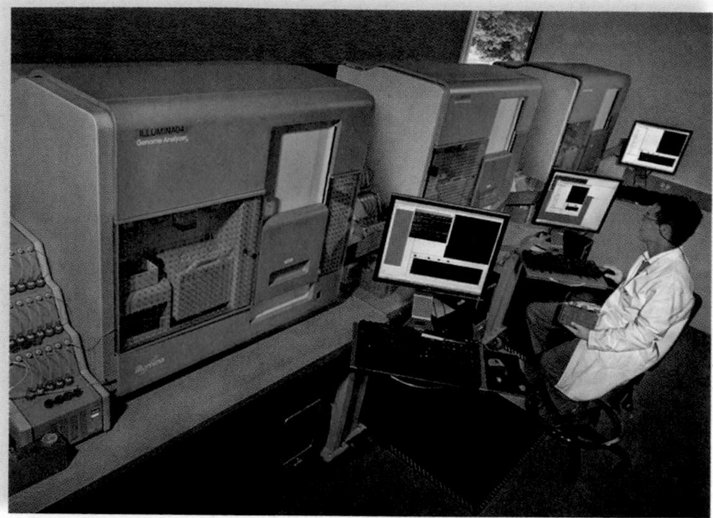

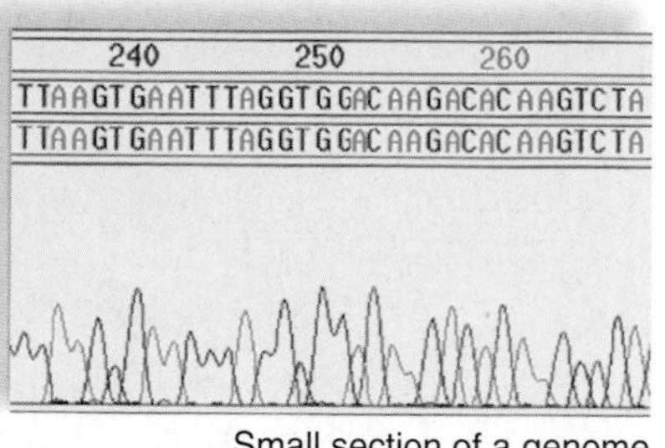

Figure 21.14 Automated DNA sequencer and an electropherogram. This typical automated DNA sequencer can sequence 1,000 base-pair sections of DNA in a matter of several hours.

DNA Sequencing

DNA sequencing is a procedure that determines the order of nucleotides in a segment of DNA, often within a specific gene. DNA sequencing allows researchers to identify specific alleles that are associated with a disease, and thus facilitate the development of medicines or treatments. This information also serves as the foundation for the study of forensic biology and even contributes to our understanding of our evolutionary history (see Chapter 22).

Animation Sanger Sequencing

When DNA technology was in its inception back in the early 1970s, this technique was performed manually using dye-terminator substances or radioactive tracer elements attached to each of the four nucleotides during DNA replication, with results being deciphered from their pattern on a gel plate. Modern day sequencing normally involves dyes attached to the nucleotides and use of a laser to detect the different dyes by an automated sequencing machine, which shows the order of nucleotides on a grid called an electropherogram (Fig. 21.14). To begin sequencing a segment of DNA, many copies of the segment are made, or replicated, using a procedure called the polymerase chain reaction.

Analyzing DNA

The **polymerase chain reaction** (**PCR**) can create billions of copies of a segment of DNA in a test tube in a matter of hours. PCR is very specific—it *amplifies* (makes copies of) a targeted DNA sequence, usually a few hundred bases in length. PCR requires the use of DNA polymerase, the enzyme that carries out DNA replication, and a supply of nucleotides for the new DNA strands. PCR involves three basic steps (Fig. 21.15) that occur repeatedly, usually for about 35 to 40 cycles: (1) a denaturation step at 95°C, where DNA is heated to become single-stranded; (2) an annealing step at a temperature usually between 50° and 60°C, where an oligonucleotide primer hybridizes to each of the single DNA strands; and (3) an extension step at 72°C, where an engineered DNA polymerase adds complementary bases to each of the single DNA strands, creating double-stranded DNA.

PCR is a chain reaction because the targeted DNA is repeatedly replicated, much in the same way natural DNA replication occurs, as long as the process continues. Figure 21.15 uses color to distinguish the old strand from the new DNA strand. Notice that the amount of DNA doubles with each replication cycle. Thus, assuming you start with only one copy of DNA, after one cycle, you will have two copies, after two cycles four copies, and so on. PCR has been in use since its development in 1985 by Kary Banks Mullis, and now almost every laboratory has automated PCR machines to carry out the procedure. Automation became possible after a temperature-insensitive (thermostable) DNA polymerase was extracted from the bacterium *Thermus aquaticus,* which lives in hot springs. The enzyme can withstand the high temperature used to denature double-stranded DNA. Therefore, replication does not have to be interrupted by the need to add more enzyme. Only a small amount of DNA is required for PCR to be effective, so it has even been possible to sequence DNA from the tiniest sample at a crime scene or from mummified human brains.

Animation Polymerase Chain Reaction

Also, following PCR, DNA can be subjected to DNA fingerprinting, also called DNA profiling. Today, DNA fingerprinting is often carried out by detecting how many times a short sequence (two to five bases) is repeated. Organisms differ by how many repeats they have at particular locations. Recall that PCR amplifies only particular portions of the DNA. Therefore, the greater the number of repeats at a location, the longer the section of DNA amplified by PCR. During a process called gel electrophoresis, these DNA fragments can be separated according to their size; the result is a pattern of distinctive bands. If two DNA patterns match, there is a high probability that the DNA came from the same source. It is customary to test for the number of repeats at several locations to further define the source.

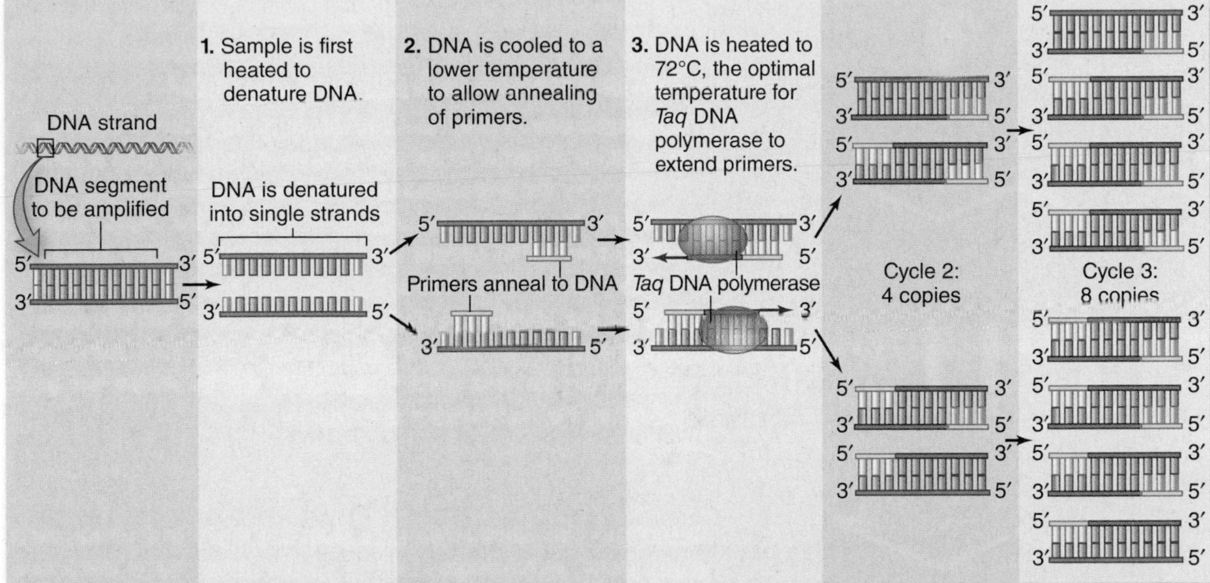

Figure 21.15 The polymerase chain reaction. The PCR process produces multiple copies of a single segment of DNA.

BIOLOGY MATTERS  **Bioethical**

DNA Fingerprinting and the Criminal Justice System

Traditional fingerprinting has been used for years to identify criminals and to exonerate those wrongly accused of crimes. The opportunity now arises to use DNA fingerprinting in the same way. DNA fingerprinting requires only a small DNA sample. This sample can come from blood left at the scene of the crime, semen from a rape case, or even a single hair root!

Advocates of DNA fingerprinting claim that identification is "beyond a reasonable doubt." But how can investigators be certain? Much of the forensic DNA fingerprinting done today uses short tandem repeats (STRs). These are stretches of noncoding DNA in our genome that contain repeated DNA sequences. Most commonly, these repeats are four bases in length; for example, CATG. You may have 11 copies of this repeat on a particular chromosome inherited from your father and only three copies on the homologous chromosome from your mother. When analyzed by electrophoresis, greater numbers of repeats correspond to increasing lengths on the DNA. Different people have unique repeat patterns, so these STRs can be used to discriminate between individuals. A particular STR pattern on a single chromosome may be shared by a number of people. However, by studying multiple STR sites, a statistically unique pattern can be developed for everyone—unless you share your DNA with an identical twin! In the United States, the FBI's Combined DNA Index System (CODIS) uses 13 STR sites (plus a marker for sex) to identify individuals.

Opponents of this technology, however, point out that it is not without its problems. Police or laboratory negligence can invalidate the evidence. For example, during the O. J. Simpson trial, the defense claimed that the DNA evidence was inadmissible because it could not be proven that the police had not "planted" O. J.'s blood at the crime scene. There have also been reported problems with sloppy laboratory procedures and the credibility of forensic experts. In one particular case, Curtis McCarty had been placed on death row three times by the same team of prosecutor and police lab analyst. After 21 years in prison, he was exonerated. The prosecutor has been accused of misconduct, and the police lab analyst was fired for falsifying laboratory data to obtain convictions.

In addition to identifying criminals, DNA fingerprinting can be used to establish paternity and maternity; determine nationality for immigration purposes; and identify victims of a national disaster, such as the terrorist attacks of September 11, 2001, the tsunamis in Indonesia in 2007, and the earthquakes in China and Haiti in 2009 and 2010.

Considering the usefulness of DNA fingerprints, perhaps everyone should be required to contribute blood to create a national DNA fingerprint databank. Some say, however, that this would constitute an unreasonable search, which is unconstitutional.

Questions to Consider

1. Would you be willing to provide your DNA for a national DNA databank? Why or why not? What types of privacy restrictions would you want on your DNA?
2. If not everyone, do you think that convicted felons, at least, should be required to provide DNA for a databank?
3. Should all defendants have access to DNA fingerprinting (at government expense) to prove they didn't commit a crime? Should this include those already convicted of crimes who want to reopen their cases using new DNA evidence?

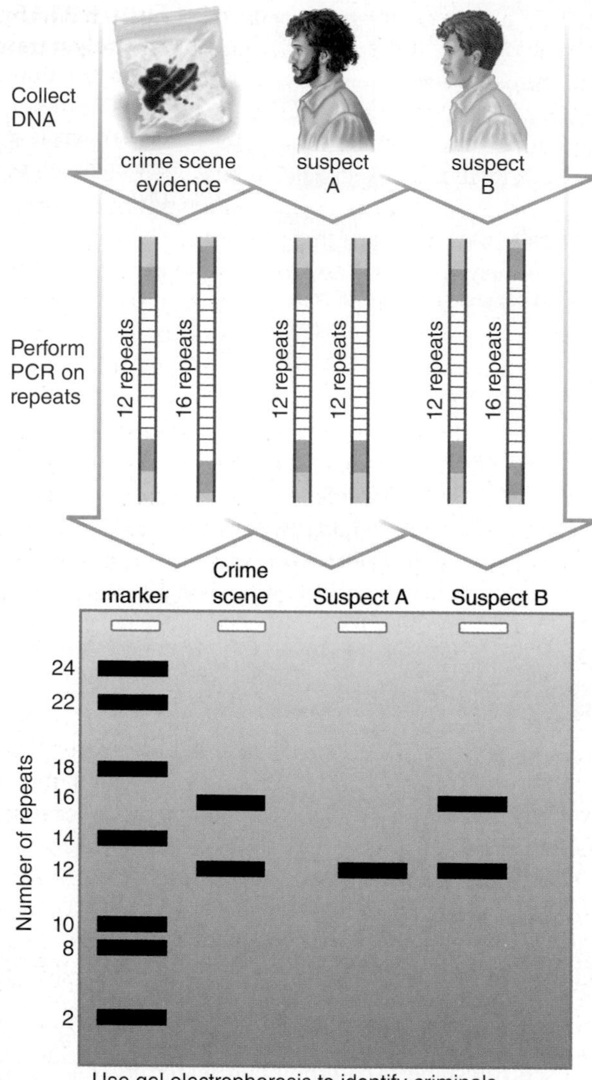

Figure 21.16 **PCR and electrophoresis used for DNA fingerprinting.**
DNA fingerprinting reveals that suspect A could not be the criminal.

DNA fingerprinting has many uses. When the DNA matches that of a virus or mutated gene, it is known that a viral infection, genetic disorder, or cancer is present. Fingerprinting DNA from a single sperm is enough to identify a suspected rapist. DNA fingerprinted from blood or tissues at a crime scene has been successfully used in convicting criminals. Figure 21.16 shows how DNA fingerprinting can be used to identify a criminal. DNA fingerprinting is also used to identify the remains of bodies. DNA fingerprinting through STR profiling was extensively used to identify the victims of the tsunamis in the past few years in Indonesia and Japan.

Applications of PCR and DNA fingerprinting are limited only by our imagination. PCR analysis has been used to identify unknown soldiers and members of the royal Russian family. Paternity suits can be settled. Environmental law enforcement authorities can identify illegally poached ivory and whale meat using these technologies. These applications have shed new light on evolutionary studies by comparing DNA from fossils with that of living organisms.

APPLICATIONS AND MISCONCEPTIONS

How do the home DNA paternity kits work?

Many laboratories now have kits allowing collection of a sample to be done at home. Collection consists of taking a buccal swab of both the child and suspected father. This is a DNA sample from inside the mouth taken by rubbing the inside of the cheeks with a sterilized swab. This sample is then sent to the lab where the DNA is replicated by PCR and DNA fingerprinting occurs. Most labs will look at 12 to 16 different chromosomal markers to identify paternity, and results are ready in about three to five days. Although efficient, private, and quick, these home test results are not approved by the court system as evidence.

Genes Can Be Isolated and Cloned

In biology, **cloning** is the production of genetically identical copies of DNA, cells, or organisms through an asexual means. **Gene cloning** can be done to produce many identical copies of the same gene. **Recombinant DNA (rDNA)**, which contains DNA from two or more different sources, allows genes to be cloned. To create recombinant DNA, a technician needs a **vector,** by which the gene of interest will be introduced into a host cell, such as a bacterium. One common vector is a plasmid. **Plasmids** are small accessory rings of DNA found in bacteria that often hold genes for antibiotic resistance. The ring is not part of the bacterial chromosome and replicates on its own. The steps for gene cloning (Fig. 21.17) follow:

① A **restriction enzyme** is used to cleave human DNA and plasmid DNA. Hundreds of restriction enzymes occur naturally in bacteria, where they cut up any viral DNA that enters the cell. They are called *restriction* enzymes because they restrict the growth of viruses. They also act as molecular scissors to cleave any piece of DNA at a specific site. For example, the restriction enzyme called *Eco*RI always cuts double-stranded DNA at this sequence of bases and in this manner:

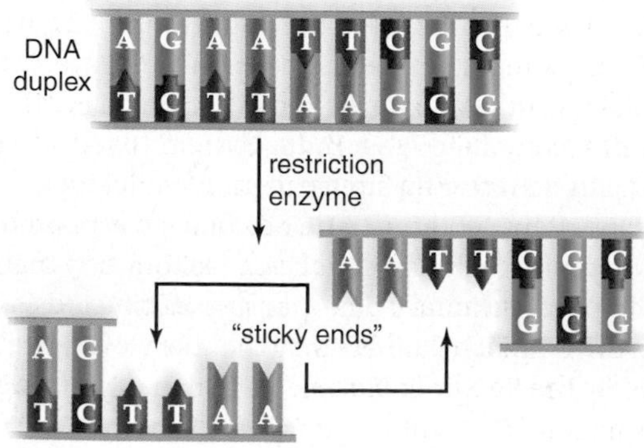

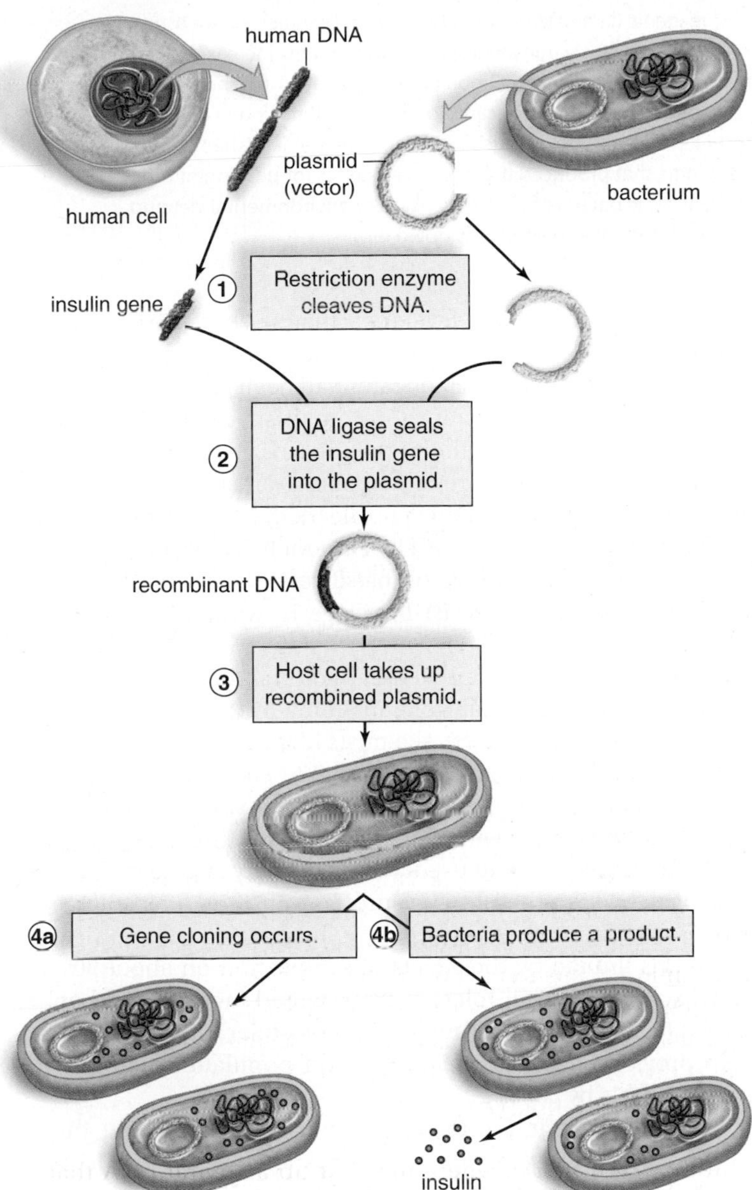

Figure 21.17 **Cloning of a human gene.**

The restriction enzyme creates a gap in plasmid DNA in which foreign DNA (possibly a human gene, such as the insulin gene) can be placed if it ends in bases complementary to those exposed by the restriction enzyme. To ensure this, it is only necessary to use the same type of restriction enzyme to cleave both human DNA containing the gene for insulin and plasmid DNA.

② The enzyme called *DNA ligase* is used to seal foreign DNA into the opening created in the plasmid. The single-stranded but complementary ends of a cleaved DNA molecule are called "sticky ends." This is because they can bind a piece of DNA by complementary base pairing. Sticky ends facilitate the sealing of the plasmid DNA with human DNA for the insulin gene. Now the vector is complete, and an rDNA molecule has been prepared.

③ Some of the bacterial cells take up a recombinant plasmid, especially if the bacteria have been treated to make them more permeable.

④ Gene cloning occurs as the plasmid replicates on its own. Scientists clone genes for a number of reasons. They might want to determine the difference in base sequence between a normal gene and a mutated gene. Or, they might use the genes to genetically modify organisms in a beneficial way.

⑤ The bacterium can also make a product (e.g., insulin) that it could not make before. For a human gene to express itself in a bacterium, the gene has to be accompanied by regulatory regions unique to bacteria. Also, the gene should not contain introns because bacteria don't have introns. However, it is possible to make a human gene that lacks introns. The enzyme called reverse transcriptase can be used to make a DNA copy of mRNA. The DNA molecule, called **complementary DNA (cDNA)**, does not contain introns. In this manner, the recombinant cells make many copies of themselves containing the new gene through mitosis, and each new cell will make the product coded for by the new gene.

Biotechnology Products

Today, bacteria, plants, and animals are genetically engineered to produce **biotechnology products.** Organisms that have had a foreign gene inserted into them are called **transgenic organisms.**

From Bacteria

Recombinant DNA technology is used to produce transgenic bacteria, grown in huge vats called bioreactors (Fig. 21.18*a*). The gene product is collected from the medium. Biotechnology products on the market produced by bacteria include insulin, clotting factor VIII, human growth hormone, tissue plasminogen activator (t-PA), and hepatitis B vaccine. Transgenic bacteria have many other uses as well. Some have been produced to promote the health of plants. For example, bacteria that normally live on plants and encourage the formation of ice crystals have been changed from frost-plus to frost-minus bacteria. As a result, new crops such as frost-resistant strawberries and oranges have been developed.

Bacteria can be selected for their ability to degrade a particular substance, and this ability can then be enhanced by **genetic engineering.** For instance, naturally occurring bacteria that eat oil can be genetically engineered to do an even better job of cleaning up beaches after oil spills (Fig. 21.18*d*). Further, these bacteria are given "suicide" genes that cause them to self-destruct when the job has been accomplished.

From Plants

Corn, potato, soybean, and cotton plants have been engineered to be resistant to either insect predation or herbicides that are widely used (Figs. 21.18*c* and 21.19). Some corn and

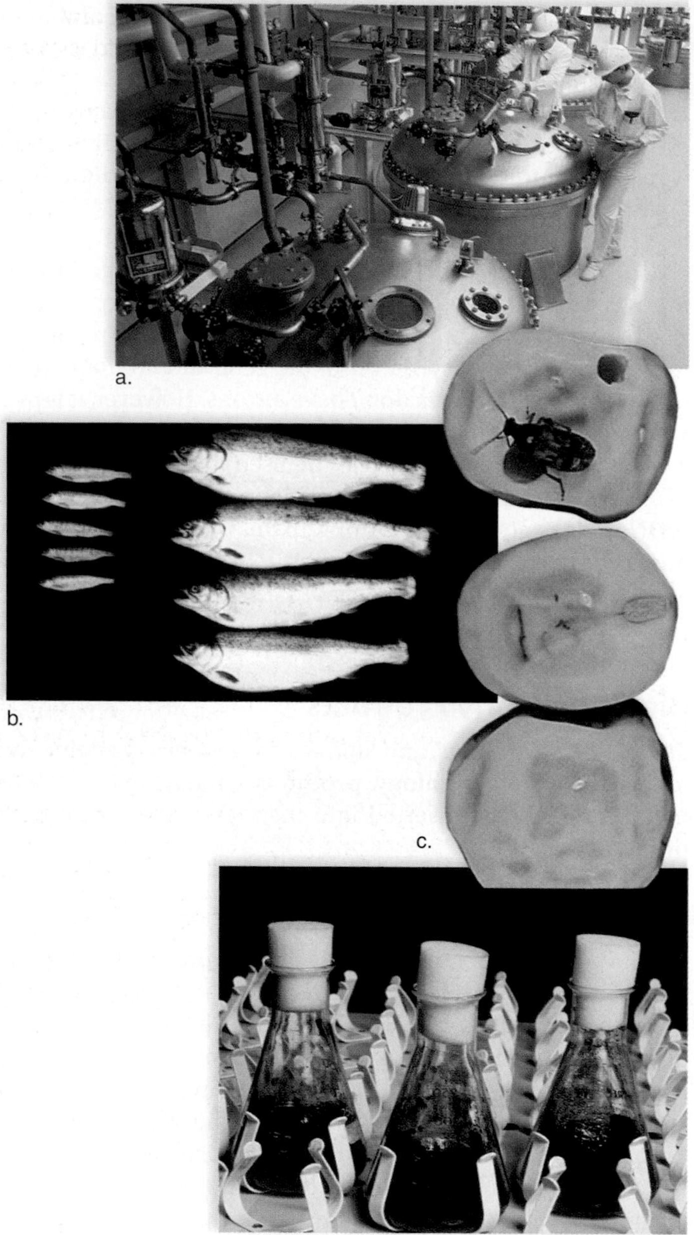

Figure 21.18 Transgenic organisms.
a. Transgenic bacteria, grown industrially in bioreactors, are used to produce medicines. Transgenic animals and plants increase the food supply. **b.** The transgenic salmon on the right are much heavier and reproduce much sooner because of the growth hormone genes they received as embryos. **c.** Pests didn't consume unblemished seeds because the plants that produced them received genes for pest inhibitors. **d.** Transgenic bacteria can also be used for environmental cleanup. These bacteria are able to break down oil.

cotton plants have been developed that are both insect- and herbicide-resistant. In 2011, 94% of the soybeans and 80% of the corn planted in the United States had been genetically engineered. If crops are resistant to a broad-spectrum herbicide and weeds are not, then the herbicide can be used to kill the weeds. When herbicide-resistant plants are planted, weeds are easily controlled, less tillage is needed, and soil erosion is minimized. The Herculex corn by Dupont has been developed with not only a large range of insect resistance but hardier stalks that resist breakage in wind and rain.

Video GM Food Safety

Crops with other improved agricultural and food quality traits are desirable. A salt-tolerant tomato has already been developed (Fig. 21.20). First, scientists identified a gene coding for a protein that transports salt ions (Na^+) across the vacuole membrane. Sequestering the salt ions in a vacuole prevents them from interfering with plant metabolism. Then, the scientists cloned the gene and used it to genetically engineer plants that overproduce the channel protein. The modified plants thrived when watered with a salty solution. Today, crop production is limited by the effects of salinization on about 50% of irrigated lands. Salt-tolerant crops would increase yield on this land. Salt- and also drought- and cold-tolerant crops might help provide enough food for a world population that may nearly double by 2050.

Potato blight is the most serious potato disease in the world. About 150 years ago, it was responsible for the Irish potato famine, which caused the death of millions of people. By placing a gene from a naturally blight-resistant wild potato

Figure 21.19 Genetically engineered plants.
Genetic engineering can produce plants resistant to herbicides (**a**) and pests (**b, c**). Drought- and salt-resistant plants can be used in dry desert areas. Plants that are richer in nutrients can also be engineered.

a. Herbicide-resistant soybean plants

b. Nonresistant potato plant

c. Pest-resistant potato plant

Transgenic Crops of the Future	
Improved Agricultural Traits	
Disease-protected	Wheat, corn, potatoes
Herbicide-resistant	Wheat, rice, sugar beets, canola
Salt-tolerant	Cereals, rice, sugarcane
Drought-tolerant	Cereals, rice, sugarcane
Cold-tolerant	Cereals, rice, sugarcane
Improved yield	Cereals, rice, corn, cotton
Modified wood pulp	Trees
Improved Food Quality Traits	
Fatty acid/oil content	Corn, soybeans
Protein/starch content	Cereals, potatoes, soybeans, rice, corn
Amino acid content	Corn, soybeans

Figure 21.20 Genetically engineered plants for desirable traits.

into a farmed variety, researchers have now made potato plants that are no longer vulnerable to a range of blight strains. Some progress has also been made to increase the food quality of crops. Soybeans have been developed that mainly produce the monounsaturated fatty acid oleic acid, a change that may improve human health.

Other types of genetically engineered plants are made to increase productivity. Leaves can be engineered to lose less water and take in more carbon dioxide. This type of modification helps a range of crops grow successfully in various climates, including those that are more drought-susceptible or have a higher average temperature than the normal growing climate. Other single-gene modifications allow plants to produce various products including human hormones, clotting factors, antibodies, and vaccines.

From Animals

Techniques have been developed to insert genes into the eggs of animals. It is possible to microinject foreign genes into eggs by hand. Another method uses vortex mixing. The eggs are placed in an agitator with DNA and silicon carbide needles, and the needles make tiny holes through which the DNA can enter. When these eggs are fertilized, the resulting offspring are transgenic animals. Using this technique, many types of animal eggs have taken up the gene for bovine growth hormone (BGH). The procedure has been used to produce larger fish, cows, pigs, rabbits, and sheep. Pigs that are bred to supply transplant organs for humans can be genetically modified in this manner also.

Gene pharming, the use of transgenic farm animals to produce pharmaceuticals, is being pursued by a number of firms. Genes that code for therapeutic and diagnostic proteins are incorporated into an animal's DNA, and the proteins appear in the animal's milk (Fig. 21.21). Plans are under way to produce drugs for the treatment of cystic fibrosis, cancer, blood diseases, and other disorders by this method. Figure 21.21 outlines the procedure for producing transgenic animals. The gene of interest is microinjected into donor eggs. Following in vitro fertilization, the zygotes are placed

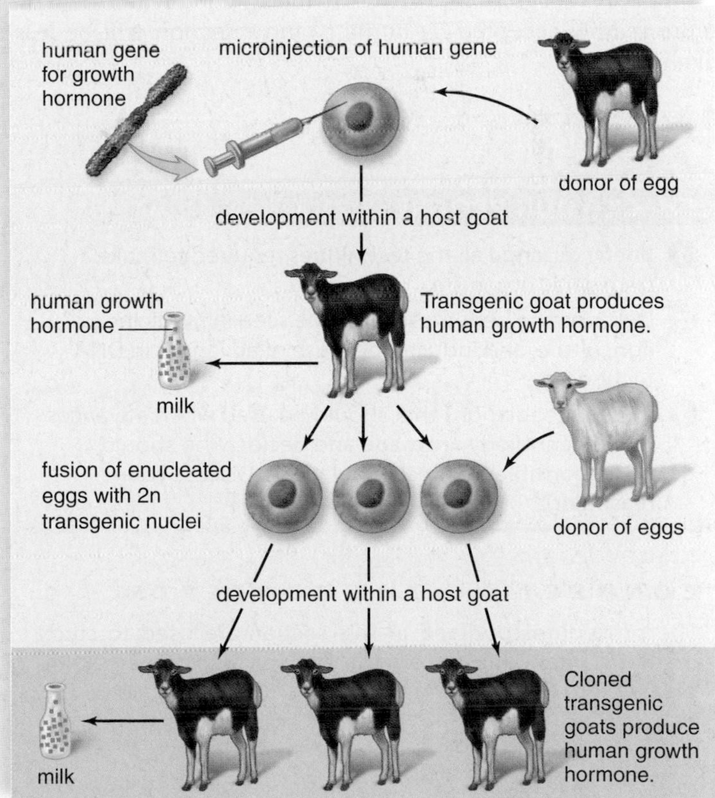

Figure 21.21 Production of a transgenic animal.
Once the desired human gene is introduced into a fertilized ovum, a single transgenic goat is produced. Using the first goat's cell nuclei and ova, a herd of transgenic goats can be developed. Each produces human growth hormone in her milk.

in host females, where they develop. After transgenic female offspring mature, the product is secreted in their milk. Then, cloning can be used to produce many animals that produce the same product. Female clones produce the same product in their milk. This technique has been beneficial in producing milk with certain antibiotics or vaccines, in addition to milk with a higher nutritional value.

Video Firefly Rx

Mouse models have also been created to study human diseases. An allele such as the one that causes cystic fibrosis can be cloned and inserted into mice embryonic stem cells. Occasionally, a mouse embryo homozygous for cystic fibrosis will result. This embryo develops into a mutant mouse that has a phenotype similar to a human with cystic fibrosis. New drugs for the treatment of cystic fibrosis can then be tested in these mice. The same scenario is seen in the OncoMouse, which carries genes for the development of cancer.

Xenotransplantation is the use of animal organs instead of human organs in transplant patients. Currently, an estimated 300,000 plus patients worldwide are in need of organ transplants. Human donors are hard to match and in most cases not available. Scientists have chosen to use the pig because knowledge of pig breeding has a long history in animal husbandry and most pig organs match the size and structure of human organs. Additionally, the ability of pigs to express genes for human-recognition proteins means that pig organs will be more readily accepted by humans and rejection will be less likely to occur.

CHECK YOUR PROGRESS 21.3

1. Briefly describe all the techniques required to make a transgenic organism.
2. Describe the benefits that can be seen in medicine, agriculture, and industry because of advances in DNA technology.
3. Identify regulations that should be used when advances in biotechnology are made and decide who should be responsible for these regulations. Discuss your reasoning.

CONNECTING THE CONCEPTS

The procedures outlined in this section are used to study many important questions in the life sciences. For a few examples, refer to the following discussions:

Chapter 17 contains a discussion on the various forms of cloning.

Chapter 24 explores how biotechnology may be used to produce food for an increasing human population.

21.4 Genomics and Gene Therapy

LEARNING OUTCOMES

Upon completion of this section, you should be able to

1. Distinguish between functional and comparative genomics.
2. Distinguish between genomics and proteomics.
3. Explain the difference between in vivo and ex vivo gene therapy.

The Human Genome Has Been Sequenced

Genetics in the twenty-first century concerns genomics, the study of genomes—our genes and the genes of other organisms. The Human Genome Project (HGP) was completed in 2003 by a coordinated effort from the U.S. government and private laboratories and contributions from many other countries like France, Japan, China, and Germany. The purpose of the HGP was to sequence the human genome. During the 13-year study, many interesting facts emerged. The human genome contains approximately 3.2 billion bases, and 99.9% of these bases are identical in sequence in all humans. The average functional gene contains only about 3,000 bases; the number of functional genes found in humans is estimated to be around 25,000, surprisingly low from the beginning estimate of 80,000 to 140,000 genes. These data were obtained through the use of advanced biotechnology including PCR, electrophoresis, and DNA sequencing. A surprising finding has been that genome size is not proportionate to the number of genes and does not correlate to the complexity of the organism. The flowering plant *Arabidopsis thaliana* has a genome size of 125 million bases and 25,500 coding genes. The common house mouse, *Mus musculus,* has a genome of 2.5 billion bases and about 25,000 coding genes. When scientists first completed the DNA sequence of the human genome, they made an estimate of the number of genes based on sequences that appeared to encode proteins. Researchers were surprised to find fewer than 25,000 genes in the human genome. Less than 2% of the entire human genome actually codes for functional proteins. Currently, the function of the majority of the genes identified in the HGP is unknown.

Functional and Comparative Genomics

Goals of continuing research in the Human Genome Project are to determine how our approximately 25,000 genes function and how together they form a human. It is hoped that this type of study, called **functional genomics,** will also discover the role of noncoding regions. Genes make up less than 2% of the human genome, and scientists in the field of genomics are working hard to find the function of the many intergenic DNA sequences.

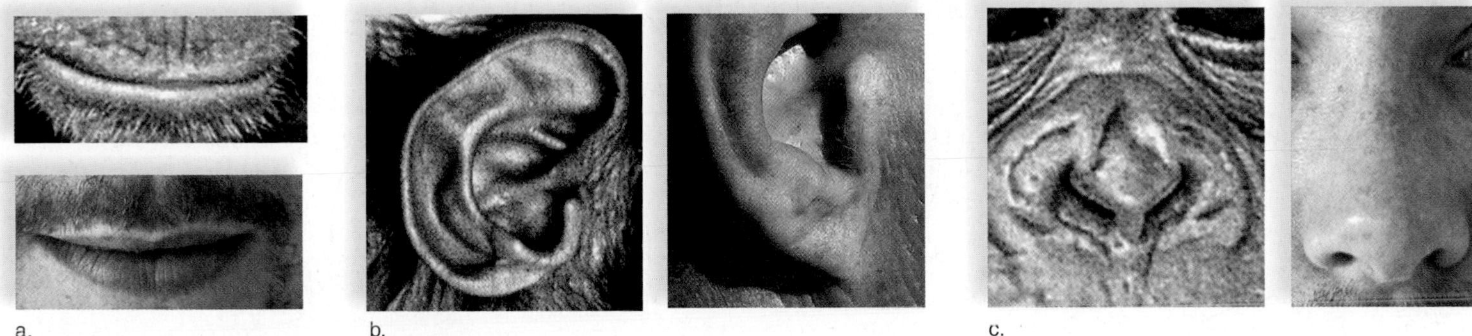

Figure 21.22 **Functional and comparative genomics between chimpanzees and humans.** Did changes in the genes for (**a**) speech, (**b**) hearing, and (**c**) smell influence the evolution of humans?

Comparative genomics is one way to determine how species have evolved and how genes and noncoding regions of the genome function. In one study, researchers compared our genome to that of chromosome 22 in chimpanzees (Fig. 21.22). They found three types of genes of particular interest: a gene for proper speech development, several for hearing, and several for smell. Genes necessary for speech development are thought to have played an important role in human evolution. You can suppose that changes in hearing may also have facilitated using language for communication between people. To explain differences in smell genes, investigators speculated that the olfaction genes may have affected dietary changes or sexual selection. Or, they may have been involved in other traits, rather than just smell. The researchers who did this study were surprised to find that many of the other genes they located and studied are known to cause human diseases. They wondered if comparing genomes would be a way of finding other genes associated with human diseases. Investigators are taking all sorts of avenues to link human base sequence differences to illnesses.

Another surprising discovery has been how very similar the genomes of all vertebrates are. Researchers weren't surprised to learn that the base sequence of chimpanzees and humans is 95–98% alike. However, they didn't expect to find that our sequence is also 85% similar to that of a mouse, for example. So, the quest is on to possibly discover how the regulation of genes explains why we have one set of traits and mice have another set, despite the similarity of our base sequences. One possibility is alternative gene splicing. We may differ from mice in the types of proteins we manufacture and/or by when and where certain proteins are present.

Proteomics and Bioinformatics

Proteomics is the study of the structure, function, and interaction of cellular proteins. Many of our genes are translated into proteins at some time, in some of our cells. The translation of all coding genes results in a collection of proteins called the human proteome. The analysis of proteomes is more challenging than the analysis of genomes. A single gene can code for more than 1,000 different proteins, and protein concentrations can differ widely in cells. Researchers have to be able to identify proteins, regardless of whether there is one or thousands of copies of a protein in a cell. Any particular protein differs minute by minute in concentration, interactions, cellular location, and chemical modifications, among other features. Yet, to understand a protein, all these features must be analyzed. Computer modeling of the three-dimensional shape of cellular proteins is an important part of proteomics. The study of cellular proteins and how they function is essential to understanding the causes of certain diseases and disorders. This study is also important in the discovery of better drugs, because most drugs are proteins or molecules that affect the function of proteins.

Bioinformatics is the application of computer technologies to the study of the genome. Specifically, it is the process of creating databases of information, then mapping and analyzing the information gained from DNA sequencing and proteomics. Genomics and proteomics produce raw data. These fields depend on computer analysis to find significant patterns in the data. As a result of bioinformatics, scientists hope to find cause-and-effect relationships between various genetic profiles and genetic disorders caused by multifactorial genes. By correlating any sequence changes with resulting phenotypes, one current focus of bioinformatics research is discovering if noncoding regions of the genome do have functions and, if so, what effect those functions may have on homeostasis.

Gene Therapy

Gene therapy is the insertion of genetic material into human cells for the treatment of a disorder. Gene therapy has been used to cure inborn errors of metabolism and also to treat more generalized disorders such as cardiovascular disease and cancer. Most recently, a 2008 clinical trial showed that gene therapy could successfully treat a type of inherited blindness. Both ex vivo (outside the body) and in vivo (inside the body) gene therapy methods are used.

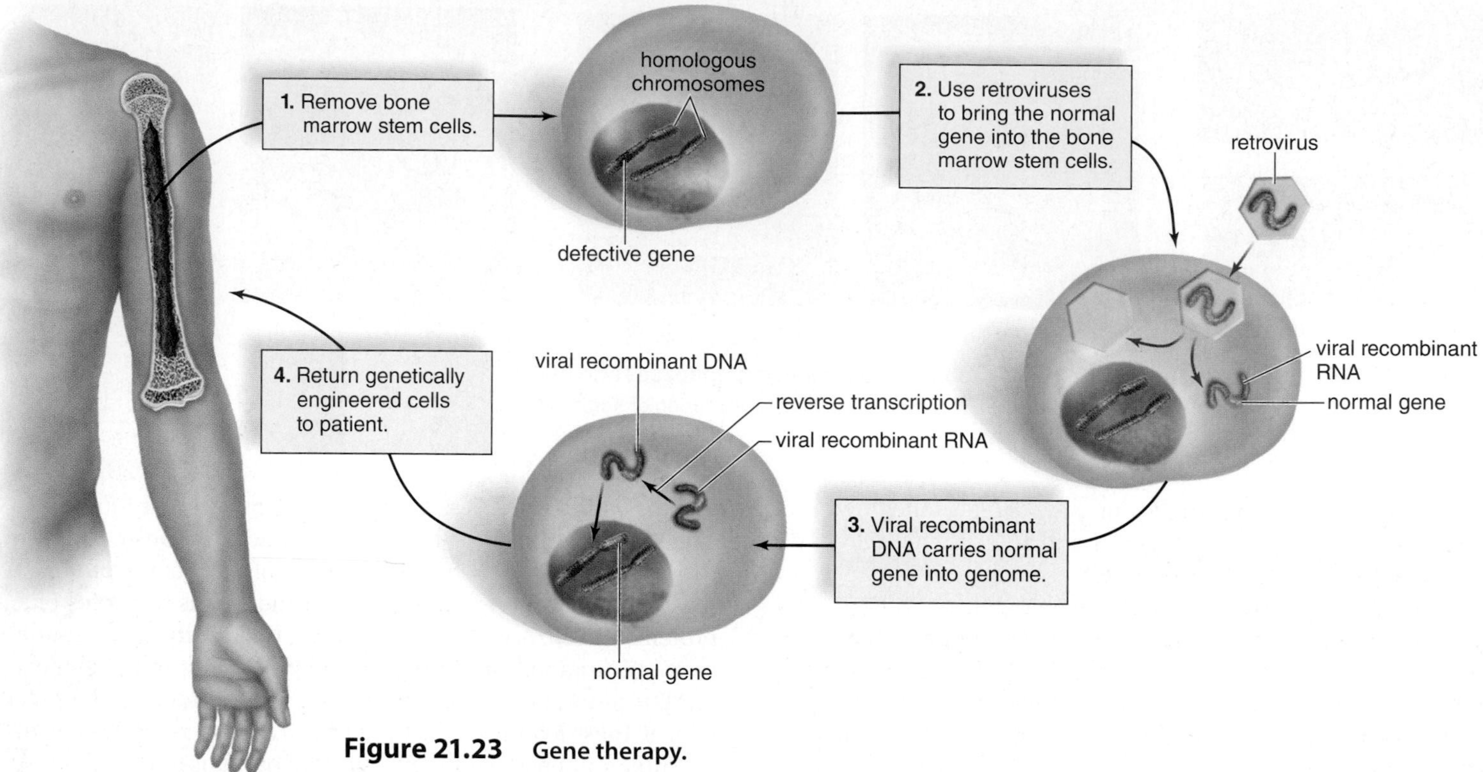

Figure 21.23 Gene therapy.

Ex Vivo Gene Therapy

Figure 21.23 describes the methodology for treating children who have severe combined immunodeficiency disease (SCID, see Chapter 7). These children lack the enzyme adenosine deaminase (ADA), which is involved in the maturation of T and B cells. To carry out ex vivo gene therapy, stem cells are removed from bone marrow. These cells are mixed with a normal gene for the enzyme that is contained in a vector, or "carrier" molecule. The most successful vectors are found to be viruses that have been altered for safety. The combined cells are then returned to the patient. Bone marrow stem cells are preferred for this procedure because they divide to produce more cells with the same genes. SCID patients who have undergone this procedure show significantly improved immune function and a sustained rise in ADA enzyme activity in the blood. Ex vivo gene therapy has also been proven effective for hemophilia A, Alzheimer disease, Parkinson disease, Crohn's disease, and certain cancers.

In Vivo Gene Therapy

Cystic fibrosis patients lack a gene that codes for the transmembrane carrier of the chloride ion. They often suffer from numerous and potentially deadly infections of the respiratory tract. For in vivo gene therapy, the therapeutic DNA is injected straight into the body cells. As in ex vivo therapy, the use of a vector, or carrier molecule, is needed. For in vivo therapy, a retrovirus or adenovirus is used as the vector that carries the corrective gene. For CF patients, the adenovirus with the functioning gene is either sprayed into the nose or injected into the lower respiratory tract. This type of gene therapy is minimally invasive and has quick results, but the effects are not as long lasting as ex vivo therapy. In vivo gene therapy has been used for cardiovascular diseases, endocrine disorders, and Huntington disease.

CHECK YOUR PROGRESS 21.4

1. Describe how an understanding of genomics influences everyday life.
2. Discuss the differences between the study of genomics and proteomics.
3. Explain why gene therapy would be used over medication and list some advantages and disadvantages.

CONNECTING THE CONCEPTS

The study of genomics and proteomics plays an important role in our understanding of evolutionary change. For more information, refer to the following discussions:

Section 22.2 examines how biochemical evidence, including the analysis of DNA, helps us to understand biological evolution.

Section 22.3 explores how DNA evidence enables a deeper understanding of human evolution.

BIOLOGY MATTERS **Science**

Testing for Genetic Disorders

Prospective parents know if either of them has an autosomal dominant disorder because the person will show it. However, genetic testing is required to detect if either is a carrier for an autosomal recessive disorder. If a woman is already pregnant, the parents may want to know if the unborn child has the disorder. If the woman is not pregnant, the parents may opt for testing of an embryo or egg before she does become pregnant. One way to detect genetic disorders is to test the DNA for mutated genes.

Testing the DNA

Two types of DNA testing are possible: testing for a genetic marker and using a DNA probe.

Genetic Markers

Testing for a genetic marker is similar to the traditional procedure for DNA fingerprinting, as discussed earlier. As an example, consider that individuals with Huntington disease have an abnormality in the sequence of their bases at a particular location on a chromosome. This abnormality in sequence is a *genetic marker.* Huntington disease, specifically, results from an STR that is so long that it actually causes a frameshift mutation within a gene even though the STR itself occurs outside, but nearby the gene. In this and similar cases, the length of the STR can be detected with PCR and analysis on an automated DNA sequencer.

DNA Microarrays

With advances in robotic technology, it is now possible to place the entire human genome onto a single microarray (Fig. 21B). The mRNA from the organism or the cell to be tested is labeled with a fluorescent dye and added to the chip. When the mRNAs bind to the microarray, a fluorescent pattern results that is recorded by a computer. Now the investigator knows what DNA is active in that cell or organism. A researcher can use this method to determine the difference in gene expression between two different cell types, such as between liver cells and muscle cells.

A mutation microarray, the most common type, can be used to generate a person's genetic profile. The microarray contains hundreds to thousands of known disease-associated mutant gene alleles. Genomic DNA from the individual to be tested is labeled with a fluorescent dye, and then added to the microarray. The spots on the microarray fluoresce if the individual's DNA binds to the mutant genes on the chip, indicating that the individual may have a particular disorder or is at risk for developing it later in life. This technique can generate a genetic profile much more quickly and inexpensively than older methods involving DNA sequencing.

DNA microarrays also promise to hasten the identification of genes associated with diseased tissues. In the first instance, mRNA derived from diseased tissue and normal tissue is labeled with different fluorescent dyes. The normal tissue serves as a control.

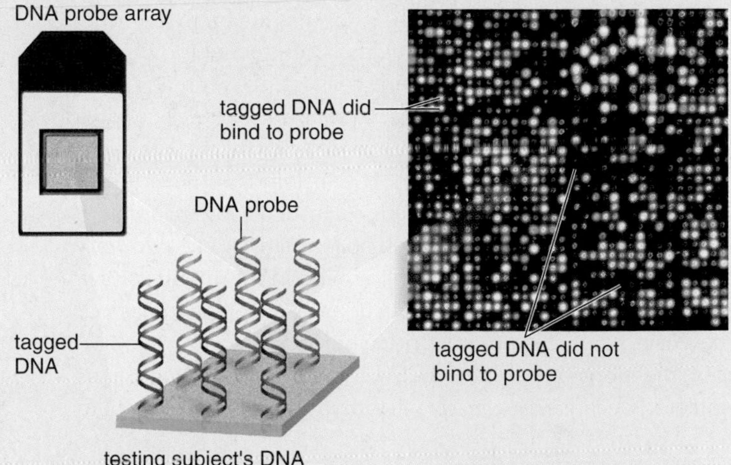

Figure 21B **Use of a DNA microarray to test for a genetic disorder.**
This DNA chip contains rows of DNA sequences for mutations that indicate the presence of particular genetic disorders. If DNA fragments derived from an individual's DNA bind to a sequence representing a mutation on the DNA chip, that sequence fluoresces, and the individual has the mutation.

The investigator applies the mRNA from both normal and abnormal tissue to the microarray. The relative intensities of fluorescence from a spot on the microarray indicate the amount of mRNA originating from that gene in the diseased tissue relative to the normal tissue. If a gene is activated in the disease, more copies of mRNA will bind to the microarray than from the control tissue, and the spot will appear more red than green.

Genomic microarrays are also used to identify links between disease and chromosomal variations. In this instance, the chip contains genomic DNA that is cut into fragments. Each spot on the microarray corresponds to a known chromosomal location. Labeled genomic DNA from diseased tissues and control tissues bind to the DNA on the chip, and the relative fluorescence from both dyes is determined. If the number of copies of any particular target DNA has increased, more sample DNA will bind to that spot on the microarray relative to the control DNA, and a difference in fluorescence of the two dyes will be detected.

Questions to Consider

1. What benefits are there when using a DNA microarray over a genetic marker such as an STR?
2. Why might a researcher want to know what genes are being expressed in different cell types?
3. How might the information from a DNA microarray be used to develop new drugs to treat disease?

CASE STUDY CONCLUSION

Dr. Weber explained that insulin for diabetes patients has been made since the late 1970s in large vats called bioreactors. Here, a non-disease-producing strain of *Escherichia coli* has been made that contains the human gene for the production of insulin through recombinant DNA technology. In the bioreactor with the recombinant cells is also a medium that contains a food source for the bacteria. The bacteria stay alive and make billions of copies of themselves, while at the same time producing human insulin. The insulin is retrieved from the medium and used for injections. Insulin lispro (Humalog) and insulin human recombinant (Humulin N) are two very common types of medical insulin made in this manner. Kaya will have to monitor her blood sugar level several times per day using a glucose meter and, when appropriate, give herself injections of insulin to help regulate her levels. Insulin is injected because if it were in pill form and used orally, it would be inactivated by the digestive system. Instead, insulin is administered by intramuscular injection so it can enter directly into the bloodstream. Advances in the treatment of diabetes have also included the use of an insulin pump. The pump is a small device worn on a belt or pocket that delivers fast-acting insulin into the body via an infusion set—a thin plastic tube ending in a small, flexible plastic cannula, or a very thin needle. The cannula is inserted beneath the skin at the infusion site, usually in the person's abdomen or upper buttocks. The infusion set is in place for two to three days (sometimes more), and then it is moved to a new location. The insulin pump is not an artificial pancreas; it is a computer-driven device that delivers fast-acting insulin in precise amounts at preprogrammed times. Under her doctor's supervision, Kaya will monitor her diabetes to discover the appropriate treatment options for her lifestyle and medical needs.

MEDIA STUDY TOOLS

Enhance your study of this chapter with media! Visit **www.mhhe.com/maderhuman13e** and go to "Media Study Tools" for this chapter to access the following:

Animations	Videos	MP3 Files
21.1 DNA Structure • Meselson and Stahl Experiment • DNA Replication **21.2** How Spliceosomes Process RNA • How Translation Works • Transcription Factors • Control of Gene Expression in Eukaryotes **21.3** Polymerase Chain Reaction • DNA Fingerprinting • Sanger Sequencing **21.4** DNA Microarray	**21.1** Tiny Genes Big Role • Drug Discovery • Halting Hepatitis • Dark Lady of DNA **21.3** World Trade Center DNA • Good Tobacco • Potato Vaccine • Cloned Milk • Firefly Rx • GM Food Safety	**21.1** DNA Replication and the Cell Cycle **21.2** Protein Synthesis

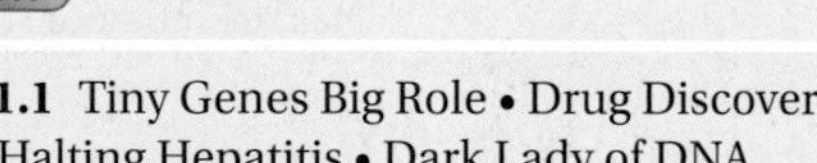
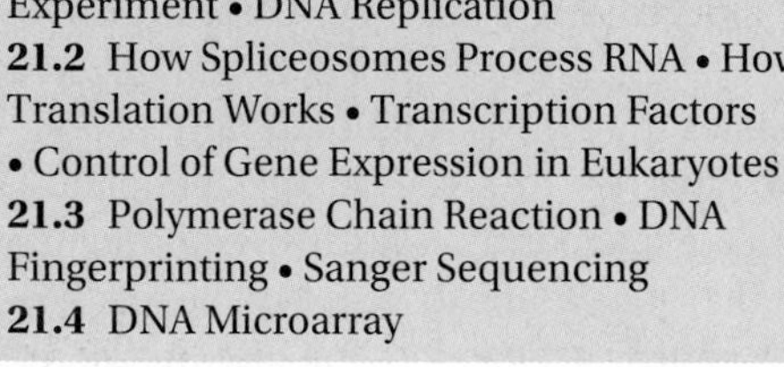

3D Animations
DNA Replication
Molecular Biology of the Gene

McGraw-Hill's 3D animations, "DNA Replication" and "Molecular Biology of the Gene," provide a dynamic exploration of the key concepts of this chapter and are available through McGraw-Hill Connect®.

SUMMARIZE

21.1 DNA and RNA Structure and Function

DNA (deoxyribonucleic acid) is the genetic material. It is organized as **genes,** located on chromosomes. DNA replicates, stores information, and mutates for genetic variability.

Structure of DNA

- DNA is a **double helix** composed of two polynucleotide strands. Each nucleotide is composed of a deoxyribose sugar, a phosphate, and a nitrogen-containing base (A, T, C, G).
- **Complementary base pairing** occurs between the strands of DNA. The base A is bonded to T, and G is bonded to C.

Replication of DNA

- During **DNA replication,** the DNA strands unzip, and a new complementary strand forms opposite each old strand (the **template**), resulting in two identical DNA molecules. **Mutations** produce variation in the genetic material.

The Structure and Function of RNA

- **RNA (ribonucleic acid)** is a single-stranded nucleic acid in which the base U (uracil) occurs instead of T (thymine).
- The four primary forms of RNA are **ribosomal RNA (rRNA), messenger RNA (mRNA), transfer RNA (tRNA),** and **small RNAs.**

21.2 Gene Expression

Gene expression leads to the formation of a product, either an RNA or a protein. Proteins differ by the sequence of their amino acids. Gene expression for proteins requires transcription and translation.

Transcription

Transcription occurs in the nucleus using an enzyme called **RNA polymerase.** The three-base DNA **codon** is passed to an mRNA that contains codons. Introns are removed from mRNA during mRNA processing.

Translation

Translation occurs in the cytoplasm at the ribosomes. The tRNA molecules bind to their amino acids, and then their **anticodons** pair with mRNA codons.

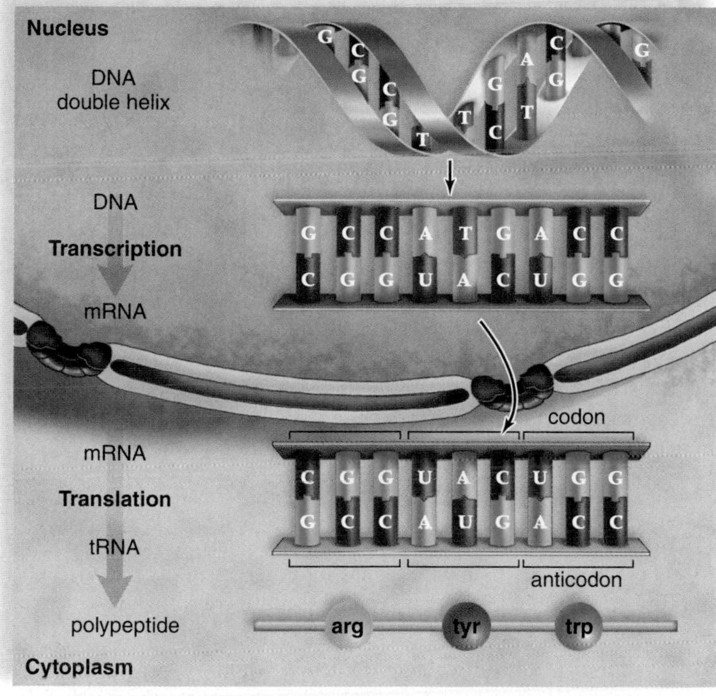

The Regulation of Gene Expression

Regulation of gene expression occurs at five levels in a human cell.

- *Pretranscriptional control:* In the nucleus; the DNA is made available to transcription factors and enzymes.
- *Transcriptional control:* In the nucleus; the degree to which a gene is transcribed into mRNA determines the amount of gene product. **Transcription factors** are involved at this stage.
- *Posttranscriptional control:* In the nucleus; involves mRNA processing and how fast mRNA leaves the nucleus.
- *Translational control:* In the cytoplasm; affects when translation begins and how long it continues. Includes inactivation and degradation of mRNA.
- *Posttranslational control:* In the cytoplasm; occurs after protein synthesis.

21.3 DNA Technology

Cloning is the production of genetically identical copies of DNA, cells, and organisms. **Gene cloning** has a variety of applications in biology.

- **Recombinant DNA (rDNA)** contains DNA from two different sources. The foreign gene and vector DNA are cut by the same **restriction enzyme** and then the foreign gene is sealed into **vector** DNA, such as bacterial **plasmids.** The reverse transcriptase enzyme may be used to make **complementary DNA (cDNA).**
- The **polymerase chain reaction (PCR)** uses DNA polymerase to make multiple copies of a specific piece of DNA. Following PCR, DNA can be subjected to DNA fingerprinting.

Some applications of DNA technology allow for **genetic engineering** and the production of transgenic organisms.

- **Transgenic organisms** (bacteria, plants, and animals that have had a foreign gene inserted into them) can produce **biotechnology products,** such as hormones and vaccines.
- Transgenic bacteria can promote plant health, remove sulfur from coal, clean up toxic waste and oil spills, extract minerals, and produce chemicals.
- Transgenic crops can resist herbicides and pests.
- Transgenic animals can be given growth hormone to produce larger animals, can supply transplant organs (**xenotransplantation**), and can produce pharmaceuticals.

21.4 Genomics and Gene Therapy

Genomics is the study of the genomes of humans and other organisms.

Functional and Comparative Genomics

- **Functional genomics** is the study of how the 25,000 genes in a human genome function.
- **Comparative genomics** is a way to determine how species have evolved and how genes and noncoding regions of the genome function.

Proteomics and Bioinformatics

- **Proteomics** is the study of the structure, function, and interaction of cellular proteins.
- **Bioinformatics** is the application of computer technologies to the study of the genome.

Gene Therapy

Gene therapy allows for the treatment of several human diseases.

- In ex vivo gene therapy, cells are removed from the body for treatment, and then reintroduced back into the body.
- In in vivo therapy, the vector is introduced directly into the body.

ASSESS

Testing Your Knowledge of the Concepts

1. Describe the general structure of a DNA molecule. (pages 484–486)
2. Summarize the process of DNA replication. (pages 486–487)
3. Describe the structure and function of the types of RNA. (pages 487–488)
4. Explain why the terms *transcription* and *translation* are appropriate for these processes. (page 489–490)
5. If a DNA strand is TAC AAT AAA CGT GTC ATT, what are the codons of mRNA, the anticodons of tRNA, and the amino acid sequence? (pages 489–491)
6. Explain the purpose of mRNA processing. (pages 490–491)
7. What are the different types of genetic control and where does each one occur in the cell? (page 494)
8. What is the polymerase chain reaction (PCR), and how is it used to produce multiple copies of a DNA segment? (page 496–497)
9. What is the methodology for producing transgenic bacteria? (pages 498–499)
10. What are some biotechnology products from bacteria, plants, and animals; what are they used for? (pages 499–501)

11. What is the difference between the study of genomics and proteomics? (pages 502–503)

12. Compare and contrast in vivo and ex vivo gene therapy. (pages 504–505)

13. The double-helix model of DNA resembles a twisted ladder in which the rungs of the ladder are
- **a.** complementary base pairs.
- **b.** A paired with G and C paired with T.
- **c.** A paired with T and G paired with C.
- **d.** a sugar-phosphate paired with a sugar-phosphate.
- **e.** Both a and c are correct.

14. The enzyme responsible for adding new nucleotides to a growing DNA chain during DNA replication is (are)
- **a.** helicase.
- **b.** RNA polymerase.
- **c.** DNA polymerase.
- **d.** a ribozyme.

15. RNA processing
- **a.** removes the introns, leaving only the exons.
- **b.** is the same as transcription.
- **c.** is an event that occurs after RNA is transcribed.
- **d.** is the rejection of old, worn-out RNA.
- **e.** Both a and c are correct.

16. During protein synthesis, an anticodon of a tRNA pairs with
- **a.** amino acids in the polypeptide.
- **b.** DNA nucleotide bases.
- **c.** rRNA nucleotide bases.
- **d.** mRNA nucleotide bases.

17. Which of the following is involved in controlling gene expression?
- **a.** the occurrence of transcription
- **b.** activity of the polypeptide product
- **c.** life expectancy of the mRNA molecule in the cell
- **d.** All of the choices are involved.

18. Restriction enzymes found in bacterial cells are ordinarily used
- **a.** during DNA replication.
- **b.** to degrade the bacterial cell's DNA.
- **c.** to degrade viral DNA that enters the cell.
- **d.** to attach pieces of DNA together.

19. Which of the following is a benefit to having insulin produced by biotechnology?
- **a.** It is just as effective.
- **b.** It can be mass-produced.
- **c.** It is nonallergenic.
- **d.** It is less expensive.
- **e.** All of the choices are correct.

20. Following is a segment of a DNA molecule. (Remember that only one strand is transcribed.) What are (a) the mRNA codons, (b) the tRNA anticodons, and (c) the sequence of amino acids?

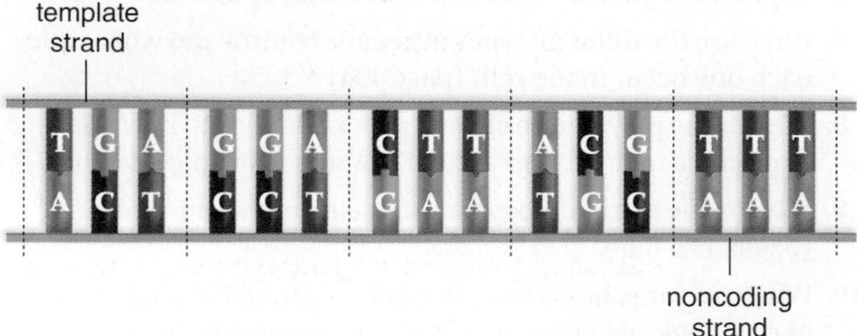

ENGAGE

Virtual Labs
Classifying Using Biotechnology, Knocking Out Genes, Gene Splicing

Several virtual labs explore the uses of biotechnology in the life sciences. "Classifying Using Biotechnology" explores how biotechnology can be used to classify bacteria. "Knocking Out Genes" examines how knocking out specific genes can create a better understanding of how plants adapt to specific environmental conditions. "Gene Splicing" uses biotechnology techniques to generate transgenic organisms.

Thinking Critically About the Concepts

The advances in biotechnology since the 1970s have made a huge impact on medicine and the treatment and management of certain diseases. The ability to combine DNA from two different organisms has given us the means to produce substances the body needs, as seen in the case study with Kaya using *E. coli*–derived human insulin. It has also given doctors the ability to insert properly functioning genes into patients lacking those genes. With certain advances in biotechnology, ethical issues arise. Think about both sides when answering the following questions:

1. What are the advantages of using a recombinant DNA human product instead of a product isolated from another organism? (For example, what advantage would there be to using human recombinant insulin versus insulin produced from cows or pigs?)
2. Are there any disadvantages to using a recombinant DNA product? (Hint: You may want to think about how the product is produced and then purified.)
3. Recombinant human growth hormone is available for children and adults with growth hormone deficiency.
   - **a.** When would use of the hormone be appropriate? Under what circumstances would hormone use be improper?
   - **b.** How would a physician know that the hormone treatment was completely effective? What further information might you need to answer this question?
4. In general, diseases caused by a single protein deficiency are more easily treated with a recombinant DNA-produced product. Yet certain cancers can be treated in this manner. What aspects of cancer could be treated with a recombinant DNA-produced product?
5. What types of regulations should be placed on advances in DNA technology and who should be responsible for these regulations?

CHAPTER

22

Human Evolution

CASE STUDY THE NEANDERTALS

For years, Neandertals have been perceived as an offshoot of human evolution, a branch of the evolutionary tree that was only distantly connected with the evolution of our species, *Homo sapiens*. This changed in 2010 when researchers in Germany obtained DNA samples from Neandertal skeletons that were between 38,000 and 45,000 years old. Upon comparing this DNA to five populations of modern humans, the research team uncovered two surprises. First, even though Neandertals and *Homo sapiens* were believed to have diverged from a common ancestor a brief 500,000 years ago, there were already 15 areas of the *Homo sapiens* genome that showed evidence of selection following the split. These included areas of the genome associated with cranial development, metabolic functions, and cognitive (thinking) abilities.

The second discovery reported by the research team was even more interesting—that Neandertals and *Homo sapiens* may have interbred, perhaps as little as 50,000 years ago. In fact, the research suggests that Neandertals are more closely related to populations of *Homo sapiens* from Europe and Asia than to African populations, meaning that Neandertals and humans might have interacted after *Homo sapiens* migrated from Africa. These discoveries have renewed an interest in discovering the genes that distinguish us from our closest ancestors. In the process, they may develop a greater understanding of the evolutionary history of our species.

As you read through the chapter, think about the following questions:

1. What are some of the known differences between Neandertals and the early Cro-Magnons?
2. What is the current thinking on the evolutionary relationship of the Neandertals to modern humans?
3. How does the evidence for evolution support the concept of human evolution?

CHAPTER CONCEPTS

22.1 Origin of Life
Chemical evolution preceded the evolution of the first cell.

22.2 Biological Evolution
Descent from a common ancestor explains the unity of life. For example, all living organisms have a cellular structure and a common chemistry from their common ancestor. Adaptation to different environments explains the great diversity of life.

22.3 Classification of Humans
The ancestry of humans, like all living organisms, indicates an evolutionary history.

22.4 Evolution of Hominins
Modern humans evolved from a group of hominins called the Australopithecines.

22.5 Evolution of Humans
Among the members of the genus *Homo*, *Homo habilis* was the first to make tools. *Homo erectus* was the first to have the use of fire. Neandertals were highly adapted to life in cold climates. Cro-Magnons resembled modern humans in appearance.

BEFORE YOU BEGIN

Before beginning this chapter, take a few moments to review the following discussions:

Section 1.1 What is the definition of the term *evolution*?

Section 2.7 What are the differences between RNA and DNA?

Section 3.1 What are the basic principles of the cell theory?

22.1 Origin of Life

LEARNING OUTCOMES

Upon completion of this section, you should be able to

1. Describe the conditions of the early Earth's atmosphere.
2. Distinguish between chemical and biological evolution.
3. Discuss how the first organic monomers and polymers may have evolved.
4. Summarize how the first true cell evolved.

In Chapter 1, we considered the characteristics shared by all living organisms. Living organisms acquire energy through metabolism, or the chemical reactions that occur within cells. Living organisms also respond and interact with their environment, self-replicate, and are subject to the forces of natural selection that drive adaptation to the environment. The molecules of life, called *biomolecules,* are organic molecules. The first living organisms on Earth would have had all of these characteristics. Yet early Earth was very different from the Earth we know today, and consisted mainly of inorganic substances. How, then, did life get started on our planet?

Advances and discoveries in chemistry, evolutionary biology, paleontology, microbiology, and other branches of science have helped to develop new, and test old, hypotheses about the origin of life. These studies contribute to an ever-growing body of scientific evidence that life originated 3.5 to 4 billion years ago (BYA) from nonliving matter in a series of four stages:

- *Stage 1: Small organic molecules.* Simple organic molecules, called *monomers,* evolved from inorganic compounds prior to the existence of cells. Amino acids, the basis of proteins, and nucleotides, the building blocks of DNA and RNA, are examples of organic monomers.
- *Stage 2: Macromolecules.* Organic monomers were joined to form larger macromolecules (*polymers*), such as DNA, RNA, and proteins.
- *Stage 3: Protocells.* Organic polymers became enclosed in a membrane to form the first cell precursors, called protocells, or probionts.
- *Stage 4: Living cells.* Probionts acquired the ability to self-replicate as well as other cellular properties.

Scientists have performed experiments to test hypotheses at each stage of the origin of life. Stages 1–3 involve the processes of **chemical evolution,** and occurred before the origin of life. Stage 4 is when life first evolved through the processes of **biological evolution** (Fig. 22.1). In this section, we examine the hypotheses and supporting scientific evidence for each stage of chemical and biological evolution, the origin of life from nonliving matter.

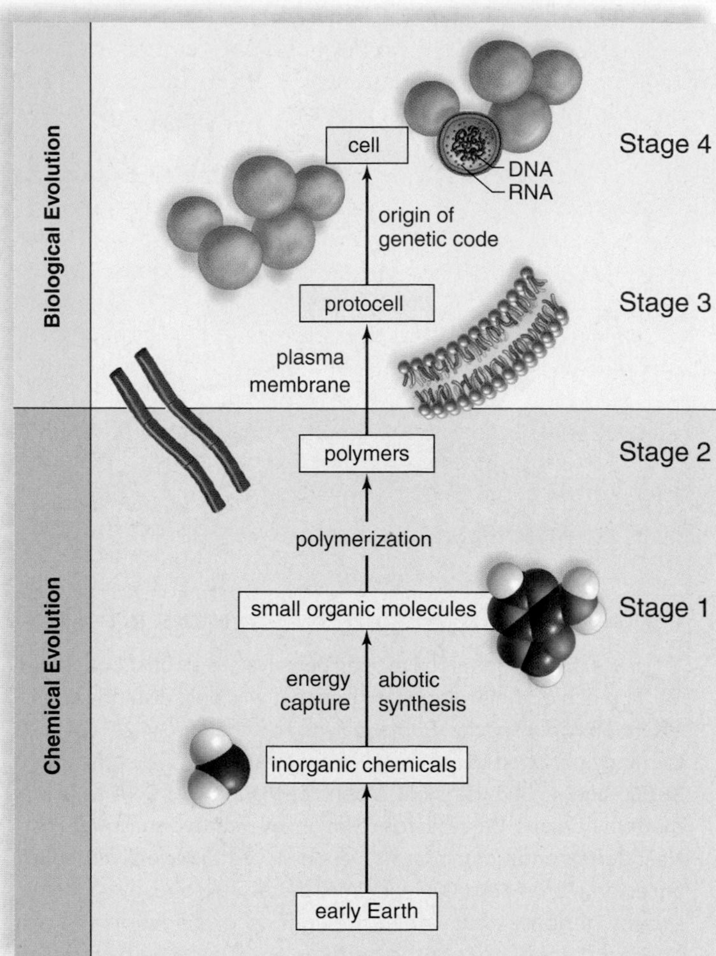

Figure 22.1 Chemical and biological evolution. The first organic molecules (*bottom*) originated from chemically altered inorganic molecules present on early Earth (Stage 1). More complex organic macromolecules were synthesized to create polymers (Stage 2) that were then enclosed in a plasma membrane to form the protocells, or probionts (Stage 3). The protocell underwent biological evolution to produce the first true, self-replicating, living cell (Stage 4).

Small Organic Molecules

The early Earth's atmosphere was not the same as today's atmosphere. Most likely, the first atmosphere was formed by gases escaping from volcanoes. If so, the primitive atmosphere would have consisted mostly of water vapor (H_2O), nitrogen (N_2), and carbon dioxide (CO_2), with only small amounts of hydrogen (H_2) and carbon monoxide (CO). The primitive atmosphere had little, if any, free oxygen.

At first, the Earth and its atmosphere were extremely hot. Water, existing only as a gas, formed dense, thick clouds. Then, as the Earth cooled, water vapor condensed to liquid water, and rain began to fall. The rain washed gases, and other chemicals, into the oceans. Rain fell in such enormous quantities over hundreds of millions of years that the oceans of the world were produced.

The primitive Earth had many sources of energy. These included volcanoes, meteorites, radioactive isotopes, lightning, and ultraviolet radiation. In the presence of so much available energy, the primitive gases may have reacted with one another. This may have produced small organic compounds, such as nucleotides and amino acids (Stage 1 of Fig. 22.1). In 1953, Stanley Miller performed an experiment (Fig. 22.2). To simulate the Earth's early environment, Miller placed the inorganic materials believed to have been present on the early Earth in a closed system, heated it, and circulated it past an electrical spark. After a week, the solution contained a variety of amino acids and organic compounds. This and other similar experiments support the hypothesis that inorganic chemicals may form organic molecules in the presence of a strong energy source, even if oxygen is not present.

Animation Miller-Urey Experiments

There are other hypotheses on how the first small organic monomers may have evolved. Researchers have proposed that thermal vents at the bottom of the Earth's oceans provided all the elements and conditions necessary to synthesize organic monomers. According to one of these hypotheses, dissolved gases emitted from thermal vents, such as carbon monoxide (CO), ammonia, and hydrogen sulfide, pass over iron and nickel sulfide minerals, also present at thermal vents. The iron and nickel sulfide molecules act as catalysts that drive the chemical evolution from inorganic to organic molecules.

It is also possible that the initial organic monomers may not have originated on this planet. Comets and meteorites have constantly pelted the Earth throughout history. In recent years, scientists have confirmed the presence of organic molecules in some meteorites. Some scientists feel that these organic molecules could have seeded the chemical origin of life on early Earth. Others even hypothesize that bacterium-like cells evolved first on another planet and then were carried to Earth. A meteorite from Mars labeled ALH84001 landed on Earth some 13,000 years ago. When examined, experts found tiny rods similar in shape to fossilized bacteria. This hypothesis continues to be investigated.

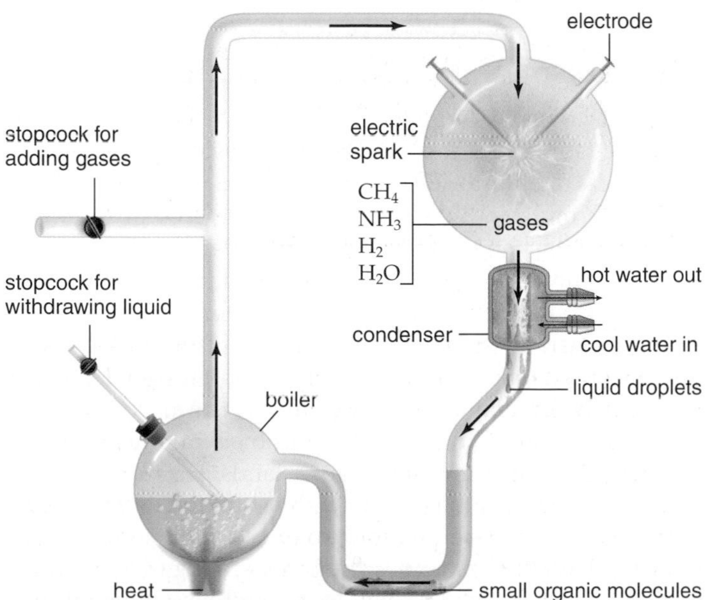

Figure 22.2 The Stanley Miller experiment.
Gases thought to be present early in the Earth's atmosphere were admitted to the apparatus, circulated past an energy source (electrical spark), and cooled to produce a liquid that could be withdrawn. Upon chemical analysis, the liquid was found to contain various small organic molecules.

Macromolecules

The newly formed small organic molecules likely joined to produce larger polymers (Stage 2 of Fig. 22.1). There are two hypotheses of special interest concerning this stage in the origin of life. One is the **RNA-first hypothesis.** This hypothesis suggests that only the macromolecule RNA was needed at this time to progress toward formation of the first cell or cells. This hypothesis was formulated after the discovery that RNA can sometimes be both a substrate and an enzyme during RNA processing (see section 21.2). At that time, the splicing of mRNA to remove introns was done by a complex composed of both RNA and protein. The RNA, not the protein, is the enzyme. RNA enzymes are called *ribozymes.* Then, too, ribosomes where protein synthesis occurs contain rRNA. Perhaps RNA, then, could have carried out the processes of life commonly associated today with DNA and proteins. Scientists who support this hypothesis are fond of saying that it was an "RNA world" some 3.5 BYA. However, DNA, being a double helix, is more stable than RNA. In the RNA-first hypothesis, RNA would have served as the first genetic material, with a transition to DNA occurring later. As far as we currently know, all organisms on the planet use DNA as their genetic material, but scientists continue to look for evidence to support the RNA-first hypothesis.

Another hypothesis is termed the **protein-first hypothesis.** Sidney Fox, an American biochemist, demonstrated that amino acids join together when exposed to dry heat. He suggested that amino acids collected in shallow puddles along the rocky shore. The heat of the sun caused them to form proteinoids, small polypeptides that have some catalytic properties. When proteinoids are returned to water, they form microspheres. Microspheres are structures composed only of protein that have many of the properties of a cell.

The Protocell

A cell has a lipid-protein membrane. Fox demonstrated that if lipids are made available to microspheres, the two tend to become associated, producing a lipid-protein membrane. A **protocell** (see Stage 3 of Fig. 22.1), which could carry on metabolism but could not reproduce, could have come into existence in this manner.

The protocell would have been able to use the still-abundant small organic molecules in the ocean as food. Therefore, the protocell was, most likely, a **heterotroph,** an organism that takes in preformed food. Further, the protocell would have been a fermenter, because there was no free oxygen.

The True Cell

A true cell can reproduce. In today's cells, DNA replicates before cell division occurs. Enzymatic proteins carry out the replication process.

How did the first cell (see Stage 4 of Fig. 22.1) acquire both DNA and enzymatic proteins? Scientists who support the RNA-first hypothesis propose a series of steps. According to this hypothesis, the first cell had RNA genes that, like messenger RNA, could have specified protein synthesis. Some of the proteins formed would have been enzymes. Perhaps one of these enzymes, such as reverse transcriptase found in retroviruses, could use RNA as a template to form DNA. Replication of DNA would then proceed normally.

By contrast, supporters of the protein-first hypothesis suggest that some of the proteins in the protocell would have evolved the enzymatic ability to synthesize DNA from nucleotides in the ocean. Then, DNA would have gone on to specify protein synthesis; in this way, the cell could have acquired all of its enzymes, even the ones that replicated DNA.

Several scientists have proposed that polypeptides and RNA evolved simultaneously. Therefore, the first true cell would have contained RNA genes that could have replicated because of the presence of proteins. This eliminates the baffling chicken-and-egg paradox: Assuming a plasma membrane, which came first, proteins or RNA? It means, however, that two unlikely events would have had to happen at the same time.

After DNA formed, the genetic code had to evolve before DNA could store genetic information. The present genetic code is subject to fewer errors than a million other possible codes. Also, the present code is among the best at minimizing the effect of mutations. A single-base change in a present codon is likely to result in the substitution of a chemically similar amino acid and, therefore, minimal changes in the final protein. This evidence suggests that the genetic code, like all other cellular processes, underwent natural selection before finalizing into today's code.

CHECK YOUR PROGRESS 22.1

1. Discuss why chemical evolution was necessary before biological evolution could occur.
2. Explain what the Miller experiments demonstrated about the formation of the first organic molecules.
3. Discuss the importance of RNA in the formation of the first protocells.

CONNECTING THE CONCEPTS

For more information on the topics presented in this section, refer to the following discussions:

Section 3.1 describes the principles of the cell theory.

Section 3.3 examines the structure of the plasma membrane of cells.

Section 21.1 describes the structure and function of RNA.

22.2 Biological Evolution

LEARNING OUTCOMES

Upon completion of this section, you should be able to

1. Explain the relationship between adaptation and the process of biological evolution.
2. Describe how the process of natural selection supports the concept of biological evolution.
3. Discuss how the fossil record, biogeography, and anatomical and biochemical evidence all support the concept of biological evolution.
4. Distinguish between homologous and analogous structure.

The first true cells were the simplest of life-forms. Therefore, they must have been **prokaryotic cells,** which lack a nucleus. Later, **eukaryotic cells** (namely, the protists), which have nuclei, evolved. Then, multicellularity and the other kingdoms (fungi, plants, and animals) evolved (see Fig. 1.5). Evidence suggests that all life on Earth has an evolutionary history and has undergone biological evolution, or change over time. Biological evolution has two important aspects: descent from a common ancestor and adaptation to the environment. Descent from the original cell or cells explains why all life has a common chemistry and a cellular structure. An **adaptation** is a characteristic that makes an organism able to survive and reproduce in its environment. Adaptations to different environments help explain the diversity of life—why there are so many different types of living organisms.

Mechanism of Biological Evolution

Charles Darwin was an English naturalist who first presented the concept that the mechanism of biological evolution was the process of **natural selection,** or descent with modification. Darwin formulated his ideas while sailing around the world as the naturalist on board the HMS *Beagle.* Between 1831 and 1836, the ship sailed in the tropics of the Southern Hemisphere.

When Darwin returned home, he spent the next 20 years gathering data to support the principle of biological evolution. Darwin's most significant contribution was to describe a mechanism for adaptation—**natural selection.** During adaptation, a species becomes suited to its environment. On his trip, Darwin visited the Galápagos Islands. He saw a number of finches that resembled one another but had different ways of life. Some were seed-eating ground finches, some cactus-eating ground finches, and some insect-eating tree finches. A warbler-type finch had a beak that could take honey from a flower. A woodpecker-type finch lacked the long tongue of a woodpecker but could use a cactus spine or twig to pull insects from cracks in the bark of a tree. Darwin thought the finches were all descended from a mainland ancestor whose offspring had spread out among the islands and become adapted to different environments.

APPLICATIONS AND MISCONCEPTIONS

What is intelligent design?

Evolution is a scientific theory. Sometimes we use the word *theory* when we mean a hunch or a guess. But in science, the term *theory* is reserved for those ideas that scientists have found to be all-encompassing because they are based on evidence (data) collected in a number of different fields. In other words, evolutionary theory has been supported by repeated scientific experiments and observations.

Some people advocate the teaching of ideas that run contrary to the theory of evolution in schools. The emphasis is currently placed on intelligent design, a belief system that maintains that the diversity of life could never have arisen without the involvement of an "intelligent agent." Many scientists, and even religions, argue that intelligent design is faith-based and not science-based. It would not be possible to test in a scientific way whether an intelligent agent exists. If it were possible to structure such an experiment, scientists would be the first to do it.

To emphasize the nature of Darwin's natural selection process, it is often contrasted with a process described by Jean-Baptiste Lamarck, another nineteenth-century naturalist. Lamarck's explanation for the long neck of the giraffe was based on the assumption that the ancestors of the modern giraffe were trying to reach into the trees to browse on high-growing vegetation (Fig. 22.3). Continual stretching of the neck caused it to become longer, and this acquired characteristic was passed on to the next generation. Lamarck's mechanism will not work because acquired characteristics cannot be inherited (Fig. 22.3).

The critical elements of the natural selection process are:

- *Variation.* Individual members of a species vary in physical characteristics. Physical variations can be passed from generation to generation. (Darwin was never aware of genes, but we know today that the inheritance of the genotype determines the phenotype.)
- *Competition for limited resources.* Even though each individual could eventually produce many descendants, the number in each generation usually stays about the same. Why? Resources are limited and competition for resources results in unequal reproduction among members of a population.
- *Adaptation.* Those members of a population with advantageous traits capture more resources and are more likely to reproduce and pass on these traits. Thus, over time, the environment "selects" for the better-adapted traits. Each subsequent generation includes more individuals adapted in the same way to the environment.

Video Finches Natural Selection

Lamarck's proposal

Originally, giraffes had short necks.

Giraffes stretched their necks in order to reach food.

With continual stretching, most giraffes now have long necks.

Darwin's proposal

Originally, giraffe neck length varied.

Competition for resources causes long-necked giraffes to have the most offspring.

Due to natural selection, most giraffes now have long necks.

Figure 22.3 The two major mechanisms for evolutionary change in the nineteenth century.
This diagram contrasts Jean-Baptiste Lamarck's process of acquired characteristics with Charles Darwin's process of natural selection.

APPLICATIONS AND MISCONCEPTIONS

When did Darwin publish his book on natural selection? What was the reaction to his work?

Darwin published his book on natural selection in 1859. It was entitled *On the Origin of Species by Means of Natural Selection, or The Preservation of Favoured Races in the Struggle for Life.* The title is usually shortened to *On the Origin of Species.* It was instantly controversial, even though he avoided the use of the word "evolution." He only alludes to humans with the statement that, by his theory, ". . . light will be thrown on the origin of man and his history."

The book is still in print and you can obtain a copy.

Darwin noted that when humans help carry out **artificial selection,** they breed selected animals with particular traits to reproduce. For example, prehistoric humans probably noted desirable variations among wolves and selected particular individuals for breeding. Therefore, the desired traits increased in frequency in the next generation. This same process was repeated many times, resulting in today's numerous varieties of dogs, all descended from the wolf. In a similar way, several varieties of vegetables can be traced to a single ancestor. Chinese cabbage, brussels sprouts, and kohlrabi are all derived from a single species, *Brassica oleracea.*

Natural selection can account for the great diversity of life. Environments differ widely; therefore, adaptations are varied. From vampire bats to sea turtles to the many finches observed by Darwin, all the different organisms are adapted to their way of life.

APPLICATIONS AND MISCONCEPTIONS

When did humans first start to practice artificial selection?

Almost all animals that are currently used in modern agriculture are the result of thousands of years of artificial selection by humans. But perhaps the longest-running experiment in artificial selection is the modern dog. Analysis of canine DNA indicates that dogs (*Canis familiaris*) are a direct descendant of the gray wolf (*Canis lupus*). This domestication, and the subsequent selection for desirable traits, appear to have begun over 130,000 years ago. Artificial selection of dogs continues to this day, with over 150 variations (or breeds) currently known.

Evidence of Evolution

Many different lines of evidence support the concept that organisms are related through descent from a common ancestor. This is significant because the more varied and abundant the evidence supporting a hypothesis, the more certain it becomes.

Fossil Evidence

Fossils remain one of the best sources of evidence for evolution. They are the actual remains of species that lived on Earth at least 10,000 years ago and up to billions of years ago. Fossils can be the traces of past life or any other direct evidence that past life existed. Traces include trails, footprints, burrows, worm casts, or even preserved droppings. Fossils can also be such items as pieces of bone, impressions of plants pressed into shale, and even insects trapped in tree resin (which we know as amber). Most fossils, however, are found embedded in or recently eroded from sedimentary rock. Sedimentation, a process that has been going on since the Earth was formed, can take place on land or in bodies of water. Weathering and erosion of rocks produce an accumulation of particles. These particles vary in size and nature and are called sediment. Sediment becomes a stratum (pl., strata), a recognizable layer in a sequence of layers. Any given stratum is older than the one above it and younger than the one immediately below it. This allows fossils to be dated.

Usually, when an organism dies, the soft parts are either consumed by scavengers or decomposed by bacteria. This means that most fossils consist only of hard parts such as shells, bones, or teeth. These are usually not consumed or destroyed. When a fossil is found encased by rock, the remains were first buried in sediment. The hard parts were preserved by a process called mineralization. Finally, the surrounding sediment hardened to form rock. Subsequently, the fossil has to be found by a human. Most estimates suggest that less than 1% of past species have been preserved as fossils. Only a small fraction of these have been found.

More and more fossils have been found because researchers, called paleontologists, and their assistants have been out in the field looking for them. Usually, paleontologists remove fossils from the strata to study them in the laboratory. Then they may decide to exhibit them. The **fossil record** is the history of life recorded by fossils. *Paleontology* is the science of discovering the fossil record. Decisions about the history of life, ancient climates, and environments can be made using the fossil record. The fossil record is the most direct evidence we have that evolution has occurred. The species found in ancient sedimentary rock are not the species we see today.

Darwin relied on fossils to formulate his theory of evolution. Today, we have a far more complete record than was available to Darwin. The record is complete enough to tell us that, in general, life has progressed from the simple to the complex. Unicellular prokaryotes are the first signs of life in the fossil record. Unicellular eukaryotes and then multicellular eukaryotes followed. Among the latter, fishes evolved before terrestrial plants and animals. On land, nonflowering plants preceded the flowering plants. Amphibians preceded the reptiles, including the dinosaurs. Dinosaurs are directly linked to the birds, but they are only indirectly linked to the evolution of mammals, including humans.

Transitional fossils are those that have characteristics of two different groups. In particular, they tell us who is related to whom and how evolution occurred. In 2004, a team of paleontologists discovered fossilized remains of *Tiktaalik roseae,* nicknamed the "fishapod" because it is the transitional form between fish and four-legged animals, the tetrapods (Fig. 22.4). *Tiktaalik* fossils are estimated to be 375 million years old, and are from a time when the transition from fish to tetrapods is likely to have occurred. As expected of an intermediate fossil, *Tiktaalik* has a mix of fishlike and tetrapod-like features that illustrate the steps in the evolution of tetrapods from a fishlike ancestor (Fig. 22.4). For example, *Tiktaalik* has a very fishlike set of gills and fins, with the exception of the pectoral, or front fins, which have the beginnings of wrist bones similar to a tetrapod. Unlike a fish, *Tiktaalik* has a flat head, flexible neck, eyes on the top of its head like a crocodile,

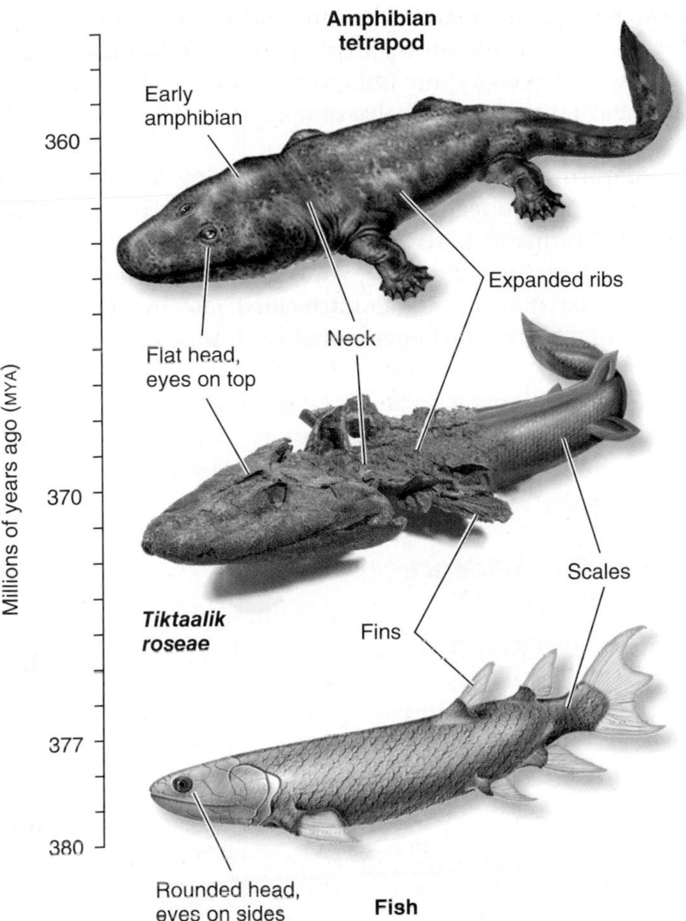

Figure 22.4 **Transitional fossils.**
Tiktaalik roseae has a mix of fishlike and tetrapod-like features.

and interlocking ribs that suggest it had lungs. These transitional features suggest that it had the ability to push itself along the bottom of shallow rivers and see above the surface of the water—features that would come in handy in the river habitat where it lived.

Even in Darwin's day, scientists knew of the *Archaeopteryx* fossils, which are intermediate to reptiles and birds. The dinosaur-like skeleton of these fossils had reptilian features, including jaws with teeth, and a long, jointed tail. But *Archaeopteryx* also had feathers and wings. Figure 22.5 not only shows a fossil of *Archaeopteryx* but also gives us an artist's representation of the animal based on the fossil remains. Another example of how transitional fossils can be used to trace the evolutionary history of an organism is the whale. It had always been thought that whales had terrestrial ancestors. Now, fossils have been discovered that support this hypothesis (Fig. 22.6). *Ambulocetus natans* ("the walking whale that swims") was the size of a large sea lion, with broad webbed feet on both fore- and hindlimbs. This animal could both walk and swim. It also had tiny hooves on its toes and the primitive skull and teeth of early whales. It is believed that *Ambulocetus* was a predator that patrolled freshwater streams looking for prey.

The origin of land mammals is also well documented. The synapsids are mammal-like reptiles whose descendants were wolflike and bearlike predators, as well as several types of piglike herbivores. Slowly, mammalian-like fossils acquired features, such as a palate, that would have enabled them to breathe and eat at the same time. They also acquired a muscular diaphragm and rib cage that would have helped them breathe efficiently. The earliest true mammals were shrew-size creatures found in fossil beds about 200 million years old.

Biogeographical Evidence

Biogeography is the study of the distribution of plants and animals in different places throughout the world. Such distributions are consistent with the hypothesis that life-forms evolved in a particular locale before they spread out. Therefore, you would expect a different mix of plants and animals whenever geography separates continents, islands, or seas. For example, Darwin noted that South America lacks a diversity of rabbits, even though the environment was suitable for them. He concluded that few rabbits lived in South America because rabbits evolved somewhere else and had no means of reaching South America. Instead, the Patagonian hare lives in South America. The Patagonian hare resembles a rabbit in anatomy and behavior, but it has the face of a guinea pig, from which it probably evolved.

As another example, both cacti and euphorbia are plants adapted to a hot, dry environment. Both are succulent, spiny, flowering plants. Why do cacti grow in North American deserts and euphorbia grow in African deserts, when each would do

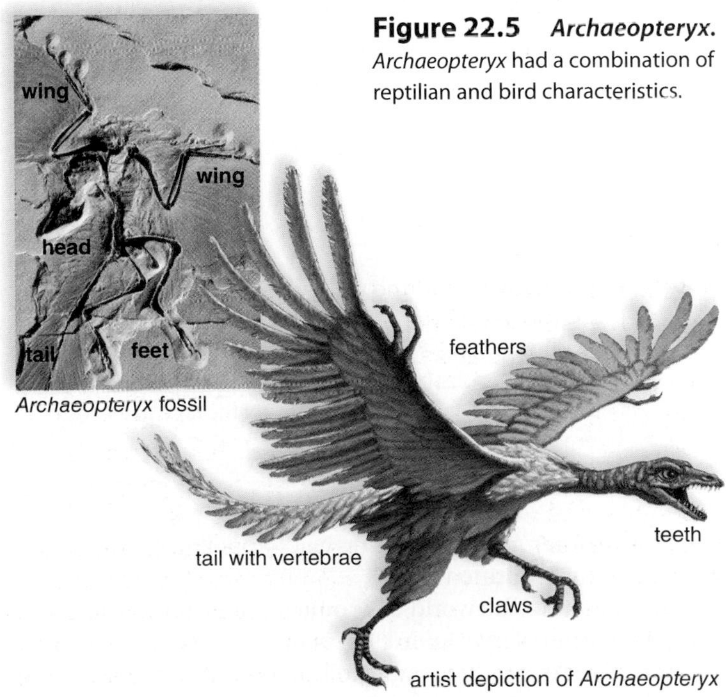

Figure 22.5 ***Archaeopteryx.***
Archaeopteryx had a combination of reptilian and bird characteristics.

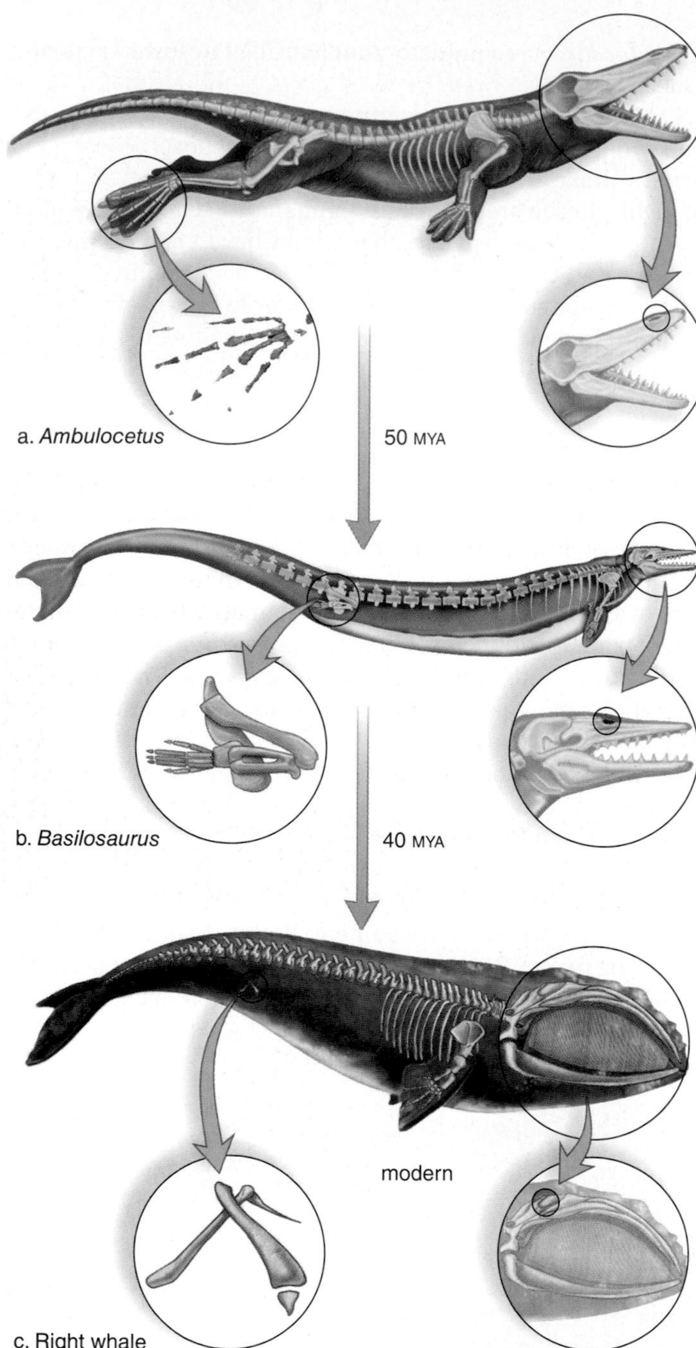

Figure 22.6 Evolution of the whales.
Transitional fossils such as *Ambulocetus* and *Basilosaurus* support the hypothesis that modern whales evolved from terrestrial ancestors that walked on four limbs. These fossils show a gradual reduction in the hindlimb and a movement of the nasal opening from the tip of the nose to the top of the head—both adaptations to living in water.

well on the other continent? They just happened to evolve on their respective continents.

The islands of the world are home to many unique species of animals and plants found no place else, even when the soil and climate are the same. Why do so many species of finches live on the Galápagos Islands, when these same species are not on the mainland? The reasonable explanation is that finches from the ancestral species migrated to all the different islands. Then, geographic isolation allowed the ancestral finches to evolve into a different species on each island.

Video Finches Adaptive Radiation

In the history of the Earth, South America, Antarctica, and Australia were originally connected. Marsupials (pouched mammals) arose at this time and today are found in both South America and Australia. But when Australia separated and drifted away, the marsupials diversified into many different forms suited to various environments of Australia (Fig. 22.7). They were free to do so because there were few, if any, placental mammals in Australia. In South America, where there are placental mammals, marsupials are not as diverse. This supports the hypothesis that evolution is influenced by the mix of plants and animals in a particular continent—by biogeography.

Anatomical Evidence

Darwin was able to show that a hypothesis that features common descent offers a plausible explanation for anatomical similarities among organisms. Vertebrate forelimbs are used for flight (birds and bats), orientation during swimming (whales and seals), running (horses), climbing (arboreal lizards), or swinging from tree branches (monkeys). Yet all vertebrate

Sugar glider, *Petaurus breviceps*, is a tree-dweller and resembles the placental flying squirrel.

The Australian wombat, *Vombatus*, is nocturnal and lives in burrows. It resembles the placental woodchuck.

Kangaroo, *Macropus*, is an herbivore that inhabits plains and forests. It resembles the placental Patagonian cavy of South America.

Figure 22.7 Biogeography.
Each type of marsupial in Australia is adapted to a different way of life. All of the marsupials in Australia presumably evolved from a common ancestor that entered Australia some 60 million years ago.

forelimbs contain the same sets of bones organized in similar ways, despite their dissimilar functions (Fig. 22.8). The most plausible explanation for this unity is that the basic forelimb plan belonged to a common ancestor. The basic plan was then modified in the succeeding groups as each continued along its own evolutionary pathway. Structures that are anatomically similar because they are inherited from a common ancestor are called **homologous structures.** In contrast, **analogous structures** serve the same function but are not constructed similarly; nor do they share a common ancestry. The wings of birds and insects and the jointed appendages of a lobster and humans are analogous structures. The presence of homology, not analogy, is evidence that organisms are related.

Vestigial structures are anatomical features that are fully developed in one group of organisms but that are reduced and may have no function in similar groups. Modern whales have a vestigial pelvic girdle and legs. The ancestors of whales walked on land, but whales are totally aquatic animals today. Most birds have well-developed wings used for flight. Some bird species (e.g., ostrich), however, have greatly reduced wings and do not fly. Similarly, snakes have no use for hindlimbs; yet some have remnants of a pelvic girdle and legs. Humans have a tailbone but no tail. The presence of vestigial structures can be explained by common descent. Vestigial structures occur because organisms inherit their anatomy from their ancestors. They are traces of an organism's evolutionary history.

The homology shared by vertebrates extends to their embryological development (Fig. 22.9). At some time during development, all vertebrates have a postanal tail and exhibit paired pharyngeal pouches. In fish and amphibian larvae, these pouches develop into functioning gills. In humans, the first pair of pouches becomes the cavity of the middle ear and the auditory tube. The second pair becomes the tonsils. The third and fourth pairs become the thymus and parathyroid glands, respectively. Why should terrestrial vertebrates develop and then modify structures like pharyngeal pouches that have lost their original function? The most likely explanation is that fish are ancestral to other vertebrate groups.

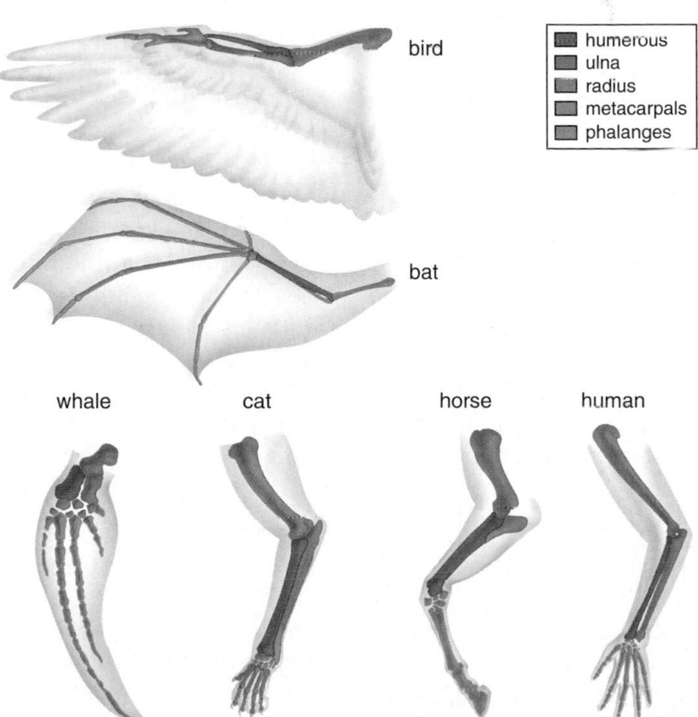

Figure 22.8 **Vertebrate forelimbs are homologous structures.** Despite differences in function, vertebrate forelimbs have the same bones.

APPLICATIONS AND MISCONCEPTIONS

What are some examples of vestigial organs in humans?

The human body is littered with vestigial organs from our evolutionary past. One example is the tiny muscles (called piloerectors) that surround each hair follicle. During times of stress, these muscles cause the hair to stand straight up—a useful protection mechanism for small mammals trying to escape predators but one that has little function in humans. Wisdom teeth are also considered to be vestigial organs, because most people now retain their original teeth for the majority of their lives.

Biochemical Evidence

Almost all living organisms use the same basic biochemical molecules, including DNA, ATP (adenosine triphosphate), and many identical or nearly identical enzymes. Further, organisms

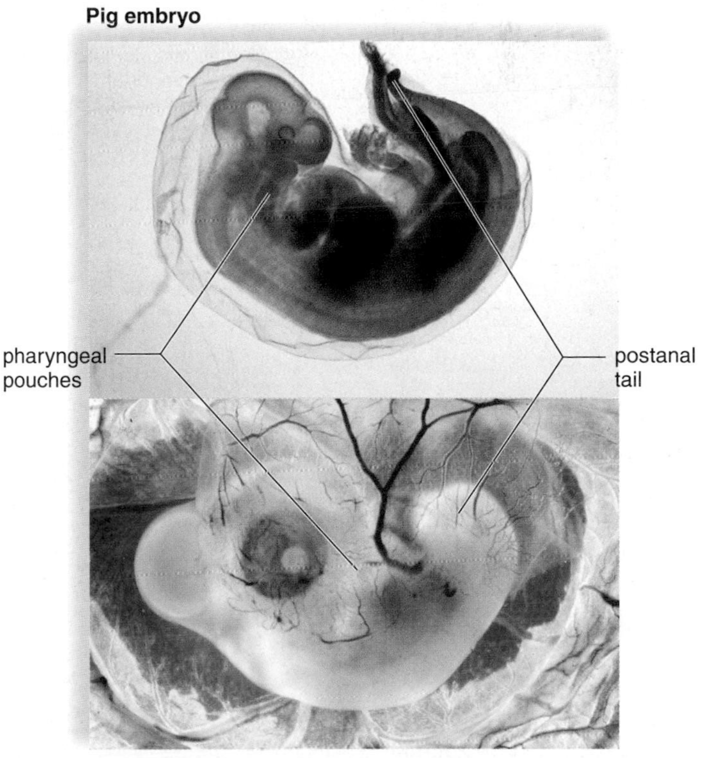

Figure 22.9 **Homologous structures in vertebrate embryos.** Vertebrate embryos have features in common, such as pharyngeal pouches, despite different ways of life as adults.

use the same DNA triplet code and the same 20 amino acids in their proteins. The sequences of DNA bases in the genomes of many organisms are now known, so it has become clear that humans share a large number of genes with much simpler organisms. Evolutionists who study development have also found that many developmental genes (called *Hox* genes) are shared in animals ranging from worms to humans. It appears that life's vast diversity has come about by only a slight difference in the regulation of genes. The result has been widely divergent types of bodies.

When the degree of similarity in DNA base sequences or the degree of similarity in amino acid sequences of proteins is examined, the data are as expected, assuming common descent. Cytochrome *c* is a molecule used in the electron transport chain of many organisms. Data regarding differences in the amino acid sequence of cytochrome *c* show that the sequence in a human differs from that in a monkey by only two amino acids. The human sequence differs from that in a duck by 11 amino acids, and from that in a yeast by 51 amino acids (Fig. 22.10). These data are consistent with other data regarding the anatomical similarities of these organisms and, therefore, their relatedness.

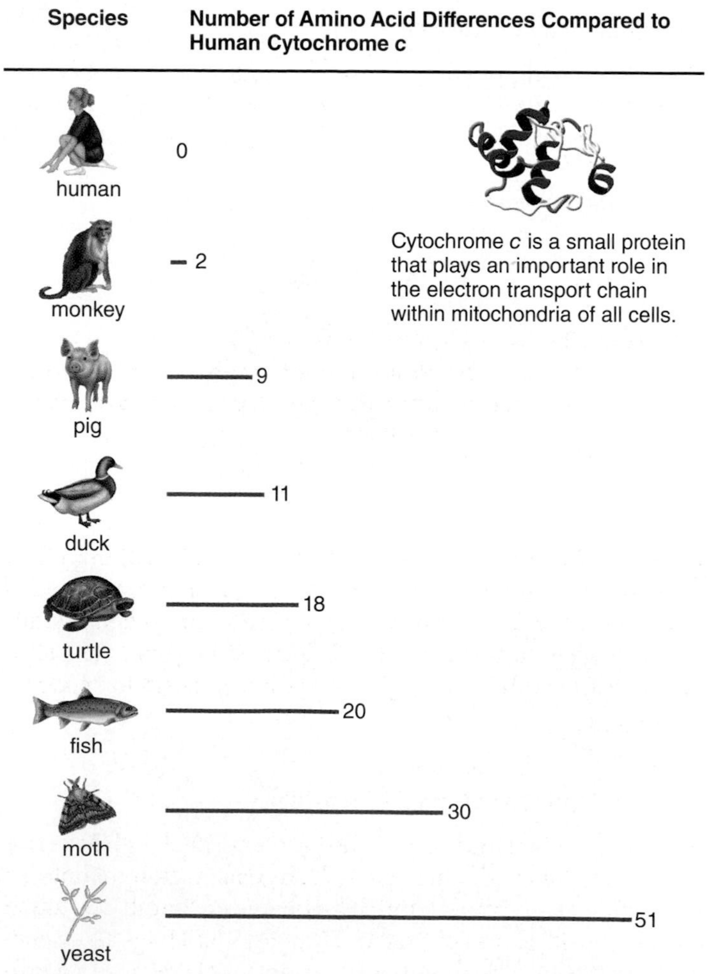

Figure 22.10 Biochemical evidence describes evolutionary relationships.
The number of amino acid differences in cytochrome *c* between humans and other species is indicated.

CHECK YOUR PROGRESS 22.2

1. Define *biological evolution,* and explain what its two most important aspects are.
2. Describe the types of evidence that support Darwin's theory of biological evolution.
3. Discuss why natural selection is the mechanism for biological evolution.

CONNECTING THE CONCEPTS

For more information on the information provided in this section, refer to the following discussions:

Section 1.1 explains why evolutionary change is the core concept of the study of biology.

Section 18.4 describes how meiosis introduces the variation that is the basis of evolutionary change.

Section 21.3 explores how scientists study changes in DNA and proteins.

22.3 Classification of Humans

LEARNING OUTCOMES

Upon completion of this section, you should be able to

1. Describe how DNA analysis is used to study primate evolution.
2. Describe the evolutionary trends that occur in the primates.
3. Compare the structure of chimpanzee and human skeletons and list the adaptations in humans that make upright walking possible.

To begin a study of human evolution, we turn to the classification of humans because biologists classify organisms according to their evolutionary relatedness. The **binomial name** of an organism gives its genus and species. Organisms in the same domain have only general characteristics in common. Those in the same genus have specific characteristics in common. Table 22.1 lists some of the characteristics that help classify humans. The dates in the first column of the table tell us when these groups of animals first appear in the fossil record.

DNA Data and Human Evolution

We are accustomed to using the characteristics given in Table 22.1 to determine evolutionary relationships, but researchers are just as apt to use DNA data. Researchers are depending more and more on DNA data to trace the history of life. DNA data are particularly useful when anatomical differences are unavailable.

For example, in the late 1970s, Carl Woese and his colleagues at the University of Illinois decided to use ribosomal RNA (rRNA) sequence data to decide how prokaryotes are

Table 22.1 Evolution and Classification of Humans

BYA/MYA*	Classification Category	Characteristics
2 BYA	Domain Eukarya	Membrane-bound nucleus
600 MYA	Kingdom Animalia	Multicellular, motile, heterotrophic
540 MYA	Phylum Chordata	Sometime in life history: dorsal tubular nerve cord, notochord, pharyngeal pouches
120 MYA	Class Mammalia	Vertebrates with hair, mammary glands
60 MYA	Order Primates	Well-developed brain, adapted to live in trees
7 MYA	Family Hominidae	Adapted to upright stance and bipedal locomotion
3 MYA	Genus *Homo*	Most developed brain, made and used tools
0.1 MYA	Species *Homo sapiens*†	Modern humans; speech centers of brain well-developed

*BYA = billions of years ago; MYA = millions of years ago.
†To specify an organism, you must use the full binomial name, such as *Homo sapiens*.

related. They knew that the DNA coding for rRNA changes slowly during evolution. Ribosomal RNA genes may change only when there is a major evolutionary event. Woese reported, on the basis of rRNA sequence data, that there are three domains of life and members of Archaea are more closely related to members of Eukarya than to those of Bacteria (Fig. 22.11). (See Fig. 1.6 for a description of these domains.) In other words, major decisions regarding the history of life are now being made on the basis of DNA/rRNA/protein sequencing data. (Fig. 22.10 gave an example of the use of protein data to show evolutionary relationships.) For example, studies of rRNA sequences indicate that among the major groups of eukaryotes, animals are more closely related to fungi than they are to plants.

Animation Three Domains

Scientists have used a variety of techniques, including DNA sequence data, to calculate when the last common ancestor for the apes and humans must have existed. While paleontologists and archeologists are getting closer to discovering this ancestor, the DNA data indicate that this ancestor must have existed about 7 MYA. For more recent events, mitochondrial DNA (mtDNA) is often used, since mtDNA changes occur more frequently than changes to the nuclear DNA. Mitochondrial DNA data indicate that humans first evolved in Africa and later migrated to Eurasia.

Humans Are Primates

In contrast to the other orders of placental mammals, **primates** are adapted to living in trees, or an arboreal life. Primates have mobile limbs; grasping hands; a flattened face; binocular vision; a large, complex brain; and a reduced reproductive rate. The order Primates may be divided into two major groups—the *prosimians,* which include lemurs, tarsiers, and lorises; and the *anthropoids,* which include monkeys, apes, and humans. This classification tells us that humans are more

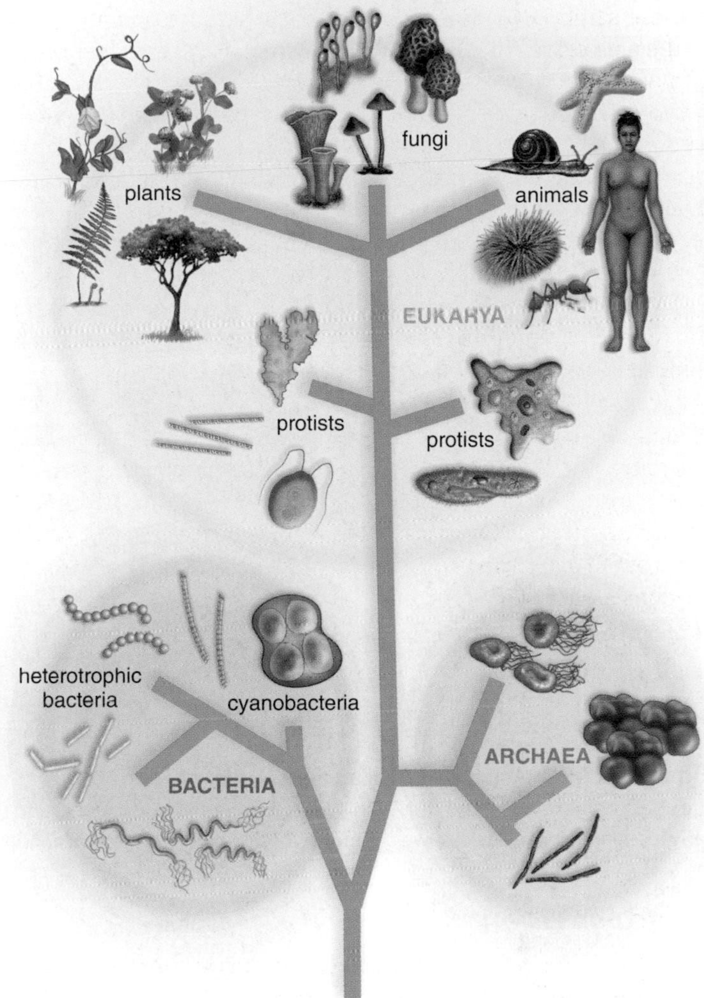

Figure 22.11 The three domains of life.
Representatives of each domain are depicted in the ovals. The evolutionary tree of life shows that domain Archaea is more closely related to domain Eukarya than either is to domain Bacteria.

closely related to the monkeys and apes (Fig. 22.12) than they are to the prosimians. After sequencing the genomes of humans and apes, geneticists have concluded that there is a 90% similarity between humans and apes. Thus, while genetically similar, there is still considerable variation present for specific adaptations.

Mobile Forelimbs and Hindlimbs

Primate limbs are mobile, and the hands and feet both have five digits each. Many primates, such as chimpanzees, have both an opposable big toe and thumb. The big toe or thumb can touch each of the other toes or fingers. Humans don't have an opposable big toe, but the thumb is opposable. This results in a grip that is both powerful and precise. The opposable thumb allows a primate to easily reach out and bring food, such as fruit, to the mouth. When locomoting, primates grasp and release tree limbs freely because nails have replaced claws.

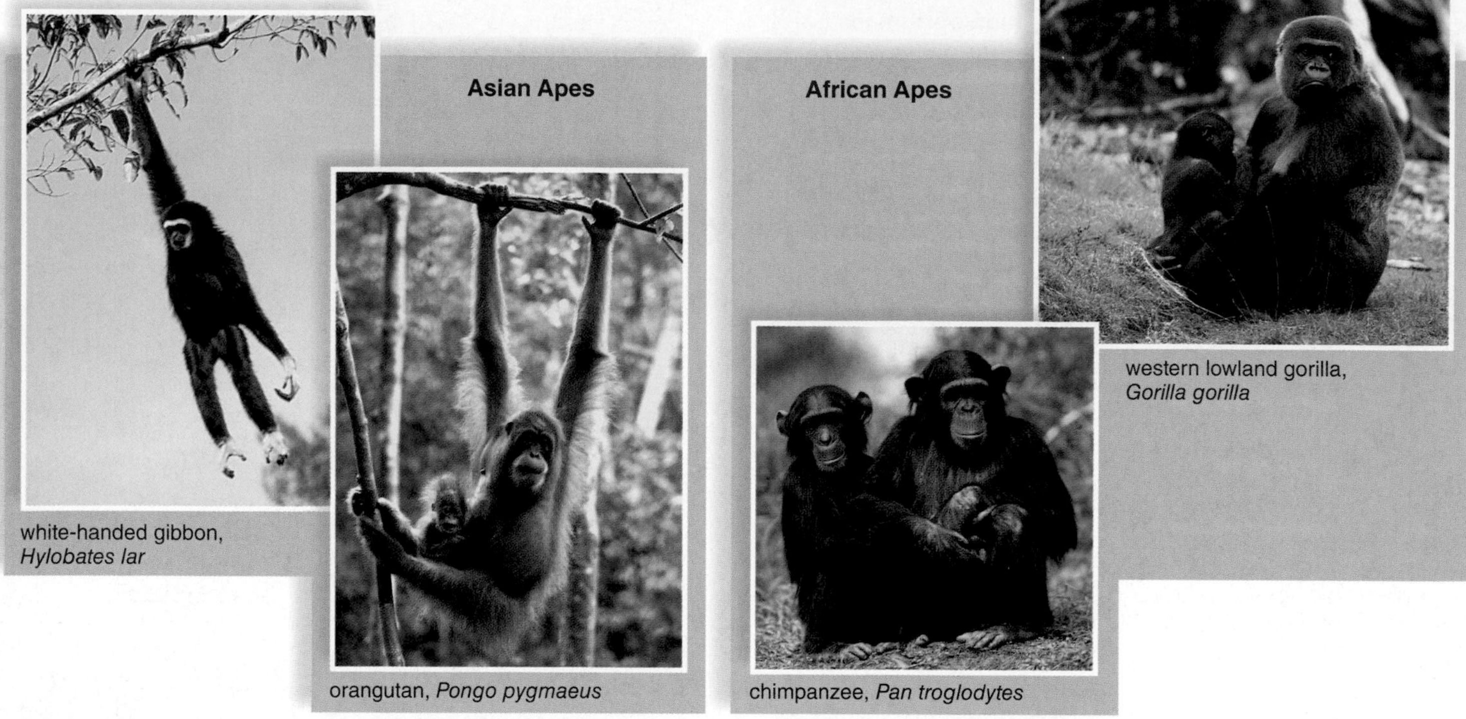

Figure 22.12 Asian and African apes.
The apes can be divided into the Asian apes (gibbons and orangutans) and the African apes (chimpanzees and gorillas). Molecular data and the location of early fossil remains tell us that we are more closely related to the African than to the Asian apes.

Binocular Vision

In chimps, like other primates, the snout is shortened considerably, allowing the eyes to move to the front of the head. The stereoscopic vision (or depth perception) that results permits primates to make accurate judgments about the distance and position of adjoining tree limbs. Humans and the apes have three different cone cells, which are able to discriminate between greens, blues, and reds (see Chapter 14 for a review). Cone cells require bright light, but the image is sharp and in color. The lens of the eye focuses light directly on the fovea, a region of the retina, where cone cells are concentrated.

Large, Complex Brain

The evolutionary trend among primates is generally toward a larger and more complex brain. The brain size is smallest in prosimians and largest in humans and Neandertals. The cerebral cortex, with many association areas, expands so much that it becomes extensively folded in humans. The portion of the brain devoted to smell is smaller. The portions devoted to sight have increased in size and complexity during primate evolution. Also, more of the brain is involved in controlling and processing information received from the hands and the thumb. The result is good hand-eye coordination in chimpanzees and humans.

Reduced Reproductive Rate

It is difficult to care for several offspring while moving from limb to limb, and one birth at a time is the norm in primates. The juvenile period of dependency is extended, and there is an emphasis on learned behavior and complex social interactions.

Comparing the Human Skeleton to the Chimpanzee Skeleton

Figure 22.13 compares anatomical differences between chimpanzees and humans, which relate to the upright stance of humans when they walk compared to the chimpanzees' practice of knuckle-walking. When chimpanzees walk, their forearms rest on their knuckles.

These differences in anatomy between chimpanzees and humans determine that humans—but not chimps—are adapted for an upright stance. (1) In humans, the spine exits inferior to the center of the skull, and this places the skull in the midline of the body. (2) The longer, S-shaped spine of humans places the trunk's center of gravity squarely over the feet. (3) The broader pelvis and hip joint of humans keep them from swaying when they walk. (4) The longer neck of the femur in humans causes the femur to angle inward at the knees. (5) The human knee joint is modified to support the body's weight; the femur is larger at the bottom, and the tibia is larger at the top. (6) Finally, the human toe is not opposable; instead, the foot has an arch. The arch enables humans to walk long distances and run with less chance of injury.

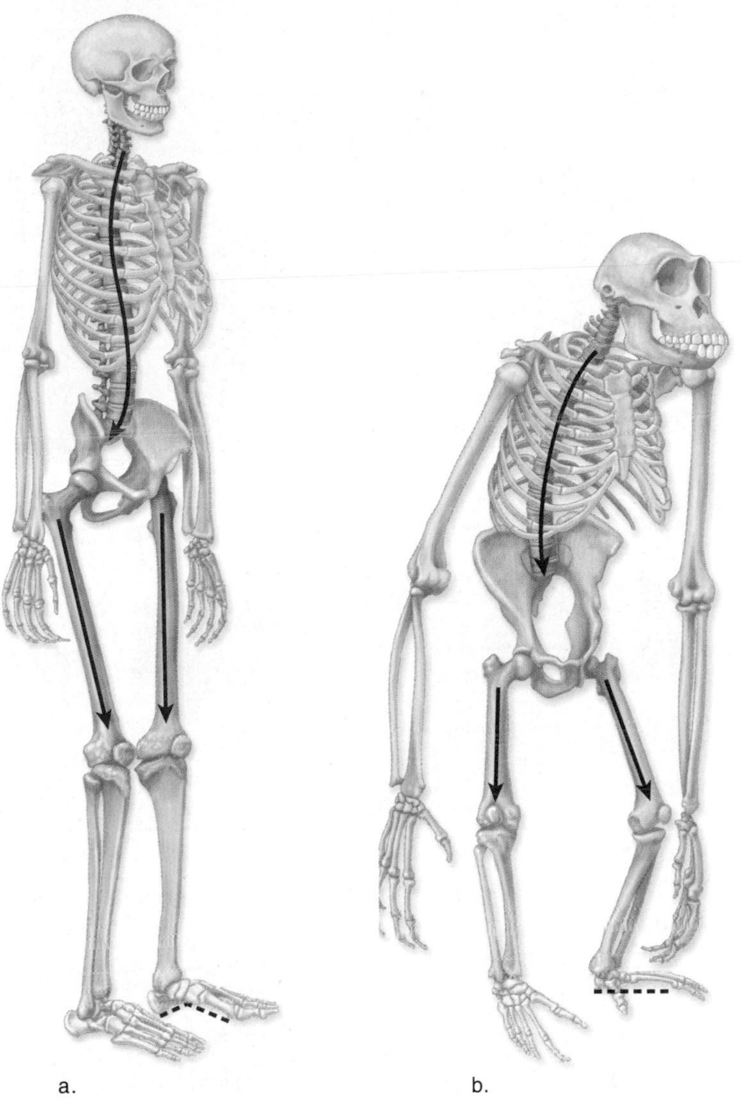

Human spine exits from the skull's center; ape spine exits from rear of skull.
Human spine is S-shaped; ape spine has a slight curve.
Human pelvis is bowl-shaped; ape pelvis is longer and more narrow.
Human femurs angle inward to the knees; ape femurs angle out a bit.
Human knee can support more weight than ape knee.
Human foot has an arch; ape foot has no arch.
a.
b.

Figure 22.13 **Adaptations in the human skeleton allow upright locomotion.**
a. Human skeleton compared to (**b**) chimpanzee skeleton.

CHECK YOUR PROGRESS 22.3

1. List the general characteristics of primates.
2. Discuss the benefits of binocular vision and a complex brain structure.
3. Summarize the major differences between the chimpanzee skeleton and the human skeleton.

CONNECTING THE CONCEPTS

For more information on the material in this section, refer to the following discussions:

Section 3.6 discusses the structure and function of the mitochondria in a eukaryotic cell.

Section 13.2 examines the function of the cerebral cortex in the human brain.

Section 21.1 explores the structure and function of DNA and RNA molecules.

22.4 Evolution of Hominins

LEARNING OUTCOMES

Upon completion of this section, you should be able to

1. Describe the major events in the evolution of the hominins.
2. Summarize the significance of the australopithecines in the study of human evolution.

Once biologists have studied the characteristics of a group of organisms, they can construct an **evolutionary tree,** which represents a working hypothesis of their past history. The evolutionary tree in Figure 22.14 shows that all primates share one common ancestor and that the other types of primates diverged from the human line of descent over time. When any two lines of descent, called a *lineage,* first diverge from a common ancestor, the genes and proteins of the two lineages

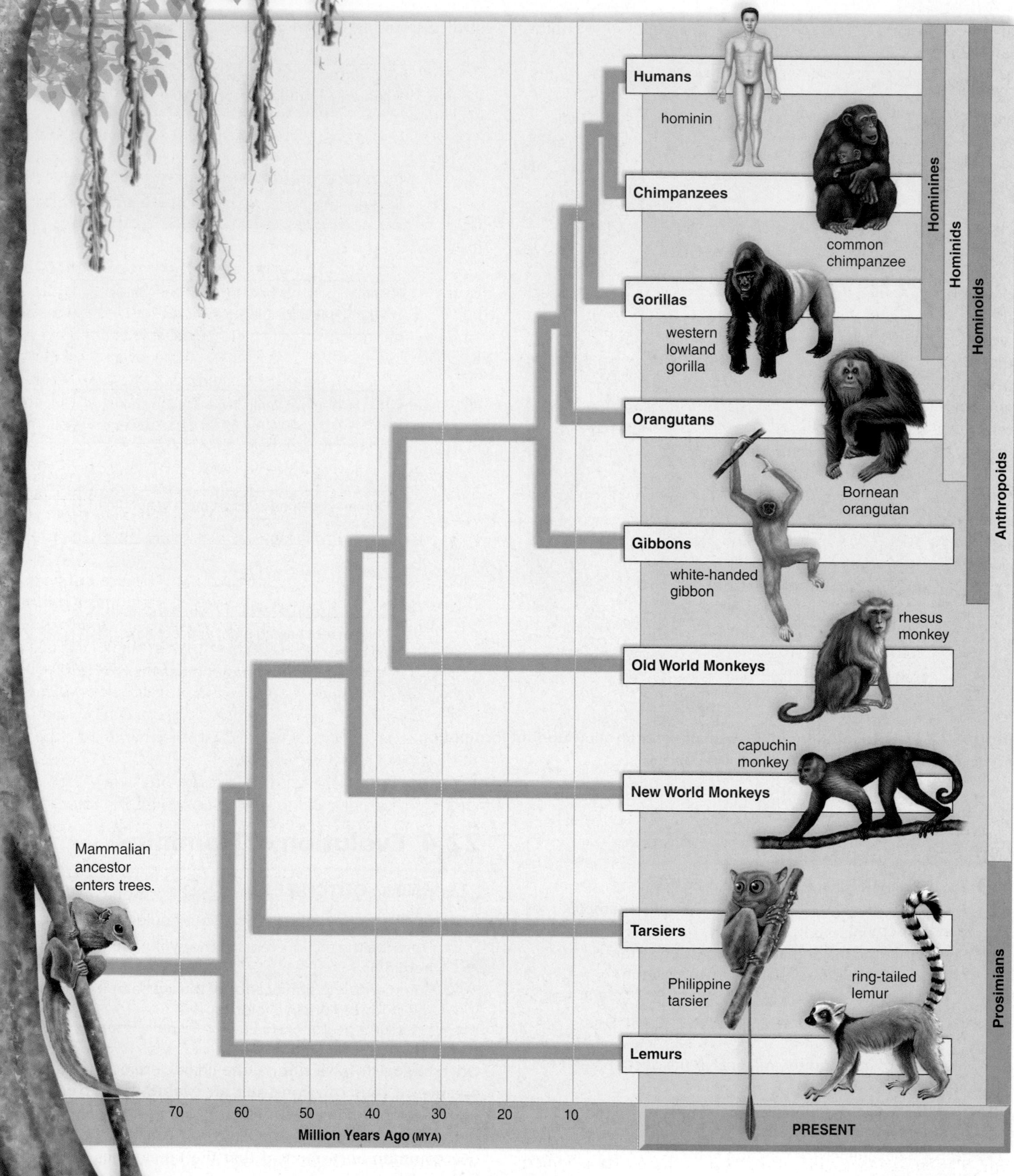

Figure 22.14 **The evolutionary tree of the primates.**
Humans are related to all other primates through a common ancestor.

are nearly identical. As time goes by, each lineage accumulates genetic changes, which lead to RNA and protein changes. Many genetic changes are neutral (not tied to adaptation) and accumulate at a fairly constant rate. Such changes can be used as a type of **molecular clock** to indicate the relatedness of two groups and when they diverged from each other. Molecular data also suggest that hominids split from the ape line of descent about 7 MYA.

The names of the various classifications of primates has changed rapidly over the past several years as new discoveries and biochemical analyses have unveiled more information on primate evolution. For example, as previously mentioned, the **prosimians** now include the lemurs, tarsiers, and lorises; the **anthropoids** include the monkeys, apes, and humans. The designation **hominid** includes the apes (gorillas and orangutans), chimpanzees, humans, and the closest extinct relatives of humans. The term **hominine** is now used to include only the gorillas, chimpanzees, and humans and their closest extinct relatives. The designation **hominin** refers to all members of the genus *Homo* and their close relatives.

One of the most unfortunate misconceptions concerning human evolution is the belief that Darwin and others suggested that humans evolved from apes. On the contrary, humans and apes are thought to have shared a common apelike ancestor. Today's apes are our distant cousins, and we couldn't have evolved from our cousins because we are contemporaries—living on Earth at the same time. Humans and apes have been evolving separately from a common ancestor for about 7 million years. Following the split between humans and apes, different environments selected for the different traits that apes and humans have now.

The First Hominins

Paleontologists use certain anatomical features when they try to determine if a fossil is a hominin. These features include **bipedal posture** (walking on two feet), the shape of the face, and brain size. Today's humans have a flatter face and a more pronounced chin than do the apes because the human jaw is shorter than that of the apes. Then, too, our teeth are generally smaller and less specialized. We don't have the sharp canines of an ape, for example. Chimpanzees have a brain size of about 400 cc, and modern humans have a brain size of about 1,300 cc.

It's hard to decide which fossils are hominins because human features evolved gradually and they didn't evolve at the same rate. Most investigators rely first and foremost on bipedal posture as the hallmark of a hominin, regardless of the size of the brain.

Earliest Fossil Hominins

Fossils have been found that can be dated at the time the ape and human lineages split. The oldest of these fossils, called *Sahelanthropus tchadensis,* dated at 7 MYA, was found in Chad, located in central Africa, far from eastern and southern Africa where other hominid fossils were excavated. The only find, a skull, appears to be that of a hominin because it has smaller canines and thicker tooth enamel than an ape. The braincase, however, is very apelike. It is impossible to tell if this hominin walked upright. Some suggest this fossil is ancestral to the gorilla.

Orrorin tugenensis, dated at 6 MYA and found in eastern Africa, is thought to be another early hominin, especially because the limb anatomy suggests a bipedal posture. However, the canine teeth are large and pointed, and the arm and finger bones retain adaptations for climbing. Some suggest this fossil is ancestral to the chimpanzee.

Two species of **ardipithicines** have been uncovered, *Ardipithecus kadabba* and *A. ramidus.* Only teeth and a few bone bits have been found for *A. kadabba,* and these have been dated to around 5.6 MYA. A more extensive collection of fossils has been collected for *A. ramidus.* To date, over 100 skeletons, all dated to 4.4 MYA, have been identified from this species; all were collected near a small town in Ethiopia, East Africa. These fossils have been reconstructed to form a female fossil specimen affectionately called Ardi.

Some of Ardi's features are primitive, like that of an ape, but others are like that of a human. Ardi was about the size of a chimpanzee, standing about 120 cm (4 ft) tall and weighing about 55 kg (110 lb). It appears that males and females were about the same size.

Ardi had a small head compared to the size of her body. The skull has the same features as *Sahelanthropus tchadensis,* but was smaller. Ardi's brain size was around 300 to 350 cc, slightly less than that of a chimpanzee brain (around 400 cc), and much smaller than that of a modern human (1,360 cc). The muzzle (area of the nose and mouth) projects forward, and the forehead is low with heavy eyebrow ridges, a combination that makes the face more primitive than that of the australopithecines (discussed next). However, the projection of the face is less than that of a chimpanzee because Ardi's teeth were small and like those of an omnivore. She lacked the strong sharp canines of a chimpanzee, and her diet probably consisted mostly of soft, rather than tough, plant material.

Ardi could walk erect, but she spent a lot of time in trees. Ardi's feet have a bone, missing in apes, that would keep her feet squarely on the ground, a sure sign that she was bipedal and not a quadruped like the apes. Nevertheless, like the apes, she has an opposable big toe. Opposable toes allow an animal's feet to grab hold of a tree limb. The wrists of Ardi's hands were flexible, and most likely she moved along tree limbs on all fours, as ancient apes did. Modern apes brachiate—use their arms to swing from limb to limb. Ardi did not do this, but her shoulders were flexible enough to allow her to reach for limbs to the side or over her head. The general conclusion is that Ardi moved carefully in trees. Although the top of her pelvis is like that of a human, and probably served for the attachment of muscles needed for walking, the bottom of the pelvis served as an attachment for the strong muscles needed for climbing trees.

Until recently, it's been suggested that bipedalism evolved when a dramatic change in climate caused the forests of East

Africa to be replaced by grassland. However, evidence suggests that Ardi lived in the woods, which questions the advantage that walking erect would have afforded her. Bipedalism does provide an advantage in caring for a helpless infant by allowing it to be carried by hand from one location to another. It is also possible that bipedalism may have benefited the males of the species as they foraged for food on the floor of the forests. More evidence is needed to better understand this mystery, but one thing is clear—the ardipithecines represent a link between our quadruped ancestors and the bipedal hominins.

Evolution of Australopithecines

The hominin line of descent begins in earnest with the **australopithecines,** a group of species that evolved and diversified in Africa. Some australopithecines were slight of frame and termed *gracile* ("slender") types. Some were *robust* ("powerful") and tended to have strong upper bodies and especially massive jaws. The gracile types most likely fed on soft fruits and leaves, whereas the robust types had a more fibrous diet that may have included hard nuts. In other words, the skull structure of australopithecines was suited to their particular diets.

Southern Africa

The first australopithecine to be discovered was unearthed in southern Africa by Raymond Dart in the 1920s. This hominin, named *Australopithecus africanus,* is a gracile type dated about 2.8 MYA. *Australopithecus robustus,* dated from 2 to 1.5 MYA, is a robust type from southern Africa. Both *A. africanus* and *A. robustus* had a brain size of about 500 cc. Their skull differences are essentially because of dental and facial adaptations to different diets.

Limb anatomy suggests these hominids walked upright. However, the proportions of the limbs are apelike. The forelimbs are longer than the hindlimbs. Some argue that *A. africanus,* with its relatively large brain, is a possible ancestral candidate for early *Homo,* whose limb proportions are similar to those of this fossil.

Eastern Africa

In the 1970s, a team led by Donald Johanson unearthed nearly 250 fossils of a hominin called *A. afarensis.* A now-famous female skeleton dated at 3.18 MYA is known worldwide by its field name, Lucy. Although her brain was small (400 cc), the shapes and relative proportions of her limbs indicate that Lucy stood upright and walked bipedally (Fig. 22.15*a*). Even better evidence of bipedal locomotion comes from a trail of footprints in Laetoli dated about 3.7 MYA. The larger prints are double, as though a smaller-sized being was stepping in the footfalls of another. There are additional small prints off to the side, within hand-holding distance (Fig. 22.15*b*).

The fact that the australopithecines were apelike above the waist (small brain) and humanlike below the waist (walked

a.

b.

Figure 22.15 ***Australopithecus afarensis.***
a. A reconstruction of Lucy on display at the St. Louis Zoo. **b.** These fossilized footprints occur in ash from a volcanic eruption some 3.7 MYA. The larger footprints are double (one followed behind the other), and a third, smaller individual was walking to the side. (A female holding the hand of a youngster may have been walking in the footprints of a male.) The footprints suggest that *A. afarensis* walked bipedally.

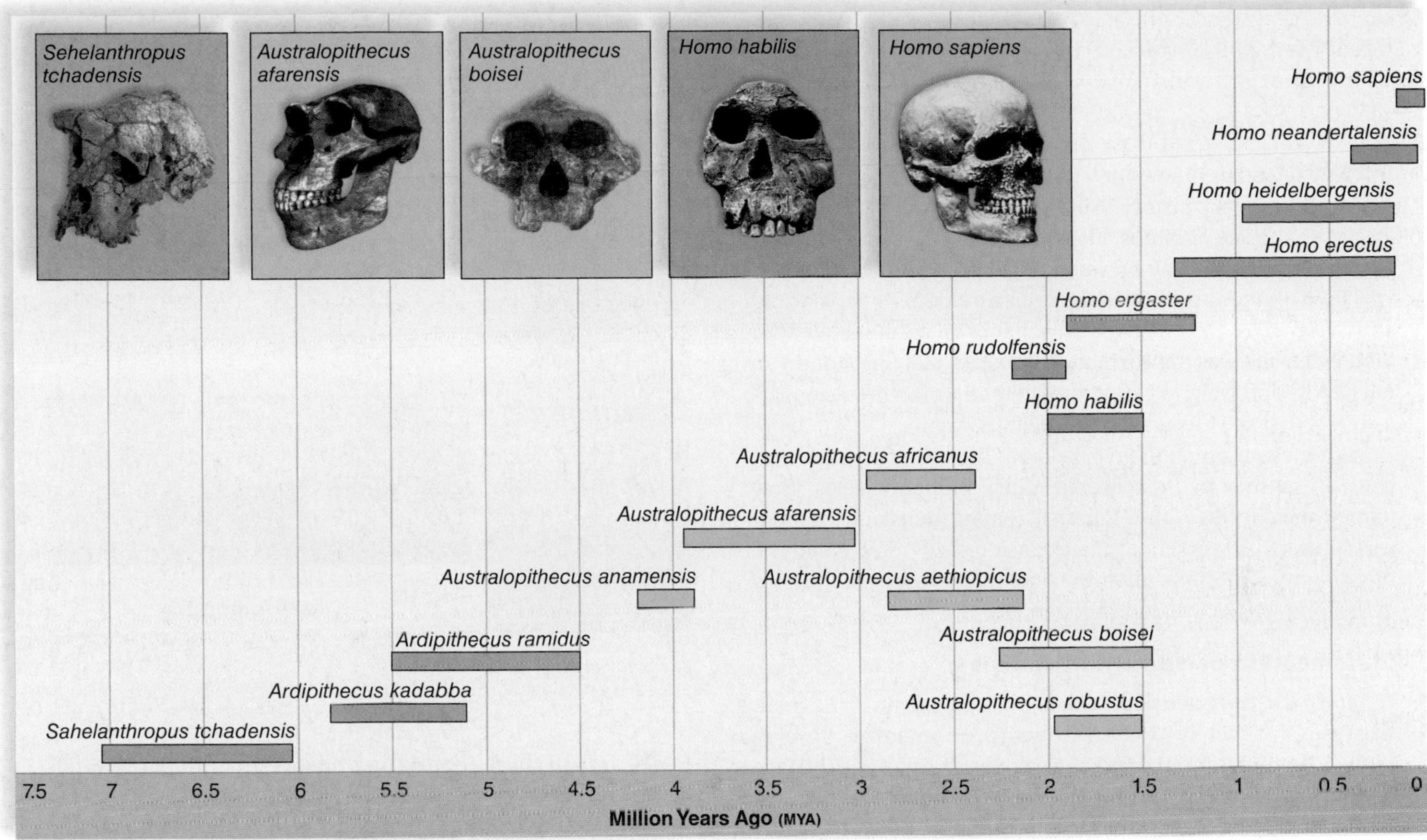

Figure 22.16 Human evolution.
Several groups of extinct hominins preceded the evolution of modern humans. The groups have been divided into the early humanlike hominins (orange), later humanlike hominins (green), early *Homo* species (lavender), and finally the later *Homo* species (blue).

erect) shows that human characteristics did not evolve all at one time. The term **mosaic evolution** is applied when different body parts change at different rates and, therefore, at different times.

Australopithecus afarensis, a gracile type, is most likely ancestral to the robust types found in eastern Africa: *A. aethiopicus* and *A. boisei. Australopithecus boisei* had a powerful upper body and the largest molars of any hominin. These robust types died out; therefore, it is possible that *A. afarensis* is ancestral to both *A. africanus* and early *Homo* (Fig. 22.16).

APPLICATIONS AND MISCONCEPTIONS

Why did Johanson name his fossil *Lucy*?

The evening of the discovery, there was a party in camp. The team gathered to celebrate the find of what appeared to be an almost complete hominid skeleton. The Beatles' song "Lucy in the Sky with Diamonds" was played over and over, and the name *Lucy* was given to the skeleton.

CHECK YOUR PROGRESS 22.4

1. Distinguish between a hominid and a hominin and give an example of each.
2. Name three features characteristic of hominins.
3. Discuss why the discovery of Lucy was an important event in the study of human evolution.

CONNECTING THE CONCEPTS

For additional information on the topics in this section, refer to the following discussions:

Section 8.2 describes the types of teeth found in modern humans.

Figure 11.2 illustrates the skeletal structure of a modern human.

Section 13.2 examines the brain structure of modern humans.

BIOLOGY MATTERS  **Science**

Homo floresiensis

In 2003, scientists made one of the most spectacular discoveries in evolutionary history. Nine skeletons were discovered in a cave on the isle of Flores. Flores is an Indonesian island east of Bali, located midway between Asia and Australia. This new species of humans grew no taller than a modern three-year-old child (Fig. 22A). One skeleton was that of an adult female who died when she was approximately 30 years old. She stood 1 m tall (3.3 ft) and weighed approximately 25 kg (55 lb). Scientists estimate that she died around 18,000 years ago.

After examining the first skeleton, the research team concluded that they had discovered a new human species. They named the species *Homo floresiensis* after the island where it was found. The workers at the excavation site nicknamed the tiny creatures "hobbits" after the fictional creatures in the *Lord of the Rings* books by J. R. R. Tolkien.

a. *Homo floresiensis*, artist's impression

b. Comparison of skulls

Figure 22A *Homo floresiensis.* **a.** Artist's re-creation of *H. floresiensis.* **b.** The *H. floresiensis* skull is smaller than that of *H. erectus* and that of *H. sapiens.*

Classification of *Homo floresiensis*

Homo floresiensis have skulls the size of a grapefruit and a brain size of approximately 417 cc. The teeth are humanlike, the eyebrow ridges are thick, the forehead slopes sharply, and the face lacks a chin. Even though the hobbits stand only 3 ft tall, they are not classified as pygmies. Despite the small body size, small brain size, and a mixture of primitive and advanced anatomical features, *H. floresiensis* is distinctly a member of genus *Homo.* The researchers believe that *Homo floresiensis* possibly evolved from a population of *H. erectus* that reached Flores approximately 840,000 years ago.

Culture of *Homo floresiensis*

Many of the habits exhibited by *H. floresiensis* are remarkably similar to those of other *Homo* species. Archaeological evidence indicates that *H. floresiensis* had the use of fire. The skeletons discovered on Flores were found in sediment deposits that also contained stone tools and the bones of dwarf elephants, giant rodents, and Komodo dragons. The dwarf elephants, or stegodons, weighed about 1,000 kg (2,200 lb) and would have posed a serious challenge to men who were only 1 m tall. Successful hunting would have required communication among members of the hunting party. The Flores diet also included fish, frogs, birds, rodents, snakes, and tortoises.

The hobbits produced sophisticated stone tools, hunted successfully in groups, and crossed at least two bodies of water to reach Flores from mainland Asia. And yet, their brain was about one-third the size of modern humans. *Homo floresiensis* is the smallest species of human ever discovered. Pound for pound, they outcompete every other member of genus *Homo.*

Further Research Needed

Researchers are interested in determining why the hobbits were so small. The first-discovered skeleton was believed to be that of a small child. There is no evidence of any other 1-m-tall adults in genus *Homo.* Modern pygmies are 1.4 to 1.5 m (4.6 to nearly 5 ft) tall. Over thousands of years, it is possible that a population of *Homo erectus* evolved into *H. floresiensis.* If so, a smaller body size would have been favored by natural selection. The members of each generation could have been smaller than the previous generation. Dwarfing of mammals on islands is a well-known process that can be seen worldwide. Islands generally have a limited food supply, few predators, and at least a few species competing for the same ecological niche. It behooves species living on islands to minimize their daily energy requirements. The smaller the body size, the fewer the calories per day required for survival. At this point, there is no substantiation for this hypothesis, but continued research may produce an answer as to why hobbits are so small.

The Extinction of *Homo floresiensis*

It appears that many of Flores' inhabitants became extinct approximately 12,000 years ago due to a major volcanic eruption. Researchers found *H. floresiensis* and pygmy stegodon remains below a 12,000-year-old volcanic ash layer. Hobbits reached the island approximately 11,000 years ago and possibly intermingled with modern humans. Rumors, myths, and legends among the indigenous tribes of Flores about "the tiny people who lived in the forest" have persisted.

Questions to Consider

1. What does the discovery of *Homo floresiensis* tell you about what we know regarding human evolution?
2. Do you think that there are other species of *Homo* yet to be discovered? Where would you look?

22.5 Evolution of Humans

LEARNING OUTCOMES

Upon completion of this section, you should be able to

1. Explain the adaptations of *Homo erectus*.
2. Distinguish between the different theories regarding *Homo sapiens* evolution.
3. Describe the differences between Neandertals and Cro-Magnons.

Fossils are assigned to the genus *Homo* if (1) the brain size is 600 cc or greater, (2) the jaw and teeth resemble those of humans, and (3) tool use is evident. In this section, we will discuss early *Homo: Homo habilis* and *Homo erectus;* and later *Homo:* the Neandertals and the Cro-Magnons, the first modern humans.

Early *Homo*

Homo habilis, dated between 2.0 and 1.9 MYA, may be ancestral to modern humans. Some of these fossils have a brain size as large as 775 cc, about 45% larger than that of *A. afarensis.* The cheek teeth are smaller than even those of the gracile australopithecines. Therefore, it is likely that these early members of the genus *Homo* were omnivores who ate meat in addition to plant material. Bones at their campsites bear cut marks, indicating that they used tools to strip them of meat.

The stone tools made by *H. habilis,* whose name means "handyman," are rather crude. It's possible that these are the cores from which they took flakes sharp enough to scrape away hide, cut tendons, and easily remove meat from bones.

Early *Homo* skulls suggest that the portions of the brain associated with speech were enlarged. We can speculate that the ability to speak may have led to hunting cooperatively. Other members of the group may have remained plant gatherers. If so, both hunters and gatherers most likely ate together and shared their food. In this way, society and culture could have begun.

Culture, which encompasses human behavior and products (e.g., technology and the arts), depends upon the capacity to speak and transmit knowledge. We can further speculate that the advantages of a culture to *H. habilis* may have hastened the extinction of the australopithecines.

Homo erectus

Homo erectus and like fossils are found in Africa, Asia, and Europe and dated between 1.9 and 0.3 MYA. A Dutch anatomist named Eugene Dubois was the first to unearth *H. erectus* bones in Java in 1891. Since that time, many other fossils have been found in the same area. Although all fossils assigned the name *H. erectus* are similar in appearance, enough discrepancy exists to suggest that several different species have been included in this group. In particular, some experts suggest that the Asian form is *Homo erectus* and the African form is *Homo ergaster* (Fig. 22.17).

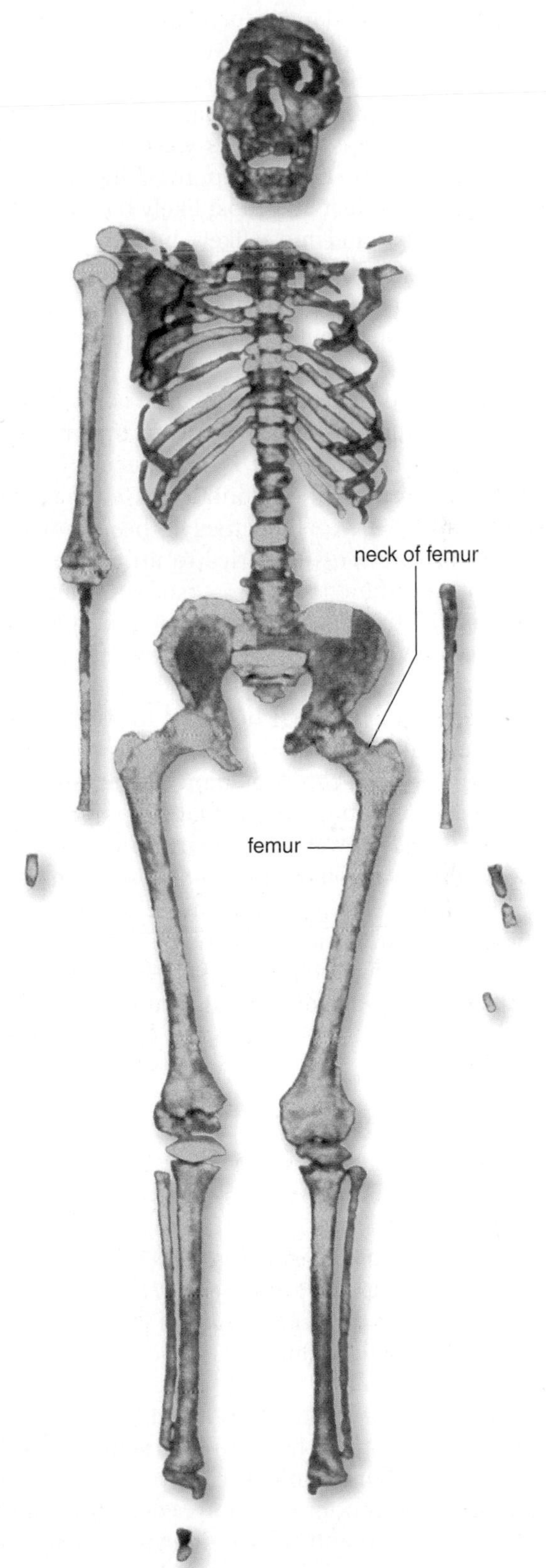

Figure 22.17 ***Homo ergaster.*** This skeleton of a ten-year-old boy who lived 1.6 MYA in eastern Africa shows angled femurs because the femur neck is quite long.

Compared with *H. habilis, H. erectus* had a larger brain (about 1,000 cc) and a flatter face. The nose projected, however. This type of nose is adaptive for a hot, dry climate because it permits water to be removed before air leaves the body. The recovery of an almost complete skeleton of a ten-year-old boy indicates that *H. ergaster* was much taller than the hominids discussed thus far. Males were 1.8 m tall (about 6 ft), and females were 1.55 m (approaching 5 ft). Indeed, these hominids were erect and most likely had a striding gait like ours. The robust and most likely heavily muscled skeleton still retained some australopithecine features. Even so, the size of the birth canal indicates that infants were born in an immature state that required an extended period of care.

Homo ergaster may have first appeared in Africa and then migrated into Asia and Europe. At one time, the migration was thought to have occurred about 1 MYA. Recently, *H. erectus* fossil remains in Java and the Republic of Georgia have been dated at 1.9 and 1.6 MYA, respectively. These remains push the evolution of *H. erectus* in Africa to an earlier date than has yet been determined. In any case, such an extensive population movement is a first in the history of humankind and a tribute to the intellectual and physical skills of the species.

Homo erectus was the first hominid to use fire and also fashioned more advanced tools than early *Homos.* These hominids used heavy, teardrop-shaped axes and cleavers. Flake tools were probably used for cutting and scraping. It could be that *H. ergaster* was a systematic hunter and brought kills to the same site over and over. In one location, researchers have found over 40,000 bones and 2,647 stones. These sites could have been "home bases," where social interaction occurred and a prolonged childhood allowed time for learning. Perhaps a language evolved and a culture more like our own developed.

Evolution of Modern Humans

Many early *Homo* species in Europe are now classified as *Homo heidelbergensis.* Just as *H. erectus* is believed to have evolved from *H. ergaster* in Asia, so *H. heidelbergensis* is believed to have evolved from *H. ergaster* in Europe. Further, for the sake of discussion, we can group together the *H. ergaster* in Africa, *H. erectus* in Asia, and *H. heidelbergensis* in Europe as a collective group of archaic humans who lived between 1.5 and 0.25 MYA.

The most widely accepted hypothesis for the evolution of modern humans from archaic humans is referred to as the **replacement model,** or out-of-Africa hypothesis (Fig. 22.18), which proposes that modern humans evolved from archaic humans only in Africa, and then modern humans migrated to Asia and Europe, where they replaced the archaic species about 100,000 years BP (before the present). However, even this hypothesis is being challenged as new genomic information becomes available on the Neandertals and Denisovans.

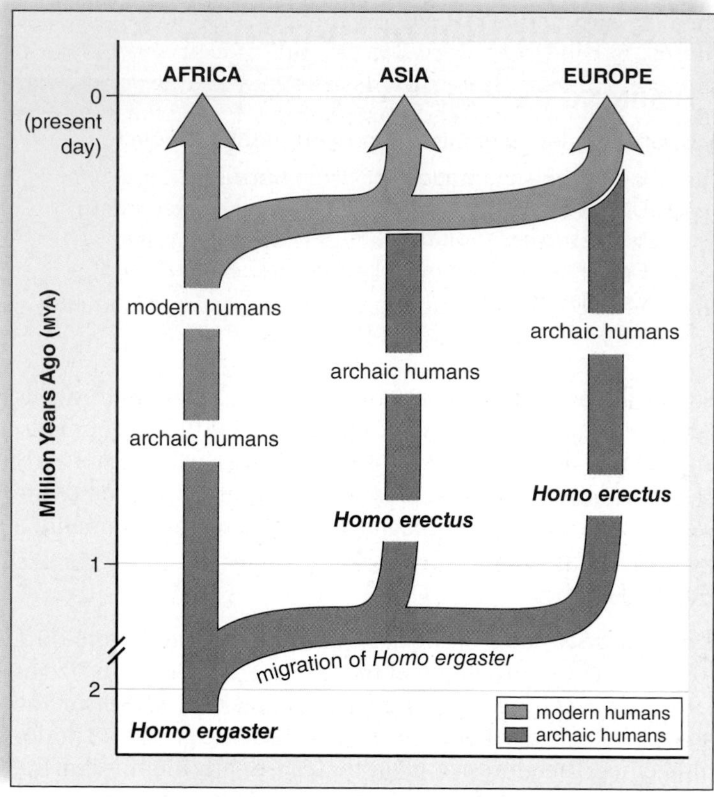

Figure 22.18 **Replacement model.**
According to the replacement model, modern humans evolved in Africa and then replaced archaic humans in Asia and Europe.

Neandertals

Neandertals (*H. neandertalensis*) take their name from Germany's Neander Valley, where one of the first Neandertal skeletons, dated some 200,000 years BP, was discovered. The Neandertals had massive brow ridges; and their nose, jaws, and teeth protruded far forward. The forehead was low and sloping, and the lower jaw lacked a chin. New fossils show that the pubic bone was long compared with ours.

According to the replacement model, Neandertals were eventually supplanted by modern humans. However, this traditional view is being challenged by studies of the Neandertal genome (completed in 2010), which suggests that not only did Neandertals interbreed with *Homo sapiens,* but between 1 and 4% of the genomes of non-African *Homo sapiens* contains remnants of the Neandertal genome. Some scientists are suggesting that Neandertals were not a separate species, but simply a race of *Homo sapiens* that was eventually absorbed into the larger population. Research continues into these and other hypotheses that explain these similarities.

Physiologically, the Neandertal brain was, on the average, slightly larger than that of *H. sapiens* (1,400 cc, compared with 1,360 cc in most modern humans). The Neandertals were heavily muscled, especially in the shoulders and neck. The bones of the limbs were shorter and thicker than those of modern humans. It is hypothesized that a larger brain than that of modern humans was required to control the extra musculature. The Neandertals lived in Europe and Asia during the last Ice Age, and their sturdy build could have helped conserve heat.

The Neandertals give evidence of being culturally advanced. Most lived in caves, but those living in the open may have built houses. They manufactured a variety of stone tools, including spear points, which could have been used for hunting. Scrapers and knives could have helped in food preparation. They most likely successfully hunted bears, woolly mammoths, rhinoceroses, reindeer, and other contemporary animals. They used and could control fire, which probably helped them cook meat and keep themselves warm. They even buried their dead with flowers and tools and may have had a religion. Perhaps they believed in life after death. If so, they were capable of thinking symbolically.

Denisovans

In 2008, a fragment of a finger bone was discovered in Denisova cave in southern Siberia. Initially, scientists thought that this might be the remains of a species of early *Homo,* possibly related to *Homo erectus.* However, mitochondrial DNA studies indicated that the fossil belonged to a species that existed around 1 million years ago, around the same time as Neandertals. The analyses suggest that the **Denisovans** and Neandertals shared a common ancestor, but did not interbreed with one another, possibly because of their geographical locations. However, what is interesting is the fact that *Homo sapiens* in the Oceania region (New Guinea and nearby islands) share around 5% of their genomes with the Denisovans. No evidence of the Denisovan genome has yet been detected in other groups of *Homo sapiens.* When coupled with the Neandertal data, this suggests that modern *Homo sapiens* did not simply replace groups of archaic humans, but rather may have assimilated them by inbreeding. Scientists are just beginning to unravel the implications of the Denisovan discovery.

Cro-Magnons

Cro-Magnons are the oldest fossils to be designated ***Homo sapiens*** and are named for a fossil location in France. Scientists recognize that Cro-Magnons were the modern humans who entered Asia and Europe from Africa around 100,000 years ago (or even earlier). Cro-Magnons had a thoroughly modern appearance (Fig. 22.19). Analysis of Neandertal DNA indicates that it is so different from Cro-Magnon DNA that these two groups of people did not interbreed. Instead, Cro-Magnons seem to have replaced the Neandertals in the Middle East and then spread to Europe 40,000 years ago. There, they lived side by side with the Neandertals for several thousand years. If so, the Neandertals are cousins and not ancestors to us.

Cro-Magnons made advanced stone tools, including compound tools, as when stone flakes were fitted to a wooden handle. They may have been the first to make knifelike blades and throw spears, enabling them to kill animals from a distance. They were such accomplished hunters that some researchers suggest they were responsible for the extinction of many larger mammals, such as the giant sloth, the mammoth, the saber-toothed tiger, and the giant ox, during the late Pleistocene epoch.

Cro-Magnons hunted cooperatively, and perhaps they were the first to have a language. Most likely, they lived in small groups, with the men hunting by day while the women remained at home with the children. It's possible that this hunting way of life among prehistoric people influences our behavior today. The Cro-Magnon culture included art. They sculpted small figurines out of reindeer bones and antlers. They also painted beautiful drawings of animals on cave walls in Spain and France (Fig. 22.19).

Figure 22.19 **The Cro-Magnons.**
Cro-Magnon people are one of the earliest groups to be designated *Homo sapiens.* Their tool-making ability and other cultural attributes, including artistic talents, are legendary.

BIOLOGY MATTERS  Bioethical

Effects of Biocultural Evolution on Population Growth

Humans today undergo biocultural evolution because culture has developed to the point that adaptation to the environment is not dependent on genes but on the passage of culture from one generation to the next.

Tool Use and Language Began

The first step toward biocultural evolution began when *Homo habilis* made primitive stone tools. *Homo erectus* continued the tradition and most likely was a hunter of sorts. It's possible that the campsites of *H. erectus* were "home bases," where the women stayed behind with the children while the men went out to hunt. Hunting was an important event in the development of culture, especially because it encouraged the development of language. If *H. erectus* didn't have the use of language, certainly Cro-Magnons did. People who have the ability to speak a language would have been able to cooperate better as they hunted and even as they sought places to gather plants. Among animals, only humans have a complex language that allows them to communicate their experiences. Words are not objects and events. They stand for objects and events that can be pictured in the mind.

Agriculture Began

About 10,000 years ago, people gave up being full-time hunter-gatherers and became at least part-time farmers. What accounts for the rise of agriculture? The answer is not known, but several explanations have been put forth. About 12,000 years ago, a warming trend occurred as the Ice Age came to a close. A variety of big-game animals became extinct, including the saber-toothed cats, mammoths, and mastodons. This may have made hunting less productive. As the weather warmed, the glaciers retreated and left fertile valleys, where rivers and streams were full of fish and the soil was good. The Fertile Crescent in Mesopotamia is one such example. Here, fishing villages may have sprung up, causing people to settle down.

The people were probably already knowledgeable about what crops to plant. Most likely, as hunter-gatherers, people had already selected seeds with desirable characteristics for propagation. Then, a chance mutation may have made these plants particularly suitable as a source of food. So now they began to till the good soil where they had settled, and they began to systematically plant crops (Fig. 22B).

As people became more sedentary, they may have had more children, especially because the men were home more often. Population increases may have tipped the scales and caused them to adopt agriculture full time, especially if agriculture could be counted on to provide food for hungry mouths. The availability of agricultural tools must have contributed to making agriculture worthwhile. The digging stick, the hoe, the sickle, and the plow were improved when iron tools replaced bronze in the stone-bronze-iron sequence of ancient tools. Irrigation began as a way to control water supply, especially in semiarid areas and regions of periodic rainfall.

Figure 22B **Primitive agriculture.** Agriculture began in several locations across the globe about 10,000 years ago.

If evolutionary success is judged by population size, agriculture was extremely beneficial to our success because it caused a rapid increase of human numbers all over the Earth. Also, agriculture ushered in civilization as we know it. When crops became bountiful, some people were freed from raising their own food. They began to specialize for other ways of life in towns and then cities. Some people became traders, shopkeepers, bakers, and teachers, to name a few possibilities. Others became the nobility, priests, and soldiers. Today, farming is highly mechanized and cities are extremely large. However, we are on a treadmill. As the human population increases, we need new innovations to produce greater amounts of food. As soon as food production increases, populations grow once again, and the demand for food becomes still greater. Will there be a point when the population is greater than the food capacity? Is that time already upon us?

Industrial Revolution Began

The Industrial Revolution began in England during the eighteenth century and with it a demand for energy in the form of coal and oil that today seems unlimited. Our ability to construct any number of tools, including high-tech computers, is not stored in our genes. We learn it from the previous generation. Our modern civilization that began due to the advent of agriculture is now altering the global environment in a way that affects the evolution of other species. Species are becoming extinct unless they are able to adapt to the presence of our civilization. It could be that biocultural evolution will be so harmful to the biosphere that the human species also will eventually be driven to extinction.

Questions to Consider

1. Can technology be used to help us not pollute the environment? How?
2. Should the extinction of other species be prevented? How can it be prevented?
3. Should the human population be reduced in size? How could this occur?

Human Variation

Humans have been widely distributed about the globe since they evolved. As with any other species that has a wide geographic distribution, phenotypic and genotypic variations are noticeable between populations. Today, we say that people have different ethnicities (Fig. 22.20).

It has been hypothesized that human variations evolved as adaptations to local environmental conditions. One obvious difference among people is skin color. A darker skin is protective against the high ultraviolet (UV) intensity of bright sunlight. On the other hand, a white skin ensures vitamin D production in the skin when the UV intensity is low. Harvard University geneticist Richard Lewontin points out, however, that this hypothesis concerning the survival value of dark and light skin has never been tested.

a.

b.

c.

Figure 22.20 **Ethnic variations in modern humans.**
a. Some of the differences between the three prevalent ethnic groups in the United States may be due to adaptations to the original environment. **b.** The Maasai live in East Africa. **c.** The Kuna live on the San Blas Islands near Panama.

Two correlations between body shape and environmental conditions have been noted since the nineteenth century. The first, Bergmann's rule, states that animals in colder regions of their range have a bulkier body build. The second, Allen's rule, states that animals in colder regions of their range have shorter limbs, digits, and ears. Both of these effects help regulate body temperature by increasing the surface area-to-volume ratio in hot climates and decreasing the ratio in cold climates. For example, the Massai of East Africa (Fig. 22.20*b*) tend to be very tall and slender, with elongated limbs. By contrast, the Eskimos, who live in northern regions, tend to have a more bulky build, with shorter limbs.

Other anatomical differences among ethnic groups, such as hair texture, a fold on the upper eyelid (common in Asian peoples), or the shape of lips, cannot be explained as adaptations to the environment. Perhaps these features became fixed in different populations due to genetic drift. As far as intelligence is concerned, no significant disparities have been found among different ethnic groups.

Genetic Evidence for a Common Ancestry

The replacement model for the evolution of humans (see Fig. 22.18) also pertains to the origin of ethnic groups. This hypothesis proposes that all modern humans have a relatively recent common ancestor (Cro-Magnon) who evolved in Africa and then spread into other regions. Paleontologists tell us that the variation among modern populations is considerably less than among human populations some 250,000 years ago. This would mean that all ethnic groups evolved from the same single, ancestral population.

A comparative study of mitochondrial DNA shows that the differences among human populations are consistent with their having a common ancestor no more than a million years ago. Lewontin, mentioned previously, found that the genotypes of different modern populations are extremely similar. He examined variations in 17 genes, including blood groups and various enzymes, among seven major geographic groups: Caucasians, black Africans, mongoloids, south Asian Aborigines, Amerinds, Oceanians, and Australian Aborigines. He found that the great majority of genetic variation—85%—occurs within ethnic groups, not among them. In other words, the amount of genetic variation between individuals of the same ethnic group is greater than the variation between ethnic groups.

CHECK YOUR PROGRESS 22.5

1. Name the fossil hominid that is the oldest of the australopithecines.
2. Name the hominin that was the first to migrate out of Africa.
3. Discuss how *Homo habilis* might have differed from the australopithecines.
4. Discuss how *Homo erectus* might have differed from *Homo habilis.*
5. Describe how the replacement model hypothesis influenced the evolution of modern humans.
6. Describe how Cro-Magnons differ from the other species in the genus *Homo.*

CONNECTING THE CONCEPTS

For more information on the topics presented in this section, refer to the following discussions:

Section 3.6 describes the structure and function of the mitochondria.

Section 23.1 examines the major terrestrial ecosystems that are occupied by humans.

CASE STUDY CONCLUSION

In a follow-up study of the Neandertal genomes, the researchers unveiled another startling discovery. The comparison of the Neandertal and human genomes suggests that the two groups may have interbred long after they had diverged from their common ancestor. The Neandertal genome shares more commonality with modern humans from outside Africa than with modern African populations. Statistical studies have suggested that almost 4% of the genomes of modern Europeans may be Neandertal in nature, suggesting that these two groups interbred freely in the past. If more samples support these findings, scientists agree that they may have to go back and revise some of their basic ideas of how modern humans evolved, and what our species' interactions may have been with the other hominins.

MEDIA STUDY TOOLS

Enhance your study of this chapter with media! Visit **www.mhhe.com/maderhuman13e** and go to "Media Study Tools" for this chapter to access the following:

Animations

22.1 Miller-Urey Experiments
22.3 Three Domains

Videos

22.2 Finches Adaptive Radiation • Finches Natural Selection

SUMMARIZE

22.1 Origin of Life

The processes of **chemical evolution** and **biological evolution** contributed to the formation of the first cells.

- Using an outside energy source, small organic molecules were produced by reactions between early Earth's atmospheric gases.
- Macromolecules evolved and interacted.
- The **RNA-first hypothesis** proposed that only macromolecule RNA was needed for the first cell or cells.
- The **protein-first hypothesis** proposed that amino acids join to form polypeptides when exposed to dry heat.
- The **protocell,** a **heterotroph,** which obtained energy by fermentation, lived on preformed organic molecules in the ocean.

The protocell eventually became a true cell once it had genes composed of DNA and could reproduce.

22.2 Biological Evolution

Biological evolution explains both the unity and diversity of life. **Prokaryotic cells** are the simplest form of life. **Eukaryotic cells** evolved from prokaryotic cells.

- Descent from a common ancestor explains the unity (sameness) of living organisms.
- **Adaptation** to different environments explains the great diversity of life.

Darwin's theory of evolution by **natural selection** explains the process of biological evolution. **Artificial selection** occurs when humans select for specific traits in a species. Natural selection is supported by the following:

- *Fossil evidence:* **Fossils,** and the **fossil record** provide the history of life in general and allow us to trace the descent of a particular group.
- *Biogeographical evidence:* **Biogeography** is the distribution of organisms on Earth, and is explainable by assuming that organisms evolved in one locale.

- *Anatomical evidence:* The common anatomies and development of a group of organisms are explainable by descent from a common ancestor. **Homologous structures** are inherited from a common ancestor. **Analogous structures** are adaptations that evolved independently. **Vestigial structures** have a function in one group of organisms but not in another, related group.
- *Biochemical evidence:* All organisms have similar biochemical molecules.

Darwin developed a mechanism for adaptation known as natural selection:

- The result of natural selection is a population adapted to its local environment.

22.3 Classification of Humans

The classification of humans can be used to trace their ancestry. Humans, like other species, are assigned a **binomial name:** *Homo sapiens.*

- Humans are **primates.**
- Primates have the characteristics of mobile limbs, binocular vision, complex brains, and a reduced reproductive rate.

22.4 Evolution of Hominins

- The evolutionary tree of humans is supported by DNA evidence, which can be used as a molecular clock to indicate when the groups diverged.
- Primates may be divided into two groups, the **prosimians** and the **anthropoids.** Humans belong to the anthropoid group.
- **Hominids** are a classification that includes apes, chimpanzees, and humans. **Hominine** refers to the gorillas, chimpanzees, and humans. **Hominin** refers only to members of the genus *Homo* and their close relatives. The first hominin (includes humans) most likely lived about 6 to 7 MYA.
- Certain features (**bipedal posture,** flat face, and brain) are used to identify fossil hominins.
- **Ardipithecines** and **australopithecines** are examples of early hominins. The features of australopithecines support the concept of **mosaic evolution.**

22.5 Evolution of Humans

Fossils are classified as *Homo* with regard to brain size (over 600 cc), jaws and teeth (resemble modern humans), and evidence of tool use.

- ***Homo habilis*** made and used tools. They were among the first to display evidence of **culture.**
- ***Homo erectus*** was the first *Homo* to have a brain size of more than 1,000 cc.
- *Homo erectus* migrated from Africa into Europe and Asia.
- *Homo erectus* used fire and may have been big-game hunters.

Evolution of Modern Humans

- The **replacement model** hypothesis says that *H. sapiens* evolved in Africa but then migrated to Asia and Europe.

Neandertals

- The **Neandertals** were already living in Europe and Asia before modern humans arrived.
- They had a culture but did not have the physical traits of modern humans.

Denisovans

- The **Denisovans** lived in the area that is now Siberia, but interacted with early *Homo sapiens.*

Cro-Magnons

- **Cro-Magnons** are the oldest fossil to be designated ***Homo sapiens.***
- Their tools were sophisticated, and they had a culture.

ASSESS

Testing Your Knowledge of the Concepts

1. Describe the differences between chemical and biological evolution. (page 510)
2. List and discuss the steps by which a chemical evolution could have produced a protocell. (pages 510–511)
3. Describe how Darwin's theory of natural selection explains biological evolution. (pages 512–514)
4. Describe the evidence that supports descent from a common ancestor. (pages 514–518)
5. List a characteristic of each category of human classification. (page 519)
6. What type of life are primates adapted to? Discuss several primate characteristics that aid them in their adaptation. (pages 519–520)
7. How is the human skeleton like that of a chimpanzee skeleton? What's the major difference? (pages 520–521)
8. Distinguish between a hominid, hominine, and hominin. (page 523)
9. Describe the characteristics of australopithecines, early *Homo, Homo erectus,* and Cro-Magnons. (pages 523–530)
10. Summarize how the replacement model explains the evolution of *Homo sapiens.* (page 528)
11. Explain the importance of the recent findings regarding Neandertals and Denisovans. (page 529)
12. Which of these did Stanley Miller place in his experimental system to show that organic molecules could have arisen from inorganic molecules on the early Earth?
    a. microspheres
    b. purines and pyrimidines
    c. gases in the atmosphere of early Earth
    d. only RNA
    e. All of the choices are correct.
13. According to Darwin,
    a. the adapted individual is the one who survives and passes on its genes to offspring.
    b. changes in phenotype are passed on by way of the genotype to the next generation.
    c. organisms are able to bring about a change in their phenotype.
    d. evolution is striving toward particular traits.
    e. All of the choices are correct.

14. If evolution occurs, we would expect different biogeographical regions with similar environments to
  a. contain the same mix of plants and animals.
  b. each have its own specific mix of plants and animals.
  c. have plants and animals with similar adaptations.
  d. have plants and animals with different adaptations.
  e. Both b and c are correct.

15. The fossil record offers direct evidence for evolution because you can
  a. see that the types of fossils change over time.
  b. sometimes find common ancestors.
  c. trace the ancestry of a particular group.
  d. trace the biological history of living organisms.
  e. All of the choices are correct.

16. Organisms such as whales and sea turtles adapted to an aquatic way of life
  a. will probably have homologous structures.
  b. will have similar adaptations but not necessarily homologous structures.
  c. may very well have analogous structures.
  d. will have the same degree of fitness.
  e. Both b and c are correct.

17. Which of these gives the correct order of divergence from the main line of descent leading to humans?
  a. prosimians, monkeys, Asian apes, African apes, humans
  b. gibbons, baboons, prosimians, monkeys, African apes, humans
  c. monkeys, gibbons, prosimians, African apes, baboons, humans
  d. African apes, gibbons, monkeys, baboons, prosimians, humans
  e. *H. habilis, H. erectus, H. neandertalensis,* Cro-Magnon

18. Lucy is a member of what species?
  a. *Homo erectus*
  b. *Australopithecus afarensis*
  c. *H. habilis*
  d. *A. robustus*
  e. *A. anamensis* and *A. afarensis* are alternative forms of Lucy.

19. A hominid includes all of the following, except
  a. chimpanzees.
  b. orangutans.
  c. prosimians.
  d. gorillas.

20. What possibly may have influenced the evolution of bipedalism?
  a. A larger brain developed.
  b. Food gathering was easier.
  c. The climate became colder.
  d. Both b and c are correct.
  e. Both a and c are correct.

21. *H. ergaster* could have been the first to
  a. use and control fire.
  b. migrate out of Africa.
  c. make tools.
  d. Both a and b are correct.
  e. All of the choices are correct.

22. The last common ancestor for African apes and hominids
  a. has been found, and it resembles a gibbon.
  b. has not yet been identified, but it is expected to be dated from about 6 to 7 MYA.
  c. has been found, and it has been dated at 30 MYA.
  d. is not expected to be found because there was no such common ancestor.
  e. most likely lived in Asia, not Africa.

23. Classify humans by filling in the missing lines.

Domain	Eukarya
Kingdom	Animalia
Phylum	a. ____
b. ____	Mammalia
c. ____	Primates
Family	d. ____
e. ____	*Homo*
Species	f. ____

In questions 24–27, match each description to a type of evolutionary evidence in the key.

Key:
  a. biogeography
  b. fossil record
  c. comparative biochemistry
  d. comparative anatomy

24. Species change over time.

25. Forms of life are variously distributed.

26. A group of related species have homologous structures.

27. The same types of molecules are found in all living organisms.

ENGAGE

Thinking Critically About the Concepts

Much of the work done in archaeology is very tedious. For example, the Science feature, "*Homo floresiensis,*" on page 526 describes the discovery of *H. floresiensis* in a limestone cave on the island of Flores. Researchers Michael Morwood, of the University of New England in Australia, and Radien Soejono, of the Indonesian Center for Archaeology in Jakarta, were in charge of the dig. They began the work in July 2001. It was not until the end of three seasons of field work that they found evidence of hominins—a single tooth. Once the first female skeleton had been found, colleague Peter Brown spent three months analyzing it.

1. In archaeology, why is it critically important to know where to dig? What clues would you look for to determine where to dig?

2. What artifacts do archaeologists use to determine a population's bioculture?

3. What types of information can scientists derive about an ancient people, using only a single human skull from that group?

CHAPTER

23

Global Ecology and Human Interferences

CASE STUDY THE POTENTIAL FUTURE OF THE EARTH

It is the year 2100 and unspoiled ecosystems, such as the one pictured here, may be a thing of the past. Scientists predict that the northeastern United States will no longer have winters that consistently get below freezing temperatures. Corn can be grown in the middle latitudes of Canada and many coastal areas will be under water. Skin cancer will reach an all-time high due to increased solar exposure. Nearly all water bodies are overgrown with algae and pond scum, and freshwater fish are disappearing. Freshwater is a limited resource and comes at a high price. Although this may seem extreme, this scenario is possible given the influence of modern humans on ecosystems and chemical cycling. Ecosystems are characterized by the interactions of biotic (living) and abiotic (nonliving) components. The biotic community depends on cycling of chemicals, such as carbon, nitrogen, and phosphorus. As we continue to burn carbon-rich fossil fuels such as oil, carbon dioxide is released into the atmosphere and acts like a blanket, trapping in solar radiation. Steady atmospheric increases in carbon dioxide contribute to global warming, which can cause climate shifts. In turn, sea levels are predicted to rise and extreme weather events, like hurricanes and cyclones, are predicted to increase in frequency. Use of detergents with phosphorus and nitrogen-containing fertilizers add excess amounts of these nutrients to water bodies, such as ponds and lakes. The result is cultural eutrophication, which causes massive algal blooms. When these blooms decompose, oxygen levels in water bodies plummet, causing widespread fish kills.

In this chapter, you will learn about how energy flows through the members of biotic communities and how abiotic chemicals and nutrients cycle through ecosystems. This knowledge will help you understand why phenomena such as global warming are a major concern.

As you read through the chapter, think about the following questions:

1. What happens to the concentration of pollutants in animals as you move from lower to higher trophic levels?
2. How do we alter biogeochemical cycles, and what are some of the environmental consequences?

CHAPTER CONCEPTS

BEFORE YOU BEGIN

Before beginning this chapter, take a few moments to review the following discussions:

Section 1.1 What is the relationship between populations and ecosystems in the levels of biological organization?

Section 1.1 Why must living organisms acquire materials and energy?

Section 1.2 What are some of the threats from humans to the biosphere?

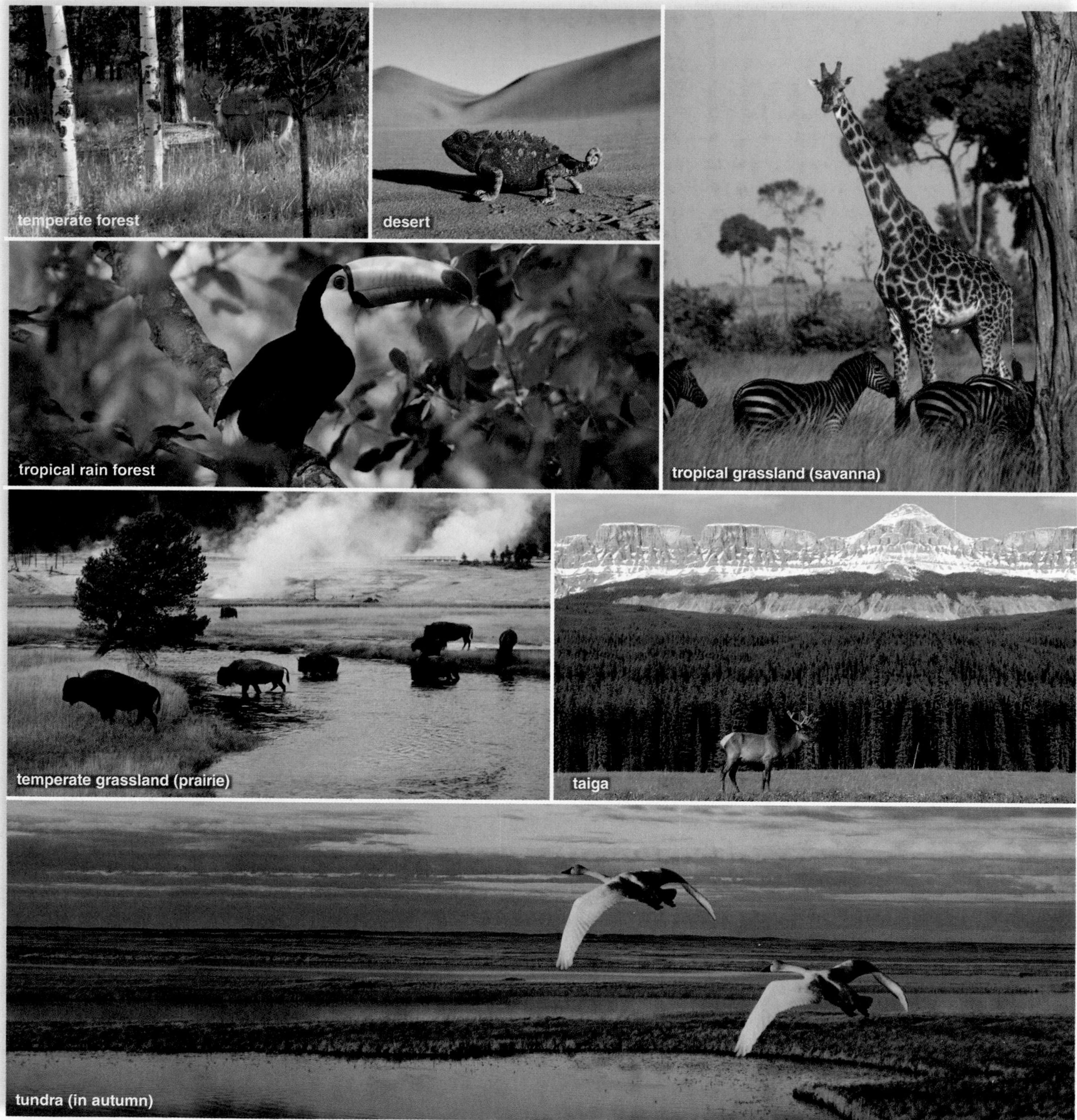

Figure 23.1 The temperature and rainfall characteristics of the major terrestrial ecosystems.
Temperate forests have mild temperatures and occur where rainfall is moderate, yet sufficient to support trees. Deserts have changeable temperatures with minimal rainfall. Tropical rain forests, which generally occur near the equator, have a high average temperature and the greatest amount of rainfall of all the terrestrial ecosystems. A tropical grassland (savanna) has high temperatures and moderate/seasonal rainfall. A temperate grassland (prairie) has low to high temperatures, with low annual rainfall. The taiga, a coniferous forest that encircles the globe, has a low average temperature but moderate rainfall. The tundra is the northernmost terrestrial ecosystem and has the lowest average temperature of all the terrestrial ecosystems, with minimal to moderate rainfall.

23.1 The Nature of Ecosystems

LEARNING OUTCOMES

Upon completion of this section, you should be able to

1. Identify the relationship between ecosystems and the biosphere.
2. Identify ways in which autotrophs and heterotrophs obtain nutrients.
3. Interpret the energy flow and biogeochemical cycling within and among ecosystems.

The **biosphere** is the portion of the Earth that contains living organisms, from the atmosphere above to the depths of the oceans below and everything in between. Specific areas of the biosphere where organisms interact among themselves and the physical and chemical environment are called **ecosystems.** The interactions that occur in an ecosystem maintain balance in that specific area, which in turn affects the balance of the biosphere. Human activities can alter the interactions between organisms and their environments in ways that reduce the abundance and diversity of life in an ecosystem. It is important to understand how ecosystems function so that we can repair past damage and predict how human activities will lead to future damage.

Types of Ecosystems

Scientists recognize several distinctive major types of terrestrial ecosystems, also called **biomes** (Fig. 23.1). Temperature and rainfall levels define the biomes. A variety of organisms adapted to the regional climate characterize each biome. The tropical rain forest, which occurs at the equator, is dominated by large, evergreen, broad-leaved trees. The savanna is a tropical grassland that supports many types of grazing animals. Temperate grasslands receive less rainfall than temperate forests (where many trees lose their leaves during the winter). Deserts receive scant rainfall and therefore lack trees. The taiga is a very cold northern forest of conifers such as pine, spruce, hemlock, and fir. Bordering the North Pole is the frigid tundra, which has long winters and a short growing season. A permafrost persists even during the summer in the tundra and prevents large plants from becoming established.

Animation Biomes

Aquatic ecosystems are divided into those composed of freshwater and those composed of salt water (marine ecosystems). The ocean is a marine ecosystem that covers 70% of the Earth's surface. Two types of freshwater ecosystems are those with standing water, such as lakes and ponds, and those with running water, such as rivers and streams. The richest marine ecosystems lie near the coasts. Coral reefs are located offshore, and salt marshes occur where rivers meet the sea (Fig. 23.2).

Video Coral Reef Ecosystems

Figure 23.2 Examples of freshwater and saltwater ecosystems.
Aquatic ecosystems are divided into those that have salt water, such as (**a**) the ocean, and (**b**) those that have freshwater, such as a river. Saltwater, or marine, ecosystems also include (**c**) coral reefs and (**d**) salt marshes.

Biotic Components of an Ecosystem

The biotic components of an ecosystem are living organisms that can be categorized according to their food source (Fig. 23.3).The abiotic components are the nonliving components, such as soil type, water, and weather. Biotic species are generally organized as autotrophs and/or heterotrophs.

Autotrophs

Autotrophs require only inorganic nutrients and an outside source of energy to produce organic nutrients for their own use and for all the other members of a community. Therefore, they are called **producers,** meaning they produce food (Fig. 23.3*a*). Photosynthetic organisms produce most of the organic nutrients for the biosphere. Algae of all types possess chlorophyll and carry on photosynthesis in freshwater and marine habitats. Green plants are the dominant photosynthesizers on land.

Video Plants

Heterotrophs

Heterotrophs need a source of organic nutrients. They are **consumers,** meaning they consume food. **Herbivores** are organisms that graze directly on plants or algae (Fig. 23.3*b*). Deer, rabbits, and caterpillars are herbivores found in terrestrial habitats. Some birds are herbivores, too. In aquatic ecosystems, some protists are herbivores. **Carnivores** feed on other animals. Spiders that feed on insects are carnivores, as are osprey that feed on fish (Fig. 23.3*c*). This example allows us to mention that there are primary consumers (e.g., plant-eating insects), secondary consumers (e.g., insect-eating spiders), and tertiary consumers (e.g., osprey, hawks). Sometimes tertiary consumers are called top predators. **Omnivores** are animals that feed on plants and animals. Humans are considered omnivores.

Video Snake Eating

Detritus feeders are organisms that feed on detritus, decomposing particles of organic matter. Marine fan worms take detritus from the water, and clams take it from the substratum. Earthworms, various beetles, termites, and ants are terrestrial

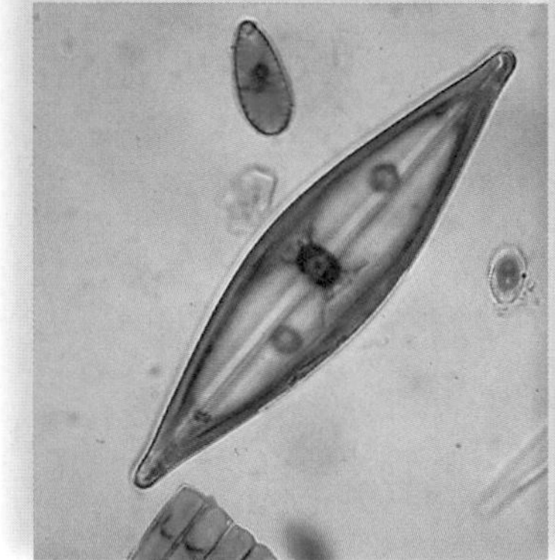

a. Producers

b. Herbivores

c. Carnivores

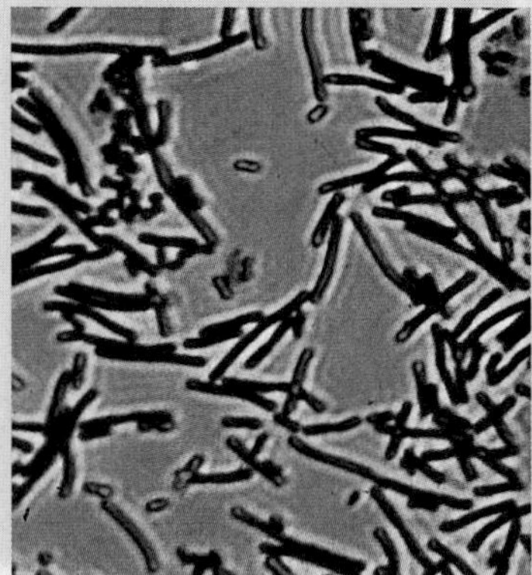

d. Decomposers

Figure 23.3 The biotic components of an ecosystem. **a.** Diatoms and green plants are autotrophs (producers). **b.** Caterpillars and rabbits are herbivores. **c.** Spiders and osprey are carnivores. **d.** Some bacteria and some mushrooms are decomposers.

APPLICATIONS AND MISCONCEPTIONS

How do producers at the bottom of the ocean floor make food without sunlight?

Producers at the bottom of the ocean floor use chemical energy instead of sunlight to make food. They are called *chemoautotrophs.* Volcanoes at the bottom of the ocean floor release hydrogen sulfide gas through cracks called hydrothermal vents. (Hydrogen sulfide is the nasty smelling gas we associate with rotten eggs.) Some chemoautotrophs split hydrogen sulfide to obtain the energy needed to link carbon atoms together to form glucose.

The glucose contained in these chemoautotrophs sustains a variety of bizarre organisms such as giant tube worms, anglerfish, and giant clams.

detritus feeders. Bacteria and fungi, including mushrooms, are decomposers. They acquire nutrients by breaking down dead organic matter, including animal wastes. Decomposers perform a valuable service because they release inorganic substances that can be taken up by plants for their growth (Fig. 23.3*d*). Otherwise, plants would be completely dependent only on physical processes, such as the release of minerals from rocks, to supply them with inorganic nutrients.

Video Decomposers

Niche

A **niche** is the role of an organism in an ecosystem. Descriptions of a niche include how an organism gets its food, how it interacts with other populations in the same community, and what eats it. Humans are also part of the cycle of life. The chemicals making up our bodies must also return to nature.

Video Cichlid Specialization

Energy Flow and Chemical Cycling

When we diagram the interactions of all the populations in an ecosystem, it is possible to illustrate two phenomena that characterize every ecosystem. One of these phenomena is energy flow, which begins when producers absorb solar (and in some cases, chemical) energy. The second, nutrient cycling, occurs when producers take in inorganic chemicals from the physical environment. Thereafter, producers make organic nutrients that can be used by them as well as by the other populations of the ecosystem. Energy flow occurs when nutrients pass from one population to another. Eventually, all the energy content is converted to heat. The heat then dissipates in the environment. Therefore, most ecosystems cannot exist without a continual supply of solar energy. Chemicals cycle when inorganic nutrients are returned to the producers from the atmosphere or soil (Fig. 23.4).

Only a portion of the organic nutrients made by autotrophs is passed on to heterotrophs due to plants using some of the organic molecules to fuel their cellular respiration (see Fig. 3.19). Similarly, only a small percentage of nutrients taken in by heterotrophs is available to higher-level consumers. Figure 23.5 shows why. Some of the food eaten by an herbivore is never digested and is eliminated as feces. Metabolic wastes are excreted in urine. Of the assimilated energy, a large portion is used during cellular respiration and thereafter becomes heat. Only the remaining food energy, converted into increased body weight (or additional offspring), becomes available to carnivores. Plants carry on cellular respiration, too, so only about 55% of the original energy absorbed by plants is available to an ecosystem. Further, as organisms feed on one another, less and less of this 55% is available in a usable form.

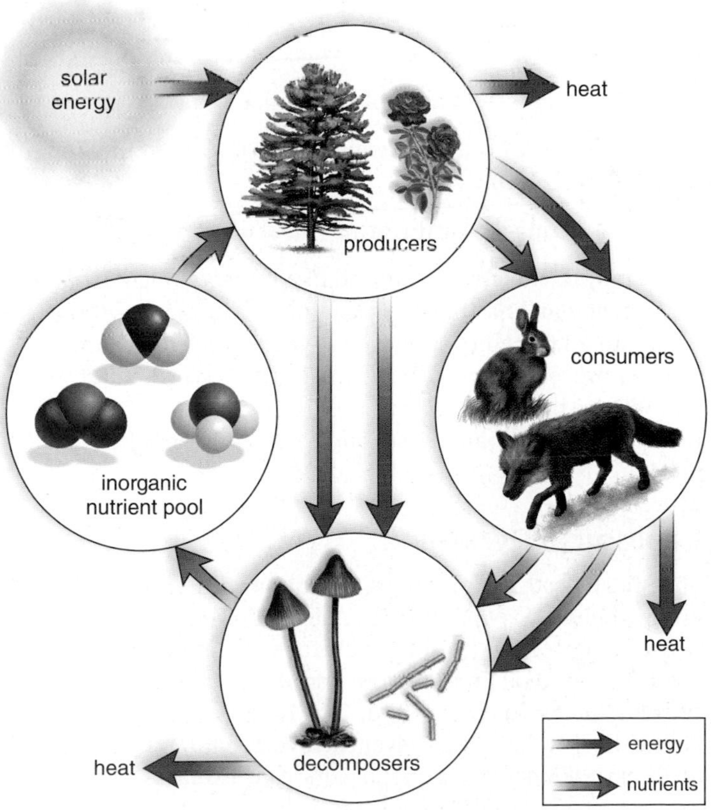

Figure 23.4 Energy flow and chemical cycling in an ecosystem.
Chemicals cycle, but energy flows through an ecosystem. As energy transformations repeatedly occur, all the energy derived from the sun eventually dissipates as heat.

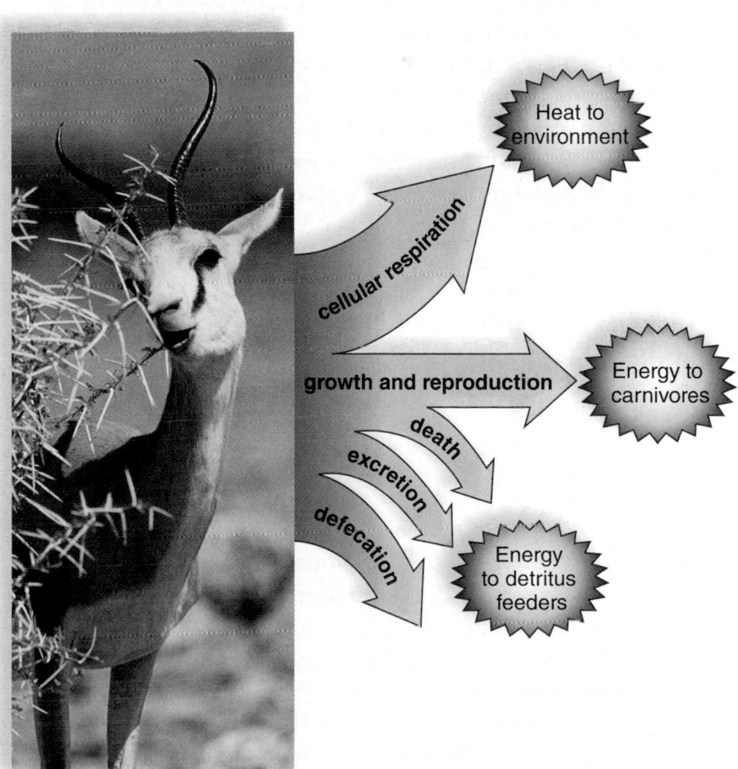

Figure 23.5 The fate of food energy taken in by an herbivore.
Only about 10% of the food energy taken in by an herbivore is passed on to carnivores. A large portion goes to detritus feeders via defecation, excretion, and death; another large portion is used for cellular respiration.

The elimination of wastes and the deaths of all organisms do not mean that substances are lost to an ecosystem. Instead they are nutrients made available to decomposers. Decomposers convert organic nutrients, such as glucose, back into inorganic chemicals, such as carbon dioxide and water. The inorganic chemicals are then released to the soil or atmosphere. Chemicals complete their cycle within an ecosystem when inorganic chemicals are absorbed by the producers from the atmosphere or soil.

CHECK YOUR PROGRESS 23.1

1. Discuss why it could be said that the biosphere is a giant ecosystem.
2. Explain why autotrophs are called producers and heterotrophs are called consumers.
3. List the different types of consumers that may be present in an ecosystem.
4. Distinguish between the movement of energy and chemicals in an ecosystem.

CONNECTING THE CONCEPTS

For more information on the topics presented in this section, refer to the following discussions:

Section 1.1 describes the relationship between populations, the ecosystems on the planet, and the biosphere.

Section 3.6 examines how organisms release the energy stored in organic nutrients by cellular respiration.

23.2 Energy Flow

LEARNING OUTCOMES

Upon completion of this section, you should be able to

1. Recognize the differences between a grazing and a detrital food web.
2. Explain the energy flow among populations through food webs and ecological pyramids.

The principles we have been discussing can now be applied using a forest ecosystem. The various interconnecting paths of energy flow are represented by a **food web,** which is a diagram that describes trophic, or feeding, relationships. Figure 23.6*a* is a **grazing food web** because it begins with an oak tree and grass. Caterpillars feed on oak leaves; mice, rabbits, and deer feed on leaves and grass at or near the ground. Various birds, chipmunks, and mice feed on seeds and nuts, but they are omnivores because they also feed on caterpillars. These herbivores and omnivores are then food for a number of different carnivores.

Figure 23.6*b* is a **detrital food web.** This type of food web begins with wastes and the remains of dead organisms. Detritus is food for decomposers and soil organisms like earthworms. Earthworms may be food for carnivorous invertebrates. In turn, salamanders and shrews may consume the carnivorous insects. The members of detrital food webs may become food for aboveground carnivores, so the detrital and grazing food webs are connected.

We naturally tend to think that aboveground plants, such as trees, are the largest storage form of organic matter and energy. This is not necessarily the case. In this particular forest, the organic matter lying on the forest floor and mixed into the soil contains over twice as much energy as the leaves of living trees. Therefore, more energy in a forest may be funneling through the detrital food web than through the grazing food web.

Video Decomposers

Trophic Levels

The arrangement of the species in Figure 23.6 suggests that organisms are linked to one another in a straight line, according to feeding or predator–prey relationships. Diagrams that show a single path of energy flow are called **food chains.** For example, in the grazing food web, we could find this **grazing food chain:**

leaves → caterpillars → mice → hawks

And in the detrital food web (Fig. 23.6*b*), we could find this **detrital food chain:**

detritus → earthworms → beetles → shrews

A **trophic level** is composed of all the organisms that feed at a particular link in a food chain. In the grazing food web in Figure 23.6*a*, going from left to right, the trees, grass, and flowers are producers (first trophic level); the first series of animals are primary consumers (second trophic level); and the next group of animals are secondary consumers (third trophic level).

Ecological Pyramids

The shortness of food chains can be attributed to the loss of energy between trophic levels. As mentioned, only about 10% of the energy of one trophic level is available to the next trophic level. Therefore, if an herbivore population consumes 1,000 kg of plant material, only about 100 kg is converted to herbivore tissue, 10 kg to first-level carnivores, and 1 kg to second-level carnivores. This general 10% rule explains why so few

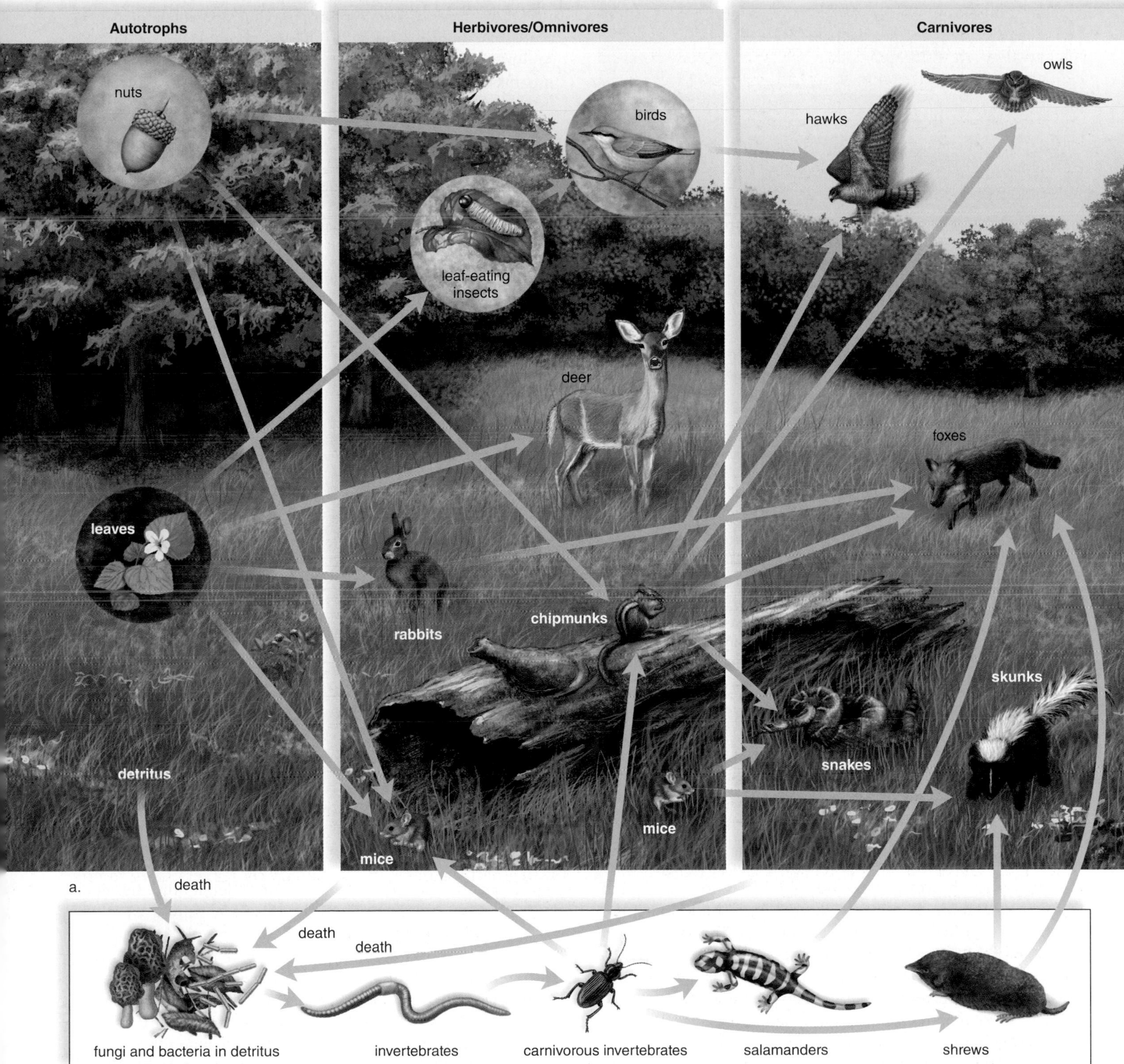

Figure 23.6 Food webs illustrate ecological relationships.
Food webs are descriptions of who eats whom. **a.** Tan arrows illustrate possible grazing food webs. The tree and other organisms that convert sun energy into food energy are producers (first trophic level). Mice and other animals that eat the producers are primary consumers (second trophic level). Carnivores that rely on primary consumer animals for energy (the hawk, fox, skunk, snake, and owl) are secondary consumers (third trophic level). **b.** Green arrows illustrate possible detrital food webs. These begin with detritus: organic waste and remains of dead organisms. Decomposers and detritus feeders recycle these organic nutrients. The organisms in the detrital food web may be prey for animals in the grazing food web, as when chipmunks feed on bugs. Thus, the grazing food web and detrital food web are interconnected.

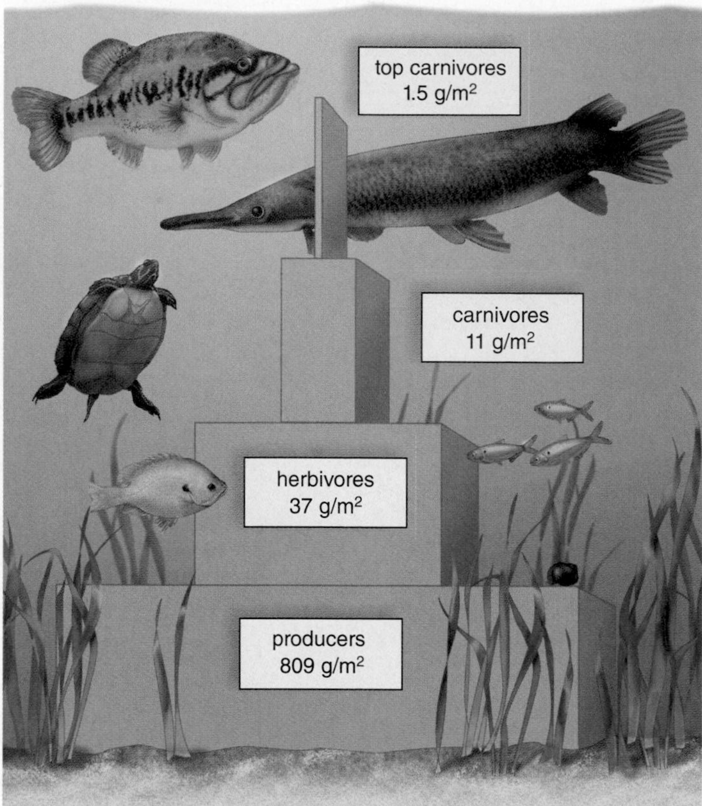

Figure 23.7 The influence of trophic level on biomass. The biomass, or dry weight (g/m²), for trophic levels in a grazing food web in a bog at Silver Springs, Florida. There is a sharp drop in biomass between the producer level and herbivore level. This is consistent with the common knowledge that the detrital food web plays a significant role in bogs.

carnivores can be supported in a food web. The flow of energy with large losses between successive trophic levels is sometimes depicted as an **ecological pyramid** (Fig. 23.7).

Energy losses between trophic levels also result in pyramids based on the number of organisms in each trophic level. When constructing a pyramid based on number of organisms, problems arise, however. For example, in Figure 23.6*a,* each tree would contain numerous caterpillars. Therefore, there would be more herbivores than autotrophs. The explanation has to do with size. An autotroph can be as tiny as a microscopic alga or as big as a beech tree. Similarly, an herbivore can be as small as a caterpillar or as large as an elephant.

Pyramids of biomass eliminate size as a factor because **biomass** is the number of organisms multiplied by the weight of organic matter contained in one organism. You would certainly expect the biomass of the producers to be greater than the biomass of the herbivores and that of the herbivores to be greater than that of the carnivores. In aquatic ecosystems, such as lakes and open seas, the herbivores may have a greater biomass than the producers. This is because algae are the only producers. Over time, algae reproduce rapidly but are also consumed at a high rate.

These types of problems are making some ecologists hesitant about using pyramids to describe ecological relationships. Another issue concerns the role played by decomposers. These organisms are rarely included in pyramids, even though a large portion of energy becomes detritus in many ecosystems.

CHECK YOUR PROGRESS 23.2

1. Explain the difference between a grazing food web and a detrital food web and give an example of each.
2. Describe the type of diagram that represents a single path of energy flow in an ecosystem.
3. Calculate the amount of biomass available to a second-level carnivore if the producer biomass in the ecosystem is 1,500 g/m².

CONNECTING THE CONCEPTS

For more information on the topics presented in this section, refer to the following discussion:

Section 24.1 examines how human population growth is threatening the structure of many ecological pyramids.

23.3 Global Biogeochemical Cycles

LEARNING OUTCOMES

Upon completion of this section, you should be able to

1. Define the term *biogeochemical cycle.*
2. Identify the steps of the water cycle, carbon cycle, nitrogen cycle, and phosphorus cycle.
3. Identify how human activities can alter each of the biogeochemical cycles.

In this section, we will examine in more detail how chemicals cycle through ecosystems. All organisms require a variety of organic and/or inorganic nutrients. For example, carbon dioxide and water are necessary nutrients for photosynthesizers. Nitrogen is a component of all the structural and functional proteins and nucleic acids that sustain living tissues. Phosphorus is essential for ATP and nucleotide production.

The pathways by which chemicals circulate through ecosystems involve both living (biotic) and nonliving (abiotic) components. Therefore, they are known as **biogeochemical cycles.** A biogeochemical cycle can be gaseous or sedimentary. In a gaseous cycle, such as the carbon and nitrogen cycles, the element returns to and is withdrawn from the atmosphere as a gas. The phosphorus cycle is a sedimentary cycle. Phosphorus is absorbed from the soil by plant roots, passed to heterotrophs, and eventually returned to the soil by decomposers.

Chemical cycling involves the components of ecosystems shown in Figure 23.8. A *reservoir* is a source of chemicals normally unavailable to producers, such as the carbon found in calcium carbonate shells on ocean bottoms. An *exchange pool* is a source from which organisms generally have the ability to take chemicals, such as the atmosphere or soil. Chemicals move along food chains in a *biotic community.*

Human activities (purple arrows) remove various chemicals from reservoirs and exchange pools and make them available to the biotic community. In this way, human activities upset the normal balance of nutrients in the environment, resulting in pollution.

Video
Salmon Farming

APPLICATIONS AND MISCONCEPTIONS

How much water is required to produce your food?

Growing a single serving of lettuce takes about 6 gallons (gal) of water. Producing an 8-oz glass of milk requires 49 gal of water. That includes the amount of water the cow drinks, the water used to grow the cow's food, and water needed to process the milk. Producing a single serving of steak consumes more than 2,600 gal of water.

Figure 23.8 The cycling of nutrients between biotic communities and biogeochemical reservoirs. Reservoirs, such as fossil fuels, minerals in rocks, and sediments in oceans, are normally relatively unavailable sources of nutrients for the biotic community. Nutrients in exchange pools, such as the atmosphere, soil, and water, are available sources of chemicals for the biotic community. When human activities (purple arrows) remove chemicals from a reservoir or an exchange pool and make them available to the biotic community, pollution can result. This is because not all the nutrients are used. For example, when humans burn fossil fuels, CO_2 increases in the atmosphere and contributes to global warming.

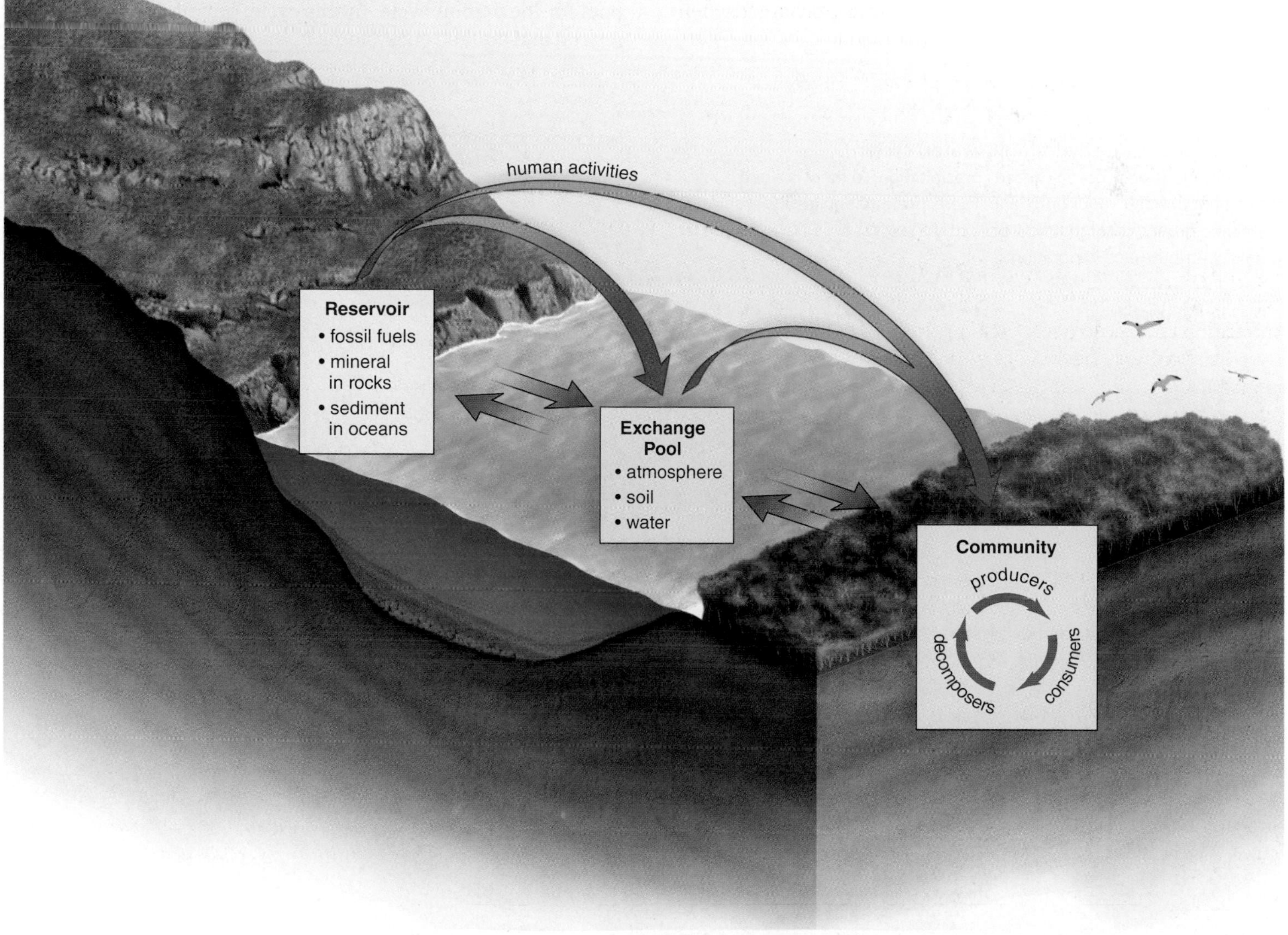

The Water Cycle

The width of the arrows in the following cycles are used to represent the approximate transfer rate of the substances between the components of an ecosystem. The **water (hydrologic) cycle** is described in Figure 23.9.

① During **evaporation** in the water cycle, the sun's rays cause freshwater to evaporate from seawater, leaving the salts behind. Net condensation occurs. During condensation, a gas is changed into a liquid. ② Vaporized freshwater rises into the atmosphere, condenses, and then falls as **precipitation** (e.g., rain, snow, sleet, hail, and fog) over the oceans and the land.

③ Water also evaporates from land and from plants (evaporation from plants is called transpiration). ④ Land lies above sea level, so gravity eventually returns all freshwater to the sea. In the meantime, much water is contained within standing waters (lakes and ponds), flowing water (streams and rivers), and groundwater. ⑤ **Runoff** is water that flows directly into nearby streams, lakes, wetlands, or the ocean.

Instead of running off, some precipitation sinks, or percolates, into the ground. This saturates the earth to a certain level. The top of the saturation zone is called the groundwater table, or the water table. ⑥ Sometimes, groundwater is also located in **aquifers,** rock layers that contain water and release it in appreciable quantities to wells or springs. Aquifers are recharged when rainfall and melted snow percolate into the soil.

Human Activities

Humans interfere with the water cycle in three ways. First, we withdraw water from aquifers. Second, we clear vegetation from land and build roads and buildings that prevent percolation and increase runoff. Third, we interfere with the natural processes that purify water and instead add pollutants like sewage and chemicals to water.

Video Thames River

In some parts of the United States, especially the arid West and southern Florida, withdrawals from aquifers exceed any possibility of recharge. This is called "groundwater mining." In these locations, the groundwater is dropping. Residents may run out of groundwater, at least for irrigation purposes, within a few short years. Freshwater, which makes up only about 3% of the world's supply of water, is called a renewable resource because a new supply is always being produced. However, it is possible to run out of freshwater when the available supply runs off instead of entering bodies of freshwater and aquifers. Freshwater may also become so polluted that it is not usable.

The Carbon Cycle

The carbon dioxide (CO_2) in the atmosphere is the exchange pool for the carbon cycle. In this cycle, organisms in both terrestrial and aquatic ecosystems exchange carbon dioxide with

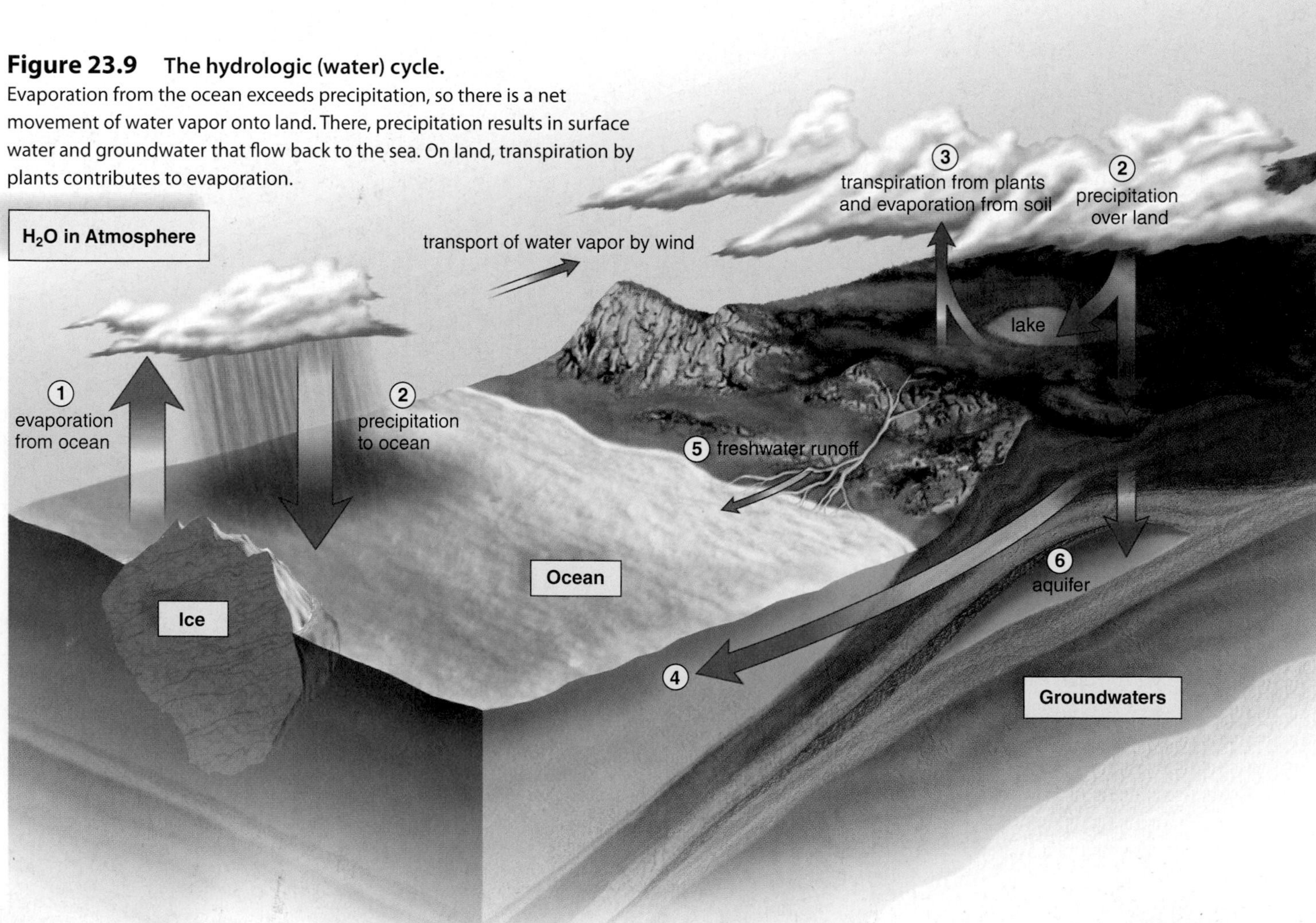

Figure 23.9 **The hydrologic (water) cycle.** Evaporation from the ocean exceeds precipitation, so there is a net movement of water vapor onto land. There, precipitation results in surface water and groundwater that flow back to the sea. On land, transpiration by plants contributes to evaporation.

the atmosphere (Fig.23.10). ① On land, plants take up carbon dioxide from the air. Through photosynthesis, they incorporate carbon into nutrients used by autotrophs and heterotrophs. ② When organisms, including plants, respire, carbon is returned to the atmosphere as carbon dioxide. Therefore, carbon dioxide recycles to plants by way of the atmosphere.

In aquatic ecosystems, the exchange of carbon dioxide with the atmosphere is indirect. ③ Carbon dioxide from the air combines with water to produce bicarbonate ion (HCO_3^-). This is a source of carbon for aquatic producers that create food for themselves and for heterotrophs. Similarly, when aquatic organisms respire, the carbon dioxide they give off becomes bicarbonate ion. ④ The amount of bicarbonate in the water is in equilibrium with the amount of carbon dioxide in the air.

Reservoirs Hold Carbon

⑤ Living and dead organisms contain organic carbon and serve as one of the reservoirs for the carbon cycle. The world's biotic components, particularly trees, contain over 800 billion tons of organic carbon. An additional 1,000–3,000 billion metric tons are estimated to be held in the remains of plants and animals in the soil. Ordinarily, decomposition of plants and animals returns CO_2 to the atmosphere.

⑥ In the history of the Earth, over the past 300 million years, plant and animal remains have been transformed into coal, oil, and natural gas. We call these materials the **fossil fuels.** Another reservoir for carbon is the inorganic carbonate that accumulates in limestone and in calcium carbonate shells. Many marine organisms have calcium carbonate shells that remain in ocean sediments long after the organisms have died. Geologic forces change these sediments into limestone.

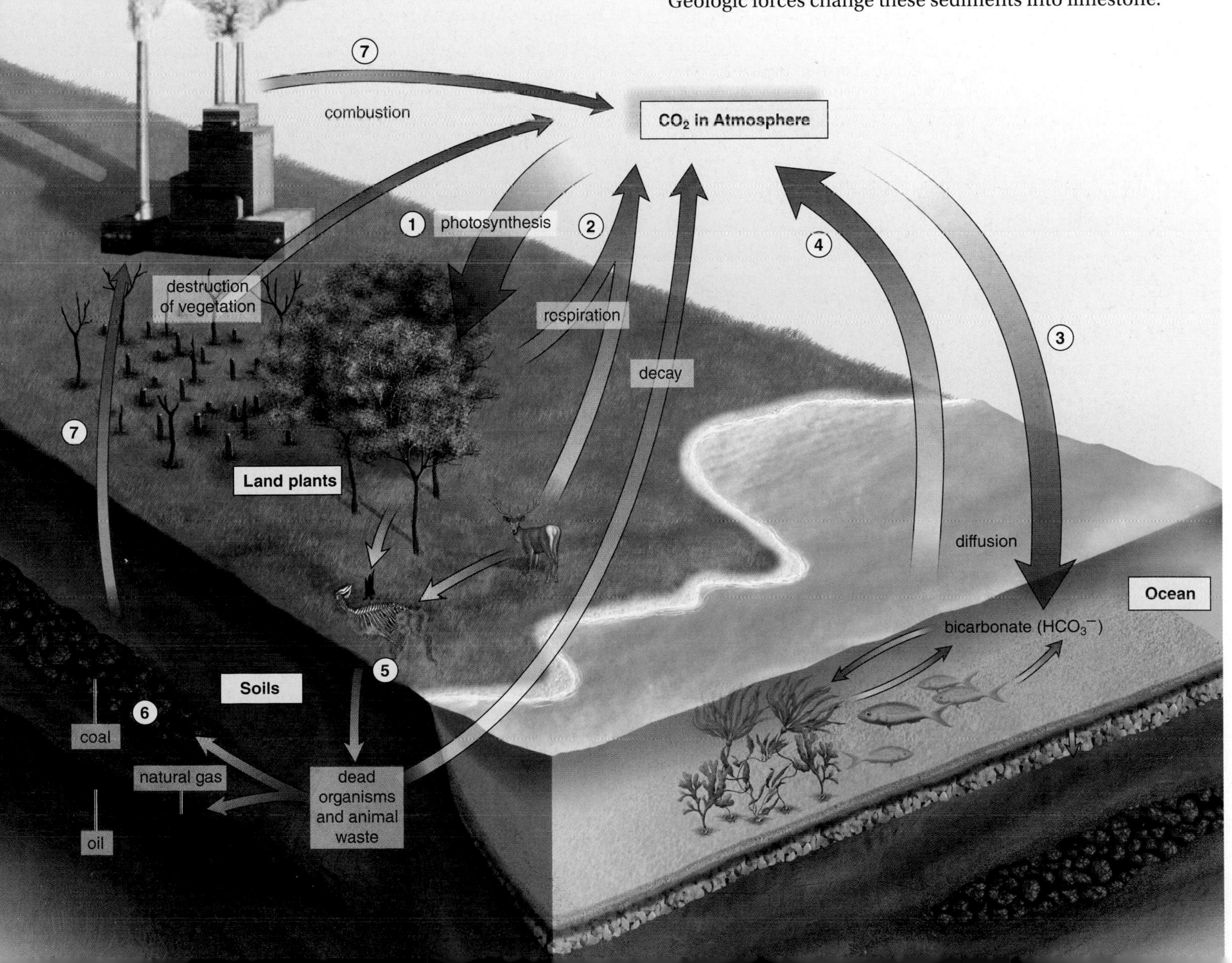

Figure 23.10 **The carbon cycle.**
The carbon cycle is a gaseous biogeochemical cycle. Producers take in carbon dioxide from the atmosphere and convert it to organic molecules that feed all organisms. The transfer rate of carbon into the atmosphere due to respiration approximately matches the rate due to withdrawal by plants for photosynthesis. Fossil fuels arise when organisms die but do not decompose. When humans burn fossil fuels and destroy vegetation (purple arrows), more carbon dioxide is added to the atmosphere than is withdrawn. This causes environmental pollution.

BIOLOGY MATTERS  Bioethical

Guaranteeing Access to Safe Drinking Water

The ability to get safe drinking water may not be something you've ever thought about. In the United States and other developed countries, safe, clean water usually pours out of your home faucets every time you turn them on. When water is plentiful, it's taken for granted and a great deal of it is wasted. However, water is a precious and often scarce resource in many parts of the world.

Over 884 million people around the world lack access to safe drinking water (Fig. 23A). Further, basic sanitation (proper disposal of human and animal waste) is unavailable to almost half. Most people lacking safe drinking water and/or basic sanitation live in China, the Middle East, or Africa. These are some of the world's poorest people. If they have to pay for water, they are often charged much more than wealthy people living in the same area. They may have to walk miles to collect water and then return to their homes.

Having safe drinking water and basic sanitation would impact people's lives and health more than any other intervention. More than 80% of diseases, including typhoid fever and cholera, are associated with foul water and improper sanitation. Each year, more than 5 million people die from water-pollution-related illnesses. This is the leading cause of death for children under age 5.

Solving the problem of making safe water available to more people is no easy task. One of the Millennium Development Goals (agreed upon by members of the United Nations in 2000) is to halve the number of people without access to safe water by 2015. The cost has been projected to be between $4 billion and $11.3 billion per year. An unusual twist to the water crisis is the interest in water as an investment and commodity to be traded or sold. The Goldman Sachs Group, Inc. (often referred to as *Goldman Sachs*), is a large global banking company involved in investment banking. Goldman Sachs estimates the value of the water industry at $425 billion. At a recent conference hosted by the company, the water shortage was listed among the world's top threats to global stability.

Each day, an average American uses 100–176 gal of water. By contrast, the typical African uses about 10 gal. We may need to consider how much of the water we use daily is necessary. More effort will be needed to ensure that all people have the ability to get safe drinking water. A more stable global community may depend on it.

Questions to Consider

1. Should developed nations assume responsibility for providing safe drinking water and sanitation to developing countries? If so, to what extent?
2. Water conservation is mandated by law in many states. Should legislation requiring water conservation be extended to include all states? Why or why not?
3. What form might this legislation take? How should it be applied?

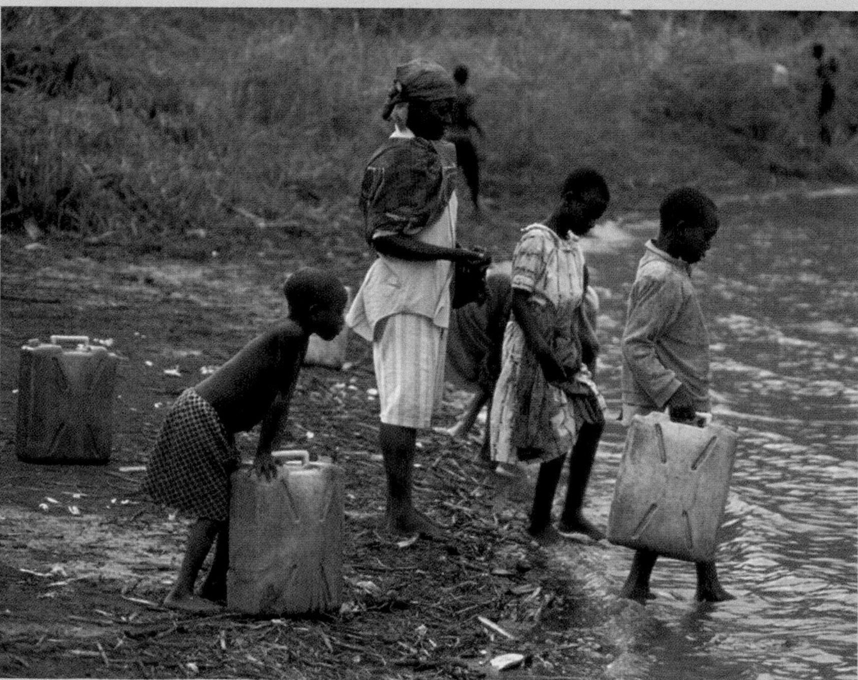

Figure 23A **Many people in developing nations lack access to safe drinking water.**
Individuals in developing countries may have little or no access to safe drinking water. Providing a universal clean water supply for all is a global necessity.

APPLICATIONS AND MISCONCEPTIONS

What are some other greenhouse gases?

In addition to carbon dioxide, these gases also play a role in the greenhouse effect:

- *Methane (CH_4):* A single molecule of methane has 21 times the warming potential of a molecule of carbon dioxide, making it a powerful greenhouse gas. Methane is a natural by-product of the decay of organic material but is also released by landfills and the production of coal, natural gas, and oil.
- *Nitrous oxide (N_2O):* Nitrous oxide is released from the combustion of fossil fuels and as gaseous waste from many industrial activities.
- *Hydrofluorocarbons:* These were initially produced to reduce the levels of ozone-depleting compounds in the upper atmosphere. Unfortunately, although present in very small quantities, they are potent greenhouse gases.

Human Activities

The transfer rates of carbon dioxide between storage of carbon during photosynthesis and the release of carbon during cellular respiration and decomposition are just about even. However, more carbon dioxide is being deposited in the atmosphere due to human activities than can be removed through photosynthesis. ⑦ This increase is largely due to the burning of fossil fuels and the destruction of forests to make way for farmland and pasture. When we do away with forests, we reduce a reservoir and lose the organisms that take up excess carbon dioxide. Today, the amount of carbon dioxide released into the atmosphere is about twice the amount that remains in the atmosphere. It's believed that much of this has been dissolving into the ocean.

CO_2 and Climate Change Large amounts of carbon dioxide and other gases are being emitted due to human activities. The other gases include nitrous oxide (N_2O) and methane (CH_4). Fertilizers and animal wastes are sources of nitrous oxide. In the digestive tracts of animals, bacterial decomposition produces methane. Decaying sediments and flooded rice paddies are methane sources, as well. In the atmosphere, nitrous oxide and methane are known as **greenhouse gases** because they act just like the panes of a greenhouse. They allow solar radiation to penetrate to Earth but hinder the escape of infrared rays (heat) back into space—a phenomenon called the **greenhouse effect.** The greenhouse gases are contributing significantly to an overall rise in the Earth's ambient temperature. This phenomenon is called **global warming,** which is contributing to the **climate change** of the Earth.

Figure 23.11 shows how the average temperature in the United States has steadily increased over the past two centuries. If the Earth's temperature continues to rise, more water will evaporate, forming more clouds. This sets up a positive feedback effect that could increase global warming still more. The global climate has already warmed about 0.6°C since the Industrial Revolution. Computer models are unable to consider all possible variables, but the Earth's temperature may rise 1.5–4.5°C by 2100 if greenhouse emissions continue at the current rates.

Video
Warming Hurts Rice

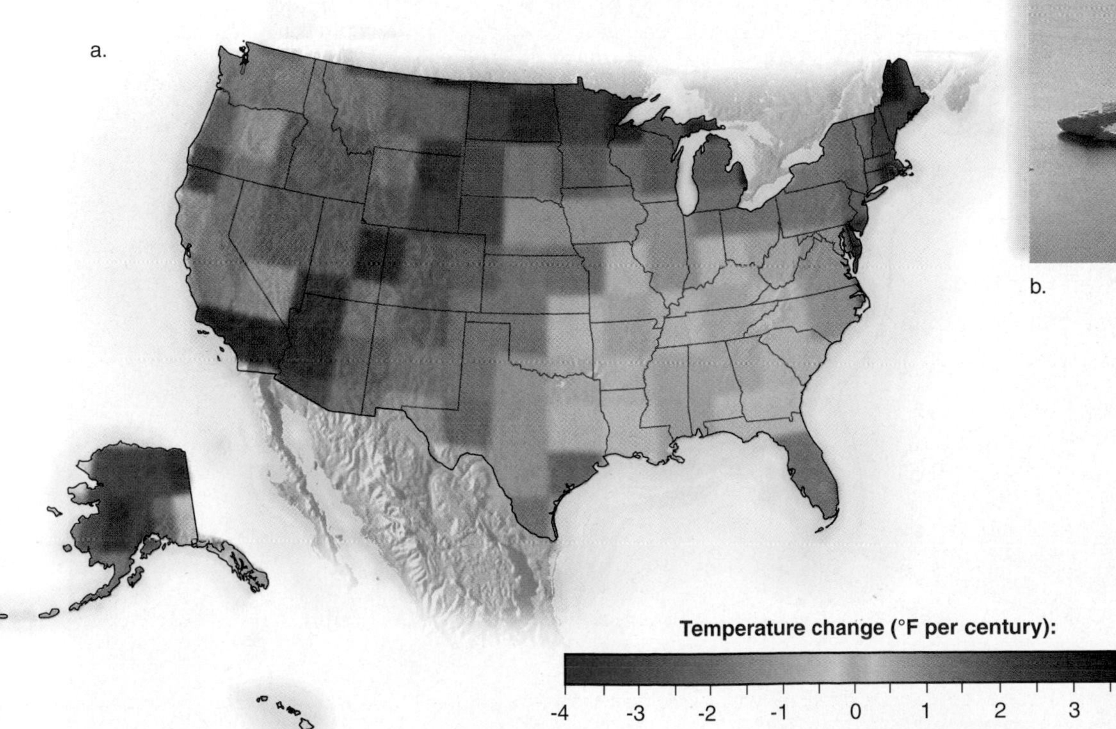

b.

Figure 23.11 Global warming and climate change. **a.** The average temperature in the United States has steadily increased over the past two centuries, leading to more severe droughts and more erratic periods of precipitation that are altering the composition of many communities. **b.** Increased extinction rates are just one of the consequences of climate change.

As climate change continues, other effects will possibly occur. It is predicted that, as the oceans warm, temperatures in the polar regions will rise to a greater degree than in other regions. As a result, sea levels will rise because glaciers will melt, and water expands as it warms. Water evaporation will increase, and most likely there will be increased rainfall along the coasts and dryer conditions inland. The occurrence of droughts will reduce agricultural yields and also cause trees to die off. Expansion of forests into Arctic areas might not offset the loss of forests in the temperate zones. Coastal agricultural lands, such as the deltas of Bangladesh and China, will be inundated with water. Billions of dollars will have to be spent to keep coastal cities such as New Orleans, New York, Boston, Miami, and Galveston from disappearing into the sea.

Video: Karoo Global Warming

The Nitrogen Cycle

Nitrogen gas (N_2) makes up about 78% of the atmosphere, in a form that is unusable by plants. Therefore, nitrogen can be a nutrient that limits the amount of growth in an ecosystem.

Video: Dung Beetles

Ammonium (NH_4^+) Formation and Use

① In the nitrogen cycle, **nitrogen fixation** occurs when nitrogen gas (N_2) is converted to ammonium (NH_4^+), a form that plants can use (Fig. 23.12). Some cyanobacteria in aquatic ecosystems and some free-living bacteria in soil are also able to fix atmospheric nitrogen in this way. Other nitrogen-fixing bacteria live in nodules on the roots of legumes, such as beans, peas,

Figure 23.12 The nitrogen cycle.

Nitrogen is primarily made available to biotic communities by internal cycling of the element. Without human activities, the amount of nitrogen returned to the atmosphere (denitrification in terrestrial and aquatic communities) exceeds withdrawal from the atmosphere (N_2 fixation and nitrification). Human activities (dark purple arrow) result in an increased amount of NO_3^- in terrestrial communities, with resultant runoff to aquatic biotic communities.

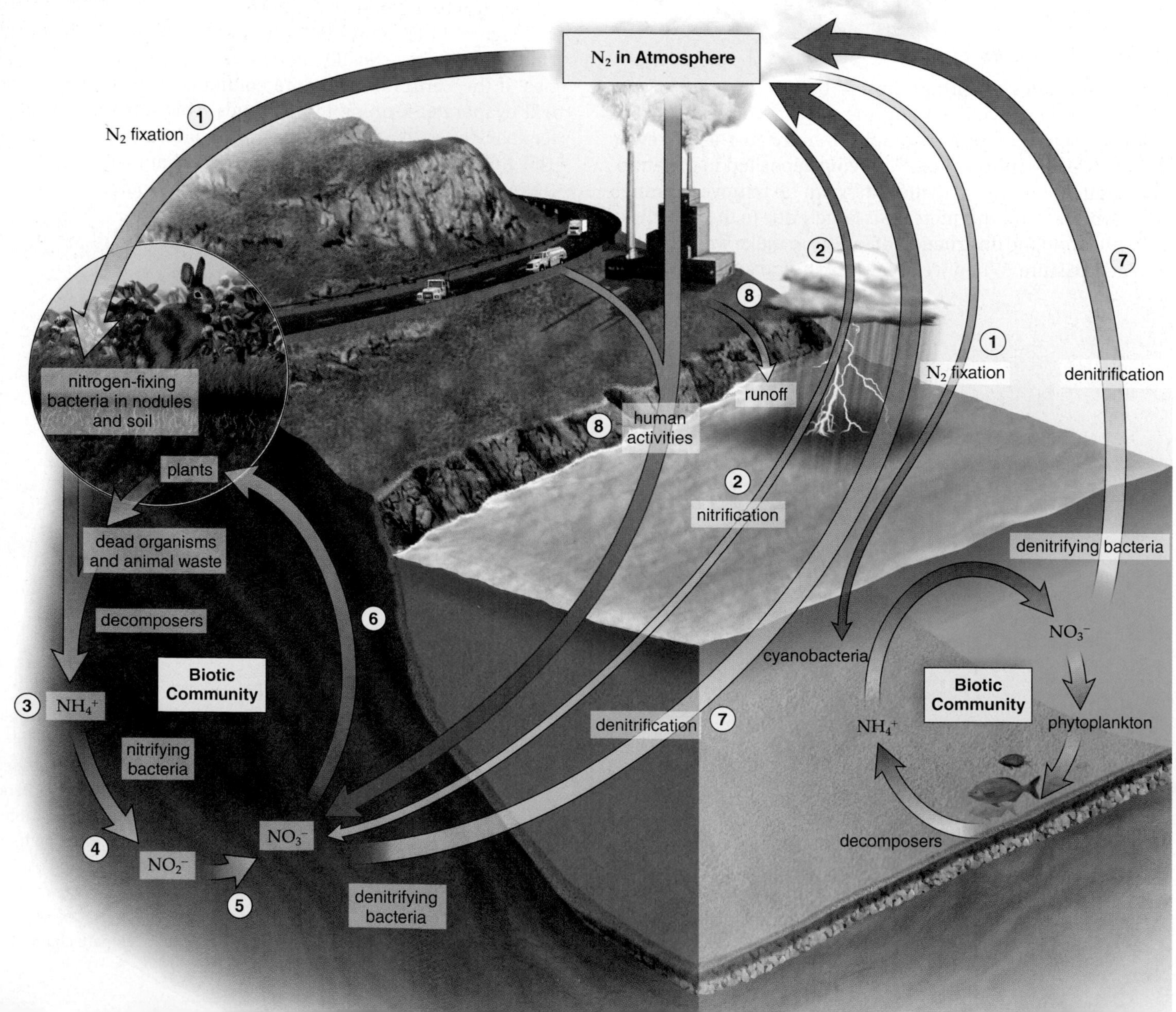

and clover. They make organic compounds containing nitrogen available to the host plants so that the plant can form proteins and nucleic acids.

Nitrate (NO_3^-) Formation and Use

Plants can also use nitrates (NO_3^-) as a source of nitrogen. The production of nitrates during the nitrogen cycle is called **nitrification.** Nitrification can occur in various ways: ② Nitrogen gas (N_2) is converted to nitrate (NO_3^-) in the atmosphere, where high energy is available for nitrogen to react with oxygen. This energy may be supplied by cosmic radiation, meteor trails, or lightning. ③ Ammonium (NH_4^+) in the soil from various sources, including decomposition of organisms and animal wastes, is converted to nitrate by soil bacteria. ④ Nitrite-producing bacteria convert ammonium to nitrite (NO_2^-). ⑤ Nitrate-producing bacteria convert nitrite to nitrate. ⑥ During the process of *assimilation,* plants take up ammonia and nitrate from the soil and use these ions to produce proteins and nucleic acids.

In Figure 23.12, observe the biotic community subcycles occurring on land and in water. Those subcycles do not depend on the presence of nitrogen gas.

Formation of Nitrogen Gas from Nitrate

⑦ **Denitrification** is the conversion of nitrate back to nitrogen gas, which enters the atmosphere. Denitrifying bacteria living in the anaerobic mud of lakes, bogs, and estuaries carry out this process during their metabolism. In the nitrogen cycle, denitrification would counterbalance nitrogen fixation if human activities were not involved.

Human Activities

⑧ Human activities significantly alter the transfer rates in the nitrogen cycle by producing fertilizers from N_2. They nearly double the fixation rate. Fertilizer, which also contains phosphate, often runs off into lakes and rivers. This results in an overgrowth of algae and rooted aquatic plants. As discussed on page 550, the result is cultural eutrophication (overenrichment), which can lead to an algal boom. When the algae die off, enlarged decomposer populations use up all the oxygen in the water. The result is a massive fish kill.

Acid deposition occurs because nitrogen oxides (NO_x) and sulfur dioxide (SO_2) enter the atmosphere from the burning of fossil fuels (Fig. 23.13*a*). Both of these gases combine with

a.

b.

c.

Figure 23.13 The effects of acid deposition.
a. Air pollution due to fossil-fuel burning in factories and modes of transportation is the major cause of acid deposition. **b.** Many forests in higher elevations of northeastern North America and northern Europe are dying due to acid deposition. **c.** Acid deposition damages architectural features.

APPLICATIONS AND MISCONCEPTIONS

How acidic is acid rain?

The pH of normal rain varies from 4.5 to 5.6 (see Fig. 2.10). That's about the acidity of tomatoes or black coffee, and neither of these is particularly corrosive or harmful to the body. Rain collected in the eastern United States during the summer has an average pH of 3.6. Vinegar has a pH similar to this rain—can you imagine watering trees or house plants with vinegar? Fog with a pH of 2 has been measured in some locations. A solution with pH 2 is only slightly less acid than stomach acid or battery acid. It's easy to imagine the damage such a solution could do to the environment.

water vapor to form acids that eventually return to the Earth in acid rain. Acid deposition has drastically affected forests and lakes in northern Europe, Canada, and the northeastern United States. The soil in these forests is naturally acidic, whereas the surface water is only mildly alkaline (basic). Soil pH is made even more acidic by acid rain. The increased acidity kills trees (Fig. 23.13*b*) and reduces agricultural yields. Marble, metal, and stonework are corroded by acid deposition, which results in the loss of architectural features (Fig. 23.13*c*).

Nitrogen oxides and hydrocarbons (HC) from the burning of fossil fuels react with one another in the presence of sunlight to produce smog. Smog contains dangerous pollutants. Warm air near the Earth usually escapes into the atmosphere, taking pollutants with it. However, during a thermal inversion, pollutants are trapped near the Earth beneath a layer of warm, stagnant air. The air does not circulate, so pollutants can build up to dangerous levels. Areas surrounded by hills are particularly susceptible to the effects of a thermal inversion. This is because the air tends to stagnate, preventing the pollution from diffusing into the atmosphere (Fig. 23.14).

The Phosphorus Cycle

In the phosphorus cycle (Fig. 23.15), ① phosphorus trapped in oceanic sediments moves onto land after a geologic upheaval. ② On land, the very slow weathering of rocks moves phosphate ions (PO_4^{3-} and HPO_4^{2-}) into the soil. ③ Some of this soil becomes available to plants, which then incorporate phosphate into a variety of molecules. Molecules requiring phosphate include phospholipids, ATP, and the nucleotides that become a part of DNA and RNA. ④ Animals eat producers and incorporate some of the phosphate into their teeth, bones, and shells. ⑤ However, the eventual death and decomposition of all organisms and their wastes do make phosphate ions available to producers once again. The amount of phosphate available within a community is generally being used within various food chains. The lack of phosphate will limit the size of populations in ecosystems.

⑥ Some phosphate naturally runs off into aquatic ecosystems, enabling the algae to acquire it before it becomes trapped in sediments. ⑦ Phosphate in marine sediments will not become available to producers on land until a geologic upheaval exposes sedimentary rocks on land, allowing the cycle to begin again. Phosphorus does not enter the atmosphere. Therefore, the phosphorus cycle is called a sedimentary cycle.

Phosphorus and Water Pollution

⑧ Humans boost the supply of phosphate by mining phosphate ores for fertilizer and detergent production. Runoff of phosphate and nitrogen into water occurs during the use of fertilizer. This contamination, as well as that from animal wastes in livestock feedlots and discharge from sewage treatment plants, results in **cultural eutrophication** (overenrichment) of waterways.

Figure 23.16 lists various sources of water pollution. *Point* sources of pollution can be traced to a specific source, whereas *nonpoint* sources are those sources that cannot be specifically identified. Industrial wastes can include heavy metals and organochlorides, such as DDT and PCBs. These materials are not readily degraded under natural conditions or in conventional sewage treatment plants. **Biological magnification** occurs as these toxic chemicals pass along a food chain. Pollutants become increasingly concentrated with each higher consumer level, because they remain in the body and are not excreted. Aquatic food chains are more likely to experience biological magnification because there are more links than in

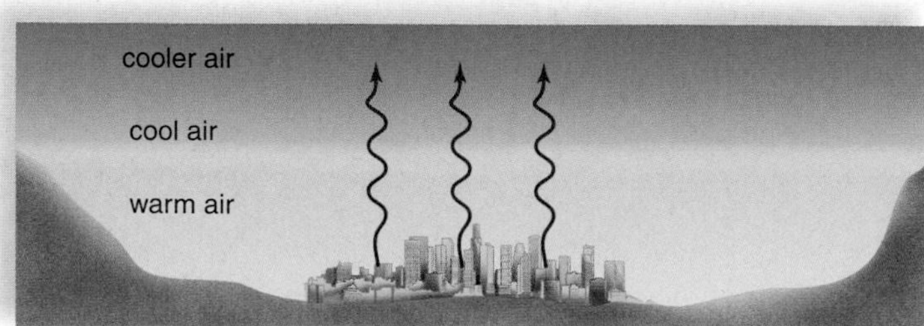

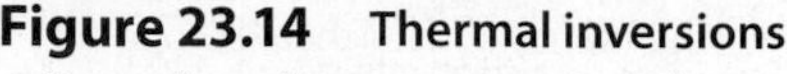
a. Normal pattern

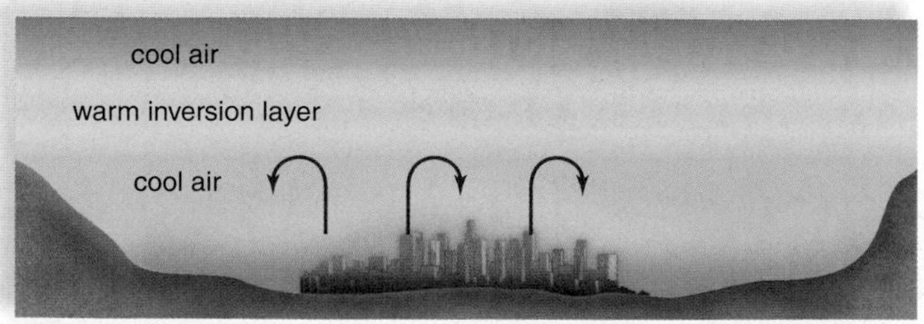

b. Thermal inversion

Figure 23.14 **Thermal inversions.**
a. Normally, pollutants escape into the atmosphere when warm air rises. **b.** During a thermal inversion, a layer of warm air (warm inversion layer) overlies and traps pollutants in cool air below.

a terrestrial food chain. Mercury levels can be high in some of the fish we eat (top consumers like shark, swordfish, and tuna) for this reason.

Coastal regions are the immediate receptors for local pollutants and the final receptors for pollutants carried by rivers that empty at a coast. Waste dumping occurs at sea. However, ocean currents sometimes transport both trash and pollutants back to shore. Offshore mining and shipping add pollutants to the oceans. Some 5 million metric tons of oil a year end up in the oceans. Large oil spills kill plankton, fish, and shellfishes, as well as birds and marine mammals.

In the last 50 years, humans have polluted the seas and exploited their resources to the point that many species are on the brink of extinction. Fisheries once rich and diverse, such as George's Bank off the coast of New England, are in severe decline. Haddock was once the most abundant species in this fishery, but now it accounts for less than 2% of the total catch. Cod and bluefin tuna have suffered a 90% reduction in population size. In warm, tropical regions, many areas of coral reefs are now overgrown with algae. This is because the fish that normally keep the algae under control have been overharvested.

Video Ocean Fishing Ban

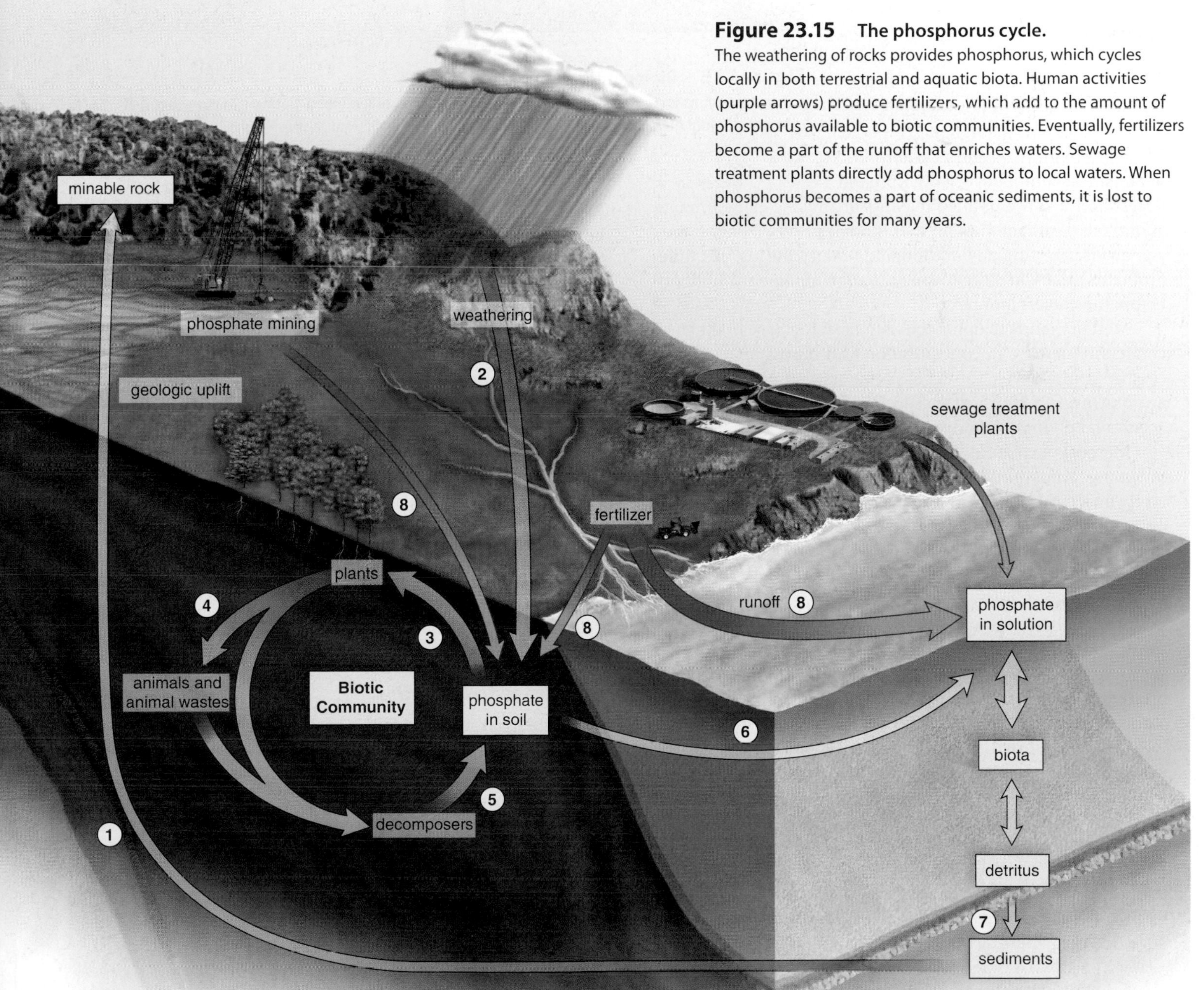

Figure 23.15 The phosphorus cycle.
The weathering of rocks provides phosphorus, which cycles locally in both terrestrial and aquatic biota. Human activities (purple arrows) produce fertilizers, which add to the amount of phosphorus available to biotic communities. Eventually, fertilizers become a part of the runoff that enriches waters. Sewage treatment plants directly add phosphorus to local waters. When phosphorus becomes a part of oceanic sediments, it is lost to biotic communities for many years.

BIOLOGY MATTERS  Science

Biomagnification of Mercury

Scientists have known since the 1950s that the emissions of mercury into the environment can lead to serious health effects for humans. Studies show that fish and wildlife exposed to mercury emissions are negatively impacted. Humans are impacted if they come into contact with affected fish and wildlife. Recent fish studies have shown a widespread contamination of mercury in streams, wetlands, reservoirs, and lakes throughout the majority of the United States.

a.

b.

Figure 23B **Biomagnification of mercury.**
a. Tuna may contain high levels of mercury due to biomagnification. **b.** Eating tuna or other fish that contain high levels of mercury can lead to problems with fetal development.

Mercury becomes a serious environmental risk when it undergoes bioaccumulation in an organism's body. Bioaccumulation occurs when an organism accumulates a contaminant faster than it can eliminate it. Most organisms can eliminate about half of the mercury in their bodies every 70 days if they can avoid ingesting any additional mercury during this time. Problems arise when organisms cannot eliminate the mercury before they ingest more.

Mercury tends to enter ecosystems at the base of the food chain and increase in concentration as it moves up each successive trophic level. Top-level predators and organisms that are long lived are the most susceptible to high levels of mercury accumulation in their body tissues.

Mercury exposure for humans generally occurs due to eating contaminated fish or breathing mercury vapor. Methylmercury is the form that leads to health problems such as sterility in men, damage to the central nervous system, and in severe cases, birth defects in infants. Developing fetuses and children can have health consequences from intake levels five to ten times lower than adults.

Studies have shown such elevated levels of mercury in sharks, tuna (Fig. 23B*a*), and swordfish that the EPA has advisories against eating these fish for women who may become or are pregnant, nursing mothers, and young children. Every state in the United States, in conjunction with federal agencies, has developed fish advisories for certain bodies of water within that state. Currently, 45 states warn pregnant women to limit their fish consumption from their waters (Fig. 23B*b*).

Mercury poisoning isn't limited to just aquatic species. Research conducted in the northeastern United States and Canada showed the presence of mercury in a variety of birds ranging from thrushes to loons to bald eagles. It is no surprise that loons and bald eagles can have high levels of mercury accumulation due to their consumption of contaminated fish. It was the presence of mercury in the songbirds that raised serious concerns among ecologists. Some speculate that northeastern songbirds are ingesting mercury when they feed upon insects that have picked up the toxin from eating smaller insects that ingested it from vegetation. This raises concerns about mercury's ability to enter food webs and bioaccumulate in previously unknown ways.

Ultimately the blame for mercury pollution falls squarely on the shoulders of humans. Every ecosystem on the planet has some degree of exposure to this pollutant, and with this exposure comes the risk of mercury contamination. Studies on species ranging from polar bears and bald eagles to whales and sharks show us that there are no limits to where mercury can be found. With coal-burning power plants being the largest human-caused source of mercury emissions , it will be up to us to find a solution to this global problem.

Questions to Consider

1. Would you support higher regulations on coal-fired power plants to reduce their mercury emissions even if it meant an increase in energy costs?
2. What is the easiest way to prevent the bioaccumulation of mercury in an organism?
3. Are there any species in an ecosystem that are not impacted by exposure to mercury?
4. Is there a way to limit the movement of mercury from one ecosystem to the next?

Sources of Water Pollution	
Leading to Cultural Eutrophication	
Oxygen-demanding waste	Biodegradable organic compounds (e.g., sewage, wastes from food-processing plants, paper mills, and tanneries)
Plant nutrients	Nitrates and phosphates from detergents, fertilizers, and sewage treatment plants
Sediments	Enriched soil in water due to soil erosion
Thermal discharges	Heated water from power plants
Health Hazards	
Disease-causing agents	Bacteria and viruses from sewage and barnyard waste (causing, for example, cholera, food poisoning, and hepatitis)
Synthetic organic compounds	Pesticides, industrial chemicals (e.g., PCBs)
Inorganic chemicals and minerals	Acids from mines and air pollution; dissolved salts; heavy metals (e.g., mercury) from industry
Radiation	Radioactive substances from nuclear power plants, medical and research facilities

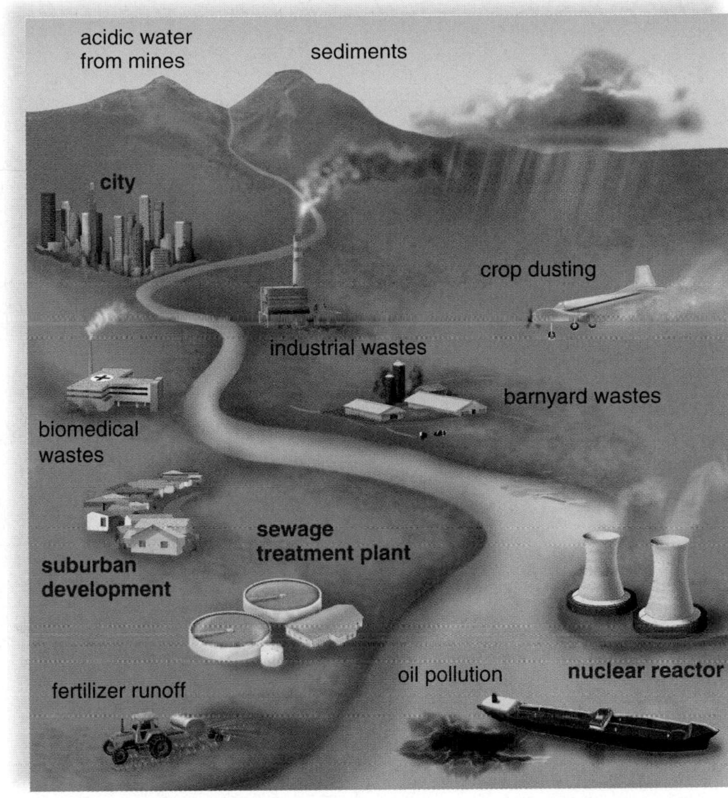

Figure 23.16 Some sources of surface water pollution.
Many bodies of water are dying due to the introduction of pollutants from point sources (in red), which are easily identifiable, and nonpoint sources (in black), which cannot be specifically identified.

CHECK YOUR PROGRESS 23.3

1. List three components that are involved in the cycling of chemicals in the biosphere.
2. Identify whether the carbon, water, nitrogen, and phosphorus cycles are examples of gaseous or sedimentary biogeochemical cycles.
3. Discuss the roles of bacteria in the nitrogen cycle.
4. Summarize the ecological problems that are associated with water, carbon, nitrogen, and phosphorus cycles.
5. Discuss the potential consequences of climate change.

CONNECTING THE CONCEPTS

For more information on the material presented in this section, refer to the following discussions:

Section 24.2 describes some of the problems associated with the overexploitation of natural resources.

Section 24.4 explores some of the ways that humans can move toward a sustainable society.

CASE STUDY CONCLUSION

Scientists predict that the Earth's ecosystems will dramatically change within the next 100 years due to human actions. Ecosystems are characterized by the interactions of living (biotic) and nonliving (abiotic) components. The biotic components of an ecosystem depend on the cycling of nutrients such as carbon, nitrogen, and phosphorus. The burning of fossil fuels and the release of excessive carbon dioxide into the atmosphere is playing a significant role in global climate change. Sea levels and extreme weather events such as hurricanes and tornados are expected to increase in frequency. As we enrich aquatic ecosystems with additional amounts of phosphates and nitrates we are causing cultural eutrophication to occur. The flow of energy through the members of biotic communities and how abiotic chemicals and nutrients cycle through ecosystems is critical to the future health of the Earth.

MEDIA STUDY TOOLS

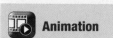

Enhance your study of this chapter with media! Visit **www.mhhe.com/maderhuman13e** and go to "Media Study Tools" for this chapter to access the following:

Animation	Videos
23.1 Biomes	**23.1** Coral Reef Ecosystems • Plants • Snake Eating • Cichlid Specialization **23.2** Decomposers **23.3** Salmon Farming • Thames River • Global Warming • Green Concrete • Warming Hurts Rice • Karoo Global Warming • Dung Beetles • Ocean Fishing Ban

SUMMARIZE

23.1 The Nature of Ecosystems

Ecology is the study of the interactions of organisms with each other and with the physical environment.

- **Ecosystems** are where living organisms interact with the physical and chemical environment.
- The sum of the ecosystems on the planet is called the **biosphere.**

Types of Ecosystems

- Terrestrial ecosystems, also called **biomes,** are forests (tropical rain forests, coniferous forests, temperate deciduous forests); grasslands (savannas and prairies); and deserts, including the tundras.
- Aquatic ecosystems are either salt water (i.e., seashores, oceans, coral reefs, estuaries) or freshwater (i.e., lakes, ponds, rivers, and streams).

Biotic Components of an Ecosystem

- In a community, each population has a habitat (place it lives) and a **niche** (its role in the community).
- **Autotrophs** (**producers**) produce organic nutrients for themselves and others from inorganic nutrients and an outside energy source.
- **Heterotrophs** (**consumers**) consume organic nutrients.
- Consumers are **herbivores** (eat plants/algae), **carnivores** (eat other animals), and **omnivores** (eat both plants/algae and animals).
- **Detritus feeders** feed on detritus, releasing inorganic substances back into the ecosystem.

Energy Flow and Chemical Cycling

Ecosystems are characterized by energy flow and chemical cycling.

- Energy flows through the populations of an ecosystem.
- Chemicals cycle within and among ecosystems.

23.2 Energy Flow

Various interconnecting paths of energy flow are called a **food web.**

- A food web is a diagram showing how various organisms are connected by eating relationships.
- **Grazing food webs** begin with vegetation eaten by an herbivore that becomes food for a carnivore.
- **Detrital food webs** begin with detritus, food for decomposers and for detritivores.
- Members of detrital food webs can be eaten by aboveground carnivores, joining the two food webs.

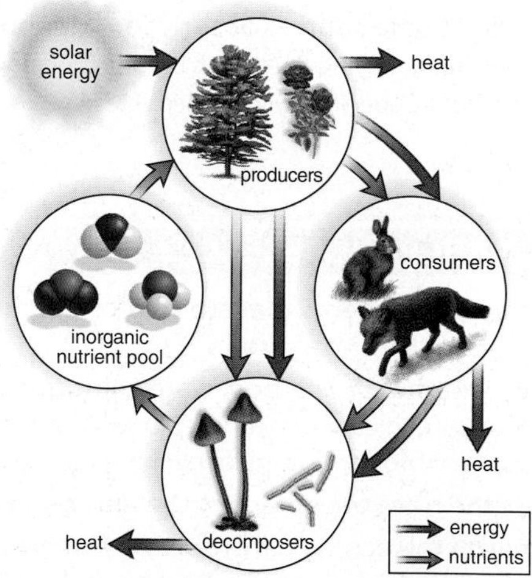

Trophic Levels

A **trophic level** is all the organisms that feed at a particular link in a **food chain.** There are two primary types of food chains: **grazing food chains** and **detrital food chains.**

Ecological Pyramids

- **Ecological pyramids** illustrate that **biomass** and energy content decrease from one trophic level to the next because of energy loss.

23.3 Global Biogeochemical Cycles

Chemicals circulate through ecosystems via **biogeochemical cycles,** pathways involving both biotic and geologic components. Biogeochemical cycles:

- can be gaseous or sedimentary; and
- have reservoirs (e.g., ocean sediments, the atmosphere, and organic matter) that contain inorganic nutrients available to living organisms on a limited basis.

Exchange pools are sources of inorganic nutrients.

- Nutrients cycle among the biotic communities (producers, consumers, decomposers) of an ecosystem.

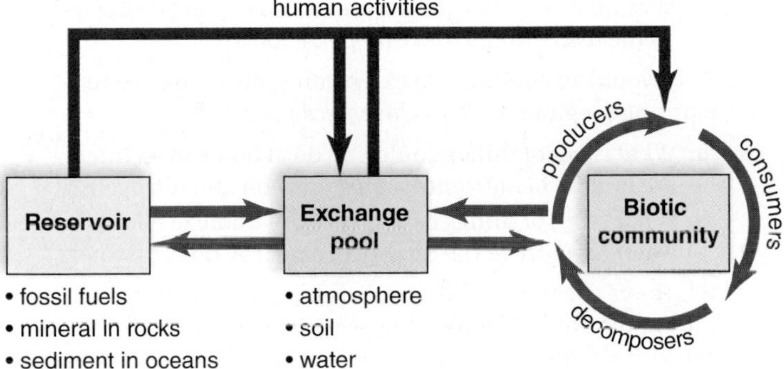

The Water Cycle

- The **water (hydrologic) cycle** is characterized by **evaporation, precipitation,** and **runoff** from the surface to lakes, rivers, and oceans.
- The primary reservoir of the water cycle is the ocean, although freshwater reserves may be located in **aquifers.**

The Carbon Cycle

- The reservoirs of the carbon cycle are organic matter (e.g., forests and dead organisms for **fossil fuels**), limestone, and the ocean (e.g., calcium carbonate shells).
- The exchange pool is the atmosphere.
- Photosynthesis removes carbon dioxide from the atmosphere. Respiration and combustion add carbon dioxide to the atmosphere.
- Imbalances in the cycling of **greenhouses gases,** such as carbon dioxide, are causing an increase in the **greenhouse effect.** These changes account for an increase in **global warming** and **climate change.**

The Nitrogen Cycle

- The reservoir of the nitrogen cycle is the atmosphere.
- Nitrogen gas must be converted to a form usable by plants (nitrates). The conversion is called **nitrification.**
- Nitrogen-fixing bacteria perform **nitrogen fixation** (in root nodules), which converts nitrogen gas to ammonium, a form producers can use.
- Nitrifying bacteria convert ammonium to nitrate.
- Some bacteria convert nitrate back to nitrogen gas (**denitrification**).
- Imbalances in the nitrogen cycle can cause **acid deposition** and acid rain.

The Phosphorus Cycle

- The reservoir of the phosphorus cycle is ocean sediments.
- Phosphate in ocean sediments becomes available through geologic upheaval, which exposes sedimentary rocks to weathering. Weathering slowly makes phosphate available to the biotic community.
- Phosphate is a limiting nutrient in ecosystems.
- Imbalances in the phosphorus cycle may lead to **cultural eutrophication** and the **biological magnification** of toxic chemicals.

ASSESS

Testing Your Knowledge of the Concepts

1. Describe the major types of terrestrial and aquatic ecosystems. (pages 536–537)
2. Name the four different types of consumers (heterotrophs) found in ecosystems. (page 538)
3. Explain why energy flows but chemicals cycle through an ecosystem. (page 539)
4. Describe the two types of food webs and two types of food chains in terrestrial ecosystems. Which of these typically moves more energy through an ecosystem? (pages 540–541)
5. What is a trophic level? An ecological pyramid? (pages 540–542)
6. What is a biochemical cycle? What is a reservoir and an exchange pool? Give an example of each. (page 543)
7. Draw a diagram to illustrate the water, carbon, nitrogen, and phosphorus biochemical cycles. Include how human activities can change a particular cycle's balance. (pages 544–551)
8. Of the total amount of energy that passes from one trophic level to another, about 10% is
   - **a.** respired and becomes heat.
   - **b.** passed out as feces or urine.
   - **c.** stored as body tissue.
   - **d.** recycled to autotrophs.
   - **e.** All of the choices are correct.
9. Nutrient cycles always involve
   - **a.** rocks as a reservoir.
   - **b.** movement of nutrients through the biotic community.
   - **c.** the atmosphere as an exchange pool.
   - **d.** loss of the nutrients from the biosphere.

For questions 10–15, match each characteristic to the cycles listed in the key. More than one answer can be used, and answers can be used more than once.

Key:

a. water cycle
b. carbon cycle
c. nitrogen cycle
d. phosphorus cycle
e. none of the cycles
f. all of the cycles

10. Occurs on land but not in the water.

11. Can occur without the participation of humans.

12. Always involves the participation of decomposers.

13. The atmosphere is involved.

14. Rocks are the reservoir in this cycle.

15. Imbalances are contributing to global climate change.

ENGAGE

The virtual lab "Model Ecosystems" provides an interactive look at some of the factors that influence the structure of an ecosystem.

Thinking Critically About the Concepts

Throughout this chapter, we have seen that human activity has a negative impact on many of the world's ecosystems. In response, many people are making efforts to reduce the environmental impact of their lifestyles. Across the country, state agencies, businesses, local organizations, and educational institutions are developing environmental projects and/or increasing their conservation methods. Many colleges and universities have initiatives that make the campuses more environmentally friendly. Our own contributions help, too. Just a few of the many ideas you can consider are water conservation, recycling trash, replacing paper or plastic containers with cloth, conserving electricity, starting a compost pile, and collecting rainwater for plants in rain barrels.

1. a. What impact might fungicides and pesticides have on detrital food webs?

b. How do you think the nutrient cycles would be affected by the use of fungicides and pesticides?

2. Why would an agricultural extension agent recommend planting a legume such as clover in a pasture?

3. a. What types of things could you do at home to facilitate the cycling of nutrients such as carbon and nitrogen?

b. What types of things could you do at home to conserve water or improve the quality of a nearby body of water?

4. If farmers use parasitoids to control flies on their farm, they may be able to stop using pesticides that contain organochlorides. What impact would this have on food chains involving wildlife around the farms?

CHAPTER 24

Human Population, Planetary Resources, and Conservation

CASE STUDY GILLS ONIONS' WASTE-TO-ENERGY PROJECT

Gills Onions is one of the largest onion producers/processors in the United States. Over 1 million lb of onions are processed every day. During processing Gills Onions generates approximately 300,000 lb of onion waste each day. Previously the waste was composted and then hauled to local farm fields and spread as fertilizer. The hauling and spreading of the onion compost led to a multitude of problems ranging from odor, runoff, and acidification of the soil, to the expense of hauling, as well as a significantly large carbon footprint. Nearly $400,000 was spent annually on the disposal of the onion waste.

Gills Onions' Advanced Energy Recovery System (AERS) has enabled them to become the first food-processing facility in the world to produce ultra-clean energy from their own waste. They are able to convert 100% of their daily onion waste into renewable energy and cattle feed.

The Advanced Energy Recovery System extracts the juice from the onion peels and uses an anaerobic reactor to produce methane-rich biogas that powers two fuel cells. This electricity is then used to power the onion processing plant, saving Gills Onions an estimated $700,000 in annual electrical cost. The remaining onion pulp is further processed into cattle feed and sold to local ranchers. Additional savings come from the elimination of $400,000 in annual cost associated with the disposal of the onion waste. Greenhouse emissions have also been reduced by the elimination of the truck traffic required to haul the compost to local fields, thus reducing Gills Onions' carbon footprint by an estimated 14,500 metric tons of CO_2 emissions per year. The AERS is expected to produce a full payback of the $10.8 million spent in less than six years.

As you read through the chapter, think about the following questions:

1. Should all major food-processing plants be required to develop and use systems similar to Gills Onions' AERS?
2. What are the potential negative consequences associated with Gills Onions' development of the AERS?
3. Can the indirect values associated with the AERS actually be measured in direct value terms?

CHAPTER CONCEPTS

24.1 Human Population Growth
The present growth rate for the world's population is currently around 1.1%. However, because of the size of the human population (over 7 billion), 83 million people are added each year.

24.2 Human Use of Resources and Pollution
Humans use land, water, food, energy, and minerals to meet their basic needs. Use of these resources leads to pollution.

24.3 Biodiversity
A biodiversity crisis is upon us because of habitat loss. The introduction of alien species, pollution, overexploitation, and disease all contribute to the crisis. Yet, wildlife has both a direct value and an indirect value for us.

24.4 Working Toward a Sustainable Society
Our present-day society is not sustainable; however, there are ways to make it more sustainable.

BEFORE YOU BEGIN

Before beginning this chapter, take a few moments to review the following discussions:

Section 21.3 How has biotechnology been applied to plants such as corn and potatoes to produce increases in food production?

Section 23.1 How does the cycling of energy and chemicals differ in an ecosystem?

Section 23.3 How are human activities influencing the water cycle?

24.1 Human Population Growth

LEARNING OUTCOMES

Upon completion of this section, you should be able to

1. Define the terms *exponential growth* and *carrying capacity* and explain how each relates to human population growth.
2. Explain the relationship between birthrate, death rate, and the annual growth rate of a population.
3. Compare and contrast the difference between more-developed countries (MDCs) and less-developed countries (LDCs) with regard to population growth.

The world's population has risen steadily to a present size of slightly over 7 billion people (Fig. 24.1). Prior to 1750, the growth of the human population was relatively slow. As more reproducing individuals were added, population growth began to rapidly increase indicating that the population was undergoing **exponential growth.** The number of people added annually to the world population peaked at about 87 million around 1990. Currently it is a little over 83 million per year.

The **growth rate** of the human population is determined by considering the difference between the number of persons born per year (birthrate, or natality) and the number who die per year (death rate, or mortality). It is customary to record these rates per 1,000 persons. For example, the world at the present time (2012) has a birthrate of 19.1 per 1,000 per year, but it has a death rate of 8.0 per 1,000 per year. This means that the world's population growth, or its growth rate, is

$$\frac{19.1 - 8.0}{1{,}000} = \frac{11.1}{1{,}000} = 0.0111 \times 100 = 1.11\%.$$

Note that whereas the birthrate and death rate are expressed in terms of 1,000 persons, the growth rate is expressed per 100 persons, or as a percentage.

After 1750, the world population growth rate steadily increased, until it peaked at 2% in 1965. It has since fallen to its present level of between 1.1 and 1.2%. Yet, the world population is still steadily growing because of its past exponential growth.

In the wild, exponential growth indicates that a population is enjoying its **biotic potential.** This is the maximum growth rate under ideal conditions. Growth begins to decline because of limiting factors such as food and space. Finally, the population levels off at the carrying capacity. The **carrying capacity** is the maximum population that the environment can support for an indefinite period. The carrying capacity of the Earth for humans has not been determined. Some authorities think the Earth may be able to sustain 50 to 100 billion people. Others think we already have more humans than the Earth can adequately support.

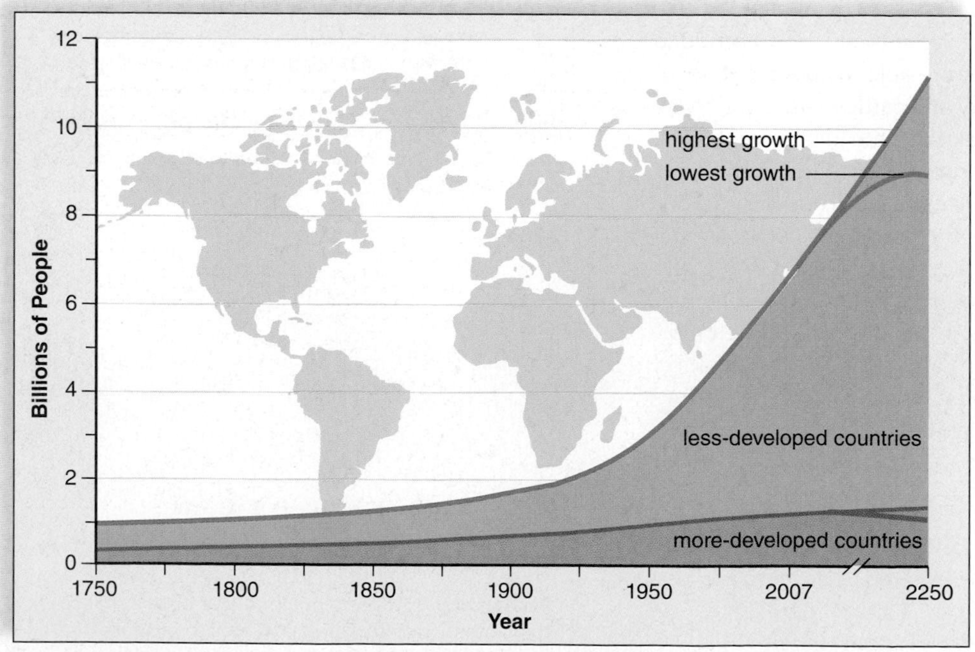

Figure 24.1 **Projections for human population growth.**
The world's population of humans is slightly over 7 billion. It is predicted that the world's population size may reach between 9 and 12 billion by 2250. Much will depend on how quickly the growth rate declines.

The MDCs Versus the LDCs

The countries of the world can be broadly divided into two groups. The more-developed countries (MDCs), typified by countries in North America and Europe, are those in which population growth is modest. The people in these countries enjoy a good standard of living. The less-developed countries (LDCs), typified by some countries in Asia, Africa, and Latin America, are those in which population growth is dramatic. The majority of people in these countries live in poverty. Many countries are in a transitional stage between being a LDC and an MDC. Population growth is decreasing and standard of living is on the rise.

The MDCs

The MDCs did not always have low population increases. Between 1850 and 1950, they doubled their populations. This was largely because of a decline in the death rate due to development of modern medicine and improvements in public health and socioeconomic conditions. The decline in the death rate was followed shortly thereafter by a decline in the birthrate. As a result, the MDCs have experienced only modest growth since 1950 (Fig. 24.1).

APPLICATIONS AND MISCONCEPTIONS

How do fertility rates vary between countries?

In 2012, the country with the highest projected fertility rate is Niger (7.52 children born per woman), followed by Uganda (6.65 children born per woman), and Mali (6.35 children born per woman), whereas Singapore was the country with lowest fertility rate (0.78 children born per woman). For death rates, South Africa had the highest rate (17.23 deaths per 1,000 people) and Qatar had one of the lowest (1.55 deaths per 1,000). For overall annual growth rates, the fastest-growing country in 2012 is projected to be Uganda (3.582%). Several countries are expected to experience declines, including Syria (–0.797%) and Jordan (–0.965%). In comparison, in the United States, the 2012 projected fertility rate is 2.06 births per woman, the death rate is 8.39 per 1,000 people, and the annual growth rate is 0.899%. *Source:* www.cia.gov

The growth rate for the MDCs as a whole is about 0.1%. In some countries the population is not increasing but instead it is decreasing in size. The MDCs are expected to increase by 52 million between 2002 and 2050, but this amount will still keep their total population at just about 1.2 billion. In contrast to the other MDCs, growth in the United States has not leveled off. The population of the United States is now greater than 300 million and continues to increase. Though the birthrate in the United States has increased slightly, much of the continued population growth is due to immigration.

The LDCs

The death rate began to decline steeply in the LDCs following World War II due to the introduction of modern medicine. However, the birthrate remained high. The growth rate of the LDCs peaked at 2.5% between 1960 and 1965. Since that time, the collective growth rate for the LDCs has declined. However, the growth rate has not declined in all LDCs. In many countries in Sub-Saharan Africa, women give birth to more than five children each.

Between 2002 and 2050, the population of the LDCs may jump from 5 billion to at least 8 billion. Some of this increase will occur in Africa, but most will occur in Asia. Many deaths from AIDS are slowing the growth of the African population. Continued growth in Asia is expected to cause acute water scarcity, a significant loss of biodiversity, and more urban pollution. Twelve of the world's 15 most polluted cities are in Asia.

Comparing Age Structure

The LDCs are experiencing a population increase because they have more women entering the reproductive years than older women leaving them. Populations have three age groups: prereproductive, reproductive, and postreproductive. This is best visualized by plotting the proportion of individuals in each group on a bar graph. This produces an age-structure diagram (Fig. 24.2).

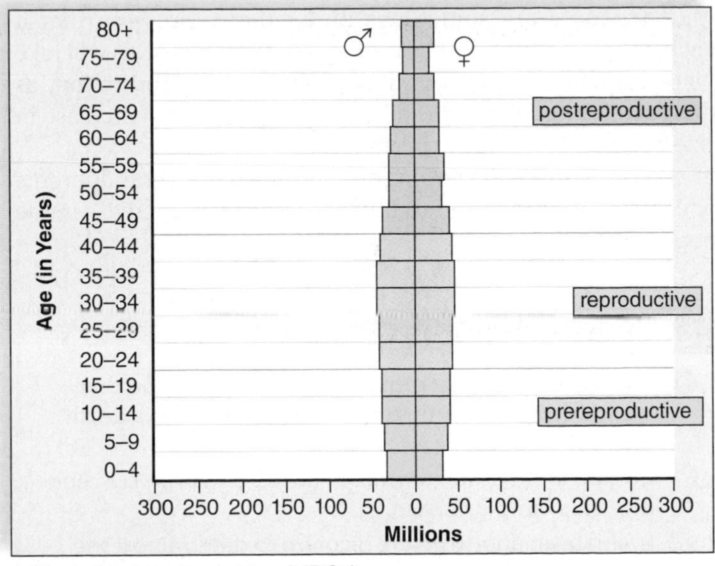

a. More-developed countries (MDCs)

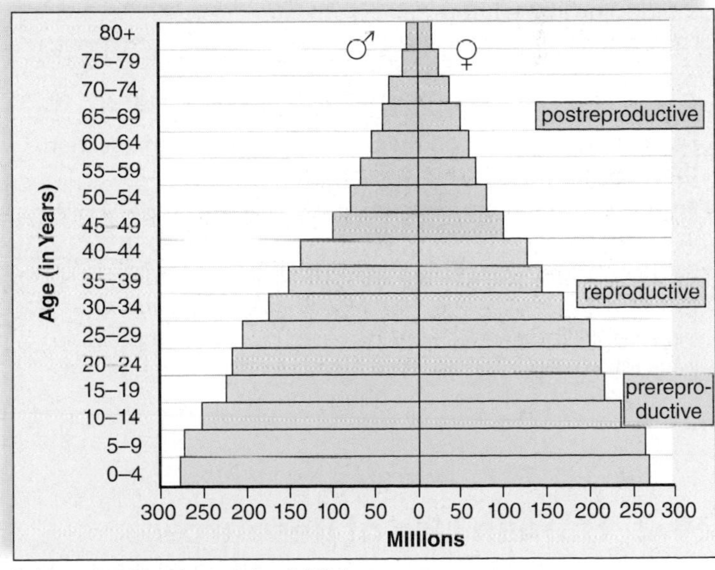

b. Less-developed countries (LDCs)

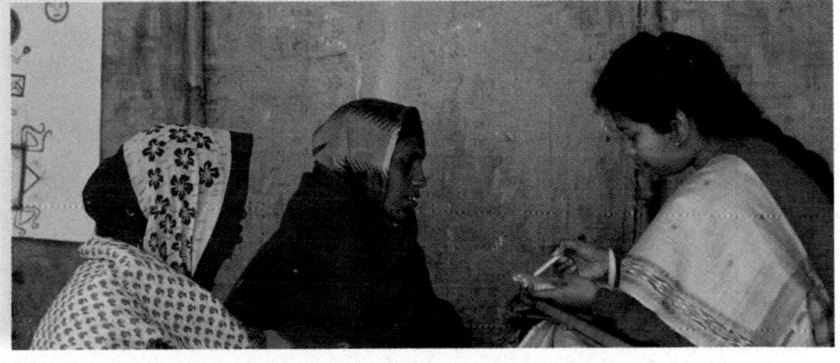

c.

Figure 24.2 **Age-structure diagrams of MDCs and LDCs.** The shape of these age-structure diagrams allows us to predict that (**a**) the populations of MDCs are approaching stabilization, and (**b**) the populations of LDCs will continue to increase for some time. (**c**) Improved women's rights and increasing contraceptive use could change this scenario. Here, a community health worker is instructing women in Bangladesh about the use of contraceptives.

Laypeople are sometimes under the impression that if each couple has two children, zero population growth will take place immediately. However, **replacement reproduction,** as this practice is called, will still cause populations to increase in size because of our life expectancy.

Most people will live long enough to become grandparents or even great-grandparents, thus resulting in an increase in the population.

CHECK YOUR PROGRESS 24.1

1. Calculate the annual growth rate of a population that is experiencing a birthrate of 20.5 per year and a death rate of 9.8 per year.
2. Compare the characteristics of an MDC with an LDC and give an example of each.
3. Evaluate an age-structure diagram to determine if the population will experience a population growth or decline in the future.

CONNECTING THE CONCEPTS

For more information on the topics presented in this section, refer to the following discussions:

Figure S.1 illustrates the number of cases worldwide of HIV in 2009.

Section 16.5 describes how assisted reproductive technologies are used to treat individuals with infertility.

Section 23.3 examines how increases in human population have interfered with natural biogeochemical cycling of water, carbon, nitrogen, and phosphorus.

24.2 Human Use of Resources and Pollution

LEARNING OUTCOMES

Upon completion of this section, you should be able to

1. Distinguish between renewable and nonrenewable resources and give an example of each.
2. Explain how human activity is influencing the natural resources of land, water, food, minerals, and energy.
3. Identify examples of biological magnification.

Humans have certain basic needs. A resource is anything from the biotic or abiotic environment that helps meet these needs. Land, water, food, energy, and minerals are the maximally used resources that will be discussed in this chapter (Fig. 24.3). The total amount of resources used by an individual to meet his or her needs is sometimes referred to as an ecological footprint. A person can make his or her ecological footprint smaller by driving an energy-efficient car, living in a smaller house, owning fewer possessions, eating vegetables as opposed to meat, and so forth.

Figure 24.3 The five classes of resources needed to meet basic human needs.
Humans use land, water, food, energy, and minerals to meet their basic needs, such as a place to live, food to eat, and products that make their lives easier.

Some resources are nonrenewable, and some are renewable. **Nonrenewable resources** are limited in supply. For example, the amount of land, fossil fuels, and minerals is finite and can be exhausted. Efficient use, recycling, or substitution can make the supply last longer, but eventually these resources will run out.

Renewable resources are capable of being naturally replenished. We can use water and certain forms of energy (e.g., solar energy) or harvest plants and animals for food. A new supply will always be forthcoming. However, even with renewable resources, we have to be careful not to overuse them before they can replenish themselves.

Unfortunately, a side effect of resource consumption can be pollution. **Pollution** is any undesired alteration of the environment and is often caused by human activities. The effect of humans on the environment is proportional to the population size and consumption levels. As the population grows, so does the need for resources and the amount of pollution caused by using these resources. Seven people adding waste to the ocean may not be alarming, but 7 billion people doing so would certainly affect its cleanliness. In modern times, the consumption of mineral and energy resources has grown faster than population size.

Land

People need a place to live. Naturally, land is also needed for a variety of uses aside from homes. Land is used for agriculture, electrical power plants, manufacturing plants, highways, hospitals, schools, and so on.

Beaches and Human Habitation

At least 40% of the world population lives within 100 km (60 mi) of a coastline, and this number is expected to increase. Living right on the coast is an unfortunate choice because it leads to beach erosion. Loss of habitat for marine organisms and loss of a buffer zone for storms also occur. Figure 24.4 shows how severe the problem can be in the United States. Coastal wetlands are also impacted when people fill them in. One reason to protect coastal wetlands is that they are spawning areas for fish and other forms of marine life. They are also habitats for certain terrestrial species, including many types of birds. Wetlands also protect coastal areas from storms. The level of storm damage to New Orleans during Hurricanes Katrina (2005) and Isaac (2012), is largely attributed to

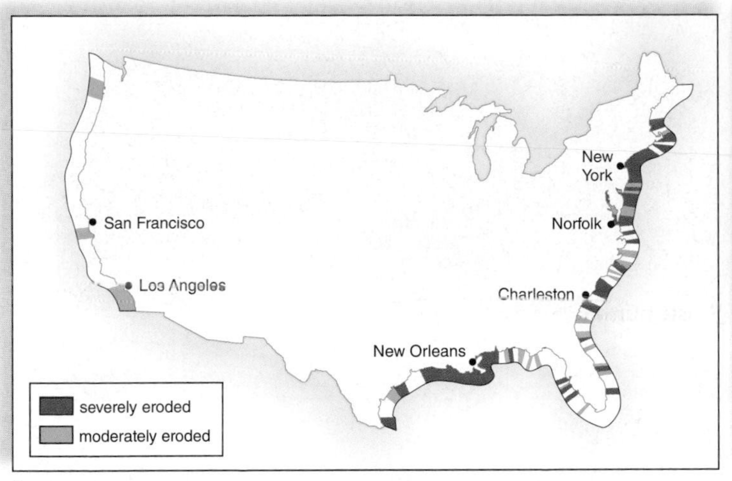

a.

b.

Figure 24.4 **Beach erosion and coastal development.**
a. Most of the U.S. coastline is subject to beach erosion. **b.** Therefore, people who choose to live near the coast may eventually lose their homes.

the lack of protective wetlands along the coast. Wetlands are also particularly subject to pollution because toxic substances placed in freshwater lakes, rivers, and streams may eventually find their way to these areas.

Semiarid Lands and Human Habitation

Forty percent of the Earth's lands are already deserts. Land adjacent to a desert is in danger of becoming unable to support human life if it is improperly managed by humans (Fig. 24.5). **Desertification** is the conversion of semiarid land to desertlike conditions.

Often, desertification begins when humans allow animals to overgraze the land. The soil can no longer hold rainwater, and it runs off instead of keeping the remaining plants alive or replenishing wells. Humans then remove whatever vegetation they can find to use as fuel or fodder for their animals. The result is a lifeless desert. That area is then abandoned as people move on to continue the process someplace else. Some estimate that nearly three-quarters of all rangelands worldwide are in danger of desertification. Many famines are due, at least in part, to degradation of the land to the point that it can no longer support humans and their livestock.

Tropical Rain Forests and Human Habitation

Deforestation, the removal of trees, has long allowed humans to live in areas where forests once covered the land (Fig. 24.6). This land, too, is subject to desertification. Soil in the tropics is often thin and nutrient-poor because all the nutrients are tied up in the trees and other vegetation. When the trees are felled and the land is used for agriculture or grazing, it quickly loses its fertility. Then it is subject to desertification.

Water

In the water-poor areas of the world, people may not have ready access to drinking water and, if they do, the water may be unsafe. It's considered a human right for people to have clean

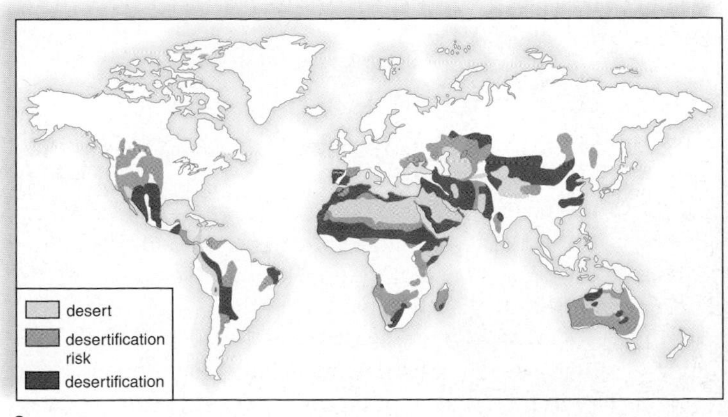

a.

b.

Figure 24.5 **Desertification.**
a. Desertification is a worldwide occurrence that (**b**) reduces the amount of land suitable for human habitation.

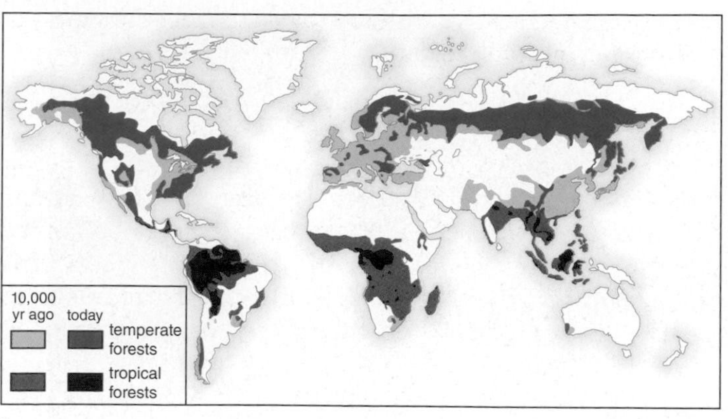

a.

b.

Figure 24.6 Deforestation.
a. Nearly half of the world's forestlands have been cleared for farming, logging, and urbanization. **b.** The soil of tropical rain forests cannot sustain long-term farming.

drinking water. In reality, most freshwater is used by industry and agriculture (Fig. 24.7). Worldwide, 70% of freshwater is used to irrigate crops! Much of a recent surge in demand for water stems from increased industrial activity and irrigation-intensive agriculture. This type of agriculture now supplies about 40% of the world's food crops. In the MDCs, more water is usually used for bathing, flushing toilets, and watering lawns than for drinking and cooking.

Increasing Water Supplies

Globally, the needs of the human population do not exceed the renewable supply of freshwater. However, this is not the case in certain regions of the United States and the world. About 40% of the world's land is desert, and deserts are bordered by semiarid land. To live in these regions humans increase the availability of freshwater by damming rivers and withdrawing water from aquifers.

Dams The world's 45,000 large dams catch 14% of all precipitation runoff and provide water for up to 40% of irrigated land. They also give some 65 countries more than half their electricity. Damming of certain rivers has been so extensive that they no longer flow as they once did. The Yellow River in China fails to reach the sea most years. The Colorado River barely makes it to the Gulf of California. Even the Rio Grande dries up before

a. Agriculture uses most of the freshwater consumed.

b. Industrial use of water is about half that of agricultural use.

c. Domestic use of water is about half that of industrial use.

Figure 24.7 Water use by agriculture, industries, and households.
a. Agriculture primarily uses water for irrigation. **b.** Industry uses water variously. **c.** Households use water to drink, shower, flush toilets, and water lawns.

it can merge with the Gulf of Mexico. The Nile in Egypt and the Ganges in India are also so overexploited they hardly make it to the ocean at some times of the year.

Dams have other drawbacks as well, not the least of which is losing water by evaporation and seepage into underlying rock beds. The amount of water lost sometimes equals the amount made available! The salt left behind by evaporation and agricultural runoff increases salinity and can make a river's water unusable farther downstream. Dams hold back less water with time because of sediment buildup. Sometimes a reservoir becomes so full of silt that it is no longer useful for storing water.

Aquifers To meet their freshwater needs, people are pumping vast amounts of water from **aquifers.** These are reservoirs found just below or as much as 1 km below the surface of the Earth. Aquifers hold about 1,000 times the amount of water that falls on land as precipitation each year. This water accumulates from rain that has fallen in far-off regions. In the past 50 years, groundwater depletion has become a problem in many areas of the world. In substantial portions of the High Plains Aquifer (stretching from South Dakota to Texas), more than half of the water has been pumped out.

Consequences of Groundwater Depletion Removal of water from aquifers is causing land **subsidence,** a settling of the soil as it dries out. An area in California's San Joaquin valley has subsided at least 30 cm due to groundwater depletion. In the worst spot, the surface of the ground has dropped more than 9 m! Subsidence damages canals, buildings, and underground pipes. Withdrawal of groundwater can cause **sinkholes.** These form when an underground cavern collapses because water no longer supports the roof (Fig. 24.8).

Saltwater intrusion is another consequence of aquifer depletion. The flow of water from streams and aquifers usually keeps them fairly free of seawater. As water is withdrawn, the water table can lower to the point that seawater backs up into streams and aquifers. Saltwater intrusion reduces the supply of freshwater along the coast.

Figure 24.8 Sinkholes may be caused by groundwater depletion.
Sinkholes occur when an underground cavern collapses after groundwater has been withdrawn.

Conservation of Water

By 2025, two-thirds of the world's population may be living in countries facing serious water shortages. Some solutions for expanding water supplies have been suggested. Planting drought- and salt-tolerant crops would help a lot. Using drip irrigation delivers more water to crops and increases crop yields as well (Fig. 24.9). Although the first drip systems were developed

a.

b.

c.

Figure 24.9 Measures that can be taken to conserve water.
a. Planting drought-resistant crops in the fields and drought-resistant plants in parks and gardens cuts down on the need to irrigate. **b.** When irrigation is necessary, drip irrigation is preferable to using sprinklers. **c.** Wastewater can be treated and reused instead of withdrawing more water from a river or aquifer.

in 1960, they're used on less than 1% of irrigated land. Most governments subsidize irrigation so heavily that farmers have little incentive to invest in drip systems or other water-saving methods. Reusing water and adopting conservation measures could help the world's industries cut their water demands by more than half.

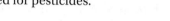

Video Thames River

Food

In 1950, the human population numbered 2.5 billion. Only enough food was produced to provide less than 2,000 calories per person per day. Now, with over 7 billion people on Earth, the world food supply provides more calories per person per day. Generally speaking, food comes from three activities: growing crops, raising animals, and fishing the seas. The increase in the food supply has largely been possible because of modern farming methods. Unfortunately, many of these methods include some harmful practices:

1. *Planting of a few genetic varieties.* The majority of farmers practice monoculture. Wheat farmers plant the same type of wheat, and corn farmers plant the same type of corn. With a *monoculture,* a single type of parasite can potentially destroy the entire crop.
2. *Heavy use of fertilizers, pesticides, and herbicides.* Fertilizer production is energy intensive, and fertilizer runoff contributes to water pollution. Pesticides reduce soil fertility because they kill off beneficial soil organisms as well as pests. Some pesticides and herbicides are linked to the development of cancer. **Agricultural runoff** places these chemicals in our water supply.
3. *Generous irrigation.* As already discussed, water is sometimes withdrawn from aquifers for crop irrigation. In the future, the water content of these aquifers may become so reduced that it may not be available to pump out any more.
4. *Excessive fuel consumption.* Irrigation pumps remove water from aquifers. Large farming machines are used to spread fertilizers, pesticides, and herbicides, as well as to sow and harvest the crops. In effect, modern farming uses large amounts of fossil fuels to produce food.

Figure 24.10 shows ways to minimize the harmful effects of modern farming practices. *Polyculture* is the planting of two or more different crops in the same area. In Figure 24.10*a*, a farmer has planted alfalfa in between strips of corn. The alfalfa replenishes the nitrogen content of the soil so that fertilizer doesn't have to be added. In Figure 24.10*b*, contour farming with no-till conserves topsoil because it reduces agricultural runoff. Contour farming is planting and plowing according to the slope of the land. No-till farming allows the previous crop to remain on the land, recycling nutrients and preventing soil erosion. Biological control (Fig. 24.10*c*) relies on the use of natural predators to destroy organisms that harm crops. This reduces the need for pesticides.

Video Needing Herbicides

Soil Loss

Land suitable for farming and grazing animals is being degraded worldwide. Topsoil is the richest in organic matter and the most capable of supporting grass and crops. When bare soil is acted on by water and wind, soil erosion occurs and topsoil is lost. As a result, marginal rangeland becomes desertized, and farmland loses its productivity.

The custom of planting the same crop in straight rows, which facilitates the use of large farming machines, has caused the

a. Polyculture

b. Contour farming

c. Biological pest control

Figure 24.10 Methods that make farming more friendly to the environment.
a. Polyculture reduces the ability of one parasite to wipe out an entire crop and reduces the need to use a herbicide to kill weeds. This farmer has planted alfalfa in between strips of corn, which also replenishes the nitrogen content of the soil (instead of adding fertilizers). Alfalfa, a legume, has root nodules that contain nitrogen-fixing bacteria. **b.** Contour farming with no-till conserves topsoil because water has less tendency to run off. **c.** Instead of pesticides, it is possible to use a natural predator. Here, ladybugs are feeding on cottony-cushion scales (insects) on citrus trees.

United States and Canada to have one of the highest rates of soil erosion in the world. Conserving soil nutrients by altering farming practices could save farmers billions of dollars annually in fertilizer costs. Much of the eroded sediment ends up in lakes and streams, where it creates problems within the aquatic ecosystem.

Green Revolutions

About 50 years ago, researchers began to breed tropical wheat and rice varieties specifically for farmers in the LDCs. The dramatic increase in yield due to the introduction of these new varieties around the world was called "the green revolution." These plants helped the world food supply keep pace with the rapid increase in world population. Unfortunately, most green revolution plants are called "high responders" because they need high levels of fertilizer, water, and pesticides to produce a high yield. They require the same subsidies and create the same ecological problems as do modern farming methods.

Genetic Engineering As discussed in Chapter 21, genetic engineering can produce transgenic plants with new and different traits. For example, resistance to both insects and herbicides are traits that can be inserted into plant DNA. When herbicide-resistant crops are planted, weeds are controlled easier, less tillage is needed, and soil erosion is minimized. Researchers also want to produce crops that tolerate salt, drought, and cold. Some progress has also been made in increasing the food quality of crops so that they will supply more of the proteins, vitamins, and minerals people need. Genetically engineered crops could result in still another green revolution.

Some are opposed to the use of genetically engineered crops. It is feared that these crops will damage the environment and lead to health problems in humans. The Health feature, "Are Genetically Engineered Foods Safe?," on page 566, discusses this issue.

Domestic Livestock

A low-protein, high-carbohydrate diet consisting only of grains such as wheat, rice, or corn can lead to malnutrition. In the LDCs, kwashiorkor, caused by a severe protein deficiency, is seen in infants and children ages one to three years old. It usually occurs after a new arrival in the family and the older children are no longer fed milk. The diet then consists of protein-poor starches. Such children are lethargic, irritable, and have bloated abdomens. Problems associated with intellectual development may also occur.

In the MDCs, many people tend to have more than enough protein in their diet. Almost two-thirds of United States cropland is devoted to producing livestock feed. This means that a large percentage of the fossil fuel, fertilizer, water, herbicides, and pesticides used are for the purpose of raising livestock. Typically, cattle are range-fed for about four months. Then they are brought to crowded feedlots where they may receive growth hormone and antibiotics. At the feedlots, they feed on grain or corn. Many animals, including cows (Fig. 24.11) and pigs, spend their entire lives cooped up in crowded pens and cages.

Figure 24.11 The effects on the environment of raising livestock.
Raising livestock requires the use of more fossil fuels and water than raising crops. Livestock waste often washes into nearby bodies of water, creating water pollution.

If livestock eat a large proportion of the crops in the United States, then raising livestock accounts for much of the pollution associated with the farming industry. Fossil fuel energy is needed not just to produce herbicides, pesticides, and to grow food but also to grow feed for livestock. Raising livestock is extremely energy-intensive. In addition, water is used to wash livestock wastes into nearby bodies of water, where it adds significantly to water pollution. Whereas human wastes are sent to sewage treatment plants, raw animal wastes are not.

For these reasons, it is prudent to recall the ecological energy pyramid (see Fig. 23.7), which shows that as you move up the food chain, not all of the energy is transferred. As a rule of thumb, for every 10 calories of energy from a plant, only 1 calorie is available for the production of animal tissue in an herbivore. A great deal of energy is wasted when the human diet contains more protein than is needed to maintain good health. It is possible to feed ten times as many people on grain as on meat.

Energy

Modern society runs on various sources of energy. Some are renewable, whereas others are nonrenewable. The consumption of nonrenewable energy supplies results in environmental degradation, which is one of the reasons why renewable energy is expected to be used more in the future.

Nonrenewable Sources

Presently, about 6% of the world's energy supply comes from nuclear power and 81% comes from fossil fuels. Both of these are finite, nonrenewable sources. It was once predicted that the nuclear power industry would fulfill a significant portion of the world's energy needs. However, this has not happened for two reasons. One is that people are very concerned about nuclear

BIOLOGY MATTERS  Health

Are Genetically Engineered Foods Safe?

A series of focus groups conducted by the Food and Drug Administration (FDA) in 2000 showed that although most participants believed that genetically engineered foods, now also called genetically modified organisms (GMOs), might offer benefits, they also feared possible unknown long-term health consequences. The discovery by activists that a type of genetically engineered corn called StarLink had inadvertently made it into the food supply triggered the recall of taco shells, tortillas, and many other corn-based foodstuffs from supermarkets. Further, the makers of StarLink were forced to buy back StarLink from farmers and to compensate food producers at an estimated cost of several hundred million dollars in late 2000.

StarLink is a type of "Bt" corn. It contains a foreign gene taken from a common soil organism, *Bacillus thuringiensis,* which makes a protein that is toxic to many insect pests. About a dozen Bt varieties, including corn, potato, and even a tomato, have now been approved for human consumption. These strains contain a gene for an insecticidal protein called CryIA. Instead, StarLink contained a gene for a related protein called Cry9C, which researchers thought might slow down the chances of pest resistance to Bt corn. In order to get FDA approval for use in foods, the makers of StarLink performed the required tests. Like the other now-approved strains, StarLink wasn't poisonous to rodents, and its biochemical structure is not similar to those of most chemicals in food that commonly cause allergic reactions in humans (called allergens). But the Cry9C protein resisted digestion longer than the other Bt proteins when it was put in simulated stomach acid and subjected to heat. Because most food allergens resist digestion in a similar fashion, StarLink was not approved for human consumption.

The scientific community is now trying to devise more tests for allergens because it has not been possible to determine conclusively whether Cry9C is or is not an allergen. Also, at this point, it is unclear how resistant to digestion a protein must be in order to be an allergen, and it is also unclear what degree of amino acid sequence similarity a potential allergen must have to a known allergen to raise concern. Dean D. Metcalfe, chief of the Laboratory of Allergic Diseases at the National Institute of Allergy and Infectious Diseases, said, "We need to understand thresholds for sensitization to food allergens and thresholds for elicitation of a reaction with food allergens."

Other scientists are concerned about the following potential drawbacks to the planting of Bt corn: (1) resistance among populations of the target pest, (2) exchange of genetic material between the transgenic crop and related plant species, and (3) Bt crops' impact on nontarget species. They feel that many more studies are needed before stating for certain that Bt corn has no ecological drawbacks.

Despite controversies, the planting of genetically engineered corn has increased annually. The USDA reports that U.S. farmers planted genetically engineered corn on between 81% and 86% of all corn acres in 2011, up from 26% in 2001 (Fig. 24A). Today, almost 90% of soybean acreage and 96% of cotton acreage is genetically modified. Some groups advocate that GMOs should be labeled as such, but this may not be easy to accomplish because, for example, most cornmeal is derived from both conventional and genetically engineered corn. So far, there has been no attempt to sort out one type of food product from the other. However, at many health food stores, foods that do not contain GMOs are now labeled.

a. 

b. 

c. 

Figure 24A Genetically engineered crops. Genetically engineered (**a**) corn, (**b**) soybeans, and (**c**) cotton crops are increasingly being planted by today's farmers.

Video GM Food Safety

Questions to Consider

1. Do you think genetically modified organisms (GMOs) should be labeled? Construct an argument both for and against labeling GMOs.
2. Some people are strongly advocating the complete removal of GMOs from the market. A few people, called ecoterrorists, are taking such drastic actions as burning crops or even setting biotechnology labs on fire. Do you think GMOs should be removed from the market? Why or why not? What further information would you need to make your decision?
3. Rice is a staple in the diet of millions of people worldwide, many of them living in less-developed countries. In some of those same countries, vitamin A deficiency is a major cause of blindness in small, malnourished children. Scientists have developed a new form of rice, called "golden rice," which is genetically modified to assist with the metabolism of vitamin A and might prevent millions of cases of blindness. The scientists who created golden rice claim it is safe for consumption, although critics say the health effects are not yet fully understood. Given its nutritional potential, should golden rice be planted and distributed on a wide scale? What information would you need to make your decision?

power dangers, such as the meltdown at the Chernobyl nuclear power plant in Russia in 1986. Further, radioactive wastes from nuclear power plants remain a threat to the environment for thousands of years. We still have not decided how best to safely store them.

Fossil fuels (oil, natural gas, and coal) are derived from the compressed remains of plants and animals that died thousands of years ago. Of the fossil fuels, oil burns more cleanly than coal, which may contain a lot of sulfur. When the use of coal releases sulfur, acid rain forms. Thus, despite the fact that coal is plentiful in the United States, imported oil is our preferred fossil fuel. Regardless of which fossil fuel is used, all contribute to environmental problems because of the pollutants released when they're burned. Current research into clean-coal technology is working to make U.S. coal less polluting.

Fossil Fuels and Global Climate Change In 1850, the level of carbon dioxide in the atmosphere was about 280 parts per million (ppm); currently, it is over 395 ppm. This increase is largely due to the burning of fossil fuels and the burning and clearing of forests. Human activities are causing the emission of other gases as well. These gases are known as **greenhouse gases.** Just like the panes of a greenhouse, they allow solar radiation to pass through but hinder the escape of infrared heat back into space.

Video Global Warming

Computer models predict the Earth may warm to temperatures never before experienced by living organisms. The global climate has already warmed about 0.6°C, and it may rise as much as 1.5–4.5°C by 2100. If so, sea levels will rise as glaciers melt and warm water expands. Major coastal cities of the United States could eventually be threatened. The present wetlands will be inundated. Great losses of aquatic habitat will occur wherever wetlands cannot move inward because of coastal development and levees. Coral reefs, which prefer shallow waters, will most likely "drown" as the waters rise.

On land, regions of suitable climate for various species will shift toward the poles and higher elevations. Plants migrate when seeds disperse and growth occurs in a new locale. The present assemblages of species in ecosystems will be disrupted as some species migrate northward faster than others. Trees, for example, cannot migrate as fast as nonwoody plants. Also, too many species of organisms are isolated to relatively small habitat patches surrounded by agricultural or urban areas. Even if such species have the capacity to disperse to new sites, suitable habitats may not be available.

Video Warming Hurts Rice

Renewable Energy Sources

Renewable types of energy include hydropower, geothermal, wind, and solar.

Hydropower Hydroelectrical plants convert the energy of falling water into electricity (Fig. 24.12*a*). Hydropower accounts for about 6% of the electrical power generated in the United States and almost 67% of the total renewable energy used. Worldwide, hydropower presently generates 20% of all electricity used. This percentage is expected to rise because of increased use in certain countries.

Much of the hydropower development in recent years has been due to the construction of enormous dams. These are known to have detrimental environmental effects (see pages 562–563). The better choice is believed to be small-scale dams that generate less power per dam but do not have the same environmental impact.

Geothermal Energy Elements such as uranium, thorium, radium, and plutonium undergo radioactive decay below the Earth's surface. This heats the surrounding rocks to hundreds of degrees Celsius. When the rocks are in contact with underground streams or lakes, huge amounts of steam and hot water are produced. This steam can be piped up to the surface to supply hot water for home heating or to run steam-driven turbogenerators. California's Geysers Recharge Project is the world's largest geothermal electricity-generating complex.

Wind Power Wind power is expected to account for a significant percentage of our energy needs in the future. A common belief is that a huge amount of land is required for the "wind farms" that produce commercial electricity. The amount of land needed for a wind farm compares favorably with the amount of land required by a coal-fired power plant or a solar thermal energy system (Fig. 24.12*b*).

A community generating its own electricity by using wind power can solve the problem of uneven energy production. Electricity can be sold to a local public utility when an excess is available. Then electricity is bought from the same facility when wind power is in short supply.

Energy and the Solar-Hydrogen Revolution Solar energy is diffuse energy that must be collected, converted to another form, and stored if it is to compete with other available forms of energy. Passive solar heating of a house is successful when the windows of the house face the sun and the building is well insulated. Successful heating also requires that heat can be stored in water tanks, rocks, bricks, or some other suitable material.

In a **photovoltaic (solar) cell,** a wafer of the electron-emitting metal is in contact with another metal that collects the electrons. Electrons are then passed along into wires in a steady stream. Spurred by the oil shocks of the 1970s, the U.S. government has been supporting the development of photovoltaics ever since. As a result, the price of buying a photovoltaic cell has dropped from about $100 per watt to around $2.43. Photovoltaic cells placed on roofs generate electricity that can be used inside a building and/or sold back to a power company (Fig. 24.12*c*).

Several types of solar power plants are now operational in California. In one type, huge reflectors focus sunlight on a pipe containing oil. The heated pipes boil water, generating steam that drives a conventional turbogenerator. In another type, 1,800 sun-tracking mirrors focus sunlight onto a molten

a.

c.

b.

d.

Figure 24.12 Sources of renewable energy.
a. Hydropower dams provide a clean form of energy but can be ecologically disastrous in other ways. **b.** Wind power requires land on which to place enough windmills to generate energy. **c.** Photovoltaic cells on rooftops and (**d**) sun-tracking mirrors on land can collect diffuse solar energy more cheaply than could be done formerly.

salt receiver mounted on a tower. The hot salt generates steam that drives a turbogenerator (Fig. 24.12*d*).

Scientists are working on the possibility of using solar energy to extract hydrogen from water via electrolysis. The hydrogen can then be used as a clean-burning fuel. When it burns, water is produced. Presently, cars have internal combustion engines that run on gasoline. In the future, vehicles are expected to be powered by fuel cells, which use hydrogen to produce electricity (Fig. 24.13). The electricity runs a motor that propels the vehicle. Fuel cells are now powering buses in Vancouver and Chicago with additional buses planned.

Hydrogen fuel can be produced locally or in central locations, using energy from photovoltaic cells. The fuel produced in central locations can be piped to filling stations using

a.

b.

Figure 24.13 Hydrogen fuel and fuel-cell vehicles.
a. This bus is powered by hydrogen fuel. **b.** Fuel-cell hybrid vehicles reduce air pollution and dependence on fossil fuels.

the natural gas pipes already plentiful in the United States. However, two major hurdles are storage and production. Advantages of a solar-hydrogen revolution are decreased dependence on oil and fewer environmental problems.

Video
Spinach Battery

Minerals

Minerals are nonrenewable raw materials in the Earth's crust that can be mined (extracted) and used by humans. Nonmetallic raw materials such as sand, gravel, and phosphate are considered minerals. Metals, such as aluminum, copper, iron, lead, and gold fall into this category, as well.

One of the greatest threats to the maintenance of ecosystems and biodiversity is surface mining, called strip mining. In the United States, huge machines can go as far as removing mountaintops to reach a mineral. The land devoid of vegetation takes on a surreal appearance, and rain washes toxic waste deposits into nearby streams and rivers.

The most dangerous metals to human health are the heavy metals. These include lead, mercury, arsenic, cadmium, tin, chromium, zinc, and copper. They are used to produce batteries, electronics, pesticides, medicines, paints, inks, and dyes. In the ionic form, they enter the body and inhibit vital enzymes. That's why these items should be discarded carefully and taken to hazardous waste sites.

Hazardous Wastes

The consumption of minerals and use of synthetic organic chemicals contribute to the buildup of hazardous waste in the environment. Every year, countries around the world discard billions of tons of solid waste on land and in freshwater and salt water. The Environmental Protection Agency (EPA) oversees the cleanup of hazardous waste disposal sites in the United States. An EPA program called the Superfund provides the funds that help pay for this cleanup. Commonly found contaminants include heavy metals (such as lead, mercury, and arsenic) and chlorine-containing organic chemicals (such as chloroform and polychlorinated biphenyls, or PCBs). Some of these contaminants interfere with hormone activity and proper endocrine system functioning.

Humanmade organic chemicals play a role in the production of plastics, pesticides, herbicides, cosmetics, and hundreds of other products. For example, the so-called *halogenated hydrocarbons* are compounds made from carbon and hydrogen and also include halogen atoms such as chlorine and fluorine. These compounds, **chlorofluorocarbons (CFCs)**, have been shown to damage the Earth's ozone shield. This shield protects terrestrial life from harmful UV radiation and has recently been depleted by the use of CFCs.

Further, these types of synthetic compounds pose a threat to the health of living organisms, including humans, because they can undergo **biological magnification.** Such chemicals are not excreted and can accumulate in an organism's body. Therefore, they become more concentrated as they pass from organism to organism along a food chain.

APPLICATIONS AND MISCONCEPTIONS

What is methylmercury and why is it dangerous?

Methylmercury is a form of the element mercury that has been bound to a methyl (CH_3) group. Because of this methyl group, methylmercury easily accumulates in the food chain by biological magnification or the concentration of chemicals. Methylmercury is released into the environment by the burning of coal, the mining of certain metals, and the incineration of medical waste. Methylmercury is a powerful neurotoxin that also inhibits the activity of the immune system. Because of this, the Food and Drug Administration (FDA) and the Environmental Protection Agency (EPA) recommend that pregnant women and small children not eat shark, swordfish, tilefish, or king mackerel and limit the amount of albacore tuna to less than 6 ounces per week. Most states have also posted warnings on eating local fish that have been caught from mercury-contaminated waters. For more information, visit the EPA website, www.epa.gov/waterscience/fish.

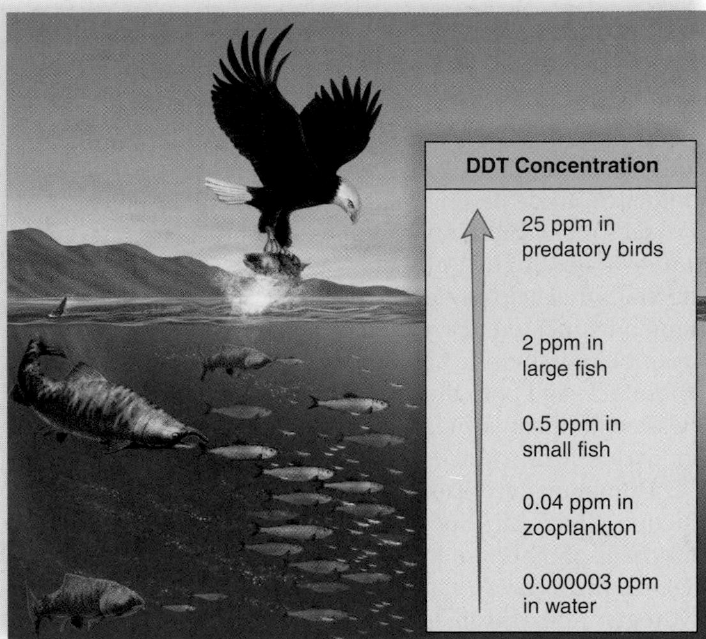

Figure 24.14 **Biological magnification concentrates chemicals in the food chain.**
Various synthetic organic chemicals, such as DDT, accumulate in animal fat. Therefore, the chemicals become increasingly concentrated at higher trophic levels. By the time DDT was banned in the United States, it had interfered with predatory bird reproduction by causing eggshell thinning.

Biological magnification is more apt to occur in aquatic food chains, which have more links than terrestrial food chains. Rachel Carson's book *Silent Spring*, published in 1962, made the public aware of the harmful effects of pesticides such as DDT. These substances accumulate in the mud of deltas and estuaries of highly polluted rivers and cause environmental problems if disturbed. After working its way up the food chain, high concentrations of DDT in predatory birds like bald eagles and pelicans interfered with their ability to reproduce (Fig. 24.14). There are health advisories about eating certain types of fish due to the high levels of mercury they contain. Humans are often the final consumers in a variety of food chains and are affected by biological magnification as well. The breast milk of humans has been found to contain significant levels of DDT, PCBs, solvents, and heavy metals.

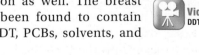

Raw sewage causes oxygen depletion in lakes and rivers. As the oxygen level decreases, the diversity of life is greatly reduced. Also, human feces can contain pathogenic microorganisms that cause cholera, typhoid fever, and dysentery. In regions of the LDCs where sewage treatment is practically nonexistent, many children die each year from these diseases. Typically, sewage treatment plants use bacteria to break down organic matter to inorganic nutrients, such as nitrates and phosphates, which then enter surface waters. The result can be cultural eutrophication (see Chapter 23).

CHECK YOUR PROGRESS 24.2

1. List the five main classes of resources.
2. Describe how humans have increased the availability of groundwater and food resources, and summarize the potential problems that these activities may be creating.
3. Summarize why is it better for humans to use renewable rather than nonrenewable sources of energy.
4. Discuss the consequences of using fossil fuels as an energy source.
5. Describe how hazardous wastes may interfere with natural environmental processes.

CONNECTING THE CONCEPTS

For more information on the topics in this section, refer to the following discussions:

Section 23.2 explores the flow of energy in an ecosystem.

Section 23.3 examines the water and carbon cycles and explores the influence of human activity on these cycles.

24.3 Biodiversity

LEARNING OUTCOMES

Upon completion of this section, you should be able to

1. Describe the factors that are contributing to the current biodiversity crisis.
2. Identify the direct values to society for conserving biodiversity.
3. Discuss the indirect values to society for conserving biodiversity.

Biodiversity can be defined as the variety of life on Earth and described in terms of the number of different species. We are presently in a biodiversity crisis. The number of extinctions (loss of species) expected to occur in the near future is unparalleled in the history of the Earth.

Loss of Biodiversity

Figure 24.15 identifies the major causes of extinction.

Habitat Loss

Human occupation and alteration of nearly every biome on Earth has contributed to the loss of biodiversity. Scientists are especially concerned about the tropical rain forests and coral reefs because they are particularly rich in species. Already, tropical rain forests have been reduced from their original 14% of landmass to the present 6%. Also, 54% of coral reefs have been destroyed or are on the verge of destruction.

Video
Coral Reef Ecosystems

Figure 24.15 Causes for the loss of biodiversity.
a. Habitat loss, alien species, pollution, overexploitation, and disease have been identified as causes of extinction of organisms. **b.** Macaws that reside in South American tropical rain forests are endangered for the reasons listed in the graph.

Alien Species

Alien species, sometimes called exotics, are nonnative members of an ecosystem. Humans have introduced alien species into new ecosystems via colonization, horticulture and agriculture, and accidental transport. For example, the pilgrims brought the dandelion to the United States as a familiar salad green. Kudzu is a vine from Japan that the U.S. Department of Agriculture thought would help prevent soil erosion. The plant now covers much landscape in the South. The zebra mussel from the Caspian Sea was accidentally introduced into the Great Lakes in 1988. It is now found in many U.S. rivers where it forms dense beds that squeeze out native mussels. Alien species that crowd out native species are termed **invasive.** One way to counteract alien species is to support native (original to the area) species.

Video Alien Invasion

Pollution

Pollution brings about environmental change that adversely affects the lives and health of living organisms. Acid deposition weakens trees and increases their susceptibility to disease and insects. This can decimate a forest. Climate change is predicted to be the cause of habitat loss due to global temperature shifts. For example, coral reefs die off as ocean temperatures increase. Coastal wetlands may be lost as sea levels increase. The depletion of the ozone shield limits crop and tree growth and kills plankton that sustain ocean life. Humanmade organic chemicals released into the environment may interfere with hormone function and affect the reproductive ability of different species.

Video Ocean Garbage

APPLICATIONS AND MISCONCEPTIONS

Why are there beagle dogs at the customs area in an international airport?

The beagles working at an airport's customs checkpoint are trained to detect agricultural items (foodstuffs, plants, animals, and the like) that may not be brought into the United States. Importing foreign species is prohibited by law. The goal of this law is to prevent the introduction of exotic life and/or diseases that might affect crops or animals in the United States. About \$2 million worth of illegal articles are seized yearly. Dogs assist in about 10% of those cases. Beagles are used because of their keen sense of smell and good-natured temperament.

Overexploitation

Overexploitation occurs when the number of individuals taken from a wild population is so great that the population cannot replace itself and becomes severely reduced. A positive feedback cycle explains overexploitation. The smaller the population, the more valuable its members and the greater the incentive to exploit the few remaining individuals.

Markets for decorative plants and exotic pets support both legal and illegal trade in wild species. Rustlers dig up rare cacti, such as the single-crested saguaro, and sell them to gardeners. Parakeets and macaws are among the birds taken from the wild for sale to pet owners. Many birds die during collection and transportation for every bird delivered alive. The same holds true for tropical fish, which often come from the coral reefs of Indonesia and the Philippines. Divers dynamite reefs or use plastic squeeze bottles of cyanide to stun them. In the process, many fish die.

Declining species of mammals are still hunted for their hides, tusks, horns, or bones. A single Siberian tiger is now worth more than $500,000 because of its rarity. Its bones are pulverized and used as a medicinal powder. The horns of rhinoceroses become ornate carved daggers, and their bones are ground up to sell as a medicine. The ivory of an elephant's tusk is used to make art objects, jewelry, or piano keys. The fur of a Bengal tiger sells for as much as $100,000 in Tokyo.

Fish are a renewable resource if harvesting does not exceed the ability of the fish to reproduce. Today, larger and more efficient fishing fleets decimate fishing stocks (Fig. 24.16). Tuna and similar fish are captured by purse seining. A very large net surrounds a school of fish, and then the net is closed in the same manner as a drawstring purse. Dolphins that accompany the tuna are killed by this type of net. Other fishing boats drag huge trawling nets, large enough to accommodate 12 jumbo jets, along the seafloor to capture bottom-dwelling fish. Trawling has been called the marine equivalent of clear-cutting trees because after the net goes by, the sea bottom is devastated (Fig. 24.16*c*). Only

APPLICATIONS AND MISCONCEPTIONS

How does one choose the best fish to eat?

Before you place your order for fish, take a moment to think about your choice. You may want to consider whether it's a sustainable species of fish, one that's being overfished, or one that contains high levels of mercury. Pacific halibut is a wild fish that is a good choice. The Pacific halibut is usually caught with bottom longlines, rather than trawling nets. This fishing technique won't damage the surrounding environment or catch unwanted fish or animals. Tilapia that is farm grown in the United States is another good choice. It's grown in closed inland systems that prevent exposure of the fish to pollutants. But avoid tilapia from Chile or Taiwan—fish there are raised in open systems where pollutants, especially mercury, can affect the fish. You might also want to consider the levels of healthy omega-3 fatty acids in the fish you eat. Farm-raised salmon or trout supply high concentrations of omega-3 acids (thought to protect against heart disease) and contain little or no mercury or other pollutants.

Shark, sole, haddock, and swordfish are poor choices. These species have been severely overfished and/or contain high levels of mercury.

Video Fishing for Trouble

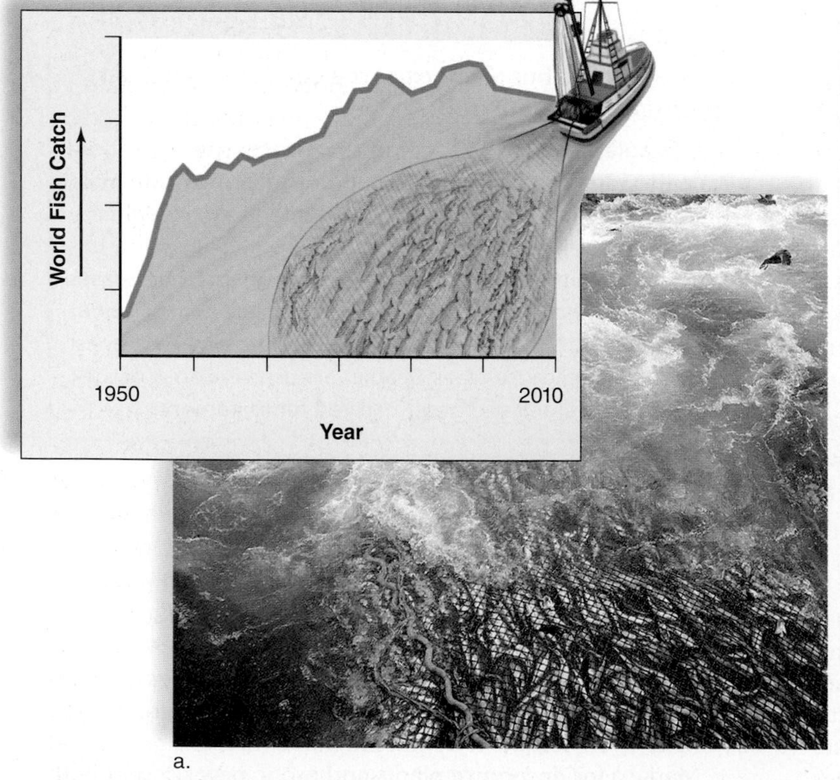

a.

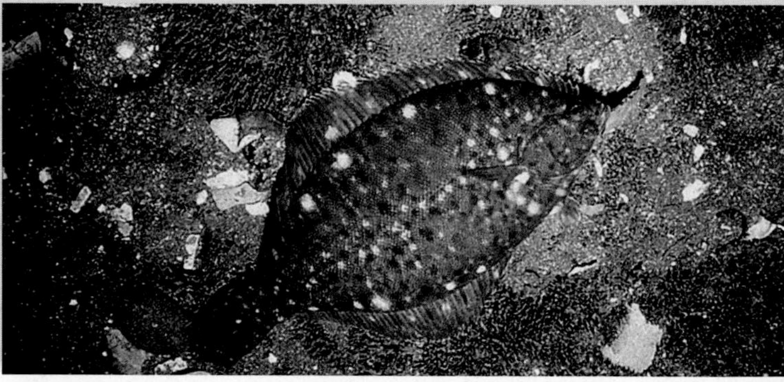

b.

c.

Figure 24.16 The impact of modern fishing practices.
a. The world fish catch has declined in recent years (inset). **b.** A flounder on a healthy seafloor. **c.** Devastation of the seafloor after trawling.

large fish are kept. Undesirable small fish and sea turtles are discarded, dying, back into the ocean. Cod and haddock were once the most abundant bottom-dwelling fish along the Northeast Coast. Now, they are often outnumbered by dogfish and skate.

A marine ecosystem can be disrupted by overfishing, as exemplified on the U.S. West Coast. When sea otters began to decline in numbers, investigators found that they were being eaten by orcas (killer whales). Usually, orcas prefer seals and sea lions to sea otters, but they began eating sea otters when seals and sea lions could not be found. The decline in seal and sea lion populations was due to the decline in the perch and herring populations as a result of overfishing. Ordinarily, sea otters keep the population of sea urchins, which feed on kelp, under control. But with fewer sea otters around, the sea urchin population exploded and decimated the kelp beds. Thus, overfishing set in motion a chain of events that detrimentally altered the food web of an ecosystem.

Video Ocean Fishing Ban

Disease

Wildlife is subject to emerging diseases just as humans are. Exposure to domestic animals and their pathogens occurs due to the encroachment of humans on wildlife habitats. Wildlife can also be infected by animals not ordinarily encountered. For example, African elephants carry a strain of herpes virus that is fatal to Asian elephants. Asian elephants can die if the two types of elephants are housed together.

The significant effect of diseases on biodiversity is underscored by a National Wildlife Health Center study. The study found that almost half of sea otter deaths along the coast of California are due to infectious diseases. Scientists tell us that the number of pathogens that cause disease are on the rise. Just as human health is threatened, so is that of wildlife. Extinctions due to disease may occur.

Direct Value of Biodiversity

Various species perform useful services for humans and contribute greatly to the value of biodiversity. The direct value of wildlife species is related to their medicinal value, agricultural value, and consumptive use value. These are discussed here and illustrated in Figure 24.17.

Medicinal Value

Most of the prescription drugs used in the United States were originally derived from living organisms. The rosy periwinkle from Madagascar is a tropical plant that has provided us with useful medicines. Potent chemicals from this plant are now used to treat leukemia and Hodgkin disease. The survival rate for childhood leukemia has gone from 10% to 90%, and Hodgkin disease is usually curable because of these drugs. Although the value of saving a life cannot be calculated, it is still sometimes easier for us to appreciate the worth of a resource if it is explained in monetary terms. Based on past success, it has been estimated that more than 300 types of drugs may yet be found in tropical rain forests. The value of this resource could be in excess of $140 billion.

The antibiotic penicillin is derived from a fungus while certain species of bacteria produce the antibiotics tetracycline and streptomycin. These drugs have proven to be indispensable in the treatment of various diseases.

Until recently leprosy was a disease for which there was no cure. The bacterium that causes leprosy will not grow in the laboratory. However, scientists discovered that it grows naturally in the nine-banded armadillo. Having a source for the bacterium made it possible to find a cure for leprosy. The blood of horseshoe crabs contains a substance called limulus amoebocyte lysate. This chemical is used to ensure that medical devices, such as pacemakers, surgical implants, and prosthetic devices, are free of bacteria. Blood is taken from 250,000 crabs a year, and then they are returned to the sea unharmed.

Video Good Poison

Agricultural Value

Crops, such as wheat, corn, and rice, are derived from wild plants that have been modified to be high producers. The same high-yield, genetically similar strains tend to be grown worldwide. At one time, rice crops in Africa were being devastated by a virus. Researchers grew wild rice plants from thousands of seed samples until they found one that contained a gene for resistance to the virus. These wild plants were then used in a breeding program to transfer the gene into high-yield rice plants. If this variety of wild rice had become extinct before being discovered, rice cultivation in Africa might have collapsed.

The use of biological pest controls (natural predators of pests) is often preferable to using chemical pesticides. When a rice pest, called the brown planthopper, became resistant to pesticides, farmers began to use natural brown planthopper enemies instead. The economic savings were calculated at well over $1 billion. Similarly, cotton growers in Cañete Valley, Peru, found that the cotton aphid was resistant to the pesticides being used. Research identified natural predators that are now being used to an ever-greater degree by cotton farmers. Again, savings have been enormous.

Most flowering plants are pollinated by animals, such as bees, wasps, butterflies, beetles, birds, and bats. The honeybee, *Apis mellifera,* has been domesticated. It now pollinates almost $10 billion worth of food crops annually in the United States. The value of wild bee pollinators to the U.S. agricultural economy has been calculated at $4.1 to $6.7 billion a year. And yet, modern agriculture often kills wild bees by spraying fields with pesticides.

Video Pollinators

Figure 24.17 The direct value of biodiversity. The direct services of wild species benefit humans immensely, and it is sometimes possible to calculate the monetary value, which is always surprisingly large.

Wild species, like the rosy periwinkle, *Catharanthus roseus*, are sources of many medicines.

Wild species, like the nine-banded armadillo, *Dasypus novemcinctus*, play a role in medical research.

Wild species, like many marine species, provide us with food.

Wild species, like the lesser long-nosed bat, *Leptonycteris curasoae*, are pollinators of agricultural and other plants.

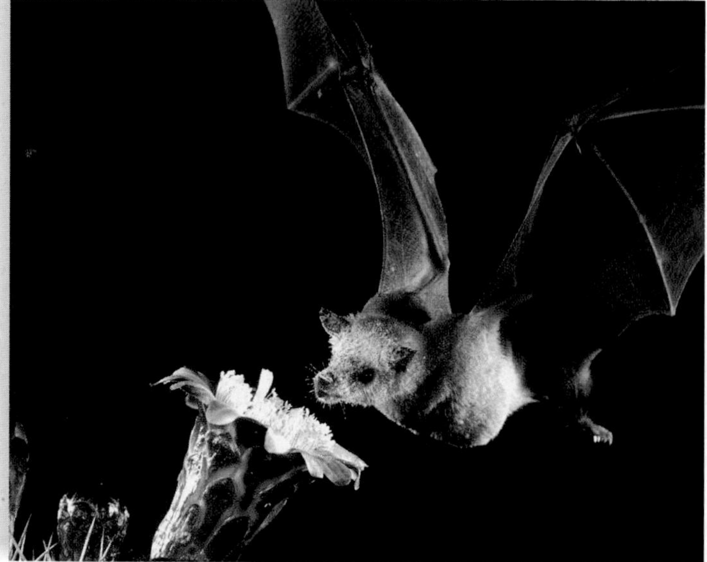

Wild species, like ladybugs, *Coccinella*, play a role in biological control of agricultural pests.

Wild species, like rubber trees, *Hevea*, can provide a product indefinitely if the forest is not destroyed.

BIOLOGY MATTERS  **Science**

Mystery of the Vanishing Bees

Imagine standing in the produce section of your supermarket. You're shocked to see that there are no apples, cucumbers, broccoli, onions, pumpkins, squash, carrots, blueberries, avocados, almonds, or cherries. This could happen at grocery stores in the future. All the crops mentioned, as well as many others, are dependent on honeybees for pollination. Your diet and $15 billion worth of crops could suffer if the honeybees aren't available to perform their important job of moving pollen. Although there are wild bee populations that pollinate crops, domestic honeybees are easily managed and transported from place to place when their pollination services are needed.

Figure 24B Parasitic mites may destroy entire bee colonies.

Colonies of honeybees have experienced a number of health problems since the 1980s. Mites—animals similar to ticks—have always been a danger for bees. Varroa mites and tracheal mites were early causes of colony stress and bee deaths (Fig. 24B). However, beekeepers were very alarmed in 2006 when entire colonies of bees began to vanish. Researchers started referring to the phenomenon as colony collapse disorder (CCD). There doesn't appear to be one single factor that causes seemingly healthy bees to vanish from their hives. Scientists now believe multiple factors may stress bees, causing them to be vulnerable to infection by a parasite or pathogen. The indiscriminate use of pesticides, the strain of being moved from place to place to pollinate crops, and/or poor nutrition (because genetically engineered plants don't provide as much food for the bees) may contribute to CCD.

CCD is also occurring in the honeybee populations in other countries. Ideally, a cause-and-effect treatment for CCD will be found soon, thanks to worldwide research dedicated to solving the problem, as well as improved funding from agricultural agencies. Until then, there are things you can do to keep bees in your area healthy. Research and then plant native plants in your yard and garden. These typically require less fertilizer and water than other plants, and they will provide more pollen and nectar for the bees. In the southern and midwestern regions of the United States, bees enjoy red clover, foxglove, bee balm, and joe-pye weed. Desert willow and manzanita will attract desert bees. Choose palms for tropical areas. In addition, native plants that flower at different times of the year will provide a constant food source. Midday is typically when bees are out foraging, so if you have to use pesticides, apply them late in the day. Plants that rely on the honeybees for pollination—as well as your body—will thank you!

Questions to Consider

1. How might the introduction of alien species contribute to CCD?
2. What other species might be suffering similar problems?

Consumptive Use Value

We have had much success cultivating crops, keeping domesticated animals, growing trees in plantations, and so forth. However, aquaculture, the growing of fish and shellfish for human consumption, has contributed only minimally to human welfare. Instead, most freshwater and marine harvests depend on the harvesting of wild fish (e.g., trout and tuna), crustaceans (e.g., shrimps and crabs), and mammals (e.g., whales). These aquatic organisms are an invaluable biodiversity resource.

The environment provides all sorts of other products that are sold in the marketplace worldwide. Wild fruits and vegetables, skins, fibers, beeswax, and seaweed are only a few examples. Also, some people obtain their meat directly from the environment. The economic value of wild pig in the diet of native hunters in Sarawak, East Malaysia, has been calculated to be approximately $40 million per year.

Similarly, many trees are still felled in the natural environment for their wood. Researchers have calculated that a species-rich forest in the Peruvian Amazon is worth far more if the forest is used for fruit and rubber production than for timber production. Fruit and the latex needed to produce rubber can be brought to market for an unlimited number of years. Once the trees are gone, fruit and latex cannot be harvested until new trees replace those that have been taken.

Indirect Value of Biodiversity

To bring about the preservation of wildlife, it is necessary to make more people aware that biodiversity is a resource of immense value. If we want to preserve wildlife, it is more economical to save ecosystems than individual species. Ecosystems perform many useful services for humans. These services are said to be indirect because they are pervasive and not easy to calculate a dollar amount associated with them. Our very survival depends on the functions that ecosystems perform for us. The indirect value of biodiversity can be associated with the following services.

Waste Disposal

Decomposers break down dead organic matter and other types of wastes to inorganic nutrients. The nutrients are then used by the producers within ecosystems. This function aids humans immensely because we dump millions of tons of waste material into natural ecosystems each year. Waste would soon cover the entire surface of our planet without decomposition. We can build expensive sewage treatment plants, but few of them break down solid wastes completely to inorganic nutrients. It is less expensive and more efficient to water plants and trees with partially treated wastewater and let soil bacteria cleanse it completely.

Video Decomposers

Biological communities are also capable of breaking down and immobilizing pollutants. These include heavy metals and pesticides that humans release into the environment.

Video Good Tobacco

Provision of Freshwater

Few terrestrial organisms are adapted to living in a salty environment. They need freshwater. The water cycle continually supplies freshwater to terrestrial ecosystems. Humans use freshwater in innumerable ways, including drinking and irrigation of crops. Freshwater ecosystems, such as rivers and lakes, also provide us with fish and other types of organisms for food.

Unlike other commodities, there is no substitute for freshwater. We can remove salt from seawater to obtain freshwater. However, the cost of desalination is about four to eight times the average cost of freshwater acquired via the water cycle.

Forests and other natural ecosystems exert a "sponge effect." They soak up water and then release it at a regular rate. When rain falls in a natural area, plant foliage and dead leaves lessen its impact. The soil then slowly absorbs it, especially if the soil has been aerated by organisms. The water-holding capacity of forests reduces the possibility of flooding. Forests release water slowly for days or weeks after the rains have ceased. Rivers flowing through forests in West Africa release between three and five times as much water at the end of the dry season, as do rivers from coffee plantations.

Prevention of Soil Erosion

Intact ecosystems naturally retain soil and prevent soil erosion. The importance of this ecosystem attribute is especially observed following deforestation. In Pakistan, the world's largest dam, the Tarbela Dam, is losing its storage capacity of 12 billion cubic meters (m^3) many years sooner than expected. Deforestation is causing silt to build up behind the dam, decreasing its storage capacity. At one time, the Republic of the Philippines was exporting $100 million worth of oysters, mussels, clams, and cockles each year. Now, silt carried down rivers following deforestation is smothering the mangrove ecosystem that serves as a nursery for these shellfish. Most coastal ecosystems are not as bountiful as they once were because of deforestation and a myriad of other environmental problems.

Biogeochemical Cycles

Chapter 23 presented the fact that ecosystems are characterized by energy flow and chemical cycling. The biodiversity within ecosystems contributes to the workings of the water, phosphorus, nitrogen, carbon, and other biogeochemical cycles. We depend on these cycles for freshwater, provision of phosphate, uptake of excess soil nitrogen, and removal of carbon dioxide from the atmosphere. When human activities upset the usual workings of biogeochemical cycles, the dire environmental consequences include the release of excess pollutants that are harmful to us. Technology is unable to substitute for any of the biogeochemical cycles.

Video Dung Beetles

Regulation of Climate

At the local level, trees provide shade and reduce the need for fans and air conditioners during the summer. In a rain forest, trees help maintain area rainfall. Forests would become arid without the trees.

Globally, forests stabilize the climate because they take up carbon dioxide. The leaves of trees use carbon dioxide when they photosynthesize, and the bodies of the trees store carbon. When trees are cut and burned, carbon dioxide is released into the atmosphere. Carbon dioxide makes a significant contribution to climate change, which is expected to be stressful for many plants and animals. Only a small percentage of wildlife may be able to move to new communities, where the weather will be suitable for them.

Ecotourism

Almost everyone prefers to vacation in the natural beauty of an ecosystem. In the United States, nearly 100 million people enjoy vacationing in a natural setting. To do so, they spend $4 billion each year on fees, travel, lodging, and food. Many tourists want to go sport fishing, whale watching, boat riding, hiking, bird watching, and the like.

CHECK YOUR PROGRESS 24.3

1. Describe the factors that are contributing to the current extinction crisis.
2. Summarize the direct and indirect benefits of preserving wildlife.
3. Discuss the importance of biodiversity to human society.

CONNECTING THE CONCEPTS

For more information on the topics presented in this section, refer to the following discussion:

Section 23.3 examines the water, carbon, nitrogen, and phosphorus cycles.

24.4 Working Toward a Sustainable Society

LEARNING OUTCOMES

Upon completion of this section, you should be able to

1. Describe the characteristics of a sustainable society.
2. Identify methods of developing sustainability within rural and urban environments.
3. List the methods of determining economic well-being and quality of life.

A **sustainable** society would always be able to provide the same amount of goods and services for future generations as it does for the current one. At the same time, biodiversity would be preserved.

To achieve a sustainable society, resources cannot be depleted and must be preserved. In particular, future generations need clean air, water, an adequate amount of food, and enough space in which to live. This goal is not possible unless we carefully regulate our consumption of resources today, taking into consideration that the human population is still increasing.

Today's Unsustainable Society

We are quick to realize that population growth in the LDCs creates an environmental burden. However, we also need to consider that the excessive resource consumption of the MDCs also stresses the environment. Sustainability is incompatible with the current level of consumption of resources as well as the level of waste produced by the MDCs. Population growth in the LDCs and overconsumption by the MDCs account for many of the problems we are facing today (Fig. 24.18).

At present, a considerable proportion of land is being used for human purposes (homes, agriculture, factories, etc.). Agriculture uses large inputs of fossil fuels, fertilizer, and pesticides that create a large amount of pollution. More freshwater is used for agriculture than in homes. Almost half of the agricultural yield in the United States goes toward feeding animals. According to the ten-to-one rule of thumb, it takes 10 lb of grain to

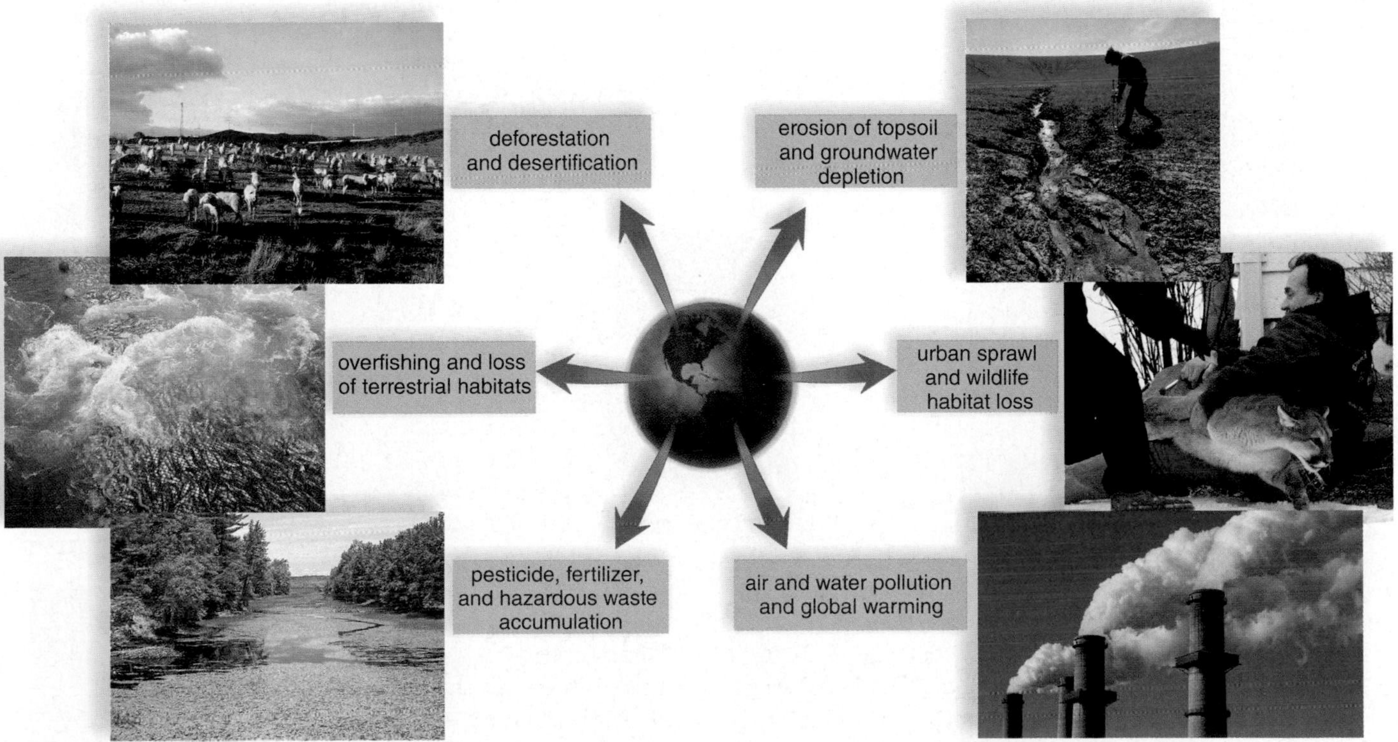

Figure 24.18 **Characteristics of an unsustainable society.**
Arrows point outward to signify that these types of activities reduce the carrying capacity of the Earth.

grow 1 lb of meat. Therefore, it is environmentally unsustainable for citizens in MDCs to eat as much meat as they do.

Farm animals and crops require freshwater from surface water and groundwater, and so do humans. Available supplies are dwindling, and the remaining groundwater is in danger of being contaminated. Sewage and animal wastes wash into bodies of surface water and cause overenrichment, which robs aquatic animals of the oxygen they need to survive.

Our society primarily uses nonrenewable fossil fuel energy, which leads to climate change, acid deposition, and smog. The result is weakened ecosystems. The demand for goods has increased to the point that facilities to meet the demand are strained. Construction of improved infrastructure to support increased transportation needs only increases the use of nonrenewable energy resources. LDCs have increased needs for energy, making it imperative for the MDCs to develop renewable energy sources.

The human population is expanding into all regions on the face of the planet, so habitats for other species are being lost, resulting in a significant extinction of wildlife.

Characteristics of a Sustainable Society

A natural ecosystem can offer clues about how to make today's society sustainable. A natural ecosystem makes use of only renewable solar energy. Its materials cycle through the various populations back to the producer once again. For example, coral reefs have been sustaining themselves for millions of years. At the same time, the reefs have provided sustenance to the LDCs. The value of coral reefs has been assessed at over $300 billion a year. Their aesthetic value is immeasurable.

It is clear that if we want to develop a sustainable society, we need to use renewable energy sources and recycle materials. We should protect natural ecosystems that help sustain our modern society. At least a quarter of the coral reefs exist close to the shores of an MDC country, and the chances are good that these coral reefs will be protected. Unfortunately, other coral reefs are threatened by unsustainable practices. The good news is that reefs are remarkably regenerative and will return to their former condition if left alone for a long enough period of time. The message of today's environmentalists is about what can be done to improve matters and use sustainable practices (Fig. 24.19). There is still time to make changes and improvements.

Sustainability should be practiced in various areas of human endeavor, from agriculture to business enterprises. Efficiency is the key to sustainability. For example, an efficient car would be ultralight and fuel efficient. Efficient cars could be just as durable and speedy as the inefficient ones of today. Only through efficiency and conservation can we meet the challenges of limited resources and finances in the future.

People generally live in either the country or the city, but the two regions depend on one another. Achieving sustainability requires that we understand how the two regions are interconnected. It would be impossible to have one sustainable and not the other because the two regions are linked.

Figure 24.19 Characteristics of a sustainable society.
Arrows point inward to signify that these types of activities increase the carrying capacity of the Earth.

What happens within one ultimately affects the other. Let's consider, therefore, the importance of both rural and urban sustainability.

Rural Sustainability

In rural areas, we must put the emphasis on preservation. We need to preserve both terrestrial ecosystems (such as forests and prairies) and aquatic ecosystems (freshwater and brackish ones along the coast). We should also preserve agricultural lands and other areas that provide us with renewable resources.

It is imperative that we take all possible steps to preserve what remains of our topsoil and replant areas with native plants. Native grasses stabilize the soil, rebuild soil nutrients, and can serve as a source of renewable biofuel. Native trees can be planted to break the wind, protect the soil from erosion, and provide consumable products. Creative solutions to today's ecological problems are very much needed.

Here are some other possible ways to help make rural areas sustainable:

- Plant *cover crops,* which often are a mixture of legumes and grasses, to stabilize the soil between rows of cash crops or between seasonal plantings of cash crops.
- Use *multiuse farming* by planting a variety of crops, and use a variety of farming techniques to increase the amount of organic matter in the soil.
- Replenish soil nutrients through composting, organic gardening, or other self-renewable methods.
- Use low-flow or trickle irrigation, retention ponds, and other water-conserving methods.
- Increase the planting of *cultivars* (plants propagated vegetatively), which are resistant to blight, rust, insect damage, salt, drought, and encroachment by noxious weeds.
- Use *precision farming (PF)* techniques that rely on accumulated knowledge to reduce habitat destruction and improve crop yields.
- Use *integrated pest management (IPM),* which encourages the growth of competitive beneficial insects and uses biological controls to reduce the abundance of pest populations.
- Plant a variety of species, including native plants, to reduce dependence on traditional crops.
- Plant *multipurpose trees*—trees with the ability to provide numerous products and perform a variety of functions, in addition to serving as windbreakers (Fig. 24.20). Remember that mature trees can provide many different types of products. For example, mature rubber trees provide us with rubber, and tagua nuts are an excellent substitute for ivory.
- Maintain and restore wetlands, especially in hurricane- or tsunami-prone areas. Protect deltas from storm damage. By protecting wetlands, we protect the spawning ground for many valuable fish nurseries.
- Use renewable forms of energy, such as wind and biofuel.
- Support local farmers to reduce the environmental impact that occurs when goods are transported long distances.

Urban Sustainability

More and more people are moving to urban environments. Much thought needs to be given about how to serve the needs of new arrivals without overexpansion of the city. Resources need to be shared in a way that will allow urban sustainability.

Here are some other possible ways to help make a city sustainable:

- Design an energy-efficient transportation system to rapidly move people about.
- Use solar or geothermal energy to heat buildings. Cool them with an air-conditioning system that uses seawater. In general, use conservation methods to regulate the temperature of buildings.
- Use *green roofs.* Grow a garden of grasses, herbs, and vegetables on the tops of buildings. This will assist temperature control, supply food, reduce the amount of rainwater runoff, and be visually appealing (Fig. 24.21).
- Improve storm-water management by using sediment traps for storm drains, artificial wetlands, and holding ponds. Increase use of porous surfaces for walking paths, parking lots, and roads. These surfaces reflect less heat and soak up rainwater runoff.
- Instead of traditional grasses, plant native species that attract bees and butterflies. These require less water and fewer fertilizers.
- Create *greenbelts* that suit the particular urban setting. Include plentiful walking and bicycle paths.
- Revitalize old sections of a city before developing new sections.
- Use lighting fixtures that hug the walls or ground and send light down. Control noise levels by designing quiet motors.

Figure 24.20 The roles of trees in a sustainable society. Trees planted by a farmer to break the wind and prevent soil erosion can also have other purposes, such as supplying nuts and fruits.

Figure 24.21 A green roof.
A green roof has plants growing on it that help control temperature, supply food, and reduce water runoff.

- Promote sustainability by encouraging recycling of business equipment. Use low-maintenance building materials rather than wood.

Assessing Economic Well-Being and Quality of Life

The gross national product (GNP) is a measure of the flow of money from consumers to businesses in the form of goods and services purchased. It can also be considered the total costs of all manufacturing, production, and services. Costs include salaries and wages, mortgage and rent, interest and loans, taxes, and profit within and outside the country. In other words, GNP pertains solely to economic activities.

When calculating GNP, economists do not necessarily consider whether an activity is environmentally or socially harmful. For example, destruction of forests due to clear-cutting, strip mining, or land development is not a part of the GNP. In the same way, the cost of medical services does not include the pain or suffering caused by illness, for example.

Measures that include noneconomic indicators are most likely better at revealing our quality of life than is the GNP. The index of sustainable economic welfare (ISEW) includes real per capita income, distributional equity, natural resources depletion, environmental damage, and the value of unpaid labor. The ISEW *does* take into account other forms of value, beyond the purely monetary value of goods and services. Another such index is called the genuine progress indicator (GPI). This indicator attempts to consider the quality of life, an attribute that does not necessarily depend on worldly goods. For example, the quality of life might depend on how much respect we give other humans. The Grameen Bank in Bangladesh decided that if women were loaned small amounts of money, they would pay it back after starting up small businesses. The loans give women the opportunity to make choices that can improve the quality of their lives. For these women, a loan is a way to sustain their lives while, in part, fulfilling their dreams. It is difficult to assign a value to well-being or happiness. However, economists are trying to devise a way to measure these values. The following criteria, among others, can be used.

Use value: actual price we pay to use or consume a resource, such as the entrance fees into national parks.

Option value: preserving options for the future, such as saving a wetland or a forest.

Existence value: saving things we might not realize exist yet. This might be flora and fauna in a tropical rain forest that, one day, could be the source of new drugs.

Aesthetic value: appreciating an area or creature for its beauty and/or contribution to biodiversity.

Cultural value: factors such as language, mythology, and history that are important for cultural identity.

Scientific and educational value: valuing the knowledge of naturalists, or even an experience of nature, as a type of rational fact.

Development of the environment will always continue. Still, we can use these values to help us direct future development. Growth creates increases in demand, but development includes the direction of growth. If we permit unbridled growth, resources will become depleted. However, if development restrains resource consumption and still promotes economic growth, perhaps a balance can be reached. We can then preserve resources for future generations.

Each person has a particular comfort level, and humans do not like to make sacrifices that reduce their particular comfort level. So, despite our knowledge of the need to protect fisheries and forests, we continue to exploit them. People from LDCs directly depend on these resources to survive and so have

APPLICATIONS AND MISCONCEPTIONS

What are some simple things you can do to conserve energy and/or water and help solve environmental problems like global warming?

A few things you could easily do include the following:

- Change the lightbulbs in your home to compact fluorescent bulbs. They use 75–80% less electricity than incandescent bulbs.
- Walk, ride your bike, carpool, or use mass transit. It will save you a lot of money, too!
- Get cloth or mesh bags for groceries and other purchases. Plastic bags may take 10–20 years to degrade. They're also dangerous to wildlife if mistaken for food and consumed.
- Turn off the water while you brush your teeth. If you don't finish a bottle of water, use it to water your plants. In dry climates, plant native plants that won't require frequent watering.

much to lose. Even so, it is difficult for them to sacrifice today for the sake of the future. Yet, there is still hope because nature is incredibly resilient. One solution to deforestation is reforestation. Costa Rica has been successfully reforesting since the early 1980s. Also, declining fisheries can be restocked and then managed for sustainability. It will take an informed citizenry, creativity, and a willingness to bring about change for the better to move toward sustainability.

CHECK YOUR PROGRESS 24.4

1. Describe the characteristics of today's society that make it unsustainable.
2. Discuss what changes are needed to convert today's society into one that is sustainable.
3. Summarize how scientists assess economic well-being and quality of life.

CONNECTING THE CONCEPTS

For more information on the material presented in this section, refer to the following discussions:

Section 23.2 describes how the 10% rule relates to energy flow in an ecosystem.

Section 23.3 examines human influence on the major biogeochemical cycles.

CASE STUDY CONCLUSION

Gills Onions processes over 1 million lb of onions every day, producing approximately 300,000 lb of waste. Previously the waste was composted and hauled to local farm fields and spread as fertilizer. This technique resulted in a variety of environmental concerns as well as a significant amount of financial loss. To help resolve the problem, Gills Onions developed an Advanced Energy Recovery System that would turn the onion waste into a methane-rich biogas that powers two fuel cells. The fuel cells in turn supply energy to the processing plant, saving an estimated $700,000 in electricity costs per year. The remaining onion waste is then sold as livestock feed, eliminating the $400,000 in expenses associated with the waste disposal. Gills Onions has become the first food-processing facility in the world to produce ultra-clean energy from their own waste. They have also reduced their carbon footprint by an estimated 14,500 metric tons of CO_2 emissions per year.

MEDIA STUDY TOOLS

Enhance your study of this chapter with media! Visit **www.mhhe.com/maderhuman13c** and go to "Media Study Tools" for this chapter to access the following:

 **Animation**	 **Videos**
24.2 World Hunger	**24.2** Thames River • Needing Herbicides • Pesticide Plants • GM Food Safety • Global Warming • Warming Hurts Rice • Karoo Global Warming • Spinach Battery • DDT **24.3** Global Reef Ecosystems • Alien Invasion • Ocean Garbage • Coral Reef Ecosystems • Ocean Fishing Ban • Good Poison • Pollinators • Decomposers • Good Tobacco • Dung Beetles • Fishing for Trouble

SUMMARIZE

24.1 Human Population Growth

- The **growth rate** of a population often displays **exponential growth,** which is represented on a graph by a steep curve.
- The **biotic potential** is the ideal growth rate for a population. The biotic potential is normally held in check by environmental resistance. This determines the **carrying capacity** of an ecosystem.
- Age-structure diagrams can be used to predict population growth. MDCs are approaching a stable population size. LDC populations will continue to increase in size even if they experience **replacement reproduction.**

24.2 Human Use of Resources and Pollution

The **ecological footprint** of an individual represents the total resources that are needed to meet his or her needs. Five resources are maximally used by humans:

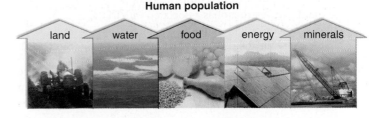

Human population

land water food energy minerals

Resources are either nonrenewable or renewable.

- **Nonrenewable resources** are not replenished and are limited in quantity (e.g., land, **fossil fuels,** minerals).
- **Renewable resources** are replenished but still are limited in quantity (e.g., water, solar energy, food).

Land

Human activities, such as habitation, farming, and mining, contribute to erosion, **pollution, desertification, deforestation,** and loss of biodiversity.

Water

Industry and agriculture use most of the freshwater supply. Water supplies are increased by damming rivers and drawing from **aquifers.** As aquifers are depleted, **subsidence, sinkhole** formation, and **saltwater intrusion** can occur. If used by industries, water conservation methods could cut world water consumption by half.

Food

Food comes from growing crops, raising animals, and fishing.

- Modern farming methods increase the food supply, but some methods harm the land, pollute water, and consume fossil fuels excessively.
- Genetically engineered plants increase the food supply and reduce the need for chemicals.
- Raising livestock contributes to water pollution and uses fossil fuel energy.
- The increased number and high efficiency of fishing boats have caused the world fish catch to decline.

Energy

Fossil fuels (oil, natural gas, coal) are nonrenewable sources. Burning fossil fuels and burning to clear land for farming cause pollutants and gases to enter the air.

- **Greenhouse gases** include CO_2 and other gases. Greenhouse gases cause global warming because solar radiation can pass through, but infrared heat cannot escape back into space.
- Renewable resources include hydropower, geothermal, wind, and solar power.

Minerals

Minerals are nonrenewable resources that can be mined. These raw materials include sand, gravel, phosphate, and metals. Mining causes destruction of the land by erosion, loss of vegetation, and toxic **agricultural runoff** into bodies of water. Some metals are dangerous to health. Land ruined by mining can take years to recover.

Hazardous Wastes

Billions of tons of solid waste are discarded on land and in water.

- Heavy metals include lead, arsenic, cadmium, and chromium.
- Synthetic organic chemicals include **chlorofluorocarbons (CFCs)**, which are involved in the production of plastics, pesticides, herbicides, and other products.
- Ozone shield destruction is associated with CFCs.
- Other synthetic organic chemicals enter the aquatic food chain, where the toxins become more concentrated (**biological magnification**).

24.3 Biodiversity

Biodiversity is the variety of life on Earth.

Loss of Biodiversity

The five major causes of biodiversity loss and extinction are

- habitat loss;
- introduction of **alien species** and **invasive** species;
- pollution;
- overexploitation of plant and animals; and
- disease.

Direct Value of Biodiversity

Direct values of biodiversity are

- medicinal value (medicines derived from living organisms);
- agricultural value (crops derived from wild plants, biological pest controls, and animal pollinators); and
- consumptive use values (food production).

Indirect Value of Biodiversity

Biodiversity in ecosystems contributes to

- waste disposal (through the action of decomposers and the ability of natural communities to purify water and take up pollutants);
- freshwater provision through the water biogeochemical cycle;
- prevention of soil erosion, which occurs naturally in intact ecosystems;
- function of biogeochemical cycles;
- climate regulation (plants take up carbon dioxide); and
- ecotourism (human enjoyment of a beautiful ecosystem).

24.4 Working Toward a Sustainable Society

A **sustainable** society would use only renewable energy sources, would reuse heat and waste materials, and would recycle almost everything. It would also provide the same goods and services presently provided and would preserve biodiversity.

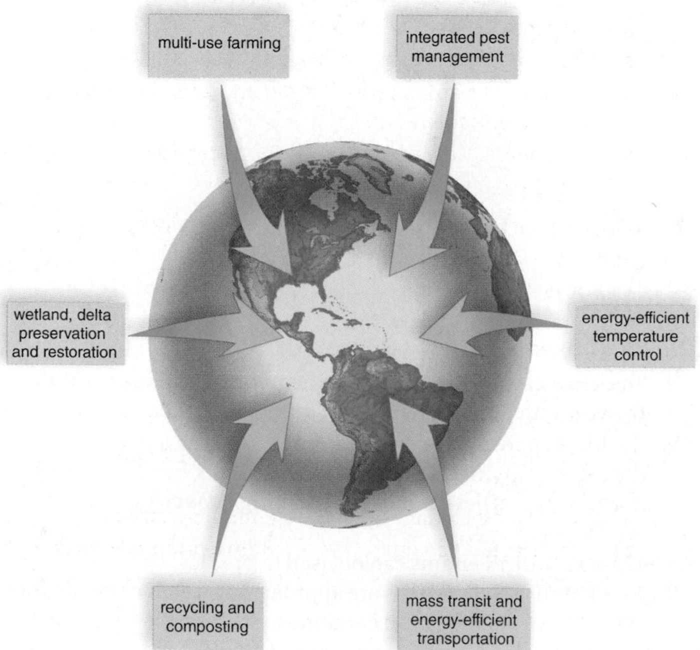

ASSESS

Testing Your Knowledge of the Concepts

1. Define carrying capacity and discuss the relevancy of this concept to the size of today's human population. (page 558)
2. Distinguish between MDCs and LDCs. Why are most LDCs, but not most MDCs, increasing in size? (pages 558–559)
3. Name three locales where humans have settled with unfortunate environmental consequences. What are those consequences? (pages 560–561)
4. Identify several nonrenewable and renewable resources. (pages 565–568)
5. Why is our society considered unsustainable? (pages 577–578)
6. Identify the values associated with assessing economic well-being and quality of life. (page 580)
7. When the carrying capacity of the environment is exceeded, the population will typically
   a. increase, but at a slower rate.
   b. stabilize at the highest level reached.
   c. decrease.
   d. die off entirely.
8. Renewable resources
   a. are in limited supply compared to nonrenewable resources.
   b. are always forthcoming but still may be inadequate for human needs.
   c. include such energy sources as wind, solar, and biomass
   d. All of these are correct.
9. Which of these are indirect values of species?
   a. participation in biogeochemical cycles
   b. participation in waste disposal
   c. provision of freshwater
   d. prevention of soil erosion
   e. All of these are indirect values.
10. Direct values of ecosystems include all of the following except
   a. discovery of a new cure for cancer.
   b. ability to raise crops on the land.
   c. discovery of a new strain of wild plants that are resistant to bacterial diseases.
   d. use of a natural predator to control crop pests.
   e. breakdown and decomposition of organic matter.
11. Which of the following is not a function that ecosystems can perform for humans?
   a. purification of water
   b. immobilization of pollutants
   c. reduction of soil erosion
   d. removal of excess soil nitrogen
   e. breakdown of heavy metals
12. In which of the following is biological magnification most pronounced?
   a. aquatic food chains
   b. terrestrial food chains
   c. long food chains
   d. energy pyramids
   e. Both a and c are correct.
13. Complete the following graph by labeling each bar with a cause of extinction, from the most influential to the least.

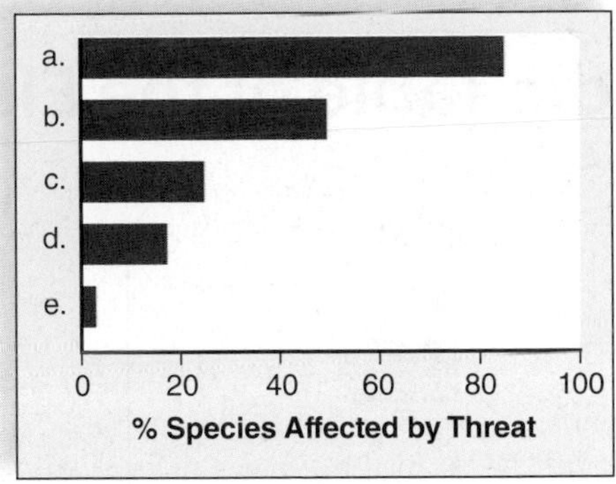

14. Which feature is part of a sustainable society?
   a. recycling and composting
   b. use of mass transit in urban environments
   c. integrated pest management to control crop damage
   d. preservation of wetlands
   e. All are features of a sustainable society.
15. Which feature is associated with the sustainability of a rural society?
   a. plant cover crops when farming
   b. use of green roofs
   c. designing energy-efficient transportation systems
   d. making areas more pedestrian- and bike-friendly
   e. All of these are features associated with the sustainability of a rural society.

ENGAGE

Thinking Critically About the Concepts

1. What environmental reasons would you give a friend for becoming a vegetarian?
2. **a.** What size footprint do you currently have on Earth? Visit www.myfootprint.org to determine the size of your footprint.
   b. What types of things could you do to decrease the size of your footprint on Earth?
3. How would failure to recycle items that can be recycled (they end up in a landfill) affect the nutrient cycles, such as the carbon cycle covered in Chapter 23?
4. **a.** What type of environmental activities/initiatives (Earth Day festivities, recycling program, etc.) are in place at your school?
   b. How could you increase awareness of environmental issues on your campus?

Periodic Table of the Elements

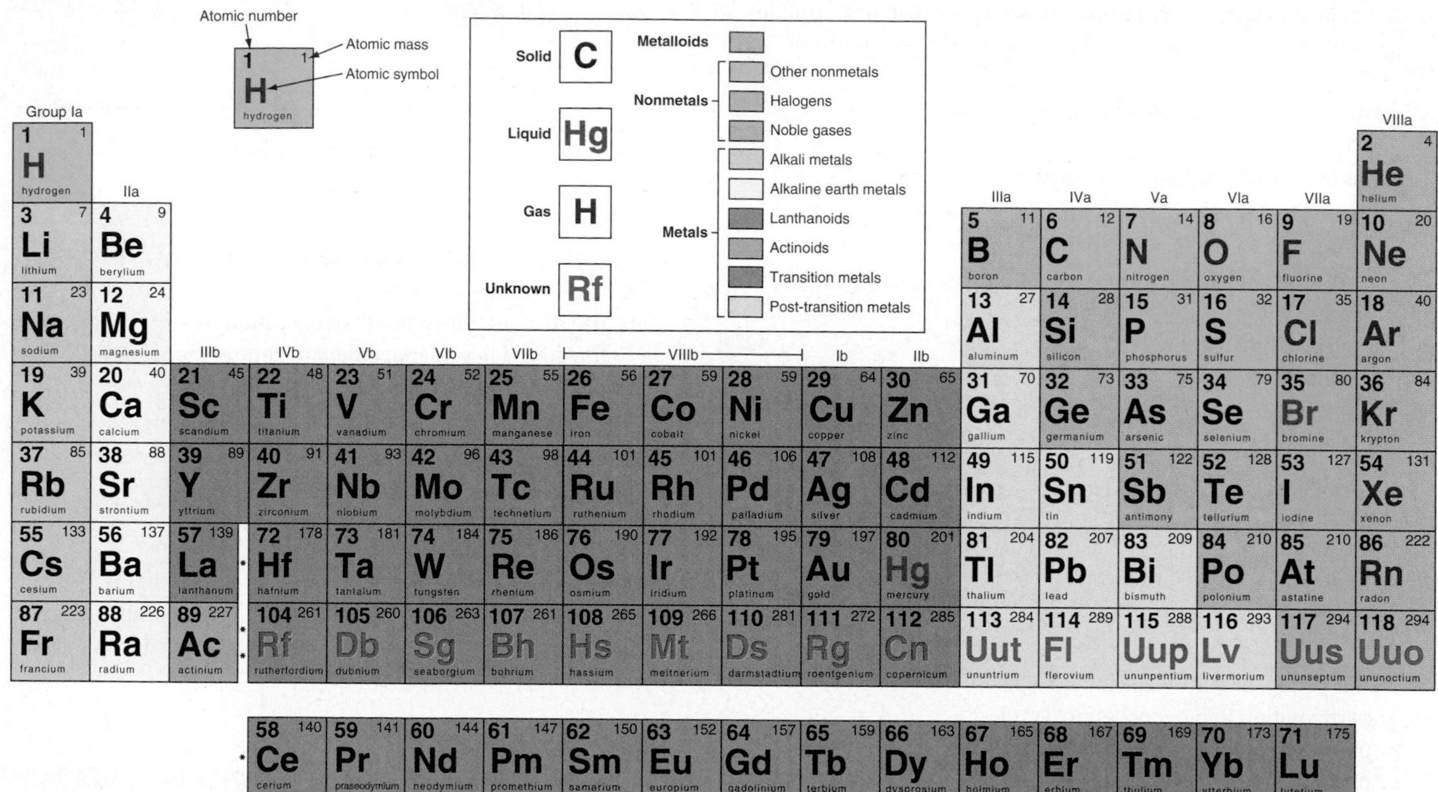

Appendix B

Metric System

Unit and Abbreviation	Metric Equivalent	Approximate English-to-Metric Equivalents
Length		
nanometer (nm)	$= 10^{-9}$ m (10^{-3} µm)	
micrometer (µm)	$= 10^{-6}$ m (10^{-3} mm)	
millimeter (mm)	= 0.001 (10^{-3}) m	
centimeter (cm)	= 0.01 (10^{-2}) m	1 inch = 2.54 cm 1 foot = 30.5 cm
meter (m)	= 100 (10^{2}) cm = 1,000 mm	1 foot = 0.30 m 1 yard = 0.91 m
kilometer (km)	= 1,000 (10^{3}) m	1 mi = 1.6 km
Weight (mass)		
nanogram (ng)	$= 10^{-9}$ g	
microgram (µg)	$= 10^{-6}$ g	
milligram (mg)	$= 10^{-3}$ g	
gram (g)	= 1,000 mg	1 ounce = 28.3 g 1 pound = 454 g
kilogram (kg)	= 1,000 (10^{3}) g	= 0.45 kg
metric ton (t)	= 1,000 kg	1 ton = 0.91 t
Volume		
microliter (µl)	$= 10^{-6}$ l (10^{-3} ml)	
milliliter (ml)	$= 10^{-3}$ liter = 1 cm^{3} (cc) = 1,000 mm^{3}	1 tsp = 5 ml 1 fl oz = 30 ml
liter (l)	= 1,000 ml	1 pint = 0.47 liter 1 quart = 0.95 liter 1 gallon = 3.79 liter
kiloliter (kl)	= 1,000 liter	

°C	°F	
100	212	Water boils at standard temperature and pressure.
71	160	Flash pasteurization of milk
57	134	Highest recorded temperature in the United States, Death Valley, July 10, 1913
41	105.8	Average body temperature of a marathon runner in hot weather
37	98.6	Human body temperature
13.7	56.66	Human survival is still possible at this temperature.
0	32.0	Water freezes at standard temperature and pressure.

Units of Temperature

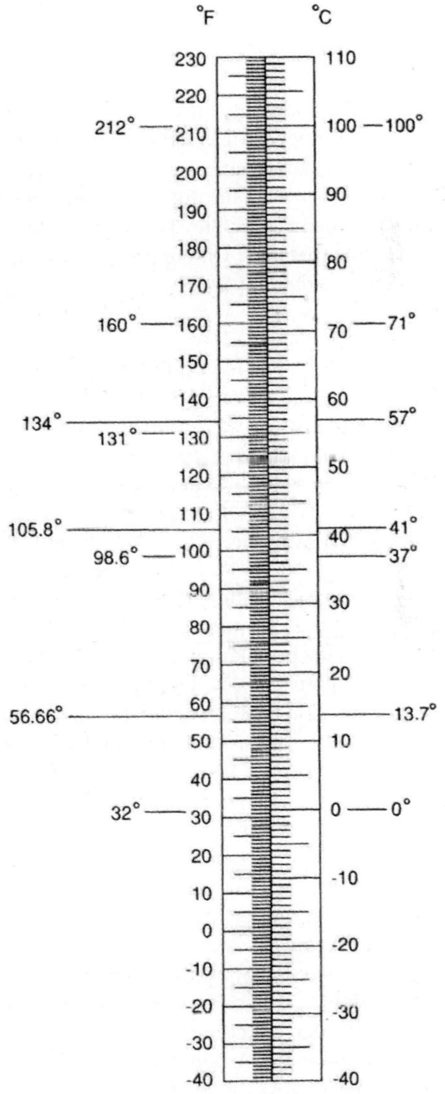

°F 212° — 100 °C
°F 160° — 71 °C
°F 134° — 57 °C
°F 131°
°F 105.8° — 41 °C
°F 98.6° — 37 °C
°F 56.66° — 13.7 °C
°F 32° — 0 °C

To convert temperature scales:

$$°C = \frac{(°F - 32)}{1.8}$$

$$°F = 1.8\,(°C) + 32$$

Appendix C

Answer Key

This appendix contains the answers to the Case Study, Testing Your Knowledge of the Concepts, and Thinking Critically About the Concepts questions, which appear at the end of each chapter, and the Check Your Progress questions, which appear within each chapter.

Chapter 1

Case Study

1. Organization, acquiring materials and energy, reproducing, growing and developing, being homeostatic, responding to stimuli, evolving; **2.** Look for any clues that correspond to the characteristics of life, for example, fossils of carbon-based life forms, or unexpected chemical deposits left behind by microscopic life; **3.** Similar environments might give us insight on varying diversity; Different environments might allow us to examine variations we do not possess.

Check Your Progress

(1.1) 1. Organization, acquiring materials and energy, reproducing, growing and developing, being homeostatic, responding to stimuli, adaptation; **2.** The smallest structural unit of all living things is the cell, and some organisms are single-celled. Organization in multicellular organisms increased from cells—tissues—organs—organ systems—organism. Organisms are further organized into populations—communities—ecosystems—the biosphere; **3.** Variations arise (usually due to random genetic mutation) that sometimes provides a survival advantage for some individuals—individuals with this advantage tend reproduce more, i.e. they have a new adaptation, causing the variation to become more common in the population. **(1.2) 1.** From the surface of the Earth through the atmosphere and down to the sea and soil; areas where life can be found; **2.** To be able to study adaptation and genetic diversity through similarities and differences; **3.** Larger populations put a strain on resources and modify ecosystems. **(1.3) 1.** Observation—watch and take note; hypothesis—form a testable statement; experiment—hypothesis is tested; conclusion—data analyzed; scientific theory—requires much support; **2.** Gives a basis of comparison for a certain variable; **3.** Pros—authored by researchers, reviewed by others; Cons—May be difficult to understand; some based on limited data. **(1.4) 1.** Anecdotal data are testimonials by individuals, and thus may be due to chance or only true in certain cases; **2.** To fully understand how the researcher obtained the data and came to the conclusion; **3.** They show relationship between quantities and summarize data in a clear, concise manner. **(1.5) 1.** Nuclear physics (cancer therapy/nuclear bomb), biotechnology (GM crops and microbes/possibly endangering biosphere), stem cell technology (treat disease/destroying embryos); **2.** Human ethics, values, and morals set guidelines for science and technology; **3.** All members.

Testing Your Knowledge of the Concepts

10. c; **11.** b; **12.** e; **13.** a; **14.** a; **15.** b; **16.** a; **17.** c

Thinking Critically About the Concepts

1. Currently the smallest living things are single-celled organisms. These organisms have a cell membrane, genetic information, and cytoplasm (cell contents). Although viruses reproduce and evolve, they lack cell characteristics and thus are considered by most people to be nonliving; **2.** It is not reliable in the terms of research data, numbers studied, or control groups, but it can be useful as participants give their opinion of the conclusion of the research through first-hand knowledge, and can provide ideas to researcher interesting in studying related phenomena; **3.** It would alter our basic definition of life if the environment was different than ours; that would also change the definition of biosphere for those places.

Chapter 2

Case Study

1. Cholesterol can cause build-ups and blockages in the cardiovascular system; **2.** No, HDLs are beneficial; **3.** Proteins function as enzymes and hormones, and are used for signaling and transport in the body. Proteins make up a certain percentage of the cholesterol molecules that relate to David's values.

Check Your Progress

(2.1) 1. An atom of carbon contains a single nucleus with six protons and six neutrons. Six electrons circle the carbon nucleus in orbitals; **2.** The first isotope has 40 - 20 = 20 neutrons. The second isotope has 48 - 20 = 28 neutrons; **3.** Isotopes are the forms of an element that differ in the number of neutrons. A radioisotope decays over time, releasing rays and subatomic particles. Radioisotopes can be used for medical purposes and for sterilizing objects, including food; **4.** The two basic types of bonding found between atoms are the ionic bond and the covalent bond. In ionic bonding, one or more atoms lose electrons and other atoms gain electrons. In this way, ions are formed. In covalent bonding, the ions are shared between atoms. **(2.2) 1.** Water is a liquid at room temperature, its temperature changes very slowly, and it requires a great deal of heat to become vapor. These characteristics keep water available for supporting life and enable it to help in cooling. Frozen water is less dense than water, so ice floats. Ice helps to insulate the water underneath, allowing marine life to survive the cold. Water molecules are cohesive yet flow freely, allowing water to fill vessels. Water is a solvent for polar molecules and facilitates chemical reactions; **2.** Acids contain greater numbers of H^+ ions than OH^- ions. Their pH will be less than 7. Bases contain greater numbers of OH^- ions than H^+ ions. Their pH will be greater than 7; **3.** The pH of soil and the body remain relatively constant because of the action of buffers. **(2.3) 1.** An organic molecule contains carbon and hydrogen, an inorganic molecule does not contain carbon; **2.** Carbohydrates—quick and short term energy storage; lipids—energy storage; proteins—support, enzymes, transport, defense, hormones, motion; nucleic acids—stores genetic information; **3.** A type of synthesis reaction called dehydration reaction where a hydroxyl group and a hydrogen group are removed as the molecule forms. **(2.4) 1.** Simple carbohydrate is used for quick and short-term energy; Polysaccharides are used for long-term energy storage; **2.** Simple carbohydrates contain low numbers of carbon atoms (3–7), complex carbohydrates contain high numbers of carbon atoms in chains of sugar (glucose) units; **3.** The human digestive tract produces enzymes that can digest types of linkages between sugars, but not between others. **(2.5) 1a.** E. Fats and oils are used for energy storage, heat insulation, as a protective cushion around organs, and to synthesize phospholipids and steroids; **2.** Phospholipids—cellular membranes; Steroids—sex hormones; **3.** Trans fats can be difficult to break down in the body, thus they may accumulate in blood vessels, blocking blood flow, for example, to the heart or brain. **(2.6) 1.** Support, enzymes, transport, defense, hormones, motion; **2.** A central carbon bonded to a hydrogen, a —NH_2 (amino group) a —COOH (acid group), and an R group; **3.** Once a protein loses its normal shape, it is no longer able to perform its usual function. **(2.7) 1.** ATP consists of an adenosine and three phosphate groups; **2.** Hydrogen bonds are strong enough to hold double-stranded DNA together, but weak enough to be easily broken to allow for DNA replication and RNA synthesis; **3.** DNA—deoxyribose sugar, phosphate, nitrogen containing bases (adenine, thymine, cytosine, guanine), double stranded; RNA—ribose sugar, phosphate, nitrogen containing bases (adenine, guanine, uracil, cytosine), single stranded. DNA is used to store genetic information and thus is more stable, and RNA is used to synthesize proteins but is relatively short-lived in the cell.

Testing Your Knowledge of the Concepts

14. b; **15.** c; **16.** b; **17.** a; **18.** b; **19.** d; **20.** d

Thinking Critically About the Concepts

1. Because cholesterol is in the food we eat and the body makes a certain amount of cholesterol as well; **2.** The different types of cholesterol include some that are more beneficial than others. Some are good and aid in transport, others build-up in the lining of the blood vessels causing blockages; **3.** Carbohydrates, lipids, oils, proteins; **4.** Altering the pH of blood will result in changes in the structures of molecules in the blood such as proteins, which will affect their function; **5.** Storage is important so the body does not have to constantly intake certain molecules.

Chapter 3

Case Study

1. Golgi apparatus; **2.** Lysosomes contain hydrolytic enzymes that digest macromolecules and cell parts; **3.** The lysosome digests excess amounts of fatty acids, if it were to malfunction, an excess of fatty acids would build up.

Check Your Progress

(3.1) 1. A cell is the basic unit of life, all living things are made up of cells, new cells arise only from preexisting cells; this is important in evolutionary history and diversity; **2.** Small cells have a greater surface-area-to-volume ratio, thus a greater ability to get material in and out of the cell; **3.** Light microscopes use light rays to magnify objects and can be used to view living specimens; electron microscopes use a stream of electrons to magnify objects and have a higher resolving power than light microscopes. **(3.2) 1a.** Nucleus, plasma membrane, cytoplasm; **2.** Eukaryotic cells have a nucleus and membrane-bounded organelles, prokaryotic cells lack these; **3.** Nucleus—through invagination of the plasma membrane. Mitochondria and chloroplast—by engulfing prokaryotic cells. **(3.3) 1.** Structure is a phospholipid bilayer with attached or embedded proteins (fluid-mosaic model), function is to keep cells intact and selectively allow passage of molecules and ions; **2.** Diffusion is the random movement of particles from the area of high to low concentration until they are equally distributed; Osmosis is the net movement of water across a semipermeable membrane from an area of high to low concentration; Facilitated transport requires a protein carrier to help molecules move through a membrane according to their concentration gradient; Active transport requires a carrier plus energy; **3.** In addition to diffusion, osmosis, facilitated transport, and active transport, materials can enter cells by invagination of the plasma membrane (endocytosis, phagocytosis, pinocytosis), and leave cells by exocytosis. **(3.4) 1.** Smooth ER—synthesizes lipids; Rough ER—synthesizes proteins and packages them in vesicles; Golgi apparatus modifies lipids and proteins from the ER, sorts and packages them in vesicles; Lysosomes—contains digestive enzymes that break down cell parts or substances entering by vesicles; **2.** Nucleus contains the DNA which directs the synthesis of proteins; ribosomes aid the mRNA in the production of proteins; Rough ER is where proteins are synthesized and folded; **3.** Endomembrane system consists of most organelles needed for the function of the cell; Transport vesicles move molecules from the ER to the Golgi. **(3.5) 1.** Microtubules, actin filaments, intermediate filaments; **2.** Cilia and flagella both contain an inner core of microtubules within a covering of plasma membrane. Cilia function by moving cells slowly in their environment and by sweeping debris. Flagella function in moving a cell very quickly; **3.** Adhesion junctions mechanically attach adjacent cells and prevent overstretching; tight junctions provide a zipperlike barrier between adjacent cells and do not allow movement of molecules between the two cells; Gap junctions are channels used for communication between adjacent cells. **(3.6) 1.** Enzymes speed up chemical reactions; **2.** Glycolysis begins with glucose and ends with two molecules of pyruvate; the citric acid cycle continues the breakdown of glucose and ends with two molecules of ATP; the electron transport chain takes NADH molecules and ends with 32 ATP molecules; **3.** Without glycolysis other, non-carbohydrate sources of energy would have to be used but cells would die quickly; If electron transport chain or citric acid cycle were missing, then glycolysis would be the only means of creating ATP, instead of more than 32 ATP molecules being made, only 2 would be made and the cell could not keep up with its energy demands.

Testing Your Knowledge of the Concepts

11. b; **12.** d; **13.** a; **14.** c; **15.** c; **16.** a; **17.** d; **18.** b; **19.** c; **20.** e; **21.** c; **22. a.** carbohydrate chain; **b.** hydrophilic heads; **c.** phospholipid bilayer; **d.** hydrophobic tails; **e.** filaments of cytoskeleton; **f.** membrane protein; **g.** membrane protein; **h.** cholesterol

Thinking Critically About the Concepts

1. Malfunctioning mitochondria would result in the malfunction of cellular respiration, thus the cell would only produce a small amount of ATP by glycolysis; **2.** Chemical reactions would not occur as quickly; **3.** Lysosomes might start digesting needed substances in the cell, or autodigesting a functioning cell.

Chapter 4

Case Study

1. All four types of tissues; **2.** Skin burns affect homeostasis of fluid levels and pH levels in the body; **3.** The dermis contains blood vessels, nerves, and glands, whereas the epidermis does not.

Check Your Progress

(4.1) 1. Connective tissue, muscular tissue, nervous tissue, and epithelial tissue; **2.** Cancers are classified according to the type of tissue from which they arise; **3.** Because the tissues from which they arise have a higher rate of cell division. **(4.2) 1.** Fibrous connective tissue (adipose tissue and dense fibrous connective tissue), supportive connective tissue (cartilage and bone), and fluid connective tissue (blood and lymph); **2.** Fibrous connective tissue—contains fibroblasts separated from each other, as well as varying numbers of collagen fibers (e.g., densely packed in tendons and ligaments); Supportive connective tissues—bone is the most compact and contains inorganic salts, while cartilage contains cells separated by a solid, yet flexible matrix; Fluid connective tissue—matrix is liquid; **3.** Blood transports nutrients and oxygen to tissue fluid. Lymph absorbs excess tissue fluids. **(4.3) 1.** Skeletal muscle—voluntary movement of body, striated cells with multiple nuclei; Smooth muscle—involuntary control of blood vessels and movement of substances in lumens of organs, spindle-shaped cells each with a single nuclei; Cardiac muscle—involuntary heart contraction, branching striated cells with a single nucleus; **2.** Skeletal muscles—attached to the skeleton; Smooth muscle—blood vessels and walls of the digestive tract; Cardiac muscle—walls of the heart; **3.** These muscles are involuntary because many of their functions are critical to sustain life—it would be difficult to constantly be aware of the array of functions taking place simultaneously in the heart, lungs, digestive and urinary tracts, etc. **(4.4) 1.** Dendrites—receive signals from sensory receptors or other neurons; Cell body—contains most of the cell's cytoplasm and the nucleus; Axon—conducts the nerve impulses; **2.** Microglia—support and nourish neurons, engulf bacterial and cellular debris; Astrocytes—produce hormones; Oligodendrocytes—form myelin sheaths; **3.** Microglia support and nourish neurons, and both cell types receive and transmit nerve impulses. **(4.5) 1.** Main functions are protection, secretion, absorption, excretion, and filtration; **2.** Simple—one layer of cells; stratified—two or more layers of cells; pseudostratified—one layer, but looks like more; squamous—thin, flat cell; cuboidal- cube shaped cell; columnar—column shape cell; transitional changes shape according to function; **3.** Simple squamous— quick diffusion (lungs, capillaries); stratified squamous—thick for protection (mouth, esophagus, vagina). **(4.6) 1.** Epidermis is composed of stratified squamous epithelium—functions in protection; dermis is beneath the dermis and contains dense fibrous connective tissue with collagen and elastic fibers—functions include adding strength to skin, sensory reception, and thermoregulation; **2.** Nails are protective covering; hair provides protection and some insulation; oil glands lubricate skin and inhibit bacteria; sweat glands aid in regulating temperature; **3.** Because the epidermis layer is routinely shed. **(4.7) 1.** Integumentary—protects body, receives sensory input, regulates temperature, synthesizes vitamin D; cardiovascular—transports nutrients and gases; lymphatic and immune—controls fluid balance, defends against infectious disease; digestive—ingests and digest food, absorbs nutrients and eliminates waste; respiratory—gas exchange, regulates pH; urinary—excretes metabolic wastes, control fluid and pH balance; skeletal—supports and protects, stores minerals, locomotion; muscular—maintains posture, moves body and internal organs, produces heat; nervous—receives, stores, and integrates sensory input, initiates motor output, coordinates organ systems; endocrine—produces hormones, coordinates organ systems, regulates metabolism; reproductive—produces and transports gametes, produces sex hormones; **2.** Dorsal cavity includes the cranial and vertebral cavities; ventral cavity contains the thoracic, abdominal, and pelvic cavities; **3.** Mucous membranes are composed of epithelium over loose fibrous connective tissue and are used for protection from bacteria and virus; serous membrane secrete a watery fluid that supports organs and lungs; synovial membranes are composed of loose connective tissue and lines cavities of freely movable joints; meninges are membranes of connective tissue in the dorsal cavity. **(4.8) 1.** The body's ability to maintain a relative constancy of its internal environment by adjusting its physiological processes—fluctuation in internal conditions can result in illness; **2.** Issue electrochemical signals, release hormones, supply oxygen, maintain body temperature, remove waste, maintain adequate nutrient levels, adjust the water-salt and acid-base balance of the blood; **3.** Negative feedback keeps a variable close to a particular value, positive feedback brings about an ever-greater change in the same direction, which assists the body in completing a process.

Testing Your Knowledge of the Concepts

11. a; **12.** d; **13.** d; **14.** c; **15.** b; **16.** a; **17.** c; **18.** d; **19.** d; **20.** d; **21.** b; **22.** b; **23.** b

Thinking Critically About the Concepts

1. c; **2.** Sweat glands, hair follicles, blood vessels and nerve endings are located in the dermis region; **3.** Protection from fluid loss and infection, and a decrease in fluid loss; **4.** Dehydration, susceptibility to infection, heat loss, and pH imbalance; **5.** Sweat glands, hair follicles, collagen, elastic fibers, sensory receptors. Hair follicles and sweat glands may never grow back once they have been destroyed; **6.** Dehydration could affect blood pressure and volume; the nervous system would not receive information from free nerve endings in the dermis.

Chapter 5

Case Study

1. Diastolic pressure occurs during relaxation of the heart ventricles; systolic pressure occurs during contraction. **2.** Blood pressure is regulated by multiple factors, such as nervous and hormonal input, all of which combine to influence the rate and strength of heart contractions, as well as the degree of constriction or dilation of blood vessels. **3.** Factors that contribute to hypertension include being overweight, dietary factors like salt intake, vascular diseases like atherosclerosis, and stress.

Check Your Progress

(5.1) 1. Heart and blood vessels; **2.** Generate blood pressure, transport blood, create exchange at the capillaries, and regulate blood flow; **3.** Lymphatic vessels collect excess tissue fluid and return it to the cardiovascular system. **(5.2) 1.** Arteries, capillaries, veins; **2.** Arteries carry blood away from the heart and have strong walls; capillaries have thin walls, where exchange can occur; veins carry blood to the heart, have thinner walls than arteries, and contain valves; **3.** Areas that need more gas exchange than others, like organs and skeletal muscles in limbs. **(5.3) 1.** Oxygen poor blood begins in the right atrium, through the tricuspid valve into the right ventricle, through the pulmonary trunk into the lungs, drops off carbon dioxide and picks up oxygen, then through pulmonary veins into left atrium, through bicuspid valve into left ventricle then out the aortic valve; **2.** "Lub" occurs when increasing pressure of blood inside a ventricle forces the cusps of the AV valve to slam shut, "dup" occurs when the ventricles relax and blood in the arteries pushes back, causing the semilunar valves to close; **3.** External is cardiac control center of the medulla oblongata and inside the heart is the conduction system of the heart. **(5.4) 1.** Heart rate; **2.** Blood pressure accounts for blood flow in arteries and capillaries, in veins it is skeletal muscle pump, respiratory pump, and valves; **3.** Valves are needed to prevent the backflow of blood in the veins. **(5.5) 1.** Right ventricle→pulmonary trunk→pulmonary arteries→pulmonary capillaries→pulmonary veins→left atrium; **2.** Left ventricle→aorta→mesenteric arteries→digestive tract capillary bed→hepatic portal vein→liver capillary bed→hepatic vein→inferior vena cava→right atrium; **3.** The pulmonary arteries carry oxygen-poor blood while the pulmonary veins carry oxygen-rich blood. **(5.6) 1.** Excess fluid goes into tissue fluid and eventually into the lymphatic vessels; **2.** Simple diffusion due to the capillary membrane; **3.** Osmotic pressure is greater than blood pressure and fluid moves back into the capillary. **(5.7) 1.** Hypertension, stroke, heart attack, aneurysm, heart failure; **2.** Common treatments include drugs to lower blood pressure, nitroglycerin given at onset of heart attack, replace diseased/damaged portion of the vessel, open clogged arteries, dissolve clots, surgery, heart transplant; **3.** Mainly due to diet, exercise, and lifestyle choices.

Testing Your Knowledge of the Concepts

15. a; **16.** e; **17.** b; **18.** d or e; **19.** c; **20.** e; **21.** d; **22.** e; **23.** c; **24. a.** jugular vein; **b.** pulmonary artery; **c.** superior vena cava; **d.** inferior vena cava; **e.** hepatic vein; **f.** hepatic portal vein; **g.** renal vein; **h.** iliac vein; **i.** carotid artery; **j.** pulmonary vein; **k.** aorta; **l.** mesenteric arteries; **m.** renal artery; **n.** iliac artery.

Thinking Critically About the Concepts

1. This is an example of the effect of extrinsic control of the heart. The brain perceives fear and this sends messages to speed the rate and strength of contraction of the heart, while the adrenal glands release stress hormones (e.g. epinephrine) which have a similar effect; 2. A. Hypertension causes the heart to have to pump against an increased load or pressure; it can also damage the fine blood vessels of the kidneys and other organs; b. Hypertension and atherosclerosis are a bad combination because the heart must pump an increased load through narrower openings; c. General factors include eating a healthier diet, getting regular exercise, stopping smoking and other drug abuse, and managing stress. He misdiagnosed his limb numbness as muscle fatigue from exercising; **2.** Diabetes; **3.** To determine the chemical composition of the blood; **4.** Aorta is the largest diameter blood vessel; if blocked no oxygenated blood gets delivered to the body tissues from the heart; **5.** Contractions are occurring too quickly and forcefully; This might cause the heart to beat irregularly and disturb blood flow to the body; **6.** Yes, lymphatic vessels have valves that prevent backflow of lymph.

Chapter 6

Case Study

1. Function in inflammation and immunity; **2.** Bone marrow is where blood cells originate and mature, and thus where most leukemias begin; **3.** The cells are not responding to normal cell cycle signals and so are dividing continuously.

Check Your Progress

(6.1) 1. Transport, defense, clotting, and regulation; **2.** Plasma proteins include albumins (contribute to osmotic pressure and transport organic molecules), globulins (transport and antibodies), and clotting factors (when activated form clots); **3.** Because it consists of cells within a matrix. **(6.2) 1.** The biconcave shape of a RBC increases cell surface area and therefore the tendency for oxygen to diffuse into and out of the cell. RBCs also contain hemoglobin chains with an iron-containing heme group that binds to oxygen; **2.** Pro—they can hold more oxygen molecules, and they are more malleable to move through vessels; Con—they cannot express genes to repair themselves or go through mitosis to divide; **3.** Anemia—iron, B_{12}, or folic acid deficiency, tired, run down feeling; sickle-cell disease— hereditary disease, hemolysis of red blood cells. **(6.3) 1.** To protect against different types of pathogens and immune system disorders; **2.** Neutrophils—granular with a multilobed nucleus, first responders in bacterial infection; Eosinophils—granular with a bilobed nucleus, increase in number during a parasitic worm infection or allergic reaction; Basophils—granular with a U-shaped nucleus, release histamine; Lymphocytes—do not have granules and have nonlobular nuclei, responsible for specific immunity to particular pathogens and their toxins; monocytes—do not have granules and are the largest of the white blood cells, phagocytize pathogens, and stimulate other white blood cells; **3.** Severe combined immunodeficiency—stem cells of white blood cells lack adenosine deaminase; Leukemia—uncontrolled white blood cell proliferation; Infectious mononucleosis—EBV infection of lymphocytes. **(6.4) 1.** Platelets—clump at the site of puncture; Thrombin—activates fibrinogen, forming long threads of fibrin; Fibrin threads—wind around platelets; **2.** Tissue damage triggers platelets to clump and release prothrombin activator, thrombin forms fibrin which winds around the platelet plug trapping RBCs forming a clot; **3.** Thrombocytopenia is low platelet count, thromboembolism is an abnormal clot that blocks a blood vessel, hemophilia is a deficiency in clotting factors. **(6.5) 1a.** Blood types A, B, AB, and O are based on the presence or absence of type A antigen and type B antigen; **2.** Type A can give to type A or AB; **3.** Once an Rh^- woman has been exposed to the Rh antigen (usually during delivery of an Rh^+ baby), she makes anti-Rh antibodies. During subsequent pregnancies these antibodies cross the placenta and destroy the Rh^+ fetal red blood cells. **(6.6) 1.** Exchanges oxygen and CO2 between the lungs and tissues, distributes nutrients from the digestive system, and removes metabolic wastes; **2.** Digestive system—provides molecules for plasma protein and blood cell formation; Urinary system—helps maintain blood pressure, pH and water-salt balance; Muscular system—heart contractions, moves blood; Nervous system—regulates heart contraction and constriction/dilation of blood vessels; Endocrine system—produces hormones that regulate blood pressure, volume, and composition; Respiratory system—gas exchange; Lymphatic system—helps maintain blood volume; Skeletal system—produces blood cells, protects heart, major vessels, and red bone marrow; **3.** Diabetes—blood sugar levels rise above normal; Decrease in blood calcium levels— problems with clotting; Kidney not producing erythropoietin—decrease in the number of red blood cells.

Testing Your Knowledge of the Concepts

10. d; **11.** c; **12.** a; **13.** e; **14.** d; **15.** b; **16.** a; **17.** d; **18.** c; **19.** b; **20.** b; **21.** e; **22.** d; **23.** e

Thinking Critically About the Concepts

1a. Carbon monoxide is found in automobile exhaust. Burning charcoal, or wood in some cases, will also give off CO. Fumes from all of these sources must be properly vented to the outside, and a CO detector should be installed; **1b.** Without oxygen, the cell mitochondria can't metabolize nutrient molecules to form ATP; **2a.** Iron, vitamin B_{12}, and folic acid; **2b.** Good sources of iron include meats, beans, nuts, spinach, and fortified foods. Foods rich in B_{12} include organ meats, fish, shellfish, and dairy products. Folic acid can be found in cereals, baked goods, leafy vegetables, fruits (bananas, melons, lemons), and organ meat; **3.** Hemoglobin is a protein, Amino acids and iron are necessary for hemoglobin synthesis; **4.** The kidneys will secrete erythropoietin after RBCs are lost or destroyed, or RBC production decreases for any reason. Another possibility is increasing demand for oxygen, such as when a marathon runner trains, or a person travels to a high altitude; **5.** Blood packed with too many red blood cells becomes too dense to travel efficiently through smaller blood vessels and capillaries; **6.** Artificial blood would need to be able to carry and release oxygen efficiently in the lungs and tissues, maintain an appropriate osmotic pressure and pH, be able to form clots only when needed, and be chemically inert enough to avoid stimulating an immune reaction.

Chapter 7

Case Study

1. Disease that results when the immune system mistakenly attacks the body's tissues; **2.** Some are common, such as fatigue, muscle soreness, and arthritis, although a butterfly-shaped rash is more specific to lupus; **3.** There appears to be a genetic predisposition but infections, stress, and hormones may play a role.

Check Your Progress

(7.1) 1. Bacteria may grow in host tissues and cause damage, and many release toxins. Viruses infect host cells, either killing them or interfering with normal functions; **2.** Capsules, flagella, fimbriae, pili, and toxins; **3.** Viruses must replicate inside a living cell. **(7.2) 1.** Lymphatic capillaries absorb excess tissue fluid and return it to the bloodstream; **2.** Primary lymphatic (lymphoid) organs are sites of lymphocyte maturation and include red bone marrow; Secondary are sites where lymphocytes react to pathogens and include lymph nodes; **3.** If lymphatic vessels were blocked, excess fluid would accumulate in the tissues. **(7.3) 1.** Skin and mucous membranes, chemical barriers, resident bacteria, inflammatory response, and protective proteins; **2.** Neutrophils and monocytes/macrophages engulf pathogens by phagocytosis; **3.** They "complement" certain immune responses, such as antibodies. **(7.4) 1.** Innate defenses act quickly but less specifically against pathogens; Adaptive defenses respond more slowly, but more specifically to antigens; **2.** Antibodies might bind to virus antigens, if present, on the cell surface; cytotoxic T cells would recognize and kill cells with viral antigens linked to MHC proteins on the surface of infected cells; **3.** Lymphocytes are mainly responsible - B cells produce plasma and memory cells, plasma cells produce antibodies, memory cells produce antibodies in the future, T cells regulate immune responses and produce cytotoxic T cells and helper T cells, cytotoxic T cells kill virus infected cells and cancer cells, helper T cells regulate immunity, and memory T cells kill in the future. **(7.5) 1.** Immunity that occurs naturally through infection or is brought about artificially by medical intervention; Active immunity—infection with a pathogen, immunization; Passive immunity—transfer of IgG antibodies across the placenta, breast feeding, gamma globulin injection; **2.** Active immunity develops after an exposure to a pathogen and passive immunity occurs when an individual is given antibodies or immune cells to combat a disease; **3.** Monoclonal antibodies can be used to target specific antigens on, for example, cancer cells, leading to their destruction; Cytokine therapy typically boosts the activity or production of white blood cells. **(7.6) 1.** Allergies—hypersensitivities to substances that ordinarily would do no harm to the body; Tissue rejection—the recipient's immune system rejects transplanted tissue is not self; Autoimmune disease—cytotoxic T cells or antibodies mistakenly attack the body's own molecules or cells, as if they are foreign antigens; **2.** When an allergen attaches to IgE antibodies on mast cells they release histamine and other chemicals that result in allergic symptoms; **3.** Because antibodies responding to bacterial or viral antigens sometimes react with self antigens, causing inflammation and tissue damage.

Testing Your Knowledge of the Concepts

16. b; **17.** d; **18.** d; **19.** b; **20.** c; **21.** c; **22.** c; **23.** a; **24.** e; **25.** d; **26.** d; **27.** c

Thinking Critically About the Concepts

1. B cells produce the IgE necessary for the allergic reaction; **2.** Anti-inflammatory drugs (e.g., antihistamines, cortisone) can help reduce the symptoms; Injection of small doses of the allergen to help a patient build immunity to the allergen; **3.** Due to the diversity of our genes giving us different immune system strengths and weaknesses, as well as different MHC molecules that bind to various parts of antigens; **4.** Barrier defenses are supposed to prevent the entrance of pathogens to someone's body, similar as a fence keeping intruders off your property; **5.** They should get another shot of anti-venom, because the first shot gave them passive immunity and perhaps prevented them from making an active, long-lasting response to the venom.

Infectious Diseases Supplement

Check Your Progress

(S.1) 1. Outbreak—when an epidemic is confined to a local area, e.g., from contaminated food; epidemic—a disease with more cases than expected in a certain area during a certain period, e.g., new influenza strains; pandemic—global epidemics, e.g., HIV/AIDS; **2.** HIV gp120 spike protein attaches to CD4 on helper T cell or macrophage, followed by fusion, entry, uncoating, reverse transcription, integration, biosynthesis and cleavage, assembly, and budding; **3.** While feeding, female mosquitos inject the parasite into humans. The parasites reproduce asexually within the human, then some develop into the sexual form that is ingested by another mosquito, to complete its life cycle; **4.** Due to mutation or new combinations after two viruses infect the same cell (see Fig. S.9); **(S.2) 1.** Diseases that are newly recognized or that have reappeared after a significant decline in incidence, e.g. avian influenza, SARS, XDR TB, MRSA; **2.** New or increased exposure to animals/insects, changes in human behavior, mutations in pathogens; **3.** Careful use of antibiotics, environmental protection, research, surveillance, and education; **(S.3) 1.** Through mutations in their genetic material, often as a result of the misuse of antibiotics; **2.** Take the entire dose of antibiotics for the entire time prescribed, do not skip doses or discontinue treatment early, do not share antibiotics, do not treat viral infections with antibiotics; **3.** Both are drug-resistant strains of pathogens that can be difficult to treat.

Chapter 8

Case Study

1. Obesity and a sedentary lifestyle contribute to many diseases including gastrointestinal disease; **2.** Heartburn is a burning sensation in the chest that is usually due to acidic stomach contents escaping into the esophagus. Sedentary lifestyle, obesity, over-eating, and gastrointestinal disease can cause heartburn; **3.** Ultrasound, X-rays, PillCam and colonoscopy can aid in the detection of polyps, cancer, IBS, diverticulitis, and IBD.

Check Your Progress

(8.1) 1. Organs—mouth, pharynx, esophagus, stomach, small intestine, large intestine, rectum, anus; accessory organs—salivary glands, liver, gallbladder, pancreas; **2.** Ingestion—the mouth takes in food; digestion—divides food into pieces and hydrolyzes food to molecular nutrients; movement—food is passed along from one organ to the next and indigestible remains are expelled; absorption—unit molecules produced by digestion cross the wall of the GI tract and enter the blood for delivery to cells, elimination—removal of indigestible wastes through the anus; **3.** Lumen, mucosa, submucosa, muscularis, serosa. **(8.2) 1.** Mouth contains a hard and soft palate, salivary glands, and a tongue; teeth have two main divisions, a crown with a layer of enamel and dentin and a root filled with dentin and pulp; pharynx is a hollow tube at the back of the throat connecting the mouth to the esophagus; esophagus is a hollow tube connecting the pharynx to the stomach; **2.** Mechanical digestion—teeth chew food into pieces convenient for swallowing and the tongue moves food around the mouth; chemical digestion—salivary amylase begins the process of digesting starch; **3.** The soft palate moves back to close off the nasal passages, and the trachea moves up under the epiglottis to cover the glottis; this prevents nasal obstructions or choking. **(8.3) 1.** The stomach stores food, initiates the digestion of protein, and controls the movement of chyme into the small intestine; the mucosa has millions of gastric pits, which lead into gastric glands that produce gastric juice; the muscularis contains an oblique layer that allows the stomach to stretch and mechanically break food down; **2.** The small intestine completes digestion using enzymes that digest all types of food and absorb the products of the digestive process; its wall contains villi that have an outer layer of columnar epithelial cells, each containing thousands of microvilli; **3.** Nutrients are not absorbed through the stomach lining because the acidity of the gastric juices (which are neutralized in the small intestine) would continuously lower the blood pH; the wall of the small intestine has microvilli to increase surface area for absorption. **(8.4) 1.** Pancreas—secretes pancreatic juice into the small intestine and insulin into the blood; liver—filters the blood, removes toxic substances, stores iron and vitamins, functions in sugar homeostasis, regulates blood cholesterol; gallbladder—stores bile; **2.** A non-functioning pancreas would not make pancreatic juices to aid in digestion; a non-functioning liver would not produce bile to emulsify fats; if the gallbladder was not functioning bile would not be stored; **3.** Regulation of digestive secretions is important to have the correct product at the correct time to digest the correct nutrient or chemical. Without these, the process of digestion could not occur properly and nutrients would not be able to get into our circulatory system and into our cells. **(8.5) 1.** Cecum and colon—absorb water and vitamins; rectum—form feces; anal canal—defecation; **2.** Large intestine contributes to homeostasis by regulating fluid balance and removing waste; **3.** Without the functions of the large intestine water and vitamin absorption would be deficient. **(8.6) 1.** Carbohydrates are simple or complex ringed structures and are used as an energy source and include products made from refined grains, beans, peas, nuts, fruits, and whole grain products; proteins are long chains of amino acids the body breaks down to make other proteins and are found in meat, eggs, and milk; lipids are fats and oils and contain either saturated or unsaturated fatty acid chains used for energy storage; minerals are elements needed either in larger quantities (calcium, phosphorus) or trace quantities (zinc, iron); vitamins are other organic compounds the body cannot synthesize, like vitamins A,D,E, and K; **2.** Carbohydrates and fats contain much energy, they are consumed in excess, and with our sedentary lifestyle are not burned off in exercise but stored; **3.** A balanced diet is needed to obtain all the chemicals and nutrients required for all cells to function correctly.

Testing Your Knowledge of the Concepts

9. d; **10.** b; **11.** c; **12.** b; **13.** d; **14.** e; **15.** c; **16.** a; **17.** c; **18.** a; **19.** b, c; **20.** a; **21.** d; **22.** e; **23.** a; **24.** c; **25.** b

Thinking Critically About the Concepts

1a. To prepare the food for enzymes to perform chemical digestion, such as chewing does. The smaller stomach from bariatric surgery no longer performs a significant amount of mechanical digestion; **1b.** Because the stomach now holds only a few ounces at a time; **2.** If stomach is overfilled, stomach contents would flow into the esophagus causing heartburn. If chronic, serious problems can result.

Chapter 9

Case Study

1. Nasal cavity, nasopharynx, oropharynx, laryngopharynx, glottis, larynx; **2.** Because the brain controls breathing and heart rate; **3.** Diet and alcohol affect sleep patterns, as well as excess weight affecting breathing rate and the functions of the upper respiratory system during sleep.

Check Your Progress

(9.1) 1. Nasal cavity→pharynx→glottis→larynx→trachea→bronchus→ bronchioles→lungs; **2.** Inspiration is inhaling air from the atmosphere into the lungs through a series of cavities, tubes and openings; expiration is exhaling air from the lungs to the atmosphere through the same structures; **3.** Ensure that oxygen enters the body and carbon dioxide leaves the body. **(9.2) 1.** Nose—filter, warm, and moisten the air; pharynx—connect nasal and oral cavities to the larynx; larynx—sound production; **2.** Nasal cavity (filters and warms incoming air), oral cavity (where food is received), and larynx (voice box); **3.** Air passage and food passage. **(9.3) 1.** Trachea—keeps lungs clean by sweeping mucus upwards and connects the larynx to the primary bronchi; bronchial tree—passage of air to the lungs; lungs—site of gas exchange between air in the alveoli and blood in the capillaries; **2.** Alveoli are composed of alveolar sacs surrounded by blood capillaries used for gas exchange; **3.** If alveoli do not function properly gas exchange cannot happen and cells will not receive oxygen or be able to remove carbon dioxide and will cease to function. **(9.4) 1.** As the volume (size) of the thoracic cavity increases (when you inhale), the pressure in the lungs decreases; as the volume decreases (when you exhale), the pressure in the lungs increases; **2.** Tidal volume is achieved through normal breathing, vital capacity is the maximum volume that can be achieved, inspiratory and expiratory reserve volume is achieved through forced breathing, and residual volume is the air that remains in the lungs after exhalation; **3.** The body would not be able to get rid of excess carbon dioxide, which would affect pH of the blood, energy metabolism, and other important functions. **(9.5) 1.** The rhythm of ventilation is controlled by a respiratory control center, located in the medulla oblongata. If you breathe any less than that, the amount of gas exchange would be insufficient to support cellular metabolism; **2.** The rhythm of ventilation is controlled by a respiratory control center located in the medulla oblongata, which automatically sends out nerve signals to the diaphragm and the external intercostal muscles of the rib cage, causing inspiration. When signals are not longer sent, the muscles relax and expiration occurs; **3.** As you hold your breath, blood CO_2 increases, which makes the blood more acidic, stimulating the respiratory center to initiate exhalation. **(9.6) 1.** External respiration refers to the exchange of gases between air in the alveoli and blood in the pulmonary capillaries; internal respiration refers to the exchange of gases between the blood in systemic capillaries and the tissue fluid; **2.** In the lungs, Hb takes up oxygen and becomes oxyhemoglobin; in the tissues Hb takes up carbon dioxide and becomes carbaminohemoglobin; **3.** Oxygen moves out of the blood into the tissues because the P_{O2} of the tissues is lower than that of the blood; in the lungs, the opposite is true, so oxygen moves from the air in alveoli into the blood. **(9.7) 1.** Common URT disorders include strep throat (sore throat and high fever), sinusitis (postnasal discharge and facial pain), otitis media (ear pain), laryngitis (sore throat and hoarseness); common LRT disorders include bronchitis (deep cough with mucus), asthma (wheezing and cough), pneumonia (chest pain, fever, and chills), and tuberculosis (nonproductive cough); pulmonary fibrosis (difficulty inflating lungs), and emphysema (ballooning of the chest); **2.** When the common respiratory infections are caused by bacteria, they are treated with antibiotics. **3.** Chronic bronchitis, emphysema, lung cancer.

Testing Your Knowledge of the Concepts

10. d; **11.** d; **12.** c; **13.** f; **14.** d; **15.** a; **16.** b; **17.** e; **18.** d; **19.** b; **20.** d; **21.** b; **22.** e; **23.** d; **24.** b; **25. a.** nasal cavity; **b.** nostril; **c.** pharynx; **d.** epiglottis; **e.** glottis; **f.** larynx; **g.** trachea; **h.** bronchus; **i.** bronchiole

Thinking Critically About the Concepts

1. The enlarged tonsils or adenoids physically obstruct the airway, inhibiting airflow. Surgical removal may be required to restore sufficient passage of air; **2.** The expression describes the movement of food particles past the epiglottis, through the glottis, and into the trachea; **3a.** Ciliated cells are damaged from smoking, so coughing prevents dust, bacteria, and other airborne contaminants from reaching the lungs; **3b.** Their ciliated epithelium is fully functioning; **4.** Diaphragm and abdominal muscles; **5.** The blood O_2 of someone who has nearly drowned is low. When hemoglobin is not bound to O_2, blood is a much darker color and appears bluish because of the diffusion of light by the skin.

Chapter 10

Case Study

1. The kidneys produce urine, excrete metabolic wastes, maintain water-salt and acid-base balance, secrete hormones, and convert vitamin D; **2.** Because urine would not be delivered from the nephrons to the renal pelvis so it could exit the kidney through the ureters; **3.** Blood cells can leak through damaged glomeruli; the kidneys are involved in regulating blood pressure through the secretion of renin.

Check Your Progress

(10.1) 1. Kidneys—primary organs of excretion, ureters—conduct urine from the kidneys to the bladder, urinary bladder—stores urine until it is expelled, urethra—small tube that extends from the urinary bladder to an external opening; **2.** Excretion of metabolic wastes, maintenance of water-salt and acid-base balance, and secretion of hormones; **3.** Waste products would build up to toxic levels, impeding the correct functioning of the cells. **(10.2) 1.** Renal cortex, renal medulla, and renal pelvis; **2.** The glomerulus is a ball of capillaries that facilitates easy passage of small molecules to the glomerular capsule; the proximal convoluted tubule has a large surface area for the reabsorption of filtrate components; the loop of the nephron facilitates reabsorption of water by the nephron; the distal convoluted tubules function in ion exchange; the collecting ducts reabsorb water and carry urine to the renal pelvis; **3.** Because various substances are reabsorbed or secreted into the blood all along the nephron. **(10.3) 1.** Glomerular filtration occurs at the glomerulus inside of the glomerular capsule; tubular reabsorption occurs at the convoluted tubules; tubular secretion occurs along the convoluted tubules; **2.** Glomerular filtration moves water, salts, nutrients, and waste molecules from the blood;. tubular reabsorption reabsorbs nutrients and salt molecules from the glomerular filtrate back into the blood;. tubular secretion secretes certain molecules into the urine; **3.** If glomerular filtration did not occur, many wastes would not be removed from the blood, although some could still be secreted. **(10.4) 1.** Compared to dilute urine, concentrated urine has less water and more solutes; **2.** Aldosterone promotes the excretion of potassium ions and the reabsorption of sodium ions, leading to a reabsorption of water; atrial natriuretic hormone promotes the excretion of sodium, leading to a decrease in water reabsorption; antidiuretic hormone increases water reabsorption; **3.** If the blood is too acidic, hydrogen ions are excreted and bicarbonate ions are reabsorbed; if the blood is too basic, hydrogen ions are not excreted and bicarbonate ions are not reabsorbed. The blood must be at the proper pH for all of the body systems to function properly; **(10.5) 1.** Infections, diabetes, hypertension, and inherited conditions cause renal disease; **2.** Hemodialysis uses an artificial kidney machine to filter the blood, removing wastes and reabsorb needed nutrients and water just like the functions of a kidney; **3.** Because wastes build up rapidly, and water-salt and acid-base balance is essential for homeostasis.

Testing Your Knowledge of the Concepts

11. d; **12.** c; **13.** a; **14.** c; **15.** c; **16.** a; **17.** c; **18.** d; **19.** a; **20.** b; **21.** d; **22.** b; **23.** a; **24.** d

Thinking Critically About the Concepts

1. In diabetes insipidus, a large amount of urine is produced to the lack of ADH-stimulated water reabsorption by the collecting ducts, whereas in diabetes mellitus a large amount of urine is produced due the osmotic effects of excess glucose in the urine; **2.** Failing kidneys often don't produce sufficient erythropoietin to stimulate normal red blood cell production; **3.** It calms the smooth muscle of the bladder, which can also slow smooth muscle contractions in the large intestine, resulting in constipation; **4.** Urinary incontinence; **5.** Damage to the kidneys may result from football tackles or blows from a hard ball (like one used in lacrosse). Players often wear pads designed to protect the ribs and kidneys from damage.

Chapter 11

Case Study

1. It provides flexibility, support, and a smooth surface for joint articulation; **2.** The epiphyses; **3.** Engaging in physical activity after surgery is important; because she is already physically active, her recovery should be easier.

Check Your Progress

(11.1) 1. Compact bone is highly organized and composed of tubular units called osteons, often surrounding a medullary cavity; spongy bone has an unorganized appearance and is composed of trabeculae, which are sometimes filled with red bone marrow; **2.** Hyaline cartilage—uniform and glassy matrix with abundant collagen fibers (the ends of long bones, in the nose, at the ends of the ribs, and in the larynx and trachea); fibrocartilage—wide rows of thick, collagenous fibers (the disks located between the vertebrae and also in the cartilage of the knee); elastic cartilage—matrix contains mostly elastin fibers (the ear flaps and the epiglottis); **3.** Without the skeletal system, movement would be difficult, organs would not be protected, blood cells would not be produced, and there would be no storage of certain materials. **(11.2) 1.** Skull, hyoid bone, vertebral column, rib cage, and the ear ossicles; **2.** The frontal bone forms the forehead, the parietal bones extend to the sides, the occipital bone curves to form the base of the skull, each temporal bone is located below the parietal bones, the sphenoid bone extends across the floor of the cranium from one side to the other, and the ethmoid bone lies in front of the sphenoid. The mandible forms the lower jaw and chin, maxillae form the upper jaw and the anterior portion of the hard palate, zygomatic bones are the cheekbone prominences, and the nasal bones form the bridge of the nose. **3.** cervical vertebrae, including C1 (atlas) and C2 (axis)—located in the neck and allow movement of the head; thoracic vertebrae—form the thoracic curvature and have long, thin, spinous processes and articular facets for the attachment of the ribs; lumbar vertebrae—form the lumbar curvature and have a large body and thick processes; sacral vertebrae—fused together, forming the pelvic curvature; coccyx—fused vertebrae that form the tailbone. **(11.3) 1.** Scapula, clavicle; humerus, radius, ulna, carpals, metacarpals, phalanges; **2.** Two coxal bones; femur, patella, tibia, fibula, tarsals, metatarsals, phalanges; **3.** In the female the iliac bones are more flared, the pelvic cavity is more shallow, but the outlet is wider. **(11.4) 1.** Fibrous, cartilaginous, synovial; **2.** Cartilaginous joints are slightly movable and found in the rib cage and intervertebral disks; fibrous joints are immovable and are found in the skull; **3.** Flexion and extension—knee; adduction and abduction—hip and shoulder; rotation and circumduction—hip and shoulder; inversion and eversion—ankle; **(11.5) 1.** Through intramembranous ossification, in which bone develops between sheets of fibrous connective tissue, and endochondral ossification, in which bone replaces a cartilage model; **2.** A hematoma is formed, next tissue repair begins, and a fibrocartilaginous callus is formed between the ends of the broken bone, then the fibrocartilaginous callus is converted into a bony callus and remodeled; **3.** When blood calcium is low, parathyroid hormone is secreted, causing osteoclasts to dissolve the bone matrix, releasing calcium into the blood. When blood calcium is high, calcitonin from the thyroid gland activates the bone-forming activity of osteoblasts.

Testing Your Knowledge of the Concepts

18. b; **19.** a; **20.** b; **21.** c; **22.** e; **23.** b; **24.** d; **25.** F; **26.** T; **27.** F; **28.** T; **29.** T; **30. a.** frontal bone; **b.** zygomatic bone; **c.** maxilla; **d.** mandible; **e.** clavicle; **f.** scapula; **g.** sternum; **h.** ribs; **i.** costal cartilages; **j.** coxal bones; **k.** sacrum; **l.** coccyx; **m.** patella; **n.** metatarsals; **o.** phalanges; **p.** temporal bone; **q.** vertebral column; **r.** humerus; **s.** ulna; **t.** radius; **u.** carpals; **v.** metacarpals; **w.** femur; **x.** fibula; **y.** tibia; **z.** tarsals

Thinking Critically About the Concepts

1. A balanced diet rich in calcium and vitamin D, and containing the proper amount of protein will speed bone repair. Weight-bearing exercise stimulates bone growth at any age; **2.** Without sunlight, a person produces less vitamin D—thus, less calcium is absorbed from the intestines, and bones become weakened. People spending more time indoors may also be less physically active; **3.** Aging effects many body systems that contribute to bone healing—the circulatory and immune systems are less able to promote inflammation, the endocrine system may produce fewer hormones involved in stimulating healing, and the musculoskeletal system itself is in a state where muscle and bone is being lost, rather than gained; **4.** The typical fast-food diet is deficient in calcium, vitamin D, and other key nutrients; **5.** Believe it or not, it is the player whose fibula is fractured. Bone cells reproduce faster than cells found in cartilage, such as the ligaments that may be torn in a sprain; **6a.** Low blood calcium triggers parathyroid hormone release; **6b.** In hyperparathyroidism, the bones lose too much calcium and become fragile. Calcium deposits form elsewhere in the body, especially in the kidneys; **7.** If the cervical vertebrae are fractured, they can no longer support the weight of the head. If the head falls forward or backward, the spinal cord will stretch and be damaged.

Chapter 12

Case Study

1. An EMG records the electrical signals that control the skeletal muscles of the body—in this case to distinguish between problems with the nerves vs. muscles; **2.** So the muscle fibers can act in unison to perform muscular functions properly; **3.** No coordinated movement, disrupted functions of organs that depend on smooth muscle contractions, heart failure.

Check Your Progress

(12.1) 1. Smooth, cardiac, and skeletal; **2.** Support the body, make bones move, produce heat, increase fluid movement in cardiovascular and lymphatic vessels, protect internal organs and stabilize joints; **3.** Orbicularis oculi, orbicularis oris, pectoralis major, serratus anterior, external oblique, quadriceps femoris, tibialis anterior, extensor digitorum longus, masseter, deltoid, biceps brachii, rectus abdominus, flexor carpi group, adductor longus, sartorius, trapezius, lattisimus dorsi, triceps brachii, extensor carpi group, extensor digitorum, gluteus maximus, biceps femoris, gastrocnemis; **4.** In opposite pairs, for example, one flexes and the other extends. **(12.2) 1.** A myofibril is a bundle of myofilaments that contract; myofilaments are actin (thin) or myosin (thick) filaments whose structure and functions account for muscle striations and contractions; a muscle fiber contains many myofibrils divided into sarcomeres which are contractile; **2.** The actin filaments (thin filaments) slide past the myosin filaments (thick filaments) toward the center. The Z lines have moved and the H zone has gotten smaller, to the point of disappearing; **3.** Calcium binds to troponin exposing myosin binding sites. Myosin uses ATP and does the work of pulling actin towards the center of the sarcomere. **(12.3) 1.** During a latent period the muscle prepares to contract; this is followed by contraction then relaxation; **2.** Stimulation of a muscle by a single electrical signal results in a simple muscle twitch (latent period, contraction, relaxation); repeated stimulation results in summation and tetanus, which creates greater force because the motor unit cannot relax between stimuli; **3.** CP pathway converts ADP to ATP while creatine phosphate is converted to creatine; during fermentation, ATP is produced while glucose is broken down to lactic acid; aerobic respiration uses cellular respiration, glycolysis, citric acid cycle, and the electron transport chain to produce ATP; **4.** Due to weight training, their muscles contain more fast twitch fibers, which fatigue more rapidly than slow twitch. **(12.4) 1.** In a strain, the muscle at a joint is stretched or torn. A sprain results in stretching and tearing of tendons and ligaments at a joint, as well as possible blood vessel and nerve damage; **2.** Myasthenia gravis is an autoimmune disease that results in impaired muscle contraction; myalgia is achy muscles which usually results from overuse; **3.** Progressive degeneration and weakening of muscles due to a loss of muscle fibers. **(12.5) 1.** Muscle movement allows body movement in response to environmental change; it is also necessary for breathing, peristalsis, moving gametes, and childbirth, and it moves fluid in blood and lymph vessels, ureters, and urinary bladder; **2.** Smooth muscles in blood vessels at body surfaces can constrict to conserve heat, also involuntary shivering can produce heat; **3.** Muscle contraction accounts for chewing and peristalsis; the digestive system absorbs nutrients needed for muscle contraction.

Testing Your Knowledge of the Concepts

10. c; **11.** b; **12.** a; **13.** d; **14.** e; **15.** a; **16** d; **17.** d; **18.** a; **19.** a, c; **20.** e; **21.** **a.** T tubule; **b.** sarcoplasmic reticulum; **c.** myofibril; **d.** Z line; **e.** sarcomere; **f.** sarcolemma

Thinking Critically About the Concepts

1a. When dystrophin is absent, calcium leaks into the cell and activates an enzyme that dissolves muscle fibers; **1b.** As muscle fibers die, fat and connective tissue take their place; **2a.** ATP is required for muscle relaxation, so rigor mortis occurs as ATP is depleted. Rapid cooling delays the progression of rigor mortis, because it slows ATP depletion; **2b.** As cold can delay the onset of rigor mortis, heat can accelerate the onset, as can extreme exercise right before death, which depletes ATP stores in the muscles; **3.** Rigor mortis diminishes when enzymes released from dying cells break bonds between actin and myosin; **4.** If the diaphragm and external intercostal muscles fail, a person can no longer breathe.

Chapter 13

Case Study

1. The myelinated nerves will no longer conduct impulses correctly; **2.** Demyelinated areas of the CNS can be seen on an MRI; an SSEP test will show problems with the speed of nerve conduction; **3.** Degeneration of the myelin sheaths covering nerves that carry impulses from the brain and spinal cord to the muscles of the legs causes problems with controlling the legs (and other parts of the body).

Check Your Progress

(13.1) 1. Sensory neurons (take nerve signals from a sensory receptor to the CNS), interneurons (lie entirely within the CNS and communicate with other neurons), and motor neurons (take nerve impulses away from the CNS to an effector); cell body, dendrites, and axon; **2.** An exchange of Na+ and K+ ions generates an action potential along the length of an axon, if the nerve is myelinated, the action potential "jumps" more quickly between nodes of Ranvier; **3.** Neurotransmitters travel across the synapse and bind to receptors. **(13.2) 1.** Provide a means of communication between the brain and the peripheral nerves, also the center for reflex actions; **2.** Cerebrum—main part of the brain, communicates and coordinates the activities of the other parts of the brain; diencephalon—contains the hypothalamus and thalamus, maintains homeostasis, receives sensory input; cerebellum—sends out motor impulses by way of the brain stem to the skeletal muscles, produces smooth, coordinated voluntary movements; brain stem—contains the midbrain, pons, and medulla oblongata, acts as a relay station and medulla has reflex centers; **3.** The RAS regulates a person's alertness, relays sensory signals to higher centers, and filters out unnecessary stimuli, which are important functions in being responsive to one's environment. **(13.3) 1.** A group of brain structures that blends primitive emotions and higher mental functions into a united whole; **2.** Amygdala—fight-or-flight, and hippocampus—learning and memory; **3.** Wernicke's area and Broca's area in the left hemisphere are related to speech, the right hemisphere is associated with more nonverbal and creative functions. **(13.4) 1.** Cranial nerves arise from the brain and there are 12 pairs. Spinal nerves arise from the spine and there are 31 pairs; **2.** A reflex action is fastest because it only passes through the spinal cord, not the brain; **3.** Without the autonomic nervous system, activities of the cardiac muscles, smooth muscles, and glands, would have to be regulated voluntarily. **(13.5) 1.** Drug therapy is used to treat a disease or disorder, drug abuse is using drugs without symptoms of disease or disorder; **2.** Alcohol is a depressant, nicotine is a stimulant;

3. Alcohol acts as a depressant by increasing the action of an inhibitory neurotransmitter called GABA, nicotine stimulates dopamine release, cocaine inhibits dopamine uptake, heroin binds to endorphin receptors, marijuana stimulates anandamide receptors.

Testing Your Knowledge of the Concepts

18. b; **19.** d; **20.** c; **21.** a; **22.** a; **23.** b; **24.** c; **25.** c; **26.** a; **27.** c; **28.** b; **29.** d; **30.** c; **31. a.** central canal; **b.** dorsal horn; **c.** white matter; **d.** cell body of interneuron; **e.** dorsal root ganglion

Thinking Critically About the Concepts

1. Myelin enables the signal to jump from node to node quickly, because the depolarization process occurs only at the node of Ranvier; **2a.** Triglycerides, phospholipids; **2b.** Unsaturated fatty acids (usually liquid at room temperature) are characterized by one or more double bonds between carbons, while saturated fatty acids (usually solid at room temperature) have all single bonds; **2c.** Animal fat, such as butter and fatty cuts of meat; **3.** Myelination enables signals to travel through axons more quickly, which helps motor skills.

Chapter 14

Case Study

1. Inner ear; **2.** Pressure waves moving through the canals cause the basilar membrane, which lies underneath the spiral organ of the cochlea, to vibrate. This causes the stereocilia in the tectorial membrane to bend, which activate nerve impulses that travel through the auditory nerve to the brain; **3.** The technology in a cochlear implant converts sound from the environment into electrical impulses that are sent to the auditory nerves. This type of technology could work on vision if detected light and sent electrical impulses to the nerves controlling vision.

Check Your Progress

(14.1) 1. Chemoreceptors respond to chemical substances, photoreceptors respond to light energy, mechanoreceptors are stimulated by mechanical forces that result in pressure, thermoreceptors are stimulated by changes in temperature; **2.** Sensation occurs when sensory receptors generate a nerve impulse that arrives at the cerebral cortex, perception is the conscious recognition of stimuli; for example, sensation is when nerve impulses from our ears reach the brain, and perception is when we recognize that sound as singing; **3.** Sensory receptors pick up changes in the internal and external environment so the body can respond to those changes and maintain homeostasis. **(14.2) 1.** By detecting the degree of muscle relaxation, the stretch of the tendons, and the movement of ligaments; **2.** Meissner corpuscles (dermis), Krause end bulbs (dermis), Merkel discs (where epidermis meets dermis), and root hair plexus (dermis) are sensitive to fine touch. Pacinian corpuscles (dermis) and Ruffini endings (dermis) are sensitive to pressure; temperature receptors are free nerve endings (epidermis); **3.** Pain receptors are sensitive to chemicals released by damaged tissues; perception of pain allows the body to avoid potential dangers in the environment. **(14.3) 1.** Taste buds on the tongue are chemoreceptors that detect food molecules; olfactory cells within the olfactory epithelium of the nasal cavity are modified neurons that detect odors; **2.** They both respond to chemical stimuli, in the tongue the stimulus is direct, in the nose it can be distant; **3.** Nerve signals generated by taste receptors go to the gustatory cortex in the parietal lobe of the brain where the sensation of taste occurs. In the nasal cavity, odor molecules stimulate olfactory cells to activate neurons in the olfactory bulb of the brain, which sends the information to the cerebral cortex where smells are perceived. **(14.4) 1.** Sclera—protects and supports eyeball; cornea—refracts light rays; pupil—admits light; choroid—absorbs stray light; ciliary body—accommodation; iris—regulates light entrance; retina—contains sensory receptors; rod cells—black and white vision; cone cells—color vision; fovea centralis—acute vision; lens—refracts and focuses light; humors—support eyeball; optic nerve—transmits impulses to brain; **2.** Rod cells function in black and white vision and cone cells function in color vision; **3.** Photoreceptors (rods and cones) synapse with bipolar cells, which synapse with ganglion cells whose axons become the optic nerve. **(14.5) 1.** The outer ear directs sound into middle ear, causing vibrations in the tympanic membrane and the ossicles that attach to the inner ear, where fluid stimulates receptors that generate impulses in nerves sending signals to the brain; **2.** The hair cells located in the spiral organ of the cochlea are mechanoreceptors, which are sensitive to the movements of fluid in the inner ear; **3.** Pressure waves move through the canals causing the basilar membrane to vibrate; this causes the stereocilia embedded in the tectorial membrane to bend, generating nerve impulses that travel to the brain. **(14.6) 1.** For rotational equlibrium—semicircular canals, ampullae, vestibular nerve, cochlea, cupula, stereocilia, hair cells, supporting cells, and endolymph; for gravitational equilibrium—utricle, saccule, otoliths, otolithic membrane, hair cells, supporting cells, and vestibular nerve; **2.** As fluid within the semicircular canal moves and displaces a cupula, the stereocilia of the hair cells bend. This causes a change in the pattern of signals sent to the brain by the vestibular nerve. **3.** Rotational equilibrium occurs when the head is moved side-to-side and gravitational equilibrium occurs when the head is moved up and down. They work together to keep the head, and body, in position according to gravity.

Testing Your Knowledge of the Concepts

14. d; **15.** a; **16.** b; **17.** c; **18.** d; **19.** c; **20.** b; **21.** b; **22.** c; **23.** d; **24.** d; **25.** c; **26.** a; **27.** d; **28. a.** retina; **b.** choroid; **c.** sclera; **d.** optic nerve; **e.** fovea centralis; **f.** ciliary body; **g.** lens; **h.** iris; **i.** pupil; **j.** cornea; **29. a.** tympanic membrane; **b.** auditory canal; **c.** stapes; **d.** incus; **e.** malleus; **f.** oval window; **g.** semicircular canals; **h.** vestibule; **i.** cochlear nerve; **j.** cochlea; **k.** auditory tube; **l.** round window

Thinking Critically About the Concepts

1. Just about the entire sensory system: taste, smell, vision (seeing your pizza!), as well as receptors for temperature and texture in your mouth; **2.** Chemoreceptors also monitor the oxygen and carbon dioxide in the blood; **3.** Vision, because the visual sensory and memory areas are found in the occipital lobe; **4.** Hearing receptors are severely damaged by continual loud noise. Without ear protection, the workers may become deaf; **5.** Both hearing and balance will be affected, sometimes severely; **6.** In addition to brain scans and other diagnostic tests, the physician might observe a loss of various brain functions, including effects on whatever senses that are perceived by the affected area of the cerebral cortex.

Chapter 15

Case Study

1. Mainly insulin and glucagon, although other hormones (e.g., growth hormone, epinephrine, cortisol) also affect blood glucose levels; **2.** Type 1 occurs when the pancreas does not produce sufficient insulin, whereas in type 2, cells become insulin-resistant; **3.** When there is not enough glucose, glucagon releases more from the body's storage areas; when there is too much glucose, insulin causes excess glucose to be stored, lowering blood levels.

Check Your Progress

(15.1) 1. A hormone is a chemical signal that affects the metabolism of a target cell; **2.** Both regulate the activities of other systems he nervous system can respond rapidly to changing stimuli, using neurotransmitters as signals, while endocrine system responses are slower but longer lasting, and use hormones as signals; **3.** Structurally, peptide hormones contain amino acids while steroid hormones are derived from cholesterol; peptide hormones act by binding to surface receptors on target cells while steroid hormones interact with receptors inside cells; **4.** Most peptide hormones can't pass through the plasma membrane, and thus work by interacting with surface receptors which in turn use second messengers to alter cell metabolism. **(15.2) 1.** Through neurotransmitters and hormones—for example, the nervous system sends input to the adrenal medullae, so that a fight-or-flight response can be triggered when needed; meanwhile, several hormones secreted by the endocrine system feed back on the hypothalamus and/or anterior pituitary; **2.** Posterior pituitary does not produce any hormones, but it stores and releases ADH—conserves water, and; oxytocin—stimulates uterine contractions and milk letdown; **3.** TSH—stimulates the thyroid to produce T3 and T4; ACTH—stimulates the adrenal cortex to produce glucocorticoids; Gonadotropic hormones—stimulate the gonads; PRL—causes breast development and milk production; MSH—skin color changes; GH—promotes growth; **4.** Stunted growth, slow metabolism, poor stress response, poorly functioning reproductive organs. **(15.3) 1.** T3 and T4 increase the metabolic rate of all cells of the body; **2.** When blood calcium is high, the thyroid gland secretes calcitonin promoting calcium uptake by the bones and lowering blood calcium; when blood calcium is low, the parathyroid gland secretes parathyroid hormone, causing bones to release calcium, the kidneys to reabsorb calcium and activate vitamin D so the intestines can absorb calcium until the blood calcium levels return to normal; **3.** Hyperthyroidism is usually an oversecretion of T3 and T4, but not calcitonin, therefore overactivity and irritability may result, along with an exophthalmic goiter in some cases; hyperparathyroidism results in osteoporosis and kidney stones due to calcium release from the bones. **(15.4) 1.** Adrenal medulla—epinephrine and norepinephrine; adrenal cortex—glucocorticoids, mineralcorticoids, and sex steroids; **2.** Short term stress response—heart rate and blood pressure increase, blood glucose rises, muscles become energized; long term stress response—increased breakdown of protein and fats instead of glucose, reduction of inflammation, sodium ions and water are reabsorbed by kidney causing blood volume and pressure to increase; **3.** Proper water-salt balance is essential to maintain blood volume and pressure, which is required for proper perfusion of tissues, lung functions, and excretion of wastes by the kidneys. **(15.5) 1.** Exocrine—produces and secretes digestive juices; endocrine—produces and secretes insulin and glucagon; **2.** When blood glucose levels are high, insulin aids in storing excess glucose; when blood glucose levels are low, glucagon releases glucose from storage and delivers to bloodstream; **3.** In type 1 diabetes the pancreas does not produce insulin; in type 2 diabetes, cells become insulin-resistant, therefore the pancreas may secrete more insulin than normal. **(15.6) 1.** Estrogen maintains the secondary sexual characteristics in the female, along with regulating the monthly uterine cycle; testosterone maintains the secondary sexual characteristics in males; **2.** Increased levels of melatonin occur at night and decrease by morning; **3.** The kidneys will secrete erythropoietin, which stimulates red blood cell production. **(15.7) 1.** By detecting input from sensory receptors, the nervous system can respond to both internal and external stimuli and maintain homeostasis; **2.** Aldosterone from the adrenal cortex acts on kidney tubules to conserve sodium and water reabsorption will follow; ADH from the anterior pituitary also increases water reabsorption by the kidneys; **3.** Because of the role of the hypothalamus in each system.

Testing Your Knowledge of the Concepts

12. b; **13.** b; **14.** d; **15.** c; **16.** c; **17.** b; **18.** e; **19.** d; **20.** d; **21.** b; **22.** c; **23.** d; **24.** a; **25.** d; **26.** e; **27.** c; **28.** b

Thinking Critically About the Concept

1. Follicle stimulating hormone and growth hormone are both protein hormones, thus they both bind to a receptor on the plasma membrane and activate a second messenger system (in this case, both use cAMP); **2.** When thyroxine is produced, negative feedback occurs to stop TSH, but when T3 and T4 are low, the anterior pituitary produces more TSH than normal; **3.** When blood glucose is too high, the excess glucose cannot be reabsorbed from the glomerular filtrate in the kidneys; water follows the glucose and an excessive amount of urine is produced, resulting in thirst; **4.** The diet would regulate the intake of glucose by favoring foods with a lower glycemic index (see chapter 8); since type 2 diabetes is associated with obesity, any diet that reduces overall caloric intake might be helpful.

Chapter 16

Case Study

1. The cervix is the opening to the uterus; **2.** Irregularities in the cells of the cervix, including precancerous and cancerous changes; **3.** The surgical removal of the uterus.

Check Your Progress

(16.1) 1. The male reproductive system functions in sperm production and insemination of the female; the female reproductive system functions in oocyte production as well as in providing a supportive environment for embryonic and fetal development; **2.** Mitosis is duplication division (number of chromosomes stays the same), meiosis is reduction division (number of chromosomes is reduced) in order to produce the gametes; **3.** Meiosis occurs in the male testes and the female ovaries. **(16.2) 1.** Organs/structures of the male reproductive system include testes, epididymis, vas deferens, seminal vesicles, prostate gland, urethra, bulbourethral glands, ejaculatory duct, penis. Sperm flows from epididymis to vas deferens to ejaculatory duct to the urethra; **2.** Spermatogonium (diploid)—primary spermatocyte—secondary spermatocyte (haploid)—spermatid—mature sperm; **3.** Testosterone promotes the development of male reproductive organs and maintains the male secondary sexual characteristics. **(16.3) 1a.** ovaries; **1b.** uterine tubes; **1c.** uterus; **1d.** vagina; **2.** The vagina receives the penis during intercourse, acts as the birth canal, and is the exit site for menstrual fluids; the uterus is the site of development of the embryo and fetus; **3.** In contrast to the male, the female needs a separate tract capable of receiving the male gamete and providing a site for embryonic and fetal development. **(16.4) 1.** Estrogen and progesterone control the uterine cycle; FSH and LH control the ovarian cycle; **2.** Corpeus luteum is maintained in the ovary and produces increasing concentrations of progesterone; progesterone shuts down the hypothalamus and anterior pituitary so that no new follicles begin in the ovary during pregnancy. The uterine lining is maintained; **3.** The hormones in birth control pills feedback to inhibit the hypothalamus and the anterior pituitary; therefore, no new follicles begin in the ovary. **(16.5) 1.** Abstinence, vasectomy, tubal ligation, IUD, contraceptive implants, contraceptive injections, birth control pills, diaphragm, condoms, morning-after pills; **2.** Vasectomy prevents sperm from entering the semen. Tubal ligation prevents the egg from getting into the fallopian tubes. In both cases the tubes are surgically closed; **3.** IVF conception occurs outside the body and the embryos are transferred to the uterus. In ICSI a single sperm is injected into an egg. **(16.6) 1.** Pelvic inflammatory disease; **2.** Cancer of the cervix; **3.** Unprotected sex.

Testing Your Knowledge of the Concepts

12a. seminal vesicle; **b.** ejaculatory duct; **c.** prostate gland; **d.** bulbourethral gland; **e.** anus; **f.** vas deferens; **g.** epididymis; **h.** testis; **i.** scrotum; **j.** foreskin; **k.** glans penis; **l.** penis; **m.** urethra; **n.** vas deferens; **o.** urinary bladder. **13.** d; **14.** b; **15. a.** uterine tube; **b.** ovary; **c.** uterus; **d.** glans clitoris;

e. labium minora; **f.** labium majora; **g.** vaginal orifice; **h.** cervix; **i.** vagina; **16.** d; **17.** c; **18.** b; **19.** c; **20.** c; **21.** d; **22.** d; **23.** e; **24.** c

Thinking Critically About the Concepts

1. When body fat composition drops too low, the body may interpret the situation as being unable to support a pregnancy, and shut down reproductive functions; **2.** One would need to collect information about various factors that can influence health, such as weight (BMI), diet, and exercise levels; in addition, one would have to survey environmental exposure to various types of chemicals, as well as other risks, like exposure to radiation or heat; **3.** Birth control pills inhibit FSH, which in turn inhibit follicle development. Fewer follicles may be the cause of a lower risk of ovarian cancer. Because all cancers arise from cells that grow out of control, fertility drugs may increase cancer risk by increasing the amount of cell division in the ovaries.

Chapter 17

Case Study

1. HCG promotes the secretion of progesterone by the corpus luteum to maintain pregnancy; **2.** Some medications have negative effects on embryonic or fetal development; **3.** Changes in size and shape of uterus, weight gain, breast swelling, increased urination, shortness of breath, difficulty sleeping.

Check Your Progress

(17.1) 1. A sperm makes its way through the corona radiata, acrosome releases digestive enzymes to digest zona pellucida, sperm and egg plasma membranes fuse, sperm enters the egg, egg and sperm nuclei fuse; **2.** The corona radiata contains cells that nourish the egg when it is a follicle in the ovary; the zona pellucida lies outside the plasma membrane and protects the egg from polyspermy; **3.** Depolarization of the egg's plasma membrane and changes in the zona pellucida as soon as one sperm touches the membrane. **(17.2) 1.** Morphogenesis refers to the shaping of the embryo; differentiation is when cells take on a specific structure and function; **2.** Chorion—develops into fetal half of the placenta; allantois—forms umbilical blood vessels; yolk sac—produces blood cells; amnion—contains fluid to cushion and protect embyo; **3.** In pre-embryonic development the zygote divides repeatedly, develops into a morula, and then a blastocyst; in embryonic development the embryo implants in the uterine wall, gastrulation occurs and primary germ layers are formed, and organ systems appear and develop. **(17.3) 1.** The path of blood (structures unique to fetus in italics) is *chorionic villi of placenta→umbilical vein→ venous ducts*→inferior vena cava→heart→blood is shunted into the left atrium by way of the oval opening and some enters the aorta by way of the arterial duct→aorta→*umbilical arteries;* **2.** Third and fourth months—skeleton becomes ossified, sex of fetus is distinguishable, fetus grows and gains weight; fifth through seventh months—fetal limbs grow, fetus gains weight; eighth through ninth months—fetus gains weight and rotates in preparation for birth; **3.** The presence of an *SRY* gene, on the Y chromosome, causes development of testes, which secrete three hormones that stimulate other male reproductive organs; without the SRY protein, ovaries and female genitals develop. **(17.4) 1.** hormones produced by the placenta include progesterone, which reduces uterine motility and inhibits the maternal immune response to the fetus; and estrogen, which increases uterine blood flow, renin-angiotensin-aldosterone activity, and liver protein synthesis; **2.** Creates a concentration gradient favorable to the flow of carbon dioxide from fetal blood to maternal blood at the placenta; **3.** Stage 1—cervix dilates; stage 2—baby is born; stage 3—placenta/afterbirth is delivered. **(17.5) 1.** Preprogrammed genetic factors (certain genes can affect life span), damage accumulation (cellular damage that accumulates over time); **2.** Integumentary—skin becomes thinner and less elastic, homeostatic adjustment to heat is limited; cardiovascular—weakening of cardiac muscle, heart valves may leak, arteries become rigid; immune—thymus involutes, B cell responses decrease; digestive—less saliva, declining liver functions; respiratory—decreased lung elasticity; excretory—decreased kidney blood supply, urinary incontinence; nervous—loss of neurons, reduced blood supply to brain; sensory—greater stimulation is needed for sense receptors to function, presbyopia; musculoskeletal—decline in bone density, loss of skeletal muscle mass; reproductive system—testosterone declines in males, females undergo menopause; **3.** Proper diet, exercise, cope effectively with stress, stay engaged mentally and socially.

Testing Your Knowledge of the Concepts

12. c; **13.** c; **14.** b; **15.** a; **16.** b; **17.** d; **18.** d; **19.** b; **20.** c; **21.** d; **22.** a; **23.** e; **24.** d; **25.** e; **26.** e; **27.** c; **28. a.** chorion; **b.** amnion; **c.** embryo; **d.** allantois; **e.** yolk sac; **f.** developing placenta; **g.** maternal blood vessels; **h.** umbilical cord

Thinking Critically About the Concepts

1. HCG is secreted by the chorion; it is filtered from the blood by the nephron during glomerular filtration and is not reabsorbed during tubular reabsorption, so it remains as a urine component; **2a.** HCG is a hormone secreted by the chorion that enters the blood; **2b.** Like other peptide or protein hormones, HCG binds to its membrane receptor and activates a second messenger inside cells; **3a.** FSH; **3b.** Decreased; **3c.** FSH stimulates follicle development and estrogen production by the follicle. Without FSH, follicle development doesn't occur and estrogen secretion does not cause the development of the uterine lining that is shed during menstruation.

Chapter 18

Case Study

1. They delay the continuation of the cell cycle until certain criteria are met to ensure proper control of cell division; **2.** During interphase, because BRCA1 normally inhibits cell division.

Check Your Progress

(18.1) 1. Chromosomes are discrete units of DNA, plus associated proteins, that assist in transmission of genetic information from one generation to the next; **2.** The karyotyping procedure causes the chromosomes of a cell to become visible, so they can be grouped into pairs for examination; **3.** They are exact copies of each other, so that they can split during mitosis, giving rise to two daughter chromosomes. **(18.2) 1.** The cell cycle includes interphase (G1, S, G2) and mitosis (prophase, metaphase, anaphase, telophase) and is how a cell divides. Checkpoints occur at G1, G2 in interphase and during mitosis between metaphase and anaphase; **2.** DNA is duplicated; **3.** Checkpoints prevent the cell from dividing unless the proper signals are present and the DNA is not damaged; **4.** To stimulate/inhibit the cell cycle according to the body's needs. **(18.3) 1.** They are the same; **2.** Prophase—chromosomes attach to the spindle; metaphase—chromosomes align at the equator; anaphase—chromatids separate and chromosomes move toward poles; telophase—nuclear envelopes form around chromosomes; **3.** During cytokinesis a contractile ring (called the cleavage furrow) pinches the cell in two; **4.** Oogenesis—during meiosis I a primary oocyte turns into a secondary oocyte and a polar body; during meiosis II a secondary oocyte, if fertilized, turns into an egg and second polar body. Spermatogenesis—during meiosis I a primary spermatocyte turns into two secondary spermatocytes; during meiosis II secondary spermatocytes turn into four haploid spermatids. **(18.4) 1.** Each daughter cell contains half as many chromosomes as the parent cell; **2.** In prophase I, crossing-over occurs, in metaphase I, independent alignment occurs;

3. Meiosis I—homologous chromosomes pair and then separate; meiosis II—sister chromatids separate, resulting in four cells with a haploid number of chromosomes. **(18.5) 1.** Meiosis I and prophase—pairing of homologous chromosomes; metaphase I—homologous duplicated chromosomes at equator; anaphase I—homologous chromosomes separate; telophase I—two haploid daughter cells. Mitosis and prophase—no pairing of chromosomes; metaphase—duplicated chromosomes at equator; anaphase—sister chromatids separate becoming daughter chromosomes that move to the poles; telophase—two daughter cells identical to the parent; **2.** Meiosis II and prophase II—no pairing of chromosomes; metaphase II—haploid number of chromosomes at equator; anaphase II—sister chromatids separate becoming daughter chromosomes that move to the poles; telophase II—four haploid daughter cells; mitosis and prophase-no pairing of chromosomes; metaphase-duplicated chromosomes at equator; anaphase—sister chromatids separate becoming daughter chromosomes that move to the poles; telophase-two daughter cells identical to parent cell; **3.** Because females must invest much energy into incubating and nourishing a fertilized egg, it would not make biological sense for females to produce as many gametes as human males, who (theoretically) have much less invested. **(18.6) 1.** Nondisjunction—failure of chromosomes to separate correctly during meiosis; **2.** Trisomy 21—three copies of chromosome 21; **3.** Turner syndrome, Klinefelter syndrome, poly-X females, Jacobs syndrome; **4.** Changes in chromosome structure, including deletions, duplications, inversions, and translocations.

Testing Your Knowledge of the Concepts

14. b; **15. a.** G_1 phase—cells grow, organelles double; **b.** S phase—DNA synthesis; **c.** G_2 phase—cell prepares to divide; **d.** mitosis—nuclear division and cytokinesis; **16.** d; **17.** b; **18.** a; **19.** d; **20.** e; **21.** a; **22.** c; **23.** d; **24.** c; **25.** d; **26.** d; **27.** b

Thinking Critically About the Concepts

1. Cell signaling genes that send a cell into the cell cycle or prevent a cell from entering the cell cycle; **2.** It could be that the body responds abnormally only to injury, but that normal mitosis functions well. Injury initiates a number of responses (e.g., inflammation) besides mitosis; **3.** Homologous chromosomes separate randomly. Siblings that look similar inherited similar homologous chromosomes; siblings that look dissimilar inherited different combinations of homologous chromosomes; **4a.** They do not undergo the phases of mitosis to form new cells that would replace the damaged cells; **4b.** They are not likely to recover fully.

Chapter 19

Case Study

1. Cancer cells lack differentiation, have abnormal nuclei, have unlimited potential to replicate, often form tumors, have no need for growth factors, undergo angiogenesis, and sometimes metastasize; **2.** They lack contact inhibition; **3.** To remove the existing tumor from the kidney and to kill any tumor cells that may exist outside the kidney.

Check Your Progress

(19.1) 1. Cancer cells do not undergo apoptosis, have unlimited replicative potential due to telomerase, do not exhibit contact inhibition, have no need for growth factors, and undergo angiogenesis and metastasis; **2.** Mutations in proto-oncogenes result in a gain-of-function that continuously promote the cell cycle, and only one copy of the gene need be mutated; mutations in tumor suppressors genes are loss-of-function mutations that can no longer inhibit the cell cycle, but both gene copies must be mutated; **3.** Lung cancer, colorectal cancer, breast cancer prostate cancer, skin cancer. **(19.2) 1.** DNA-linkage studies have revealed breast cancer genes (*BRCA1* and *BRCA2*), a tumor suppressor gene has been associated with retinoblastoma, and an abnormal RET gene predisposes an individual to thyroid cancer; **2.** Ionizing radiation, tobacco smoke, pollutants, certain viruses; **3.** Don't smoke, eat a healthy, low-fat diet, exercise, use sunscreen. **(19.3) 1.** Change in bowel or bladder habits; a sore that does not heal; unusual bleeding or discharge; thickening or lump in breast or elsewhere; indigestion or difficulty in swallowing; obvious change in wart or mole; nagging cough or hoarseness; **2.** Self-examination, mammography, Pap test, fecal occult blood test, colonoscopy, digital rectal exam, PSA test; **3.** Tumor marker tests detect cancer when in the very early stages when it is treatable; genetic tests alert the patient to the predisposition to develop cancer and so indicate earlier and more frequent screening and preventative behaviors. **(19.4) 1.** Surgery (removal of tumor), radiation therapy (localized exposure to ionizing radiation), chemotherapy (whole body treatment with drugs that inhibit DNA synthesis or damage DNA); **2.** To remove an existing tumor, inhibit growth of new tumors, and make sure no other tumor cells have developed outside of tumor origin; **3.** p53 triggers cell death in many cancer cells but not in normal cells; an adenovirus has been engineered to kill cells that lack a p53 gene; **4.** Antiangiogenic drugs confine and reduce tumors by breaking up the network of new capillaries that are normally recruited by a tumor.

Testing Your Knowledge of the Concepts

9. d; **10.** d; **11.** c; **12.** b; **13.** a; **14.** c; **15.** e; **16.** b; **17.** b; **18.** c; **19.** b; **20.** d; **21.** d; **22.** a; **23.** d; **24.** b;

Thinking Critically About the Concepts

1. Chemotherapy and radiation therapy attack all rapidly reproducing cells, including immune cells, thus Cody's immune responses will be diminished; **2.** These cancers are more likely to spread, because the cells are immature (undifferentiated). In general, the more differentiated cancer cells are, the less likely they are to metastasize; **3.** The renal artery, renal vein, and/or any of the arteries within the kidney; **4.** Low-grade fever, abdominal pain, and "fussiness"—irritability, possibly indicating pain; **5.** To prevent the virus from ever infecting the girl's body; **6.** Active, artificial immunity; **7.** Abstinence, form and remain in a monogamous relationship, and practice safer sex using a condom; **8.** The lymphatic system is often a route used by cancer cells as they spread, so removing lymph nodes will remove cancer cells that may have spread there; **9.** A well-balanced diet, exercise, avoid smoking and excessive alcohol, practice safer sex in a monogamous relationship; **10.** The consequences can be serious, or even fatal, if cancer cells have metastasized to vital organs such as the brain, lungs, or liver, prior to being detected, unless chemotherapy or radiation can successfully eradicate the cells.

Chapter 20

Case Study

1. One that is caused by two recessive alleles. Autosomal recessive, autosomal dominant, sex-linked; **2.** They can be carriers (heterozygous) for a recessive disorder; **3.** In many ways—for example, the genetic disorder phenylketonuria is manifested as the disease phenotype only when an affected (homozygous) individual ingests foods that contain relatively high levels of the amino acid phenylalanine.

Check Your Progress

(20.1) 1. Gene—DNA segment on chromosome that codes for a trait; allele—alternative form of a gene; locus—particular site where a gene is found on a chromosome; chromosome—chromatin condensed into a compact structure; dominant—displays phenotype with only one allele copy; recessive—needs both alleles to express phenotype; heterozygous—having one dominant and one recessive allele; **2.** Genotype refers to the genes of an individual; phenotype is the expressed characteristics of an individual; **3.** The three genotypes are homozygous dominant, heterozygous (which both display the dominant phenotype) and homozygous recessive which displays the recessive phenotype. **(20.2) 1.** Homolo-gous chromosomes separate independently during meiosis I, so each gamete has an equal chance of receiving either allele, therefore all possible combinations of chromosomes and alleles occur in the gametes; **2.** Each child will receive one *e* and one *W* from one parent and one *E* and one *w* from the other parent, thus all children will have the genotype *EeWw*—i.e., they are heterozygous for both traits; **3.** Assuming simple dominance, the child is *ddff,* the father is *DdFf,* and the mother is *ddff;* **(20.3) 1a.** *aa;* **1b.** *AA* or *Aa;* **2.** The chance is nearly 0%, except for the very rare possibility of a spontaneous mutation in the CF gene early in a child's development; **3.** Due to cultural and/or religious beliefs governing marriage and procreating with members of the same culture or religion. **(20.4) 1.** Control of trait by several sets of alleles with an additive effect; **2.** Height—genetics controls the potential height, but without proper diet and nutrition the full potential height would not be reached; **3.** Diversity in alleles allows for more flexible adaptations to the environment. **(20.5) 1.** Mother is X^BX^b and father is X^BY; **2.** Yes, assuming she is heterozygous, she can give her normal X chromosome to her offspring; **3.** Y chromosome has very few genes, mostly ones that determine male gender.

Testing Your Knowledge of the Concepts

11. e; **12.** b; **13.** b; **14.** a; **15.** a; **16.** a; **17.** d; **18.** d; **19.** e; **20.** a

Thinking Critically About the Concepts

1. Yes, unless both parents are homozygous recessive, because affected children can have unaffected parents and heterozygotes have an unaffected phenotype; **2.** No; **3.** X-linked would occur more often than autosomal due to the fact that males only have one X chromosome and will thus exhibit the disorder every time they inherit the X chromosome containing the defective allele; **4.** Call flies *with* the trait "A"; call those *without* the trait "B." Cross "A" flies with "A" flies; cross "B" flies with "B" flies. If the trait is dominant then the "A" parent flies would be either AA or Aa, and the ratio of offspring with the trait to without the trait would be 3:1, and all the offspring of "B" x "B" would lack the trait (because the parents were all "BB"). If the trait was recessive, these ratios would be reversed; **5.** Drosphilia reproduces much faster than humans do—a single pair can produce hundreds of offspring in a couple of weeks, therefore experiments can progress much more rapidly. Additionally, experiments with flies do not raise the types of ethical questions inherent in human experimentation!

Figure Questions

Fig. 20.8: Because she passed on the *a* allele to her first two children. **Fig. 20.9:** Because he passed on the *a* allele to his third child. **Fig. 20.18:** No child could have type O blood.

Chapter 21

Case Study

1. Recombinant DNA technology, PCR; **2.** Gene expression is the production of a protein from DNA which involves the processes of transcription and translation; **3.** Genes are cloned by recombinant DNA techniques, which often involve cutting out DNA sequences with restriction enzymes, then splicing the DNA into a plasmid that is inserted into a bacterium. This is done to produce large amounts of a desired gene product in relatively pure form.

Check Your Progress

(21.1) 1. DNA stores genetic information; RNA aids DNA in the production of proteins. Structure of both include a sugar phosphate backbone and nitrogen containing bases. Differences include double stranded (DNA) or single stranded (RNA), type of sugar can be deoxyribose (DNA) or ribose (RNA), both contain the nitrogen containing bases cytosine, guanine, and adenine, but DNA contains thymine, and RNA contains uracil; **2.** rRNA makes up part of the ribosomal subunits, tRNA transfers amino acids to the ribosomes during translation, mRNA is the complementary copy of DNA produced by transcription, and small RNAs are involved in several processes including splicing and regulation; **3.** During the S phase of the cell cycle, DNA helicase unwinds and "unzips" double-stranded DNA, allowing DNA polymerase to add complimentary nucleotides to each original strand, and ligase to seal any gaps, forming two identical DNA molecules. **(21.2) 1.** Transcription occurs in the nucleus and results in an RNA strand complementary to the DNA gene; translation occurs in the cytoplasm and is the process where ribosomes and tRNA molecules bind to the mRNA strand and make a protein; **2.** The genetic code is how each three-base sequence combines to form 64 possible codons, which code for the 20 amino acids (plus three stop codons) which are used to make proteins. The tRNA molecule will base pair its anticodon region to the codon region of the mRNA when bringing in its correct amino acid; **3.** The universal characteristic of the genetic code demonstrates an evolutionary relationship history among different organisms; **4.** Types of control—pretranscriptional, transcriptional, posttranscriptional, translational, and posttranslational; these are important to homeostasis because they regulate protein production in different types of cells and aid in diversity of proteins produced. **(21.3) 1.** A desired gene is identified and isolated; it is then introduced into a fertilized ovum producing an offspring that synthesizes the desired gene product; **2.** Disease protection, herbicide resistance, salt tolerance, drought tolerance, cold tolerance, improved yield, modified wood pulp, fatty acid/oil content, protein/starch content, amino acid content; **3.** Safety regulation for population health and well-being should be instituted and regulation is the responsibility of all citizens. **(21.4) 1.** Researchers are studying the functions of our 20,500 genes and how they form a human being, as well as what goes wrong in certain disorders. Proteomics and bioinformatics are involved; **2.** Genomics is the study of genomes (DNA sequences), whereas proteomics is the study of proteins (expressed genes); **3.** Gene therapy has a longer lasting result than medication but is limited in capability and very expensive.

Testing Your Knowledge of the Concepts

13. e; **14.** c; **15.** e; **16.** d; **17.** d; **18.** c; **19.** e; **20a.** ACU'CCU'GAA'UGC'AAA; **b.** UGA'GGA'CUU'ACG'UUU; **c.** thr-pro-glu-cys-lys

Thinking Critically About the Concepts

1. It would have exactly the same amino acid sequence as the natural human product; **2.** The product could be contaminated by the growing process, in other words, with bacteria or bacterial products if grown in bacteria. Also, the organism in which it is grown may modify the product in a manner that the human body does not; **3a.** It would be appropriate to use growth hormone in those who are deficient in the natural production. It would be inappropriate to use growth hormone to help normal people grow taller; **3b.** Depending on how early a growth hormone deficiency was diagnosed, it might be difficult to know whether a patient receiving growth hormone has reached his or her maximum growth potential—knowing the height of all immediate family members would allow an estimation of what the patient's natural height should be; **4.** Recombinant proteins that attack the tumor or enhance the immune system, products that inhibit the growth of the tumor in various ways (such as preventing the tumor from developing a blood supply), vaccines to prevent infection by agents that cause cancer; **5.** Regulations in ethics, health and safety should be considered by all citizens.

Chapter 22

Case Study

1. Neandertals were muscular and had a slightly larger brain than modern humans. They manufactured a variety of stone tools. Cro-Magnons had a modern appearance, made advanced tools, and may have been the first to make knife-like blades and spears; **2.** Some recent studies suggest that modern humans have some Neandertal DNA (1-4%). but others indicate that Neandertals are cousins and not ancestors to us; **3.** Out-of-Africa hypothesis proposes that modern humans originated only in Africa then migrated out of Africa and supplanted populations of early *Homo* in Asia and Europe about 100,000 years ago. The multiregional continuity hypothesis proposes that evolution to modern humans was essentially similar in several different places with each region showing continuity of its own anatomical characteristics.

Check Your Progress

(22.1) 1. Newly formed small organic molecules had to join to form macromolecules that had enzymatic activity, which in turn would lead to self-replication; self-replicating molecules could become surrounded by membranes, forming cells; **2.** Miller's experiment demonstrated that inorganic molecules believed to be present in primitive Earth's environment could combine to form organic molecules; **3.** RNA may have been the only macromolecule needed at that time to progress towards formation of the first cell. **(22.2) 1.** Change in a population or species over time. Its two most important aspects are descent from a common ancestor and adaptation to the environment; **2.** Fossil, biogeographical, anatomical, and biochemical; **3.** Because variation, competition, and adaptation account for the great diversity of living things. **(22.3) 1.** Mobile forelimbs and hindlimbs, binocular vision, large complex brains, reduced reproductive rate, grasping hands, flattened face; **2.** Binocular vision allows for depth perception and the ability to make accurate judgments about distance and position. Complex brain with many foldings allows for many association areas; **3.** Human spine exits from the skull's center while ape spine exits from rear of skull; human spine is S-shaped while ape spine has a slight curve; human pelvis is bowl-shaped while ape pelvis is longer and more narrow; human femurs angle inward to the knees while ape femurs angle out; human knee can support more weight than ape knee; human foot has an arch while ape foot does not; **(22.4) 1.** Hominids include apes, chimpanzees, humans, and all extinct close relatives; hominins include all members of genus *Homo* and their close relative. **2.** Bipedal posture, shape of face and size of brain; **3.** The study of Lucy provided important information regarding structure and function of Australopithecines, the earliest known hominins. **(22.5) 1.** *Australopithecus afarensis* (Lucy); **2.** *Homo ergaster*; **3.** Larger brain size, smaller teeth, enlarged portions of the brain associated with speech, culture; **4.** Larger brain, flatter face, used fire, fashioned advanced tools; **5.** *H. sapiens* evolved from *H. ergaster* only in Africa and thereafter *H. sapiens* migrated to Europe and Asia. These findings suggest that we all descend from a few individuals and are more genetically similar; **6.** A thoroughly modern appearance, advanced stone tools, possibly first with a language, and culture that included art.

Testing Your Knowledge of the Concepts

12. c; **13.** b; **14.** e; **15.** e; **16.** e; **17.** a; **18.** b; **19.** c; **20.** b; **21.** d; **22.** b; **23. a.** chordata; **b.** class; **c.** order; **d.** hominidae; **e.** genus; **f.** *Homo sapiens;* **24.** b; **25.** a; **26.** d; **27.** c

Thinking Critically About the Concepts

1. There are many places to dig on the surface of the Earth where people have inhabited, and without some clues on the surface, it would be difficult to determine where to begin. Human habitations usually are built near water sources, near food sources, on high places (for protection), or near shelter (like a cave). Archaeologists look for unusual dirt structures, such as mounds or evidence of digging/holes. Sometimes aerial surveys and infrared photographs provide evidence of unusual structures that are not visible from the ground; **2.** Type of pottery, decorative items, jewelry, burial artifacts, monuments and memorials, seeds, tools; **3.** Brain case size can help determine intellectual capacity, location of the spine's exit from the skull determines whether upright stance, location of the eye sockets determines whether forward vision, size and shape of the jaw determine food source.

Chapter 23

Case Study

1. The concentration of pollutants is increased due to biological magnification, as many of these toxins accumulate in tissues; **2.** Examples of human interference include the withdrawal of water from aquifers, clearing vegetation from land, and adding of pollutants, all of which interfere with the water cycle, as well as interference with carbon cycling by the burning of fossil fuels and destruction of forests, resulting in increased greenhouse gases in the atmosphere.

Check Your Progress

(23.1) 1. Because the entire biosphere is a place where organisms interact among themselves and with the physical and chemical environment; **2.** Autotrophs require only inorganic nutrients and energy to produce food; heterotrophs need a source of organic nutrients and they must consume food; **3.** Herbivores, carnivores, omnivores, detritus feeders; **4.** Energy flow—as nutrients pass from one population to another, all the energy is eventually converted into heat; chemical cycling—inorganic nutrients are returned to the producers from the atmosphere or soil. **(23.2) 1.** A grazing food web begins with photosynthetic producers such as trees and grass; a detrital food web begins with decomposers such as bacteria, fungi, and earthworms; **2.** Food chain; **3.** 1.5 g/m^2. **(23.3) 1.** Reservoir, exchange pool, biotic community; **2.** Carbon, water, and nitrogen are examples of gaseous biogeochemical cycles and phosphorous is an example of a sedimentary biogeochemical cycle; **3.** Certain bacteria convert (fix) nitrogen gas to ammonium; some bacteria can carry out nitrification, meaning they convert ammonium or nitrate in soil to nitrate, which plants can use; **4.** Ground water shortage, climate change, acid deposition, cultural eutrophication; **5.** Increased climate warming, sea levels increase, glaciers melt, increased rainfall along coastlines, dryer inland conditions.

Testing Your Knowledge of the Concepts

8. c; **9.** b; **10.** e; **11.** f; **12.** b, c, d; **13.** a, b, c; **14.** d; **15.** b

Thinking Critically About the Concepts

1a. Fungicides will kill the fungi and pesticides will kill the insects that are part of the detritus food webs; **1b.** Fungicides kill fungi that decompose dead/discarded tissues and pesticides kill many insects that are detritivores that recycle dead/discarded tissues. The nutrients in the dead/discarded tissues will not be released without the activities of the fungi and insects causing adverse effects to the cycles; **2.** Legumes like clover and soybeans have nitrogen-fixing bacteria living in their roots. These bacteria supply the plant with a form of nitrogen the plant can use. Nitrogen that is not needed by the plants (extra) is added to the soil around the plant, which enriches the soil for other plants. Fertilizer also adds nitrogen to the soil, but it may run off during heavy rainfalls, which contributes to overgrowth in aquatic habitats; **3a.** Answers will vary, but may include recycling newspapers, cans, and so forth or starting a compost pile for food scraps. The compost can be used to enrich the soil used for gardens; **3b.** Answers will vary, but may include turning off the water while brushing your teeth or running the dishwasher/washing machine with full loads only. Others might include minimizing the use of fertilizer and pesticides that can run off into bodies of water; **4.** Organochlorides accumulate in organisms higher up on a food chain, discontinuing the use of pesticides that contain organochlorides would improve the health of all organisms in food chains.

Chapter 24

Case Study

1. An argument can be made that not only would this be better for long-term sustainability, but it would also be profitable for the businesses, as well; **2.** One possible criticism of this system is that some of the wastes is sold as cattle feed, but raising livestock under intensively managed conditions is one of the most damaging practices for the environment; **3.** One problem with efforts to encourage businesses to develop more sustainable practices is that it can be difficult to measure the value of these practices in short-term, financial units.

Check Your Progress

(24.1) 1. 1.07%; **2.** MDC (United States)—increased medicine, public health, socioeconomic conditions; LDC (Bangladesh)—increased growth, birth rates, deaths, urban populations, decreased availability of water and decreased biodiversity; **3.** Compare the number of young women entering the reproductive years to older women leaving them. If the number of young women is greater, the population will grow. If the number of older women is greater, the population will decline. **(24.2) 1.** Land, water, food, energy, and minerals; **2.** Groundwater—damming rivers and withdrawing water from aquifers; leads to land subsidence and saltwater intrusion; Food resources—planting a few genetic varieties, heavy use of fertilizers, pesticides and herbicides, generous irrigation, and excessive fuel consumption; leads to oil loss, agricultural runoff, energy consumption, susceptibility to pathogens; **3.** Renewable means more can be produced over time, thus benefitting future generations; nonrenewable resources cannot be replaced once used; **4.** Fossil fuels contribute to environmental problems because of the pollutants released when they are burned; **5.** Hazardous wastes may damage the Earth's ozone shield, pose a threat to the health of living things, and raw sewage causes oxygen depletion in lakes and rivers. **(24.3) 1.** Habitat loss, alien species, pollution, overexploitation, and disease; **2.** Direct value—medicinal value, agricultural value, consumptive use value; indirect value—waste disposal, provision of fresh water, prevention of soil erosion, biogeochemical cycles, regulation of climate, ecotourism; **3.** Biodiversity is important to maintain a large variety of alleles present in a gene pool to maintain populations, for human use as well as enjoyment. **(24.4) 1.** Large portion of land used for human purposes, agriculture uses large amounts of nonrenewable energy and creates pollution, more freshwater is used in agriculture than used in homes, almost half of the agriculture yield goes toward feeding animals, decrease in surface water, use of nonrenewable fossil energy, expansion of the population into all regions of the planet; **2.** To make rural areas sustainable: plant cover crops, plant multiuse crops, use low flow irrigation, use precision farming to reduce habitat destruction, use integrated pest management, plant multipurpose trees, restore wetlands, use renewable forms of energy. To make urban areas sustainable: use energy-efficient modes of transport, use solar or geothermal energy to heat buildings, use green roofs, improve storm-water management, plant native grasses for lawns, create greenbelts, revitalize old sections of cities, use more efficient light fixtures, recycle business equipment, use low-maintenance building materials; **3.** Criteria include use value, option value, existence value, aesthetic value, cultural value, scientific and educational value.

Testing Your Knowledge of the Concepts

7. c; **8.** d; **9.** e; **10.** e; **11.** b; **12.** e; **13. a.** habitat loss; **b.** alien species; **c.** pollution; **d.** overexploitation; **e.** disease **14.** e; **15.** a

Thinking Critically About the Concepts

1. Less energy is needed to grow plant material than animals for human consumption; **2a.** Answers will vary; **2b.** Answers will vary. Example: drive less or drive an energy efficient vehicle, use alternative means of heating/cooling a home, purchase locally grown food; **3.** Most items in a landfill are not exposed to decomposers, so the nutrients do not cycle back; **4a.** Answers will vary; **4b.** Answers will vary. Examples: recycling programs, ride sharing/bus use, water/electricity use, events associated with Earth Day.

Glossary

The terms listed below represent the key terms presented in the text. The majority of these are boldfaced when first defined in the text and in the end-of-chapter summaries. For specific page numbers, refer to the Index.

A

absorption Taking in of substances by cells or membranes.

acetylcholine Neurotransmitter active in both the peripheral and central nervous systems.

acetylcholinesterase (AChE) Enzyme that breaks down acetylcholine bound to postsynaptic receptors within a synapse.

acid Molecules that raise the hydrogen ion concentration in a solution and lower its pH numerically.

acid deposition The return to Earth in rain or snow of sulfate or nitrate salts of acids produced by commercial and industrial activities.

acidosis Excessive accumulation of acids in body fluids.

acquired immunodeficiency syndrome (AIDS) Disease caused by HIV and transmitted via body fluids; characterized by failure of the immune system.

acromegaly Condition resulting from an increase in growth hormone production after adult height has been achieved.

acrosome Cap at the anterior end of a sperm that partially covers the nucleus and contains enzymes that help the sperm penetrate the egg.

actin One of two major proteins of muscle; makes up thin filaments in myofibrils of muscle fibers. See also myosin.

actin filament Cytoskeletal filaments of eukaryotic cells composed of the protein actin; also refers to the thin filaments of muscle cells.

action potential Electrochemical changes that take place across the axon membrane; the nerve impulse.

active immunity Resistance to disease due to the immune system's response to a microorganism or a vaccine.

active site Region of an enzyme where the substrate binds and where the reaction occurs.

active transport Movement of a molecule across a plasma membrane from an area of lower concentraition to one of higher concentration; uses a carrier protein and energy.

acute bronchitis Infection of the primary and secondary bronchi.

adaptation Organism's modification in structure, function, or behavior suitable to the environment.

Addison disease Condition resulting from a deficiency of adrenal cortex hormones; characterized by low blood glucose, weight loss, and weakness.

adenine (A) One of four nitrogen bases in nucleotides composing the structure of DNA and RNA.

adipocyte Cell that stores fat

adipose tissue Connective tissue in which fat is stored.

ADP (adenosine diphosphate) Nucleotide with two phosphate groups that can accept another phosphate group and become ATP.

adrenal cortex Outer portion of the adrenal gland; secretes mineralocorticoids, such as aldosterone, and glucocorticoids, such as cortisol.

adrenal gland An endocrine gland that lies atop a kidney; consisting of the inner adrenal medulla and the outer adrenal cortex.

adrenal medulla Inner portion of the adrenal gland; secretes the hormones epinephrine and norepinephrine.

adrenocorticotropic hormone (ACTH) Hormone secreted by the anterior lobe of the pituitary gland that stimulates activity in the adrenal cortex.

aerobic Chemical process that requires oxygen.

afterbirth Placenta and the extraembryonic membranes, which are delivered (expelled) during the third stage of parturition.

agglutination Clumping of red blood cells due to a reaction between antigens on red blood cell plasma membranes and antibodies in the plasma

aging Progressive changes over time, leading to loss of physiological function and eventual death.

agranular leukocyte White blood cell that does not contain distinctive granules.

agricultural runoff Water from precipitation and irrigation that flows over fields into bodies of water or aquifers.

albumin Plasma protein of the blood having transport and osmotic functions.

aldosterone Hormone secreted by the adrenal cortex that decreases sodium and increases potassium excretion; raises blood volume and pressure.

alien species Nonnative species that migrate or are introduced by humans into a new ecosystem; also called exotics.

alkalosis Excessive accumulation of bases in body fluids.

allantois Extraembryonic membrane that contributes to the formation of umbilical blood vessels in humans.

allele Alternative form of a gene; alleles occur at the same locus on homologous chromosomes.

allergen Foreign substance capable of stimulating an allergic response.

allergy Immune response to substances that usually are not recognized as foreign.

alveoli (sing., **alveolus**) Microscopic air sacs in a lung; site of gas exchange.

Alzheimer disease Brain disorder characterized by a general loss of mental abilities.

amino acid Organic molecule having an amino group and an acid group, which covalently bonds to produce peptide molecules.

amnion Extraembryonic membrane that forms an enclosing, fluid-filled sac.

ampulla Base of a semicircular canal in the inner ear.

amygdala Portion of the limbic system that functions to add emotional overtones to memories.

amylase Enzyme that breaks down carbohydrates.

amyotrophic lateral sclerosis (ALS) Chronic, progressive motor neuron disease characterized by the gradual degeneration of the nerve cells resulting in death.

anabolic steroid Synthetic steroid that mimics the effect of testosterone.

anaerobic Chemical reaction that occurs in the absence of oxygen; example are the fermentation reactions.

analogous structure Structure that has a similar function in separate lineages but differs in anatomy and ancestry.

anaphase Mitotic phase during which daughter chromosomes move toward the poles of the spindle.

anaphylactic shock Severe systemic form of allergic reaction involving bronchiolar constriction, impaired breathing, vasodilation, and a rapid drop in blood pressure with a threat of circulatory failure.

androgen General term for a male sex hormone (e.g., testosterone).

anemia Inefficiency in the oxygen-carrying ability of blood due to a shortage of hemoglobin.

aneurysm Saclike expansion of a blood vessel wall.

angina pectoris Condition characterized by thoracic pain resulting from occluded coronary arteries; precedes a heart attack.
angiogenesis Formation of new blood vessels; one characteristic of cancer.
angioplasty Surgical procedure for treating clogged arteries in which a plastic tube is threaded through a major blood vessel toward the heart and then a balloon at the end of the tube is inflated, forcing open the vessel. A stent is then placed in the vessel.
anorexia nervosa Eating disorder characterized by a morbid fear of gaining weight.
anterior pituitary Portion of the pituitary gland controlled by the hypothalamus and that produces seven types of hormones, some of which control other endocrine glands.
anthropoid Group of primates that includes monkeys, apes, and humans.
antibiotic resistance A characteristic of pathogens that causes them to survive treatment with chemicals (antibiotics) that normally would kill them.
antibody Protein produced in response to the presence of an antigen; each antibody combines with a specific antigen.
antibody-mediated immunity Specific mechanism of defense in which plasma cells derived from B cells produce antibodies that combine with antigens.
anticodon Three-base sequence in a tRNA molecule base that pairs with a complementary codon in mRNA.
antidiuretic hormone (ADH) Hormone secreted by the posterior pituitary that increases the permeability of the collecting ducts in a kidney.
antigen Foreign substance, usually a protein or a polysaccharide, that stimulates the immune system to produce antibodies.
antigen-presenting cell (APC) Cell that displays the antigen to the cells of the immune system so they can defend the body against that particular antigen.
anus Outlet of the digestive tract.
aorta Major systemic artery that receives blood from the left ventricle.
apoptosis Programmed cell death involving a cascade of specific cellular events leading to death and destruction of the cell.
appendicular skeleton Portion of the skeleton forming the pectoral girdles and upper extremities and the pelvic girdle and lower extremities.
appendix In humans, small, tubular appendage that extends outward from the cecum of the large intestine; composed of lymphoid tissue.
aquaporin Protein membrane channel through which water can diffuse.
aqueous humor Clear, watery fluid between the cornea and lens of the eye.
aquifer Rock layers that contain water released in appreciable quantities to wells or springs.
ardipithicines Common name for species in the genus *Ardipithicus*, an early hominin.
arteriole Vessel that takes blood from an artery to capillaries.
arteriovenous shunt A pathway, usually abnormal, that connects an artery directly to a vein.
artery Vessel that transports blood from the heart.
articular cartilage Hyaline cartilaginous covering over the articulating surface of the bones of synovial joints.
artificial selection The intentional breeding of organisms for the selection of certain traits.
association area One of several regions of the cerebral cortex related to memory, reasoning, judgment, and emotional feelings.
aster Short, radiating fibers about the centrioles at the poles of a spindle.
asthma Condition in which bronchioles constrict and cause difficulty in breathing.
astigmatism Blurred vision due to an irregular curvature of the cornea or the lens.
atherosclerosis Condition in which fatty substances accumulate abnormally beneath the inner linings of the arteries.
atom Smallest particle of an element that displays the properties of the element.
atomic mass Mass of an atom equal to the number of protons plus the number of neutrons with the nucleus.
atomic number Number of protons within the nucleus of an atom.
ATP (adenosine triphosphate) Nucleotide with three phosphate groups. The breakdown of ATP into ADP + Ⓟ makes energy available for energy-requiring processes in cells.
atrial natriuretic hormone (ANH) Hormone secreted by the heart that increases sodium excretion and, therefore, lowers blood volume and pressure.
atrioventricular (AV) bundle Group of specialized fibers that conduct impulses from the atrioventricular node to the ventricles of the heart; also called AV bundle.
atrioventricular (AV) node Small region of neuromuscular tissue that transmits impulses received from the sinoatrial node to the ventricles.
atrioventricular (AV) valve Valve located between the atrium and the ventricle.
atrium One of the upper chambers of the heart, either the left atrium or the right atrium, that receives blood.
auditory canal Curved tube extending from the pinna to the tympanic membrane.
auditory tube Extension from the middle ear to the nasopharynx that equalizes air pressure on the eardrum; also called the eustachian tube.
australopithecine Any of the early hominins; classified into several species of *Australopithecus.*
autoimmune disease Disease that results when the immune system mistakenly attacks the body's tissues.
autonomic system Branch of the peripheral nervous system that has control over the internal organs; consists of the sympathetic and parasympathetic systems.
autosome Any chromosome other than the sex chromosomes.
autotroph Organism that can capture energy and synthesize organic nutrients from inorganic nutrients.
axial skeleton Portion of the skeleton that supports and protects the organs of the head, the neck, and the trunk.
axon Elongated portion of a neuron that conducts nerve impulses typically from the cell body to the synapse.
axon terminal Small swelling at the tip of one of many endings of the axon.

B

bacillus A rod-shaped bacterial cell.
bacteria Members of one of three domains of life; prokaryotic cells other than archaea with unique genetic, biochemical, and physiological characteristics.
Barr body Dark-staining body (discovered by M. Barr) in the nuclei of female mammals that contains a condensed, inactive X chromosome.
basal nuclei Nerve cells that integrate motor commands to ensure balance and coordination.
base Molecules that lower the hydrogen ion concentration in a solution and raise the pH numerically.
basement membrane Layer of nonliving material that anchors epithelial tissue to underlying connective tissue.
basophil White blood cell with granular cytoplasm; able to be stained with a basic dye.

B cell (B lymphocyte) Lymphocyte that matures in the bone marrow and, when stimulated by the presence of a specific antigen, gives rise to antibody-producing plasma cells.

B-cell receptor (BCR) Molecule on the surface of a B lymphocyte to which an antigen binds.

bicarbonate ion Ion that participates in buffering the blood; the form in which carbon dioxide is transported in the bloodstream.

bile Secretion of the liver temporarily stored and concentrated in the gallbladder before being released into the small intestine, where it emulsifies fat.

bilirubin The yellow-orange bile pigment produced from the breakdown of hemoglobin.

binge-eating disorder Condition characterized by overeating episodes that are not followed by purging.

binomial name Two-part scientific name of an organism. The first part designates the genus, the second part the specific epithet.

biodiversity Total number of species, the variability of their genes, and the communities in which they live.

biogeochemical cycle Circulating pathway of elements such as carbon and nitrogen, involving exchange pools, storage areas, and biotic communities.

biogeography Study of the geographic distribution of organisms.

bioinformatics Computer technologies used to study the genome.

biological evolution Change in life that has taken place in the past and will take place in the future; includes descent from a common ancestor and adaptation to the environment.

biological magnification Process by which substances become more concentrated in organisms in the higher trophic levels of a food web.

biology Scientific study of life.

biomass The number of organisms multiplied by their weight.

biome A major regional community of organisms, characterized primarily by climate and the types of plant species that are present.

biosphere Zone of air, land, and water at the surface of the Earth in which living organisms are found.

biotechnology product Product created by using biotechnology techniques.

biotic potential Maximum reproductive rate of an organism, given unlimited resources and ideal environmental conditions.

bipedal posture Ability to walk upright on two feet; characteristic of the primates.

birth control method Prevents either fertilization or implantation of an embryo in the uterine lining.

birth control pill Oral contraceptive containing a combination of estrogen and/or progesterone.

blastocyst Early stage of human embryonic development that consists of a hollow, fluid-filled ball of cells.

blind spot Region of the retina lacking rods or cones where the optic nerve leaves the eye.

blood Type of connective tissue in which cells are separated by a liquid called plasma.

blood doping Practice of boosting the number of red blood cells in the blood to enhance athletic performance.

blood pressure Force of blood pushing against the inside wall of a vessel; measured as a ratio of systolic to diastolic pressure.

blood transfusion Introduction of whole blood or a blood component directly into the bloodstream.

body mass index (BMI) Calculation that uses an individual's height and weight to determine whether the individual is overweight or obese.

bolus Small lump of food that has been chewed and swallowed.

bone marrow transplant Process that involves the transfer of bone marrow from one individual to another; often used in cancer treatments following chemotherapy.

bone remodeling Ongoing mineral deposits and withdrawals from bone that adjust bone strength and maintain levels of calcium and phosphorus in blood.

brain Enlarged superior portion of the central nervous system located in the cranial cavity of the skull.

brain stem Portion of the brain consisting of the medulla oblongata, pons, and midbrain.

Braxton Hicks contraction Strong, late-term uterine contractions prior to cervical dilation; also called false labor.

Broca's area Region of the frontal lobe that coordinates complex muscular actions of the mouth, tongue, and larynx, making speech possible.

bronchi (sing., **bronchus**) The two major divisions of the trachea leading to the lungs.

bronchiole Smaller air passages in the lungs that begin at the bronchi and terminate in alveoli.

buffer Substance or group of substances that tend to resist pH changes of a solution, thus stabilizing its relative acidity and basicity.

bulbourethral gland Either of two small structures located below the prostate gland in males; each adds secretions to semen.

bulimia nervosa Eating disorder characterized by binge eating followed by purging via self-induced vomiting or use of a laxative.

bursae Saclike, fluid-filled structures, lined with synovial membrane, that occurs near a joint.

bursitis Inflammation of any of the friction-easing sacs called bursae within the knee joint.

C

calcitonin Hormone secreted by the thyroid gland that increases the blood calcium level.

calorie Amount of heat energy required to raise the temperature of 1 g of water 1°C.

cancer Malignant tumor whose nondifferentiated cells exhibit loss of contact inhibition, uncontrolled growth, and the ability to invade tissue and metastasize.

capsule Gelatinous layer surrounding the cells of blue-green algae and certain bacteria.

carbaminohemoglobin Hemoglobin carrying carbon dioxide.

carbohydrate Class of organic compounds that includes monosaccharides, disaccharides, and polysaccharides.

carbonic anhydrase Enzyme in red blood cells that speeds the formation of carbonic acid from the reactants water and carbon dioxide.

carcinogen Environmental agent that causes mutations leading to the development of cancer.

carcinogenesis Development of cancer.

carcinoma Cancer arising in epithelial tissue.

cardiac cycle One complete cycle of systole and diastole for all heart chambers.

cardiac muscle Striated, involuntary muscle found only in the heart.

cardiovascular system Organ system in which blood vessels distribute blood powered by the pumping action of the heart.

carnivore Consumer in a food chain that eats other animals.

carrier Heterozygous individual who has no apparent abnormality but can pass on an allele for a recessively inherited genetic disorder.

carrying capacity Maximum number of individuals of any species that can be supported by a particular ecosystem on a long-term basis.

cartilage Connective tissue in which the cells lie within lacunae separated by a flexible proteinaceous matrix.

cecum Small pouch that lies below the entrance of the small intestine and is the blind end of the large intestine.

cell Smallest unit that displays the properties of life; always contains cytoplasm surrounded by a plasma membrane.

cell body Portion of a neuron that contains a nucleus and from which dendrites and an axon extend.
cell cycle Repeating sequence of cellular events that consists of interphase, mitosis, and cytokinesis.
cell-mediated immunity Specific mechanism of defense in which T cells destroy antigen-bearing cells.
cell theory One of the major theories of biology; states that all organisms are made up of cells and cells come only from pre-existing cells.
cellular respiration Metabolic reactions that use the stored energy in carbohydrates, but also from fats or amino acids, to produce ATP molecules.
cellulose Polysaccharide that is the major complex carbohydrate in plant cell walls.
central nervous system (CNS) Portion of the nervous system consisting of the brain and spinal cord.
centriole Cellular structure, existing in pairs, that possibly organizes the mitotic spindle for chromosomal movement during mitosis and meiosis.
centromere Area of a chromosome where sister chromatids are held together.
centrosome Central microtubule organizing center of cells. In animal cells, it contains two centrioles.
cerebellum Part of the brain located posterior to the medulla oblongata and pons that coordinates skeletal muscles to produce smooth, graceful motions.
cerebral cortex Outer layer of cerebral hemispheres; receives sensory information and controls motor activities.
cerebral hemisphere One of the large, paired structures that together constitute the cerebrum of the brain.
cerebrospinal fluid Fluid found in the ventricles of the brain, in the central canal of the spinal cord, and in association with the meninges.
cerebrum Main part of the brain consisting of two large masses, or cerebral hemispheres; the largest part of the brain in mammals.
cervix Narrow end of the uterus, which projects into the vagina.
chancre Sore that appears on the skin; first sign of syphilis.
checkpoint Regulatory location in the cell cycle at which the cell assesses whether to proceed with cell division.
chemical evolution Increase in the complexity of chemicals over time that could have led to the first protocells.
chemical signal Molecule that brings about a change in a cell, tissue, organ, or individual when it binds to a specific receptor.
chemoreceptor Sensory receptor sensitive to chemical stimuli—for example, receptors for taste and smell.
chlamydia Sexually transmitted disease caused by the bacterium *Chlamydia trachomatis;* can lead to pelvic inflammatory disease.
chlorofluorocarbons (CFCs) Organic compounds containing carbon, chlorine, and fluorine atoms. CFCs, such as Freon, can deplete the ozone shield by releasing chlorine atoms in the upper atmosphere.
cholesterol Form of lipid; structural component of plasma membrane and precursor for steroid hormones.
chondrocyte Type of cell found in the lacunae of cartilage.
chordae tendineae Tough bands of connective tissue that attach the papillary muscles to the atrioventricular valves within the heart.
chorion Extraembryonic membrane that contributes to placenta formation.
chorionic villi Treelike extensions of the chorion that project into the maternal tissues at the placenta.
choroid Vascular, pigmented middle layer of the eye.
chromatin Network of fine threads in the nucleus composed of DNA and proteins.
chromosome Chromatin condensed into a compact structure.
chronic bronchitis Obstructive pulmonary disorder that tends to recur; marked by inflamed airways filled with mucus and degenerative changes in the bronchi, including loss of cilia.
chyme Thick, semiliquid food material that passes from the stomach to the small intestine.
cilia (sing., **cilium**) Short, hairlike projections from the plasma membrane; may be involved in movement.
ciliary body Structure associated with the choroid layer that contains ciliary muscle and controls the shape of the lens of the eye.
circadian rhythm Biological rhythm with a 24-hour cycle.
circumcision Removal of the prepuce (foreskin) of the penis.
cirrhosis Chronic, irreversible injury to liver tissue; commonly caused by frequent alcohol consumption.
citric acid cycle Cycle of reactions in mitochondria that begins with citric acid; it breaks down an acetyl group as CO_2, ATP, NADH, and $FADH_2$ are given off; also called the Krebs cycle.
cleavage Cell division without cytoplasmic addition or enlargement; occurs during the first stage of animal development.
cleavage furrow Indentation that begins the process of cleavage, by which human cells undergo cytokinesis.
climate change Changes in weather patterns over periods of time; currently used to indicate changes in climate (global warming, dessertification, etc.) associated with human activities on the planet.
clonal selection model Concept that an antigen selects which lymphocyte will undergo clonal expansion and produce more lymphocytes bearing the same type of antigen receptor.
cloning Production of identical copies; can be either the production of identical individuals or, in genetic engineering, the production of identical copies of a gene.
clotting Process of blood coagulation, usually when injury occurs.
coagulation Homeostatic mechanism that forms blood clots in response to injury.
coccus Spherical bacterial cell.
cochlea Portion of the inner ear that resembles a snail's shell and contains the spiral organ, the sense organ for hearing.
cochlear nerve Either of two cranial nerves that carry nerve impulses from the spiral organ to the brain; also called the auditory nerve.
codominance Inheritance pattern in which both alleles of a gene are equally expressed.
codon Three-base sequence in mRNA that causes the insertion of a particular amino acid into a protein or termination of translation.
coenzyme Nonprotein organic molecule that aids the action of the enzyme to which it is loosely bound.
collagen fiber White fiber in the matrix of connective tissue; gives flexibility and strength.
collecting duct Duct within the kidney that receives fluid from several nephrons; the reabsorption of water occurs here.
colon Major portion of the large intestine, consisting of the ascending colon, the transverse colon, and the descending colon.
colony-stimulating factor (CSF) Protein that stimulates differentiation and maturation of white blood cells.
color blindness Deficiency in one or more of the three types of cone cells responsible for color vision.
color vision Ability to detect the color of an object; dependent on three types of cone cells.
columnar epithelium Type of epithelial tissue with cylindrical cells.
community Assemblage of populations interacting with one another within the same environment.
compact bone Type of bone that contains osteons consisting of concentric layers of matrix and osteocytes in lacunae.
comparative genomics Study of the evolutionary relationship of species based on differences in the structure of their genomes.
complementary base pairing Hydrogen bonding between particular bases; in DNA, thymine (T) pairs with adenine (A), and guanine (G) pairs with cytosine (C); in RNA, uracil (U) pairs with A, and G pairs with C.
complementary DNA (cDNA) DNA that has been synthesized from mRNA by the action of reverse transcriptase.

complement system Series of proteins in plasma that form a nonspecific defense mechanism against a microbe invasion; it complements the antigen–antibody reaction.

compound Substance having two or more different elements united chemically in a fixed ratio.

conclusion Statement made following an experiment as to whether the results support the hypothesis.

condom For females, a large polyurethane tube with a flexible ring that fits onto the cervix. For males, a sheath used to cover the penis; used as a contraceptive and, if latex, to minimize the risk of transmitting infection

cone cell Photoreceptor in retina of eye that responds to bright light; detects color and provides visual acuity.

congenital hypothyroidism Condition resulting from improper development of the thyroid in an infant; characterized by stunted growth and intellectual disability.

connective tissue Type of tissue that binds structures together, provides support and protection, fills spaces, stores fat, and forms blood cells; adipose tissue, cartilage, bone, and blood are types of connective tissue.

constipation Delayed and difficult defecation caused by insufficient water in the feces.

consumer Organism that feeds on another organism in a food chain; primary consumers eat plants, and secondary consumers eat animals.

contraceptive Medication or device used to reduce the chance of pregnancy.

control group Sample that goes through all the steps of an experiment but lacks the factor or is not exposed to the factor being tested; a standard against which results of an experiment are checked.

convulsion Sudden attack characterized by a loss of consciousness and severe, sustained, rhythmic contractions of some or all voluntary muscles.

cornea Transparent, anterior portion of the outer layer of the eye.

coronary artery Artery that supplies blood to the wall of the heart.

coronary bypass operation Therapy for blocked coronary arteries in which part of a blood vessel from another part of the body is grafted around the obstructed artery.

corpus callosum Bridge of nerve tracts that connects the two cerebral hemispheres.

corpus luteum Yellow body that forms in the ovary from a follicle that has discharged its secondary oocyte; it secretes progesterone and some estrogen.

cortisol Glucocorticoid secreted by the adrenal cortex that responds to stress on a long-term basis; reduces inflammation and promotes protein and fat metabolism.

covalent bond Chemical bond in which atoms share one pair of electrons.

cramp Muscle contraction that causes pain.

cranial nerve Nerve that arises from the brain.

creatinine Nitrogenous waste; the end product of creatine phosphate metabolism.

Creutzfeldt-Jakob disease (CJD) Prion related disease of the nervous system of humans.

Cro-Magnon Common name for first fossils to be designated *Homo sapiens.*

crossing-over Exchange of genetic material between nonsister chromatids of a tetrad during meiosis.

cuboidal epithelium Type of epithelial tissue with cube-shaped cells.

cultural eutrophication Enrichment of water by inorganic nutrients used by phytoplankton. Often, overenrichment caused by human activities leads to excessive bacterial growth and oxygen depletion.

culture Total pattern of human behavior; includes technology and the arts, and depends upon the capacity to speak and transmit knowledge.

Cushing syndrome Condition resulting from hypersecretion of glucocorticoids; characterized by thin arms and legs and a "moon face," and accompanied by high blood glucose and sodium levels.

cutaneous receptor Sensory receptors for pressure and touch found in the dermis of the skin.

cyclic adenosine monophosphate (cAMP) ATP-related compound that acts as the second messenger in peptide hormone transduction; it initiates activity of the metabolic machinery.

cyclin Protein that regularly increases and decreases in concentration during the cell cycle.

cystic fibrosis (CF) A generalized, autosomal recessive disorder of infants and children in which there is widespread dysfunction of the exocrine glands.

cystitis Inflammation of the urinary bladder.

cytokine Type of protein secreted by a T cell that stimulates cells of the immune system to perform their various functions.

cytokinesis Division of the cytoplasm following mitosis and meiosis.

cytoplasm Contents of a cell between the nucleus and the plasma membrane that contains the organelles.

cytosine (C) One of four nitrogen bases in nucleotides composing the structure of DNA and RNA.

cytoskeleton Internal framework of the cell, consisting of microtubules, actin filaments, and intermediate filaments.

cytotoxic T cell T cell that attacks and kills antigen-bearing cells.

D

data Facts or pieces of information collected through observation and/or experimentation.

daughter cell Cell that arises from a parent cell by mitosis or meiosis.

dead air space Volume of inspired air that cannot be exchanged with blood.

defecation Discharge of feces from the rectum through the anus.

deforestation Removal of trees from a forest in a way that continuously reduces the size of the forest.

dehydration reaction Chemical reaction resulting in a covalent bond with the accompanying loss of a water molecule.

dehydroepiandrosterone (DHEA) Precursor to testosterone, the male sex hormone.

delayed allergic response Allergic response initiated at the site of the allergen by sensitized T cells, involving macrophages and regulated by cytokines.

deletion Change in chromosome structure in which the end of a chromosome breaks off or two simultaneous breaks lead to the loss of an internal segment; often causes abnormalities (e.g., cri du chat syndrome).

denaturation Loss of normal shape by an enzyme so that it no longer functions; caused by a less-than-optimal pH or temperature.

dendrite Branched ending of a neuron that conducts signals toward the cell body.

Denisovans Recently discovered species of hominins that existed at about the same time as the Neandertals.

denitrification Conversion of nitrate or nitrite to nitrogen gas by bacteria in soil.

dense fibrous connective tissue Type of connective tissue containing many collagen fibers packed together; found in tendons and ligaments, for example.

dental caries Tooth decay that occurs when bacteria within the mouth metabolize sugar and give off acids that erode teeth; a cavity.

deoxyhemoglobin Hemoglobin not carrying oxygen.

depolarization When the charge inside the axon changes from positive to negative.

dermis Region of skin that lies beneath the epidermis.

desertification Denuding and degrading a once-fertile land, initiating a desert-producing cycle that feeds on itself and causes long-term changes in the soil, climate, and biota of an area.

desmosome Includes arrangement of protein fibers that tightly hold the membranes of adjacent cells together and prevent overstretching.
detrital food chain Straight-line linking of organisms according to who eats whom, beginning with detritus.
detrital food web Complex pattern of interlocking and crisscrossing food chains, beginning with detritus.
detritus feeder Any organism that obtains most of its nutrients from the detritus in an ecosystem.
development Group of stages by which a zygote becomes an organism or by which an organism changes during its life span; includes puberty and aging, for example.
diabetes mellitus Condition characterized by a high blood glucose level and the appearance of glucose in the urine, due to a deficiency of insulin production and failure of cells to take up glucose.
dialysate Material that passes through the membrane in dialysis.
diaphragm 1. Dome-shaped horizontal sheet of muscle and connective tissue that divides the thoracic cavity from the abdominal cavity. 2. A birth control device consisting of a soft rubber or latex cup that fits over the cervix.
diarrhea Excessively frequent bowel movements.
diastole Relaxation period of a heart chamber during the cardiac cycle.
diastolic pressure Arterial blood pressure during the diastolic phase of the cardiac cycle.
diencephalon Portion of the brain in the region of the third ventricle that includes the thalamus and hypothalamus.
differentiation Development of an unspecialized cell into one with a more specialized structure and function.
diffusion Movement of molecules or ions from a region of higher to lower concentration; it requires no energy and stops when the distribution is equal.
digestion Breaking down of large nutrient molecules into smaller molecules that can be absorbed.
digestive system Organ system including the mouth, esophagus, stomach, small intestine, and large intestine (colon) that receives food and digests it into nutrient molecules. Also has associated organs: teeth, tongue, salivary glands, liver, gallbladder, and pancreas.
dihybrid cross Genetic cross that usually involves individuals who are different for two traits; often uses two are heterozygoustraits, producing a 9:3:3:1 phenotypic ratio in their offspring.
diploid (2n) Cell condition in which two of each type of chromosome are present in the nucleus.
disaccharide Sugar that contains two units of a monosaccharide (e.g., maltose).
distal convoluted tubule Final portion of a nephron that joins with a collecting duct; associated with tubular secretion.
diuretic Drug used to counteract hypertension by causing the excretion of water.
diverticulosis A condition in which portions of the digestive tract mucosa have pushed through other layers of the tract forming pouches where food may collect.
DNA (deoxyribonucleic acid) Nucleic acid polymer produced from covalent bonding of nucleotide monomers that contain the sugar deoxyribose; the genetic material of nearly all organisms.
DNA ligase Enzyme that links DNA fragments; used during production of rDNA to join foreign DNA to vector DNA.
DNA replication Synthesis of a new DNA double helix prior to mitosis and meiosis in eukaryotic cells and during prokaryotic fission in prokaryotic cells.
domain The primary taxonomic group above the kingdom level; all living organisms may be placed in one of three domains.
dominant allele Allele that exerts its phenotypic effect in the heterozygote; it masks the expression of the recessive allele.
dopamine Neurotransmitter in the central nervous system.
dorsal-root ganglion Mass of sensory neuron cell bodies located in the dorsal root of a spinal nerve.
double helix Double spiral; describes the three-dimensional shape of DNA.
drug abuse Dependence on a drug, which assumes an "essential" biochemical role in the body following habituation and tolerance.
Duchenne muscular dystrophy Chronic progressive disease affecting the shoulder and pelvic girdles, commencing in early childhood. Characterized by increasing weakness of the muscles, followed by atrophy and a peculiar swaying gait with the legs kept wide apart. Transmitted as an X-linked trait; affected individuals, predominantly males, rarely survive to maturity. Death is usually due to respiratory weakness or heart failure.
duodenum First part of the small intestine where chyme enters from the stomach.
duplication Change in chromosome structure in which a particular segment is present more than once in the same chromosome.

E

ecological footprint A measure of an individual's demand on the Earth's resources and ecosystems.
ecological pyramid Pictorial graph based on the biomass, number of organisms, or energy content of various trophic levels in a food web—from the producer to the final consumer populations.
ecosystem Biological community together with the associated abiotic environment; characterized by energy flow and chemical cycling.
ectopic pregnancy Implantation of the embryo in a location other than the uterus, most often in an uterine tube.
effacement During the first stage of labor, the uterine contractions of labor occur in such a way that the cervical canal slowly disappears as the lower part of the uterus is pulled upward toward the baby's head.
effector Muscle or gland that responds to stimulation.
egg Female gamete having the haploid number of chromosomes fertilized by a sperm, the male gamete.
elastic cartilage Type of cartilage composed of elastic fibers, allowing greater flexibility.
elastic fiber Yellow fiber in the matrix of connective tissue, providing flexibility.
electrocardiogram (ECG) Recording of the electrical activity associated with the heartbeat.
electron Negative subatomic particle, moving about in an energy level around the nucleus of an atom.
electron shell Average location, or energy level, of an electron in an atom.
electron transport chain Passage of electrons along a series of membrane-bound carrier molecules from a higher to lower energy level; the energy released is used for the synthesis of ATP.
element Substance that cannot be broken down into substances with different properties; composed of only one type of atom.
elimination Process of expelling substances from the body.
embolus Moving blood clot that is carried through the bloodstream.
embryo Immature developmental stage not recognizable as a human being.
embryonic development Period of development from the second through eighth weeks.
embryonic disk Stage of embryonic development following the blastocyst stage that has two layers; one layer will be endoderm, and the other will be ectoderm.
emerging diseases Diseases caused by pathogens that are newly recognized in the last 20 years.
emphysema Degenerative lung disorder in which the bursting of alveolar walls reduces the total surface area for gas exchange.

emulsification Breaking up of fat globules into smaller droplets by the action of bile salts or any other emulsifier.

endochondral ossification Ossification that begins as hyaline cartilage and subsequently replaced by bone tissue.

endocrine gland Ductless organ that secretes a hormone or hormones into the bloodstream.

endocrine system Organ system involved in the coordination of body activities; uses hormones as chemical signals secreted into the bloodstream.

endomembrane system A collection of membranous structures involved in transport within the cell.

endometriosis Condition in which cells of the endometrium (uterus) migrate outside of the reproductive tract.

endometrium Mucous membrane lining the interior surface of the uterus.

endoplasmic reticulum (ER) System of membranous saccules and channels in the cytoplasm, often with attached ribosomes.

enzymes Metabolic assistants, usually proteins, that accelerate the rate of chemical reactions.

eosinophil White blood cell containing cytoplasmic granules that stain with acidic dye.

epidemic More cases of a disease than expected in a certain area for a certain period of time.

epidemiology The study of diseases in populations; includes the causes, distribution, and control of these diseases.

epidermis Region of skin that lies above the dermis.

epididymis Coiled tubule next to the testes where sperm mature and may be stored for a short time.

epiglottis Structure that covers the glottis during the process of swallowing.

epinephrine Hormone secreted by the adrenal medulla in times of stress; adrenaline.

epiphyseal plate Growth plate of a long bone; allows for increase in the length of a long bone during childhood

episiotomy Surgical procedure performed during childbirth in which the opening of the vagina is enlarged to avoid tearing.

episodic memory Capacity of brain to store and retrieve information with regard to persons and events.

epithelial tissue Type of tissue that lines hollow organs and covers surfaces; also called epithelium.

erectile dysfunction Failure of the penis to achieve or maintain erection.

erythropoietin (EPO) Hormone, produced by the kidneys, that speeds red blood cell formation.

esophagus Muscular tube for moving swallowed food from the pharynx to the stomach.

essential amino acids Amino acids required in the human diet because the body cannot make them.

essential fatty acid Fatty acid required in the human diet because the body cannot make it.

estradiol Sex hormone; form of estrogen.

estrogen Female sex hormone that helps maintain sex organs and secondary sex characteristics.

eukaryotic cell Type of cell that has a membrane-bound nucleus and membranous organelles.

evaporation Conversion of a liquid or a solid into a gas.

evolution Descent of organisms from common ancestors with the development of genetic and phenotypic changes over time that make them more suited to the environment.

evolutionary tree Diagram that describes the evolutionary relationship of groups of organisms; a common ancestor is presumed to have been present at points of divergence.

excretion Removal of metabolic wastes from the body.

exocrine gland Gland that secretes its product to an epithelial surface directly or through ducts.

exophthalmic goiter Enlargement of the thyroid gland accompanied by an abnormal protrusion of the eyes.

experiment Artificial situation devised to test a hypothesis.

experimental variable Value expected to change as a result of an experiment; represents the factor being tested by the experiment.

expiration Act of expelling air from the lungs; also called exhalation.

expiratory reserve volume Volume of air that can be forcibly exhaled after normal exhalation.

exponential growth Growth at a constant rate of increase per unit of time; can be expressed as a constant fraction or exponent.

external respiration Exchange of oxygen and carbon dioxide between alveoli and blood.

exteroceptor Sensory receptor that detects stimuli from outside the body (e.g., taste, smell, vision, hearing, and equilibrium).

extinction Total disappearance of a species or higher group.

extraembryonic membrane Membrane that is not a part of the embryo but is necessary to the continued existence and health of the embryo.

F

facial tic Involuntary muscle movement of the face.

facilitated transport Use of a plasma membrane carrier to move a substance into or out of a cell from higher to lower concentration; no energy required.

familial hypercholesterolemia Genetic disorder characterized by an inability to remove cholesterol from the bloodstream; predisposes individual to heart attack.

farsighted Vision abnormality due to a shortened shape of the eye causing light rays to focus in back of retina when viewing close objects.

fat Organic molecule that contains glycerol and fatty acids; found in adipose tissue.

fatty acid Molecule that contains a hydrocarbon chain and ends with an acid group.

fermentation Anaerobic breakdown of glucose that results in a gain of two ATP and end products, such as alcohol and lactate.

fertilization Union of a sperm nucleus and an egg nucleus, which creates a zygote.

fetal development Period of development from the ninth week through birth.

fiber Structure resembling a thread; also, plant material that is nondigestible.

fibrin Insoluble protein threads formed from fibrinogen during blood clotting.

fibrinogen Plasma protein that is converted into fibrin threads during blood clotting.

fibroblast Cell in connective tissues that produces fibers and other substances.

fibrocartilage Cartilage with a matrix of strong collagenous fibers.

fibromyalgia Chronic, widespread pain in muscles and soft tissues surrounding joints.

fibrous connective tissue Tissue composed mainly of closely packed collagenous fibers and found in tendons and ligaments.

fimbria (pl., **fimbriae**) Small, bristlelike fiber on the surface of a bacterial cell, which attaches bacteria to a surface.

first messenger Chemical signal, such as a peptide hormone, that binds to a plasma membrane receptor protein and alters the metabolism of a cell because a second messenger is activated.

flagella (sing., **flagellum**) Long extensions of the cytoskeleton (microtubules) that serve to propel a cell through a fluid medium.

fluid-mosaic model Model for the plasma membrance based on the changing location and pattern of protein molecules in a fluid phospholipid bilayer.
focus Bending of light rays by the cornea, lens, and humors so that they converge and create an image on the retina.
follicle Structure in the ovary that produces a secondary oocyte and the hormones estrogen and progesterone.
follicle-stimulating hormone (FSH) Hormone secreted by the anterior pituitary gland that stimulates the development of an ovarian follicle in a female or the production of sperm in a male.
fontanel Membranous region located between certain cranial bones in the skull of a fetus or infant.
food chain Order in which one population feeds on another in an ecosystem, from detritus (detrital food chain) or producer (grazing food chain) to final consumer.
food web In ecosystems, complex pattern of interlocking and crisscrossing food chains.
foramen magnum Opening in the occipital bone of the vertebrate skull through which the spinal cord passes.
formed element Constituent of blood that is either cellular (red blood cells and white blood cells) or at least cellular in origin (platelets).
fossil Any past evidence of an organism that has been preserved in the Earth's crust.
fossil fuel Fuel, such as oil, coal, or natural gas, that is the result of partial decomposition of plants and animals coupled with exposure to heat and pressure for millions of years.
fossil record History of life recorded from remains from the past.
fovea centralis Region of the retina consisting of densely packed cones; responsible for the greatest visual acuity.
fragile X syndrome Most common form of inherited intellecutal disability; results from mutation to a single gene and results in deficiency of a protein critical to brain development.
functional genomics Study of all the nucleotide sequences, including structural genes, regulatory sequences, and noncoding DNA segments, in the chromosomes of an organism.

G

GABA (gamma aminobutyric acid) Major inhibitory neurotransmitter in the CNS.
gallbladder Organ attached to the liver that serves to store and concentrate bile.
gallstone Crystalline bodies formed by concentration of normal and abnormal bile components within the gallbladder.
gamete Haploid sex cell; the egg or sperm, which join in fertilization to form a zygote.
gamma globulin Large proteins found in the blood plasma and on the surface of immune cells, functioning as antibodies (IgG).
ganglia (sing., **ganglion**) Collections of nerve cell bodies found outside the central nervous system in the peripheral nervous system.
gap junction Junction between cells formed by the joining of two adjacent plasma membranes; it lends strength and allows ions, sugars, and small molecules to pass between cells.
gastric gland Gland within the stomach wall that secretes gastric juice.
gastrulation Stage of animal development during which germ layers form, at least in part, by invagination.
gene Unit of heredity existing as alleles on the chromosomes; in diploid organisms, typically two alleles are inherited—one from each parent.
gene cloning Production of one or more copies of the same gene.
genetic engineering Alteration of DNA for medical or industrial purposes.
genotype Genes of an individual for a particular trait or traits; often designated by letters, for example, *BB* or *Aa*.
gerontology Study of aging.
gingivitis Inflammation of the gums in the oral cavity (mouth).
gland Epithelial cell or group of epithelial cells specialized to secrete a substance.
glaucoma Increasing loss of field of vision; caused by blockage of the ducts that drain the aqueous humor, creating pressure buildup and nerve damage.
global warming Predicted increase in the Earth's temperature, due to human activities that promote the greenhouse effect.
globulin Type of protein in blood plasma. There are alpha, beta, and gamma globulins.
glomerular capsule Double-walled cup that surrounds the glomerulus at the beginning of the nephron.
glomerular filtrate Filtered portion of blood contained within the glomerular capsule.
glomerular filtration Movement of small molecules from the glomerulus into the glomerular capsule due to the action of blood pressure.
glomerulus Cluster; for example, the cluster of capillaries surrounded by the glomerular capsule in a nephron, where glomerular filtration takes place.
glottis Opening for airflow in the larynx.
glucagon Hormone secreted by the pancreas that causes the liver to break down glycogen and raises the blood glucose level.
glucocorticoid Type of hormone secreted by the adrenal cortex that influences carbohydrate, fat, and protein metabolism; see cortisol.
glucose Six-carbon sugar that organisms degrade as a source of energy during cellular respiration.
glutamate Major excitatory CNS neurotransmitter.
glycemic index (GI) Blood glucose response of a given food.
glycogen Storage polysaccharide composed of glucose molecules joined in a linear fashion but having numerous branches.
glycolysis Anaerobic breakdown of glucose that results in a gain of two ATP molecules.
Golgi apparatus Organelle, consisting of saccules and vesicles, that processes, packages, and distributes molecules about or from the cell.
gonad Organ that produces gametes; the ovary produces eggs, and the testis produces sperm.
gonadotropic hormone Chemical signal secreted by the anterior pituitary that regulates the activity of the ovaries and testes; principally, follicle-stimulating hormone (FSH) and luteinizing hormone (LH).
gonadotropin-releasing hormone (GnRH) Hormone secreted by the hypothalamus that stimulates the anterior pituitary to secrete follicle-stimulating hormone (FSH) and luteinizing hormone (LH).
gout Joint inflammation caused by accumulation of uric acid.
granular leukocyte White blood cell with prominent granules in the cytoplasm.
gravitational equilibrium Maintenance of balance when the head and body are motionless.
gray matter Nonmyelinated axons and cell bodies in the central nervous system.
grazing food chain Straight-line linking of organisms according to who eats whom, beginning with a producer.
grazing food web Complex pattern of interlocking and crisscrossing food chains that begins with populations of autotrophs serving as producers.
greenhouse effect Reradiation of solar heat toward the Earth because gases, such as carbon dioxide, methane, nitrous oxide, and water vapor, allow solar energy to pass through toward the Earth but block the escape of heat back into space.
greenhouse gases Gases involved in the greenhouse effect.
growth Increase in the number of cells and/or the size of these cells.
growth factor Chemical signal that regulates mitosis and differentiation of cells that have receptors for it; important in such processes as

fetal development, tissue maintenance and repair, and hematopoiesis; sometimes a contributing factor in cancer.

growth hormone (GH) Substance secreted by the anterior pituitary; controls size of individual by promoting cell division, protein synthesis, and bone growth.

growth plate Cartilaginous layer within an epiphysis of a long bone that permits growth of bone to occur.

growth rate A percentage that reflects the difference between the number of persons in a population who are born and the number who die each year.

guanine (G) One of four nitrogen-containing bases in nucleotides composing the structure of DNA and RNA; pairs with cytosine.

H

hair cell Cell with stereocilia (long microvilli) that is sensitive to mechanical stimulation; mechanoreceptor for hearing and equilibrium in the inner ear.

hair follicle Tubelike depression in the skin in which a hair develops.

haploid (n) The n number of chromosomes—half the diploid number; the number characteristic of gametes that contain only one set of chromosomes.

hard palate Bony, anterior portion of the roof of the mouth.

hay fever Seasonal variety of allergic reaction to a specific allergen. Characterized by sudden attacks of sneezing, swelling of nasal mucosa, and often asthmatic symptoms.

heart Muscular organ located in the thoracic cavity whose rhythmic contractions maintain blood circulation.

heart attack Damage to the myocardium due to blocked circulation in the coronary arteries; also called a myocardial infarction.

heartburn Burning pain in the chest that occurs when part of the stomach contents escape into the esophagus.

heart failure Syndrome characterized by distinctive symptoms and signs resulting from disturbances in cardiac output or from increased pressure in the veins.

helper T cell T cell that secretes cytokines that stimulate all types of immune system cells.

hemodialysis Cleansing of blood by using an artificial membrane that causes substances to diffuse from blood into a dialysis fluid.

hemoglobin Iron-containing pigment in red blood cells that combines with and transports oxygen.

hemolysis Rupture of red blood cells accompanied by the release of hemoglobin.

hemophilia Genetic disorder in which the affected individual is subject to uncontrollable bleeding.

hemorrhoid Abnormally dilated blood vessels of the rectum.

hepatic portal vein Vein leading to the liver and formed by the merging blood vessels leaving the small intestine.

hepatic vein Vein that runs between the liver and the inferior vena cava.

hepatitis Inflammation of the liver. Viral hepatitis occurs in several forms.

herbivore Primary consumer in a grazing food chain; a plant eater.

heterotroph Organism that cannot synthesize organic molecules from inorganic nutrients and therefore must take in organic nutrients (food).

heterozygous Possessing unlike alleles for a particular trait.

hexose Six-carbon sugar.

hippocampus Portion of the limbic system where memories are stored.

histamine Substance, produced by basophils in blood and mast cells in connective tissue, that causes capillaries to dilate.

homeostasis Maintenance of normal internal conditions in a cell or an organism by means of self-regulating mechanisms.

hominid Member of the family Hominidae, which contains australopithecines and humans.

hominin An extinct or modern species of humans.

hominine Classification of primates that includes the humans, gorillas, and chimpanzees.

Homo erectus Hominid who used fire and migrated out of Africa to Europe and Asia.

Homo habilis Hominid of 2 MYA who is believed to have been the first tool user.

homologous chromosome Member of a pair of chromosomes that are alike and come together in synapsis during prophase of the first meiotic division.

homologous structure Structure similar in two or more species because of common ancestry.

homologue Member of a homologous pair of chromosomes.

Homo sapiens Modern humans.

homozygous dominant Possessing two identical alleles, such as *AA*, for a particular trait.

homozygous recessive Possessing two identical recessive alleles, such as *aa*, for a particular trait.

hormone A protein or steroid produced by a cell that affects a different cell, the so-called target cell

human chorionic gonadotropin (HCG) Hormone produced by the chorion that functions to maintain the uterine lining.

human immunodeficiency virus (HIV) Virus responsible for AIDS.

human leukocyte antigen (HLA) Protein in a plasma membrane that identifies the cell as belonging to a particular individual and acts as an antigen in other organisms.

Huntington disease Genetic disease marked by progressive deterioration of the nervous system due to deficiency of a neurotransmitter.

hyaline cartilage Cartilage whose cells lie in lacunae separated by a white, translucent matrix containing very fine collagen fibers.

hydrogen bond Weak bond that arises between a slightly positive hydrogen atom of one molecule and a slightly negative atom of another, or between parts of the same molecule.

hydrolysis reaction Splitting of a compound by the addition of water, with the H^+ being incorporated in one fragment and the OH^- in the other.

hydrolyze To break a chemical bond between molecules by insertion of a water molecule.

hydrophilic Type of molecule that interacts with water by dissolving in water and/or forming hydrogen bonds with water molecules.

hydrophobic Type of molecule that does not interact with water because it is nonpolar.

hypertension Elevated blood pressure, particularly the diastolic pressure.

hypothalamic-inhibiting hormone One of many hormones produced by the hypothalamus that inhibits the secretion of an anterior pituitary hormone.

hypothalamic-releasing hormone One of many hormones produced by the hypothalamus that stimulates the secretion of an anterior pituitary hormone.

hypothalamus Part of the brain located below the thalamus that helps regulate the internal environment of the body and produces releasing factors that control the anterior pituitary.

hypothesis Supposition that is formulated after making an observation; it can be tested by obtaining more data, often by experimentation.

I

immediate allergic response Allergic response that occurs within seconds of contact with an allergen, caused by the attachment of the allergen to IgE antibodies.

immune system White blood cells and lymphatic organs that protect the body against foreign organisms and substances and also cancerous cells.

immunity Ability of the body to protect itself from foreign substances and cells, including disease-causing agents.

immunization Use of a vaccine to protect the body against specific disease-causing agents.

immunosuppressive Inactivating the immune system to prevent organ rejection, usually via a drug.

implantation Attachment and penetration of the embryo into the lining of the uterus (endometrium).

incomplete dominance Inheritance pattern in which the offspring has an intermediate phenotype, as when a red-flowered plant and a white-flowered plant produce pink-flowered offspring.

incontinence The inability of the body to control urination or defecation.

incus The middle of three ossicles of the ear that serve to conduct vibrations from the tympanic membrane to the oval window of the inner ear.

infant respiratory distress syndrome Condition in newborns, especially premature ones, in which the lungs collapse because of a lack of surfactant lining the alveoli.

infectious diseases Diseases caused by pathogens such as bacteria, viruses, fungi, parasites, protozoans, and prions.

infectious mononucleosis Acute, self-limited infectious disease of the lymphatic system caused by the Epstein–Barr virus and characterized by fever, sore throat, swelling of lymph nodes and spleen, and the proliferation of monocytes and abnormal lymphocytes.

infertility Inability to have as many children as desired.

inflammatory response Tissue response to injury that is characterized by redness, swelling, pain, and heat.

ingestion The taking of food or liquid into the body by way of the mouth.

initiation Mutation of a single cell that may lead to cancer development.

inner cell mass An aggregation of cells at one pole of the blastocyst, destined to form the embryo proper.

inner ear Portion of the ear consisting of a vestibule, semicircular canals, and the cochlea, where equilibrium is maintained and sound is transmitted.

insertion End of a muscle attached to a movable bone.

inspiration Act of taking air into the lungs; also called inhalation.

inspiratory reserve volume Volume of air that can be forcibly inhaled after normal inhalation.

insulin Hormone secreted by the pancreas that lowers the blood glucose level by promoting the uptake of glucose by cells, and the conversion of glucose to glycogen by the liver and skeletal muscles.

integrase Viral enzyme that enables the integration of viral genetic material into a host cell's DNA.

integration Summing up of excitatory and inhibitory signals by a neuron or by some part of the brain.

integumentary system Organ system consisting of skin and various organs, such as hair, found in skin.

intercalated disk Region that holds adjacent cardiac muscle cells together; disks appear as dense bands at right angles to the muscle striations.

interferon Antiviral agent produced by an infected cell that blocks the infection of another cell.

interkinesis Period between meiosis I and meiosis II, during which no DNA replication takes place.

interleukin Cytokine produced by macrophages and T cells that functions as a metabolic regulator of the immune response.

intermediate filament Ropelike assemblies of fibrous polypeptides in the cytoskeleton that provide support and strength to cells; so called because they are intermediate in size between actin filaments and microtubules.

internal respiration Exchange of oxygen and carbon dioxide between blood and tissue fluid.

interneuron Neuron located within the central nervous system that conveys messages between parts of the central nervous system.

interoceptor Sensory receptor that detects stimuli from inside the body (e.g., pressoreceptors, osmoreceptors, and chemoreceptors).

interphase Cell cycle stage during which growth and DNA synthesis occur when the nucleus is not actively dividing.

interstitial cell Hormone-secreting cell located between the seminiferous tubules of the testes.

intervertebral disk Layer of cartilage located between adjacent vertebrae.

intramembranous ossification Ossification that forms from membranelike layers of primitive connective tissue.

intrauterine device (IUD) Birth control device consisting of a small piece of molded plastic (and sometimes copper) that is inserted into the uterus; believed to alter the uterine environment so that fertilization does not occur.

invasive Describes cells, such as tumor cells, that invade normal cells.

inversion Change in chromosome structure in which a segment of a chromosome is turned around 180°; this reversed sequence of genes can lead to altered gene activity and abnormalities.

invertebrate Animal without a vertebral column or backbone.

in vitro fertilization (IVF) Form of assisted reproductive technology in which the egg is fertilized outside of the body, and then placed within the uterus.

ion Charged particle that carries a negative or positive charge.

ionic bond Chemical bond in which ions are attracted to one another by opposite charges.

iris Muscular ring that surrounds the pupil and regulates the passage of light through this opening.

isotope Atoms with the same atomic number but a different atomic mass due to a differentnumber of neutrons.

J

jaundice Yellowish tint to the skin caused by an abnormal amount of bilirubin (bile pigment) in the blood, indicating liver malfunction.

joint Articulation between two bones of a skeleton.

juxtaglomerular apparatus Structure located in the walls of arterioles near the glomerulus; regulates renal blood flow.

K

kidney Organ in the urinary system that produces and excretes urine.

kingdom One of the categories used to classify organisms; the category above phylum.

kinocilium The largest stereocilium.

L

lacteal Lymphatic vessel in an intestinal villus; it aids in the absorption of lipids.

lactose intolerance Inability to digest lactose because of an enzyme deficiency.

lacunae (sing., **lacuna**) Small pit or hollow cavity, as in bone or cartilage, where a cell or cells are located.

Langerhans cell Specialized epidermal cell that assists the immune system.

lanugo Short, fine hair that is present during the later portion of fetal development.

large intestine Last major portion of the digestive tract, extending from the small intestine to the anus and consisting of the cecum, the colon, the rectum, and the anal canal.

laryngitis Infection of the larynx with accompanying hoarseness.

larynx Cartilaginous organ located between the pharynx and the trachea that contains the vocal cords; also called the voice box.

learning Relatively permanent change in behavior that results from practice and experience.

lens Clear, membranelike structure found in the eye behind the iris; brings objects into focus.

leptin Hormone produced by adipose tissue; acts on the hypothalamus to signal satiety.

leukemia Cancer of the blood-forming tissues leading to the overproduction of abnormal white blood cells.

ligament Tough cord or band of dense fibrous connective tissue that joins bone to bone at a joint.

limbic system Associates various brain centers, including the amygdala and hippocampus; governs learning, memory, and various emotions such as pleasure, fear, and happiness.

lineage Evolutionary line of descent.

lipase Enzyme that digests fats and lipids; secreted by the pancreas.

lipid Class of organic compounds that tends to be soluble only in nonpolar solvents such as alcohol; includes fats and oils.

liver Large, dark red internal organ that produces urea and bile, detoxifies the blood, stores glycogen, and produces the plasma proteins, among other functions.

locus (pl., **loci**) Particular site where a gene is found on a chromosome. Homologous chromosomes have corresponding gene loci.

long-term memory Retention of information that lasts longer than a few minutes.

loop of the nephron Portion of the nephron lying between the proximal convoluted tubule and the distal convoluted tubule that functions in water reabsorption; also called the loop of Henle.

loose fibrous connective tissue Tissue composed mainly of fibroblasts widely separated by a matrix containing collagen and elastic fibers.

lumen Cavity inside any tubular structure, such as the lumen of the digestive tract.

lung cancer Malignant growth that often begins in the bronchi.

lungs Paired, cone-shaped organs within the thoracic cavity; function in internal respiration and contain moist surfaces for gas exchange.

luteinizing hormone (LH) Hormone that controls the production of testosterone by interstitial cells in males and promotes the development of the corpus luteum in females.

lymph Fluid, derived from tissue fluid, that is carried in lymphatic vessels.

lymphatic organ Organ other than a lymphatic vessel that is part of the lymphatic system; includes lymph nodes, tonsils, spleen, thymus, and bone marrow.

lymphatic system Organ system consisting of lymphatic vessels and lymphatic organs that transport lymph and lipids; aids the immune system.

lymph node Mass of lymphatic tissue located along the course of a lymphatic vessel.

lymphocyte Specialized white blood cell that functions in specific defense; occurs in two forms—T cell and B cell.

lymphoma Cancer of lymphatic tissue.

lysosome Membrane-bound vesicle that contains hydrolytic enzymes for digesting macromolecules.

lysozyme Enzyme found in tears, milk, saliva, mucus, and other body fluids that destroys bacteria by digesting their cell walls.

M

macromolecule Extremely large biological molecule; refers specifically to proteins, nucleic acids, polysaccharides, lipids, and complexes of these.

macrophage Large phagocytic cell derived from a monocyte that ingests microbes and debris.

mad cow disease Slowly progressive fatal disease affecting the central nervous system of cattle; transmissible to humans; scientific name is bovine spongiform encephalopathy.

major histocompatibility complex (MHC) Cluster of genes on chromosome 6 concerned with self-antigen production; matching these is critical to success of organ transplants. The MHC includes the human leukocyte antigen (HLA) genes.

malleus The first of three ossicles of the ear that serve to conduct vibrations from the tympanic membrane to the oval window of the inner ear.

Marfan syndrome Congenital disorder of connective tissue characterized by abnormal length of the extremities and weakness of the aorta.

mass An atom's quantity of matter.

mass number The sum of the number of protons and neutrons in the nucleus of an atom.

mast cell Cell to which antibodies, formed in response to allergens, attach, causing it to release histamine, thus producing allergic symptoms.

mastoiditis Inflammation of the mastoid sinuses of the skull.

matrix Unstructured semifluid substance that fills the space between cells in connective tissues or inside organelles.

matter Anything that takes up space and has mass.

mechanoreceptor Sensory receptor that responds to mechanical stimuli, such as that from pressure, sound waves, and gravity.

medulla oblongata Part of the brain stem that is continuous with the spinal cord; controls heartbeat, blood pressure, breathing, and other vital functions.

medullary cavity Cavity within the diaphysis of a long bone containing marrow.

megakaryocyte Large cell that gives rise to blood platelets.

meiosis Type of nuclear division that occurs as part of sexual reproduction in which the daughter cells receive the haploid number of chromosomes in varied combinations.

melanocyte Melanin-producing cell found in skin.

melanocyte-stimulating hormone Substance that causes melanocytes to secrete melanin in lower vertebrates.

melatonin Hormone secreted by the pineal gland that is involved in biorhythms.

memory Capacity of the brain to store and retrieve information about past sensations and perceptions; essential to learning.

memory T cell T cell that differentiates during an initial infection and responds rapidly during subsequent exposure to the same antigen.

meninges (sing., **meninx**) Protective membranous coverings about the central nervous system.

meningitis Condition that refers to inflammation of the brain or spinal cord meninges (membranes).

menopause Termination of the ovarian and uterine cycles in older women.

menstruation Loss of blood and tissue from the uterus at the end of a uterine cycle.

messenger RNA (mRNA) Type of RNA formed from a DNA template that bears coded information for the amino acid sequence of a polypeptide.

metabolism All of the chemical reactions that occur in a cell.

metaphase Mitotic phase during which chromosomes are aligned at the equator of the mitotic spindle.

metastasis Spread of cancer from the place of origin throughout the body; caused by the ability of cancer cells to migrate and invade tissues.

microtubule Small cylindrical structure that contains 13 rows of the protein tubulin around an empty central core; component of the cytoskeleton; present in the cytoplasm, centrioles, cilia, and flagella.

micturition Emptying of the bladder; urination.

midbrain Part of the brain located below the thalamus and above the pons; contains reflex centers and tracts.

middle ear Portion of the ear consisting of the tympanic membrane, the oval and round windows, and the ossicles; where sound is amplified.

mineral Naturally occurring inorganic substance containing two or more elements; certain minerals are needed in the diet.

mineralocorticoid Type of hormone secreted by the adrenal cortex that regulates water–salt balance, leading to increases in blood volume and blood pressure.

mitochondria (sing., **mitochondrion**) Membrane-bound organelle in which ATP molecules are produced during the process of cellular respiration.

mitosis Type of cell division in which daughter cells receive the exact chromosomal and genetic makeup of the parent cell; occurs during growth and repair.

mitotic spindle Microtubule structure that brings about chromosomal movement during nuclear division.

mole A unit of scientific measurement for atoms, ions, and molecules.

molecular clock Mutational changes that accumulate at a presumed constant rate in regions of DNA not involved in adaptation to the environment.

molecule Union of two or more atoms of the same element; also, the smallest part of a compound that retains the properties of the compound.

monoclonal antibody One of many antibodies produced by a clone of hybridoma cells that all bind to the same antigen.

monocyte Type of agranular white blood cell that functions as a phagocyte and an antigen-presenting cell.

monohybrid cross Genetic cross that involves individuals that are heterozygous for one trait; shows the phenotype of the dominant allele but carries the recessive allele.

monosaccharide Simple sugar; a carbohydrate that cannot be decomposed by hydrolysis (e.g., glucose).

monosomy One less chromosome than usual.

morphogenesis Emergence of shape in tissues, organs, or entire embryo during development.

morula Spherical mass of cells resulting from cleavage during animal development, prior to the blastula stage.

mosaic evolution Concept that human characteristics did not evolve at the same rate; for example, some body parts are more humanlike than others in early hominids.

motor neuron Nerve cell that conducts nerve impulses away from the central nervous system and innervates effectors (muscles and glands).

motor unit Motor neuron and all the muscle fibers it innervates.

mouth Oral cavity; location where mechanical and chemical digestion begin.

movement Motion.

MRSA Methicillin-resistant *Staphylococcus aureus*, a type of bacterium that causes "staph" infections that is no longer susceptible to certain antibiotics including methicillin.

mucosa Membrane that lines tubes and body cavities that open to the outside of the body; mucous membrane.

mucous membrane Membrane lining a cavity or tube that opens to the outside of the body; also called mucosa.

multicellular Referring to organism composed of many cells; usually has organized tissues, organs, and organ systems.

multifactorial trait Controlled by several allelic pairs; each dominant allele contributes to the phenotype in an additive and like manner.

multiple allele Inheritance pattern in which there are more than two alleles for a particular trait; each individual has only two of all possible alleles.

multiple sclerosis (MS) Disease in which the outer myelin layer of nerve fiber insulation becomes scarred, interfering with normal conduction of nerve impulses.

multiregional continuity hypothesis Proposal that modern humans evolved independently in at least three different places: Asia, Africa, and Europe.

muscle dysmorphia Mental state in which a person considers his or her body to be underdeveloped and becomes preoccupied with bodybuilding and diet; affects more men than women.

muscle fiber Muscle cell.

muscle tone Continuous, partial contraction of muscle.

muscle twitch Contraction of a whole muscle in response to a single stimulus.

muscular dystrophy Progressive muscle weakness and atrophy caused by deficient dystrophin protein.

muscularis Two layers of muscle in the gastrointestinal tract.

muscular system System of muscles that produces movement, both within the body and of its limbs; principal components are skeletal, smooth, and cardiac muscle.

muscular tissue Type of tissue composed of fibers that can shorten and thicken.

mutagen Agent, such as radiation or a chemical, that brings about a mutation.

mutation Alteration in chromosome structure or number and also an alteration in a gene due to a change in DNA composition.

myalgia Muscular pain.

myasthenia gravis Chronic disease characterized by muscles that are weak and easily fatigued. It results from the immune system's attack on neuromuscular junctions so that stimuli are not transmitted from motor neurons to muscle fibers.

myelin sheath White, fatty material, derived from the membrane of Schwann cells; forms a covering for nerve fibers.

myocardium The middle of the three layers of the wall of the heart; composed of cardiac muscle.

myofibril Contractile portion of muscle cells that contains a linear arrangement of sarcomeres and shortens to produce muscle contraction.

myoglobin Pigmented molecule in muscle tissue that stores oxygen.

myosin One of two major proteins of muscle; makes up thick filaments in myofibrils of muscle fibers. See also actin.

myxedema Condition resulting from a deficiency of thyroid hormone in an adult.

N

NAD$^+$ (nicotinamide adenine dinucleotide) Coenzyme that functions as a carrier of electrons and hydrogen ions, especially in cellular respiration.

nail Protective covering of the distal part of fingers and toes.

nasal cavity One of two canals in the nose, separated by a septum.

natural selection Mechanism of evolutionary change resulting in adaptation to the environment.

Neandertal Hominin with a sturdy build who lived during the last Ice Age in Europe and the Middle East; hunted large game and left evidence of being culturally advanced.

nearsighted Vision abnormality due to an elongated eyeball from front to back; light rays focus in front of retina when viewing distant objects.

negative feedback Mechanism of homeostatic response in which a stimulus initiates reactions that reduce the stimulus.

nephron Microscopic kidney unit that regulates blood composition by glomerular filtration, tubular reabsorption, and tubular secretion.

nerve Bundle of long axons outside the central nervous system.

nerve signal Action potential (electrochemical change) traveling along a neuron.

nervous system Organ system consisting of the brain, spinal cord, and associated nerves that coordinates the other organ systems of the body.

nervous tissue Tissue that contains nerve cells (neurons), which conduct impulses, and neuroglia, which support, protect, and provide nutrients to neurons.

neuroglia Nonconducting nerve cells that are intimately associated with neurons and function in a supportive capacity.

neuromuscular junction Region where an axon terminal approaches a muscle fiber; the synaptic cleft separates the axon terminal from the sarcolemma of a muscle fiber.

neuron Nerve cell that characteristically has three parts: dendrites, cell body, and axon.

neurotransmitter Chemical stored at the ends of axons that is responsible for transmission across a synapse.

neutron Neutral subatomic particle, located in the nucleus and having a weight of approximately one atomic mass unit.

neutrophil Granular leukocyte that is the most abundant of the white blood cells; first to respond to infection.

niche Role an organism plays in its community, including its habitat and its interactions with other organisms.

nitrification Process by which nitrogen in ammonia and organic molecules is oxidized to nitrites and nitrates by soil bacteria.

nitrogen fixation Process whereby free atmospheric nitrogen is converted into compounds, such as ammonium and nitrates, usually by bacteria.

nociceptor Sensory receptor that is sensitive to chemicals released by damaged tissues or excess stimuli of heat or pressure; pain receptors.

node of Ranvier Gap in the myelin sheath around a nerve fiber.

nondisjunction Failure of homologous chromosomes or daughter chromosomes to separate during meiosis I and meiosis II, respectively.

nonrenewable resource Minerals, fossil fuels, and other materials present in essentially fixed amounts (within human timescales) in our environment.

norepinephrine Neurotransmitter of the postganglionic fibers in the sympathetic division of the autonomic system; also, a hormone produced by the adrenal medulla.

nuclear envelope Double membrane that surrounds the nucleus and is connected to the endoplasmic reticulum; has pores that allow substances to pass between the nucleus and the cytoplasm.

nuclear pore Opening in the nuclear envelope that permits the passage of proteins into the nucleus and ribosomal subunits out of the nucleus.

nucleic acid Polymer of nucleotides; DNA and RNA are both examples.

nucleolus Dark-staining, spherical body in the cell nucleus that produces ribosomal subunits.

nucleoplasm Semifluid medium of the nucleus, containing chromatin.

nucleotide Monomer of DNA and RNA consisting of a 5-carbon sugar bonded to a nitrogen-containing base and a phosphate group.

nucleus Membrane-bounded organelle that contains chromosomes and controls the structure and function of the cell.

nutrient Chemical substance in foods that is essential to the diet and contributes to good health.

O

obesity Excess adipose tissue; exceeding ideal weight by more than 20%.

oblique At an angle between horizontal and vertical.

oblique layer Muscle layer of the stomach that is responsible for mixing the food with gastic juice and mechanical digestion.

oil Substance, usually of plant origin and liquid at room temperature, formed when a glycerol molecule reacts with three fatty acid molecules.

oil gland Gland of the skin associated with hair follicle; secretes sebum; also called sebaceous gland.

olfactory cell Modified neuron that is a sensory receptor for the sense of smell.

omnivore Organism in a food chain that feeds on both plants and animals.

oncogene Gene that disrupts the normal cell cycle, causing cancer.

oncology The study of cancer.

oogenesis Production of an egg in females by the process of meiosis and maturation.

opportunistic infection Infection that has an opportunity to occur because the immune system has been weakened.

optic chiasma X-shaped structure on the underside of the brain formed by a partial crossing-over of optic nerve fibers.

optic nerve Either of two cranial nerves that carry nerve impulses from the retina of the eye to the brain, thereby contributing to the sense of sight.

optic tract Groups of neurons from the optic nerve that sweep around the hypothalamus. Most fibers synapse with neurons in nuclei within the thalamus.

orbitals Pathways in which electrons travel around the nucleus of an atom.

organ Combination of two or more different tissues performing a common function.

organelle Small membranous structure in the cytoplasm of eukaryotic cells having a specific structure and function.

organic Refers to a molecule that always contains carbon and hydrogen, and often contains oxygen as well; organic molecules are associated with living organisms.

organic molecule Type of molecule that contains carbon and hydrogen—and often contains oxygen also.

organism Individual living thing.

organ system Group of related organs working together.

origin End of a muscle attached to a relatively immovable bone.

osmosis Diffusion of water from an area of high concentration to low concentration through a selectively permeable membrane.

osmotic pressure Measure of the tendency of water to move across a selectively permeable membrane; visible as an increase in liquid on the side of the membrane with higher solute concentration.

ossicle One of the small bones of the middle ear—malleus, incus, and stapes.

ossification Formation of bone tissue.

osteoblast Bone-forming cell.

osteoclast Cell that causes the breakdown of bone.

osteocyte Mature bone cell located within the lacunae of bone.

osteogenesis imperfecta Genetic condition that results in very weak bones.

osteoporosis Condition in which bones break easily because calcium is removed from them faster than it is replaced.

otitis media Infection of the middle ear, characterized by pain and possibly by a sense of fullness, hearing loss, vertigo, and fever.

otolith Calcium carbonate granule associated with ciliated cells in the utricle and the saccule.

outbreak A disease epidemic that is confined to a local area.

outer ear Portion of ear consisting of the pinna and auditory canal.

oval window Membrane-covered opening between the stapes and the inner ear.

ovarian cycle Monthly follicle changes occurring in the ovary that control the level of sex hormones in the blood and the uterine cycle.

ovaries Female gonads that produces eggs and the female sex hormones.

ovulation Release of a secondary oocyte from the ovary; if fertilization occurs, the secondary oocyte becomes an egg.

oxygen debt Amount of oxygen needed to metabolize lactate, a compound that accumulates during vigorous exercise.

oxyhemoglobin Compound formed when oxygen combines with hemoglobin.

oxytocin Hormone released by the posterior pituitary that causes contraction of uterus and milk letdown.

ozone hole Seasonal thinning of the ozone shield in the lower stratosphere at the North and South poles.

ozone shield Accumulation of O_3, formed from oxygen in the upper atmosphere; a filtering layer that protects the Earth from ultraviolet radiation.

P

pacemaker Another name for the SA (sinoatrial) node; establishes the electrical rythym of the heart.

pancreas Internal organ that produces digestive enzymes and the hormones insulin and glucagon.

pancreatic amylase Enzyme secreted by the pancreas that digests starch to maltose in the small intestine.

pancreatic islets (islets of Langerhans) Masses of cells that constitute the endocrine portion of the pancreas.

pandemic An increase in the occurrence of a disease within a large and geographically widespread population (often refers to a worldwide epidemic).

Pap test Analysis done on cervical cells for detection of cancer.

parasympathetic division That part of the autonomic system that is active under normal conditions; uses acetylcholine as a neurotransmitter.

parathyroid gland Gland embedded in the posterior surface of the thyroid gland; it produces parathyroid hormone.

parathyroid hormone (PTH) Hormone secreted by the four parathyroid glands that increases the blood calcium level and decreases the blood phosphate level.

parent cell Cell that divides so as to form daughter cells.

Parkinson disease Progressive deterioration of the central nervous system due to a deficiency in the neurotransmitter dopamine.

parturition Processes that lead to and include birth and the expulsion of the afterbirth.

passive immunity Protection against infection acquired by transfer of antibodies to a susceptible individual.

pathogen Disease-causing agent.

pectoral girdle Portion of the skeleton that provides support and attachment for an arm; consists of a scapula and a clavicle.

pelvic girdle Portion of the skeleton to which the legs are attached; consists of the coxal bones.

pelvis Bony ring formed by the sacrum and coxae.

penis External organ in males through which the urethra passes; also serves as the organ of sexual intercourse.

pentose Five-carbon sugar. Deoxyribose is the pentose sugar found in DNA; ribose is a pentose sugar found in RNA.

pepsin Enzyme secreted by gastric glands that digests proteins to peptides.

peptide bond Type of covalent bond that joins two amino acids.

peptide hormone Type of hormone that is a protein, a peptide, or derived from an amino acid.

pericardium Protective serous membrane that surrounds the heart.

periodontitis Inflammation of the periodontal membrane that lines tooth sockets, causing loss of bone and loosening of teeth.

periosteum Fibrous connective tissue covering the surface of bone.

peripheral nervous system (PNS) Nerves and ganglia that lie outside the central nervous system.

peristalsis Wavelike contractions that propel substances along a tubular structure such as the esophagus.

peritonitis Generalized infection of the lining of the abdominal cavity.

peritubular capillary network Capillary network that surrounds a nephron and functions in reabsorption during urine formation.

Peyer's patches Lymphatic organs located in the small intestine.

phagocytosis Process by which amoeboid-type cells engulf large substances, forming an intracellular vacuole.

pharynx Portion of the digestive tract between the mouth and the esophagus that serves as a passageway for food and also for air on its way to the trachea.

phenotype Visible expression of a genotype—for example, brown eyes or attached earlobes.

pheromone Chemical signal released by an organism that affects the metabolism or influences the behavior of another individual of the same species.

phospholipid Lipid molecule that forms the bilayer of the cell's membranes; has a polar, hydrophilic head bonded to two nonpolar, hydrophobic tails.

photoreceptor Sensory receptor in retina that responds to light stimuli.

photosynthesis Process by which plants, algae, and some bacteria harvest the energy of the sun and convert it to chemical energy.

pH scale Measurement scale for hydrogen ion concentration; logarithmic scale.

pilus Elongated, hollow appendage on bacteria used to transfer DNA from one cell to another.

pineal gland Endocrine gland located in the third ventricle of the brain; produces melatonin.

pinna Part of the ear that projects on the outside of the head.

pituitary dwarfism Condition in which an affected individual has normal proportions but small stature; caused by inadequate growth hormone.

pituitary gland Endocrine gland that lies just inferior to the hypothalamus; consists of the anterior pituitary and posterior pituitary.

placebo Treatment that is an inactive substance (pill, liquid, etc.) administered as if it were a therapy in an experiment but that has no therapeutic value.

placenta Structure that forms from the chorion and the uterine wall and allows the embryo and then the fetus to acquire nutrients and rid itself of wastes.

plaque Accumulation of soft masses of fatty material, particularly cholesterol, beneath the inner linings of the arteries.

plasma Liquid portion of blood; contains nutrients, wastes, salts, and proteins.

plasma cell Cell derived from a B lymphocyte specialized to mass-produce antibodies.

plasma membrane Membrane surrounding the cytoplasm that consists of a phospholipid bilayer with embedded proteins; functions to regulate the entrance and exit of molecules from the cell.

plasma protein Protein dissolved in blood plasma.

plasmid Self-replicating ring of accessory DNA in the cytoplasm of bacteria.

platelet (thrombocyte) Component of blood necessary to blood clotting; also called a thrombocyte.

pleura Serous membrane that encloses the lungs.

pneumonectomy Surgical removal of all or part of a lung.

pneumonia Infection of the lungs that causes alveoli to fill with mucus and pus.

polar Combination of atoms in which the electrical charge is not distributed symmetrically.

polar body In oogenesis, a nonfunctional product; two to three meiotic products are of this type.
pollution Any environmental change that adversely affects the lives and health of living organisms.
polygenic trait Trait is controlled by several allelic pairs; each dominant allele contributes to the phenotype in an additive and like manner.
polymerase chain reaction (PCR) Technique that uses the enzyme DNA polymerase to produce millions of copies of a particular piece of DNA.
polyp Small, abnormal growth that arises from the epithelial lining.
polypeptide Polymer of many amino acids linked by peptide bonds.
polyribosome String of ribosomes simultaneously translating regions of the same mRNA strand during protein synthesis.
polysaccharide Polymer made from sugar monomers; the polysaccharides starch and glycogen are polymers of glucose monomers.
pons Portion of the brain stem above the medulla oblongata and below the midbrain; assists the medulla oblongata in regulating the breathing rate.
population Organisms of the same species occupying a certain area.
positive feedback Mechanism in which the stimulus initiates reactions that lead to an increase in the stimulus.
posterior pituitary Portion of the pituitary gland that stores and secretes oxytocin and antidiuretic hormone, which are produced by the hypothalamus.
precapillary sphincter Smooth muscle ring that controls blood flow through a capillary bed.
precipitation Water deposited on the Earth in the form of rain, snow, sleet, hail, or fog.
pre-embryonic development Development of the zygote in the first week, including fertilization, the beginning of cell division, and the appearance of the chorion.
prefrontal area Association area in the frontal lobe that receives information from other association areas and uses it to reason and plan actions.
primary germ layer One of the three layers (ectoderm, mesoderm, and endoderm) of embryonic cells that develop into specific tissues and organs.
primary motor area Area in the frontal lobe where voluntary commands begin; each section controls a part of the body.
primary somatosensory area Area dorsal to the central sulcus where sensory information arrives from skin and skeletal muscles.
primate Animal that belongs to the order Primates; includes prosimians, monkeys, apes, and humans, all of whom have adaptations for living in trees.
principle Theory generally accepted by an overwhelming number of scientists; a law.
prion An infectious particle that is the cause of diseases, such as scrapie in sheep, mad cow disease, and Creutzfeldt–Jakob disease in humans; it has a protein component, but no nucleic acid has been detected.
producer Photosynthetic organism at the start of a grazing food chain that makes its own food (e.g., green plants on land and algae in water).
product Substance that forms as a result of a reaction.
progesterone Female sex hormone that helps maintain sex organs and secondary sex characteristics.
progression In cancer, a second mutation that allows cells to invade surrounding tissues.
prokaryotic cell Type of cell that lacks a membrane-bounded nucleus and organelles.
prolactin Hormone secreted by the anterior pituitary that stimulates the production of milk from the mammary glands.
promotion In cancer, development of a group of cells from a single mutated cell.
prophase Phase of nuclear division phase during which chromatin condenses so that chromosomes appear and the nuclear envelope dissolves.
proprioceptor Sensory receptor in skeletal muscles and joints that assists the brain in knowing the position of the limbs.
prosimian Member of a group of primates that includes lemur and tarsiers and may resemble the first primates to have evolved.
prostaglandin Hormone that has various and powerful local effects.
prostate gland Gland located around the male urethra below the urinary bladder; adds secretions to semen.
protease Enzyme capable of breaking peptide bonds in a protein or polypeptide.
protein Molecule consisting of one or more polypeptides.
protein-first hypothesis In chemical evolution, the proposal that protein originated before other macromolecules and allowed the formation of protocells.
proteomics The study of all proteins in an organism.
prothrombin Plasma protein converted to thrombin during the steps of blood clotting.
prothrombin activator Enzyme that catalyzes the transformation of the precursor prothrombin to the active enzyme thrombin.
protocell In biological evolution, a possible cell forerunner that became a cell once it could reproduce.
proton Positive subatomic particle, located in the nucleus and having a weight of approximately one atomic mass unit.
proto-oncogene Normal gene that can become an oncogene through mutation; involved in the regulation of the cell cycle.
provirus Latent form of a virus in which the viral DNA is incorporated into the chromosome of the host.
proximal convoluted tubule Highly coiled region of a nephron near the glomerular capsule, where tubular reabsorption takes place.
pseudostratified columnar epithelium Appearance of layering in some epithelial cells when, actually, each cell touches a baseline and true layers do not exist.
pulmonary artery Blood vessel that takes blood away from the heart to the lungs.
pulmonary circuit Circulatory pathway that consists of the pulmonary trunk, the pulmonary arteries, and the pulmonary veins; takes oxygen-poor blood from the heart to the lungs and oxygen-rich blood from the lungs to the heart.
pulmonary fibrosis Accumulation of fibrous connective tissue in the lungs; caused by inhaling irritating particles, such as silica, coal dust, or asbestos.
pulmonary vein Blood vessel that takes blood from the lungs to the heart.
pulse Vibration felt in arterial walls due to expansion of the aorta following ventricle contraction.
Punnett square Gridlike guide used to calculate the expected results of simple genetic crosses.
pupil Opening in the center of the iris of the eye.
Purkinje fibers Specialized muscle fibers that conduct the cardiac impulse from the AV bundle into the ventricles.
pus Thick, yellowish fluid composed of dead phagocytes, dead tissue, and bacteria.
pyelonephritis Inflammation of the kidney due to bacterial infection.

R

radioisotope Unstable form of an atom that spontaneously emits radiation in the form of radioactive particles or radiant energy.
reactant Substance that participates in a reaction.

recessive allele Allele that exerts its phenotypic effect only in the homozygote; its expression is masked by a dominant allele.
recombinant DNA (rRNA) DNA that contains genes from more than one source.
rectum Terminal end of the digestive tube between the sigmoid colon and the anus.
red blood cell (erythrocyte) Formed element that contains hemoglobin and carries oxygen from the lungs to the tissues; also called erythrocyte.
red bone marrow Blood-cell-forming tissue located in the spaces within spongy bone.
reduced hemoglobin Hemoglobin carrying hydrogen ions.
referred pain Pain perceived as having come from a site other than that of its actual origin.
reflex Automatic, involuntary response of an organism to a stimulus.
refractory period Time following an action potential when a neuron is unable to conduct another nerve impulse.
renal artery Vessel that originates from the aorta and delivers blood to the kidney.
renal cortex Outer portion of the kidney that appears granular.
renal medulla Inner portion of the kidney that consists of renal pyramids.
renal pelvis Hollow chamber in the kidney that lies inside the renal medulla and receives freshly prepared urine from the collecting ducts.
renal vein Vessel that takes blood from the kidney to the inferior vena cava.
renewable resource Resources normally replaced or replenished by natural processes; resources not depleted by moderate use. Examples include solar energy, biological resources such as forests and fisheries, biological organisms, and some biogeochemical cycles.
renin Enzyme released by kidneys that leads to the secretion of aldosterone and a rise in blood pressure.
replacement model Proposal that modern humans originated only in Africa; then migrated out of Africa and supplanted populations of early *Homo* in Asia and Europe about 100,000 years ago; also called the Out-of-Africa hypothesis.
replacement reproduction Population in which each person is replaced by only one child.
repolarization When the charge inside the axon resumes a negative charge.
reproduce To produce a new individual of the same type.
reproductive system Organ system that contains male or female organs and specializes in the production of offspring.
residual volume Amount of air remaining in the lungs after a forceful expiration.
respiratory control center Group of nerve cells in the medulla oblongata that sends out nerve impulses on a rhythmic basis, resulting in involuntary inspiration on an ongoing basis.
respiratory system Organ system consisting of the lungs and tubes that bring oxygen into the lungs and take carbon dioxide out.
resting potential Polarity across the plasma membrane of a resting neuron due to an unequal distribution of ions.
restriction enzyme Bacterial enzyme that stops viral reproduction by cleaving viral DNA; used to cut DNA at specific points during production of recombinant DNA.
reticular fiber Very thin collagen fibers in the matrix of connective tissue, highly branched and forming delicate supporting networks.
reticular formation Complex network of nerve fibers within the central nervous system that arouses the cerebrum.
retina Innermost layer of the eye that contains the rod cells and the cone cells.
retinal Light-absorbing molecule that is a derivative of vitamin A and a component of rhodopsin.
retrovirus RNA virus containing the enzyme reverse transcriptase that carries out RNA to DNA transcription.
reverse transcriptase Enzyme that speeds the conversion of viral RNA to viral DNA.
rheumatic fever Disease caused by bacterial infection, characterized by fever, swelling and pain in the joints, sore throat, and cardiac involvement.
rheumatoid arthritis Persistent inflammation of synovial joints, often causing cartilage destruction, bone erosion, and joint deformities.
rhodopsin Light-absorbing molecule in rod cells and cone cells that contains a pigment and the protein opsin.
ribosomal RNA (rRNA) Type of RNA found in ribosomes where protein synthesis occurs.
ribosome RNA and protein in two subunits; site of protein synthesis in the cytoplasm.
rigor mortis Contraction of muscles at death due to lack of ATP.
RNA (ribonucleic acid) Nucleic acid produced from covalent bonding of nucleotide monomers that contain the sugar ribose; occurs in three major forms: messenger RNA, ribosomal RNA, and transfer RNA.
RNA-first hypothesis In chemical evolution, the proposal that RNA originated before other macromolecules and allowed the formation of the first cell or cells.
RNA polymerase During transcription, an enzyme that joins nucleotides complementary to a DNA template.
rod cell Photoreceptor in retina of eyes that responds to dim light.
rotational equilibrium Maintenance of balance when the head and body are suddenly moved or rotated.
rotator cuff Tendons that encircle and help form a socket for the humerus and also help reinforce the shoulder joint.
round window Membrane-covered opening between the inner ear and the middle ear.
rugae Deep folds, as in the wall of the stomach.
runoff Water—from rain, snowmelt, or other sources—that flows over the land surface, adding to the water cycle.

S

SA (sinoatrial) node Small region of neuromuscular tissue that initiates the heartbeat; also called the pacemaker.
saccule Saclike cavity in the vestibule of the inner ear; contains sensory receptors for gravitational equilibrium.
salivary amylase Secreted from the salivary glands; the first enzyme to act on starch.
salivary gland Gland associated with the mouth that secretes saliva.
saltatory conduction Movement of nerve impulses from one neurofibral node to another along a myelinated axon.
saltwater intrusion Movement of salt water into freshwater aquifers in coastal areas where groundwater is withdrawn faster than it is replenished.
sarcolemma Plasma membrane of a muscle fiber; also forms the tubules of the T system involved in muscular contraction.
sarcoma Cancer that arises in muscles and connective tissues.
sarcomere One of many units arranged linearly within a myofibril whose contraction produces muscle contraction.
sarcoplasmic reticulum Smooth endoplasmic reticulum of skeletal muscle cells; surrounds the myofibrils and stores calcium ions.
saturated fatty acid Fatty-acid molecule that lacks double bonds between the atoms of its carbon chain.
Schwann cell Cell that surrounds a fiber of a peripheral nerve and forms the myelin sheath.
science Development of concepts about the natural world, often by using the scientific method.
scientific method Process of attaining knowledge by making observations, testing hypotheses, and coming to conclusions.
scientific theory Concept supported by a broad range of observations, experiments, and conclusions.

sclera White, fibrous, outer layer of the eyeball.
scoliosis Abnormal, laterial (side-to-side) curvature of the vertebral column.
scrotum Pouch of skin that encloses the testes.
secondary oocyte In oogenesis, the functional product of meiosis I; becomes the egg.
second messenger Chemical signal such as cyclic AMP that causes the cell to respond to the first messenger—a hormone bound to a receptor protein in the plasma membrane.
selectively permeable Having degrees of permeability; the cell is impermeable to some substances and allows others to pass through at varying rates.
semantic memory Capacity of the brain to store and retrieve information with regard to words or numbers.
semen Thick, whitish fluid consisting of sperm and secretions from several glands of the male reproductive tract.
semicircular canal One of three tubular structures within the inner ear that contain sensory receptors responsible for the sense of rotational equilibrium.
semilunar valve Valve resembling a half moon located between the ventricles and their attached vessels.
seminal vesicle Convoluted structure attached to the vas deferens near the base of the urinary bladder in males; adds secretions to semen.
seminiferous tubule Long, coiled structure contained within chambers of the testis; where sperm are produced.
sensation Conscious awareness of a stimulus due to nerve impulses sent to the brain from a sensory receptor by way of sensory neurons.
sensory adaptation Phenomenon of a sensation becoming less noticeable once it has been recognized by constant repeated stimulation.
sensory neuron Nerve cell that transmits nerve impulses to the central nervous system after a sensory receptor has been stimulated.
sensory receptor Structure that receives either external or internal environmental stimuli and is a part of a sensory neuron or transmits signals to a sensory neuron.
septum Wall between two cavities; in the human heart, a septum separates the right side from the left side.
serosa Membrane that covers internal organs and lines cavities without an opening to the outside of the body.
serotonin Neurotransmitter derived from the amino acid tryptophan.
serous membrane Membrane that covers internal organs and lines cavities without an opening to the outside of the body; also called serosa.
Sertoli cell Cell associated with developing germ cells in seminiferous tubule; secretes fluid into seminiferous tubule and mediates hormonal effects on tubule.
serum Light yellow liquid left after clotting of blood.
severe combined immunodeficiency disease (SCID) Congenital illness in which both antibody- and cell-mediated immunity are lacking or inadequate.
sex chromosome Chromosome that determines the sex of an individual; in humans, females have two X chromosomes and males have both an X and Y chromosome.
sex-linked Refers to allele that occurs on the sex chromosomes but may control a trait that has nothing to do with the sex characteristics of an individual.
short-term memory Retention of information for only a few minutes, such as remembering a telephone number.
sickle-cell disease Genetic disorder in which the affected individual has sickle-shaped red blood cells subject to hemolysis.
simple goiter Condition in which an enlarged thyroid produces low levels of thyroxine.
sinkhole Large surface crater caused by the collapse of an underground channel or cavern; often triggered by groundwater withdrawal.
sinus Cavity or hollow space in an organ such as the skull.
sinusitis Infection of the sinuses, caused by blockage of the openings to the sinuses and characterized by postnasal discharge and facial pain.
sister chromatids One of two genetically identical chromosomal units that are the result of DNA replication and are attached to each other at the centromere.
skeletal muscle Striated, voluntary muscle tissue found in muscles that move the bones.
skeletal system System of bones, cartilage, and ligaments that works with the muscular system to protect the body and provide support for locomotion and movement.
skill memory Capacity of the brain to store and retrieve information necessary to perform motor activities, such as riding a bike.
skin Outer covering of the body; can be called the integumentary system because it contains organs such as sense organs.
skull Bony framework of the head, composed of cranial bones and the bones of the face.
sliding filament model An explanation for muscle contraction based on the movement of actin filaments in relation to myosin filaments.
small intestine Long, tubelike chamber of the digestive tract between the stomach and large intestine.
small RNAs Short RNA molecules that help to regulate gene expression.
smooth muscle Nonstriated, involuntary muscle tissue found in the walls of internal organs; also called visceral muscle.
sodium-potassium pump Carrier protein in the plasma membrane that moves sodium ions out of and potassium ions into cells; important in nerve and muscle cells.
soft palate Entirely muscular posterior portion of the roof of the mouth.
somatic system That portion of the peripheral nervous system containing motor neurons that control skeletal muscles.
spasm Sudden, involuntary contraction of one or more muscles.
species Group of similarly constructed organisms capable of interbreeding and producing fertile offspring; organisms that share a common gene pool.
sperm Male gamete having a haploid number of chromosomes and the ability to fertilize an egg, the female gamete.
spermatogenesis Production of sperm in males by the process of meiosis and maturation.
sphincter Muscle that surrounds a tube and closes or opens the tube by contracting and relaxing.
spinal cord Part of the central nervous system; the nerve cord that is continuous with the base of the brain plus the vertebral column that protects the nerve cord.
spinal nerve Nerve that arises from the spinal cord.
spiral organ Organ in the cochlear duct of the inner ear responsible for hearing; also called the organ of Corti.
spirillum Group of bacteria that exhibit a variety of spiral or undulating shapes or are comma-shaped.
spleen Large, glandular organ located in the upper left region of the abdomen; stores and purifies blood.
spongy bone Porous bone found at the ends of long bones where red bone marrow is sometimes located.
sprain Injury to a ligament caused by abnormal force applied to a joint.
squamous epithelium Type of epithelial tissue that contains flat cells.
standard error Number used in evaluating statistical data to show the range of variation in the data.
stapes The last of three ossicles of the ear that serve to conduct vibrations from the tympanic membrane to the oval window of the inner ear.
starch Storage polysaccharide found in plants; composed of glucose molecules joined in a linear fashion with few side chains.
stereocilia (sing., **stereocilium**) Long, flexible microvilli that superficially resemble cilia. Within the inner ear, these signal changes in body position and help to maintain balance and equilibrium.

steroid Type of lipid molecule having a complex of four carbon rings; examples are cholesterol, progesterone, and testosterone.

steroid hormone One of a group of hormones derived from cholesterol.

stimulus Change in the internal or external environment that a sensory receptor can detect, leading to nerve impulses in sensory neurons.

stomach Muscular sac that mixes food with gastric juices to form chyme, which enters the small intestine.

strain Injury to a muscle resulting from overuse or improper use.

striae gravidarum Linear, depressed, scarlike lesions occurring on the abdomen, breasts, buttocks, and thighs due to the weakening of the elastic tissues during pregnancy.

striated Having bands; in cardiac and skeletal muscle, alternating light and dark crossbands produced by the distribution of contractile proteins.

stroke Condition resulting when an arteriole in the brain bursts or becomes blocked by an embolism; also called cerebrovascular accident.

subcutaneous layer Tissue layer that lies just beneath the skin and contains adipose tissue.

submucosa Layer of connective tissue underneath a mucous membrane.

subsidence Occurs when a portion of the Earth's surface gradually settles downward.

substrate Reactant in a reaction controlled by an enzyme.

sudden infant death syndrome (SIDS) Any sudden and unexplained death of an apparently healthy infant aged one month to one year.

summation Combination of signals at a synapse.

surfactant Agent that reduces the surface tension of water; in the lungs, a surfactant prevents the alveoli from collapsing.

sustainable Ability of a society or ecosystem to maintain itself while also providing services to human beings.

suture Type of immovable joint articulation found between bones of the skull.

sweat gland Skin gland that secretes a fluid substance for evaporative cooling; also called sudoriferous gland.

sympathetic division The part of the autonomic system that usually promotes activities associated with emergency (fight-or-flight) situations; uses norepinephrine as a neurotransmitter.

synapse Junction between neurons consisting of the presynaptic (axon) membrane, the synaptic cleft, and the postsynaptic (usually dendrite) membrane.

synapsis Pairing of homologous chromosomes during prophase I of meiosis I; allows for crossing-over to occur.

synaptic cleft Small gap between presynaptic and postsynaptic membranes of a synapse.

syndrome Group of symptoms that appear together and tend to indicate the presence of a particular disorder.

synovial joint Freely movable joint having a cavity filled with synovial fluid.

synovial membrane Membrane that forms the inner lining of the capsule of a freely movable joint.

systemic circuit Blood vessels that transport blood from the left ventricle and back to the right atrium of the heart.

systemic lupus erythematosus Syndrome involving the connective tissues and various organs, including kidney.

systole Contraction period of the heart during the cardiac cycle.

systolic pressure Arterial blood pressure during the systolic phase of the cardiac cycle.

T

taste bud Sense organ containing the receptors associated with the sense of taste.

Tay–Sachs disease Lethal genetic disease in which the newborn has a faulty lysosomal digestive enzyme.

T cell (T lymphocyte) Lymphocyte that matures in the thymus. Cytotoxic T cells kill antigen-bearing cells outright; helper T cells release cytokines that stimulate other immune system cells.

T-cell receptor (TCR) Molecule on the surface of a T lymphocyte to which an antigen binds.

technology The science or study of the practical or industrial arts.

tectorial membrane Membrane that lies above and makes contact with the hair cells in the spiral organ.

telomere Tip of the end of a chromosome.

telophase Mitotic phase during which daughter chromosomes are located at each pole.

template Pattern or guide used to make copies; parental strand of DNA serves as a guide for the production of daughter DNA strands, and DNA also serves as a guide for the production of messenger RNA.

tendinitis An inflammation of muscle tendons and their attachments.

tendon Strap of fibrous connective tissue that connects skeletal muscle to bone.

testes (sing., **testis**) Male gonads that produce sperm and the male sex hormones.

test group Group exposed to the experimental variable in an experiment; compare to control group.

testosterone Male sex hormone that helps maintain sexual organs and secondary sex characteristics.

tetanus Sustained muscle contraction without relaxation.

tetany Severe twitching caused by involuntary contraction of the skeletal muscles due to a calcium imbalance.

thalamus Part of the brain located in the lateral walls of the third ventricle that serves as the integrating center for sensory input; it plays a role in arousing the cerebral cortex.

thermoreceptor Sensory receptor that is sensitive to changes in temperature.

threshold Electrical potential level (voltage) at which an action potential or nerve impulse is produced.

thrombin Enzyme that converts fibrinogen to fibrin threads during blood clotting.

thrombocytopenia Insufficient number of platelets in the blood.

thromboembolism Obstruction of a blood vessel by a thrombus that has dislodged from the site of its formation.

thrombus Blood clot that remains in the blood vessel where it formed.

thymine (T) One of four nitrogen-containing bases in nucleotides composing the structure of DNA; pairs with adenine.

thymosin Peptide secreted by the thymus that increases production of certain types of white blood cells.

thymus Lymphatic organ, located along the trachea behind the sternum, involved in the maturation of T lymphocytes in the thymus gland. Secretes hormones called thymosins, which aid the maturation of T cells and perhaps stimulate immune cells in general.

thyroid gland Endocrine gland in the neck that produces several important hormones, including thyroxine, triiodothyronine, and calcitonin.

thyroid-stimulating hormone (TSH) Substance produced by the anterior pituitary that causes the thyroid to secrete thyroxine and triiodothyronine.

thyroxine (T_4) Hormone secreted from the thyroid gland that promotes growth and development; in general, it increases the metabolic rate in cells.

tidal volume Amount of air normally moved in the human body during an inspiration or expiration.

tight juntion Junction between cells when adjacent plasma membrane proteins join to form an impermeable barrier.

tissue Group of similar cells that perform a common function.

tissue fluid Fluid that surrounds the body's cells; consists of dissolved substances that leave the blood capillaries by filtration and diffusion.

tonicity Osmolarity of a solution compared with that of a cell. If the solution is isotonic to the cell, there is no net movement of water; if the solution is hypotonic, the cell gains water; and if the solution is hypertonic, the cell loses water.

tonsillectomy Surgical removal of the tonsils.

tonsillitis Infection of the tonsils that causes inflammation and can spread to the middle ears.

tonsils Partially encapsulated lymph nodules located in the pharynx.

total artificial heart (TAH) A mechanical replacement for the heart, as opposed to a partial replacement.

toxin Poisonous substance produced by living cells or organisms. Toxins are nearly always proteins that are capable of causing disease on contact or absorption with body tissues.

tracer Substance having an attached radioisotope that allows a researcher to track its whereabouts in a biological system.

trachea Passageway that conveys air from the larynx to the bronchi; also called the windpipe.

tracheostomy Creation of an artificial airway by incision of the trachea and insertion of a tube.

tract Bundle of myelinated axons in the central nervous system.

transcription Process whereby a DNA strand serves as a template for the formation of mRNA.

transcription factor In eukaryotes, protein required for the initiation of transcription by RNA polymerase.

trans fat Fats, which occur naturally in meat and dairy products of ruminants, that are also industrially created through partial hydrogenation of plant oils and animal fats.

transfer RNA (tRNA) Type of RNA that transfers a particular amino acid to a ribosome during protein synthesis; at one end, it binds to the amino acid, and at the other end it has an anticodon that binds to an mRNA codon.

transgenic organism Free-living organism in the environment that has a foreign gene in its cells.

translation Process whereby ribosomes use the sequence of codons in mRNA to produce a polypeptide with a particular sequence of amino acids.

translocation Movement of a chromosomal segment from one chromosome to another nonhomologous chromosome, leading to abnormalities (e.g., Down syndrome).

triglyceride Neutral fat composed of glycerol and three fatty acids.

triiodothyronine (T_3) Hormone produced by the thyroid gland that contains three iodine atoms; the metabolically active form of the thyroid hormones.

trisomy Three copies of a chromosome in a cell.

trophic level Feeding level of one or more populations in a food web.

trophic relationship In ecosystems, feeding relationships such as grazing food webs or detrital food webs.

tropomyosin Protein that functions with troponin to block muscle contraction until calcium ions are present.

troponin Protein that functions with tropomyosin to block muscle contraction until calcium ions are present.

trypsin Protein-digesting enzyme secreted by the pancreas.

T (transverse) tubule Membranous channel that extends inward.

tubal ligation Method for preventing pregnancy in which the uterine tubes are cut and sealed.

tuberculosis infection of the lungs, caused by the bacteria *Mycobacterium tuberculosis.*

tubular reabsorption Movement of primarily nutrient molecules and water from the contents of the nephron into blood at the proximal convoluted tubule.

tubular secretion Movement of certain molecules from blood into the distal convoluted tubule of a nephron so that they are added to urine.

tumor Cells derived from a single mutated cell that has repeatedly undergone cell division; benign tumors remain at the site of origin, and malignant tumors metastasize.

tumor suppressor gene Gene that codes for a protein that ordinarily suppresses cell division; inactivity can lead to a tumor.

tympanic membrane Located between the outer and middle ear where it receives sound waves; also called the eardrum.

U

umbilical cord Cord connecting the fetus to the placenta through which blood vessels pass.

unsaturated fatty acid Fatty acid molecule that has one or more double bonds between the atoms of its carbon chain.

uracil (U) The base in RNA that replaces thymine found in DNA; pairs with adenine.

urea Primary nitrogenous waste of humans derived from amino acid breakdown.

uremia High level of urea nitrogen in the blood.

ureter One of two tubes that take urine from the kidneys to the urinary bladder.

urethra Tubular structure that receives urine from the bladder and carries it to the outside of the body.

urethritis Inflammation of the urethra.

uric acid Waste product of nucleotide metabolism.

urinary bladder Organ where urine is stored before being discharged by way of the urethra.

urinary system Organ system consisting of the kidneys and urinary bladder; rids the body of nitrogenous wastes and helps regulate the water–salt balance of the blood.

uterine cycle Monthly occurring changes in the characteristics of the uterine lining (endometrium).

uterine tube Tube that transports eggs to the uterus; also called oviducts or Fallopian tubes; location of fertilization.

uterus Organ located in the female pelvis where the fetus develops; also called the womb.

utricle Saclike cavity in the vestibule of the inner ear that contains sensory receptors for gravitational equilibrium.

V

vaccine Antigens prepared in such a way that they can promote active immunity without causing disease.

vagina Organ that leads from the uterus to the vestibule and serves as the birth canal and organ of sexual intercourse in females.

valve Membranous extension of a vessel of the heart wall that opens and closes, ensuring one-way flow.

vas deferens Tube that leads from the epididymis to the urethra in males.

vasectomy Method for preventing pregnancy in which the vasa deferentia are cut and sealed.

vector In genetic engineering, a means to transfer foreign genetic material into a cell (e.g., a plasmid).

vein Vessel that transports blood back to the heart; often has valves due to the low pressure of the blood.

vena cava Major vein of the body; returns blood from the systemtic circuit to the right atrium.

ventilation Process of moving air into and out of the lungs; also called breathing.

ventricle Cavity in an organ, such as a lower chamber of the heart or the ventricles of the brain.

venule Vessel that takes blood from capillaries to a vein.

vermiform appendix Small, tubular appendage that extends outward from the cecum of the large intestine.

vernix caseosa Cheeselike substance covering the skin of the fetus.

vertebral column Series of joined vertebrae that extends from the skull to the pelvis.

vertebrate An animal with a vertebral column.

vesicle Small, membrance-bounded sac that stores substances within a cell.

vestibule Space or cavity at the entrance of a canal, such as the cavity that lies between the semicircular canals and the cochlea.

vestigial structure Remains of a structure that was functional in some ancestor but is no longer functional in the organism in question.

villi (pl., **villus**) Small, fingerlike projection of the inner small intestinal wall.

virus Noncellular, parasitic agent consisting of an outer capsid and an inner core of nucleic acid.

visual accommodation Ability of the eye to focus at different distances by changing the curvature of the lens.

vital capacity Maximum amount of air moved into or out of the human body with each breathing cycle.

vitamin Essential requirement in the diet, needed in small amounts. Vitamins are often part of coenzymes.

vitamin D Required for proper bone growth.

vitreous humor Clear, gelatinous material between the lens of the eye and the retina.

vocal cord Fold of tissue within the larynx; creates vocal sounds when it vibrates.

vulva External genitals of the female that surround the opening of the vagina.

W

water (hydrologic) cycle Interdependent and continuous circulation of water from the ocean to the atmosphere, to the land, and back to the ocean.

Wernicke's area Brain area involved in language comprehension.

white blood cell (leukocyte) Type of blood cell that is transparent without staining and protects the body from invasion by foreign substances and organisms; also called a leukocyte.

white matter Myelinated axons in the central nervous system.

X

XDR TB Extensively drug-resistant tuberculosis, a type of bacterium that causes tuberculosis (TB) that is no longer susceptible to almost all of the drugs normally used to treat TB.

xenotransplantation Use of animal organs, instead of human organs, in human transplant patients.

X-linked Refers to allele located on an X chromosome, but may control a trait that has nothing to do with the sex characteristics of an individual.

Y

yolk sac Extraembryonic membrane that encloses the yolk of birds; in humans, it is the first site of blood cell formation.

Z

zygote Diploid cell formed by the union of sperm and egg; the product of fertilization.

Credits

Photo Credits

Chapter 1

Opener: Courtesy NASA/JPL/University of Arizona; 1.1 (leech): © St. Bartholomews Hospital/Photo Researchers; 1.1 (mushrooms): © IT Stock/age fotostock RF; 1.1 (bacteria): © Science Source/Photo Researchers; 1.1 (meerkat): © Jami Tarris/Getty Images; 1.1 (sunflower): © Photodisc Green/Getty RF; 1.1 (*Giardia*): Courtesy CDC/Dr. Stan Erlandsen; 1.3a; © Vol. 124/Corbis RF; 1.3b: © John Cancalosi/Getty Images; 1.4a (seedling): © Herman Eisenbeiss/Photo Researchers; 1.4a (tree): Courtesy Paul Wray, Iowa State University; 1.4b (fertilization): © Dr. David Phillips/Visuals Unlimited; 1.4b (fetus): © Brand X Pictures/Punchstock RF; 1A: © Michael Freeman/Corbis RF; 1.6 (protist): © Michael Abby/Visuals Unlimited; 1.6 (fungi): © Ingram Publishing RF; 1.6 (plant): © Pat Pendarvis; 1.6 (animal): © Corbis RF; 1.6 (Archaea): © Ralph Robinson/Visuals Unlimited; 1.6 (Bacteria): © A.B. Dowsett/SPL/Photo Researchers; 1.7a: © Scenics of America/PhotoLink/Getty RF; 1.7b: © K. Tumanowicz/Photo Researchers; 1.8: © Ryan McVay/Getty RF; 1.9 (main): © Tony McDonough/epa/Corbis; 1.9 (inset): © Eye of Science/Photo Researchers; 1.10a (both): © blickwinkel/Alamy; 1.10c: © Phanie/Photo Researchers; 1.12a: © Comstock Images/Getty RF; 1.12b: © Photo by Ron Nichols, USDA Natural Resources Conservation Service.

Chapter 2

Opener: © liquidlibrary/PictureQuest RF; 2.3a: © Biomed Commun./Custom Medical Stock Photo; 2.3b (patient): Courtesy National Institutes of Health; 2.3b (brain scan): © Mazzlota et al./Photo Researchers; 2.4a(both): © Tony Freeman/PhotoEdit; 2.4b: © Geoff Tompkinson/SPL/Photo Researchers; 2.5b (crystals, shaker): © Evelyn Jo Johnson; 2.8b: © Clerkenwell/Getty RF; 2.13: © Jeremy Burgess/SPL/Photo Researchers; 2.14: © Don W. Fawcett/Photo Researchers; 2.15: © Science/Visuals Unlimited; 2.2 (both): © Purestock/Superstock RF; p. 36 (hair): © Comstock/PunchStock RF; p. 36 (hemoglobin): © P. Motta & S. Correr/Photo Researchers; p. 36 (muscle): © Suza Scalora/Getty RF.

Chapter 3

Opener: © Bruce Dale/National Geographic/Getty Images; 3.1 (red blood cells): © Prof. P. Motta, Dept. of Anatomy, Univ. LaSapienza Rome/SPL/Photo Researchers; 3.1 (nerve cells): © Dr. Dennis Kunkel/Visuals Unlimited; 3.1 (bone cells): © The McGraw-Hill Companies, Inc.; 3.3a: © David M. Phillips/Visuals Unlimited; 3.3b: © Alfred Pasieka/Photo Researchers; 3.3c: © 2013/Warren Rosenberg/Biological Photo Service; 3Aa: © David Wrobel/Visuals Unlimited; 3Ab: © M. Schliwa/Visuals Unlimited; 3.4a: © Dennis Kunkel/Visuals Unlimited; 3.9 (all): © Dennis Kunkel/Phototake; 3.13 (nuclear pores): Courtesy E.G. Pollock; 3.13 (ER): © R. Bolender & D. Fawcett/Visuals Unlimited; 3.15b: © Y. Nikas/Photo Researchers; 3.15c: © David M. Phillips/Photo Researchers; 3.17: © Dr. Don W. Fawcett/Visuals Unlimited; 3B: © C Squared Studios/Getty RF.

Chapter 4

Opener: © PHANIE/Photo Researchers; 4.2 (all), 4.5a, c: © Ed Reschke; 4.5b: © The McGraw-Hill Higher Education, Dennis Strete, photographer; 4.6, 4.8 (all): © Ed Reschke; 4.10: © John D. Cunningham/Visuals Unlimited; 4.11a: © Ken Greer/Visuals Unlimited; 4.11b: © James Stevenson/SPL/Photo Researchers; 4B (top, both): © AFP/Getty Images; 4B (bottom, both): © Paul Cooper Photography, Paris.

Chapter 5

Opener: © Adam Gault/Getty RF; 5.2 (left): © Ed Reschke; 5.2 (right): © Biophoto Associates/Photo Researchers; 5.3b: © SIU/Visuals Unlimited; 5.4 (top): © Dr. Don W. Fawcett/Visuals Unlimited; 5.5d: © Biophoto Associates/Photo Researchers; 5.6b–c: © Ed Reschke; 5.6d: © Lester Lefkowitz/Corbis; 5.7: © Comstock Images/PictureQuest; 5A: © Ryan McVay/Getty RF; 5B: © Biophoto Associates/Photo Researchers; 5.14a: © Pascal Goethgheluck/SPL/Photo Researchers; 5.15 (right): Courtesy SynCardia Systems, Inc.

Chapter 6

Openers (main): © SPL/Photo Researchers and (inset): © Andrew Syred/Photo Researchers; 6.1: © Doug Menuez/Getty RF; 6.3a: © Andrew Syred/Photo Researchers; 6.3c: © Ed Reschke/Getty Images; p. 118 (sickled cell): © Phototake, Inc./Alamy; 6.7b: © Eye of Science/Photo Researchers; 6A: © The McGraw-Hill Companies, Inc., Mark Dierker, photographer; 6B: © Keith Brofsky/Getty RF.

Chapter 7

Opener: © ISM/Phototake; 7.1b: © Dr. David M. Phillips/Visuals Unlimited; 7.1c: © Dr. Dennis Kunkel/Visuals Unlimited; 7.1d: © Dr. Gary D. Gaugler/Phototake; 7.2: © CNRI/SPL/Photo Researchers; 7.4a: © Dr. Hans Gelderblom/Visuals Unlimited; 7.4b: © K.G. Murti/Visuals Unlimited; 7.6a: © R. Calentine/Visuals Unlimited; 7.6c: © Ed Reschke; 7.6b: © Ed Reschke/Getty Images; 7.6d: © Fred E. Hossler/Visuals Unlimited; 7.12b: Courtesy Dr. Arthur J. Olson, Scripps Institute; 7.15b: © Steve Gschmeissner/Photo Researchers; 7.17a: © John Lund/Drew Kelly/Blend Images/Corbis RF; 7.17b: © Digital Vision/Getty RF; 7.17c: © Photodisc Collection/Getty RF; 7.18: © Southern Illinois University/Photo Researchers.

Infectious Diseases Supplement

Opener: © Bettmann/Corbis; S.6: © SPL/Photo Researchers; SA: © 2000–2006 Custom Medical Stock Photo; S.7: © ISM/Phototake; S.10: © China Photo/Reuters/Corbis; S.12: © David McCarthy/Photo Researchers.

Chapter 8

Opener: © Bubbles Photolibrary/Alamy; 8.5c: © Steve Gschmeissner/Getty RF; 8.6 (villi): © Kage Mikrofotografie/Phototake; 8.6 (microvilli): Reprinted from Medical Cell Biology, Charles Flickinger, copyright 1979, with permission from Elsevier; 8B (all): Courtesy Given Imaging, Ltd; p. 184 (people): © BananaStock/age fotostock RF; p. 184 (scale): © Photodisc/Getty RF; p. 184 (foods): © Photolink/Getty RF; 8.12: © Cole Group/Getty RF; 8.13: © Volume 20/Photodisc/Getty RF; 8C: © Evelyn Jo Johnson; 8.15: USDA, ChooseMyPlate.gov website; 8.16a: © Ted Foxx/Alamy RF; 8.16b: © Donna Day/Stone/Getty Images; 8.16c: © Corbis RF.

Chapter 9

Opener: © Image Source/Getty RF; 9.3 (center): © Science Photo Library/Getty RF; 9.4 (left): © CNRI/Phototake; 9.5b (bottom half): © Ed Reschke; 9A: Courtesy The Wyss Institute, Harvard University; 9.9: © Veronique Burger/Photo Researchers; 9.13a: © Matt Meadows/Getty Images; 9.13b: © Biophoto Associates/Photo Researchers.

Chapter 10

Openers: (polycystic, right): © Biophoto Associates/Photo Researchers and (normal, left): © SIU/Visuals Unlimited; 10A: © AP Photo/Brian Walker; 10.3b: © James Cavallini/Photo Researchers; 10.3c: © Ralph T. Hutchings/Visuals Unlimited; 10.4 (both): © Science Photo Library/Getty RF; 10.5a: © Joseph F. Gennaro Jr./Photo Researchers; 10C: © Ian Hooton/Photo Researchers; 10.11: © AJPhoto/Photo Researchers.

Chapter 11

Opener: © Southern Stock Corp/Corbis; 11.1 (hyaline, compact bone): © Ed Reschke; 11.1 (osteocyte): © Biophoto Associates/Photo Researchers; 11.4b: © Corbis RF; 11A: © Michael Donne/Photo Researchers; 11.9a: © Gerard Vandystadt/Photo Researchers; 11Ba–b (both): © Scott Camazine/Photo Researchers; 11C (woman, left): © Corbis RF; 11Ca (normal bone): © Susumu Nishinaga/Photo Researchers; 11Cb (osteoporosis): © Alan Boyde/Visuals Unlimited; 11C (woman, right): © Bill Aaron/PhotoEdit; 11.14b: © Tony Freeman/PhotoEdit.

Chapter 12

Opener: © Erproductions Ltd/Blend Images/Corbis; 12.1 (smooth): © The McGraw-Hill Companies, Inc. Dennis Strete, photographer; 12.1 (cardiac, skeletal): © Ed Reschke; 12.4 (both): © The McGraw-Hill Companies, Inc. J.W. Ramsey, photographer; 12.6 (gymnast): © Corbis RF; 12.6 (myofibril): © Biology Media/Photo Researchers; 12.7 (top right): © Victor B. Eichler; 12A: © Thinkstock/Getty RF; 12.12 (man): © Lawrence Manning/Corbis; 12.12 (muscle fibers): © G.W. Willis/Visuals Unlimited; 12.12 (woman): © Corbis RF; 12A (left): © ImageState/PunchStock RF; 12A (center): © Nancy Ney/Photodisc/Getty RF; 12A (right): © Getty Images/Stockbyte RF; 12B: © Justin Sullivan/Getty Images; 12C: © Focus on Sport/Getty Images.

Chapter 13

Opener: © Peter Duddek/Visum/The Image Works; 13.2 (myelin): © 2013/M.B. Bunge/Biological Photo Service; 13.2 (cell body): © doc-stock/Visuals Unlimited; 13.5a: © Science VU/Lewis-Everhart-Zeevi/Visuals Unlimited; 13Aa: Drawing is from "Histologie due Systeme Nerveux de l'Homme et des Vertebres." Courtesy of Instituto Cajal, Madrid, Spain; 13Ab: © David Becker/Photo Researchers; 13.7a: © Karl E. Deckart/Phototake; 13.7d: © The McGraw-Hill Companies, Inc. Rebecca Gray, photographer, Don Kincaid, dissections; 13.8b: © Colin Chumbley/Science Source/Photo Researchers; 13.13 (all): © Marcus Raichle; 13.14: © Dr. Richard Kessel & Dr. Randy Kardon/Visuals Unlimited; 13.18 (all): © Science VU/Visuals Unlimited.

Chapter 14

Openers (both): © AP Images/Gene J. Puskar; 14.4b (both): © Omikron/SPL/Photo Researchers; 14.8a: © Lennart Nilsson/Scanpix; 14.9b: © Biophoto Associates/Photo Researchers; 14A: © Pascal Goetgheluck/Photo Researchers; 14.13 (stereocilia): © P. Motta/SPL/Photo Researchers; 14B (both): Courtesy Dr. Yeohash Raphael, the University of Michigan, Ann Arbor.

Chapter 15

Opener: © Russ Curtis/Photo Researchers; 15.8a: © AP/Wide World Photos; 15.8b: © General Photographic Agency/Getty Images; 15.9 (all): From Clinical Pathological Conference, "Acromegaly, Diabetes, Hypermetabolism, Proteinura and Heart Failure," American Journal of Medicine 20 (1956) 133, with permission from Excerpta Medica, Inc.; 15.10a: © Bruce Coleman, Inc./Alamy; 15.10b: © Medical-on-Line/Alamy; 15.10c: © Dr. P. Marazzi/Photo Researchers; 15.15a: © Custom Medical Stock Photo; 15.15b: © NMSB/Custom Medical Stock Photo; 15.16 (both): Courtesy Shannon Halverson; 15.17: © Peter Arnold, Inc./Alamy; 15.18 (both): © Victor P. Eroschenko; 15A: © The McGraw-Hill Companies, Inc. Jill Braaten, photographer; 15.21: © Evelyn Jo Johnson.

Chapter 16

Opener: © UpperCut Images/Getty RF; 16.3b: © Anatomical Travelogue/Photo Researchers; 16.4b: © Ed Reschke; 16A: © Getty Images; 16.8 (Graafian follicle): © Ed Reschke; 16.14a: © Saturn Stills/Photo Researchers; 16.14b: © Keith Brofsky/Getty RF; 16.14c: © The McGraw-Hill Companies, Inc. Lars A. Niki, photographer; 16B: © Nancy R. Cohen/Getty RF; 16.16: © CC Studio/SPL/Photo Researchers; 16.17: © Dr. Olivier Schwartz/Institut Pasteur/Photo Researchers; 16.18a–b (both): Courtesy Centers for Disease Control, Atlanta, GA. (Crooks, R. and Baur, K.) *Our Sexuality,* 8/e, Wadsworth 2000 17.4, p. 489; 16.18c: © G. W. Willis/Visuals Unlimited; 16.19: © Biomedical Imaging Unit, Southampton General Hospital/Photo Researchers; 16C (left): © Vol. 161/Corbis RF; 16C (right): © David

Raymer/Corbis; 16D: © Brand X Pictures RF; 16.20: © Melba Photo Agency/Alamy RF.

Chapter 17

Opener: © John Slater/Getty RF; 17.1: © David M. Phillips/Visuals Unlimited; 17.2 (fertilization): © Don W. Fawcett/Photo Researchers; 17.2 (2-cell): © Rawlins-CMSP/Getty Images; 17.2 (morula): © RBM Online/epa/Corbis; 17.2 (implantation): © Bettmann/Corbis; 17.6a: © Lennart Nilsson/Scanpix; 17.8: © Neil Bromhall/Getty Images; 17.11: Courtesy Dr. Howard Jones, Eastern VA Medical School; 17.13c: © Rune Hellestad/Corbis; 17.14 (all): © Image Source/Getty RF; 17.15: © Ronnie Kaufman/Corbis; 17C: © Associated Press.

Chapter 18

Opener: © Kevin Mazur/Contributor/WireImage/Getty Images; 18.1: © CNRI/SPL/Photo Researchers; 18.5 (top): © Scott Camazine/Photo Researchers; 18.5 (bottom): © Edward Kinsman/Photo Researchers; 18.8 (early prophase, prophase, metaphase, anaphase, telophase): © Ed Reschke; 18.8 (early metaphase): © Michael Abbey/Photo Researchers; 18.16a: © Scott Wintrow/Getty Images; 18.16b (left): © CNRI/SPL/Photo Researchers; 18.19b: Courtesy The Williams Syndrome Association; 18.20b: © Fred, Herbert, and Hendrik van Dijk. "Images of Memorable Cases, 50 Years at the Bedside: Case 133."

Chapter 19

Opener: © Strauss/Curtis/Corbis; 19.1a: © Biology Media/Photo Researchers; 19.1b: © SPL/Photo Researchers; 19.1c: © Martin M. Rotker/Photo Researchers; 19.8A: © Custom Medical Stock Photo; 19.8B: © ISM/Phototake; 19.8C: © Dr. P. Marazzi/Photo Researchers; 19.8D: © Custom Medical Stock Photo; 19.8E: © Dr. Ken Greer/Visuals Unlimited; 19.9 (both): © UHB Trust/Getty Images; 19.10: © BSIP/Phototake; 19.11: © Simon Fraser/Royal Victoria Infirmary/Photo Researchers; 19D: © Obstetrics & Gynecology/SPL/Photo Researchers; 19E: © Visuals Unlimited/Corbis.

Chapter 20

Opener: © Don Tremain/Getty RF; 20.2a: © HFPA, 63rd Golden Globe Awards; 20.2b: © Dynamic Graphics/PictureQuest RF; 20.2c–f: © The McGraw-Hill Companies, Inc. Bob Coyle, photographer; 20.2g: © Jupiterimages/Getty RF; 20.2h: © Creatas/PunchStock RF; 20A (both): © Brand X/SuperStock RF; 20.11b; © Pat Pendarvis; 20.12 (both): Courtesy Dr. Hemachandra Reddy, The Neurological Science Institute, Oregon Health & Science University; 20.13: Courtesy University of Connecticut/Peter Morenus, photographer; 20.15: © Jane Burton/Photoshot; 20.16 (top right): © Ed Reschke; 20.17: © Mediscan/Medical-On-Line; 20.21 (left, right): Courtesy Dr. Rabi Tawil, Director, Neuromuscular Pathology Laboratory, University of Rochester Medical Center; 20.21 (center): Courtesy Muscular Dystrophy Association; 20B (queen): © Stapleton Collection/Corbis; 20B (prince): © Huton Archive/Getty Images.

Chapter 21

Opener: © Blend Images/Alamy RF; 21A (both): © Science Source/Photo Researchers; 21.11: Courtesy Alexander Rich; 21.14 (sequencer): © Lawrence Berkeley Nat'l Lab/Roy Kaltschmidt, photographer; 21.18a: © Maximilian Stock Ltd./Photo Researchers; 21.18b: Courtesy Robert H. Devlin, Fisheries and Oceans Canada; 21.18c: © Richard Shade; 21.18d: © Jerry Mason/Photo Researchers; 21.19 (all): Courtesy Monsanto; 21.20 (both): Courtesy Eduardo Blumwald; 21.22a (top): © Digital Vision/Getty RF; 21.22a (bottom): © Getty RF; 21.22b (left): © Digital Vision/Getty RF; 21.22b (right): © The McGraw-Hill Companies, Inc. Bob Coyle, photographer; 21.22c (left): © Digital Vision/Getty RF; 21.22c (right): © Getty RF; 21B: © Deco/Alamy RF.

Chapter 22

Opener: © Tom McHugh/Photo Researchers; 22.5 (fossil, left): © Jean-Claude Carton/PhotoShot; 22.5 (drawing, right): © Joe Tucciarone; 22.6a (fossil *Ambulocetus* foot): © J.G.M. Thewissen, Northeastern Ohio Universities College of Medicine; 22.7 (sugar glider): © ANT Photo Library/Photo Researchers; 22.7 (wombat): © Photodisc Blue/Getty RF; 22.7 (kangaroo): © George Holton/Photo Researchers; 22.9 (both): © Carolina Biological Supply/Phototake; 22.12 (gibbon): © Hans & Judy Beste/Animals Animals; 22.12 (orangutan, chimps, gorillas): © Creatas/PunchStock RF; 22.15a: © Dan Dreyfus and Associates; 22.15b: © John Reader/Photo Researchers; 22.16 (*Sehalanthropus tchadensis*): © HO/AFP/Newscom; 22.16 (*A. afarensis*): © Friedrich Saurer/Alamy; 22.16 (*A. boisei*): © Science VU/NMK/Visuals Unlimited; 22.16 (*H. habilis*): © Herve Conge/ISM/Phototake; 22.16 (*H. sapiens*): © 2007 Educational Images Ltd./Custom Medical Stock Photo; 22Aa: © Peter Schouten/National Geographic Society/Reuters/Corbis; 22Ab: © AP Photo/Richard Lewis; 22.17: © Science VU/Visuals Unlimited; 22.19: Image # 608, American Museum of Natural History, Library; 22B: © Peter Bowater/Photo Researchers; 22.20a: © PhotoDisc/Getty RF; 22.20b: © Sylvia S. Mader; 22.20c: © Adam Crowley/Getty Images.

Chapter 23

Opener: © Jeremy Woodhouse 2008/Getty Images; 23.1 (temperate): © Chris Johnson; 23.1 (desert): © Martin Harvey/Getty RF; 23.1 (savanna): © Gregory Dimijan/Photo Researchers; 23.1 (rain forest): © Steve Bloom Images/Alamy; 23.1 (prairie): © Vol. 86/Corbis RF; 23.1 (taiga): © Image Source/Alamy RF; 23.1 (tundra): © Johnny Johnson/Getty Images; 23.2a: © Francois Gohier/Photo Researchers; 23.2b: © The McGraw-Hill Companies, Inc. Carlyn Iverson, photographer; 23.2c: © David Hall/Photo Researchers; 23.2d: © Vol. 121/Corbis RF; 23.3a (left): © Ed Reschke; 23.3a (right): © Authors Image/PunchStock RF; 23.3b (left): © Corbis RF; 23.3b (right): © Gerald C. Kelley/Photo Researchers; 23.3c (left): © Bill Beatty/Visuals Unlimited; 23.3c (right): © Joe McDonald/Visuals Unlimited; 23.3d (left): © SciMAT/Photo Researchers; 23.3d (right): © IT Stock/age fotostock RF; 23.5: © George D. Lepp/Photo Researchers; 23A (left): © igor kisselev, www.close-up.biz/Getty RF; 23A (right): © Howard Davies/Corbis; 23.11b: © Jan Martin Will/Shutterstock Images; 23.13a: © Larry Lee Photography/Corbis RF; 23.13b: © Oxford Scientific/Photolibrary/Getty Images; 23.13c: © Sue Baker/Photo Researchers; 23Ba: © Brian J. Skerry/Getty Images; 23Bb: © Jose Luis Pelaez Inc./Blend Images LLC RF.

Chapter 24

Opener: © J. Emilio Flores/La Opinion/Newscom; 24.2c: © Tim Graham/Getty Images; 24.3 (land): © Vol. 39/PhotoDisc/Getty RF; 24.3 (water): © Evelyn Jo Johnson; 24.3 (food): © The McGraw-Hill Companies, Inc. John Thoeming, photographer; 24.3 (energy): © Photo Link/Getty RF; 24.3 (minerals): © T. O'Keefe/PhotoLink/Getty RF; 24.4b: © Melvin Zucker/Visuals Unlimited; 24.5b: © Carlos Dominguez/Photo Researchers; 24.6b: © L. Hobbs/PhotoLink/Getty RF; 24.7a: © Comstock Images/Alamy RF; 24.7b: © Jeremy Samuelson/FoodPix/Getty Images; 24.7c: © Stockbyte/PunchStock RF; 24.8: © AP Photo/Mark Gormus; 24.9a: © Bruno Barbier/Getty Images; 24.9b: © Inga Spence/Visuals Unlimited; 24.9c: © View Stock/Alamy; 24.10a: © Laish Briston/Visuals Unlimited; 24.10b: © Inga Spence/Visuals Unlimited; 24.10c: Courtesy V. Jane Windsor, Division of Plant Industry, Florida Department of Agriculture & Consumer Services; 24.11: Photo by Tim McCabe, USDA Natural Resources Conservation Service; 24Aa: Courtesy USDA/Doug Wilson, photographer; 24Ab: © Norm Thomas/Photo Researchers; 24Ac: Courtesy Carolyn Merchant/UC Berkeley; 24.12a: © Corbis RF; 24.12b: © Glen Allison/Getty RF; 24.12c: © Danita Delimont/Getty Images; 24.12d: © Photo Link/Getty RF; 24.13a: Courtesy DaimlerChrysler; 24.13b: © Warren Gretz/DOE/NREL; 24.15b: © IT Stock/PunchStock RF; 24.16a: © Shane Moore/Animals Animals; 24.16b (both): © Peter Auster/University of Connecticut; 24.17 (periwinkle): © Steven P. Lynch; 24.17 (armadillo): © PhotoDisc/Getty RF; 24.17 (fishing trawler): © Tim Laman/Getty RF; 24.17 (bat): © Merlin D. Tuttle/Bat Conservation International; 24.17 (lady bug): © Martin Ruegner/Masterfile RF; 24.17 (rubber): © Bryn Campbell/Stone/Getty Images; 24B: Courtesy USDA/Lila De Guzman, photographer; 24.18 (deforestation): © Carlos Dominguez/Photo Researchers; 24.18 (overfishing): © Shane Moore/Animals Animals; 24.18 (water pollution): © Michael Gadomski/Animals Animals; 24.18 (air pollution): © Larry Lee Photography/Corbis RF; 24.18 (habitat loss): © Bill Zimmer/Associated Press; 24.18 (erosion): © USDA/Nature Source/Photo Researchers; 24.19 (farming): © Inga Spence/Visuals Unlimited; 24.19 (delta preservation): © Peter DeJong/Associated Press; 24.19 (recycling): © Jeffrey Greenberg/Photo Researchers; 24.19 (mass transit): Courtesy DaimlerChrysler; 24.19 (temperature control): © Reuters/Corbis; 24.19 (integrated pest management): Courtesy V. Jane Windsor, Division of Plant Industry, Florida Department of Agriculture & Consumer Services; 24.20: © Photo by Erwin Cole, USDA Natural Resources Conservation Service; 24.21: © Diane Cook and Len Jenshel/Getty Images; p. 581 (land): © Vol. 39/PhotoDisc/Getty RF; p. 581 (water): © Evelyn Jo Johnson; p. 581 (food): © The McGraw-Hill Companies, Inc. John Thoeming, photographer; p. 581 (energy): © Photo Link/Getty RF; p. 581 (minerals): © T. O'Keefe/PhotoLink/Getty RF.

Line Art

Chapter 5

5.14c: Source: American Heart Association; 5.15: Figure from www.texasheart.org AB/OMED Replacement Heart. Reprinted with permission from ABIOMED, Inc.

Index

Note: Page references followed by *f* and *t* refer to figures and tables, respectively.

A

B

C

D

E

H

I

J

K

L

M

N

O

P

T

U

V

W

X

Y

Z